3rd International Conference on Applied Materials and Manufacturing Technology (ICAMMT 2017)

AA006464

IOP Conference Series: Materials Science and Engineering
Volume 242

Changsha, China
23 – 25 June 2017

ISBN: 978-1-5108-4893-1
ISSN: 1757-8981

Printed from e-media with permission by:

Curran Associates, Inc.
57 Morehouse Lane
Red Hook, NY 12571

Some format issues inherent in the e-media version may also appear in this print version.

Copyright© (2017) by the Institute of Physics
All rights reserved. The material featured in this book is subject to
IOP copyright protection, unless otherwise indicated.

Printed by Curran Associates, Inc. (2017)

For permission requests, please contact the Institute of Physics
at the address below.

Institute of Physics
Dirac House, Temple Back
Bristol BS1 6BE UK

Phone: 44 1 17 929 7481
Fax: 44 1 17 920 0979

techtracking@iop.org

Additional copies of this publication are available from:

Curran Associates, Inc.
57 Morehouse Lane
Red Hook, NY 12571 USA
Phone: 845-758-0400
Fax: 845-758-2633
Email: curran@proceedings.com
Web: www.proceedings.com

Table of contents

Volume 242

2017 3rd International Conference on Applied Materials and Manufacturing Technology (ICAMMT 2017)

23–25 June 2017, Changsha, China

Accepted papers received: 12 September 2017
Published online: 28 September 2017

Preface

Composite Materials and Materials Science

Advanced Manufacturing Technology and Mechanical Technology

Industrial Robots, Control and Automation

IOP Conf. Series: Materials Science and Engineering **242** (2017) 011001 doi:10.1088/1757-899X/242/1/011001

Preface

The 2017 3rd International Conference on Applied Materials and Manufacturing Technology (ICAMMT 2017) has been held on June 23-25, 2017 in Changsha, China. The conference addressed recent development in the following,
(1) Composite Materials and Materials Science
(2) Advanced Manufacturing Technology and Mechanical Technology
(3) Industrial Robots, Control and Automation

This volume of IOP Conference Series: Materials Science and Engineering (MSE) is a compilation of the accepted papers in ICAMMT 2017 and represents important results that were presented in the conference. The electronic submission and handling of the manuscripts via the conference website, including the selection of reviewers and evaluation of manuscripts, were identical to the procedures applied to manuscripts submitted as regular contributions for publication in peer reviewed journals.

The organization of this conference and the preparation of proceedings volume would have been impossible without the tremendous efforts and dedication of many individuals. Our two conference chairs oversaw the organization of the Conference. We would like to acknowledge a large team of reviewers for their timely submission of quality reports. Without this support, the conference could not have been successful. We sincerely hope that ICAMMT 2017 will serve as an ideal forum for academia, industrial practitioners and government agents.

Prof. Shiming Chen, Tongji University, China
General Chairman of ICAMMT 2017

Content from this work may be used under the terms of the Creative Commons Attribution 3.0 licence. Any further distribution of this work must maintain attribution to the author(s) and the title of the work, journal citation and DOI.
Published under licence by IOP Publishing Ltd

IOP Conf. Series: Materials Science and Engineering **242** (2017) 011001 doi:10.1088/1757-899X/242/1/011001

Organizing Committees

Conference Chair
Prof. Shiming Chen, Tongji University, China
Prof. Yeping Wang, Tongji University, China

Chair of International Technical Committee
Prof. Peijiang Zhou, Wuhan University, China

International Technical Committee
Xinli Guo, Southeast university, China
Zhangxu Chen, Putian University, China
Fei Yu, Shanghai Institute of Technology, China
Jinhe Wang, Shanghai University, China
Zhenrong Yan, Shanghai Boiler Works Co, Ltd
Qingsheng Wang, Russian St. Petersburg State Polytechnic University, China
Yanhua Zhao, Chongqing University of Arts and Science, China
Tao Zhu, China University of Mining & Technology, China
Ş. Hakan Atapek, Kocaeli University, Turkey
Meilinda Nurbanasari, Institut Teknologi Nasional, Bandung, Indonesia
Jagannath Pal, CSIR National Metallurgical Laboratory, India
Binsheng Hu, North China University of Science and Technology, China
Yi Zhang, Chinese Academy of Sciences, China
Zaiyu Zhang, Guizhou University, China
Haofeng Zhao, Nanjing University of information technology, China
Ritayan Chatterjee, Haldia Institute of Technology, India
M A GafurM A Gafur, Pilot Plant and Process Development Centre, Bangladesh
Ming Dong, Shanghai Jiao Tong University, China
Carlos Caceres, The University of Queensland, Australia
Shupei Cheng, Nanjing University, China
Haozhong Cheng, Shanghai Jiao Tong University, China
Lam Kin Che, The Chinese University of Hongkong, HK
Zhenqian Chen, Southeast University, China
Wenbin Cai, Fudan University, China
Heinz-Gunter, Brokmeier, Technische Universitat Clausthal,Germany
Jinglong Bu, Hebei United University, China
Shahrum Abdullah, University Kebangsaan Malaysia, Malaysia

Peer review statement

All papers published in this volume of *IOP Conference Series: Materials Science and Engineering* have been peer reviewed through processes administered by the proceedings Editors. Reviews were conducted by expert referees to the professional and scientific standards expected of a proceedings journal published by IOP Publishing.

Content from this work may be used under the terms of the Creative Commons Attribution 3.0 licence. Any further distribution of this work must maintain attribution to the author(s) and the title of the work, journal citation and DOI.
Published under licence by IOP Publishing Ltd

Inverse gas chromatography as a method for determination of surface properties of binding materials

Jihai Yu[1], Xiaolei Lu[1], Chunxia Yang[2], Baoli Du[3], Shuxian Wang[4], Zhengmao Ye[1, 4*]

[1] Shandong Provincial Key Laboratory of Preparation and Measurement of Building Materials, University of Jinan, Jinan, Shandong Province, China

[2] School of Chemical and Chemical Engineering, University of Jinan, Jinan, Shandong Province, China

[3] Dong Guan Chuang Jie New Materials CO., LED.

[4] School of Material Science and Engineering, University of Jinan, Jinan, ShandongProvince, China

*E-mail address: mse_yezm@ujn.edu.cn

Abstract. Inverse gas chromatography (IGC) is a promising measurement technique for investigating the surface properties of binding materials, which are the major influence element for the adsorption performance of superplasticizer. In this work, using the IGC method, blast furnace slag (BFS), sulphoaluminate cement (SAC) and portland cement (P·O) are employed to systematically evaluate the corresponding dispersive component (γ_s^d), specific surface free energy (γ_s^{ab}), and acid-base properties. The obtained results show that γ_s^d contributes to a major section of the surface free energy in the three binding materials, suggesting they are of a relatively low polarity. Compared to the two kinds of cements, the BFS possesses the highest dispersive and specific surface free energies (the values are 45.01 mJ/m^2 and 11.68 mJ/m^2, respectively), and also exhibits a wider distribution range of γ_s^d, indicating their surfaces are heterogeneous. For acid-base properties, the results indicate the surfaces of three samples are basic in nature. In addition, the adsorption investigation shows that per unit surface of BFS adsorbs the most superplasticizer molecules, which indicates the higher surface free energies is beneficial to the superplasticizer adsorption.

1. Introduction

The Binding materials cooperated by superplasticizer are widely used in building industry. The adsorption capability of binding materials surfaces would influence the water reducing effect of the superplasticizer. QU *et al.*, reported that the adsorption of cement particles was influenced by many factors such as mineral composition of cement, alkali sulphate and mineral admixture in cement [1]. S.K. Agarwal *et al.*, found that different grade of P·O showed different compatibility with superplasticizer [2]. Erdogdu. S. showed that the mineral composition of cement will influenced the adsorbed capability of cement particles and the effect of a superplasticizing admixture depended on the composition of cement rather than the amount used [3]. Surface free energy is also an important influential element for the performance of binding materials, there are few reports about the relationship between the adsorption capability and surface free energy nevertheless. Although the traditional thin-layer wicking technique has been employed by some researchers to study the dynamic

Content from this work may be used under the terms of the Creative Commons Attribution 3.0 licence. Any further distribution of this work must maintain attribution to the author(s) and the title of the work, journal citation and DOI.
Published under licence by IOP Publishing Ltd

IOP Conf. Series: Materials Science and Engineering **242** (2017) 012001 doi:10.1088/1757-899X/242/1/012001

contact angles and surface free energy of cement binding materials [4], the experimental conditions are usually difficult to be controlled.

As an alternative to classic contact angle measurement, IGC has proved to be a sensitive and easy-operating technique for the characterization of materials surface properties of [5-7], and are widely applied in the fields of food detection and composite materials. Surface energy analysis technique based on IGC has been found to be very effective for characterizing wetting phenomena on powders and fibers [8-10]. Lubomír Lapčík used the IGC for the measurement of the surface energy of demineralised whey and skimmed milk and showed the surface energy profiles [11]. Jeong et al., studied the surface properties of $CaCO_3$ by IGC [12]. Naveed Arsalan et al., introduced the IGC to characterize the surface energy of sandstones (Ottawa sand and Berea sandstone) [13], and they indicated that the use of IGC enables an accurate picture of the surface at different temperatures and other physical conditions by taking into account surface heterogeneity and the interaction forces responsible for the adsorption[14]. Therefore, based on the studies as mentioned above, the IGC method may be also suitable for analyzing the surface properties of binding materials.

In this study, we investigate the surface properties by the IGC method. The adsorption weight of naphthalene based superplasticizer on the surface of binding materials is also measured. Combining the surface energy test and adsorption capacity measurement, there is a positive relevance between surface energy and adsorption capacity of binding materials particles. The relevance between surface properties and superplasticizer adsorption on binding material surfaces should be further studied by the IGC method.

2. Materials and methods

2.1. Materials

Some SAC, P·O and BFS were produced by factories in Shandong Province. The chemical components and physical properties of the materials are listed in Table 1.

Table 1 Chemical composition and physical properties of cement (wt %).

Material	SiO_2	Fe_2O_3	Al_2O_3	CaO	MgO	SO_3	Specific Surface Area ($m^2 \cdot g^{-1}$)
SAC	10.96	3.71	28.9	45.25	1.45	8.88	2.2
P·O	19.95	2.90	4.71	60.58	1.41	—	2.0
BFS	32.31	0.29	16.36	35.53	10.41	2.74	1.26

* Specific surface area of binding materials is measured by a commercial inverse gas chromatograph SMS IGC-SEA (surface measurement systems, UK).

Values of Gutmann's donor and modified acceptor numbers for the polar probes are displayed in Table 2

Table 2 Physical constants of the probes used in the IGC measurements [15-17].

Probes	a($\times 10^{-19} m^2$)	γ_l^d (mJ/m^2)	$a(\gamma_l^d)^{0.5}$ [$m^2(mJ/m^2)^{0.5}$]	Modified acceptor numbers(kJ/mol)	Gutmann's donor (kJ/mol)	Specific characteristic
n-Heptane	5.73	20.3	2.58×10^{-18}	-	-	Neutral
n-Octane	6.28	21.3	2.89×10^{-18}	-	-	Neutral
n-Nonane	6.89	22.7	3.28×10^{-18}	-	-	Neutral

n-Hexane	5.10	18.4	2.19×10^{-18}	-	-	Neutral
Ethyl acetate	3.30	19.6	1.46×10^{-18}	6.3	71.8	Amphoteric
Dichloromethane	2.45	24.5	3.83×10^{-18}	16.4	0.0	Acid
Acetonitrile	2.14	27.5	1.12×10^{-18}	19.4	58.3	Amphoteric

2.2. Methods

2.2.1 IGC method. The IGC test was carried out using a commercial inverse as chromatograph SMS IGC-SEA (surface measurement systems, UK). The injection system used by the SMS IGC-SEA allows the precise control of the injection size, and therefore different amounts of probe vapor can be chosen to pass through the sample column achieving different surfaces coverage (n/nm) If a series of probe vapors is injected at the same surface coverage, the surface free energy and Gibbs specific free energy values can be determined. Consequently, the injections of probe vapors at different surface coverage will referred to as a surface free energy profile. The surface chemistry of the samples was assessed using the Gutmann acid (Ka) and base (Kb) numbers, determined based on the Gutmann approach using the following polar probes: Dichloromethane, Ethyl acetate, Acetonitrile, Chloroform. Ka and Kb values of the samples were calculated using the γ_s^{ab} values of polar probes at that particular surface coverage.

In IGC, the term "inverse" indicates that the solid state materials to be characterized was packed into the chromatographic column and this material was probed by known gas mixtures which were injected into the column [18]. For the preparation of columns, 250mg of the powder was filled into a salinized standard column (3mm ID, 300mm long) using the SMS sample-packing device. All the columns have been prepared using the same batch of material and using the same frequency and duration of tapping during packing.

The stationary phase characterization was achieved by portioning the sample between the mobile phase and the stationary phase, indicated by the time taken to elute the samples. The molecular probes were injected at infinite dilution in order to rule out lateral probe-probe interactions and favor probe-stationary phase interactions only [19]. The dead volume was determined by injecting methane, a probe molecule that does not interact significantly with low energy materials

The total surface free energy of solid materials is often divided into dispersive γ_S^d and specific (γ_S^{ab}) parts, as is shown in Equation 1. Dispersive (apolar) interactions, also known as Lifshitz-van der Waals interactions, consist of London, Keesom and Debye interactions. Specific (polar) interactions explain all other types of interactions [20].

$$\gamma_s = \gamma_s^d + \gamma_s^{ab} \quad (1)$$

A standard IGC method for determining γ_S^d of the stationary phase relies on a series of linear alkane molecular probes. The increasing carbon numbers in the chain lead to the increase of adsorption energy. The n-alkane line is obtained by plotting the adsorption free energy of probes against the carbon number n of the injected alkane probes. In this method, a plot of $RT\ln(V_{N,n})$ versus the carbon number (of alkanes) should produce a linear correlation. Where R is the gas constant 8.31 J mol^{-1} K^{-1} T is the temperature; $V_{N,n}$ is the retention volumes of the n-alkanes probes with carbon number n. The dispersive component of the solid sample can be determined from the slope of the correlation.

$$-\Delta G_{CH_2} = 2(\gamma_{CH_2}\gamma_s^d)^{1/2} N_A a_{CH_2} \quad (2)$$

Where γ_S^d is the dispersive component of the solid surface free energy, a_{CH_2} is the cross sectional area of the methylene group, γ_{CH_2} is the surface dispersive free energy of a solid material constituted solely of methylene groups. N_A is Avogadro's number, ΔG_{CH_2} is the slope corresponds to the increment of free energy per methylene groups [21].

Specific surface free energy can be examined from the probes plot of $RT\ln V_N$ against $a(\gamma_l^d)^{0.5}$. The distance between the ordinate values of the polar probe datum point and the n-alkane reference line gives the specific component of the surface free energy.

$$\frac{\gamma_s^{ab}}{AN^*} = \frac{DN}{AN^*} K_a + K_b \quad (3)$$

DN and *AN** are Gutmann's donor and modified acceptor numbers, respectively. K_a and K_b are obtained from a plot of $-\frac{\gamma_s^{ab}}{AN^*}$ versus $\frac{DN}{AN^*}$ with K_a as the slope and K_b as the intercept.

The overall acid-base character of the samples' surface can be evaluated from the K_b/K_a ratio [22].

2.2.2 Adsorption test. Naphthalene based superplasticizer (NBS) was used as adsorbent. Before this test, NBS solutions with different concentration were taken for ultraviolet spectrophotometry to find the characteristic absorption peak, as is shown in the Figure 1. A curve of concentration to absorbance can be obtained by linear fitting method at the wavelength of 229 nm, which was shown in Figure 2. In the test, 2 g of cement were put into 200 g NBS agent solution with the concentration of 6mg/L to get the suspension of cement particles. A certain amount of suspension was taken out for centrifuging and the supernatant was taken for ultraviolet spectrophotometry. Then the absorbance of the supernatant can be got. The weight of NBS adsorbed by cement particles could be calculated referring to the standard adsorption curve of the NBS.

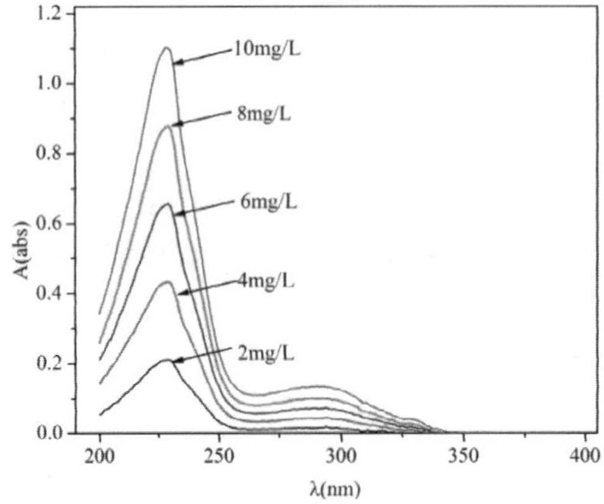

Figure 1 Adsorption spectra of NBS standard solution.

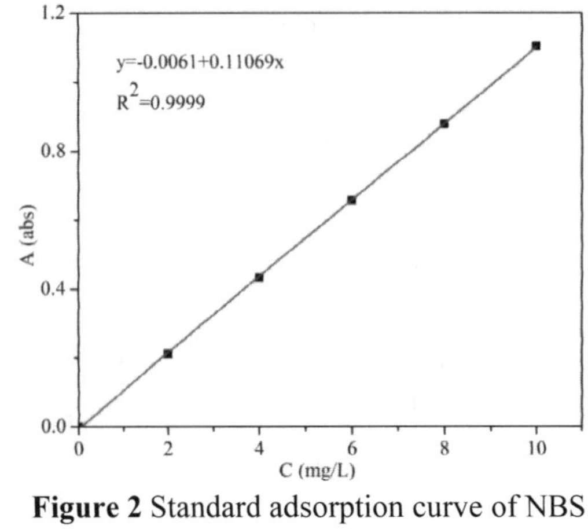

Figure 2 Standard adsorption curve of NBS.

3. Results and discussion

3.1 Dispersive surface free energy profiles and distribution

The dispersive surface free energy profiles (γ_S^d) of cement and BFS are displayed in Figure 3. The calculate γ_S^d of SAC, P·O and BFS fell into the ranges of 41.35-51.99 mJ/m^2, 40.5-53.01 mJ/m^2 and 41.89-58.56 mJ/m^2, respectively. The finally adoptive γ_S^d of the three materials are 43.34, 42.55 and 45.01 mJ/m^2 respectively.

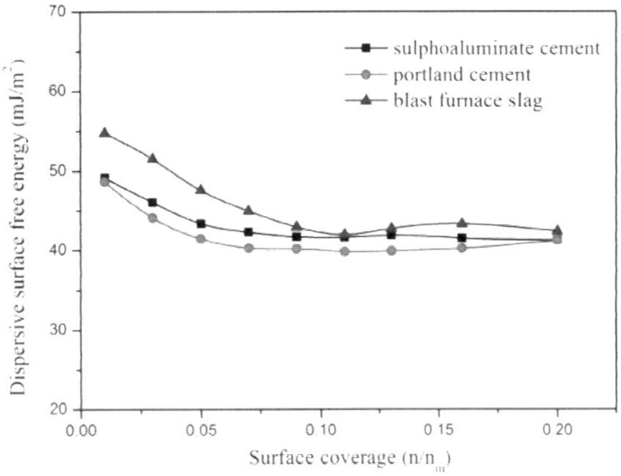

Figure 3 Dispersive surface free energy profiles of binding materials

As is shown in Figure 3, it can be seen that the γ_S^d values for SAC, P·O and BFS showed a decreasing trend with an increase in the surface coverage. The γ_S^d showed the highest value when the coverage is 0.01%. The retention volume of samples decreased with increasing surface coverage, because the interaction between the alkane probes and less energetic sites at high surface coverage was weaker [23-25]. The γ_S^d values for samples decreased remarkably as the surface coverage increased from 0.01% to 0.13% then increased slightly with still greater surface coverage. This may be because the samples surfaces are heterogeneous and rough [20].

The Figure 4 displayed the distribution of dispersive surface free energy of three samples. It showed a characteristically broader distribution for BFS (42.39 mJ/m2 to 54.78 mJ/m2). The dispersive surface free energy distribution of SAC and P·O were narrower, ranging from 41.30 mJ/m2 to 49.20 mJ/m2 and 41.27 mJ/m2 to 48.63 mJ/m2 respectively with relatively the same area increment occupancy of 2%.

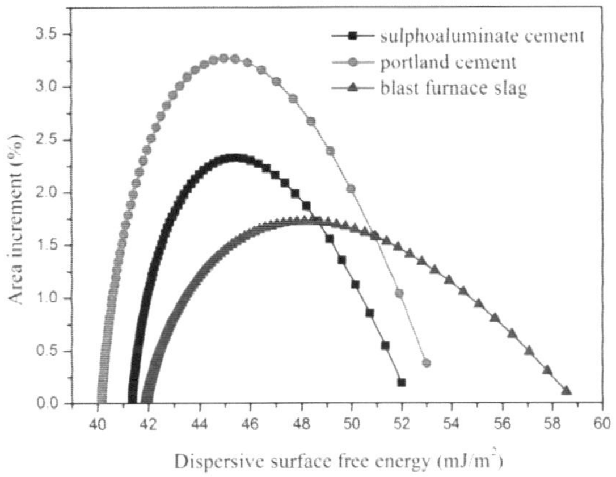

Figure 4 Dispersive surface free energy distributions of binding materials

Surface energy values are inherently independent of surface area, particle size, and particle shape. For some materials (such as silica and fiber) high surface area is associated with high surface free energy [17, 26]. However, in some cases it is not true, because a larger surface area implies a higher concentration of active sites or even a change in composition per unit surface. In theory, the surface free energy values will not be affected by changes in surface roughness. However, the orientation and interaction of IGC probe molecule can vary slightly if surface roughness is of the same magnitude as the molecular probe diameters [21]. Thus the surface free energy is not only associated with structural variations, but also with the density or type of functional groups orientated on the surface. In this study, the three samples have different composite and surface rough, so the dispersive surface free energy was affected by various factors.

3.2 Specific surface free energy and acid-base surface characteristics

Specific surface free energy can be examined from the probes plot of $RT \ln V_N$ against $a(\gamma_l^d)^{0.5}$. The distance between the ordinate values of the polar probe datum point and the n-alkane reference line gives the specific component of the surface free energy. The results show that the specific surface free energy (γ_S^{ab}) of the three binding materials are 7.43, 8.20 and 11.68 mJ/m^2 respectively.

Acid-base surface characteristics of samples depend on the ability to participate in specific interactions. The K_a and K_b were used for characterization of the acidity or basicity of the surface of samples. The surface chemistry of samples is routinely characterized by Gutmann's acid-base theory. If $K_b/K_a > 1$, the surface can be considered to be basic, meaning the surface of the solid prefers to donate electrons, while $K_b/K_a < 1$, the surface can be considered to be acidic, indicating that it prefers to accept electrons. The K_b/K_a ratios were 5.20, 4.49 and 3.01 for SAC, P·O and BFS respectively. This indicated that the surface of samples is basic in nature. The basic character of SAC is stronger than the others.

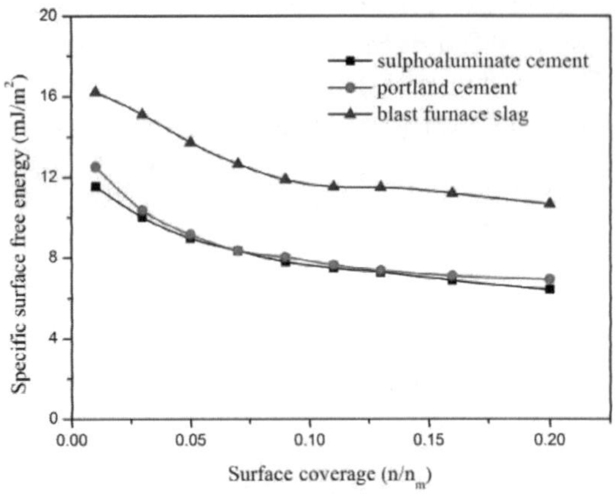

Figure 5 Specific surface free energy profiles of binding materials

As is shown in the Figure 5 the specific surface free energy of BFS was still the highest which is the same case for dispersive surface free energy. It is similar for the specific surface free energy of SAC and P·O.

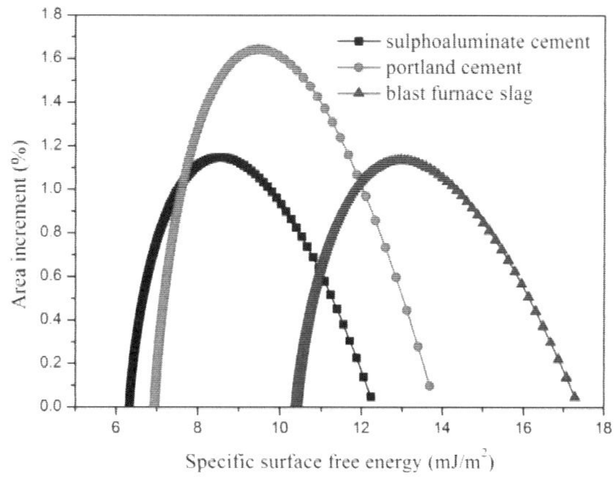

Figure 6 Specific surface free energies distributions of binding materials

Figure 6 displayed the specific surface free energy distribution, it can be found from the picture that the specific surface free energy distribution range of three samples were similar. This was different from the dispersive surface free energy distribution. It shows that the dispersive surface free energy and specific surface free energy were influenced by different factors.

3.3 Adsorption properties

The experiment results of ultraviolet spectrophotometry listed in Table 3 showed that the adsorption capacity of BFS was the highest. The mass of NSP in the suspension can be calculated by the standard absorption spectrum of naphthalene based SP then the absorbance of NSP by binding materials can be calculated. The weight of NSP adsorbed by unit surface of SAC, P·O and BFS were 0.161mg, 0.155mg and 0.175mg respectively. The result was corresponding to the laboratory finding of LIU et al., [27]. They found that the adsorbing capacity of SAC is higher than the adsorbing capacity of P·O.

To counteract the effect of specific surface area on the weight of NSP adsorbed. The weight of superlasticizer adsorbed by unit surface of samples was measured. The weight of NSP adsorbed on unit surface of three samples is 0.161 mg, 0.155mg, and 0.175mg respectively. The NSP adsorption capacity of BFS is the strongest.

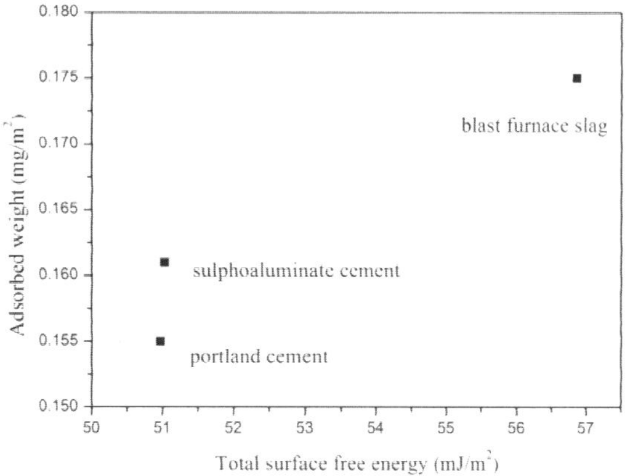

Figure 7 Adsorption weight by unit surface versus total surface free energy

As is shown in Figure 7 there is a positive correlation between surface free energy and adsorption capacity of samples. However the influencing mechanism of surface free energy on adsorption capacity of binding material is complicated and many works are needed to be done to establish the accurate relationship between surface free energy and adsorption. The IGC method will be an important technology in the study.

4. Conclusion

The γ_S^d value significantly decreased with the increasing of surface coverage from 0.01% to 0.13% then slightly increased during the increasing of surface coverage from 0.13% to 0.2%. This maybe because that the samples are heterogeneous and rough. The γ_S^d values of SAC, P·O and BFS fell into 41.5-49.3 mJ/m^2, 41.2-48.7 mJ/m^2 and 42.4-54.8 mJ/m^2 respectively. The BFS particles exhibits relatively wider range of γ_S^d distribution so its surface is more heterogeneous. Both the three binding materials are basic in nature. BFS showed higher values in both dispersive surface free energy and specific surface free energy. The unit surface of BFS adsorbed the most NSP molecules. Combining the measurement results of surface free energy and adsorption capacity of the samples it showed that the adsorption capacity of sample particles increases with higher surface free energy. IGC method is alternative for investigation of binding materials, and has potential advantages for the study of binding materials surface. The application of IGC in binding materials industry should be further explored.

Acknowledgements

This work is supported by National Natural Science Foundation of China (No. 51272092 No. 51302105) and Shandong Provincial Science and Technology Major Project (New Industry) (No. 2015ZDXX0702B01). Meanwhile, this work is supported by Program for Scientific Research Innovation Team in Colleges and Universities of Shandong Province.

References

[1] Qu JD, Peng JH, Chen MF, Zhang JX, Wan TZ, (2005) The research progress of adsorption properties of water reducers on the cement particles Journal of Building Materials 8: 410-417.
[2] Agarwal SK, Masood I, Malhotra SK, (2000) Compatibility of superplasticizers with different cements Construction and Building Materials 14: 253 -257.
[3] Erdogdu S, (2000) Compatibility of superplasticizers with cements different in composition Cement and Concrete Research 30: 767 -773.
[4] Liu JP, Li H, Tian Q, Wang YJ, (2013) Determination of dynamic contact angles of cement and mineral admixtures based on thin-layer wicking technique Journal Of Southeast University(Natural Science Edition) 43: 1074-1079.
[5] Gamelas JAF, Ferraz E, Rocha F, (2014) An insight into the surface properties of calcined kaolinitic clays: the grinding effect Colloid. Surface A 455: 49–57.
[6] Ylä-Mäihäniemi PP, Heng JYY, Thielmann F, Williams DR, (2008) Inverse gas chromatographic method for measuring the dispersive surface energy distribution for particulates, Langmuir 24: 9551–9557.
[7] Riedl B, Matuana LM, (2006) Inverse gas chromatography of fibers and polymers, Encyclopedia of Surface and Colloid Science, 3018–3031.
[8] Lapčík L, Kubíček P, Lapčíková B, Zbořil R, Nevěčná T, (2014) Study of penetration kinetics of sodium hydroxide aqueous solution into wood samples. BioResources 9: 881–893.
[9] Martinez-Padilla LP, Garcia-Mena V, Casas-Alencaster NB, Sosa-Herrera MG, (2014) Foaming properties of skim milk powder fortified with milk proteins. International Dairy Journal 36: 21–28.
[10] Lazar P, Otyepková E, Banáš P, Fargašová A, Šafářová K, Lapčík L, (2014) The nature of high surface energy sites in graphene and graphite. Carbon 73: 448–453.
[11] Lapčík JR, Lapčík L, De Smedt S, Demeester J, Chabreček P, (1998) Hyaluronan: Preparation, structure, properties and applications. Chemical Reviews 8: 2663–2684.
[12] Jeong SB, Yang YC, Chae YB, Kim BG, (2009) Characteristics of the treated ground calcium carbonate powder with stearic acid using the dry process coating system, Mater. Trans. 50: 409–414.
[13] Arsalan N, Palayangoda SS, Burnett DJ, Buiting JJ, Nguyen QP, (2013) Surface energy characterization of sandstone rocks, Journal of Physics and Chemistry of Solids, 74: 1069-1077.
[14] Arsalan N, Palayangoda SS, Burnett DJ, Buiting JJ, Nguyen QP, (2013) Surface energy

characterization of carbonate rocks, Colloids and Surfaces A: Physicochemical and Engineering Aspects, 436: 139-147.
[15] Karaoglan GK, Sakar D, (2011) Surface characterization of a novel D-p-D type ligand by IGC at infinite dilution Chromatographia 73: 93–98.
[16] Steele DF, Moreton RC, Staniforth JN, Young PM, Tobyn MJ, Edge S, (2008) Surface energy of microcrystalline cellulose determined by capillary intrusion and inverse gas chromatography AAPS J. 10: 494–503.
[17] Cordeiro N, Gouveia C, Moraes AGO, Amico SC, (2011) Natural fibers characterization by inverse gas chromatography. Carbohyd. Polym 84: 110–117.
[18] Perruchot C, Chehimi MM, Vaulay MJ, Benzarti K, (2006) Characterisation of the surface thermodynamic properties of cement components by inverse gas chromatography at infinite dilution. Cem. Concr. Res. 36: 305–319.
[19] Voelkel A, (2004) Inverse gas chromatography in characterization of surface. Chemometr. Intell. Lab. 72: 205–207.
[20] Yao ZT, Ge LQ, Ji XS, Tang JH, Xia MS, Xi YQ, (2015) Surface properties studies of bivalve shell waste by the IGC technique: Probing its significant potential application in the polymer industry. Journal of Alloys and Compounds 621: 389-395.
[21] Thielmann F, Butler D, Determination of the dispersive surface energy of Paracetamol by iGC at infinite dilution, SMS Application Note 202
[22] Gutmann V, The Donor–Acceptor Approach to Molecular Interaction Plenum Press, 1978.
[23] Ho R, Hinder SJ, Watts JF, Dilworth SE, Williams DR, Heng JYY, (2010) Determination of surface heterogeneity of d-mannitol by sessile drop contact angle and finite concentration inverse gas chromatography. Int. J. Pharmaceut 387: 79–86.
[24] Burnett DJ, Khoo J, Naderi M, Heng JYY, Wang GD, Thielmann F, (2012) Effect of processing route on the surface properties of amorphous indomethacin measured by inverse gas chromatography. AAPS PharmSciTech. 13: 1511–1517.
[25] Lapčík L, Otyepková E, Lapčíková B, Otyepka M, (2013) Surface energy analysis (SEA) study of hyaluronan powders. Colloid. Surface A 436: 1170–1174.
[26] Zhang WJ, Leonov AI, (2001) IGC study of filler–filler and filler–rubber interactions in silica-filled compounds. J. Appl. Polym. Sci. 81: 2517–2530.
[27] LIU T, LI YJ, CHEN J, YANG W, WU X, (2013) The research of adsorption of naphthalene-based superplasticizer on alite-sulphoaluminate cement. Concrete 20: 73-76.

IOP Conf. Series: Materials Science and Engineering **242** (2017) 012002 doi:10.1088/1757-899X/242/1/012002

Synthesis of silver/silver chloride/graphene oxide composite and its surface-enhanced Raman scattering activity and self-cleaning property

Nan Zhao[a, b], Xiao Fei[a], Xiaonong Cheng[b], Juan Yang[b, c]

[a] Jingjiang College of Jiangsu University, Zhenjiang, 212013, P.R. China

[b] School of Materials Science and Engineering, Jiangsu University, Zhenjiang, 212013, P.R. China

[c] Key Laboratory of Soft Chemistry and Functional Materials, Nanjing University of Science and Technology, Ministry of Education, Nanjing, 210094, P.R. China

Corresponding author e-mail address: yangjuan6347@mail.ujs.edu.cn

Abstract. Recently, silver nanoparticles decorated with graphene and graphene oxide (GO) sheets can be employed as surface-enhanced Raman scattering (SERS) substrates. However, their SERS activity on macromolecular compound detection is all one-time process. In order to solve this issue and decrease the cost of routine SERS detection, silver/silver chloride (Ag/AgCl) with photocatalytic activity under visible light was introduced. In this study, a novel, simple and clean approach is carried out for synthesis of the Ag/AgCl/GO composite. The Ag/AgCl colloidal solution is obtained by hydrothermal method and then mixed with GO solution to obtain the Ag/AgCl/GO composite using a facile electrostatic self-assembly method. Results showed that the Ag/AgCl/GO composite has the optimized SERS activity to Rhodamine 6G molecules with the maximum enhancement factor value of 3.8×10^7. Furthermore, the Ag/AgCl particles with high efficient and stable photocatalytic activity under visible light lead to an outstanding self-cleaning property of the Ag/AgCl/GO composite.

1. Introduction

In recent years, graphene (GE) and its derivatives are the hot and novel carbonaceous material study for a significant research endeavor since their excellent and uniquely mechanical, electrical as well as chemical properties [1-3]. The important theme in GE-based study is to survey their surface-enhanced Raman scattering (SERS) properties, particular in the noble metal–GE or graphene oxide (GO) composite (such as silver, gold) [4]. The SERS technique allows for highly sensitive detection and structural analysis with ultra-low concentrations, and has exhibited potential applications in different areas, such as chemistry, biomedicine, energy saving, and environment science [5]. Zhao et al. have prepared uniform silver nanoparticles (Ag NPs) with small sizes on GE nanosheets for active SERS substrate by an etching copper template strategy [3]. Cheng et al. reported that a SERS substrate of Ag NPs/GO can be fabricated by covalent assembly of Ag NPs and GO [5]. Our previous work [4, 6] has successfully synthesized Ag NPs–GE and gold nanoparticle/graphene oxide composite films by electrostatic self-assembly method. Due to the good SERS effect, the Raman signals of Rhodamine 6G (R6G) adsorbed on Ag NPs-GE substrate have been enhanced obviously and the mechanism was also discussed.

Content from this work may be used under the terms of the Creative Commons Attribution 3.0 licence. Any further distribution of this work must maintain attribution to the author(s) and the title of the work, journal citation and DOI.
Published under licence by IOP Publishing Ltd

However, to the best of our knowledge, almost all of the SERS detection process of the GE or GO-based nanocomposite materials on molecules are not recyclable. It is believed that the Raman detections are always on macromolecular compounds, if the macromolecules can be degraded after detection, which means apart from SERS sensitivity, the composite material also has a good photocatalytic activity, and then the substrate material used in Raman detection can be reused. To address this challenge, silver chloride (AgCl) was introduced into the composite materials in this work. The AgCl particles are well-known as a photosensitive material and a semiconducting material with direct/indirect band gap of 5.6 eV and 3.25 eV, respectively [7]. A semiconductor photocatalyst has good optical performance that can transfer the light energy into chemically available charge carriers. And then, the redox reactions will be driving on the surface of photocatalyst when the absorbed incident photons energy is higher than the band gap of the photocatalyst [8]. Lately, many reports [7, 9] have illustrated that the Ag/AgX (X=Cl, Br) compounds possess outstanding photocatalytic performance and high stability owing to the surface plasmon resonance (SPR) effect of the noble metal nanoparticles at diverse light irradiations, especially the Ag/AgCl has lured arresting attention as a rising photocatalyst work for pollutant degradation under visible light irradiation. So the optimized SERS activity and excellent self-cleaning property could be integrated into the designed Ag/AgCl/GO composite substrate.

Herein, self-assembly process is utilized to prepare Ag/AgCl/GO composite. Ag/AgCl particles are firstly obtained by hydrothermal method followed by mixing with GO solution by electrostatic self-assembly method to obtain Ag/AgCl/GO composite. The R6G molecules, as the probing molecules, were served for testing the SERS sensitivity of the obtained composite and the self-cleaning property was characterized by UV-vis and Raman experiments. This kind of composite is promising for the routine SERS detection.

2. Methods

2.1. Preparation of GO solution and Ag/AgCl

GO was prepared according to the reported Hummers method [10]. 10 mg of GO powders were firstly dispersed in 100 mL deionized (DI) water with stirring for 1 h and sonication for 2 h. Then, the prepared GO solution was preserved and used for the following experiment.

The Ag/AgCl was prepared by a hydrothermal method. As a typical synthesis recipe, the milky AgCl colloidal solution was prepared by mixing $AgNO_3$ (20 mL, 0.02 M), polyelectrolyte-poly (diallyldimethylammonium chloride) (PDDA) (20 mL, 0.2 M) and NaCl solution (20 mL, 0.02 M), followed by adding and mixing glucose (60 mL 0.84 mM) with vigorous stirring for 2 h. The molar ratio of $[AgNO_3]/[NaCl]$ was fixed at 1:1 in our experiment. Then the colloidal solution was transferred into a pressure-resistant vessel, and was heated to 180 °C in vacuum for 24 h. After the reaction, the resulting dispersion was washed and centrifuged three times. Then the final pellet was dissolved in 20 mL of deionized water and preserved in a vial.

2.2. Preparation of Ag/AgCl/GO composite substrate

The glass substrates were firstly cleaned by immersing in ethanol with sonication for 12 h. The 10 mL of freshly prepared GO solution was mixed with 10 mL of Ag/AgCl suspension slowly with stirring for 12 h. Afterwards, the pretreated glass substrates were immerged in the Ag/AgCl/GO solution for 30 min and then rinsed with DI water followed by drying in flowing nitrogen. The samples were disposed by identical process for four times to prepare Ag/AgCl/GO composite substrate at room temperature.

2.3. Preparation of sample for SERS measurements

The prepared SERS substrates were firstly immersed in 20 mL, 10^{-5} M R6G ethanol solution for 6 h. The R6G molecules can be adsorbed on the surface of substrates. After that, the unabsorbed molecules were removed by DI water rinsing for three times. Finally, the above-mentioned substrates were dried under room temperature for SERS measurements. The whole preparation process was carried out in a dark room.

2.4. Characterization

The crystal structure of the composite was conducted on a Philips 1730 powder X-ray diffractometer (XRD) with Cu Kα radiation (λ=1.5406 Å). The morphologies of the samples were observed by the JEOL-6460 scanning electron microscope (SEM) equipped with an Oxford INCA Energy SEM 350 EDS system. A thermos ESCALAB 250 spectrophotometer of the X-ray photoelectron spectroscopy (XPS) measurements with Al-Kα radiation were used to analyze the constitution of the substrates. The absorption spectra were performed on a Shimadzu UV-2550 spectrophotometer. The Raman spectra were recorded by a MXR Raman system operating at 532 nm with irradiation power of 1 mW and the laser spot size is ca. 2.1 μm^2. The acquisition time of each measurement is 10 s. The relative standard deviation (RSD) is within 5% based on the strongest band intensity of R6G on the Raman spectrum.

3. Results and discussion

Fig. 1(a, b) show the SEM images of the Ag/AgCl samples synthesized using chemical transformation of $AgNO_3$ by hydrothermal method. In Fig.1 (a), it can be seen that the products were consist of many homogeneous particles with special polyhedral morphology. Under the microscope, most of the particles show cube-like morphology with the particle size of about 100 nm. Fig. 1(b) is the high-resolution SEM picture of the Ag/AgCl products and the structure presented distinct facets and slippy exterior surfaces. Fig. 1(c) is the SEM image of the surface of Ag/AgCl/GO composite substrate. SEM precisely presented the microcosmic morphology and detailed structure of the Ag/AgCl/GO products. It is well-known that the oxygen-containing groups of GO carry negative charges. The Ag/AgCl colloidal solution was positively charged for the participation of PDDA [9]. The Ag/AgCl particles combined with the GO faultlessly by electrostatic self-assembly method as shown in Fig. 1(c). Fig. 1(d) shows the XRD pattern of the Ag/AgCl products. The distinct diffraction peaks at 2θ of 38.1° (111), 44.3° (200), 64.4° (220), 77.4° (311), corresponding to the representative cubic phase of Ag (JCPDS card No. 65-2357), could be unambiguously detected, which reveals the triumphant phase transformation from AgCl to cubic Ag. Furthermore, the accessional peaks at 27.8° (111), 32.2° (200), 46.2° (220), 54.8° (311), 57.4° (222), 67.5° (400) and 74.5° (331) in Fig.1 (d) could match with the classically cubic phase of AgCl (JCPDS card No. 35-0270). The relatively low and weak intensity of the peaks of AgCl indicates the low content or poor crystallinity of AgCl in the sample. The XRD results prove that the coexistence of metallic Ag and AgCl is in the Ag/AgCl.

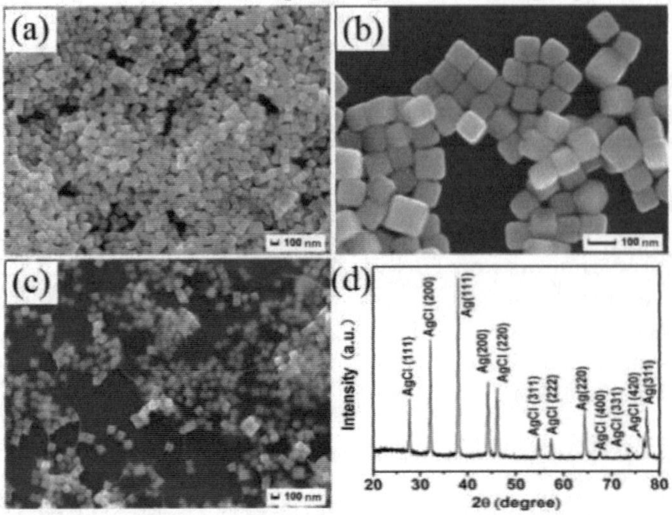

Fig 1 (a, b) SEM images of Ag/AgCl and (c) Ag/AgCl/GO composite substrate (d) XRD pattern of Ag/AgCl.

XPS is also employed to characterize the surface chemical states of the sample. In Fig. 2 (a), the characteristic peaks in the Cl 2p spectrum are at approximately 197.6 eV and 199.2 eV, corresponding well with Cl 2p 3/2 and Cl 2p 1/2, which indicates the presence of chlorine as Cl^-. Fig. 2 (b) displays the typical Ag 3d spectrum of the prepared Ag/AgCl/GO composite. The characteristic peaks for the Ag 3d at around 367.34 eV and 373.0 eV are attributed to Ag (+) 3d 5/2 and Ag (+) 3d 3/2, respectively. The peaks at 368.73 eV and 374.2 eV are assigned to the binding energies of Ag (0) 3d 5/2 and Ag (0) 3d 3/2, respectively. The XPS analysis indicates that the Ag presented as Ag^+ in AgCl and as Ag^0 in metallic Ag [11]. The C 1s XPS spectrum of Ag/AgCl/GO composite in Fig. 2(c) can be deconvoluted into four peaks arising from C=C/C–C (284.8eV) in the aromatic rings, C–O (286.3eV) of epoxy and alkoxy, C=O (287.7eV) and O–C=O (288.8eV) groups, respectively [6]. The result illustrates that the Ag/AgCl nanoparticles have been decorated on the surface of GO, which is accordant with the results of SEM and XRD.

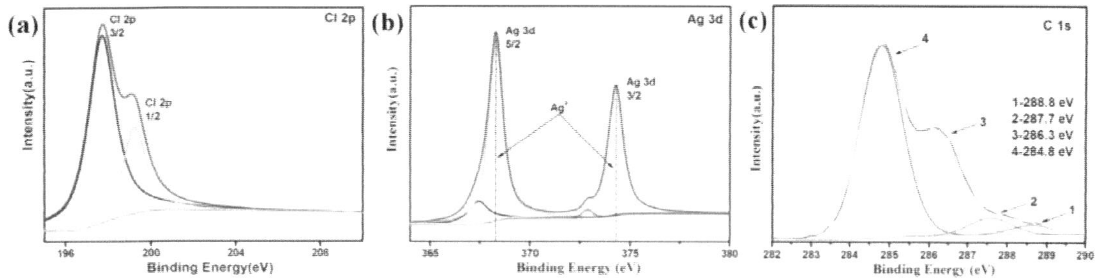

Fig 2 XPS spectra of (a) Cl 2p (b) Ag 3d and (c) C 1s of the Ag/AgCl/GO composite.

The SERS property of the obtained Ag/AgCl/GO substrate was studied in detail by using R6G as probing molecule. First of all, Ag/AgCl/GO substrate was immersed in 20 mL R6G ethanol solution (10^{-5} M) for 6 h. The tested SERS spectrum for the above substrate is shown in Fig. 3(A)-(a). The Raman spectra of 0.02 M R6G solution measured at 532 nm is also shown in Fig. 3(A)-(c). In the normal Raman spectrum of R6G, the absorbed bands are all relatively weak, but the R6G molecules adsorbed on Ag/AgCl/GO substrate exhibits the stiffening of characteristic bands for explicit excitation energy, and the five strong bands are observed at 611, 773, 1184, 1363 and 1649 cm^{-1} (Fig. 3(A)-(a)). The band at 611 cm^{-1} is ascribed to the C–C–C ring in-plane vibration mode, and the band at 773 cm^{-1} is assigned to the C–H out-of-plane bend mode. The two bands at 1129 and 1184 cm^{-1} are both attributed to the C–H in-plane bend modes. The bands at 1311 as well as 1574 cm^{-1} are assigned to the N–H in-plane bend modes, while the bands at 1363, 1509 and 1649 cm^{-1} are ascribed to the C–C stretching modes. These results illuminate that the Raman signals of R6G molecules adsorbed on Ag/AgCl/GO substrate are enhanced remarkably. Experiment was designed to explore the effect of GO on the SERS activity of the composite substrate. In the comparison of Fig. 3(A)-(a) and Fig. 3(A)-(b), it can be seen that Ag/AgCl/GO has the more powerful enhancement effect on R6G molecules than that of Ag/AgCl, which indicates that the GO sheet indeed amplified the SERS signal of the R6G probe molecules. Fig. 3(B) shows the SERS spectra of R6G with diverse concentrations on Ag/AgCl/GO substrate. As the concentration of R6G decreases, the characteristic bands of R6G such as 611, 1363 and 1649 cm^{-1} are quite sensitive to the concentration, so these vibration modes were used as sign for detecting the presence of R6G. The obtained detection limit of R6G in our experiment is as low as 1×10^{-7} M.

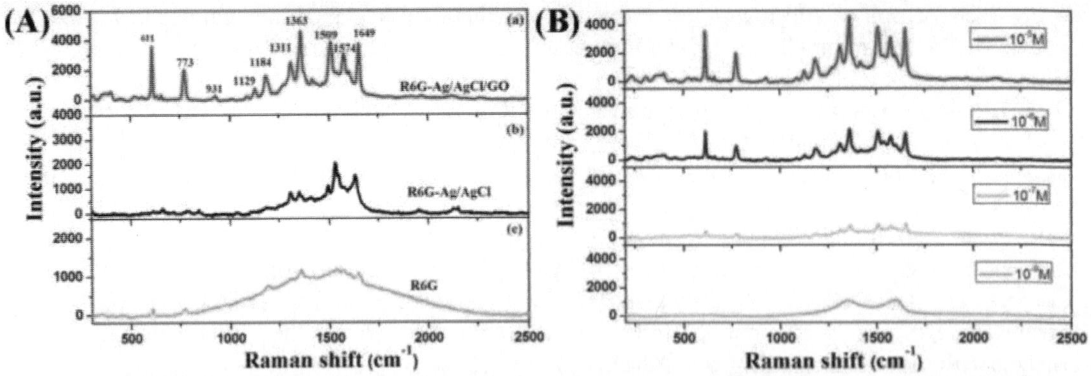

Fig 3 (A) SERS spectra of (a) R6G (10^{-5} M) adsorbed on Ag/AgCl/GO substrate (b) R6G (10^{-5} M) adsorbed on Ag/AgCl substrate and (c) 0.2 M R6G solution measured at 532 nm. (B) SERS spectra of R6G adsorbed on Ag/AgCl/GO substrate with different concentrations.

The self-cleaning property of the Ag/AgCl/GO composite film was firstly investigated by visible-light degradation. Ag/AgCl/GO film after SERS detection was illuminated under visible-light and tested by UV-vis spectrometer and the results are shown in Fig. 4(A). After SERS detection, there is an obvious absorption band around 532 nm, indicating the existence of R6G on the composite substrate [4]. The obvious decrease of absorption peak with irradiation time indicates that the R6G absorbed on the composite substrate could be degraded under visible light and R6G could be almost completely decomposed after irradiation for 12 h. For comparison, the Ag/AgCl/GO composite substrate after absorption of R6G was put in dark and it was found that the intensity of UV-vis spectra maintained even after 30 h. Ag/GO composite without AgCl was also prepared for comparison, which could not lead to the photocatalytic decomposition of R6G shown in the inset of Fig. 4(A). The results suggest that the degradation of R6G was caused by the photocatalytic reactions of AgCl on Ag/AgCl/GO sample under visible-light.

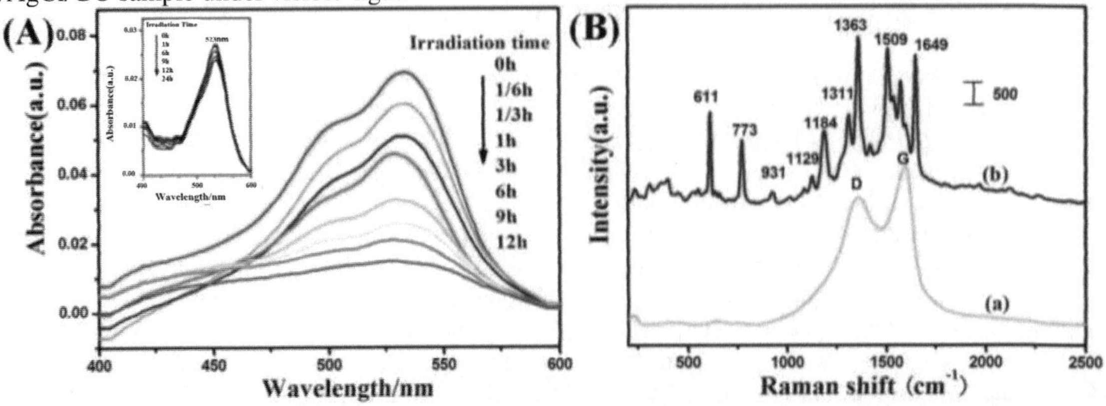

Fig 4 (A) Changes in UV–vis absorption spectra of R6G adsorbed on Ag/AgCl/GO substrate, the inset is changes in UV–vis absorption spectra of R6G adsorbed on Ag/GO substrate as a function of irradiation time. (B) The SERS spectra of samples (a) the Ag/AgCl/GO substrate after SERS detection and then irradiated under visible light for 12 h (b) the photo-degradated Ag/AgCl/GO substrate adsorbed R6G again.

Raman experiments were also carried out to detect the self-cleaning property. As shown in Fig. 3(A)-(a), the Raman signals of R6G adsorbed on Ag/AgCl/GO composite substrate are enhanced prominently. After the Raman detection, the film was irradiated under visible-light for 12 h and characterized by Raman again. It is clearly seen in Fig. 4(B)-(a) that the signals of R6G disappeared, only characteristic peaks of GO can be found, the D band (1355 cm^{-1}) and the G band (1600 cm^{-1}), which represent the amount of sp3 carbons in the surroundings and the E2g phonon of sp2 C atoms, respectively [6]. Results indicated that the R6G can really be decomposed under the photocatalytic degradation. After the photocatalytic degradation, the Ag/AgCl/GO substrate was immersed into 20 mL 10^{-5} M R6G ethanol solution again and the Raman spectrum was recorded in Fig. 4(B)-(b), which states clearly the SERS effect. This is the first cycle and we repeated the procedure for five cycles and the results are shown in Fig. 5, the enhancement factor (*EF*) values are also given. The *EF* value can be estimated as follows: $EF=(I_{SERS}/I_{bulk})/(N_{ads}/N_{bulk})$ [4]. Where ISERS is the intensity of a vibrational mode in the SERS spectrum and Ibulk is the intensity of the same mode in the Raman (non-SERS) spectrum. The I_{SERS}/I_{bulk} can be obtained from the SERS spectra directly [6]. The N_{bulk} and N_{ads} are estimated by approximate calculation using monolayer-absorbed mode [4].

In Fig. 5(a), it can be seen that the SERS intensity decrease gradually with the growth of cycling numbers and the intensity decrease obviously after 5 times. In Fig. 5(b), the tendency can be seen clearly by the *EF* values. The *EF* values from 0 to 5 times are 3.8×10^7, 3.5×10^7, 2.9×10^7, 2.1×10^7, 7.4×10^6 and 7.2×10^6, 6.7×10^6, 3.6×10^6, 3.4×10^6, 1.5×10^6 for C-C-C ring in-plane vibration (ca. 611 cm^{-1}) and C–C stretching (ca.1363 cm^{-1}) respectively. The decay of the SERS property during the recycle might be caused by the oxidation of Ag^0 on the surface. The mechanism study and property improvement are carrying out in our Lab.

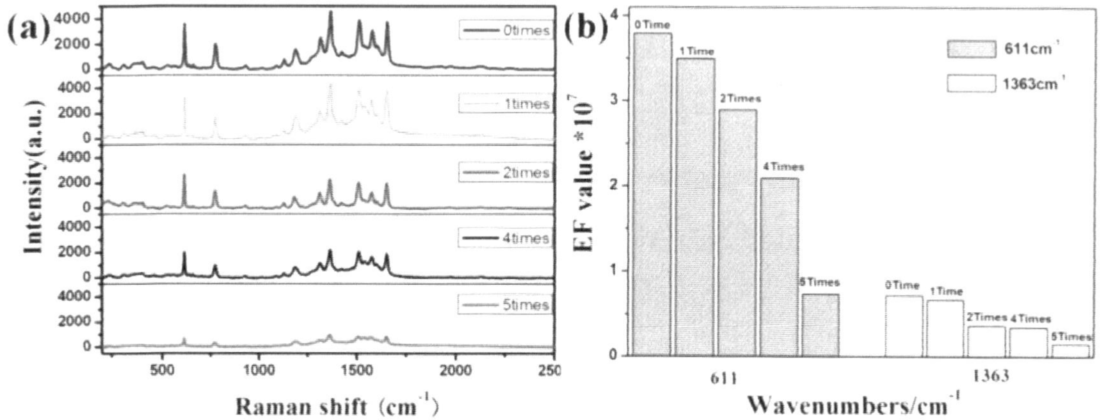

Fig 5 (a) The SERS spectra of R6G adsorbed on Ag/AgCl/GO substrate for different self-cleaning cycles; (b) Raman EF values at 611 cm^{-1} and 1363 cm^{-1} with different self-cleaning cycles.

4. Conclusion

The Ag/AgCl/GO composite with superior SERS activity and self-cleaning property has been prepared through the hydrothermal method followed by electrostatic self-assembly process. All the preparation processes are fast, green and simple. The prepared materials exhibit enhanced SERS activity for R6G and high visible-light photocatalytic performance for the degradation of R6G. The composite is recyclable, readily prepared and disposable, which possesses a huge convenience for the conventional SERS detection.

Acknowledgements

This work was supported by National Science Foundation of China (51572114), Key University Science Research Project of Jiangsu Province (16KJB430009), Fundamental Research Funds for the Central University (No.30920140122003), Undergraduate Students of Jiangsu Province (201513986003Y) and Senior Talent Foundation of Jiangsu University (15JDG078).

References

[1] K.S. Novoselov, A.K. Geim, S.V. Morozov, D. Jiang , M.I. Katsnelson, I.V. Grigorieva, S.V. Dubonos, A.A. Firsov, Two-dimensional gas of massless dirac fermions in graphene, Nature. 438 (2005) 197-200.

[2] Y. Li, J. Yang, S.H. Zheng, W.W. Zeng, N. Zhao, M. Shen, One-pot synthesis of 3D TiO2-reduced graphene oxide aerogels with superior adsorption capacity and enhanced visible-light photocatalytic performance, Ceram. Int. 42 (2016) 19091-19096.
[3] H. Zhao, H.G. Fu, T.S. Zhao, L. Wang, T.X. Tan, Fabrication of small-sized silver NPs/graphene sheets for high-quality surface-enhanced Raman scattering, J. Colloid. Interface. Sci. 375 (2012) 30-34.
[4] Y. Li, J. Yang, Y.Z. Zhou, N. Zhao, W.W. Zeng, W.W. Wang, Fabrication of gold nanoparticles/graphene oxide films with surface-enhanced Raman scattering activity by a simple electrostatic self-assembly method, Colloid. Surface A 512 (2017) 93-100.
[5] Z.J. Qian, Y.C. Cheng, X.F. Zhou, J.H. Wu, G.J. Xu, Fabrication of graphene oxide/Ag hybrids and their surface-enhanced Raman scattering characteristics, J. Colloid. Interface. Sci. 397 (2013) 103-107.
[6] Y.Z. Zhou, X.N. Cheng, D. Du, J. Yang, N. Zhao, S.B. Ma, T. Zhong, Y.H. Lin, Graphene–silver nanohybrids for ultrasensitive surface enhanced Raman spectroscopy:size dependence of silver nanoparticles, J. Mater. Chem. C. 2 (2014) 6850-6858.
[7] P. Wang, B. Huang, X. Qin, X. Zhang, Y. Dai, J. Wei, M.H. Whangbo, Ag/AgCl: a highly efficient and stable photocatalyst active under visible light, Angew. Chem. Int. Ed. 47 (2008) 7931-7933.
[8] S.M. Liu, J.L. Zhu, Q. Yang, P.P. Xu, J.H. Ge, X.T. Guo, Facile fabrication and enhanced photocatalytic performance of Ag/AgCl-activated sepiolite heterostructure photocatalyst, J. Mater. Sci. 51 (2016) 2565-2572.
[9] Y.Z. Zhou, R. Chen, T.T. He, K. Xu, D. Du, N. Zhao, X.N. Cheng, J. Yang, H.F. Shi, Y.H. Lin, Biomedical potential of ultrafine Ag/AgCl nanoparticles coated on graphene with special reference to antimicrobial performances and burn wound healing, ACS. Appl. Mater. Interfaces. 8 (2016) 15067-15075.
[10] W.S. Hummers, R.E. Offeman, Preparation of graphene oxide, J. Am. Chem. Soc. 80 (1958) 1339-1339.
[11] Y.Z. Zhou, J. Yang, T.T. He, H.F. Shi, X.N. Cheng, Y.X. Lu, Highly stable and dispersive silver nanoparticle–graphene composites by a simple and low-energy-consuming approach and their antimicrobial activity, Small 9 (2013) 3445-3454.

Research and application of high performance GPES rigid foam composite plastic insulation boards

Hongming sun[1,a], Hongsheng xu[1,b], Feifei Han[1,c]

[1]Shandong Provincial Academy of Building Research, Jinan, Shandong, Post: 250031

E-mail: [a] jzjns106@sina.com, [b] jick_ki@163.com, [c] feifeihan1984@163.com

Abstract:A new type of heat insulation board named GPES was prepared by several polymers and modified nano-graphite particles, injecting high-pressure supercritical CO2. Compared with the traditional thermal insulation material, GPES insulation board has higher roundness bubble and thinner bubble wall. Repeatability and reproducibility tests show that melting knot, dimensional stability, strength and other physical properties are significantly better than traditional organic heat insulation materials. Especially the lower and more stable thermal conductivity of GPES can significantly reduce thermal insulation layer thickness. Obviously GPES is the best choice of insulation materials with the implement of 75% and higher energy efficiency standard.

1. Introduction

With the implementation of 75% building energy efficiency in china, the thickness of insulation layer is increasing, therefore, the issue of structural safety in exterior insulation is more prominent. Since the enforcement of GB50016-2014 code for fire protection design of buildings（The Chinese national standard）, the contradiction between exterior wall insulation and fire prevention has been more sharp than before. Firstly, building energy efficiency design standards has been raised from 65% to 75%, the thickness of insulation layer has been almost more than double. Especially for the passive building in our province, the thickness of EPS could be 25-30cm. The increased thickness of insulation layer, raises the proportion of combustible per unit area and total in external insulation system by two times, increases the risk of flame spreading during the fire.

At the same time, the insulation system's ability of resisting gravity (the outer protective layer) to prevent shearing by itself has fallen sharply, the length of anchor bolt has been almost doubled as well, so it is inevitable that the structure safety of external thermal insulation system has high risks because of the low resistant capacity for stretching, torsion and shear.

In view of the above problems, project team developed GPES rigid foam plastic composite insulation board (hereinafter referred to as the GPES insulation board), which uses high barrier property materials as a primary structure, prepared by several polymers, modified nano-graphite and so on, via 30MPa high pressure and high temperature, injected high pressure supercritical carbon dioxide fluid, finally released from low temperature and high pressure. Its density is between 32 and 38 kg/m^3 .After a large number of repeatability and reproducibility tests, various property of the GPES are excellent. GPES can meet the requirement of 75% energy saving standards and higher passive building construction technology standard by combining with the different forms of insulation structure. It costs lower than other thermal insulation material such as EPS, XPS, PU etc..

Content from this work may be used under the terms of the Creative Commons Attribution 3.0 licence. Any further distribution of this work must maintain attribution to the author(s) and the title of the work, journal citation and DOI.
Published under licence by IOP Publishing Ltd

2. GPES insulation boards

GPES rigid foam plastic insulation board was prepared by several polymers, modified nano-graphite and additives, via 30MPa high pressure and high temperature, injected high pressure supercritical CO_2 fluid, finally released from low temperature and high pressure. It called GPES insulation board for short. Insulation board mixing modified material and various of high-molecular polymers, it adopt the foaming of high pressure supercritical carbon dioxide fluid rather than the blowing agent such as tetrachloroethane, pentane, or Freon etc. In this way, the bubble become more balance, shorter diameter(less than 50um), higher roundness bubble, most of

Closed bubbles has lower heat conductive coefficient. Thus the thermal insulation material with better performance and lower heat conductive coefficient in friendly environment has been obtained.

2.1. Classification of GPES insulation board

GPES insulation board is divided into GPES-I and GPES-IIaccording to the thermal conductivity, the range of its thermal conductivity are shown in table 1

Table 1 Classification of GPES insulation board

Category	Thermal Conductivity W/(m·K)
GPES-I	≤0.020
GPES-II	≤0.025 ~ >0.020

2.2. Research and development of GPES-I insulation board

After continuous efforts of three years, the second generation of GPES insulation board (GEPS-I) with better performance was developed by the project team. Its theory is based on the first generation of GPES insulation boards (GPES-II), and the technical formula was continued improving through 49 batch insulation board production experiment test. The insulation board has excellent physical and combustion performance as the first generation of GPES, but lower coefficient of thermal conductivity and more excellent long-term stability that its 28d heat conduction coefficient is less than 0.020 W/(m·K).

Table 2 The results of orthogonal experiment of GPES- I insulation board

No.	Density kg/m^3	Compressive strength MPa	Water absorption %	Dimensi-onal stability %	Tensile Strength MPa	Bending load N	Bending Deforma-tion mm	Oxygen -index %	Classifi ca-tion for burning behavior	28d Thermal Conductivity W/(m·K)
35	34.4	0.34	0.4	0.2	0.35	41	40	30.5	B1	0.0198
43	34.5	0.36	0.3	0.3	0.35	40	52	30.2	B1	0.0190
2	34.6	0.35	0.4	0.4	0.38	42	41	31.7	B1	0.0175
29	34.5	0.34	0.2	0.5	0.36	43	39	31.6	B1	0.0183
48	34.8	0.37	0.5	0.6	0.39	45	54	31.0	B1	0.0176
38	33.9	0.34	0.3	0.4	0.33	37	50	30.2	B1	0.0191
49	34.0	0.34	0.4	0.7	0.34	39	46	30.8	B1	0.0187
32	34.6	0.37	0.4	0.3	0.38	43	42	30.5	B1	0.0189

2.3. Morphology of GPES- I insulation plate under electron microscope

Figure 1 is a SEM photo of GPES-I insulation board bubble hole section, it shows the microcellular structural characteristics of GPES-I insulation board. Figure 2 is a SEM photo of the GPES-I insulation board bubble wall. It can be seen that the ultra-thin bubble wall is less than 0.5um. Microporous structure, ultra-thin bubble wall, high barrier properties to fundamentally ensure that the GPES-Iexcellent thermal barrier .The composite plastic with high performance was processed by the special mixing, so that its thermal performance and physical properties are significantly better than the traditional organic insulation materials.

FIG.1 SEM of GPES-Iinsulation board cell

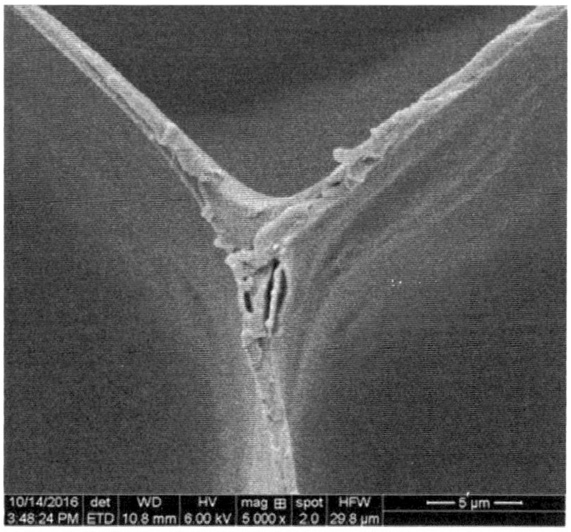

FIG.2 SEM of GPES-Iinsulation board

3. Performance analysis of GPES thermal insulation boards

3.1. Coefficient of thermal conductivity

3.1.1. GPES-I coefficient of thermal conductivity long-term stability. Coefficient of thermal condu ctivity is the main indicator of thermal properties of materials and is also a key index to ensure its hea t preservation performance. GPES adopt supercritical CO_2 foaming by a specific process, which makes high obturator rate of block foam obturator structure with bubbles of uniform size and thinner wall. GPES - II coefficient of thermal conductivity is stable within 0.025 W/(m·K) for long-term. Based on this, after many times of technological formula adjustments, GPES-Icoeff icient of thermal conductivity is stable within 0.020 W/(m·K) for long-term .GPES-Iwas maintenan ced according to GB/T 30595-2014, and aged 14d, 28d, 45d, 60d, 75d and 90d respectively. we test coefficient of thermal conductivity of some lots of GPES -I insulation board that aged 0-90d. As can be seen from table 3 the test result, 28d thermal conductivity coefficient of 4 8# of the samples can reach 0.0176 W/(m·K), 90d coefficient of thermal conductivity is 0.017 8 W/(m·K).

Table 3 The results of orthogonal experiment of GPES- I insulation boards (Aged 0-90d) W/(m·K)

No.	Thermal Conductivity W/(m·K)					
	14d	28d	45d	60d	75d	90d
35	0.0192	0.0198	0.0195	0.0196	0.0199	0.0198
2	0.0180	0.0175	0.0183	0.0190	0.0194	0.0195
29	0.0183	0.0183	0.0190	0.0185	0.0184	0.0185
48	0.0158	0.0176	0.0173	0.0179	0.0180	0.0178
49	0.0178	0.0187	0.0185	0.0190	0.0193	0.0196
32	0.0187	0.0189	0.0192	0.0193	0.0197	0.0196

3.1.2. GPES-II coefficient of thermal conductivity long-term stability. GPES-II was maintained accordingly to GB/T 30595-2014, and aged 90 d.The coefficient of thermal conductivity was tested according to GB/T 10294-2008. Table 4 and figure 3 are respectively coefficient of thermal conductivity and change curve of GPES-I and GPES-II that aged 90 d.

Table 4 GPES-Iand GPES-II's coefficient of thermal conductivity（Aged 0-90d）W/(m·K)

	Thermal Conductivity							
	3d	7d	14d	28d	45d	60d	75d	90d
GPES-I	0.0190	0.0182	0.0174	0.0192	0.0190	0.0192	0.0195	0.0194
GPES-II	—	—	0.0222	0.0230	0.0235	0.0239	0.0242	0.0243

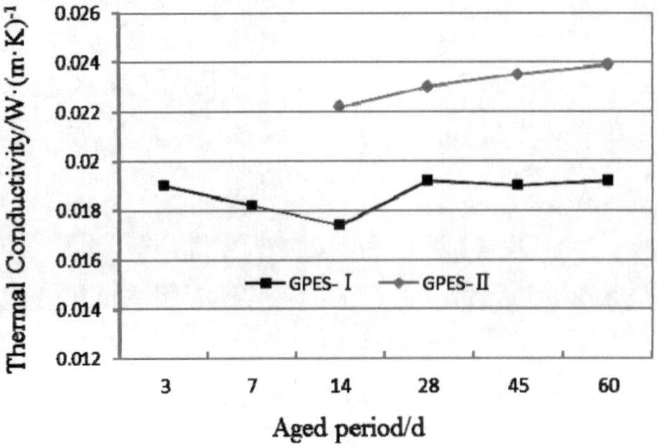

FIG.3 Curves of thermal conductivity of GPES- I and GPES- II （Aged0~90d）

From the experimental results, coefficient of thermal conductivity within 90 d basically remain unc hanged of GPES-I, while that of GPES-IIincreases slightly. Tendency curves of thermal conductivity of GPES-Iand GPES-II illustrates that GPES-I has more excellent long-term stability of thermal performance.

3.2. The durability of GPES

In order to investigate the durability of GPES insulation board's thermal performance，contrast experiment was carried on 5 groups of samples taken from GPES-I, GPES-II and PU plate, which has already aged 90d. The method is shown below:

Test method: (1) raise its temperature up to (70 + 5)°C within 1 h, relative humidity (90 + 5)%, and keep it for 3 h; (2) cool down its temperature to (-20±5)°C within 1 h, and keep it for 3 h; Temperature (23 + 2)°C, relative humidity (50 + 5) %, keep it for 16h,and then 15, 30, 45 and 60 times of durability test was adopted respectively, each cycle is 24h. The results of synchronized test can be seen in below table and figure.

Table 5 Endurance test curves of GPES- I and GPES- II's thermal performance W/(m·K)

Number of cycles		0	15	30	45	60	Loss rate
Thermal conductivit y coefficient	GPES-I	0.0194	0.0198	0.0207	0.0211	0.0210	8.25%
	GPES-II	0.0244	0.0256	0.0263	0.0267	0.0265	8.60%
	PU	0.0237	0.0261	0.0268	0.0272	0.0272	14.7%

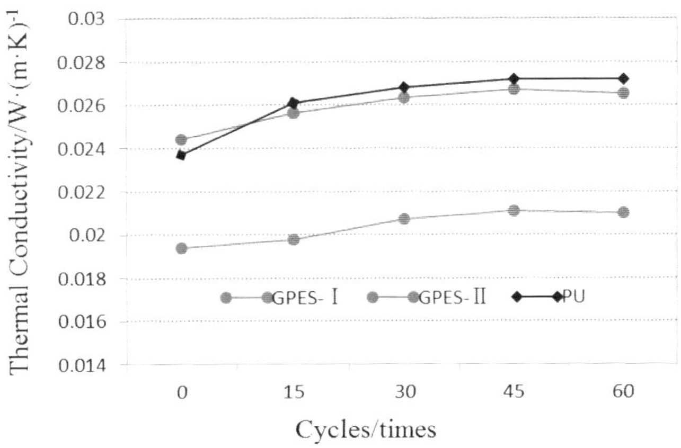

FIG.4 Endurance test curves of GPES- I and GPES- II 's thermal performance

After 60 times of humid-hot-cold cycles, GPES-I and GPES-II's thermal conductivity are 0.0210 and 0.0265W/(m·K), PU is 0.0272W/(m·K), increased by 8.25%, 8.6% and 14.7% than their initial thermal conductivity coefficient. As you can see, after a long-term of humid hot and cold cycle, the structure of GPES tends to be stable. It means that being used over the years, GPE-Iand the loss thermal conductivity of GPES-II could reach 8.25% and 8.6%, but far less than that of the PU 14.7%. The conclusion is GPES has more excellent durability of thermal performance.

3.3. Mechanical property

3.3.1. Tensile strength. As the main mechanical performance indexes, tensile strength of insulation board is to ensure that building external thermal insulation engineering maintain its normal function. Every day we sliced same size of sample from the GPES insulation board production line after steady production. We sorted the sample and aged for 28 days, then tested its tensile strength perpendicular to the direction of the plate. Test results are shown in table 6.

Table 6 Tensile strength of GPES perpendicular to the plate surface MPa

Items		Sample number			Average	GB/T 30595-2014
		1	2	3		
Tensile strength perpendicular to the direction of the plate	GPES-I	0.33	0.32	0.34	0.33	≥0.20
	GPES-II	0.34	0.33	0.34	0.34	

It can be seen from the results in the table, the GPES-I and GPES-II insulation board's tensile strength perpendicular to the direction of the plate were 0.33 and 0.34 MPa, which is very excellent. Considering the smooth surface and high hydrophobicity of GPES may cause bonding interface with mortar is not strong, a special interface agent applicable to GPES insulation board has been developed. The interface agent has excellent adhesion and water resistance that the bonding strength of GPES

IOP Conf. Series: Materials Science and Engineering **242** (2017) 012003 doi:10.1088/1757-899X/242/1/012003

insulation board and mortar remain above 0.40 MPa after the application of the agent.

Table 7 Bonding tensile strength of GPES and mortars treated with special interface agent MPa

Items	28d original intensity		Water resistant to 48h, dry 2h		Water resistant to 48h,dry 7d	
	GPES-I	GPES-II	GPES-I	GPES-II	GPES-I	GPES-II
Bonding mortar	0.42	0.44	0.40	0.41	0.43	0.45
GB/T 30595-2014 standard	≥0.20		≥0.10		≥0.20	
Rendering coat mortar	0.45	0.46	0.42	0.42	0.45	0.45
GB/T 30595-2014 standard	≥0.20		≥0.10		≥0.20	

XPS's tensile strength perpendicular to its surface and bond strength with mortar are below 0.25 MPa, while GPES's tensile strength perpendicular to its surface is higher than 0.30 MPa. After treatment the interface by special interface agent, GPES's bonding strength with mortar could be higher than 0.40 MPa. Main mechanical performance of GPES insulation board is more than 1.5 times superior to the XPS.

3.3.2. Other mechanical properties. Compression strength, bending load and bending deformation are also important indicators,. The GPES-I and GPES-II insulation board's compression strength and bending load, bending deformation has been tested respectively, the results are shown in table 8.

Table 8 Other mechanical properties of GPES- I and GPES- II insulation board

Items	Category		GB/T 30595-2014
	GPES-I	GPES-II	
Bending load of fracture /N	43	41	/
Bending deformation /mm	50	48	≥20
Compression strength /MPa	0.33	0.32	≥0.20

The results indicate that the bending deformation of GPES is excellent and more than 2 times superior to standard of GB/T 30595-2014. Its compression strength is more than 0.30 MPa, which is superior to XPS by more than 50%

The internal structure of insulation board is the important factors influencing its compression strength and bending deformation, the higher its value, the stronger of durability and impact resistance of insulation board. The mixed grafting technology used in the process of forming the GPES insulation board, makes high degree of polymer crosslinking, compact structure, and high obturator rate, which contributed to the excellent bending properties and compression strength of GPES.

3.4. Water absorption

Water absorption is an important indicator for the performance of insulation materials, high water absorption will lead to gradually reduce the thermal insulation properties. This result also shows that in the following four kinds of insulation materials, the smallest water absorption rate insulation plate of GPES-Iand GPES-IIis 0.3%, compared with the board water absorption of XPS decreased by 60% and more. According to the heat transfer theory of porous materials, the lower the water absorption of the material, the more stable the insulation effect. The very low water absorption for GPES plate is to ensure that its thermal insulation performance to the perfection.

Table 9 Water absorption rate of several thermal insulation materials %

Performance parameters	GPES-I	GPES-II	XPS	EPS	PU
Water absorption per volume	0.3	0.3	0.8	1.5	2.0
GB/T 30595-2014			≤1.5%		

3.5. Dimensional stability

Insulation board with poor Dimension stability performance is easy to bending deformation, causing the wall to crack or fall off, thereby reducing the external thermal insulation system security. In regard to GPES-I and GPES-II dimension stability, prescriptive indicator of XPS board's dimension stability used for external thermal insulation system is: 1.2% or less which according to GB/T 30595-2014 standard. The dimensional change rate of the GPES-Iand GPES-II along length, width and thickness direction are tested after 48h at the temperature 70°C. As the test results, the dimension stability of GPES-Iand GPES-II are better than the prescriptive indicator.

Table10 The dimensional change rate of GPES- I and GPES- II

Items	Model		GB/T 30595-2014
	GPES-I	GPES-II	
Length	0.6	0.1	
Width	0.6	0.4	≤1.2%
Thickness	0.4	0.2	

3.6. Combustion Performance

The combustion performance of the insulation board concerns the safety of the entire external wall insulation project, therefore it attracts much attention. GPES insulation board adopted CO_2 fluid foaming, which not only reduce VOC emissions, but also play a role in flame retardant. Through the test of combustion experiments, GPES-Iand GPES-II insulation boards are up to B_1 level.

4. GPES insulation board exterior insulation system

The system was mainly consist of high-performance GPES insulation board, fixed material (adhesive mortar, anchoring parts), fireproof screed-coat (composite sizing agent coat), plaster coat and finish coat. It contains two systems, namely high-performance insulation board paste- compound slurry (the basic structure can be seen in Table 11) and the system of high-performance insulation board double-deck break joint paste -compound slurry (the basic structure can be seen in Table 12).

The engineering design of high- performance thermal insulation board can select a appropriate external insulation system according to the building energy efficiency requirements. High-performance insulation board paste - composite slurry system has been applied for public buildings in 65% energy saving and residential buildings in 75% energy saving , high-performance insulation board double-deck break joint paste -compound slurry system has been applied for public buildings in 65% energy saving, residential buildings in 75% energy saving and the passive green construction with ultra-low energy consumption.

IOP Conf. Series: Materials Science and Engineering **242** (2017) 012003 doi:10.1088/1757-899X/242/1/012003

Table11 The basic structure of high-performance insulation board paste- compound slurry system

Base wall	System basic structure						Structure diagram
	Bonding layer ①	Thermal insulation layer ②	Fire screed-coat ③	Protective layer: Rendering coat④	Protective layer: Finish coat⑤	Plate seam⑥	
Concrete wall and all kinds of masonry walls	Bond-ing morta-r	Special interface agent + GPES + Special interface agent	20mm composite layer of insulation coating	Rendering coat mortar + Alkali resistant fiberglass mesh	Coating materials	10mm Joints with embedded full of composite insulation coating	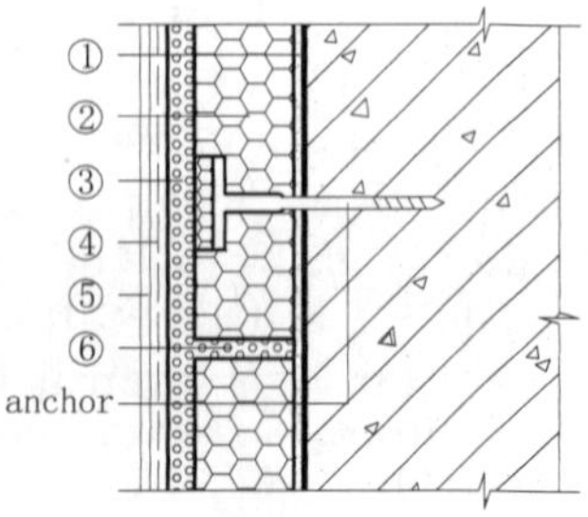

Table12 The basic structure of high-performance insulation board double-deck break joint paste-compound slurry system

Base wall	System basic structure									Structure diagram
	Bonding layer ①	Thermal insulation layer②	Bonding layer ③	Thermal insulation layer ④	Fire screed-coat ⑤	Protective layer: Rendering coat⑥	Protective layer: Finish coat ⑦	The inner seam ⑧	The outer seams⑨	
Concrete wall and all kinds of masonry walls	Bonding mortar	Special interface agent + GPES insulation board + Special interface agent	Bondi-ng mortar	Special interface agent + GPES insulation board + Special interface agent	20mm composite layer of insulation coating	Rendering coat mortar + Alkali resistant fiberglass mesh	Coating materials	5mm Joints with embedded full of bonding mortar	10mm Joints with embedded full of composite insulation coating	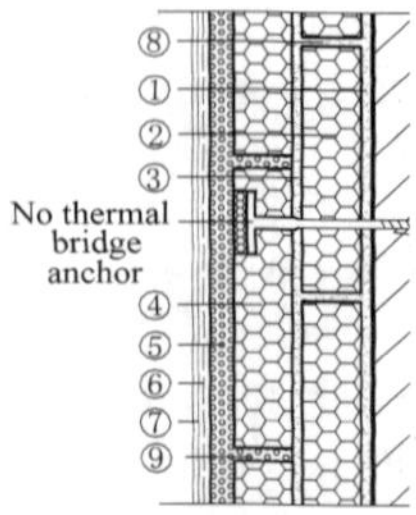

5. Conclusion

Comparing with current insulation board, this new type of insulation board has many advantages such as lower thermal conductivity, relatively larger apparent density and high compressive strength and tensile strength, lower water absorption, better moisture resistance and dimensional stability etc.. Under the same energy efficiency standards, its thickness can be 30% thinner than EPS and XPS board, it has more advantages than polyurethane insulation board—high cost-effective, the price is 1/3 of polyurethane insulation board, it is better than other insulation materials in current marketing.

The development and popularization of the GPES insulation board can impel a high-level step for

IOP Conf. Series: Materials Science and Engineering **242** (2017) 012003 doi:10.1088/1757-899X/242/1/012003

energy efficiency standards of building energy-saving work, and provide reliable product and technical support. It has the huge practical significance for popularization and implementation of high-level energy efficiency standards such as 75% power-saving and passive ultra-low energy consumption, and at the same time, it can provide huge economic benefits for the building energy-saving industry and it has a broad application prospects.

References

[1] GB/T 30595-2014,*Exernal thermal insulation composite systems based on extruded polystyrene*,S.
[2] GB/T 10801.2-2002,*Rigid extruded polystyrene foam board for thermal insulation（XPS）*,S.
[3] GB/T 10801.1-2002 ,*Moulded extruded polystyrene foam board for thermal insulation（EPS）*,S.
[4] GB 8624-2012.*Classification for burning behavior of building materials and products*.S.
[5].GB/T.10294-2008.*Thermal insulation-Determination of steady-state thermal resistance and related properties-Guarded hot plate apparatus*.S.

The research of the grow speed of $AgGa_{1-x}In_xSe_2$ monocrystalline synthesized by Vertical Bridgman method

Shukai Zhang

College of materials and Chemistry & Chenical Engineering,Chengdu University of Technology,Chengdu,Sichuan,China

Zhangshukai0602@163.com

Abstract. $AgGa_{1-x}In_xSe_2$ is an infrared non-linear optical crystal which has been widely used in laser radar, infrared laser distant control and atmosphere environment monitoring area. The monocrystalline of $AgGa_{1-x}In_xSe_2$ usually synthesize by vertical Bridgman method. However, the melting and crystallization point of crystal can be changed by different x value of $AgGa_{1-x}In_xSe_2$. To synthesize high quality monocrystalline of $AgGa_{1-x}In_xSe_2$ with one sure x value, we should measure the melting point and crystallization point, calculate the degree of super-cooling, to design suitable temperature field distribution for this crystal.

1. Introduction

$AgGa_{1-x}In_xSe_2$, an infrared non-linear optical crystal with good performance, of which the crystallization and melting point are varied with the change of x value [1-3]. So far, several methods are normally used to synthesize $AgGa_{1-x}In_xSe_2$ monocrystalline, including Czochralski (CZ), Heat Exchange Method (HEM), Temperature Gradient Technique (TGT), kyropoulos (KY), Top Seeded Solution Growth (TSSG), Horizontal Bridgman (HB) and Vertical Bridgman (VB) method [4-6]. Due to the previous reason, we need to design suitable temperature field distribution for $AgGa_{1-x}In_xSe_2$ with one sure x value to match the following two requirements when using vertical Bridgman method to synthesize monocrystalline:

(1) $AgGa_{1-x}In_xSe_2$ have a big degree of super-cooling, thus a proper high temperature gradient is required to control the seed crystal into a short length to improve the quality of monocrystalline;

(2) Too high temperature gradient leads to the increasing of thermal stress, this may result in crystal defect or even cracking. To obtain good quality monocrystalline, we should avoid this situation [1,7].

In our work, we synthesized the $AgGa_{0.7}In_{0.3}Se_2$ ($AgGa_{1-x}In_xSe_2$, x=0.3) polycrystalline powder, which is the raw material of $AgGa_{0.7}In_{0.3}Se_2$ monocrystalline, then used Differential Thermal Analysis (DTA) to measure the melting point and crystallization point in different cooling speed, calculated degree of super-cooling of each speed, meanwhile we also measured the temperature distribution curve of growth furnace which was used to synthesizing $AgGa_{0.8}In_{0.2}Se_2$ ($AgGa_{1-x}In_xSe_2$, x=0.2). Finally, we combine DTA data with the temperature distribution curve to discuss how to change the temperature distribution to match the requirement of synthesizing $AgGa_{0.7}In_{0.3}Se_2$.

2. Experiment

2.1. Synthesize $AgGa_{0.7}In_{0.3}Se_2$ polycrystalline

Content from this work may be used under the terms of the Creative Commons Attribution 3.0 licence. Any further distribution of this work must maintain attribution to the author(s) and the title of the work, journal citation and DOI.
Published under licence by IOP Publishing Ltd

IOP Conf. Series: Materials Science and Engineering **242** (2017) 012004 doi:10.1088/1757-899X/242/1/012004

We use high-purity In, Ga, Ag, Se elementary substance as raw material, mixture them as stoichiometric ratio with 0.3% surplus Se as a result of Se volatilize. Put Se, Ga, In, Ag into clean quartz ampoule in sequence and dry them in 100°C then sealing quartz ampoule after vacuuming to 1×10^{-3}Pa. Use vertical synthesis furnace to heat them to 870°C with the speed of 100°C/h, then stay in 870°C for 6 hours, continue to heat with the speed of 50°C/h till 1070°C.finally cool quartz ampoule down to 20°C.Take the sample out of quartz ampoule and pulverize it.

2.2. DTA of $AgGa_{0.7}In_{0.3}Se_2$ polycrystalline
1. We put a little sample powder into crucible, gather it at the bottom of crucible.

2. Prepare another crucible without anything, then put both of two crucibles into SDT-Q600 differential thermal analyzer. The powder is protected by N_2 gas.

3. Select the heating speed of SDT-Q600 differential thermal analyzer as 10°C/min, the top temperature as 850°C， the cooling speed as 5°C/min, then waiting until SDT-Q600 operation finish, we obtain the heat flow-temperature data under 5°C/min cooling speed.

4. Then we select another two cooling-speed of 10°C/min and 15°C/min, repeat previous process.

2.3. The measure of growth furnace thermal field
The measure object is $AgGa_{0.8}In_{0.2}Se_2$ monocrystalline growth furnace made by lab, with Japannese shimada company's FP93 type precision digital temperature director. We use Pt-Pt&Rh alloy heat thermocouple to measure the thermal field.

1. Select three heat resistance wire valus of fusant temperature field as 950°C, 905°C, 800°C, then preheat 12 hours.

2.Put thermocouple into furnace, measure the temperature of furnace from 10cm under the opening to 45cm. at the area of 10-32cm, descend 2cm then stay 1min; at the area of 32-45cm, descend 1cm then stay 1min. Record the data after the thermodetector have a stable display each time.

3. Select three heat resistance wire valus of fusion temperature field as 930°C, 910°C, 660°C, then preheat 12 hours.

4. Measure the temperature of furnace from 39cm under the opening to 59cm. at the area of 39-59cm, descend 1cm then stay 1min. Record the data after the thermodetector have a stable display each time.

5. Select three heat resistance wire valus of fusion temperature field as750°C, 715°C, 690°C, then preheat 12 hours.

6. Measure the temperature of furnace from 59cm to 39cm by rise thermocouple at the area of 59-39cm, descend 1cm then stay 1min. Record the data after the thermodetector have a stable display each time.

3. Result and Discussion

3.1. Representation of $AgGa_{0.7}In_{0.3}Se_2$ polycrystalline
Use XRD to analysis the powder sample, we obtain XRD pattern as Figure 1 and it agrees with standard PDF card (No35-1096). So we can guarantee the sample is $AgGa_{0.7}In_{0.3}Se_2$.

IOP Conf. Series: Materials Science and Engineering **242** (2017) 012004 doi:10.1088/1757-899X/242/1/012004

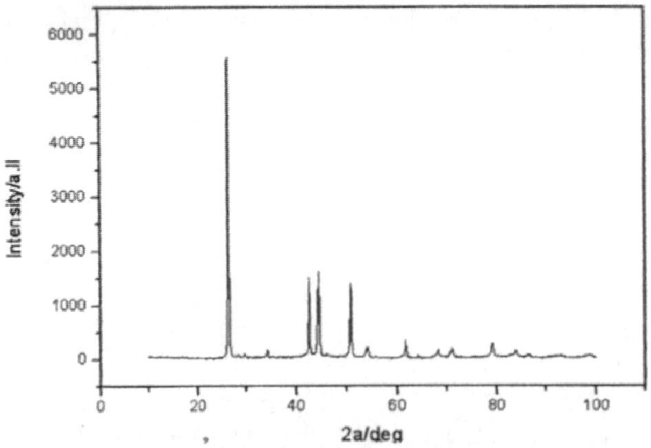

Figure 1. The image of XRD of $AgGa_{0.7}In_{0.3}Se_2$.

3.2. The heat flow-temperature influenced by cooling speed

Actually, the temperature of molten mass (T) is lower than melting point(T_m) in melt system,the difference between two value is the degree of super-cooling (ΔT):

$$\Delta T = T_m - T$$

The degree of super-cooling provide a driving force of phase change, enable crystal to grow.

We obtain data and heat flow-temperature curve as table 1 and Figure 2.

Table 1. The relationship of cooling-speed and melting point

Cooling-speed	Melting point	Crystallization point	Degree of super cooling
5°C/min	811°C	793°C	18°C
10°C/min	810°C	790°C	20°C
15°C/min	811°C	788°C	23°C

Heat speed: 10°C/min

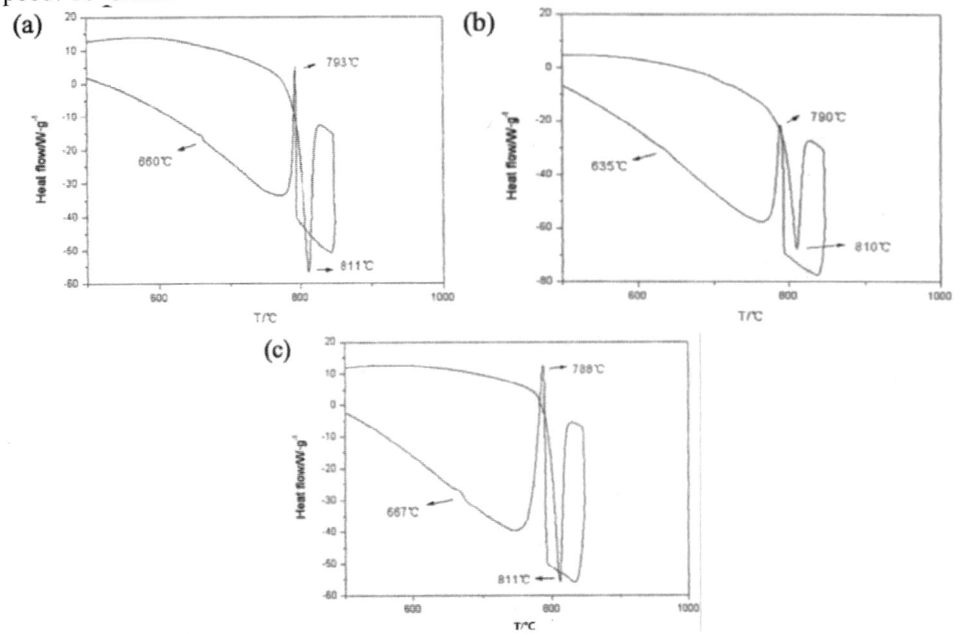

Figure 2. The relationship of heat flow and temperature in different cooling speed. (a) 5°C/min cooling speed. (b) 10°C/min cooling speed. (c) 15°C/min cooling speed (b)

It is obvious that the crystallization point and the degree of super-cooling decrease with the improving of cooling speed.

It is obviously that each of three heat flow-temperature curve has two sharp phase change peaks. The peak of 810°C is endothermic peak. This moment the sample melt, the solid-liquid phase change takes place and endotherm. In the following cooling process, there are three exothermic peaks of 793°C, 790°C, 788°C at each heat flow-temperature curve, this moment the sample crystalize, the liquid-solid phase change take place and exotherm. Meanwhile, we should note there is a weak exothermic peak near 660°C in each curve. Consider the melt point of InSe is 660°C, this exothermic peak may be related to the InSe phase change resulted from the crystalline inner component segregation [1].

In differential thermal analysis, we can regard the speed of cooling as the degree of sample's temperature difference. The change trend of super-cooling degree under different cooling speed can reflect the change trend of temperature difference between crystal and fusant in the solid liquid interphase under different vertical temperature gradient distribution. They are positive correlation.

The growth process of monocrystalline is complex in the crucible. We use simple model to discuss. Assume the grow speed of crystal as the conduction of heat in one- dimensional space. Base on the heat conduction equation:

$$v=\Delta T\,(KS-KF)/\,(\rho_m L)$$

In the equation, ΔT is the temperature difference of solid liquid interphase. K_S and K_F is the heat conductivity of crystal and fusant. ρ_m is the fusant desity near the melting point. L is the crystallization of latent heat released from grow unit mass of crystal.

It is obviously that ΔT have positive correlation with crystal grow speed. So we are able to consider that the sample's change trend of super-cooling degree in DTA are consistent with the grow speed of $AgGa_{1-x}In_xSe_2$ crystal in grow furnace.

3.3. The temperature distribution situation of temperature field

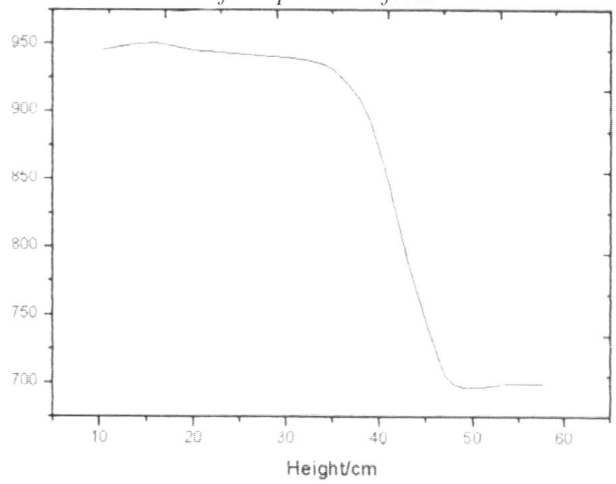

Figure 3. The curve of furnace temperature field

From the curve of furnace temperature field, we can see, the area from 10cm to 36cm belongs to fusant temperature field. There is a little difference in temperature distribution, ensure raw material fully melt. The curve fluctuates slightly because of outside environment temperature interfering. Area from 36 cm to 47 cm is crystal growth temperature field. This area has a large temperature degree, provide grow driving force to the crystallize of fusant. The temperature of 42-43 cm is close to the melting and crystallization point. The crystallization of $AgGa_{1-x}In_xSe_2$ monocrystalline mainly take place in this area. From 47 cm to 59 cm is the area of post processing temperature field. The temperature distribution is equally, and lower than the crystallization point but higher than 660°C, which may segregate InSe independent phase. Enable crystal mold further and avoid the component segregation phenomenon, which have effect on the quality of monocrystalline.

IOP Conf. Series: Materials Science and Engineering **242** (2017) 012004 doi:10.1088/1757-899X/242/1/012004

3.4. Discussing of grow furnace temperature curve

3.4.1. Discussing of fusant temperature field curve. The fusant area from 10cm to 32cm is theoretically a constant temperature field. The temperature of this area is higher than raw material. But there exist following interfere factors:

1. The outer insulating layer of grow furnace can't completely prevent the thermal exchange between grow furnace and out environment.
2. A lot of heat run away from the opening area of furnace.
3. Temperature director feedback control mechanism.

So, the temperature is not equality everywhere in this area. Meanwhile, the fusant temperature field should meet following requires:

1. This area dynamic temperature should not lower than 811°C to make sure $AgGa_{0.7}In_{0.3}Se_2$ raw material completely melt.
2. Consider that the volatilizing temperature of Se is 684°C, enable to reduce the loss of Se volatilize, the dynamic temperature should not too high.
3. Because of the upside and downside curve of growth temperature field connect with the fusant temperature field and the post-processing temperature field smoothly. The fusant temperature should not too low, enable the growth temperature has a suitable temperature gradient.

Consider these situation, we should let the temperature properly high.

3.4.2. Discussing of growth temperature field. The area of 32cm to 36cm is a transition between fusant and growth temperature field. This area near the fusant area, The temperature gradient in this area is small, the growth of crystal is not in this area, so we don't discuss any more. From the 32cm to 47cm area, this part is the main growth area of crystal. From heat flow-temperature curve we can see that the crystallization point is located in the 41cm-43cm area, $AgGa_{0.8}In_{0.2}Se_2$ monocrystalline grow in this area. For a suitable fast grow speed and control the size of other direction seed crystal, the temperature gradient of this area should be the largest.

3.4.3. Discussing post-processing temperature field. From the area 47cm to 59 cm is the post-processing temperature field, this area the distribution of temperature is equality. The temperature is lower than crystallization point, but higher than 660°C. The main function of Post- processing temperature field is enable crystallize further more, and avoid possibly InSe phase component segregation.

3.4.4. Possibly change ways of growth temperature field curve. In order to reform the furnace temperature distribution of $AgGa_{0.8}In_{0.2}Se_2$ to suit the the synthesize of $AgGa_{0.7}In_{0.3}Se_2$ monocrystalline, first we should consider measures of change temperature gradient, then combine the different melt and crystallization point of two crystalline with measures to discuss witch measure is better.

The influence of increase the growth temperature gradient:

1. Remain the post processing temperature, increase the fusant area temperature to rise the temperature gradient. That can increase the growth driving force, and decrease the risk of break after crystallization. But the shortage are that may case to the decomposing and volatilization of raw material. Meanwhile, improve the fusant temperature will lead to the temperature increase in same position. The crystallization will happen under original situation; this will short the post-processing temperature field distance.
2. Remain the fusant temperature, Decrease the post processing temperature to improve the temperature gradient. This measure can ensure the monocrystalline has enough length to get post processing. But that will product large thermal stress, which may break the crystal structure. Meanwhile, crystallization will take place above the original situation, the raw material may not fully melt.

The influence of decrease the temperature gradient:

1. Remain the post processing temperature, decrease the fusant temperature to reduce the temperature gradient. This measure can reduce decomposing and volatilization. But reducing the temperature gradient equals to slowing down the grow speed of monocrystalline, the size of other direction seed crystal will bigger than that of not reduce the fusant temperature. Moreover, the raw material will crystallize above original situation.
2. Remain the fusant temperature, increase the post processing to reduce the temperature gradient. That can decrease thermal stress, but short the length of post processing.

3.4.5. Comprehensive discussion. According to those situation, the temperature curve of $AgGa_{0.7}In_{0.3}Se_2$ growth furnace should match the following requires:

1. In order to avoid drastically volatilize of Se, the value of $AgGa_{0.7}In_{0.3}Se_2$ fusant temperature field should not higher than that of $AgGa_{0.8}In_{0.2}Se_2$.
2. Consider that the $AgGa_{0.8}In_{0.2}Se_2$ has a crystallization point of 814°C and a melting point of 843°C, with a super-cooling degree of 29°C [2], each of them are higher than that of $AgGa_{0.7}In_{0.3}Se_2$. Enable to provide equally growth driving force, we can properly improve the temperature gradient of growth temperature field.
3. The temperature gradient leads to thermal stress. The max temperature gradient crystal allowed can be calculate through the following formula [3]:

$$G_{max}=\frac{2\varepsilon}{\alpha R^{3/2}}\left(\frac{2}{h}\right)^{1/2}$$

ε is critical fracture strain value, $\boldsymbol{\alpha}$ is crystal coefficient of thermal expansion, $\boldsymbol{R}$ is crystal radius, $\boldsymbol{h}$ is crystal heat exchange coefficient. Considering that different of each valus between $AgGa_{0.8}In_{0.2}Se_2$ and $AgGa_{0.7}In_{0.3}Se_2$ is much less than reality experiment error, so we can regard both G_{max} as equality.

4. The post processing temperature should higher than 660°C.

4. Conclusion

In conclusion, we can change the temperature field curve of $AgGa_{0.8}In_{0.2}Se_2$ to that of $AgGa_{0.7}In_{0.3}Se_2$ in the way that we remain the fusant and post processing temperature, improve the temperature gradient. But to solve the problem of making accurate curve, we should do more research about fusant volatilize speed, the exothermic peak at 660°C and calculate G_{max}.

References

[1] A L Zhao, BJ Zhao, SF Zhu, BJ Chen, ZY He. Study on two-phase precipitates of $AgGa_{1-x}In_xSe_2$ [J]. Journal of synthetic crystals, 2013, 42(5): 785.
[2] CF Xu, BJ Zhao, SF Zhu, Y Huang, WL Zhu. Study on growth temperature field of $AgGa_{1-x}In_xSe_2$ [J]. Journal of Sichuan university, 2006, 43(5): 1062.
[3] KC Zhang, LH Zhang. Crystal growth science and technology [M]. Shanghai Science Press, 1997, 482.
[4] J Liu, BJ Zhao, SF Zhu, Y Huang, JJ Zhang, WL Zhu. Synthesis of $AgGa_{1-x}In_xSe_2$ polycrystalline materials [J]. Journal of Rare Earth, 2006, 24: 269.
[5] VV Badikov, GM Kuz'Micheva, VL Panyutin, VB Rybakov, VI Chizhikov. Preparation and structure of $AgGa_{1-x}In_xSe_2$ single crystals [J]. Inorganic Materials, 2003, 39 (10): 1028-1034.
[6] Y Huang, BJ Zhao, SF Zhu, WL Zhu, CF Xu. Properties of $AgGa_{1-x}In_xSe_2$ single crystals grown by Bridgman method [J]. Crystal Research & Technology, 2010, 42 (3): 227-230.
[7] R Santos-Ortiz, E Tupitsyn, I Nieves, P Bhattacharya, A Burger. Growth improvement and characterization of $AgGa_{1-x}In_xSe_2$ chalcopyrite crystals using the horizontal Bridgman technique [J]. Journal of Crystal Growth, 2011, 314: 293-297.

IOP Conf. Series: Materials Science and Engineering **242** (2017) 012005 doi:10.1088/1757-899X/242/1/012005

Preparation of Janus Ag nanoparticles by liquid/liquid interfacial reaction

Zhicheng Bian, Yongdong Xiang, Chengliang Hu, Zhihong Luo[1*], Kun Luo[2*]

[1] Corresponding authors: Zhihong Luo, Email: luozhihong615@163.com
[2] Kun Luo: luokun@glut.edu.cn

Guangxi Key Laboratory of Universities for Clean Metallurgy Comprehensive Utilization of Nonferrous Metal Resource, College of Materials Science and Engineering, Guilin University of Technology, Guilin, 541004, PR China

Abstract. Janus silver (Ag) nanoparticles have been successfully synthesized by intense agitation at the interface of toluene and water which contains Ag $(PPh_3)_2NO_3$ and THPC. Contact angle of the film prepared by L-B technology is measured which indicates that the Ag nanoparticle is amphipathicity because they have hydrophobic region and hydrophilic region separately. TEM image showed the assembled morphology of Ag instead of monodispersed particles. After purification, analysis of FT-IR and XPS showed the Janus Ag nanoparticles are protected by ligands of THPC and PPh_3 with a ratio of 1:1.

1. Introduction

In 1978, Lehn came up with "supermolecular chemistry", which investigate the self-assembled molecular based on the interaction of molecular including hydrogen bond, electrostatic force, Van der Waals' force, hydrophilic and hydrophobic interaction. By using these interactions, various supermolecular are prepared. Janus particles is a typical structure of these supermolecular, which attracted intensive interest in the fields of self-assembling and molecular recognition[1], catalysts[2], optical biosensor [3]and etc[4,5,]due to their anisotropy.

Synthesis of amphiphilic Janus nanoparticles requires creating two distinct compartments on the surface in a cost-effective andreliable way. Masking was one of the techniques initially applied on larger Janus particles and scaled down to the nanoscale involving the protection of one side and modification of another, followed with the removal of the protection. However, the use of gas/liquid, liquid/liquid, and liquid/solid interfaces that trap homogeneous nanoparticles is much more popular[6]. Previous researches have reported the preparation of Janus metal nanoparticles or Janus alloy nanoparticles by using metallo-organic compound as precursor, THPC as reductant, and the reaction mechanism and application were investigated. In this paper, we prepared ligands modified Ag by using ultrasonic, which is a rapid method. Janus Ag particles were prepared by using liquid/liquid interfacial reaction. The structure, morphology and the distribution of ligands on the surface of Janus Ag particles were measured.

2. Experimental

2.1Chemicals

Triphenylphosphine (PPh_3), Tetramethylolphosphonium chloride (THPC), Sinopharm Chemical Reagent Co., Ltd. (China).Other chemical agents werepurchased from Wuhan Jinlin Chemical

Content from this work may be used under the terms of the Creative Commons Attribution 3.0 licence. Any further distribution of this work must maintain attribution to the author(s) and the title of the work, journal citation and DOI.
Published under licence by IOP Publishing Ltd

Technology Co., Ltd.(China).Deionized water was used in all experiments.

2.2 Preparation of Ag (PPh3) xNO3
0.1065 g $AgNO_3$ and 0.6560 g PPh_3 was dissolved in 1 mL H_2O and 10 mL toluene, respectively. The obtained solutions were mixed under ultrasonic for 10 min and the white precipitate was obtained in water phase, which was washed with ethyl ether three times and dried at 65°C for 1 h. 3 mL CH_2Cl_2 was added to dissolved the precipitate and then the solution was placed in refrigirate at -15°C for 15 min. 18 mL n-pentane was added into the above solution to separate crystal out, which was washed by n-pentane for 3 times, and dried at 65°C for 1 h to obtain white powder for FTIR and XRD measurement.

2.3 Preparation of Janus Ag nanopartiles
$AgNO_3$ solution (0.6 mL, 25 mM) and PPh_3 solution (1.2 mL, 50 mM) were mixed under ultrasonic for 10 min to obtain white precipitate for use. NaOH (0.09 mL, 1M) , 14.01 mL H_2O and 13.8 mL toluene were mixed by emulsifier under 4000 r/min, and then THPC solution (0.3 mL, 50 mM) and the oboved white precipitate was added in sequence, the mixture was kept stirring for another 2 h to obtain self-assembled Janus Ag film. The assembled Janus Ag film on quartz was prepared in Langmiur film balnce (JML04C1-P).

2.4 Characterization
Janus Ag film was characterized by a contact angle goniometer (JWA-360, Xiamen Chongda Intelligent Technology Co., Ltd., China), an XPS (ESCALAB250Xi, ThermoFisher Scientific), FTIR (Thermo Nexus 470), XRD (Xpert-Pro), TEM (JEM-2100F) and ICP-OES(Optima 8000, PerkinElmer).

3. Results and discussions

3.1 Characterization of Ag (PPh3) xNO3
The structure of Ag $(PPh_3)_xNO_3$was measured by FTIR and XRD. In Fig.1a, the band at 3055 cm^{-1} is assigned to vibration of C-H, as well as the peaks at 745 $cm^{-1}$and 694 cm^{-1}. The peaks at 1435 $cm^{-1}$and 513 cm^{-1} are attributed to vibration of benzene. The peaks at 1385 $cm^{-1}$and 1310 cm^{-1} belong to the vibration of $-NO_2$. These vibrations prove the existence of PPh_3.

Fig.1b shows the XRD pattern of Ag $(PPh_3)_xNO_3$. The red line is the measured XRD pattern which is similar to the simulated results. The sharp XRD peaks indicate high crystallinity. According to the XRD pattern, the obtained Ag $(PPh_3)_xNO_3$ belongs to orthorhombic system (Aba2, a=14.846Å, b=25.440Å, c=10.318Å). These results indicate that Ag $(PPh_3)_xNO_3$can be prepared by untrasonic

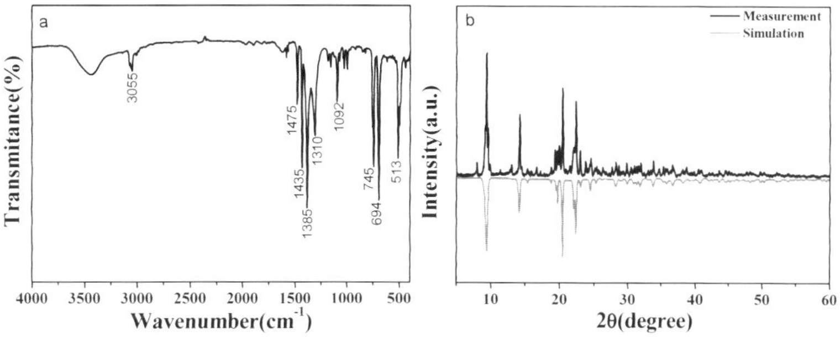

Fig1. FTIR (a) and XRD(b) spectra of $Ag(PPh_3)_xNO_3$

3.2 Characterization of Janus Ag nanoparticles
The morphology of Janus Ag nanoparticles were measured by TEM which are showed in Fig.2a and 2b. No monodispersed Janus Ag nanoparticles are observed, indicating the Janus Ag nanoparticles are

self-assembled to form film. Ag nanoparticles have three major lattice fringe which can be assigned to (111), (200) and (311) with the interplanar spacing of 0.236 nm, 0.204 nm and 0.123 nm, respectively. In Fig.2b, only (200) facet of Ag nanoparticles can be observed.

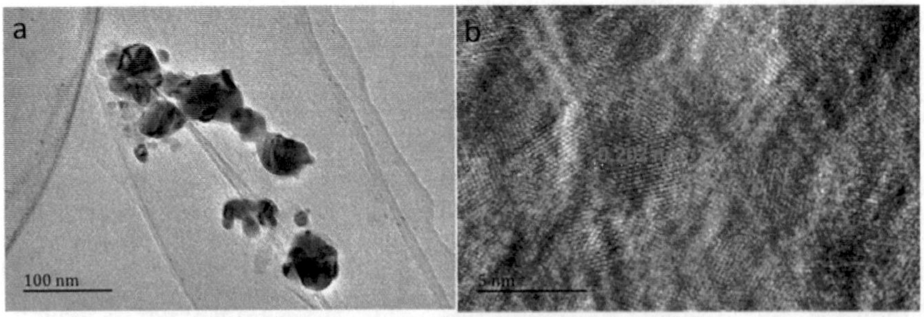

Fig.2 TEM images of Janus Ag nanoparticles

The structure of Janus Ag nanoparticles were measured by UV-vis and FTIR analysis. Janus Ag nanoparticles were dispersed in DMF for UV-vis analysis (inset of Fig.2a). The absorption peak at about 443 nm manifests the existence of Ag nanoparticles (Fig.2a). Fig.2b shows the FTIR of Janus Ag nanoparticles, the peaks at about 3443 $cm^{-1}$and 1400 cm^{-1} are assigned to the stretch vibration and transformation vibration of –OH. The peaks at about 1629 $cm^{-1}$and 611cm^{-1} are attributed to the vibrations of benzene and C-H, which indicates the existence of PPh_3.

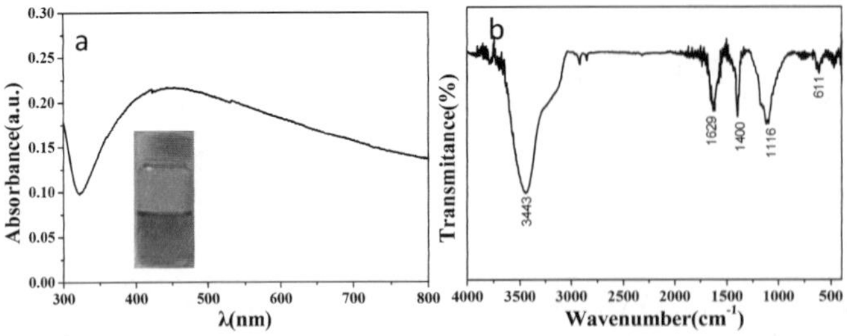

Fig.3 UV-vis (a) and FTIR (b) spectrum of Janus Ag nanoparticles

The contact angle of H_2O is measured to investigate the amphiphilic property of Janus Ag particles. Before measurement, Janus Ag nanoparticles were deposited onto the hydrophilic and hydrophobic quartz substrate through Langmuir-Blodgett method, respectively. Quartz substrate was treated by piranha solution (98% H_2SO_4:30%H_2O_2=3:1) to introduce hydrophilic property, and then immersed in trimethylchlorosilane/DMF solution for 20 min to introduce hydrophobic property. The contact angles of H_2O on Janus Ag nopoaprticles deposited on hydrophilic and hydrophobic quartz substrate are showed in Fig.4.

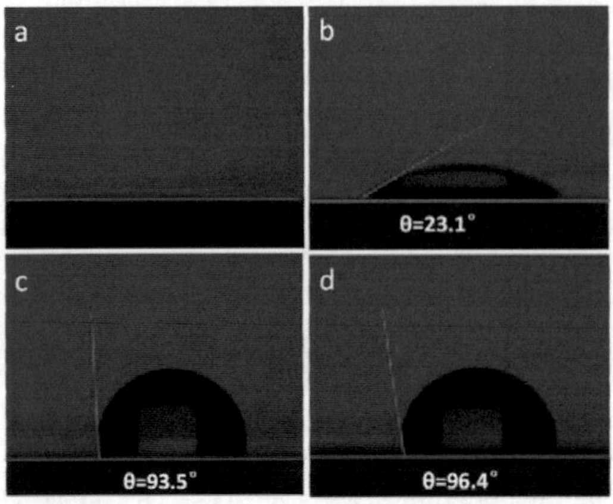

Fig.4 Contact angles of H_2O on hydrophilic (a) and hydrophobic (c) quartz substrate, on Janus Ag nanoparticles deposited on hydrophilic (b) and hydrophobic (d) quartz substrate.

Before deposition of Janus Ag nanoparticles, the contact angle of H_2O on hydrophilic and hydrophobic quartz substrate is 0° and 93.5°. After deposition of Janus Ag nanoparticles, the contact angles increase, the contact angle on Janus Ag nanoparticles is 23.1° and 96.4°, exhibiting hydrophilic and hydrophobic properties.

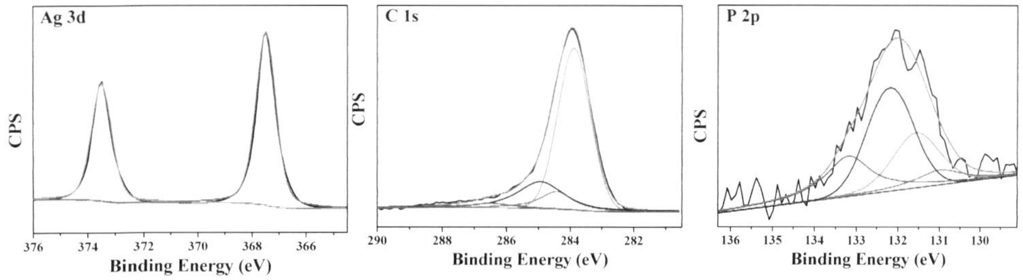

Fig.5 C 1s, P 2p and Ag 3d spectrum of Janus Ag nanoparticles.

Table1 XPS data of the C 1s, P 2p, Ag 3dof the sample

Elements	BE(eV)	Fwhm(eV)	Atomic percentage[a](%)	Possible chemistry
	283.88	1.22	65.77	C-C[7]
C 1s	284.43	1.22	10.42	C-OH[8]
	284.94	1.7	20.41	C-P[9]
	286.51	1.42	3.40	C=O[10]
	131	1.33	9.05	PPh_3[11]
P 2p	131.56	1.33	25.05	THP
	132.19	1.33	41.68	PPh_3O[12]
	133.20	1.33	24.22	THPO[13]
Ag 3d	367.50	0.87	59.86	Ag^0 $3d_{5/2}$[14]
	373.5	0.87	40.14	Ag^0 $3d_{3/2}$[15]

Atomic percentage[a] = A_{peak} / A_{total} for a selected element

The surface property of Janus Ag nanoparticles are further investigated by XPS analysis, the results are showed in Fig. 5 and Table 1. From Ag 3d spectrum, the ratio of $3d_{3/2}$ and $3d_{5/2}$ is about 2:3, indicating the existence of Ag^0. The atomic percentage of C-C, C-OH, C-P and C=O is about 65.77 at.%, 10.42 at.%, 20.41 at.% and 3.40at.%. According to the structure of PPh_3/PPh_3O (P is connected with three benzene) and THP/THPO (P is connected with three C-OH), the ratio of C-P to C-OH is calculated to be 1.05:1. According to P 2p spectrum, the ratio of PPh_3/PPh_3O and THP/THPO is about 1.03:1, which is in line with the results obtained from C 1s spectrum. We can conclude that Ag has both hydrophilic and hydrophobic ligands with the ratio about 1:1.

4. Conclusions

In conclusion, we find that Ag $(PPh_3)_xNO_3$ precusorcan be prepared by ultrasonic, XRD and FTIR confirm the crystal structure of Ag $(PPh_3)_xNO_3$ which has the PPh_3 ligands. Janus Ag nanoparticles can be prepared by vigorous stirring which exhibits the morphology of self-assembled film. The Janus Ag nanoparticles has hydrophilic and hydrophobic properties due to the existence of THP/THPO and PPh_3/PPh_3O with the ratio about 1:1.

Acknowledgement

Z.C. Bian and Y.D. Xiang contributed equal to this work. This work was supported by Natural Science Foundation of Guangxi Province (No.2015GXNSFBA139220and 2016GXNSFAA380107) and the funding from Collaborative Innovation Center for Exploration of Hidden Nonferrous Metal Deposits and Development of New Materials in Guangxi, Guangxi Key Laboratory of Processing for Non-ferrous Metal and Featured Materials.

IOP Conf. Series: Materials Science and Engineering **242** (2017) 012005 doi:10.1088/1757-899X/242/1/012005

References

[1] Dhananjay D, Alan TH and Doyle PS 2001 *J. Langmuir.* **23** 4669-74
[2] Cole-Hamilton DJ 2010 *J. Sci.* **327** 41-42
[3] Himmelhaus M and Takei H 2000 *J. Sens. Actuators, B. Chemical* **63** 24-30
[4] Nisisako T, Torii T,Takahashi T and Takizawa Y 2006 *J. Adv. Mater.* **18** 1152-56
[5] Yuet KP, Hwang DK, Haghgoole R and Doyle PS 2010 *J. Langmuir.* **26** 4281-87
[6] Luo K, Xiang YD, Wang HM and Luo ZH 2016 *J. J. Mater. Sci. Technol.* **32** 733-7.
[7] Blass PM. 1989 *J.Surf. Sci.***21** 785-102.
[8] Rebholz M, Matolin V, Prins R and Kruse N 1991 *J. Surf. Sci.* **251** 1117-22
[9] Larsson R, Malek A and Folkesson B 1980 *J. Chem. Scr.* **16** 47-51
[10] Beamson G and Briggs D 1992 *J. Mol. Phys.***76** 919-36
[11] Clark DT, Woolsey IS, Robinson SD, Laing KR and Wingfield JN 1977 *J. ChemInform.***8**
[12] Stec WJ, Morgan WE, Albridge RG and Wazer JRV 1972 *J. Inorg. Chem.* **11**
[13] Kouba JK, Pierce JL and Walton RA 1980*J. J. Organomet. Chem.* **202**C 105-7.
[14] Hoflund GB, Weaver JF and Epling WS 1994 *J. Surf. Sci. Spectra.* **3** 163-8
[15] LiuY, Jordan RG and Qiu SL 1994*J.Phys. Rev. B: Condens. Matter.* **50** 4478-84

IOP Conf. Series: Materials Science and Engineering **242** (2017) 012006 doi:10.1088/1757-899X/242/1/012006

High conductive and scalable Ag nanowires flexible transparent electrode by nanowelding with physical methods

W W He[1,2,a], X H Yan[1], Y F Long[1], Y M Liang[1], C Pan[1], J L Zhao[1] and Q X Liu[3,b]

[1] City College, Wuhan University of Science and Technology, Wuhan 430083, China.

[2] State Key Laboratory of Materials Processing and Die & Mould Technology, Huazhong University of Science and Technology, Wuhan 430074, China.

[3] Wuhan University of Science and Technology, Wuhan, 430080, China.

[a]heww1987@126.com, b liuqiuxin1956@sohu.com

Abstract. Transparent electrodes (TEs) are very important for electronic devices. At present, ITO is gaining the largest market share but will be reduced. Ag nanowires (AgNWs) TEs is acknowledged as one of the most potential alternative to ITO. However, AgNWs TEs still have electrical problems because of the low contact between the AgNWs. In this paper, we report three physics methods to increase the conductivity of AgNWs TEs by nanowelding the contact of nanowires. For heat-resistant materials, 200 ^{0}C heat-nanowelding can help to reduce the sheet resistance by 96.7%. For pressure resistant materials, 20MPa pressure-nanowelding can help to increase the conductivity by 98.7%. And the transmittance (>90%) remains constant during the above process. Yet, both of these methods cannot improve the adhesion between nanowires and the substrates. Luckily, tight adhesion can be obtained by overcoating a PEDOT: PSS lalyer on AgNWs film which can reduce the sheet resistance by 87.8%. This means that things are usually not perfect, and they have their own advantages and lay the foundation for the popularization and application of AgNWs TEs. In a word, these three nano-welding methods are all suit for manufacture on a large scale for high conductive AgNWs TEs.

1. Introduction

By 2019, transparent electrode (TE) market will reach $4.6 billion. That's because TEs are the essential components to a wide variety of electronic devices, such as transparent heaters, liquid crystal displays, organic light emitting diodes (OLEDs), touch screens, organic photovoltaics (OPVs), microchips, and smart windows.[1-4] And Indium tin oxide (ITO) is the most wildly used TE that account for 93 % of the market.[5] However, there are many problems which limit its further development. Due to the vapor-phase coating processes, the cost of ITO (10 Ω/sq.) is $ 26 m^{-2} which is correspondingly high.[6] And we can see yellow for low-sheet resistance (< 60 Ω/sq.) ITO. In addition, the cost of indium is $ 600 kg^{-1} which is a little high because of the low abundance (0.05 ppm).[7] What's more, ITO would not work on flexible devices in principle. For these reasons, the search for alternatives to ITO is motivated. Professional assessment that China will become the largest winner of transparent electrodes. That's because Chinese government controls the supply chain of indium/ITO, on the other hand, China will be the main consumer of TEs. Obviously, the ultimate market opportunity for alternative TEs is in fully flexible devices.

So far, the alternatives include Transparent Conduct oxides (TCOs), conducting polymers, carbon nanotube (CNT), graphene, and metal nanowires.[8-11] It seems that the closer approach to ITO, the more likely to be recognized as a substitute. Unfortunately, all TCOs failed when encountering a

Content from this work may be used under the terms of the Creative Commons Attribution 3.0 licence. Any further distribution of this work must maintain attribution to the author(s) and the title of the work, journal citation and DOI.
Published under licence by IOP Publishing Ltd

flexible device. At present the commonly used conducting polymers including Poly (3, 4-ethylenedioxythiophene) poly (styrenesulfonate) (PEDOT: PSS) and similar polythiophenes which performance is not really good (90 % 100 Ω/sq.).[12] Conducting polymers have been widely used for anti-electrostatic layers, and PEDOT: PSS has been used as an alternatives in commercial products. For years, CNTs has received a high degree of attention. In many ways, the performance of CNT is not as good as ITO, and it needs to be improved. Due to the large contact resistance (*Rc*) of CNTs, the sheet resistance (*Rsh*) is 400 Ω/sq. at 90 % transmittance.[13] But if exfoliated/doped with chlorosulfonic superacid, the *Rsh* will be 60 Ω/sq. at 90.9 % transmittance which is still not good enough.[14] Now, graphene is getting the most attention and considered to be the ultimate win because of its potential low cost. Graphene based on chemical vapor deposition (CVD) process has excellent performance but is not practical for production. And the graphene based on solution-phase coating is relatively poor performance (80 % 3 kΩ/sq.).[15] There is still a long way before graphene really replaces ITO. The performance of metal grids fabricated by the nanopatterned method is better than ITO (90 % 0.8 kΩ/sq.).[16] But nanopatterned method will lead to additional cost and lower yields.[17] Fortunately, metal nanowires can meet the conditions of alternatives. At present, metal nanowires include gold (Au), copper (Cu), and silver (Ag) nanowires. As we all know Au is too expensive and Cu is 6 % less conductive than silver. Furthermore, Cu is oxidized quickly than Ag at ambient conditions. The performance of Ag nanowires (AgNWs) TE is even better than ITO (95 % 20 kΩ/sq.).[18] In a word, AgNWs TE is acknowledged as the proverbial "low-hanging fruit" for potential alternative to ITO.

However, AgNWs TEs still have electrical problems because of the low contact between the AgNWs. As the experimental results, *Rc* is the main source of *Rsh*. There are many methods to reduce *Rc*, such as laser nano-welding, graphene coating, chemical annealing, plasmonic welding, deposition of Au, TiO_2 or ZnO particles.[19-25] Lee et al. have demonstrated that chemical annealing can be applied to the roll-to-roll production.[26]

In this work, we report three physics methods to increase the conductivity of AgNWs TEs by nanowelding the contact of nanowires for high conductive AgNWs TEs. First of all, heat-nanowelding is a simple and effective way to reduce the film resistance. For heat-resistant materials, 200 ^{0}C heat-nanowelding can help to reduce the sheet resistance by 96.7% while the transmittance increases slightly to 92%. Secondly, pressure-nanowelding is an extremely effective way to even form the perfect contact at the junction. For pressure resistant materials, 20MPa pressure-nanowelding can help to increase the conductivity by 98.7%. And the transmittance remains 93% which is good enough for most of the devices. Yet, both of these methods cannot improve the adhesion between nanowires and the substrates. Luckily, tight adhesion can be obtained by overcoating a PEDOT: PSS lalyer on AgNWs film which can reduce the sheet resistance by 87.8%. Unsatisfactorily, PEDOT-nanowelding will cause a 10% reduction in transmittance. In a word, things are usually not perfect. This means that things are usually not perfect, and they have their own advantages and lay the foundation for the popularization and application of AgNWs TEs. In a word, these three nano-welding methods are all suit for manufacture on a large scale for high conductive AgNWs TEs.

2. Experiments

2.1. Formation of transparent and conductive film using colloid of Ag nanowires

Colloid of Ag nanowires was first prepared by dispersing the purified Ag nanowires into isopropanol solution with a concentration of 10 mg/mL. After ultrasonication for 2 min, the colloid was ready for film formation by spin-coating or doctor-blading. Substrates of sodium-line glass (SLG) or flexible poly (ethylene terephthalate) (PET) were then cleaned by sequential ultrasonication in acetone, ethanol and deionized water for 30 min, respectively. After cleaning, substrates were submitted to O_2 plasma treating to make the surface hydrophilic. Colloid was immediately spun or rolled on treated substrates to form a thin film. After drying, the film was heated in air at 150-200°C for 20 min and cooled naturally. To enhance the contact at the junction, pressure of 20 MPa was applied on Ag film between underlying PET substrate and a clean but hydrophobic PET. Tight adhesion of Ag nanowires on

IOP Conf. Series: Materials Science and Engineering **242** (2017) 012006 doi:10.1088/1757-899X/242/1/012006

underlying substrate was realized by overcoating a PEDOT layer on Ag film. Overcoating was realized by spinning or rolling isopropanol solution of PEDOT (PEDOT:isopropanol = 1:4 in volume) on Ag film and drying in air at 120°C for 5 min.

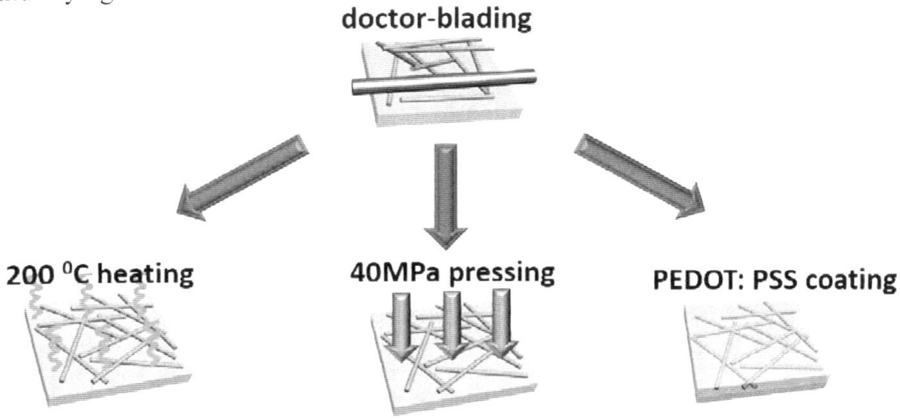

Figure 1. Schematic diagram of the experimental processes.

2.2. *Characterizations*

Optical reflection and transmission spectra were obtained on a UV-vis-NIR spectrometer (Shimadzu SolidSpec-3700) using an integral sphere detector. The sheet resistance values were measured by a four-point probe technique.

3. Results and discussions

3.1. *Conductivity*

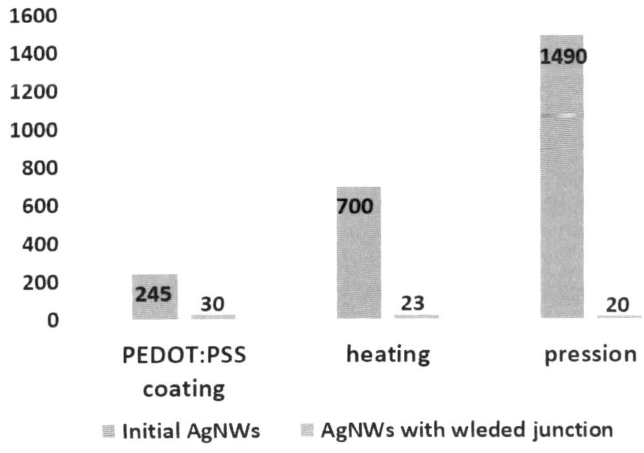

Figure 2. Conductivity of initial AgNWs TEs and AgNWs TEs by nanowelding with physical method.

Some text.

In fact, the conductivity of AgNWs TE is not always very well when they formed by doctor-blading. And the poor conductivity is due to poor contact between silver nanowires which forming a large contact resistance. The best way to solve this problem is the nanowelding between the nanowires. Inventive methods have been tried, such as PEDOT-nanowelding, heating- nanowelding or pressing-nanowelding. Figure 2a shows the conductivity of initial AgNWs TEs and AgNWs TEs by nanowelding with physical method. First of all, heat-nanowelding is a simple and effective way to reduce the film resistance. For heat-resistant materials, 200 ^{0}C heat-nanowelding can help to reduce the sheet resistance from 700 down to 23 Ω/sq. and decreased by 96.7%. Secondly, pressure-nanowelding is another extremely effective way to even form the perfect contact at the junction. For pressure resistant materials, 20MPa pressure-nanowelding can help to increase the conductivity from 1490 to 20 Ω/sq., an increase of 98.7%. Thirdly, overcoating a PEDOT: PSS lalyer on AgNWs film which can reduce the sheet resistance by 87.8%. Unsatisfactorily, PEDOT-nanowelding will cause a 10% reduction in transmittance. From the perspective of improved conductivity, the pressure method is more advantageous than the other two methods.

3.2. *Morphology*

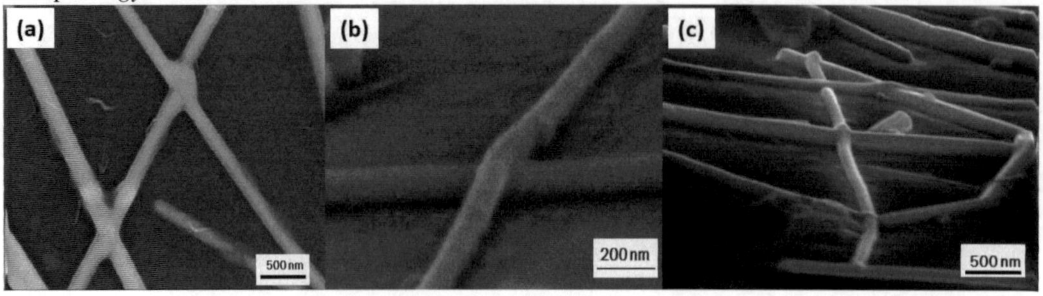

Figure 3. SEM image of the junction of AgNWs. (a) PEDOT: PSS coating; (b) heating; (c) pressing.

Figure 3 is the SEM image of the junction of AgNWs progressed by three method: PEDOT: PSS coating, heating, and pressing. We can see that there is a common characteristic of the above three methods that they all assisted joining at the nanowire junctions. That is the real reason why they reduce the sheet resistance. As seen in Figure 3a, the nanowires are welded together under the action of PEDOT: PSS and there is a layer of organic matter on the contact. So we can speculate that the principle of PEDOT-nanowelding is that electronically transmitted via PEDOT: PSS material. As seen in Figure 3b, the nanowires are slightly welded together under the effect of heat-welding. This means that nanowires at the intersection of nanowires have a small amount of melting and welding together. By heat-welding the nanowires, electrons will be transported within the nanowire. As seen in Figure 3c, under the action of pressure, the nanowires form a perfect combination, even as there is just only one nanowire. As we all know, the pressure can make the metal to achieve perfect welding. So the transmission of electrons is certainly carried out within the nanowires. Contact the conductivity of the electrode, we know that the closer the nanowires are, the better the conductivity of the electrodes. And the transmission of electrons within the nanowires is better than the transmission of external matter.

3.3. *Optical properties*

In addition to conductivity, the transparency of the transparent electrode is also a very important parameter. Figure 4 is the optical properties of AgNWs TEs. We can see that, PEDOT: PSS layer coverage will reduce the transmittance by 10%, while the other two methods are maintained at about 92% transmittance, almost no impact. This is because PEDOT this dark blue semiconductor material that will enhance the absorption and scattering of light. Heat welding and pressure welding are only slightly changed the contact of the nanowires without other materials. It is not difficult to understand that the above two modes do not cause significant changes in transmittance. Taking into account the change in the optical transmittance, heating and pressing may be better than coating a PEDOT: PSS layer.

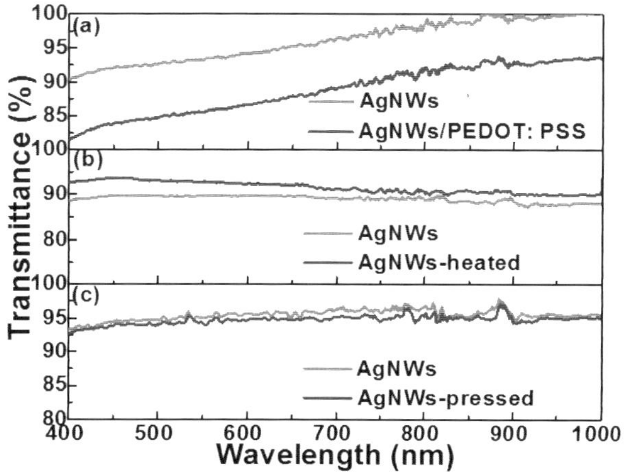

Figure 4. Optical properties of the original and the welded AgNWs TEs. (a) PEDOT: PSS coating; (b) heating; (c) pressing.

3.4. Adhesion with the substrate

In order to investigate the adhesion strength of the transparent electrode, a 90 °peeling test was conducted. A 12-mm-wide piece of 3M scotch tape was attached to an electrode sample and then pulled off the tape. After coating a PEDOT: PSS layer, it was clearly shown that the adhesion between the substrate and the thin film was significantly increased as shown in Figure 5. While pristine AgNWs network film could be easily removed from the substrate during the 3M tape testing (Figure 5a), after the PEDOT: PSS treatment, the AgNW/PEDOT: PSS hybrid film showed very strong adhesion to the substrate while maintaining its original conductivity even after tape testing (Figure 5b). This strong and stable substrate adhesion might come from the film uniformity after the filling the gap between nanowires network with conducting polymer. Unfortunately, the film of the other two methods did not successfully pass the tape pasting test. This is due to the slightness van der Waals force of the nanowires with the substrate. From the perspective of improved adhesion with the substrate, the PEDOT-welding method is more advantageous than the other two methods.

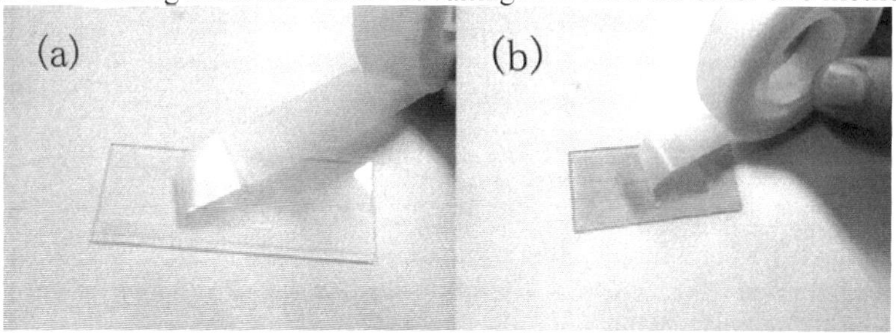

Figure 5. Adhesion force test (3M tape test) of AgNWs electrodes of a) pristine AgNWs network before nanosoldering and b) AgNW/PEDOT:PSS hybrid film after nanosoldering.

4. Conclusions

In conclusion, we report three physics methods to increase the conductivity of AgNWs TEs by nanowelding the contact of nanowires: heating-nanowelding, pressure-nanowelding, and PEDOT-nanowelding. There is a common characteristic of the above three methods that they all assisted joining at the nanowire junctions. That is the real reason why they reduce the sheet resistance. From the perspective of improved conductivity, the pressure method is more advantageous than the other two methods. 20MPa pressure-nanowelding can help to increase the conductivity from 1490 to 20 Ω/sq., an increase of 98.7% while other two methods providing 87.8% and 96.7% growth. Contact the

the morphology of the junction of the junction, we know that the closer the nanowires are, the better the conductivity of the electrodes. And the transmission of electrons within the nanowires is better than the transmission of external matter. Taking into account the change in the optical transmittance, PEDOT: PSS layer coverage will reduce the transmittance by 10%, while the other two methods are maintained at about 92% transmittance, almost no impact. Therefore, taking into account the transmittance, heating and pressing may be better than coating a PEDOT: PSS layer. However, from the perspective of improved adhesion with the substrate, the PEDOT-welding method is more advantageous than the other two methods. This is due to the slightness van der Waals force of the nanowires with the substrate that the film of the other two methods did not successfully pass the tape pasting test. This means that things are usually not perfect, and they have their own advantages and lay the foundation for the popularization and application of AgNWs TEs. In a word, these three nano-welding methods are all suit for manufacture on a large scale for high conductive AgNWs TEs.

Acknowledgements
This work was financially supported by the Hubei Natural Science Foundation (2016CFB266), State Key Laboratory of Materials Processing and Die & Mould Technology, Huazhong University of Science and Technology (P2018-012), and Science and technology research project of Hubei Provincial Education Department (B2016442), and Doctoral independent research talent fund (2015CYBSKY005).

References

[1] He W W, Wu K J and Wang K 2014 *Sci. Rep.* **4** 3715.
[2] Kim H J, Kim D I and Kim S S 2017 *Aip Adv.* **7** 3685.
[3] Boundy T, Boyton M, Taylor P. 2017 *J. Clean. Prod.*
[4] Lee J H, Kim A. 2017 *Org. Electron.* **47** 147-151.
[5] Ghaffarzadeh K and Das K 2014 *Transparent Conductive Films (TCF) 2014-2024: Forecasts, Markets, Technologies*, IDTechEx.
[6] Ito K 2015 *8. Sulfurization of Physical Vapor-Deposited Precursor Layers [M],* Copper Zinc Tin Sulfide-Based Thin-Film Solar Cells, John Wiley & Sons Ltd p 181.
[7] U S Geological Survey 2014 *Mineral Commodity Summaries, Indium, U.S.* Department of the Interior, Washington, D.C., USA, 74.
[8] Kim J, Lee G H and Yun G H 2017 *Sci. Adv. Mater.*
[9] Kirihara K, Wei Q and Mukaida M 2017 *Synthetic Met.* **225** 41-48.
[10] Amirighadi S, Raoof J B and Chekin F 2017 *Mat. Sci. Eng. C.*
[11] Kumar R, Jahan K and Nagarale R K 2017 *Acs Appl. Mater. Interfaces* **7** 593-601.
[12] Schaubroeck D, Smet J D and Willems W 2016 *Appl. Surf. Sci.* **376** 151.
[13] Geng H-Z, Kim K K and So K P 2007 *J. Am. Chem. Soc.* **129** 7758.
[14] Hecht D S, Heintz A M and Lee R 2011 *Nanotechnology* **22** 075201.
[15] Lee J H, Shin D W and Makotchenko V G 2009 *Adv. Mater.* **21** 4383.
[16] Catrysse P B and Fan S H 2010 *Nano Lett.* **10** 2944.
[17] Woerle J and Rost H 2011 *MRS Bull* **36** 789.
[18] Song M, You D S and Lim K 2013 *Adv. Funct. Mater.* **50** 4177.
[19] Lee S, Shin S and Lee 2015 *Adv. Funct. Mater.* **25**, 3114.
[20] Lee D, Lee H and Ahn Y 2015 *Carbon* **81** 439.
[21] Shin D Y, Yi G R and Lee D 2013 *Nanoscale* **5** 5043.
[22] Garnett E C, Cai W and Cha J J 2012 *Nat Mater.* **11** 241.
[23] Hu L, Kim H S and Lee J Y 2010 *ACS Nano* **4** 2955.
[24] Zhu R, Chung C H and Cha K C 2011 *ACS Nano* **5** 9877.
[25] Morgenstern F S F, Kabra D and Massip S 2011 *Appl. Phys. Lett.* **99** 183307.
[26] Lee S J, Kim Y-H and Kim J K 2014 *Nanoscale* **6** 11828.

IOP Conf. Series: Materials Science and Engineering **242** (2017) 012007 doi:10.1088/1757-899X/242/1/012007

Finite element analysis of bending performance on polyurethane composite panel

Minli Jia[1], Hongqiao Li[1] and Xiaoming Wang[2]

[1]School of Civil Engineering, Yantai University, 264005, China

[2]Wanhua Energy-saving Building Materials Co., LTD, 264003, China

Abstract: The finite element analysis model of polyurethane composite panel (simply named PCP) is established by using ABAQUS software. In view of the PCPs made of different thickness of surface board, their bending performance is carried out on finite element analysis, and the load-deflection curves which come from it are compared with the experimental results. The results show that the values between finite element analysis and experiment agree well with each other. It can be deduced that the established finite element model is fit to simulate the bending test of PCPs. The simulation not only has certain reference significance to the optimal design for the bending performance of PCPs, but also to the choice of PCPs in the practical project.

1. Introduction

Polyurethane composite panel is a kind of sandwich board which polyurethane rigid foam is firmly bonded between two layers of surface board made of metal or non-metal which is in good fire resistance. As a type of new building material in engineering, polyurethane composite panels are widely used for external building envelope and internal partitions in prefabricated houses. They provide good function performance for building, for instance, sound insulation, thermal insulation, heat preservation, fire prevention and so on. In addition, polyurethane composite panel not only responses to the policy of housing industrialization in China, realizing mass production in the factory and high degree of prefabrication assembly in construction site, but also become a important development direction of the construction industry, beneficially promoting the construction industrialization in our country.

At present, the application of composite panel is mainly being focused on the sandwich plate of metal surface, which of non-metal surface is still in the stage of development in our country. The main research methods of polyurethane composite board are classical analytical method, approximation method and finite element method. Based on sandwich panels of the metal surface plate, Qin Peicheng made a research on constitutive relation of polyurethane foam material. Yu Min studied deformation mechanism and design formula of sandwich panels with finite element method[1-2]. The formula for calculating deflection of sandwich panels with metal surface plate was put forward by Davies[3]. On the basis of the predecessor's work, the bending performance of sandwich plate which made of non-metal surface studied more are Structural Insulated Panels with wood surface board[4-5]. Li Hongqiao analyzed the compressive performance of polyurethane composite board with fiber cement board[6].The finite element method was widely used in the research of composite plates by Styles and others, which provides a reference for the finite element study of sandwich panel[7].

In conclusion, the research of sandwich panels with non-metal surfaces has got a certain achievement. However, there are few researches based on the composite panel made of fiber cement

Content from this work may be used under the terms of the Creative Commons Attribution 3.0 licence. Any further distribution of this work must maintain attribution to the author(s) and the title of the work, journal citation and DOI.
Published under licence by IOP Publishing Ltd

surface board and polyurethane core, and there are extremely few studies using finite element method on comparing the bending performance of metal surface sandwich panels with non-metal surface sandwich panels. Therefore, it is pretty necessary to carry out studying on the polyurethane composite panels.

This paper based on the bending experiments[8], which polyurethane composite panels were applied a concentrated load in the middle, establishes research model for the bending performance of polyurethane composite panels with different thickness and materials of surface board by means of the finite element software ABAQUS, and compares calculating results with the test results, which not only has certain reference significance to the optimal design for the flexural performance of polyurethane composite panel, but also to the choice of panels in the practical engineering projects.

2. Experiment

2.1. Introduction of Experiment

The polyurethane composite panels involved in this paper are provided by Wanhua Energy-saving Building Materials Co., LTD. Polyurethane composite panels make full use of the good cohesion of polyurethane foam firmly bonded to the surface board which is fiber cement surface board or color steel plate. According to the test scheme[8], the bending test of sandwich panels is carried out on the YZS-500 hydraulic testing machine. The specimen is supported by two high strength steel columns and the load is applied to the middle of the composite panel with cuboid steel beam. The moment in the middle of specimen is M=PL/4, the values of the load applied and the mid-span deflection of the panels are measured by the sensor and the displacement meters respectively. Finally, the curve of load-deflection is plotted. Figure.1 shows the simplified experiment device.

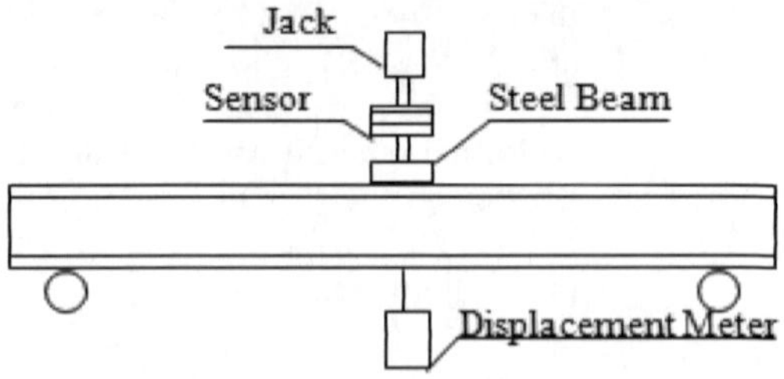

Figure 1. Experiment device.

2.2. Experiment results

Figure.2 shows mid-span load-deflection curves of bending test for the polyurethane composite panels in four groups. It can be seen from the figure that the stiffness of PCP-2 is higher and bearing capacity is stronger than PCP-1, because of the fiber cement board of PCP-2 being thicker than PCP-1. Similarly, compared with PCP-3, PCP-4 has thicker color steel surface plate, so it accordingly presents higher stiffness and stronger bearing capacity. Comparing PCP-3 with PCP-2, it is found that the bearing capacity of PCP whose lower surface board is color steel plate is stronger than PCP whose lower surface board is fiber cement board.

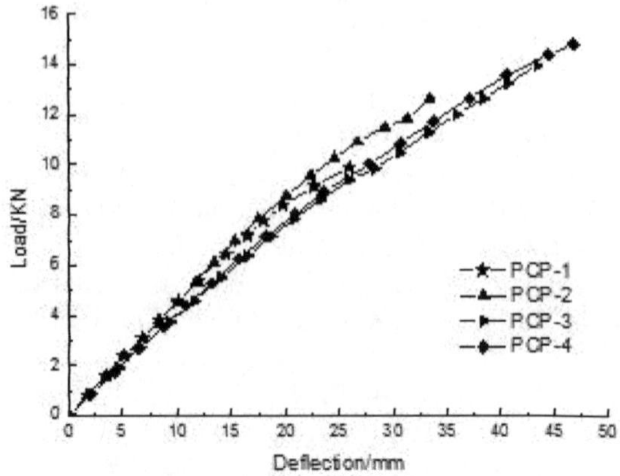

Figure 2. Load-deflection curve.

In conclusion, the thicker the surface plate of sandwich boards is, the higher the stiffness and the stronger bearing capacity become. On the same conditions, the color steel plate could be used to improve the ductility of polyurethane composite panels.

IOP Conf. Series: Materials Science and Engineering **242** (2017) 012007 doi:10.1088/1757-899X/242/1/012007

3. Finite element model

3.1. Three dimensional finite element model

Table1 shows the size on four groups of specimens. Although thickness of surface boards or polyurethane composite panels is different from each other, the length and the width are same. Figure.3 shows the finite element model established in accordance with the practical specimen.

Table 1. Specimens size.

PCP Groups	Plane Size /mm	Span /mm	Thickness of Upper Surface Board/mm	Thickness of Lower Surface Board/mm	Thickness of Polyurethane /mm
PCP-1	2000×1000	1800	6board A	6board A	68
PCP-2	2000×1000	1800	8board A	8board A	64
PCP-3	2000×1000	1800	8board A	0.5board B	71.5
PCP-4	2000×1000	1800	10board A	0.5board B	69.5

Note: board A means fiber cement board, board B means color steel plate.

The bearings are 15mm diameter circular columns, and the steel beam applying load is a 15mm×15mm×1000mm cuboid. Solid element of C3D8R is used to establish the model of surface board and polyurethane core. The bearings and the steel beam applying load in the finite element model are set as rigid body. The displacement load is applied by steel beam at the rate of 2mm/min. Except U3 and rotational displacement, the rigid body displacement are restrained (the direction should is shown in Figure.3). Boundary conditions of finite element model correspond to experiments. The model is simplified as a simply supported beam, which is depicted as Figure.4. Because there is no degumming during experiments, tie as an ideal contact is used between panel and core material. The contact between steel beam and sandwich board as well as between bearings and composite panel is surface to surface, and finite sliding is also considered. Grid of the model established in finite element software is divided into hexahedron.

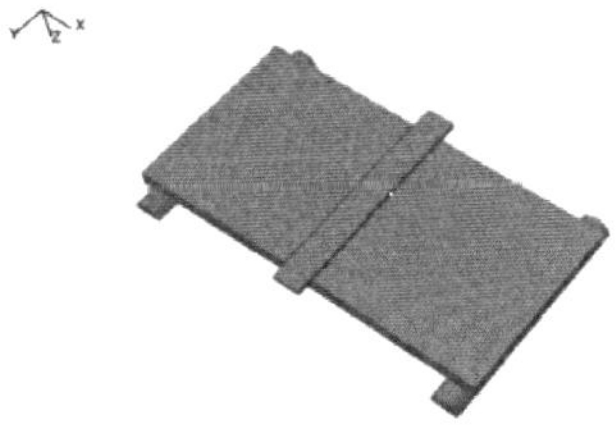

Figure 3. Finite element model.

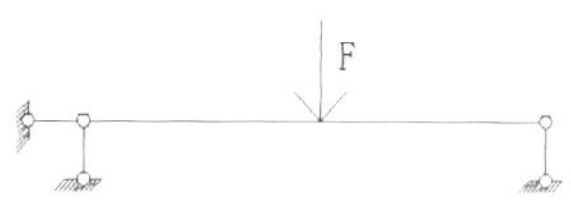

Figure 4. Boundary conditions.

According to design standard in China[9], the mid-span deflection of PCPs is not allowed to exceed L/200, the letter of "L" standing for the span of PCPs. It can be concluded from table 2 that the failure load is three or four times as large as the load corresponding with f=L/200. Figure. 2 illustrates the curves of load-deflection performed in elastic stage before deflection reaching L/200. Therefore, this paper only study the elastic stage of PCPS.

Table 2. Experiment Results of Polyurethane Composite Panel.

Number of PCPs	Load(f=L/200)/KN	Destruction Load/KN
PCP-1	3.22	9.83
PCP-2	3.63	12.39
PCP-3	3.12	13.94
PCP-4	3.27	14.65

3.2. Materials description

Materials parameters involved in this paper are based on the performance test report on wall materials produced by Wanhua Energy-saving Building Materials Co., LTD. Because the materials used in the experiment were not the same batch, the material parameters used as analyzing are different slightly. It should be noted that the problem mentioned above have little influence on the accuracy of the analysis. The specific parameters are shown in table 3.

Table 3. Material Parameters.

Materials	Elasticity Modulus E/Mpa	Yield Strength f_y/Mpa	Poisson Ratio μ
Polyurethane PCP1/2	10	0.142	0.3
Polyurethane PCP3/4	8	0.142	0.3
Fiber cement board	3200	—	0.22
color steel plate	210000	345	0.3

Note: "-"means not applicable.

4. Results and discussion

4.1. Stress nephogram

Figure.5 shows the strain nephogram for PCP1/PCP2 in finite element method. It's found that the model established can simulate the deformation of specimens well. Figure. (a) is the stress cloud of a whole PCP. It can be seen that the tensile stress of lower surface plate is largest. The stress of core material is so smaller that it's not obvious in figure (a). In order to make the stress of polyurethane to be more intuitive, figure (b) specially shows stress nephogram of core material. It can be seen from figure (b) that the phenomenon of stress concentration occurs on both sides of the cuboid steel beam in a certain angle, which corresponds with failure locations of specimen shown in Figure.(c) . Since PC1/PC2 made of brittle materials belong to brittle members, compressive property of which is much better than tensile property. Therefore, with the increase of load applied, though compressive stress concentration occurs in the finite element model, there is no phenomenon of compression failure occurring. On the contrary, when the tensile stress of lower surface board which is fiber cement board exceeds the tensile strength, lower surface board occurs brittle fracture failure. Almost at the same time, core polyurethane occurs similar fracture.

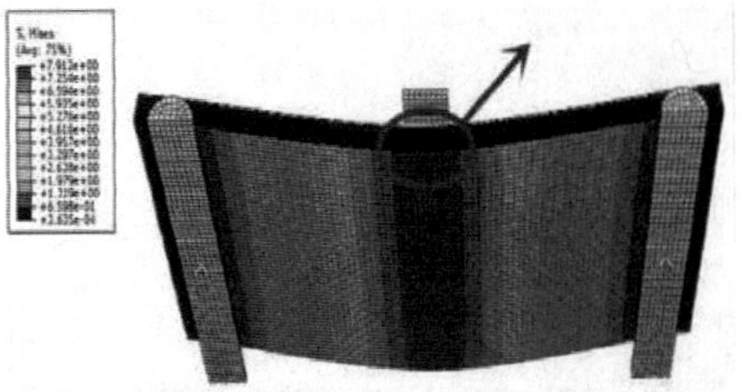

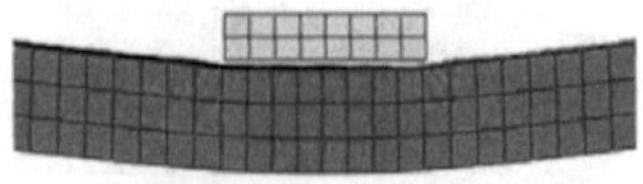

(b) Core material stress nephogram.

Figure 5. (a) PCPs stress nephogram. (c) Specimen failure.

The stress nephogram of PCP3/PCP4 which is established with finite element method is shown in figure 6. The meaning of Figure.(a) to Figure.(c) is as same as PCP1/PCP2 mentioned above. It is found from figure (a) that the stress under steel beam is maximum. However, it's not the main cause of fracture because it's compressive stress. There is a little difference of fracture path between Figure.(b) and Figure.(c). The potential causes of error between simulation and test are as follows: on the one hand, the material and the interface in finite element model is assumed as homogeneous and ideal respectively. On the other hand, there may be initial imperfection in the specimen, then, the weak part in PCPs may crack during the test.

IOP Conf. Series: Materials Science and Engineering **242** (2017) 012007 doi:10.1088/1757-899X/242/1/012007

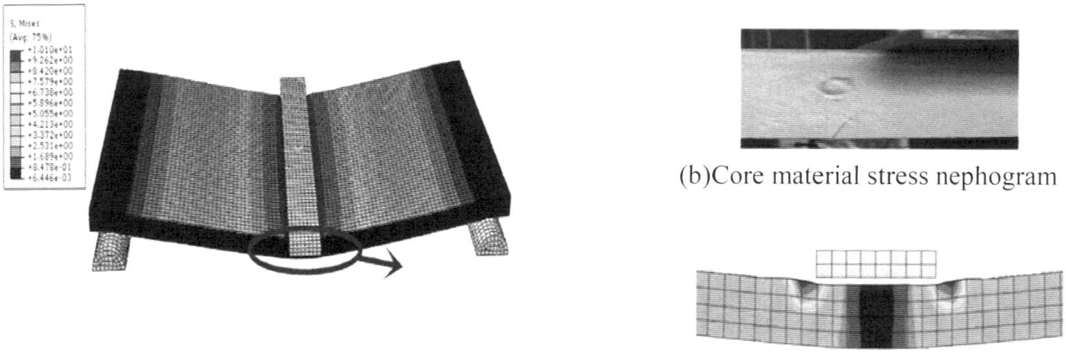

(b)Core material stress nephogram

Figure 6.(a) PCPs stress nephogram (c) Specimen failure

4.2. Load-deflection curve

Load-deflection curves of FEA and Experiment are shown in Figure.7-Figure.10. FEA represents the results of finite element analysis, and experiment curves are the results of tests. From these results, we can find that the load-deflection curves of banding performance for PCPs are linearly distributed, which means the specimens are in elastic stage before destruction. The potential reasons existing more obviously inaccuracy between two lines of curve in figure 7 may be as follows. Firstly, there is initial imperfection in PCPs. Secondly, it's impossible to conduct an experiment in ideal conditions, maybe there is a certain friction between high strength steel columns and PCPs. Thirdly, during the test, the stiffness of PCPs is increased because of bearings influenced by artificial factors or external factors which are not ideal hinge bearings. However, the results of simulation coincide well with test in general. It presents important reference significance for the choice of polyurethane PCPs in practical projects.

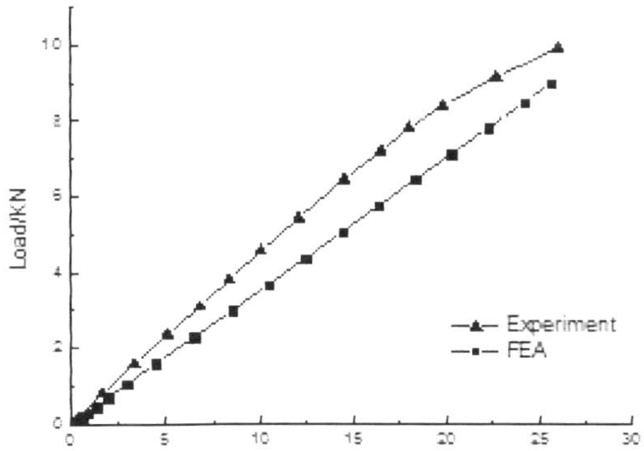

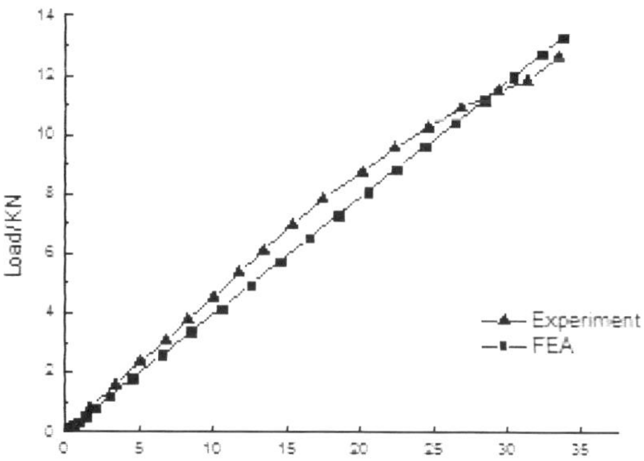

Figure 7. Mid-span load-deflection curve of PCP1. **Figure 8.** Mid-span load-deflection curve of PCP2.

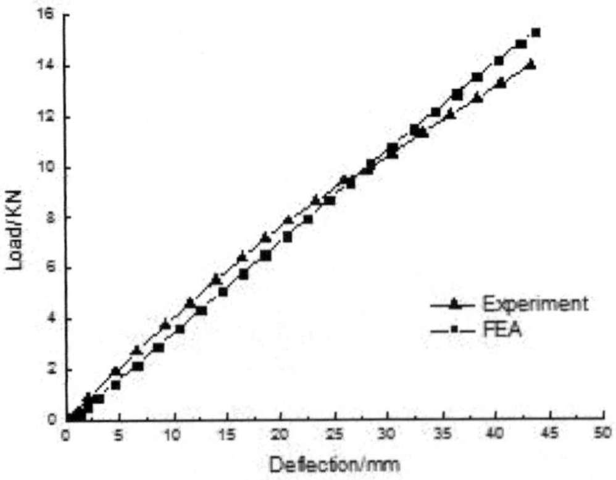

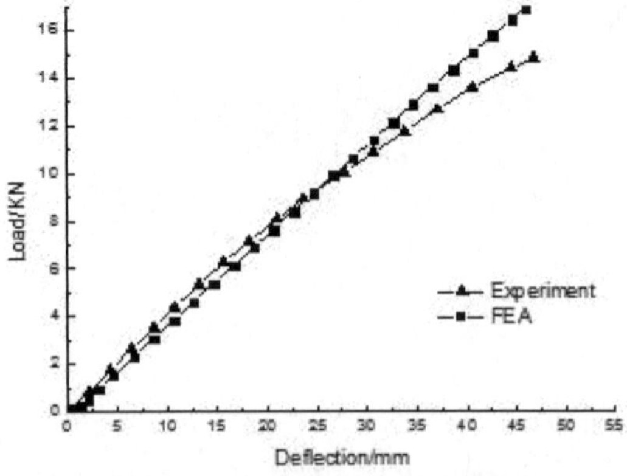

Figure 9. Mid-span load-deflection curve of PCP3. **Figure 10.** Mid-span load-deflection curve of PCP4.

5. Conclusions

(1)Finite element models for the bending performance of polyurethane composite panel are established. Comparing specimens experiment data with stress nephogram coming from ABAQUS, it's found that the failure feature of PCPs whose lower structural surface board is fiber cement board presents brittle fracture under vertical load, failure beginning with lower surface board. The failure of PCPs whose lower structural surface board is color steel plate also shows the feature of brittle fracture, but failure beginning with the bottom of core material. Both kinds of failures what have been mentioned above are tension failures. And the specimen failure coincides well with stress cloud.

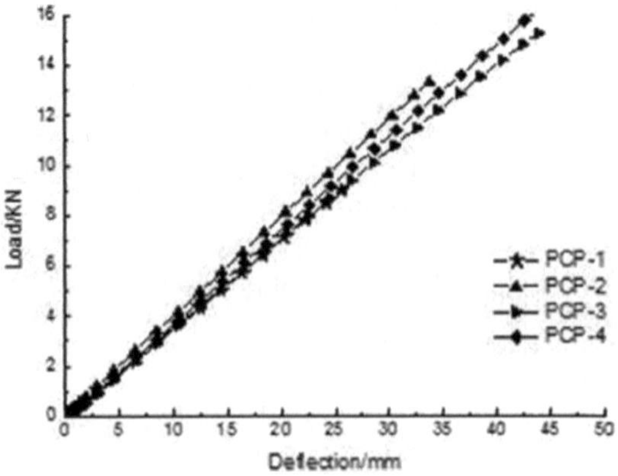

Figure11. Mid-span load-deflection curve of FEA.

(2) In view of bending performance of polyurethane composite panels with different thickness and materials of surface board, comparing the simulation results got through finite element method (shown in Figure.11) with the results from experiments (shown in Figure.2), it can be seen that their tendencies are similar. Both the simulation and the experiment results show that the thicker the surface plate of PCPs is, the higher their stiffness and the stronger their bearing capacity become. On the same conditions, the color steel plate could be used to improve the ductility of polyurethane composite board.

Acknowledgments

The article was financially supported by Wanhua Energy-saving Building Materials Co.,LTD. I am also grateful for the assistance from Mr Li and Mr Wang.

References

[1] Zha X X, Tang Z R and Feng L. 2014. Bending properties of sandwich panels with calcium-silicate faces and polyurethane cores [J]. *J. Har. Ins. Technol* **46** (6) 1-7

[2] Yu M, Zha X X, Yang S, et al. 2008. Finite element analysis on mechanical properties of polyurethane and XPS sandwich panel[J]. *J. Har. Ins. Technol* **40** (12) 1914-18

IOP Conf. Series: Materials Science and Engineering **242** (2017) 012007 doi:10.1088/1757-899X/242/1/012007

[3] Davies J M. 2001. Lightweight Sandwich Construction[J]
[4] Kawasaki T, Zhang M, Wang Q, et al. 2006 *J. Wood. Sci* **52**(4) 302-10
[5] Morley, M. 2000. Building with structural insulated panels (SIPs): Strength and Energy Efficiency Through Structural Panel Construction [M]. *Newtown.Taunton Pres.*
[6] Li H Q, Yue T H, Guo Z H. 2013. Experimental Study on Compressive Performance of Polyurethane Composite Panel[J]. *Advanced Materials Research* **638** 2693-96
[7] Styles M, Compston P, Kalyanasundaram S. 2008. *Composite Structures* **86**(1–3) 227-32
[8] Li H Q, Wang X M, Sun Z G. 2012.Experimental Study on the Bending Capability of Polyurethane Composite Panel[J]. *Applied Mechanics & Materials* **174-177** 1042-46
[9] GB/T23932-2009 Double skin metal faced insulating panels for building[S]. *Beijing. Standards Press of China*

IOP Conf. Series: Materials Science and Engineering **242** (2017) 012008 doi:10.1088/1757-899X/242/1/012008

Thermal stress analysis on 2D C/SiC composites of thin-walled structure of anti-heat head cone

Wei Zhang, Shengli Lv

Science and Technology on UAV Laboratory, Northwestern Polytechnical University, Xi'an, 710072, China

Corresponding author: zw15836085@nwpu.edu.cn

Abstract. The thermal protection performance of C/SiC composites can be improved by coating the thermal barrier coating on the surface of C/SiC composites.The wall heat flux density distribution of the 2D C/SiC composite thermal insulation structure at the front end of the hyper-sonic vehicle was calculated by steady state heat transfer at different Mach numbers in the atmosphere. In the thermal stress calculation, an effective finite element numerical model is established by the method of partitioning and based on the flexible gradient density method. The model was treated as heterogeneous materials which consists of C/SiC composite region, the coating zone and transition zone, and the results show that the maximum thermal stress appears near the end of the head cone.

1. Introduction

The thermal barrier coating on the surface of C/SiC composites can greatly improve the high temperature performance of the material and has been applied in the anti-heat structures and engines of the spacecrafts. The C/SiC composite structure with thermal barrier coating can be considered as a structural system composed of multi-layer composite structure. The structure system is usually composed of substrate, bonding layer and thermal barrier coating [1].The finite element numerical simulation can effectively reduce the number of tests. However, many scholars have not considered the influence of coating when calculating the mechanical properties of C/SiC composites, which makes the calculation model not consistent with the actual structure. Of course, the reliability of the calculation results is poor [2].The effect of thermal barrier coating on the damping, mass and stiffness of 2D-C/SiC composites should not be neglected [3]. Recently, the minimum net cross-section images of 2D-C/SiC composites were observed, and the distribution of the thermal barrier coating, the macroscopic holes and the density distribution gradient were obtained. The results of the finite element prediction were in agreement with the experimental results better[4].In this study, the finite element model of 2D-C/SiC composites included the thermal barrier coating , and the thin-walled structure response under the action of aerodynamic heat was calculated.

2. Numerical calculation model of aerodynamic shape and finite element model

The hypersonic high-speed cruise missile front-end anti-heat cone structure was taken as the object of study [5]. To ensure a better aerodynamic shape, the geometric parameters has referred to a model of hypersonic cruise missile pneumatic structure. The anti-heat head cone structure is mainly composed of internal metal material separator and frame, as well as external high temperature resistant C/SiC composites with thermal barrier coating with a thickness of 6 mm. The top of the C/SiC composite head cone is spherical and the inner diameter R is 28mm.The rear is the cone, the rear end diameter D

Content from this work may be used under the terms of the Creative Commons Attribution 3.0 licence. Any further distribution of this work must maintain attribution to the author(s) and the title of the work, journal citation and DOI.
Published under licence by IOP Publishing Ltd

is 366mm; the total length of the cone structure is 450mm (Figure 1).

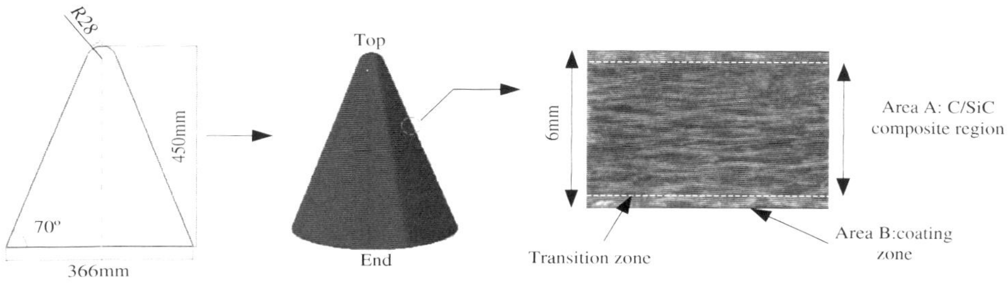

Figure 1. 2D-C/SiC composites thin-walled structure.

It can be seen from the sectional view that the material has a significant partition (Figure 1).The zones are divided into zones A and B, representing the C/SiC composite region and the coating zone respectively, which have different density and mechanical properties. In the zones A, there are a lot of sparse holes, and zones B is more dense, almost no holes exist. Transition zone is very narrow between A and B. The apparent partition of the specimen is mainly due to the fact that the C/SiC composite prepared by the CVI process will form a dense layer with very low porosity on the surface.There is a significant density gradient from the inside to the outside, and the density values of the materials in zones A and B are very different. There is a significant transition between the two zones (White dotted line in Figure 1). Zone A, transition zone and zone B accounts for 84%, 2%, 14% of the thickness respectively. The mechanical properties are shown in Table 1, in which properties of transition zone and zone B were predicted by linear extrapolation based on the density.

Table 1. Properties of coated 2D-C / SiC composites at room temperature. Unit: GPa

District	$E_1 = E_2$	E_3	$\mu_{23}=\mu_{13}$	μ_{12}
C/SiC zone[2]	102.2	42.4	0.23	0.15
Transition zone	122.64	50.88	0.23	0.15
SiC coating zone	480	--	--	0.17

3. Calculation of thin-walled structure of thermal head cone

3.1. Aerothermal results

In order to reduce the computational complexity and improve the computational efficiency, combined with the time of heat transfer effect, every 10 seconds was selected as a time period in total reentry time (130 seconds) in the Fluent software. The SST K-ω turbulence model was used to calculate the relationship between the wall heat flux density and the reentry phase. The steady state heat transfer was used to calculate the heat flux distribution of the wall under different Mach number (Figure 2) [6], and the value under steady state was taken as the wall heat flux at different time. The distribution of the heat flux along the axis of the wall was calculated as follows: The extreme points appear at the top of the head cone, and the heat flux fluctuates between 0 and 10mm. The heat flux between 10 and 200mm has little fluctuation, between 200-450mm heat flux tends to be gentle (Figure 3). Based on the analysis of the distribution law of the wall heat flux density and the characteristics of the structure, the head cone structure was divided into nine regions along the axis: 0~10mm was divided into a region, 10~200mm equidistant divided into three regions, 200~450mm equidistant divided into five regions.

IOP Conf. Series: Materials Science and Engineering **242** (2017) 012008 doi:10.1088/1757-899X/242/1/012008

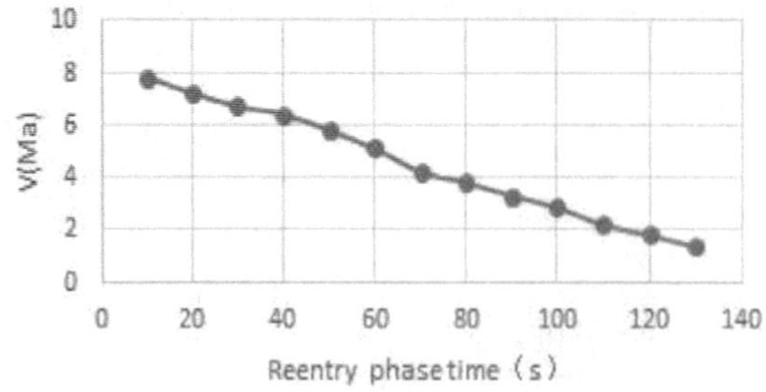

Figure 2. Relationship between velocity and reentry times.

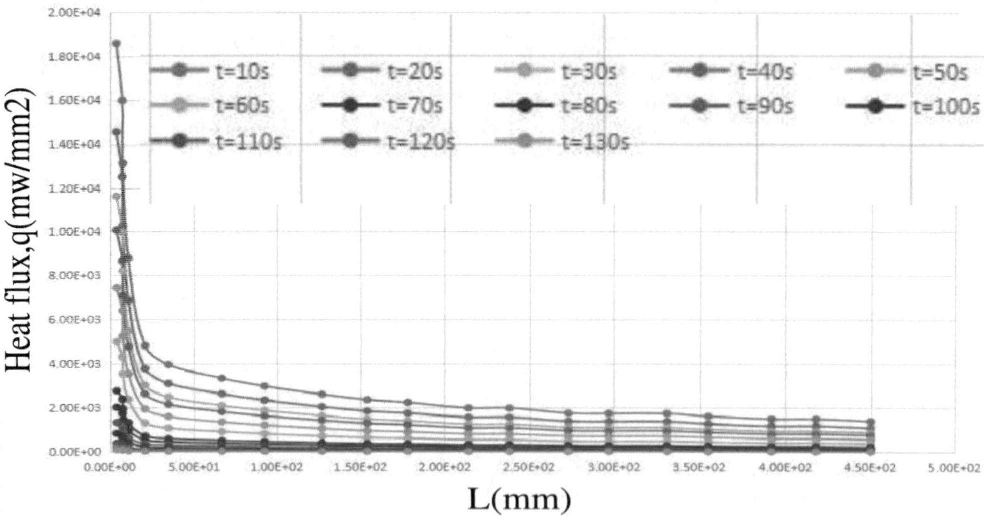

Figure 3. The heat flux in different zones.

3.2. Thermal stress calculation results

The thermodynamic coupling calculation can be divided into 12 intervals(Figure 4).The change of heat flux in each interval was obtained by fitting and amplitude loading. The initial value was taken as 1.As the time increases, the amplitude changes from 1 to 0.009, and the changing relation is as the same as the relation between maximum heat flux and time. In order to apply the heat flux more accurately, analytic field was applied to the structure in the ABAQUS. Select the different moments and the changing relationship between the heat flux and the function along the axis were adopted. The relationship between elastic modulus, shear modulus, thermal expansion coefficient, specific heat capacity, heat transfer coefficient and temperature were used according to the literature [7,8]. In the thermodynamic coupling calculation, the temperature field and the force field coexisted and interact with each other, and need to be solved at the same time [9]. The S4RT shell element was used to calculate the stress in each numerical point along the thickness direction (integral points) in the numerical calculation. The trend of temperature change in 12 regions of the cone is almost the same, and with the increase of reentry time, the temperature rise trend of each region becomes gentle (Figure 5), and the max temperature is 2104°C.As the time increases, the gradual decrease in the velocity of the flight causes the surface heat flux to become smaller, while the increase in the structural temperature caused by the aerodynamic heating becomes smaller. But there is a temperature difference between the different regions that will lead to differences in the distribution of structural thermal stress and thermal deformation. The thermal stress of the structure increases greatly in the first 20 s, and the increase range is less in 40~120s. The maximum stress was 118.3MPa below the ultimate strength at 80s (Figure 5) in the vicinity of the end of the cone, not in the region with the highest temperature, and

IOP Conf. Series: Materials Science and Engineering **242** (2017) 012008 doi:10.1088/1757-899X/242/1/012008

the smallest stress appeared at the top of the cone.

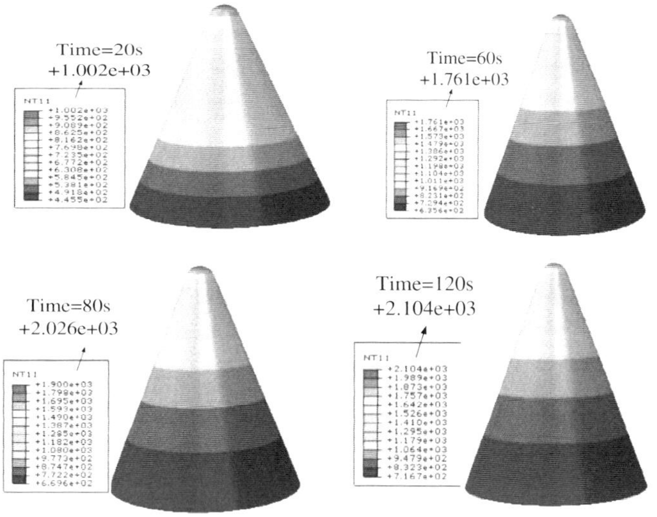

Figure 4. Temperature Distribution.

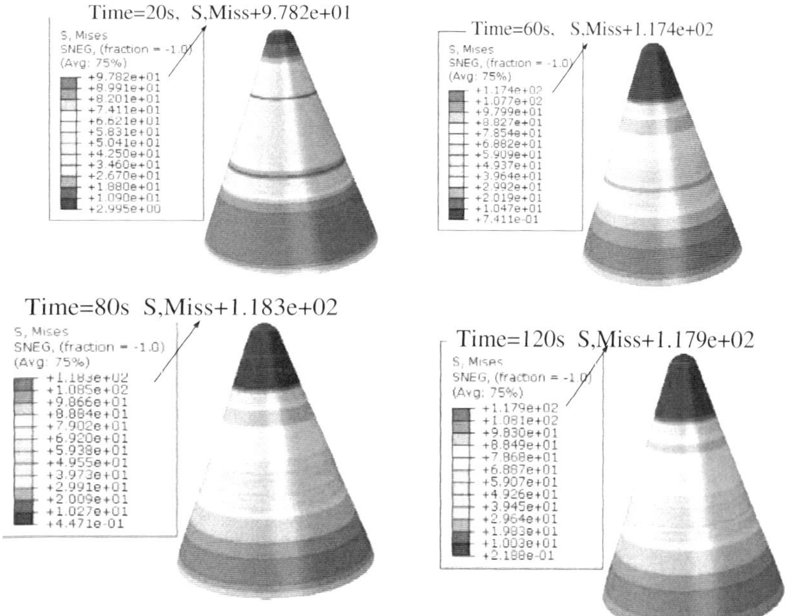

Figure 5. Thermal stress distribution.

4. Conclusion

The structural force is entirely transmitted to the end of the cone, which is the main bearing zone. In the high temperature region of the top, after the peak time the unit heat absorbed from the outer surface decreases with the increase of the reentry time. This is because the thickness of the thin-walled structure is only 6mm, the temperature difference between the inside and outside the wall gradually becomes smaller, which reduces the thermal stress. The maximum thermal stress of the structure occurs at a time of 3/4 near the reentry time due to the fact that the temperature difference between the inside and the outside of the wall is the largest, and the temperature difference of the thin-walled structure becomes smaller after that time. The above-mentioned reason causes the resultant force to be transmitted to the end of the head cone becoming smaller.

IOP Conf. Series: Materials Science and Engineering **242** (2017) 012008 doi:10.1088/1757-899X/242/1/012008

References

[1] Yichun Zhou,Qixing Liu,Li Yang,et al. Failure Mechanisms And Life Prediction Of Thermal Barrier Coatings.*Chinese Journal Of Solid Mechanics*. 2010, 31(5):504-31.

[2] Lv Y, Lv S L, Lv G Z. Integrated predication of macroscopic properties of composites based on finite element computational micromechanics. ICAS 2008, Anchorage, USA, SEP 10-14, 2008.

[3] Gregory N, Morscher.Modeling the elastic modulus of 2D woven CVI SiC composites. *Composites Science and Technology*. (2006)66:2804-14.

[4] Guo HongBao.FEM calculation about tensile strength of 2D-C/SiC composites with circular holes.*Acta materiae Compositae Sinica*. 2014,31(2):448-455.

[5] Yao Enliang. Numerical Computation and Research on Aerodynamic Heating over Aircraft of Ball-cone Shape at Supersonic flux. Nanjing University of Science&Technology in China.2015.

[6] Guo HaiFeng,Huang ChangQiang,et al.Re-entry Trajectory Optimization For Supersonic Missile Considering multiple Constraints.*Journal of ballistic(in China)*.2013,25(3):10-15.

[7] LI Kaiyuan,XU Yongdong,Zhang Litong,et al. Effects Of Fabric Architectures On The Thermal Expansion Coefficient And The Thermal Diffusivity Of Carbon Fiber Reinforced Silicon Carbide Compostes. *Journal Of The Chinese Ceramic Society*. 2008,11:1564-69.

[8] Zhang Qing. Coefficients Of Thermal Expansion Of C/SiC Carbide Composites[D]. Northwestern Polytechnical University of China.2004.

[9] Zhang Bi-feng. Original And Service Properties Of C/SIC Composites By Pip Method [D].National University of Defense Technology of China 2010.

Application of lightweight materials in structure concept design of large-scale solar energy unmanned aerial vehicle

Wei Zhang[1] Shengli Lv[1] XiQi Guan[2]

[1]Science and Technology of UAV Laboratory, Northwestern Polytechnical University, Xi' an, China

[2]College of Aeronautics,Northwestern Polytechnical University, Xi'an, China

Corresponding author: zw15836085@nwpu.edu.cn

Abstract: Carbon fiber composites and film materials can be effectively used in light aircraft structures, especially for solar unmanned aerial vehicles. The use of light materials can reduce the weight of the aircraft, but also can effectively improve the aircraft's strength and stiffness. The structure of the large aspect ratio solar energy UAV was analyzed in detail, taking Solar-impulse solar aircraft as an example. The solar energy UAV has a wing aspect ratio greater than 20, and the detailed digital model of the wing structure including beam, ribs and skin was built, also the Finite Element Method was applied to analyze the static and dynamic performance of the structure. The upper skin of the wing is covered with silicon solar cells, while the lower skin is light and transparent film. The single beam truss form of carbon fiber lightweight material is used in the wing structure. The wing beam is a box beam with rectangular cross sections. The box beam connected the front parts and after parts of the ribs together. The fuselage of the aircraft was built by space truss structure. According to the static and dynamic analysis with Finite Element method, it was found that the aircraft has a small wingtip deflection relative to the wingspan in the level flight state. The first natural frequency of the wing structure is pretty low, which is closed to the gust load.

1. Introduction

Setting solar cells on the aircraft and manufacturing solar powered aircraft is the most creative application of solar cells. Transforming solar radiation energy into electric energy to drive the electric propeller for thrust, the solar powered aircraft can stay at an altitude of 10-30km for several months even years, contributing itself a bright prospect on environmental monitoring, border patrol, data link relay, etc[1,2]. Apart from related conception research, many researchers have completed the engineering validation study of High-altitude Long-endurance Unmanned Aerial Vehicle. As one of the famous solar powered drone, American "Helios" [3] lasted for 4 hours at an altitude of 29,531m in a region which is 18 degrees north latitude of Hawaii on August 13, 2001, breaking the record of maximum ceiling of propeller powered fixed-wing aircraft. As for the most well-known solar powered aircraft, "solar-impulse" adopts conventional wing layout with a span (63m) approximately equal to A380 of AIRBUS, but surprisingly, it merely weighs like a family car. What's more, there are totally 11.628 thousand solar cells covering an area of 200m2 on its straight wing and horizontal tail. Currently "Solar-impulse" can't fly at night and above 10 thousand meters. In Feb, 2014, the advanced "Solar-impulse II" finished its first flight. Compared with the former, its weight of structure has increased from 1600kg to 2300kg, its span from 63m to 72m. The aircraft carried out a global flight in 2015 and passed through China. In this study, the application of lightweight materials in the

Content from this work may be used under the terms of the Creative Commons Attribution 3.0 licence. Any further distribution of this work must maintain attribution to the author(s) and the title of the work, journal citation and DOI.
Published under licence by IOP Publishing Ltd

conceptual design of solar energy unmanned aerial vehicles was carried out including the static analysis and modal analysis of the wing structure.

2. Structure layout

The structure of the wing has a large surface ratio, large aspect ratio and large span characteristics. The main materials of wing are carbon fiber honeycomb composites, and the upper skin is covered with a layer of solar cells, while the lower skin is covered with a layer of lightweight transparent film. Wing ribs are installed between the two layers to support an form the aerodynamic shape. Like "Solar-impulse", low speed thick airfoil with large chord length was used in wing design, and also with slender wing trailing edge to expand the wing area for more solar cells. The performance of the "Solar-impulse" aircraft as a reference [4], the overall size parameters are as follows.

Table 1. Summary of Solar Impulse HB-SIA features [4].

Wing span	63.4m	Cruising speed	70km/h
Fuselage length	21.85m	Max. Altitude	8500m
Fuselage height	6.4m	Empty weight	1600kg
Takeoff speed	35km/h		

According to reference [4,5], the detailed measurements of wing structure were designed. The partial measurements of the wing and some parts are as follows: The wing is divided into inner and outer segments. The unilateral inner wing measures about 20.25m long, outer segment about 11.45m long including winglet. The chord length of inner wing is approximately 3.15m, and the dihedral of the outer segment is about 8 degrees. The whole wing has 120 wing ribs when 80 wing ribs are in inner wing and the other are in both left and right outer wings. The measurements of wing ribs of outer wing segment decrease linearly. And in the inner segment the largest wing rib is 1.5 times bigger than the smallest one.

The wing rib is composite structure, and the front and rear portions of it are joined together by a rectangular beam. The ribs can bear the local aerodynamic force transferred from the skin to rectangular beam then, which will produce bending moment on wing ribs, so it's effective for bearing the unexpected bending moment to add some support structure in wing ribs. The bending moment at wing root becomes incredibly large with the wing span extending, hence, it's important to decrease the weight of wing structure. New materials with high specific stiffness and lightening hole are efficiently designed for decreasing weight.

The wing beam is rectangular in shape and composed of closed box sections when the rectangular beam support all the wing structure, which solves the problem about the large bending moment at wing root for aircraft with large aspect ratio. The 7th wing rib from the plane of wing symmetry is installed with electric propeller, and it only contributes a little of moment on account of the short distance away from wing root, so its impact on the wing is not pivotal. Moreover, there is also a electric propeller at where the inner and outer wing segments joint and it can balance wing lift when the aircraft flies. The biggest rib is a truss structure with height of 0.42m, and it is divided into 6 sections, the length of each section is 0.8m, 0.7m, 0.6m, 0.3m, 0.3m and 0.45m from the back of the ribs (Figure 1).

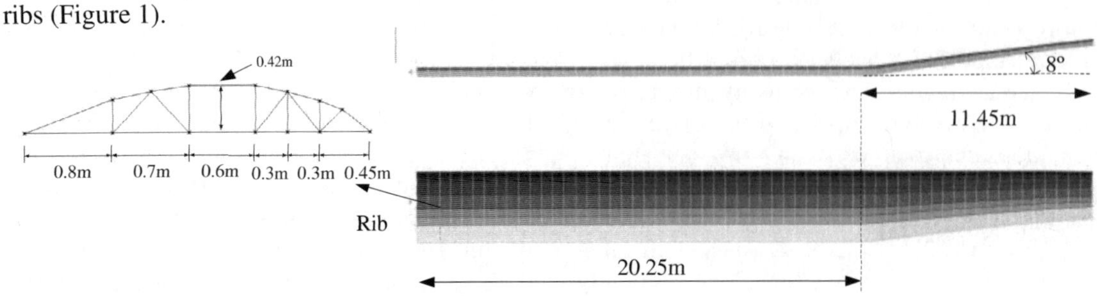

Figure 1. The wing and ribs layout.

The fuselage length is 21m, accounting for 1/3 of wing length, and the fuselage width is 0.4m (Figure 2). The nose which is truss structure which is located below the plane of wing symmetry, and there is a equipment compartment in the head, which can accommodate other equipment such as satellite antenna. The space truss structure is the main form of the aircraft fuselage, and 4 longitudinal flanges make up the main structure of the aircraft fuselage when vertical and oblique carbon fiber brackets support the adjacent flanges.

The horizontal tail is connected with fuselage likes upper wing layout, and horizontal stabilizer and elevator constitute the horizontal tail which has 24 ribs. The horizontal tail consists of a horizontal stabilizer and a rudder, containing 24 ribs, and uses approximately symmetrical airfoil with lightening hole on each rib which joints together with a rectangular beam. The horizontal tail measures 9.12m long and 0.33m wide, while the vertical tail is vertically embedded in the fuselage tail. Similarly the vertical tail uses symmetrical airfoil whose chord length is 1.65m, maximum thickness 0.33m, span 6.4m, and it has 13 ribs (Figure 2).

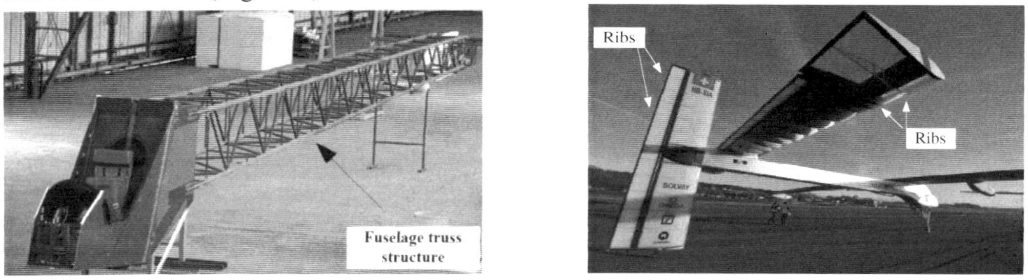

Figure 2. Fuselage detailed structures, horizontal and vertical tail detailed structures [5].

3. Static analysis of wing structure

The main material of the wing beam is carbon fiber which is quite suitable for aircraft with large aspect ratio because of its small density, large stiffness and large elastic modulus. However, decreasing the weight of wing ribs matters in contrast to the requirements of stiffness, and it means that carbon fiber honeycomb sandwich can be used. The skin is the primary part to support the aerodynamic force, so the small-density polyester film which doesn't damage easily when suffering huge deformation is suitable. The material properties are shown in the following table.

Table 2. Mechanical properties of lightweight materials [6].

Material	Density(kg/m^3)	Elastic Modulus	Poisson
Carbon-Fiber	1400	200	0.3
Carbon Fiber	350	100	0.3
PET Film	1450	3.495	0.3

The measurements of the hollow rectangular beam are as follows: width (0.5~0.6m), height (0.2~0.3m), wall thickness: 1~2cm. The weight (400N) of motor and corresponding propulsion device are respectively imposed on the 7th wing rib and the 40th rib from the fuselage in the form of concentrated force [5]. What's more, the weight of solar batteries and structure are imposed on the wing beam in the form of approximative distributed force. The structural finite element calculations were performed with the MSC. Patran/Nastran software under the action of gravity. After calculating and analyzing, the maximum displacement of wing is -0.808m showing at the wing tip, and at the wing root comes the maximum stress shown as 2.64MPa.In level flight state where the wing lift equals the aircraft weight. It is assumed that the whole lift is generated by the wing, and the lift on the wing is distributed according to the ellipse. The maximum displacement of the wing is 0.371m at wing tip, the maximum stress is 1.54MPa far below the ultimate strength at the wing root (Figure 3).

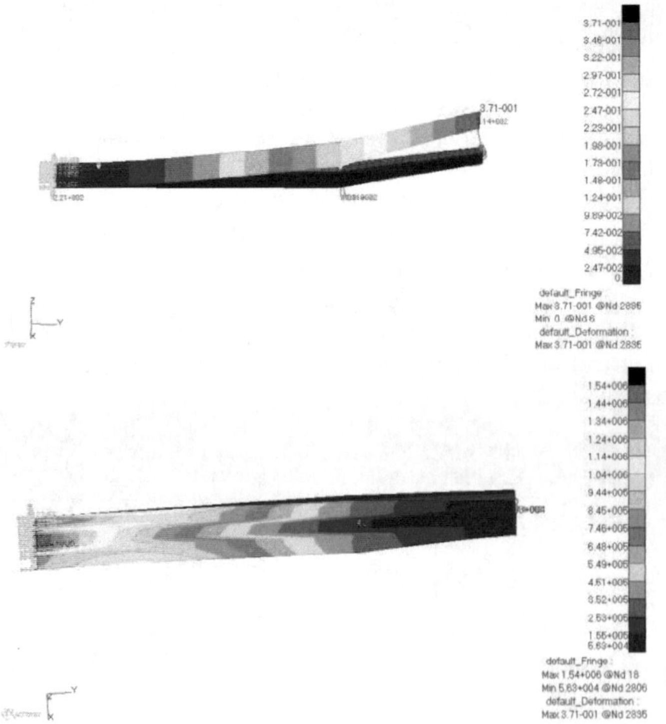

Figure 3. Displacement and stressdistribution in level flight state.

4. Natural modal analysis of the wing structure

The modal frequency of wing is found low through modal analysis, and its main modal frequency is in the range of 0-20Hz, basically consistent with the conclusion of reference [5]. The low order mode of wing is close to the frequency of gust load. The first-order modal frequency is about 0.6Hz and the vibration mode is a bending mode on the vertical direction.

Table3. Modal vibration frequency.

Modal	Frequency	Modal	Frequency
First	0.64	Third	3.98
Second	1.22	Fourth	7.11

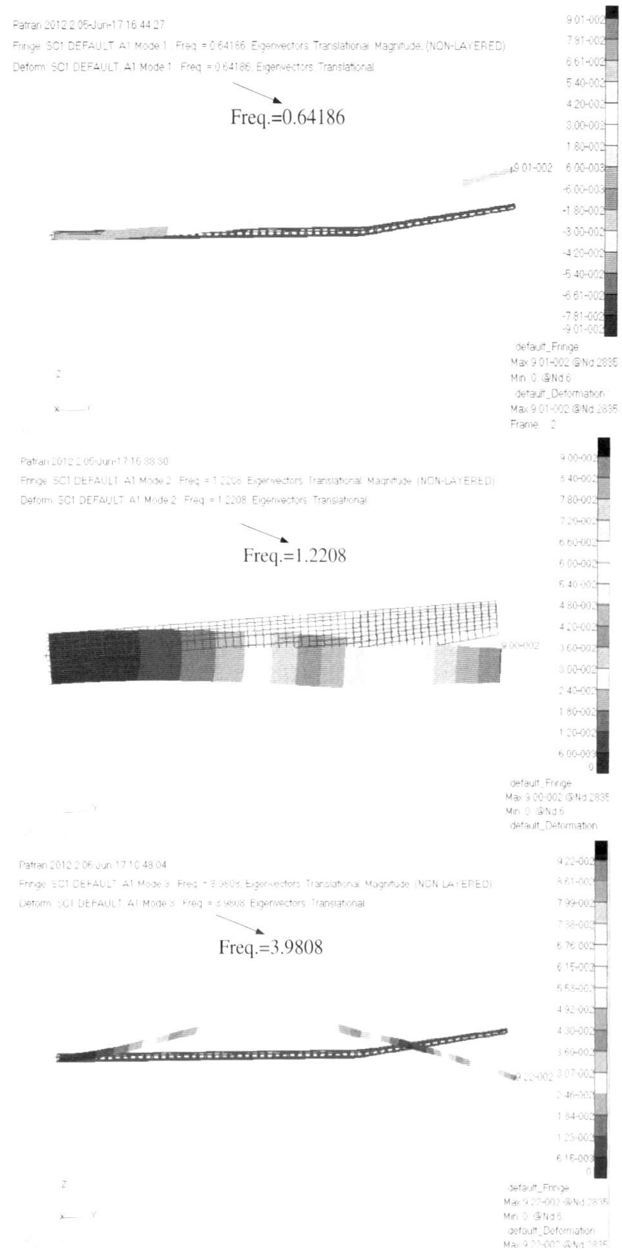
Freq.=0.64186
Freq.=1.2208
Freq.=3.9808

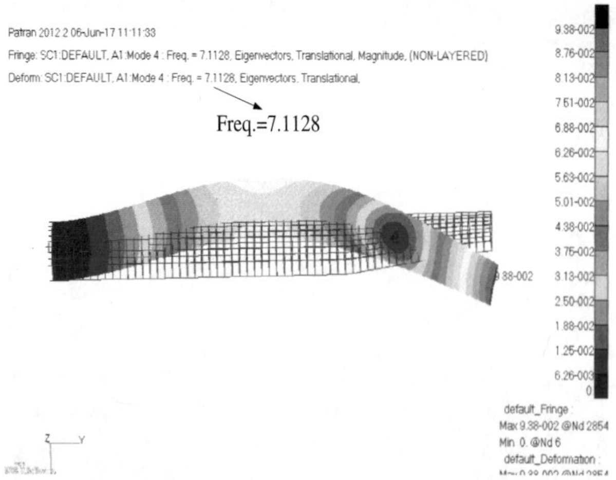

Figure 4. Modal vibration shapes (First-Fourth).

5. Conclusion

Lightweight materials were used in the Solar UAV structure design, which took the "Solar-impulse" aircraft as a reference. The aspect ratio of the wing is more than 20, resulting in low wing structure stiffness. If encountering gale and gust load, the wing structure will resonate with a huge bending deflection appearing, that may result in structural damage of solar batteries. The fragile mechanical properties of solar silicon cells taken into consideration, in order to achieve practical purposes, double fuselage structure layout will effectively increase the structural stiffness of such aircraft in future.

References

[1] Yan Yan-fa. Status and Challenges of Thin-Film Solar Cells.*Optics & Optoelectronic Technology*.2015.13(6):1-4.

[2] Chang min,Zhou Zhou,Wang Rui. Primary Parameters Determination for Year-round Solar-powered Aircraft of Wing-sail Type at Higher Latitudes.*Acta Aeronautica et Astronautica Sinica*.2013.34(1):1-11.

[3] Stanley R. Herwitz Orchestrating A near-real-time imaging mission in national airspace using a Solar-Powered UAV *AIAA* 2003-6617:1-10.

[4] Marc Boeswald, Arne Vollan, Yves Govers, Peter Frei3.Solar Impulse-Ground Vibration testing And Finite Element Model Validation Of LightWeight Aircraft . IFASD-2011-132.

[5] Around the word in a solar airplane. [EB/OL]. [2015-03-24].http//:www.solar-impulse.com.

[6] Yin ChengYue,Gao PuYun,Guo jian,et,al.Mechanical performance analysis on wing's of Zephyr7 UAV. *Structure & Environment Engineering*.2009,39(3):19-25.

Influences of powder granularity on crystallizing characteristics in mica-contained glass ceramic

L N Xu[1, 2, 4], D Y Kong[2], Q B Tian[2], Z J Lv[3]

[1] Shandong Co-innovation Center of Green Building, Shandong Jianzhu University, Jinan 250101, Shandong, China

[2] School of Materials Science and Engineering, Shandong Jianzhu University, Jinan 250101, Shandong, China

[3] School of Mechanical and Electrical Engineering, Shandong Jianzhu University, Jinan 250101, Shandong, China

[4] Author to whom any correspondence should be addressed, xln@sdjzu.edu.cn

Abstract. A machinable mica-contained glass ceramic in the SiO_2-Al_2O_3-MgO-F glassy system was prepared by ball milling and hot pressed sintering. Three kinds of powder sizes of base glass were chosen and the effects of the glass powder sizes on the crystallization were explored by x-ray diffraction and scanning electron microscopy techniques. The results indicate that mica crystal as a major phase and $KFeSi_2O_6$ and mullite as minor phases are crystallized. Applying pressure at 670°C has little influences on the types of crystal precipitated and the preferential growth of crystal. The powder sizes, however, have obvious effects on the morphology of precipitated mica crystals. In the glass sample with a powder size of d_{50}=16.4 μm, the plate-shaped mica phase is precipitated. As the powder size decrease to 9.9 μm and 3.3 μm, however, the particle-shaped mica is formed instead of the plate-shaped crystals.

1. Introduction

Mica-contained glass ceramics based on SiO_2-Al_2O_3-MgO-F glassy system has gained increasing attention owing to high heat shock resistance, high electrical insulation, and especially the good machinability. The properties of glass ceramic are dependent on microstructure, the characteristics of mica phase precipitated and the ratio of residue glassy phase [1-5]. Recently, the synthesis of glass-ceramic by powder sintering has been carried out to improve the properties of mica-based glass-ceramics for broadening the application fields of the glass ceramics [4, 6-10]. The crystallization by powder sintering has several advantages compared with the traditional bulk devitrification: 1) it is less limited for the glass forming ranges due to the efficient surface crystallization; 2) it is convenient to produce glass-ceramic-based composites by mixing fibers, ceramic particles or mineral phases with the glass powders; and 3) the complex shapes and the accuracy are easily controlled by accurately designing the mold and forming method [7, 8, 10-13]. During the powder sintering, the characteristics of glass powders have greatly influences on the crystallization of glass ceramics [10, 11, 14]. In this paper, the mica-composition glass powders with various sizes are to fabricate machinable glass ceramics by hot pressed sintering. The properties of the glass powders and the crystallizing characteristics of glass ceramics obtained are investigated.

Content from this work may be used under the terms of the Creative Commons Attribution 3.0 licence. Any further distribution of this work must maintain attribution to the author(s) and the title of the work, journal citation and DOI.
Published under licence by IOP Publishing Ltd

IOP Conf. Series: Materials Science and Engineering **242** (2017) 012010 doi:10.1088/1757-899X/242/1/012010

2. Experimental

The base glass composition is listed in Table 1. All the raw materials are pure chemical agent. The base materials were accurately weighed and mixed homogeneously in corundum crucible. The mixtures were melted at 1430°C for 2 h and quenched into cooling water to get the base glass particles. Then the glass particles were ball milled for various times in the high pure-Al_2O_3 tank. After ball milling, the particle distribution of glass powders was measured by Laser Particle Size Analyzer (LPSA: Winner 2000s, Winner Ltd, China). The obtained glassy powders were filled into the graphite die, then the temperature was heated to 670 °C, and a pressure of 40 MPa was applied at the same time. The pressure was unloaded after the holding time for 1 h and the temperature was heated to 950 °C for 1h to obtain glass ceramics. The crystalline phases precipitated in the glass ceramics were identified by X-ray Diffraction (XRD: D/MAX-YB, Japan). The microstructures of the samples were examined in Scanning Electronic Microscopy (SEM: JCXA-733, JEOL Ltd, Japan).

Table 1. Ingredients of base glass (mass %)

Sample	SiO_2	B_2O_3	Al_2O_3	MgO	F	K_2O	ZrO_2	Fe_2O_3
%	35.46	6.6	16.49	13.47	16.49	8.52	2.47	0.5

3. Results and discussion

Figure1 shows the cumulative distribution of the glass powders with various ball-milling times. From the curves of the cumulative distribution, it can be seen that the particle sizes of glass powders gradually decrease with the lengths of milling time. The particle sizes of glass powders by 1 h, 5 h and

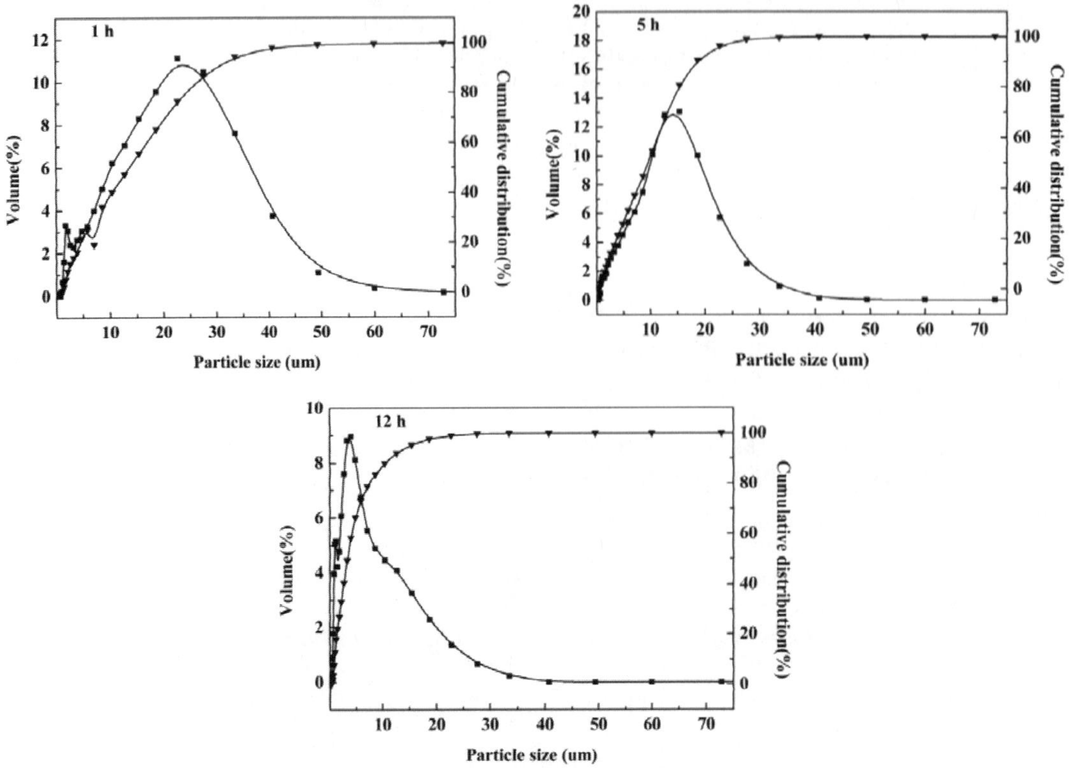

Figure1. Particle size distribution of different samples milling for various times

12 h ball milling are d_{50}=16.4 μm, 9.6 μm and 3.3 μm, d_{90}=30.1 μm, 19.2 μm and 10.2 μm, respectively. The XRD pattern after 12 h milling is shown in figure 2. According to this pattern, there is no obvious diffraction peak after 12 h by ball milling treatment indicating the glass powders show a typical amorphous characteristic and there are no impurities mixed.

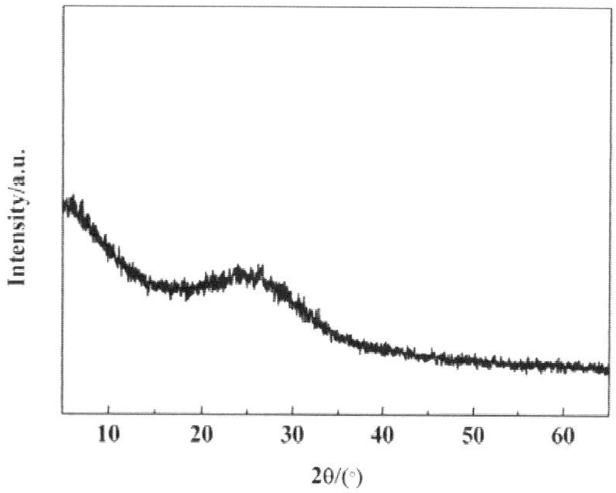

Figure 2. XRD pattern of the base glass powders by ball-milling for 12 h

Figure 3 shows the XRD results of glass ceramic samples by hot-press sintering. From the XRD results, it can be concluded that the diffraction peak intensities decrease with the fineness of the glass powders. In the sample with glass powder size of d_{50}=16.4 μm, the diffraction intensities of crystals are higher than that the powder size is d_{50}=9.6 μm (see figure 3(1h-1,2) and figure 3(5h-1,2)). Furthermore, the diffraction intensities of crystals further lowered in the sample with glass powder size of d_{50}=3.3 μm (see figure 3(12h-1, 2)). However, the kinds of phase precipitated in the samples with different powder sizes are same. There are mica crystal as major phase, and mullite and $KFeSi_2O_6$ crystals as minor phases. In the vertical and parallel orientation of the pressure direction, there is no obvious difference for the diffraction peak intensity of crystals. This indicates that the precipitation of crystal in the different directions is not affected by the pressure, and the orientation of precipitation crystal trends to be uniform. This is contrary to several reports [2, 15], in which, the preferred precipitation of crystals is seemed to tend to be formed by the hot pressed processes, suggesting that applying the pressure at 670°C has little influences on the preferential growth of crystals.

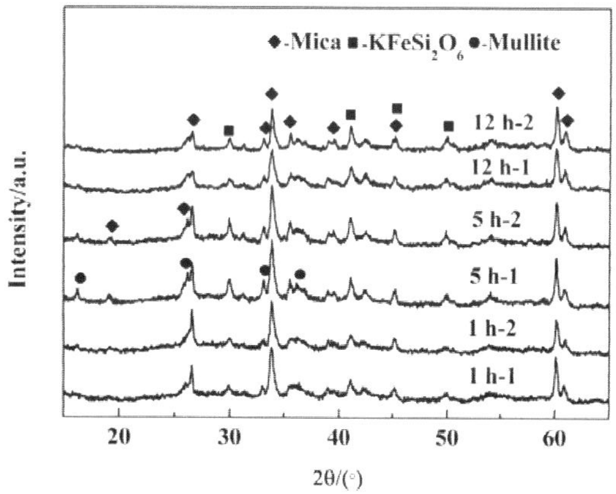

Figure 3. XRD patterns of glass samples with powder size and observed perpendicular (1) or parallel (2) to press direction

Figure 4 shows the SEM micrographs the sintered glass ceramics for various glass powders sizes. In the sample with glass powder size of d_{50}=16.4 μm, the plate shaped crystals with about 5.0 μm are precipitated, and this crystal phase is interlocked and has the typical characters of mica crystal. Therefore, the plate crystals should be mica phases. In the sample with glass powder size of d_{50}=9.6 μm, it is found that the typical plate-shaped mica is disappeared (see figure 4b), in which the particle-like mica phase is observed. There is no obvious boundary among glass powders, which induces the homogeneous distribution of mica crystal. This tendency is more obvious with the decrease of glass powder size. SEM image of glass powder sintered with the size of d_{50}=3.3 μm are shown in Figure 4c. According to the image, the precipitation of particle-like crystal shows a more homogeneity in the base glass. Therefore, the sizes of glass powders have a great effect on sintering of the glass-ceramic and on microstructure of precipitation crystal. With the decrease of size of glass powders, the increase of surface energy can accelerate the nucleation of glass powders. The boundaries of glass powders vanish and the microstructure tends to be consistent, which is in favor of improving machinability of glass-ceramic [8, 12]. Meanwhile, more crystal nucleus are formed with finer of glass powders. The more crystal nucleus form, the less growth spaces before they impinge each other. Therefore, mica crystal precipitated becomes smaller and this may the reason that lower intensities of diffraction peaks are observed in finer powder in XRD.

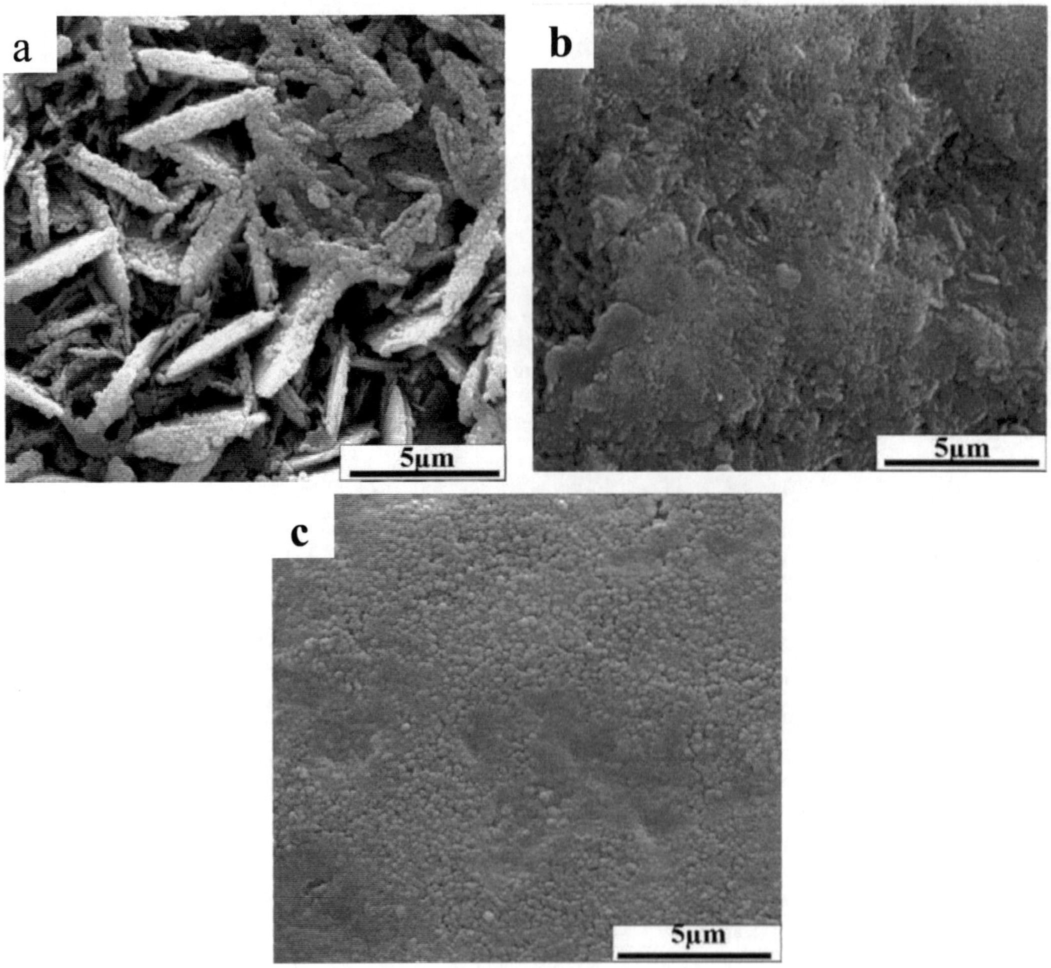

Figure 4. SEM micrographs of glass samples with powder size of d_{50}= (a) 16.4 μm, (b) 9.6 μm and (c) 3.3 μm

4. Conclusions

In the $MgO-Al_2O_3-SiO_2-F$ glass system fabricated by hot pressed sintering, the glass-ceramic are composed of mica crystal and little mullite and $KFeSi_2O_6$ crystal precipitated. The plate-like mica crystal is precipitated homogeneously in glass matrix, and was transformed to particle-shaped phase with the diminution of the glass powders. The decrease of size of the glass powders can hinder the precipitation of crystals but is favorable to form the homogeneous and uniform microstructures.

References

[1] Casasola R, Pérez J M and Romero M 2016 *J. Am. Ceram. Soc.* **99** 484-91
[2] Cheng K G, Wan J L and Liang K M 1999 Mater. Lett. **39** 350-53
[3] Baik D S, No K S, Chun J S and Cho H Y 1997 *J. Mater. Proc.Tech.* **67** 50-54
[4] Rad B A and Alizadeh P 2009 *Ceram. Inter.* **35** 2775-80
[5] Faeghi-Nia A and Ebadzadeh T 2012 *Ceram. Inter.* **38** 2653-58

[6] Alizadeh P, Yekta B E and Javadi T 2008 *J. Europ. Ceram. Soc.* **28** 1569-73
[7] Uno T, Kasuga T, Nakayama S and Ikushimal A J 1993 *J. Am. Ceram. Soc.* **76** 539-541.
[8] Taruta S, Hayashi T and Kitajima K 2004 *J. Europ. Ceram. Soc.* **24** 3149-54
[9] Suzuki S S, Taruta S, and Takusagawa N 1998 *Korean J. Ceram.* **4** 363-67.
[10] Montazerian M, Alizadeh P and Yekta B E 2008 *J. Eur. Ceram. Soc.* **28** 2693- 99
[11] Tian Q B, Kong D Y and Zhai T 2016 *Sci. Eng.Comp. Mater.* **23** 573-77
[12] Taruta S, Yonekura H, Suzuki S S, Kitajima K and Takusagawa N. 2002 *J. Ceram. Soc. Jap.* **110** 208-10
[13] Habelitz S, Carl G and Rüssel C 2001 *Mater. Sci. Eng.* A **307** 1-14.
[14] Xu J and Gao F 2009 *J. Chin. Ceram. Soc.* **37** 409-13(in Chinese)
[15] Habelitz S, Carl G, Rüssel C, Thiel S, Gerth U, Schnapp J D, Jordanov A and Knake H 1997 *J. Non-Crys. Solids* **220** 291-98

IOP Conf. Series: Materials Science and Engineering **242** (2017) 012011 doi:10.1088/1757-899X/242/1/012011

Heterogeneous Fenton-like degradation of methylene blue (MB) by magnetic nanoparticles $Fe_3O_4@TiO_2$ in neutral condition

Junfeng Hua and Mei Huang *

College of Chemical & Biochemical Engineering, Zhejiang University, Hangzhou, 310027, China

huangm@zju.edu.cn

Abstract. The traditional Fenton system (Fe^{2+}–H_2O_2) only works in an acidic environment and produces a large quantity of sludge. In this study, magnetic nanoparticles $Fe_3O_4@TiO_2$ were used as a high-performance Fenton-like catalyst for the degradation of MB. The surface morphology was characterized by Transmission Electron Microscopy (TEM). And the Fenton-like catalytic activity of $Fe_3O_4@TiO_2$ was evaluated under different pH condition and various H_2O_2 in feed concentration, respectively. The results indicated that nanoparticles $Fe_3O_4@TiO_2$ show high efficiency for methylene blue (MB) degradation under the reaction condition of H_2O_2 40mM, $Fe_3O_4@TiO_2$ 20mM, MB 300mg/L, pH 7 and 30℃, the mineralization of persistent MB can achieve 72% COD removal ratio.

1. Introduction

Wastewater produced by chemical industry, textile industry and leather manufacturing, often contains considerable amounts of organic pollutants, which can cause negative impacts for ecosystems and humans for their toxicity, carcinogenic or mutagenic properties. Therefore, it is significant to degrade the organics in wastewater before discharge [1]. Fenton oxidation is one of the simplest and most efficient AOPs in environment remediation. However, one disadvantage of this method is that the reaction achieves its high activity at relatively low pH range, usually at approximate pH=3, which is not beneficial for the chemical equipment and operation. In addition, the wastewater after treatment is preferred to be neutralized before discharge, so a large amount of slurry will be produced, which causes the problem of solid waste for further treatment and greatly increasing the running cost.

In recent years, heterogeneous Fenton-like catalysts, such as iron-based clays and zeolites, and iron oxide minerals [2, 3], have received considerable interests due to their advantages of facile recovery and recycling as well as dramatically decreased slurry. And among these catalysts, Fe_3O_4 nanoparticles not only have unique characteristic of magnetic feature but also showed higher reactivity in neutral condition. For example, when the concentrations of Fe_3O_4 and H_2O_2 were 5g/L and 1.2M, phenol and aniline could be removed completely after 6h at 308K. However, it is found that the total organic carbon abatement efficiency for phenol and aniline were only 42.79% and 40.38%, respectively [4].

Recently, TiO_2 coupled with iron oxide has been reported to have a good magnetic response and displayed high photocatalytic efficiency [5-7]. In this work, we have successfully synthesized the $Fe_3O_4@TiO_2$ NPs, which were applied to the Fenton-like degradation of methylene blue (MB) as a model of recalcitrant contaminants. The effects of operational parameters as concentrations of catalyst and H_2O_2, and pH value on the degradation efficiency of TC were systematically evaluated.

Content from this work may be used under the terms of the Creative Commons Attribution 3.0 licence. Any further distribution of this work must maintain attribution to the author(s) and the title of the work, journal citation and DOI.
Published under licence by IOP Publishing Ltd

Extraordinarily high and stable catalytic activity has been achieved on Fe_3O_4@TiO_2 with a wide working pH range of 5–9.

2. Materials and methods

2.1. Reagents
All reagents used in the experiments were of chemical reagent grade and used without further purification. 30% (w/w) hydrogen peroxide, ferrous sulfate ($FeSO_4 \cdot 7H_2O$), titanium tetrachloride ($TiCl_4$), $NH_3 \cdot H_2O$ (25%, w/w), H_2SO_4 and NaOH were obtained from Beijing Chemicals Corporation (Beijing, China).

2.2. Fe_3O_4@TiO_2 MNPs preparation
Fe_3O_4@TiO_2 MNPs were prepared by coprecipitation of Fe^{2+} and Ti^{4+} in aqueous solutions. In a typical procedure, 0.06mol/L $FeCl_3$ and $TiCl_4$ solutions were mixed uniformly, and then dilute alkaline solution was added drop wise to the solution. After 2h vigorous stirring under an Ar stream, the black particles were deposited and separated from solution by a powerful magnet, washed with de-ionized water and ethanol to neutral pH. Then Fe_3O_4@TiO_2 MNPs were dried under vacuum and stored in the desiccator for further experiments.

2.3. Experimental procedure
Fe_3O_4@TiO_2 MNPs was dispersed in 100mL MB solution. And the specific pH of MB solution was adjusted with solutions of H_2SO_4 or NaOH. Degradation of MB was initiated by adding a desired dosage of H_2O_2. Then the mixture was vibrated 250rpm at 30℃. At selected time intervals, Ailquots were collected and sediment were separated by magnetic separation. The concentration of MB was determined after Na_2SO_3 was added to quench excess H_2O_2.

2.4. Characterization of Fe_3O_4@TiO_2 MNPs
TEM images were conducted with a CM200 transmission electron microscope.

3. Results and discussion

3.1. Characterization of Fe_3O_4@TiO_2 MNPs

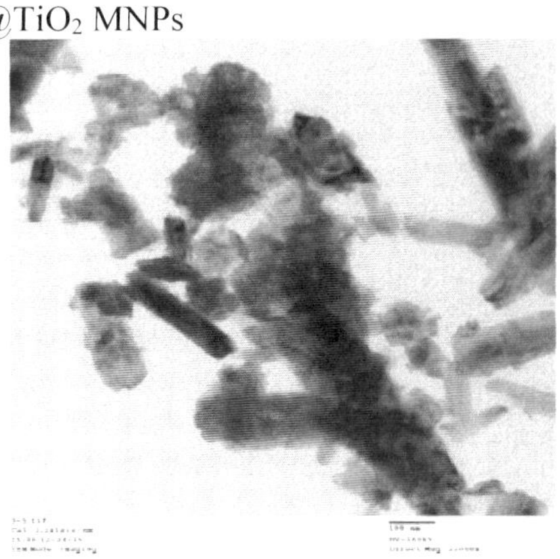

Fig.1. TEM image of synthesized Fe_3O_4@TiO_2 MNPs

Fig. 1 displays TEM images of Fe_3O_4@TiO_2 MNPs samples. It can be seen that the obtained Fe_3O_4@TiO_2 MNPs are plane particles, which show great difference from uniform spherical of Fe_3O_4 MNPs. And the particle size is about 20nm, which has good consistency with samples' XRD results.

3.2. Effect of Fe_3O_4@TiO_2 MNPs amounts

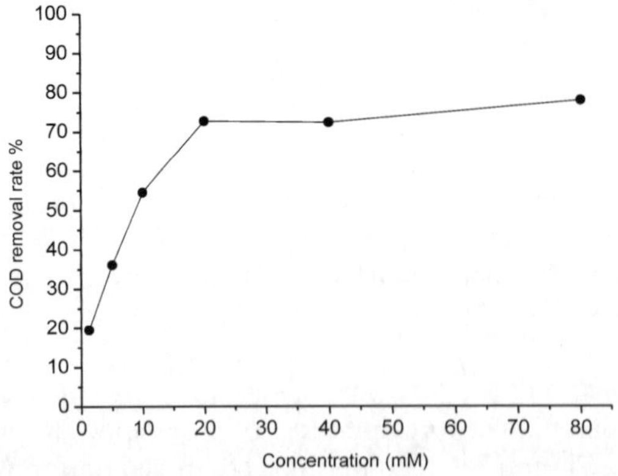

Fig.2. Effect of Fe_3O_4@TiO_2 MNPs on COD removal efficiency at pH7.

Fig. 2 illustrates the effect of Fe_3O_4@TiO_2 MNPs amounts on the COD removal rate with initial MB 300mg/L and H_2O_2 dosage 40mM at pH=7 and T=30℃. It is found that with the increased addition of Fe_3O_4@TiO_2 MNPs from 1.25mM to 80mM, COD abatement efficiency also increased from 20% to 78%. The extra catalytic activity by Fe_3O_4 MNPs is related to the increased amount of active sites on Fe_3O_4@TiO_2 surface. Because more active sites in reaction system can accelerate more production of HO•, MB can be degraded with improved efficiency.

3.3. Effect of H_2O_2 concentration

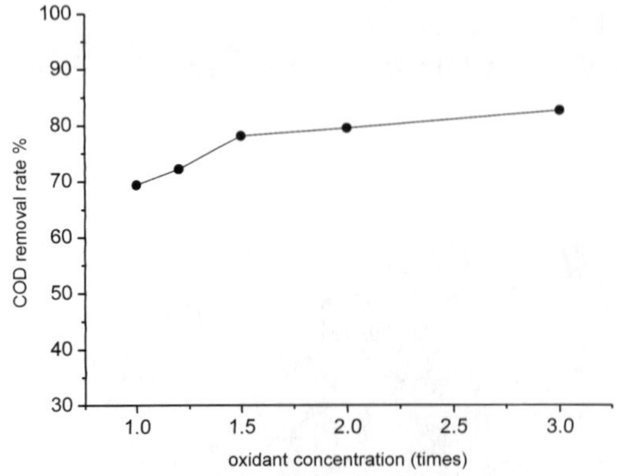

Fig.3. Effect of H_2O_2 dosage on COD removal efficiency at pH7

Fig. 3 indicates that the optimal H_2O_2 concentration is about 40 mM for the MB degradation on Fe_3O_4@TiO_2 MNPs at pH=7 and T=30 ℃. It is clear that COD removal rate changed slowly from 69% to 83% with increased H_2O_2 concentration in the range of 40-120mM. And the existence of optimum H_2O_2 concentration is related to the effect of excessive H_2O_2. Because of its scavenging effect on OH radicals, efficiency of OH radicals decreased in the solution so that H_2O_2 oxidation efficiency also decreased.

3.4. Effect of pH

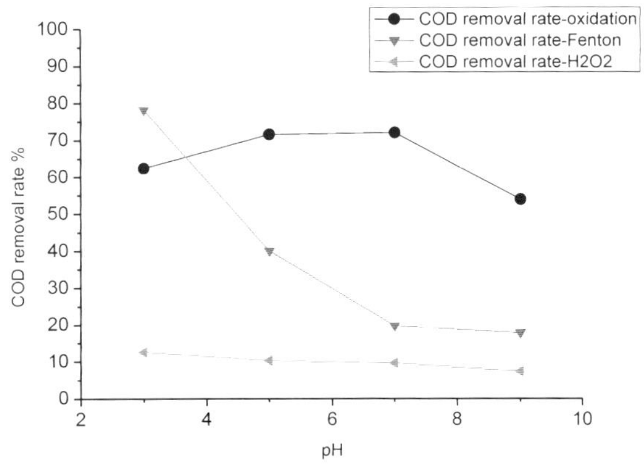

Fig.4. Catalytic oxidation effect of MB analogues at different pH

The degradation of MB under different pH values was evaluated as shown in Fig.4. It is found that the COD abatement efficiency was less than 10% with only 40mM H_2O_2, which is ascribed to the weak oxidation potential of H_2O_2 compared with hydroxyl radicals. In the typical Fenton reaction, the maximum78% COD removal rate was observed with the initial parameters of pH=3. But the efficiency decreased with the increased pH, which is consistent with the usual Fenton reaction. However, the highest degradation of MB, about 72% COD removal rate, was achieved in the Fenton-like system catalyzed by Fe_3O_4@TiO_2 MNPs at pH=7. Therefore, it is reasonable to consider that SiO_2 coated on Fe_3O_4 could effectively enhance the catalytic activity of Fe_3O_4.

4. Conclusion

Highly active Fe_3O_4@TiO_2 MNPs were successfully synthesized by co-precipitation method, and used as heterogeneous Fenton-like catalysts that can degrade MB from aqueous solution efficiently. Unlike the traditional Fenton process, the COD removal efficiency remained relatively high at a wide pH range from 5~9 for MB.

Acknowledgements

We acknowledge financial support from Science & Technology Department of Zhejiang Province (Program no.2016F50029), China.

References

[1] A.N. Soon, B.H. Hameed, Heterogeneous catalytic treatment of synthetic dyes in aqueous media using Fenton and photo-assisted Fenton process, Desalination 269 (2011) 1–16

[2] R. Gonzalez-Olmos, F. Holzer, F.D. Kopinke, A. Georgi, Indications of the reactive species in a heterogeneous Fenton-like reaction using Fe-containing zeolites, Applied Catalysis A 398(2011) 44–53

[3] G.B. Ortiz de la Plata, O.M. Alfano, A.E. Cassano, Decomposition of 2-chlorophenol employing goethite as Fenton catalyst. I. Proposal of a feasible, combined reaction scheme of heterogeneous and homogeneous reactions, Applied Catalysis B: Environmental 95 (2010) 1–13

[4] S. Zhang, X. Zhao, H. Niu, Y. Shi, Y. Cai, G. Jiang, Superparamagentic Fe_3O_4 nanoparticles as catalysts for the catalytic oxidation of phenolic and aniline compounds, J. Hazard. Mater. 167 (2009) 560-566.

[5] M.M. Mohameda, W.A. Bayoumya, M.E. Goherb, M.H. Abdob, T.Y. Mansour El-Ashkarb, Optimization of α-Fe_2O_3@Fe_3O_4 incorporated N-TiO_2 as super effective photocatalysts under visible light irradiation, Applied Surface Science 412 (2017) 668–682.

[6] M.A. Sarvin, S. Bijan, Tajoddin, N. Nosrat, Recyclable $Fe_3O_4/SiO_2/TiO_2$/Cu nanocomposites:

synthesis, characterization and investigation of the photocatalytic and magnetic property, Journal of Materials Science-Materials in Electronics 28 (2017) 9456-9463
[7] K.H. Choi, S.Y. Park, B.J. Park, J.S. Jung, Recyclable Ag-coated Fe_3O_4@TiO_2 for efficient photocatalytic oxidation of chlorophenol, Surface & Coatings Technology 320 (2017) 240–245

IOP Conf. Series: Materials Science and Engineering **242** (2017) 012012 doi:10.1088/1757-899X/242/1/012012

The soften mechanisms of martensitic heat resistant steels during hot deformation

Wenfeng Zhang [1], Yunqiang Liu [1], Dong Ye [1], Xiaogang Liu [1], Wei YAN [*2]

1. Guangxi Colleges and Universities Key Laboratory of Robot & Welding, Guilin University of Aerospace Technology, Guilin 541004, China.

2. Institute of metal research, CAS, Shenyang 110016, China

*conresspongding author: wfzhang11b@alum.imr.ac.cn

Abstract. Hot deformations for the experimental steels were carried out in the temperature range of 900-1200 ℃ and the strain rate range 10-3-100 s-1. In this study, the critical condition for dynamic recrystallization (DRX) was determined by a cooperation of the minimum value of $-(\partial\theta/\partial\sigma)$ and the deviation point of the linear slope of θ - σ curve. From the analysis, it was found that only one slope ratio of critical strain, ε_c versus critical stress, σ_c appeared in the 9Cr and 10Cr steels , whilst two different slopes in P92 and NS steels due to the augmentation of auxiliary softening effect of the dynamic strain-induced transformation (DSIT) . The results also exhibited that $ln(Z)$ kept an approximate positive linear relationship with $ln(\varepsilon_c)$ below the value of 40, but barely changed beyond it. However, the $ln(\sigma_c)$ value increased continuously with the rise of the $ln(Z)$ value at a smaller growth rate after the value of $ln(Z)$ reached 40 than that of the initial part.

1. Introduction

Thermomechanical processing (TMP) has become a crucial tool to advance the structure and enhance the mechanical properties of steels, including the heat resistant steels. During a TMP process, recrystallization plays the main role in the associated microstructure evolution [1] which mainly determines the deformation characteristics of the steels during the hot working processes. Besides of the recrystallization referring to dynamic recrystallization (DRX), metadynamic recrystallization (MDRX), static recrystallization (SRX), other softening mechanisms may also be strongly auxiliary to have some effects on the deformation behaviour, such as dynamic recovery (DRV), static recovery (SRV) and dynamic strain induced transformation (DSIT). Which softening mechanisms execute the microstructure evolution depends on chemical compositions, prior machining procedures and the deformation parameters.

Actually, each of the softening mechanisms has to satisfy its own prerequisite in order to take place, i. e, the DRX only occurs when a critical strain is reached, and at the same time, the minimum rate of energy dissipating is received [2]. To control the microstructure during deformation, the softening mechanisms should be well analysed. Generally, whether a softening procedure occurs or not during a specific hot deformation, it could be identified through both analysing the stress-strain curves up to the peak and the microstructure observation [3].

The heat resistant steels being used in the ultra-supercritcal (USC) power plants not only bear extrusion at high temperature but also experience a creep deformation under a constant stress at high temperature due to the service situation. The creep deformation could also be analysed through the constitutive equation, and the parameters calculated for the hot deformation are suitable for the creep deformation. Therefore, it is of essential importance to study the deformation characteristics and the

Content from this work may be used under the terms of the Creative Commons Attribution 3.0 licence. Any further distribution of this work must maintain attribution to the author(s) and the title of the work, journal citation and DOI.
Published under licence by IOP Publishing Ltd

microstructure evolution of the experimental steels during hot deformation. The martensitic heat resistant steels are assumed to be the most competitive candidates for the USC application at 650 °C thanks to their high heat conductivity, nice thermal fatigue resistance and high stress corrosion resistance [6]. Thus, the main objective of this study is to establish the soften mechanisms that took place during the hot deformation, to predict the creep behaviour and the failure mechanism during service.

2. Experimental

This study was conducted on four types of martensitic heat resistant steels. The chemical compositions of the experimental steels are listed in Table 1. The steels were machined into samples of 8 mm in diameter and 12 mm in gauge length.

The rob samples were homogenized in vacuum at 1200 °C for 5 min, and then cooled down to the deformation temperature of 900-1200 °C at a cooling rate of 10 °C/s. After stabilizing holding for 1 min at the deforming temperature, the compression testing samples were deformed to 60% at the strain rate range of 10^{-3}-100 s^{-1}. The samples were quenched to room temperature as soon as the compressions were finished.

The microstructure of deformed samples under different compression parameters was observed through an optical microscope, a scanning electron microscope (SEM).

Table 1. Chemical compositions of the experimental steels, wt%

Steel	C	Si	Mn	Cr	Mo	W	V	Nb	B	Co	N
NS	0.021	0.09	1.25	9.37	-	1.42	0.15	0.06	-	-	0.037
P92	0.11	0.37	0.46	8.77	0.42	1.73	0.17	0.057	0.0028	-	0.048
9Cr	0.089	0.31	0.50	8.58	0.40	1.65	0.18	0.060	0.0022	1.64	0.040
10Cr	0.088	0.31	0.50	10.42	0.40	2.55	0.18	0.056	0.0022	2.19	0.058

3. Results

3.1 Stress-strain curves critical conditions for DRX

The stress-strain curves of the experimental steels obtained from four different strain rates labelled beside their corresponding curves are shown in Figure 1, in which the critical strain, ε_c, and the peak strain, ε_p, are marked on the curves. All the curves showed an initial work hardening but only a few out of them developed to a clear stress peak, indicating a DRX taking place. It was found that the distinct peaks only appeared in the curves that obtained under high temperature and low strain rate deformation situations (low *Z* values). With the increase of *Z* value in the beginning, the peak became broad and the critical strain climb up correspondingly. But when the *Z* value reached a higher level, the stress curve showed a 'flat-top' shape with no peak. This flat-top behaviour of stress curve was traditionally thought to be an indication of no occurrence of DRX and implied that the dynamic recovery was the only softening mechanism operated during the deformation process [4].

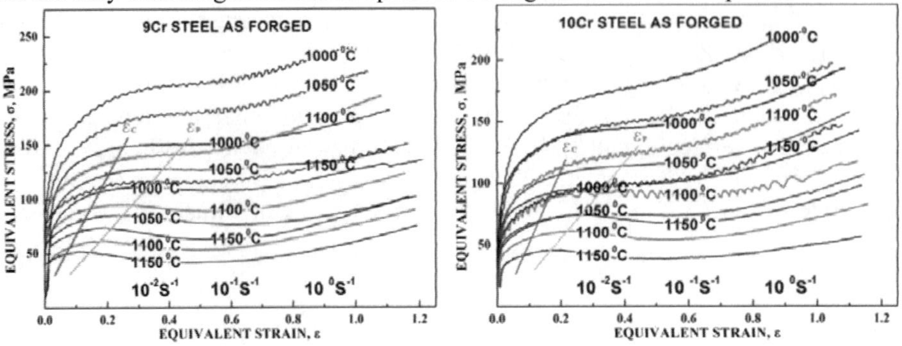

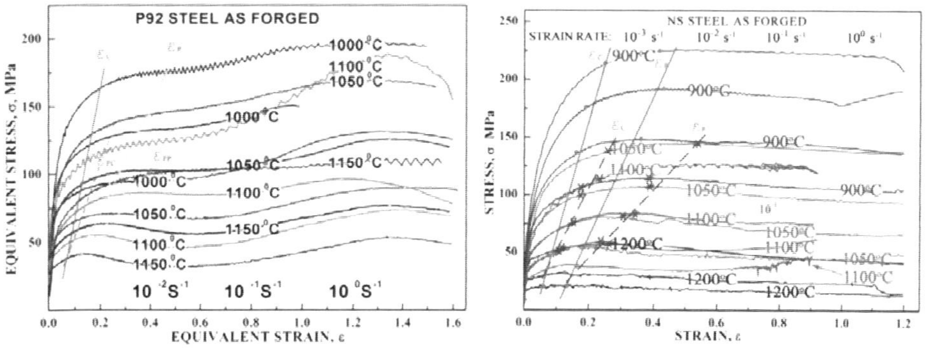

Figure 1. Equivalent stress-strain curves with indication of ε_c and ε_p of 9Cr steel, 10Cr steel, P92 steel and NS steel in sequence.

The existing approach to the onset of DRX was based mainly on energetic considerations. The critical strain for the initiation of DRX can be deduced from either the direct microstructure observation or the analysis of the flow curve. The former is a more complicated and time consuming process.

3.2 The change of dominate soften mechanisms with different Z parameters

Generally speaking, the increase of Z value prompted the critical strain for the initiation of DRX correspondingly. The theory was confirmed by the approximate linear growth of the third polynomial order fitting curves of $ln(\varepsilon_c)$ plots at the beginning, as shown in Figure 2 for all the experimental steels. However, the increase of $ln(\varepsilon_c)$ value slowed down and then levelled off at Z value beyond 40. This resulting hump recorded by Stewart et al. was explained by the segregation of substitutional impurities (P in particular) to the subboundaries of newly nucleated DRX grains [3]. But Nanba acclaimed that the critical strain was expected to slightly depend on the temperature and have no apparent dependence on either the strain rate or the initial grain size for plain carbon steels [5]. That is to say, at low deformation temperature, the higher dislocation density would prompt the initiation of DRX[6] and thus slowed down the increase of the critical strain leading to a barely varying $ln(\varepsilon_c)$ value.

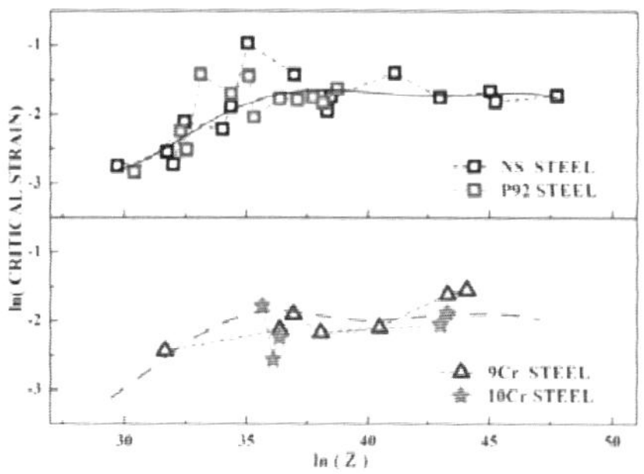

Figure 2. Curves of $ln(\varepsilon_p)$ and $ln(\varepsilon_c)$ values versus $ln(Z)$ value for the experimental steels

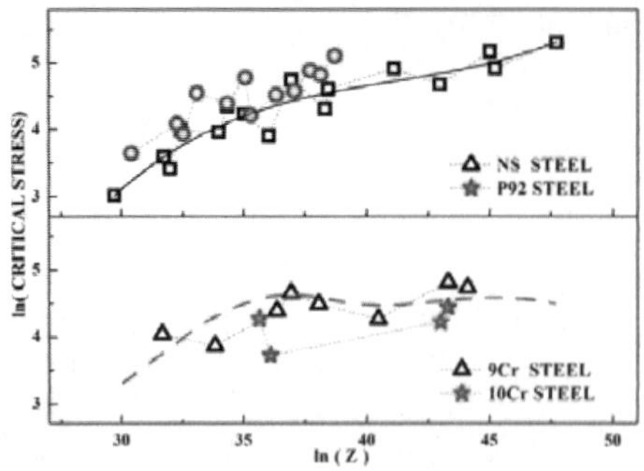

Figure 3. Relationship between ε_c and ε_p values for the experimental steels.

Unlike *ln*(ε_c) values, the *ln*(σ_c) values exhibited two different trends with the increase of *ln*(*Z*) value, as shown in Figure 3. For the 9Cr and 10Cr steels, the trends of *ln*(σ_c) versus *ln*(*Z*) were similar to that of *ln*(ε_c) versus *ln*(*Z*). But as for the NS and P92 steels, the *ln*(σ_c) values still kept linearly rising rather than barely varied when the *ln*(*Z*) values were beyond 40. This phenomenon gave a possible explanation to the trends of *ln*(σ_c) curves: the slope ratio of σ_c and ε_c got augmented at high *Z* value, which ensured the continuous growth of *ln*(σ_c) after the *ln*(*Z*) value of 40.

3.3 Microstructure evolution

The evolution of microstructure during hot deformation was closely related to DRV, DRX, MDRX and DSIT in the experimental steels [5]. With the change of *Z* value, the steels exhibited various microstructure characteristics, as demonstrated in Figure 4.

An analysis on the stress curve suggested that DRV took place before the critical strain for initiation of DRX under all investigation conditions [6,8]. At high temperature, the DRV got prompted, and the DRX started at a lower strain and proceeded faster [7]. Meanwhile, the fine and medium size particles that reduced the recovery process stabilized the substructure which would slow down the nucleation and pin the grain boundaries diminished to rare [8]. All the above factors gave explanations to the isometric martensitic grains deformed at 1200 °C/10^{-3} s^{-1}. However, the deformation temperature was high enough for the nucleated DSIT ferrite to grow at a high speed. That was why almost 50% of the grains were composed by the equiaxed ferrite.

As denoted by other studiers [13-15], the decrease of temperature would promote the DSIT nucleation and consequently resulted in relatively finer ferrite grain size. The ferrite grains formed in the experimental steels tended to merge into allotriomorphic rather than equiaxial grains at the relatively low temperature and low strain rate (*Z* value around 40). This was because the nucleation size of the DSIT ferrite was small and the growth of DSIT ferrite was sensitive to both temperature and strain rate. But the DSIT ferrite exhibited an elongated shape which implied no DRX occurring in DSIT ferrite, while the martensitic grains showed approximately equiaxed shape indicating that DRX happened in the austenite during deformation.

The MDRX process only involved the growth of previously nucleated dynamic grains and thus enlarged the grain size in a certain extent. Therefore, the MDRX grain size depended only on the *Z* parameter [15,16] and was finer with the *Z* value increasing. This is why when the *Z* value rose up to 60, there was rarely distinct DSIT ferrite left, replaced by a certain volume of ferrite-martensite blend and why the martensitic microstructure of high *Z* value showed finer grains compared with that of low *Z* value.

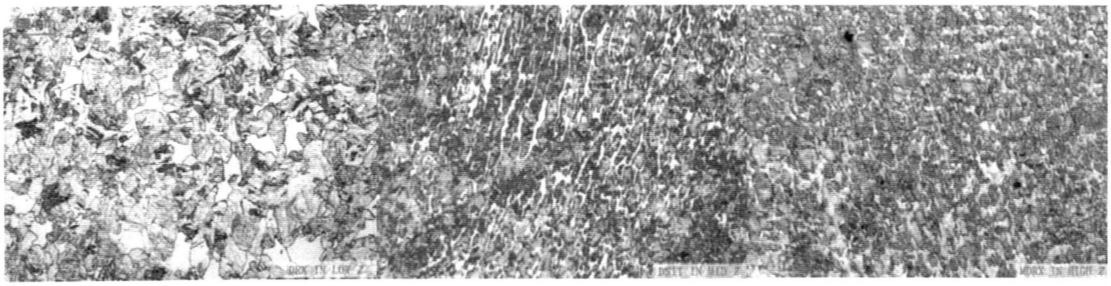

Figure 4. Microstructure evolution and precipitation behavior of NS steel during hot deformation.

4. Conclusion

Beneath the diversity of microstructures and hot deformation processes, there are some fundamental characteristics of four types of martensitic heat resistant steels as follows.

- The value of $ln(\varepsilon_c)$ of all the experimental steels rose linearly with $ln(Z)$ increasing at the beginning but this acceleration slowed down and levelled off at high Z conditions. The $ln(\sigma_c)$ curve changed with $ln(Z)$ in a similar manner to the trend of $ln(\varepsilon_c)$ versus $ln(Z)$ for 9Cr and 10Cr steels while it still kept a smaller linear growth when $ln(Z)$ value beyond 40 for NS and P92 steels.
- The high temperature provoked the starts of DRX at lower strain and prompted its proceeding, leading to equiaxial grains. DSIT ferrite was fond to grow along the prior austenite grain boundaries, which formed the stripe structure. The MDRX and DSIT processes were responsible for the ferrite-martensite blend structure.

Acknowledgements

This work was financially supported by the National Science Foundation for Young Scientists of China (Grant No. 51601044), Guangxi Natural Science Foundation (NO. 2015GXNSFBA139225), Natural Science Foundation of Guilin University Aerospace of Technology (No. YJ1405), the Opening Project of Guangxi Colleges and Universities Key Laboratory of robot & welding: Research of microstructure and mechanical properties on CLAM steel after EBW and the Dr. Start-up fund for "Study on improving the microstructure stability of the heat resistant steel by modifying the precipitates" from Guilin University Aerospace of Technology.

References

[1] W. Yan, W. Wang, Y.-Y. Shan, et al. Microstructural stability of 9–12%Cr ferrite/martensite heat-resistant steels [J]. Frontiers of Materials Science, 2013, 7: 1-27.

[2] L. Milović, T. Vuherer, I. Blačić, et al. Microstructures and mechanical properties of creep resistant steel for application at elevated temperatures [J]. Materials & Design, 2013, 46: 660-667.

[3] W.-F. Zhang, X.-L. Li, W. Sha, et al. Hot deformation characteristics of a nitride strengthened martensitic heat resistant steel [J]. Materials Science and Engineering: A, 2014, 590: 199-208.

[4] Wenfeng Zhang, Qingyong Su, Wei Yan, et al. Precipitation behavior in a nitride-strengthened martensitic heat resistant steel during hot deformation [J]. Materials Science & Engineering A, 2015, 639: 173-180.

[5] B. Dutta, E.V. 1, C.M. Sellars. Mechanism and kinetics of strain induced precipitation of Nb(C, N) in austenite [J]. Acta metall, mater., 1992, 40: 653--662.

[6] W.F. Zhang, W. Sha, W. Yan, et al. Analysis of deformation behavior and workability of advanced 9Cr-Nb-V ferritic heat resistant steels [J]. Materials Science and Engineering A, 2014,

[7] W.-F. Zhang, W. Sha, W. Yan, et al. Constitutive modeling, microstructure evolution and processing map for a nitride strengthened heat resistant steel [J]. Journal of Materials Engineering and Performance, 2014,

[8] A. Momeni, K. Dehghani. Characterization of hot deformation behavior of 410 martensitic stainless steel using constitutive equations and processing maps [J]. Materials Science and Engineering: A, 2010, 527: 5467-5473.

[9] A. Marchattiwar, A. Sarkar, J.K. Chakravartty, et al. Dynamic Recrystallization during Hot Deformation of 304 Austenitic Stainless Steel [J]. Journal of Materials Engineering and Performance, 2013,

[10] A. A. Khamei, K. Dehghani. Modeling the hot-deformation behavior of Ni 60 wt%–Ti 40 wt% intermetallic alloy [J]. Journal of Alloys and Compounds, 2010, 490: 377-381.

[11] Wenfeng Zhang, Qingyong Su, MiXua, et al. Precipitation behavior in a nitride-strengthened martensitic heat resistant steel during hot deformation [J]. Data in Brief, 2015, 4: 395-399.

[12] S. Allain, O. Bouaziz, T. Lebedkina, et al. Relationship between relaxation mechanisms and strain aging in an austenitic FeMnC steel [J]. Scripta Mater, 2011, 64: 741-744.

[13] K. Kimura, Y. Toda, H. Kushima, et al. Creep strength of high chromium steel with ferrite matrix [J]. International Journal of Pressure Vessels and Piping, 2010, 87: 282-288.

[14] A. Nagode, L. Kosec, B. Ulet, et al. Review of creep resistant [J]. Metabk, 2011, 50: 45-48.

[15] W.F. Zhang, P. Hu, Q. G. Zhou, et al. Effect of Heat Treatment on the Mechanical Properties and the Carbide Characteristics of A High Strength Low Alloy Steel [J]. J Iron Steel Res. Int. supplement 1-1, 2011, 18: 143-147.

[16] W.F. Zhang, W. Yan, W. Sha, et al. The impact toughness of a nitride-strengthened martensitic heat resistant steel [J]. Science China Technological Sciences, 2012, 55: 1858-1862.

IOP Conf. Series: Materials Science and Engineering **242** (2017) 012013 doi:10.1088/1757-899X/242/1/012013

Enhanced electrochemical performance of a LTO/carbon nanotubes/graphene composite as an anode material for Li-ion batteries

Aijia Wei[1, 2], Wen Li[1, 2], Lihui Zhang[1, 2] and Zhenfa Liu[1, 2*]

[1] Institute of Energy Resources, Hebei Academy of Science, Shijiazhuang, China

[2] Hebei Engineering Research Center for Water Saving in Industry, Shijiazhuang, Hebei 050081, China

E-mail: lzf63@sohu.com

Abstract. A $Li_4Ti_5O_{12}$/carbon nanotubes/graphene composite has been successfully prepared by a solid-state method. For comparison, pure LTO and $Li_4Ti_5O_{12}$/graphene composite were also synthesized using the same method. The materials were characterized by X-ray diffraction (XRD) and scanning electron microscopy (SEM) to confirm the structure and morphology. The results reveal that LTO particles are well dispersed and wrapped in the graphene sheets with cross-linked carbon nanotubes. The electrochemical results show that the $Li_4Ti_5O_{12}$/carbon nanotubes/graphene composite exhibits the best rate capacity, which lead to a charge capacity of 169.0, 168.5, 167.1, 153.2, 144.5, 131.5 mAh g^{-1} at 0.2, 0.5, 1, 3, 5 and 10 C, respectively between 1 and 3 V (1 C = 160 mAh g^{-1}). The synergistic effect of graphene and carbon nanotubes constructing 3D networks could enhance the electronic conductivity of $Li_4Ti_5O_{12}$/carbon nanotubes/graphene composite.

1. Introduction

In recent years, rechargeable lithium-ion batteries (LIBs) have been considered as a promising power system for energy storage. Lithium ion batteries are rapidly replacing conventional lead-acid batteries in our daily life from personal portable equipment to large-scale electric energy storage applications such as electric vehicles (EVs) and hybrid electric vehicles (HEVs) [1, 2]. Spinel $Li_4Ti_5O_{12}$ (LTO) has been regarded as an ideal anode material for LIBs because it possesses some unique advantages, such as the negligible volume variation during insertion/extraction of the lithium ion and the high redox plateaus at around 1.55 V (vs Li/Li^+) [3, 4]. However, the LTO also has some inherent disadvantages. The poor electronic conductivity (< 10^{-13} S cm^{-1}) an sluggish Li^+ diffusion coefficient (10^{-13} – 10^{-9} cm^2 s^{-1}) of LTO, which results in low capacities at high current rates [5]. In order to overcome the drawbacks, many effective strategies have been executed, such as designing LTO nanostructures, doping with ions and surface coating with some conductive components (carbon, graphene, carbon nanotubes) [6-8]. Notably, the surface coating with some conductive components has been proved to be an effective and facile way to improve the electronic conductivity and Li^+ diffusion ability of LTO.

To our best knowledge, graphene and carbon nanotubes are two new type nano-carbon materials, which have been widely used in LIBs [9, 10]. However, most studies are limited to graphene or carbon nanotubes modifying electrode materials alone. It is meaningful to combine graphene with carbon nanotubes together to modify electrode materials, which could construct a three dimensional (3D) conducting networks and improve the electrochemical performances. For example, Lei et al.

Content from this work may be used under the terms of the Creative Commons Attribution 3.0 licence. Any further distribution of this work must maintain attribution to the author(s) and the title of the work, journal citation and DOI.
Published under licence by IOP Publishing Ltd

synthesized a three-dimensional $LiFePO_4$/carbon nanotubes/graphene composite by wet ball milling followed by a two-step heat treatment process [11]. In our paper, we have prepared a LTO/carbon nanotubes/graphene composite by a two step solid-state method follow by sintering in a box furnace under air atmosphere. By contrast, the LTO/graphene composite and pure LTO are also synthesized by the same method, and the electrochemical performances of these three materials are discussed.

2. Experiments and Methods

2.1 Synthesis and characterization

Pure LTO powders were synthesized by solid-state reaction. In brief, a stoichiometric amount of anatase-TiO_2 (Chongqing Xinhua Chemical Co., Ltd.), Li_2CO_3 (Sichuan Tianqi Lithium Industries Inc., China) and anhydrous ethanol were mixed by ball milling at 300 rpm for 15 h using zirconia as the milling media, the mass ratio of the materials to zirconia balls was selected to be 1:20. The mixture was subsequently dried at 85°C for 5 h. Finally, the mixed solid powders were calcined at 800°C for 10 h in an air atmosphere to obtain pure LTO. The molar ratio of Li:Ti was 4:5 and excessive Li (5%) was added to compensate for the volatilization of Li during the calcined process.

Graphene (Qingdao hi tech Polytron Technologies Inc) and carbon nanotubes (Beijing Cnano Technology Limited) powders were purchased for this study. For fabricating LTO/carbon nanotubes/graphene composite: 5% wt. graphene and 5% wt. carbon nanotubes were dispersed in anhydrous ethanol with sonication for 0.5 h. Then, the carbon nanotubes and graphene suspension and LTO powders were mixed by ball milling (the weight percentage ratio of the total carbon nanotubes and graphene to LTO is 10%) at 500 rpm for 5 h. Subsequently, the mixture was dried at 85°C for 4 h, followed by calcining at 500°C for 5 h under N_2 atmosphere to obtain the homogeneous LTO/carbon nanotubes/graphene composite (abbreviated for LTO-CNT-G). For comparison, a LTO/graphene composite (the weight percentage ratio of graphene to LTO is 10%) was also prepared by the same process. This composite was abbreviated LTO-G.

Powder XRD (Ultima IV, Riguku) with Cu K_α radiation was used to identify the phase composition of all samples. The diffraction patterns were collected at room temperature by scanning stepwise in the range of 10–90° at a scanning rate of 0.02° per 10 s. The morphology of the materials was characterized by SEM (Tescan Maia3 MHL).

2.2 Electrochemical measurements

The electrochemical properties of all samples were tested by adopting a CR2032 coin-type half-cell with Li foil as the counter electrode. The cathode slurry was prepared by homogeneously mixing the active materials, Super-P, and polyvinylidene fluoride (PVDF) in a mass ratio of 85:10:5 in N-methyl-2-pyrrolidone (NMP) solvent. Then, the slurry was cast onto a Cu foil and dried for 12 h in vacuum at 105°C. Finally, the electrode laminate was punched into disks (10 mm in diameter) and dried in a vacuum oven at 105°C for 24 h. The coin cell was assembled entirely in an argon-filled glovebox. The loading density of the active materials on each electrode was about 1.52 mg cm^{-2}. The separator was Celgard 2400 (Celgard Inc., USA), and the electrolyte (Capchem Technology (Shenzhen) Co., Ltd.) was a solution of 1 mol L^{-1} $LiPF_6$ in ethylene carbonate, dimethyl carbonate, and diethyl carbonate (1:1:1, volume ratio).

Galvanostatic charge-discharge tests were carried out on an automatic galvanostatic charge-discharge unit (Land 2001A, Wuhan, China) between 1 and 3 V at C-rates of 0.2 C, 0.5 C, 1 C, 3 C, 5 C and 10 C (1 C = 250 mAh g^{-1}) at 25 °C.

3. Results and discussion

3.1 Material characterization

Figure1 shows the XRD patterns of pure LTO, LTO-G, and LTO-CNT-G composites. All the XRD patterns are indexed to the standard spinel structure of LTO (card NO. 49-0207) with *Fd*-3m space group. The absence of graphite diffraction peaks from LTO-G and LTO-CNT-G composites may be ascribe to the low quantity and amorphous structure. The XRD patterns prove that the structures of LTO-G and LTO-CNT-G composites are not destroyed after the graphene and carbon nanotubes incorporated with LTO particles.

Figure 2 shows the SEM images of pure LTO, LTO-G, and LTO-CNT-G composites. It can be seen that the LTO particles are well dispersed and wrapped in the graphene sheets with cross-linked carbon nanotubes to form 3D conducting networks, which can offer more conductive pathways for electron transfer during the charge/discharge process.

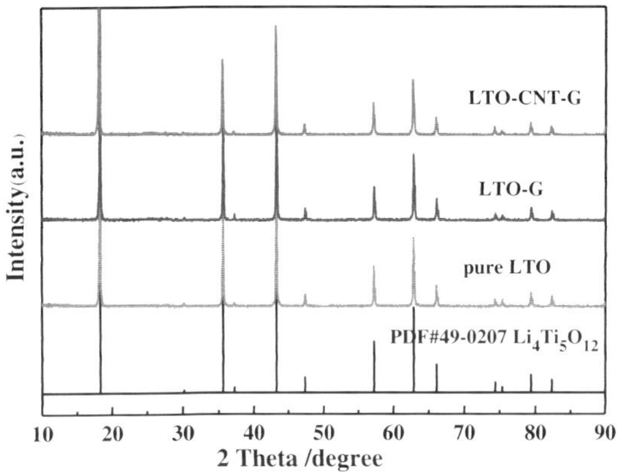

Figure 1: XRD patterns of pure LTO, LTO-G composite, LTO-CNT-G composites and the standard XRD patterns of LTO.

Figure 2: SEM images of (a) pure LTO, (b) LTO-G and (c) LTO-CNT-G composites.

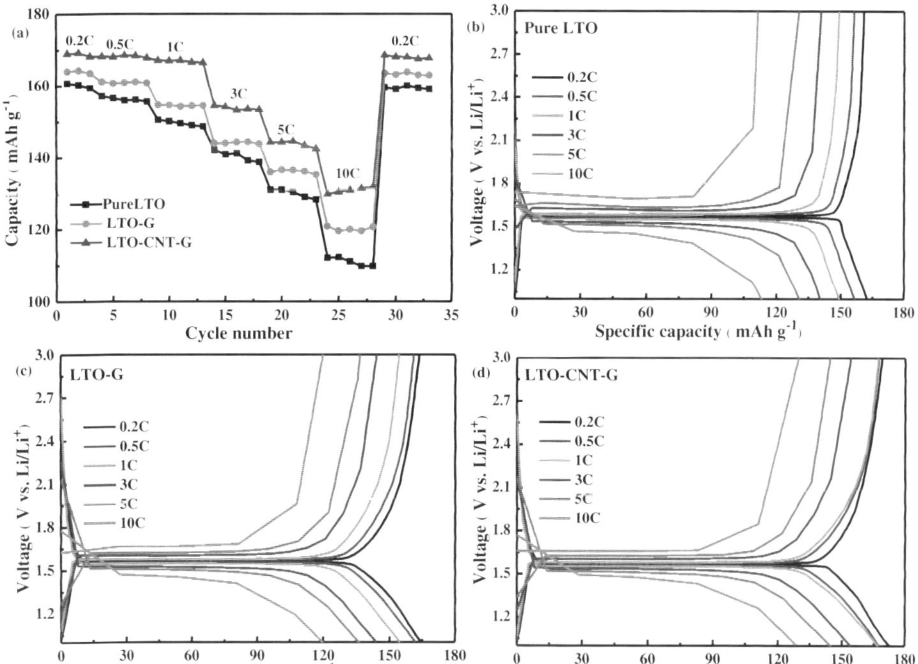

Figure 3: (a) Rate capacity of pure LTO, LTO-G and LTO-CNT-G composites from 0.2 C to 10 C, and (b-d) galvanostatic charge-discharge curves at different C-rates for pure LTO, LTO-G composite, LTO-CNT-G composites between 1 and 3 V (1 C = 160 mAh g^{-1}).

IOP Conf. Series: Materials Science and Engineering **242** (2017) 012013 doi:10.1088/1757-899X/242/1/012013

3.2 Electrochemical properties

Figure 3a displays the rate capacity of pure LTO, LTO-G, and LTO-CNT-G composites at C-rates from 0.2–10 C in the range of 1 and 3 V. The LTO-CNT-G composite shows a superior rate capacity than LTO-G composite and pure LTO. The galvanostatic charge-discharge curves of pure LTO, LTO-G and LTO-CNT-G composites are shown in Figure 3b–d, respectively. The LTO-CNT-G composite exhibits a smallest potential difference between the charge and discharge platforms compared with LTO-G composite and pure LTO at high C-rates. This is attributed to that LTO particles incorporated with graphene and carbon nanotubes could improve electronic conductivity and reduce electrode polarization , thereby enhancing the rate capacity of LTO. The charge capacity values of the LTO-CNT-G composite are 169.0, 168.5, 167.1, 153.2, 144.5, 131.5 mAh g^{-1} at 0.2, 0.5, 1, 3, 5 and 10 C, (1 C = 160 mAh g^{-1}), respectively. However, the charge capacity values of the LTO-G composite are 164.3, 160.9, 154.6, 144.3, 136.2, 120.0 mAh g^{-1}, respectively, and the charge capacity values of pure LTO are 160.3, 155.9, 149.5, 139.2, 130.4, 111.1 mAh g^{-1}, respectively.

4. Conclusions

In summary, the LTO-CNT-G composite has been successfully synthesized by a solid-state method. The characterization results show that LTO particles are well dispersed and wrapped in the graphene sheets with cross-linked carbon nanotube. The electrochemical results indicate that the LTO-CNT-G composite displays a superior rate capability than the LTO-G composite and pure LTO, which may be attributed to that the synergistic effect of graphene and carbon nanotubes could enhance the electronic conductivity and reduce electrode polarization. Therefore, the LTO-CNT-G composite is a promising anode material for lithium ion batteries.

Acknowledgments

This work was financially supported by Science-technology Support Plan Projects of Hebei Province (17394411D) and (17964407D).

References

[1] Cao F F, Guo Y G and Wan L J 2011 *Energy Environ. Sci.* **4** 1634.
[2] Etacheri V, Marom R, Elazari R, Salitra G and Aurbach D 2011 *Energy Environ. Sci.* **4** 3243.
[3] Liao J Y, Chabot V, Gu M, Wang C, Xiao X and Chen Z 2014 Nano Energy **9** 383.
[4] Wu N, Yang Z Z, Yao H R, Yin Y X, Gu L and Guo Y G 2015 *Angew. Chem., Int. Ed.* **54** 5757.
[5] Zhu G N, Liu H J, Zhuang J H, Wang C X, Wang Y G and Xia Y Y 2011 *Energy Environ. Sci.* **4** 4016.
[6] Yu L, Wu H B and Lou X W D 2013 *Adv. Mater.* **25** 2296.
[7] Qi Y L, Huang Y D, Jia D Z, Bao S J and Guo Z P 2009 *Electrochim. Acta* **54** 4772.
[8] Li X, Qu M Z, Huai Y J and Z L Yu 2010 *Electrochim. Acta* **55** 2978.
[9] Li Y, Zhang H, Wang S, Chen Y, Shi Z and Li N J 2016 *J Mater Chem A* **4** 11247.
[10] Yang J, Wang J, Li X, Wang D, Liu J and Liang G 2012 *J Mater Chem* **22** 7537.
[11]Lei X L, Chen Y M, Wang W G , Ye Y P, Zheng C C, Deng P, Shi Z C and Zhang H Y 2015 *J. Alloys Comp.* **626** 280.

Study of the Curie temperature in a transverse Ising film with the next nearest-neighbor interactions

Z P Li, Z X Lu

School of Mechanical & Vehicle Engineering, Linyi University, Linyi 276005, China

E-mail: lizhipeng@lyu.edu.cn

Abstract. The mean-field approximation is used to study the Curie temperature in a transverse Ising film with the next nearest-neighbor interactions. The main emphasis of our research is focused on the dependence of Curie temperature on the decaying exponent δ. In the meanwhile, how the effect of film thickness N on the character of phase transition is also calculated.

1. Introduction

The transverse Ising film has been concerned for long time in theory and experiment [1-10]. Theoretically, the transverse Ising model, from the microcosmic point of view, has been used to study the phase transition properties of ferroelectric thin films or ferroelectric superlattices [2-10]. By taking into account the nearest-neighbor interactions, Wang *et al.* [2] and Sy [4], respectively, investigated the surface modification effect on the Curie temperature Tc of a ferroelectric thin film with the mean-field approximation. In fact, the long-range interaction dominates in a ferroelectric thin film. As a consequence, the next nearest-neighbor interactions could not be neglected. In some literatures [5-8], by considering different ranges of exchange interactions, the mean-field approximation and effective-field theory with correlations are used to calculate the dependence of Curie temperature Tc on various exchange interaction parameters. Recently, we studied the phase transitions in a transverse Ising film with the next nearest-neighbor interactions by using the mean-field approximation [9]. Very recently, the nearest-neighbor interactions and the next nearest-neighbor interactions as well as the third nearest-neighbor interactions have been included when we calculated the phase diagram of a ferroelectric thin film by the use of mean-field approximation [10].

In the present paper, the Curie temperature in a transverse Ising film with the next nearest-neighbor interactions has been studied by using the mean-field approximation. The dependence of Curie temperature Tc on the film thickness N is shown. Meanwhile, the influence of decaying exponent δ on the phase transition properties is also obtained.

2. Model and formulas

Figure 1 is the three-dimension structure diagram for the transverse Ising film with a thickness N. The Hamiltonian of this system is [5-10]

$$H = -\frac{1}{2}\sum_{ij}\frac{J_{ij}}{r_{ij}^{\delta}}S_i^z S_j^z - \Omega_b \sum_{i\in b} S_i^x - \Omega_s \sum_{i\in s} S_i^x , \tag{1}$$

where S_i^x and S_i^z, respectively, is the spin-1/2 component along the x and z direction at the site i. J_{ij} denotes the interchange interaction between the two pseudo-spins. J_{ij}=Js while the two pseudo-spins at the surface layer, otherwise J_{ij}=Jb. Ω_i denotes the transverse field in the i-th layer. $\Omega_i=\Omega_s$ while the transverse field on the surface layer, however $\Omega_i=\Omega_b$ while the transverse field in the bulk layer. δ

Content from this work may be used under the terms of the Creative Commons Attribution 3.0 licence. Any further distribution of this work must maintain attribution to the author(s) and the title of the work, journal citation and DOI.
Published under licence by IOP Publishing Ltd

represents the decaying exponent.

With the mean-field approximation [5, 6, 9, 10], the polarization in the i-th layer of the transverse Ising film is

$$R_i = \langle S_{iz} \rangle = \left(\langle H_{iz} \rangle / 2|H_i|\right) / \tanh\left(|H_i| / 2k_B T\right), \tag{2}$$

where

$$\langle H_{iz} \rangle = \sum_j \frac{J_{ij}}{r_{ij}^{\delta}} S_j^z , \tag{3}$$

$$|H_i| = \sqrt{\Omega_i^2 + \langle H_{iz} \rangle^2} . \tag{4}$$

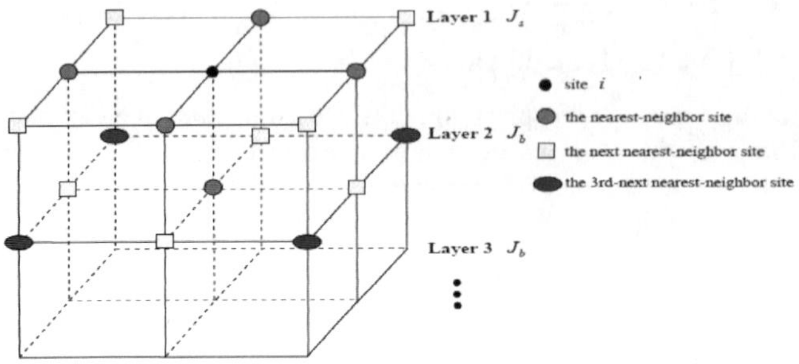

Figure 1. The 3-D structure diagram for a transverse Ising film.

For an N-layer transverse Ising film, the equation (2) will be expanded to N linear equations [9, 10]. For the layer 1,

$$\tau_s R_1 = 4J_s R_1 + J_b R_2 + 4\frac{J_s}{\sqrt{2}^{\delta}} R_1 + 4\frac{J_b}{\sqrt{2}^{\delta}} R_2 , \tag{5}$$

for the layer N,

$$\tau_s R_N = 4J_s R_N + J_b R_{N-1} + 4\frac{J_s}{\sqrt{2}^{\delta}} R_N + 4\frac{J_b}{\sqrt{2}^{\delta}} R_{N-1} , \tag{6}$$

for the other layers,

$$\tau_b R_i = 4J_b R_i + J_b R_{i-1} + J_b R_{i+1} + 4\frac{J_b}{\sqrt{2}^{\delta}} R_i + 4\frac{J_b}{\sqrt{2}^{\delta}} R_{i-1} + 4\frac{J_b}{\sqrt{2}^{\delta}} R_{i+1} . \tag{7}$$

Where the parameters τ_s and τ_b is respectively defined as follows:

$$\tau_s = 2\Omega_s \coth\left(\Omega_s / 2k_B T\right), \tag{8}$$

$$\tau_b = 2\Omega_b \coth\left(\Omega_b / 2k_B T\right). \tag{9}$$

According to [9], the Curie temperature Tc for an N-layer transverse Ising film is calculated by

$$\begin{vmatrix} X_s & -1 & & & & & \\ -1 & X_b & -1 & & & & \\ & -1 & X_b & & & & \\ & & & \ddots & & & \\ & & & & X_b & -1 & \\ & & & & -1 & X_b & -1 \\ & & & & & -1 & X_s \end{vmatrix}_{N\times N} = 0 , \tag{10}$$

where

$$X_s = \frac{\tau_s - 4J_s - 4\dfrac{J_s}{\sqrt{2}^{\delta}}}{J_b + 4\dfrac{J_b}{\sqrt{2}^{\delta}}}, \quad X_b = \frac{\tau_b - 4J_b - 4\dfrac{J_b}{\sqrt{2}^{\delta}}}{J_b + 4\dfrac{J_b}{\sqrt{2}^{\delta}}}. \tag{11}$$

3. Numerical results

Within the framework of the mean-field approximation [9, 10], the Curie temperature of a transverse Ising film with the next nearest-neighbor interaction can be calculated from equations (5-11). In the following numerical calculations, the transverse field $\Omega_s=\Omega_b=\Omega$ is supposed for convenience.

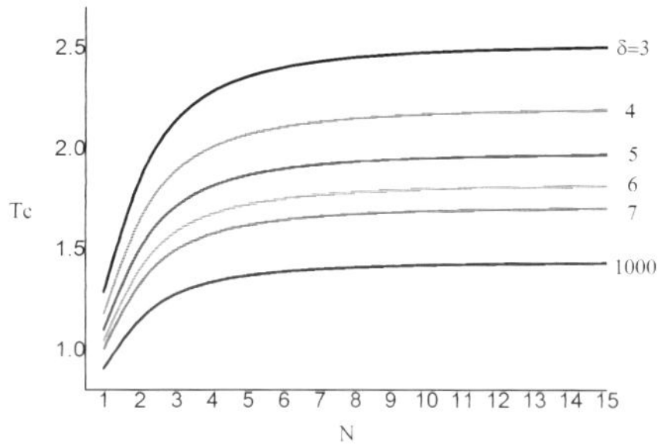

Figure 2. The dependence of Curie temperature Tc on the film thickness N for Js=Jb.

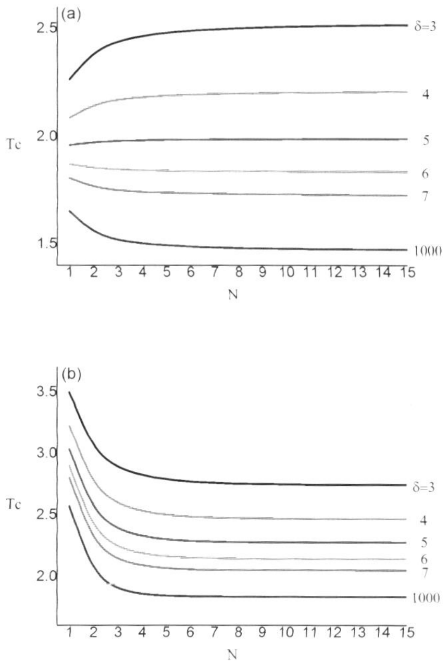

Figure 3. The dependence of Curie temperature Tc on the film thickness N for Js>Jb. All curves are for Jb=1.0, Ω=1.0. (a) Js=1.35, (b) Js=1.80.

Figure 2 shows the dependence of Curie temperature Tc on the film thickness N for different values of decaying exponent δ when Js=Jb. In Figure 2, Js=Jb=1.0 and Ω=1.0 are selected. For a certain decaying exponent δ, it is obvious that the Curie temperature Tc increases with the film thickness N increases. On the other hand, for the transverse Ising thin film with a certain thickness N, with the increase of decaying exponent δ, however, the Curie temperature Tc decreases. In addition, the larger the value of decaying exponent δ, the closer to each other the value of Curie temperature Tc. These characters are similar to those of in [9]. However, Figure 2 is more intuitionistic and explicit.

Figure 3 shows the dependence of Curie temperature Tc on the film thickness N for different values of decaying exponent δ when Js>Jb. In Figure 3(a), Js=1.35, Jb=1.0 and Ω=1.0. It is obvious that the surface interchange interaction Js has an important impact on the Curie temperature Tc. For a different surface interchange interaction Js, all the curves of Tc~N have some significant changes. The curvature and direction of every curve is obvious differently. However, the larger the decaying exponent δ, the closer the Curie temperature Tc. These features still remain.

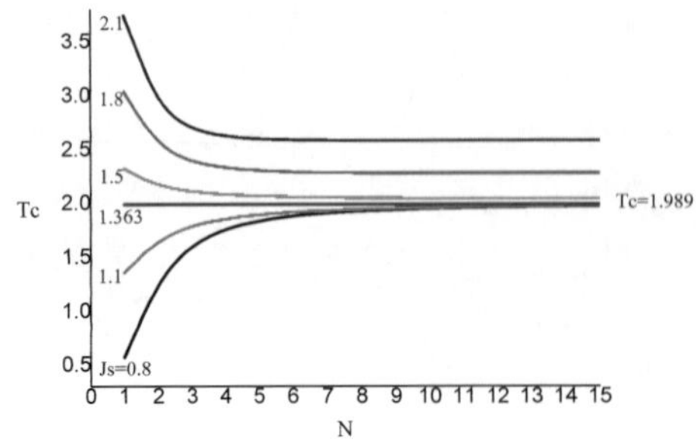

Figure 4. The dependence of Curie temperature Tc on the film thickness N for different Js. All curves are for Jb=1.0, Ω=1.0, δ=5.0.

Figure 4 shows the dependence of Curie temperature Tc on the film thickness N for different values of surface interchange interaction Js when the decaying exponent δ is fixed. Obviously, for a certain Js which is smaller than Js=1.363, the Curie temperature Tc increases with the film thickness N increases. Eventually, the curves of Tc~N for different values of Js would be close to the bulk transition temperature Tc=1.989 with film thickness N increases. These features are similar to those of in [10].

4. Conclusions

In conclusion, the mean-field approximation has been used to calculate the Curie temperature for a transverse Ising film by taking into account the next nearest-neighbor interaction. The dependence of the Curie temperature Tc on the film thickness N for different values of decaying exponent δ is discussed. Meanwhile, the effect of surface interchange interaction Js on the Curie temperature Tc is also executed. Our results are also compared with those of the third nearest-neighbor interaction [10].

References

[1] Scott J F 2007 *Science* **315** 954
[2] Wang C L, Zhong W L and Zhang P L 1992 *J. Phys.: Condens. Matter* **4** 4743
[3] Wang C L, Smith S R P and Tilley D R 1994 *J. Phys.: Condens. Matter* **6** 9633
[4] Sy H K 1993 *J. Phys.: Condens. Matter* **5** 1213
[5] Wang Y G, Zhong W L and Zhang P L 1997 *Solid State Commun.* **101** 807
[6] Xin Y, Wang C L, Zhong W L and Zhang P L 1999 *Phys. Lett. A* **260** 411
[7] Yao D L, Wu Y Z and Li Z Y 2002 *Phys. Status Solidi B* **231** 3
[8] Tabyaoui A, Madani M, Ainane A and Saber M 2005 *Physica A* **358** 150
[9] Li Z P 2015 *IFEEA* 364
[10] Li Z P 2016 *Phase Transitions* **89** 1119

IOP Conf. Series: Materials Science and Engineering **242** (2017) 012015 doi:10.1088/1757-899X/242/1/012015

Simulation of microcrack initiation in nanocomposite ceramic tool materials

T Zhou [1, *] and C Huang[2]

[1] School of Mechanical and Automotive Engineering, Qilu University of Technology, Jinan 250353, P. R. China

[2] Center for Advanced Jet Engineering Technologies (CaJET), School of Mechanical Engineering, Key Laboratory of High-efficiency and Clean Mechanical Manufacture (Ministry of Education), Shandong University, Jinan 250061, P. R. China

[*]Corresponding author. E-mail address: zhoutingting506@163.com

Abstract. In this work, the microcrack initiation behavior of nanocomposite ceramic tool materials caused by the residual stress initiated during the cooling stage is simulated by means of a micromechanical model based on the Voronoi tessellation and the cohesive element theory. Influences of microstructure morphologies such as nanoparticle size, nanoparticle volume content and microstructure types on the microcrack distribution and material properties are analyzed respectively. The conclusions are useful for providing theoretical basis for the development of nanocomposite ceramic tool materials.

1. Introduction

The brittle fracture is known as the main fracture mode of ceramic tool materials. Cracks usually initiate from internal defects like pores and microcracks in ceramic tool materials. On one hand, a crack grows under external load and is affected by microstructural heterogeneities such as grain boundaries, porosities, impurities and second phase particles. On the other hand, residual stress forms during the cooling stage after sintering procedure due to the mismatch of thermal expansion coefficients and elastic modulus between matrix and second phases in composite ceramic tool materials, which also has an important effect on the crack path. If the superimposition of the residual stress on the external stress caused by applied load is above the fracture strength of materials, microcracks are liable to initiate in the vicinity of the main crack tip. Microcracking consumes extra fracture energy, releases the stress concentration around the crack tip and is useful for the material toughening. However, if the residual stress is too large, the strength of materials tends to be reduced.

Nanocomposite ceramic tool materials possess more excellent mechanical properties compared with microcomposite ceramic tool materials[1, 2]. Due to the size differences between nanoparticles and microsized matrix grains in nanocomposite ceramic tool materials, in which nanoparticles distribute both along GBs and in grains, the residual stress field and the microcrack initiation caused by the residual stress are different from that in the microstructure of microcomposite ceramic tool materials, as well as the toughening mechanism of the residual stress and the microcrack initiation. Therefore, studying the residual stress distribution and the microcrack initiation behavior of nanocomposite ceramic tool materials is able to provide useful theoretical basis for material development.

In this work, the microcrack initiation in nanocomposite ceramic tool materials caused by the residual stress initiated during the cooling stage will be simulated. And influences of nanoparticle size,

Content from this work may be used under the terms of the Creative Commons Attribution 3.0 licence. Any further distribution of this work must maintain attribution to the author(s) and the title of the work, journal citation and DOI.
Published under licence by IOP Publishing Ltd

nanoparticle volume content and microstructure types on microcrack initiation and material properties will be discussed as well.

2. Simulation models

The simulation of microcrack initiation and cracking behavior will be carried out by means of a micromechanical model based on the Voronoi tessellation and the cohesive element theory. The microstructure models are developed by programming with MATLAB[3]. The Voronoi tessellation is utilized to characterize microstructures of nanocomposite ceramic tool materials. And nanoparticles generation are implemented until the size and the volume content of nanoparticles reaching the set value. Then cohesive elements with fracture criteria should be placed along all triangular volumetric element boundaries in order to form the arbitrary cracking paths. Finally, an inp file is compiled to introduce the cohesive model to ABAQUS for FEM calculation after the nodes and elements that lie in matrix grains, matrix GBs, nanoparticles and interfaces are determined.

Figure 1 (a) is a microstructure model represented by the Voronoi tessellation. And Figure 1(b) is the Voronoi tessellation diagram after generating cohesive elements and triangular elements. In Figure 1(b), each Voronoi polygon stands for a matrix grain, while white triangles indicate volumetric elements of matrix grains and blue triangles stand for volumetric elements of nanoparticles. Red and yellow quadrilaterals indicate cohesive elements along matrix GBs and those in matrix grains respectively. Black and green quadrilaterals characterize cohesive elements along interfaces and those in nanoparticles respectively. For easy display reason, the thickness of cohesive elements is magnified and coarser finite element mesh is adopted in Figure 1(b). It is obvious that cohesive elements with fracture criteria distribute among arbitrary adjacent volumetric elements, no matter in grains or along GBs of matrix phases and nano phases. Therefore both intergranular and transgranular cracking can be simulated with our model.

It should be noted that the addition of cohesive elements around volumetric finite elements easily induces the stiffness reduction of a numerical model. However, this issue can be ignored by carefully choosing the cohesive element strength and the finite element size[4].

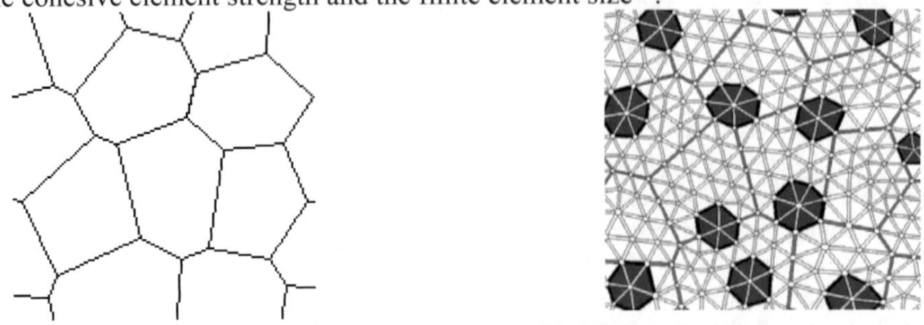

(a) Before embedding cohesive elements (b) After embedding cohesive elements

Figure 1. Cohesive element model.

3. FEM boundary conditions and parameters setting

Residual stress and microcracks tend to initiate in materials during the cooling stage. In this work, 1500°C is taken as the temperature at which the residual stress initiates during the cooling stage and the room temperature is set as 20°C. Therefore, the initiation of residual stress and microcracks can be modeled and the influence of microstructure morphologies on the microcrack distribution can be analyzed. Free boundary conditions are adopted in this step. Then external load is applied and the microscopic cracking behavior in the presence of residual stress and microcracks can be modeled. Explicit integration is employed to conduct the cracking simulation and constant velocity load is imposed symmetrically at the top and bottom edges of the model. The corresponding velocity load can be obtained by the expression $v = \dot{\varepsilon} \cdot h$, while the constant strain rate $\dot{\varepsilon} = 5 \times 10^3$ is adopted in the simulation and h is the height of the model. Bilinear traction-separation law for brittle fracture is used to describe the properties of cohesive elements.

Al_2O_3/SiC_n nanocomposite ceramic tool materials are taken for the research subject. The thermal expansion coefficient for Al_2O_3 matrix is 8×10^{-6} and that for SiC is 4.35×10^{-6}. The other material parameters are shown in Table 1.

Table 1. Material properties for the microstructure of Al_2O_3/SiC_n.

Element type	E (GPa)	ν	T_{max} (MPa)	Γ (J/m^2)
Triangular elements in Al_2O_3	390	0.23	—	—
Cohesive elements in Al_2O_3	390	0.23	644	2.3
Cohesive elements in GBs	78	0.25	186	1
Triangular elements in SiC	430	0.2	—	—
Cohesive elements in SiC	430	0.2	950	5.8
Cohesive elements in interfaces between Al_2O_3 and SiC	250	0.25	712	2.5

4. Simulation results and discussions

4.1 Influences of nanoparticle size

Due to the size differences between nanoparticles and matrix grains, nanoparticles can distribute in matrix grains as well as along matrix GBs in nanocomposite ceramic tool materials. Figure 2 and Figure 3 respectively show intergranular (all the nanoparticles are located along matrix GBs) and intragranular (all the nanoparticles are located in matrix grains) microstructure models for Al_2O_3/SiC_n with different sizes of nano SiC particles. The microstructure models have the same morphology with the average Al_2O_3 matrix size $d_{matrix}=1\ \mu m$ and volume content of nano SiC V_{nano}=10%. The nanoparticle sizes in Figure 2 and Figure 3 are 70*nm* and 150*nm* respectively.

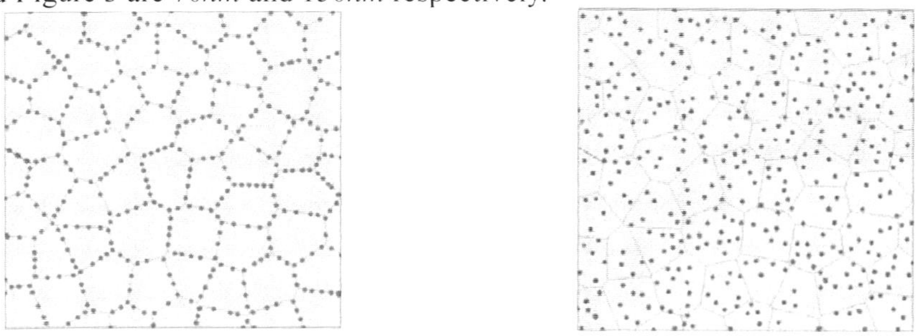

(a) Intergranular model (b) Intragranular model

Figure 2. Microstructure models with $d_{nano}=70nm$.

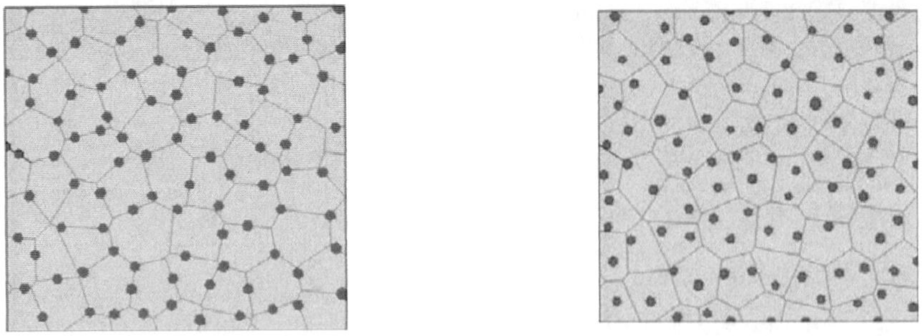

(a) Intergranular model (b) Intragranular model

Figure 3. Microstructure models with $d_{nano} = 150nm$.

The numerical results of residual stress field in the microstructures with $d_{nano} = 70nm$ after FEM calculations are represented in Figure 4. Because of the fact that the thermal expansion coefficient of SiC is smaller than that of Al_2O_3 matrix, the shrinkage of nano SiC particles is smaller than that of Al_2O_3 matrix grains during the cooling stage. Thus compressive stress acts in nano SiC particles, whereas tensile stress prevails in Al_2O_3 matrix grains. More complicated stress condition exists between nanoparticles and matrix grains, which is affected by the location of nanoparticles. It can be observed from Figure 4 that the residual stress distributing around nanoparticles acts as compressive stress in the radial direction and acts as tensile stress in the tangential direction, no matter what kind of the microstructure types is.

The numerical results of microcrack initiation in the microstructures are shown in Figure 5 ($d_{nano} = 70nm$) and Figure 6 ($d_{nano} = 150nm$). It can be observed that the microcrack length is equivalent with the nanoparticle size and shorter microcracks initiate in the microstructure models with smaller nanoparticles, which indicates that the fracture strength of materials increases as the nanoparticle size decreases.

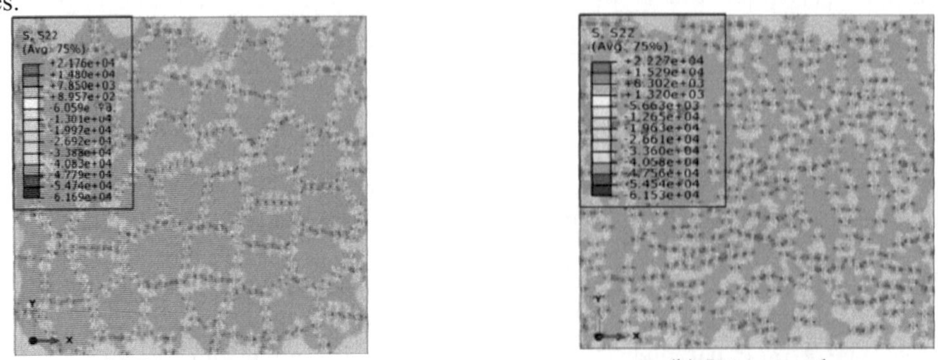

(a) Intergranular (b) Intragranular

Figure 4. Residual stress field in microstructures with $d_{nano} = 70nm$.

The radial residual stress and the tangential residual stress around nanoparticles can be expressed as[5],

$$\begin{cases} \sigma_r = P(\frac{R}{r})^3 \\ \sigma_t = -\frac{P}{2}(\frac{R}{r})^3 \end{cases} \quad (1)$$

Where, r is the distance between adjacent nanoparticles. R is the radius of nanoparticles.

It can be deduced from equation (1) that lower tangential tensile stress initiates in microstructures with smaller nanoparticles, which produces smaller microcracks in size. Besides, the pinning effect of smaller nanoparticles on the GBs is stronger[6], which leads to smaller matrix size after the sintering procedure. Therefore, conclusions can be drawn that nanocomposite ceramic tool materials with smaller nanoparticles possess higher fracture strength. This is one of the reason why the microcrack size is smaller in the microstructure of ceramic tool materials with smaller nanoparticles.

Levin et al.[7] regarded that the fracture strength of matrix is liable to be reduced after the addition of nano SiC to ceramic tool materials, which contributes to transgranular cracking. However, the fracture toughness of materials tends to decrease if the fracture strength of matrix is reduced too much. The effect of tensile stress area in the matrix on the fracture toughness of materials can be expressed as[8],

$$K_{IC} = K_{I0} - 2q\sqrt{\frac{2D_i}{\pi}} \qquad (2)$$

Where, K_{I0} is the critical fracture strength factor of matrix. q is the local residual tensile stress. D_i is the length of tensile stress area. In terms of nanocomposite ceramic tool materials with the same volume content of nanoparticles, reducing the nanoparticle size is liable to increase the number of nanoparticles and decrease the distance between nanoparticles. The length of the tensile stress area D_i is relevant with the distance between nanoparticles, therefore D_i is liable to be reduced, as well as the weakening effect on the matrix strength. Accordingly, it is beneficial for improving the fracture toughness of ceramic tool materials by reducing nanoparticle size based on equation (2).

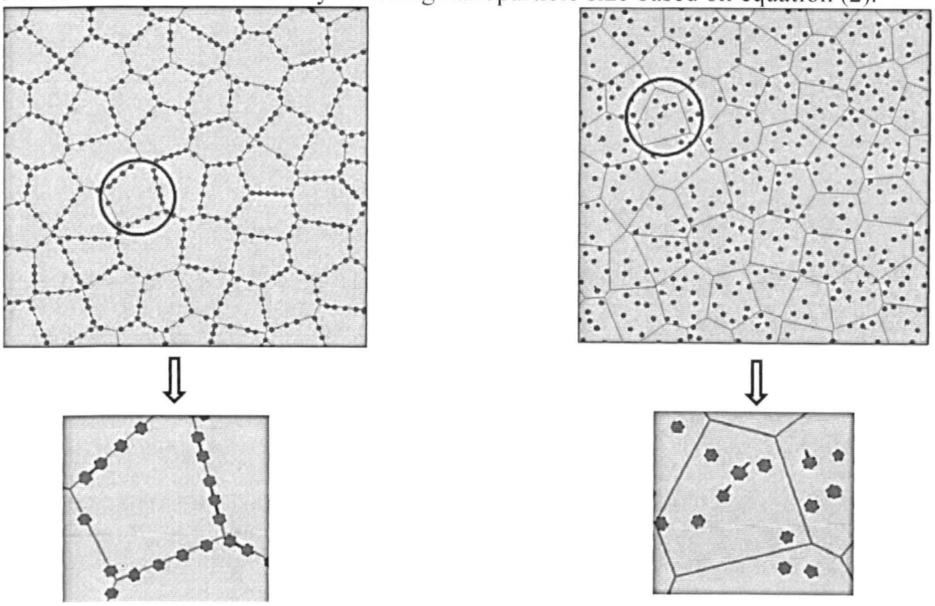

(a) Intergranular (b) Intragranular

Figure 5. Microcracks initiated in microstructures with $d_{nano} = 70nm$.

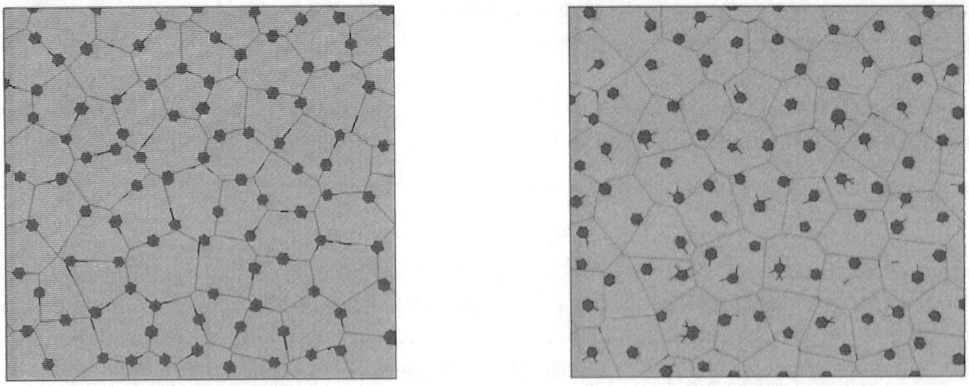

(a) Intergranular (b) Intragranular

Figure 6. Microcracks initiated in microstructures with $d_{nano} = 150nm$.

4.2 Influence of nanoparticle volume content

The microstructure models for nanocomposite ceramic tool materials with the volume content of nano SiC as 5%, 10%, 15% and 30% respectively are established. The microstructures are all the intragranular/intergranular structures with the average matrix grain size $d_{matrix} = 1\ \mu m$ and the average nanoparticle size $d_{nano} = 70nm$.

Figure 7 represents the simulation results of microcrack initiation. With the increase of nanoparticle volume content, the distances between partial nanoparticles decrease and microcracks connect with each other, which is disadvantageous for the improvement of material strength. When the volume content of nanoparticles is too high (30%), some nanoparticles agglomerate and the residual tensile stress increases. Accordingly, microcracks are more liable to initiate around agglomerated nanoparticles. Therefore, it should take measures to control the nanoparticle dispersion during the actual sintering procedure in order to avoid the nanoparticle agglomeration.

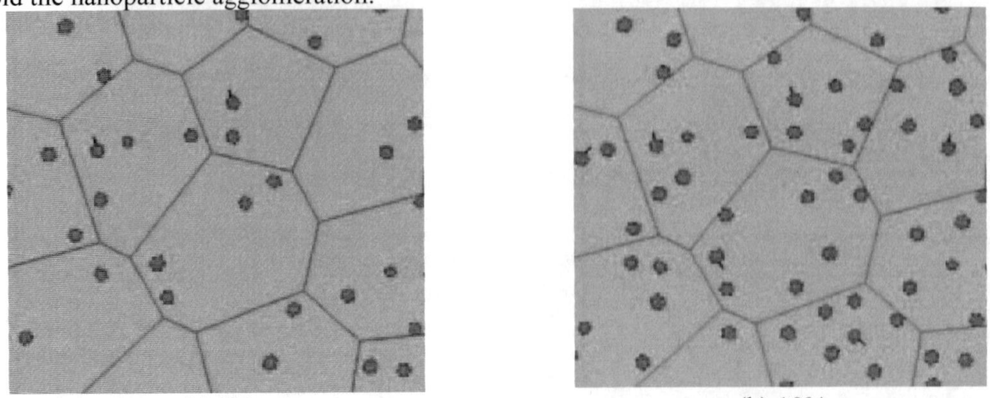

(a) 5% (b) 10%

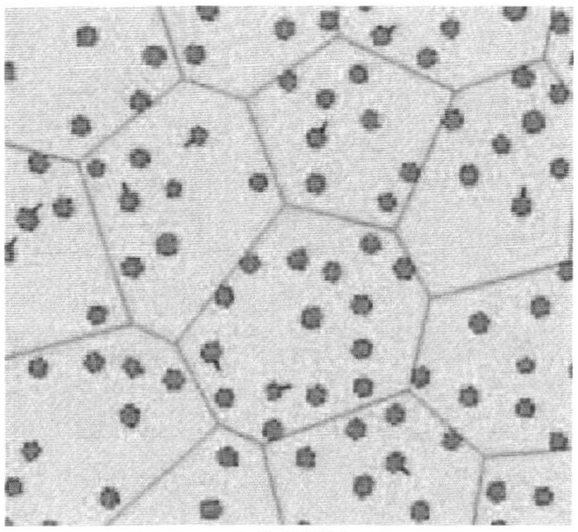

(c) 15%

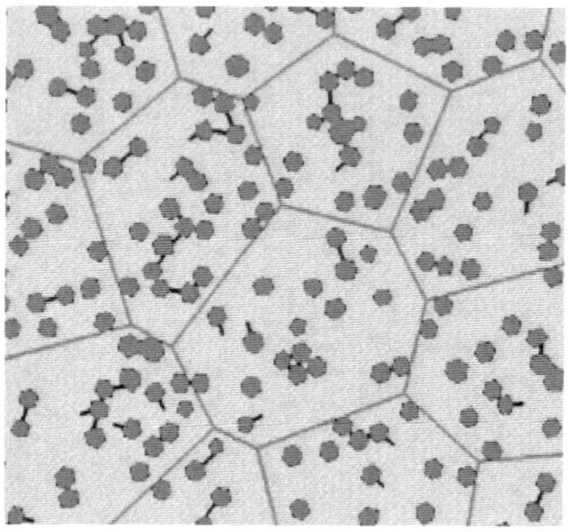

(d) 30%

Figure 7. Microcrack initiation in microstructures with different nanoparticle volume content.

4.3 Influence of microstructure types

In order to explore the influence of nanoparticles distribution on mechanical properties, intragranular and intergranular models for Al_2O_3/SiC_n are established, as shown in Figure 3. The average size of Al_2O_3 matrix grains equals to $1\mu m$, while the average size and the volume content of nano SiC are 150nm and 10% respectively.

4.3.1 Simulation of microcrack initiation. The simulation results of microcrack initiation in intergranular and intragranular microstructures are shown in Figure 6. It can be found that microcracks tend to initiate at matrix GBs around nanoparticles in the intergranular microstructure, since the tangential tensile stress forms around nanoparticles located along matrix GBs and the fracture strength of GBs is weaker than that of grains. As to the intragranular microstructure, in which nanoparticles are all located in the matrix, microcraks produce in the radial direction of some in-matrix nanoparticles under the tangential tensile stress.

4.3.2 Simulation of microscopic cracking behavior under the external load. External load is applied to the microstructure models with microcracks in order to study the effects of intragranular and intergranular microcracks on the fracture mode and the critical fracture energy release rate of materials. Figure 8 is the cracking paths in the intergranular microstructure. It can be observed that transgranular cracking is more evident in the microstructure without microcracks in comparison with the microstructure with microcracks. It indicates that intergranular cracking tends to happen due to the radial microcrakcs around nanoparticles that are located at GBs, which is disadvantageous for improving the fracture toughness of materials.

Figure 9 represents the cracking paths in the intragranular microstructure. It can be observed that transgranular cracking happens in the intragranular microstructure with microcracks under the applied load (Figure 9 (a)), while cracks tend to propagate along GBs in the intragranular microstructure without microcracks Figure 9 (b)). It implies that microcracks around in-matrix nanoparticles in the intragranular microstructure play an important role in the fracture mode transition.

The results of critical fracture energy release rate during the cracking process for the intergranular and intragranular microstructures are shown in Figure 10. For the intergranular microstructures, the critical fracture energy release rate for the model with mcirocracks is less than that for the model without microcracks. As to the intragranular microstructures, the critical fracture energy release rate for the model with mcirocracks is close to that for the model without microcracks. Microcracks induced by residual stress tend to cause stress concentration and have a negative effect on the fracture strength of materials. Besides, the fracture surface energy is liable to be reduced since the existence of microcracks tends to decrease the actual fracture surfaces during the propagation process of main crack. Although microcracks around nanoparticles located along GBs is liable to induce the transgranular cracking in materials to some extent, it is still disadvantageous for the fracture toughness improvement of materials.

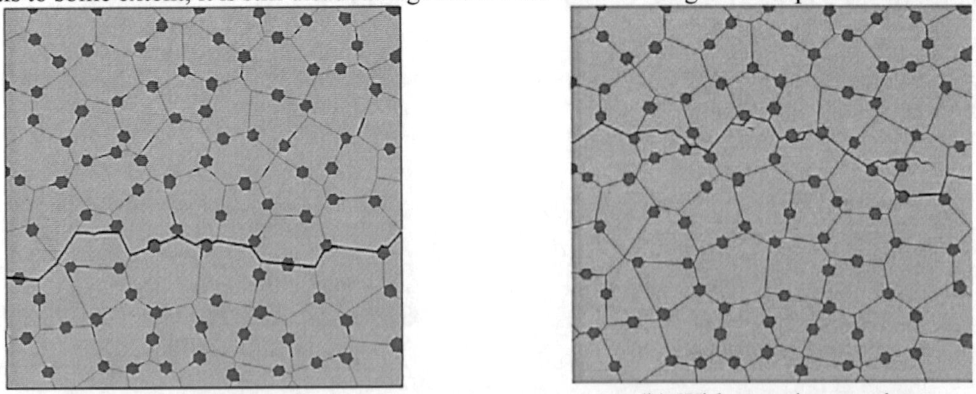

(a) With microcracks (b) Without microcracks

Figure 8. Cracking paths of intergranular microstructures under external load.

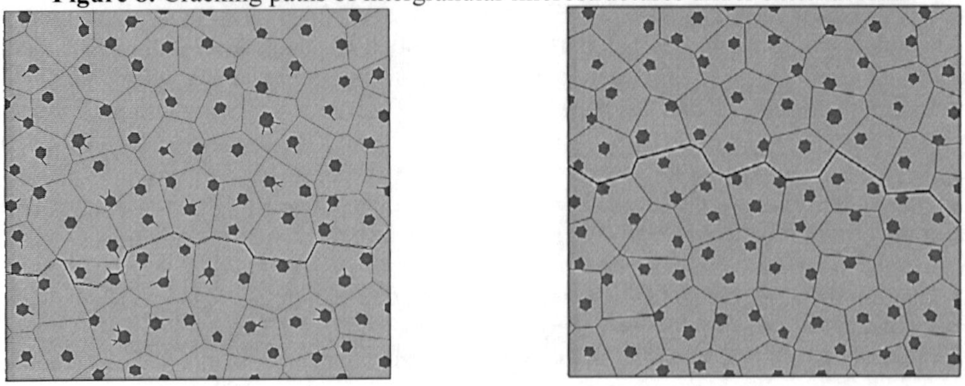

(a) With microcracks (b) Without microcracks

Figure 9. Cracking paths of intragranular microstructures under external load.

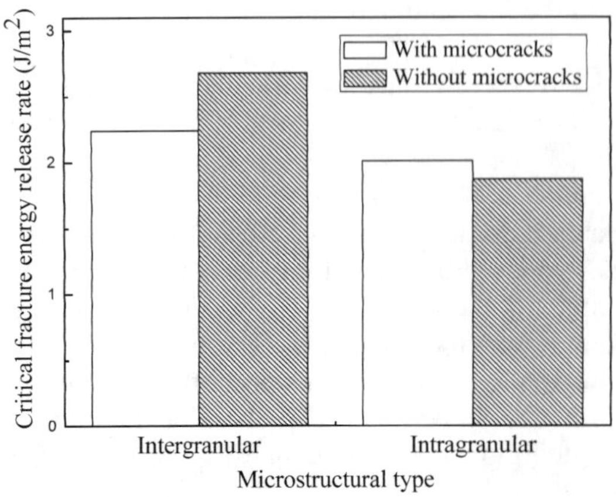

Figure 10. Critical fracture energy release rate of microstructures with and without microcracks.

5. Conclusions

In this paper, the numerical simulation of microcrack initiation behavior in nanocomposite ceramic tool materials caused by the residual stress is carried out by means of a micromechanical model based on the Voronoi tessellation and the cohesive element theory. Influences of microstructure morphologies such as nanoparticle size, nanoparticle volume content and microstructure types on the microcrack initiation and material properties are analyzed respectively. The following conclusions can be obtained.

Shorter microcracks initiate in microstructures with smaller nanoparticles, which is good for the material strength. The microcrack density increases as the nanoparticle volume content rises. When the nanoparticle volume content increases to some extent, microcracks located among adjacent nanoparticles are liable to connect with each other, which results in the decrease of fracture toughness. For the intergranular microstructure, microcracks mainly form at GBs around nanoparticles. As to the intragranular microstructure, microcracks primarily produce in the vicinity of nanoparticles that are located in matrix grains. And it can be concluded from the numerical simulation under the applied load that the presence of microcracks in materials is liable to reduce fracture surface areas in the process of crack propagation and decrease the fracture surface energy of materials, which has a negative effect on the fracture toughness of materials.

Acknowledgements

The work is supported by National Natural Science Foundation of China (51405253).

References

[1] Nakahira A, Niihara K and Hirai T 1986 Microstructure and mechanical properties of Al_2O_3-SiC composites *J. Ceram. Soc. Jpn.* 94(8) p 767
[2] Karch J, Birringer R and Gleiter H 1987 Ceramics ductile at low temperature *Nature*, 330 p 556-558
[3] Zhou T and Huang C 2015 Simulation of crack propagation in single phase ceramic tool materials *Comp. Mater. Sci.*, 104 p 177-184
[4] Tomar V Zhai J and Zhou M 2004 Bounds for element size in a variable stiffness cohesive finite element model *Int. J. Numer. Meth. Eng.* 61 p 1894-1920
[5] Xu Z 1982 *Elastic Mechanics* (Beijing: Higher Education Press) p 279
[6] Smith C 1948 Grains, phases and interphases: an interpretation of the microstructure *T. Metall. Soc. AIME* 175 p 15-51.
[7] Levin I Kaplan W Brandon D and Layyous A 1995 Effect of SiC submicrometer particle size and content on fracture toughness of alumina–SiC "nanocomposites" *J. Am. Ceram. Soc.* 78(1) p 254-256.
[8] Zhang G and Jin Z 1994 Toughening mechanism of particle toughened ceramics *J. Chinese Ceram. Soc.* 22(3) p 259-269.

IOP Conf. Series: Materials Science and Engineering **242** (2017) 012016 doi:10.1088/1757-899X/242/1/012016

Preparation of Diatomite Supported Nano Zinc Oxide Composite Photocatalytic Material and Study on its Formaldehyde Degradation

Liguang Xiao[1] Bo Pang[2]

School of Materials Science and Engineering, Jilin Jianzhu university, Changchun 130118, China

1)Corresponding author: xlg627@163.com 2)1004625664@qq.com

Abstract. This experiment used zinc nitrate as precursor, ethanol as solvent and polyethylene glycol as dispersant, diatomite as carrier, diatomite loaded nano Zinc Oxide was prepared by sol-gel method, in addition, the formaldehyde degradation was studied by two kinds of experimental methods: preparation and loading, preparation and post loading, The samples were characterized by SEM, XRD, BET and IR. Experimental results showed that: Diatomite based nano Zinc Oxide had a continuous adsorption and degradation of formaldehyde, formaldehyde gas with initial concentration was $0.7mg/m^3$, after 36h degradation, the concentration reached $0.238mg/m^3$, the degradation rate reached to 66%.

1. Introduction

Nano Zinc Oxide was a n type semiconductor[1]. Because of the separation efficiency of photogenerated electrons and holes of nano Zinc Oxide was higher than TiO_2, therefore, nano Zinc Oxide had better photocatalytic activity than TiO_2[2], the formaldehyde could be degraded into harmless gas at normal temperature and pressure, and no pollution could be caused by two times, moreover, the temperature range required by photocatalysis was wide and the cost was low, it had some potential applications in air purification, and had a better prospect in environmental control[3], But nano Zinc Oxide because of nano particles was poor dispersion and difficult to fix and so on, if it was directly applied to the paint coating interior walls, it was easy to fall off, Diatomite which had large particle and strong adsorbability, could be used as the carrier of nano Zinc Oxide, it could can overcome the defects of uneven dispersion and have better photocatalytic effect[4-5].

2. Experiment process

2.1 Main experimental materials and equipments

Materials included: diatomite (Changbai County in Jilin province), six zinc nitrate (AR) and citric acid (AR); polyethylene glycol 6000; ethanol; equipments included: HY-1600 used high temperature energy-saving furnace, DF-101S heat constant temperature type magnetic stirrer, WD-9415B type ultrasonic cleaner, TM3030 SEM、X-Ray Diffractometer System 554800、VC-Sorb 2800TP specific surface and aperture analyzer、IRAffinity-1，the formaldehyde degradation of the environment the cabin was made of organic glass laminated, fully enclosed, volume was about $1m^3$, the carrier coated composite powder for frosted glass, effective area was $0.25m^2$.

Content from this work may be used under the terms of the Creative Commons Attribution 3.0 licence. Any further distribution of this work must maintain attribution to the author(s) and the title of the work, journal citation and DOI.
Published under licence by IOP Publishing Ltd

2.2 Preparation of diatomite supported nano Zinc Oxide

Zinc nitrate and citric acid in the mixture of 20:1 and 100mL in deionized water, stirred evenly, ethanol of 10ml was put slowly, diatomite was added under the 2:1 ratio of zinc nitrate with diatomite then ultrasonic treatment for 30min; the suspension was obtained in flask, added trace amount of polyvinyl alcohol, magnetic mixed 6h, static 6h, filtration, dried, grinded, under 600°C roasted 2h, preparation of diatomite supported nano composite photocatalytic materials Zinc Oxide powder called A; Zinc nitrate and citric acid by 20:1 mixed with 100mL deionized water, evenly stirring, ethanol of 10ml was put slowly, then ultrasonic treatment for 30min, the suspension into the flask, magnetic stirring after 5h by zinc nitrate and diatomite was added under the 2:1 ratio of zinc nitrate with diatomite and trace polyvinyl alcohol, continue stirred 1h static 6h, filtration, dried, grinded, under 600°C roasted 2h, preparation of diatomite supported nano composite photocatalytic materials Zinc Oxide powder called B.

2.3 Formaldehyde degradation

Formaldehyde degradation experiment: firstly, the formaldehyde adsorption experiment in the reaction cabin (such as Figure 1) was carried out to overcome the error of the following experiment. Then, the diatomite based nano Zinc Oxide sample coated on the glass. Formaldehyde initial concentration of 0.7mg/m^3 was injected into reaction chamber, residual formaldehyde gas concentration was measured every 5h, adsorption degradation time-formaldehyde concentration curve was made .

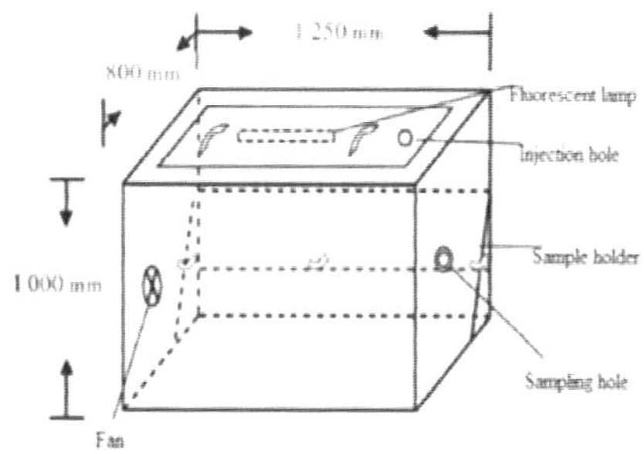

Figure 1. Degradation of formaldehyde gas reaction chamber.

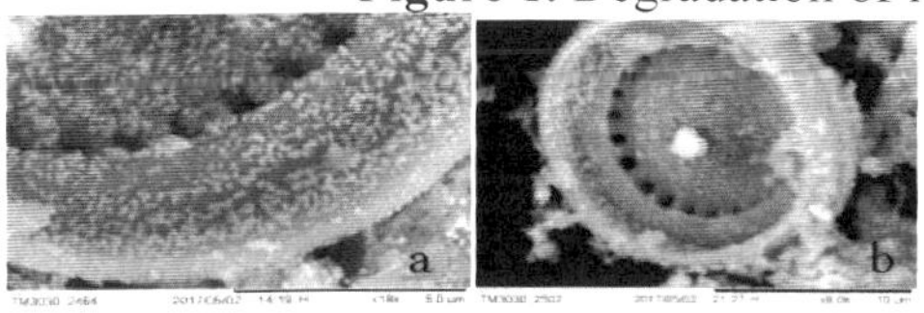

Figure 2. SEM of composite photocatalytic material powder A (a、b).

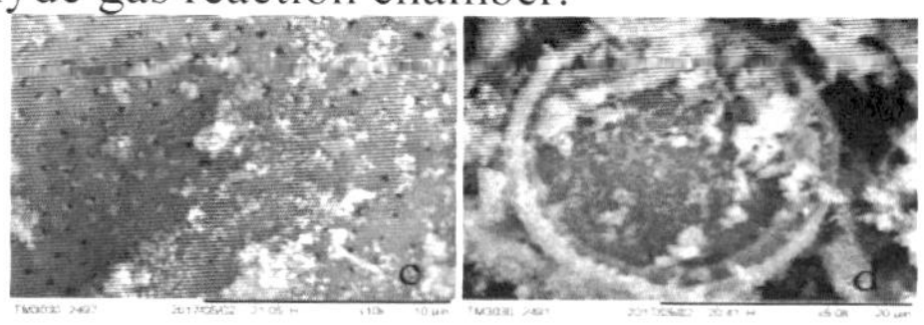

Figure 3. SEM of composite photocatalytic material powder B (c、d).

3. Experimental results and analysis

3.1 Morphology analysis of the sample

Figure 2 and 3 were SEM of composite photocatalytic material powder A and B. the distribution of nano Zinc Oxide on diatomite was observed, part of the grain size distribution was dense, the existence of large granular aggregate amount of nano particles, founded that Zinc Oxide of composite photocatalytic material powder A was dispersed, the Zinc Oxide dispersion of composite photocatalytic material powder A was better than the nano particles in Figure 3. As can be seen from the SEM, there was no obvious change in the morphology of the nano Zinc Oxide of composite photocatalytic material powder A compared with B, just the dispersion was better and the particle size was uniform. The preparation method of diatomite supported nano Zinc Oxide should be prepared by the method of preparation and loading.

IOP Conf. Series: Materials Science and Engineering **242** (2017) 012016 doi:10.1088/1757-899X/242/1/012016

3.2 X ray diffraction analysis

Figure 4 and 5 were the XRD spectra of composite photocatalytic material powder A and B, the composite photocatalytic material powder A had a certain change compared with the composite photocatalytic material powder B. In Figure 3 and 4, the diffraction peak at 2 theta to 28° was SiO_2, 2 theta was 31°, 38° and 56°, the diffraction peak was ZnO, and the peaks in the XRD diagram were very strong and sharp, and there was no other impurity peak, which showed that the purity of the sample was very high. X ray diffraction peak intensity of samples was strong, due to the different preparation methods of diatomite supported nano Zinc Oxide to internal lattice changes, the intensity of diffraction peak that load after the preparation higher than preparating and loading.

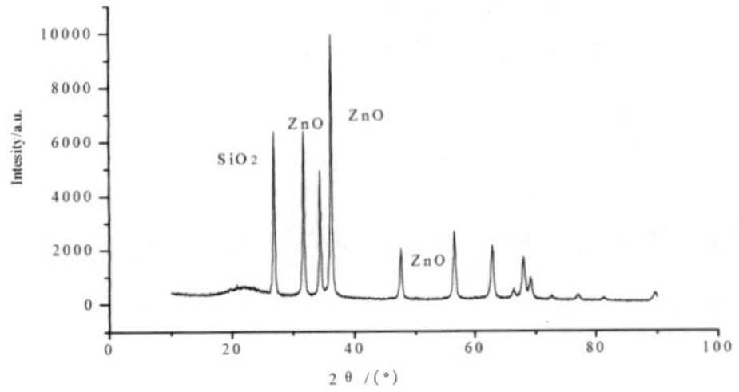

Figure 4. XRD of Composite Photocatalytic Material Powder A.

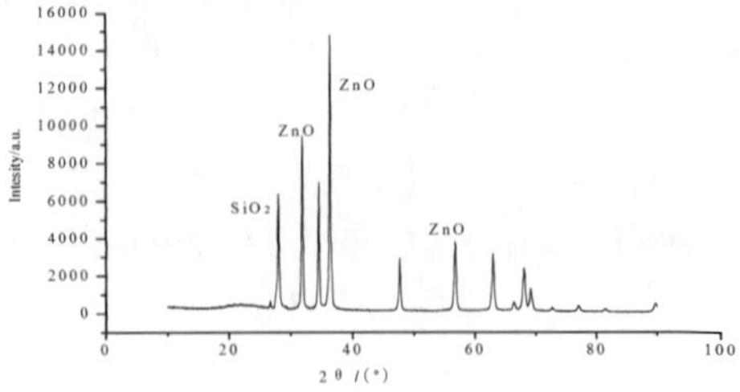

Figure 5. XRD of Composite Photocatalytic Material Powder B.

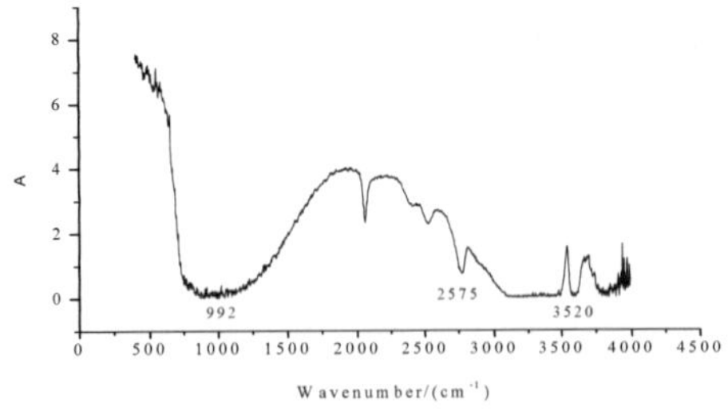

Figure 6. IR of Composite Photocatalytic Material Powder A.

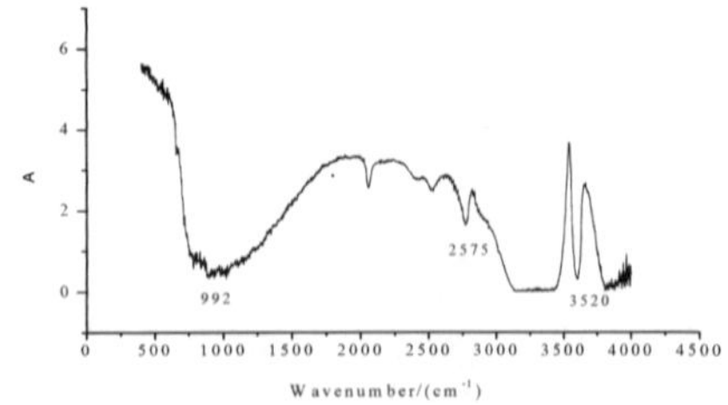

Figure 7. IR of Composite Photocatalytic Material Powder B.

3.3 Study on formaldehyde degradation of samples

Table 1 was the specific surface area and average particle size of the diatomite supported nano Zinc Oxide composite photocatalyst composite powder A and B, from that: because the nano particles increased, the surface area was greatly increased, average particle size decreased, diatomite as a carrier of nano Zinc Oxide particles, in photocatalytic process, and the contact area with formaldehyde molecule will be bigger, so as to improve the formaldehyde degradation.

Table 1. Specific surface area and average particle size of sample.

Sample	Specific surface area□m^2/g□	Average particle size□nm□
Diatomite	14.56	9.2
Composite powder A	45.56	5.1
Composite powder B	40.23	5.3

3.4. Infrared analysis of samples

The infrared spectra of the sample were shown in Figure 6 and 7, From that: there were O-Si-O symmetric stretching vibration bands and O-Si-O asymmetric stretching vibration bands in the range of 900~1200cm-1. There was an obvious absorption peak at 992cm-1. It was the asymmetric stretching vibration of Si-O-Zn. It showed that the bonding between Zinc Oxide and diatomite occurs during the preparation process, formed Si-O-Zn bond. In the range of 3500~3600cm-1, it was the position of hydroxyl vibration, indicated the expansion of the structure water and the adsorbed water, and the vibration peak of the composite photocatalytic material powder B was obviously increased, indicated that the hydroxyl content was very high in the sample. With XRD analysis and SEM images, it can be concluded that most of the Zinc Oxide particle load to the surface and pores of diatomite.

3.5. Study on formaldehyde degradation of samples

From figure 8, the initial concentration was 0.7mg/m^3 formaldehyde gas in the reaction chamber after 3h can be reduced to 0.6mg/m^3. From the curve: the reaction tank on adsorption of formaldehyde gas saturation concentration was about 0.1mg/m^3, so during the formal experiment the concentration of formaldehyde gas were increased in 0.1mg/m^3 based.

From Figure 9, during 9h, formaldehyde concentration decreased rapidly; then, the formaldehyde concentration decreased slowly, because the initial concentration of formaldehyde was higher, the formaldehyde was rapidly adsorbed by sub micropores of diatomite; With decreasing concentration of formaldehyde in the reaction chamber, adsorption of diatomite to formaldehyde reached equilibrium. In fluorescent light irradiation, inspired Zinc Oxide to produce highly active nano hole electron pairs, the high activity of electron hole pair would take electronics of O_2 and H_2O around and its generated strong oxidation ·OH、·O_2^-, reached the purpose of formaldehyde degradation. Diatomite composite photocatalyst powder A and B after the adsorption and degradation of 36h, the formaldehyde degradation rate were 29%, 64.6%, 51.7%, indicated that the adsorption of formaldehyde on single diatomite belonged to physical adsorption, while the preparation and loading than the loaded after the preparation of formaldehyde degradation rate was 12.9%. Because Zinc Oxide nano composite photocatalyst B powder agglomerate phenomenon, generated ·OH、·$O_2^-$was less, so the formaldehyde degradation rate was low.

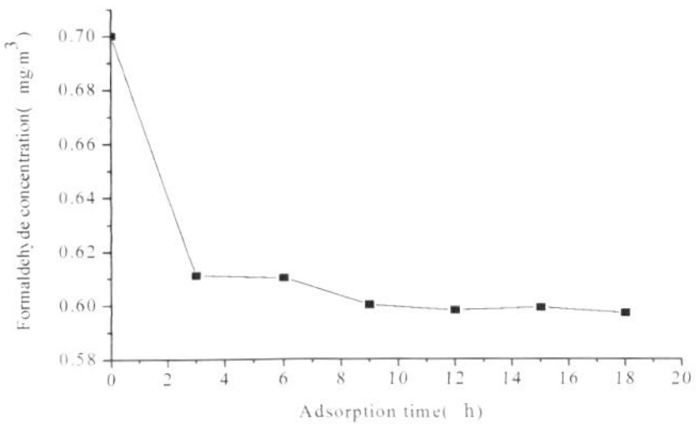

Figure 8. Formaldehyde adsorption of reactor.

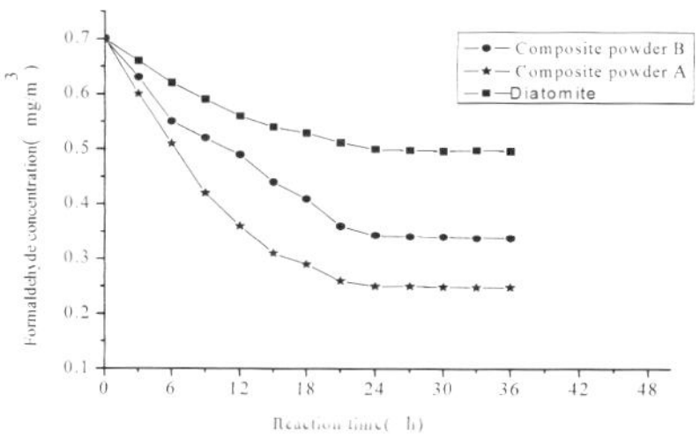

Figure 9. Adsorption and degradation of formaldehyde of sample.

4. Conclusion

Composite photocatalyst material powder was prepared by sol-gel method using diatomite as a carrier, Zinc Oxide as photocatalytic materials, the experimental method was by preparing and loading, composite powders were dispersed uniformly and the specific surface area was 45.56 m^2/g, the average particle size was 5.1nm. After 36h the degradation of formaldehyde rate reached to 64.6%, 37% higher than the single diatomite adsorption.

Acknowledgments

This work was financially supported by the National Key Technologies Research and Development Program of China (No. 2016YFC0701002).

References

[1] Huarong Huang, Research progress in photocatalytic properties of nanometer Zinc Oxide[J]. Guangzhou chemical industry, 2016,(17):50-60.
[2] Y Z Li, W Xie, X L Hu. Comparison of dye photodegradation and its coupling with light-to-electricity conversion over TiO_2 and ZnO[J]. Langmuir, 2010,26: 591-597.
[3] Chenglin Yu, Hongen Quan, Yong Kang, Experimental study on degradation of formaldehyde by

diatomite nano TiO_2[J]. Journal of Environmental Science, 2012,(01):116-122.
[4]bAo C H, Lee S C, Yu J Z. Photodegradation of formaldehyde by photocatalyst TiO_2: effects on the presences of NO, SO_2 and VOCs[J]. Applied Catalysis B, 2004,54:41-50.
[5] Ding Z, Chen X D, Lin P. Photocatalytic degradation of formaldehyde in air by nanometer TiO_2[J]. Research of Environmental Science, 2006,19(4):74-79.

IOP Conf. Series: Materials Science and Engineering **242** (2017) 012017 doi:10.1088/1757-899X/242/1/012017

Study on the infrared radiation performance of Tourmaline composite and its effect on the diesel characteristic

Jian Bin LIAO[1,2,3,a] ; Hong Liang YU[2,3*]; Di SUN[2,3]; Fong Yuan MA[4]

1. Marine Engineering College, Dalian Maritime University, Dalian 116026, China
2. Marine Engineering College, Jimei University, Xiamen 361021, China
3. Provincial Key Laboratory of Naval Architecture & Ocean Engineering, Xiamen 361021, China
4. National Taiwan Ocean University, Keelung City, Taiwan

a:ljb558@126.com; *Corresponding author: yu1202@hotmail.com

Abstract. The black tourmaline, magnesium tourmaline, and spinel were ground into powder, and the infrared radiation material was prepared by adding the ceramic powder, clay and the other material into the tourmaline powder according to a certain proportion. The infrared radiation property was tested and analyzed, the diesel was infrared radiation activated by the composite material, and the physicochemical property of fuel oil was analyzed pre-test and post-test. The result shows that the infrared absorption spectrum of the black tourmaline of different particle size is stable. After the diesel oil was infrared radiation activated by tourmaline composite materials, the physicochemical property of diesel oil was changed, the activation energy decreased, the viscosity and surface tension of fuel oil were reduced.

1. Introduction

The research of improving the combustion efficiency of diesel engine fuel, reducing the fuel consumption and the discharge of pollutants is quite wide range at present. Patents abroad show that it is feasible to activate fuel molecules, improve fuel combustion efficiency and the emission performance by using far infrared radiation. The current domestic research of infrared radiation materials mainly focus on agriculture, water treatment, air purification, biological medicine health industry, textile industry, etc. Infrared is part of the electromagnetic spectrum, and the infrared of 2.5 ~ 50 μm wavelength can enhance the vibration of groups and atoms of the molecules and strengthen the molecular rotation, thus the internal energy of the radiated molecules were increased.

Tourmaline is characterized by the electromagnetic field. When the environment temperature varies, tourmaline crystal would produce surface charge of the same number and opposite polarity on both ends of the symmetry axis of the third times, and the voltage is produced on both ends of the crystals and the far infrared radiation is released as the temperature changes [1].Studies shows that the infrared radiant material can be used to activate engine fuel hence to improve the combustion efficiency of fuel and reduce the harmful gas emissions [2-5]. The research of the effect of tourmaline composite materials on boiler diesel combustion emissions has been carried out [6]. Research mainly focuses on the high radiation materials of 3~14μm band in Domestic, however, the infrared emissivity is low in 3~5μm wavelength [7].

In order to improve the activation efficiency of far infrared fuel, the research of far infrared radiation properties of the tourmaline, spinel and other composite material were studied. The nature tourmaline was composite modified by using the spinel material of high infrared emissivity in the

Content from this work may be used under the terms of the Creative Commons Attribution 3.0 licence. Any further distribution of this work must maintain attribution to the author(s) and the title of the work, journal citation and DOI.
Published under licence by IOP Publishing Ltd

3~5μm band, and the activation of the far-infrared negative ions of tourmaline composite materials for the fuel and the changes of physical and chemical properties were analyzed.

2. Mechanism of infrared radiation

Infrared radiation, also known as infrared light, is firstly discovered by the British astronomer F.W. Herschel in 1800. Infrared light is an electromagnetic wave located in the out end of the visible red light and will radiate infrared radiation energy over absolute zero (- 273 °C) [8, 9]. Infrared radiation can be divided into near infrared wavelength (0.76~1.5μm), infrared (1.5~5.6μm) and far infrared wave (8~14μm) and super infrared (15~1000μm).

The infrared radiation can be depicted by emissivity ε, namely

$$\varepsilon = M/Mb \qquad (1)$$

Where M is the radiant exitance of objects, Mb is the radiant exitance of blackbodies, and ε reflects the radiation ability of objects. Blackbody emissivity (or absorption)is 1, namely α=1,therefore, the α ratio of the other object whose temperature is more than absolute zero degree and the ideal blackbodies is less than 1.The emissivity changes over the material of permittivity, surface roughness, temperature, wavelength, observation direction, the radiant energy wavelength and observation angle. The infrared radiation of blackbody follows the below rule:

(1) Stefan-Boltzmann's law

$$M_b = \varepsilon\sigma T^4 \qquad (2)$$

Where, Mb is the total power of blackbody radiation, σ is Stefan-Boltzmann constantσ=5.699×10-8W·m-2K-4 , ε is the blackbody radiation coefficient, T is the absolute temperature of blackbody. The Stefan - Boltzmann's law depicts that the total radiation power of various wavelengths is proportional to the four times of blackbody temperature (Kelvin) T.

(2) Wayne displacement law

$$\lambda_m T = C \qquad (3)$$

Where, λm is the radiation wavelength of peak, T is the blackbody absolute temperature, C is constant, the experimental measured value is about 2898μm·K. Wayne's law reflects the product of the peak wavelength and the blackbody temperature T is constant. This means that the short wave is the direction of motion as the T increases , on the contrary, long wave is the direction of motion as T drops.

(3) Planck's law

Planck's radiation law, a basic thermal conduction rule between objects, describes that the radiation power per unit time and area is proportional to the four times of blackbody temperature, the mathematical expression is as follows

$$M_{b\lambda} = \left(2\pi c^2 h / \lambda^5\right)\left(e^{hv/kt} - 1\right) = \left(c_1 / \lambda^5\right)\left(e^{c_2/\lambda T} - 1\right) \qquad (4)$$

Where, $M_{b\lambda}$ is radiant exitance of spectrum; λ is the infrared radiation wavelength, μm; C is the light speed in vacuum; h is Planck's constant, h=6.6256×10-34J·s; .V is infrared radiation frequency, k is the Boltzmann constant; T is the thermodynamics temperature of blackbody; c1 is the first radiation constant, c2 is the second radiation constant.

3. Mechanism of infrared radiation

3.1. Tourmaline and composite materials preparation and characterization

Tourmaline chemical general formula for NaR3Al6 [Si6O18] [BO3] 3 (OH, F) 4, is containing boron with additional anionic ring silicate minerals, the main chemical composition is SiO2, FeO, Fe2O3, B2O3, Al2O3, Na2O, MgO style, Li2O and MnO2 [10];Belong to the trigonal system, the basic unit for (SiO4) 6 after the three ring [11].Metallic elements such as sodium, calcium and magnesium after filling in the gaps between the three sides ring [12].R on behalf of the metal cation in the chemical formula, and when R for Ca2 + magnesium 2 +, Fe2 +, or Li + Al3 +, respectively form calcium dravite, black tourmaline tourmaline and lithium mineral [13], tourmaline columnar crystals were nearly triangle, its two side have positive and negative polarities of the natural, piezoelectric and pyroelectric property (focal electrical) [14].

Tourmaline, chemical general formula NaR3Al6[Si6O18][BO3]3(OH,F)4, is cycle silicate minerals containing boron and additional anionic. The chemical composition of tourmaline is SiO2、FeO、Fe2O3、B2O3、Al2O3、Na2O、MgO、Li2O、MnO2 ,etc.[10], belongs to the trigonal system, and is the composite three ring with basic unit (SiO4)6 [11]. The metallic elements such as sodium, calcium and magnesium fills in the gaps of the composite three ring [12]. In the chemical formula R represents the metal cation, and when R is Ca2+Mg2+,Fe2+,Li+ Al3+, and the tourmaline forms calcium tourmaline, dravite, black tourmaline and lithium tourmaline, respectively [13]. Tourmaline crystal is nearly triangle columnar, and its crystallization ends have natural positive and negative polarities, piezoelectric and pyroelectric property [14].

According to certain quality percentage, the black tourmaline, dravite and spinel material were ground into below 200 mesh by ball mill after broken. The far infrared radiation compound material was prepared by adding a percentage of the ceramic powder, clay, CaO and other composite materials into the tourmaline powder, and then sintered into the spherical composite particles A with diameter of 5 mm and B with diameter of 2 mm under 950 ℃,and the particles are shown in Figure.1.

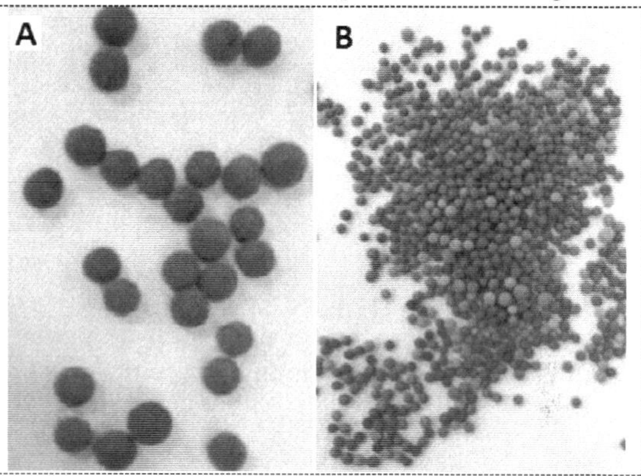

Figure 1. Particles A and B of composite

The microstructure of tourmaline and the composite material were observed with XL30W/TMP SEM of PHILIPS Company, and the microscopic structure of rough tourmaline and its powder are as shown in Figure 2 and Figure 3. Because of the tourmaline natural electric polarity, particles are adsorbed with each other, and small particles are chained connected to each other.

Figure 2. The SEM image of rough Tourmaline

Figure 3. The SEM image of Tourmaline powder

It can be seen that the rough tourmaline is characterized by compact structure and rough surface, however the powder is characterized by smooth surface and section lamellar structure. The composite material has obvious polymerization phenomenon, and the tourmaline particles distributed uniformly.

Figure 4.The SEM image of Tourmaline particle A

Figure 5.The SEM image of Tourmaline particle B

3.2. Infrared characteristic research of tourmaline and composite material

The black tourmaline from Xinjiang was taken as the sample and the infrared characteristics of different size tourmaline were researched. The sample 1 is rough tourmaline, the sample 2 is tourmaline powder, the sample 3 is tourmaline composite particles A with diameter of 5 mm, and the sample 4 is tourmaline composite particles B with diameter of 3 mm. These samples were infrared property tested, and its infrared spectrum diagram are as shown in Fig. 6.

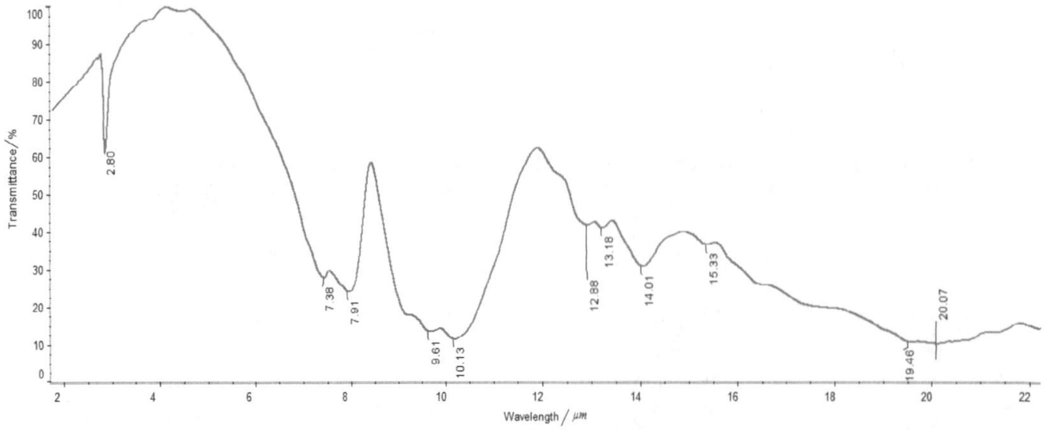

a.rough tourmaline spectrum

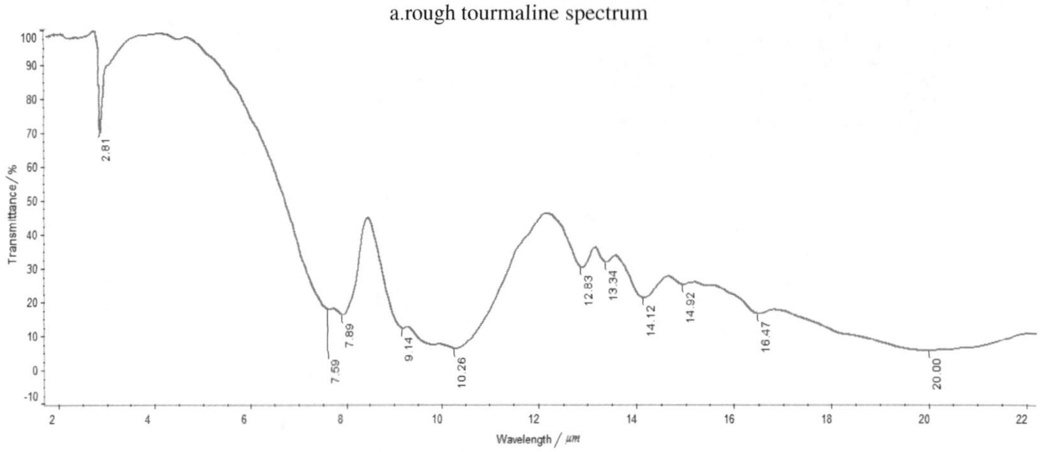

b. tourmaline powder spectrum

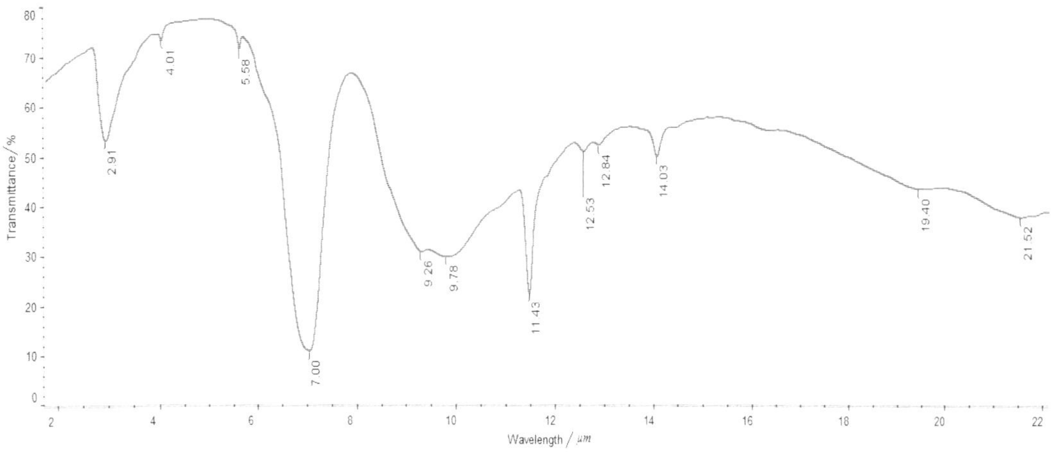

c. composite large particle A spectrum

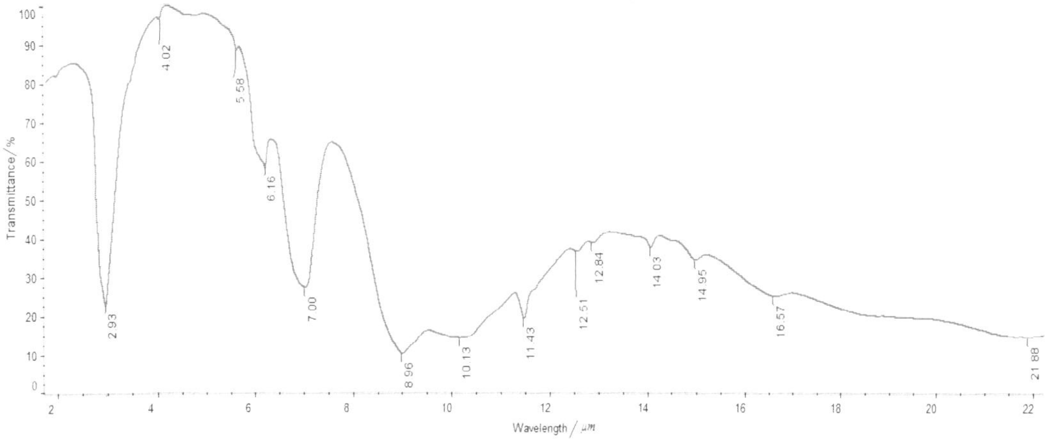

d. composite small particle B spectrum

Figure 6. Infrared absorption characteristic analysis of tourmaline curve

Kirchhoff's law shows that the monochromatic radiation rate is equal to the monochromatic absorption values under the same temperature and corresponding wavelength for different objects. Therefore, the infrared radiation capacity is strong in absorption peak. It can be seen from Fig.6 that the absorption is not obvious in 4~7μm wavelength(2500-1429cm^{-1})and the absorption peak is obvious in 2.8μm(3571cm^{-1}),7.5μm(1333cm^{-1}),9.0μm(1111cm^{-1}),19.4μm(515cm^{-1}). The infrared absorption spectrum of the black tourmaline of different size particle is stable, and the absorption spectrum column characteristics are the similar under the room temperature. The absorption peak is obvious near 9.0μm(1111cm^{-1}), according to Kirchhoff's law that the infrared absorption rate of A λ equals emissivity e λ, the normal emissivity near 9.0μm is strong. As is seen from Fig.6 that the infrared absorption effect for tourmaline composite materials with 5mm particle diameter in range of 2.83μm~21.89μm is better.

3.3. The effect of infrared radiation property on fuel oil physicochemical properties

The high-energy infrared in 2.5~22μm wavelength of infrared radiant material can be used to activate fuel oil, resonance excite fuel molecules bond energy, improve the fuel combustion efficiency, reduce harmful gas emissions, and eventually catalyze combustion. Take two same oil samples, the first sample is A and the second sample is B that was carried out radiation activation treatment by using the prepared infrared radiation composites above. The oil samples were observed with microscope respectively, and the results are as follows:

- The diagram of fuel unactivated with far-infrared radiation is as shown in Fig. 7

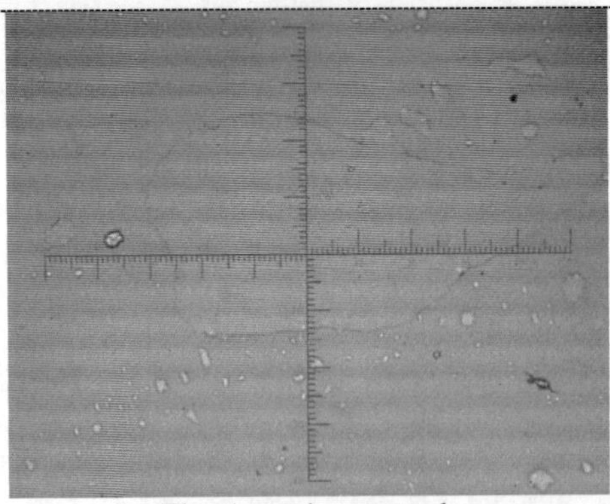

a. The view using cover glass

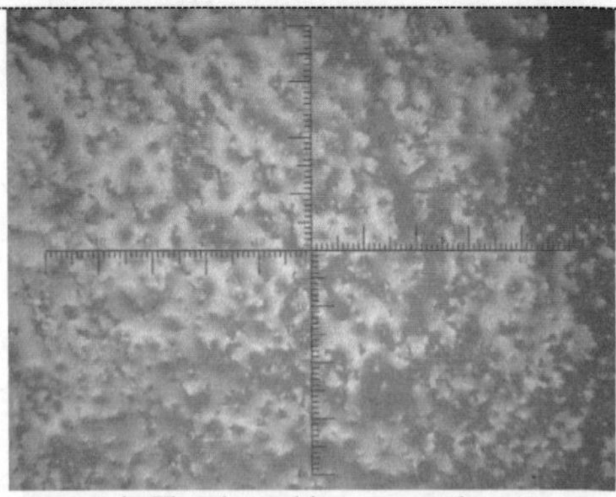

b. The view without cover glass

Figure 7. The diesel oil unactivated by far infrared radiation

- The diagram of fuel activated with far-infrared radiation is as shown in Fig. 8

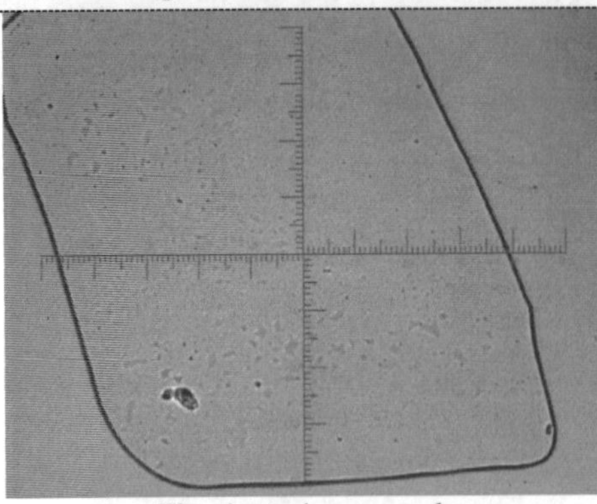

a. The view using cover glass

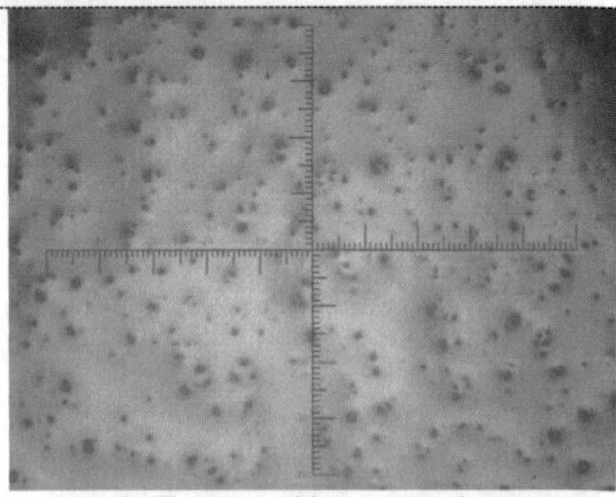

b. The view without cover glass

Figure 8. The diesel oil activated by far infrared radiation

It can be seen from Fig.7 and 8 that the particles in A sample appear scaly and the interface of oil and water is not obvious, however the particles in B sample appear broken, fuel gap structure is uniform, and the interface of oil and water is obvious. The structure characteristic of fuel oil in B sample contributes to the decrease of the fuel oil viscosity, which further increases the atomization effect and combustion efficiency. The physicochemical property of two oil samples were tested respectively, and the results are shown as table 1.

Table 1. The test results of two sample

sample	closed flash point (°C)	Gross Calorific Value (MJ/kg)	Net Calorific Value (MJ/kg)	interfacial tension @25°C (mN/m)	density @25°C （Kg/m3）	viscosity @20°C (mm2/s)
A	79.5	46.16	43.03	19.5	836.8	4.685
B	78.5	45.96	42.89	17.0	836.2	4.629

It is clear in table 1 that the interfacial tension changes obviously, drops from 19.5 to 17.0 mN/m after the diesel oil was activated by composite infrared radiation under room temperature, and the interfacial tension reduces 12.82%. Namely, the composite has significant influence on the interfacial tension of fuel oil. In fact, fuel oil is characterized by strong clustering and most of the fuel molecule are in the form of double dimer, trimer or polymer. Therefore, the oil droplets are bigger and atomization quality is poor, which leads to the incomplete combustion, the fuel consumption and flue gas emissions increase.

In addition, the interfacial tension of oil has close relationship with the diameter of the fuel jet [15].When the interfacial tension of fuel oil reduces, the diameter of the oil droplets also decreases. Moreover, the infrared ray can be C-H and C-C key resonance absorbed, which makes it easier disconnect, restructure into carbon dioxide and water, and release energy [16]. Therefore ,the reduction of the oil interfacial tension can achieve fuel sufficient burning, further energy conservation and emissions reduction.

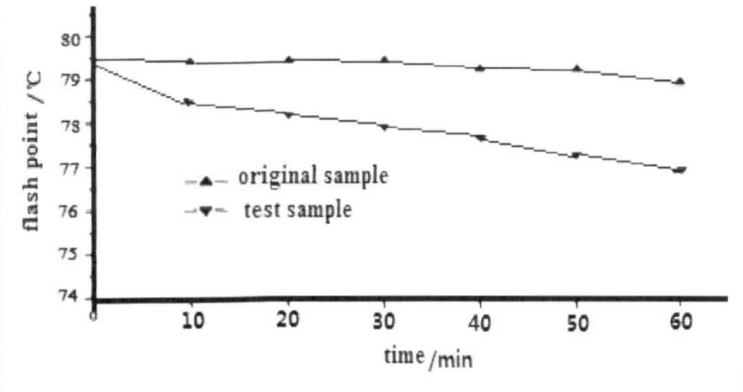

Figure 9. Diesel oil flash point change over infrared radiation time.

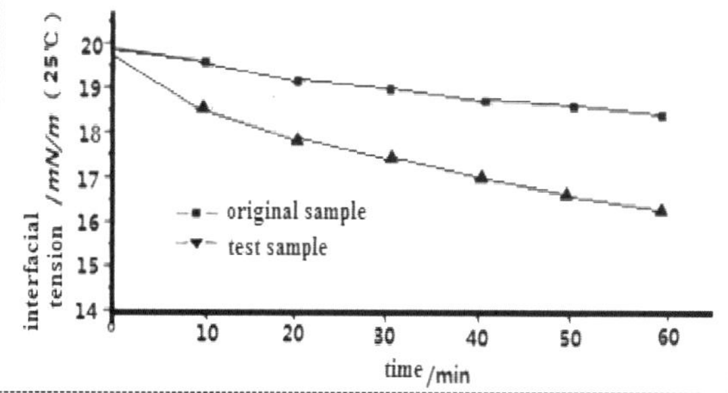

Figure 10. Diesel surface tension change over infrared radiation time.

Furthermore, the flash point drops by 1.26% , the gross calorific value and net calorific value reduce by 0.43% and 0.33% respectively after the diesel oil was activated by the infrared radiation of composite material. The results show that the safety of saving and transportation and the calorific value of sample are not significant damaged after infrared radiation activation.

4. Conclusion

(1) The infrared radiation properties of natural black tourmaline in different particle size were researched and analyzed ,and the result shows that the infrared absorption spectrum of the black tourmaline is stable and the infrared radiation characteristic is similar under room temperature.

(2) The infrared radiation composite was prepared by a certain percentage of black tourmaline, tourmaline powder, drivate, spinel, ceramic powder, clay, and other composite materials, and then the diesel oil was activated with the infrared radiation composite. The result shows that the interfacial tension and the viscosity of diesel significantly reduce of and the flash point, gross calorific value and net calorific value of sample have no substantial change. The infrared radiation activation would not damage the oil sample and can effectively improve the physicochemical property of fuel oil.

Acknowledgments

This research was financially supported by:

1.Science and technology project of Fujian Province（No. 2015Y0038;2015H0006;2016H0067）

2. Fujian Sci-tech Innovation Platform Construction Project (2014H2001)

3. Xiamen science and technology project (No. 3502Z20151231)

4. Xiamen Southern Oceanographic Center (14GZB66NF30)

References

[1] Yang, R.Z, Yang, M.Z, Liao, Z T & Zhou, Z.Y 2002 Natural Black Tourmalines' character of infrared radiation. Journal of Tongji University. Vol 30(2)pp183-188

[2] MIZUTANI T 2005 Device for saving fuel. *Japan Patent*, 2005106043

[3] HASHIMOTO K, NAKAGAWA H. 2005 Apparatus for enhancing combustion efficiency of liquid fuel. *Japan Patent*, 2005241626

[4] Xu G.K, Liang J.S, Wang L.J, Meng J.P & Liang G.C 2005 Effects on combustion of gasoline of rare earth composite materials for far infrared radiation. *Journal of Rare Earths*, 23(S1) pp247-250.

[5] Wang L.J, Liang J.S, Liang G.C, Xu G.K & Meng J.P 2004 Effect of rare earth on fuel activation of far infrared ceramic. *Journal of Rare Earths*,22(S3) pp255-257.

[6] Li F.P, Liang J.S, Meng J.P, Ding Y, Liang, G.C, & Xu G.K 2007 Effect of tourmaline/resin composite materials on the combustion of diesel oil for oil-burning boiler. *Journal of the*

Chinese Ceramic Society, Vol 35(4)pp517-521.
[7] Li H.T & Liu J.X 2005Application and Present Development of High Efficiency Far Infrared Radiation Ceramics.*Advanced Caremics*,2005(2) pp24-29.
[8] Acharjee S, Ray J, Dey P, Bhattacharyya D, Banerjee M, & Chattopadhyay, B 2016 Mineral chemistry of tourmaline from mashak pahar, south purulia shear zone (spsz), eastern indian shield.*Journal of Earth System Science*, 125(8)pp1681-1696.
[9] Berryman B, Plessen E, Rhede B, Koch-Müller D, Heinrich M, & Wilhelm 2015 Synthetic and natural ammonium-bearing tourmaline. *American Mineralogist*, 100(1)pp250-256.
[10] Meng J.P, Liang J.S, Liang G.C, Yu J.M & Pan Y.F 2006 Effects of tourmaline on microstructures and photo catalytic activity of TiO2 /SiO2 composite powders. *Trans Nonferrous Met Soc China*, 2006(16)pp547-550.
[11] Setkova T, Shapovalov Y, & Balitsky V 2011 Growth of tourmaline single crystals containing transition metal elements in hydrothermal solutions. *Journal of Crystal Growth*, 318(1)pp904-907.
[12] Desbois G & Ingrin J 2007 Anisotropy of hydrogen diffusion in tourmaline. *Geochimica Et Cosmochimica Acta*, 71(21)pp5233-5243.
[13] Gujar A.R, Ambre N.V, Mislankar P.G, & Iyer S.D 2010 Ilmenite, magnetite and chromite beach placers from south maharashtra, central west coast of india. *Resource Geology*, Vol 60(1)pp71-86.
[14] Mo Z.L, Sun Y, Guo R.B & Feng C 2011 Research Progress in the Performance and Application of Tourmaline. *Bulletin of The Chinese Cermic Society*, Vol 30(4)pp822-826
[15] Wang D.H,Li D.M, Xu G.P, Wu G.X & Yang J.Y 1999 The Effect of Surface Tension and Viscosity on the Atomization of Fuel. *Journal Shanghai Normal University*, Vol28(4)pp47–50.
[16] Wei Q.T, Sun J, Wang M, Sun Y, & Liu Y 2002 Infrared ray used for automobile energy saving and pollution reduction. *Municipal Administration & Technology*. Vol 4(3)pp25–28

A study of tensile residual strength of composite laminates under different patch-repaired series

M.H. Ding [1,2,3*],S.zhan[3], Y.H. Tang [3*] , L. Wang [3], D.Q. Ma [1] ,R.G.Wang[2]

1 Hafei Aviation Industry Company Limited, Harbin 150027, PR China.

2 Harbin Institute of Technology, Harbin 150001, PR China.

3 College of Material Science and Chemical Engineering, Harbin Engineering University, Harbin 150001, PR China.

E-mail: ding197392@163.com; zhanshuai2012@163.com; wl_880330@126.com; 670217424@qq.com; 863829942@qq.com

Abstract. The tensile behavior of composite laminate structures repaired by bonding external patches was studied in the paper. Two different types of patches including wedge patches and inverted wedge patches were used and failure mechanisms, failure load and strength predictions were studied. A convenient and fast method of building 2-D finite element modeling (FEM) of laminate structure repaired was proposed and the strength of repaired laminate structures was calculated by FEM. The results showed that more than 80% tensile strength of the undamaged laminate could be recovered by bonding patch repairs. Moreover, the results indicated that the strength of inverted wedge patches repair were higher than that of wedge patches repair. FEM simulation results indicated that high stress concentration was found along the edges of invert patches and the most weakness part located in the adhesive bondline. FEM analysis results showed that the strength predicted matched well with the test strength.

1. Introduction

The composite has been widely used in aerospace so far. However, there are some repair questions to be solved due to damage caused by accidental impact, hailstones, bird strike and lightning strike. [1-4] Thus, the better repair techniques and processes for composite components are needed to be studied. Many different composite repair techniques have been used to repair different types of damages in composite components with different geometric construction. And one of those was bonding repair. For bonding repair, a load carrying repair must be adopted for intact aerodynamic contour of composite components in a permanent repair of structure damaged.[5-7] Relevant experiments about composite material repair were in progress. Campiho *et al*[8] developed interface finite elements model based on the indirect use of fracture mechanics and studied the effects of the single and double-lap repairs on the bearing load strength of composite materials by using the ABAQUS software. The reported results indicated that the strength of the adhesives affects immensely on the failure mode, whereas its fracture toughness exhibit a little influence. Oztelcan *et al*[9] focused on developing appropriate test specimens to evaluate computationally and experimentally the effects of static and cyclic loads on the strength of a composite blade with a repair site and a local finite element model with the appropriate boundary conditions was proposed. Kumar *et al* [10] analysed the failure modes

Content from this work may be used under the terms of the Creative Commons Attribution 3.0 licence. Any further distribution of this work must maintain attribution to the author(s) and the title of the work, journal citation and DOI.
Published under licence by IOP Publishing Ltd

IOP Conf. Series: Materials Science and Engineering **242** (2017) 012018 doi:10.1088/1757-899X/242/1/012018

of bonding repaired laminate structure under tensile loading. Test results revealed that failure occurred into two modes including fracture and pull-out of fibre and cohesive shear failure of the adhesive film. Breitzman *et al* [11] examined the strength of the composite repaired under tensile loading with and without overlay plies. The results showed that the patch stacking layers affected greatly the strengh recover of the laminate and the strength could exceed 80% of the undamaged laminate by using overlay piles. The performance of a repaired structure using adhesive bonding depends not only on the adhensive properties, bonding mode and fracture mode, *et al*, but also on the patch shape design and adequate shape of bonding patch was in need of designing in order to ensure maximum repair efficiency. [12-13] To our knowledge, there were few systemic reports on effects of patch shape design on strength recover in bonding repair. In addition, the FEM models proposed by previous researcher were complex, which was not suitable for fast calculation with high precision.

In the study, bonding repair with two different patches including inverted wedge and wedge patches was used to repair composite laminate damage and effects of patch types on the tensile residual strength were discussed. In addition, a convenient and fast 2-D FEM was also proposed to predict the strength of composite components repaired. These researches favour to increase the repair efficiency and minmize the repair cost for composite components.

2. Experiments

In the external patch technique, considering bending strains due to eccentric load path, the patch must be capable of withstanding the high peel and shear stresses which develop at the edges of the overlap of patches. [14-15] In the experiments, two types of patches made up of prepreg were shown in **Fig. 1.**

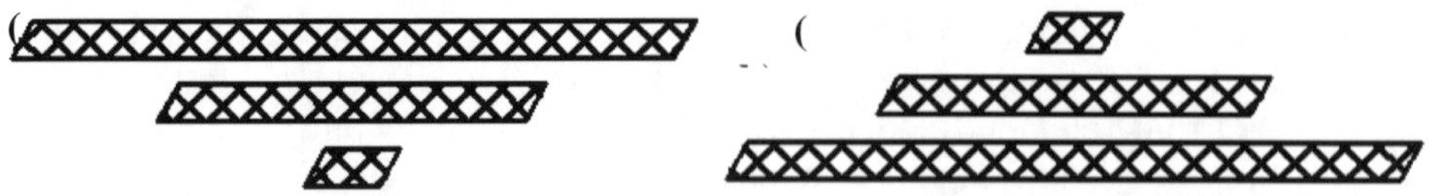

Fig. 1 The diagram of two different patches with three layers: (a) SeriesI:Wedge prepreg patch; (b) SeriesII:Inverted wedge prepreg patch

A useful data base on specimens repaired by wedge prepreg patch under tensile loading has been prepared in previous work. [16-19] It showed that the required mechanical performance could be achieved by the repair method of using wedge prepreg patch. This study extended to repair surface damage by inverted wedge prepreg patch and analyze the repairing effect. The stacking sequences of laminates used in this work were followed: [45/0/0/45], [45/0/45/0/45] and [45/0/45/0/45/0/45]. The schematic of repaired specimens was shown in **Fig. 2**.

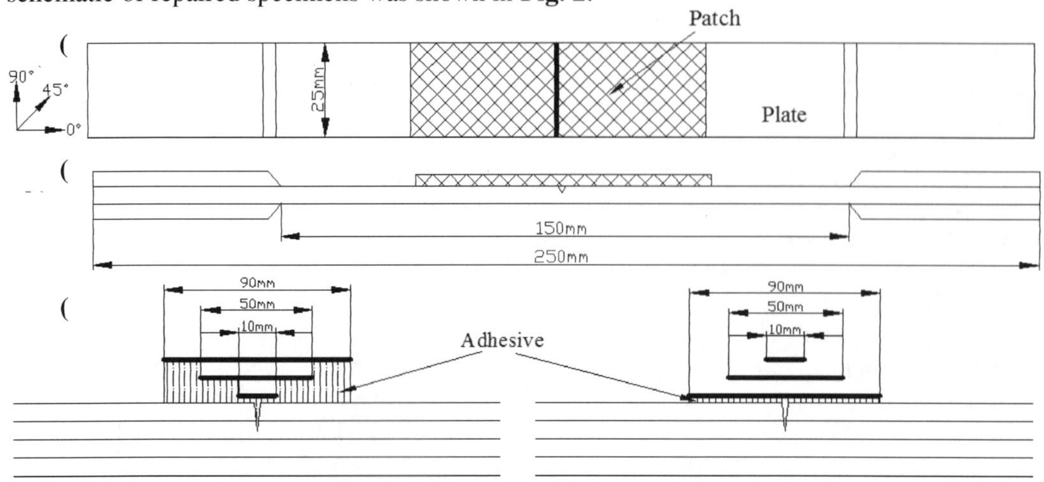

Fig. 2 Test specimens: (a) Front view; (b) Side view; (c) Two different of repaired details

IOP Conf. Series: Materials Science and Engineering **242** (2017) 012018 doi:10.1088/1757-899X/242/1/012018

Fig. 2(a) and (b) showed two layers scratch were made in the middle of the specimens (250×25×0.285n, n means layers) and then the rectangle patches were used to bond one-sided of the specimen. Beveled Al plates (25×50×2mm) were bonded to each end of the test specimens. All patches for repairing specimens were designed according to following stacking sequence: [0/0/45] and outermost 45° layer increased the impact resistance performance and minimized cracking. After repairing, C-Scans were performed on all of the repaired specimens to evaluate the quality of component repaired. The tensile tests were operated at room temperature on a MTS universal test machine and five specimens were tested for each configuration specimens. The beginning load and the maximum load were set at 0kN and 100kN, respectively. Load speed was 2% of maximum load.

In this work, a convenient and fast 2-D modeling was built in order to analyze the strength of repaired plates by using MSC.Patran with MSC.Nastran solver. The mesh pattern around the patch was kept very fine in order to capture the high stress and the failure concentration. Boundary conditions were simulated by fixing one end and applying a continuing stage loading in the longitudinal direction at the other. The adhesive bondline was simulated as being isotropic. The patches and the plate were meshed using the Quad-4 node elements. With [45/0/0/45] stacking laminate as example, its symmetrical structure was divided vertically. As shown in Fig. 3(a), A, B, C and D four parts were designed as different stacking layers and the finite element grid connected in turn. The entire model built was shown in Fig. 3(b).

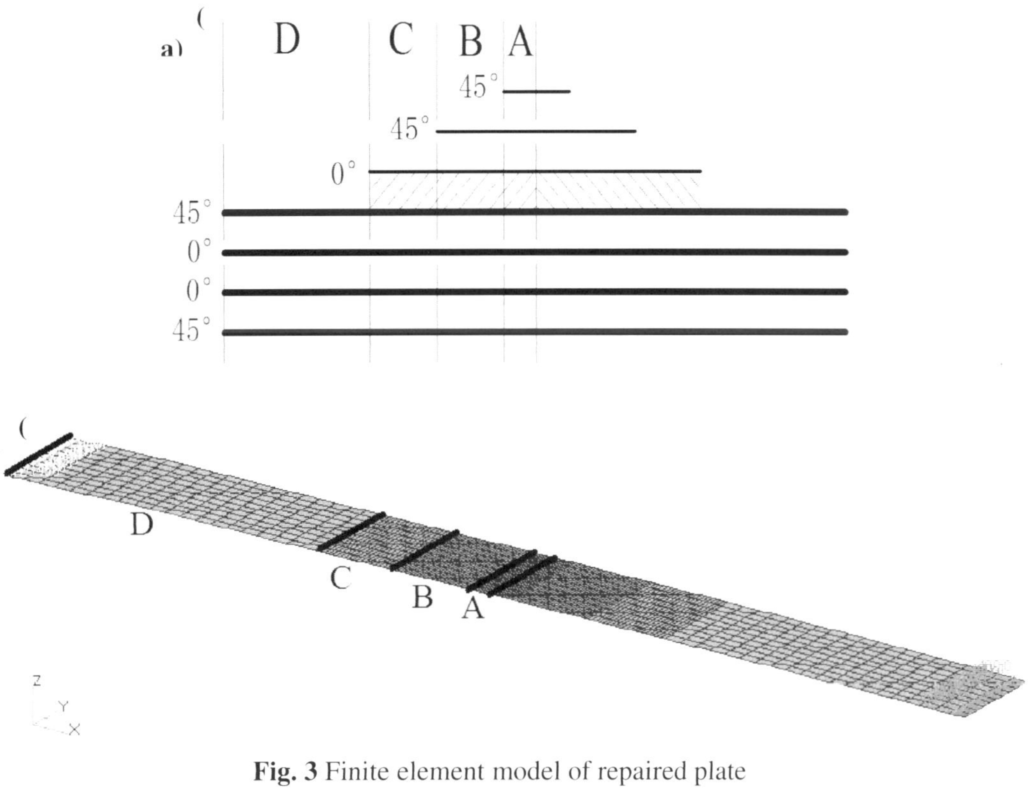

Fig. 3 Finite element model of repaired plate

3. Results and Discussions

3.1 Fracture surface observation

In the experiments, the sound of fiber broken off could be heard. The fracture surfaces were examined by low magnification photography in order to analyze the failure mechanisms of patch repairs. For the undamaged specimens, a large amount of fibre breakage and delamination could be seen near the fracture **Fig. 4(a)** . In addition, it could be observed that fracture located in the weak section of the plates. The specimens repaired with two different patches were shown in **Fig. 4(b)** and **Fig. 4(c)**. The fibre breakage, delamination and patches separations can be observed in the location of damage plate. Failure process was as follows. [10] Firstly, the adhesive was particularly damaged, and then whole patch separated from parent plate under bearing the tensile load which exceeded the maximum shear load of the adhesive. Meanwhile the stress concentrating formed in the location of scratch and scratch extended into deeper section of the laminate, and then final failure occurred.

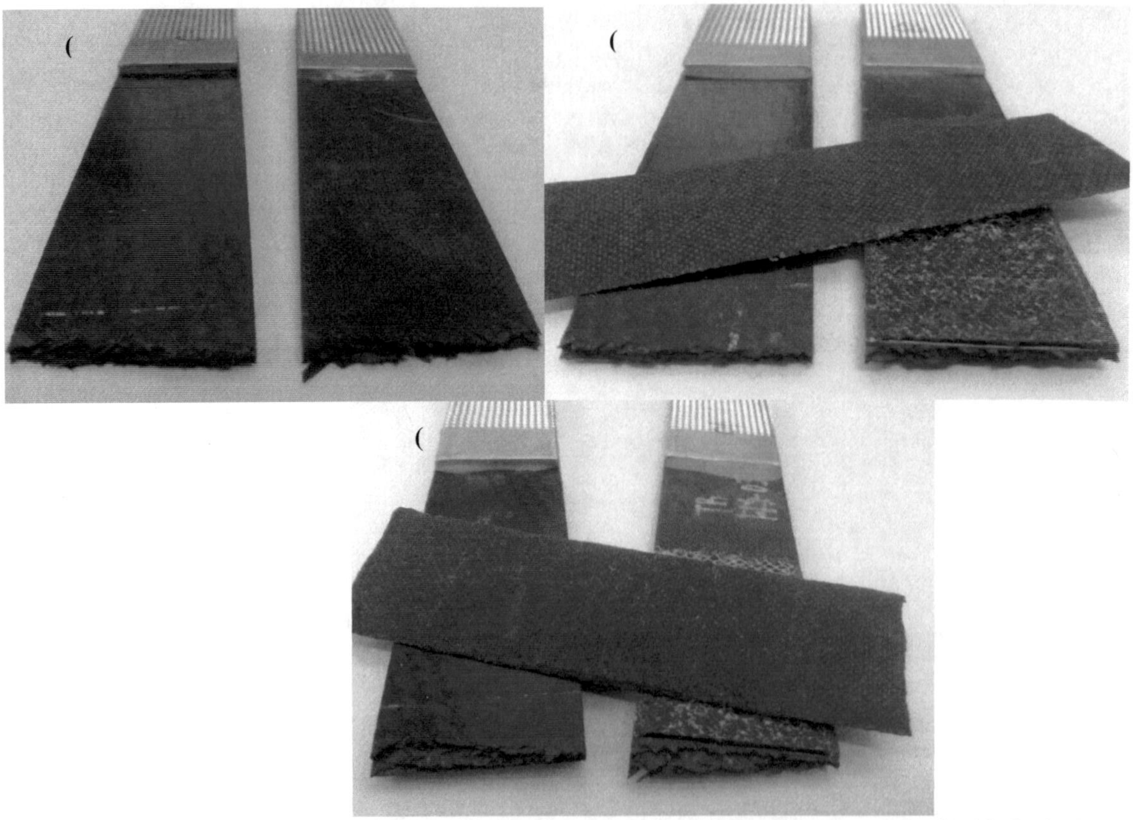

Fig. 4 Photographs of failed specimens: (a) Good specimens; (b) Specimens repaired with SeriesI; (c) Specimens repaired with SeriesII

3.2Analysis of load and displacement curves

Load and displacement curves of specimens with three types of stacking sequences were shown in **Fig. 6** and tensile loads of failure were listed in Table. 2. It could be observed from **Fig. 6** that the strength of unrepaired specimens changed in a linear fashion at different loading and their failure loading were about 13kN, 17kN and 24kN, respectively. For repaired specimens, the similar linear behavior could also be seen before failure. For laminate structure repaired by wedge patch, the upper large patch stacked on the lower small patch, which resulted in the forming of a wedge angle (**Fig. 6**) at the end of pre-patch. When repaired laminate structure bore load, the patch fractured layer by layer, not fractured as a whole, which was fit for bearing large load. However, there was not a wedge angle in invert wedge patch and the loaded repaired laminate structure with invert wedge patch fractured integrally. Thus, the laminate structure repaired by Series II could bear a little higher tensile loading than that of series I. The data listed in **Table. 1** showed the tensile strength of composite laminate repaired by series I and series II methods could recover 80% strength of the undamaged laminate, which indicated that Series I and Series II were used practically in bonding repair.

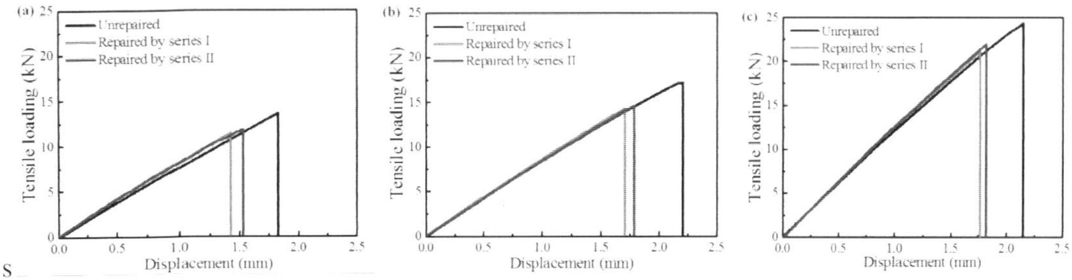

A s

Fig. 5 Load and displacement of specimens: (a) [45/0/0/45]; (b) [45/0/45/0/45]; (c)

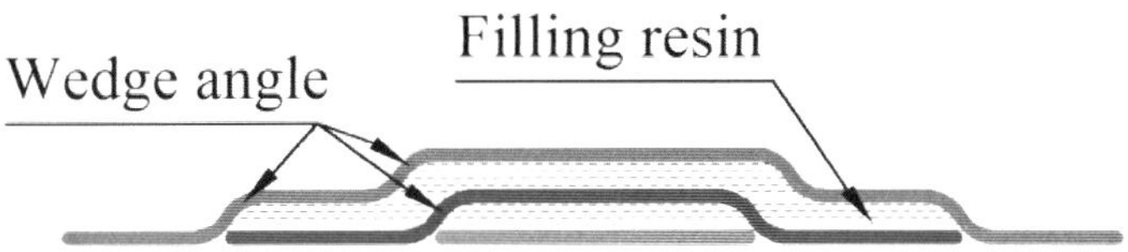

[45/0/45/0/45/0/45]

Fig. 6 Wedge angle in wedge patch

Table 1 The tensile failure load (KN)

Specimens	[45/0/0/45]	[45/0/45/0/45]	[45/0/45/0/45/0/45]
Undamaged	13.61	17.17	24.37
Repaired by SeriesI	11.48	14.30	21.12
Repaired by SeriesII	11.77	14.63	21.96

3. 3 Strength prediction of patch repairs

As for the composite components, Maximum stress criterion was applied to predict Maximum load of damage initiation as defined in the following equation:

$$\sigma_1 = X_t,\ \sigma_2 = Y_t \text{ and } |\tau_{12}| = S \qquad (1)$$

If $\sigma_1 > X_t$, $\sigma_2 > Y_t$ or $|\tau_{12}| > S$, damage will occur.

For repaired laminate, this criterion could get maximum load of damage initiation, but the real damage modes of repaired structure was not described clearly by it.

The finite element analysis was carried out by applying a 2-D modeling. **Table. 2** gave the prediction of strength value in each modeling. It could be seen that the strength values of repaired modeling were lower than those of the undamaged parent material. According to the phenomenon of the test results, it could be concluded that the adhesive was the weakest part in the repaired structure. For laminate structure repaired, the adhesive strength should be analyzed used the FEM. **Fig. 7** and **Fig. 8** gave the FEM results of the maximum stress of the adhesive bondline.

Table 2 The prediction of strength value (KN)

Modeling	[45/0/0/45]	[45/0/45/0/45]	[45/0/45/0/45/0/45]
Undamaged	12.36	15.63	23.60
Repaired by SeriesI	11.14	14.06	20.83
Repaired by SeriesII	11.46	14.37	21.61

IOP Conf. Series: Materials Science and Engineering **242** (2017) 012018 doi:10.1088/1757-899X/242/1/012018

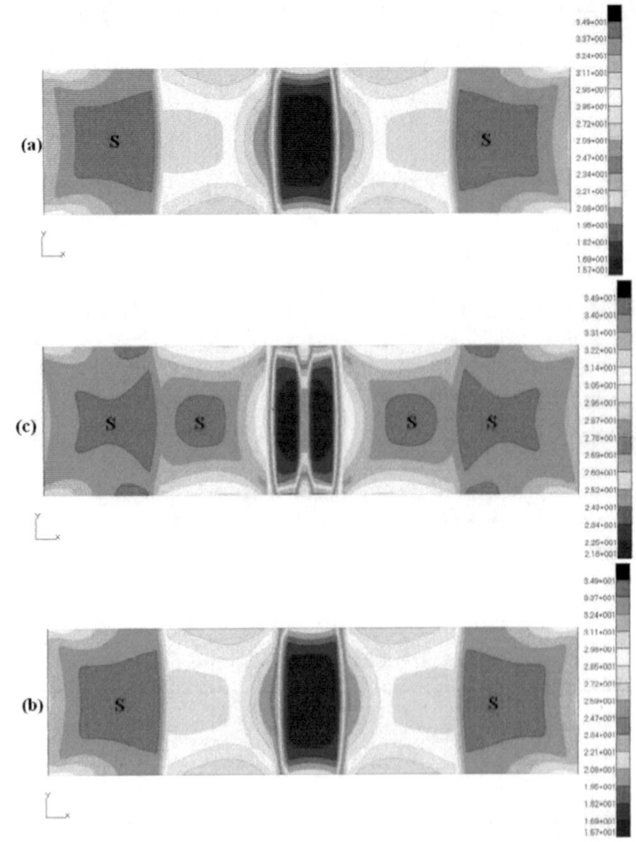

Fig. 7 The maximum stress nephogram of the laminate repaired with wedge patch: (a) 4 layers; (b) 5 layers; (c) 7 layers

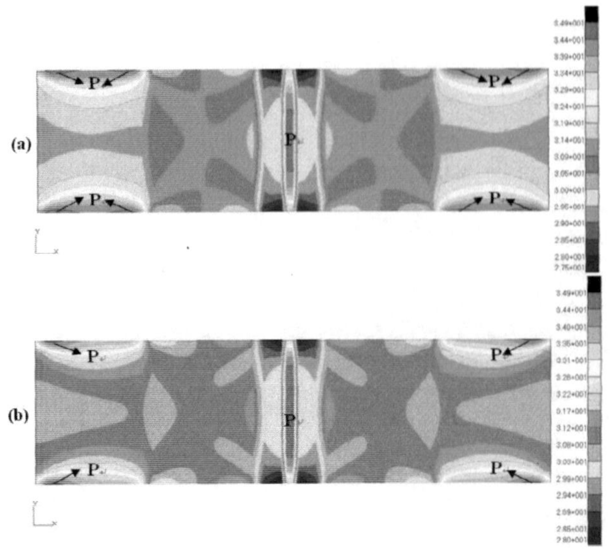

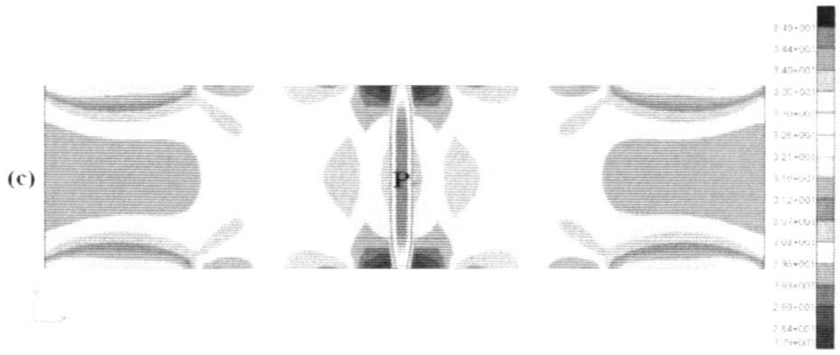

Fig. 8 The maximum stress nephogram of the laminate repaired with inverted wedge patch: (a) 4 layers; (b) 5 layers; (c) 7 layers

It could be seen from **Fig. 7** that both ends of adhesive bondline (S area in Fig. 8) were the weakest location in the laminate repaired with wedge prepreg patch. The bearing stress extended to the middle part with the increasing in layer thickness and the stress concentration was formed near the damage position. The repaired structure failed when the adhesive bondline reached its maximum bearing stress. It could be seen from **Fig. 8** that the weakest part of the repaired structure with inverted wedge prepreg patch located in the edge of adhesive bondline (P area in **Fig. 8**). The edge stress concentration could be observed in four layers laminate and it decreased with the increasing in layer thickness. The repaired structure failed when the adhesive bondline reached its maximum stress. In a similar study, Liu [20] studied the effects of patch stacking sequences on the fracture process and showed that cohesive damage initiated in the bondline close to the longitudinal patch edges where the stress focused. In addition, patch stacking sequences had a great impact on the strength of laminate. [21]

The FEM results and the test results are shown in **Fig. 9**. It could be seen that the strength were on the rise with the increasing layer stacking thickness, and the FEM results matched the test results well.

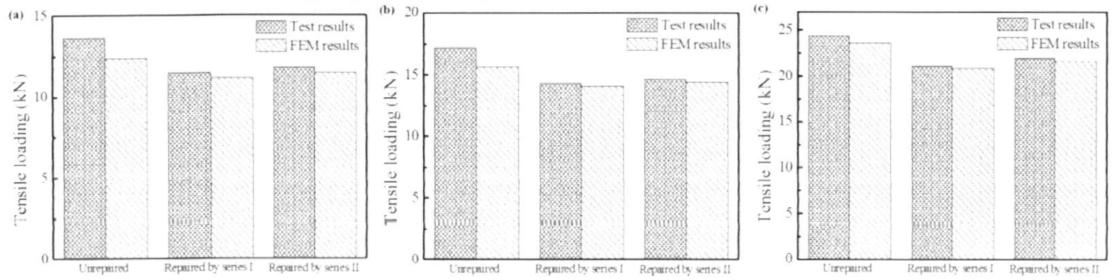

Fig. 9 *Compare of the test results with the FEM results: (a) [45/0/0/45]; (b) [45/0/45/0/45]; (c) [45/0/45/0/45/0/45]*

4. Conclusion

In the paper, the scratch laminates were repaired by patch-bonding with wedge patch and inverted wedge patch. The effect of composite patch type and layer thickness on the strength of laminate repaired was studied and a 2-D model was purposed to predict the repaired strength results. The conclusions were summarized as follow.

(1) The tensile strength of laminate plate after repair could recover more than 80% with wedge patch and inverted wedge patch. In addition, the tensile strengths were 11.46KN, 14.37KN and 21.61KN by using inverted wedge patch and they were a little bit higher than that of wedge patch repair.

(2) By analyzing the stress nephogram, patch type affected stress distribution and stress concentration place. The stress focused on the edge of patch by inverted wedge patch repair and it concentrated in the center area along the vertical direction of patch by wedge patch repair.

(3) The repaired FEM was built and results showed that the simulation was in good agreement with experimental data, which could be used in quick calculate of repaired structure in an emergency.A subsubsection. The paragraph text follows on from the subsubsection heading but should not be in italic.

Acknowledgements
This work was financially supported by the Application Technology Research and Development Projects of Heilongjiang Province (GC13A103), the Postdoctoral Sustentation Fund for Heilongjiang Province (LBH-Z12272) and Science and Technology Research Projects of Harbin City (2014AB4BG061).

References

[1]. R. Jones, W. K. Chiu, and R. Smith: 'Airworthiness of composite repairs: failure mechanisms', *Eng. Fail. Anal.*, 1995, 2, (2), 117-128.
[2]. J. Denney and S. Mall: 'Characterization of disband effects on fatigue crack growth behavior in aluminium plate with bonded composite patch', *Eng. Fract. Mech.*, 1997, 57, (5), 507-525.
[3]. H. J. Park: 'Bearing failure analysis of mechanically fastened joints in composite laminates', *Compos. Struct.*, 2001, 53, (2), 199-211.
[4]. M. L. Dano, G. Gendron, and A. Picard: 'Stress and failure analysis of mechanically fastened joints in composite laminates', *Compos. Struct.*, 2000, 50, (3), 287-296.
[5]. A. Baker: 'Bonded composite repair of fatigue-cracked primary aircraft structure', *Compos. Struct.*, 1999, 47, (1-4), 431-443.
[6]. P. Chalkley and A. Baker: 'Development of a generic repair joint for certification of bonded composite repairs', *Int. J. Adhes. Adhes.*, 1999, 19, (2-3), 121-132.
[7]. V. Turage and S. Ripudaman: 'Modeling of a patch repair to a thin cracked sheet', *Eng. Fract. Mech.*, 1999, 62, (2-3), 267-289.
[8]. R. D. S. G. Campilho, M. F. S. F. de Moura, and J. J. M. S. Domingues: 'Modelling single and double-lap repairs on composite materials', *Compos. Sci. Tech.*, 2005, (65), 1948-1958.
[9]. C. Oztelcan, O. O. Ochoab, and J. Martin: 'Design and analysis of test coupons for composite blade repairs', *Compos. Struct.*, 1997, (37), 185-193.
[10].S. B. Kumar, I. Sridhar, and S. Sivashanker: 'Tensile failure of adhesively bonded CFRP composite scarf joints', *Mater. Sci. Eng.*, 2006, (132), 113-120.
[11].T. D. Breitzman, E. V. Iarve, and B. M. Cook: 'Optimization of a composite scarf repair patch under tensile loading', *Compos.*, 2009, (40), 1921-1930.
[12].M. Davis and D. Bond: 'Principles and practices of adhesive bonded structural joints and repairs', *Int. J. Adhes. Adhes.*, 1999, (19), 91-105.
[13].P. Camanho and M. Lambert: 'A design methodology for mechanically fastened joints in laminated composite materials', *Compos. Sci. Tech.*, 2006, 66, (15), 3004-3020.
[14].C. Soutis, D. M. Duan, and P. Goutas: 'Compressive behaviour of CFRP laminates repaired with adhesively bonded external patches', *Compos. Struct.*, 1999, (45), 289-301.
[15].J. Xiong and R. A. Shenoi: 'Integrated experimental screening of bonded composites patch repair schemes to notched aluminium-alloy panels based on static and fatigue strength concepts', *Compos. Struct.* 2008, 83, (3), 266-272.
[16].C. C. Peng, X. J. Gong, and D. Hearn: 'Tensile behaviour of patch-repaired CFRP laminates', *Compos. Struct.*, 2001, (93), 582-589.
[17].R. D. S. G. Campilho, M. F. S. F. de Moura, and J. J. M. S. Domingues: 'Numerical prediction on the tensile residual strength of repaired CFRP under different geometric changes', *Int. J. Adhes. Adhes.*, 2009, (29), 195-205.
[18].Olivier Sicot, X. L. Gong, and X. J. Gong: 'Dependence between aging treatments and residual stresses on composite laminate', *Mater. Sci. Forum.*, 2006, (524 – 525), 813-817.
[19].C. Soutis and F. Z. Hu: 'Design and performance of bonded patch repairs of composite structure', *P. I. Mech. Eng.*, 1997, 4,(211), 263-271.
[20].X. Liu and G. Wang: 'Progressive failure analysis of bonded composite repairs', *Compos. Struct.*, 2007, (81), 331-340.
[21].R. Asokan, T. Manikandan, adn C. Ganesan: 'Evaluation of tensile strength on repaired composite adhesive bonded joint', *Procedia. Engineering* 2012, (38), 2337-2346

Assessment of Atomic Mobility for the Bcc Phase of the Ti-Al-Mo System

Yi Chen, Liang Cheng, Yunlong Zhang

Key Construction Laboratory of Green Forming and Equipment from Jiangsu Province, Jiangsu University of Technology, Changzhou, 213001, China
E-mail: webchenyi@hotmail.com

Abstract. The atomic mobility for the bcc phase of the Ti-Al-Mo ternary system was developed by critical assessment of the experimental diffusion data by using the DICTRA software. The results shown the calculated diffusion coefficients presented the good agreements by the comprehensive comparisons with the experimental diffusion coefficients. The developed Ti-Al-Mo ternary atomic mobility was then subsequently validated by appropriate prediction of the interdiffusion behavior with comparison of the diffusion-couple experiments in the literature.

1. Introduction
β titanium alloy exhibits excellent formability thanks to its sustainability of 100% β when quenched from β phase area [1], yet offers significantly higher strengths by separating out the α phase when aged at α+β phase area [2, 3], which are widely used in aerospace and automotive industries [1-3]. The microstructure of the β-Ti alloys are largely depended on the recrystallization, phase transformation and grain growth process, while much of these processes are governed by the diffusion interaction [4-6] during thermomechanical processing and the follow heat treatment. Therefore, knowledge of diffusion kinetics is critical to understand the influence of alloying additions, schedule and optimize the heat treatment parameters and control of microstructural development for new type Ti-based alloys.

Nowadays, by combined with computational thermodynamics, the computational diffusion kinetics [7, 8] can predict the kinetic process of high quality for many commercially important alloy systems by using the DICTRA software, in which the atomic mobility database of the alloy system should be used [8]. More importantly, the mobility database can be applied for simulation the full diffusion picture of titanium alloys of interest without extra experimental measurements by combining the CALPHAD-base (Calculation of Phase Diagram) thermodynamic database [11-14].

The ternary Ti-Al-Mo system is an important base system of many titanium alloys such as Ti-5553 [15] and β 21S [16]. Due to the missing of the mobility database, the aim of this work is to assess the atomic mobility for the bcc phase of the Ti-Al-Mo ternary system by using the DICTRA software, and finally obtain the diffusion characteristic of the Ti-Al-Mo system with simulation of some diffusion couples by applying the acquired atomic mobility parameters.

2. Model Description
The atomic mobility Mi is suggested by Andersson and Ågren first [16] and later modified by Jönsson[17], which is expressed as

$$M_i = M_i^0 \exp\left(\frac{-Q_i^S}{RT}\right)\frac{1}{RT}{}^{mg}\Gamma, \quad (1)$$

Content from this work may be used under the terms of the Creative Commons Attribution 3.0 licence. Any further distribution of this work must maintain attribution to the author(s) and the title of the work, journal citation and DOI.
Published under licence by IOP Publishing Ltd

where Q_i^S is the activation energy, M_i^0 is the frequency factor, T is the temperature and R is the gas constant, and ${}^{mg}\Gamma$ is the effect of ferromagnetic ordering. If there is no ferromagnetic effect, we combined the first two parts of the mobility parameter into a single parameter, i.e., $Q_i = -Q_i^S + RTlnM_i^0$. This parameter Q_i is approximated to be of composition dependence, which can be similar treated in the phenomenological CALPHAD approach by using the Redlich-Kister polynomial [18], i.e,

$$Q_i = \sum_p x_p Q_i^p + \sum_p \sum_{q>p} x_p x_q \left[\sum_{r=0,1,2,\ldots} {}^rQ_i^{p,q} (x_p - x_q)^r \right] + \sum_p \sum_{q>p} \sum_{v>q} x_p x_q x_v \left[v_{pqv}^s \, {}^sQ_i^{p,q,v} \right], \ (S = p, q, v), \ (2)$$

where the Q_i^p is the value Qi of species i in pure species p and the x_p is the mole fraction of species p, while the ${}^rQ_i^{p,q}$ and ${}^rQ_i^{p,q,v}$ are the binary and ternary interaction parameters.

All the diffusivities can be expressed to the atomic mobility Mi, e.g., the tracer diffusivity D_i^* is directly related to the atomic mobility by the follow equation,

$$D_i^* = RTM_i, \ (3)$$

and the interdiffusion coefficient $\tilde{D}_{pq}^n$ can be expressed as,

$$\tilde{D}_{pq}^n = \sum_{i=1}^{n-1} (\delta_{ip} - x_p) x_i M_i \left(\frac{\partial \mu_i}{\partial x_q} - \frac{\partial \mu_i}{\partial x_n} \right), \qquad (4)$$

where the δ_{ip} is equal to 1 when i=p and otherwise equal to 0, and in which μi is the chemical potential of species i. By fitting to experimental obtained diffusion coefficients, those binary and ternary mobility parameters, Q_i^p, ${}^rQ_i^{p,q}$ and ${}^rQ_i^{p,q,v}$, can be numerically assessed upon the proposed relations.

In the diffusion couple, the composition is varying with the diffusion time t in the diffusion zone, which can be expressed as [19],

$$\frac{1}{V_m} \frac{\partial x_i}{\partial t} + \nabla \cdot \tilde{J}_i = 0, \ (5)$$

where $\tilde{J}_i$ is the interdiffusion flux and the molar volume Vm is generally treated as a constant. By adding the different boundary conditions and initial conditions, this equation can be solved numerically to present the form of the concentration profile.

3. Evaluation of experimental diffusion data

By applying the Matano-Kirkaldy and the modified Hall methods, the interdiffusion and impurity diffusion coefficients of the bcc phase of the Ti-Al-Mo ternary alloys at the Ti-rich corner have been recently determined at the temperature of 1523 K [20]. Before using those to optimize the ternary interaction parameters in this work, all those interdiffusion coefficients were examined and obey the thermodynamic constraints [21].

4. Optimization procedure

During the optimization procedure, the thermodynamic factor of diffusion should be obtained first by using CALPHAD-base thermodynamic database. In this paper, the thermodynamic parameters of the Ti-Al-Mo ternary were provided from CompuTherm [22] for calculation.

The mobility parameters of the bcc phase of Ti-Al binary was taken from the works of Chen et al. [23] was accepted by this work whereas the descriptions of the Ti-Mo binary assessed by Huang et al. [24] was used in this work. Then, by using DICTRA software [25], all the other mobility parameters were optimized by fitting to the selected experimental diffusion coefficients [20]. Table 1 shows the accessed mobility parameters of the Ti-Al-Mo ternary system.

Table 1. Assessed atomic mobilities for the BCC phase of the Ti-Al-Mo ternary system.

Mobility	Parameter (J/mole)	Reference
Mobility of Al		
Q_{Al}^{Al}	-215000-80.2*T	[23]
Q_{Al}^{Ti}	-154855.1839-129.012*T	Present work
Q_{Al}^{Mo}	-215000-80.2*T	Present work
$^{0}Q_{Al}^{Al,Mo}$	0	Present work
$^{0}Q_{Al}^{Ti,Al}$	-488445.3+78.3*T	Present work
$^{1}Q_{Al}^{Ti,Al}$	-262410.72-182.65*T	Present work
$^{0}Q_{Al}^{Ti,Mo}$	-389841	Present work
Mobility of Mo		
Q_{Mo}^{Al}	-448004.86-74.69*T	Present work
Q_{Mo}^{Ti}	-196255.4-105.21*T	Present work
Q_{Mo}^{Mo}	-448004.86-74.69*T	Present work
$^{0}Q_{Mo}^{Al,Mo}$	0	Present work
$^{0}Q_{Mo}^{Ti,Mo}$	172696.72-101.96*T	[24]
$^{1}Q_{Mo}^{Ti,Mo}$	-162707.90	[24]
$^{0}Q_{Mo}^{Ti,Al}$	46667	Present work
Mobility of Ti		
Q_{Ti}^{Al}	-199404.05-90.78*T	[23]
Q_{Ti}^{Ti}	-151989.95-127.37*T	[23]
Q_{Ti}^{Mo}	-207422.23-178.30*T	[24]
$^{0}Q_{Ti}^{Ti,Mo}$	247023.39-161.01*T	[24]
$^{1}Q_{Ti}^{Ti,Mo}$	189952.94	[24]
$^{0}Q_{Ti}^{Ti,Al}$	-223061.6+184.84*T	Present work
$^{1}Q_{Ti}^{Ti,Al}$	304789.2-194.43*T	Present work

5. Results and discussions

5.1 The Ti-Al-Mo ternary

In order to validation of the assessed mobility parameters, the diffusion coefficients are calculated and then compared with the experimental data. Figure 1 shows the calculated main ternary interdiffusion coefficients, $\widetilde{D}_{AlAl}^{Ti}$ and $\widetilde{D}_{MoMo}^{Ti}$, and the cross coefficients, $\widetilde{D}_{AlMo}^{Ti}$ and $\widetilde{D}_{MoAl}^{Ti}$, which are compared with the experimental data [20] at 1523 K. It can be seen that the agreements are very reasonable, which reviews particularly good for the main coefficients. However, it shows some minor discrepancy for the cross interdiffusion coefficients, which is mainly due to the usual scatter and relatively larger uncertainty of the experimental cross interdiffusion coefficients. In this work, the calculated cross interdiffusion coefficients $\widetilde{D}_{AlMo}^{Ti}$ and $\widetilde{D}_{MoAl}^{Ti}$ reviews the regularly change rather than abnormally changing with the variation of the composition, which means our calculations allow the variation of interdiffusion coefficients in the concentration to be appropriately predicted, i.e., the main interdiffusion coefficients of Al increase with increasing the Al content or decreasing the Mo content; however, decreasing the Al or Mo content both leads to a rise of the main interdiffusion coefficients of Mo. The calculated results also show the cross interdiffusion coefficients of Al and Mo increase with increasing both the Al and Mo content.

5.2 Simulation of diffusion couple experiments

The accuracy of the assessed mobility parameters can be further validated by comparing the predicted in-depth diffusion behavior arising from interdiffusion with the available diffusion experiment. By combining the obtained mobility parameters of the Ti-Al-Mo ternary the thermodynamic parameters, the diffusion couple experiment can be predicted by solving Eq. (5) numerically. Figure 2 shows the examples of the Ti-Al-Mo ternary diffusion couples for the concentration profile of Ti/Ti-22.83Al-5.94Mo in Fig. 2 (a) and Ti-5Mo/Ti-15.09Al in Fig. 2 (b) both at 1523 K with the experimental data [20] for comparison. The results show the agreement is apparently very satisfactory. Figure 3 compares the simulated diffusion paths with the experimental, the curves on the ternary isotherm mapping the locus of the compositions in planes parallel to the couple interface throughout the diffusion zone, with the experimental points measured from the diffusion couples at 1523 K [20]. The results also show the good agreement for a majority of the diffusion couples.

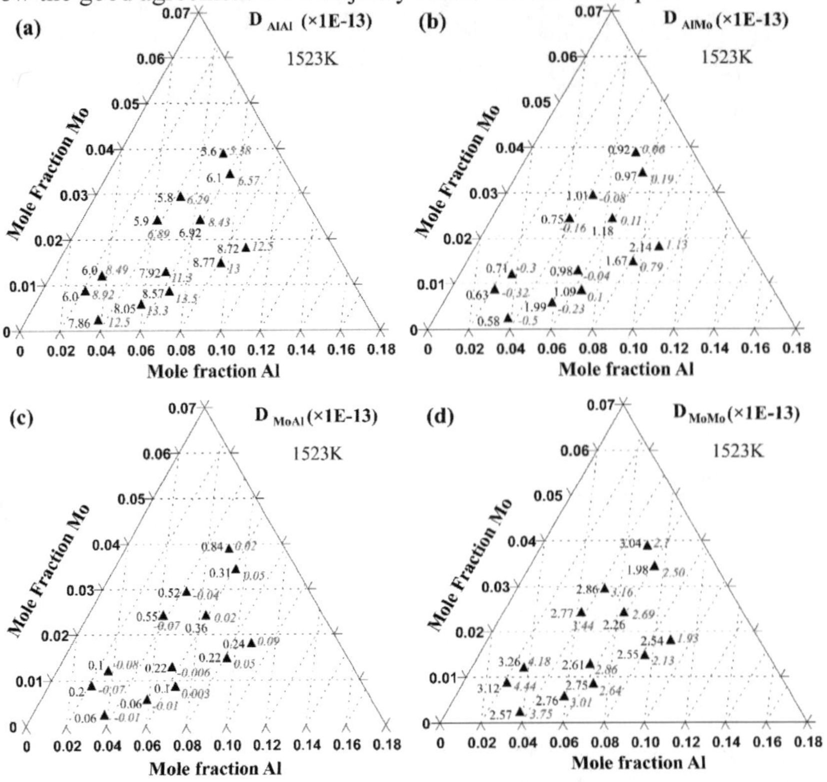

Figure 1. Calculated interdiffusion coefficients (in italic font) of the Ti-Al-Mo bcc ternary alloys at 1523 K in a comparison with the experimental measurement [20] (a) $\tilde{D}^{Ti}_{AlAl}$, (b) $\tilde{D}^{Ti}_{AlMo}$, (c) $\tilde{D}^{Ti}_{MoAl}$, (d) $\tilde{D}^{Ti}_{MoMo}$.

6. Conclusion

The DICTRA software was used for developing the diffusion mobility for the bcc phase of the Ti-Al-Mo ternary by assessing the experimental diffusion data. The comprehensive comparisons were made between the calculated diffusion coefficients and the available experimental data, which show the general satisfactory agreements. By combining with the CALPHAD-base thermodynamic parameters, the developed atomic mobility database has been successfully validated through further predicting of a number of ternary diffusion experiments.

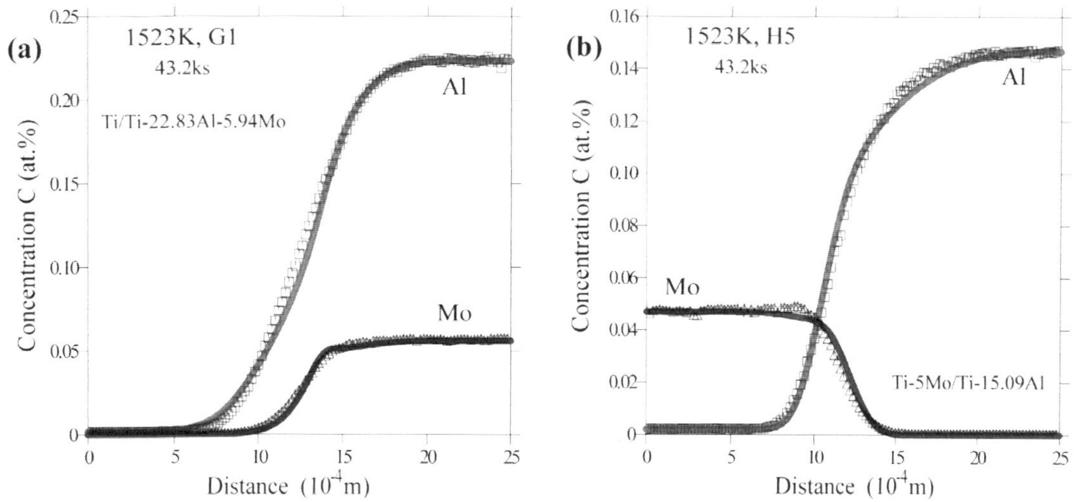

Figure 2. Simulated concentration profiles for (a) Ti/Ti-22.83Al-5.94Mo and (b) Ti-5Mo/Ti-15.09Al diffusion couples at 1523K in a comparison with the experimental measurement [20].

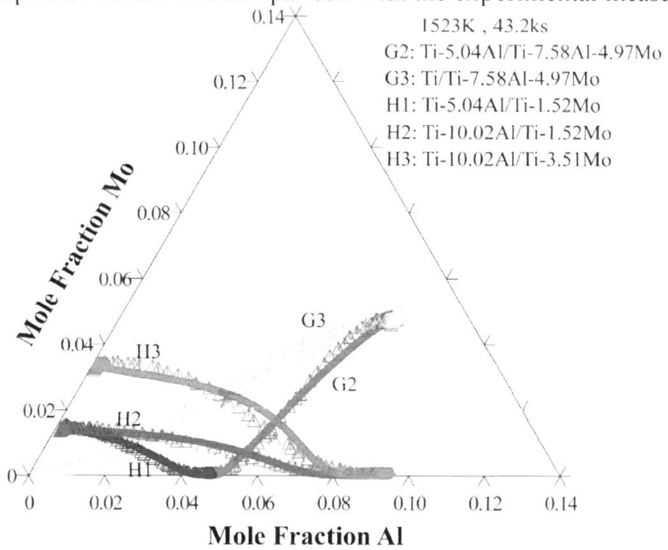

Figure 3. Simulated diffusion paths for ternary couples at the temperatures of 1523K compared with the experimental measurement (symbols) [20].

Acknowledgements
The authors would like to thank the National Natural Science Foundation of China (Grant No. 51601077) and the Natural Science Foundation of Jiangsu Province (Grant No. BK20160291) for the financial support.

References
[1] Boyer R R 1996 *Mater. Sci. Eng., A* **213** 103.
[2] Brewer W D, Bird R K and Wallace T A 1998 *Mater. Sci. Eng., A* **243** 299.
[3] Ivasishin O M, Markovsky P E, Matviychuk Y V, Semiatin S L, Ward C H and Fox S 2008 *J. Alloys Compd.* **457** 296.
[4] Cui Y W, Xu G, Kato R, Lu X G, Kainuma R and Ishida K 2013 *Metall. Mater. Trans. A* **44** 1621.
[5] Cui Y W, Tang B, Kato R, Kainuma R and Ishida K 2011 *Metall. Mater. Trans. A* **42** 2542.
[6] Cui Y W, Kato R, Omori T, Ohnuma I, Oikawa K, Kainuma R and Ishida K 2010 *Scripta Mater.* **62**

171.
[7] Andersson J, Helander T, Höglund L, Shi P and Bo S 2002 *Calphad* **26** 273.
[8] Borgenstam A, Höglund L, Ågren J and Engström A 2000 *J. Phase Equilib.* **21** 269.
[9] Saunders N and Miodownik A P 1998 *CALPHAD-A Comprehensive Guide* ed Cahn R W (Oxford: Pergamon Press) chapter 10.
[10] PanTitanium http://computherm.com/databases.html
[11] Huang L, Cui Y, Chang H, Zhong H, Li J S and Zhou L 2010 *J. Phase Equilib. Diff.* **31** 135.
[12] Li W, Tang B, Cui Y W, Hu R, Chang H, Li J S and Zhou L 2011 *Calphad* **35** 384.
[13] Liu Y, Pan T, Zhang L, Yu D and Ge Y 2009 *J. Alloys Compd.* **476** 429.
[14] Liu Y, Zhang L and Yu D 2009 *J. Phase Equilib. Diff.* **30** 334.
[15] Nag S, Banerjee R, Hwang J, Harper M and Fraser H 2009 *Philos. Mag.* **89** 535.
[16] Chaudhuri K and Perepezko J H 1994 *Metall. Mater. Trans. A* **25** 1109.
[17] Andersson J O and Ågren J 1992 *J. Appl. Phys.* **72** 1350.
[18] Jönsson B 1994 *Z. Metallkd.* **85** 498.
[19] Redlich O and Kister A T 1948 *Ind. Eng. Chem.* **40** 345.
[20] Chen Y, Tang B, Xu G L, Wang C Y, Kou H C, Li J S and Cui Y W. *Metall. Mater. Trans. A* **45** 1647.
[21] Kirkaldy J S, Weichert D and Haq Z U 1963 *Can. J. Phys.* **41** 2166.
[22] http://www.computherm.com/index.php?route=common/home
[23] Chen Y, Li J S, Tang B, Kou H C, Segurado J and Cui Y W 2014 *Calphad* **46** 205.
[24] Huang L 2010 *Master's Thesis of the Northwestern Polytechnical University* p 46.
[25] http://www.thermocalc.com/products-services/software/dictra/

IOP Conf. Series: Materials Science and Engineering **242** (2017) 012020 doi:10.1088/1757-899X/242/1/012020

Advances of Researches on Improving the Stability of Foams by Nanoparticles

G Wang, K L Wang and C J Lu

Key Laboratory of Enhanced Oil Recovery of Ministry Education, Northeast Petroleum University, Daqing, 99 university street, China

E-mail: wg_0459@163.com

Abstract. Recently, nano-tech made a change of traditional oil-gas exploration. Considering that foam fluid had a poor stability, investigators proposed to add nanoparticles to stabilize the foam fluid system. This paper described the mechanism of particles to improve the stability of the foam fluid in detail; and emphasized the synergistic effect between nanoparticles and surfactants and its effect on the foaming and foam stability of dispersions; and reviewed the latest applications of foam fluid that was stabilized by nanoparticle in enhancing oil-gas recovery, in which there are analysis that showed that the nanoparticles not only greatly increase the stability of the foam fluid, but also improve the efficiency of foam fluid; and lastly, forecasted the development of nanotechnology in petroleum areas.

1. Introduction

In recent years, among all the researches on enhancing oil recovery efficiency, the foam flooding has attracted more and more attention because of its unique seepage and oil displacement performance[1-5]. On the one hand, the apparent viscosity of foam is larger, which means that it can significantly improve the oil-water mobility ratio, and greatly enhances the macroscopic swept volume; on the other hand, due to the decrease of interfacial tension, the residual oil saturation can be reduced and the microcosmic oil displacement efficiency can be improved [6]. But at the same time, foam fluids also face some challenges technically and economically, such as the source of gas, the cost of compression, and the adsorption of surfactants on rocks, and the biggest one, the stability of foam fluid. Foam fluid is a thermodynamically unstable system [7-10]. Under the reservoir conditions, the surfactant will pyrolyze at high temperature and high pressure, while the ionic surfactant will precipitate at high salinity, and the foam fluid will defoam when encounter in-place oil [11-13]. In order to get a more stable foam system, the leading way at present is to add polyacrylamide and other foam stabilizer, but the macromolecular foam stabilizers will always cause additional damage to the strata, and their prices is expensive, and they are easy to break down at high temperatures. Therefore, based on the shortcomings mentioned above, we need to find a new alternative to stabilize the foam, and to further expand the scope of application of foam fluid.

As early as the beginning of the twentieth century, Ramsden found that colloidal size solid particles could be used to stabilize immiscible phases [14]. Pickering and other researchers further explored this phenomenon, they researched the function mechanism granular emulsifier, and found that particles can be adsorbed to the surface of gas and liquid and stabilize the foam [15]. Foams whose particle is stabilized have not received widespread attention and application in the past for a long period of time, and the reasons are that, firstly, the conventional surfactant foaming agent can basically adapt to the general foaming problem; secondly, the research and application of particle foaming agents require the

Content from this work may be used under the terms of the Creative Commons Attribution 3.0 licence. Any further distribution of this work must maintain attribution to the author(s) and the title of the work, journal citation and DOI.
Published under licence by IOP Publishing Ltd

cooperation of the preparation of ultrafine particles and surface modification technology. With the deepening of foam research and the improvement of the preparation of ultrafine particles and the surface modification technology, as a nano-sized particle that possesses special physical and chemical properties, nanoparticle has great application prospects and advantages in stabilizing foam. In this paper, the mechanism of nanoparticles stabilizing foams is introduced in detail, and the application of foam stabilized by nanoparticle in enhancing oil recovery is reviewed, and the future research direction and application prospect are discussed.

2. Mechanism of nanoparticles stabilizing foam

The mechanism of particle stabilizing emulsion is that nanoparticles or micronparticles accumulate at the gas-water interface and form a dense wrap, effectively preventing the dispersed phase droplets from coalescing or merging due to electrostatic attraction, interfacial energy change, spatial structure and other factors . Unlike conventional surfactant foaming agents, the particles form a particulate solid layer on the gas-water surface and the surfactant is formed typical dissolved molecular agglomerate on the gas / water surface.

When the particle-wrapped droplets infinitely close to each other, the state of thin layer of particles in the interface, can be mainly divided into the following two ways [16]:

1) Single-layer arrangement of bridged particles (cf. fig. 1). The particles are arranged in the interface with only one single layer. At this time, the particles are wetted by the two kind of liquid phases, but most of the particles are still immersed in the continuous aqueous phase. Compared with the double-layer arrangement of particles, the stability of the single-layer structure depends mainly on the influence of the spatial resistance, preventing the particles from leaving the bridge layer.

2) The particles are tightly warped and arranged in a stratified layer at the gas-water interface (cf. fig. 2). In this case, the particles prevent droplet coalescence as a physical barrier, and enhance the stability of the aqueous film between the droplets and the continuous phase, which depends on the capillary pressure and interfacial rheological properties among all the particles. The location of the particles in the droplet interface depends on the wettability of the particles themselves.

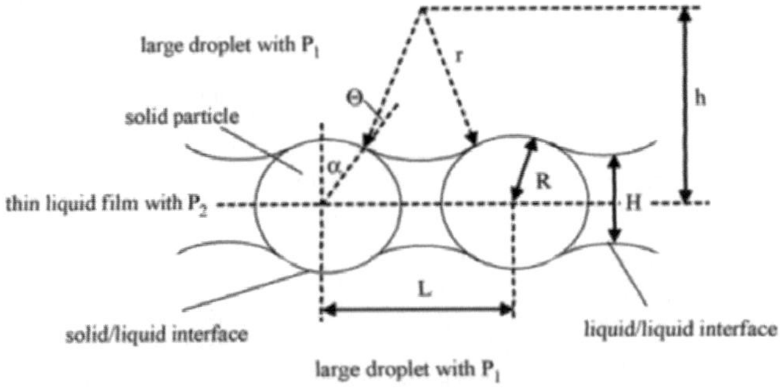

Fig.1. Schematic of the hexagonally packed single layer of spherical particles, stabilizing the thin liquid film between two large droplets [17].

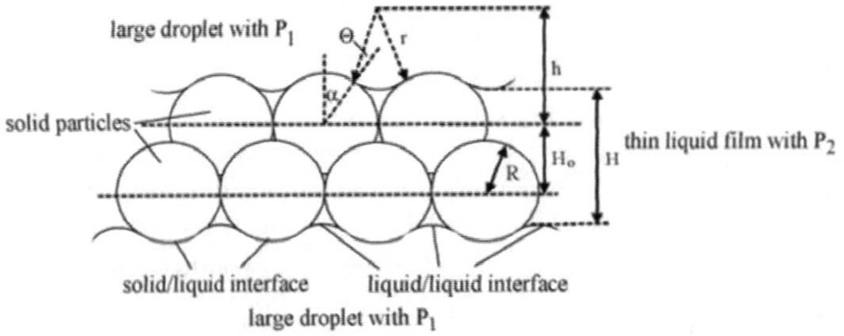

Fig.2. Schematic of the closely packed double layer of spherical particles, stabilizing the thin liquid film between two large droples(Ho=1.633R) [17].

Particle foaming agent is characterized by its greater adsorption energy at the gas-water interface, in other words, its adsorption heat is greater. Thus, the adsorption at the interface can be considered irreversible to adsorb or achieving a balance of adsorption, that is to say, the stability of the particle-stabilized foam is better, and once the foam is formed, it is not easy to be demulsified, and the conventional surfactant foaming agent may desorbs from the interface after a while, leading to foam collapse.

3. Foam flooding under nanoparticles stabilization to enhanced oil recovery

Although stabilizing foam with nanoparticles is not an emerging technology, it is still in the indoor research stage in oil industry areas and most of the research directions are to improve the oil-gas recovery efficiency. Since the foam has a high viscosity in the stratum, it is possible to avoid the phenomenon of "viscous fingering", and blocking water but not oil, it thus is particularly suitable for heterogeneous stratum. Foam flooding can improve both the sweep efficiency and displacement efficiency, which makes it a promising method to enhance recovery efficiency [18]. But the stability of the foam has always been a challenge to hinder the foam from further application, but indoor experiments have shown that nanoparticles can solve this problem, and now there are some regions that are conditional enough to provide support for small-scale experiments, so that they can provide guidance for future field applications.

CO_2 foam flooding can effectively control the mobility of CO_2, which can solve the problems of common viscous fingering phenomenon of CO_2, and decreasing of sweep efficiency due to low density upward along the stratum. Without using surfactant, Worthen, et al. [19] stabilized the foam at 1200 ~ 3000psi (1psi = 6.895 kPa) with supercritical CO_2 fluid and 50% silanized coating of partially hydrophobic SiO_2 nanoparticles, from which the CO_2 foam can maintain transparent and stable for 23 hours, which can be effectively used for CO_2 foam flooding.

Yu, et al. [20,21] injected SiO_2 nanoparticles dispersion and supercritical CO_2 together into a sandstone core in 2013, from which there's CO_2 foam produced. This experiment measured the mobility of foam and retention of particles, and came to the conclusion of that the nanoparticle-stabilized CO_2 foam can improve the recovery efficiency of sandstone after water flooding, meanwhile, the permeability of the core is almost constant, which indicated that nanoparticles do not clog the stratum and can migrate in the pores. Moreover, San, et al. [22] studied the effects of different mineralization degree on nanoparticles-stabilized CO_2 foam. It was found that, under a certain consistency, NaCl and $CaCl_2$ promoted the foaming, foam stabilizing and mobility controlling, but the polyvalent ions have an inhibitory effect on the stability of CO_2 foam.

Throughout the researches above, it is concluded that, nanoparticles can not only stabilize the CO_2 foam well, but also increase the viscosity of the CO_2 foam to control the mobility in the reservoir conditions.

Compared to CO_2, the properties of N_2 are more stable. It will not corrode the ground equipment and down-hole string, and it is of low density, low hydrostatic head, so it's suitable for shallow depth of oil-gas reservoirs. In order to solve the instability of N_2 foam, Sun, et al. [23] injected partly hydrophobic SiO_2 and anionic surfactant SDS into the micro-model and sandstone filling model to flood oil, in order to evaluate the stability of the foam and the efficiency of oil-gas recovery, and the results showed that SiO_2 / SDS foam are more thermally stable than SDS foam and it has a good quality of oil flooding for homogeneous or heterogeneous strata.

Singh, et al. [24] observed nanoparticles exist in the Plateau boundary of foam under fluorescence microscope, so it effectively delay the foam of bleeding and coarsening. After water flooding, the particles that produced by the injection of SiO_2 nanoparticles and surfactants co-produced foam, can increase the recovery efficiency by about 10%.

In order to further saving costs, Eftekhari, et al. [25] injected coal by-products of nano-scale fly ash particles and anionic surface-active AOS together into the porous medium, the study of N_2 foam foaming and foam. It was found that the addition of nano-fly ash resulted in better foam foaming, more oil flooded, and better stability in the presence of oil than the foam produced by AOS alone.

In addition to conventional N_2 foam flooding, N_2 vapor foam flooding technology is an effective method for improving oil recovery when it is for heavy oil reservoirs with high gravity and high viscosity. In order to overcome the foam stability and a large number of additive losses and other issues, Khajehpour, et al. [26] in the reservoir conditions, the preferred concentration of SiO_2 nanoparticles and the type of surfactant, and the co-generation of the foam used for core flooding experiments. Studies have shown that the addition of nanoparticles can better reduce the flow and control the vapor foam.

Regardless of CO_2 foam or N_2 foam, indoor experiments have shown that foam foaming, foam stabilization; flow control effect and enhanced oil recovery efficiency can be effectively improved after adding nanoparticles. How to combine nanoparticles and oilfield chemical technology, and further applied to the scene, is a step must be taken. In addition, extraction in the field of oil and gas from the conventional oil and gas to unconventional oil and gas across in China, and the stability of nanoparticles in the application of unconventional oil and gas resources, the nanoparticles in the formation of the flow and retention will be low porosity and low permeability It is a question that must be solved and answered.

4. Problems and prospects of stabilizing foam by nanoparticles

While low oil prices have led to a significant reduction in R&D spending by some oil giants, many colleges and universities are currently working on nanotechnology in oil and gas exploration, where nanoparticle stabilization is one of the most popular research directions. As for the nanoparticles-stabilized foam becoming practical, I think that the facing challenges and prospects of future research are of the following aspects.

(1) The mechanism and influencing factors of nanoparticles stabilizing foams need further study. For example, after the addition of nanoparticles, the interaction between the interface changes in the law is not perfect. In addition, some scholars have concluded that the addition of nonionic surfactants can also be synergistic with the nanoparticles, indicating that in addition to electrostatic adsorption to change the surface wetting of particles, there are other stable foam mechanisms [27]. Only when these problems are solved, the concentration of nanoparticles and surfactants can be optimized, which will be different performance of the foam applied to different areas of oil and gas exploitation.

(2) Some of the laboratory materials used in the study is partially hydrophobic nanoparticles that must be obtained by surface modification, which further increases the cost. But the current downturn in energy prices needs to further reduce the cost of nanotechnology. Optimize the concentration of nanoparticles and consider the use of hydrophilic nanoparticles and surfactants together to stabilize the foam is the idea of response.

(3) At present, the application of nanoparticles-stable foam in oil and gas exploration is still only in the laboratory experimental stage. With regard to the seepage of the foam in the formation of nanoparticles-stabilization, the retention of the nanoparticles in the pores, the relative numerical simulation of the settling of the proppant in the nanoparticles foam fracturing fluid is rare, and the small tests on the site are almost none. However, the combination of experimental data and simulation results, applied to the field of the test is necessary to be carried out. This can also provide a basis for solving the potential damage of the nanoparticles to the formation.

5. Conclusion

Nanoparticles-stable foam can solve many long-standing problems in traditional oil and gas exploration methods. Whether the global research hotspots of shale gas and the national "thirteenth five" plan to vigorously promote the development of coalbed methane, nanoparticles stable foam has a wide range of prospects. In the near future, nanotechnology will write a new chapter in oil and gas exploration.

References

[1] Yuan X, Wang K and Chen J 2010 *Acta. Petro. Sin.* **01** 87-90

[2] Lee S and Kam S I 2013 *EOR. Field. Case. Stu.* **12** 23-63
[3] Li H, Hou J and Li W 2014 *Petro. Ge. Reco. Effic.* **04** 93-96
[4] Pang Z, Liu H and Zhu L 2015 *J. Petro. Sci. Eng.* **128** 184-193
[5] Sheng J J 2013 *EOR. Field. Case. Stu.* **14** 251
[6] Bayat A E, Rajaei K and Junin R 2016 *Colloid. Surface. A.* **511** 222-231
[7] Da C, Xue Z and Worthen A J 2016 *Viscosity and stability of dry CO_2 foams for improved oil recovery*(SPE Improved Oil Recovery Conference) p 11
[8] Gu S, Li W and Luo W 2016 *Chem. Tech. Fuels. Oil.* **52** 386-395
[9] Carpenter C 2016 *J. Petrol. Technol.* **68** 87-88
[10] Yang W, Wang T and Fan Z 2017 *Energ. Fuel.* **31** 4721-4730
[11] Farajzadeh R, Andrianov A and Krastev R 2012 *Adv.Colloid.Interfac.* **183** 1-13
[12] Andrianov A, Farajzadeh R and Mahmoodi M 2012 *Ind. Eng. Chem. Res.* **51** 2214-2226
[13] Li R F, Yan W and Liu S 2010 *SPE. J.* **15** 928-942
[14] Ramsden W 1903 *Proc. R. Soc. London.* **72** 156
[15] Pickering S U 1907 *J.Chem. Sci.* **91** 2001
[16] Dickinson E 2010 *Curr. Opin. Colloid. In.* **15** 40-49
[17] Kaptay G 2006 *Colloid. Surface. A.* **282** 387-401
[18] Li Zhaomin 2010 *Application of foams in oil and gas production*(Beijing: Petroleum Industry Press) p 246
[19] Worthen A,Bagaria H and Chen Y 2012 *Nanoparticle stabilized carbon dioxide in water foams for enhanced oil recovery* (SPE Improved Oil Recovery Symposium) p 1-7
[20] Yu J, Mo D and Liu N 2013 *The application of nanoparticle-stabilized CO_2 foam for oil recovery* (SPE International Symposium on Oilfield Chemistry) p 1-9
[21] Yu J, Khalil M and Liu N 2014 *Fuel* **126** 104-108
[22] San J, Wang S and Yu J 2016 *Nanoparticle stabilized CO_2 foam: effect of different ions* (SPE Improved Oil Recovery Conference) p 1-12
[23] Sun Q, Li Z and Li S 2014 *Energ. Fuel.* **28** 2384-2394
[24] Singh R 2015 *Energ. Fuel.* **29** 467-479
[25] Eftekhari A, Krastev R and Farajzadeh R 2015 *Ind. Eng. Chem. Res.* **54** 12482-12491
[26] Khajehpour M, Etminan S R and Goldman J 2016 *Nanoparticles as foam stabilizer for steam-foam process* (SPE EOR Conference at Oil and Gas West Asia) p1-14
[27] Zhang S, Sun D and Dong X 2008 *Colloid. Surface. A.* **32** **4**1-8

IOP Conf. Series: Materials Science and Engineering **242** (2017) 012021 doi:10.1088/1757-899X/242/1/012021

The Particle Distribution in Liquid Metal with Ceramic Particles Mould Filling Process

Qi Dong, Shu-ming Xing

School of Mechanical, Electronic and Control Engineering, Beijing Jiaotong University,Beijing 100080,China

E-mail: 15116360@bjtu.edu.cn

Abstract. Adding ceramic particles in the plate hammer is an effective method to increase the wear resistance of the hammer. The liquid phase method is based on the "with the flow of mixed liquid forging composite preparation of ZTA ceramic particle reinforced high chromium cast iron hammer. Preparation method for this system is using CFD simulation analysis the particles distribution of flow mixing and filling process. Taking the 30% volume fraction of ZTA ceramic composite of high chromium cast iron hammer as example, by changing the speed of liquid metal viscosity to control and make reasonable predictions of particles distribution before solidification.

1. Introduction

Adding ceramic particles (radius=1mm) in high chromium cast iron can effectively enhance the performance of wear parts to improve the wear resistance of the plate hammer [1-3]. Kaihong Zheng [4] used the infiltration method to prepare matrix composite of ZTA ceramic particles (1~3 mm) with high chromium cast iron. After heat treatment the composite abrasive wear performance is 5.9 times of Cr20 high chromium cast iron.

2. Particles distribution model

The establishment of the theoretical model of particles distribution in a two-phase mixture during the filling process. The parameters influence the particles distribution in the two-phase flow of granular particles and metal melt [5]:

$$\frac{6\sqrt{2}v_p\emptyset}{d_p}=\frac{\rho_f}{\rho_p}\left(\frac{18v_f}{{d_p}^2}+\frac{2.7{v_f}^{0.313}v_t^{0.687}}{{d_p}^{1.313}}\right) \tag{1}$$

With the particle volume concentration increasing, the interaction frequency between particles increasing too, the particles are prone to collisions and agglomeration. When the fluid and particle density difference is small, the ability of the particle following fluid pulsation will be enhanced, the agglomeration of particles will be decreased. With the increase of particle diameter, it is not conducive to the interaction between fluid and particles, the aggregation easy to happen. In addition, with the grain terminal (fluid flow) increasing, agglomeration and separation will be destroyed in hydrodynamic shear.

3. Filling simulation

The convective mixing filling process involving liquid metal, particles and air phase. The calculation using Eulerian multiphase flow model, because of the particles volume ratio is greater than 10% in the field of wide distribution. The particles as discrete systems for processing, it needs to be as quasi

Content from this work may be used under the terms of the Creative Commons Attribution 3.0 licence. Any further distribution of this work must maintain attribution to the author(s) and the title of the work, journal citation and DOI.
Published under licence by IOP Publishing Ltd

continuous medium particles or quasi fluid.

3.1Volume fraction[7]

The ZTA ceramic particle was selected as the reinforced phase and the physical parameters are shown in Table 1.The particles will be simplified for spherical particles in the simulation process.

Table 1. Physical parameters of ZTA ceramic particles.

	Particle diameter (mm)	density ($g\cdot cm^{-3}$)	porosity (λ)
Ceramic Particles Zr40(40%ZrO_2)	2-4	4.2-4.6	0.3-0.5

The matrix of plate hammer selection of high chromium cast iron, which has characteristics of low thermal conductivity, wide range of crystallization temperature, alloy liquid viscosity, flow resistance[6]. The physical parameters see table 2:

Table 2. Physical parameters of Cr20.

temperature /K	density /Kg/m^3	liquidus temperature /K	solidus temperature /K	kinetic viscosity /Pa·s	surface tension /N/m
1573	7620	1563	1493	0.006	1.88
1623	7620	1563	1493	0.005	1.88

3.2 Volume fraction[7]

$$V_q = \int_V \alpha_q \, Dv \tag{2}$$

$$\sum_{q=1}^{n} \alpha_q = 1 \tag{3}$$

3.3 Fundamental equations

3.3.1The conservation equations of continuity equation Q[8]

$$\frac{\partial}{\partial t}(\alpha_q) + \nabla \cdot (\alpha_q \vec{v}_q) = \frac{1}{\rho_q}\left(\sum_{p=1}^{n} \dot{m}_{pq} - \alpha_q \frac{d_q \rho_q}{dt}\right) \tag{4}$$

The momentum equation of liquid phase:

$$\begin{aligned} \frac{\partial}{\partial t}(\alpha_l \rho_l \vec{v}_l) + \nabla \cdot (\alpha_l \rho_l \vec{v}_l \vec{v}_l)2 \\ = -\alpha_l \nabla p + \nabla \cdot \bar{\bar{\tau}}_l + \alpha_l \rho_l \vec{g} + \alpha_l \rho_l (\vec{F}_l + \vec{F}_{lift\,l} + \vec{F}_{V_m,l}) \\ + \sum_{p=1}^{n} [K_{pl}(\vec{v}_p - \vec{v}_l) + m_{pl} \vec{v}_{pl}] \end{aligned} \tag{5}$$

Momentum conservation equation of particle phase:

$$\frac{\partial}{\partial t}(\alpha_s\rho_s\vec{v}_s) + \nabla\cdot(\alpha_s\rho_s\vec{v}_s\vec{v}_s) = -\alpha_s\nabla p - \nabla p_s + \nabla\cdot\overline{\overline{\tau_s}} + \alpha_s\rho_s\vec{g} + \alpha_s\rho_s(\vec{F}_s + \vec{F}_{lift\,s} + \vec{F}_{V_m,s}) + \sum_{p=1}^{n}[K_{ps}(\vec{v}_p - \vec{v}_s) + m_{ps}\vec{v}_{ps}] \tag{6}$$

Momentum conservation equation of gas:

$$\frac{\partial}{\partial t}(\alpha_g\rho_g\vec{v}_g) + \nabla\cdot(\alpha_g\rho_g\vec{v}_g\vec{v}_g) = -\alpha_g\nabla p - \nabla p_g + \nabla\cdot\overline{\overline{\tau_g}} + \alpha_g\rho_g\vec{g} + \alpha_g\rho_g(\vec{F}_g + \vec{F}_{lift\,g} + \vec{F}_{V_m,g}) + \sum_{p=1}^{n}[K_{pg}(\vec{v}_p - \vec{v}_g) + m_{pg}\vec{v}_{pg}] \tag{7}$$

3.3.2 Turbulence model[9]. In the standard k-$\mathcal{E}$ turbulence model, the relationship between turbulent kinetic energy K, dissipation rate $\mathcal{E}$ and turbulent viscosity μ expressed as:

$$\mu = \rho C_\mu \frac{k^2}{\mathcal{E}} \tag{8}$$

The turbulent kinetic energy equation K:

$$\rho\frac{\partial k}{\partial t} + \rho\mu\frac{\partial k}{\partial x} = \frac{\partial}{\partial x}\left[\left(\mu + \frac{\mu}{\sigma}\right)\frac{\partial k}{\partial x}\right] + \psi - \rho\mathcal{E} \tag{9}$$

Equation of turbulent kinetic energy dissipation rate $\mathcal{E}$:

$$\rho\frac{\partial \mathcal{E}}{\partial t} + \rho\mu\frac{\partial \mathcal{E}}{\partial x} = \frac{\partial}{\partial x}\left[\left(\mu + \frac{\mu}{\sigma}\right)\frac{\partial \mathcal{E}}{\partial x}\right] + \frac{\mathcal{E}}{k}(c_1\psi - c_2\rho\mathcal{E}) \tag{10}$$

The turbulent phase:

$$\psi = \mu\frac{\partial \mu}{\partial x}\left(\frac{\partial \mu}{\partial x} + \frac{\partial \mu}{\partial x}\right) \tag{11}$$

C_μ=0.09, c_1=1.44, c_2,1.92, σ_k=1, $\sigma_\mathcal{E}$=1.3

4. Results and discussion

The distribution of particles in the straight runner and the cavity of the workpiece have been found after simulation, and the particle distribution expressed by volume fraction. The particles distribution on the centre line of the model are shown in the figures to the below:

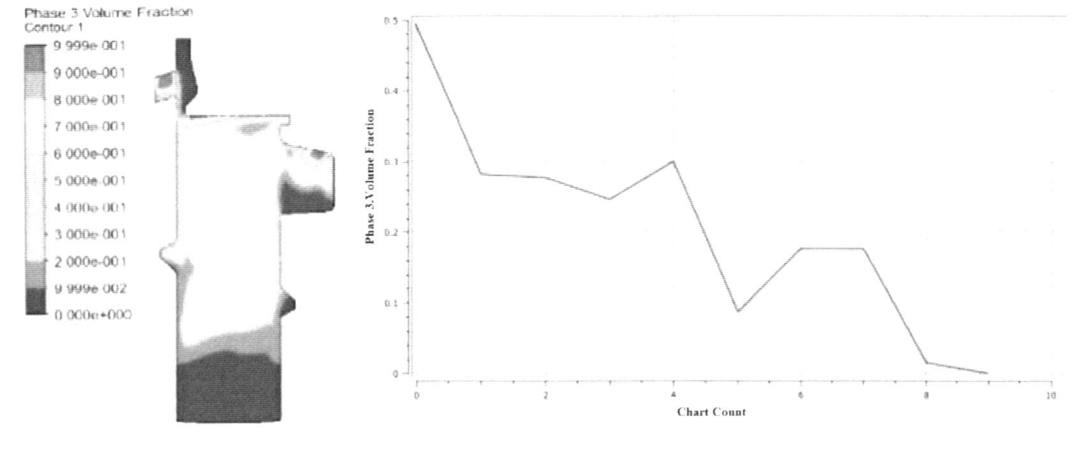

Figure 1. Distribution of particle volume fraction along the centre of hammer when the molten metal with a velocity of 0.5m/s

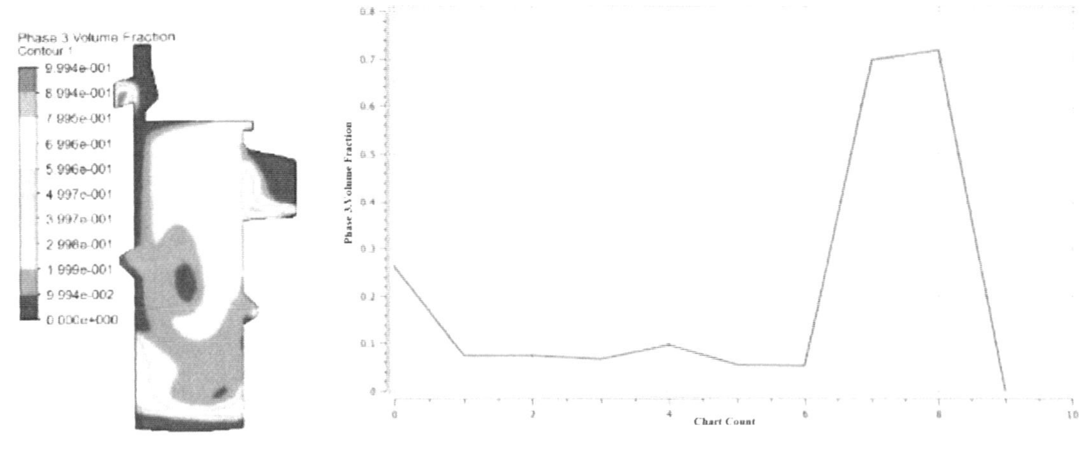

Figure 2. Distribution of particle volume fraction along the centre of hammer when the molten metal with a velocity of 1m/s

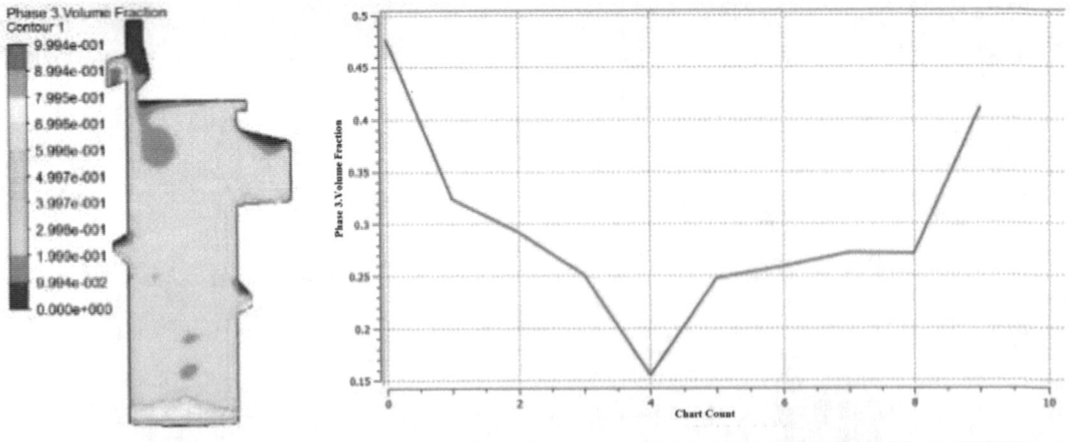

Figure 3. Distribution of particle volume fraction along the centre of hammer when the molten metal with a velocity of 2m/s

According to the particle volume fraction set of 30%, make curve y=0.3 linear, the curve and straight line to form uniform area to characterize the degree of particle distribution. The greater area shows that the volume fraction of the particles deviates from the set value is greater, the distribution of particles are uneven. The smaller particle volume fraction means that the distribution is close to the set value, the particle distribution are uniform. By calculation, the molten metal with a velocity of 2m/s was obtained particle distribution more uniform than other groups.

The influences of liquid metal speed on the particle distribution include: when the speed is 0.5 m/s, 1 m/s, 2 m/s, the time needed for the workpiece cavity filled with respectively 3.58s, 1.79s, 0.89s. In the different viscosity, when the filling time long, will be easy to cause the particles float. In simulation results, the particle volume fraction is very low at the bottom of the hammer .When the filling time is short, the floating particles is not obvious. The greater liquid metal flow velocity lead to large flow shear stress relatively, it can be promote particle dispersion to reducing the particle gathered. The bottom of the plate hammer is the turning point of the workpiece cavity, and the side the wall have no arc, rectangular region on both sides were the primary causes of liquid metal flow slow. The particle retention in the workpiece at the bottom lead to particle volume fraction significantly larger is the result of slow molten metal velocity.

5. Conclusions

(1) At a certain speed in the two-phase mixing and flow process, the possibility of particles aggregation can be reduced;

(2) The uniform particle distribution was obtained when the liquid metal flow velocity is 2m/s;

(3) Through the analysis of the simulation results, it is concluded that the greater the flow velocity of metal liquid is the greater shearing force of fluid flow. The high-speed is the promotion of particle phase dispersion. This result is similar with the forecast of the mathematical model.

References

[1] Yingjie Liu,Keqiang Cheng.The analysis of Wear failure [M]. Beijing:China Machine Press,1991,3-7

[2] Jiwen Li,Guoshang Zhang,Shizhong Wei. Study on technology of bimetal liquid-liquid composite wear-resistant plate hammer [J]. Foundry,2010,59(3):300-303

[3] Jun Du,XiaoYu Chong,YeHua Jiang.Numerical simulation of mold filling process for high chromium cast ironmatrix composite reinforced by ZTA ceramic particles.International Journal of Heat and Mass Transfer.2015,89: 872-883

[4] Kaihong Zheng,Sanmei Zhao,Juan Wang. Particle reinforced high chromium cast iron matrix composites preparation, microstructure and properties [J]. Foundry.2012, 61(2): 165-168

IOP Conf. Series: Materials Science and Engineering **242** (2017) 012021 doi:10.1088/1757-899X/242/1/012021

[5] S.L.Soo. "Gas-solid Systems",in Handbook of Mulitphase Systems, Edited by Gad Hetsrom,MoGraw-Hill.1982,PP.1-15
[6] Binshi Huang,Tang Xi. Casting process and properties of high chromium cast iron [J]. Modern cast iron,1988.(2):36
[7] Stoehr R,Wang C.Coupled heat transfer and fluid flow in the filling of castings[J], AFS transactions. 1988, 96:733-740
[8] BROWN R. E. Magnesium at NADCA Congress and Exposition [J]. Light Metal Age,2000,8(3-4): 103-105.
[9] Ik-Tae Im W-S K,Kwan-Soo Lee.A unified analysis of filling and solidification in casting with natural convection[J]. International Journal of Heat and Mass Transfer. 2001, 44; 1507-1515

IOP Conf. Series: Materials Science and Engineering **242** (2017) 012022 doi:10.1088/1757-899X/242/1/012022

Research on Innovation Model of Manufacturing Materials from the Perspective of Green Development

Xu Jie

Sichuan Academy of Social Sciences, Chengdu, Sichuan, China, 610071

E-mail: jie_xuscu@126.com

Abstract. From the perspective of green development, this paper constructs the innovation system of manufacturing materials on the basis of relevant literature at home and abroad. After that, this paper analyzes the dimensions of manufacturing material innovation and on this basis, puts forward the innovation models of manufacturing materials, and finally explores the upgrading paths of manufacturing material innovation models.

1. Introduction

Innovation is the motive force of economic development. At present, the manufacturing materials technology is gradually seeking to reduce pollution, reduce consumption and improve the ecology. Therefore, the innovation of manufacturing materials based on green development has attracted much attention. For manufacturing materials research, innovation is not a linear or mechanical process, but a dynamic and multi factor system, that is, manufacturing material innovation system. The innovation system contains a variety of innovative models, through interaction and transformation, to generate innovation power, and ultimately achieve sustainable economic development.

2. Literature review

This article mainly summarizes the domestic and foreign scholars' research from two aspects of green manufacturing and technological innovation.

2.1. Green manufacturing

Since 1980s, scholars have begun to study green manufacturing. With regard to the concept and connotation of green manufacturing, some scholars believe that in the manufacturing process, green product production management, pollution prevention and innovative investment in clean technology are three dimensions of green manufacturing connotation (Hart, 1997). About the composition of green manufacturing, some scholars believe that from the system and integrated idea studying, construction of green supply chain model is to resolve the conflict between manufacturing industry and the environment, to achieve optimization of environment and sustainable development of the manufacturing industry (Wang Yingluo, 2003). With regard to the system structure of green manufacturing, some scholars believe that green manufacturing is a comprehensive manufacturing system including product life cycle, resource optimization and environmental impact (Liu Fei, 1997,2004).

2.2. Technological innovation

Schumpeter believed that the fundamental force to promote economic development lies in innovation. On the technical innovation, some scholars believe that the innovation of ecological technology, can

Content from this work may be used under the terms of the Creative Commons Attribution 3.0 licence. Any further distribution of this work must maintain attribution to the author(s) and the title of the work, journal citation and DOI.
Published under licence by IOP Publishing Ltd

IOP Conf. Series: Materials Science and Engineering **242** (2017) 012022 doi:10.1088/1757-899X/242/1/012022

promote the strong innovation ability of enterprises to develop new markets, and to have the weak innovation ability of enterprises to further reduce costs (Grubb, Uiph, 2002). On the path of green innovation, some scholars believe that the product development, supply chain evaluation, new product and market portfolio, emission rights and emission reduction internal transformation, process optimization is the realization path of green innovations (Kolk, Pinkse, 2005). On the path of innovation, some scholars believe that the technology innovation and development, mature industrial organization structure reform and innovation, is the innovation path of industrial transformation and upgrading (Li Zhaoyou, 2015).

2.3. Review of related research
According to the existing research, we find that there are more achievements in the field of natural science, and less research results in the field of social sciences. Much research has been done on materials, manufacturing, machinery and environment, and the study of management and economics has just started. The existing research focuses on green manufacturing, with more connotation and technology, and little research on innovation and path of green manufacturing.

3. Innovation system of manufacturing materials from the perspective of green development
Under the background of green economy, manufacturing industry needs "green" material with high technology content, low resource consumption and less environmental pollution.

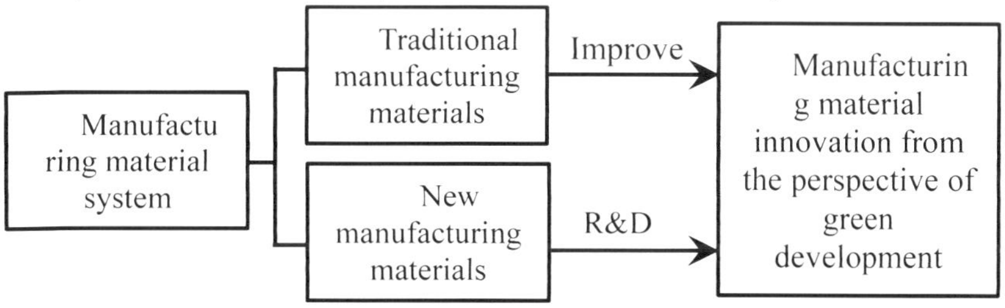

Figure 1. Innovation system of manufacturing materials

As shown in Figure 1, the research and development of new materials mainly refers to the development of new green manufacturing materials. The improvement of traditional materials mainly refers to the green transformation of traditional manufacturing materials.

4. Innovation model of manufacturing materials from the perspective of green development
According to the influence degree of factors, the innovation of manufacturing materials is divided into Technological innovation dimension (T) and Institutional change dimension (I). The two dimensions can form a two-dimensional space. Due to different circumstances, the dimensions of technological innovation and institutional change are different from each other, so different types of innovation models will be formed.

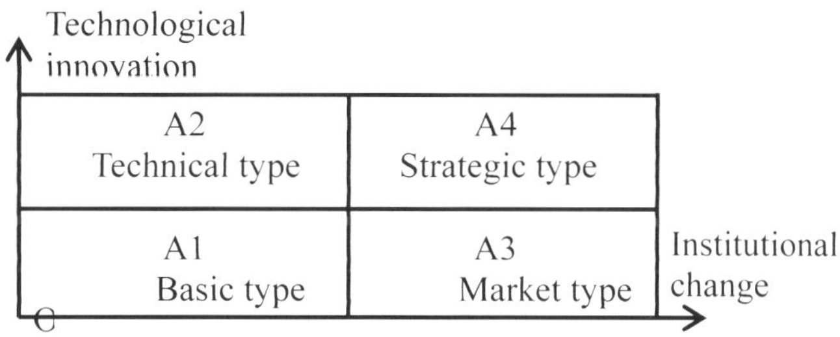

Figure 2. Innovation model of manufacturing materials

As shown in Figure 2, the dimensions of technological innovation and institutional change constitute four innovative models of A1, A2, A3 and A4.

Basic type (A1): The characteristics of this model are lack of technological innovation and lack of institutional change.

Technical type (A2): the characteristics of this model are strong technological innovation and weak system transformation.

Market type (A3): the prominent feature of this model is weak technological innovation and strong institutional change.

Strategic type (A4): the prominent feature of this model is strong technological innovation and strong institutional change.

5. Promotion of manufacturing material innovation model from the perspective of green development

According to the dimension of technological innovation and the dimension of institutional change, the upgrading paths of A1, A2, A3 and A4 are as follows.

Basic type (A1): eliminate non green materials. Upgrading technological innovation ability and transforming to technological type (A2). Or enhance the ability of institutional change, transforming to market type (A3). As shown in figure 3.

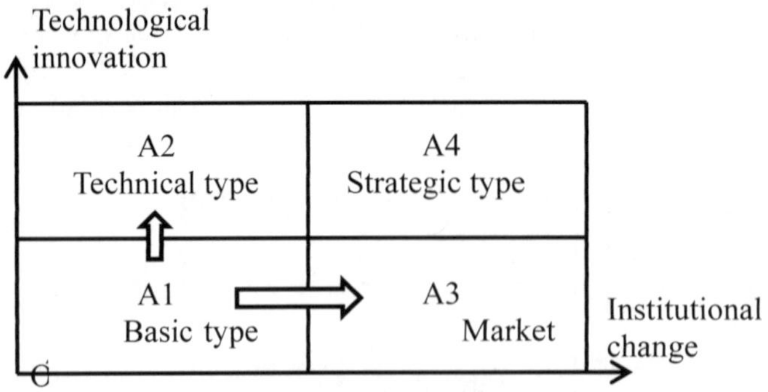

Figure 3. The improving path of Basic type

Technology type (A2): green manufacturing materials research and development. Upgrading the capability of institutional change and transforming to strategic type (A4). As shown in figure 4.

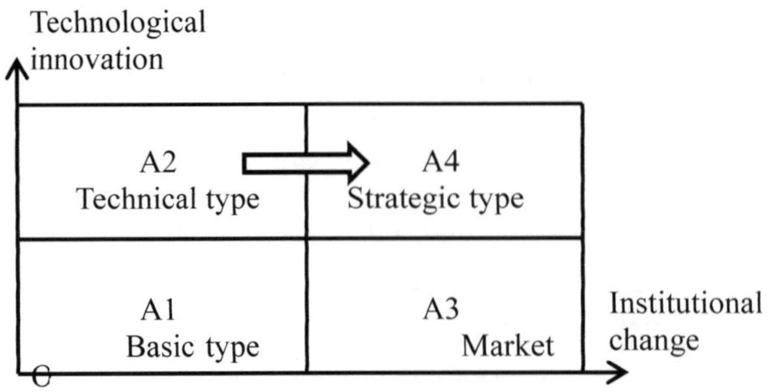

Figure 4. The improving path of Technical type

Market type (A3): Green improvement of traditional manufacturing materials. Upgrading technological innovation ability and transforming to strategic type (A4). As shown in figure 5.

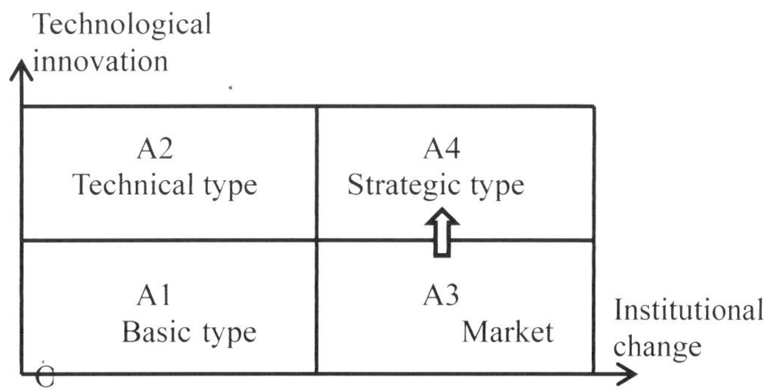

Figure 5. The improving path of Market type

Strategic type (A4): green manufacturing, materials research and development, green improvement of traditional manufacturing materials. Promoted from A4 to A4'. As shown in figure 6.

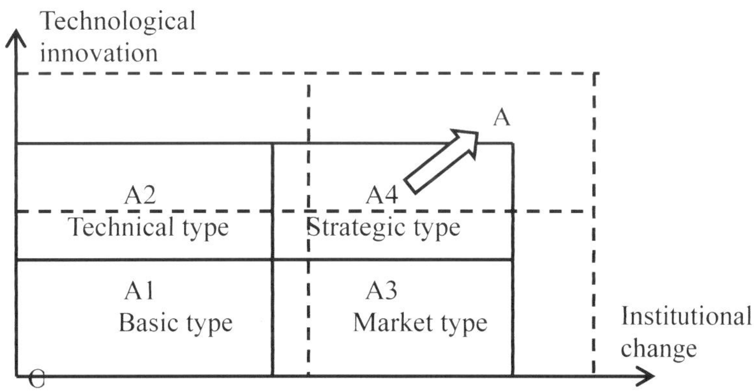

Figure 6. The improving path of Strategic type

6. Conclusion

This paper constructed the innovation system of manufacturing materials, which could be divided into research and development of new materials and improvement of traditional materials. According to the influence degree of factors, the innovation of manufacturing materials was divided into technological innovation dimension and institutional change dimension. These two dimensions constituted four innovative models, such as: Basic type (A1), Technical type (A2), Market type (A3) and Strategic type (A4). These four innovation models could be effectively promoted by upgrading the specific capabilities contained in technological innovation dimension and institutional change dimension.

References

[1] Hart. Beyond Greening Strategies for a Sustainable World[J].Harvard Business Review,1997.
[2] Melngk S A， Smith R T.Green Manufacturing. Dearborn,USA： Society of Manufacturing Engineers， 1996.
[3] C Freeman. The Economics of Industrial Innovation[J].Social Science Electronic Publishing, 1997, 1(2):215-219.
[4] Baoshan Li. Integration-Management Innovation in Modern Time [M].Beijing: China Renmin University Press.1998, P56-69.
[5] Robert A. Burgelman, etc..Strategic Management of Technology and Innovation[M].China Machine Press.1998.8.
[6] Michael Porter.Green and Competitive:ending the stalemate. Harvard Business Review, 1995.

IOP Conf. Series: Materials Science and Engineering **242** (2017) 012023 doi:10.1088/1757-899X/242/1/012023

Solid-state Chemical Reaction Synthesis and Characterization of Lanthanum Tartrate Nanocrystallites Under Ultrasonication Spectra

LI Dao-hua*, HE Shao-fen, CHEN Jie, JIANG Cheng-yan, YANG Cheng

College of Chemistry and Chemical Engineering, Neijiang Normal University, Neijiang 641112, China

E-mail: daohl@163.com

Abstract. Under near room temperature, by using sodium tartrate, lanthanum trichloride, lanthanum acetate, and lanthanum nitrate as reactants, we synthesize the lanthanum tartrate nanocrystals in one step under the condition that the infrared lamp is irradiated and the polyethylene glycol 400 is used as the surfactant, where the synthesis method of nanoparticles by solid phase chemical reaction is used. Deionized water after washing with ethanol washes. Tartaric acid lanthanum crystalline powder is gotten, respectively. Analyze solid products phase using X-ray diffraction (XRD) and electron diffraction; characterize the functional groups structure with infrared spectra, and analyzethe change of functional groups in drugs before and after reaction; observe the size, shape, and size distribution of particlesby transmission electron microscopy(TEM). Testing 3 types of tartaric acid crystal powder of lanthanum are nano-crystallization, solid products are granular and relatively uniform in size, with an average particle diameter of about 40 nm, and the yield rate is approximately 92.3%. Furthermore, during the synthesis, the solid-state reaction conditions including raw materials, matching proportion of reactants, additions of inert substance, addition of trace solvents, surfactants and porphyrization time, etc, all have some influence on the morphology, particle size and size distribution of the final products. During the synthesis of the lanthanum tartrate nanocrystallites, the solid state reaction conditions such as changing reactant, matching proportion of reactant, adding inert substance, joining a little solvent or surface active solvent and grinding at different times may influence morphology, particle size and the size distribution of final products.

1. Introduction

Nanocrystalline material has small size effect, quantum size effect, volume effect, surface effect and macroscopic quantum tunnel effect. It exhibits excellent characteristics in mechanics, catalysis, optics, electrics, magnetics, acoustics, calorifics, superconducting technology, chemical and biological activity, and so on. It also has high practical value in national defense, electronics, chemical industry, nuclear technology, metallurgy, astronautics, light industry, biological and medical industry. Recently it has become a foundation for developing special materials and has been active in the fields of physical, chemical and materials science research [1, 2, 3].

In the continuous extensive research on nanocrystallite, it has become especially important to obtain highly pure, super fine and uniform nanocrystallite with optional physical chemical characteristics, and produce new nanomaterials. The key to solving the demand lies in the research and development of new synthesis techniques. The study on the solid-state chemical reaction under ambient temperature and near ambient temperature has made great progress in recent years. Applying

Content from this work may be used under the terms of the Creative Commons Attribution 3.0 licence. Any further distribution of this work must maintain attribution to the author(s) and the title of the work, journal citation and DOI.
Published under licence by IOP Publishing Ltd

solid-state chemical reaction to produce nanomaterials is a new method which was developed recently. The preparation of nanocrystallite generally requires: clean surface, controllable particle shape, size distribution that prevents particle agglomeration, easy collection, better stability and high productivity. The solid-state chemical reaction method can meet these needs and it is easy to operate with advantages of simple processes such as synthesis, high yield, high selectivity, uniform particle size distribution, controllable size and less pollution. It can prevent or reduce the phenomenon of agglomeration of the liquid phase and the phenomenon of particle agglomeration caused by intermediate step and high-temperature reaction[4,5,6]. The solid-state chemical reaction process consists of four steps: diffusion-reaction-nucleation-growth. When the nucleation rate is greater than the nucleus's growth speed, it is beneficial to the formation of nanocrystallite. But if the nucleus's growing rate is greater than the nucleation rate, a lump crystal forms [7, 8].

Lanthanum tartrate is an important rare earth coordination compound with practical mechanical properties and luminescent properties. The nanometer solid material composed of lanthanum tartrate nanoparticles has specific quantum size effect, volume effect, surface and interfacial effect, and presents a unique mechanics, optics, electrical, acoustics, magnetism, superconductivity, chemical and biological activity that differ from macroscopic objects and from single molecules. Lanthanum tartrate nanocrystals have very important research and use value in defense, electronics, chemical industry, metallurgy, aerospace and medicine. This paper discusses the application or the solid-state chemical reaction to synthesize the nanocrystalline lanthanum tartrate under near ambient temperature and ultrasonication. With characterization and observation of the solid phase, particle size and morphology performed through powder X-ray diffraction (XRD), transmission electron microscopy (TEM) and electron diffraction (ED). It also discussed, the factors which the influence to the final product such as types of reactant, changing the proportions of reactant, addition of inert substances, mixing trace solvent or surfactant and time of porphyrizing [9, 10].

2. Experimental

Composing experiment regents are all the analytical purity baking 10.0 mmol $La(Ac)_3 \cdot 3H_2O$ and 15.0 mmol $Na_2C_4H_4O_6 \cdot 2H_2O$ to grind in the agate pot with intrared light during grinding, the temperature is about 35 ~ 40° C, with the operate of ultrosonic, to wash the lomposing 3 times with distilled water, and then wash 2 times with alcohol, drying, finally the lanthanum tartrate nanocrystlline. The processes was repeated three separate times using $LaCl_3 \cdot 7H_2O$, $La(NO_3)_3 \cdot 6H_2O$, $La_2(SO_4)_3 \cdot 9H_2O$ as the reactant in place of $La(Ac)_3 \cdot 3H_2O$. During each synthesis, the types of factors influencing the final product under different solid-state reaction conditions such as reactant, changing the proportions of reactant, addition of inert substances, mixing of trace solvent or surfactant and time of porphyrizing was also observed. The solid phase was characterized by powder X-ray diffraction (XRD) and electron diffraction (ED). The particle size and its distribution and morphology of the prepared nanocrystallite were observed by transmission electron microscopy (TEM). The TEM and ED patterns were obtained by utilizing a Japanese electronic type of JEM-200 CX transmission electric microscope at an accelerated voltage of 160 kV and an amplification of one hundred thousand times.

3. Results and Discussion

Table 1 shows the XRD patterns of lanthanum tartrate nanocrystallite synthesized according to the conditions of No.16 in table 2. Results of the synthesized lanthanum tartrate appears the XRD pattern of analytically pure lanthanum tartrate and the standard XRD pattern of lanthanum tartrate from JCPDS card, and which has no impurity peaks. However, the diffraction peaks of the synthesized product were significantly widened and conspicuously weaker. According to X-ray polycrystal diffraction theory, the widening of the peak diffraction is because of the superrefining of the particle size. This indicates that the particle size of the synthesized lanthanum tartrate is a single phase, ultrafine particle. According to the width of the peaks of the XRD pattern, and utilizing the formula of scherrer, the crystal's average particle size was calculated to be 40 nm. The particle size of the product is determined by the relative magnitude of the nucleation speed and the crystal nucleus's growing

IOP Conf. Series: Materials Science and Engineering **242** (2017) 012023 doi:10.1088/1757-899X/242/1/012023

speed. Results indicate that this condition enables the attainment of smaller nanocrystallites at ambient temperature, the nucleation speed is quicker than that of the nucleus's growth.

Table 1. X-ray diffraction result of the $La_2(C_4H_4O_6)_3$ nanocrystallite.

Reaction system	the liquid phase synthesis $La_2(C_4H_4O_6)_3$						Reaction system	$La(Ac)_3 \cdot 3H_2O$ + $Na_2C_4H_4O_6 \cdot 2H_2O$ $\rightarrow La_2(C_4H_4O_6)_3$					
2θ/(°)	15.9	16.2	21.7	36.1	45.3	55.1	2θ/(°)	15.6	17.8	20.2	36.1	45.1	53.2
d/Å	2.06	3.11	1.36	3.11	1.91	1.02	d/Å	2.07	3.21	1.30	3.21	2.01	1.03
Reaction system	$LaCl_3 \cdot 7H_2O$ + $Na_2C_4H_4O_6 \cdot 2H_2O$ $\rightarrow$ $La_2(C_4H_4O_6)_3$						Reaction system	$La(NO_3)_3 \cdot 6H_2O + Na_2C_4H_4O_6 \cdot 2H_2O \rightarrow La_2(C_4H_4O_6)_3$					
2θ/(°)	14.7	17.7	20.1	35.8	45.2	54.9	2θ/(°)	15.3	16.9	20.3	36.3	46.2	53.8
d/Å	2.12	3.07	1.31	3.11	2.21	1.05	d/Å	2.08	3.23	1.32	3.32	2.11	1.12
Reaction system	$La_2(SO_4)_3 \cdot 9H_2O$ + $Na_2C_4H_4O_6 \cdot 2H_2O$ $\rightarrow$ $La_2(C_4H_4O_6)_3$						Reaction system	the purity with analyses $La_2(C_4H_4O_6)_3$					
2θ/(°)	15.5	17.5	20.2	35.9	44.8	53.6	2θ/(°)	15.8	18.5	20.6	37.1	46.2	55.1
d/Å	2.09	3.08	1.27	3.22	2.17	1.05	d/Å	2.06	3.23	1.35	3.17	2.12	1.08

Figure 1 shows the ED pattern of this same sample (synthesized according to No.16 in table 2). The result shows that the round line of the diffraction is clear and is the result of multi-crystal diffraction. Calculation of the d value (table 1) of the corresponding diffraction, it shows that the product is the multi-crystal $La_2(C_4H_4O_6)_3$. Figure 2 shows the TEM result of the sample. The shape of the nanocrystalline $La_2(C_4H_4O_6)_3$ is approximately spherical in shape. The majority of the nanocrystallite's size is approximately 40 nm. This is consistent with the X-ray diffraction result.

Figure 1. The ED pattern of the $La_2(C_4H_4O_6)_3$ nanocrystallite

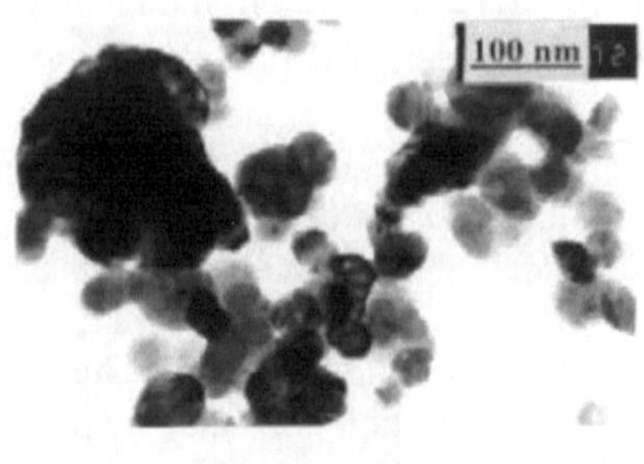

Figure 2. TEM photograph of the $La_2(C_4H_4O_6)_3$ nanocrystallite

Results indicate that separately using $LaCl_3 \cdot 7H_2O$, $La(NO_3)_3 \cdot 6H_2O$ and $La_2(SO_4)_3 \cdot 9H_2O$ in place of $La(Ac)_3 \cdot 3H_2O$ as the reactant as synthesized according to the conditions of No.1 to 4 in table 2, yield similar XRD patterns, TEM results and the ED patterns as when using the latter. All these experiments resulted in the synthesis of single nanocrystalline $La_2(C_4H_4O_6)_3$. The X-ray diffraction analysis results are indicated in table 1. But it can be seen from the TEM graph that the particle for nanocrystalline $La_2(C_4H_4O_6)_3$ of different reactants are eavenly dispersed and the order of granular sizes is: $La(Ac)_3 \cdot 3H_2O < La_2(SO_4)_3 \cdot 9H_2O < LaCl_3 \cdot 7H_2O < La(NO_3)_3 \cdot 6H_2O$ (table 2), among which the particle size of nanocrystallite $La(Ac)_3 \cdot 3H_2O$ is the smallest. During experimentation, observing from changing rate of the mixture's color after reactants are mixed and porphyrized indicates the speeds of reaction are different. Using $La(Ac)_3 \cdot 3H_2O$ as the reactant, yielded the fastest rate of reaction and the faster the reaction speed is, the faster the speed of nucleation growth.

IOP Conf. Series: Materials Science and Engineering **242** (2017) 012023 doi:10.1088/1757-899X/242/1/012023

Table 2. Morphology and particle size of nanocrystalline $La_2(C_4H_4O_6)_3$ under different conditions by solid-state reaction.

No.	Mole ratio	Inert substance and ratio of mixture (between La)	Solvent (1 mL)	Reaction time/min	Iorphology	Particle size/nm	Yield %
1	$LaCl_3$•$7H_2O$,$Na_2C_4H_4O_6$(1:1.5)			30	Approx ball shape	80	89.1
2	$La(NO_3)_3$•$6H_2O$,$Na_2C_4H_4O_6$(1:1.5)			30	Approx ball shape	90	90.2
3	$La_2(SO_4)_3$•$9H_2O$,$Na_2C_4H_4O_6$(1:1.5)			30	Rhombus shape	70	90.2
4	$La(Ac)_3$•$3H_2O$, $Na_2C_4H_4O_6$ (1:1.5)			30	Rhombus shape	60	92.6
5	$La(Ac)_3$•$3H_2O$, $Na_2C_4H_4O_6$ (1:2)			30	Rhombus shape	60	93.8
6	$La(Ac)_3.3H_2O$, $Na_2C_4H_4O_6$ (1:3)			30	Rhombus shape	80	91.2
7	$La(Ac)_3$•$3H_2O$, $Na_2C_4H_4O_6$ (1:3)	NaCl (8:1)		30	Rhombus shape	70	92.2
8	$La(Ac)_3$•$3H_2O$, $Na_2C_4H_4O_6$ (1:2)	NaAc(1:1)		30	Rhombus shape	80	91.3
9	$La(Ac)_3$•$3H_2O$, $Na_2C_4H_4O_6$ (1:2)	NaAc(2:1)		30	Rhombus shape	70	90.2
10	$La(Ac)_3$•$3H_2O$, $Na_2C_4H_4O_6$ (1:2)	NaAc(4:1)		30	Rhombus shape	80	91.0
11	$La(Ac)_3$•$3H_2O$, $Na_2C_4H_4O_6$ (1:2)	NaAc(8:1)		30	Rhombus shape	80	91.6
12	$La(Ac)_3$•$3H_2O$, $Na_2C_4H_4O_6$ (1:2)	NaAc(16:1)		30	Rhombus shape	70	91.8
13	$La(Ac)_3$•$3H_2O$, $Na_2C_4H_4O_6$ (1:2)	NaAc(32:1)		30	Rhombus shape	80	92.9
14	$La(Ac)_3$•$3H_2O$, $Na_2C_4H_4O_6$ (1:2)	NaAc(8:1)	Alcohol	30	Approx ball shape	70	93.2
15	$La(Ac)_3$•$3H_2O$, $Na_2C_4H_4O_6$ (1:2)	NaAc(8:1)	Acetonitrile	20	Approx ball shape	40	89.2
16	$La(Ac)_3$•$3H_2O$, $Na_2C_4H_4O_6$ (1:2)	NaAc(8:1)	Acetonitrile	30	Approx ball shape	40	92.3
17	$La(Ac)_3$•$3H_2O$, $Na_2C_4H_4O_6$ (1:2)	NaAc(8:1)	Acetonitrile	60	Rhombus shape	50	92.2
18	$La(Ac)_3$•$3H_2O$, $Na_2C_4H_4O_6$ (1:2)	NaAc(8:1)	Phenthiol	30	Rhombus shape	50	92.7
19	$La(Ac)_3$•$3H_2O$, $Na_2C_4H_4O_6$ (1:2)	NaAc(8:1)	OP	30	Approx ball shape	60	91.2
20	$La(Ac)_3$•$3H_2O$, $Na_2C_4H_4O_6$ (1:2)	NaAc(8:1)	H_2O	30	Approx ball shape	90	90.1

OP: poly ethylene glycol dioctyl ether

The XRD patterns of nanocrystalline $La_2(C_4H_4O_6)_3$ synthesized according to different matching proportions of reactants (synthesized according to No.4 to 6 in table 2) are similar. It is shown in the TEM pattern that by increasing the anion proportion of the reactant, the granularity of particles is reduced and its dispersity is improved. This is because with increasing density of the reactant, the nucleation speed is accelerated therefore decreasing particle size. Results indicate that by adding inert substances like NaCl and NaAc can produce smaller particle sizes, X-ray powder diffraction (XRD) of the products with the latter yielding results slightly better. X-ray powder diffraction (XRD) results of different matching proportions (synthesized according to No.7 to 13 in table 2) shows that the XRD pattern of $La_2(C_4H_4O_6)_3$ is similar to the standard XRD pattern of $La_2(C_4H_4O_6)_3$, and that the product is a nanocrystalite. It can be seen from the TEM photograph that the particle of $La_2(C_4H_4O_6)_3$ is in the shape of a rhombus. Moreover, the larger the matching proportion of NaAc is, the smaller the produced particle size of $La_2(C_4H_4O_6)_3$ becomes. When the matching proportion reaches 16:1, the size of the particle diameter will be about 70 nm. In solid-state reaction those substances that do not take part in the reaction all belong to inert substances, and the appropriate amount of inert substances, the ones which can be added in order to change the speed of nucleation and the nucleus's growth. On one hand, the added insert substances can make the reactant be more dispersive and mix more evenly, and make the growing speed of the product particle slow down and the produced particle becomes less. On the other hand, by adding insert, the reaction speed of the system and the speed of nucleation would slow down. Therefore, the matching proportion should be appropriately controlled so as to achieve the best result.

The XRD pattern of $La_2(C_4H_4O_6)_3$ synthesized by adding various kinds of trace solvents or surfactants (synthesized according to No.14 to 19 in table 2) all yielded similar results as well. All the products from these reactions were single phase nanocrystalline $La_2(C_4H_4O_6)_3$. It can be seen from the TEM graph that by adding trace solvent or surfactant, the product particle size becomes slightly smaller and the particle diameters are more evenly dispersed. Solvents or surfactants can promote the contact and dispersion of the reactants in solid-state reaction system, and quicken the reaction speed.

Among the solvents and surfactants tested the best result yielded acetonitrile.

Different time spent on porphyrizing the reactant,the XRD pattern of $La_2(C_4H_4O_6)_3$ are similar.However, the size differences of the $La_2(C_4H_4O_6)_3$ nanocrystallite particles are minimal between a porphyrization time of 20 min and 30 min. However, after porphyrizing the reactants for 60 min, the nanocrystallite particle sizes becomes larger (synthesized according to No.15, 16, 17 in table 2). After 30 min of porphyrizing, the chemical reactions is basically complete and if porphyrization continues, the growth of the nanocrystalline nucleus becomes the main reaction process. Therefore, resulting in a larger diameter of the nanocrystalline particle. Table 2 shows the TEM graph of the nanocrystalline $La_2(C_4H_4O_6)_3$ (synthesized according to No.17 in table 2). The graph indicates that the produced nanocrystalline $La_2(C_4H_4O_6)_3$ is rhombus spherical, most of which are about 50 nm in size.

4. Conclusions

In conclusion, under near ambient temperature and ultrasonication, nanocrystalline $La_2(C_4H_4O_6)_3$ can be synthesized by solid-state reaction. The resulting particle sizes are relatively uniform, the morphology of the crystal is spherical, and the average particle diameter is about 40 nm.

Furthermore, during the synthesis of nanocrystalline $La_2(C_4H_4O_6)_3$, varying the solid-state reaction conditions such as the types of reactants, matching proportion of reactants, additions of inert substance, addition of trace solvents or surfactants and porphyrization time may influence the morphology, particle size and size distribution of the final products. The best condition for solid-state synthesis nanocrystalline $La_2(C_4H_4O_6)_3$ involves the utilization of $La(Ac)_3 \cdot 3H_2O$ and $Na_2C_4H_4O_6 \cdot 2H_2O$ (1:1.5) as reactants, addition of inert substance at a ratio of 8 to 1 acetonitrile (1 mL). The synthesis is achieved during a porphyrization period of 30 min under heatred conditions. Finally, the yield rate of nanocrystalline $La_2(C_4H_4O_6)_3$ is approximately 92.3%.

Acknowledgements

This work was financially Supported by Students Innovation and Entrepreneurship Training Program in Neijiang Normal University (X201635,X201633), and "Basic Chemistry Teaching Team"-Undergraduate Teaching Project by Neijiang Normal University (〔2015〕96), and Patent Project of Science & Technology Department in Neijiang Normal University (P016062).

References

[1] Zhuang J, Chi Y H and Wang D 2007 *Preparation of Different Topography $NiCo_2O_4$ with*

Surfactant by Solid State Reaction, Journal of Inorganic Materials, vol 22, chapter 1 p40.
[2] Wang Y 2004 *Nano and Submicron-structured Sulfide Self-lubricating Coatings*, Tribology Letters, vol17, chapter 2 p165.
[3] Tong W P, Tao N R and Wang Z B, et al 2003 *Nitriding Iron at Lower Temperatures*, Science, vol 299, p686.
[4] Si W Y, Pei X M , and Jia H S, et al2005 *Preparation of Nanosized α-Al_2O_3*. Diamond and Abrasives Engineering, vol 2, chapter 146 p33.
[5] Qiang J B, Wang Y M and Wang D H, et al 2004 *Quasicrystals in the Ti-Zr-Ni Alloy Systerm*. J. Non-Cry. Sol., vol 334/335, p223.
[6] Zhao X, Huang L and Li H R, et al 2016 *Promotional effects of zirconium doped $CeVO_4$ or the low-temperature selective catalytic reduction of NO_x with NH_3.* Applied Catalysis B:Environmental, vol 183, p269.
[7] Nikolay K, Jorge G and Freek K, et al 2016 *Recent developments in zeolite membranes for gas separation*. Journal of Membrane Science vol 499, p65.
[8] Bai C, Liu M. 2013 *From chemistry to nanoscience: not just a matter of size*. Angewandte Chemie International Edition vol 52, chapter 10, p2678.
[9] Ma J Q and Shen Y 2014 *Al-doping chitosan-Fe (III) hydrogel for removal of fluoride from aqueous solutions* Chemical Engineering Journal, vol 248, p98.
[10] Hewage D, Roudjane M and Yang D S, et al 2015 *Lanthanum–mediated C–H Bond Activation of Propyne and Identification of La(C_3H_2) Isomers* Journal of Physical Chemistry A, vol 119, chapter 12, p2857.

Effect of Continuous DC Longitudinal Magnetic Field Heat Treatment on Soft Magnetic Properties of $Fe_{78}Si_9B_{13}$ Amorphous Cores

Xu Yuhua, Zhu Zhenghou, Yin Lei and Liu Jilei

School of Materials Science & Engineering, Nanchang University, Jiangxi, China

E-mail: z00708@sina.com

Abstract. The effect of continuous DC longitudinal magnetic field heat treatment on soft magnetic properties of $Fe_{78}Si_9B_{13}$ amorphous cores was studied. The results showed that comparing with the amorphous cores without magnetic field heat treatment, the permeability increases and the coercivity decreases after the longitudinal magnetic field heat treatment. At the same time, the rectangular ratio Br/Bm of the amorphous cores increases from 0.58 to 0.94, and the rectangular hysteresis loop is obtained, the loss value of Ps the amorphous cores is significantly reduced at low frequency, while it is almost the same at the high frequency. At f =50Hz and Bm=1.3T, the loss value of Ps is 0.278W/kg in the heat treatment without magnetic field, and the loss value of Ps was 0.153W/kg in the longitudinal magnetic field heat treatment, which was 44.9% lower than that of heat treatment without magnetic field. The effect of structural relaxation on magnetic properties of the magnetic cores in longitudinal magnetic heat treatment is not significant, and the magnetic properties exhibit good stability.

1. Introduction

FeSiB amorphous alloy is widely used in the preparation of distribution transformer cores owing to excellent soft magnetic properties [1-3]. It can greatly reduce the no-load loss and no-load current of the transformers [4-5]. With the advancement of energy saving and consumption reduction in power grid work, amorphous alloy distribution transformers will be widely promoted and applied.

Amorphous alloy transformer cores are firstly annealed under the protection of nitrogen, and then in longitudinal magnetic field with the flowing nitrogen atmosphere, which is called "two-step heat treatment" [6-7]. The so-called longitudinal magnetic field heat treatment refers to the large magnetic field which is applied in the same direction as the actual magnetic flux of the magnetic cores during the heat treatment. It can be used to increase the rectangular ratio of the magnetic cores. However, the two-step heat treatment not only consumes large energy and time, but also damages the magnetic cores due to tranfer of large mass core.

In this paper, the two-step heat treatment method is improved with the combination of two heat treatment steps. At the same time, the magnetic field is changed into DC longitudinal magnetic field, which is applied to the whole process of magnetic core heat treatment. This paper mainly studies the soft magnetic properties of the amorphous cores under the DC longitudinal magnetic field heat treatment, and it also evaluates the application value of the magnetic core.

2. Experimental

$Fe_{78}Si_9B_{13}$ amorphous alloy strip with bright and clean surface was prepared by single roll rapid quenching method, the width is 50±0.2 mm and the thickness is 27±0.2μm. The amorphous alloy strip

Content from this work may be used under the terms of the Creative Commons Attribution 3.0 licence. Any further distribution of this work must maintain attribution to the author(s) and the title of the work, journal citation and DOI.
Published under licence by IOP Publishing Ltd

is made into a ring amorphous core with an outer diameter of 100mm and an inner diameter of 60mm by using a belt winding machine. $Fe_{78}Si_9B_{13}$ amorphous cores were subsequently heat treated at 380°C for 90min in nitrogen gas atmosphere. Heat treatment is in two ways: one is the heat-treatment without magnetic field in nitrogen gas atmosphere(short for "heat treatment without magnetic field"); the other is the longitudinal magnetic field heat treatment that the magnetic field is 800 A/m in nitrogen gas atmosphere (short for " longitudinal magnetic field heat treatment ").

The phase structure of $Fe_{78}Si_9B_{13}$ amorphous alloy strip was analyzed by X-ray diffraction (XRD, Bruker D8 Advance) with Cu-Kα radiation at 40Kv and 40mA. The microstructure of amorphous alloy strip was observed by 3D-microscopy with large depth of field (VHX-100). The soft magnetic properties of the $Fe_{78}Si_9B_{13}$ amorphous cores were measured by MATS-2010SD test device.

Table.1 Chemical composition of amorphous alloy strip

Chemical element	Fe	Si	B	C	S
Content(mass fraction) /%	91.14	4.98	2.92	0.085	0.011

3. Results and discussion

3. 1 Structural analysis of $Fe_{78}Si_9B_{13}$ amorphous alloy strip

Fig.1 shows the XRD pattern of the $Fe_{78}Si_9B_{13}$ amorphous alloy strip, From the XRD results of the $Fe_{78}Si_9B_{13}$ amorphous alloy strip, the diffraction peak is a diffuse peak and there is no sharp diffraction peak corresponding to the crystal, which showed that the $Fe_{78}Si_9B_{13}$ alloy strip is amorphous structure.

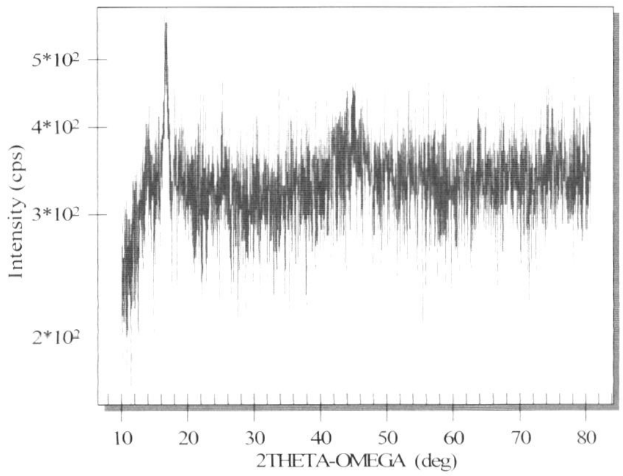

Fig.1 XRD pattern of the $Fe_{78}Si_9B_{13}$ amorphous alloy strip

The microstructure of amorphous alloy strip is shown in figure 2. We can see that a small amount of pores, impurities and pits on the surface of the amorphous alloy strip.

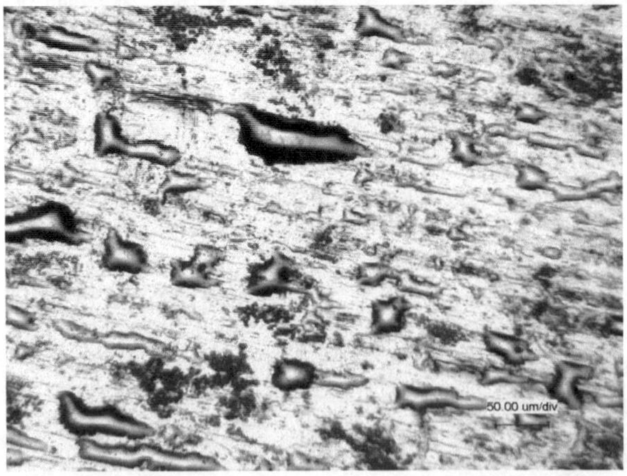

Fig.2 Microstructure of the $Fe_{78}Si_9B_{13}$ alloy strip

3. 2 The effect of longitudinal magnetic field heat treatment on the soft magnetic properties of the amorphous cores

Figure 3 shows the dynamic hysteresis loop of the amorphous cores in the different states at *f*=50Hz.

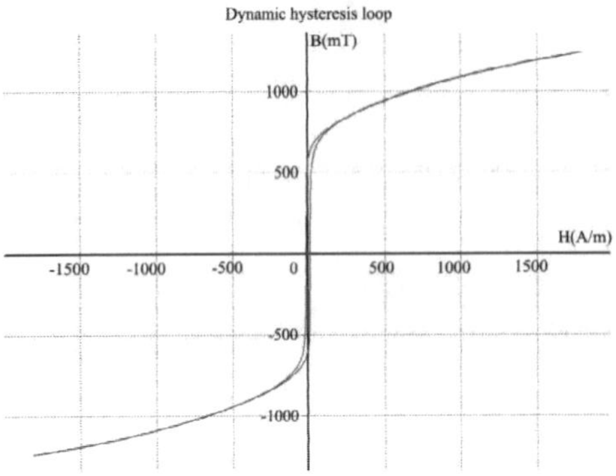

(a) casted core

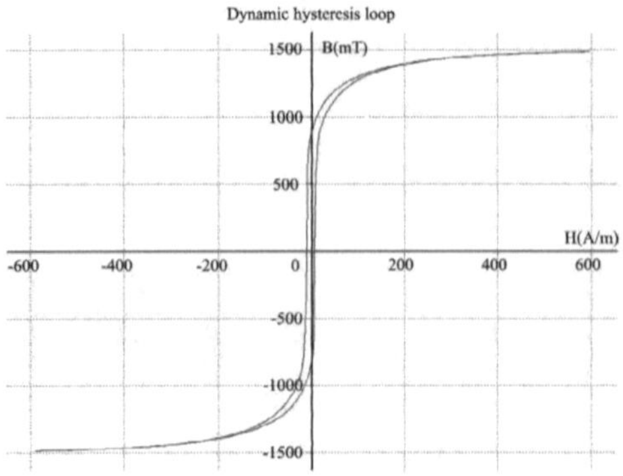

(b) heat treatment without magnetic field

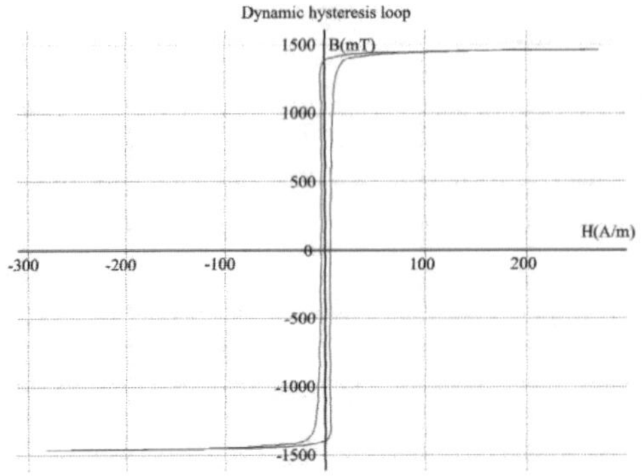

(c) longitudinal magnetic field heat treatment

Fig.3 The dynamic hysteresis loop of the amorphous cores in different state (*f*=50Hz)

It can be seen from Fig.3 (a), (b) and (c) that the amorphous cores are more easily saturated after heat treatment. When *Hm*=800 A/m, the magnetic induction Bm of the amorphous cores that in casted core, heat treatment without magnetic field and longitudinal magnetic field heat treatment is 1.04, 1.48 and 1.47, respectively. This is mainly due to the stress caused by the rapid cooling and winding of the

IOP Conf. Series: Materials Science and Engineering **242** (2017) 012024 doi:10.1088/1757-899X/242/1/012024

amorphous cores is removed after heat treatment, which makes it easier to magnetize and obtain good magnetic properties. The magnetic parameters of the amorphous cores in different states are given by Table 2.

Table.2 Magnetic parameters of the amorphous cores under different condition(*f*=50Hz)

	B_{800}/T	Hc/(A/m)	μ_m	Br/T
casted core	1.04	11.52	34800	0.59
heat treatment without magnetic field	1.48	8.12	73300	0.86
longitudinal magnetic field heat treatment	1.47	5.23	212700	1.39

(B_{800} represents the magnetic induction B value which was measured at H = 800 A/m)

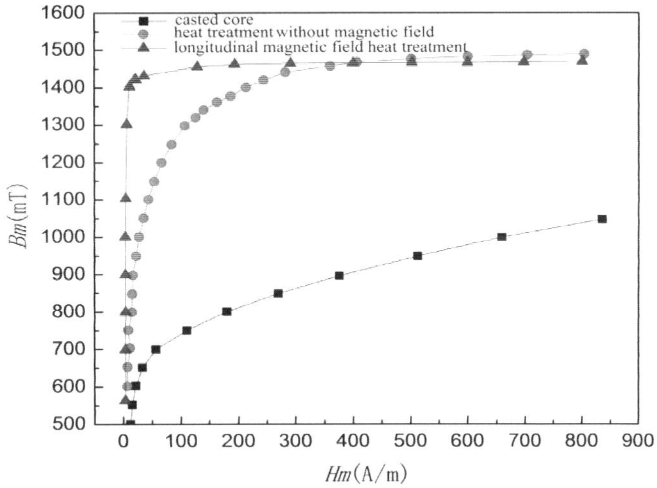

Fig.4 The magnetization curve of the amorphous cores in different states (*f*=50Hz)

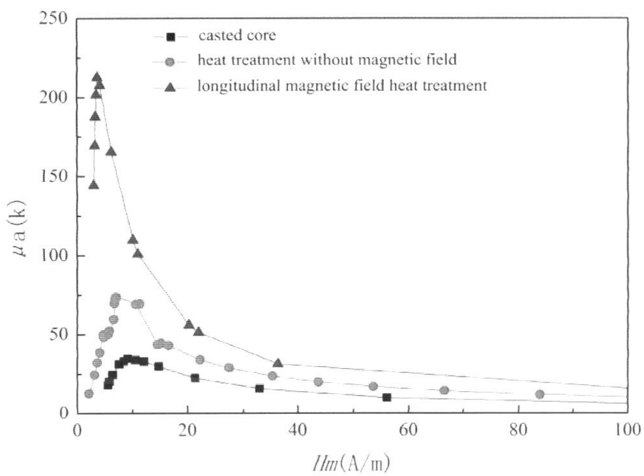

Fig.5 The permeability curve of the amorphous cores in different states (*f*=50 Hz)

The magnetization curve and permeability curve of the amorphous cores in different states are shown in Fig. 4 and Fig. 5 respectively. It can be seen from Fig.4, the magnetic induction of the amorphous cores under the casted state is difficult to saturate and the magnetic properties are poor;

The saturation magnetic induction Bs of the amorphous cores in heat treatment without magnetic field and longitudinal magnetic field heat treatment are 1.48 T and 1.47 T, respectively, and the Bs are almost the same, This is because the composition of the magnetic material plays a dominant role in the saturation magnetic induction unless the structure has a greater change. But the saturation magnetic induction of the amorphous cores after longitudinal magnetic field heat treatment is close to saturation, when the magnetic field is less than 50A/m. And when the magnetic field is 200A/m, the saturation magnetic induction of the amorphous cores after heat treatment without magnetic field is close to saturation. It can be seen that the magnetizing current can be reduced when the magnetic core works, this can reduce the no-load current for amorphous distribution transformers. The reduction of excitation current is mainly due to the uniaxial anisotropy of the system after longitudinal magnetic field heat treatment increases permeability.

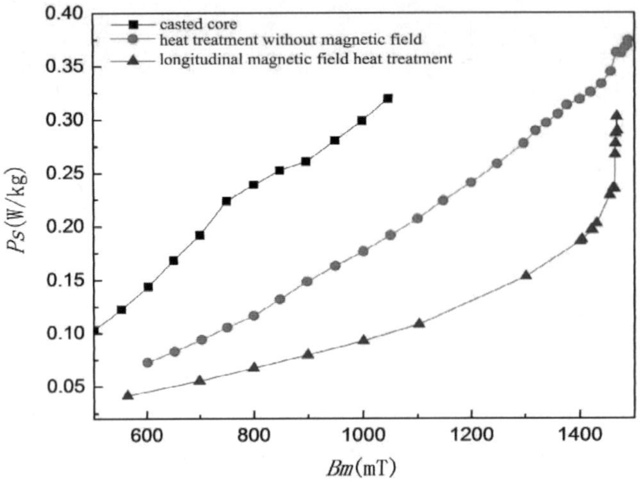

Fig.6 The loss curve of the amorphous cores in different state (f=50Hz)

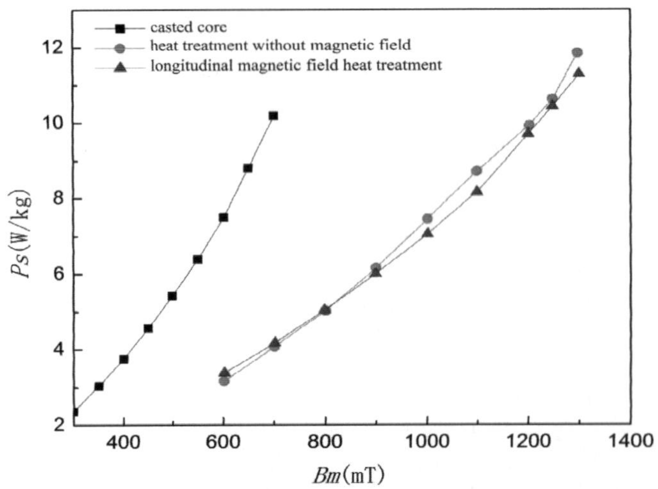

Fig.7 The loss curve of the amorphous cores in different state (f=1kHz)

The loss curves of the amorphous cores at f=50Hz and f=1KHz are shown the Fig.6, Fig.7 respectively. Comparison of the loss curves of the amorphous cores in heat treatment and casted state. When f and Bm are fixed, the loss of the amorphous cores after heat treatment is significantly lower than that of the casted state. At f =50Hz and Bm=1.3T, the loss value of Ps is 0.278W/kg in the heat treatment without magnetic field, and the loss value of Ps was 0.153W/kg in longitudinal magnetic field heat treatment, which was 44.9% lower than that of heat treatment without magnetic field. At f =1KHz and B_m=1.3T, the loss value of Ps was 11.84W/kg in heat treatment without magnetic field , and the loss value of Ps is 11.28W/kg in longitudinal magnetic field heat treatment, which is 4.7% lower than that of heat treatment without magnetic field. Compared with heat treatment without magnetic field, the loss value Ps of the amorphous cores in longitudinal magnetic field heat treatment is significantly reduced at low frequency, while it is almost the same at the high frequency. The reason is that the magnetic core loss is composed of hysteresis loss, eddy current loss and residual loss. In low frequency, the loss of magnetic core is mainly hysteresis loss. In high frequency, eddy current loss and residual loss are the main parts.

3.3 The effect of structural relaxation on magnetic properties of the amorphous cores in longitudinal magnetic field heat treatment

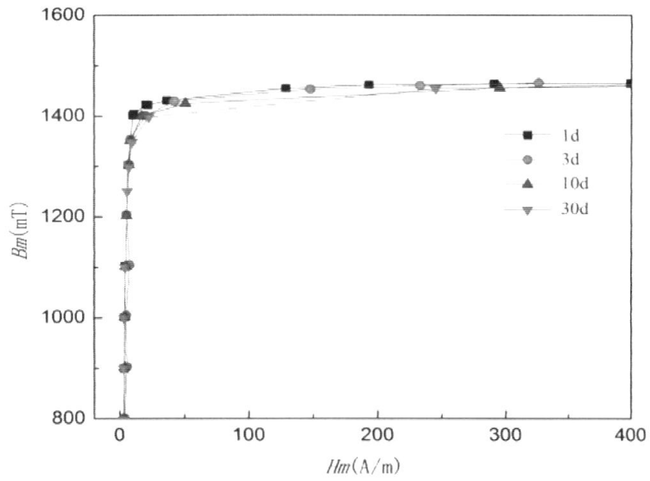

Fig.8 Stability of the magnetization curve for the amorphous coresin longitudinal magnetic field heat treatment (f=50Hz)

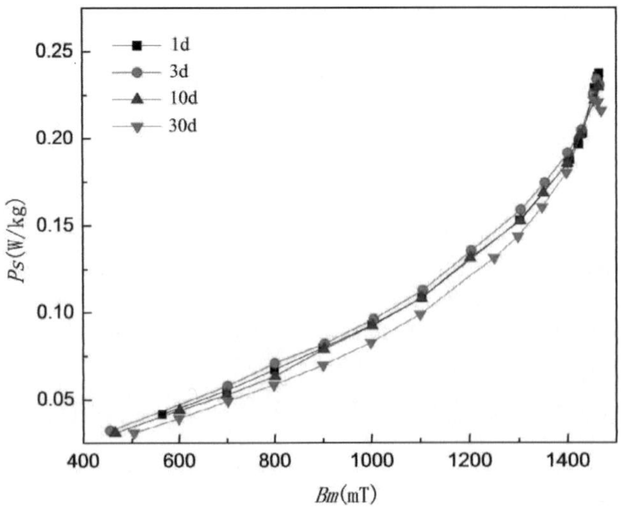

Fig.9 Stability of the loss curve for the amorphous coresin longitudinal magnetic field heat treatment (f=50Hz)

Figure 8 and figure 9 show the stability of the magnetization curves and the loss curves for the amorphous cores respectively. Structural relaxation phenomenon of iron-base amorphous occurs due to it is a metastable state. It can be seen from Fig. 8 and Fig. 9 that the magnetization curve and the loss curve of the amorphous cores after the longitudinal magnetic field heat treatment are measured at different times are basically identical. This shows that the effect of structural relaxation on magnetic properties of the amorphous cores after longitudinal magnetic field heat treatment is not significant, and the magnetic properties exhibit good stability.

4. Conclusions

(1) The amorphous cores after longitudinal magnetic heat treatment can be obtained rectangular hysteresis loop. The rectangular ratio Br/Bm increases from 0.58 in heat treatment without magnetic field to 0.94.

(2) Comparing with the amorphous cores without magnetic field heat treatment, the loss value of Ps the amorphous cores is significantly reduced at low frequency, while it is almost the same at the high frequency. At f =50Hz and Bm=1.3T, the loss value of Ps is 0.278W/kg in heat treatment without magnetic field and 0.153W/kg in the longitudinal magnetic field heat treatment, which was 44.9% lower than that of heat treatment without magnetic field.

(3) The effect of structural relaxation on magnetic properties of the amorphous cores in the longitudinal magnetic field heat treatment was not significant, and the magnetic properties exhibit good stability.

Acknowledgments

This study was supported by National Natural Science Foundation of China (No.: 61361008) and Graduate Innovation Special Fund of Jiangxi Province (No. YC2016-B007).

References

[1] Xia Xiaoge, Zhu Zhenghou, Ma Guangbin. Infiunce of anneal processes on piezomagnetic properties of FeCuNbSiB amorphous alloy's strip[J]. Journal of Functional Materials,2006, 37(12):1881-1883. (in chinese)

[2] Ma Guangbin, Zhu Zhenghou, Xia Xiaoge. Influence of annealing conditions on stress impedance effect of FeCuNbSiB amorphous alloy strip[J]. Hot Working Technology, 2006, 35(20):1-3.(in Chinese)

[3] Yoshizawa Y, Oguma S, Yamauchi K. New Fe-based soft magnetic alloys composed of ultrafine grain structure[J]. Journal of Applied Physics, 1988, 64(10):6044-6046.
[4] Luciano B A, Kiminami C S. An amorphous core transformer: design and experimental performance[J]. Materials Science & Engineering A, 1997, s226–228(96):1079-1082.
[5] Liu Yan, Tang Jinquan, Zhang Jianjun. Some key problems of dry- type transformer with amorphous cores[J]. Transformer, 2007, 44(6):7-10.(in Chinese)
[6] Zhang Wanli, Jiang Hongchuan, Yang Guoning, et al. Influence of annealing temperature on stress impedance effect of FeCoSiB films and FeCoSiB/Cu/FeCoSiB sandwich layers[J]. Journal of Functional Materials, 2005, 4(36):516-518. (in Chinese)
[7] Yao Zhisong,Yao Lei.Structure, Principle and Application of New Distribution Transformer[M]. Beijing:China Machine Press, 2008.(in Chinese)

IOP Conf. Series: Materials Science and Engineering **242** (2017) 012025 doi:10.1088/1757-899X/242/1/012025

Study on the Application Mode and Legal Protection of Green Materials in Medical-Nursing Combined Building

Xian Zhiyong

Sichuan Academy of Social Sciences, Chengdu, Sichuan, China, 610071

E-mail: xianzy2018@hotmail.com

Abstract. In the context of green development, green materials are the future trend of Medical-Nursing Combined building. This paper summarizes the concept and types of green building materials. Then, on the basis of existing research, it constructs the green material system framework of Medical-Nursing Combined building, puts forward the application mode of green building materials, and studies the policy and legal protection of green material application.

1. Introduction

Medical-Nursing Combined mode is a new mode of providing for the aged, and is an important measure to solve the problem of old-age support caused by the aging of the population. At present, the standard of Medical-Nursing Combined building is not perfect, and the accidents of safety and quality are frequent. In the construction industry "green" trend, to explore the application of green materials in Medical-Nursing Combined building, is of great significance to create a safe and comfortable environment and promote the old-age care standardization.

From the existing literature, in 1986, Britain began to adopt the classification method of residential buildings proposed by HTA. The United States implements the communities that can continue to live and care. Japan vigorously promotes small-scale home care. In order to improve the living quality and the dementia care, Swedish creating a "shared home".

The research of Chinese scholars in this field is mainly embodied in two aspects. First, analyze the construction of pension facilities. Chen Yang (2013) put forward the concept of "Medical-Nursing Combined Building", and discussed its construction strategy and method under the mode of community support for the aged. Luo Deqi(2004) explained the design principles and suggestions for the residential environment and facilities of the elderly in the context of human settlements. Second, based on the needs and characteristics of the elderly, put forward the design strategy suitable for the elderly. Zhou Yanxia (2013) divided the pension facilities into living space, public activity space, public service space and traffic space. On the basis of the activities and life demands of the elderly, she put forward the general plan of each space. Zhou Bo (2009) conducted a field investigation of the nursing home, classified the life behavior and the corresponding spatial factors of the elderly, and explored the relationship between them in the process of architectural design. Zhao Xiaozheng (2010) studied the whole construction process of the pension facilities, and combed the whole process from the pre project planning to the project operation management.

At present, the existing related research has some academic and application foundation, and it has positive significance to solve the problem of the building for the aged. However, most of the current studies mainly focus on improving the efficiency of nursing, and lack the deep care for the elderly. In view of the social background, the economic situation, the population distribution and the cultural

Content from this work may be used under the terms of the Creative Commons Attribution 3.0 licence. Any further distribution of this work must maintain attribution to the author(s) and the title of the work, journal citation and DOI.
Published under licence by IOP Publishing Ltd

idea, the Medical-Nursing Combined mode of any country can not be copied directly. In the future research, we can find a breakthrough from the advanced experience abroad, and try to structure a suitable Medical-Nursing Combined building.

2. The concept and categories of green building materials

In the context of green development, green building materials are the future trend of Medical-Nursing Combined building.

2.1. The concept of green building materials

Green building materials, also known as ecological building materials or environmentally friendly building materials. It refers to the building materials which use clean production technology, less consume or no consume the natural resources and energy, and more conducive to health and environmental protection. The green building material has the characteristics of light regulation, temperature regulation, noise elimination, heat insulation, antistatic, etc., which can minimize pollution, protect the environment and provide a healthy environment for the elderly.

2.2. Categories of green building materials

The sustained and rapid economic development has provided opportunities for the development of the Medical-Nursing Combined building materials industry. The wide application of green building materials is an important means to develop the Medical-Nursing Combined building with the concept of national green development.

Saving type of green building materials. Traditional building materials are purely natural, and the manufacture of building materials requires the consumption of mineral resources as a price. Green building materials through the substitution of raw materials to achieve resource conservation, to avoid damage to the natural environment.

Environmental type of green building materials. Pollution-free, non-toxic and radioactive green materials, such as diatom mud, meet the goal of green living environment for the elderly, and have become the first choice for more and more people to buy building materials.

Space type of green building materials. Temperature control building materials, such as sun protection materials, heat insulation materials and heat radiation materials, optical building materials, such as light absorbing materials and reflective materials, etc., can be applied to the design of the Medical-Nursing Combined building.

3. Green material system framework of Medical-Nursing Combined building

Based on the product life cycle, this paper presents a structure of four layers of green material system framework of Medical-Nursing Combined building, including the product life cycle process layer, green technology layer, green evaluation and monitoring layer, green support layer.

The product life cycle process layer mainly includes materials, green design, manufacturing, green packaging, use, green recycling, green remanufacturing.

Green technology layer mainly includes: green design technology, green manufacturing technology, green packaging technology, green recycling technology, green remanufacturing technology.

The green evaluation and monitoring layer mainly includes: evaluation, data collection and process monitoring.

The green support layer mainly includes database and knowledge base, planning and standard, information support.

As shown in figure 1.

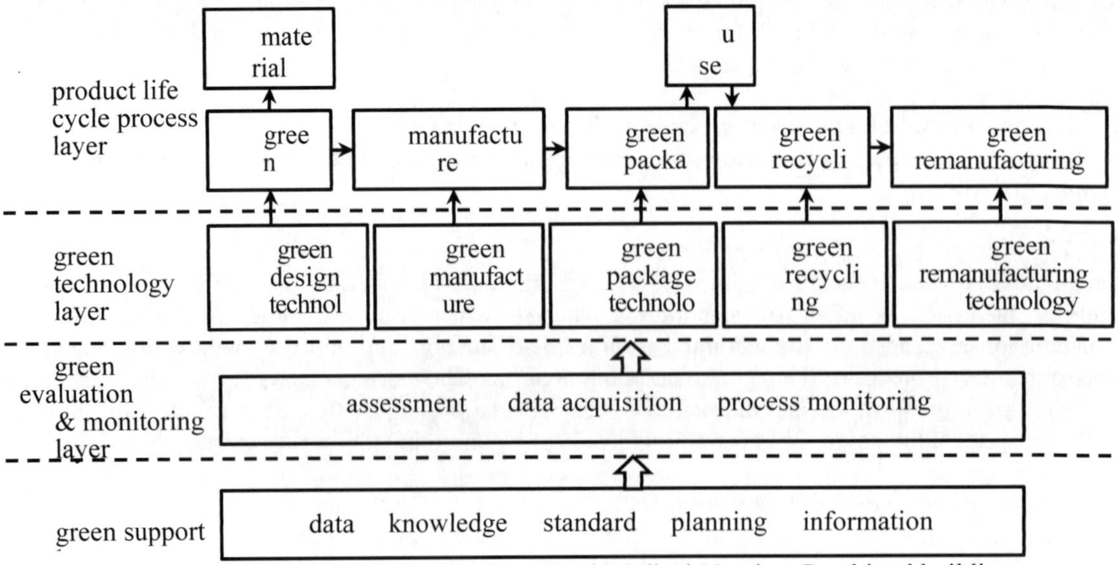

Figure 1. Green material system framework of Medical-Nursing Combined building

4. Application mode of green materials in Medical-Nursing Combined building

On the basis of the previous parts, this paper proposes a five level application mode of green materials in Medical-Nursing Combined building. As shown in figure 2.

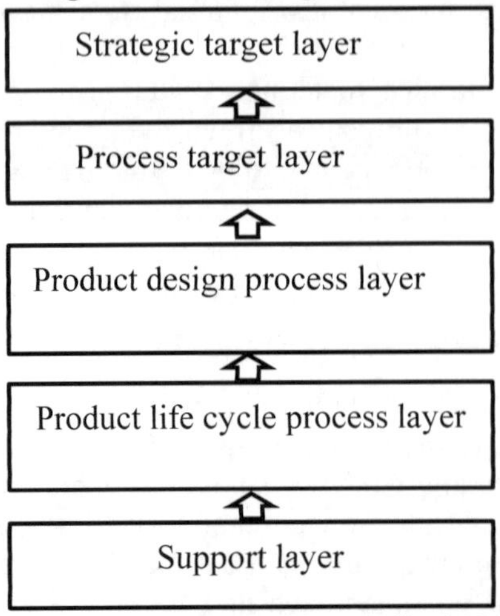

Figure 2. Application mode of green materials in Medical-Nursing Combined building

The first layer is the strategic target layer, that is to say, green building materials should pursue economic benefits and maximize the efficiency of sustainable development in the application.

The second layer is the process target layer, including six goals, such as time, quality, cost, service, resource consumption, and environmental impact. The main goals are: short production cycle, high product quality, low product cost, good pre-sales and after-sales service, less consumption of resources and small the environmental impact.

The third layer is the product design process layer, which mainly includes product design, material selection, manufacturing environment design, process design, packaging design, product recycling design and remanufacturing design.

The fourth layer is the product life cycle process layer, mainly including raw material acquisition, manufacturing process, product assembly, product packaging, product use and maintenance, product recycling, processing and remanufacturing.

The fifth layer is the support system layer, which mainly includes: green design support system, clean production support system, management information system and environmental impact assessment system.

5. Policy norm and legal guidance

Policies and laws provide a guarantee for the application of green materials in Medical-Nursing Combined building.

Firstly, the government construction supervision departments should perfect policies on green building materials standards for specific quantification and standardization to medical-nursing combined building. The NPC will eliminate the fuzziness of definition and standard of green building materials through legal formulation and guide the construction enterprises and the public to consciously use green building materials.

Secondly, the government should also set up a reward system through policy and law, in order to improve the enthusiasm on green building materials development of production enterprises and green building materials use of construction enterprises.

6. Conclusion

Green building materials are effective methods and approaches to solve the problems of safety and quality of Medical-Nursing Combined building. This paper constructed the green material system framework of Medical-Nursing Combined building, put forward the application mode of green building materials, and studied the policy and legal protection of green material application.

References

[1] Zhang Huanan, Fang Jin, Bi Hongxia. Discussion on the New Pension Model of Community Medical-Nursing Combined for the Empty Nest Elderly [J]. Economic Forum, 2015, 08:82-85.

[2] Gong Xun. Study on the Management Mode of Medical-Nursing Combined [J]. Medical and Social Sciences, 2015, 09:1.

[3] Cheng Liang. Medical-Nursing Combined: Explore to the New Development Path of Pension Institutions [J]. Journal of Zhongzhou, 2015, 04.

[4] Fang HongWei. The path Choice of Medical-Nursing Combined in the Pension Service Industry [J]. China of Party and Government Cadres Forum, 2015, 09:83-86.

[5] Gong Xun, Wang Zheng. the Service System of Medical-Nursing Combined in City Circle of Hubei province [J]. China Gerontology Journal, 2015, 21:6288-6290.

Two main and a new type rare earth elements in Mg alloys: A review

Linghang Kong

Department of Mechanical Engineering, Harbin University of Science and Technology Rongcheng Campus, China

Abstract. Magnesium (Mg) alloys stand for the lightest structure engineering materials. Moreover, the strengthening of Mg alloys in ductility, toughness and corrosion predominates their wide applications. With adding rare earth elements in Mg, the mechanical properties will be improved remarkably, especially their plasticity and strength. A brief overview of the addition of rare earth elements for Mg alloys is shown. The basic mechanisms of strengthening Mg alloys with rare earth elements are reviewed, including the solid solution strengthening, grain refinement and long period stacking ordered (LPSO) phase. Furthermore, the available rare earth elements are summarized by type, chemical or physical effects and other unique properties. Finally, some challenge problems that the research is facing and future expectations of rare-earth Mg alloys are stated and discussed.

1. Introduction

Magnesium (Mg) alloys are considered to be utilized in automobile, electronics and aerospace industry due to their high specific strength, casting formability, performance of resistance shielding, damping shock absorption and etc. However, in contrast to the traditional steel or aluminum materials, they have not been widely accepted by consumer. Their commercial products are mainly fabricated by die casting. Compared with cast products, the wrought Mg alloys only have a small market which the proportion is less than 5%, because of their low ductility or toughness and poor corrosion which will limit their wide applications in several fields [1]. So, it is extremely significant for researchers to improve the formation and corrosion resistance of Mg alloys.

The properties and applications of Mg alloy and pure Mg are perplexed by these two main problems above. It is necessary for researchers to find some effective methods to solve these problems. Usually, there are two kinds of methods: i) fabrication and processing; ii) alloying, which can achieve these goals. In the past centuries, researchers have investigated many different kinds of methods for fabrication and processing of Mg alloys, such as powder metallurgy [2], friction stir processing [3], equal channel angular pressing (ECAP) [4 and etc. Except of the traditional casting method, almost all the fabrication processing which mentioned above can refine the microstructure grain, and lead the Mg alloy to high ductility, elongation, yield strength in the Hull-Petch equation, even the superplasticity [5]. On the other hand, the common adding alloy elements are Zn, Zr, Al, Li, Ca, Ag, Mn, and rare-earth elements [6] and etc. Recently, a large number of researchers have investigated many different kinds of rare-earth elements and find out effects of these elements which can refine the crystal particles, enhance the creep and corrosion resistance at elevated temperature and improve the mechanical properties due to the presence of second phase in microstructure. Rare-earth elements are located at the third subgroup (IIIB) in the periodic table, and their chemical properties are similar because of the common structures of peripheral electrons, as well as the similar structures of secondary electrons. So,

Content from this work may be used under the terms of the Creative Commons Attribution 3.0 licence. Any further distribution of this work must maintain attribution to the author(s) and the title of the work, journal citation and DOI.
Published under licence by IOP Publishing Ltd

the solubility of rare earth element in Mg matrix is higher than any other normal element, which can lead the formation of more second phases, and then the mechanical properties of Mg alloy can be improved by leaps and bounds. For instance, in Mg-Zn-Y-Zr system alloys with different processing methods, Bae et al. reported a maximum elongation of 780%in a hot rolled Mg-3.0Zn-0.5Y-1.5Zr (wt.%) alloy at 450°C and a strain rate of $5\times10^{-4}s^{-1}$ [7]. Moreover, Tang et al. reported that an ECAP Mg-5.8Zn-1.0Y-0.48Zr alloy exhibited an elongation of 800% at 350°C and $1.7\times10^{-3}s^{-1}$ [8]. Because of the presence of icosahedral quasi-crystal Mg_3Zn_6Y (I phase) and $Mg_3Zn_3Y_2$ (W phase) which can result in refinement of grains, both of them have achieved the superplasticity on rare-earth Mg alloys [7]. Xu et al. produced Mg-8.2Gd-3.8Y-1.0Zn-0.4Zr alloy with ultra-high yield strength of 426 MPa and ultimate strength of 517 MPa by large-strain hot rolling and aging [9]. However, disadvantages of the rare-earth Mg alloys is that the high cost and natural resource scarcity of rare-earth elements, especially for heavy rare-earth elements, make their large-quantity usage (usually ~10 wt.%) unpractical in industry applications. In contrast, the advantage of high specific strength of Mg alloys compared with Al alloys gradually losses due to the increased density of the rare-earth Mg alloys [10]. In this regard, the developments of low-cost, rare-earth Mg alloys with high strength and ductility are strongly desired.

In this review, the strengthening effects of common rare-earth elements on Mg alloy were critically stated. Furthermore, the future and development of rare-earth Mg alloys was proposed.

2. Types and affections of rare-earth elements in magnesium alloys

2.1 Gadolinium

Gd is used to increase the elevated temperature strength of Mg-alloy and improve the microstructure of alloys. It is worth noting that Gd shows higher solubility in Mg compared with other rare earth elements. This can provide simultaneous solid solution hardening and precipitation strengthening, and the later could enhance thermal stability of the microstructure in Mg alloys.

According to the research of Du et al. [11], the α-Mg and Al_2Gd phase will be formed in AZ21-Gd alloy. With the addition of Gd increasing, the size and number of α-Mg phase will be decreased while the size and number of Al_2Gd phase will be increased. When the addition of Gd approaches to 3.5 wt.%, the number of intermediate phase will also be increased, especially around grain boundaries or grain boundaries. Also the elevated tensile strength and yield stress of the alloy increases with increasing addition of Gd. Then, Reba Mahmuch et al. [12] investigated Mg-9Gd-4Y-0.4Zr alloy in the temperature range of 573-773K by shear punch testing which shows that m value, strain rate sensitivity index, increases with increasing temperature from 0.22 at 623K to 0.4 at 723K and then decreases to 0.32 with a further increase in temperature. And the lattice diffusion accommodated grain boundary sliding can be considered as the dominant deformation mechanism of the material in the superplastic region. In addition, M. Sarebanzadeh et al. [13] investigated Mg-3Gd-1Zn alloy processed by ECAP (equal-channel angular pressing) with SPT method, which demonstrated that the microstructure was refined with the average grain size of 1.7 μm and the Mg_5Gd and $Mg_3Gd_2Zn_3$ particles were detected which can act as grain growth inhibitors during superplastic deformation along with particle strengthening. And the microstructure is showed by Fig.1.

However, Yang et al. [14] pointed out that Gd acts as grain refinement strengthening and grain boundary strengthening, which improves tensile properties of AM50 Mg alloy. But, Gd improves the plastic deformation properties and reduces the toughness at the same time. AM50Gd1 Mg alloy is also quasi-cleavage fracture. So, it could draw a conclusion that Gd improves the tensile strength, but reduces the deformation capacity of AM50 Mg alloy.

Apart from improving the strength and ductility, some literatures point that the creep and corrosion resistance [15] of Mg alloy also can be increased remarkably with adding Gd.

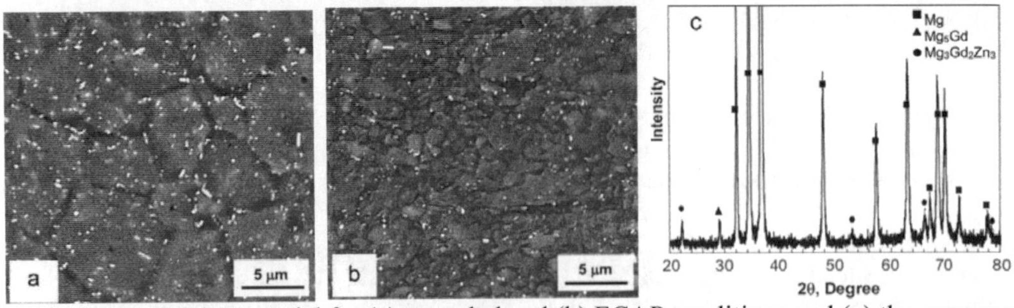

Fig. 1. SEM images of the material for (a) extruded and (b) ECAP conditions and (c) the corresponding XRD pattern indicated in the image [13]

2.2 Yttrium

According to the result of simulation, the lattice structure of Y is similar to that of Mg, which meets the similar dissolve mutually theory. So, the effect of Y is to improve elevated temperature plasticity and the creep resistance of Mg alloys. For ZK60 alloy, the addition of rare-earth element Yttrium (Y) into it can remarkably improve the mechanism properties at elevated temperature and ductility, because of the formation of ternary Mg-Zn-Y phase with high thermal stability [16]. Li et al. [17] investigated Mg-4Zn-1Y alloy for discovering the tensile properties and evolution of microstructure. Although it is not similar to that of Xie et al. [16], the tensile strength and elongation of as-cast alloy were 125.22MPa and 8.2% respectively and those of extruded alloy were 254.94 MPa and 17.9%, which means the tensile properties including strength and elasticity and plasticity were all improved greatly, which microstructure is showed in Fig.2 [17]. With addition of Y element, especially in the Mg-Zn-RE alloy as ZK60, it is significant to form two kinds of phase, *e.g.*, I-phase (Mg_3Zn_6Y) and W-phase (Mg_3Zn_3Y), and the eutectic temperature are 450°C and 510°C, respectively [18]. Rosalie et al. [19] suggested that Mg-Zn-Y alloy can be quasicrystal-strengthened by I-Phase which is icosahedral quasicrystal structure with high hardness, high strength and low surface energy. I-phase can refine the microstructure and increase elongation by dynamic recrystallization and the dissevered effect between matrix and I-Phases decreases as I-Phases are spheroidized through hot extrusion.

Recently, mixing Y and Gd into Mg alloy has been the mainstream measure of improving the mechanism properties, such as Mg–10Gd–3Y–0.5Zr alloy [20] with maximum strain value of 18% while that of traditional Mg alloy is ~12.3%. It can be explained that the reciprocal action of Y and Gd affect significantly the deformation behavior of the extruded Mg alloys by predominantly suppressing the twinning due to the finer grain size, the relatively more random crystallographic texture, and the presence of RE-containing precipitates.

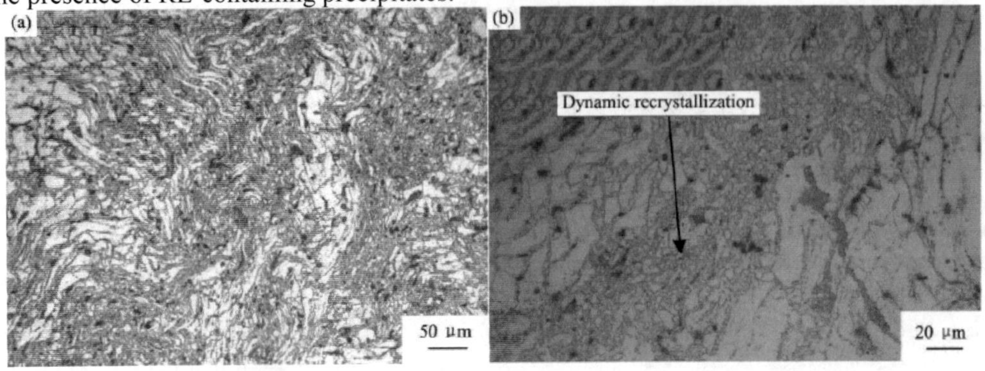

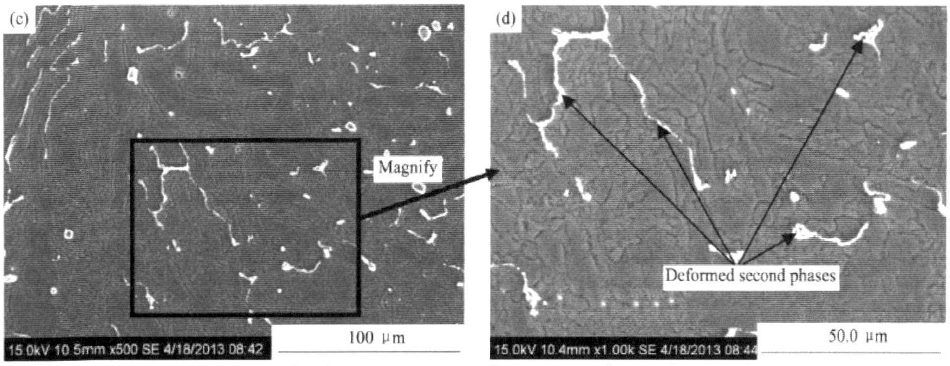

Fig. 2. OM (a, b) and SEM (c, d) of microstructure of cross section of extruded Mg-4Zn-1Y alloy[17]

2.3 Dysprosium

Up to now, the addition of Dysprosium (Dy) is reported few. But, its effects are also significant and have special properties in common, because of the similar size between Dy and Y in atom order of magnitudes. Both of them will have the similar effecting mechanism. The group of Yang et al. [21] have investigated that the tensile strength of the alloy with the addition of Y decreased while increased in the alloys with the single addition of Dy. Compound add two RE elements made the increasing of the tensile strength, yield strength and elongation of alloy, which increase 5.3%, 17.1% and 177.4% compared ZK60 alloy without RE element. And Guangli Bi et al. [22] studied the deformation of an extruded Mg-2Dy-0.5Zn (at%) alloy under uniaxial tensile test at 300°C with 3×10^{-5}~$3\times10^{-1}s^{-1}$, which exhibited a quasi-superplastic deformation with an elongation of 105% at 300 °C with 3×10^{-5} s^{-1} due to the formation of the LPSO phase which can improve the deformation and dislocation of the grain boundary slip.

3. Summary and future perspectives

Except of the rare earth elements above, there are also many other kinds of alloying elements, such as neodymium (Nd), promethium (Pm), ytterbium (Yb) and etc., which are also investigated in other kinds of Mg alloys [23]. Three difficulties towards application are still remaining to be solved. Firstly, the high cost of the rare earth elements, like Dy, Pm and Yb, which can only be used in research and not be utilized in engineering. Relatively low-cost rare element, *e.g.*, scandium (Sc), needs to be developed with the similar effect for replacement of the high-cost ones needs to be developed with the similar effect for replacement of the high-cost ones [24]. Secondly, in order to realize some specific properties, the content of rare earth element needs to be paid attention for mass producing. Thus, the main target is to improve the mechanical properties of Mg alloys with little content (~1wt.%) of rare earth element.

It is necessary to search for the alloying mechanism between rare earth element and Mg alloy in order to develop the most optimal ratio of alloy in Mg alloys. Researchers have also turned to investigate some effective processing method, such as equal-channel angular pressing (ECAP), rapidly solidified powder metallurgy (RSPM), hydrostatic extrusion (HE), friction stir processing (FSP), severe torsion straining (STS) and etc. Each of them has its own exotic point which can be utilized in different conditions. In consequence, there are three main problems waiting to be improved: i) high cost of elements; ii) processing method of Mg-RE alloy and iii) changing the role of rare element in Mg.

Recently, with the development rare earth elements adding Mg alloy, they have been widely accepted by more and more people, and their application will be expanded. The high strength, heat resistant and corrosion resistant of rare earth Mg alloys will not only increase the applications in recent automobile, communications, electron industry and etc. but also can promote and exploit new fields for rare earth Mg alloys. It is significant for researchers to complete the existing rare earth Mg alloys which mentioned above. For Mg alloys and rare earth Mg alloys, it will be enormously facilitate the development of these materials.

Reference

[1] Kainer KU, Kaiser F (2003) Magnesium alloys and technology. Deutsche Gesellschaft für Materialkunde, New York
[2] Rashad M, Pan F, Asif M et al (2015) Enhanced ductility of Mg-3Al-1Zn alloy reinforced with short length multi-walled carbon nanotubes using a powder metallurgy method, Prog Nat Sci 25:276-281.
[3] Freeney TA, Mishra RS (2010) Effect of Friction Stir Processing on Microstructure and Mechanical Properties of a Cast-Magnesium–Rare Earth Alloy, Metall Mater Trans A 41:73-84.
[4] Mostaed E, Fabrizi A, Dellasega D et al (2015) Microstructure, mechanical behavior and low temperature superplasticity of ECAP processed ZM21 Mg alloy, J Alloys Compd 638:267-276.
[5] Yu WB, Liu Z, Cheng N et al (2006) Research and Development of Wrought Magnesium Alloy Containing Rare Earth Elements, Mater Rev 20:65-68.
[6] Zucchi F, Grassi V, Frignani A et al (2006) Electrochemical behaviour of a magnesium alloy containing rare earth elements, J Appl Electrochem 36:195-204.
[7] Bae DH, Kim Y, Kim IJ (2016) Thermally stable quasicrystalline phase in a superplastic Mg-Zn-Y-Zr alloy, Mater Lett 60:2190-2193.
[8] Xu C, Zheng MY, Xu SW et al (2013) Microstructure and mechanical properties of Mg–Gd–Y–Zn–Zr alloy sheets processed by combined processes of extrusion, hot rolling and ageing, Mater Sci Eng A 559:844-851.
[9] Gao L, Chen RS, Han EH (2009) Effects of rare-earth elements Gd and Y on the solid solution strengthening of Mg alloys, J Alloys Comp 481:379-384.
[10] Jin G, Du W, Li J et al (2007) Effects of Gadolinium Addition on High Temperature Mechanical Properties of Mg-2Al-1Zn Alloy. J Chin RE Soc 45:1111-1121.
[11] Alizadeh R, Mahmudi R, Langdon TG (2014) Superplasticity of a fine-grained Mg-9Gd-4Y-0.4Zr alloy evaluated using shear punch testing, J Mater Res Tech 3:228-232.
[12] Sarebanzadeh M, Roumina R, Mahmudi R et al (2015) Enhancement of superplasticity in a fine-grained Mg–3Gd–1Zn alloy processed by equal-channel angular pressing, Mater Sci Eng A 646:249-253.
[13] YANG M, Liu YH, Liu JA et al (2014) Corrosion and mechanical properties of AM50 magnesium alloy after being modified by 1 wt.% rare earth element gadolinium, J RE Ear 32:558-563.
[14] Shi LL, Huang Y, Yang L et al (2015) Mechanical properties and corrosion behavior of Mg-Gd-Ca-Zr alloys for medical applications, J Mech Behav Biomed 47:38-48.
[15] Yang Q, Xiao BL, Ma ZY et al (2011) Achieving high strain rate superplasticity in Mg–Zn–Y–Zr alloy produced by friction stir processing, Scripta Mater 65:335-338.
[16] LI JB, Wang F, Mao PL et al (2014) Evolution of microstructure and tensile properties of extruded Mg-4Zn-1Y alloy, J RE Ear 32:1189-1195.
[17] Xie GM, Ma ZY, Geng L (2009) Effect of Y Addition on Microstructure and Mechanical Properties of Friction Stir Welded ZK60 Alloy, J Mater Sci Tech 25:351-355.
[18] Xie GM, Luo ZA, Ma ZY et al (1993) Superplastic Behavior of Friction Stir Processed ZK60 Magnesium Alloy, Instrum Exp Tech 30:339.
[19] Rosalie JM, Somekawa H, Singh A et al (2013) Effect of precipitation on strength and ductility in a Mg-Zn-Y alloy, J Alloys Comp 550:114-123.
[20] Mirza FA, Chen DL, Li DJ et al (2013) Effect of rare earth elements on deformation behavior of an extruded Mg-10Gd-3Y-0.5Zr alloy during compression, Mater Deg 46:411-418.
[21] Wu A, Xia C, Dong L. Distributing (2007) Evolvement and Effect on the Mechanical Properties of the Rare-Earth Ce, Y in ZK60 Alloys, RE Met Mater Eng 36:1955-1959.
[22] Yang L, Huang Y, Peng Q et al (2011) Mechanical and corrosion properties of binary Mg-Dy alloys for medical applications, Mater Sci Eng B 176:1827-1834.

[23] Bi G, Li Y, Huang X et al (2015) Deformation behavior of an extruded Mg–Dy–Zn alloy with long period stacking ordered phase, Mater Sci Eng A 622:52-60.
[24] Mordike BL, Stulíková I, Smola B (2005) Mechanisms of creep deformation in Mg-Sc-based alloys, Metall Mater Tran A 36:1729-1736.

Influence of length-to-diameter ratio on shrinkage of basalt fiber concrete

Ruijie MA [1, a)], Jiansen Yang[1, b)], Yuan Liu[1, c)] and Xiaojun Zheng[1, d)]

[1] School of Civil and Hydraulic Engineering, Ningxia University, Yinchuan 750021, China

E-mail: [a)]578131988@qq.com, [b)]Corresponding author: yjs508@163.com, [c)] 18795390588@163.com, [d)] 547774399@qq.com

Abstract. In order to study the shrinkage performance of basalt concrete, using the shrinkage rate as index, the work not only studied the influence of different length-to-diameter ratio (LDR) on plastic shrinkage and drying shrinkage of basalt fiber concrete, but also analyzed the action mechanism. The results show that when the fiber content is 0.1%, the LDR of 800 and 1200 take better effects on reducing plastic shrinkage, however the fiber content is 0.3%, that of LDR 600 is better. To improve drying shrinkage, the fiber of LDR 800 takes best effect. In the concrete structure, the adding basalt fibers form a uniform and chaotic supporting system, optimize the pore and the void structure of concrete, make the material further compacted, reduce the water loss, so as to decrease the shrinkage of concrete effectively.

1. Introduction

Concrete shrinkage refers to the phenomenon of concrete volume shrinkage in the initial condensation or during hardening. Generally including plastic shrinkage, chemical shrinkage, drying shrinkage and carbonation shrinkage, large shrinkage will cause the concrete cracking, which has the most significant impact on the plastic shrinkage and drying shrinkage of concrete. Shrinkage is mainly related to the properties of raw materials, mix ratio, curing methods and so on. Excessive and uneven shrinkage will produce internal stress and even cracks in the product and component, which will affect the quality and durability of concrete [1]. As a new material of inorganic fiber, basalt fiber has many strong points, including high temperature resistance, corrosion resistance, alkali resistance, strong chemical resistance, good thermal stability and other advantages, therefore, in the construction industry. Basalt fiber plays a unique role in enhancing the shrinkage of concrete [2]. At present, there are a lot of researches on the influence of basalt content on the shrinkage of concrete, but for the influence of the length to diameter ratio (LDR) is relatively few. So this paper studied and analyzed the drying shrinkage properties and plastic shrinkage properties of basalt fiber, in order to figure out the influence of different LDR of basalt fiber on dry shrinkage properties and plastic shrinkage properties of concrete and provide reference for its engineering application.

2. Raw Materials and Experiments

2.1 Raw Materials

(1) Cement: P • O 42.5 cement produced by Ningxia Sai Ma Cement Factory.

(2) Sand: fineness modulus 2.32, meet the II zone level, mud content of 2.0%, clod content of 0.8%.

Content from this work may be used under the terms of the Creative Commons Attribution 3.0 licence. Any further distribution of this work must maintain attribution to the author(s) and the title of the work, journal citation and DOI.
Published under licence by IOP Publishing Ltd

(3) Coarse aggregate: gravel, continuous gradation, maximum particle size is 20 mm, mud content of 0.45%, clod content of 0.3%, crushing index of 9.8%, needle sheet content of 7.3%.

(4) Fiber: Nanjing Mandecard Technology Co., Ltd. Producted the basalt fiber, length is 12 mm, the actual fiber density is 2.65 g/cm3.

(5) Gelatine Powder: VINNAPAS RE5044N, Dispersible vinyl acetate-vinyl copolymer powder, white powder, produced by Germany Warker, its solid content of 99%, apparent density of 490kg/ m3, needle sheet content of 7.3%.

(6) Water reducer: naphthalene series superplasticizer, brown yellow powder, solid content of >92%, 0.3% of the amount of cementitious material, the recommended dosage of 0.75%.

2.2. *Test Plan*

Considering the performance requirements of concrete, the slump is controlled at 80mm to 120mm. Using single-factor test program, after several trial mixing, determined the basalt fiber concrete mix ratio (Table 1).

TABLE 1. Mix proportions design of basalt fiber concrete.

	Material consumption per unit volume /(kg·m^{-3})						
number	**cement**	**gelatine powder**	**sand**	**gravel**	**LDR(volume content)**	**water**	**water reducer /%**
XW-1	400	16	638	1186	0	160	0.75
XW2-1	400	16	637	1184	400(0.1%)	160	0.75
XW3-1	400	16	637	1182	600(0.1%)	160	0.75
XW4-1	400	16	636	1180	800(0.1%)	160	0.75
XW5-1	400	16	635	1179	1000(0.1%)	160	0.75
XW6-1	400	16	634	1177	1200(0.1%)	160	0.75
XW2-2	400	16	637	1184	400(0.3%)	160	0.75
XW3-2	400	16	637	1182	600(0.3%)	160	0.75
XW4-2	400	16	636	1180	800(0.3%)	160	0.75
XW5-2	400	16	635	1179	1000(0.3%)	160	0.75
XW6-2	400	16	634	1177	1200(0.3%)	160	0.75

Note: (1) the amount of fiber in the brackets is the percentage of the total volume of the fiber;
(2) The amount of the water-reducing agent is the mass percentage of the cementitious material.

2.3. *Experimental Method*

Accordance with the standard test methods of long-term performance and durability of ordinary concrete(GB/T 50082-2009) on shrinkage test contact method, preparation of 100mm ×100mm×515mm shrinkage specimens, after 3d standard curing temperature, removed the products in a constant temperature and humidity environment, which temperature is (20±2)°Cand relative humidity of (60±5)%. Measured the initial length immediately, thereafter, according to 1d, 3d, 7d 14d, 28d, 45d, 60d, 90d of time interval (contraction age), measure each length of products, and calculate the shrinkage rate respectively. The plastic shrinkage was carried out using non-contact method, and the deformation was measured continuously for 72 hours after molding.

3 Test results and analysis

3.1. *Effect of LDR on Plastic Shrinkage of Basalt Concrete*

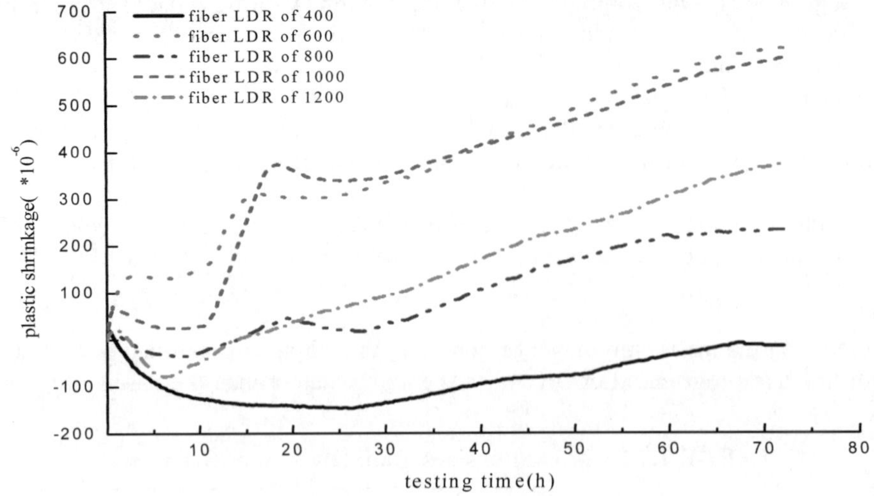

FIGURE 1. At 0.1% when different dosage of fiber LDR on basalt plastic shrinkage of concrete

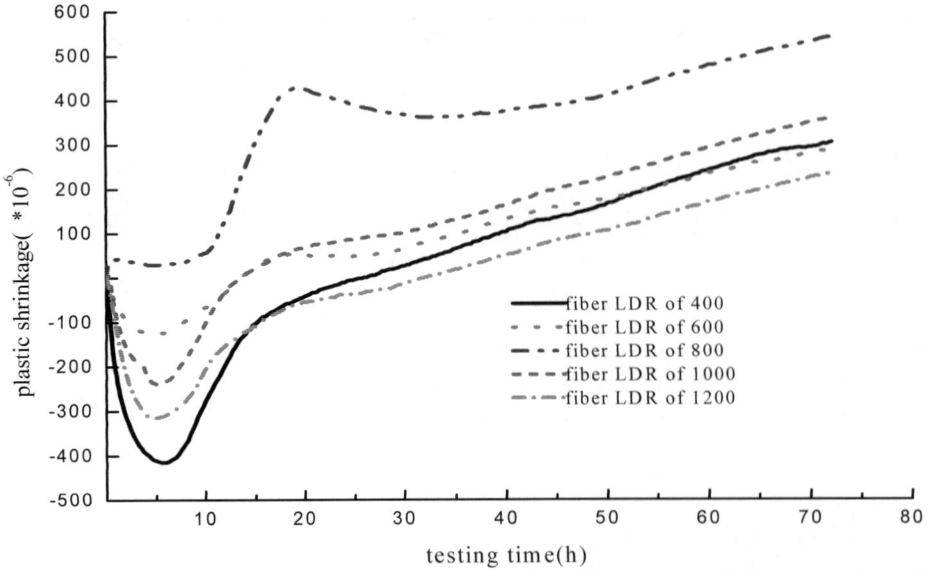

FIGURE 2. At 0.3% when different dosage of fiber LDR on basalt plastic shrinkage of concrete

It can be seen from Figure 1 and Figure 2 that the effect of basalt fiber on plastic shrinkage crack is better when the ratio of fiber is 0.1% and the LDR are 800 and 1200.When the aspect ratio is 800 of basalt fiber, the effect is the best, the plastic shrinkage of concrete is only 182×10^{-6}. When the fiber content is relatively much at 0.3%, the shrinkage age is within 11h, the basalt fiber with the LDR of 800 has the best effect, and the shrinkage rate is kept between $(0\text{-}50)\times10^{-6}$, but overall, when the LDR is 600 of the basalt fiber, effect is relatively good, the shrinkage rate at 72h is 290×10^{-6}. According to the study by Banthia [3]. The fracture area and the maximum crack width are inversely proportional to the aspect ratio of the fiber:

$$A_c=\alpha\left(\frac{l}{d}\right) \quad (1) \qquad W_{max}=\beta\left(\frac{l}{d}\right) \quad (2)$$

But the proportion coefficient α and β vary with the fiber content. The total surface area of fiber per unit volume (S_f) can also be used as a parameter to reflect the crack area and the maximum crack width. But the relationship between them are exponential [4].

$$A_c = \alpha e^{-bS_f} \quad (3) \qquad W_{max} = Ae^{-BS_f} \quad (4)$$

It can be seen that the effect of LDR of 1200 and at 0.3% when aspect ratio of 400 and 600 is relatively good, which is basically consistent with the conclusion of this experiment.

3.2. Effect of LDR on Drying Shrinkage of Basalt Concrete

As can be seen from Figure 3 and figure 4. For the drying shrinkage, the shrinkage of concrete is the most obvious when the LDR is 800. This is due to the three-dimensional chaotic distribution of basalt fiber in the specimen. The length of the basalt fiber with the LDR of 1000 and 1200 is relatively large, LDR 600 and 400 are kind of short, the side wall effect [5] is relatively strong (The side wall effect is that fiber in between the mold and the aggregate or next to the surface of side wall, due to the restriction of the edge wall, the fiber tends to parallel to the direction of the side wall). Basalt fiber along the coarse aggregate interface cannot resist cracks. However, the side wall effect of basalt fiber with the aspect ratio of 800 is relatively weak, and there is a certain amount of basalt fiber intersects with the interface of cemented aggregate. Once the initial microcracks that parallel to the interface have a trend to develop, this part of basalt fiber can prevent its development very well, delayed the formation of initial micro cracks, and effectively restrained the shrinkage of concrete.

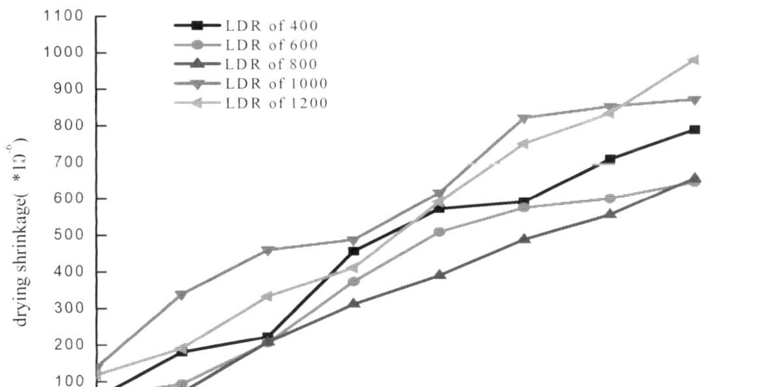

FIGURE 3. At 0.1% when different dosage of fiber LDR on basalt drying shrinkage of concrete

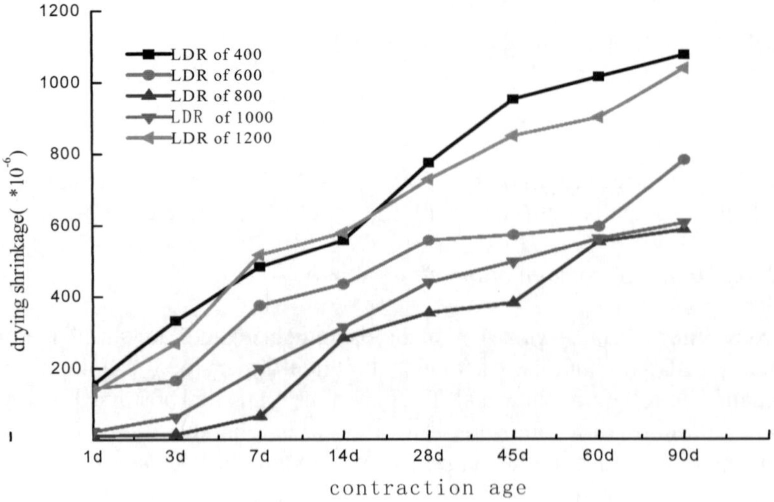

FIGURE 4. At 0.3% when different dosage of fiber LDR on basalt drying shrinkage of concrete

4 Mechanism analysis

There are two kinds of theories about the crack resistance mechanism of fiber, from different angles to explain this mechanism. Firstly, Romuadi from the United States proposed "fiber spacing theory"[6], according to the fracture mechanics theory to explain that fiber plays a role in the crack resistance for cracks of concrete, to improve the strength of concrete, must reduce the number and scale of cracks and the original defects in concrete, when the fiber spacing less than a certain value, the tensile strength of concrete will be increased. Secondly, Swamy from the United Kingdom proposed "composite theory"[7], starting from the composite principle that constituted by composite materials, considering fiber as a concrete strengthening system, applying the composite principle to deduce the tensile strength of fiber reinforced concrete, put forward a relationship among the tensile strength of fiber concrete and the amount, the direction, the LDR and the bond force of fiber.

Fiber spacing is more suitable to explain the plastic crack resistance mechanism of fiber concrete. Before the concrete is not hardened, the fiber has the effect of crack resistance. There are tens of millions to hundreds of millions of fibers per unit volume of concrete, the effect of crack resistance of fiber is mainly reflected in the number of fibers, the smaller the spacing between fibers, the more obvious the effect of the early anti-cracking. When the fiber spacing is large enough to a certain value, the fiber has little effect on the early crack resistance of concrete. When the fiber content is same, in the premise of ensuring the full dispersion of fiber, from the point of view of anti-cracking, it is appropriate to use fine fibers. When the fiber content is certain, the fiber content more thinner and the number of fibers in the unit volume concrete is numerous, the effect of crack resistance is more remarkable. The fiber length should be adapted to the size of coarse aggregate, for the larger diameter of coarse aggregate, we should select longer fiber.

In the concrete structure, the addition of fiber in the concrete to form a uniform and chaotic supporting system, optimizing the concrete pore and the void structure, reducing the number of capillary and internal porosity, making the material more compact, reducing the water loss, so as to reduce the shrinkage of concrete effectively.

At the same time, because of the elastic modulus of the fiber is big, it counteracted the pressure of capillary, thereby reducing the total capillary pressure. So the addition of fiber significantly reduced the shrinkage of concrete.

5 Conclusion

(1) The effect of basalt fiber on plastic shrinkage crack is better when the ratio of fiber is 0.1% and the LDR are 800 and 1200. When the LDR is 800 of basalt fiber, the effect is the best, the plastic shrinkage of concrete is only 182×10^{-6}. When the fiber content is relatively at 0.3% and the LDR is 600 of the basalt fiber, effect is relatively good, the shrinkage rate at 72h is 290×10^{-6}.

(2) For the drying shrinkage performance of concrete, the LDR of 800 basalt fiber takes good effect on suppressing shrinkage.

(3) In the concrete structure, the addition of fiber in the concrete formed a uniform and chaotic supporting system, optimizing the pore and the void structure of concrete, making the material more compact, reducing the water loss, so as to reduce the shrinkage of concrete effectively.

Acknowledgments

The authors gratefully acknowledge the financial supports provided by National Natural Science Foundation of China (51168041 and 51568056) and the Graduate Innovation Project of Ningxia University (GIP201637).

References

[1] Li Shihua. Influence of Admixtures on Shrinkage Property and Cracking Behavior of Concrete [J]. Zhengzhou: Zhengzhou University, 2012.
[2] Wang Xingzhou. Research on the plastic shrinkage of fiber concrete [D]. Jilin: Jiling University, 2006.
[3] Nemkumar Banthia and Cheng Yan,Shrinkage Cracking in Polyolefin Fiber-Reinforced Concrete.ACI Materials Journal,V.97, No.4, July-August 2000.
[4] Kang Jingfu. Shrinkage of concrete at early age and approach on the anti-cracking mechanism of polypropylene fiber. Concrete, 2013, 159 (1):10-12.
[5] Xu Ping, Cui Jian. Effect of fiber - aspect ratio and content on compressive strength of steel fiber reinforced concrete [J]. Manufacturing· material, 2004, 42(482）: 60-61.
[6] Hu Xiaoqing, Wang Yongsheng. Defect Analysis of Theory of Fiber Spacing [J].An Hui Construct, 2011, (02): 73-74.
[7] Li Jianqiang. Experimental research on Ultra High Toughness Cementitious Composites [D]. Chang An university, 2012.

Numerical investigation of wrinkle for multi-point thermoforming of Polymethylmethacrylate sheet

Heli Peng[1, a)], Haijian Liu[1, b)] and Xiaolong Zhang[1, c)]
[1] Shanghai Spaceflight Precision Machinery Institute, Shanghai 201600, China.

E-mail: [a)] Corresponding author: phl12616040811@126.com, [b)]623513470@qq.com, [c)]hangtian402@163.com

Abstract. The multi-point thermoforming (MPTF) machine and process of resin sheet were introduced, the glass transition temperature and stress relaxation curves of Polymethylmethacrylate (PMMA) sheet were tested, and the material parameters of PMMA sheet were got. Lots of numerical simulation of wrinkle for spherical part with MPTF of PMMA sheet were put forward, the effects of thickness, radius and forming temperature on the wrinkle were investigated, and the non-wrinkle forming limit diagram are determined. Then, MPTF experiments of spherical part with PMMA sheet was done to verify the numerical results. Consequently, the spherical part has good shape accuracy, which confirms that PMMA sheet formed by MPTF is feasible and the non-wrinkle forming limit diagram obtained by numerical simulation is practicable.

1. Introduction

Resin sheet has characters of low density, high specific strength, corrosion preventive, good ductility and good insulation, which is used in many fields, such as aerospace, automobile, architecture, medical instruments, transport and so on. There are three kinds of main forming methods for resin sheet, which are compression moulding, thermal diaphragm forming and bulging forming. All of these methods need specific mould, and it is hard to achieve multi-species, small batch and personal manufacture for its long manufacturing cycle and high cost, so it is urgent to demand flexible methods for forming resin sheet.

In the past two decades, considerable progresses have been made to study flexible forming technology of resin sheet. Gertner [1] proposed a rolling method to form polymer sheet after heating. Franzen [2] used single point incremental method to form the polymer sheet. Silva [3] and Martins [4] studied the single point incremental forming theory, optimized the forming process and carried out a series of experiments through polymer plate with good ductility. Walczyk [5] designed and developed flexible experimental machine by combining the reconfigurable tooling and vacuum thermoforming to form composite aircraft parts. Su [6] proposed the multi-point thermoforming to manufacture freeform panels by combining the multi-point forming and thermoforming. Peng [7] have developed multi-point thermoforming (MPTF) machine and studied the feasibility and practicability of MPTF.

Due to the manufacturing trends of individuation and flexibility for resin sheet, MPTF has broad application prospects. While the MPTF of resin sheet is freeform, and wrinkle is easily generated, which affect the forming quality. So, it is necessary to study the wrinkle and its influencing factors in order to obtain good profile shape.

Content from this work may be used under the terms of the Creative Commons Attribution 3.0 licence. Any further distribution of this work must maintain attribution to the author(s) and the title of the work, journal citation and DOI.
Published under licence by IOP Publishing Ltd

2. MPTF Machine and Process

As is known, the resin sheet has three kinds of aggregate states, which are glassy state, high-elastic state and viscous flowing state. Generally, the resin sheet is formed at high-elastic state for good plasticity and ductility. Similar to sheet metal, the resin sheet can be formed by multi-point press forming or multi-point stretch forming, where the heating equipment is fabricate and the forming machine is expensive. So, the MPTF machine was developed by combining the single multi-point die and heating pressure box, as shown in Fig.1.

Fig.2 shows the MPTF process of resin sheet, which contain four steps. The first step: the shape of multi-point die was formed by calculating the height and adjusting the position of punch element based objective part. The second step: the resin sheet was placed above the multi-point die, heat was generated by lots of heating rods placed above the resin sheet until the temperature of resin sheet above its glass transition temperature, and the temperature is kept for several minutes. The third step: the compressed gas is pour into forming box through air pimp until the resin sheet accords fully with the multi-point die. The fourth step: the heating rods and air pump are closed, then the resin product is removed when it backs to room temperature.

FIGURE 1. MPTF Machine

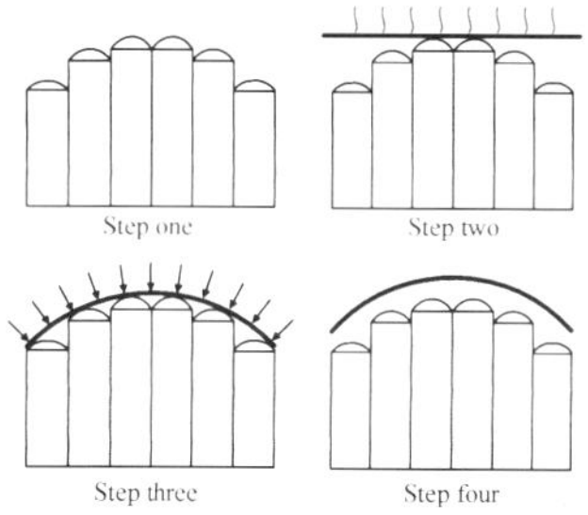

FIGURE 2. MPTF Process

3. Finite element simulation of MPTF for PMMA sheet

3.1 Material Parameters

PMMA is selected as the study object for its good comprehensive performance. The glass transition temperature of PMMA is measured as 105°C by dynamic mechanical analysis. The tests of stress relaxation are put forward in universal electronic testing machine, and the real stress relaxation curves of PMMA at different temperatures are shown in Fig.3.

Generalized Maxwell model can be selected to describe the viscoelastic behaviour of resin sheet, its relaxation property can be described by Prony series in ABAQUS, and the Prony series of generalized Maxwell model can be expressed by the following equation:

$$e(t) = 1 - \sum_{i=1}^{n} e_i(1\text{-}\exp(-\frac{t}{\tau_i})) \tag{1}$$

Where e_i and τ_i are coefficients of Prony equation.

The coefficients of Prony equation for PMMA are listed in Table1. The effects of temperature and time on the mechanical properties of PMMA sheet can be transformed into each other under certain conditions based on equivalent conversion principle of time and temperature, which indicate that the viscoelastic behaviour of resin sheet in different temperature can be forecasted by displacement factor. The displacement factor is the function of temperature, and it can be calculate by the Williams-Landel-Ferry (WLF) [8] given by the eq.2.

TABLE 1. Material parameters of PMMA

parameter	i_1	i_2	i_3	i_4	i_5
e_i	0.2788	0.1589	0.2729	0.2506	0.0403
τ_i	0.0136	0.1049	1.3188	11.379	569.62

$$\lg a_T = -\frac{C_1(T-T_0)}{C_2+(T-T_0)} \tag{2}$$

Where $C_1$and C_2 are the constant, T_0 is the reference temperature, T is the objective temperature, and the a_T is displacement factor, which can be expressed bythe eq.3.

$$a_T = \frac{\tau}{\tau_0} \tag{3}$$

Where τ and τ_0 represent the time of stress relaxation when the temperature of T and T_0, respectively. The constants of $C_1$and C_2 for PMMA sheet are 6.71 and 65.01, respectively.

3.2 Finite Element Model

Numerical simulation of MPTF for PMMA sheet is carried out by ABAQUS. Spherical part is selected as the objective shape of formed part. Due to the symmetry, only a quarter of finite element model is analysed, as shown in the Fig.4. The FEM is composed of multi-point die, elastic cushion and resin sheet. The size of multi-point die is 400×400mm, punch number is 20×20 and punch radius is 20 mm. To save computing time of CPU, multi-point die only retain the hemispheric end; it is simplified to rigid shell surface and modelled with bilinear quadrilateral three dimensional rigid element R3D4. The thermostable silica is selected as the elastic cushion, its size is 400×400×20mm, and it is modeled with hexahedral solid element C3D8R. The resin sheet is modeled with quadrilateral shell element S4R. Effective hourglass control techniques are used to avoid the spurious deformation modes during finite element simulations. On the two symmetrical planes, the displacements normal to the planes and the rotations around the planes are constrained in the finite element simulations.

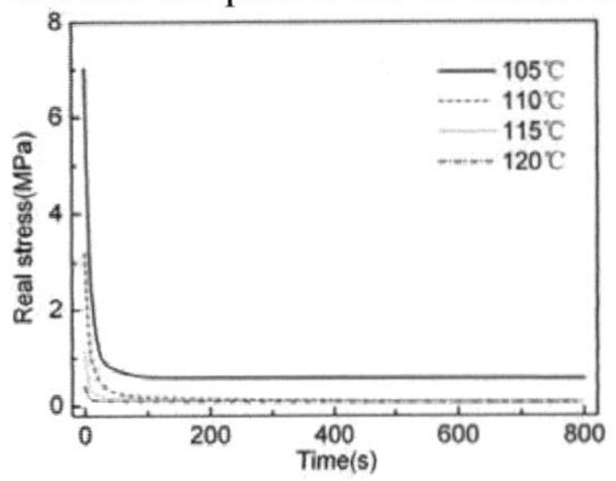

FIGURE 3. Stress relaxation curves of PMMA

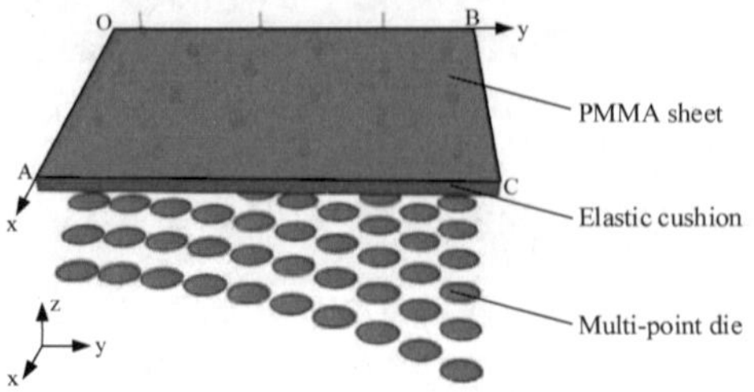

FIGURE 4. The FEM of MPTF for PMMA sheet

4. Numerical simulation of wrinkle in MPTF of PMMA sheet

When the radius of spherical part is 300mm, the forming is 110°C, four kinds of thicknesses are investigated. The z coordinate of line AC on spherical part with different thicknesses is shown in Fig.5(a). When the thickness is 1mm, the maximum value of wrinkle is 9.3mm. With the thickness increase to 2mm, the maximum value of wrinkle decreases to 5.4mm. When the thickness increases to 4mm, no wrinkles appear in spherical part. It can be found that the thicker the thickness is, the smaller the wrinkle will be for resin sheet.

When the thickness of PMMA sheet is 2mm, the forming temperature is 110°C, three kinds of radii are investigated. Fig.5(b) shows the z coordinate of line AC on spherical part with different radii. The following features can be found: the larger the forming radius is, the smaller the wrinkle will be. These results illustrate that the larger the forming radius is, the smaller the wrinkle will be, since a smaller forming radius leads to a larger deformation.

Fig.5(c) presents the z coordinate of line AC on spherical part with different forming temperatures when the thickness of resin sheet is 2mm and forming radius is 500mm. Obviously, there is no wrinkle on spherical part when the forming temperature is 110°C. With the forming temperature increase, the maximum wrinkling values of spherical part are 1.0mm and 3.1mm when the temperature are 115°Cand 120°C respectively. These results illustrate that the higher the forming temperature is, the larger the average shape error will be. Because a higher forming temperature leads to a more soft resin sheet, which leads to a larger value of wrinkle.

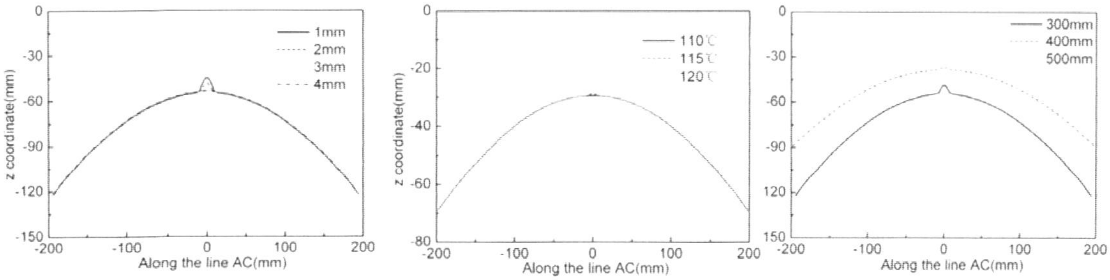

(a) Different thickness (b) Different radius (c) Different temperature

FIGURE 5. The z coordinate of spherical parts along the line AC with different forming parameters

The wrinkle of spherical part with MPTF of resin sheet is influenced by many factors such as the thickness, radius and temperature. It is difficult to control wrinkling by considering all of these complex factors. However, the limit curves of spherical part without wrinkle are beneficial for choosing forming parameters, which can be got by lots of numerical simulation and computations. When the size of multi-point die is 400×400mm, punch number is 20×20 and punch radius is 20 mm, the non-wrinkle forming limit curves of spherical parts are shown as Fig.6, which are plotted by getting the minimum radius of spherical when the temperature and thickness are fixed. The increasing amplitude spherical radius is chosen as 50mm in order to save computing time. It can be seen that if the radius above the limit curve, the spherical part can be formed well.

5. MPTF experiment of spherical part

The forming experiment of spherical part is carried out in the MPTF machine. The effective forming size of multi-point die is 400×400mm, punch number is 20×20 and punch radius is 20 mm. The experimental conditions are set as follows: the size of PMMA sheet is 400×400×2 mm, spherical radius is 500mm and the forming temperature is 110 °C. Fig.7 shows the experimental photo of spherical part with MPTF of PMMA sheet. It can be found that the spherical part has good profile shape and no wrinkle appear, which verifies the correction of numerical results.

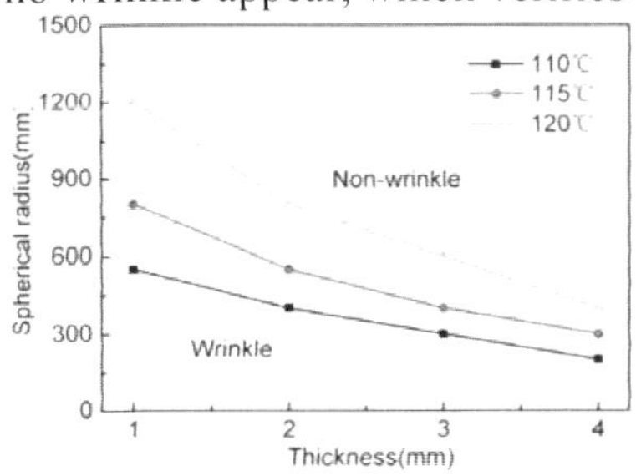

FIGURE 6. The non-wrinkle forming limit

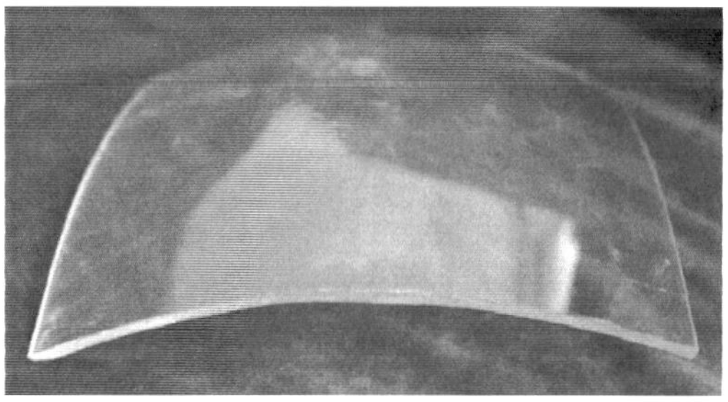

FIGURE 7. The photo of spherical part with MPTF

6. Conclusions

The machine and process of MPTF for resin sheet were introduced, and the material parameters of PMMA sheet were got. Numerical simulations are carried out to investigate the effects of thickness, radius and temperature on the MPTF of PMMA sheet. The results show that the thicker the sheet thickness or the larger the forming radius or the lower the forming temperature, the smaller the wrinkle will be. The non-wrinkle limit diagrams of spherical part with MPTF of PMMA sheet are got by lots of numerical simulations, and then the MPTF experiment is done. The result shows that the spherical part has no wrinkle and good surface quality, which illustrate that the MPTF of PMMA sheet is feasible and the forming limit diagram is practical.

Acknowledgments

This project was supported by National Science Foundation of China (51505278).

IOP Conf. Series: Materials Science and Engineering **242** (2017) 012028 doi:10.1088/1757-899X/242/1/012028

References

[1] Y. Gertner and A. K. Miller, Journal of Thermoplastic Composite Materials **9**, 151-182(1996).
[2] V. Franzen and L. Kwiatkowski, Journal of materials processing technology **209**, 462-469(2009).
[3] P. A. F. Martins and L. Kwiatkowski, CIRP Annals-Manufacturing Technology **58**, 229-232(2009).
[4] M. B. Silva and L. M. Alves, European Journal of Mechanics-A/Solids **29**, 557-566(2010).
[5] D. Simon and L. Kern, Procedia CIRP **17**, 2014, 853-858(2014).
[6] S. Z. Su and M. Z. Li, Key Engineering Materials **504**, 839-844(2012).
[7] H. L. Peng and M. Z. Li, Journal of Advanced Manufacturing Technology **67**, 2811-2817(2013).
[8] M. L. Williams and R. F. Landel, Journal of the American Chemical Society **77**, 3701-3707(1955).

Nitrogen doped CNT/Li4Ti5O12 composite for the improved high-rate electrochemical performance of lithium-ion batteries

Bin Ren[1, 2], Wen Li[1, 2], Aijia Wei[1,2], Rui He, Lihui Zhang[1,2] and Zhenfa Liu[1,2,a)]

[1] Institute of Energy Resources, Hebei Academy of Sciences, Shijiazhuang, Hebei Province 050081, PR China

[2] Hebei Engineer Research Center for Water Saving in Industry, Shijiazhuang, Hebei Province 050081, PR China

[a)] Corresponding author: lzf63@sohu.com

Abstract. A novel Li4Ti5O12 (LTO) composite with nitrogen doped multi-walled carbon nanotubes (CNTs), denoted N-C-LTO, was successfully prepared via simple thermal annealing of CNTs in the presence of melamine. For comparison, LTO, C-LTO (Li4Ti5O12/CNT) were also synthesized. N-C-LTO demonstrated the best electrochemical performance among the samples. Even at a high charge/discharge rate of 20 C, the reversible capacity was maintained at the high level of 100 mAhg-1. Moreover, after 150 cycles at 3.0 C, 90.7 % of the capacity was retained with negligible capacity fading. The excellent electrochemical performance was possibly due to the nitrogen in the doped CNTs, which maintained the benefits of the nitrogen as good electron donating.

1. Introduction

In the last decade, spinel $Li_4Ti_5O_{12}$ (LTO) has been considered as a promising anode material due to its outstanding safety performance and ultra-long lifetime [1-3]. Despite these advantages, drawbacks, such as poor electronic conductivity and sluggish Li-ion diffusion in the bulk, are still present [4-6]. To overcome these drawbacks, two typical approaches have been developed. The first approach is to synthesize nanostructured $Li_4Ti_5O_{12}$ materials with various morphologies to facilitate lithium-ion transport by shortening the diffusion path and to improve the intercalation kinetics by providing a large electrode/electrolyte contact area [7-9]. The other approach is to improve the electronic conductivity of $Li_4Ti_5O_{12}$ by doping with metal ions or through surface modification by coating or mixing with conductive species, such as Ag [10], carbon materials [11-13], etc. According to the literature, the synthesis of carbon-modified composites presents a simple but more effective way to improve the electronic conductivity. As a novel carbon material, carbon nanotubes (CNTs) are useful for improving the performance of $Li_4Ti_5O_{12}$ due to their high length/diameter ratio and unique conductivity. Several studies have demonstrated LTO/CNT composites with highly improved performance [14,15]. Therefore, a novel approach to modify LTO/CNT composites to further enhance the performance is highly desired.

The addition of dopants (N) in the CNT lattice affords a better electron transfer rate than that of ordinary CNTs. According to the literature, CNTs show enhanced conductivity upon nitrogen doping, as nitrogen atoms act as electron donors and decrease the band gap of the material [16,17] To the best of our knowledge, LTO/ N-CNT composites have been rarely reported. Therefore, we examined the

Content from this work may be used under the terms of the Creative Commons Attribution 3.0 licence. Any further distribution of this work must maintain attribution to the author(s) and the title of the work, journal citation and DOI.
Published under licence by IOP Publishing Ltd

effect of N-CNT on the electrochemical properties of LTO. In this paper, we successfully prepared the LTO/N-CNT composite and obtained excellent electrochemical performance. The effects of N doping was studied. In addition, characterizations of the LTO/ N-CNT composite, such as the electrochemical performance, phase purity, morphology, are also discussed in this paper.

2. Experimental section

2.1 Preparation of Li4Ti5O12
$Li_4Ti_5O_{12}$ (LTO) was prepared from TiO_2 (anatase structure) and Li_2CO_3 via a solid-state method. The starting materials, TiO_2 and Li_2CO_3 in a Li:Ti molar ratio of 0.85:1, were mixed in water and ball-milled for 6 h. The obtained binary precursor slurry was treated by the spray-drying method to form a dry powder. The dry powder was then calcinated at 750 °C for 12 h under air atmosphere with a temperature ramp rate of 5 °C/min.

2.2 Preparation of N-CNT
A commercial multi-walled carbon nanotube (CNT) sample (Cnano Technology 9100) was used as the starting material. According to the supplier, these nanotubes had an average diameter of 10-15 nm and a carbon purity higher than 97%. For the preparation of N-CNT, 1 g of the CNT sample was mixed with 0.075 g of N using melamine as the nitrogen precursor (5 wt% of N in N-CNT), and the mixture was ball-milled for 4 h. Then, the resulting material was heated under N_2 flow to 600 °C and subjected to thermal treatment at this temperature for 2 h.

2.3 Preparation of LTO/CNT, LTO/N-CNT
Then, 4 g LTO (300 mesh) and 0.2 g CNT, N-CNT were mixed, and the mixture was then ball-milled for 5 h and finally dried at 60 °C for 12 h. After that, the dried mixture was calcined at 500 °C for 5 h under nitrogen flow to obtain LTO/CNT, LTO/N-CNT. The corresponding products are denoted C-LTO (LTO/CNT), N,C-LTO (LTO/N-CNT)

2.4 Electrochemical measurements
The electrochemical properties of all samples were measured using a CR2032 coin-type half-cell in which the cathode and Li metal anode were separated by a porous polypropylene film (Celgard 2400, Celgard Inc., USA). The cathode slurry was prepared by the homogeneous mixing of the active material (LTO material), Super-P, and polyvinylidene fluoride (PVDF) in a mass ratio of 85:10:5 in N-methyl-2-pyrrolidone (NMP). The resulting slurry was cast onto Cu foil and dried under vacuum for 12 h at 90 °C. Finally, the electrode laminate was punched into disks (10 mm in diameter) and dried in a vacuum oven at 105 °C for 24 h. The coin cells were assembled in an argon-filled glove box. The electrolyte (Capchem Technology (Shenzhen) Co., Ltd.) was a 1 mol L^{-1} solution of $LiPF_6$ in ethylene carbonate, dimethyl carbonate, and diethyl carbonate (volume ratio = 1:1:1).

Galvanostatic charge-discharge tests were carried out using an automatic galvanostatic charge-discharge unit (Land 2001A, Wuhan, China) between 1 and 3 V at C-rates of 0.2, 1, 3, 10 and 20 C (1 C = 250 mAh g^{-1}) at 25 °C. Electrochemical impedance spectroscopy (EIS) was performed between 10 mHz and 100 kHz before and after the cycling tests using an electrochemical workstation (Interface 1000, Gamry, USA).

2.5 Characterization of the materials
Powder X-ray diffraction (XRD, Ultima IV, Rigaku) utilizing Cu-Kα radiation was used to characterize the phases and crystal structures of all samples. The diffraction patterns were collected at room temperature by step scanning over a range of 10-90° at a scanning rate of 5°/min. The morphologies of the materials were characterized by scanning electron microscopy (SEM, Hitachi S-4800, Japan) and transmission electron microscopy (TEM, JEOL, JEM-2010F). The surface

compositions of the LTO materials were examined by X-ray photoelectron spectroscopy (XPS, PHI5600 Physical Electronics).

3. Results AND Discussion

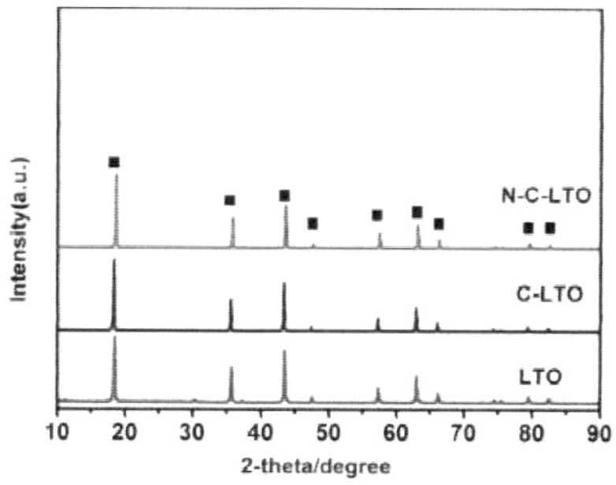

Fig. 1 XRD patterns of LTO, C-LTO, N,C-LTO.

Fig. 1 shows the XRD patterns of LTO, C-LTO, N,C-LTO. The five as-prepared materials show intense, sharp peaks that can be assigned to spinel $Li_4Ti_5O_{12}$ with good crystallinity in accordance with JCPDS card no. 49-0207. No extra phases are detected, confirming the high phase purity. Additionally, no diffraction peaks corresponding to carbon are present, presumably due to the low CNT content in these composite materials.

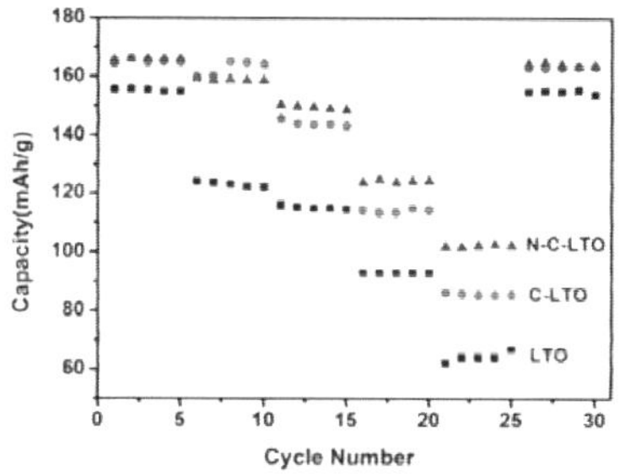

Fig. 2 Cyclic performances of the LTO, C-LTO, B,C-LTO, N,C-LTO, and N-B-C-LTO samples at different

rates: 1st–5th cycles at 0.2 C, 6th–10th at 1C, 11th–15th at 3 C, 16th–20th at 10C, and 21th–25th at 20 C.

Fig. 2 shows the cyclic performances of the LTO, C-LTO, N,C-LTO samples over 5 cycles at different rates of 0.2, 1.0, 3.0, 10 and 20 C. As shown in Fig. 4, the discharge capacity of all the samples gradually decreased with increased rate. However, the two CNT-LTO composite samples manifested a higher reversible capacity than LTO at all rates, and both the CNT-LTO composite samples showed almost the same reversible capacities at 0.2, 1 and 3 C. However, at the higher rate of 20 C, the capacity of the doped CNT-LTO composite showed a much higher reversible capacity than the non-doped CNT-LTO composite. N-C-LTO demonstrated the highest capacity at 20 C. These results indicate that the N -C-LTO composite exhibits the best rate capability among the samples of LTO, C-LTO, N,C-LTO possibly due to the CNTs show enhanced conductivity upon nitrogen doping, as nitrogen atoms act as electron donors and decrease the band gap of the material.

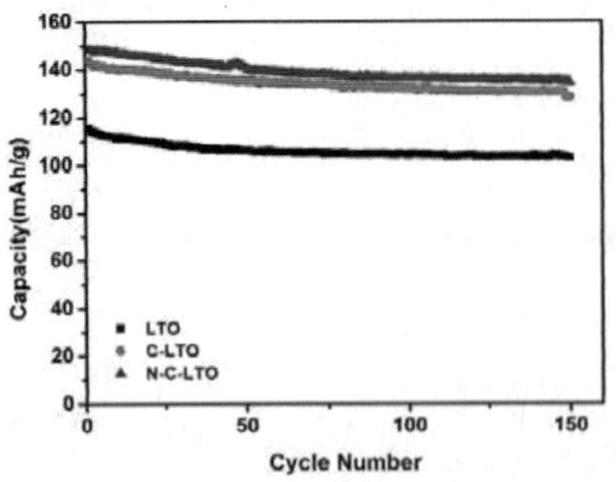

Fig. 3 The cycling performance of LTO, C-LTO, N,C-LTO over 150 cycles at 3 C.

The cycling performance of various samples at 3 C is shown in Fig. 5. In the initial cycle, the discharge capacity of LTO, C-LTO, N,C-LTO is 115.8, 141.8, 148.4 mAh/g, respectively. The discharge capacity decreases to 103.2, 128.5, 137.5mAh/g after 150 cycles of charging/discharging, maintaining 89.5%, 90.7%, 92.7% of the capacity relative to the initial cycle. The N-C-LTO sample exhibits superior electrochemical performance. As shown above, N-doping of the CNTs not only improved the discharge capacity at high C-rates but also enhanced the cycling stability.

4. Conclusion

We successfully synthesized N doped CNT-LTO composite materials using a simple ball-milling process followed by calcination. The results show that the electronic conductivity and the electrochemical performance can be improved through nitrogen doping. The N-B-C-LTO demonstrated a much better electrochemical performance. At a high charge/discharge rate of 20 C, the reversible capacity reached the high level of 100 mAhg-1. Moreover, after 150 cycles at 3.0 C, 92.7% of the capacity was retained with negligible capacity fading.

Acknowledgments

This work was supported by the Major State Basic Research Developmen Program of Hebei Province (17964407D), Science and Technology Program of Hebei Province (16294403D) and Project of Hebei Academy of Sciences (17705).

References

[1] H.G. Jung and J. Kim, J. Power Sources **196**, 7763–7766 (2011)..
[2] H. Ni and W.L. Song, Electrochem. Commun. **40**, 1–4 (2014).
[3] A. Nugroho and S.J. Kim, Electrochem. Commun. **13**, 650-653 (2011)
[4] N. Wu and Z. Z. Yang, Angew. Chem., Int. Ed. **54**, 5757-5761 (2015)
[5] D. Wang and N. Ding, J. Mater. Sci. **44**, 198-203 (2008)
[6] L. Wang and H. Zhang, Electrochim. Acta **142**, 202-207 (2014)
[7] B. Zhao and X. Deng, Adv. Energy Mater. **6** 1500924 (2016)
[8] L. Yu and H. B. Wu, Adv. Mater. **25** 2296-2300 (2013)
[9] A.S. Prakash and P. Manikan, Chem. Mater. **22**, 2857–2863(2010)
[10] J.G. Kim and D. Q. Shi, Nano Res. **6**, 365–372 (2013)
[11] H.W. Bai and X. L. Zan, Chemelectrochem **3**, 91-97 (2016)
[12] H. Ge and T. Hao, Acs Appl. Mater. Interfaces **8**, 9162–9169 (2016)
[13] Y. Ding and G. R. Li, Electrochim. Acta, **102,** 282-289 (2013)
[14] J. J. Huang and Z.Y. Jiang, Electrochim. Acta **53,** 7756–7759 (2008)
[15] H.F. Ni and L.Z. Fan, J. Power Sources **214**, 195-199 (2012)
[16] Y.X. Zhang and J. Zhang, Chemsuschem **45,** 1240-1250 (2014)

IOP Conf. Series: Materials Science and Engineering **242** (2017) 012030 doi:10.1088/1757-899X/242/1/012030

Research progress of polyolefin fibrillating fiber

Shu Hu[1,a)],Hui Guo[1,2,b)] ,Xiangchao Chen[1,c)] and Peng Li[1,d)]

[1]Science and Technology Innovation Center,Shenzhen Selen Science & Technology Co., Ltd, Shenzhen 518132,China.

[2]Shenzhen Selen Advanced Material Technology Research Institute Ltd, Shenzhen 518052, China.

E-mail: [a)]Corresponding author:hushu@szselen.com, [b)]guohui@szselen.com, [c)]chenxiangchao@szselen.com, [d)]lipeng@szselen.com

Abstract: The properties and producing methods of polyolefin fibrillating fiber were introduced in this paper. The research progress of polyolefin fibrillating fiber and the fibrillating device were reviewed. At the same time, the disparity of fibrillating device at home and abroad was discussed. The problems existing in the domestic polyolefin fibrillating fiber were pointed out, and the development trend of f polyolefin fibrillating fiber was discussed.

1. Introduction

The fibrillating fiber is a new type of synthetic fiber which has developed rapidly in 1870s. At present, the most widely studied and used industrial applications are mainly polytetrafluoroethylene fibrillating fiber and polypropylene as typical polyolefin fibrillating fiber. The PTFE material has excellent thermal stability and chemical resistance, but its raw materials and processing equipment are expensive. Considering the complex processing technology and difficult melt processing, large shrinkage of products and relatively low production efficiency, the PTFE materials are only used in some special occasions [1]. With lighter in density, lower price and material requirement, simple process and high automation of equipments, high in tensile strength and resistant to chemical corrosion, the polyolefin fibrillating fiber is more suitable for industrial production and application[2]. The section of fibrillating fiber is in a flat shape. It contains a lot of superfine fiber, which has high degree of fibrillation, high specific surface area, and strong adsorbing ability. It can be used as the main fiber filter material, as well as small amount of "large size fiber"-skeleton to support network. When made into non-woven air filter materials, it has a fluffy structure, high filtration efficiency and large dust holding capacity but very low air resistance[3]. When polarized with high voltage, the filtration efficiency is further improved, which make it a appropriate material for high efficiency and low resistance air filter production.

2. Process for producing polyolefin fibrillating fiber

The fibrillating fiber is made from polyolefin material with high crystallization and low liquidity. First, the polyolefin is prepared into a thin film. Through full longitudinal stretching and orienting, the longitudinal strength of film is greatly increased, correspondingly the transverse strength is extremely low. Thus the film has a tendency to burst into slim fiber along the longitudinal. After thermal forming and fibrillation process, the polyolefin fibrillating fiber can be obtained as long fiber, or cut into a certain length of short fiber[4].

Long fiber: mixing, extrusion granulation, feeding, extrusion molding, Film preparation, thermal

Content from this work may be used under the terms of the Creative Commons Attribution 3.0 licence. Any further distribution of this work must maintain attribution to the author(s) and the title of the work, journal citation and DOI.
Published under licence by IOP Publishing Ltd

IOP Conf. Series: Materials Science and Engineering **242** (2017) 012030 doi:10.1088/1757-899X/242/1/012030

stretching (orienting), heat-setting, fibrillating (fibrils collection/twisting), and rolling.

Short fiber: mixing, extrusion granulation, feeding, extrusion molding, film preparation, thermal stretching (orienting), heat-setting, fibrillating, ibrils collection, (traction-threading, steam heating, drag, curl), short cut, pack spare.

The manufacturing methods of polyolefin fibrillating fiber are mainly composed of three core technologies:

(1) Preparation of the polyolefin film

Blown film process has advantages of low price of production equipment, high film production, comfortable hand feel of the product, which is mainly used for low-end products such as reticular fibrillated fiber substrate in buildings, bind rope, packaging bags, carpet and decorative fabric etc. As the higher requirements of fibrillated fiber products on quality and stability, the large foreign equipments for flat wire and fibrillation fiber are mainly based on flat film method. For example, the Supertex production line (as shown in figure 1)designed by Farley Brothers company , can produce fibrillation fiber used as lining materials in bulk container, low shrinkage fibrillated fiber used in carpet base cloth and high strength fibrillated fiber used in weaving geotextile. The Supertex production line can produce 300-3300dtex fibrillated fiber, meeting kinds of specific requirements[5].

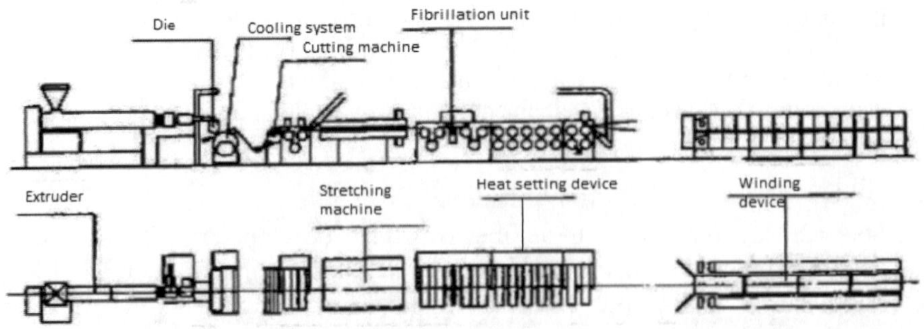

FIGURE 1. Supertex extrusion line for high strength fibrillating fiber.

(2) Film drawing

After decades of application in production, in order to get more fine fiber, after film forming the film is usually sliced into 5 ~ 30 mm ribbon online for hot drawing. Besides, cutting back materials can be recycled.

Cutting usually adopts coaxial blade cutting method. The cutter can be square, diamond, long hole type and can also be a sharp alien blade. The blade material can choose stainless steel, tungsten steel, hard alloy and etc. In order to prolong the service life of blade, coating process is usually used. The ideal thickness of the blade is 0.12 ~ 0.25 mm. According to the requirement of denier of the fiber, a variety of different spacing specifications are formed.

In order to precisely cut, the cutting device should be installed in the position where the film is in tension (as shown in figure 2 and figure 3).

FIGURE 2. The Cutting unit of Starlinger Corp in Austria.

FIGURE 3. The Cutting unit for polyolefin network fiber production line

Thermal film drafting device generally adopts five and seven stretching rollers, making a ideal stretching ratio, as well as finer fibers. The heating methods for stretching mainly divided into electric heating and steam heating and oil heating. Steam heating uses too much energy and is eliminated at this stage. Electric heating mostly uses far-infrared electric thermal roasting, which saves electricity and is convenient for automatic control.

The oil heating is heated by the double helix connecten with the stainless steel multi-layer high pressure hose in the drawing roll, and the temperature of the roller is controlled by thermal oil with high boiling point.

In order to prevent the film from non-uniform heating and fast cooling, the preheating and heat insulation device can be added before stretching roller or between two stretching rollers.

Preheating device/insulation device can be divided into three kinds-hot water (steam) tank, contact device (arched plate) and hot oven. The hot air circulation device with heating oven are mainly used abroad. How to maintain uniform temperature distribution inside the oven without impacting the flatness of the narrow strip needs special attention.

Thermal setting device generally adopts the same as the draft process, but the setting temperature is slightly a little higher than the drawing temperature about 5~ 10°Ctechnically.

(3) Fibrillation of film[6]

Film fibrillating is the core process in fibrillating fiber production. By highly uniaxial tensile (8 ~ 13 times), the longitudinal strength of the film reaches high, while transverse strength reduces to the limit. Through mechanical fibrillation or chemical methods, the film stretches and forms a mesh, then breaks up into fiber, winds for spare. The fibrillation method can be summarized into three categories: the random mechanical fibrillation, the chemical mechanical fibrillation and the adjustable mechanism.

Because the fiber produced by the first two methods is uneven structurally, and the fiber diameter and denier are uncontrolled, random mechanical method and chemical mechanical method is not applied to textile fibers.

Adjustable mechanical method uses mechanical means such as knifing and embossing to fabricate regular fibrillating fiber (reticular fiber) in early times. It develops later that one or two specially designed high-speed roller (as shown in figure 4 and 5) is used. After stretching along the longitude, the banded film passes through the roller fixed with blade or steel needles, and is cut into uniform fibrous mesh.

FIGURE 4. Fibrillation device of Taian Haidai Rope Machinery Science Technology Co.,Ltd.

FIGURE 5. Stretching machine and fibrillation for tape systems fb9 of oerlikon corporate.

Until the 1990s, due to the limited processing capacity and precision of the mechanical equipment, the film thickness of the flat fabrication method is not as thin as the blown method's, which cannot meet the processing requirements of the superfine fiber. With development of electrical automation, mechanical instrument system, precision mold processing and plastic modification, the use of cooling water and cooling roller in flat method come out. This makes flat method better than blown method in film thickness, hand feeling and production efficiency. Film prepared with flat method is more uniform and stable in size, which contributes to film fabrication. Besides, it needs less floor space. Since the mass production at home and abroad in 1890s, the single screw extruder is used, with the screw diameter Φ90~120mm and the draw ratio 30. For example, SJ-L starEX 800 /TM and starEX 1500/TM from starlinger company (see figure 6), with ultimate melting capacity 330kg/h and 600 kg/h, maximum production rate 420m/min and 420kg/min respectively, is mainly used for bag, FIBC weaving cloth, geotextile and agricultural cloth, carpet yarn, man-made lawn film crack silk back etc[7]. Oerlikon company[8] has similar plastic film Tape systems FB9 series extrusion line (as shown in figure 7). Flat wire machine or film line equipped with transverse adjustable high-precision fibrillating device can be used for producing superfine fibrillating fiber as the air filter material.

FIGURE 6. Flat wire drawing machine of oerlikon corporate.

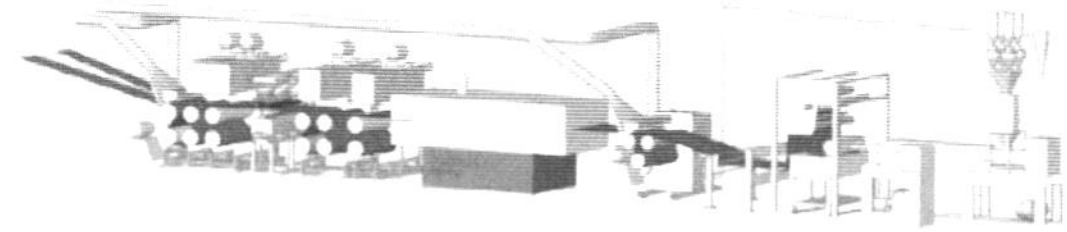

FIGURE 7. Schematic diagram of ape systems FB production line of oerlikon corporate.

3. Research progress of polyyolefin fibrillating fiber

The study of polyolefin fibrillating fiber abroad started earlier. In the 1830s, the Germans Heinrich Jaeque[9] discovered the potential of tape yarn(Slit) and fibrillating fiber (Split) production in the study of tensile phenomenon of polyvinyl chloride (PVC) and polystyrene membrane. Most of the early studies were confined to the laboratory, with no practical value. In the 1860s, with the industrialization of the polypropylene, it opened a world for thermoplastic polyolefins fibrillating fiber. Our country has been researching on the slitting tape yarn and film fibrillating since the 1970s. In domestic Beijing Plastics Factory pioneered the use of cutting tools on the pilot and expanded production of polypropylene fibrillating fiber. They made detail trial test and theoretical analysis on the fibrillation device, technological conditions and the influence factors of fibrillation, especially on basis of forming fibrillated network, average length, denier of the fiber and uniformity, and the cause of broken filament[10]. The fibrillation fiber was first applied to filter material in cigarette. At that time the cigarette filter materials acetate mainly depended on import, thus general attention had been paid on polypropylene fibrillation fibers,for its bonded mesh structure,not easy to loose hair, hardness, and high rate of mouth and so on.

Guojun Liu and others[11] blended PP with LDPE and HDPE respectively in extrusion and blow tube membrane as the filter material of cigarette. LDPE has good film forming stability, and is more soft than PP, for which the wrinkling problem of membrane in the herringbone plate can be solved. However, HDPE and PP have significant difference in characteristics of thermal fluidity, and the thickness contribution of the film is heterogeneous, which is not favorable to thermal stretch and fibrillation. When titanium dioxide (TiO_2) is added in the polyolefin as nucleating agent, the crystallinity and orientation degree of membrane increased. It not only benefits preparation of fibrillating fibers, but also improves whiteness of the product[12]. Haitao Wang and Huanfang Zhang[13-14] respectively used two-step grafting method and high energy electron beam radiation to graft the surface of polypropylene fiber, fabricating polypropylene filter with high absorption. The intercept efficiency of filter fiber after grafting modified was significantly increased, considering total particle phase of mainstream smoke, nicotine and tar content of phenolic compounds.

The raw material for preparing polyolefin fibrillating film requires low liquidity, high tensile strength and high elongation at break. Hence in the practice PP needs to be modified to increase its strength and toughness, in order to improve the effect of fibrillation. Shu Hu and others[15] combined the surface graft modification of hyperbranched polymer nano powder with POE to improve PP. The fibrillating fiber they prepared has characteristics of high fibrillation degree, high fracture strength and initial modulus. Crystalline ethylene chain segments in POE have good compatibility with PP, so it's easy to get smaller dispersed phase particle and narrow particle size distribution in PP matrix. The compatibility and dispersity of hyperbranched modified polymer nano powder in the polymer base material are obviously improved, which to a certain extent expands and reinforces the PP/POE. This not only improves the tensile strength and toughness of fibrillating fiber, but also plays an important role in reducing breakage in the production process[16]. After curled up and short cut, the film fibrillating fiber can be made into air filtration material with high efficiency and low resistance by means of non-spinning needle-strengthening and high voltage polarization.

4. Progress in research on polyolefin fibrillation equipment

In the 1870s, Yazawa Masahide and others[17] set small regular interval protrusions in the surface of the rotating device, in which the high oriented undirectional film slits itsself into fibrillating fiber. The shell company[18] adopts tube membrane method to prepare the polypropylene film. The film is stretched along longitudinal direction in a hot oven, and then goes through a roller with more than one circular needle. The top of the needle is higher than the roller's surface, and the needle has greater linear velocity than the film has. The rotating direction of needle roller accords with the film movement, while the needles tilts towards the opposite direction of the film movement in a certain angle. When the thermoplastic oriented film goes through the needle roller surface, the film fibrillation appears under the tension of needle line, which forms reticular fiber of average size 10 d. After further processing and it can be made into fabric yarn for the carpet backing, and decorative fabric.

CB Olson and others[19] found that the folding of the membrane leads to vibration during the stretching process and jagged fissures during the cutting period. By locating the serrated roller and pressure roller on the parallel surface of the cylinder, and seperating each other, foldings can be avoided. 3M Company[20~21] treated the oriented film with fluid energy such as ultrasound, high-presssure jet of water organic solvent liquid, air or nitrogen or carbon dioxide, to make its microporous surface fibrillated, and get the fibrillation products with minimal diameter, high strength and high modulus. 3M Company was the first company who applied fibrillation fiber in air filtration material with high efficiency and low resistance. They charged polymer film with static electricity, polarized the fibrillation fiber, and then sticked them together by ultrasonic as non-woven filter. By these means the filtration efficiency and dust capacity of the filter material improved significantly, and the air resistance loweredat the same time[22]. Oricon company[23~24] coated PTFE composite between battens of fibrillation roller surface to reduce friction, ensuring smooth pass of the fiber on fibrillating godet. Xueyue Zhao and others[25] installed triangular blades on the connecting parts, and fixed the connecting parts in dovetail groove of fibrillation roller. However the preblem is it took a lot of time to replace and install the blade. Burckhardt Company[26] fixed the needles on a special needle holder, which simplified the installation of the needles on the needle bed and greatly reduces cost of material and time in replacing the fibrillating mechanism. Chunfeng Liang and others[27] holded that net-like stripes bump in the surface of roller between needle roller, combined with flower pattern in the outer surface, can increase the friction between the fiber bundle and peripheral surface of roller. It made the ribbon or fibrillated film wind on the fibrillation godet in a whole, and could prevent the fiber bundle or film deviating from fibrillation or roller in the process of movement.

At present, the gap in fibrillation device between domestic and foreigner is mainly on needle roller and the needle materials, materials processing, and processing precision, etc.

Specific reasons exist in the following aspects:

(1) Inferior equipment in processing.

In order to save the cost, inferior bits are used, leading to uneven hole size in the needle roller. The aperture is different in size, and the hole is a wave. Poor precision of drilling machine, perpendicular hole and vibrating machine, causing nonuniform hole size and even supression deformation.

(2) Unreasonable materials chosen.

The domestic needle roller is mainly made of brass pipe and Japanese bronze pipe. The pipe is too hard to force uniformly in compression moulding and the needle roller is easy to distort. Fiber roll steel plated hard chrome is mainly used abroad, which is durable and not easy to rust.

(3)The precision of domestic fibrillaring needle is rather poor, and the material and heat treatment are not qualified, causing not enough hardness, easy deformation of long use. While the hardness of imported steel needle exceeds the HRC60. Jiangsu Textile Equipment Research Institute the LTD co. owns the fibrillar mechanism with the highest rows of needle roller and needle density in domestic, which can achieve 10 single needle/cm. India BASANT Company and Sitaram Products Company also achieve the same level, while India Rktexparts Company can make 16 needle/cm in single row.

At present, the world's top level is Switzerland Burckhardt, 25 needle/cm single row, while in stagger way of multi line distribution the needle density can reach 100 cm/row needle, or even 125 needle/cm. This attributes to the needle with minimum diameter 1.4 mm, the accuracy about 0.3 mm, needlepoint exceeding 2 mm in domestic. While imports needle can make minimum diameter 0.3 ~ 0.4 mm, accuracy 0.1 mm, needlepoint exceed 0.5 mm.

(4) The tessellation (assembly) of the fibrillation needle is a delicate work, for the deformation of needle tip caused in installation is irreparable. The needle roller manufacturer uses the internal polish to make repair of the inner diameter, but it is of little effect. To circular polishing inside the roller, it actually aggravates the decentration situation between the needle tip diameter Ds and installing diameter d. For the brass has been suppressed with deformation, it is difficult to find concentric point, to lacate the jig. The inner polishing turns to produce the speaker hole, one side loose and the other tight in assembling, leading the pinpoint of diffrerent height, and seriously affecting the quality of the fibrillation roller. Special craft has been used in needle roller products abroad. It is no need to polish the inner side repeatedly, and batch of needle roller can keep the needlepoint beating degrees within 0.1 mm, 0.03 mm within d size error in the installation. Assembling standard shaft with needle roller, all is sliding bearing assembly effect, not too loose and not too tight. It's no need to polish the inner hole repeatedly, with high coaxial degree and easy assembly.

(5)According to the experience of foreign companies, there should be strict requirements for dynamic balance and radial vibration during the manufacture and installation of fibrillation equipment. The spacing of needles should be identical. The axis of needle must be perpendicular to the needle roller when installation. It cannot tilt, and the tip height should be consistent. In order to improve the service life of equipment, hard chromium is usually coated in the roll surface, which makes it more difficult to deal with the needles.

5. Summary and prospect

With the rapid growth of economy, China's chemical fiber industry has kept nearly 20% of high-speed increase for seven years in a row under the rapidly expansive demand of textile market at home and abroad. However, for a long time, there has been a large gap between domestic and foreign manufacturers in plastic film extrusion mould, processing precision, stretching equipment and stability of the cooling system. The fact is that no fibrillating rollers and needles is specially designed for producing superfine fibrillating fiber. Although domestic film fiber production is very high, they are only used in plastic woven bag, jumbo bag, rope, membrane crack yarn, concrete reinfocing fiber in construction and other low-end products. But it is never seen in high-end field, which has more qualification in denier and evenness of fibrillating fiber.

In recent years, the chemical fiber industry in China is still developing at high speed, along with the increasing production. What's more, the special fiber is one of the key points of the 13th Five-year Plan. Although the polyolefin film fibrillating fiber and its products have many advantages, there is a large gap between China and overseas in polyolefin film fibrillating fiber at present, from the product categories and functionality, stitch density of fibrillating roller, the accuracy of machanism and process stability. It restricts the applications of polyolefin film fibrillating fiber in our country and its special use of air filtration material.

First of all we should study advanced foreign technology of related areas, absorb the core knowledge of equipment and process, and the essence of design idea. Especially we should upgrade the equipment and technology of fibrillation, enrich the theory of fiber preparation, solve the problem of nonuniform denier, fiber broken, and improve the stability of our products and production efficiency. Secondly according to the practical application, through modifying raw materials, we need to take a active part in developing high functional fibrillating fiber, such as antibacterial, green flame retardant, resistance to ultraviolet, far-infrared modification, electret masterbatch, etc. Last but not least, update its variety and quality level to meet the needs of the market, in order to expand its application range. For the nationwide haze presents currently and gets worse and worse, it is hard to solve this problem effectively in the short term. In this aspect, it is urgent to break the foreign monopoly of film fibrillating fiber technology in the field of air filtration material. The research on film fibrillation fiber in the use of air filtration material with high efficiency and low resistance, can not only satisfy the the increasingly stringent requirements to air filter material and equipment, but also provide more effective protection for public health. Besides, it is also in accord with the national industrial policy, energy saving and emission reduction.

Acknowledgments

Fund Project: Project of special fund for development of strategic emerging industries in Shenzhen （JSGG20153024152817） --"High efficiency and low resistivity air filter material development project"
Corresponding Author: Shu Hu, Mainly engaged in conductive/antistatic polymer composite materials and functional fiber research, E-mail:hushu@szselen.com. Mobile phone: +86 18682350972.

References

[1] Tian Zhang, Zuming Hu and JunRong Yu, Synthesis technology and application of polyphenylene sulfide (China synthetic fiber industry, Yueyang, 2012), pp. 36–43.
[2] Zhilian Pan, Shaoqi Zhou , Research on the process and equipment of fibrillating fiber(China synthetic fiber industry, Yueyang, 1979), pp. 26–34.
[3] Lanlan Chen, Chuyang Zhang, Baopu Yin, CN Patent,No. 105,040,127(11 November 2015)
[4] Ming Yi, Split fiber, (Synthetic fiber news, Shanghai, 1974), pp.38.
[5] Shuochen Huang, Synthetic fiber processing machinery for the thermoplastic resin processed into short fibers, split fiber, fibrillated yarn and multifilament (Textile technology overseas--Handbook of chemical fiber, dyeing and finishing, environmental protection, Shanghai, 1992), pp. 11-12, 15.
[6] Zhilian Pan, The development of fibrillating fiber abroad (Synthetic fiber industry,Yueyang, 1980), pp. 65-70.
[7] Changchun Lu, Development of high-performance monofilament technology (Synthetic fiber industry,Yueyang, 2007), pp. 40-43.
[8] Shuyun Chen, Yimin Wang, Oerlikon Barmag:EvoTape- the revolution of tape production (International Textile Guide, Shanghai, 2016), pp. 46-47.
[9] Heinrich Jacgue. thread and fiber of oegantic thermoplastic materials and process of producing the same, U.S. patent,No.2,185,789(2 January 1940)
[10] Beijing plastic products factory, Preliminary summary of the manufacture of polypropylene fibrillating by cutting tools-"fibrillating"method section(Plastic,Beijing,1977), pp. 19-29.
[11] Guojun Liu, Youde Sun,Ming Xie, Study on process of split fiber for cigarette filter tips(Guangdong Chemical Fiber,Guangzhou,1992) ,pp. 9-14.
[12] Guojun Liu, Youde Sun, Ming Xie, Study on process of split fiber for cigarette filter tips (Guangdong Chemical Chemical,Guangzhou, 1993) ,pp. 20-22,39.
[13] Haitao Wang， Junfu Wei，Ao Wang，A novel filtration of cigarette smole prepared by two steps grafting of AA and AMPS monomer onto PP matrix （ Journal of Functional Materials,Chongqin,2013） ,pp. 573-576,580.

[14] Huanfang Zhang, Long Dian, Junfu Wei, Study on Preparation and properties of a new type of polypropylene fiber for cigarette by one-step grafting(New Chemical Materials,Beijing,2016), pp. 161-163,167.
[15] Shu Hu, Hui Guo, CN Patent No. 105,926,079(7 September 2016)
[16] Jingwei Chen, “”Study on the Preparation, Properties and Structure of Hard Elastic Polypropylene Membrane, Master degree thesis, South China University of Technology, 2012.
[17] Masahide Yazawa, Tokio Okada, U.S. Patent No.3,662,935(26 September 1972)
[18] Samuel McMeekin, Drumbeg, Belfast, U.S. Patent No. 3,658,221(5 April 1972)
[19] Carl B.Olson, Carl S.Weisner,Phillio H.Parker, U.S. Patent No.4,129,632(12 December 1978)
[20] MD.Swan, S. Damodaran, M.A. Perez, CN Patent No.1,352,663(5 June 2002)
[21] M.A. Perez, MD.Swan, J.W. Klose, CN Patent No.1,334,886(6 February 2002)
[22] Hendrik.Boeter, CN Patent No. 1,469,769 (21 January 2004)
[23] J. Weinhold, CN Patent No. 102,869,820(9 January 2013)
[24] J. Weinhold, B. Wallner, M. Neubert, CN Patent No. 102,884,229(16 January 2013
[25] Xueyue Zhao, Guocheng Zong, CN Patent No. 2,889,653(18 April 2007)
[26] Theodore.Burkhard, Hayner.Hausdorff, Thomas.Gallman, CN Patent No.103,476,980(9 March 2016)
[27] Chunfeng Liang, CN Patent No. 205,741,405(30 November 2016)

IOP Conf. Series: Materials Science and Engineering **242** (2017) 012031 doi:10.1088/1757-899X/242/1/012031

Removal of phosphate from aqueous solution using modified zeolite clays

Fangqun Gan, Pinzhu Qin, Guan ying, Rong Tang & Jien Hu

College of Environment and Ecology, Jiangsu Open University, Nanjing, 210017, China

Abstract: Phosphate adsorption capabilities of different treated zeolite clays were assessed from aqueous solution. Natural zeolites were treated by thermal activation over 120–800 °C for 2 h, and by hydrochloric ac-id solution from 1%-9%, respectively. The mechanisms of phosphate adsorption of these modified products were also studied through adsorption isotherms and adsorption kinetics experiments. The acid activation in-creased the phosphate sorption capacity and the highest phosphate sorption capacity occurred at AZT9 (zeo-lite activated by 9% hydrochloric acid), while thermal activation did not have the same effect. AZT9 showed higher adsorption rate than natural zeolite (ZT). The phosphate adsorption isotherms of ZT, H400 and AZT9 are fitted well with Langmuir equation in terms of R2 values. It suggested that the adsorption of phosphate on zeolite was homogeneous multilayer adsorption. Based on R2 values, the kinetics of phosphate adsorption on these three zeolites can be satisfactorily described by pseudo-second-order kinetic equation. Results of this study suggested that acid modified zeolite could be potentially used as a promising adsorption media for phosphate removal.

1 Introduction

Zeolite is a kind of aluminosilicate with tetrahedral skeleton structure existing widely in nature. It's chemical composition is [M_2(I),M(II)$OAl_2O_3nSi_2 \cdot mH_2O$, M_2(I) and M(II) are monovalent and bivalence metal ions respectively, usually sodium, potassium, strontium, calcium, barium and so on. There are many holes and channels with uniform size inside the crystal lattice. The surface area is very large, reaching 400~800 m^2 g^{-1}. The zeolites also have great dispersion force and large static electricity, so it has good adsorption performance.

Xiao Juqiang et al. (2003) studied the adsorption of phosphorus by natural clinoptilolite from Gansu. The results showed that the adsorption equilibrium could be reached in 10 minutes. The removal rate of phosphorus is 83.7%. Meng Hong (2001) discussed adsorption effect of the activated zeolite, which had better adsorption effect in the pH range of 4~12, and adsorption capacity could be above 4 mg g^{-1}. The phosphorus removal rate of city sewage reached 84.8%, but the removal efficiency for poly phosphide was not ideal. Zhao Guiyu et al. (2007) used the natural clinoptilolite from Jinyun in Zhejiang province to remove phosphorus from sewage, and the adsorption reaction accorded with the Langmuir equation. At 25 °C, the maximum saturated adsorption capacity of zeolite with 0.3 μm for phosphorus was 101.01 mg kg^{-1}. Sun Jiashou (1997) applied modified zeolite on phosphorus wastewater and found that phosphate adsorption capacity could reach 24.57 mg g^{-1}, the removal rate of phosphorus is greater than 98%, the effluent phosphorus concentration could also be reduced to 0.46 mg L^{-1}. While further experiments showed that less adsorption capacities were tested for phosphoric acid and poly phosphoric acid.

Content from this work may be used under the terms of the Creative Commons Attribution 3.0 licence. Any further distribution of this work must maintain attribution to the author(s) and the title of the work, journal citation and DOI.
Published under licence by IOP Publishing Ltd

Therefore, this study was designed to: (1) modify natural zeolite with different simple methods to form zeolite products, then (2) to evaluate the potentials of using the zeolite for phosphate removal from wastewater in batch adsorption experiments, and finally (3) to ascertain the practical applicability of the zeolite products for real wastewaters.

2 Materials and methods

2.1 *Adsorbents*

Natural zeolite clay (ZT) used was from Dantu County (China). The natural sample was manually ground and selected for particles < 160 μm. The bulk chemical analysis of the ZT was listed in Table 1.

Thermal activated zeolites were prepared by heating over a range of temperatures (120, 200, 300, 400, 500, 600, 700 and 800 °C) at a period of 2 h. They were labeled as H120, H200, etc.

Thirty grams of natural zeolite were dispersed into 250 mL of hydrochloric acid solution (volume fraction of 1%, 3% and 9%) at 70°C. After a mixing time of 2 h, the clay suspension was centrifuged after cooling, then the solid sediments was washed with distilled water until no Cl^- was tested by $AgNO_3$ test solution. The products were dried at 105 °C, ground and stored in polyethylene bottles before use. These were labeled AZT1, AZT3 and AZT9.

Table 1. Bulk chemical analysis of the zeolites.

Chemical analysis (wt.%)	ZT	H400	AZT9
Al_2O_3	13.72	14.55	13.66
Fe_2O_3	1.93	2.13	1.73
CaO	2.03	2.21	0.98
MgO	2.20	2.33	1.99
MnO_2	0.037	0.039	0.03
K_2O	1.07	1.42	1.26
Na_2O	1.20	1.66	1.42
TiO_2	0.21	0.24	0.23
P_2O_5	0.10	0.11	0.01
SiO_2	65.92	67.57	68.68
LOI[a]	11.54	7.34	9.87

[a] Loss on ignition at 950 °C

2.2 *Batch adsorption experiments*

2.2.1 Single point adsorption experiments. To evaluate the impact of heated zeolites on the phosphate adsorption capacity, natural and modified zeolite with different temperature were used. A defined volume of the phosphate stock solution, with concentrations of 10 mg P/L was added to 0.5 g of adsorbents. The mixture was stirred at 200 rpm in thermostatic shaker for a defined period, then the orthophosphate removal was calculated and the optimum material was selected. To evaluate the impact of acid activated zeolite on the phosphate adsorption capacity, natural and modified zeolite with different acid volume were used. A defined volume of the phosphate stock solution, with concentrations of 20 mg P/L was added to 0.5 g of adsorbents. The remaining steps were the same as described above.

2.2.2 Phosphate adsorption isotherms. Phosphate adsorption isotherms were carried out with different initial concentrations of phosphate and a fixed concentration of the adsorbents at room temperature (25 °C). 0.5 g of the sample was loaded in 100 mL conical flasks, and 50 mL of PO_4^{3-} solution was then added. Thirteen levels of initial phosphate concentrations (0, 3, 6, 10, 30, 50, 80, 100, 120, 140, 160, 180 and 200 mg P/L) were used. The flask was capped and stirred magnetically at 200 rpm for 24 h to ensure approximate equilibrium. After phosphate adsorption, the solution was filtered through a 0.22 μm membrane filter and then analyzed for PO_4^{3-}. The quantity of adsorbed phosphate (adsorption capacity) was calculated from the decrease of the phosphate concentration in solution.

2.2.3 Phosphate adsorption kinetics. Phosphate adsorption kinetics was evaluated at room temperature (25 °C) and at original pH value. The initial phosphate concentration was 10 mg P/L while the adsorbent quantity was 0.5 g. Reaction solutions were sampled at intervals between 0 and 24 h of adsorption.

2.2.4 pH effect on phosphate adsorption. The effect of pH on phosphate removal by H400 was examined in the pH region between 2 and 10. The tests were carried out at constant experimental conditions with an initial solution concentration 10 mg P/L and 0.5 g adsorbent. All the tests were carried out at constant retention time, fixed at 24 h.

2.2.5 Sample Analysis. XRD patterns of the prepared samples were acquired with a D/MAX2200 X-ray diffractometer using CuKα radiation (40 kV, 40 mA) and a Ni filter. All XRD patterns were obtained from 3.0° to 60.0° with a scan speed of 4.0 °/min. Chemical analyses of the adsorbents were determined by $LiBO_2$-ICP-AES method (Soil Science Society of China, 2002), using an induced coupled plasma atomic emission spectrophotometer (ICP-AES) (Thermo IRIS Advantage). The analysis of phosphate (as phosphorous) was done by the molybdenum-blue ascorbic acid method with a UV-Vis spectrophotometer (UV/VIS 721 model) (Soil Science Society of China, 2002).

3 Results and discussion

3.1 *Phosphate adsorption capacity of activated zeolites*

The phosphate adsorption capacities of eight thermal treated zeolites are given in Figure 1(a). The curves showed that zeolite heated at different temperature had different phosphate adsorption capacity. Compared with ZT, thermal treated samples did not behave better phosphate adsorption ability. When the heating temperature increases to 400 °C, the adsorption capacity increases with temperature, and then the phosphate adsorption capacity no longer increases, even reduces. In order to clarify the phosphate adsorption mechanism, so H400 was still chosen as the next batch adsorption experimental material. Figure 1(b) showed the phosphate adsorption capacities of three acid treated zeolites, compared with natural sample. While the acid solution was set at 9%, the phosphate adsorption capacity reaches the maximum, followed by 3% and 1%. In comprehensive consideration of removal efficiency and economic costs, AZT9 was used as subsequent test material.

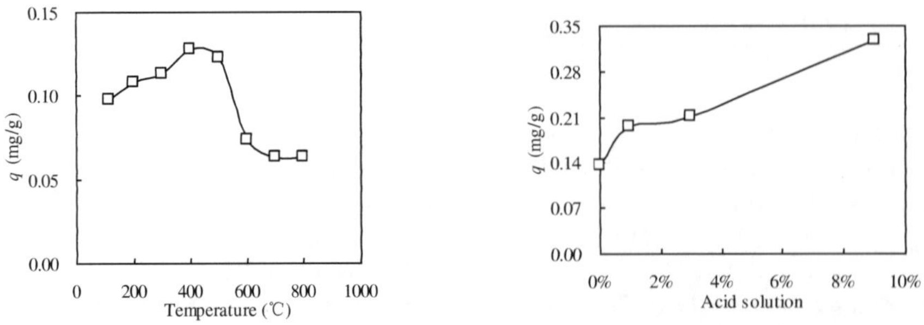

(a) heated zeolites (b) acid activated zeolites

Fig. 1 Variation of the phosphate adsorption capacity of heated zeolites and acid activated zeolites

3.2 *Maximum adsorption capacity*

The isothermal adsorption data of ZT, H400 and AZT9 are summarized in Figure 2. According to Fig. 2, AZT9 had highest adsorption capacities than the H400 and AT, which were 1.18 mg/g, 0.44 mg/g and 0.39 mg/g, respectively, at the phosphate initial concentration of 200 mg/L. It is noted here that H400 had higher adsorption capacities than ZT when the initial phosphate concentration was above 10 mg/L. The adsorption data were fitted with the Langmuir isotherm:

$$q = q_m bC/(1+bC)$$

Where q (mg/g) and C (mg/L) are the equilibrium concentrations of an adsorbate on the zeolite and in the aqueous solution, respectively, b is adsorption isothermal constant (Table 2). The adsorption data agree with the Langmuir isotherm reasonably. With reference to phosphate adsorption tests, parameters such as size of adsorbate, retention time and phosphate concentration have varied, and these factors also influence the outcomes of experiments. And there are no researchers who have carried out the investigations exactly in the same way and data should not be compared (Westholm, 2006). Therefore, data on P adsorption capacity should only be compared when similar experimental conditions have been employed (Gan et al., 2011).

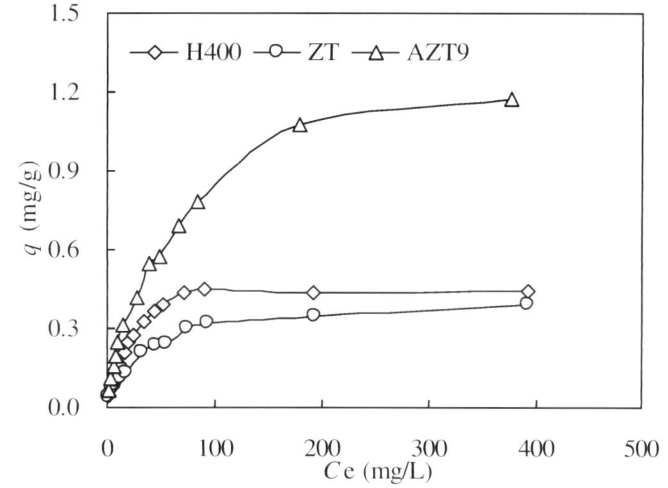

Fig. 2 Adsorption isotherms of natural zeolite (ZT), H400 and AZT9 plotted as equilibrium adsorbed amount (q) vs. equilibrium concentration (Ce) at 25 °C

Table 2 Estimated isotherm and kinetic model parameters for phosphate adsorption.

T (25 °C)	Langmuir isotherm			Second-order equation	
	q_m (mg/g)	b	R^2	k_2 g/(mg min)	R^2
ZT	0.285	0.067	0.974	62.31	0.9977
H400	0.503	0.044	0.996	49.69	0.9995
AZT9	0.735	0.055	0.977	306.58	0.9999

3.3 *Phosphate adsorption kinetic results*

In Figure 3, the results of phosphate adsorption kinetic data for ZT, H400 and AZT9 at 25 °C are presented. The AZT9 had fast kinetics than AT and H400, the majority of phosphate adsorption on the AZT9 was completed in 0.5 h. Four typical kinetic equations were used for fitting the experimental data, while estimated parameters of pseudo second-order equation were listed based on R^2. All the adsorption system of phosphate/ZT, H400 and AZT9 obey the pseudo second-order kinetic model and thus supports the assumption behind the model that the adsorption is due to chemisorption (Demirbas, 2008).

3.4 *pH effect on phosphate adsorption*

The impact of pH on the phosphate adsorption on the H400 is shown in Figure 4. As seen from Figure 4, the pH value of the phosphate solution plays an important role in the whole adsorption process and particularly on the adsorption capacity. The phosphate adsorption tends to decrease with the increase of pH for the ZT, from 0.26 mg P/g at pH 3.0 to 0.02 mg P/g at pH 9.0. Similar trends were also observed for phosphate adsorption on fly ash (Agyei et al., 2002) and other adsorption materials. The low phosphate uptake in alkaline pH range can be attributed to the competition of hydroxide ions with phosphate ions for adsorption sites. The pH of the aqueous solution is an important variable that influences the adsorption of anions and cations at the solid-liquid interfaces.

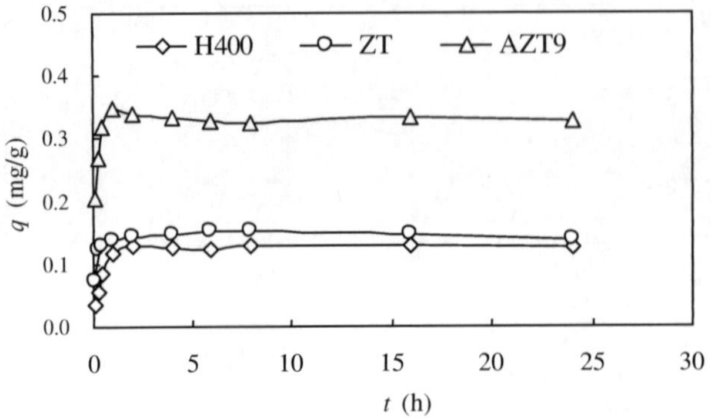

Fig. 3 Phosphate adsorption kinetic data for ZT, H400 and AZT9 (C_0=10 mg/L)

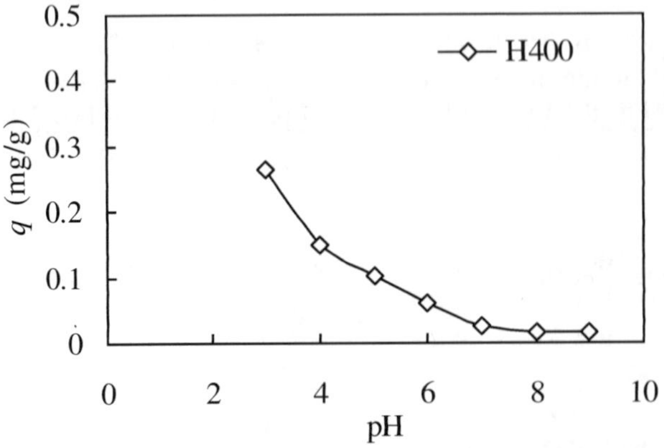

Fig. 4 Impact of pH on phosphate adsorption capacity of AT and H400 (C_0=200 mg/L)

4 Conclusions

The performance characteristics were investigated to remove phosphate for heated zeolites and acid activated zeolites. It has been demonstrated that the natural zeolite has phosphate adsorption ability and can be used as a low cost adsorbent. Hydrochloric acid activation is an effective method to enhance the adsorption capacity of the natural clay, and is also an easily handling operation compared with other chemical activation.

References

[1] Agyei N., Strydomb C., Potgieter J. (2002). The removal of phosphate ions from aqueous solution by fly ash, slag, ordinary Portland cement and related blends. *Cement Concrete Research*. 32: 1889-1897.

[2] Gan F., Zhou J., Wang H., *et al.* (2011). Phosphate adsorption on granular palygorskite: Batch and column studies. *Water Environment Research*. 83(2): 147-153.
[3] Meng H., Peng C., Xiao J., *et al.* (2001). Study on removal of phosphide activated zeolite. *Comprehensive Utilization of Mineral Resources*. 4: 10-14 (in Chinese).
[4] Westholm, L. (2006). Substrates for phosphorus removal potential benefits for on-site wastewater treatment? *Water Research*. 40: 23-36.
[5] Soil Science Society of China. (2002). Soil Agrochemistry Analysis Method (in Chinese). China Agricultural Press, Beijing, China.
[6] Sun J., Liu Y. (1997). Study on the adsorption characteristics of cross-linked clay minerals (II) - selective adsorption of phosphate ions by aluminum pillared montmorillonite]. *Journal of Wuhan Institute of Chemical Technology*. 19(1): 34-37 (in Chinese).
[7] Xiao J., Li Yujin, Zheng Hui. (2003). Study on phosphorus removal performance of zeolite. *Gansu Environmental Research and Monitoring*. 16 (3): 252-254 (in Chinese).
[8] Zhao G., Zhou Q. (2007). Adsorption of phosphorus from wastewater onto zeolite. *Technology of Water Treatment*. 33(2): 34-37 (in Chinese).

IOP Conf. Series: Materials Science and Engineering **242** (2017) 012032 doi:10.1088/1757-899X/242/1/012032

Progress on preparation of graphene and its application

Zhenyu Wang

Wuhan University of Technology, Wuhan 430000, China

1300334274@qq.com

Abstract: Graphene is a kind of new material with special chemical structure and excellent phy sical properties, demonstrating great application prospects in the fields of electronics, physics and chemistry. In this paper, some preparation methods of graphene were reviewed, and the advantages and disadvantages of these methods were analyzed, and the newest applications and development prospects of graphene were reviewed.

1. Introduction

Graphene, the parent of all graphitic forms (Figure 1), has become one of the most exciting topics of research in the last three to four years. Graphene is a one-atom-thick planar sheet of sp2-bonded carbon atoms densely packed in a honeycomb crystal lattice. It is the mother element of some carbon allotropes, including graphite, carbon nanotubes, and fullerenes [1]. Research on graphene has been successfully capitalized after its discovery in 2004 by Novoselov et al. at The University of Manchester [2]. This revolutionary discovery has added a new dimension of research in the fields of physics, chemistry, biotechnology, and materials science. The ''thinnest'' known material graphene exhibits excellent electrical conductivity, mechanical flexibility, optical transparency, thermal conductivity, and low coefficient of thermal expansion behavior [3]. The unique features of graphene have attracted tremendous interest both in academics and industry.

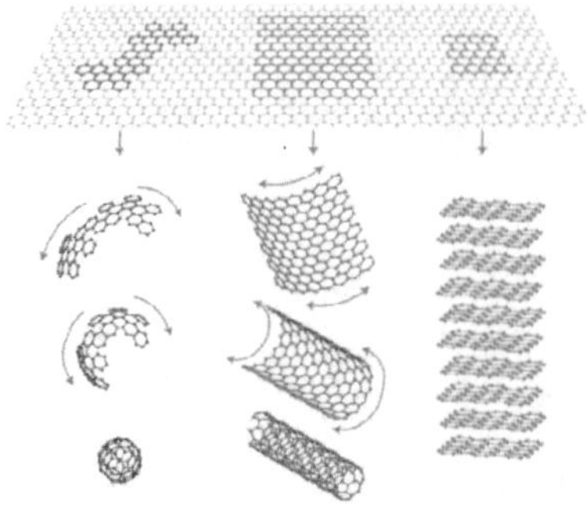

FIGURE 1 Graphene: the parent of all graphitic forms [4]

2. Graphene Synthesis Methods

To date, several techniques have been established for graphene synthesis. However, mechanical cleaving (exfoliation) (Novoselov et al. 2004) [2], chemical exfoliation (Allen et al. 2010; Viculis et al. 2003)[5, 6], chemical synthesis (Park and Ruoff 2009)[7], and thermal chemical vapor deposition (CVD) (Reina et al. 2009) synthesis are the most commonly used methods today[8]. Some other techniques are also reported such as unzipping nanotube (Jiao etal. 2010; Kosynkin et al. 2009; Jiao et al. 2009) and microwave synthesis (Xin et al. 2010)[9-12]; however, those techniques need to be

Content from this work may be used under the terms of the Creative Commons Attribution 3.0 licence. Any further distribution of this work must maintain attribution to the author(s) and the title of the work, journal citation and DOI.
Published under licence by IOP Publishing Ltd

explored more extensively. An overview of graphene synthesis techniques is shown in the flow chart in Figure 2.

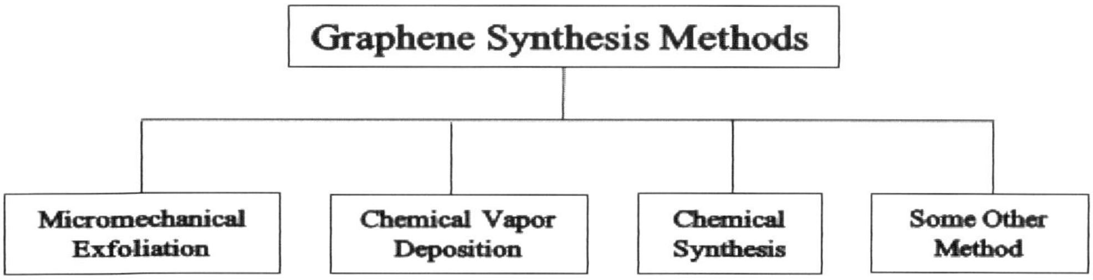

FIGURE 2 Different graphene synthesis methods

2.1 Micromechanical Exfoliation

Micromechanical exfoliation is the first recognized method of graphene synthesis. Novoselov et al. (2004) used adhesive tape to produce a single graphene layer by a mechanical cleaving technique from 1-mm-thick HOPG [2]. Micromechanical exfoliation can be used to generate "high quality" graphene that is electrically isolated for fundamental studies of transport physics and other properties, but does not yet appear to be scalable to large area. It typically produces graphene "particles" with lateral dimensions on the order of tens to hundreds of micrometers.

While mechanical cleavage of graphene layers from a graphite crystal has afforded the study of the properties of single-layer graphene or bilayer graphene, the method is not suitable for large scale synthesis of single-layer graphene or of few-layer graphene.

2.2 Chemical Vapor Deposition (CVD)

Chemical vapor deposition is a chemical process by which a substrate is exposed to thermally decomposed precursors and the desired product deposited onto the substrate surface at high temperature Because the high temperature is not desired in many cases, plasma-assisted decomposition and reaction may lower the process temperature[13].

When a thermal CVD process involves chemical reactions of the reacting gases by generating plasma* inside a vacuum chamber, which leads to the deposition of thin film on the substrate surface, the process is known as plasma-enhanced chemical vapor deposition (PECVD). The plasma can be generated inside a PECVD system using RF (AC frequency), microwave, and inductive coupling (electrical currents produced by electromagnetic induction) PECVD has a wide range of advantages over other conventional CVD methods. By this technique, aprocess can be carried out at relatively low temperature compared to the other thermal CVD processes; hence, it is more feasible for industrial-scale applications [14].

2.3 Chemical synthesis: graphene from reduced graphene oxide

Chemical synthesis is the first method that demonstrated graphene synthesis by a chemical route. In the year 1962, Boehm et al first demonstrated monolayer flakes of reduced graphene oxide, which was recently acknowledged by the graphene inventor Andre Geim. The method involves the synthesis of a graphite oxide (GO) by oxidation of graphite, dispersing the flakes by sonication, and reducing it back to graphene[15]. There are three popular methods available for GO synthesis: the Brodie method (Brodie 1860) [15], Staudenmaier method (Staudenmaier 1898) [16], and Hummers and Offeman method (Hummers and Offeman 1958) [17]. All three methods involve oxidation of graphite using strong acids and oxidants. Brodie and Staudenmaier used a combination of potassium chlorate ($KClO_3$) with nitric acid (HNO_3) to oxidize graphite, and the Hummers method involves treatment of graphite with potassium permanganate ($KMnO_4$) and sulfuric acid (H_2SO_4). Graphite salts made by intercalating graphite with strong acids such as H_2SO_4, HNO_3 or $HClO_4$ have also been used as precursors for the subsequent oxidation to GO.

2.4 Organic Synthesis

Synthetic organic chemists attempt to prepare graphene with full controllability by another strategy, in which certain type of graphene sheets can be synthesized by organic reactions. In organic chemistry, polycyclic aromatic hydrocarbons are considered as two-dimensional graphene segments composed of all-sp2 carbons. Pioneering contributions to the synthesis of polycyclic aromatic hydrocarbons were made by Scholl and Clar in the first half of the 20th century [18].

3. Applications of graphene

The fascinating of properties graphene and their derivatives enable them to be used in many potential applications [15-19].

3.1 Hydrogen storage material

Hydrogen has high energy density by mass. Therefore, it is considered as a key energy carriers. For practical applications, storage of hydrogen requires high gravimetric and volumetric densities, fast reaction kinetics, a low adsorption temperature, good reversibility, and low cost. The most common methods of storing hydrogen are in the form of gas in pressurized cylinders, cryogenic liquid, or as adsorbed gas in solid materials. Currently, the high temperature dissolution of hydrogen in metal hydrides poses a problem. Therefore, storage of hydrogen in materials by chemical process is indeed a safer method. Graphene can be an ideal material for hydrogen storage owing to the light weight, high surface area, and chemical stability. Hydrogen can be chemically stored in graphene by physisorption or chemisorption.

3.2 Energy storage devices

Graphene is a promising electrode material due to its high theoretical surface area and electron transferability along its 2-D surface. Graphene based electrodes are used as rechargeable lithiumion batteries and electrochemical double-layer capacitors. Graphene nanocomposites can form a conducting 3-D network due to the uniform dispersion of silicon particles and reconstitution of graphene platelets, which are important aspects for the high storage capacities

3.3 Memory and Photovoltaic Devices

Graphene-related nanomaterials have also been used in memory devices, transparent electrodes, electron acceptors, and light adsorbers because of their good electronic properties, transparency, and large specific surface area.

3.4 Transparent Conductive Films

Graphene materials are promising candidates for transparent conductive films because of their high carrier mobility, electrical conductivity, and optical transmittance in the visible range of the spectrum. Graphene-base transparent conductive films have been used as electrodes for dye-sensitized solar cells, liquid crystal devices, and organic light-emitting diodes.

4. Conclusions and Perspective

In this article, the preparationand applications of graphene have been reviewed. Preparation of high quality graphene materials in a costeffective manner and on the desired scale is essential for many applications. Since new methods to synthesize graphene keep getting more and more interesting, we hope to see a solution that provides not just a fair balance between scalability and quality, but the best of both. The aim of this article is to provide an insight into graphene synthesis technology by providing an introduction and comparative study. Future efforts for graphene and n-layer graphene such as achieving desired surface functionalization and e.g., the "cutting" or preparation into desired shapes, could generate novel structures with many applications.

References

[1] Sur U K. Graphene: A Rising Star on the Horizon of Materials Science [J]. International Journal of Electrochemistry, 2012, (2012-9-20), 2012, 2012.
[2] Novoselov K S, Geim A K, Morozov S V, et al. Electric field effect in atomically thin carbon films [J]. science, 2004, 306(5696): 666-669.
[3] Hansora D P, Shimpi N G, Mishra S. Graphite to Graphene via Graphene Oxide: An Overview on Synthesis, Properties, and Applications [J]. JOM, 2015, 67(12):2855-2868.
[4] Geim A K, Novoselov K S. The rise of grapheme [J]. Nature materials, 2007, 6(3): 183-191.
[5] Allen Mattew J, Tung Vincent C, Kaner Richard B. Honeycomb carbon a review of graphene[J]. Chem. Rev, 2010, 110(1): 132-145.
[6] Viculis L M, Mack J J, Kaner R B. A chemical route to carbon nanoscrolls [J]. Science, 2003, 299(5611): 1361-1361.
[7] Park S, Ruoff R S. Chemical methods for the production of graphenes [J]. Nature nanotechnology, 2009, 4(4): 217-224.
[8] Reina A, Thiele S, Jia X, et al. Growth of large-area single-and bi-layer graphene by controlled carbon precipitation on polycrystalline Ni surfaces [J]. Nano Research, 2009, 2(6): 509-516.
[9] Jiao N, Herndl G J, Hansell D A, et al. Microbial production of recalcitrant dissolved organic matter: long-term carbon storage in the global ocean [J]. Nature Reviews Microbiology, 2010, 8(8): 593-599.
[10] Kosynkin D V, Higginbotham A L, Sinitskii A, et al. Longitudinal unzipping of carbon nanotubes to form graphene nanoribbons[J]. Nature, 2009, 458(7240): 872-876.
[11] Jiao L, Zhang L, Wang X, et al. Narrow graphene nanoribbons from carbon nanotubes [J]. Nature, 2009, 458(7240): 877-880.
[12] Xin H S, Schaefer D M, Liu Q P, et al. Effects of polyurethane coated urea supplement on in vitro ruminal fermentation, ammonia release dynamics and lactating performance of Holstein dairy cows fed a steam-flaked corn-based diet[J]. Asian-Aust. J. Anim. Sci, 2010, 23: 491-500.
[13] Chiu J, Yeh C, Chiu P. Method for growing graphene by chemical vapor deposition: U.S. Patent 20,170,144,888[P]. 2017-5-25.
[14] Mathur S, Gonullu Y, Fischer T. Interfacial Modification of Heterojunction Metal Oxide Photo Anodes for Efficient Solar Water Splitting[C]//Meeting Abstracts. The Electrochemical Society, 2017 (25): 1215-1215.
[15] Brodie B C. Sur le poids atomique du graphite [J]. Ann. Chim. Phys, 1860, 59(466): e472.
[16] Staudenmaier L. Verfahren zur darstellung der graphitsäure [J]. European Journal of Inorganic Chemistry, 1898, 31(2): 1481-1487.
[17] Hummers Jr W S, Offeman R E. Preparation of graphitic oxide [J]. Journal of the American Chemical Society, 1958, 80 (6): 1339-1339.
[18] Kissel P, Erni R, Schweizer W B, et al. A two-dimensional polymer prepared by organic synthesis [J]. Nature chemistry, 2012, 4(4): 287-291.
[19] Hansora D P, Shimpi N G, Mishra S. Graphite to graphene via graphene oxide: an overview on synthesis, properties, and applications[J]. JOM, 2015, 67(12): 2855-2868.

IOP Conf. Series: Materials Science and Engineering **242** (2017) 012033 doi:10.1088/1757-899X/242/1/012033

Effect of sieving polymer concentration on separation of 100 bp DNA Ladder by capillary gel electrophoresis

T Nakazumi[1] and Y Hara[1]*

[1]Research Institute for Sustainable Chemistry, ISC, National Institute of Advanced Industrial Science and Technology, AIST, Central 5-2, 1-1-1 Higashi, Tsukuba 305-8565, Japan; E-Mail: nakazumi.t@aist.go.jp

*Correspondence: y-hara@aist.go.jp; Tel.: +81-29-81-6236

Abstract. We studied the effect of sieving polymer concentration on separation of a 100 bp DNA Ladder by capillary gel electrophoresis (CGE) using hydroxyethyl cellulose (HEC) with a molecular size of 1000 k. For measurement purposes, we selected a fused silica capillary with total length of 15 cm and effective length of 7.5 cm; this was applied to compact CGE equipment for a Point-Care-Testing (POCT) system. Measurement results of the 100 bp DNA Ladder sample indicated that small DNA separation was significantly affected by HEC sieving polymer concentration. This was due to the level of entanglement between small DNA molecules and the sieving polymer chain significantly influencing migration time, mobility, and resolution length of the CGE process. We concluded that 1.0 w/v % HEC sieving polymer concentration was optimal for CGE separation of DNA ≥ 1000 bp in the 100 bp DNA Ladder (100–1500 bp) when using the short-length capillary.

1. Introduction

Capillary gel electrophoresis (CGE) is a very useful tool for nucleic acid separation [1-4] and is especially suitable for separation of small DNAs produced by the polymerase chain reaction (PCR). For separation of PCR products by the CGE method, polymer solutions filled in capillary are needed to serve as the sieving polymer [5, 6]. Furthermore, in order to increase the accuracy of CGE separation analysis, many different aspects need to be considered, such as the type of sieving polymer solution, buffer solution, and migration process, among others [12-22]. There have been many previous studies regarding different sieving polymer solutions, such as poly(acrylamide) [7,8], poly(ethylene oxide) (PEO) [9], cellulose and its derivatives [10,11], and poly(vinyl pyrrolidine) (PVP) [12]. Linear polymer chains are basically used as the sieving polymer solution for separation of small DNAs during the CGE process; the linear polymer solution can be easily changed to the new sieving polymer solution with each CGE measurement. During the CGE separation process, DNA samples are entangled with the sieving polymer chain. Samples with different lengths interact differently with the sieving polymer chain and the migration time of the different DNA samples therefore also differs. Migration time is thus influenced by the level of entanglement between the sieving polymer chain and DNA samples [13, 14]. In this study, to prevent electroosmotic flow (EOF) during the CGE DNA separation process, a linear polyacrylamide chain was selected as the coating polymer for the fused silica capillary wall [15, 16]. The direction of EOF in the fused silica capillary is opposite to the direction of flow of DNA samples [17-19]. If conducting CGE without polymer coating on the capillary wall, DNA sample peaks cannot be detected because samples cannot be inserted into the capillary due to the prevention of EOF in the opposite direction to DNA samples. Because of this,

Content from this work may be used under the terms of the Creative Commons Attribution 3.0 licence. Any further distribution of this work must maintain attribution to the author(s) and the title of the work, journal citation and DOI.
Published under licence by IOP Publishing Ltd

there have been many discussion of how to manage EOF by using coating polymers for the capillary wall; examples include cross-linked polyacrylamide [20], poly(N-acryloyl aminoethoxyethanol) [21], and poly(N-acryloyl aminopropanol) [22], among others.

In this study, we investigated the effect of sieving polymer concentration on separation of a 100 bp DNA Ladder. Hydroxyethyl cellulose (HEC), with a molecular size of 1000 k, was selected as the sieving polymer chain. The capillary had total length of 15 cm and effective length of 7.5 cm; these dimensions were selected to apply a short-size capillary to a compact CGE system that we are now developing. The compact CGE equipment is expected to be applicable for a Point-Care-Testing (POCT) system. Compared to slab gel electrophoresis (SGE), the CGE method is faster and requires smaller volume of DNA samples, with high-resolution in real time for analysis of PCR products. To develop the compact CGE system for POCT, in this investigation, we studied the influence of sieving polymer concentration on the separation process when using the short-size capillary to increase separation accuracy. Through measurement of small DNA samples, we show that the results of DNA separation were significantly affected by the sieving polymer concentration, due to entanglement between DNA molecules and sieving polymer chains.

2. Materials and Methods

2.1. Procedure for coating of short-size fused silica capillary

Firstly, the capillary was washed by using 1 N NaOH (15 min), water (15 min), and methanol (15 min). Next, an aqueous solution totaling 20 ml (containing 3-Methacryloxypropyltrimethoxysilane (Shin-Etsu chemical) (80 μL) and methanol (1 ml), with the addition of one drop of acetic acid) was sucked up into the capillary for 2 h at room temperature. Through this process, the silane coupling agent was covalently bonded to the glass wall of the fused-silica capillary. After washing with methanol and water, the monomer solutions were sucked up into the fused silica capillary for 2 h at room temperature. The monomer aqueous solutions were prepared by dissolving the acrylamide monomer (0.7 g), ammonium persulphate (APS) (20 mg), and N,N,N',N'-tetramethylethylenediamine (TEMED) (20 μL) in 20 ml aqueous solution. Before use, the monomer solutions were degassed for 30 min using nitrogen. The fused silica capillaries were then rinsed with water.

2.2. Self-constructed instrument for capillary gel electrophoresis (CGE)

All CGE measurements were performed with a self-constructed instrument composed of a high-voltage power supply (HJPQ-10P3, Matsusada) and a microscope with epi-illumination (IX73, Olympus, Tokyo, Japan) (Fig. 1). To detect the conjugate of SYBR Green II and DNA, a mercury lamp was used through the optical filter (U-FBWA, Olympus, Tokyo, Japan). By passing through the filter, a wavelength of 460–495 nm was utilized as the excitation light source. Fluorescence originating in the DNA samples was collected using a × 60 objective lens (UPlanFLN, Olympus) and the fluorescence signal was detected by a photomultiplier tube (PMT) (H8249-101, Hamamatsu Photonics, Hamamatsu, Japan). In addition, the signal received at the PMT was converted to a digital signal using National Instrument NI USB-6341. LabVIEW software (National Instrument) was used to control the applied voltage and acquire digital data. Fused silica capillaries with 75 μm circles (Polymicro Technologies, Phoenix, AZ, USA) were cut to 15 cm. As noted above, the effective length of the 15 cm capillary was 7.5 cm. To inject the DNA samples into the capillary, 1.5 kV was applied for 1 s; separation was then performed at 100 V/cm.

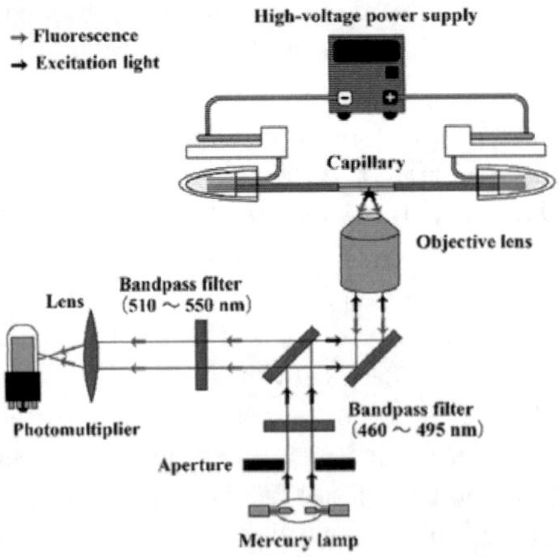

Figure 1. Schematic illustration of the capillary gel electrophoresis (CGE) instrument

2.3. CGE chemicals.

To prepare the running buffer for this experiment, 5 × TBE buffer (44.5 mM Tris, 44.5 mM boric acid, 1.0 mM EDTA, pH 8.3) (Nippon Gene) was diluted to 0.5 × with ultrapure water. The sieving polymer selected was hydroxyethyl cellulose (HEC), with a molecular size of 1000 k. The composition of the HEC polymer solution was adjusted to 0.5%, 0.8% and 1.0 w/v %, including 0.5 × TBE buffer and 2 × SYBR Green II (TAKARA BIO). The 100 bp DNA Ladder (TAKARA BIO), comprised of 11 double-stranded fragments with 100 bp, 200 bp, 300 bp, 400 bp, 500 bp, 600 bp, 700 bp, 800 bp, 900 bp, 1000 bp, and 1500 bp, was selected as the measurement target. The 130 μg/mL 100 bp DNA Ladder was diluted 10 times and was used in the CGE injection sample.

3. Results and Discussion

In this study, the fused silica capillary was coated with the polyacrylamide linear polymer chain, coupled with the silane coupling agent. The acrylamide linear polymer chain was synthesized after washing the capillary and chemically bonded to the silane coupling agent. When we measured small DNA samples using the capillary without coating, we did not detect the DNA peaks; this is because the DNA samples do not insert into the capillary due to EOF with opposite direction to that of the DNA samples. The capillary coating procedure is therefore very important for analysis of small DNAs by CGE.

Fig. 2 shows CGE measurement results for the 100 bp DNA Ladder using different concentrations of the sieving polymer chain (HEC, 1000 k). At > 1.0 w/v % HEC polymer solution, it was difficult to insert the polymer solution due to its high viscosity. In this study, we therefore measured CGE using concentrations ≤ 1.0 w/v %. As shown in Fig. 2, the fluorescence intensity at 1.0 w/v % was lower than at 0.5 w/v % and 0.8 w/v %. We considered this to be related to the insertion process of small DNAs. When the sieving polymer concentration increased, the mesh density of the sieving polymer chain increased. The introduction amount of small DNA samples therefore decreased and fluorescence intensity also decreased, as fluorescence intensity is determined by the amount of small DNA samples introduced.

To aid in evaluation of the electropherograms in Fig. 2, Fig. 3 summarizes the relationship between migration time and HEC concentration during the CGE process. As shown, migration time increased with an increase in sieving polymer concentration at all lengths of the small DNA samples. As the sieving polymer concentration increased, entanglement between the sieving polymer chain and DNA samples increased. With high concentrations of sieving polymer, DNA sample detection time was therefore longer than with low polymer concentrations. In addition, Fig. 4 shows the relationship between mobility and concentration of the HEC solution. As shown, when HEC concentration decreased, mobility increased, because this is determined by the velocity of DNA samples in the capillary filled with HEC. As the HEC polymer concentration increases, electrophoresis of the DNA samples is inhibited. We therefore considered the mobility of DNA samples to be significantly affected by the concentration of the sieving polymer chain under constant applied voltage conditions.

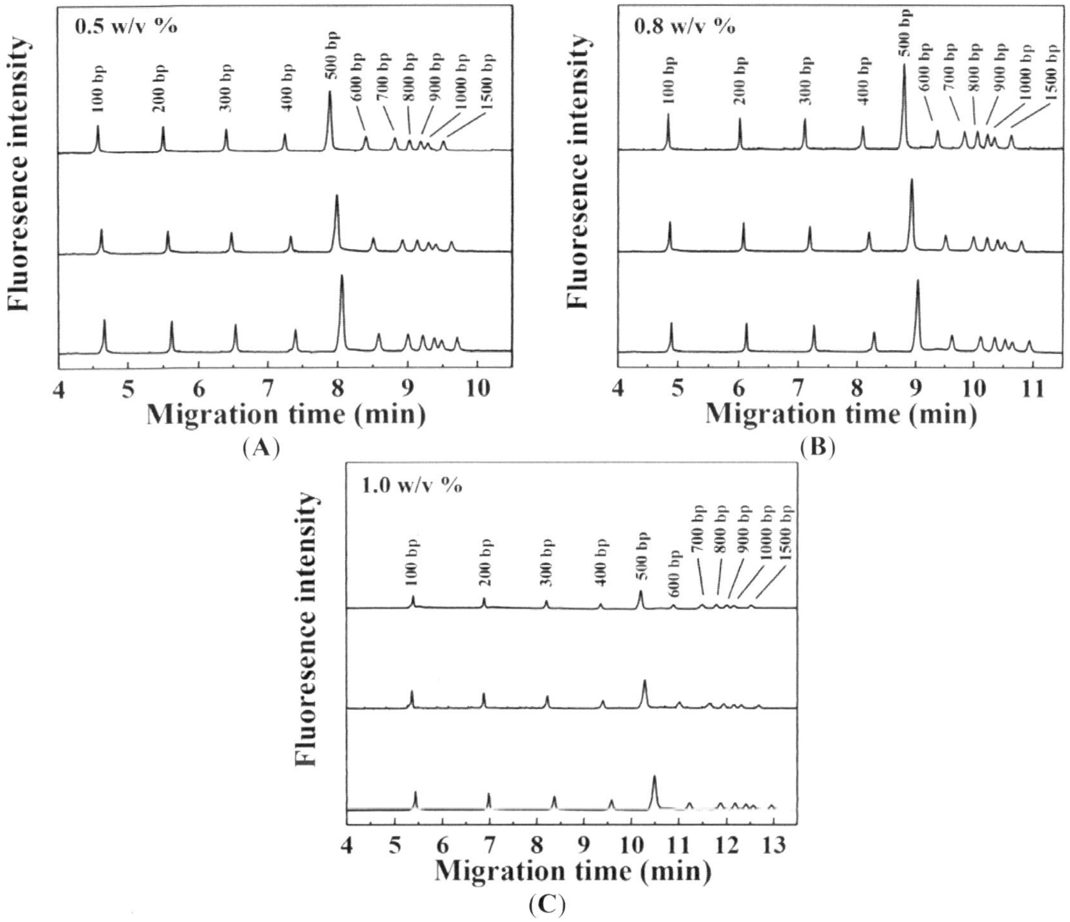

Figure 2. Separation of 100 bp DNA Ladder using different sieving polymer concentrations; (A) 0.5 w/v %, (B) 0.8 w/v %, and (C) 1.0 w/v %

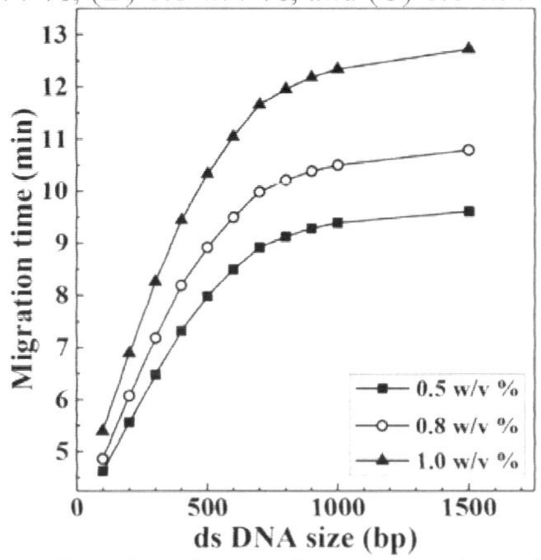

Figure 3. Relationship between migration time and concentration of HEC (as the sieving polymer) during CGE

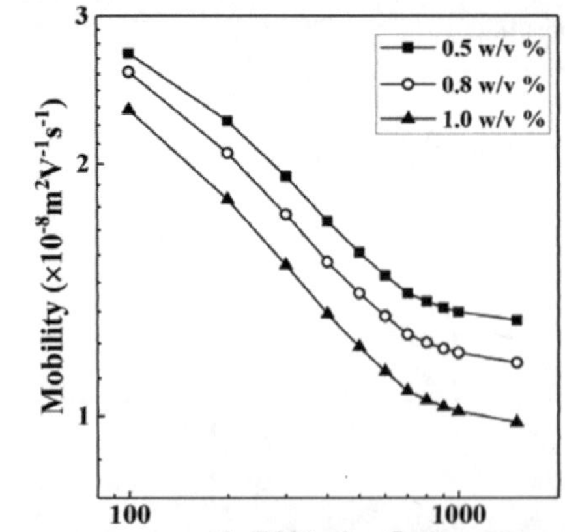

Figure 4. Relationship between mobility and concentration of HEC (as the sieving polymer) during CGE

Figure 5 shows the resolution length results of the 100 bp DNA Ladder samples. Resolution (Rs) was calculated using the following equation:

$$Rs = 1.18\ \Delta t / (w_{0.5}1 + w_{0.5}2) \tag{1}$$

Δt is the difference in migration time of the adjacent two peaks in the electropherograms. $w_{0.5}$ represents full width at half maximum of the peak in the graph. The resolution length (RSL) was determined by the following equation:

$$RSL = \Delta n / Rs \tag{2}$$

where Δn is the DNA length-difference of adjacent peaks in the graph, and Rs is the resolution value. As shown in Figure 5, the resolution length of the 100 bp DNA Ladder was almost the same ≤ 900 bp. At values ≥ 1000 bp, the difference in the resolution length increased. When the sieving polymer concentration was 1.0 w/v %, the RSL value was notably lower than at other concentrations ≥ 1000 bp. Lower RSL values indicate higher separation performance. Under these CGE conditions, for clear separation of the 1000–1500 bp DNA samples with the short-length capillary, we therefore conclude that 1.0 w/v % HEC (1000 k) sieving polymer concentration is optimal.

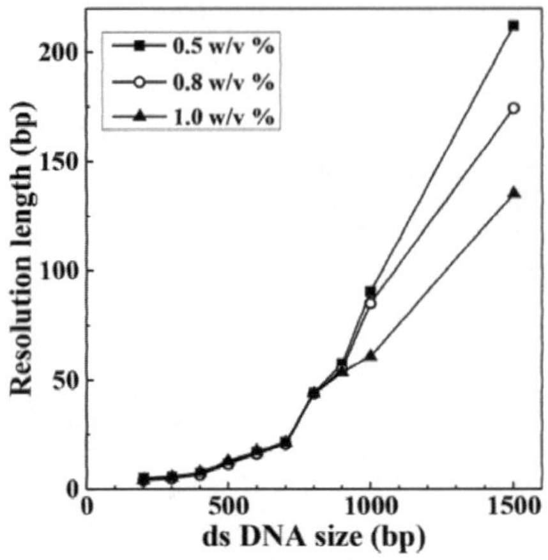

Figure 5. Relationship between resolution length and concentration of HEC (as the sieving polymer) during CGE

4. Conclusions

In this study, we investigated the influence of the concentration of sieving polymer (HEC, molecular size 1000 k) on separation of a 100 bp DNA Ladder sample with a short-length capillary (total length 15 cm and effective length 7.5 cm). When the concentration of sieving polymer increased, migration time increased. In addition, mobility decreased as the sieving polymer concentration increased. This clearly indicates that the strength of entanglement between DNA samples and the sieving polymer was determined by sieving polymer concentration. This is because resistance to DNA flow increased in the capillary as sieving polymer concentration increased. In addition, the resolution length was almost the same for DNA length $\leq$ 900 bp. On the other hand, under conditions $\geq$ 1000 bp in the 100 bp DNA Ladder sample, the resolution length at 1.0 w/v % was lower than at other concentrations. The optimal sieving polymer concentration for separation of the DNA $\geq$ 1000 bp in the 100 bp DNA Ladder sample (100–1500 bp) with the short-length capillary was thus 1.0 w/v %.

Acknowledgments

This work was supported by Grants-in-Aid (KAKENHI) for Young Scientists (B) (16K17493, 15K18014) and Scientific Research (B) (17H03209, 15H03827).

References

[1] Grossbach U 1965 Acrlyamide gel electrophoresis in capillary columns *Biochim. Biophys. Acta* **107**180-182

[2] Righetti P G (ed.) 1996 *Capillary Electrophoresis in Analytical Biotechnology* (Boca Raton, Florida: CRC Press)

[3] Cohen A S, Najarian D R, Paulus A, Guttman A, Smith J A and Karger B L 1988 *Proc. Natl. Acad. Sci.* USA **85** 9660

[4] Mitchelson K R and Chen J 2004 *Capillary Electrophoresis of Nucleic Acid: Introduction to the capillary electrophoresis of nucleic acids* (Springer)

[5] Righetti P G and Gelfi C 1997 Electrophoresis gel media: the state of the art *J. Chromatogr. B* **699** 63-75

[6] Guttman A 2003 Gel and polymer-solution mediated separation of biopolymers by capillary electrophoresis *J Chromatogr Sci* **41** 449-459

[7] Heiger D N, Cohen A S and Karger B L 1990 Separation of DNA restriction fragments by high performance capillary electrophoresis with low and zero crosslinked polyacrylamide using continuous and pulsed electric fields *J. Chromatogr.* **516** 33–48

[8] Best N, Arriaga E, Chem D Y and Dovichi N J 1994 Separation of fragments up to 570 bases in length by use of 6% T non-Cross-linked polyacrylamide for DNA sequencing in capillary electrophoresis *Anal. Chem.* **66** 4063

[9] Fung E N and Yeung E S 1995 High-speed DNA sequencing by using mixed poly(ethylene oxide) solutions in uncoated capillary columns *Anal. Chem.* **67** 1913

[10] Dolnik V and Gurske W A 1999 Capillary electrophoresis in sieving matrices: selectivity per base, mobility slope, and inflection slope *Electrophoresis* **20** 3373

[11] De Carmejane O, Yamaguchi Y, Todorov T I and Morris M D 2001 Three-dimensional observation of electrophoretic migration of dsDNA in semidilute hydroxyethylcellulose solution *Electrophoresis* **22** 2433

[12] Gao Q and Yeung E S 1998 A matrix for DNA separation: genotyping and sequencing using poly(vinylpyrrolidone) solution in uncoated capillaries *Anal. Chem.* **70** 1382

[13] Masubuchi Y, Oana H, Matsumoto M, Doi M and Yoshikawa K 1996 Conformational dynamics of DNA during biased sinusoidal field gel electrophoresis *Electrophoresis* **17** 1065-1074

[14] Oana H, Masubuchi Y, Matsumoto M, Doi M, Matsuzawa Y and Yoshikawa K 1994 Periodic motion of large DNA molecules during steady field gel electrophoresis *Macromolecules* **27** 6061-6067

[15] Cifuentes A, Canalejas P and Diez-Masa J C 1999 Preparation of linear polyacrylamide-coated capillaries: study of the polymerization process and its effect on capillary electrophoresis performance *J. Chromatogr. A* **830** 423-438
[16] Schmalzing D, Piggee C A, Foret F, Carrilho E and Karger B L 1993 Characterization and performance of a neutral hydrophilic coating for the capillary electrophoretic separation of biopolymers *J. Chromatogr. A* **652** 149-159
[17] Horvath J and Dolník V 2001 Polymer wall coating for capillary electrophoresis *Electrophoresis* **22** 644–655
[18] Lucy C A, MacDonald A M and Gulcev M D 2008 Non-covalent capillary coatings for protein separations in capillary electrophoresis *J Chromatogr A* **1184** 81-105
[19] Robb C S 2007 Applications of physically adsorbed polymer coatings in capillary electrophoresis *Journal of Liquid Chromatography & Related Technologies* **30** 729–759
[20] Cifuentes A, de Frutos M, Santos J M and Diez-Masa J C 1993 Separation of basic proteins by capillary electrophoresis using cross-linked polyacrylamide-coated capillaries and cationic buffer additives *J. Chromatogr. A* **655** 63-72
[21] Simò-Alfonso E, Gelfi C, Sebastiano R, Citterio A and Righetti P G 1996 Novel acrylamido monomers with higher hydrophilicity and improved hydrolytic stability: I. synthetic route and product characterization *Electrophoresis* **17** 723
[22] Chiari M, Dell' Ort N and Gelain A 1996 Synthesis and characterization of capillary columns coated with glycoside-bearing polymer *Anal Chem* **68** 2731–2736

IOP Conf. Series: Materials Science and Engineering **242** (2017) 012034 doi:10.1088/1757-899X/242/1/012034

Influence of thickness of alkyl-silane coupling agent coating on separation of small DNA fragments in capillary gel electrophoresis

T Nakazumi and Y Hara*

Research Institute for Sustainable Chemistry, ISC, National Institute of Advanced Industrial Science and Technology, AIST, Central 5-2, 1-1-1 Higashi, Tsukuba 305-8565, Japan; E-Mail: nakazumi.t@aist.go.jp

*Correspondence: y-hara@aist.go.jp; Tel.: +81-29-81-6236

Abstract. To simplify the process of coating capillaries with fused silica, we herein set out to develop a one-step procedure for coating capillaries to prevent electro-osmotic flow (EOF) during the separation of small DNA fragments. We selected a short capillary (total length = 15 cm; effective length = 7.5 cm) for use in a compact capillary gel electrophoresis (CGE) system. To develop a one-step coating procedure, we employed alkyltrimethoxysilane agents because they are cheap and can be easily acquired, in contrast to polyethylene glycol (PEG) silane coupling agents. We examined a 100-bp DNA Ladder sample using fused silica capillaries, which were coated with alkyltrimethoxysilane agents of five different molecular lengths (C4, C6, C8, C12, and C16). We found that a fused-silica capillary with C8 alkyltrimethoxysilane is optimal for separating small DNA samples.

1. Introduction

High-resolution capillary gel electrophoresis (CGE) has proven to be a reliable tool for separating small DNA samples in a short time [1–4]. CGE can clearly separate single- and double-stranded DNA fragments by utilizing a sieving polymer [5–12]. CGE has many advantages over slab gel electrophoresis (SGE): CGE can separate PCR products despite the use of smaller sample volumes, and in lesser time. In addition, by using CGE, we can detect the separation of small DNA fragments with high resolution and in real time more successfully than with SGE. However, the use of CGE to separate a DNA sample would lead to electro-osmotic flow (EOF) because the flow direction of the EOF in the capillary is opposite to that of the DNA samples [13–20]. Some studies have reported the use of capillaries coated with a non-ionic polymer chain to prevent EOF [21–23]. Unfortunately, the synthesis of polymer chains on the capillary walls is a complex and cumbersome procedure. Therefore, this process is not realistic for the mass production of capillaries coated with linear polymer chains.

In the present study, we set out to develop a simple method for coating the capillaries used in the separation of small DNA fragments. That is, we attempt to develop a one-step procedure to use capillary coating for preventing the EOF. At present, we are developing a compact analytical device for application to a CGE system for incorporation into a point-of-care-testing (POCT) system. To miniaturize the CGE equipment, the capillaries should be as short as possible. Therefore, in the present study, we adopted a short capillary (total length = 15 cm; effective length = 7.5 cm) that would be used with the developed compact analytical device. For our developed POCT apparatus, we planned to eliminate both the washing process and the need for injecting a sieving polymer into the

Content from this work may be used under the terms of the Creative Commons Attribution 3.0 licence. Any further distribution of this work must maintain attribution to the author(s) and the title of the work, journal citation and DOI.
Published under licence by IOP Publishing Ltd

capillary when separating a DNA sample. To achieve this, we proposed the use of disposable short capillaries that can remove the need for both the washing process and sieving polymer solution filling work. To mass produce disposable capillaries, a simple and cost-effective means should be used for reducing the production cost and time. Therefore, we believe that the one-step coating of the capillary will make a significant contribution to realizing mass production, especially for short capillaries that can be used in a compact POCT device.

In the present study, to develop a one-step coating procedure for the fused silica capillary wall, we used alkyltrimethoxysilane agents with five different molecular lengths because they are cheap and easy to acquire, in contrast to polyethylene glycol (PEG) silane coupling agents. To optimize the lengths of the alkyltrimethoxysilane agents, we measured the CGE by using the fused silica capillaries with five different lengths of coating because the EOF strength is greatly affected by the length of the silane coupling agents. These CGE measurements allowed us to determine the optimal length of alkyltrimethoxysilane for separating small DNA samples with short fused-silica capillaries.

2. Materials and methods

2.1. Butyltrimethoxysilane (BTS) coating for 15-cm long fused-silica capillary

After washing the capillary with 1 N NaOH (15 min), water (15 min), and then methanol (15 min), a total of 20 mL of methanol solution containing butyltrimethoxysilane (BTS) (80 μL) and one drop of acetic acid was passed through the capillary for 2 h at room temperature, so that it covalently bonded to the glass wall of the fused-silica capillary. Next, the BTS-coated capillary was washed with methanol (15 min) and water (15 min) at room temperature.

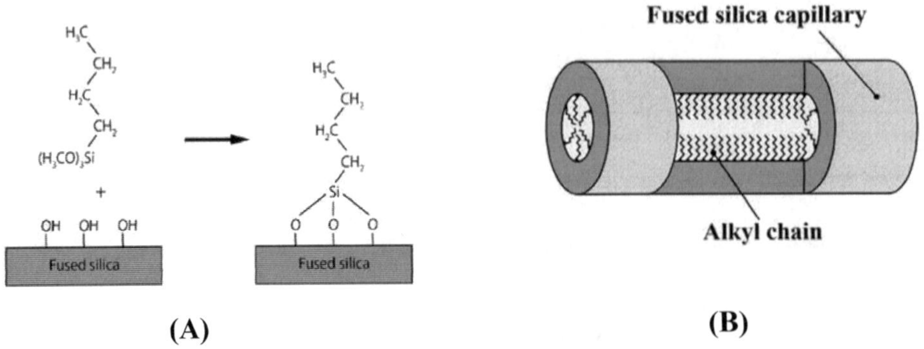

Figure 1. Schematic of (A) butyltrimethoxysilane coating procedure and (B) inner surface of fused-silica capillary, modified with alkyltrimethoxysilane.

2.2. Coating of 15-cm fused-silica capillary with four types of alkyltrimethoxysilane agents

First, the fused-silica capillary was washed with 1 N NaOH (15 min), water (15 min), and ethanol (15 min). Then, 20 mL of hexadecane solution containing 80 μL of alkyltrimethoxysilane coupling agent (hexyltrimethoxysilane (HTS), octyltrimethoxysilane (OTS), dodecyltrimethoxysilane (DDS), hexadecyltrimethoxysilane (HDS), or octadecyltrimethoxysilane (ODS)) was passed through the capillary for 2 h at room temperature, such that it covalently bonded to the glass wall of the fused-silica capillary. Next, the capillaries coated with the alkyltrimethoxysilanes were flushed with ethanol and water at room temperature.

Table 1. Alkyltrimethoxysilane coupling agents and their molecular chain lengths.

Silane coupling agent	Abbreviation	Length (nm)
Butyltrimethoxysilane	BTS (C4)	0.54
Hexyltrimethoxysilane	HTS (C6)	0.80
Octyltrimethoxysilane	OTS (C8)	1.05
Dodecyltrimethoxysilane	DDS (C12)	1.56
Hexadecyltrimethoxysilane	HDS (C16)	2.08

2.3. Equipment developed for capillary gel electrophoresis (CGE)

We used capillary gel electrophoresis (CGE) equipment developed in house for all the measurements. This equipment consisted of a microscope with epi-illumination (IX73, Olympus, Tokyo, Japan) and a high-voltage power supply (HJPQ-10P3, Matsusada, Kusatsu, Japan). To enable detection using this equipment, the DNA was conjugated using SYBR Green II. We used a mercury lamp with an optical filter (U-FBWA, Olympus, Tokyo, Japan), which allowed the passage of light of wavelength 460–495 nm. This was done because the DNA conjugated with SYBR Green II has an excitation wavelength in this region. The fluorescence of the DNA conjugated by SYBR Green II was collected by a 60× objective lens (UPlanFLN, Olympus, Tokyo, Japan). The corrected fluorescence signal could be detected by using a photomultiplier tube (PMT) (H8249-101, Hamamatsu Photonics, Hamamatsu, Japan). The detected signal was converted to a digital signal by utilizing a multifunction I/O device (NI USB-6341, National Instrument, Austin, TX, USA). For every measurement, we used the fused silica capillaries with a 75-µm circle (Polymicro Technologies, Phoenix, AZ, USA). To inject the 100-bp DNA Ladder sample, we applied 1.5 kV to the capillary for 1 s. The separation of the DNA samples was performed at 100 V/cm.

2.4. Chemicals

We used a 0.5× TBE buffer (44.5-mM Tris, 44.5-mM boric acid, 1.0-mM EDTA, pH 8.3) (Nippon Gene, Toyama, Japan) diluted with ultrapure water to act as the running buffer for all the measurements. Hydroxyethyl cellulose (HEC) (molecular size: 1300 k) was utilized as the sieving polymer for the CGE process. The HEC was adjusted to 0.5 w/v % and contained 0.5× TBE buffer and 2× SYBR Green II (Takara Bio Inc., Kusatsu, Japan). As the DNA samples, we selected a 100-bp DNA Ladder (Takara Bio Inc., Kusatsu, Japan) consisting of 11 double-stranded fragments with 100, 200, 300, 400, 500, 600, 700, 800, 900, 1000, and 1500 bp as the measurement targets for all these measurements. The 130-µg/mL 100-bp DNA Ladder sample was used as the injection sample.

3. Results and discussion

To develop a simple and fast coating method, we used alkyltrimethoxysilane agents because they can coat a fused silica capillary wall in a single step after the completion of the washing process (see Figure 1). In addition, the alkyltrimethoxysilane agents can be acquired at a low cost compared to silane-coupling agents with a polyethylene glycol (PEO) chemical structure. To evaluate the effect of the molecular length of the alkyltrimethoxysilane agents on the separation of small DNA fragments, we measured five silane-coupling agents with different molecular lengths, namely, BTS (C4), HTS (C6), OTS (C8), DDS (C12), and HDS (C16) (see Table 1). Figure 2 shows the separation results for the 100-bp DNA Ladder sample as obtained by CGE. As shown in Figure 2(A), there was no peak in the separation graph. The calculated thickness of the electronic double layer was 1.44 nm, while that of BTS (C4) was about 0.54 nm, implying that the BTS (C4) coating cannot effectively prevent EOF. The flow direction of the EOF is opposite to that of the DNA samples. Therefore, the DNA samples cannot be injected into the capillary by disturbing the EOF. In contrast, for alkyltrimethoxysilane lengths of ≥C6, we measured the peak of the small DNA samples during CGE. The calculated molecular length of C6 is about 0.8 nm, implying that the C8 coating can prevent the EOF opposite flow more effectively than C6. As shown in Figure 2(C), we can obtain clear peaks for every DNA sample length in the 100-bp DNA Ladder sample. The calculated molecular length of the OTS (C8) is about 1.05 nm. This trend indicates that the slightly different molecular length of the alkyltrimethoxysilane agent significantly affects the result of the separation for the small DNA samples in the CGE process. In contrast, when DDS (C12) and HDS (C16) are used (see Figure 2(D) and 2(E)), we cannot obtain a clear peak in the CGE graph again. We believe that this occurred because of the hydrophobic nature of the alkyltrimethoxysilane agents on the capillary wall. Because the strength of hydrophobicity increases with increase in the long chains of the alkyltrimethoxysilane agent, we believe that the flow in the capillary during CGE changed from plug flow (OTS (C8)) to laminar flow (DDS (C12), with HDS (C16)) (see Figure 3) owing to the change in the contact angle between the solution and the capillary wall. The laminar flow affected the measurement of small DNAs during CGE because DNA fragments of the same length scattered in the flow. That is, the density of DNA of the same length in the laminar flow was lower than that in the plug flow. When the density of the DNA of the same length increased, we obtained clear separation peaks in the electrophoregram.

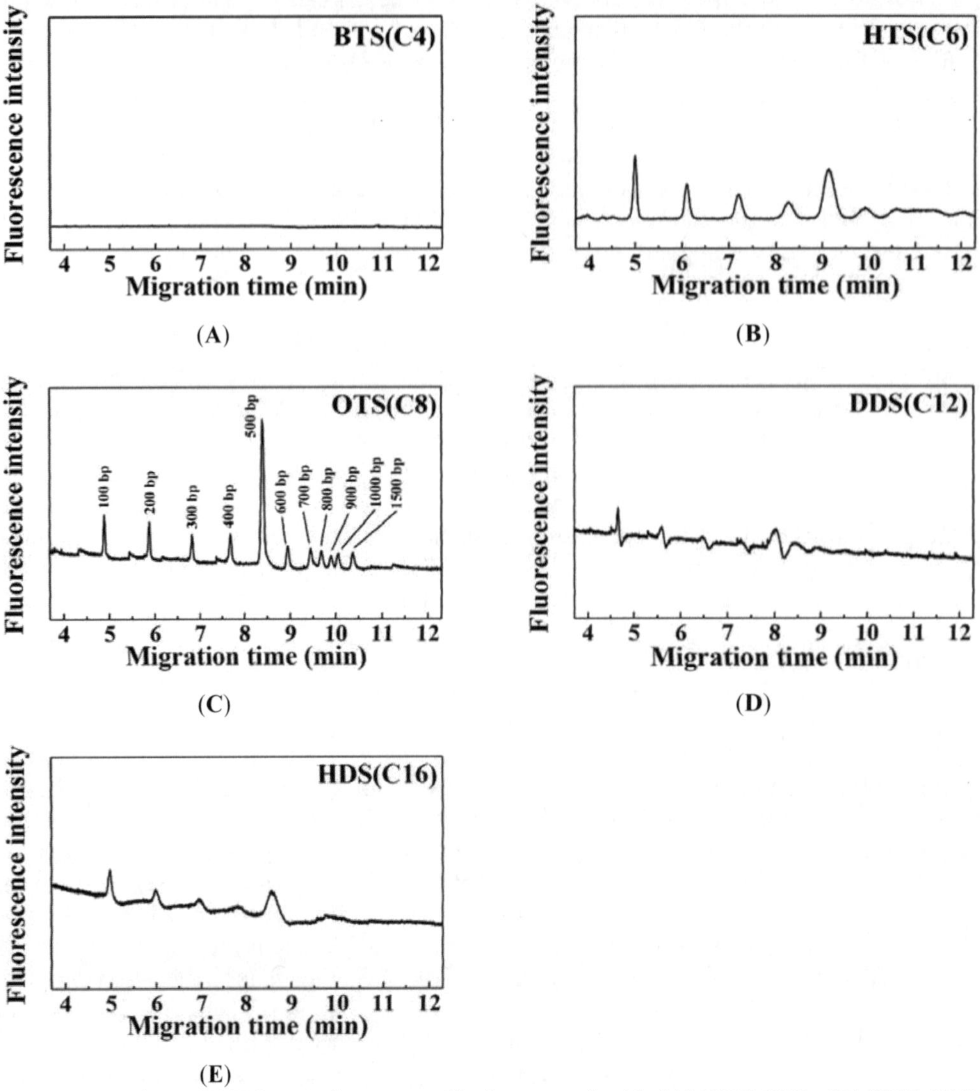

Figure 2 Separation of DNA samples on capillaries coated with (A) BTS (C4), (B) HTS (C6), (C) OTS (C8), (D) DDS (C12), and (E) HDS (C16).

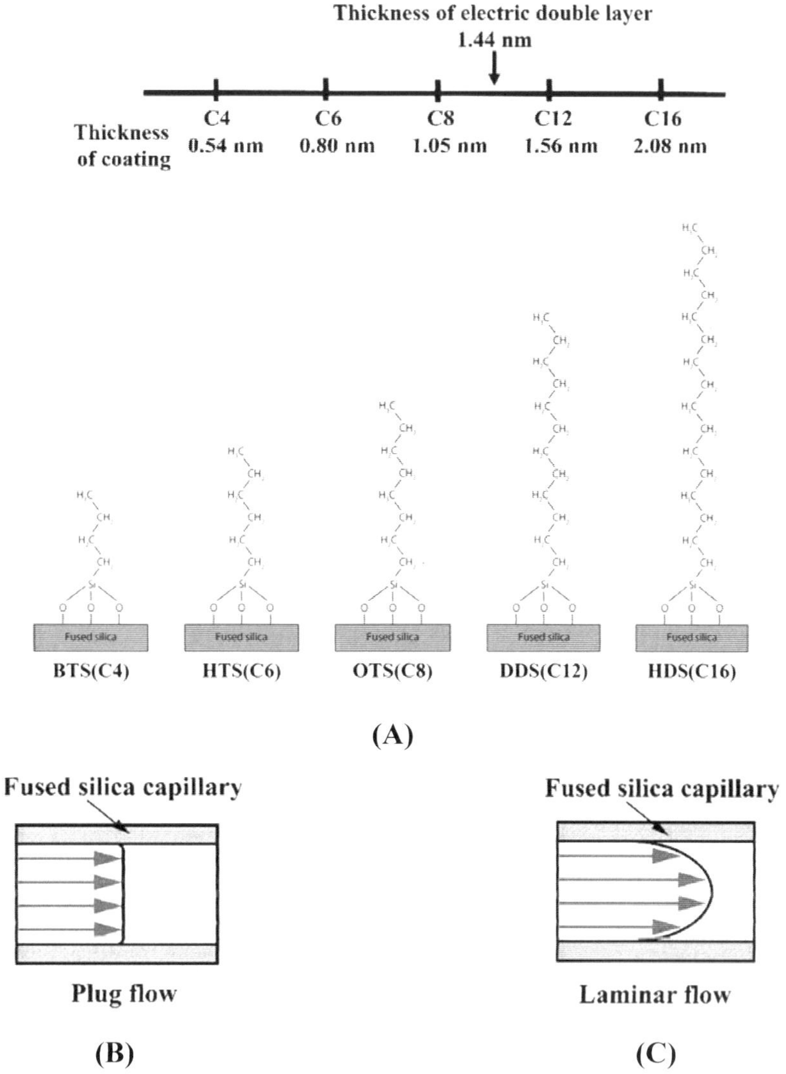

Figure 3. Schematic of capillaries coated with (A) OTS (C 8), (B) HDS (C12), and ODS (C16).

4. Conclusions

In the present study, to develop a simple method for coating a fused silica capillary, we selected alkyltrimethoxysilane agents with five different molecular lengths. CGE measurement revealed that the OTS (C8) can be detected in all the DNA samples in the 100-bp DNA Ladder sample. In the case of BTS (C4), there is no peak in the CGE graph because the DNA samples cannot be injected into the capillary due to EOF with a flow in the opposite direction to that of the DNA samples. As we used capillaries coated with DDS (C12) and HDS (C16), we could not separate the DNA fragments from the 100-bp DNA Ladder again. This trend indicates that the flow in the capillary changed from a plug flow (OTS (C8)) to a laminar flow (DDS (C12) and HDS (C16)) due to an increase in the hydrophobicity of alkyltrimethoxysilane agents on the capillary wall.

Acknowledgments
This work was supported by Grants-in-Aid (KAKENHI) for Young Scientists (B) (16K17493, 15K18014) and Scientific Research (B) (17H03209, 15H03827).

References

[1] Mitchelson K R and Cheng J 2001 *Capillary Electrophoresis of Nucleic Acids: Introduction to the Capillary Electrophoresis of Nucleic Acids* (Springer Science & Business Media)
[2] Mitchelson K R and Cheng J 2001 *Capillary Electrophoresis of Nucleic Acids Volume Ⅱ: Practical Application of Capillary Electrophoresis* (Springer Science & Business Media)
[3] Paulus A, Gassmann E, Field M J 1990 *Electrophoresis* **11** 702
[4] Yin H F, Lux J A, Shomburg G 1990 *J. High Resolut. Chromatogr.* **13** 624
[5] Chu B J, Liang D H 2002 *J. Chromatogr. A* **966** 1–13
[6] Manabe T, Chen N, Terabe S, Yohda M, Endo I 1994 *Anal. Chem.* **66** 4243
[7] Cohen A S, Najarian D R, Paulus A, Guttman A, Smith J A, Karger B L 1988 *Proc. Natl. Acad. Sci. USA* **85** 9660
[8] Wei W and Yeung E S 2000 *J. Chromatogr. B* **745** 221
[9] Kim Y and Yeung E S 1997 *J. Chromatogr. A* **781** 315
[10] Albarghouthi M N, Buchholz B A, Doherty E A, Bogdan F M, Zhou H, Barron A E 2001 *Electrophoresis* **22** 737
[11] Song J M and Yeung E S 2001 *Electrophoresis* **22** 748
[12] Barta C, Ronai Z, Sasvari-Szekely M, Guttman A 2001 *Electrophoresis* **22** 779
[13] Horvath J, Dolník V 2001 *Electrophoresis* **22** 644–655
[14] Lucy C A, MacDonald A M, Gulcev M D 2008 *J. Chromatogr. A* **1184** 81–105
[15] Robb C S 2007 J. *Liq. Chromatogr. Relat. Technol.* **30** 729–759
[16] Liu C-Y 2001 *Electrophoresis* **22** 612–628
[17] Huhn C, Ramautar R, Wuhrer M, Somsen G W 2010 *Anal. Bioanal. Chem.* **396** 00297
[18] Cifuentes A, Canalejas P, Diez-Masa J C 1999 *J. Chromatogr. A* **830** 423–438
[19] Schmalzing D, Piggee C A, Foret F, Carrilho E, Karger B L 1993 *J. Chromatogr. A* **652** 149–159
[20] Cifuentes A, de Frutos M, Santos J M, Diez-Masa J C 1993 *J. Chromatogr. A* **655** 63–72
[21] Simò-Alfonso E, Gelfi C, Sebastiano R, Citterio A, Righetti P G 1996 *Electrophoresis* **17** 723
[22] Chiari M, Dell'Ort N, Gelain A 1996 *Anal. Chem.* **68** 2731–2736

Study on penetration-induced initiation of energetic fragment

Xiangxin Qiao, Heyang Xu

Shenyang Ligong University, Shenyang China 110159

xxq2002@163.com, 390671553@qq.com

Abstract. In order to investigate penetration-induced initiation of energetic fragment penetrating target, PTFE/Al (mass ratio 73.5/26.5) pressed and sintered into a Φ8mm × 8mm cylinder. To form energetic fragment, the cylinder was put into a closed container made by 35CrMnSiA. The container is 12mm long, 2mm thick. Energetic fragments were launched by a 14.5mm ballistic gun with a series of velocities and the penetrate process was simulated by AUTODYN-3D. The results show that the stress peak of energetic material exceed the initiation threshold, and energetic material will deflagrate, when energetic fragments impact velocity more than 800 m/s. The research results can provide reference for designs of energetic warhead.

1. Introduction

Energetic fragments, also known as reactive material enhanced penetrator, which usually consists of energetic materials as core and is wrapped with high strength metal[1]. Because of strong penetrating ability and "class detonation" damage effect caused by large released chemical energy, the energetic fragment attracts both domestic and foreign scholar attention. Energetic fragment filled with different energetic material impact steel target was studied by Jun-feng Shuai[1] et al. The experiment results show that the deflagration effect of PTFE/Ti is better than PTFE/Al, damage capability of fragment increases with the shell thickness, and the target aperture by energetic fragment is 40% higher than that by inert fragment. JIANG Jian-wei[2] et al. Studied the strength model and state equation of PTFE/Al and verified the validity of the model by AUTODYN. Yin Yi-feng[3] analyzed dynamic behavior of reactive material enhanced penetrator impact target, and studied the structure parameters, projectile/target materials and impact conditions in detail.

In this paper, using AUTODYN to simulate energetic fragment penetrate target, and verified the validity of simulation by range test. Analyzed fragment velocity variation and internal pressure of energetic material when energetic penetrate target in different velocities and angles. To provide guidance for engineering application of energetic materials in terminal ballistics.

2. Range Test

Before experiment, preparing energetic material. Mixing Al powder and PTFE by mass ratio 26.5/73.5, and pressed into Φ8mm × 8mm cylinder by press machine, then sintered in vacuum furnace. Sintering temperature in 365 ℃~385 ℃, the heating rate was about 10 ℃/min, density of sintered material is about 2.4 g/cm^3. Finally, the energetic material was filled into 35CrMnSiA shell and confined with an end cap. Figure 1 shows the test site layout, target is 10 m away from ballistic gun, placed high-speed camera (model:

Content from this work may be used under the terms of the Creative Commons Attribution 3.0 licence. Any further distribution of this work must maintain attribution to the author(s) and the title of the work, journal citation and DOI.
Published under licence by IOP Publishing Ltd

IOP Conf. Series: Materials Science and Engineering **242** (2017) 012035 doi:10.1088/1757-899X/242/1/012035

PHANTOM220) 10 m away from target's left side, shooting frequency is 3000fps. using 14.5mm ballistic gun launching the energetic fragment with sabot, see Figure 2, adjust velocities by changing charge, target material is 4340 steel in 6mm thick. Test consumes 10 rounds of projectiles, and launch velocities is 600 m/s~1500 m/s, interval is about 100 m/s. Part of results are shown in Table 1.

The results show that the energetic fragments will deflagrate, when velocity is higher than 800 m/s, contrarily, there is no deflagration, when velocity below 800 m/s.

Figure 1. Test site layout

Figure 2. Energetic fragment and propellant cartridges

Table 1 typical fragment penetrating the target

Velocity(m/s)	t=0.20 ms	t=0.23 ms	t=0.26 ms
706			
812	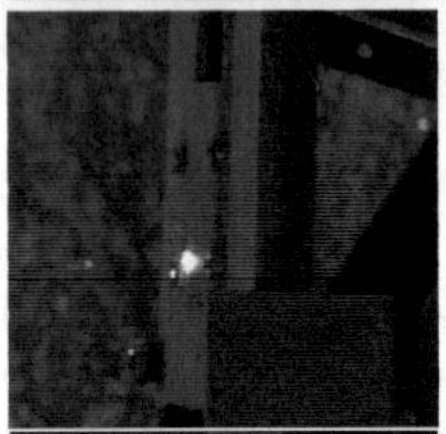		
902			

3 Numerical Simulation

3.1 Computational model

According the size and metal of energetic fragments used in experiment to build up fragment model, and target model is a 6 mm thick circular plate made by 4340 steel. Considering the symmetry of structure, a 1/2 model will be established with symmetry constraint by AUTODYN-3D, using mm-mg-μs unit system, Lagrange algorithm and hexahedron mesh. In order to clearly describe penetration, a schematic sketch for fragment impact target in different angles is depicted, V is penetration velocity, α is target angle. Finite element model and geometric model are shown in figure 3.

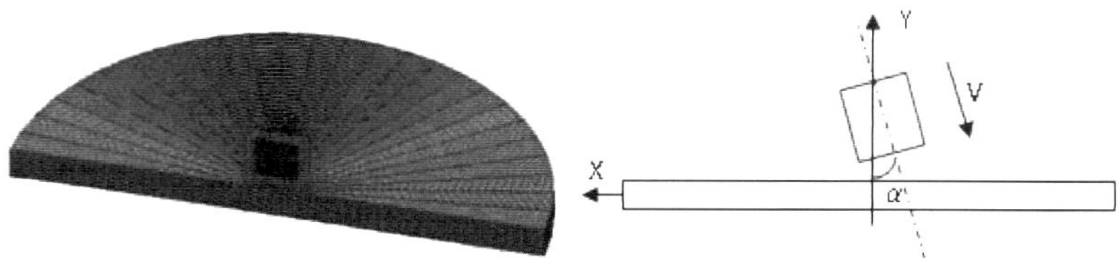

Figure.3 The finite element model of energetic fragment penetrating target and geometrical model

3.2 Material model

The shock EOS for most metals and the Johnson–Cook plasticity model are adopted to describe the dynamic response of fragment shell, energetic material and target. To prevent the mesh distortion during calculation, all materials added erosion algorithm. The strain values of PTFE/Al, 35CrMnSiA and 4340 steels are 0.8, 1.5 and 1.5, respectively. The material parameters used in simulation are given in Table 2.

Table 2 Material Parameters

Parameters	PTFE/Al	35CrMnSiA	4340 steel	Parameters	PTFE/Al	35CrMnSiA	4340 steel
A/MPa	8.044	1500	792000	G/MPa	666	——	——
B/MPa	250.6	3000	510000	γ	0.9	——	——
N	1.8	0.2	0.26	c_0(m·s^{-1})	1450	——	——
C	0.4	0.014	0.014	S	2.258	——	——
θ_m (k)	463	1793	1793	$\dot{\varepsilon}$ / s^{-1}	——	206	——
θ_r (k)	299	300	299	E/GPa	——	206	200
ρ (kg·m^{-3})	2.40	7.83	7.8	m	——	1.03	0.94

3.3 Effect of velocity on internal stress distribution of energetic materials

In simulation, the energetic fragments velocities is 600 m/s~1500 m/s and interval is 100 m/s. Due to small strain rate of energetic materials, the reference [4] shows that energetic material internal stress peak in posterior good agreement with theoretical calculations through theoretical calculation and numerical simulation. Therefore, setting observation points in posterior of energetic material before calculation. Plot the calculate results into line, and then compared with experimental results in reference [5] and the initiation threshold σ_c in reference [6]. The relationship between impact velocity and internal stress is shown in figure 4.

From figure 4(a), we can notice that the internal stress peak of energetic material is lower than initiation threshold and the material didn't deflagrate when penetrate velocity is lower than 800 m/s, which is conformed to experimental result. And the figure 4(b) shows that the stress peak of energetic material increases with impact velocities. Because of steel shell used in this paper, the energetic material is subjected to large stress. Therefore the stress is higher than that in reference [6], while the trend of calculated results are consistent with the experimental results[6], combining above experimental results can prove the validity of the computational model.

IOP Conf. Series: Materials Science and Engineering **242** (2017) 012035 doi:10.1088/1757-899X/242/1/012035

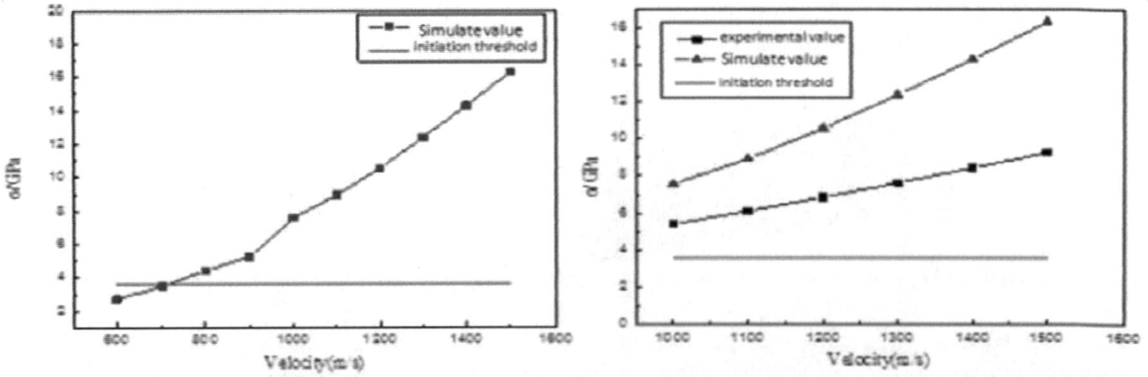

(a) relation of penetration rate and initiation threshold (b) experimental/simulate value and initiation threshold

Figure 4. The penetration rate on the influence of the internal stress distribution in energetic materials

3.4 Explosive Behavior of Targets with Different Angle of Energetic Fragment

When PTFE/Al subjected to shock wave, it may ignite and deflagrate, so its chemical reaction rate is related to impact condition closely. If the overpressure of energetic material reaches or exceeds the threshold pressure σ_c=3.6GPa for deflagration[6], the energetic material will deflagrate. Penetration-induced initiation behavior of energetic fragments, means its penetration ability and deflagration performance, now analysis velocity decay and internal stress of fragments penetration 6 mm thick 4340 steel target in 800 m/s, and determine whether deflagration occurs. In order to clearly observe the internal stress peak of energetic materials, nine observation points are uniformly set on energetic materials, as shown in figure 5.

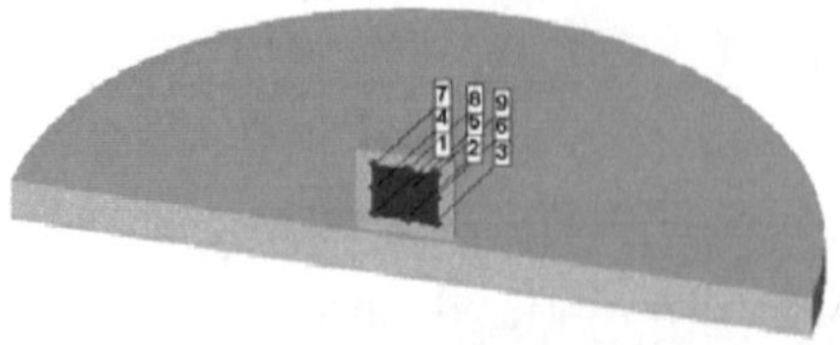

Figure 5. The internal observation points of energetic materials

Figure 6 shows the velocity decay curve of energetic fragments penetrating the target at different angles and the distribution of internal stress peak of the energetic materials. From figure 6(a), we can see all fragments penetrated through the target effectively. When impact angles is 0°, 5°and 10°, the velocity fluctuates is obviously, velocity decreases rapidly at 0~15μs, and becomes slowly after 15μs, all fragments' residual velocities are more than 300 m/s. In 0~18μs, the velocity curves of 15°, 20°and 25° fluctuates is not obvious, compared with the velocity curves of 0°, 5°and 10°, It is attributed the great stress generated instantaneously during the fragments impacting the target with relatively large impact angle. With the penetration, compressive stress gradually unloading, acceleration similar to the linear changes so that velocity curve is similar to a straight line. The change of velocity curve decreases and trends to be gentle after 20μs. Residual velocity is between 200 m/s to 260 m/s, lower than the velocity for 5°, 10°and 15°, which indicates that the bigger impact angle, the greater consume energy of fragment penetrating target, so that the fragment leave less kinetic energy and lower velocity. From figure 6(b), the trends of stress is consistent except individual jumping points of energetic materials at 5°. The stress peak of energetic material exceeds the initiation threshold and occur deflagration when impact angle is 0°, 5°and 10°, while the internal stress peak is below initiation threshold and can't deflagration at 15°, 20°and 25°.

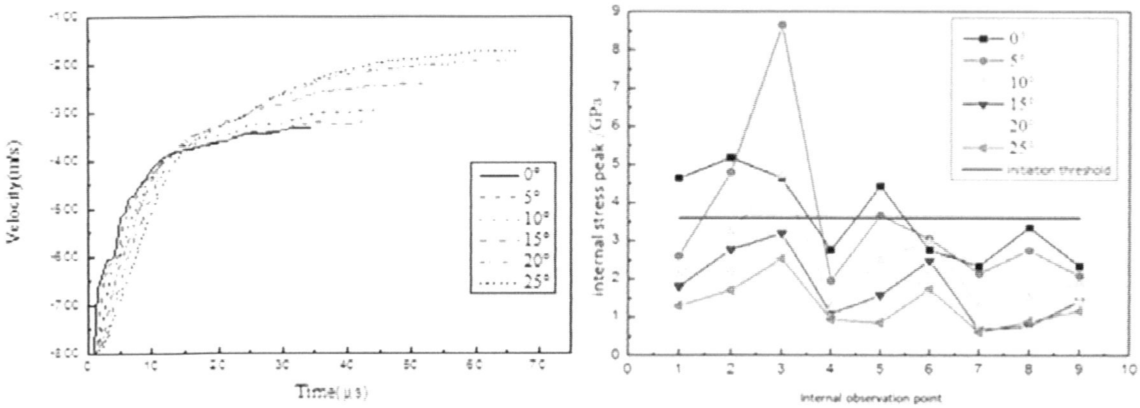

(a) curves of speed versus time (b) internal stress peak

Figure.6 Penetrating Target at different angles of internal stress peak in energetic materials distribution

4. Conclusions

In this paper, through the experiment and the simulation of energetic fragment impact, we received following conclusions:

1) The energetic material will deflagrate when the penetrate velocity exceeding 800m / s.

2) The residual velocity of the fragment decreases with the increase of angle. When the angle is 0°~10°, the internal stress peak of energetic material exceeds the initiation threshold, and when the angle is 15~20°, the internal stress peak is below the initiation threshold.

3) The results of numerical simulations of the energetic fragments penetrating target have good agreement with the experimental results.

4) The results of this study can provide reference for the design of energetic fragment warheads and damage research.

References

[1] SHUAI Jun-feng, JIANG Jian-wei, WANG Shu-you et al. Compound reactive fragment penetrating steel target[J]. Chinese Journal of Energetic Material, 2009, 17(6): 722-723.

[2] JIANG Jian-wei, WANG Shu-you, ZHANG Mou et al. Modeling and simulation of JWL equation of state for reactive Al / PTFE mixture[J]. Journal of Beijing Institute of Technology, 2012, 21(2): 143-149.150-155.

[3] YING Yi-feng. Research on terminal penetration-induced initiation and damage effect of reactive material enhanced penetrator[D]. Beijing: Beijing Institute of Technology, 2015.

[4] Raftenberg M N, Willis M J, Kirby G C. Modeling the impact deformation of rods of a pressed PTFE/Al composite mixture[J]. International Journal of Impact Engineering, 2008, 35:1735-1744.

[5] WANG Hai-fu, ZHENG Yuan-feng, YU Qing-bo et al. Initiation behavior of covered explosive subjected to reactive fragment[J].Journal of Beijing Institute of Technology, 2012,21(2): 143-149.

[6] Ames R G. Energy Release Characteristics of Impact-Initiated Energetic Materials[C]. Mater. Res. Soc. Symp. Proc.Vol.896:2006.

IOP Conf. Series: Materials Science and Engineering **242** (2017) 012036 doi:10.1088/1757-899X/242/1/012036

Effect of microalloying elements on microstructure and properties of quenched and tempered constructional steel

Qingshen Ma[a,b], Leqing Huang[a,b], Guobiao Di[a,b], Yanfeng Wang[a,b], Yongda Yang[a,b], Changwen Ma[a,b]

[a] Shougang research institute of technology, Beijing 100041, China
[b] Beijing key Laboratory of Green Recyclable Process for Iron & steel Production Technology,Bei Jing 100043

huangleqing0716@126.com

Abstract: The effects of microalloying elements Nb, V and Ti on microstructure and properties of quenched and tempered steel were studied. Results showed that the addition of microalloying elements led to the formation of bainite and increased strength, while the austenization and ferrite transformation temperature was barely affected, i.e. 10°C. Microalloying elements shortened the incubation time for bainite transformation by refinement of austenite grain, and decreased the hardenability by forming carbides and therefore reducing the carbon content of super-cooled austenite. Either of them promoted the bainite transformation. The better tempering stability was ascribed to the as hot-rolled bainite microstructure and secondary carbide precipitation during tempering.

1. Introduction

For the high-strength low-alloy steels with quenched and tempered delivery conditions, it is necessary to add Cr, Ni and Mo elements to ensure the hardenability of the steel plate. The addition of these elements increases the carbon equivalent and welding crack sensitivity. Therefore, in order to ensure welding performance, the amount of these elements would be restrained. In this paper, a 450MPa grade quenched and tempered steel was selected as the research object. Two chemistry, i.e. with and without addition of Nb, V and Ti, were designed to investigate the effects of microalloying elements on the microstructure and mechanical properties.

2. Experimental

The experimental material is 450MPa grade quenched and tempered steel. The chemical composition of the experimental steels is listed in Table 1. Two chemical compositions are proposed, one of which is referred to as non-microalloyed, and the other is microalloyed.

Table 1 Chemical composition of the investigated steels (wt %)

Elements	C	Si	Mn	P	S	Ni+Cr+Mo	Nb + Ti+V
Non-microalloyed	0.079	0.24	1.02	0.006	0.004	1.9	-
Microalloyed	0.082	0.29	0.97	0.006	0.004	1.9	0.087

The ingots were prepared by vacuum melting, and forged into 220mm ×170mm×175mm in size. After homogenization at 1200°C for 150min, the ingots were rolled into 60 mm in recrystallization region. After holding, they were rolled into 20mm in non-recrystallization region and finishing rolling temperature of 840°C. Then, the hot-rolled plates were air cooled down to room temperature.

Content from this work may be used under the terms of the Creative Commons Attribution 3.0 licence. Any further distribution of this work must maintain attribution to the author(s) and the title of the work, journal citation and DOI.
Published under licence by IOP Publishing Ltd

Subsequently, the hot-rolled samples were quenched and tempered. After holding at 900°C for 50min, they were immediately quenched into water, then reheated to 650 °C, 670°C and 700 °C, respectively, with a holding time of 60min.

Cylindrical specimens with 5 mm diameter and 25 mm length were cut from hot-rolled plates for the measurement of phase transformation temperature using dilatometer. According to GB/T 225-2006, bar samples with a dimension of Ø25mm×100 mm were machined from forged ingots for JOMINY test. Specimens with the dimension of Ø10 mm×50 mm and 10 mm×10 mm×55 mm were prepared after heat treatment. Tensile tests and CVN tests were carried out at room temperature and -40°C respectively. Microstructure characterizations were made by optical microscopy (OM) and scanning electron microscopy (SEM) with samples polished and etched for 13s in the mixture of 4% nitric acid and alcohol.

3. Results

3.1 Effect of microalloying elements on microstructure and mechanical property of hot-rolled sample

The microstructures and mechanical properties of hot-rolled samples were shown in Fig.1 and Table 2, respectively. As shown in Fig. 1, the microstructure of non-microalloyed sample is "P +F". Because of the solute microalloying elements in austenite which delay the ferrite transformation, and promote bainite transformation [2], the microstructure of microalloyed sample is "B+F+P". The differences in microstructures lead to better properties of microalloyed sample when compared with non-microalloyed one. The lower Impact energy was due to the process of air-cooling after rolling.

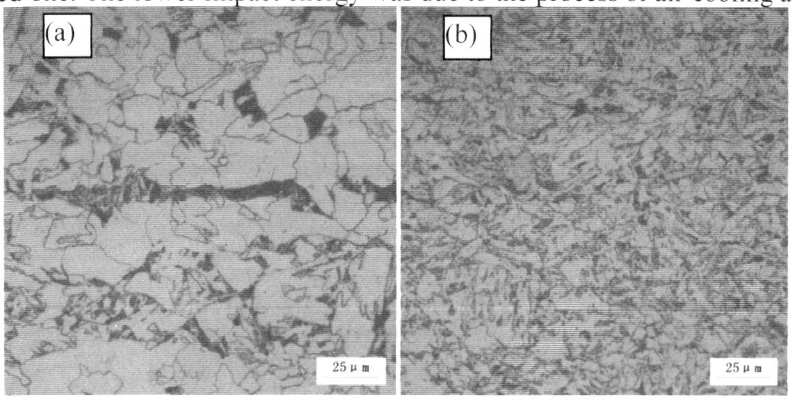

Fig.1 OM images of hot-rolled samples (a)non-microalloyed; (b)microalloyed

Table 2 Mechanical properties of hot-rolled samples.

Elements	Transverse				Longitudinal				-40°C
	Rel/MPa	Rm/MPa	A/%	Z/%	Rel/MPa	Rm /MPa	A /%	Z/%	AKv/J
Non-microalloyed	350	530	23.0	80	335	570	20.0	79	41 46 62 50
Microalloyed	501	738	18.0	54	507	790	21.5	66	24 31 48 34

3.2 Effect of microalloying elements on hardenability

The effect of microalloying elements on the transformation temperature was measured by dilatometer. Results were shown in Table 3. It is shown that the addition of microalloying elements, Nb, V, Ti, has little effect on the phase transformation temperature. The temperature of fully austenization for the microalloyed and non-microalloyed steels is 857.6 °C and 864.8°C, respectively. Thus, the quenching temperature was chosen as 900 °C.

Table 3 Effect of microalloying elements on transformation temperature

Elements	Heating		Cooling	
	Start transformation temperature/° t	Finish transformation temperature /° t	Start transformation temperature /° t	Finish transformation temperature /° t
Non-microalloyed	726.5	857.6	733.1	671.9
Microalloyed	729.7	864.8	745.5	678.5

JOMINY tests were carried out with the specimen Ø25 mm×100 mm in size. The sample was heated to 900 °C, holding for 30 min, followed by water-quenching immediately. Variation in hardness as a function of the distance to the quenching surface is shown in Fig.2. It can be seen that, within 0 ~ 15mm, the hardness of non-microalloyed specimen is higher than that of microalloyed one, because the corresponding microstructure is martensite and bainite, respectively. However, for >15mm, the microstructures of both steels were bainite, while the hardness of microalloyed specimen was higher than that of microalloyed one, due to precipitation strengthening microalloying carbides. The JOMINY tests showed that microalloying elements contribute to the decreased hardenability.

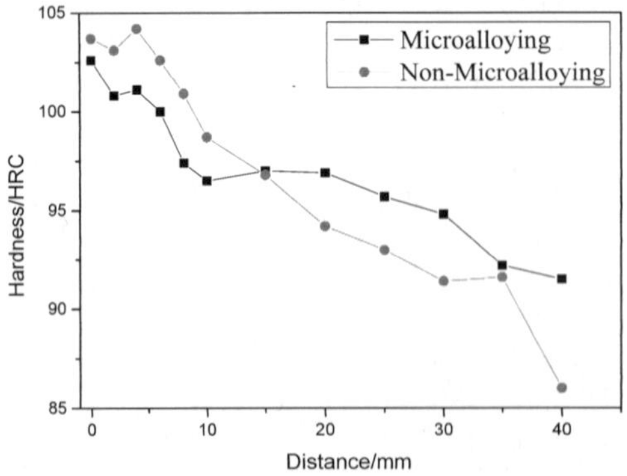

Fig.2 Effect of microalloying elements on hardenability

The microstructures and mechanical properties of quenched specimens were shown in Fig.3 and Table 4, respectively. The microstructure of non-microalloyed steel is martensite (Fig.3(a)), and that of microalloyed steel is "bainite + small quantity of martensite"(Fig.3(b)). As shown in Table 4, the non-microalloyed steel achieved a higher yield strength but a lower plasticity.

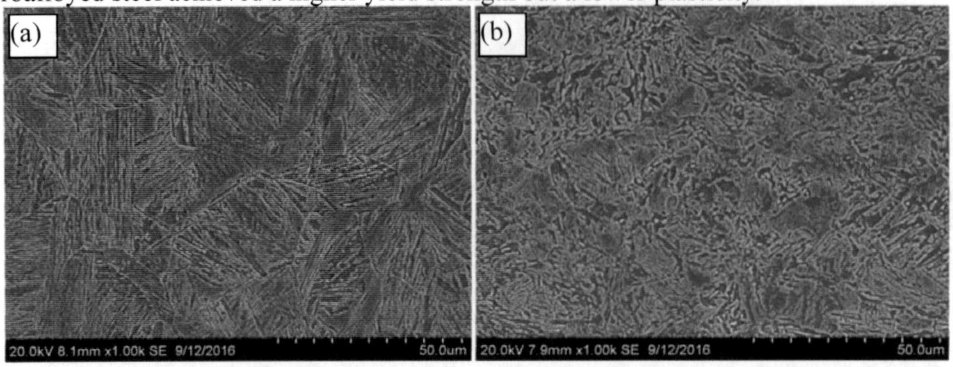

Fig.3 Microstructure of two kinds of quenched and tempered steels (a) non-microalloyed; (b) microalloyed

Table 4 Effect of microalloying elements on properties of quenched specimens

Elements	Transverse				Longitudinal				-40°C AKv/J
	Rel/MPa	Rm/MPa	A/%	Z/%	Rel/MPa	Rm /MPa	A /%	Z/%	
Non-microalloying	676	855	14.0	73	754	956	13.0	72	198 207 217 207
Microalloying	596	857	17.0	71	610	846	17.0	71	133 152 158 148

3.3 Effect of microalloying elements on microstructure and properties of quenched and tempered steel

Effect of microalloying elements on microstructure and mechanical properties of quenched and tempered steel were studied. As shown in Fig.4, the microstructure of the non-microalloyed and microalloyed steels is tempered martensite and "bainite+massive ferrite+small amount of tempered martensite", respectively. In Fig.5, it is suggested that the strength of the tempered specimen is lower than that of the quenched state. However, with the increasing of tempering temperature, the strength of non-microalloyed steel decreased and the plasticity increased. However, when tempered in the range of 650 °C -700 °C, the strength and plasticity of microalloyed steel had little change and showed good tempering stability.

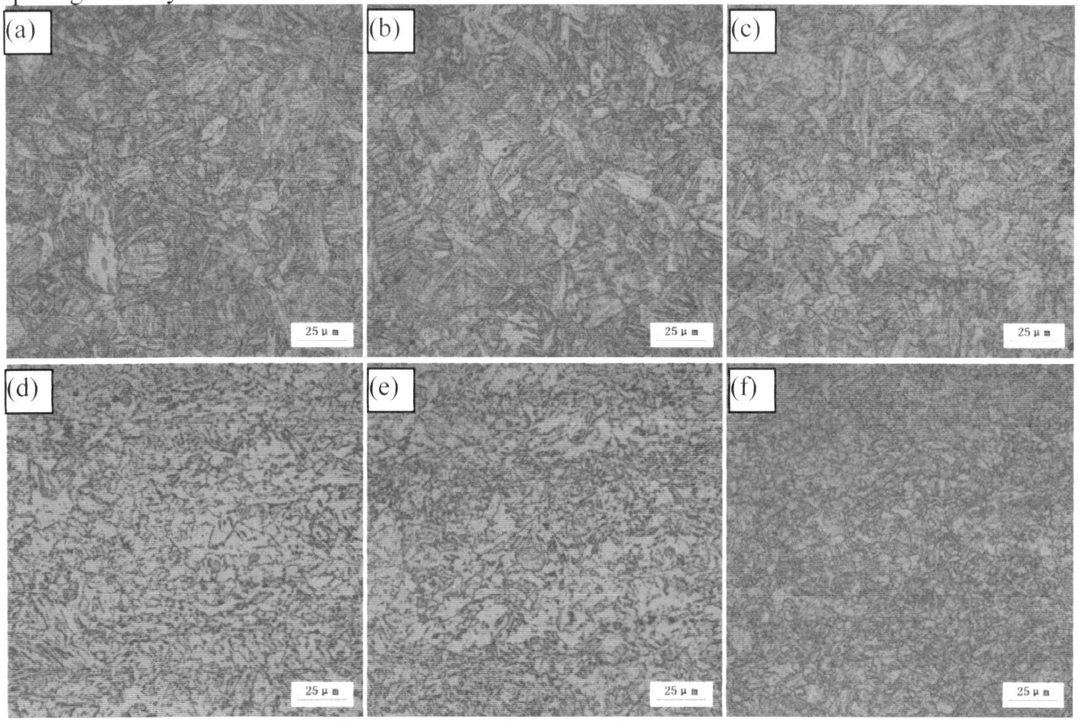

Fig.4 Microstructure of quenched and tempered steel

Non- microalloyed: (a)650°C;(b)670°C;(c)700°C; Microalloyed;(e) 650°C;(f) 670°C;(g) 700°C

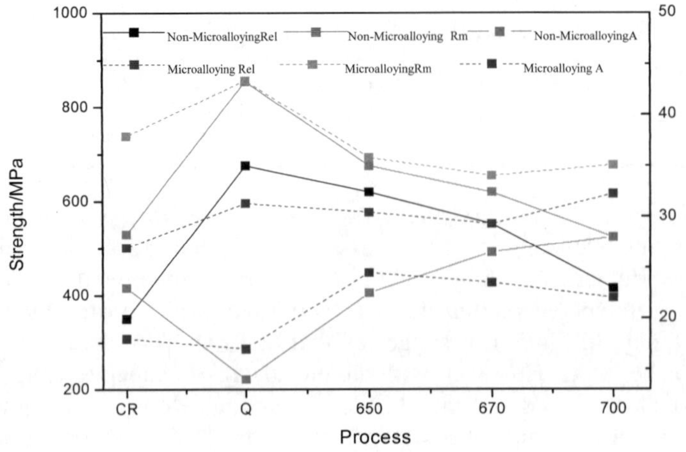

Fig.5 Effect of microalloying elements on property of quenched and tempered steel.

4. Discussion

The influence of microalloying elements, Nb, V and Ti, on the microstructures and properties of quenched and tempered steel mainly includes two aspects. One is the effect of microalloying elements on the transformation during quenching, the other is the effect of precipitation of microalloying elements on properties during tempering .The latter has been researched systematically[3-4]. Thus, the former would be discussed in detail.

It can be seen from Table 3 that the addition of Nb, V, Ti microalloying elements has little effect on the austenization temperature, with the same austenization temperature and cooling rate, the effect of microalloying elements on the microstructure transformation during quenching was explained by empirical formula designed for the heat treatment process. The critical cooling rate formula is shown in Formula 1. Formula2 shows the critical cooling rate of martensite transformation when quenched in the oil. Formula 3 shows the expression for the calculation of the bainite transformation incubation period[5]. It can be seen from the formula that the effect of microalloying elements is not present. The microalloying elements precipitate in the form of carbides during the rolling and air-cooling, resulting in a decreased carbon content in austenite during quenching[6]. The additions of microalloying elements increase the transformation temperature of super-cooled austenite to martensite and critical cooling rate. On the other hand, it has already been confirmed that microalloying elements play a role on grain refinement during steel rolling. Due to the genetic function of the austenite grain size in the austenization process[7], the addition of microalloying elements decreases the grain size of the supercooled austenite during quenching. According to the incubation period of bainite transformation, with the other chemical composition of the same, the addition of microalloying elements reduces the carbon content in the supercooled austenite and decreasing the grain size. The combination of these two factors shortens the incubation period of bainite transformation, promoting the transformation from supercooled austenite to bainite. Thus, the addition of microalloying elements promotes bainite transformation and inhibits martensite transformation.

$$Ms=561-474C-33Mn-17Cr-17Ni-21Mo \quad (1)$$

Where Ms represents martensite start temperature;

$$LogvM=9.81-4.62C+1.10Mn+0.54Ni+0.50Cr+0.60Mo+0.00183PA \quad (2)$$

Where vM presents critical cooling rate, PA represents austenization condition

$$\tau_B = \frac{(2.34 + 10.1C + 3.8Cr + 19Mo) * 10^{-4}}{\beta * \exp\left(-\frac{27500}{RT}\right) * (\Delta T)^2 * 2^{(G-1)/2}} \int_0^X \frac{\exp[X^2(1.9C + 2.5Mn + 0.9Ni + 1.7Cr + 4Mo - 2.6)]}{X^{\frac{2(1-X)}{3}}(1-X)^{\frac{2X}{3}}} \quad (3)$$

Where τ represents incubation period, T represents transformation temperature, β represents correction factor, X represents transformation, △T represents degree of supercooling, G represents grain size.

5. Conclusions

1. Microalloying elements have little effect on the thermodynamics of phase transformation in the quenched and tempered steel.

2. Microalloying elements can increase critical cooling rate and shorten the incubation period of bainite transformation, therefore promote bainite transformation during quenching process.

3. The better tempering stability was ascribed to the hot-rolled bainite microstructure and secondary carbide precipitation during tempering in microalloyed steel.

Refenrces

[1] ZHAO Yan-ling. Effect of Micro-alloying elements on the properties of quenching and tempered low alloy high strength steel[J]. Wide and Heavy Plate, 2004, 10(5): 14-19.
[2] CAO Yong-qing. Effect of Nb on hardenability of austenite steel[N]. World Meatal.2002 /12/31 /007 /
[3] ZHANG Ke,YONG Qi-long,etc.Effect of tempering temperature on microstructure and mechanical properties of high Timicroalloyed directed quenching high structure steel[J]. ACTA METALLURGICA SINICA, 2014, 50(8): 913-920.
[4] XU Li-shan,YU Wei,etc. Microstructure and mechanical properties of low carbon bainite steel treated by quenching and tempering[J]. Journal of University of Science and Technology Beijing, 2012, 34(2): 125-131.
[5] HUANG Chun-feng. Emerprical formulas of mechanical properties of steels and heat treatment processing[J]. heat treatment, 1998 (4): 12-17.
[6] SUN Man-li,JIANG Bo,etc. Effect of microalloying Nb on the microstructure and mechanical properties of high carbon steel[J]. Anhui metallurgy, 2015 (2): 36-39.
[7] FU Han-guang,CHEN Xiang,etc.Prograss on study of high silicon wear resistant casting steel[J]. Research on iron and steel, 2005, 33(1): 59-63.

Production of gauge 1.25Cr-0.5Mo and carbide distribution in heat treatment structures

Lv YanChun, Zou Yang, Zhao Nan, Qin LiYe

Shougang Group Technology Research Institute, China

Abstract: In order to research into the structural features of heat treated guage $1^{1/4}$Cr-1/2Mo steel, relevant constituents are designed and smelted, and tests inclusive of heating, rolling, heat treating and PWHT are conducted. The tested steel plate resists the relevant processes, PWHT test and step cooling. Handled in different temperatures, the steel plate demonstrates regularity in its carbide distribution formed in corresponding heat treated structures.

1. Introduction

Cr-Mo steel applied in petrochemical industry and heat-engine plants permanently confronts great heat, high pressure and hydro attack. Composed of relatively high amount of alloy, Cr-Mo steel can produce bainite structure in air cooling conditions. As a heat-resisting metal, high temperature-chemical stability and mechanical stability are its most basic features. Gauge steel plates are characterized by weightiness and thickness, which may well lead to uneven temperature in cross section or a lack of synchronization of structural shift. Based on pressure and temperature, current hydroreformer adds to the tempering embrittlement of $2^{1/4}$Cr-1Mo steel as amount of Cr and Mo increase, and in long term brittleness and material malleability decrease inextricably surfaces. Research shows that tempering emrbrittlement is fundamentally when alloy steel was being tempered or subsequently being cooled the gathering of inimical elements like P、Sn、Sb、As in austenite matrix which undermines the cohesive force of matrix in grain boundary. When tension or shock applies, grain boundary is subject to breakage and causes Intergranular fracture[1-7]. In consequence, the study of gauge and high-alloy$1^{1/4}$Cr-1/2Mo steel's heat treatment process and uniformity plays a vital role in material's long-term performance.

2. Experiment method

(1) The smelting of molten steel

The hydroreformer was designed in line with the chemical composition of $1^{1/4}$Cr-0.5Mo steel plate. Our major concern touches upon the following points: Firstly, the impurity elements including P、Sb、Sn、As. Secondly, the content of Mn is around 0.5%, considering Mn's contribution to overall intensity.

Table 1 Chemical composition of hydroreformer-oriented steel plate

Speci-men	C%	Si%	Mn%	P%≤	S%≤	Alt%	Cr%	Mo%	Sb	Sn	As	J	X
1	0.145	0.55	0.49	0.007	0.002	0.034	1.217	0.498	-	-	0.006	71.8	7.45
2	0.126	0.56	0.45	0.007	0.002	0.034	1.182	0.486	0.002	0.002	0.009	87.9	9.5

Molten iron desulphurization technique and converter double slag technique were employed to get

Content from this work may be used under the terms of the Creative Commons Attribution 3.0 licence. Any further distribution of this work must maintain attribution to the author(s) and the title of the work, journal citation and DOI.
Published under licence by IOP Publishing Ltd

rid of S and P. Deoxidation alloying was conducted out of converter. S was further cleared through LF refining process and H through RH process. Heavy plate casting followed.

(2) Heating method

Temperature control within the heating furnace kept the tapping temperature of the steel plate. The steel plate here is 400mm thick. The heating duration is illustrated in the following table.

Table 2 The furnacing duration of the hydroreformed $1^{1/4}$Cr-1/2Mo steel plate

	thickness/mm	heating duration/min
Cold billet	250	250~300
	300	300~450
	400	400~500

(3) Rolling of steel plates

For hydroreformed $1^{1/4}$Cr-1/2Mo steel plates thinner than 90mm, normally hot rolling is preferred. Rolling passes were not separated and in rough rolling and pass reduction was heightened. In finish rolling multiple methods were available in rolling shape control, leaving no holding time between passes. The finish cooling temperature was around 650°C. For hydroreformed $1^{1/4}$Cr-1/2Mo steel plates thicker than 90mm, rolling passes were not separated and in rough rolling, and its finish cooling temperature was around 670°C.

(4) Heat treatment process

Hydroreformed $1^{1/4}$Cr-1/2Mo steel plates thicker than 90mm were normalized in our effort to retrieve bainite structure, with normalization temperature fixed at 930℃. Corresponding holding time was 60±1min after plate core are heated to expected temperature. Tempering temperature was initially settled at 700℃.

3.Results and discussion

3.1 composition design

(1) Impact of Mn

Like Si, Mn can drive tempering embrittlement. That said, Mn is different from Si as it exerts bigger influence on steel's hardenability. When hardenability is required and ferrite is undesirable, the amount of Mn should remain low but not too low.

(2) Impact of Sn, Sb, P and As

Sn, Sb, P and As are all impurity elements that cause marked embrittlement, therefore should be kept in low levels in smelting. Besides, the brittleness sensitivity coefficient X（X=(l0P+5Sb+4Sn+As)*10-2≤15ppm）according as both national and American standards is supposed to be minimum, and this echoes with the need to be technically curbed.

(3) Impact of Cr and Mo

Cr gives a evident lift of steel's high temperature performance by improving steel's anti-oxidation, which prevents graphitization, and corrosion resistance (hydro attack). Properly heated, Cr-Mo steel can acquire tensile strength and comprehensive mechanical properties. But as Cr increases, susceptibility to tempering embrittlement will soar; especially when Cr accounts for 2.0～3.0%, susceptibility to embrittlement is higher.

Mo is the pivotal source of steel's high temperature strength. With its strong tendency to form carbide, Mo can prevent pearlite transformation and make steel astoundingly subject to bainite, thereby promotes the comprehensive mechanic performance of Cr-Mo steel. In the meantime, Mo can make steel tempering more stable, eliminate or reduce tempering embrittlement of Cr-containing steel.

3.2 Rolling process

95mm thick hydroformed 1.25Cr-0.5Mo steel mainly adopted direct converter steelmaking, at the same time combining rough rolling of double passes width spread and longitudinal rolling. The pass reduction and pass reduction rate is illustrated in figure 1. From figure 1 we can note that rough rolling of steel plate remained steady, with 4-5 pass reduction rates above 35mm each, while the maximum pass reduction rate nearing 25%. The innermost part of the thick plate was sufficiently deformed, thereby adding to the mechanic homogeneity in the thickness of steel plate, and thus the great level of stability in the rolling process.

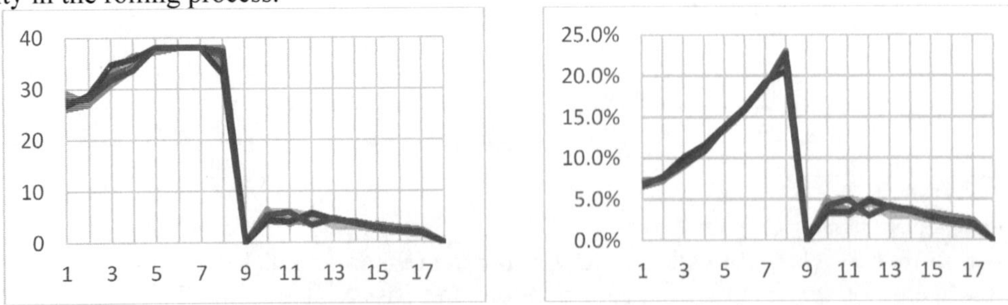

95mm-sized plate's pass reduction 95mm-sized plate's pass reduction rate

Figure 1 distribution of pass reduction and corresponding rate of 95mm thick steel

60mm thick hydroreformed 1.25Cr-0.5Mo steel mainly adopted forming pass plus 6-9 pass width spread plus longitudinal rolling, whereas 2 steel plates adopted forming pass and full cross-rolling technique, whose impact on performance pending further research. As shown in the following Figure 2, the overall pass reduction stood fewer than 30mm, and the low level was due to the a lengthier width of the steel plate. That means pass reduction concentrated at the width spread passes, and after longitudinal rolling the rolling force and torque were relatively high, thus a low level of pass reduction. The width of the steel is a major reason for a meager single-pass reduction and reduction rate in rough rolling.

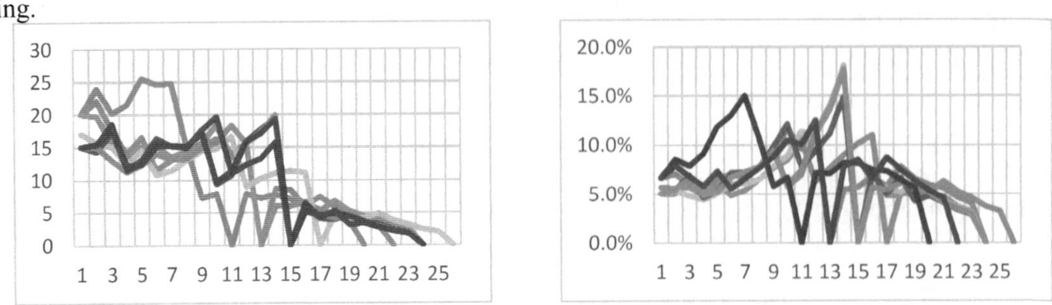

60mm-sized plate's pass reduction 60mm-sized plate's pass reduction rate

Figure 2 distribution of pass reduction and pass reduction rate of 60mm thick steel

3.3 Heat treatment process

For 60-95mm thick hydroreformed steel plate, the normalizing temperature was 930±10℃, holding time afterwards was 95min. Chilled water flow rate following normalization was shown in Figure 3.

After weak cooling water treatment, plate shape of 95mm thick plate remained steady. The self-tempering temperature ranged between 160℃ and 180℃. Temperature difference of the whole plate was less than 20℃, reaching the foreseen goal.

Table 3 Amount of water used on 60-95mm sized hydroreformed steel plate

Specimen	**1**	**2**	**3**	**4**	**5**	**6**	**7**	**8**	**9**	**10**	roll speed m/min
60mm	120	485	120	200	580	200	90	325	90	81	14
	11	**12**	**13**	**14**	**15**	**16**	**17**	**18**	**19**	**20**	

	398	81	480	580	200	400	200	400	200	400	
	1	**2**	**3**	**4**	**5**	**6**	**7**	**8**	**9**	**10**	
	120	485	120	200	580	200	90	325	90	81	
95mm	**11**	**12**	**13**	**14**	**15**	**16**	**17**	**18**	**19**	**20**	5
	398	81	480	580	200	400	200	400	200	400	

3.4 tempering performance and carbide distribution

95 mm thick hydroreformed 1.25Cr-0.5Mo steel plate has gone through post-tempering performance test, as shown in Figure 4. Holding post-tempering mechanic parameters held as a reference, we can see that all mechanic performance fits into requirements of the protocol, and residual amount of intensity created room for the improvement of mechanic performance for Max.PWHTand Min.PWHT.

Table 4 Post-temp performance of 95mm gauge hydroreformed 1.25Cr-0.5Mo steel plate

specimen	Rel /MPa	Rm /MPa	A /%	Z /%	Shock temperature	Impact energy/J			cold bending d=3a
1	443	589	27.5	72.2	-20℃	169	195	241	up to par
2	436	584	25.5	65.1	-20℃	245	239	261	up to par
3	440	580	27.0	72.1	-20℃	177	162	220	up to par
4	448	597	26.5	62.1	-20℃	177	172	180	up to par
5	448	592	23.5	63.5	-20℃	179	220	229	up to par
6	435	590	25.5	62.7	-20℃	198	214	213	up to par
7	438	586	23.5	65.0	-20℃	105	188	114	up to par
8	444	590	26.5	66.2	-20℃	205	249	186	up to par

Hydroreformed 1.25Cr-0.5Mo steel plates turned predominantly lath bainite after being tempered in 700℃, and the carbide mostly took on beam current distribution in the form of bainite. Part of the carbide was rod-shaped about 1μm long, part sphere-shaped with a diameter between 200nm and 500nm, and some other smaller sphere-shaped carbide distributed therein.

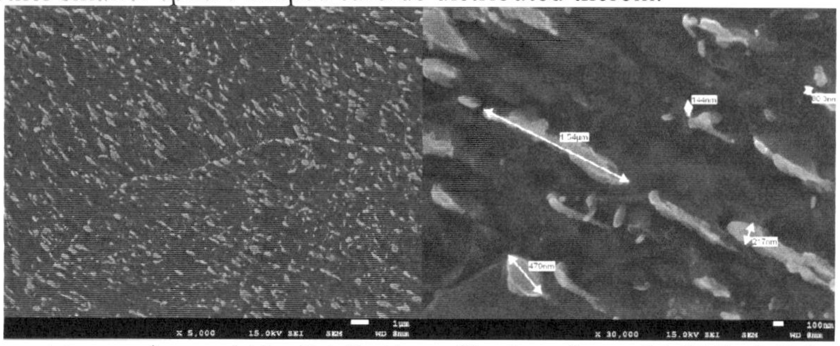

Figure 3 post-tempering carbide distribution of hydroreformed 1.25Cr-0.5Mo steel

EDS analysis indicates that the majority of the rod-shaped carbide was alloy cementite and the sphere-shaped carbide was carbide of Cr, which was inferred to be Cr7C3, pending TEM phase arrangement differentiation. The still less sizable sphere-shaped carbide was too limited in size for EDS to yield a report of its composition, thereby pending TEM phase arrangement differentiation. On austenite matrix mainly distributed alloy cementite as shown in Figure 4.

IOP Conf. Series: Materials Science and Engineering **242** (2017) 012037 doi:10.1088/1757-899X/242/1/012037

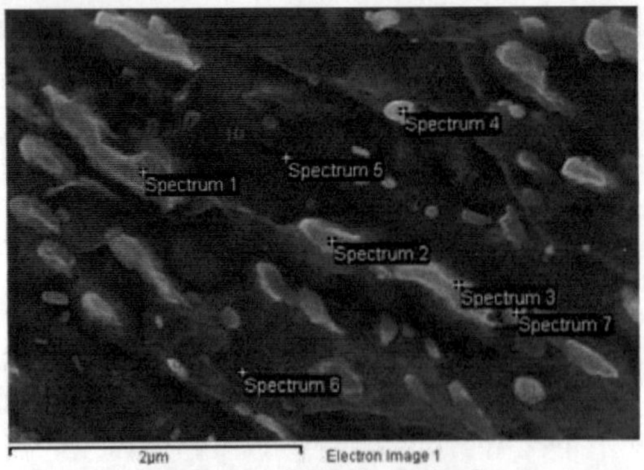

Figure 4 EDS analysis

Table 5 Composition of each point position

	C	Si	Cr	Mn	Fe
1	7.38	0.52	2.30		89.79
2	6.31	0.65	1.58		91.46
3	6.66	0.57	1.58		91.19
4	11.35	0.61	4.75		83.29
5	3.98	0.69	1.54		93.79
6	3.82	0.61	1.56		94.01
7	13.45	0.60	7.01	1.37	77.57
8	7.61		2.62		89.77
9	9.72	0.72	3.61		85.99
10	4.46		1.66		93.88
11	10.90	0.82	3.14		85.14

3.5 Post-PWHT carbide morphological distribution

3.5.1 Min.PWHT Following 700℃ tempering of 1.25Cr-0.5Mo steel, the post-Min.PWHT carbide morphological distribution took on the form shown in Figure 5. Compared with conditions during tempering, the 1.25Cr-0.5Mo steel tempered in 700℃ demonstrated such features after Min.PWHT procession: Alloy cementite disappeared in huge amount while Cr's carbide rose in proportion; in some spots Mo's carbide together with MnS precipitated, though in meager amount.

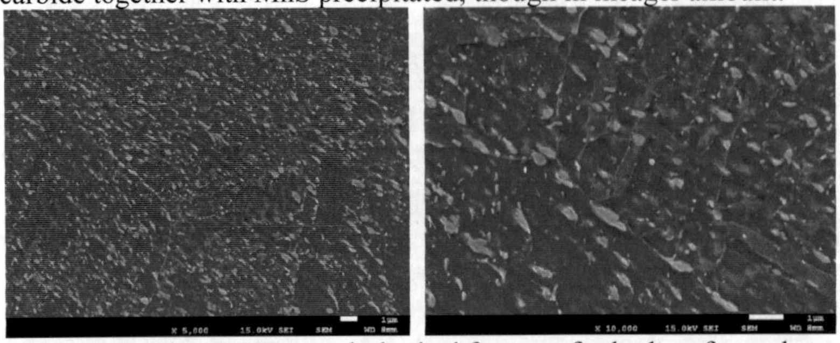

Figure 5 The carbide's post-Min.PWHT morphological features for hydroreformed steel tempered in 700℃

3.5.2 Max.PWHT Following 730℃ tempering of 1.25Cr-0.5Mo steel, the post-Min.PWHT carbide morphological distribution took on the form shown in Figure 6. Post-Min.PWHT conditions didn't vary much from tempering conditions, i.e. Cr's carbide remained the bulk.

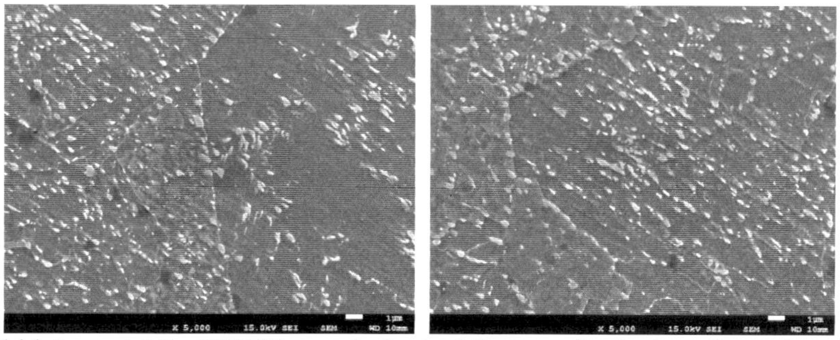

Figure 6 The carbide's post-Min.PWHT morphological features for hydroreformed steel tempered in 730℃

3.6 Post-step-cooling carbide's morphological distribution within steel plate

Following 700℃ tempering of 1.25Cr-0.5Mo steel, the post-step-cooling carbide morphological distribution took on the form shown in Figure 7. The result after step cooling resembles that after Min.PWHT, as sphere-shaped Cr's carbide still took the majority. As far as EDS's results are concerned, some spots yielded composite carbide of Cr and Mo, though in meager proportion. Stable carbide distribution both prior to and after the step cooling contributes to lower level of ΔvTr55, bolstering post-tempering anti-brittleness performance of steel plates that are to be hydroreformed.

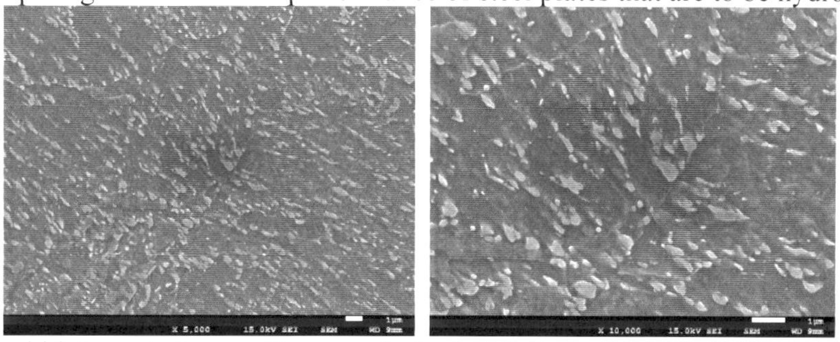

Figure 7 The carbide's post-step-cooling morphological features for hydroreformed steel tempered in 700℃

Following 730℃ tempering of 1.25Cr-0.5Mo steel, the post-step-cooling carbide morphological distribution resembles that following Min PWHT procession, as sphere-shaped Cr carbide dominated in amount. Though some spots yielded Mo's carbide, it was low in proportion. Stable carbide distribution both prior to and after the step cooling contributes to lower level of ΔvTr55, bolstering post-tempering anti-brittleness performance of steel plates that are to be hydroreformed.

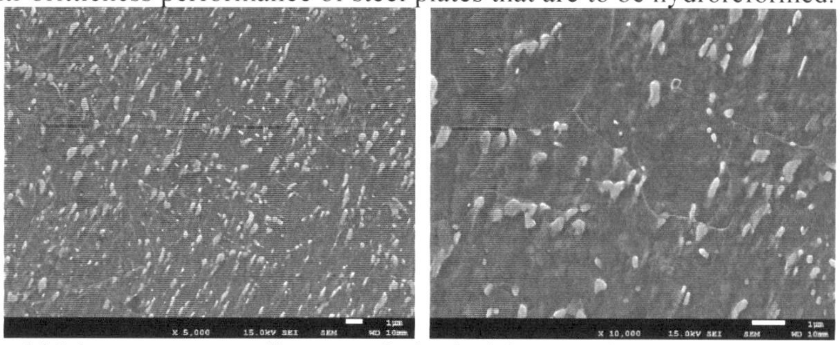

Figure 8 The carbide's post-step-cooling morphological features for hydroreformed steel tempered in 730℃

For tempered 1.25Cr-0.5Mo steel, the post-step-cooling carbide morphological distribution was characterized by higher proportion of pearlite. Within ferrite matrix, carbide's majority was still Cr's carbide, whose formation needs to be observed and analyzed through TEM.

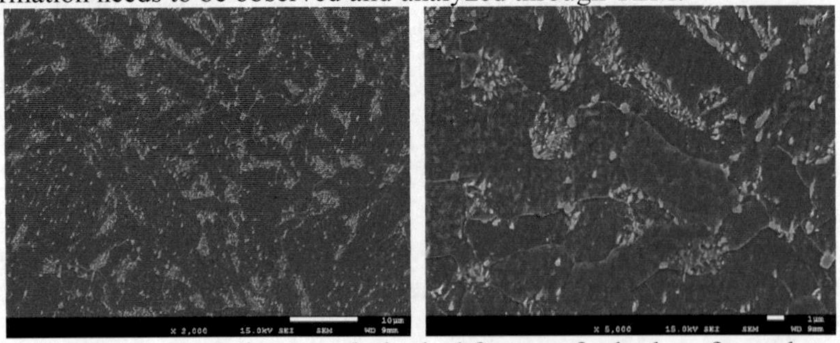

Figure 9 The carbide's post-step-cooling morphological features for hydroreformed steel tempered in 760℃

Following 700℃and 730℃ tempering of 1.25Cr-0.5Mo steel, the post-step-cooling carbide morphological size, distribution and positions largely remained the same, with its bulk Cr's carbide. Following 760℃ tempering of 1.25Cr-0.5Mo steel, the post-step-cooling carbide demonstrated higher proportions of pearlite, formation mechanism of which pending TEM analysis.

4. Conclusion

1) The smelting of steel plate met set standards of the test, and both coefficient J,X met the requirements.

2) After tempering, all mechanic performance all fitted before-set requirements. Meanwhile the residual amount of intensity was relatively high thereby created room for the improvement of mechanic performance for Max.PWHTand Min.PWHT.

3) Compared with tempering conditions, post Min.PWHT features disappearance of alloying cementite and increase of Cr's carbide in proportion; in some spots Mo's carbide together with MnS precipitated, though in meager amount.

4) Following 700℃and 730℃ tempering of 1.25Cr-0.5Mo steel, the post-step-cooling carbide morphological size, distribution and positions largely remained the same, with its bulk Cr's carbide; post 760℃ tempering the steel yielded greater amount of pearlite.

Reference

[1] Foldyna V, Jakobova A, Vodarek V, Kubon Z. In: Coutsouradis D, et al, editor. Materials for advanced power engineering.Kluwer Academic Publisher, I, 1994. p. 453–464.

[2] Spiradek K, Bauer R, Zeiler G. In: Coutsouradis D, et al, editor. Materials for advanced power engineering.Kluwer Academic Publisher, I, 1994. p. 251–262.

[3] Tsuchida Y, Okamoto K, Tokunaga Y. ISIJ Int. 1995;35:309– 316. w4x Hamada K, Tokuno K, Tomita Y, Mabuchi H, Okamoto K. ISIJ Int. 1995;35:86–91.

[4] R.B. Scarlin, in: A. Strang, W.M. Banks, R.D. Conroy, M.J. Goulette (Eds.), Advances in Turbine Materials, Design and Manufacturing, The Institute of Materials, London, 1997, p. 242.

[5] R.W. Vanstone, in: J. Lecomte-Beckers, F. Schubert, P.J. Ennis (Eds.), Materials for Advanced Power Engineering, Forschungscentrum Ju ¨lich GmbH, p, Julich, 1998, p. 139.

[6] P. Polcik, PhD Thesis, Shaker Verlag, Aachen, 1999.

[7] J. Hald, S. Straub, in: J. Lecomte-Beckers, F. Schubert, P.J. Ennis (Eds.), Materials for Advanced Power Engineering, Forschungszentrum Julich GmbH, Ju ¨lich, 1998, p. 155.

IOP Conf. Series: Materials Science and Engineering **242** (2017) 012038 doi:10.1088/1757-899X/242/1/012038

Study on slag forming mechanism of hot metal containing titanium in converter

H B Wang, Y C Lv, L Y Qin, Y Liu, C W Ma, X T Kong

[1] Shougang Research Institute of Technology, Beijing, 100043
[2] Beijing Key Laboratory of Green Recyclable Process for Iron & steel Production Technology, Beijing, 100043
[3] Beijing Engineering Research Center of Energy Steel, Beijing, 100043

wanghaibao1982@shougang.com.cn

Abstract. The use of titanium containing molten iron can expand the sources of raw materials, reduce the cost, but the process produce serious foaming slag, low dephosphorization proportion and the consumption of lime is high. The technicians have improved the operation process through experiments and solved the problem of efficient and smooth blowing of hot metal containing titanium. Through mine phase composition analysis of slag using SEM and EDS, the distribution of titanium and the regularity of phosphorus retention are found, and measures to prevent phosphorus recovery in the later stage of converter are put forward.

1 Introduction

In order to ease the tension of the ordinary iron ore resources, reduce raw material procurement costs, a certain amount of ore which containing titanium and vanadium is added in the blast furnace in the ironmaking plant and about 0.4% trace elements appear in molten iron. Its characteristics are that the content of titanium in molten iron is high, and the content of titanium is 0.15% ~ 0.45%, results in low melting point and poor slag splashing effect in the steelmaking process. There will be a series of problems by the use of high titanium and vanadium contented hot metal, such as poor dephosphorization effect, serious foaming slag, serious splash, higher lime consumption, lower furnace age and so on [1-3].

In view of the above problems, the technical staff of the steel plant started the experiment and improved the original "double slag" steelmaking method and achieved good results.

The new BOF steelmaking process is divided into three periods, detitanium period dephosphorization period and decarburization period. With oxygen supply to the converter detitanium period begin and end after three minutes. In detitanium period the basicity should be low and the oxygen lance keep high, with a small batch of ore added.

Liquid slag in the converter mainly from Ti, Si, Mn, Fe oxidation during the early period of blowing, the slag is removed from the converter before the severe carbon oxidation period, and the titanium rich slag is poured out. The ore is added into the converter and the lime is used to make the dephosphorization slag. Then, the traditional "double slag" process is carried on until tapping [4-8]. Through the observation of slag micro morphology and analysis of the mine phases in the process of converter steelmaking, the process of steelmaking can be further optimized to lay a foundation for reducing the smelting cost and increasing the efficiency.

Content from this work may be used under the terms of the Creative Commons Attribution 3.0 licence. Any further distribution of this work must maintain attribution to the author(s) and the title of the work, journal citation and DOI.
Published under licence by IOP Publishing Ltd

2 Experimental materials and methods

2.1 Experimental materials

The sample is taken from a 100t top and bottom combined blown converter. Chemical components of the slag are shown in table 1.

Table1. Chemical composition of slag for experiment (wt %)

Sample No.	TFe	FeO	SiO2	Al_2O3	CaO	MgO	MnO	P2O5	TiO2	R
detitanium	11.18	11.7	20.22	1.26	37.91	3.16	7.91	2.43	8.84	1.87
dephosphorization	8.36	5.14	14.92	2.26	53.28	4.31	3.55	3.52	4.03	3.57
tapping	14.26	10.8	12.77	2.1	48.73	4.24	3.41	2.29	3.03	3.81

As can be seen from table 1, the main components of slag used in this experiment are Ca, Fe, Si, Mn, Mg and Ti, and the slag is basic.

2.2 Experimental methods

The slag is broken and inlaid, and the sample is prepared by grinding and polishing with a polishing machine. The experimental equipment is S-3400N scanning electron microscope. The acceleration voltage is from 0.3 to 30KV, the magnification is 50~300 thousand times.

Using SEM backscatter low vacuum mode, micro morphology of the specimen was observed. The backscattered electron energy is higher and sensitive to the atomic number of the sample material. The larger the average atomic number is, the more the backscattered electron appeared which make the image brighter. Through material qualitative analysis; and using X-ray spectrometer to determine the components of micro elements, the mineral compositions of slag obtained finally.

3 Experimental results

3.1 Morphology observation and energy spectrum analysis of detitanium slag

Using SEM backscatter low vacuum mode, micro morphology of the specimen was observed. As shown in Figure 1, the morphology of the main minerals can be roughly divided into 3 phases: white phase (A1 and A2), gray phase (B1 and B2) and black and white stripe phase (C1, C2, C3).

Figure1. Morphology and energy spectrum position of detitanium slag

The matrix of the converter slag is grey phase, and the irregular white phase is distributed intermittently in the matrix. Some of the white phase boundaries are clear and rectangular, and some of the white phases have black and white stripes in the middle.

The results of energy spectrum analysis are shown in table 2.

Table2. Composition analysis of energy spectrum of detitanium slag (wt %)

Position	Color	O	Mg	Al	Si	P	Ca	Ti	V	Mn	Fe
A1	white	36.98	3.19	1.06	5.04		9.17	11.32	4.52	10.92	17.8
A2		41.68	2.55	1.12	4.74	0.62	8.8	9.65	2.39	9.55	18.9
B1	gray	45.97	1.48	1.34	13.22	1.08	20.37	3.25		6.11	7.18
B2		43.49	2.25	0.5	12.96	1.08	20.05	3.1		8.21	8.38
C1	stripes	40.59	2.27	0.46	8.78	1.36	19.16	6.68	1.64	7.63	11.44
C2		43.28	1.93	0.88	8.82	1.34	17.62	5.77	1.57	8.08	10.72
C3		40.91	2.1	0.52	9.32	1.51	20.47	5.14	1.87	7.28	10.89

Through spectrum analysis showed that the white phase mainly contains Fe, Ti, Mn, Ca and other elements, the content of Fe is about 18%, the content of Ti is about 10%, the content of Mn is about 10%, the content of Ca in about 9%, and a small amount of Si and Mg elements, it can be deduced that the white phase consists mainly of $CaO \cdot Fe_2O_3$, $TiO_2 \cdot Fe_2O_3$ and manganese oxide, as well as a small amount of silicate.

The grey matrix was mainly composed of Ca, Si, Fe and Mn elements, the content of Ca is about 20%, the content of Si is about 13%, the content of Fe is about 8%, the content of Mn is about 8%, it can be inferred that the grey phase consists mainly of $2CaO \cdot SiO_2$, $CaO \cdot Fe_2O_3$, and manganese oxide.

The black and white combined phase was mainly composed of Ca, Fe, Si, Mn and Ti elements, the content of Ca is 18% ~ 21%, the content of Fe is about 11%, the content of Si is about 9%, the content of Mn is about 8%, the content of Ti is about 6%, it can be inferred that black and white combined phase consists mainly of $CaO \cdot Fe_2O_3$,$CaO \cdot SiO_2$, manganese oxide and $TiO_2 \cdot Fe_2O_3$. All of the three phases contained Ti and a small amount of phosphorus.

3.2 Morphology observation and energy spectrum analysis of dephosphorization slag

Using SEM backscatter low vacuum mode, micro morphology of the specimen was observed. As shown in Figure 2, the morphology of the main minerals can be roughly divided into 3 phases: white phase (A1 and A2), gray phase (B1 and B2) and light gray phase (C1, C2, C3). The distribution of white phase and light gray phase is uneven. The shape of the two phase distribution is consistent, and the gray matrix phase is divided into many elliptical shapes.

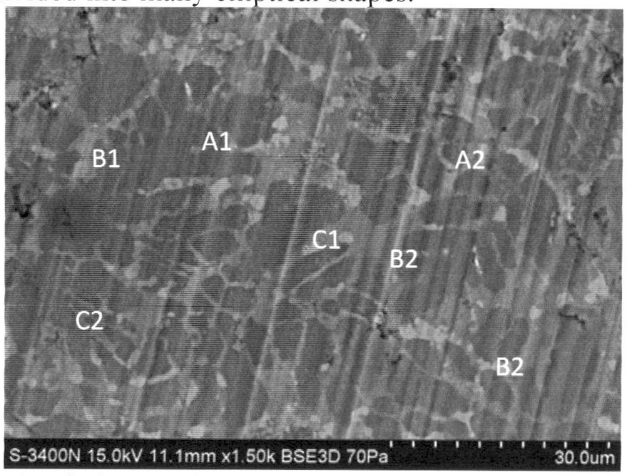

Figure2. Morphology and energy spectrum position of dephosphorization slag

The results of energy spectrum analysis are shown in table 3.

Table3. Composition analysis of energy spectrum of dephosphorization slag (wt %)

Position	Color	O	Mg	Al	Si	P	Ca	Ti	V	Mn	Fe
A1	white	36.7	1.27	0.64	2.85	0.84	15.51	0.88		9.02	32.28
A2		36.68	1.49	0.51	2.48	0.57	15.49	0.91		9.18	33.7
B1	gray	42.82	0.73	1	9.36	2.45	35.95	1.43	1.06	0.85	4.35
B2		45.86	1.08	0.61	9.66	2.48	35.33	1.55			3.43
B3		44.79	1.13	0.86	9.49	2.45	34.97	1.37		1.13	3.81
C1	light gray	40.78	1	5.59	2.71	0.71	30.39	2.54	1.72	1.31	13.25
C2		38.97	0.88	5.18	2.85	0.99	30.11	2.64	1.74	2.2	14.44

Through spectrum analysis showed that the white phase mainly contains Fe, Ca, Mn elements, the content of Fe is about 33%, the content of Ca is about 16%, the content of Mn is about 9%, a small amount of Si and Mg elements, it can be inferred that the white phase is mainly composed of $CaO \cdot Fe_2O_3$, $3CaO \cdot SiO_2$ and manganese oxide .

The gray matrix was mainly composed of Ca and Si elements, the content of Ca is about 35%, the content of Si is about 10%, also contains a small amount of Fe elements, it can be inferred that the grey phase consists mainly of $2CaO \cdot SiO_2$ and a small amount of $CaO \cdot Fe_2O_3$.

The light gray phase mainly contains Ca, Fe and Al elements, the content of Ca is about 30%, the content of Fe is about 14%, the content of Al is about 5%, also contains a small amount of Ti element, it can be inferred that the light gray phase consists mainly of $CaO \cdot Fe_2O_3$, $CaO \cdot Al_2O_3$, and a small amount of TiO_2. In the gray phase, the content of phosphorus is highest, about 2.5%, in white or light gray phase the content of phosphorus is below 1%.

3.3 Morphology observation and energy spectrum analysis of tapping slag

Using SEM backscatter low vacuum mode, micro morphology of the specimen was observed. As shown in Figure 3.

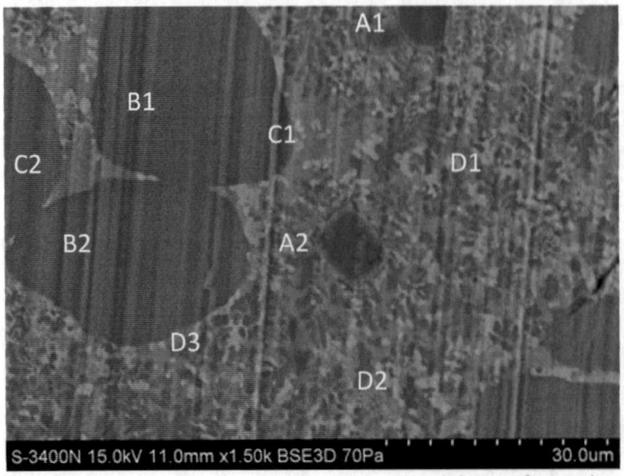

Figure3. Morphology and energy spectrum position of tapping slag

the morphology of the main minerals can be roughly divided into 4 phases: dark gray phase (A1 and A2), gray phase (B1 and B2), and light gray phase (C1 and C2) and white phase (D1, D2, D3).Dark gray phase embedded in the matrix, the boundary is clear, shape is rectangular. Light gray and white phase morphology consistent, distributed in the grey matrix and the grey matrix is mostly all covered by white and light grey phase in some area.

The results of energy spectrum analysis are shown in table 4.

Table4. Composition analysis of energy spectrum of tapping slag (wt %)

Position	Color	O	Mg	Al	Si	P	Ca	Ti	Cr	Mn	Fe
A1	dark	42.77	24.14		2.16	0.62	11.6	0.66	0.78	4.39	12.88
A2	gray	43.06	22.36	0.6	1.98	0.57	12.25		1.11	3.88	14.18
B1	gray	46.54	0.65	0.61	8.98	2.51	34	0.91			5.81
B2		45.09	0.98	0.51	9.21	2.87	34.45	1.09			5.8
C1	light	36.86	1.13	2.55	2.88	0.89	29.33	1.63		2.32	20.84
C2	gray	40.55	1.32	2.13	2.95	0.67	30.22	2.15		1.49	19.5
D1	white	36.39	3.56	0.65	3.07	0.78	17.79	0.94		6.49	30.34
D2		35.15	3.19	0.46	2.11	0.58	18.17	0.8		6.41	32.71
D3		39.13	3.17	0.43	2.67	0.69	16.06	1.03	0.78	5.49	30.54

Through the energy spectrum analysis shows that the dark gray phase mainly contains Mg, Fe, Ca elements, Mg content is about 23%, the content of Fe is about 13%, the content of Ca in about 12%, also contains a small amount of Mn element, it can be inferred that the dark gray phase is mainly composed of MgO and $CaO \cdot Fe_2O_3$ and a small amount of $CaO \cdot SiO_2$.

the gray phase mainly contains Ca, Si and Fe elements, the content of Ca is about 34%, the content of Si is about 9%, the content of Fe is about 6%, it can be inferred that

the gray phase is mainly composed of $2CaO \cdot SiO_2$ and $CaO \cdot Fe_2O_3$.

The light gray phase mainly contains Ca, Fe elements, the content of Ca is about 30%, the content of Fe is about 20%, also contains a small amount of Si element and Al element, it can be inferred that the light gray phase is mainly composed of $CaO \cdot Fe_2O_3$ and a small amount of $3CaO \cdot SiO_2$ and $CaO \cdot Al_2O_3$.

The white phase mainly contains Fe, Ca and Mn elements, the content of Fe is about 30%, the content of Ca is about 17%, the content of Mn in about 5%, also contains a small amount of Mg and Si, it can be inferred that the white phase is mainly composed of $CaO \cdot Fe_2O_3$ manganese oxide and a small amount of $CaO \cdot SiO_2$. All of the four phases contain a certain amount of phosphorus.

4 Analysis and discussion

4.1 The distribution and regularity of Ti element

(1)Through the observation of slag micro morphology and analysis of the mine phase shows that all of the three phases contain Ti elements. The reason is that TiO_2 have relatively stable performance and poor surface activity so the binding ability with other oxides is not strong, as a result the distribution is dispersed in the slag.

(2)There is a significant correlation between Ti content and Fe content, this is because the Fe element level is a measure of oxidizability of the slag. With a Sufficient oxidation in converter, the content of TiO_2 is higher. When the Fe element is about 30% and higher temperature in the converter, Ti elements decrease in the content of mineral phase.

(3)There is a significant correlation between the Ti element and the content of Al, that because of the Al_2O_3 surface tension in molten steel is large, easily combined into aggregates particle groups and adsorption of TiO_2 increased.

(4)With the increase of slag oxidation, the phase colored with white and black will gradually change into white phase with higher Ti content, which will be beneficial to the removal of Ti elements. All of the three phase in the detitanium slag contain P elements, which is widely distributed in the slag, but the content of phosphorus in each phase is low.

4.2 Analysis of control measures in dephosphorization period

In detitanium period all the Si element is oxidized and the SiO_2 content is low in dephosphorization period after the detitanium slag poured out. Therefore, in order to increase the dephosphorization capacity of the slag, silica was added to increase the SiO_2 content of the slag and lime was added to adjust the basicity.

Dephosphorization period is concentrated in about 3 ~ 6min, At this stage, in order to better remove phosphorus in molten iron, the blowing temperature should not be too high, so as to avoid the rephosphorization. In addition, the high temperature will lead to the decomposition of ferric phosphate, so be sure to increase the basicity before the temperature rises and turn the ferric phosphate into calcium phosphate.

4.3 Analysis of phosphorus recovery during decarburization period

Through the observation of decarbonized slag by micro morphology and analysis of the mine phase shows that the highest phosphorus contented phase is $2CaO \cdot SiO_2$ and $3CaO \cdot SiO_2$ mixing phase in converter. When the FeO in slag increases, the phase of $2CaO \cdot SiO_2$ and $3CaO \cdot SiO_2$ translated to low melting point phase, such as calcium ferrite, ferric phosphate and ferric silicate. At the same time, the solid phase of phosphorus is decomposed into small particles, and the phosphorus elements return to the liquid steel during the decomposition of solid phosphorus, and the phenomenon of phosphorus recovery occurs. Therefore, the FeO content should be reduced in the later period of converter blowing.

5 Conclusion

Through the study following conclusions can be obtained.

(1) In detitanium period, there is a significant correlation between Ti content and Fe content, this is because the Fe element level is a measure of oxidizability of the slag. With a Sufficient oxidation in converter, the content of TiO_2 is higher. When the Fe element is about 30% and higher temperature in the converter, Ti elements decrease in the content of mineral phase. All of the three phase in detitanium slag contain P element, which is widely distributed in the slag, but the content of phosphorus in each phase is low.

(2)In the dephosphorization period be sure to increase the basicity before the temperature rises and turn the $Fe_2O_3 \cdot P_2O_5$ into $3CaO \cdot 2P_2O_5$.

(3) In the decarburization period, when the FeO in slag increases, the phase of $2CaO \cdot SiO_2$ and $3CaO \cdot SiO_2$ translated to low melting point phase, such as $CaO \cdot Fe_2O_3$, and $Fe_2O_3 \cdot SiO_2$. At the same time, the phenomenon of phosphorus recovery occurs.

References

[1] Cvijovic Z and Radenkovic G 2006 *Corros. Sci.* **48** 3887.
[2] Chen T H and Yang J R 2001 *Mater. Sci. Eng., A.***A311** 28.
[3] Ahn Y S and Kang J P 2000 *Mater. Sci. Technol.***16** 382.
[4] Iacoviello F, Casari F and Gialanella S 2005 *Corros. Sci.* **47** 909.
[5] Tsai W T and Chen M S 2000 *Corros. Sci.* **42** 545.
[6] Rondelli G, Vicentini B and Sivieri E 1997 *Corros. Sci.* **39** 1037.
[7] Feng H, Zhou X Y, Liu H and Song Z G 2015 *Journal of Iron and Steel Research* **27** 1.
[8] Wang Y X, Chen Y M, Liu Z Y, Fang Y T and Wang G D 2009 *Journal of Iron and Steel Research* **21** 51.

IOP Conf. Series: Materials Science and Engineering **242** (2017) 012039 doi:10.1088/1757-899X/242/1/012039

Study on the surface constitute properties of high-speed end milling aluminum alloy

Huang Xiaoming [1,*], **Li Hongwei** [1] , **Ma Yumeng** [2]

[1] Mechatronics Engineering Department, Binzhou University;
[2] College of Aeronautical Engineering, Binzhou University

* Corresponding author's e-mail address: hxm2552@163.com

Abstract. The physical and mechanical properties of the metal surface will change after the metal cutting processing. The comprehensive study of the influence of machining parameters on surface constitute properties are necessary. A high-speed milling experiment by means of orthogonal method with four factors was conducted for aluminum alloy7050-T7451. The surface constitutive properties of the Al-Alloy surface were measured using SSM-B4000TM stress-strain microprobe system. Based on all the load-depth curves obtained, the characteristics parameters such as strain hardening exponent n and yield strength σ_y of the milling surface are calculated. The effect of cutting speed, feed rate, and width and depth of cut on n and σ_y was investigated using the ANOVA techniques. The affecting degree of milling parameters on n and σ_y was $v > f_z > a_p > a_e$. The influence of milling parameters on n and σ y was described and discussed.

1. Introduction

7050-T7451 aluminium alloy has been widely used in aircraft structures due to their perfect strength, fracture toughness, and high stress/density ratio after supersaturated solid solution and prestretching treatment [1, 2]. High-speed milling is one of the most common processes to produce structural parts out of aluminum alloy in the aerospace industry. Under the effect of cutting force, cutting heat and so on, the physical and mechanical properties of metal changed greatly, which was different from the matrix material [3, 4]. The stress-strain state can significantly influence the fatigue strength, creep, and stress-corrosion-cracking resistance of metals. Surface rolling for aeronautical monolithic component distortion correction was a new kind theory and process method [5]. To establish the surface rolling correction analysis model, accurate milling surface constitutive properties and the influence of milling parameters on them were pre-requisite.

Material constitutive model reflects the physical and mechanics behavior, which described the relationship between material flow stress, strain, strain rate and temperature, etc. Many scholars have studied the constitutive relation of 7050 aluminum alloy materials. Wu etc. [6] studied the flow stress and strain characteristics of 7050 aluminum alloy within the scope of the strain rate of 0.01-20s-1 and temperature of 593-743K. In the study, constitutive equations were obtained by isothermal compression experiment and the dynamic analysis, which can be used to predict the stress-strain relationship of higher temperature. Yongzhi, H etc. investigated the strain rate enhancement parameters and thermal softening parameters at room temperature (400-2500 s-1) and strain rate of 2500 s-1 under different temperature by split hopkinson pressure bar (SHPB) experiments [7, 8]. The modified Johnson-Cook constitutive relation model of 7050-T7451 aluminum alloy was established.

Content from this work may be used under the terms of the Creative Commons Attribution 3.0 licence. Any further distribution of this work must maintain attribution to the author(s) and the title of the work, journal citation and DOI.
Published under licence by IOP Publishing Ltd

IOP Conf. Series: Materials Science and Engineering **242** (2017) 012039 doi:10.1088/1757-899X/242/1/012039

The above research can study the constitutive relationship of the basis material at different temperature and strain rate, which reflect the mechanical behavior of the whole materials.

The physical and mechanical properties of aluminum alloy surface will change along the depth direction after the metal cutting processing, and the impact depth was about 0.1mm [9]. It is very critical to predict surface constitutive properties in a high-speed milled aluminium alloy, and a comprehensive study of the influence of machining conditions on surface strain hardening exponent n and yield strength σ_y is necessary. Indentation experiment is one of the effective methods to obtain the metal surface properties, which is a non-destructive testing technology[10,11]. In this article, orthogonal milling test of 7050-T7451 alloy were set up firstly. B4000TM stress-strain microprobe system was used to obtain the load-depth data of aluminum alloy, and the stress strain relations of the milled surface were calculated based on the load-depth data. The mathematical model for milling surface constitutive model was established and the effect of machining parameters on constitutive model was discussed.

2. Experimental setup and procedure

2.1. Work material

The 7050-T7451 aluminum alloy block used in this experiment was manufactured by Kaiser Aluminum and Chemical Corp., USA. The mechanical properties of the material are given in Table 1.

Table 1. Material properties of aluminum alloy 7050-T7451

Density (g/cm3)	Tensile Strength, Ultimate (MPa)	Tensile Strength, Yield (MPa)	Modulus of Elasticity (GPa)	Poisson Ratio
2.83	524	469	71.7	0.33

2.2. Milling experimental design

Experiments based on orthogonal methods were performed to reduce variations in the output response. Factorial design was applied with 4 factors and 4 levels (44 designs), factors consist of cutting speed (v), feed rate (f_z), width of cut (a_p), and depth of cut (a_e). The details are shown in Table 2. Values of cutting speed above 300m/min were chosen so that it falls in the range of high-speed machining.

Table 2. Orthogonal experiment parameters

level	v(m/min)	f_z (mm)	a_p (mm)	a_e (mm)
1	314	0.06	3	6
2	502	0.1	4	8
3	628	0.14	5	10
4	942	0.18	6	12

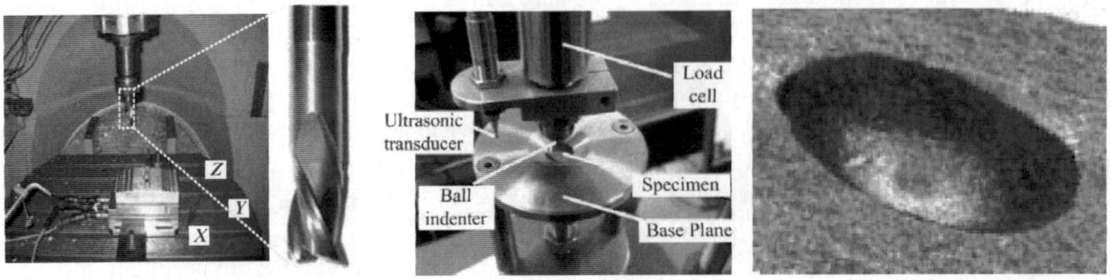

(a) (b) (c)

Figure 1. The experiment setup: (a) milling experiment; (b) Indentation experiment; (c) Indentation appearance.

Milling test was performed on a DECKEL MAHO DMU 70V 5-axis universal machining center, and the tool was end mill without coating. Number of teeth was 3; the rake angle was 14°; the relief angle was 12°; the corner radius was 1mm. The experiment setup is shown in Figure 1(a).

2.3. Indentation experiment
B4000TM stress-strain microprobe system was used to obtain the load-depth data of aluminum alloy. Indentation experiment test was conducted at room temperature (18.3°C). The test process contains multiple loading and unloading phase, the test system is shown in Figure 1(b). Spherical indenters made of tungsten carbide were used with diameters 0.45mm. Seven kinds of loading and unloading stages were set up to obtain load-depth curve data respectively. Indenting velocity of ball indenter speed was set at 0.005334 mm/s for a quasi-static state.

Large field depth microscope VHX-600E was used to observe indentation morphology, as shown in Figure 1(c).

3. Surface constitutive model construction
Plastic flow of materials is generally represented by the equation[12]:

$$\sigma_t = K\varepsilon_p^{\ n} \tag{1}$$

Where σ_t is the true stress, ε_p is the true plastic strain, n is the strain hardening exponent and K is the strength coefficient.

A computer program is used to solve the following equations to determine the true stress and true strain values. From the plastic depth h_p, the plastic diameter d_p is calculated by iterating the following equation,

$$d_p = \left\{ \frac{2.753P \times D(1/E_{spec} + 1/E_{ind})\left(4h_p^{\ 2} + d_p^{\ 2}\right)}{4h_p^{\ 2} + d_p^{\ 2} - 4h_p D} \right\}^{1/3} \tag{2}$$

Where P is the measured load, D is the indenter diameter, E_{spec} and E_{ind} are the elastic modulus of the specimen and indenter (6.41×10^5 MPa), respectively. The true stress and plastic strain values are then calculated using the following equations,

$$\varepsilon_p = 0.2d_p/D \tag{3}$$

$$\sigma_t = 4P/\pi d_p^{\ 2}\delta \tag{4}$$

Where δ is a constant related to the constraint effect for plastic deformation. δ is calculated by iterating the following set of equations:

$$\delta = \begin{cases} 1.12 & \Phi \le 1 \\ 1.12 + \tau \ln\Phi & 1 < \Phi \le 27 \\ 2.87\alpha_m & \Phi > 27 \end{cases} \tag{5}$$

Where $\tau = (2.87\alpha m - 1.12)/\ln(27)$, $\Phi = \varepsilon_p E_{spec}/0.43\sigma_t$ .The α_m parameter is material dependent and is held at a constant value of 1.0 for 7050-T7451 alloy [13].

Values of n and K are determined by regression analysis of the data fitted to equation (1). Data points from all loading cycles are fit by linear regression analysis to the following relationship:

$$P/d_t^{\ 2} = A(d_t/D)^{m-2} \tag{6}$$

Where d_t is the total indentation diameter, m is the Meyer's coefficient and A is a material parameter obtained from the regression analysis.

For each loading cycle, the total penetration depth (h_t) is measured while the load is applied and the depth is converted to a total indentation diameter (d_t) using the following equation:

$$d_t = 2\left(h_t D - h_t^{\ 2}\right)^{1/2} \tag{7}$$

The yield strength (σ_y) is calculated from the expression:

$$\sigma_y = \beta_m A \tag{8}$$

Where β_m is a constant (0.219) for 7050-T7451 alloy[14].

4. Results and discussion

4.1. Regression equation and ANOVA significance testing

The profile of load-depth curve was obtained by B4000TM stress-strain microprobe system. The maximum indentation depth was 0.1mm. Figure 2 represents a typical load-depth curve obtained for a cutting speed of 628m/min, feed rate of 0.06mm/z, and depth of cut and width of cut were 5mm and 12mm. The yield strength analysis is carried out by taking into account simultaneous occurrence of yielding and strain hardening of the material under conditions of multiaxial compression. There were seven consecutive processes of work hardening of surface materials.

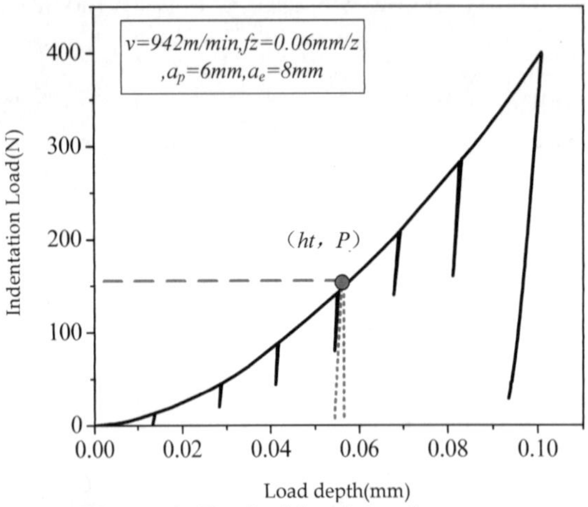

Figure 2. Typical load-depth curves

Table 3 shows the orthogonal experiment schemes and experiment results. It can be found that there is a complex relationship between cutting parameters and surface constitutive model in the milling process. Under the conditions of the present experiment, the yield strength of 7050-T7451 surface layer changed from 351MPa to 533MPa, and the strain hardening exponent changed from 1.12 to 1.67. It is important to put the relation between cutting parameters and surface constitutive model in a mathematical form by the obtained results. Therefore, a mathematical model between machining parameters and n, σ_y was as follows:

$$n = C_1 v^{b1} f_z^{b2} a_p^{b3} a_e^{b4}$$
$$\sigma_y = C_2 v^{b5} f_z^{b6} a_p^{b7} a_e^{b8} \tag{9}$$

Where C is a coefficient factor depending on workpiece material, machine tool, and cutting condition; b_1 to b_8 are exponents of v, f_z, a_p, a_e, respectively. The empirical equations of surface strain hardening exponent n and yield strength σ_y could be achieved based on multiple linear regression analysis, as shown (10).

$$n = 8.35 v^{-0.221} f_Z^{0.359} a_p^{0.112} a_e^{0.151}$$
$$\sigma_y = 3870 v^{-0.307} f_Z^{0.268} a_p^{0.141} a_e^{0.098} \tag{10}$$

An analysis of variance (ANOVA) for surface constitutive models was shown in Table 4. The ANOVA results indicated that the most significant factor was cutting speed. The F value were 38.5 and 31.1 for n and σ_y at the level of 0.05. The affecting degree of the parameters was described as $v > f_z > a_p > a_e$ in order.

Table 3. Orthogonal experiment schemes and experiment results

No	v(m/min)	f_z (mm)	a_p (mm)	a_e (mm)	σ_y(MPa)	n
1	942	0.06	6	8	351.25	1.12
2	942	0.1	5	6	381.63	1.26
3	942	0.14	4	12	433.21	1.54
4	942	0.18	3	10	437.08	1.59
5	628	0.06	5	12	403.47	1.27
6	628	0.1	6	10	466.29	1.52
7	628	0.14	3	8	452.74	1.53
8	628	0.18	4	6	490.29	1.66
9	502	0.06	4	10	411.36	1.27
10	502	0.1	3	12	461.15	1.52
11	502	0.14	6	6	519.86	1.67
12	502	0.18	5	8	517.50	1.67
13	314	0.06	3	6	433.89	1.26
14	314	0.1	4	8	533.00	1.63
15	314	0.14	5	10	515.27	1.46
16	314	0.18	6	12	487.48	1.25

Table 4: Significance testing of the machining parameters on n and σ_y

Factors	Sum of squares		DF	Mean of squares		F		Major-minor order of factors
	n	σ_y		n	σ_y	n	σ_y	
v	19.2	428.6	3	23.1	435.9	38.5	31.12	$v>f_z>a_p>a_e$
f_z	10.4	182.5	3	13.1	196.9	20.24	15.37	
a_p	2.89	56.4	3	2.8	63.5	6.39	7.72	
a_e	1.6	39.6	3	1.2	42.4	3.48	2.53	
Error	9.4	65.3	3	10.1	56.7			
Total	43.5	772.4	15					

4.2. Influence of the machining parameters on n and σy

Figure 3 shows the influence of machining parameters on the n and σ_y values by the range analysis taking into account the four parameters reported above. It is possible to describe how the n and σ_y are affected by the cutting parameters. The horizontal axis represents the cutting parameters, and the left and right vertical axis represents the surface yield strength σ_y and strain hardening exponent n.

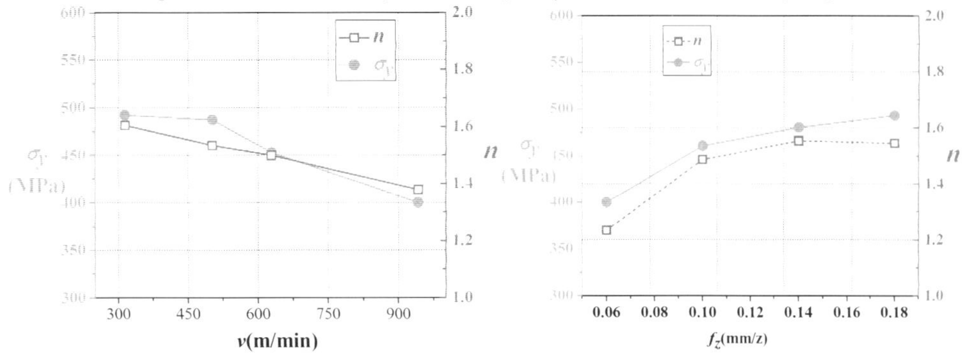

(a) Effect of cutting speed (b) Effect of feed rate

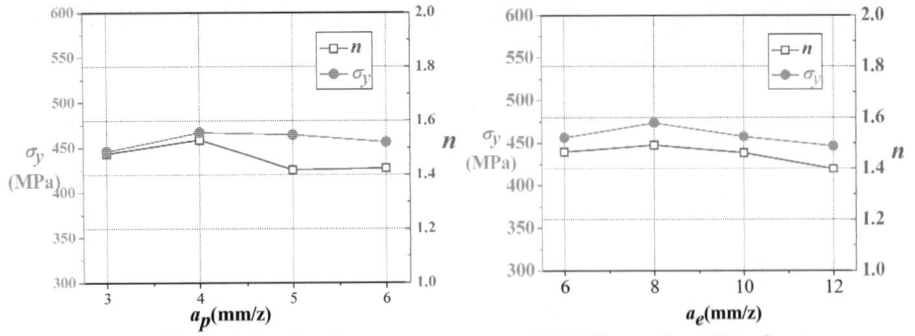

(c) Effect of depth of cut (d) Effect of width of cut

Figure 3. Range analysis of on n and σ_y

Figure 3(a) shows that the increase of the cutting speed from v=314m/min to v=628m/min causes a consecutive reduction of the compressive residual stresses. The surface constitute properties were associated with the previous milling plastic deformation. The observed effect may be explained by an understanding of the material strain rate of cutting that determines the nature of plastic deformation in the machining. At the lowest cutting speed of 314m/min, the chips stay in the machining zone for a relatively longer duration and less heat dissipates into the machined surface, which leads to increase of σ_y and n on the machined surface.

As seen in Figure 3(b), an increase of the feed rate f_z from 0.06mm/tooth to 0.1mm/tooth leads to a pronounced effect on the surface constitutes properties. But, further increase in the feed rate to 0.18mm/tooth shows a less pronounced increase of σ_y and n. The increased cutting force and more plastic deformation were generated with the increasing of the feed rate.

Figure 3(c) shows that a slight change of σ_y and n was observed when the depth of cut increased from 3mm to 6mm. Further, the chart of Figure 3(d) shows that the width of cut does not present an obvious influence on the surface constitutes properties. A variation of the cutting depth and width causes a direct proportional increase of the forces. At the same time, cutting temperature increases with more cutting heat. It is envisaged that the effect of cutting force goes hand-in-hand with the cutting temperature change, which offsets the influence on the 7050-T7451 alloy milling surface plastic deformation.

5. Conclusions

The influence of machining parameters on the surface constitutive properties of 7050-T7451 aluminum alloy was invested by orthogonal cutting and indentation experiment. The important conclusion drawn from the present research was summarized as follows.

(1) Plastic deformation caused by milling will change the surface constitute properties of 7050-T7451 alloy. Under the conditions of the present experiment, the yield strength of 7050-T7451 surface layer changed from 351MPa to 533MPa, and the strain hardening exponent changed from 1.12 to 1.67.

(2) The surface stresses could be predicted effectively by applying cutting speed, feed rate, depth of cut, width of cut, and their interactions in multiple regression exponent models. Their affecting degree diminishes in order, described as $v > f_z > a_p > a_e$ based on the ANOVA analysis of orthogonal experiment.

(3) It has been shown that the decrease of the cutting speed and the increase of the feed rate lead to significant increase of the yield strength and strain hardening exponent on 7050-T7451 finished surface.

Acknowledgments

The authors acknowledge the support from the Shandong Provincial Natural Science Foundation, China (Grant no. ZR2016EEP04).

References

[1] Heinz, A.; Haszler, A.; Keidel, C.; Moldenhauer, S.; Benedictus, R.; Miller, W. S. Recent development in aluminium alloys for aerospace applications. Materials Science and Engineering A, 2000, 280: 102–107.

[2] Bao, R.; Zhang, X.; Fatigue crack growth behaviour and life prediction for 2324-T39 and 7050-T7451 aluminium alloys under truncated load spectra. International Journal of Fatigue, 2011, 32:1180–1189.

[3] Haikuo, M.; Xinda, H.; Yichu, Y.; Xiaoming, Z.; Han, D.; Investigation of surface integrity of aluminum lithium alloy in high speed machining. Scientia Sinica Technologica, 2014, 44:89-98.

[4] Aatakhov, V.P. Surface integrity-definition and importance in functional performance. London: Springer, 2010.

[5] Zhongqiu, W.; Jianfeng, L.; Jie, S.; Guiyu, L. FEM analysis of deformation correction by side-wall rolling of aircraft monolithic components. China Mechanical Engineering, 2009, 20: 612-616.

[6] Wu, B.; Li, M.Q.; Ma, D.W. The flow behavior and constitutive equations in isothermal compression of 7050 aluminum alloy. Materials Science and Engineering: A, 2012, 542:79-87.

[7] Yongzhi, H.; Liwen, G.; Xin-jun, L.; Hailong, C. Research and revise on constitutive equation of 7050-T7451 aluminum alloy in high strain rate and high temperature condition. Journal of Materials Engineering, 2012, 12:7-13.

[8] Xiuli, F.; Xing, A.; Yi, W.; Song, Z. Flow stress characteristics and constitutive equation at high temperature for 7050 aluminum alloy. Journal of Wuhan University of Technology, 2006, Volume 28, pp. 113-116.

[9] Zhitao, T.; Yi, W.; Xing, A. The influence of tool flank wear on residual stresses induced by milling aluminum alloy. Journal of Materials Processing Technology, 2009, 209:4502–4508.

[10] Tabor, D. The hardness of metals. Oxford: Clarendon Press, 1951.

[11] Haggag, F. M. Field indentation microprobe for structural integrity evaluation: US, 4852397, 1989.

[12] Xin, Y.; Jie, S.; Qingchun, X.; Xiong, H. Milling surface properties of 7050-T7451 aluminum alloy and establishment of constitutive model.The Chinese Journal of Nonferrous Metals, 2015, 25:2982-2989.

[13] Weiguo, G.; Hongwei, T. Strain rate sensitivity and constitutive models of several typical aluminum alloys. The Chinese Journal of Nonferrous Metals, 2009, 19: 56-61.

[14] Haggag, F. M.; Bell, G. E. C. Measurement of yield strength and flow properties in spot welds and their HAZs at various strain. The ASM 3rd international conference on trends in welding research, USA, Gatlinburg, 1992 :637-642.

IOP Conf. Series: Materials Science and Engineering **242** (2017) 012040 doi:10.1088/1757-899X/242/1/012040

Numerical simulation study on rolling-chemical milling process of aluminum-lithium alloy skin panel

Z B Huang[1], Z G Sun[1], X F Sun[1] and X Q Li[2]

[1]Shanghai Aircraft Manufacturing Co., Ltd., COMAC, Shanghai 201325, P. R. China
[2]School of Mechanical Engineering and Automation, Beihang University, Beijing 100191, P. R. China

E-mail: zbhuang126@126.com, sunzhonggang@comac.cc

Abstract. Single curvature parts such as aircraft fuselage skin panels are usually manufactured by rolling-chemical milling process, which is usually faced with the problem of geometric accuracy caused by springback. In most cases, the methods of manual adjustment and multiple roll bending are used to control or eliminate the springback. However, these methods can cause the increase of product cost and cycle, and lead to material performance degradation. Therefore, it is of significance to precisely control the springback of rolling-chemical milling process. In this paper, using the method of experiment and numerical simulation on rolling-chemical milling process, the simulation model for rolling-chemical milling process of 2060-T8 aluminum-lithium alloy skin was established and testified by the comparison between numerical simulation and experiment results for the validity. Then, based on the numerical simulation model, the relative technological parameters which influence on the curvature of the skin panel were analyzed. Finally, the prediction of springback and the compensation can be realized by controlling the process parameters.

1. Introduction

Single curvature parts such as fuselage skin panels are important parts of the aircraft, and this kind of single curvature parts are usually formed by rolling-chemical milling process. Through the rolling technology, parts can be formed. Moreover, through the chemical milling process, a part of material can be removed to meet the structural requirement and reduce the weight [1]. This sort of technology is usually faced with the problem of geometric accuracy caused by springback. In most cases, the methods of manual adjustment and multiple roll bending are used to control or eliminate the springback. However, these methods can lead to the increase of product cost and cycle, and also cause the degradation of the material performance. Therefore, precisely controlling the springback of rolling-chemical milling process is of important significance.

2060-T8 is the newly developed third-generation aluminum alloy material [2]. So far, domestic and abroad civil aviation companies have not applied it into the airplane parts in quantity. Besides, there are few studies on the numerical simulation of rolling-chemical milling process [3]. In this paper, through using experiment and numerical simulation on rolling-chemical milling process, the simulation model for 2060-T8 aluminum-lithium alloy skin rolling-chemical milling process is established and testified by the comparison between numerical simulation and experiment results for the validity. Then, based on the numerical simulation model, the relative process parameters influencing the rolling results of the skin panel is analyzed. Finally, the prediction of springback and springback compensation are realized by controlling the process parameters.

Content from this work may be used under the terms of the Creative Commons Attribution 3.0 licence. Any further distribution of this work must maintain attribution to the author(s) and the title of the work, journal citation and DOI.
Published under licence by IOP Publishing Ltd

IOP Conf. Series: Materials Science and Engineering **242** (2017) 012040 doi:10.1088/1757-899X/242/1/012040

2. Establishment of the rolling-chemical milling model

The experimental material used in this paper is the 2060-T8 aluminum-lithium alloy rolled sheet whose chemical composition is shown as Table 1 [4]. The size of the experimental material is 2 mm × 100 mm × 500 mm. According to the feature of fuselage skin material, the roll bending direction is perpendicular to the rolling direction. Figure 1 shows the schematic diagram of the roll bending process.

Table 1. Chemical composition of 2060-T8 aluminum-lithium alloy

Material	Cu	Li	Zn	Mg	Mn	Zr	Ag	Si	Fe	Al
2060-T8	3.9	0.8	0.32	0.7	0.29	0.1	0.34	0.02	0.02	Bal.

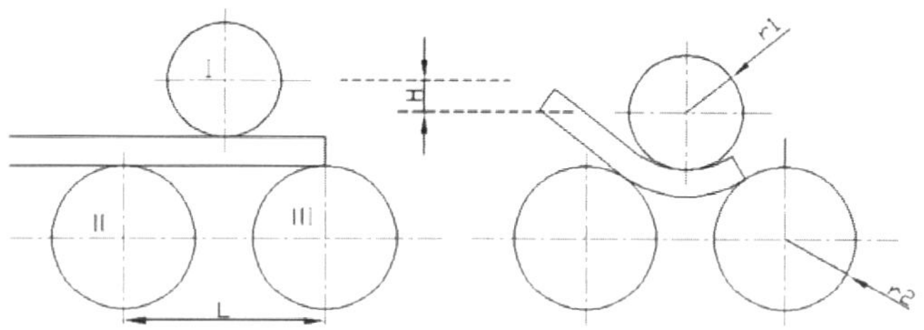

Figure 1. The symmetrical tri-axial roll bending process

A symmetrical tri-axial roll bending machine was selected in the experimental rolling phase. The radius of the upper roller is r1 = 40 mm, the radius of the two-bottom roller is r2 = 50 mm, and the spacing L = 200 mm. During the experimental process, the bending radius was controlled by adjusting the roller stroke H, and the specific process is shown as follows: First, upper roller presses down to the specified amount of stroke. Then, the bottom rollers start to rotate and drive the upper roller and metal sheet through friction. When the rolling comes to the boundary part, the bottom rollers begin the reverse rotation, so the motion is repeated to obtain the required curvature radius. In the end, the rollers are hoisted and the sheet discharges springback. After the roll-bending process finished, the experiment comes to the chemical milling phase. Besides, the size of the chemical-milling area is 400mm × 80mm, and the depth of chemical milling is 0.8 mm.

The roll bending and chemical-milling process of the skin part involve two working procedures. Thus, in order to accurately describe the springback caused by the stress redistribution, the finite element model needs to transmit the residual stress after the roll-bending process as well as the springback into chemical-milling model to realize the stress release after the material has been removed. As a result, the numerical simulation model was divided into two stages according to the rolling-chemical milling process.

Roll bending stage: the metal sheet with the size of 2 mm × 100 mm × 500 mm was used to conduct the experiment of the skin roll bending, and the numerical simulation model is presented in Figure 2.

Chemical-milling stage: The simulation was performed with the sub-model method of the software ABAQUS [5]. Through adjusting the unit thickness of plate, the change of the parts' whole structure during the milling process was stimulated, destroying the balance condition of the original stress inside the material and redistributing the internal stress to make the specimen changed. Finally, the implicit algorithm was selected to compute the springback after chemical milling.

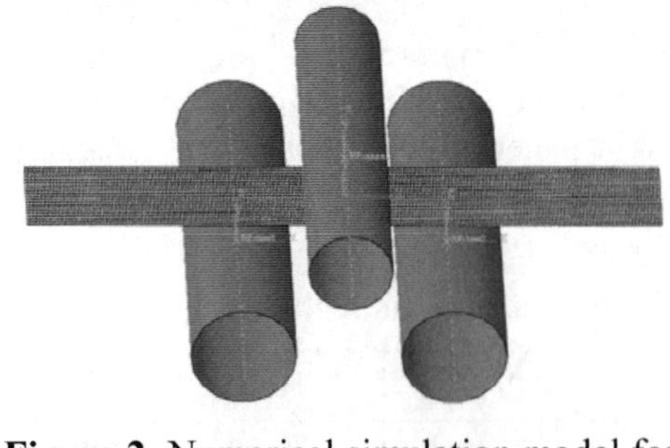

Figure 2. Numerical simulation model for the roll-bending

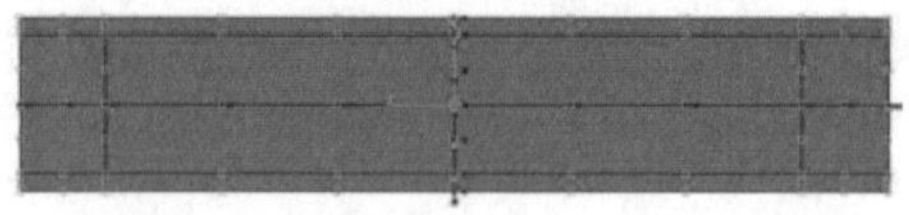

Figure 3. Numerical simulation model for the chemical-milling process

Figure 4 shows the test specimen of the roll-bending and chemical-milling process. And it can be seen the comparison of the simulation and the experimental results, as shown in Figure 5. The overall error is 3.8%, and the biggest error is 7.9%, appearing in the springback change after the chemical-milling process with the roller stroke of 28mm. Thus, in this paper, the finite element model of the whole process of the roll-bending and chemical-milling is accurate, and it can meet the production demand.

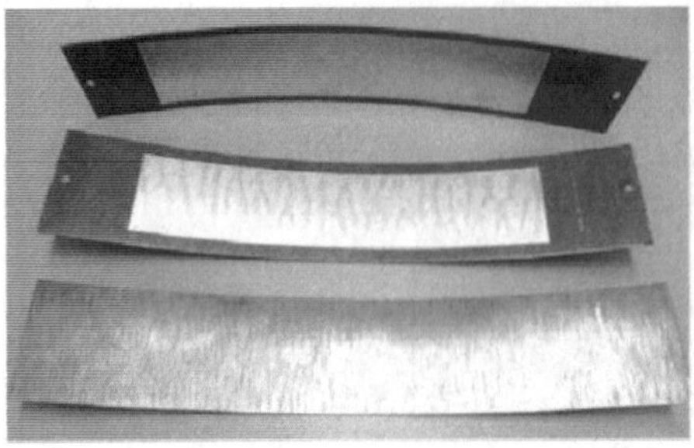

Figure 4. Roll bended-chemical milled workpieces

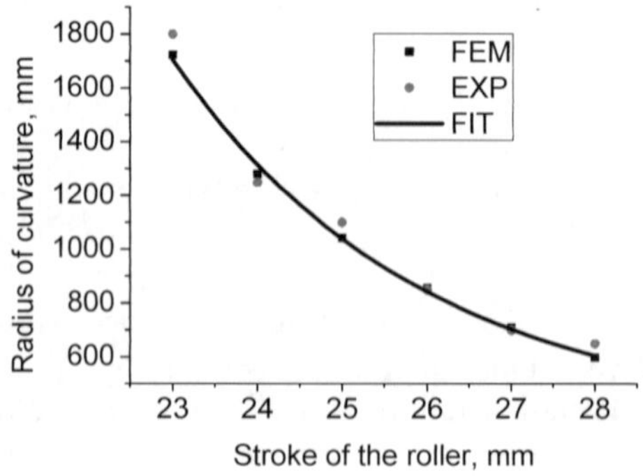

Figure 5. Comparison of numerical simulation results and experiment results

3. Results and discussions

3.1. The effect of the roller stroke

In this paper, five sets of different roller stroke were selected to understand the influence on the forming curvature radius. The simulation results are shown in Figure 5. It can be seen that the relationship between the part curvature radius and the centre roller stroke is close to a power exponential function. Additionally, the roller stroke volume for other formed curvature radius can be quantitative analyzed with the fitted result data. Thus, this paper obtains the approximate map relationship between central roller and radius through substituting experiments based on numerical simulation, and basically meets the requirement of actual production.

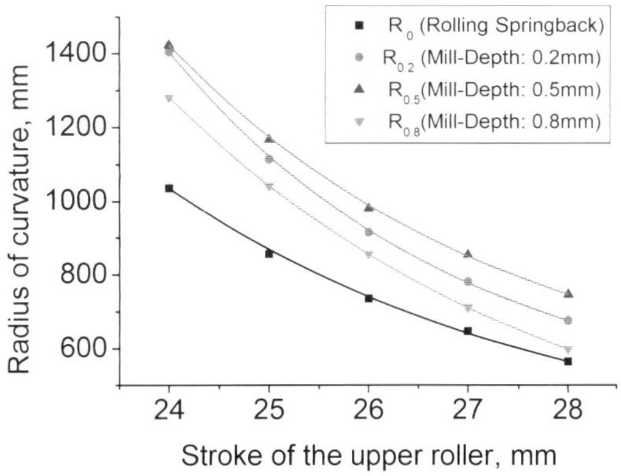

Figure 6. Effect of chemical-milling depth on curvature radius

3.2. The effect of the chemical-milling depth

The numerical simulation result of parts with different chemical-milling depth is shown in Figure 6. According to the figure, the impact of the chemical-milling process on the springback of the formed parts caused by the roll bending is relatively large, and the deformation of the curvature radius is up to 40%. This is because that there exists milling stress in the interior of the part after the chemical-milling process. And the release of the stress leads to the increase of the curvature radius. Thus, certain radius compensation should be considered during roll-bending process of this kind of parts. In addition, with the increase of the milling depth, the curvature radius of the parts increases first and then decreases. This is caused by the following conditions: the stress distributes unevenly along the thickness direction after the roll-bending process; the difference of the chemical-milling depth directly causes the parts to generate internal stress after the chemical-milling process; besides, the difference of the milling stress directly influences the change of the curvature radius after the stress has been released.

Figure 7 presents the distribution of internal stress after the roll bending process. When it reaches to the milling depth of area which is distributed tensile stress, the part curvature radius increases. Besides, the main reason is that the stress balance of the sheet is broken when the stress area has undergone the chemical-milling process, which causes the crushing stress being greater than the tensile stress. Therefore, to achieve the rebalanced state, the sheet leads to the release of the crushing stress, and then causes the enlargement of the curvature radius. With the increase of milling depth, it turns to the crushing stress distribution area, which gradually lessens the crushing stress inside the material. Therefore, the material releases the tensile stress to gradually reduce the internal stress, resulting in the enlargement of the curvature radius.

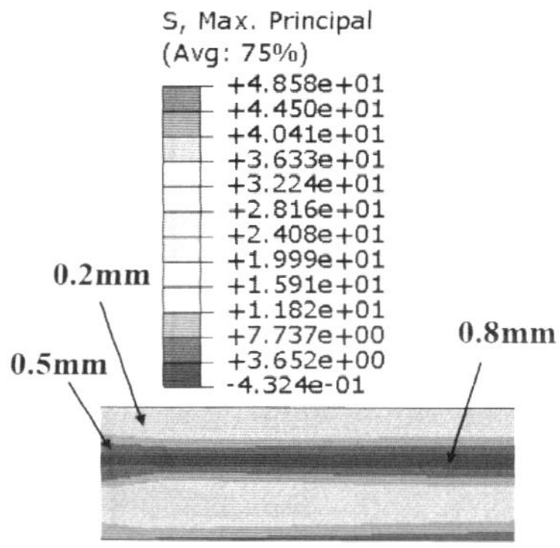

Figure 7. Distribution of the internal stress

4. Summary

This paper analyzes the technology of the roll-bending and chemical-milling process. Combined with the features of the two kind of processing technology, a numerical simulation model of the whole roll-bending and chemical-milling process was established based on the ABAQUS software. Through the contrastive analysis between the simulation results and experimental results, the correctness of the finite element model was verified and the following conclusion can be drawn:

Firstly, An approximate map relationship between central roller and part radius was obtained through substituting experiments by numerical simulation, and basically meets the demand of actual production, providing guiding significance for the determination of roll bending process parameters. Secondly, based on the established model, the effect of chemical-milling depth on the part springback was discussed. That is to say, with the increase of the milling depth, instead of increasing progressively, the springback volume increases firstly and decreases later.

Acknowledgments
This work is financially supported by the Fund of Shanghai Rising-star Program (No. 15QB1401000).

References

[1] He L and Wang J 2010 Forming Process of Chemical Milling for X Aircraft *Hongdu Science & Technology* **144** 13
[2] Zhang X, Li X and Li D 2015 Experimental Study on Milling of New Al-Li Alloy 2060T8 Sheet *Aeronautical Manufacturing Technology* **472** 46
[3] Dang H, Zhou M and Gong G 2014 Experimental Study on Roll Forming of Al-Li Alloy Aircraft Skin *Journal of Netshape Forming Engineering* **6** (6) 94
[4] Zhang Y, Tao W, Chen Y, Chen J, Liu H and Jiang M 2013 Shaping Control of Welds for Fiber Laser Welding T-Section of Fuselage Panels *Chinese Journal of Lasers* **40** (12) 79
[5] Shi Y and Zhou Y 2006 *Examples for the Finite Element Commercial Code ABAQUS* (Beijing: China Machine Press) p 198-200

Study on acoustic emission source localization of 16Mn structural steel of high temperature deformation

Yubo Zhang[1,2,a], Muhan Deng[1,b], Rui Yang[1,c,*], Feixiang Jin[2,d]

[1] China Waterborne Transport Research Institute, Beijing, 100088, China
[2] School of Materials Science and Engineering, Beijing University of Aeronautics and Astronautics, Beijing, 100191, China

[a]zyb@wti.ac.cn, [b]dengmh@wti.ac.cn,[c]orangez80@163.com,[d]59024400@qq.com,
*Corresponding author

Abstract. The location technique of acoustic emission (AE) source for deformation damage of 16Mn steel in high temperature environment is studied by using linear time-difference-of-arrival (TDOA) location method. The distribution characteristics of strain induced acoustic emission source signals at 20℃ and 400℃ of tensile specimens were investigated. It is found that the near fault has the location signal of the cluster, which can judge the stress concentration and cause the fracture.

1. Introduction
Because of its excellent characteristics 16Mn steel has been widely used in construction, transportation, petrochemical and other engineering fields. However, due to its shortcomings such as high thermal conductivity and poor fire resistance, the safety of the steel structure in the fire or high temperature environment has attracted much attention.

In the course of service, the steel structure will deform or even break under the action of the external load. The process of generation and change of internal defects will release energy in the form of elastic wave and propagate through the material. The defects in this process will become the source of acoustic emission (AE) .The acoustic emission can be used to locate the acoustic emission source, determine the location of the material defect, and localize the defect in the space. There are two methods for AE localization: area location method and time-difference-of-arrival (TDOA) location method. Compared with regional positioning method, TDOA location method is widely used because of its wide range of application and higher positioning accuracy. The research and application of AE technology in material science mainly focus on at normal atmospheric temperature, and there are few literatures about high temperature monitoring. C J Feng has studied the characteristics of acoustic emission signals during the high temperature tensile process of Q345R steel. It is found that the use of waveguide bars plays an important role in the acquisition of acoustic emission signals in high temperature environments. Sound emission is used to monitor the change of internal structure defects in special environment, so as to provide experimental and theoretical basis for safe operation and evaluation of steel structure.

2. Experimental procedure

2.1. Linear TDOA location method

Content from this work may be used under the terms of the Creative Commons Attribution 3.0 licence. Any further distribution of this work must maintain attribution to the author(s) and the title of the work, journal citation and DOI.
Published under licence by IOP Publishing Ltd

IOP Conf. Series: Materials Science and Engineering **242** (2017) 012041 doi:10.1088/1757-899X/242/1/012041

The time difference method measures a number of sensors into a certain geometry to determine the relative moveout of the stress wave from the acoustic emission source to each sensor. The location of the acoustic emission source is obtained by geometric relation calculation and the acoustic emission signal is directly related to it. When the ratio of the length to radius of the object to be detected is very large, a Linear TDOA location method should be used to locate the AE source, such as pipe, bar, steel beam, steel plate and so on. Figure 1 is the distribution model of TODA linear sensor array, two or more sensors are distributed in one dimension X axis coordinate system.

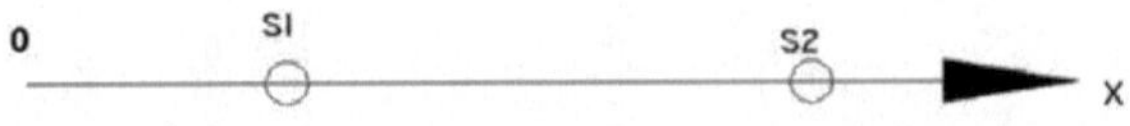

Figure 1. The distribution model of TODA linear sensor array

If there's an acoustic emission signal between the S1 and S2 to produce acoustic emission signals, the signal is reached at S1, the probe time is t_1, the signal reaches S2, and the probe time is t_2, therefore, the arrival of the two sensors is $\Delta t=t_2-t_1$. The distance of AE source from S1 is ：

$$d = \frac{1}{2}(D - \Delta t V) \qquad (1)$$

Where D is the distance between the two probes, and V is the propagation velocity of sound waves in the sample.

2.2. *Specimens*

The material used in this paper is low alloy structural steel 16Mn steel. The mechanical properties are shown in Table 1.

Table 1. The mechanical parameters of 16Mn steel

parameters	Yield strength σ_s(MPa)	Tensile strength σ_b(MPa)	Modulus of elasticity E(MPa)	Impact energy AK/J(20C°)	Elongation η (%)
value	345	560	206000	≥34	≥22

The standard plate tensile samples conforming to national standards GB/T228-2002 were prepared with the shape and dimensions as shown in Figure 2. The circular hole with Φ8mm at both ends is used to clamp the specimen together with the pins of the test machine. The acoustic emission guide bar is welded at a circular hole of Φ4mm as shown in Figure 3.

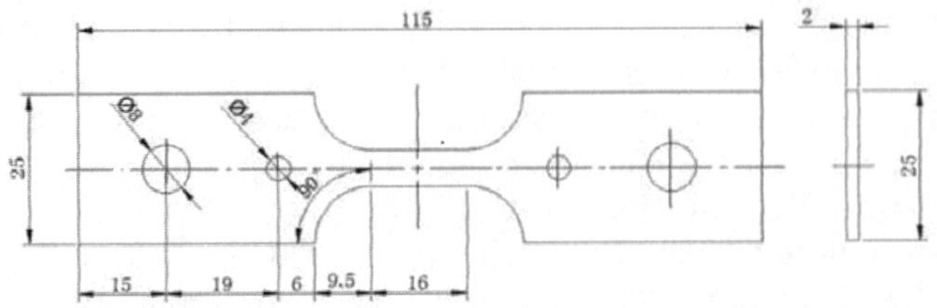

Figure 2. Details of the plate tensile specimens (mm)

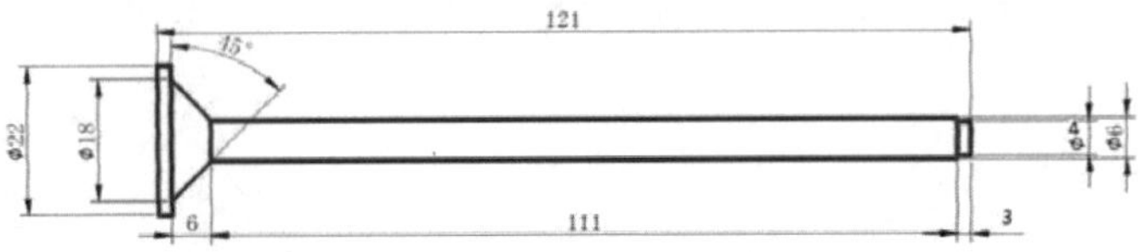

Figure 3. Details of the waveguide bars (mm)

2.3. *Test procedure*

High temperature tensile test using MTS880 material testing machine and equipped with high temperature control system. Acoustic emission instrument is used to monitor the acoustic emission source signal during deformation process. The acoustic emission instrument is a DISP with AEwin v2.19 acoustic emission instrument developed by PAC company. The sensor is fixed at the end of the wave guide bar by magnetic ring.

In this experiment, the loading of the sample is loaded by the method of displacement loading control and the tensile rate of 2mm/min. Meanwhile, the acoustic emission signal is monitored and the source location of the transmitter is carried out. When the sample and instrument are installed, the sample is heated to the set temperature (20℃,400℃) and the heat preservation is 15min. Loading process: preloading (below the yield strength) - unloading - reloading, the Kaiser experiment is used to determine whether the electromagnetic noise generated by the device is excluded.

3. Results and discussion

3.1. *Characterization of deformation damage of steel 16Mn by acoustic emission*

The amplitude of acoustic emission source localization signal varies with strain during the deformation damage of 16Mn steel at normal atmospheric temperature and high temperature can be seen in Figure 4. (a) , (b). At normal temperature (20℃) the tensile curve of the sample has an obvious yield plateau, at high temperature (400℃) the strength of the material decreases and the yielding platform disappears. The curve of the plastic deformation stress rise stage becomes smoother.

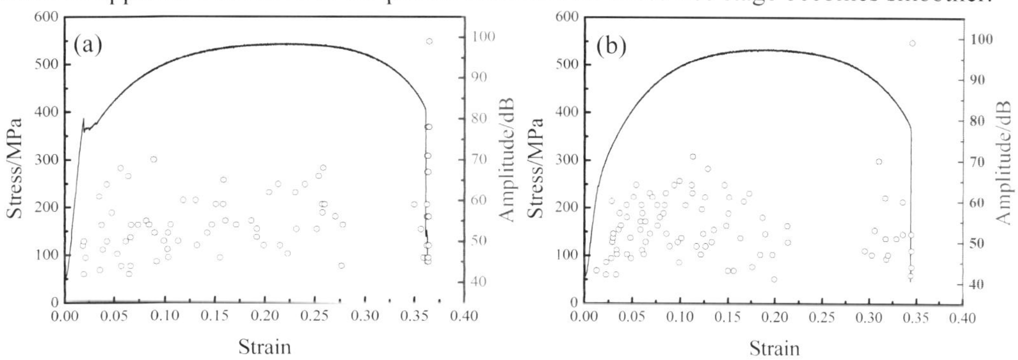

Figure 4. The amplitude of acoustic emission source localization signal varies with strain during the deformation damage of 16Mn steel (a)20℃， (b)400℃

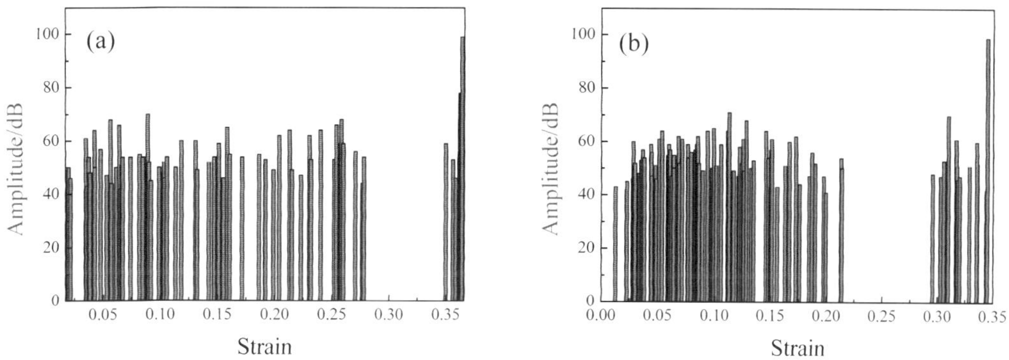

Figure 5. The amplitude of acoustic emission source localization signal of 16Mn steel (a)20℃， (b)400℃

Figure 5. (a) , (b) show the amplitude of acoustic emission source localization signal of 16Mn steel at different temperatures. Under the influence of high temperature, the bonding force between crystal atoms decreases, which accelerates the manager deformation and grain boundary sliding. There are more acoustic emission locating signals near the fracture.

3.2. Analysis of test results of acoustic emission source localization

In order to further explore the law of AE localization signals in high temperature deformation damage, Figure 6. shows the spatial distribution of AE signals with strain at different temperatures.

At the yield stage, the positioning signal is the least and the signal amplitude is below 50dB. The main reason is that the signal produced in the yield stage is mainly continuous signal, corresponding to dislocation movement. There is no obvious structural damage and cannot form position signals. There are more signals in the strain hardening stage, and the whole signal amplitude is between 45-75dB. In the rapidly rising stage of stress, signals with higher amplitude (65-75dB) appear, which may be related to rapid deformation. A cluster signal appears near the fracture show that the local stress concentration caused the material shrinkage neck appeared, began to appear microporous, microporous aggregation leads to the final fracture, a close 98dB signal corresponding to the material fracture.

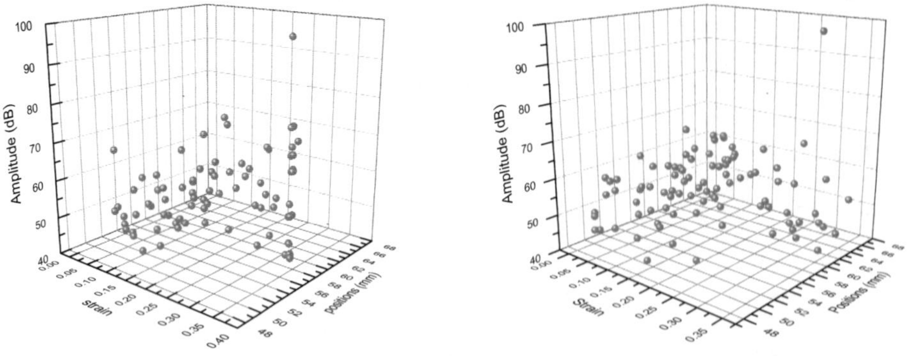

Figure 6. The spatial distribution of AE source localization signal amplitude with strain (a)20℃，(b)400℃

In order to verify the accuracy of AE location signals for the determination of necking and fracture, the actual measurements of the stretched sample were carried out, as shown in Table 2.

Table 2. Measured results of acoustic emission source location

Temperature(°C)	Sample breaking point(mm)	AE location breaking point(mm)
20	56	56.2
400	61	60.47

From the distribution law of acoustic emission signals at 2 temperatures, it can be seen that the location method of high temperature acoustic emission source using guided wave bar is feasible, and the AE location signals are all distributed in the range of the sample gauge. By comparing and analyzing the location signals, it is found that the AE localization signals have higher accuracy in judging the severely deformed region and the breaking point. From this we can deduce that the acoustic emission location signal given in the whole strain deformation damage stage is accurate.

4. Conclusions

The experimental results show that it is feasible to locate the high temperature acoustic emission source using the guided wave auxiliary method.

The tensile deformation damage process of 16Mn steel at high temperature can be reflected in the structural signals. The spatial distribution and time distribution of acoustic emission source agree with tensile deformation damage. The distribution of acoustic emission source before necking is uniform in the gauge interval. After entering the contraction stage, the acoustic emission signals are concentrated around the fracture point, and the appearance of the necking of the material corresponding to the local serious stress concentration appears. The location of microstructure damage resulting from the change of high temperature mechanical properties of 16Mn steel can be understood by acoustic emission source localization. The results will serve as a reference for real-time on-line monitoring and early warning of high temperature acoustic emission.

Acknowledgments
This work was financially supported by National Key Technology R&D Program of China (2015BAG20B04).

References

[1] Yubo Zhang, Research on deformation of structural steel applying the acoustic emission method at complex service conditions D. Beijing: Beihang University, 2017
[2] Jian Xu, En-Hou Han, Xinqiang Wu. Acoustic emission response of 304 stainless steel during constant load test in high temperature aqueous environment J. Corrosion Science, 2012(63): 91–99
[3] G T Shen, R S Geng, S F Liu, Acoustic emission source location, J. Nondestructive Testing, 2002(24)3:114–117
[4] Bazant, ZP, Zhou, Y, Why did the World Trade Center collapse? - Simple analysis,J. Journal of engineering mechanics-asce, 2002, 28(1):2–6
[5] Z Y Han, H Y Luo, Y B Zhang, Effects of micro-structure on fatigue crack propagation and acoustic emission, J. Materials Science & Engineering A 559(2013)534 - 542
[6] Z M Yuan, Y K Ma, Z Y He, The acoustic emission technique and its application ,M. Beijing: Machinery Industry Press, 1985
[7] Dunegan H, Modal analysis of acoustic emission signals, J. Journal of Acoustic Emission, 1998(15):1–4.
[8] Masoud Rabiei, Mohammad Modarres. Quantitative methods for structural health management using in situ acoustic emission monitoring, J. International Journal of Fatigue, 2013(39):81–89.
[9] Zhang Yubo, Luo Hongyun, Li Junrong, et al. An Integrated processing method for fatigue damage identification in a steel structure based on acoustic emission signals J. Journal of Materials Engineering and Performance (2017)26:1784-1791
[10] Yubo Zhang, Yiqin Li, Hongyun Luo. The Probabilistic Safety Assessment of the Rail Mounted Gantry Cranes J. Materials Science Forum, 2013, 749:450-455

IOP Conf. Series: Materials Science and Engineering **242** (2017) 012042 doi:10.1088/1757-899X/242/1/012042

Simulation analyzing the influence of cutting HT250 by self-prepared Si3N4 insert at different feed rate

Bo Wang[1,4], Yang Zhang[2] and Jian-nong Liang[3]

[1,2,3] Zhuhai College of Jilin University, Zhuhai, 519041, China

[4]School of Electro-mechanical Engineering, Guangdong University of Technology, Guangzhou 510006, CHINA

Abstract. Metal cutting process is an extremely complicated problem. Using the commercial finite element software (Deform), the process has been reliably simulated. In this paper, the process of cutting HT250 by self-prepared Si3N4 insert was simulated at different feed rate base on Finite Element Method (FEM). After simulation, the results of cutting force and temperature were availably obtained. The influence of cutting properties at different feed rate was effectively predicted through analysing these results.

1. Introduction

Metal cutting process is a very complicated nonlinear problem. The traditional cutting test method is difficult to get accurate and stable results. However, using the commercial finite element software, metal cutting process has been able to be reliably simulated and predicted. The finite element simulation by computer can achieve the required accuracy and the high reliability, avoid actual material cutting experiments, and save time, manpower and material resources.

Childs [1,2] considers that the success and reliability models are dependent upon work material flow stress. Trent et al. [3,4,5] considers that the simulation models are also dependent upon friction parameters between the tool and workpiece interfaces for accurate analysing the metal machining process. This paper studies the dynamic process of cutting HT250 by Si3N4 inserts using FEM. The numerical models (geometric model, material behaviour model, contact and friction model, heat transfer model, etc.) were created before simulation. The simulation tests were carried out at different cutting condition (V_c=300m/min, a_p=0.5mm, f=0.1/0.15/0.2mm/rev). The results of cutting force and tool temperature were availably obtained after simulation. The cutting properties of Si3N4 insert were effectively predicted through analysis these results.

2. Finite element models

2.1. Geometries and mesh generation

The geometric model of Si3N4 inserts was created in Deform software according to the type of MCGNR. And then the workpiece and insert were respectively meshed with 45419 and 26994 isoparametric quadrilateral elements [6]. The tool nose geometry mesh was generated with element size ratio of 6 for improving the accuracy of analysis. The geometric models and mesh generations of Si3N4 insert and HT250 workpiece were respectively shown a and b in fig.1

Content from this work may be used under the terms of the Creative Commons Attribution 3.0 licence. Any further distribution of this work must maintain attribution to the author(s) and the title of the work, journal citation and DOI.
Published under licence by IOP Publishing Ltd

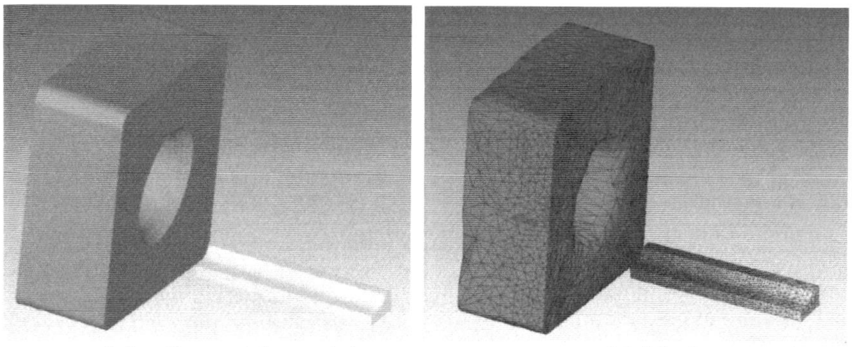

(a) Geometric models (b) Mesh generations

Fig.1 Geometries and mesh generation

2.2. Material behaviour model

（1）Workpiece material behaviour model

Accurate flow stress models are considered highly necessary to represent work material constitutive behavior under high strain rate deformation conditions [7].The constitutive law proposed by Johnson and Cook [8] provides a good description of material behavior subjected to large strains, high strain-rates and thermal softening. The Johnson-Cook model, provided in Deform software, was used in the simulation process. The law was described by Eq.1. The elastic and thermos-mechanical properties of HT250 were shown in Table2.

$$\bar{\sigma} = [A + B(\bar{\varepsilon})^n]\left[1 + C\ln\left(\frac{\dot{\bar{\varepsilon}}}{\dot{\bar{\varepsilon}}_0}\right)\right]\left[1 - \left(\frac{T - T_{room}}{T_{melt} - T_{room}}\right)^m\right] \quad (1)$$

In Eq.1, σ is the material flow stress, $\bar{\varepsilon}$ is the equivalent plastic strain, $\dot{\bar{\varepsilon}}_0$ is the reference plastic strain rate, $\dot{\bar{\varepsilon}}$ is the plastic strain rate, T_{room} and T_{melt} are respectively the room and the melting temperatures. The coefficients (A, B C, n and m), listed in Table 3, are obtained from tensile curves using a fitting program [9].

Table 1 Elastic and thermal properties parameters of HT250

Density (Kg/m3)	Poisson's ratio	Yong's modulus 10^3MPa	Heat conductivity (W/(m·K))	Thermal expansion (10^{-6}/℃)
7190	0.26	126	100℃ 49.1	−100~20℃ 10.0
			200℃ 48.1	20~200℃ 11.0
			300℃ 47.1	200~400℃ 12.5
			400℃ 46.1	
			500℃ 45.1	

Table 2 HT250 material constants for J-C model:

A	B	C	n	m	T_{room}	T_{melt}	$\dot{\bar{\varepsilon}}_0$	$\dot{\bar{\varepsilon}}$
573	380	0.034	0.17	0.12	20	1250	1	1

（2）Insert material behaviour model

Deform software provides different methods of defining the flow stress. In materials database of Deform software, the Al2O3 ceramic is defined by the Power Law [6], shown below Eq.2. In this study, Si3N4 ceramic was also defined by the Power law.

$$\bar{\sigma} = c \times \bar{\varepsilon}^n \times \dot{\bar{\varepsilon}}^m + y \quad (2)$$

In Eq.2, $\bar{\sigma}$ is Flow stress, $\bar{\varepsilon}$ is effective plastic strain, $\dot{\bar{\varepsilon}}$ is effective strain rate, c is material constant, n is strain exponent, m is Strain rate exponent, y is initial yield value. The properties of Si3N4

IOP Conf. Series: Materials Science and Engineering **242** (2017) 012042 doi:10.1088/1757-899X/242/1/012042

inserts were shown in Table 3.

Table 3 Properties of Si3N4 ceramic inserts

Bending strength (MPa)	HV hardness (GPa)	Fracture toughness (MPa·m1/2)	Poisson's ratio	Yong's modulus 10^3MPa	Heat conductivity (W/(m·K))	Heat capacity (J/(kg·K))
905	14.3	9.02	0.25	310	31	550

2.3. Contact and friction model

Zorev [10] proposed the most realistic description of normal and frictional stress distribution at tool-chip interface. The author assumed that the tool-chip interface was subdivided into two zones: sticking zone and sliding zone [11]. In the sticking zone, the shear stress reach the saturation values τ_{max}. In the sliding zone, the frictional shear stress does not reach the saturation values τ_{max}.The frictional stress can be expressed as Eq3, $\mu = 0.45$ [12].

$$\begin{cases} \tau_f = \mu \times \sigma_n & \text{if } \mu \times \sigma_n < \tau_{max} \text{ (Sliding)} \\ \tau_f = \tau_{max} & \text{if } \mu \times \sigma_n \geq \tau_{max} \text{ (Sticking)} \end{cases} \quad (3)$$

2.4. Heat transfer model

In machining process, the temperature increases in cutting zone results from material plastic deformation and friction at tool-workpiece interface. Most of the heat is taken away by chip, parts of the heat is transferred into the air. In order to improve the accuracy of analysis, the heat transferred into air is expressed as Eq.4[13].

$$Q_h = h \times (T_w - T_0) \quad (4)$$

In Eq.4, $h = 0.02$ and is convection coefficient, T_w is the surface temperature of insert and workpiece, $T_0 = 20°C$ and is environment temperature [6].

3. Finite element simulation and discussions

Using the aforementioned models, the simulation tests were carried out with specific cutting parameters. The cutting parameters were listed in Table 4. The number of simulation steps was defined to 500. The simulation process was shown in Fig.2.

Table 4 Cutting parameters

Cutting speed V_c (m/min)	Depth of cut a_p (mm)	Feed rate f (mm/rev)
300	0.5	0.1/0.15/0.2

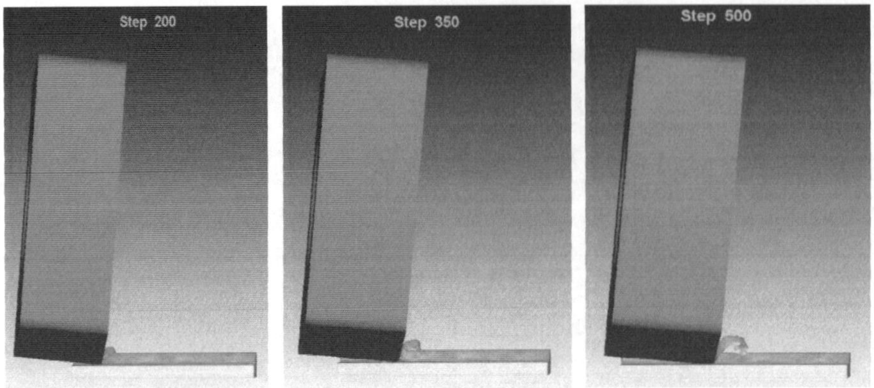

(a) 200 step (b) 350 step (c) 500 step

Fig.2 Simulation process

IOP Conf. Series: Materials Science and Engineering **242** (2017) 012042 doi:10.1088/1757-899X/242/1/012042

3.1. Cutting force analysis

Cutting force is main analysis parameter, which is related to cutting heat generation, tool wear, machining accuracy. The simulation results of cutting force were shown in Fig. 3.The cutting force were gradually rising from zero up to the maximum, stabilize within a certain range finally. Accompany with chip separation, when node number was increasing, cutting force was increasing. On the contrary, cutting force is reduced. So cutting force appeared wave.

The max cutting force was respectively 403/648/535(N) while feed rate was respectively 0.1/0.15/0.2 (mm/rev). The cutting force was the maximum at f=0.15 than others.

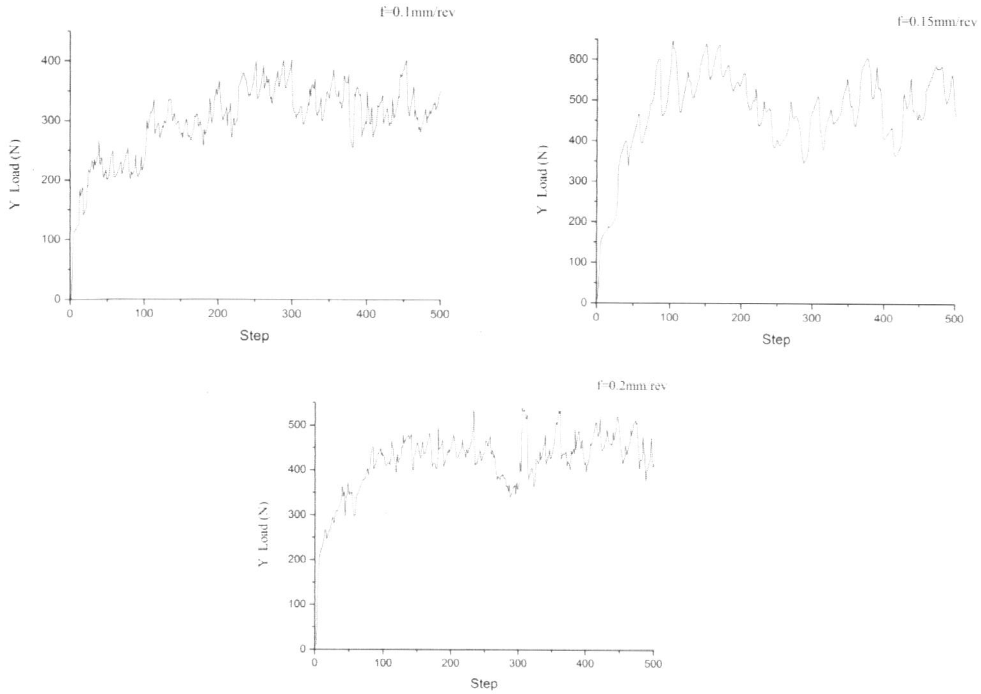

Fig.3 Results of cutting force

3.2. Temperature fields analysis

Temperature is another main analysis parameter. Heat generation is mainly caused by metal deformation in cutting process. The results of temperature analysis were shown in Fig. 4. The highest temperature focused on the nose of insert, which was the interface of insert and workpiece, where was the most severe shear zone of materials. However, Fig. 8 showed that the temperature of Si3N4 insert was only more than 23℃. The most of heat generated caused by deformation was taken away by chip. The low thermal conductivity of ceramic material prevented heat from passing on the insert, which effectively ensured tool life.

The highest temperature was respectively 23.1/24.9/24.4(℃) while feed rate was respectively 0.1/0.15/0.2 (mm/rev). The tool temperature was the highest at f=0.15 than others.

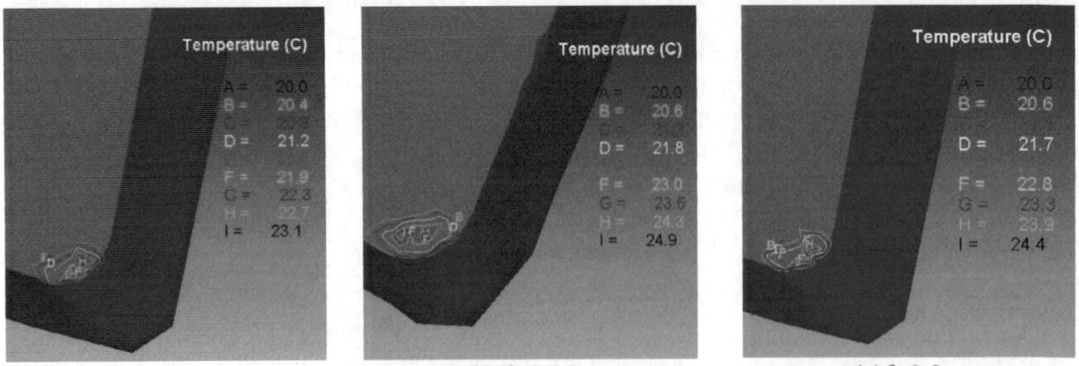

(a) f=0.1 (b) f=0.15 (c)f=0.2

Fig.4 Temperature fields analysis

4. Conclusions

This paper simulated the cutting HT250 process by self-prepared Si3N4 ceramic insert at different feed rate. Depended on analyzing the results of cutting force and tool temperature after simulation, the cutting properties of self-prepared Si3N4 insert were predict. The simulation results concluded:

(1) The cutting force was the maximum at f=0.15 mm/rev.

(2) The tool temperature was the highest at f=0.15 mm/rev.

(3) Other factors (cutting speed, cutting depth) should be taken into account for predicting accurately the cutting properties of self-prepared Si3N4 insert.

Acknowledgements

This work was financially supported by Premier-Discipline Enhancement Scheme Supported by Zhuhai Government Fund. (Fund code: 2015YXXK14)

References

[1] Childs, T.H.C., 1998. Material property needs in modelling metal machining. In: Proceedings of the CIRP International Workshop on Modeling of Machining Operations, Atlanta, Georgia, 19 May, p. 193–202.

[2] Childs, T.H.C., Maekawa, K., Obinakawa, T., Yamane, Y., 2000. *Metal Machining Theory and Applications*. Elsevier Ltd, p. 408.

[3] Trent, E.M., Wrigth, P.K., 2000. *Metal Cutting*, 4th ed. B-Heinemann, p. 434.

[4] Shaw, M.C., 2005. *Metal Cutting Principles*, 2nd ed. Oxford Science Publications, p. 651.

[5] Astakhov, V.P., 1999. *Metal Cutting Mechanics*. CRC Press, p. 297.

[6] Deform-user Manual 2006. SFTC-DeformV9.0.1, Scientific Forming Technologies Corporation Ed. Columbus, OH, USA.

[7] Pedro J. Arrazola, Tugrul Ozel, 2010. Investigations on the effects of friction modeling in finite element simulation of machining. International Journal of Mechanical Sciences 52, 31–42

[8] Johnson GR, Cook WH. A constitutive model and data for metals subjected to large strains, high strain rates and high temperatures. In: Proceedings of the seventh international symposium on ballistics, The Hague, The Netherlands, 1983. p. 541–547.

[9] Jiang-feng Luo, 2008. Study on Numerical Simulation of Dynamic Compress and High Speed Cutting for Cast Iron. Coal Mine Machinery, Sep.2008, p.45-47. In Chinese.

[10] Zorev NN. Inter-relationship between shear processes occurring along tool face and shear plane in metal cutting. International research in production engineering. New York: ASME; 1963 42–49.

[11] N. Ben Moussa, H. Sidhom, C. Braham, 2012. Numerical and experimental analysis of residual stress and plastic strain distributions in machined stainless steel. International Journal of Mechanical Sciences, 64, 82–93.

[12] Xing Ai, Hong Xiao. Ceramic Tool Cutting Machining. China Machine Press,1988. In Chinese.

[13] Huang J, Shivpuri R, Cheng X, Bedekar V, Matsumoto Y, Hashimoto F, Watkins TR. Effect of feed rate, workpiece hardness and cutting edge on subsurface residual stress in the hard turning of bearing steel using chamfer-hone cutting edge geometry. Mater Sci Eng 2005; 394:238–48.

Study on influence of gas content of coal and gas outburst

Xianzhan Chen[1, 2, a)]

[1] School of Mining and Safety Engineering, Anhui University of Science and Technology, Huainan 232001, China.
[2] Anhui Coal Mine Safety Supervision Bureau,Hefei230088, China.

E-mail: [a)] xianzhanchen@126.com

Abstract. In order to study the influence of gas content on the strength of coal and gas outburst, a set of device for simulating coal and gas outburst is improved by adding flowmeter and gas pressure sensors. The simulation experiments of coal and gas outburst were carried out with gas content of 7.5m3/t, 8.0m3/t, 8.5m3/t and 9.0m3/t respectively. The test results show that with the increase of gas content, the weight of outburst coal and the outburst distance increases gradually, the variation trend of crushing is not obvious either.Introduction

1. Introduction
Coal and gas outburst is a dynamic phenomenon that coal body in the front of mining face bursts into mining space with plenty of gas suddenly[1]. It is influenced by several factors, such as ground stress, gas pressure, coal property and gas conten[t[2-4]]. Some scholars research it by method of simulating tests[5]. In this paper, the simulation tests of coal and Gas Outburst under different gas content are carried out to study the influence of gas content on the strength of coal and gas outburst.

2. Experiment device

2.1 Function and structure of testing device
As shown in Fig.1, The gas pressure sensor and temperature sensor are added on the basis of the original device.

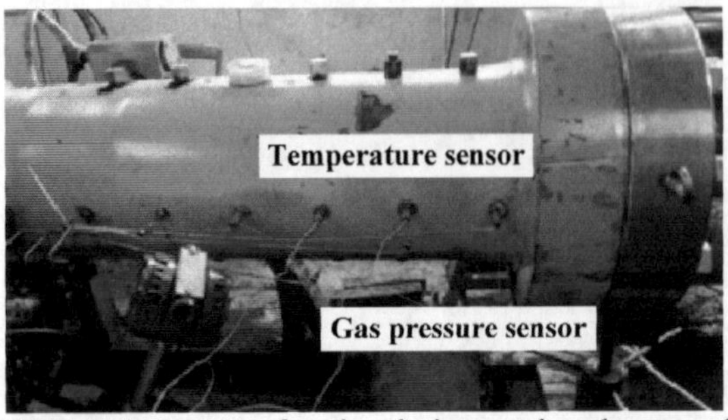

FIGURE 1. Apparatus for simulating coal and gas outburst

Content from this work may be used under the terms of the Creative Commons Attribution 3.0 licence. Any further distribution of this work must maintain attribution to the author(s) and the title of the work, journal citation and DOI.
Published under licence by IOP Publishing Ltd

2.1.1 Compacting organization. The compacting organization is plunger type and the plunger is installed into side of sample cylinder. The liquid chamber is formed by plunger, cylinder wall and end cap. The coal seam is compacted by advance of plunger pushed by external high-pressure pump and its distance of travel is 400mm. As shown in Fig 2.

FIGURE 2. Compacting organization

2.1.2 High speed photography. This high speed 24bit CMOS camera system combines with advanced CMOS and electronic technique. The process of coal and gas outburst is transient which is less than 1s. So, high speed photography is adopted to record the outburst phenomenon. In the test, it is initiated before the outburst is induced and thus outburst process can be recorded, which is shown in Fig 3.

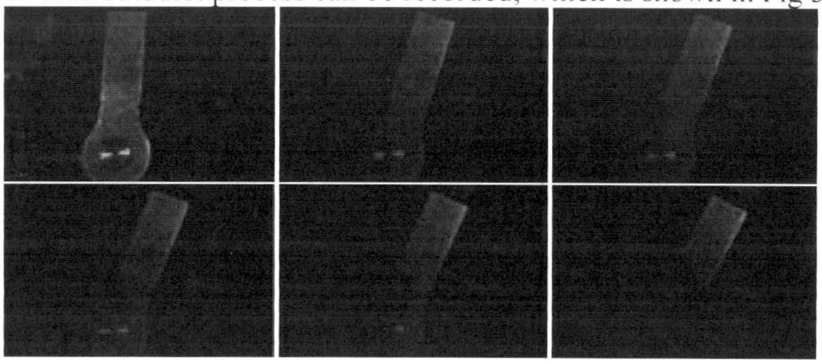

FIGURE 3. Coal and gas outburst

2.1.3 Gas meter and gas flowmeter. The gas flow is measured by flowmeter. However, its usage range is smaller than 0.3MPa. Thus, the flow of gas is calculated according to the volume of gas bottle multiplying change of gas pressure.

$$V_0 = {}^{P_1 V_1 T_0}/_{P_0 T_1} \quad (1)$$

Where：

——Pressure under standard atmospheric condition(P0=101.325KPa=0.101325MPa)；

——Temperature under standard atmospheric condition(T0=20℃=293.15K)；

—— Gas volume under standard atmospheric condition(m^3)；

——Pressure under current condition(MPa)；

——Temperature under current condition(K)；

——Gas volume undercurrent condition(m^3)。

According to equation(1), the gas volumes under standard atmospheric condition before and after coal's adsorption of gas are obtained and the gas is exactly gas content of coal. Besides, while the pressure is 0.25MPa, its error rate is 0.2% compared to the value measured by flowmeter, demonstrating it feasibility.

FIGURE 4. pressure gauge and flowmeter

3. Simulating test of coal and gas outburst

3.1 Experiment scheme

Grain size distribution of coal specimen is displayed in table 1 below.

TABLE 1. Grain size distribution of coal specimen

	<0.75mm	0.75mm~1.6mm	1.6mm~2.5mm	2.5mm~5.0mm
Mass percent/%	**9**	**38**	**29**	**24**

3.2 Experiment

The testing results is shown in table 2.

TABLE 2. Parameters in influence of ground stress on outburst experiment

NO	Ground stress/MPa	Gas content/m^3/t	Outburst caliber/mm	Moisture content/%	Total coal weight/Kg	Outburst condition	Outburst intensity/Kg
1	15	7.5	50	1.5	64.4	Outburst	25.9
2	15	8.0	50	1.5	64.2	Outburst	26.0
3	15	8.5	50	1.5	64.5	Outburst	26.5
4	15	9.0	50	1.5	64.7	Outburst	27.0

3.3 regional and particle size distribution of coal

The collecting apparatus of outburst coal sample divides outburst zone into 5 parts.

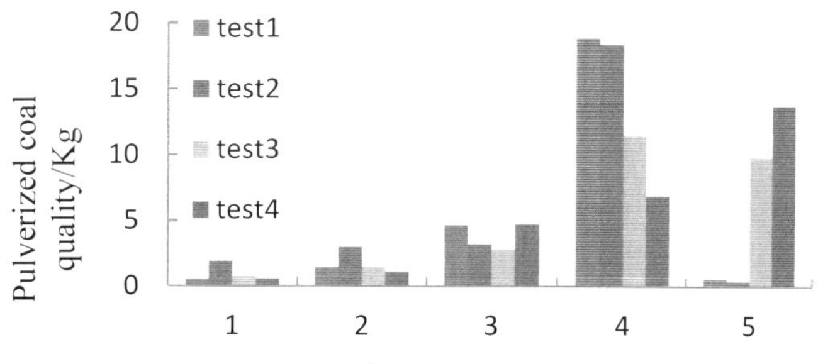

FIGURE 5. Weight distribution after coal and gas outburst

From fig.5, the gas content has obvious influence on the intensity of coal and gas outburst.The weights of outburst coal in tests 1～4 are 25.9Kg、26.0Kg、26.5Kg、27.0Kg respectively. With the increase of gas content, the outburst strength is increasing, and the influence of gas content on the intensity of coal and gas outburst is nonlinear.

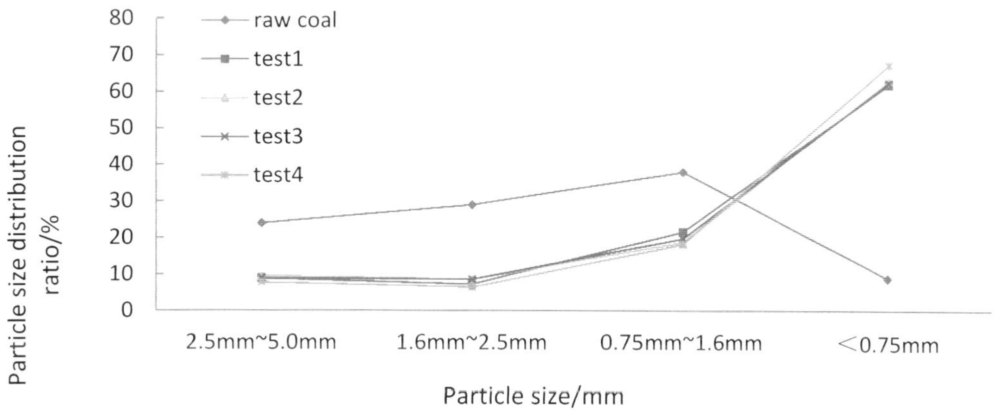

FIGURE 6. Comparison of grain size before and after outburst

Comparison of coal's grain size before and after outburst is shown in fig.6. Generally speaking, the proportion of bigger grain size decreases in coal sample after outburst while the smaller one rises up correspondingly,showing strong crushing property. The reason of coal's crushing during outburst is that the absorbed gas in porous particle desorbs into free gas, contributing to a pressure gap of gas inside and outside particle during swelling process. Besides, the crushing effect is stronger while the effect of gas swelling on coal sample is more evident[8-10]. Under conditions that underground stress and outburst caliber remain the same, the outburst distance and intensity rise up with increase of gas content. However, the change is not prominent and the variation trend of crushing is not obvious.

4. Conclusion

1) Other test conditions remain unchanged, With the increase of gas content, the quality of outburst coal body increases with the increase of gas pressure, but the quality of outburst coal body is relatively close.

2) With the increase of gas content, the outburst distance of coal sample increases and the outburst strength increases, but the change is not obvious, and the tendency of pulverization is not obvious.

About the author:
Xianzhan Chen (1983.6-),male, Current Ph.D Candidates , Engineer, Study on Gas Control in Coal Mine.Tel:17775388927, E-mail:xianzhanchen@126.com

References

[1] Xu Jiang,Tao Yunqi,Yin Guangzhi,et al.Development and applicationof coal and gas outburst simulation test device ［J］.ChineseJournal of Rock Mechanics and Engineering,2008,27(11):2354－2362.

[2] Hu Qianting,Zhou Shining,Zhou Xinquan.Mechanical mechanismof coal and gas outburst process ［J］.Journal of China Coal Society,2008,33(12):1368－1372.

[3] Guo Chenye,Xian Xuefu,Yao Weijing,et al.Relationship betweenthe fracture zone and cave of coal or gas outburst caving in coal androck seams ［J］.Journal of China University of Mining & Technology,2010,39(6):802－807.

[4] WANG Changming,WANG Zhen ,LIANG Yunpei,et al. Formation mechanism coal and gas outburst cave[J]. Mining safety and Environmental Protection,2008,35(4):60–62.(in Chinese) .

[5] Xu Jiang,Liu Dong,Peng Shoujian,et al.Experimental research oninfluence of particle diameter on coal and gas outburst ［J］.ChineseJournal of Rock Mechanics and Engineering,2010,29(6) : 1231 －1237.

Research on stress corrosion behavior of CCSE40 welded by underwater wet welding with austenitic welding rod in seawater

Y Zou[1], Q Bai[1,2,*], S Dong[2], Z L Yang[1] and Y Gao[1]

[1] Shandong Provincial Key Laboratory of Ocean Environment Monitoring Technology, Institute of Oceanographic Instrumentation, Shandong Academy of Science. 28 Zhejiang Road, Qingdao, China.

[2] Ocean University of China, Qingdao, China

Abstract. The stress corrosion behavior of CCSE40 welded by underwater wet welding with austenitic welding rod in seawater was studied. Microstructure, mechanical property and stress corrosion cracking susceptibility of the underwater wet welding joint were analyzed by metallographic observation, tensile and bending tests, slow strain rate test (SSRT) and SEM. The results indicated that the weld zone (WZ) and the heat affected zone (HAZ) were all sensitive to the stress corrosion, and the WZ was more sensitive than the HAZ.

1. Introduction

Affected by the water environment, the arc of underwater wet welding for water thermal decomposition can cause hydrogen content in the weld bath in water much larger than that in air (up to 62-68%). The hydrogen partial pressure is higher, so that there is a lot of hydrogen in the weld tissue, about 30-40 ml/100 g, maximum up to 60-70 ml/100 g. This is several more times than that in air when welding with acidic electrodes. The high hydrogen content and fast cooling rate lead to the poor plasticity and toughness of the underwater wet welding joints, which exacerbates the cold crack formation and seriously affects the service safety of the underwater wet welding on ocean engineering structures [1, 2].

Hydrogen plays a major role in the stress corrosion. The content of hydrogen in the metal exceeding the allowable value can generate cracks and pores, which cause hydrogen embrittlement in the weld. It can result in hydrogen cracking, and eventually destroy the structure.

In the hydrogen-induced cracking study, the accurate determination of hydrogen concentration is significant. The hydrogen-induced cracking sensitivity of high-strength steel is closely related to the hydrogen concentration in the steel. The higher the hydrogen concentration, the stronger the sensitivity of hydrogen-induced cracking. Takagi et al. [4] have evaluated the hydrogen-induced cracking sensitivity of high strength steel materials with critical hydrogen diffusion concentrations. Nevertheless, due to the difficulty of the accurate determination of the hydrogen concentration in material, there is little report on the quantitative relationship between hydrogen concentration and material strength index. Akhurst [5] used the local hydrogen pressure to replace the hydrogen concentration in the material. Zielinski [6] studied the effect of the cathodic polarization on the mechanical properties of carbon steel, high strength steel and high strength low alloy steels in artificial seawater, and used linear regression analysis to indicate the relationship between the hydrogen-induced cracking sensitivity and the amount of hydrogen related. In recent years, with the rapid development of detection and measurement

Content from this work may be used under the terms of the Creative Commons Attribution 3.0 licence. Any further distribution of this work must maintain attribution to the author(s) and the title of the work, journal citation and DOI.
Published under licence by IOP Publishing Ltd

technology, many new test methods have been applied, such as TGA-MS technology and TDS (Thermal Desorption Spectrometry) hydrogen measurement technology. TDS technology can improve the measurement accuracy of hydrogen concentration in high strength steel up to 10^{-8} (0.01 wppm), which plays a very important role in the quantitative study of hydrogen induced cracking of high strength steel and the mechanism of hydrogen induced cracking [7-9]. Wang [10-12] used TDS technology and SSRT methods to study the quantitative relationship between the notch tensile strength of the high strength steel and the hydrogen content. The results showed that the diffusible hydrogen led to the decrease of the tensile strength of the high strength steel, and the degree of decrease was exponentially related to the content of the diffusible hydrogen, while the non-diffusible hydrogen had no effect on the hydrogen cracking. Jayalakshmi [13] used TGA-MS technique to accurately derive the nonlinear relationship between the hydrogen concentration and the hydrogen charging time.

There is no consensus on the mechanism of hydrogen embrittlement. The common view of the various theories is that the hydrogen atoms are induced by stress-induced diffusion to the high stress site. When the hydrogen is concentrated to the critical value, the fracture stress of the material decreases and breaks. It is not clear how the aggregated hydrogen works. Hydrogen embrittlement is a complex process of the interaction of multiple mechanisms. Researchers used different mechanisms to explain the observed phenomena. There is no mechanism to explain the hydrogen induced cracking in all environments.

In this paper, the microstructure and mechanical properties of CCSE40 welded by underwater wet welding with austenitic welding rod were studied. The sensitivity of stress corrosion was evaluated by slow strain rate test (SSRT) and scanning electron microscopy (SEM). The fracture mechanism of welded joint was analyzed.

2. Experimental materials and methods

The material under investigation was CCSE40 high strength low-alloy steel plate with 45° V-shape of the weld pool by underwater manual welding. The filler material was austenitic stainless steel electrode. The chemical composition of base metal (BM) and weld zone (WZ) are given in Tables 1. Multi-pass welding was applied to complete the joint. The underwater welding was carried out in a tank, of which the size was 2 m×3 m×3 m. The tank was filled with seawater, the depth of which was 2 m.

Table 1. Chemical compositions of base metal and WZ of CCSE40 steel (mass fraction / %)

Position	C	Mn	Si	S	P	Mo	Ni	Cr	Fe
BM	0.18	1.2	0.50	0.035	0.035	0.08	0.40	0.20	Bal.
WZ	0.10	1.6	0.78	0.020	0.020	4.5	22.5	20.6	Bal.

Metallographic observations were carried out using GX51 Olympus metallurgical microscopy. A 10×30×10 mm specimen was obtained from the welded plate which composed of the BM, the WZ and the heat affected zone (HAZ). After ground with the SiC abrasive paper and polished to a mirror finish (1μm), this weldment sample was immersed into the etching solution. Because of different corrosion resistance, two etchants were used to reveal the microstructure of the joint. The etching of BM and HAZ was carried out using 3% Nital solutions (3 ml nitric acid (HNO_3) 97 ml ethanol) then the microstructure of WZ was etched by the solution made of 5g $CuSO_4$, 20ml hydrochloric acid and 20ml distilled water.

The fracture morphologies of the WZ, HAZ and BM for the underwater wet welding joint after SSRT were carried out using Philips XL30 scanning electron microscope.

The tensile and bending tests were carried out by universal testing machine. A cuboid specimen with 220×12×10 mm was taken along the vertical direction of the weld for the tensile test, while a cuboid specimen with 150×12×10 mm for the bending test. After cutting, the specimen was polished with metallurgical paper up to No.800 grade. The speeds of the tensile and the bending tests were 1 mm/min and 2 mm/min separately.

The size of SSRT sample is shown in Figure 1. The unit is mm. The coupons were polished with metallurgical paper up to No.1000 grade and cleaned with acetone. The slow strain rate tensile testing machine was used for stress corrosion test. The rate was 0.002 mm/min. The test solution was natural seawater.

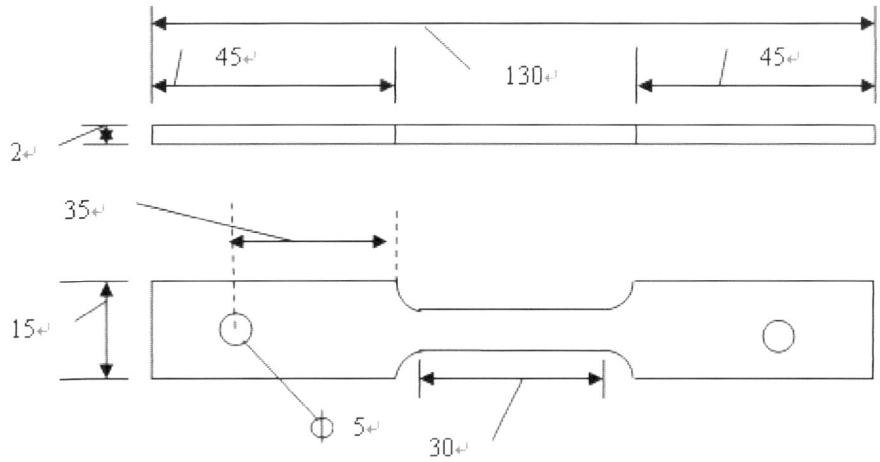

Figure 1. SSRT sample

3. Results and discussion

3.1 Microstructure analysis

Figure 2 shows the microstructures of the fusion zone and the WZ. The top and the bottom of the picture are the near-fused line of the weld and the coarse zone of the HAZ. The coarse zone of the HAZ had a martensite structure, while the near-fused line of the weld had a sorbite and ferrite structure. As can be seen from Figure 2a, there was a crack parallel to the fused line in the coarse region of the HAZ, which was a hydrogen delayed crack. Under the effect of large heat input, the coarse zone of the HAZ overheating produced coarse grain. Under the condition of water cooling, the hardened martensite structure was formed. In combination of the constraints of the bottom plate and the hydrogen diffusion enrichment, the delayed crack formed. It was cold crack. Figure 2b shows a weld metal picture, and a crystal crack formed along a columnar grain boundary can be observed.

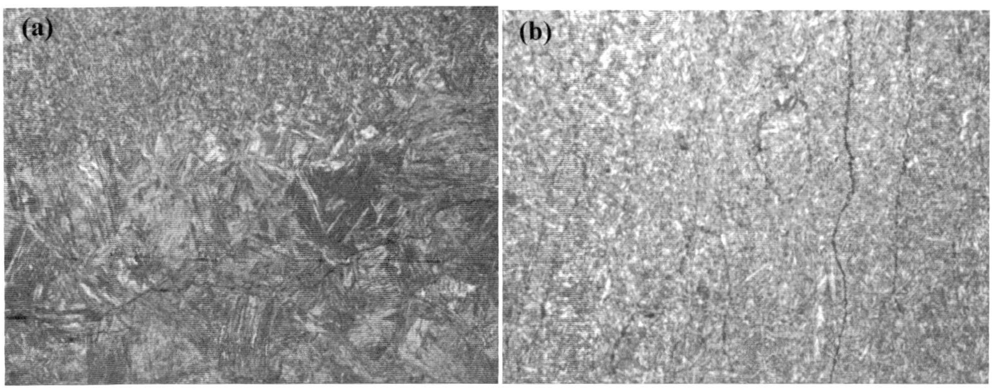

Figure 2. Microstructures of the weld joint (a) the fusion zone, (b) the WZ

3.2 Mechanical property test

The mechanical properties of the underwater wet welding joint of austenitic electrode are shown in Table 2. The face bending of the coupon did not break, while the back bending cracked along the zone near the HAZ and the fusion line. The tensile coupon, without exception, spread along the HAZ and the fused line to the weld (Figure 3). The tensile strength of the weld is higher than that of the base metal. The reason is that the hydrogen-induced cracking of the heat-affected zone leads to a decrease in the mechanical properties near the heat-affected zone.

Table 2. Mechanical properties of the underwater wet welding joint

Tensile properties of the welding joint			Tensile properties of the weld metal		Bending properties of the welding joint
Average tensile strength (MPa)	Average yield strength (MPa)	Fracture position	Tensile strength (MPa)	Elongation (%)	
509.9	401.4	Near the HAZ and the fused line	512.4	1.7	Face bending passed through. Back bending began to crack at 20°

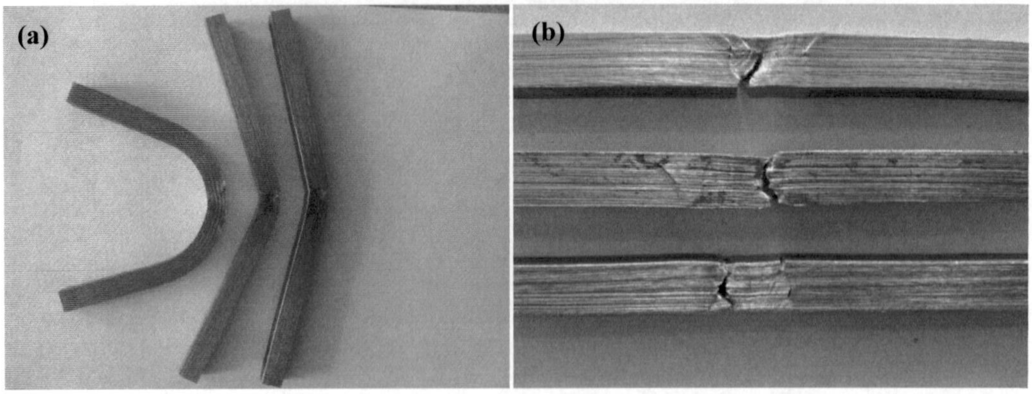

Figure 3. Morphologies of the underwater wet welding with austenitic welding rod after mechanical property test (a) bending test, (b) tensile test

3.3 Slow strain rate test (SSRT)

SSRT is of great significance in evaluating the stress corrosion cracking susceptibility of materials. As the test is in the indoor environment, the effects of other factors such as temperature, electrode potential and pH on the corrosion process can be studied simultaneously during SSRT. When the crack-free samples are slowly tensile failure in a specific medium and inert medium separately, according to the elongation δ, the contraction of area ψ, the fracture time t_f, the absorbing energy, the fracture morphology, the characteristics of the secondary crack and other parameters to evaluate the materials stress corrosion cracking susceptibility.

The severity of stress corrosion cracking is usually related to the strain rate. If the strain rate is very large, the ductile fracture occurs before the material producing the effective stress corrosion cracking. On the contrary, if the strain rate is very small, the film breaks and repassivation follows, so that the crack tip exposed metal is too late to produce corrosion reaction. That is to say, it cannot produce stress corrosion cracking, but can produce a ductile fracture. The lower the strain rate, the lower the elongation, the longer the breaking time requires, the longer the time it need in the medium.

The relative difference (I) between the elongation (δ) in the inertia medium and that in the corrosive medium or between the contraction of area (ψ) in these two mediums are taken as a measurement of the stress corrosion susceptibility. That is $I_\delta = (\delta_a - \delta_c)/\delta_a$ 或 $I_\psi = (\psi_a - \psi_c)/\psi_a$. Where "a" denotes the inert medium and "c" denotes the corrosive medium. The larger the I_δ or I_ψ, the more sensitive to the stress corrosion.

The elongations for the BM, WZ and HAZ by SSRT in seawater and glycerin as well as the brittleness index which reflects the stress corrosion susceptibility are shown in Figure 4. As can be seen from the picture that the WZ and the HAZ were all sensitive to the stress corrosion. The weld was the most sensitive one, while the base metal was the most insensitive one. In seawater, the weld was the most susceptible to the stress corrosion.

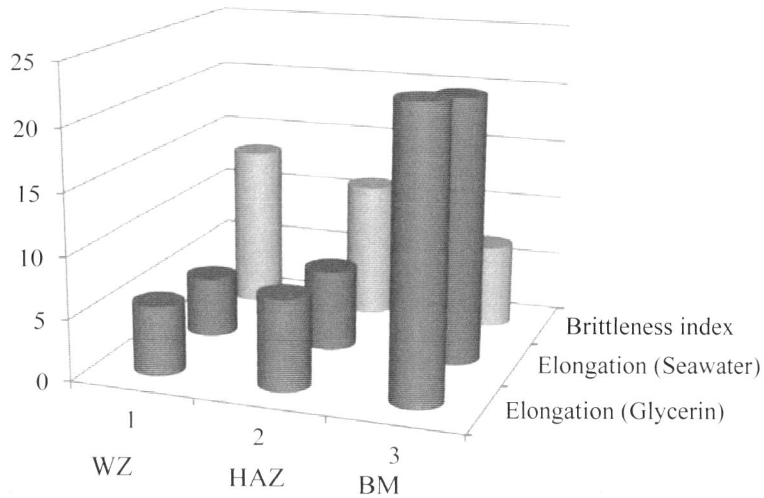

Figure 4. SSRT results for the underwater wet welding in seawater

3.4 Scanning electron microscopy (SEM)

Figure 5 shows the fracture morphologies of the WZ, HAZ and BM for the underwater wet welding joint in seawater after SSRT. It can be seen from the picture that the WZ and the BM were ductile fracture, while the HAZ was quasi-cleavage fracture. The dimple center of the weld had a mixed phase. There were a large number of typical tectonic-shaped secondary cracks on the main fracture surface, which indicated that there was a hydrogen embrittlement mechanism for the underwater wet welding joints in seawater.

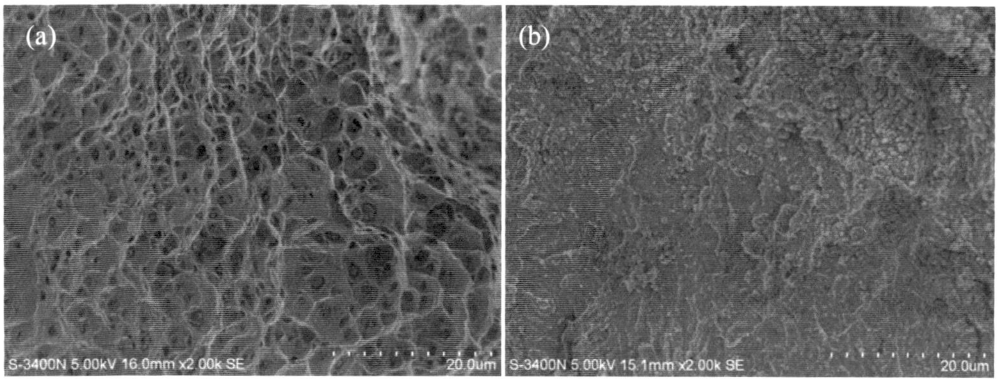

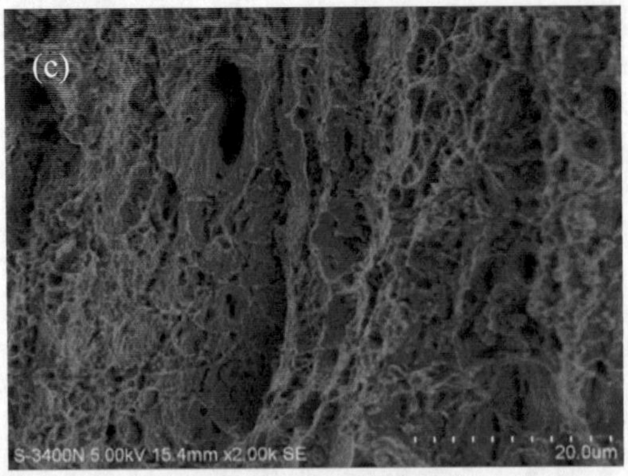

Figure 5. Fracture morphologies of the WZ, HAZ and BM for the underwater wet welding joint in seawater after SSRT (a) WZ, (b) HAZ, (c) BM

4. Conclusion

(1)The coarse zone of the HAZ had a martensite structure, while the near-fused line of the weld had a sorbite and ferrite structure. There was a crack parallel to the fused line in the coarse region of the HAZ, which was a hydrogen delayed crack.

(2)The mechanical properties of the underwater wet welding joint of austenitic electrode showed that the back bending cracked along the zone near the HAZ and the fusion line. The tensile coupon spread along the HAZ and the fused line to the weld.

(3)From the SSRT results, the weld was the most sensitive to the stress corrosion, while the base metal was the most insensitive.

(4)The fracture morphologies showed that the WZ and the BM were ductile fracture, while the HAZ was quasi-cleavage fracture.

Acknowledgement
The authors appreciate the financial support by National Natural Science Foundation of China (No. 50971118) and National High-Tech Research & Development Program of China (No. 2008AA092901)

References

[1] Yu J R, Zhang Y L and Jiang L P 2001 *Weld. Technol.* **30** 2
[2] Ma C Y, Zhao J W and Song W Q 2007 *Proceedings of the 13th China Ocean (Shore) Engineering Symposium* 659
[3] Li H L, Yu Z H, Li Y, Chu W Y and Qiao L J 2009 *Corros. Prot.* **30** 678
[4] Takagi S, Inoue T, Hara T and Takahashi T 2000 *Tetsu-to-Hagane* **86** 689
[5] Akhurst K N and Baker T J 1985 *Metall. Trans. A* **12** 1059
[6] Zielinski A and Domzalicki P 2003 *J. Mater. Process. Tech.* **133** 230
[7] Wei F G, Hara T, Tsuchida T and Tsuzaki K 2003 *ISIJ International* **43** 539
[8] Nagumo M 1985 *Mater. Sci.Technol.* **1** 711
[9] Wang M Q, Akiyama E and Tsuzaki K 2006 *Mater. Sci. Forum* **512** 55
[10] Wang M Q, Akiyama E and Suzaki K 2005 *Scripta Mater.* **53** 713
[11] Wang M Q, Akiyama E and Tsuzaki K 2005 *Scripta Mater.* **2** 403
[12] Wang M Q 2006 *Trans. Mater. and Heat Treat.* **27** 57
[13] Jayalakshmi S, Kim K B and Fleury E 2006 *J. Alloys Compd.* **417** 195
[14] Yau T L 1982 *Corrosion* **38** 615
[15] Lyle F F and Norris E B 1978 *Corrosion* **34** 193
[16] Turn J C, Wilde B E and Troianos C A 1983 *Corrosion* **39** 364

IOP Conf. Series: Materials Science and Engineering **242** (2017) 012045 doi:10.1088/1757-899X/242/1/012045

Generating elastic band gaps in square lattice grid by periodical cut-off operations

Lin Han[1,*], **Xiao-mei Li**[2] **and Yan Zhang**[1]

[1] College of Mechanics and Materials, Hohai University, 211100, Nanjing, P. R. China

[2] Department of Geotechnical Engineering, Nanjing Hydraulic Research Institute, 210024, Nanjing, P. R. China

[*]corresponding author, hanlin@hhu.edu.cn.

Abstract. From periodically cutting off rods in square lattice grids, we give an easy and feasible way to generate elastic band gaps (BGs) in it. We define the model of the periodical cut-off (PCO) square lattice grid, and study the frequency dispersion relation and the motion modes of rods, as well as the frequency responses. On this basis, we confirm that the PCO operation is feasible to generate BGs in square lattice grids. Furthermore, the BGs generated in PCO square lattice grids are basically locally resonant BGs with low frequency ranges which attributes to the motions of exposed cantilevers.

1. Introduction

Lattice grids are cellular structures composed of periodical cells, and usually used in solar panels, interstage of spacecraft launcher [1], aircraft fuselage and wing [2], for the merits of high specific stiffness and specific strength. Several types, including square lattice grid, triangular lattice grid, hexagonal lattice grid, kagome lattice grid, reentrant lattice grid, etc. are often adopted [3]. For now, the static properties of lattice grids are studied a lot, including various material properties depending on periodical structures [4], buckling [5], fracture [5], etc. Besides, the versatility of lattice grids in vibration insulation, energy absorption [7], heat dissipation, etc. also has been borne in upon researchers recently, for wide prospect in applications. However, the study on the versatility of lattice grids is still rare, and should be paid more attention. We mainly focus the characteristic of controlling elastic waves.

Recently, the propagation of elastic waves in periodical structures or materials has been received much attention [8-14]. In specific periodical structures or materials, elastic waves fall into some frequency ranges, which are called band gaps (BGs), cannot propagate. The periodical structures or materials are the so-called phononic crystals (PCs). Due to the BG characteristics, PCs could have many applications such as high-precision mechanical systems [9], noise control [15], vibration control [16, 17], vibration reduction [18], transducers and wireless communications [19]. Clearly, lattice grids with periodical structures could bring BG characteristics for elastic wave control. Based on the PC concept, the propagation characteristics of elastic waves in lattice grids has been studied, including the BG range [20], BG control [3, 21], as well as directional propagation of elastic waves [22, 23]. However, the current studies are still not enough to support actual applications, for the BG ranges are hard to meet the actual requirements. So the BGs regulation and control method urgently need breakthrough.

Content from this work may be used under the terms of the Creative Commons Attribution 3.0 licence. Any further distribution of this work must maintain attribution to the author(s) and the title of the work, journal citation and DOI.
Published under licence by IOP Publishing Ltd

Actually, for general PCs, BGs regulation and control have some measures, which are mainly based on the changes of material properties and distribution, lattice type and size, etc. Specifically, by means of some materials which could easily control properties, such as piezoelectric material [24], electrorheological material [25], magnetostriction material [26], BGs could be controlled. Furthermore, rotating the scatterers or adding scatterers in cells [28] could help to changing the material distribution. Additionally, changing the deformation form by external effects could change the lattice size or even lattice type, so the BGs could be changed [29, 30]. While the most of these measures are hard to implement in lattice grids, because one kind of actual lattice grid is generally made by just one material, and the BGs characteristics are just determined by the periodical geometry. So when a kind of certain lattice grid is given, how can we change the propagation characteristics of elastic waves in it? Specifically, BGs are hard to show for square lattice grids. And how can we do to obtain BGs in square lattice grids by using simple methods? We think periodically cutting off particular rods in square lattice grids is an easy and feasible way to realize. Cutting off rods in square lattice grids in periodically pattern, could not only reduce the stiffness of structure, but also change the mass distribution of structure due to the exposed cantilevers, so the frequency dispersion relation would be changed, and BGs might be obtained.

In this paper, firstly, we define the model of the periodical cut-off (PCO) square lattice grid, and study the elastic BGs characteristics in it. Then we determine the BGs mechanism from analysing the frequency dispersion relation and motion modes in it. The influence of the cantilever length on BGs is also illustrated. Finally, we give the frequency response analysis of the corresponding finite specimen to support our conclusions.

2. Model

Figure 1 shows the original square lattice grid and the PCO square lattice grid. The PCO operation is implemented according to the square lattice type, so the PCO square lattice grid still maintains the square lattice type. The original square lattice grid has the lattice constant of a, while the lattice constant of PCO square lattice grid is twice as that of the original square lattice grid. Actually, there could be several different kinds of PCO operations which just need to meet at least two requirements: (1) the PCO lattice grids should still maintain some kinds of periodicity, but not necessarily the same as that of the original lattice grids; (2) the PCO lattice grids should not disintegrate into pieces. Here we just consider one kind of PCO square lattice to study the effect and mechanism of PCO operations on square lattice grids.

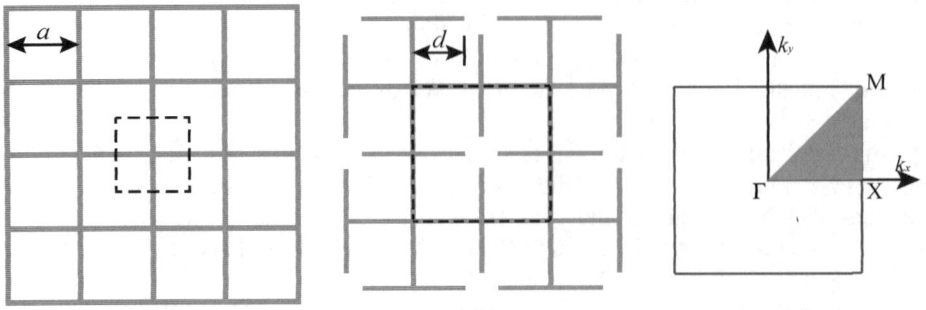

(a) (b) (c)

Figure 1. (a) The square lattice grid, (b) PCO square lattice grid and (c) irreducible Brillouin zone

Clearly, the PCO operation generates several cantilevers which vertically stretch from the middle position of each rod. We can change the length of the cantilever d by adjusting the size of cut-off parts. Because the exposed cantilevers lose the load-carrying ability, thus the stiffness of PCO square lattice grids must be lower than the original square lattice grids. In spite of this, the cantilevers still have the mass effect. Moreover, the lattice constant increases after the PCO operation. These three factors make it possible to change the propagation characteristic of elastic waves in it. Cantilevers must be the key to generate the elastic BGs in PCO square lattice grids.

3. Elastic BGs

The frequency dispersion relations of lattice grids could be calculated by the finite element method, the details could be seen in [31]. We just consider the propagation of in-plane waves. Here, we use the normalized frequency f/f_0 to describe the frequency dispersion relation and BGs ranges of PCO square lattice grids, where f_0 is the basic frequency of the simply supported beam constructed by the rod in lattice grid. $f_0=\pi\ (EI/\rho_l)^{1/2}/2l^2$, where l is the length of the rod, EI is the flexural stiffness of the rod, $\rho_l=\rho S$ is the line density, ρ is the material density, S is the cross section area of the rod. Because of using the normalized frequency, the results below could be independent on the material type. In other words, the PCO square lattice grids with identical construction have the same frequency dispersion relations; even they are composed of different materials. So it is more convenient to study the influence of geometry construction on BGs. Here we choose the cross section of rod as the square shape, which is easy to manufacture. Actually, the cross section of the rod could be any shape, and just need to be the same at any position of rod. Let the square cross section width be b , so the slenderness ratio of rod cold be defined as $\gamma=l/b$. In this paper, we just analyze the case of $\gamma=10$. Specifically, we set l=0.07 m and d=0.06 m, in order to more conveniently find the influence of cantilevers and the mechanism based on the case of relatively long cantilevers. The cells and the corresponding Brillouin zone are shown in figure1.

In order to comprehensively illustrate the changes of frequency dispersion relation and BGs after cut, we calculate the frequency dispersion relations of the PCO square lattice grids with partial cantilevers left (d=0.06 m) and no cantilevers left (d=0), as well as the original square lattice grid. The comparison is shown in figure 2. Clearly, the original square lattice grid and the PCO square lattice grids with no cantilevers left both have no BGs in the concerned frequency range, which also verifies the fact that the square lattice grid is hard to obtain BGs. The difference is the latter one has bigger lattice constant, thus its frequency dispersion curves are compressed to low frequency direction. While the PCO square lattice grid with partial cantilevers generates two BGs, which cover the ranges of 0.39~0.40 and 0.48~0.53.

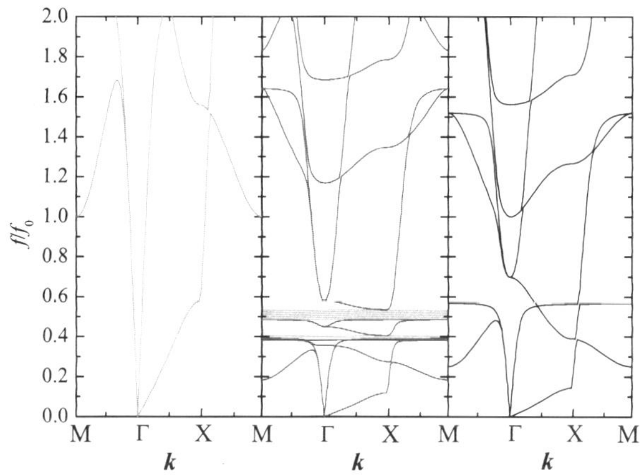

Figure 2. Frequency dispersion relations of the original square lattice grid (left), PCO square lattice grids with partial (middle) and no (right) cantilevers. The shadow zones are the BGs ranges.

Definitely, the PCO square lattice grids with or without partial cantilevers have the similar features in frequency dispersion curves. Because the two cases have the same lattice constant, as well as the load-carrying skeleton and material. But, several new relatively flat curves occur in the case of PCO square lattice grid with partial cantilevers, which could only be attributed to the existence of cantilevers. The motion modes of a cell of the PCO square lattice grid with cantilevers at certain wave vectors on frequency dispersion curves (table 1) support the conclusion. The 1st BG locates between the 4th and 5th order curves, and the frequencies corresponding to wave vector X determine the 1st BG range. The longitudinal motions of cantilevers and flexural motions of skeleton are shown at wave vector X in 4th order curve. While the flexural motions of cantilevers are shown at wave vector X in 5th order curve. The discrepancy of cantilever motions leads to the 1st BG. Similarly, the 2nd BG is generated. While the motions which determine the 2nd BG are both flexural modes of cantilevers, which but behaves different types. However, the 1st BG is too narrow to work. Actually, when we cut off longer parts, the 1st BG disappears. So the 2nd BG is really worth considering.

Table 1. Motion modes of a cell of the PCO square lattice grid with cantilevers at wave vector X

4th order	5th order	6th order	7th order
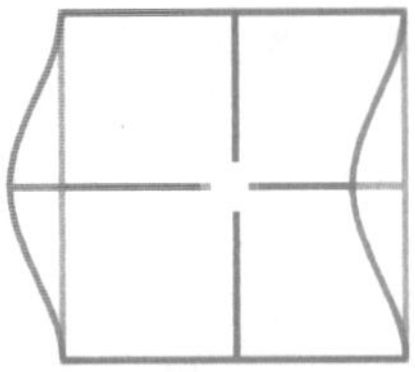	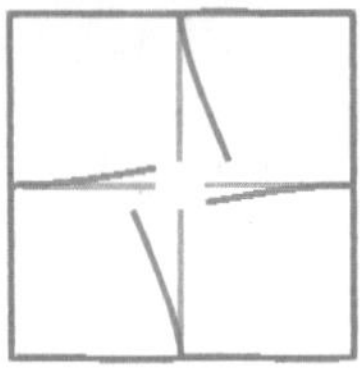		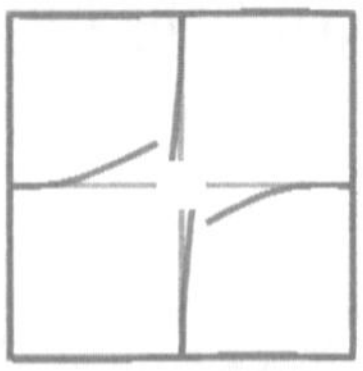

Due to the cantilevers motion at the lower limit of the 2nd BG are the pure first flexural mode in the same direction, we think the first flexural motion of cantilevers determines the lower limit of the 2nd BG. The natural frequency corresponding to the first flexural motion of cantilever is given as $f_1=1.875^2(EI/\rho_l)^{1/2}/2\pi l^2$. So we get the normalized frequency f_1/f_0 is 0.48, which matches the lower limit of the 2nd BG. Evidently, the cantilevers indeed determine the lower limit of the 2nd BG. The conclusion also support that the 2nd BG is generated by the localized resonance of cantilevers. So the PCO square lattice grid is actually a localized resonant PC, although it is composed of just one material.

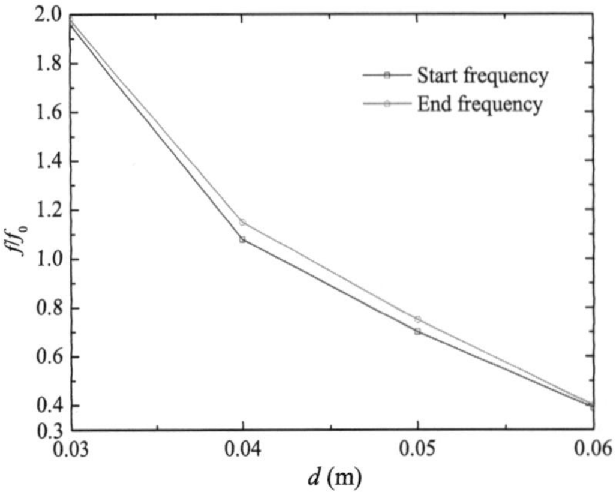

Figure 3. The start (square) and end (circle) frequencies of the BG of the PCO square lattice grid with different lengths of cantilevers.

For verifying the above conclusions, we also calculate other three cases with cantilevers lengths of 0.05 m, 0.04 m and 0.03 m, and give the corresponding BG ranges with different cantilevers lengths, which are shown in figure 3. Surely, the lower limit of the BG is determined by the first flexural mode of cantilevers. The longer cantilever leads to the lower natural frequency. So the longer cantilever case gives the lower BG frequency. Additionally, the BG range changes with the change of cantilever length, while the difference is little.

After the theoretical analysis of infinite PCO square lattice grids, then we study the practicable finite cases, to find the feasibility of BG regulation and control from PCO operations. We choose the specimen with 8×6 cells which is shown in figure 4 (a). We just focus the wave propagation along the ΓX direction which corresponds to the frequency dispersion curves between ΓX. We apply the harmonic displacement impulses which sweep over the normalized frequency range of 0~0.6 to the left end of specimen, and then get the frequency responses at point A on the other end. Two kinds of impulses which correspond to the in-plane longitudinal and transverse vibrations are applied and marked as I and II in figure 4 (a). Figure 4 (b) illustrates the frequency responses at point A. The shadow area shows the both distinct attenuation areas of the two kinds of impulses, which is the actual BG range. Clearly, just the above mentioned 2nd BG exists in frequency response results. The 1st BG is too narrow to show. The actual BG covers the range of 0.47~0.54, which matches the theoretical BG result very well. The results verify that the square lattice grid indeed generates BGs after PCO operations. Additionally, the PCO square lattice grids could give relatively low frequency BGs based on the locally resonant mechanism, which could be used in many potential applications.

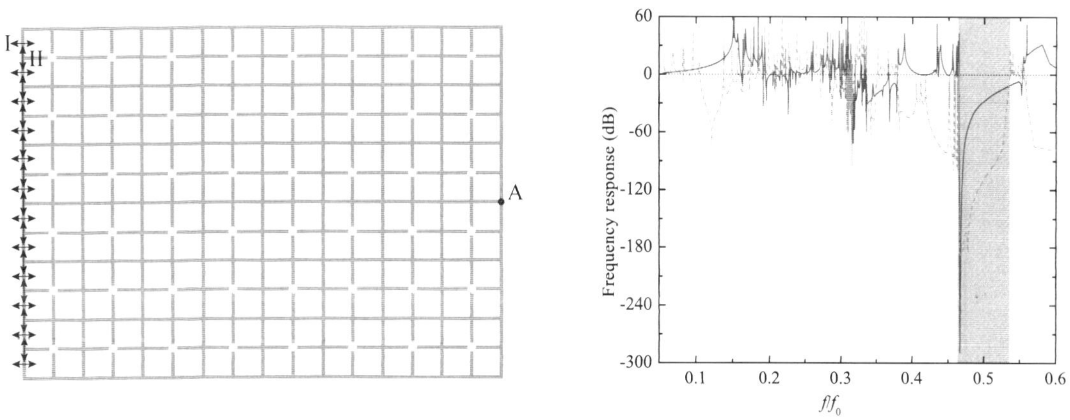

(a) (b)

Figure 4. (a) Finite specimen of the PCO square lattice grid. The in-plane longitudinal (I) and transverse (II) vibration impulses are applied. (b) The frequency response is obtained at point A. The overlapping of curves of transverse (dash) and longitudinal (solid) vibrations gives the actual BG.

4. Conclusions

In summary, we mainly define the model of PCO square lattice grid, and study the mechanism and properties of elastic BGs in it. We confirm that the PCO operation is feasible to generate BGs in square lattice grids. Furthermore, the BGs generated in PCO square lattice grids are basically locally resonant BGs with low frequency ranges which attributes to the motions of exposed cantilevers. The lower edge of the locally resonant BG could be predicted by the natural frequency of first flexural mode of cantilevers. One can easily control the BG location from adjusting the size of cut-off parts. All the features make the BGs in PCO square lattice grids more feasible in applications.

Acknowledgement
The authors would like to thank the support of the National Natural Science Foundation of China (Grant No. 51608173), Natural Science Foundation of Jiangsu Province (Grant No. BK20150796) and

Fundamental Research Funds for the Central public welfare scientific research institutes (Grant No. Y316022).

References:

[1] Vasiliev V V and Razin A F 2006 *Compos. Struct.* **76** 182
[2] Noor A K, Venneri S L, Paul D B and Hopkins M A 2000 *Comput. Struct.* **74** 507
[3] Phani A S, Woodhouse J and Fleck N A 2006 *J. Acoust. Soc. Am.* **119** 1995
[4] Chen Y, Liu X and Hu G 2014 *Comptes Rendus Mécanique* **342** 273
[5] Vigliotti A, Deshpande V S and Pasini D 2014 *J. Mech. Phys. Solids* **64** 44
[6] Cui X, Xue Z, Pei Y and Fang D 2011 *Int. J. Solids Struct.* **48** 3453
[7] Gonella S, To A C and Liu W K 2009 *J. Mech. Phys. Solids* **57** 621
[8] Sigalas M M and Economou E N 1992 *J. Sound Vib.* **158** 377
[9] Kushwaha M S, Halevi P, Dobrzynski L and Djafari-Rouhani B 1993 *Phys. Rev. Lett.* **71** 2022
[10] Liu Z, Zhang X, Mao Y, Zhu Y Y, Yang Z, Chan C T and Sheng P 2000 *Science* **289** 1734
[11] Shu H, Shi X, Li S, Liu S, Wang W and Shi D 2014 *Int. J. Mod. Phys. B* **28** 1450231
[12] Han L, Zhang Y, Ni Z, Zhang Z and Jiang L 2012 *Physica B* **407** 4579
[13] Zhang Y, Han L and Jiang L 2013 *Phys. Status Solidi B* **250** 1245
[14] Han L, Zhang Y, Jiang L and Zhang Z 2013 *Phys. Status Solidi - R.* **7** 514
[15] Martinezsala R, Sancho J, Sanchez J V, Gomez V, Llinares J, Meseguer F 1995 *Nature* **378** 241
[16] Yu D, Wen J, Zhao H, Liu Y and Wen X 2008 *J. Sound Vib.* **318** 193
[17] Yu D, Liu Y, Zhao H, Wang G and Qiu J 2006 *Phys. Rev. B* **73** 64301
[18] Farhat M, Guenneau S and Enoch S 2009 *Phys. Rev. Lett.* **103** 24301
[19] Mohammadi S, Eftekhar A A, Hunt W D and Adibi A 2009 *Appl. Phys. Lett.* **94** 51906
[20] Liebold-Ribeiro Y and Körner C 2014 *Adv. Eng. Mater.* **16** 328
[21] Nolde E, Craster R V and Kaplunov J 2011 *J. Mech. Phys. Solids* **59** 651
[22] Gonella S and Massimo R 2008 *J. Sound Vib.* **312** 125
[23] Spadoni A, Ruzzene M, Gonella S and Scarpa F 2009 *Wave Motion* **46** 435
[24] Dai L, Jiang S, Lian Z, Hu H and Chen X 2015 *J. Sound Vib.* **337** 150-160
[25] Yeh J 2007 *Physica B* **400** 137
[26] Robillard J, Matar O B, Vasseur J O, Deymier P A, Stippinger M, Hladky-Hennion A, Pennec Y and Djafari-Rouhani B 2009 *Appl. Phys. Lett.* **95** 124104
[27] Wu F, Liu Z and Liu Y 2002 *Phys. Rev. E* **66** 46628
[28] Yao Y, Hou Z and Liu Y 2007 *Phys. Lett. A* **362** 494
[29] Wang L and Bertoldi K 2012 *Int. J. Solids Struct.* **49** 2881
[30] Bertoldi K and Boyce M 2008 *Phys. Rev. B* **77** 52105
[31] Han L, Zhang Y, Li X, Jiang L and Chen D 2016 *Crystals* **6** 11

IOP Conf. Series: Materials Science and Engineering **242** (2017) 012046 doi:10.1088/1757-899X/242/1/012046

Comparison of a high lithium-ion conductivity solid electrolyte prepared by two methods

Shang Xuefu[1], Cheng Si, Zhang Jinhao, Wang Yawei

Department of Physics, Faculty of Science, Jiangsu University, Zhenjiang, Jiangsu, 212013

[1] Corresponding author.E-mail address: shangxuefu@yeah.net

Abstract. A NASICON-type structured solid electrolyte of precursor for $Li_{1.3}Al_{0.5}Ti_{1.3}Nb_{0.2}(PO_4)_3$ with high lithium-ion conductivity was prepared with solid-state reaction. Two samples have same structures without impurity phases. The highest total electrical conductivities of sintered pellets made with two methods of press method and tape casting calculated at 25°C were 4.65×10^{-4} S/cm and 7.1×10^{-4} S/cm, respectively. The conductivity of sample synthesized with tape casting is significantly increased, but relative density is 90% which is lower than relative density of 96% for sintered pellets synthesized with press method.

1. Introduction

In recent years, the rechargeable lithium-ion battery is widely studied as power equipment of mobile electronic product and electrically driven vehicle [1]. Currently, the commercial lithium-ion battery exists safety issue due to leakage and combustibility. That the solid state electrolyte possesses higher stability excellent mechanical properties and non-flammable can optimize safety issue. Hence many solid electrolyte have been studied such as monocrystal Li_3N, sulfide-base glasses, LISICON, NASICON [2].

1976, Goodenough designed a new solid electrolyte formed with the three-dimensional network structure and named Na super ionic conductor (NASICON) [3]. The NASICON-type structure consist of MO_6 octahedra and PO_4 tetrahedra and is linked with corner to form a $[Ti_2(PO_4)_3]^-$ rigid three-dimensional skeleton. The two different sites for mobile lithium ions are A_1 and A_2, respectively. The A_1 sites are fully occupied by lithium ions and lie an elongated octahedral oxygen environment, but the A_2 sites are completely vacant [4]. Anon [2][5] has reported $LiTi_2(PO_4)_3$ with poor ionic conductivity and high porosity, if the Ti^{4+} sites are partially substituted by some trivalent M^{3+} (M=Al or Sc) in the $LiTi_2(PO_4)_3$ systems, the conductivity has effective enhancement and the maximum can reach 7.0×10^{-4} s/cm at 298 k for $Li_{1.3}Al_{0.3}Ti_{1.7}(PO_4)_3$. The Ti^{4+} partially replaced by Al^{3+}+Li^+ and the Li^+ insertion onto the A_2 sites are not effect for the Li^+ mobility. The conductivity enhancement is mainly attributed to the grain boundary but no correlation with bulk. So the grain boundary is main factors to improve the conductivity in $LiM_2(PO_4)_3$ systems [6].

In general, the LATP precursor is synthesized by conventional melting-quenching [7][8], sol-gel [9] or solid-state method. 1996, Fu reported LATP specimens obtained by using melting-quenching method, the maximum conductivity of 1.3×10^{-3} s/cm was measured at room temperature [7]. The LATP nano-powder reported by Y. Yoon et al sintered at 900°C for 3 hours by using a sol-gel method has a room temperature conductivity of 7.8×10^{-5}s/cm [9]. However, the melting-quenching inevitably cause a lose of Lithium due to sintered pelleted need heating 1450°C [7] and the sol-gel method need higher cost coprecipitator [10]. But the solid-state is a method of simple, low cost and suitable for mass

Content from this work may be used under the terms of the Creative Commons Attribution 3.0 licence. Any further distribution of this work must maintain attribution to the author(s) and the title of the work, journal citation and DOI.
Published under licence by IOP Publishing Ltd

production. E. Kazakevicius et al reported a new material $Li_{1.3}Ge_{1.4}Ti_{0.3}Al_{0.3}(PO_4)_3$ by Ge^{4+} partial substitution Ti^{4+} in the LATP NASICON structure, which exhibit good electrolytic property [11]. Then, Zhang et al reported the highest total electrical conductivity of 1.29×10^{-3}s/cm at 25°C from $Li_{1.4}Al_{0.4}Ti_{1.4}Ge_{0.2}(PO_4)_3$ [14], but the study about $Li_{1.4}Fe_xAl_{0.4-x}Ti_{1.6}(PO_4)_3$ was not significant break contrasted with $Li_{1.4}Al_{0.4}Ti_{1.4}Ge_{0.2}(PO_4)_3$ [15]. The water-stable lithium ion conducting solid electrolyte can be used as an interlayer to insulate air and water [12] and the LATP-Epoxy hybrid sheets obtained by pouring 2,2-bis(4-glycidyloxyphneyl)propane and 1,3-phenylenediamine in tetrahydrofuran into the LATP sheets have the conductivity of 4×10^{-4}s/cm and watertightness [13]. In this study, we attempted to research LANTP by pentavalent Nb partial substitution Ti sites, and prepared $Li_{1.3}Al_{0.5}Ti_{1.3}Nb_{0.2}$ $(PO_4)_3$ sheets by tape casting and press method, and discussed the performance.

2. Experimental

The NASICON-type $Li_{1.3}Al_{0.5}Ti_{1.3}Nb_{0.2}(PO_4)_3$ precursor obtained by using solid-state reaction[16]. Stoichiometric amounts of Li_2CO_3, TiO_2, Nb_2O_5, Al_2O_3, and $NH_4H_2PO_4$ were ball-milled with Zirconia balls in a zirconia vessel for 2h at 400 rpm using high-energy mechanical milling (HEMM) with a planetary micro mill, and the mixed powder calcined at 600°C for 4 h. The obtained powders again was ball-milled using HEMM before pressed into pellets at 150 MPa, then calcined at 850°C for 5 h. The sintered pellets were reground and thirdly ball-milled for 10h. The fine precursor powder was made into $Li_{1.3}Al_{0.5}Ti_{1.3}Nb_{0.2}(PO_4)_3$ pellets by isostatically pressing at 150 MPa and tape casting reported by P. Zhang [17]. Briefly, the fine $Li_{1.3}Al_{0.5}Ti_{1.3}Nb_{0.2}(PO_4)_3$ powder dispersed in a mixed solution of ethanol and toluene (1:1) using menhaden fish oil (2 wt% to LANTP) as a dispersant. Then the mixed slurry was ball-milled at 400 rpm using a ZrO_2 vessel and ZrO_2 ball for 12h with HEMM. Then, the Polyvinyl butyl alcohol (8 wt% to LANTP) as a binder and butyl benzyl phthalate (7 wt% to LANTP) as a plasticizer was poured into the mixed slurry and ball-milled using the HEMM at 150 rpm for another 12 h. Then the slurry was be removed air bubbles by using a planetary vacuum mixer (Thinky, Japan) for 9 min. After tape casting, the green sheets were kept in a sealed box with a small amount of ethanol in a refrigerator to slowly dry at 5°C for 24 h. Several green sheets were hot pressed at 90°C for 10 min and then sintered at 900°C for 7 h. The diameter of obtained pellets with thickness of approximately 0.3 mm is 7 mm.

The phases analysis was operated by a Rigaku Rint 2500 diffractometer with Cu Kα radiation in the 2θ range from 10° to 90° at a scanning step rate of 0.02°/s as X-ray diffraction (XRD) analysis. The samples both side were sputtered a thin gold layer as electrodes, then the electrical conductivity was measured by using an impedance phase analyzer in the frequency range of 0.1 Hz- 1 MHz with the bias voltage at 10mV. The bulk and grain boundray conductivity of samples were calculated from complex plots using Zview 2. Three-point bending strength of LANTP samples was measured at room temperature using a materials tester (Shimadzu EX-SX 500 N). Then the dates of the samples were obtained by using two method of tape casting and press method were compared and discussed.

3. Results and Discussion

Figure.1 shows the XRD patterns of the $Li_{1.3}Al_{0.5}Ti_{1.3}Nb_{0.2}(PO_4)_3$ pellets made with two methods of press method and tape casting and sintered in the temperature 900°C for 7h. The two samples prepared by using two methods were compared about the impurities phases and the lattice constants, and the analysis indicates the two samples of the NASICON-type structure $Li_{1.3}Al_{0.5}Ti_{1.3}Nb_{0.2}(PO_4)_3$ without impurity phases such as $AlPO_4$, Li_3PO_4, and monoclinic NASICON-type $Li_2AlTi(PO_4)_3$. Previously, Shang reported the XRD patterns of the $L_{i1+x-y}Al_xNb_yTi_{2-x-y}(PO_4)_3$(x=0-0.6, y=0-0.3) systems by press method and the impurity

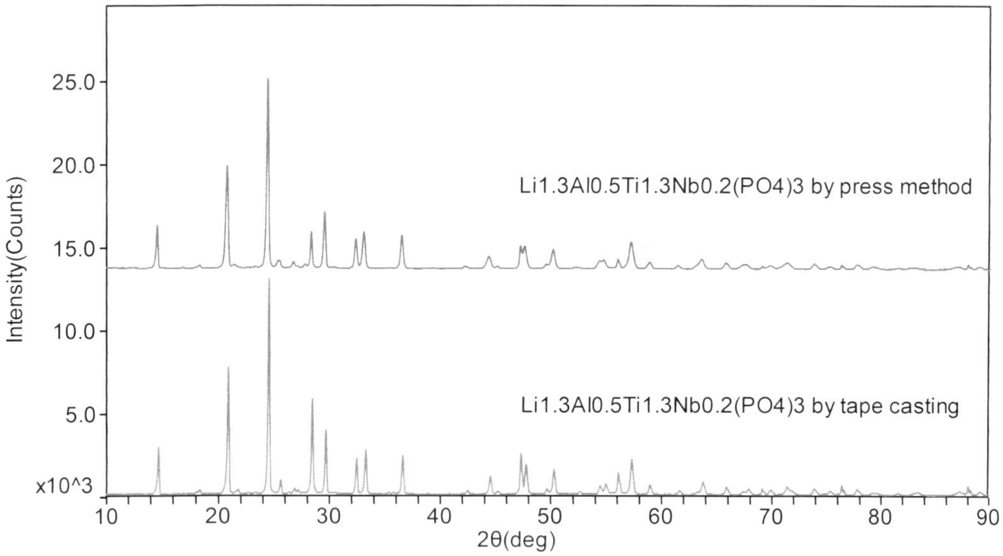

Figure 1. XRD patterns for $Li_{1.3}Al_{0.5}Ti_{1.3}Nb_{0.2}(PO_4)_3$ by two methods sintered at 900 °C for 7 h.

phases were not detected in the XRD patterns of $Li_{1.3}Al_{0.5}Ti_{1.3}Nb_{0.2}(PO_4)_3$, but the impurity phases of $Li_2AlTi(PO_4)_3$ and $Al(PO_4)_3$ were observed in the other type $Li_{1+x-y}Al_xNb_yTi_{2-x-y}(PO_4)_3$ [18]. The lattice parameters calculated by the hexagonal unit cell of $Li_{1.3}Al_{0.5}Ti_{1.3}Nb_{0.2}(PO_4)_3$ prepared with two methods are same and there are a=0.8513 Å and c=2.0878 Å. The relative densities of $Li_{1.3}Al_{0.5}Ti_{1.3}Nb_{0.2}(PO_4)_3$ pellets sintered at 900°C for 7h were calculated from the ratio of theoretical densities and actual measurement densities, the relative density of 96% was obtained for $Li_{1.3}Al_{0.5}Ti_{1.3}Nb_{0.2}(PO_4)_3$ pellet synthesized by using press method, which is larger than the relative density of 90% for $Li_{1.3}Al_{0.5}Ti_{1.3}Nb_{0.2}(PO_4)_3$ pellet synthesized by using tape casting. The two different manufacture processes have insignificant effect for the lattice parameters, but the relative density is related with two different methods.

Figure.2 shows the impedance profiles of samples for $Li_{1.3}Al_{0.5}Ti_{1.3}Nb_{0.2}(PO_4)_3$ pellets synthesi zed with two methods of tape casting and press method. Only one large semicircle was observed and a straight line follow it. The complex impedance plots show similar behavior with LATP. T he intercept of the semicircle on the real axis at high frequency represents the bulk resistance, an d the diameter of semicircle in high frequency range was attributed to the grain boundary resistan ce [7]. The total conductivity of $Li_{1.3}Al_{0.5}Ti_{1.3}Nb_{0.2}(PO_4)_3$ is the maximum compared with other p atterns of $Li_{1+x-y}Al_xNb_yTi_{2-x-y}(PO_4)_3$ [18]. The highest grain boundary conductivity and bulk cond uctivity of $Li_{1.3}Al_{0.5}Ti_{1.3}Nb_{0.2}(PO_4)_3$ prepared with press method sintered at 900°C for 7h were fo und, which are 1.35×10^{-3} S/cm and 7.1×10^{-4} S/cm at 25 °C, respectively. The lager grain boundar y conductivity and bulk conductivity of sample prepared with tape casting method are 2.38×10^{-3} S/cm and 1.08×10-3 S/cm at 25 °C, the total conductivity of 7.4×10-4

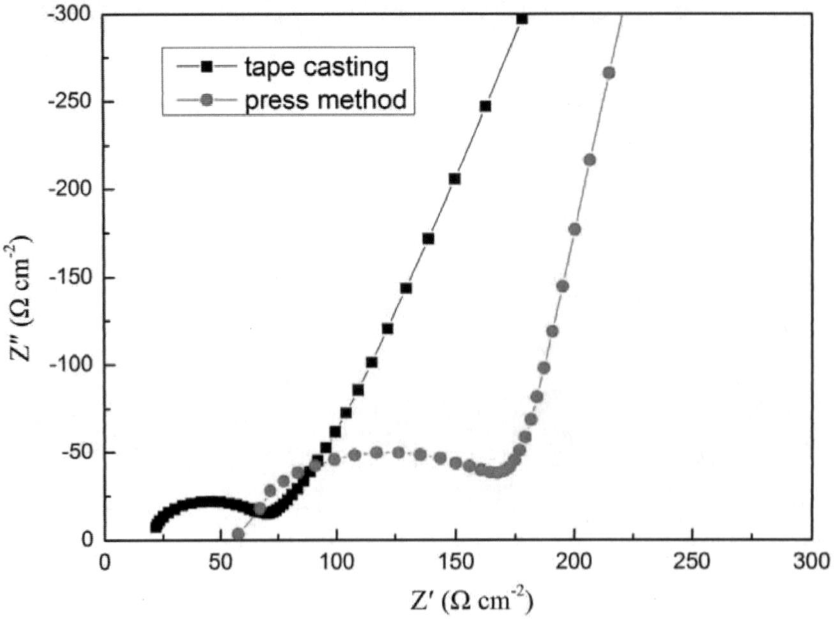

Figure 2. Impedance profiles of Au/$Li_{1.3}Al_{0.5}Ti_{1.3}Nb_{0.2}(PO_4)_3$/Au by two methods measured at 25 °C

S/cm is also larger than the total conductivity of 4.65×10^{-4} S/cm for sample made by using press method. The larger grain boundary conductivity of samples made by using tape casting method may be contributed to the particle distribution of sample more uniform than samples made by using press method. The total conductivities of two samples for $Li_{1.3}Al_{0.5}Ti_{1.3}Nb_{0.2}(PO_4)_3$ are higher than $Li_{1.3}Al_{0.3}Ti_{1.7}(PO_4)_3$($2.3\times10^{-4}$ S/cm at 25 °C) [19]. The dates of three-point bending strength for two samples prepared by using press method and tape casting are 50 N mm^{-2} and 90 N mm^{-2}, respectively.

4. Conclusion

The precursor of NASICON-type lithium-ion conducting solid electrolyte for $Li_{1.3}Al_{0.5}Ti_{1.3}Nb_{0.2}(PO_4)_3$ was prepared by using solid-state reaction, the sintered pellets were fabricated at 900°C for 7 h by using two method of press method and tape casting. Two samples were the purity phase by XRD pattern analysis, the a lattice parameter and c lattice parameter are all 0.8513 Å and 2.0878 Å. The relative densities of samples made with press method and tape casting are 96% and 90%, respectively. The conductivity of 7.4×10^{-4} S/cm for samples made by using tape casting is larger than the conductivity of 4.65×10^{-4} S/cm for sample made by using press method. The three-point bending strengths are 50 N mm^{-2} and 90 N mm^{-2}, respectively. The NASICON-type lithium-ion conducting structured solid electrolyte of LATP by partially adding pentavalent Nb ion can increase the total conductivity. The two samples all have highlight performance for promising candidate of solid electrolyte.

Acknowledgment

The authors gratefully acknowledge the financial supports from the National Natural Science Foundation of China (11304125) and the Research Foundation for Advanced Talents, Jiangsu University (10JDG077).

References

[1]Armand M and Tarascon J M, 2008, *Nature* 451-652.
[2]Asachi G, Imanaka N and Aono H, 1996, *Adv Mater* 8 No.2 127-35.
[3]Goodenough J B, Hong H Y-P and Kafalas J A, 1976, *Mat. Res. Bull* 11 203-20.

IOP Conf. Series: Materials Science and Engineering **242** (2017) 012046 doi:10.1088/1757-899X/242/1/012046

[4]Martinez-Juarez A, Pecharroman C, Iglesias J E and Rojo J M, 1998, *J. Phys. Chem.* B 102 372-75.
[5]Aono H, Sugimoto E, Sadaoka Y, Imanaka N and Adachi G, 1990, *J. Electrochem. Soc* No.4 Vol.137 1023-27.
[6]Aono H, Sugimoto E, Sadaoka Y, Imanaka N and Adachi G, 1990, *Chem. Lett* 1825-28.
[7]J Fu, 1996, *Solid State Ionics* 96 195-200.
[8]Thokchom J S and Kumar B, 2006, *Solid state Ionics* 177 727-32.
[9]Yoon Y, Kim J, Park C and Shin D, 2013, *J. Ceram Process Res* Vol.14 No.4 563-66.
[10] Ma F, Zhao E, Zhu S, Yan W, Sun D, Jin Y and Nan C, 2016, *Solid State Ionics*, 29 7-12.
[11] Kazakevicius E, Urcinskas A, Keziionis A, Dindune A, Kanepe Z and Ronis J, 2006, *Electrochimica Acta* 51 6199-202.
[12] Imanishi N, Zhang T, Shimonishi Y, Hasegawa S, Hirano A, Takeda Y, Yamamoto O and Sammas N, 2010, *Chem. Commun.* 46 1661-3.
[13] Takahashi K, Johnson P, Imanishi N, Sammes N, Takeda Y and Yamamoto O, 2012, *J.Electrochem.Soc* 159(7) A1065-69.
[14] Zhang P, Matsui M, Hirano A, Takeda Y, Yamamoto O and Imanishi N, 2013, *Solid State Ionics* 253 175-80.
[15] Zhang P, Matsui M, Takeda Y, Yamamoto O and Imanishi N, 2014, *Solid State Ionics,* 263 27-32.
[16] Shang X, Nemori H, Mitsuoka S, Xu P, Matsui M, Takeda Y, Yamamoto O and Imanishi N, 2016, *Frontiers in Energy Research* Vol.4 Article.12.
[17] Zhang P, Wang H, Lee Y, Matsui M, Takeda Y, Yamamoto O and Imanishi N 2015 *J.Electrochem. Soc* 162 A1265-71.
[18] Shang X, Nemori H, Mitsuoka S, Matsui M, Takeda Y, Yamamoto O and Imanishi N, 2016, *Solid State Ionics* 297 43-8.
[19] Xu X, Wen Z, Wu J and Yang X, 2007, *Solid State Ionics* 178 29-34.

IOP Conf. Series: Materials Science and Engineering **242** (2017) 012047 doi:10.1088/1757-899X/242/1/012047

Study on bonded slip of reinforced concrete based on extended finite element

Hailin Wu, Fubing Cui, Liangcai Zhu, Zhiqiang Zheng

College of Hydraulic & Environmental Engineering, Three Gorges University, Yichang 443002，China

wuhailinwhl@163.com

Abstract. In this paper, for the tensile test of reinforced concrete specimens, the numerical simulation model is established to simulate the whole process of concrete specimen from crack generation and extension to final fracture based on the extended finite element method. Based on the numerical simulation results and the tensile test results of concrete specimens, the differences of the numerical simulation results with the bond slip relationship and those without the bond slip relationship are analysed. The results show that the numerical analysis results of bearing ratio of steel bar with bond slip relationship are close to those of the test, and its numerical analysis results of crack width are also closer to the measured values of the test. And there is great deviation between the numerical analysis results without bond slip relationship and the test results, at the same time, the feasibility of solving the continuous - discontinuous problem based on extended finite element method is verified.

1. Introduction

The bond and slip characteristics between reinforced concrete and concrete are one of the hot issues in the numerical simulation of reinforced concrete structures. With the continuous development of numerical methods, numerical simulation has become an important mean to study the reinforced concrete structure. For a long time, the simulation of the bond relationship between the reinforcement and the concrete is often simplified, in other words, both are considered to be well bonded [1]. In the study on numerical simulation of reinforced concrete, it is very few to take both the cracking of concrete and the bond slip relationship between the reinforcement and the concrete into account. Therefore, based on the extended finite element method, the numerical analysis model is established to simulate the bearing process of reinforced concrete specimen, the stress and the crack width of the reinforced concrete specimen are compared with the results of the test, and the differences of the results with the bond slip relationship and those without the bond slip relationship are analysed, at the same time, the feasibility of solving the continuous - discontinuous problem based on extended finite element method is verified.

2. Bond slip characteristics of reinforced concrete

In the reinforced concrete components, these two materials of different properties can be effectively combined to force together. The main reason is that reinforced concrete has a bond slip characteristics, and it produces cohesive force between the reinforcement and the concrete. Through the interaction, the mutual transmission of force and compatible deformation can occur between the two materials.

Content from this work may be used under the terms of the Creative Commons Attribution 3.0 licence. Any further distribution of this work must maintain attribution to the author(s) and the title of the work, journal citation and DOI.
Published under licence by IOP Publishing Ltd

2.1. bond-slip constitutive relations

The bond-slip constitutive relation used in this paper is the bond-slip relationship proposed by CEB-FIP [2]. The constitutive relation is expressed as follows

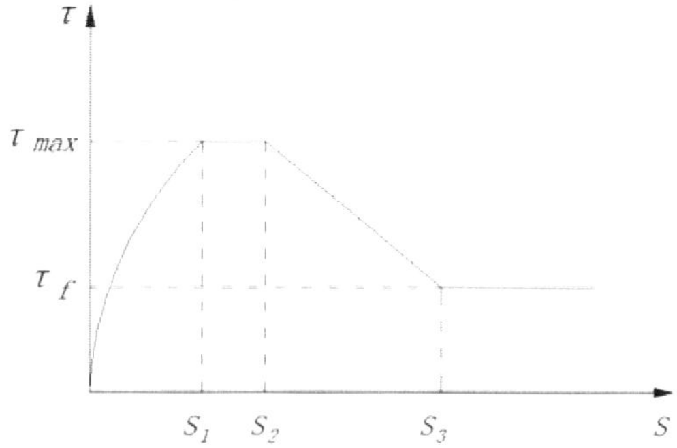

Figure 1. Bonded slip constitutive relation of reinforced concrete.

$$\begin{cases} \text{Nonlinear section: } \tau=\tau_{max}\left(\frac{s}{s_1}\right)^{\alpha} \ ; 0\leq s\leq s_1 \\ \text{Platform section: } \tau=\tau_{max} \ ; \ s_1<s\leq s_2 \\ \text{Descent section: } \tau=\tau_{max}-(\tau_{max}-\tau_f)\left(\frac{s-s_2}{s_3-s_2}\right) \ ; \ s_2\leq s\leq s_3 \\ \text{Residual segment: } \tau=\tau_f \ ; \ s_3<s \end{cases} \tag{1}$$

Where. τ is the bond stress between the concrete and steel bar; s is the bond-slip between the concrete and steel bar; τ_f is the bond residual stress between the concrete and steel bar.

In consideration of the bond between concrete and steel bar is better in test specimen, the relevant parameters are as follows:

$s_1=s_2=0.6mm$、$s_3=1mm$、$\alpha=0.4mm$、$\tau_f=0.15\tau_{max}$、$\tau_{max}=2.0\sqrt{f_c}$ 、$f_c=0.8f_{cu}$、f_{cu} is the compressive strength of the concrete cube.

2.2. Bond slip simulation method

In the extended finite element simulation of reinforced concrete structures, the way of establishing models of steel bar can be divided into three types: integral type model, composite type model and separate type model [3]. In the monolithic model, the reinforcement is dispersed throughout the elements, and the element is treated as a continuous and uniform material. The contribution of steel bars in the whole structure is reflected by adjusting the mechanical properties of the elements. This method can only get the overall results of the structure under load, and the effect of reinforcement and concrete cannot be obtained separately. When it is assumed that the bond between the reinforcement and the concrete in the model is intact and there is no slip between the two, the modular model will be used. In a separate model, the reinforcement and concrete are treated as different units, and they are divided into different units separately. In the plane, the concrete can be divided into triangular or quadrilateral elements, steel can also be divided into triangular or quadrilateral unit. Taking the reinforced as an elongated material into account, we usually ignore its lateral shear effect, so that the steel can be regarded as a linear unit. This method can greatly reduce the number of units, and can avoid the interface in the steel and concrete interface with a lot of transition unit due to the unit is too dense [4]. Therefore, when using the extended finite element method to analyse the bond slip relationship, we should use the separate type model.

When the separate type model is used, the reinforcement and the concrete in the component are mutually constrained and produce a relative slip after the member is stressed. In order to simulate this effect, it is necessary to add a coupling element in the reinforcement and concrete. The coupling unit includes spring unit [5], no thickness quadrilateral bond unit [6], bond area unit [7]. In this paper, spring element is used to simulate the bond slip behavior between steel bar and concrete.

3. Study on Simulation of Bonded Slip of Reinforcement and Concrete

In this paper, the extended finite element method is used to simulate the axial tensile test of reinforced concrete specimen [8] which has been completed by our team. The size of the specimen is shown in the figure below, and a reinforcement which is a diameter of 10 mm is located at the center of the section.

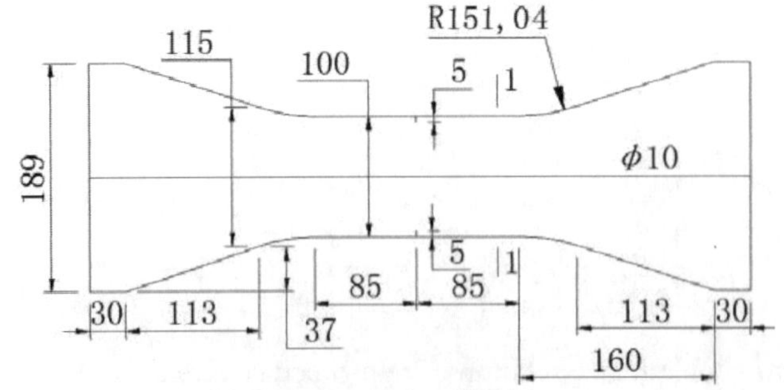

(a) Specimen size diagram.

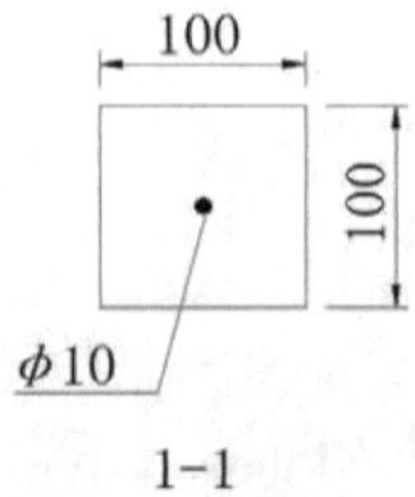

(b) Reinforcement diagram.

Figure 2. Specimen diagram.

Concrete strength grade is C25, and steel strength grade is HRB400 under test. Concrete elastic modulus, ultimate tensile strength and fracture energy are the experimental measured values. The material parameters are shown in the following table.

Table 1. Material parameters of reinforced concrete specimens.

Name	Bulk density (KN/m3)	Elastic modulus (GPa)	Poisson ratio	Ultimate tensile strength (MPa)	Fracture energy (N/m)
Steel bar	78.0	200	0.3	—	—
Concrete	25.0	28.59	0.167	1.54	116.3

3.1. A brief introduction of models

In this paper, the models are established by the separate type model, and reinforcement and concrete are divided into grids separately. The cracking criterion of concrete is the maximum principal stress criterion. The concrete model adopts the eight-node isoparametric element, and steel bar model adopts two-node bar element. The specimen model has a total of 6485 elements, including 6396 concrete elements and 89 steel bar elements.

(a) Concrete grid.

(b) Steel grid.

Figure 3. Reinforced concrete specimen grid.

During the test, the specimen was clamped by the fixture. When the clamp is clamped, the displacement of the clamping end of the specimen is almost the same as that of the clamp. In order to simulate the constraint and force of the specimen, two reference points are established in the model and a coupling constraint is established between the clamping end and the reference point. At the side of the reference point, another reference point is applied a full constraint, and at the other side, concentrated force load is applied, as shown in figure 4. In order to compare the numerical analysis results and test results, the concentrated force is step loading.

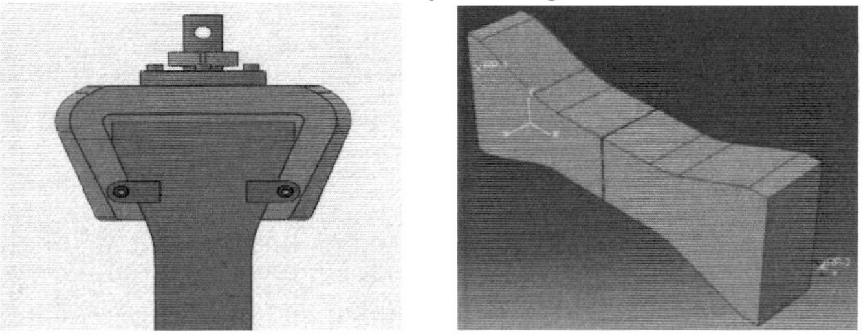

Figure 4. Boundary condition of reinforced concrete specimen.

In this paper, under the condition of not including the bond slip relationship of reinforced concrete, the steel bar is embedded in the concrete. According to the elastic modulus of steel and concrete and cross-sectional size to share the quantity of the load, and it makes them coordinated deformation. But, under the condition of including the bond slip relationship of reinforced concrete, the three-spring model is established, and the three springs elements which are orthogonal to each other are built between corresponding reinforcement node and concrete node. The spring which is perpendicular to the axis of the steel bar is used to simulate the vertical force between the reinforcement and the concrete. Since the mutual compression between the reinforcement and the concrete is not taken into account in this paper, its stiffness take the maximum value is 1.0×10^9N/m; and the spring which is parallel to the axial direction of reinforcement simulate the bonded slip relationship between the reinforcement and concrete, and the relationship is decided by formula 1.

3.2. Bearing Analysis of Reinforcement in Reinforced Concrete Specimen

The results of steel stress of the specimen model under the loading are shown in the following figure, and the feasibility of the numerical simulation method is analyzed by comparing the steel stress of the model with that of the test under similar load.

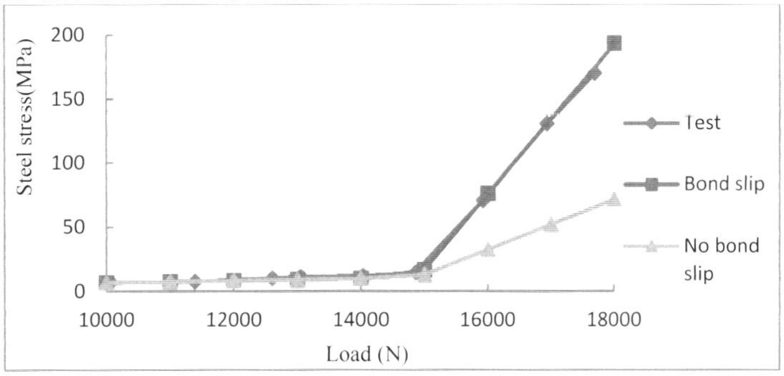

Figure 5. The curve of steel stress and load.

From the above figure, the results of steel stress of the model including bonded slip, the model excluding bonded slip and the test are close under different grade before crack. After the concrete cracking, the results of the model including bonded slip and those of the test are approximately equal. However, there is a great deviation between the numerical simulation results without bond slip and the test.

Table 2. Bearing ration of steel bar (%).

	Numerical simulation		Test	
Load (N)	Bearing ratio of steel bar with bond slip relationship	Bearing ratio of steel bar without bond slip relationship	Load (N)	Bearing ratio of steel bar
18000	71.3	26.5	17682	63.9
17000	52.4	20.4	16938	51.4
16000	31.5	13.5	15924	29.5
15000	7.3	5.7	14855	6.5
14000	4.6	4.6	14030	5.7
13000	4.5	4.6	13045	5.8
12000	4.5	4.5	12599	5.4
11000	4.5	4.5	11391	4.4
10000	4.5	4.5	10066	4.6

From the above figure, the bearing ratio of steel bar of the model with bonded slip, the model without bonded slip and the test are close under different grade before crack. After the concrete cracking, the results of the model including bonded slip and those of the test are approximately equal. However, there is a great deviation between the numerical simulation results without bond slip and the test. Before the cracking of the concrete, the concrete in the specimen takes most of the load. When the tensile load of the specimen is increased to 15000N, the cracks are generated at both sides of the concrete specimen. As the load increases, the crack gradually extends to the axial of the specimen. The quantity of bearing of the specimen is gradually reduced, and the numerical analysis results with bond slip are in accordance with those of the test. Therefore, the numerical results with bond slip are closer to those of the test.

3.3. Analysis of Crack Width of Reinforced Concrete Specimen

In this paper, the crack width of reinforced concrete specimen under different bearing capacity is studied, and the calculated value, numerical analysis value and the measured value of crack width are compared and analyzed. Since the model is only subjected to axial tensile load, it is type I cracking. The numerical analysis values are obtained by using the displacement difference between nodes on both sides of the crack at the cracking unit, the formula calculated using the "concrete structure design specifications" (GB 50010-2010) in the calculation of the maximum crack width formula, The formula is as follows:

$$W_{max}=a_{cr}\varphi\frac{\sigma_s}{E_s}(1.9c_s+0.08\frac{d_{eq}}{\rho_{te}}) \quad (2)$$

$$\varphi=1.1-0.65\frac{f_{tk}}{\rho_{te}\sigma_s} \quad (3)$$

Where, W_{max} is the maximum crack width, mm; a_{cr} is component force characteristic coefficient; σ_S is steel stress, N/mm^2; E_s is elastic modulus of steel bars, N/mm^2; c_s is the distance from the outer edge of the outermost longitudinal reinforcement to the bottom of the tension zone, mm; d_{eq} is equivalent diameter of longitudinal reinforcement in tension zone, mm; ρ_{te} is ratio of reinforcement.

The calculated values of crack width, numerical analysis and shaft tensile test of reinforced concrete specimen are shown in the following table.

Table 3. Crack width (mm).

Formula calculated value	Numerical value		The measured value
	Bonding slip	No bonding slip	
0.203	0.096	0.019	0.120
0.067	0.066	0.014	0.098
0.032	0.035	0.008	0.051
0.007	0.005	0.002	0.000

As can be seen from the above table data, the results of numerical analysis of crack width considering bond slip are closer to those of test and formula calculation value. The measured value of the crack width is greater than the calculated value of the bond slip and the calculated value of the formula, and the formula calculation is the smallest. Most of the fractal width numerical analysis value deviation is smaller than the calculated value of the formula. Therefore, it can be concluded from the table that the numerical value of crack width is closer to the experimental value.

4. Conclusions

In this paper, based on the extended finite element method, the numerical model of the tensile test of the reinforced concrete specimen is established to simulate the specimen from the crack to the specimen fracture damage under the action of axial tensile load. And compared with the test results, the differences of the results with the bond slip relationship and those without the bond slip relationship are analysed, and the reliability and feasibility of the numerical calculation method are verified, the results show that:

1. Before the cracking of the concrete, the steel bearing ratio of including the bond slip, excluding the bond slip, and the model test under the different level of loading is close. However, after the cracking of the concrete, the results of steel bearing ratio of numerical analysis with the bond slip relationship are close to the results of the tensile test, and there is great deviation between the numerical analysis result without bond slip relationship and the test result.

2. Before the concrete cracking, the concrete in the specimen takes up most of the load. When the tensile load of the specimen is increased to 15000N, the cracks are generated at both sides of the concrete specimen, and as the load increases, the cracks gradually expand toward the central position. Then, concrete unloading slowly, the bearing ratio is also gradually reduced, the specimen bearing gradually reduced by the steel to bear, the numerical analysis results with bond slip relationship is closer to the results of the test.

3. The numerical analysis results of the crack width with the bond slip are closer to the measured results of test and the calculated results. The relationship of the crack width of the calculated results of the formula, the numerical analysis results with the bond slip and the measured values is both the calculated results and the numerical analysis results with bond slip relationship are close to the measured value of the test, and the numerical analysis results is closer to the measured value of test.

Acknowledgements

This work was financially supported by the National Natural Science Foundation of China (No. 51379107)

References

[1]Su Xiaoyong 2007 Nonlinear Computer Simulation for the Reinforced Concrete Structure (Sichuan: Southwest Jiaotong University)

[2]Yang Tao,Zou Daoqin 2013 Numerical Simulation of Crack Growth of Reinforced Concrete Beam based on XFEM vol 47 (Zhejiang: Journal of Zhejiang University) pp 495-501

[3]Chen Jianping,Xu Xunqian and Bao Hua 2004 Research Status of the Bond-Slip between

Reinforcing Steel Bar and Concrete vol 3 (Jiang Su: Journal of Nantong Institute of Technology) pp 51-56

[4]Chen Qiang 2011 Numerical Simulation on Bond-slip Relationship of Reinforced Concrete and Crack in Reinforced Concrete Beam (Zhejiang: Zhejiang University)

[5]Dong Zheren 1998 Design and Nonlinear Analysis of Steel Lined Reinforced Concrete Penstock (Beijing: China Water Power Press)

[6]Shen Jumin,Wang Chuanzhi and Jiang Jianjing 1993 Finite Element and Plate Shell Limit Analysis of Reinforced Concrete (Beijing: Tsinghua University Press)

[7]Lu Xilin,Jin Guofang and Wu Xiaohan 1996 Nonlinear Finite Element Theory and Its Application of Reinforced Concrete Structure (Shanghai: Tongji University Press)

[8]WU Hailin, RAN Hongzhou and WU Long 2015 Research on the Methods and Devices for Uniaxial Tension Test of Reinforced Concrete vol 15 (Beijing: Science Technology and Engineering) pp 275-278

IOP Conf. Series: Materials Science and Engineering **242** (2017) 012048 doi:10.1088/1757-899X/242/1/012048

Model test research on effect of flow accelerating-board in a pumping station

Yuan Ding[1], TongChun Li[1&2,*], XiaoQing Liu[1], Yun Guo[3], MinZhe Zhou[1]

[1]College of Water Conservancy and Hydropower, Hohai University, NO.1 Xikang Road, Gulou District Nanjing, China

[2]National Engineering Research Center of Water Resources Efficient Utilization and Engineering Safety, NO.1 Xikang Road, Gulou District, Nanjing, China

[3]Shenzhen Longhua district environmental protection and water resources bureau, Room 119A, Unit 3, Guohong Building, NO.98 Meilong Road, LongHua District, Shenzhen, China

*ltchhu@163.com

Abstract. Generally, the sedimentation in the forebay of pumping station may result in bad flow patterns, which will decrease efficiency of pump device and cause the vibration of pump house and units, or other safety problems. To research the improvement of this impact in an actual project, a physical model test was established for the original scheme of one pumping station. One part of results show that the flow velocity in the channel of regulating-pool is low under the high-water level condition, and it's easy to cause the sedimentation in the regulating-pool. According to this problem, we propose a flow accelerating-board scheme for the regulating-pool. The final results show that this scheme could effectively increase the flow velocity at the bottom and reduce the sedimentation in the regulating-pool. Although the hydraulic loss of regulating-pool increased, it could be able to satisfy the design requirements.

1. Introduction

The forebay of pumping station is an important hydraulic structure which connects the channel and the inlet channel. In order to provide good flow conditions for the water pump, the flow in the forebay should be such as that the velocity distribution is uniform, the water flow is smooth and there are no vortices in the forebay[1][2][3]. The actual observation and later theoretical studies have shown that the bad forebay caused recirculation, vortex, disturbance of water intake and deterioration of water pump suction conditions, the efficiency of the pump units decreased, the vibration of the pumping station and the units start-up difficulties, cavitation and noise[4]. In the pumping station with large sediment concentration, the adverse hydraulic conditions in forebay can also cause the sedimentation or erosion at the bottom of forebay. Therefore, it is important to improve the efficiency of pump and prevent the sedimentation by improving the flow patterns in the forebay of pumping station through some necessary measurement[5].

In this paper, there is a regulation-pool in front of the forebay, the partition wall in the regulation-pool is setting up to guide the water to enter the forebay easily. The flow velocity in the channel flow is low under the high water level condition, and it's easy to cause the sedimentation in the regulating-pool. So, it's necessary to make further modification of the regulating-pool under the condition of ensuring the head loss to meet the design requirements, to increase the flow velocity at the bottom of the channel

Content from this work may be used under the terms of the Creative Commons Attribution 3.0 licence. Any further distribution of this work must maintain attribution to the author(s) and the title of the work, journal citation and DOI.
Published under licence by IOP Publishing Ltd

IOP Conf. Series: Materials Science and Engineering **242** (2017) 012048 doi:10.1088/1757-899X/242/1/012048

flow and inhibit the sedimentation in the regulating-pool. According to the model test results of the pumping station, the scheme of adding the flow accelerating-board in the regulating-pool was put forward to verify the feasibility of increasing flow velocity at the bottom of the channel flow and suppressing the sedimentation in the regulating-pool.

2. Model foundation

2.1. General situation of the pumping station

A pumping station is an important part of the upper reaches Huangpu River water source connected pipe project. The original design: through two water pipes to supply water to two regulating-pools, the effective volume of a single pool is about 10000m3, the elevation of the bottom is 1.00m (the elevation of the ground outside the pumping station is 4.50m), set up partition wall in the regulating-pool. Each regulating-pool corresponding to a forebay, and supply water to four pumps. Intake sump is setting up to each pump alone which is separated by partition wall, the width and length of intake sump are 5.0m and 17.8m respectively, the elevation of the Intake sump bottom is 7.00m, the intake sump link up with the regulating-pool through inclined floor of the forebay. The planar graph of the pumping station is shown in Fig.1.

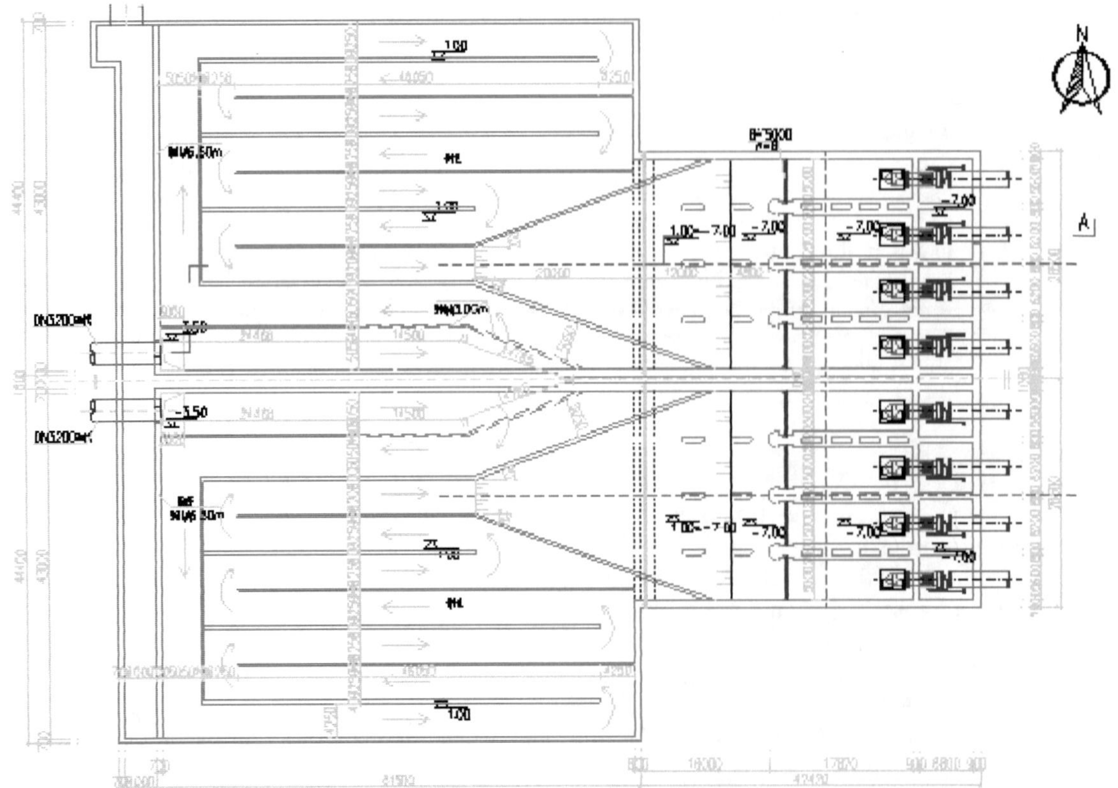

Fig.1 The planar graph of the pumping station

2.2. Model design

The physical model of the pump station adopts the whole normal hydraulic model, and it's designed according to gravity similarity criterion. Considering the requirements of the flow at drag square area and the choice of the model pump, choose $\lambda_l = 10$ as the linear scale for the model. The similarity scales of other physical quantities are shown in Table 1.

Table 1. The scale of the model.

Gradient scale	Angle scale	Velocity scale	Flow scale	Roughness scale
1	1	3.16	316.23	1.47

The forebay and intake sump of the pumping station are made of concrete, and the roughness is 0.013～0.014, the roughness of the model is 0.009～0.010. The sidewall of the model is made of pure cement, and the roughness is 0.010. The regulating-pool and intake sump of the model are made of organic glass and plastic, and the roughness is about 0.009. The roughness of the model generally satisfies the similar requirements of roughness.

3. Analysis of test results.

3.1. The original scheme.

The original scheme selects designed operating conditions of pumping station as the test condition （the water supply scale of the pumping station is 2.4 million m3 / d）, six units are running, and the flow of each pump is 4.954m^3 / s. The test conditions are shown in Table 2.

Table 2. The test conditions.

Condition	The water level in the forebay (m)	The number of running units	The flow of forebay (m^3/s)	Remarks
1	3.80		14.862	Designed normal water level
2	1.80	3	14.862	Designed lowest water level
3	5.80		14.862	Designed highest water level

The test results of condition 1 show that the water flow from the regulating-pool into the forebay has to turn 9 times, due to the large number of channel flow and the narrow width, the flow velocity in the channel flow is large and the distribution is uneven, the main flow is on one side, and the local hydraulic loss is large at the corner. There is a very clear flow separation and a long narrow recirculation zone at the corner. In the regulating-pool, the partition wall area which is connected with the diffusion section of the forebay has a large range of retention region due to the large flow area. The flow patterns in the regulating-pool of condition 1 is shown in Fig.2, the velocity distribution in the regulating-pool is shown in Fig.3.

Fig.2 The flow patterns in the regulating-pool of condition 1

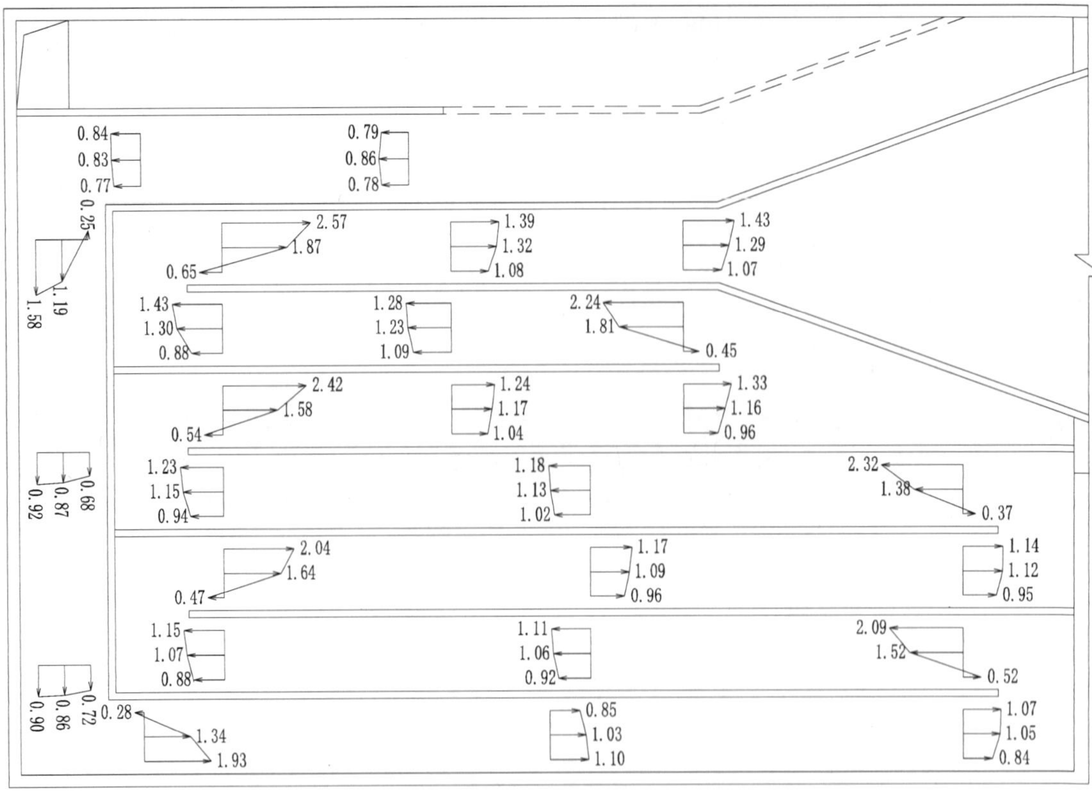

Fig.3 The velocity distribution in the regulating-pool of condition 1

In the test of condition 2, it is observed that when the minimum water level is 1.80m in the forebay, the flow velocity in the channel flow is large due to the shallow water depth at the forebay intake.

The test results of condition 3 show that the hydraulic loss of the regulating-pool is reduced due to the increase of the water depth and the decrease of the flow velocity. The flow patterns in the regulating-pool and forebay is similar to condition 1.

The test results of the original scheme show that there is a large range of retention region at the corner, the distribution of the flow velocity in the channel flow is uneven, the main flow is on one side. It's easy to cause the sedimentation in regulating-pool at the partition wall area which is connected with the diffusion section of the forebay due to the large flow area.

3.2. The flow accelerating-board scheme.

When the pumping station is running at the high water level, the flow velocity in the regulating-pool is significantly reduced. When the pumping station is running 3 units, the average flow velocity in the channel flow is about 0.34m / s, which is easy to cause the sedimentation in regulating-pool. Therefore, it is necessary to make further modification of the regulating pool under the condition of ensuring the hydraulic loss to meet the design requirements, to increase the flow velocity at the bottom of the channel flow and inhibit the sedimentation in regulating-pool.

The flow accelerating-board scheme is about to add boards in the regulating-pool, a expected bottom flow velocity can be obtained by arranging a sufficient number of boards in the channel flows and selecting a suitable bottom elevation. At the same time, the addition of flow accelerating-board will increase the local hydraulic loss in regulating-pool, the more number of the boards and the lower bottom elevation will increase the hydraulic loss. In the test, control the largest water surface falls is 0.20m, through the different orifice height, different number of boards and different layout, multi-program comparison, get a more reasonable layout of the flow accelerating-board. The layout of the flow accelerating-board in the regulating-pool is shown in Fig.4 and Fig.5 (The elevation of the bottom of each board is 3.50m).

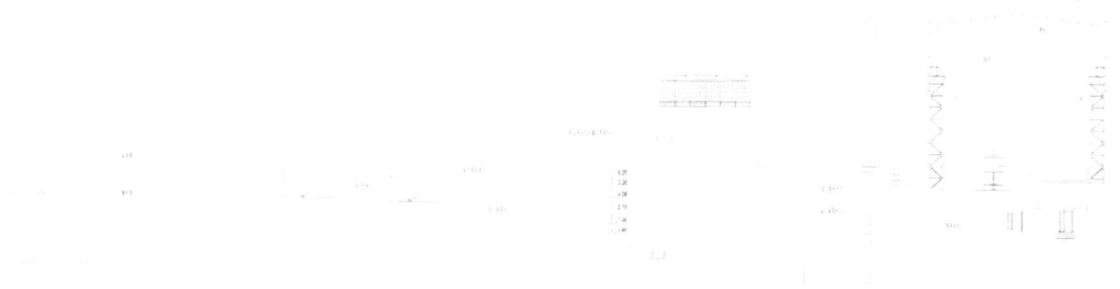

Fig.4 The sectional drawing of the layout of the flow accelerating-board in the regulating-pool

Fig.5 The planar graph of the layout of the flow accelerating-board in the regulating-pool

The flow accelerating-board scheme is plan to solve the sedimentation problem in the regulating-pool under the high water level operating condition, so we choose condition 3 to test. The results of the test show that when the boards are set up, the upstream flow shrink and dive before the board, and the bottom flow velocity gradually increased. In the area 5m downstream of the board, the main flow is in the lower part of the channel flow. As the increase of distance from the board, the main flow gradually spread up from the bottom, the bottom flow velocity has decreased and the upper flow velocity has increased. Until get to the next board, the upstream flow shrink and dive again, and the bottom flow velocity increased.

The velocity distribution in the regulating-pool of condition 3 is shown in Fig.6~Fig.8, It can be seen from the figures that the flow velocity at the bottom of each channel flow is large, and the flow velocity at the bottom of the channel flow is 0.45 m / s ~ 0.60m / s. The velocity is larger than the velocity without the flow accelerating-board (about 0.34 m / s), but the average velocity in the channel flow is slightly lower than the velocity under the designed normal water level condition.

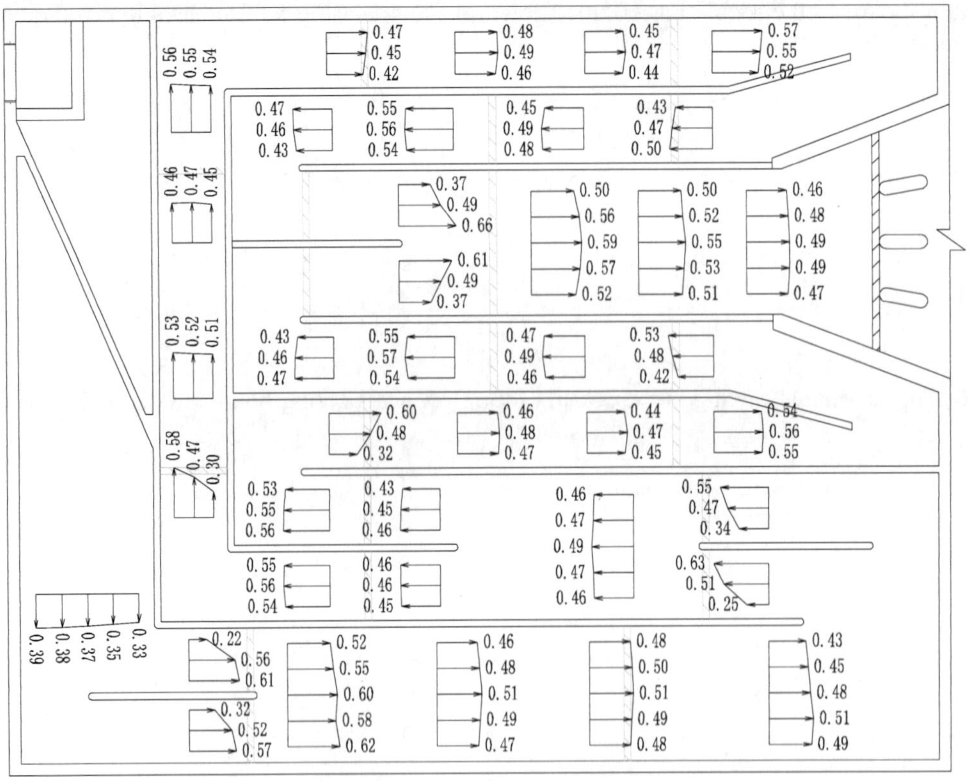

Fig.6 The velocity distribution at the bottom of condition 3

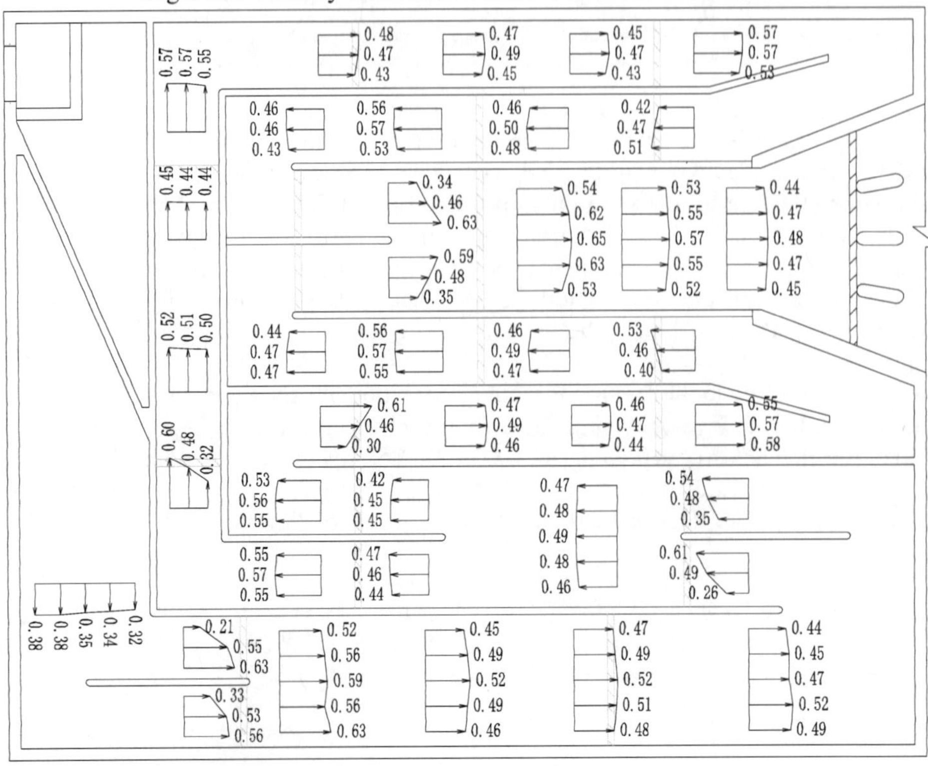

Fig.7 The velocity distribution at the surface 1.5m distance from the bottom of condition 3

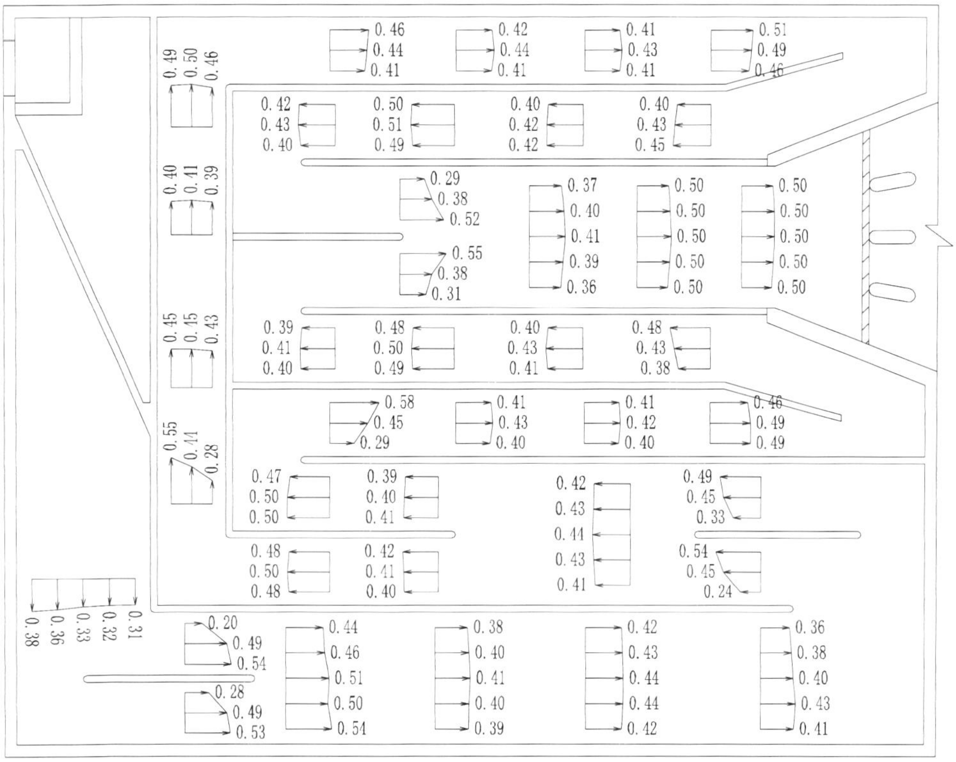

Fig.8 The velocity distribution at the surface 3m distance from the bottom of condition 3

After setting the flow accelerating-board in the regulating-pool, because of the increase of the velocity at the bottom of the channel flow, the frictional hydraulic loss and the local hydraulic loss of the channel flow increase. The water surface has fallen 0.21m from the regulating-pool to the forebay of condition 3, basically meets the design requirements.

4. Conclusion

In this paper, a physical model test is established for a pumping station. The test results of the original scheme show that the flow in the regulating-pool is affected by the partition wall, and there is a recirculation zone at the corner. The distribution of velocity in the channel flow is uneven, and the main flow is on one side. In the regulating-pool, the partition wall area which is connected with the diffusion section of the forebay has a large range of retention region due to the large flow area, where the sedimentation is easy to occur.

According to the test results view of the original scheme, we propose a rectification scheme: the flow accelerating-board scheme. The test results show that this scheme can effectively increase the flow velocity at the bottom of the regulating-pool, and effectively reduce the sedimentation in the channel flow when the pumping station is running at high water level. Although the hydraulic loss of the regulating-pool is increased, but also meets the design requirements.

Acknowledgement

This research was financially supported by the Ministry of Water Resources of China (201501033) and the Research on key technology of safety monitoring and remote cooperation for large pumping station group in South-to-North Water Diversion Project (BE2015734).

References

[1] Jiashan T 1988 *The Harm of Turbulent Flow in Pumping Station and Countermeasures* vol 02 (Nanjing: Journal of Hohai University (Natural Sciences)) p 10-19.

IOP Conf. Series: Materials Science and Engineering **242** (2017) 012048 doi:10.1088/1757-899X/242/1/012048

[2] Jianggang F and Jie *L* 2010 *Flow patterns in forebays of large urban water source pumping stations and countermeasures* vol 30(Nanjing: Advances in Science and Technology of Water Resources) chapter 2 p 70–74.
[3] Jianggang F and Xiaosheng W and Hongwei T 2010 *The countermeasures of flow patterns in forebays of large urban water source pumping stations* vol 12(Beijing: Water & Wastewater Engineering) chapter 36 p 51-54.
[4] Chao L 2003 *Water pump and water pump station* (Beijing: Peking university press).
[5] Xiaohuan S 2015 *Numerical simulation and model test research on optimization for flow patterns in the forebay of pumping station*

Static Stiffness Optimization of Rubber Absorber Based on Taguchi Method

Quanshan Xiao[1,2,a], Yinglong Zhao[1,2], Jin Zhu[1,2] and Jian Liu[1,2]

[1] Institute of Noise & Vibration, Naval University of Engineering, Wuhan 430033, China

[2] National Key Laboratory on Ship Vibration & Noise, Wuhan 430033, China.

[a] Corresponding author: 970042415@qq.com

Abstract. Static stiffness is an important property to ensure good bearings capacity performance of rubber absorber. Taking BE-300 type rubber absorber as an example, in order to obtain desirable static stiffness, Taguchi orthogonal design is proposed in this study based on the finite element method(FEM) replacing the practical test. Firstly, the FEM model of original rubber absorber which is consistent with practical test is established through constitutive test of rubber material. Dimensional parameters are one of the key factors affecting static stiffness, regarding bottom clearance height, rubber support layer thickness, width of support bearings and distance between large radius and small radius as controllable factors to design orthogonal FEM model by using Taguchi orthogonal methods. 25s FEM models with different level of controllable factors combinations are built based on the established FEM analysis technique. For each controllable factors combination scheme, the mean sensitivity parameters are calculated to evaluate the incidence of each controllable factor. The result show that the width of support bearings is a dominant factor affecting static stiffness. Finally, the optimization scheme can be proposed based on sensitivity analysis. Compared with the result from orthogonal optimization simulation and practical test, the static stiffness is much closer to the desired value. The study demonstrate that the orthogonal simulation is a useful approach for designing rubber absorber to specified static stiffness.

1. Introduction

Rubber absorbers have been widely used in modern ships because of its superior design and vibration reduction performance. BE-300 type rubber absorber, which are with features of low intrinsic frequency, excellent damping capacities, stable performance and compact construction, and can be installed in confined space, became one of superior absorber using in various ship. Therefore, it is essential to study its mechanical properties to design products that fulfil with performance requirements.

The structural optimization of rubber absorber includes topology optimization, shape optimization and dimension optimization[1-3]. Orthogonal design is a conventional technique method during the manufacture of rubber products. Taguchi orthogonal design method, with multiple controllable factors, treat sensitivity parameter as an indicator to evaluate which level of controllable factors are the key to the response results. The purpose of improving test efficiency and reducing cost can be achieved with high reliability. In this paper, the optimization design of rubber absorber is replaced by FEM. Using sensitivity parameters to evaluate the magnitude of the influence of the dimensional parameters on the static stiffness, Finally, an optimal scheme of controllable factor combination satisfying the desired

Content from this work may be used under the terms of the Creative Commons Attribution 3.0 licence. Any further distribution of this work must maintain attribution to the author(s) and the title of the work, journal citation and DOI.
Published under licence by IOP Publishing Ltd

value is determined.

2. FEM simulation of rubber absorber

2.1 Material Parameters Obtain

The constitutive relation of rubber material is essential for the FEM simulation of rubber absorber[4], Usually, the constitutive equations are derived according to the differential energy function by the description of the strain energy function, The strain energy function of the 2 order polynomial model define as follows[5]:

$$W(I_1,I_2)=\sum_{i,j=0}^{2}C_{ij}(I_1-3)^i(I_2-3)^j+\sum_{i=1}^{N}\frac{1}{D_i}(J-1)^{2i} \tag{1}$$

In the formula (1), C_{ij} is the Rivlin coefficient, C_{00}=0, I_1 and I_2 are the first and second Green strain invariants, and J is the volume ratio of the rubber before and after deformation, and the parameter of D_i determines whether the rubber material is compressible. 2 is the polynomial order, when treats the rubber material as incompressible material, J=1[6], so 2 order polynomial model define as follows:

$$W(I_1,I_2)=\sum_{i,j=0}^{2}C_{ij}(I_1-3)^i(I_2-3)^j \tag{2}$$

Strain energy function of rubber materials generally tested by uniaxial tensile and uniaxial compression. As shown in figure 1, uniaxial tensile is refer with GB/T 528-2009[7], the rubber specimen is dumbbell shaped. The thickness of the specimen is 2±0.2mm, the length of the test is 25±0.5mm, and the width of the narrow part is 6±0.4mm. As shown in figure 2, the tensile test is carried out on the SANS testing machine. The elongation of large deformation part is measured with extensometer. Uniaxial tensile tests of rubber materials are designed to obtain complete tensile stress-strain data. Because of the Mullins effect and hysteresis effect in the filled rubber, the loading and displacement are not synchronized in the test. In order to reduce the experimental error caused by the material itself, the tensile test of the rubber specimen was carried out by quasi-static tensile rate. figure 3 is the stress-strain curve obtained by uniaxial tensile test fitting with the 2 order polynomial model.

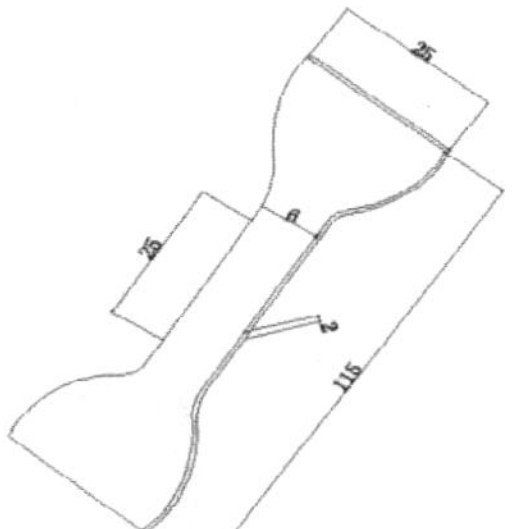

Figure 1. rubber specimen of uniaxial tensile

Figure 2. uniaxial tensile test

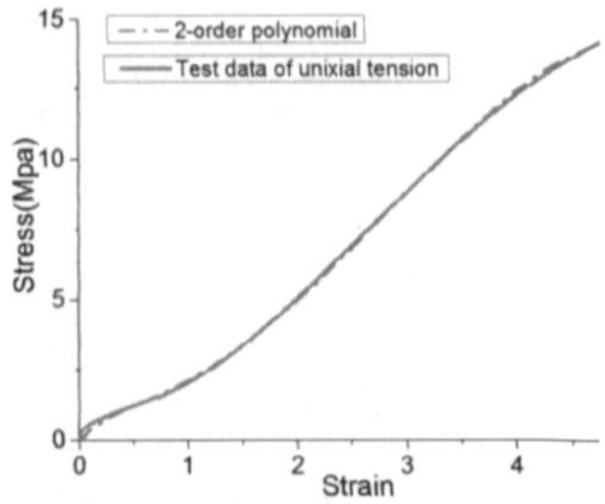

Figure 3. 2 order polynomial fitting with uniaxial tensile.

The tensile and compressive strain of rubber is very different, only have the tensile test data can not describe the constitutive model of rubber material well. Study shows that the relationship between uniaxial compression and equiaxial tension is describe as[8]:

$$\begin{cases}\sigma_{Eb}=\dfrac{\sigma_{Ec}}{\sqrt{(1+\varepsilon_{Ec})^3}}\\ \varepsilon_{Eb}=\sqrt{\varepsilon_{Ec}+1}-1\end{cases} \tag{3}$$

Where σ_{Ec} is compressive stress, ε_{Ec} is compressive strain. σ_{Eb} is equivalent tension stress , ε_{Eb} is equivalent equiaxial tension strain. Therefore, equiaxial tension can be replaced by uniaxial compression to obtain pure compression stress and strain of rubber. With reference with GB/T 528-2009[9], the uniaxial compression specimen is cylinder and carried out on the SANS testing

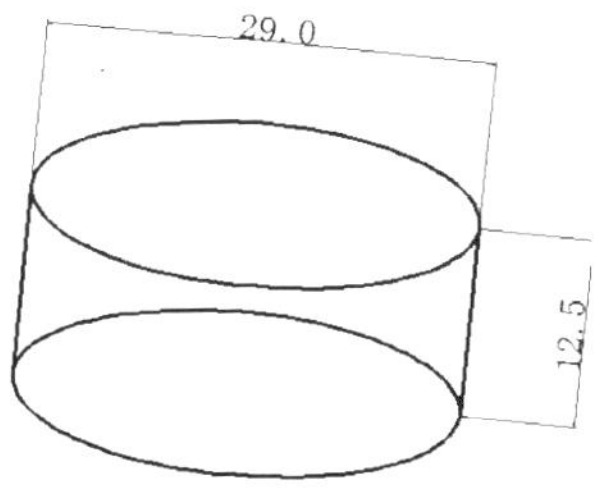

Figure 4. rubber specimen of uniaxial compression

Figure 5. uniaxial compression test.

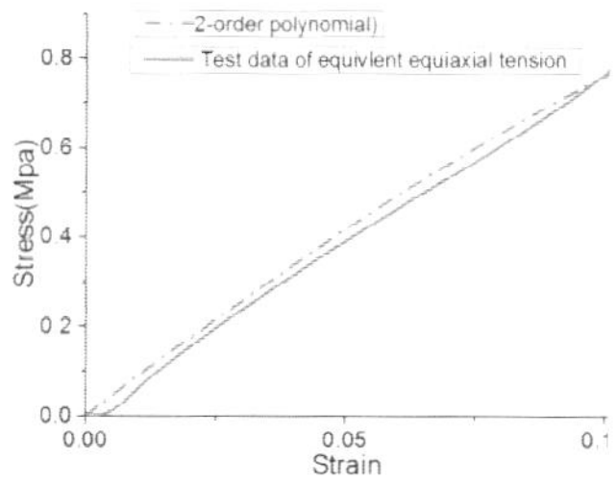

Figure 6. 2 order polynomial fitting with equivalent equiaxial tension result.

machine(as shown in figure 4& 5), firstly, compressing the specimen at 10mm/min. and then relaxing the specimen at the same speed, repeat 4 times, take the last compression data as valid compressing data of rubber, based on formula (3), the fitting curve of quivalent stress and strain with 2 order polynomial model are shown in figure 6.

As can be seen from figure3 and figure 6, 2 order polynomial model and experimental data are all fitting well, so it is reasonable to select the 2 order polynomial model as a hyperelastic constitutive model for the absorber rubber material. The fitting results are obtained with 2 order polynomial model parameters are:C_{10}=0.27832Mpa, C_{01}=0.46764Mpa, C_{20}=-5.1424E-02Mpa, C_{11}=0.38994Mpa, C_{02}=-0.46408Mpa, the metal part is carbon steel of Q235 which the modulus is E=2.1E6Mpa, the poisson's ratio is μ =0.3.

2.2 FEM Analysis and Test

The deformation of rubber absorber under 3000N downward force is analysed by FEM based on ABAQUS software[4]. The rubber part is simulated by hybrid C3D8H units, and the metal part is simulated by reduced C3D8R units. The static compression test was carry out on the MTS test machine (as shown in figure 7). Because of the typical hysteresis and Mullins effects of rubber materials, the loading and unloading paths do not coincide. Therefore, this test circularly loading and unloading 4 times, and takes the hysteretic line as an index to measure the static deformation of the rubber absorber. The testing result of static stiffness of the rubber absorber is 786.44N/mm. The static simulation results of rubber absorber are obtained under 3000N downward axial force, and the deformation nephogram as shown in figure 8.

From the FEM analysis results in figure 8, the deformation of the rubber absorber is 3.710mm, so the static stiffness is calculated to be 808.63N/mm, which is about 3% higher than the experimental result. The test calculation results and simulation results plotted in figure 9, it can be seen that the FEM analysis results and experimental results are quite close, which illustrate that the analytic results

Figure 7. Static test diagram of rubber absorber

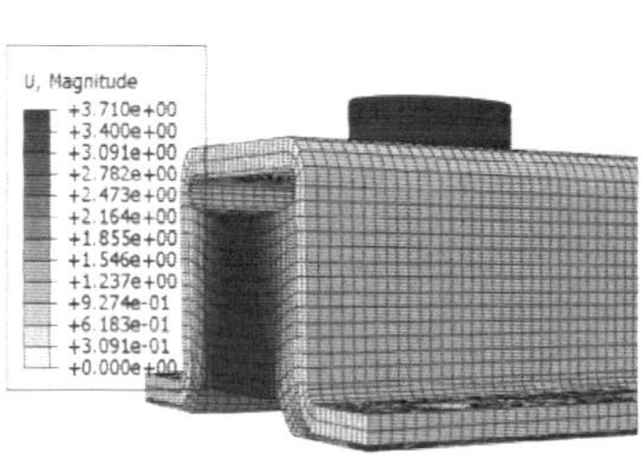

Figure 8. Simulation results under 3000N vertical pressure

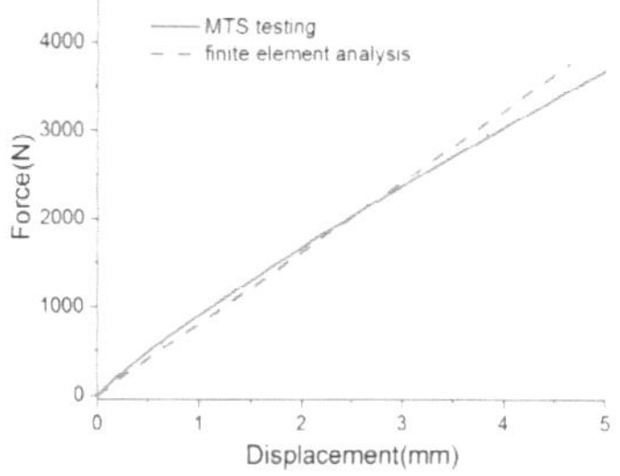

Figure 9. comparison of MTS testing and static simulation result rubber absorber

of FEM simulation instead of rubber absorber test is reasonable.

3. Orthogonal design and optimization

3.1 Taguchi Orthogonal Design theory

The basic idea of Taguchi parameter design method is to establish orthogonal test table which contain various controllable factors that affect properties[10]. By analysing each controllable factor combination of the products, the optimum combination scheme can be determined. Orthogonal design is an important means of Taguchi methods[11]. Mean sensitivity parameter of each level refers to the influence of controllable factors affect the properties[12]. The greater the sensitivity is, the greater the influence are. When determining the optimization scheme, the level of the controllable factor which with higher sensitivity is preferentially determined.

Assuming there exist adjustment factors in the multiple tests, the average of the experimental results can be adjusted to the target properties of m, with n test data:$y_1, y_2, \cdots, y_n$, the total fluctuation S_T is calculated as [13, 14]:

$$S_m = \frac{1}{n}\left(\sum_{i-1}^{n} y_i\right)^2 \tag{4}$$

and the error variance V_e is calculated as: $$V_e = \frac{\sum_{i-1}^{n} y_i^2 - \frac{1}{n}\left(\sum_{i-1}^{n} y_i\right)^2}{n-1} \tag{5}$$

Sensitivity is an indicator of the average product quality response. the sensitivity is defined as:

$$S = 10\lg\frac{1}{n}(S_m - V_e) \tag{6}$$

Where S is sensitivity, S_m and V_e can be obtained by formula (4) and formula (5).

3.2 Orthogonal Simulation Design

According to the requirements of installation condition and BE-300 type rubber absorber existing design conclusion, 4 factors which essential to the static stiffness are selected as controllable factors of Taguchi orthogonal design plan. As shown in figure 10, this 4 factors respectively are bottom clearance height h_1 (factor A), rubber support layer thickness h_2 (factor B), the width of support bearings l (factor C), the distance between large radius and small radius axis of D_1 (factor D). Sensitivity is an indicator of the average product quality characteristics, so in a test orthogonal table, each scheme must have two or more experimental data, otherwise the sensitivity will be meaningless.

The orthogonal test in this study is completed by computer simulation, so there is no external interference, so the external environment influence can be excluded. Two kinds of mesh size of rubber material are selected as noise factors (factor Y), the one approximate global mesh size is 3mm and the another is 4mm, two attributes of mesh are C3D8H. The level of the 3 controllable factors is determined by referring to the magnitude parameters of the existing absorber. Interference factors and

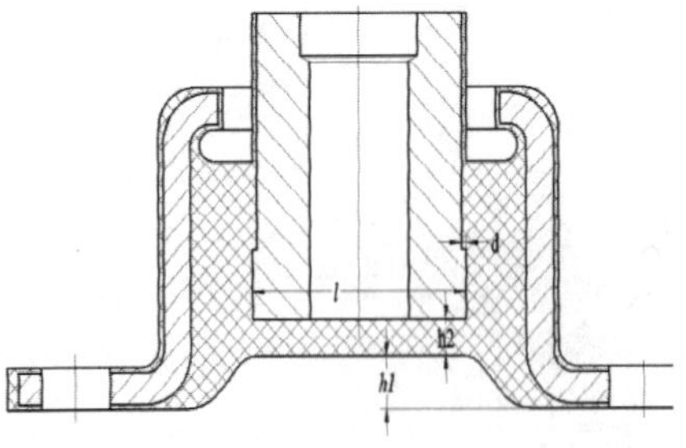

Figure 10. The geometry parameters of rubber absorber

Table 1. Interference factors and its level

factors		level				
controllable factors	A($h1$/mm)	8	10	12	14	16
	B($h2$/mm)	5	7	9	11	13
	C(l/mm)	44	46	48	50	52
	D(d/mm)	1	2	3	4	5
noise factors	Y	mesh size is 3mm			mesh size is 4mm	

its level arrangement are as shown in table 1. According to the selection of controllable and noise factors of each level, orthogonal analysis table of L25 (5^4) is established. The orthogonal table ensure the times of repetition of one level is the same in each column, and each level collocation between two rows arbitrarily in horizontal is balanced. Completely FEM model is established in the *ABAQUS* software in accordance with the dimension combination scheme in table 2. For the same model, the rubber absorber model is divided into two mesh sizes referring to table 1. A total of 25×2 times FEM analysis were carried out, and the obtained simulation results as shown in table 2.

Table 2. Orthogonal arrangement with FEM analysis results

NO	Level			Static stiffness N/mm		Sensitivity S	NO	Level			Static stiffness N/mm		Sensitivity S
	A	B	C	Y_1	Y_2			A	B	C	Y_1	Y_2	
1	8	5	44	749.63	753.58	57.520	14	14	5	48	844.12	848.90	58.553
2	10	7	46	815.00	820.79	58.254	5	16	7	50	916.31	930.52	59.308
3	12	9	48	878.48	885.22	58.908	16	8	11	46	844.36	849.86	58.559
4	14	11	50	929.08	933.13	59.380	17	10	13	48	898.74	900.90	59.083
5	16	13	52	964.32	978.79	59.749	18	12	5	50	940.44	945.18	59.488
6	8	7	48	893.92	896.59	59.039	19	14	7	52	1024.24	1036.99	60.262
7	10	9	50	966.81	973.08	59.735	20	16	9	44	704.72	714.63	57.021
8	12	11	2	1044.20	1047.49	60.389	21	8	13	50	1000.33	1003.34	60.016
9	14	13	44	719.94	727.98	57.194	22	10	5	52	1034.13	1040.94	60.320
10	16	5	46	759.11	763.16	57.629	23	12	7	44	737.10	740.38	57.370
11	8	9	52	1076.43	1085.38	60.676	24	14	9	46	788.85	792.39	57.959
12	10	11	44	768.25	774.39	57.745	25	16	11	48	826.22	834.26	58.384
13	12	13	46	803.00	809.50	58.129							

3.3 Optimization Analysis

The absorber specification usage request that the vertical deformation of BE-300 rubber absorber under rated load (3000N) is 3.5~5.0*mm*. Taking the vertical deformation of 3.5*mm* as an example, the static stiffness is 857.1N/mm. The simulation results of BE-300 rubber absorber are about 3% higher than the experimental results, therefore, the static stiffness targets of the simulation result should be m=857.1×(1+0.03) =882.8*N/mm*.

The mean sensitivity of one controllable factor for a level are respectively calculated. Taking the controllable factor of A's 1 level as an example, the simulation scheme contains A's 1 level include No.1, No.6, No.11, No.16 and No.21. Then the mean sensitivity of A's 1 level can be calculated by

Table 3. Mean sensitivity analysis results

level	A/*N/mm*	B/*N/mm*	C/*N/mm*	D/*N/mm*
M1	59.162	58.702	57.370	58.852
M2	59.027	58.846	58.106	58.799
M3	58.857	58.860	58.793	58.912
M4	58.670	58.891	59.585	58.778
M5	58.418	58.834	60.279	58.793
Δ	0.744	0.189	2.909	0.134
Rand	2	3	1	4
$\overline{M}$				46.85

Table 4. Mean static stiffness analysis results

level	A/*N/mm*	B/*N/mm*	C/*N/mm*	D/*N/mm*
M1	915.3	867.9	739.1	882.3
M2	899.3	881.2	804.6	879.2
M3	883.1	886.6	870.7	889.4
M4	864.6	885.1	953.8	876
M5	839.2	880.7	1033.3	874.8
$\overline{M}$			880.3	

$$M1_A = \frac{S_1 + S_6 + S_{11} + S_{16} + S_{21}}{5} \tag{7}$$

$M1_A$ is the mean sensitivity of A's 1 level, S_1 is sensitivity of No.1. The mean sensitivity of the remaining controllable factors can be obtained in turn, and the results are shown in table 3.

The mean static stiffness of one controllable factor for a level also can be calculated. Firstly, calculating the mean static stiffness of each experiments, then based on the calculation methods which similar with mean sensitivity to obtain the mean static stiffness of each level of controllable factors. The results shown in table 4.

As can be seen in table 4, *M1~M5* is 5 different levels of controllable factors, $\overline{M}$ is the mean value of 25 times FEM analysis results. Δ is the maximum mean static stiffness minus the minimum of one controllable factor at different levels, and Δ indicate the impact significance of controllable factors. As can be seen from table 3, the impact significance of sensitivity of S is $\Delta_C > \Delta_A > \Delta_B > \Delta_D$, show that the most effective factor is the width of support bearings, the distance between large radius and small radius only have slightly influence on the static stiffness. Therefore, the first step of the optimization scheme should select the level of C which closest to the desired value m. As can be seen from table 4, the mean static stiffness of C3 equal to 870.7*N/mm*, closer to the desired value m=882.8*N/mm*. Then selecting the level of controllable factor A, from the table 4, A3 is closer to the desired value. After the level of C and A is determined, according to table 4, the difference between optimization value and the desired value m is ∇_1=(870.7+883.1)- 2×882.8=-11.8 *N/mm*. Therefore, the mean static stiffness of B and D should slightly higher than the desired value m, so that the overall static stiffness can be approached to the desired value m as possible. The mean static stiffness of B3 and D3 combinations is slightly higher than the desired value m, the difference ∇_2=(889.4+886.6)- 2×882.8=10.4 *N/mm*, and the total difference $\nabla=\nabla_1+\nabla_2=-1.4$N/mm. So the optimization scheme can be determined as A3, C3, B3, and D3. This combination is the scheme No.3 which in the orthogonal table 2.

4. The optimization results

The FEM analysis results of optimization scheme of No.3 are shown in figure 11 and in figure 12.

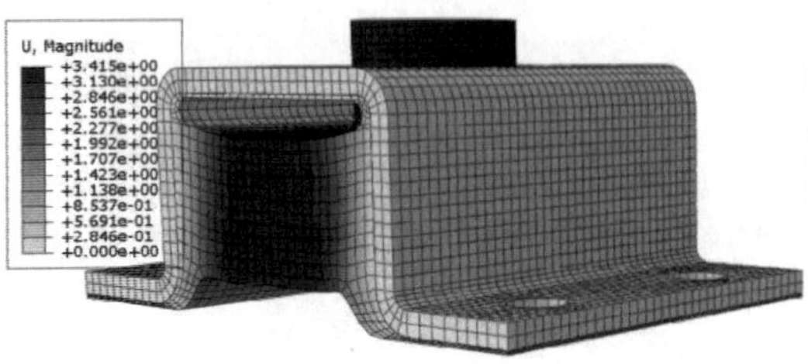

Figure 11. Optimization result of mesh size is 3*mm*

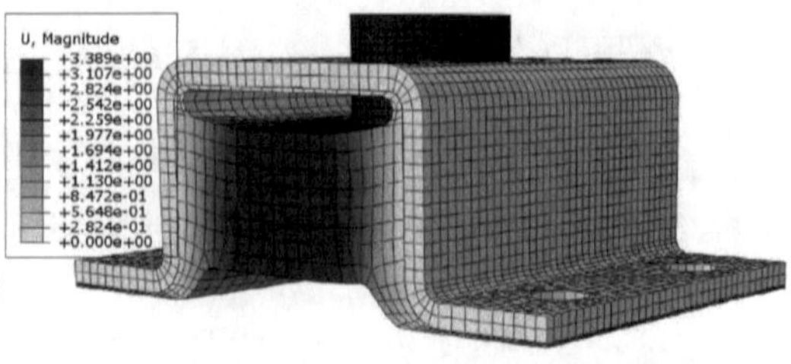

Figure 12. Optimization result of mesh size is 4*mm*

Figure 11 and figure 12 are the static stiffness FEM analysis results of approximate global mesh size is 3*mm* and 4*mm*. It can be seen that the calculated mean static stiffness is 881.85N/mm, which is closer to the desired value *m*=882.8*N/mm*, this indicate that the optimization scheme can improve the static stiffness of rubber absorber effectively.

5. Conclusion

The performance of static stiffness is the key to ensure bearing capacity of rubber absorber. Four factors which are essential to the static stiffness are selected as controllable factors to optimize the static stiffness. Optimization scheme indicated that the combination of A3, C3, B3, and D3 can improve the static stiffness to the desired value effectively. This calculation results show that using sensitivity parameter as evaluation index in Taguchi orthogonal methods is reasonable, and the Taguchi orthogonal methods can help to design static stiffness of rubber absorber approach to the desired value utmost.

Acknowledgments

The authors are grateful to National Key Laboratory on Ship Vibration & Noise for the testing facilities that have been used to conducted this investigation. The authors would like to thank Mr. Luo for the design of drawings.

References

[1] Dimitrovová Z, Rodrigues HC. 2010. Optimization of passive vibration isolators mechanical characteristics. *Structural & Multidisciplinary Optimization* 42:325-340.
[2] Hopkins C ed. 2009. Introduction to Designing Elastomeric Vibration Isolators.
[3] Lee W, Lee W, Youn S, Youn S. 2003. Topology optimization of rubbers in the vibration isolators.
[4] Iavornic C, Praisach ZI, Vasile O, Gillich GR, Iancu V. 2011. Study of stress and deformation in elastomeric isolation systems using the Finite Element Method., *Wseas International Conference on Signal Processing, Computational Geometry and Artificial Vision, Iscgav'11, Wseas International Conference on Systems Theory and Scientific Computation, Istasc*, pp 239-244.
[5] R WL, H LZ. 2003. Modeling method of constitutive law of rubber hyperelasticity based on finite element simulations. *RUBBER CHEM TECHNOL* 76:271-285.
[6] J B, D WR. 1997. Nonlinear continuum mechanics for Finite Element Analysis. Cambridge University Press, Cambridge.
[7] 2009. Determinaion of tensile sress strian of vulcanized rubber or thermoplastic rubber., *GB/T 528-2009*. China Standard Press.
[8] Lihong S. 2012. Study on the generalized Mechanical model and creep mechanism of Polyurethane isolator. Naval University of engineering, Wuhan.
[9] 2009. Determinaion of compressive strian properties of vulcanized rubber or thermoplastic rubber., *GB/T 7757-2009*. China Standard Press.
[10] Mao M. 2015. Optimization of the Ijection Molding Process Parsmeters of Water segregator Using Taguchi Method. Anhui Polytechnic University, Hefei.
[11] Tang SH, Tan YJ, Sapuan SM, Sulaiman S, Ismail N, Samin R. 2007. The use of Taguchi method in the design of plastic injection mould for reducing warpage. *J MATER PROCESS TECH*:418-426.
[12] Dezfuli FH, Alam M. 2016. Experiment-Based Sensitivity Analysis of Scaled Carbon-Fiber-Reinforced Elastomeric Isolators in Bonded Applications. *Fibers* 4.
[13] Zhang Z. 2008. The reducing of quality characteristic value fluctuation of pusher kiln furniture (slab) by Taguchi Method. Shanghai Jiaotong University, Shanghai.
[14] Liu Y. 2015. Optimization on Quenching Process of Large Backup Rolls Based on Taguchi Method. Yanshan University, Yanshan.

IOP Conf. Series: Materials Science and Engineering **242** (2017) 012050 doi:10.1088/1757-899X/242/1/012050

The research of suspen-dome structure

Shengyuan Gong[1, a)]

[1] College of civil and safety engineering, Dalian Jiaotong University, Dalian 116028, China

[a)]arrow1994@foxmail.com

Abstract. After overcoming the shortcomings of single-layer latticed shell and cable dome structure, the suspen-dome was developed by inheriting the advantages of them, and it was recognized and applied as a new type of prestressed force large span space structure. Based on the analysis of the background and mechanical principle, the researches of suspen-dome are reviewed, including form-finding analysis, the analysis of static force and stability, the dynamic behaviors and the earthquake resistant behavior, the analysis of prestressing force and optimization design, and the research status of the design of the fir-resistant performance etc. This thesis summarizes the methods of various researches, being a reference for further structural performance research and structural engineering application.

1. Introduction

With the improvement of society and of the economy, the demand of large span space structure has seen a drastically increase. The single-layer latticed shell has graceful mould, proven technique, and advantageous construction technique, therefore it was used extensively in small and medium sized span project. However, the structure is very sensitive to the initial imperfection, which makes its stability become the governing factor; the double layer shell overcomes the shortcomings of the weakness of the shell, being no longer sensitive to the initial imperfection, but it possesses dense rod pieces. With the enlargement of the span, the dead load will lead to larger tensile force in the surrounding ring beams. As a result, the usage of iron and project cost are higher. The newer cable dome structure vastly influences the non-cable structural component because of the cable in the system, the cost is high and the construction is difficult.

To solve the problems in the large span space structures mentioned above, professor Mamoru Kawaguchi [1] in Hosei University in Japan put forward the idea of suspen-dome structure system in 1993, it includes the superior single latticed shell and the lower whole stretch-draw part, the whole stretch-draw part includes vertical stay bars, radial pull rods and circumferential lassoes.

The suspen-dome fused the advantages of the single latticed shell structures and the dome structure as one of the newly composite structures. On the one hand, it brings the whole stretch-draw part into the lower single latticed shell structure, which improves the stability of the single latticed shell structure; On the other hand, it reduces the difficulties of the design, analysis, and construction because the upper lasso of the cable dome was replaced by the single latticed shell structure. Since the idea of suspen-dome was proposed, it has been applied to a dozen of large projects at home and abroad, such as the Beijing Olympic badminton gymnasium in 2008, the gymnasium of Anhui University, the dome of Mitsuoka in Japan, and the gymnasium of Dalian, which displays the application prospect of it.

Content from this work may be used under the terms of the Creative Commons Attribution 3.0 licence. Any further distribution of this work must maintain attribution to the author(s) and the title of the work, journal citation and DOI.
Published under licence by IOP Publishing Ltd

2. The research development of the suspen-dome structure

2.1 Static and Buckling Analysis

After Professor Mamoru Kawaguchi [1] put forward the idea of suspen-dome structure, he made a series of theoretical analysis and experimental research of the system. Model tests on a small-scale suspen-dome with a diameter of 3.0m and a rise of 0.45 m have also been carried out [1], in which the load-carrying capacity of stability and the rigidity were proved to be increased greatly compared with the single-layer dome. Finite element (FE) analysis results also showed good agreement with results from experiments. A design example that is used is a suspen-dome structure, of which span and rise are 200 and 30 m, respectively [2]. The static response of them on the pretension cables were compared. Here is the conclusion of the experiment and the finite element analysis: The internal force of radial bars in suspen-dome structure is only half of the single-layer latticed shell, the outer ring commutator bars also have significant force reductions, however the force of inner ring bars have little difference.

The stability of it was showed thorough the research of the literature [3-5], and they drew the following conclusions: (1) The buckling strength of the suspen-dome structure and it sensibility to the initial imperfection is superior to the relative single-layer latticed shell structure; (2) Prestressed force, the length of the stay bars and the sections of the lassoes have a great influence on the stability of the structure. The material nonlinearity was motioned in the literature [6] and the cable-strut joint structure have a larger influence on the global stability of the suspen-dome structure. The research of the literature [7] showed that the construction deviation probabilistic method can perfectly illustrate the impact on the stability of the suspen-dome structure of the initial geometric imperfections. The experiment [8] shows that the ultimate load under half-span load was obviously inferior to that under the full-span load.

2.2 Dynamic and Seismic Behavior Analysis

Chen Zhihua [5] used the finite element software ANSYS to analyze the self-vibration characteristics of the suspen-dome structure. He analyzed the impact of the loads, rise, span, external-prestressing loop cable, bearing restraint on the structure's self-vibration, he gave a proposal value to the structural design as well; The time-history method was used to analyze the suspen-dome structure's response to the earthquake, and the influence of the parameters like the prestressing force, rise span ratio to its response. And followed are the conclusions: The frequency spectrum of the suspen-dome structure is intensive; the structure have a complicated vibration; the first few frequencies have an apparently enlargement than the single latticed shell structure, but their vibration modes have little differences; The suspen-dome structure's respond to the horizontal wave is obviously larger than the vertical wave, and the internal force is reduced the single latticed shell structure; and it conducted a consistent conclusion with the theoretical analysis through pulsating excitation and impact excitation in exiting actual projects. Li Meng [9] did a research about the dynamic performance of the suspen-dome structure under the geological process, the results indicate that: when a structural seismic analysis was conducted, the role of multi-dimensional earthquake should be considered; Border suspen-dome structure has a larger seismic coefficient and a stronger the structure boundary is needed. Li Li [10] researched the self-vibration of the suspen-dome structure through changing many kinds of parameters, and she concluded the distribution law of the seismic internal force, seismic coefficient of the suspdome structure elements. Then, she gave the recommend values for structural design as well. Zhang Ailin [11] made following conclusion through the dynamic experiment of the Olympic badminton gym: the frequency of the suspen-dome structure is intensive-the period is long; the vibration centered on the vertical direction; the symmetric form and the asymmetric form take the major parts of the vibration, and the structural damping is between 0.02 and 0.03. Tian Shixuan [12] established the structure failure load of cellular automata (CA) model based on energy balance equation for structural energy response analysis and the incremental dynamic analysis conclusion of the suspen-dome structure under the ground motion.

2.3 Prestressing Force Analysis

The mechanical principle of the suspen-dome structure is enhancing the single latticed shell structure through bringing in prestressing force to bear the lower tensegrity structure to improve its capacity. Therefore, the design of the lower tensegrity structure is of vital importance in structure designing. At present, the designing principle of prestressing force is minimizing the peak values of the supports and the latticed shell members. Professor Mamoru Kawaguchi [1] took the method of trial and error, and the successive approximation method to design the party dome. These methods are simple in concept, but they are lacking in efficiency and effect. Chen Zhihua [13] put forward the simplex algorithm of the prestressed cable support system based on the mechanical balance principle. This method is convenient, but its precision is not high. Zhang Mingshan [14] used the genetic algorithm to conduct a secondary majorization of the suspen-dome structure. This method takes a single-target genetic algorithm, which means only one target can be optimized every time. When the secondary majorization is being conducted, the first level optimization structure has already changed, so this method cannot guarantee the accuracy of the majorization. After that, Zhang Mingshan [14] used the local analysis method to confirm the initial prestressing force distribution of the suspen-dome structure based on the balance matrix theory. The simplified calculation is the simplest method of the above with low construction cost, however its accuracy is the lowest as well. On the contrary, the genetic algorithm method is the most complicated method, but it has the highest accuracy and a wide application. In practical engineering, engineers can select these methods based on physical condition.

2.4 Fire Resistance and Temperature Analysis

In the aspect of the fire resistance design of the suspen-dome structure, Huang Wenqin [15] simulated the unitary fire-resistant performance of the suspen-dome structure under large enclosure fire using software ANSYS. He analyzed the structure behavior and its failure mechanism under fire. Then the sensitive of parameters like temperature field distribution, load ratio, and geometry and boundary conditions were analyzed by numerical simulation, and the critical temperature of the structure under fire was obtained. Based on this, the evaluation method of the fire resistance conditions of the suspen-dome structure was put forward.

Liu Hongbo [16] analyzed the construction processes of bearing frame-fastener type steel tube supporting frame, considered the influence of sliding friction and temperature change to the construction control theory. He also researched the influence of the solar radiation and structural behavior under temperature load, by using the temperature field numerical simulation method. Then he concluded that the environmental temperature changes under the tensioning construction stage of the suspen-dome structure had a great influence on the tensioning construction. He suggests that the influence of temperature change on the tensioning construction should be considered when designing the tension control values of the suspen-dome.

3 Conclusion

(1) The suspen-dome structure is kind of a highly efficient form of prestressed spatial structure. It possesses a superior static and dynamic performance, while its construction cost is low. The suspen-dome structure has already been applied in many large projects at home and abroad, and it has shown great superiority in many aspects.

(2) Current studies of the suspen-dome structure are mainly concentrated in the static, dynamic and stability analysis, prestressing force design and optimization design of the structure, etc.

(3) Researches on the fire resistance design and temperature analysis of the suspen-dome structure are relatively few, and further researches are needed.

References

[1] Mamoru Kawaguchi, Masaru Abe, Ikuo Tatemichi, et al. Design, tests and realization of “suspen-dome” system[J]. Journal of the IAAS, 1999, 40 (3):179-192.

[2] Mamoru Kawaguchi, Masaru Abe, Tatsuo Hatato, et al. Structural tests on the "suspen-dome" system[C]. Proceedings of IASS Symposium. Marid, Spain: IASS, 1994: 384-392.
[3] Cui Xiaoqiang, Guo Yanlin. Research of Elastic Limit Bearing Capacity of the Kiewitt Suspen-dome Structure [J]. Journal of Building Structures, 2003, 24(1): 74-79.
[4] Li Yongmei, Zhang Yigang. Stability and Static Analysis of Cable-Supported Lattice Shells [J]. International Journal of Space Structure, 2003, 9(1): 25-30.
[5] Chen Zhihua, Li Yang. Parameter analysis on stability of a suspen-dome [J]. International Journal of Space Structure, 2005, 20(2): 115-124.
[6] Ge Jiaqi, Zhang Guojun, Wang Shu, etc. The Overall Stability Analysis of the Suspen-dome Structure System of the Badminton Gymnasium for 2008 Olympic Games [J]. Journal of Building Structures, 2007, 28(6): 22-30.
[7] Liu Xuechun, Zhang Ailin, Ge Jiaqi, etc. Study on the influence of construction deviation random distribution on the integral stability of suspen-dome. [J]. Journal of Building Structures, 2007, 28(6): 76-82.
[8] Zhang Ailin, Liu Xuechun, Wang Dongmei, etc. Seismic Behaviors Study of the 2008 Olympic Games New Type Suspen-dome Structure [J]. Journal of Building Structures, 2007, 28(6): 58-67.
[9] Li Meng. Research of Static and Dynamic Performance of the Chord Dome Considering Supporting Structure [D]. Xi'an: Xi'an University of Architecture and Technology, 2010.
[10] Li Li. The Anti-Seismic Property Analysis of the Suspen-dome Structure [D]. Harbin: Harbin Institute of Technology. 2008.
[11] Zhang Ailin, Liu Xuechun, Wang Dongmei, etc. Seismic Behaviors Study of the 2008 Olympic Games New Type Suspen-dome Structure [J]. Journal of Building Structures, 2007, 28(6): 68-75.
[12] Tian Shixuan. Seismic Response Analysis of Suspen-dome Based on Energy Method [D] Harbin: Harbin Institute of Technology. 2013.
[13] Kang Wenjiang, Chen Zhihua, Heung-Fai Lam, et al. Analysis and design of the general and outmost-ring stiffed suspen-dome structures [J]. Engineering Structures, 2003, 25(13):1685-1695.
[14] Zhang Mingshan, Dong Shilin, Zhang Zhihong, Distribution of Initial Pre-stress and Stability Analysis of Suspen-dome [J]. International Journal of Space Structure, 2004, 10(2): 9-12.
[15] Huang Wenqin, Study of Numerical Analysis and Fire Resistance Conditions of the Suspen-dome Structure Under Fire [J]. 2013.
[16] Liu Hongbo, Study on the Construction Control Theory and Temperature Effect of Suspen-dome Structure [J]. 2011.

IOP Conf. Series: Materials Science and Engineering **242** (2017) 012051 doi:10.1088/1757-899X/242/1/012051

Study on influence of three kinds of stress on crack propagation in butt welds of spiral coil waterwall for ultra supercritical boiler

Zhenrong Yan[1,2,a)],Jun Si[2,b)]

[1]Shanghai Boiler Works Co.,Ltd. Shanghai,200245,China

[2]Shanghai Institute of Special Equipment Inspection and Technical Research.Shanghai, 200062, China

E-mail: [a)]Corresponding author: yanzr2010@163.com, [b)]sijun@ssei.cn

Abstract. The spiral coil waterwall is the main pressure parts and the core functional components of Ultra Supercritical Boiler. In the process of operation, the spiral coil waterwall is under the combined action of welding residual stress, installation defects stress and working fluid stress, Cracks and crack propagation are easy to occur in butt welds with defects. In view of the early cracks in the butt welds of more T23 water cooled walls, in this paper, the influence of various stresses on the crack propagation in the butt welds of spiral coil waterwall was studied by numerical simulation. Firstly, the welding process of T23 water cooled wall tube was simulated, and the welding residual stress field was obtained. Then,on the basis, put the working medium load on the spiral coil waterwall, the supercoated stress distribution of the welding residual stress and the stress of the working medium is obtained. Considering the bending moment formed by stagger joint which is the most common installation defects, the stress field distribution of butt welds in T23 water-cooled wall tubes was obtained by applying bending moment on the basis of the stress field of the welding residual stress and the working medium stress. The results show that, the welding residual stress is small, the effect of T23 heat treatment after welding to improve the weld quality is not obvious; The working medium load plays a great role in the hoop stress of the water cooled wall tube, and promotes the cracks in the butt welds; The axial stress on the water cooled wall tube produced by the installation defect stress is obvious, the stagger joint ,and other installation defects are the main reason of crack propagation of spiral coil waterwall. It is recommended that the control the bending moment resulting from the stagger joint not exceed 756.5 NM.

1. Introduction

The spiral coil waterwall is the main pressure parts of ultra supercritical boiler. The manufacturing workload is large and the production process is complicated [1]. The spiral coil waterwall is the key functional component of ultra supercritical boiler for variable pressure operation. Therefore, In the manufacture, installation and operation of the spiral coil waterwalll, all kinds of stresses occur in the water cooled wall tube, of which the welding residual stress, thermal expansion stress and installation stress are the main factors [2]. The water cooled wall tube will also bear the working fluid stress in the variable pressure operation. This puts forward very high requirements for the performance of T23 material which be used to manufacture the spiral coil waterwall.

Foreign papers [3] reported that T23 joints which the wall thickness is not more than 10 mm, welding not only without preheating, but also cancel the welding heat treatment. Especially, the

Content from this work may be used under the terms of the Creative Commons Attribution 3.0 licence. Any further distribution of this work must maintain attribution to the author(s) and the title of the work, journal citation and DOI.
Published under licence by IOP Publishing Ltd

simplification of welding process makes T23 an ideal material for producing the spiral coil waterwall for super supercritical boiler.T23 steel is widely used in the spiral coil waterwall of 1000 MW ultra supercritical tower boiler in china.

However, in recent years, it has been reported [4,5] that T23 butt welds of the spiral coilwaterwall are prone to crack during operation, leading to leakage accidents and frequent outages of the units. Many scholars have carried out a lot of research on the early failure of T23 butt welds. The failure reasons are classified into three categories: over temperature pipe burst, high temperature corrosion, combined effects of welding stress, thermal stress and working medium stress [6]. The document [7] also pointed out that superimposed effect will produce between the welding residual stress with the working medium stress during the process of operation, start and stop. It is possible that cracks occur in the welded joint and its region nearby under the combined action of welding stress, thermal stress and working medium stress, which eventually results in cracking and leakage of water-cooled wall tubes.

Previous studies on cracks of butt welds in water-cooled wall tubes are rarely involve installation defect stress. In fact, the butt welds of the spiral coil waterwall is divided into the manufacture butt welds and the installation butt welds. The literature [8] pointed out that the cracks in the butt welds of T23 water cooled wall tubes generally appear in the installation butt welds, rarely appeared in the manufacture butt welds which have been made heat treatment after welding. In the process of welding of the spiral coil waterwall on field installation, geometric defects, such as misalignment, discount and angular deformation often occur. These defects cause great stress at the welded joint, which is called installation defect stress, which intensifies the crack propagation in the butt welds of T23 water-cooled wall tubes. Therefore, this paper uses ABAQUS finite element numerical simulation method, to investigate the influence of welding residual stress, installation defects stress and working fluid stress on crack propagation in butt weld of T23 water-cooled wall tubes, so that find out the main factors leading to the failure of butt welds of water-cooled wall tubes.

2. Physical model and numerical simulation

2.1 Establishment of Physical Model

Take a the T23 butt weld crack of spiral coil waterwall for 1000MW ultra supercritical boiler in a power plant as a example, as shown in Figure 1. When ultra supercritical boiler is the rated load of the working medium pressure P in water-cooled wall tubes is 27.9MPa,tube wall temperature is about 400 °C~550°C,the tube specification is 38.1 mm ×6.8 mm, take length 100mm which include the butt weld as the test sample. The physical parameters of T23 steel at various temperatures are shown in Table 1. Relationship between wall temperature and allowable stress of T23 steel tube are shown in Table 2. In the meshing of finite element mesh, the simulation of the welding temperature field uses the unit type DC3D8, the simulation of the stress field uses the unit type C3D8R. The mesh is divided into a cube with 1mm sides, it can not only guarantee the accuracy of calculation, but also greatly reduce the time required for calculation and improve the computational efficiency.

FIGURE 1. Inner wall crack in the weldment of the T23 tube

TABLE 1. The physical parameters of T23 steel at various temperatures

Temperature/°C	20	50	100	200	300	400	500	600	650
Elastic modulus /GPa	208	206	203	196	189	181	171	160	154
Thermal conductivity /W/m.K	34.4	34.8	35.8	36.8	36.6	35.8	34.6	33.1	32.2
Expansion coefficient /10^{-6}//K	-	11.3	11.6	12.2	12.6	13.1	13.5	13.9	14.0
Specific heat /J/(kg.k)	440	460	480	510	550	630	660	770	860
yield limit /MPa	400	380	365	355	345	325	300	255	205

TABLE 2. Relationship between wall temperature and allowable stress of T23 steel tube

T23 wall temperature /°C	450	500	525	550
allowable stress /MPa	117	111	105	87

2.2 Numerical Simulation of Welding Process

The welding process is sequentially coupled, that is, the simulation of the welding temperature field is first carried out, and then the calculated temperature field is introduced into the stress calculation model to calculate the stress field. Combined with birth and death element method, the Finite Element Method (FEM) is applied to simulation of welding temperature field. The welding process of finite element simulation will need to remove the weld part (kill), and then the welding step by step to add (reborn).

The environmental temperature during welding has some influence on the welding stress. In this paper, the predefined temperature field is defined as the ambient temperature of 20 °C. When the welding is defined in Interaction, the convection coefficient between the inner and outer walls of the T23 tube and ambient air is $20w/m^2\cdot°C$. According to the actual welding process of T23 water-cooled wall tubes, T23 water-cooled wall tubes after welding is cooled to room temperature by the air. During the simulation of welding temperature field, the heat source is imposed to body in the form of the thermal flow.Firstly, needs to definite the collection" set" which can impose load to body , then each "set " is subjected to a load of body heat flow, The magnitude of the load can be calculated from the welding process. According to the actual welding process parameters:I=100A, U=10V, welding speed 20mm/min , it is calculated that appropriate heat flux density is $6.25\times10^{10}w/m^2$, select the heat transfer efficiency is 40%, then each load unit set is 2.5×10^{10}.

3. Numerical simulation analysis

3.1 Influence of Welding Residual Stress

The residual stress field is obtained by finite element numerical simulation. It is found that the maximum residual stress 89.6MPa, which is located in the outer wall of the heat affected zone. According to the ASME standard, the relation between allowable stress and wall temperature of T23 material is shown in Table 2. In the ultra supercritical boiler, the working temperature of T23 water-cooled wall tubes is about 450°C~550°C, apparently, the residual stress of single layer welding is very high, and it is close to or higher than the allowable temperature under the highest equivalent temperature. The local stress distribution in the weld is shown in Figure 2. The residual stress in the heat affected zone is the largest. It shows that the heat affected zone is easy to cause cracks in the fluctuating load of the ultra supercritical boiler under variable pressure operation. Many documents have mentioned that cracks occur in the heat affected zone, probably due to welding residual stresses.

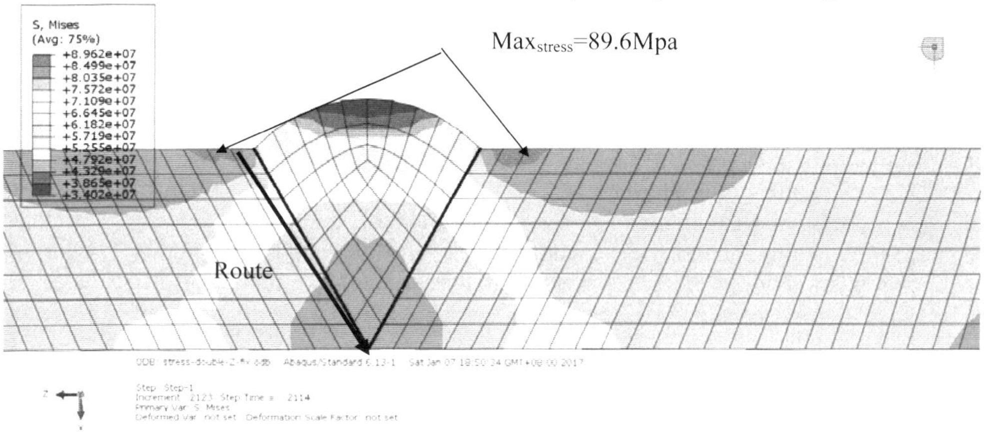

FIGURE. 2 local residual stress field distribution in weld

As shown in Figure 2, the stress variations along the fusion line to the weld root from the outer heat affected zone. The stress from the outer heat affected zone of the tube to the weld root is reduced first and then increased, and the minimum stress is 72MPa, appearing at about 2.3mm from the outer wall. The maximum stress is in the heat affected zone of the outer wall, the stress at the welding root is 84.2MPa, and the stress values are relatively high.

3.2 Influence of Working Fluid Stress

The rated working pressure of ultra supercritical tower boiler is 27.9MPa, so the maximum internal pressure of T23 water wall tube is 27.9MPa, and the external operating temperature is about 500°C. According to the internal pressure 27.9MPa, the calculated axial tensile stress of the tube is 16.8MPa. The welding residual stress field is loaded into the model as the initial stress field, and set the ambient temperature is 500 °C. The stress distribution of the T23 water-cooled wall tubes after superposition of the working load and the welding residual stress is shown in Figure 3. As can be seen from Figure 3, the maximum stress of the T23 water-cooled wall tube is distributed on the inner wall of the tube, where the stress at the welding root is concentrated, and the maximum stress is 113MPa. According to the established column coordinates, the radial, circumferential and axial stress field distributions of the work load and the welding residual stress are superimposed, can calculate radial forces, circumferential forces, and axial forces, respectively. The largest of the three is the circumferentia stress, and where the welding root it reaches 65MPa.

IOP Conf. Series: Materials Science and Engineering **242** (2017) 012051 doi:10.1088/1757-899X/242/1/012051

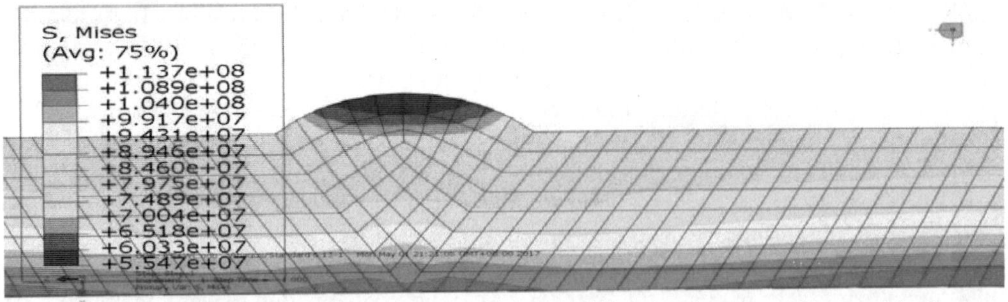

FIGURE. 3 Stress distribution after superposition of welding load and welding residual stress

3.3 Influence of Installation Defect Stress

In the installation of T23 water-cooled wall tubes, it is used brute force to butt joint when the stagger, the discount and the angle shift occur frequently. The welding process of T23 water-cooledwall tubes in the installation:

When welding parts correspond, the inner wall should be flush, and the wrong boundary value of the tube shall not exceed 0.5mm.When meet the wrong boundary value of the tube, can only be opened on both sides of fin by cutting, and then force counterparts, after the butt weld be done, the fin on the both sides will again weld between the T23 water-cooledwall tubes. The length of the fin being cut is not able to the same, and the force required to correct the T23 tube is different, and the bending moment is different for the weld.

The effect of bending moment remains when the T23 water-cooled wall tube is welded well. The influence of bending moment on stress distribution in T23 tubes by using ABAQUS. The stress field superimposed on the welding residual stress and the work load is taken as the initial stress field into the model, and then the bending moment load is applied at one end, The change of the three principal stresses of the T23 tube and the bending moment is obtained, as shown in Figure 4. With the increase of bending moment, radial stress, hoop stress and axial stress increase, but radial stress and hoop stress increase less, axial stress increases obviously.

According to table 2, it is found that the allowable stress is between 117-87MPa at the working temperature (450°C~550°C) of the T23 water-cooled wall tubes. In order to ensure the safety of use, choose 87MPa as the reference of the allowable stress. In order to avoid the axial stress beyond the allowable stress of the material, the maximum bending moment under the allowable stress is calculated by the finite element software. When the axial stress maximum allowable stress 87MPa, the maximum bending moment allowed is 756.5NM, the stress distribution is shown in Figure 5.The maximum stress value 238MPa is located in the outer heat affected zone of the pipe inside the bend. At this time, the radial stress is very small between the three, only 12.2Mpa.the hoop stress is larger than 74Mpa, and it is located in the inner part of the bend.

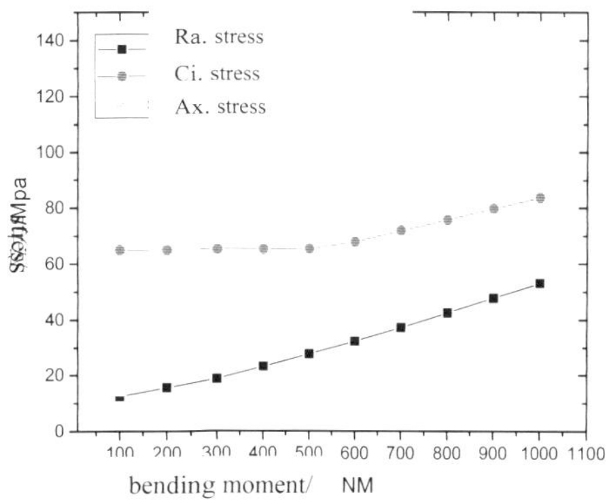

FIGURE 4. The change of the three principal stresses of the T23 tube and the bending moment

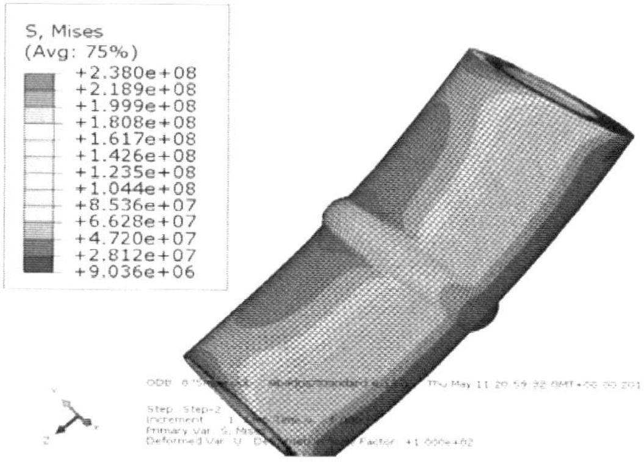

FIGURE 5. Stress distribution at bending moment of 756.5NM

4. Conclusions

The welding residual stress, especially the main stress in the three directions, is small, which the tensile stress is not enough to cause the damage of the material. Therefore, the effect of post welding heat treatment on reducing welding residual stress is not obvious. The working load has great effect on the hoop stress of T23 water-cooled wall tubes. The stress of T23 water-cooled wall tubes is mainly determined by the working load without considering the tension of the fin. The bending moment caused by the stagger mouth has obvious effect on the axial stress of the T23 water-cooled wall tubes. In order to ensure the safe and reliable operation of the T23 water-cooled wall tubes, it is recommended that the bending moment not exceeding 756.5NM be used.

Acknowledgments
This work is supported by Program of Shanghai Subject Chief Scientist (B type) (No.15XD1522700).

References

[1] Ahmad J, Purbolaksono J, Beng L C, et al. Failure investigation on rear water wall tube of boiler [J]. Engineering Failure Analysis. 2009, 16(7): 2325-2332.
[2] Yang Fu, Zhang Yinlin, Ren Yongning, et al.Welding of new type heat-resistant steels [M].Beijing: China Electric Power Press, 2006: 79-86(in Chinese).
[3] Vaillant J C, Vandenberghe B, Hanhn B, et al.T/P23, 24, 911 and 92: new grades for advanced coal-fired power plants-Properties and experience [J].International Journal of Pressure Vessels and Piping, 2008, 85(12): 38-46.
[4] Dhooge A, Vekeman J.New generation 2.25Cr steels T/P23 and T/P24 weldability and high temperature properties [J].Welding in the World, 2005, 49(9/10): 31-49.
[5] Ji Xianwu,Duan Peng,Li Ju, et al.Application of T23 steel to 1 000 MW ultra-supercritical units[J].East China Electric Power, 2009, 37(12): 2097-2100(in Chinese).
[6] Wang Changwen,He Peng.Analysis of the leakage of T23water wall tube of 1000 MW ultra supercritical boiler[J].Journal of Shanghai University of Electric Power, 2014, 30(4): 339-345(in Chinese).
[7] Almer J D, Cohen J B, Moran B. The effects of residual macrostresses and microstresses on fatigue crack initiation [J]. Materials Science & Engineering A. 2000, 284(01): 268-279.
[8] WANG Xue, ZHU Di, HU Lei, et al. Analysis on Short-term Failure of T23 Steel Welded Joints in Water Wall of USC Tower Boilers[J]. Proceedings of the CSEE, 2015, 35(Supplement): 154-161.

IOP Conf. Series: Materials Science and Engineering **242** (2017) 012052 doi:10.1088/1757-899X/242/1/012052

Study on loading path optimization of internal high pressure forming process

Shufeng Jiang, Hengda Zhu and Fusheng Gao

School of Mechatronics Engineering, Qiqihar University, Qiqihar, 161006, China

E-mail: 474118867@qq.com

Abstract. In the process of internal high pressure forming, there is no formula to describe the process parameters and forming results. The article use numerical simulation to obtain several input parameters and corresponding output result, use the BP neural network to found their mapping relationship, and with weighted summing method make each evaluating parameters to set up a formula which can evaluate quality. Then put the training BP neural network into the particle swarm optimization, and take the evaluating formula of the quality as adapting formula of particle swarm optimization, finally do the optimization and research at the range of each parameters. The results show that the parameters obtained by the BP neural network algorithm and the particle swarm optimization algorithm can meet the practical requirements. The method can solve the optimization of the process parameters in the internal high pressure forming process.

1. Introduction

Because of its light weight, good rigidity and less forming process, internal high pressure forming parts have been widely used in the production of automobile parts and components [1]. According to statistics, about 30% of the parts of the car body can be manufactured by internal high-pressure forming process. In the early 1980s, western countries began to study the forming technology of internal high pressure. By 1990s, internal pressure forming technology has begun to be widely used in the automotive field. Internal high pressure forming is a new plastic forming technology for tube forming under this background [2].

The quality of internal high pressure forming is related to many factors, such as the material and size of the pipe, the quality of the mould, the friction coefficient between the tube and the die and the loading path of the forming process [3]. In order to obtain qualified parts, researchers have carried out a great deal of research on the factors that affect forming [4]. Studies have shown that, after solving the material, mold and friction and other factors, there must be a suitable loading path to get high-quality forming parts [5]. Therefore, it is important to find a method to calculate and optimize the loading path to improve the forming quality, reduce the reject rate and reduce the production cost of internal high pressure forming parts [6].

Loading path has great impact on the quality of parts during the processing. How to get the optimal loading path is a central issue which always been researched by domestic scholars and foreign scholars [7]. Now the common method has theoretical calculations and numerical simulation. Internal high pressure forming process is so complicated that the result which is required by theoretical calculation is inaccurate. Because theoretical calculations need certain assumption and predigestion. The theory of numerical simulation is that use finite element software to simulate forming process [8]. It needs to adjust the input variables constantly to require the optimal loading path, and the results finally

Content from this work may be used under the terms of the Creative Commons Attribution 3.0 licence. Any further distribution of this work must maintain attribution to the author(s) and the title of the work, journal citation and DOI.
Published under licence by IOP Publishing Ltd

IOP Conf. Series: Materials Science and Engineering **242** (2017) 012052 doi:10.1088/1757-899X/242/1/012052

obtained are not an accurate result but a similar result. In response to these problems, the article propose a new method which gets the input parameters and the output by finite element software firstly, then calculates the mapping relationship between the input parameters and the output by BP neural network, finally uses the particle swarm optimization to optimize and get the optimal loading path.

2. Basic theories

Neural network is an algorithm which can simulate brain's physical function, and it has strong ability of logical operation and linear approximation. It includes computer, physics, mathematics, biology and other disciplines, and has broader prospect compared with mathematical modeling. BP neural network is a multilayer feed forward networks which is trained by error back propagation algorithm. Its basic theory is that input layer neurons receive input information, the middle layer deal with and convert message, and the output layer neurons calculate and export processing results. When the actual output has larger mistake compared with the expected result, BP neural network begin doing back-propagation. The algorithm adjusts constantly the network's weights and threshold to adjust the mistake to the appropriate extent [9].

Particle swarm optimization (PSO) is an intelligent algorithm which is proposed by simulating the phenomenon of bird predation. It is proposed by J. Kennedy and R. C. Eberhart in 1995. For introducing simple speed-die placement method and global searching tactics, particle swarm optimization avoids complex genetic manipulation and has higher searching efficiency and convenient performance. In recent years, particle swarm optimization has been widely used machinery, electricity, chemical industry and other fields to solve global extreme, integer programming, multi-objective optimization and other issues [10].

3. Test and simulation

3.1. Select design variables

Tube's internal high pressure forming process is affected by many factors, yielding pressure P_1, forming pressure P_2, finishing pressure P_3 and axial feeding Δl is the four most important parameters. These parameters can be calculated by such formula.

$$P_1 = \frac{2 \cdot t_0}{D_0 \text{-} t_0} \cdot \sigma_s \tag{1}$$

$$P_2 = \frac{2 \cdot t_0}{D_p \text{-} t_0} \cdot \sigma_b \tag{2}$$

$$P_3 = \frac{2}{\sqrt{3}} \cdot \sigma_f \cdot \left[\ln \frac{r_b}{r_b \text{-} t_0} \right] \tag{3}$$

$$\Delta l = \frac{\int_0^{l_1} f(x)dx}{R} \text{-} \Delta H \tag{4}$$

In the formula, D_0 is tube's diameter, t_0 is the tube's initial thickness, D_p is the tube's forming thickness, r_b is the mode's minimum fillet, $f(x)$ is the forming tube's profile curve, R is the tube's initial radius, ΔH is the forming tube's height of branch tube, σ_s is the tube's yielding limit, σ_b is the tube's strength limit, σ_f is the tube's flowing stress.

The subject of this test is a stainless steel tube. Its material is SS304, length is 200 mm, thickness is 1.5 mm, diameter is 20 mm. we want to get a four-way tube whose diameter is 20 mm and branch's height is 20 mm through internal high pressure forming. By the formula (1) to (4) and parameter in the

Table 1, we can get the curve about the relationship between the axial feeding and time (Figure 1) and the curve about the relationship between the internal pressure and time (Figure 2).

Table 1. Parameter list

t_0	D_0	D_p	r_b
1.5 mm	20 mm	20 mm	5 mm
σ_s	σ_b	σ_f	ΔH
205 MPa	520 MPa	625 MPa	20 mm

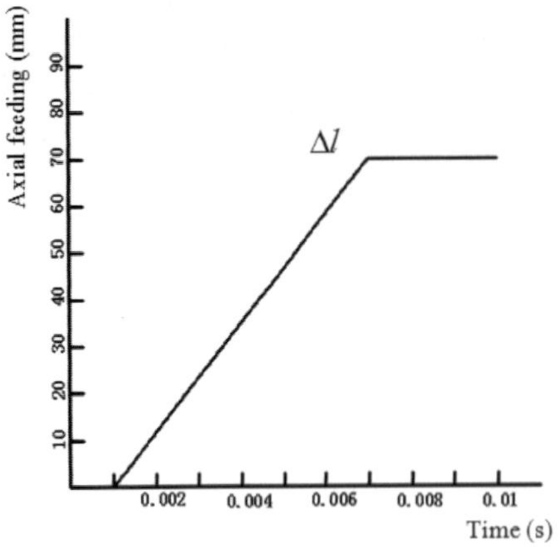

Figure 1. Relationship between the axial feeding and time

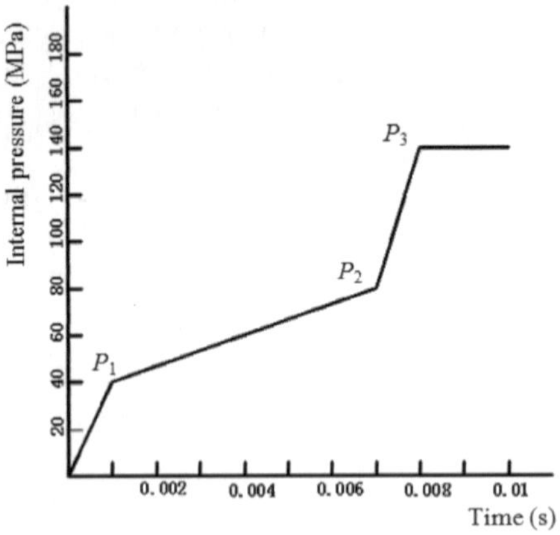

Figure 2. Relationship between internal pressure and time

This article make yielding pressure P_1, forming pressure P_2, finishing pressure P_3 and axial feeding Δl as the design variables, and according to the calculation result set each variable's range: $30 \le P_1 \le 50$, $50 \le P_2 \le 90$, $100 \le P_3 \le 200$, $50 \le \Delta l \le 70$.

3.2. Training BP neural network

When we want to evaluate a four-way tube's quality by internal high pressure forming, we often analyze branch tube's height ΔH, the minimum thickness of the tube $T_{\min}$ and the maximum thickness of the tube $T_{\max}$. The article takes these parameters as BP neural network's output variables and takes yielding pressure P_1, forming pressure P_2, finishing pressure P_3 and axial feeding Δl as the input variables. According to Empirical formula $m = \sqrt{m_1 + m_2} + a$, we set hidden layer's number of the nodes to 5. In the formula, m_1 is the number of input layer and m_2 is the number of the output layer. a is a constant whose range is from 1 to 10. According to these factors, we can get a model of BP neural network which has 3 layers.

We random select 40 sets of data from the range of input parameters. Then we use finite element software-DYNAFORM to simulate these parameters and get the corresponding output data. We make 21 sets of data of them as BP neural network's training sample, make others as BP neural network's training sample examining sample. The training result is as shown in Figure 3 and Figure 4. From Figure 3 and Figure 4, we can know that this BP neural network's forecasting accuracy is very well.

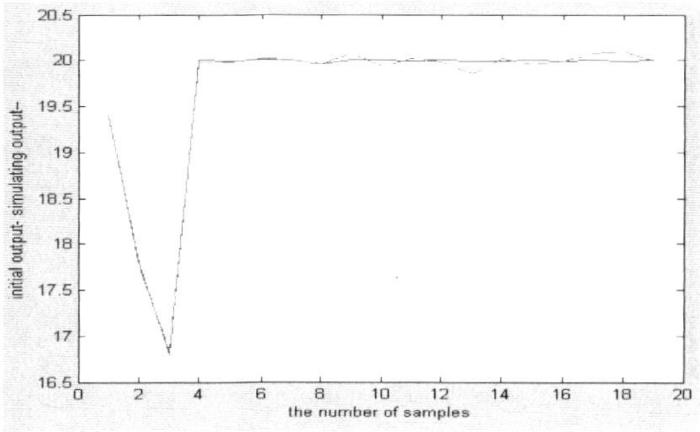

Figure 3. Comparison figure of ΔH 's training result

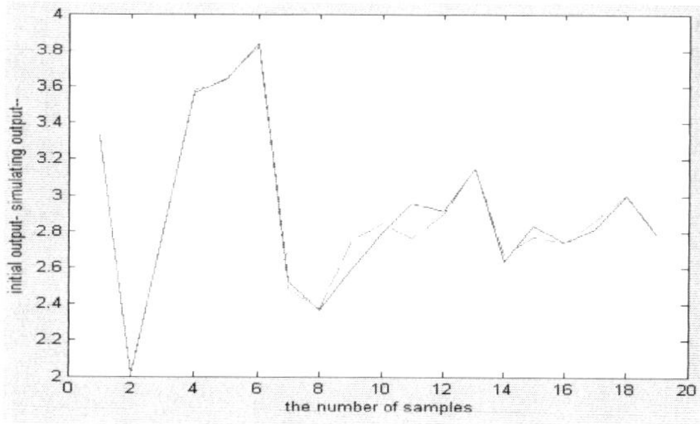

Figure 4. Comparison figure of T_{min} 's training result

4. Simulate the loading path

In this article we will simulate branch tube's height ΔH , the minimum thickness of the tube T_{min} and the maximum thickness of the tube T_{max} , it is a Multi-objective optimization problem. In order to simulate these three parameters same time, we need to make them into an entirety. In this article we describe them with formula (5).

$$f(x) = 0.1\Delta H + t_{min} - t_{max} \tag{5}$$

The value of the ΔH is 10 times compared with the value of the T_{min} and T_{max} , so we set the modulus of ΔH to 0.1 in order to make these parameters in the same order of magnitude. The value of ΔH and T_{min} is the smaller the better, but the value of T_{max} is the bigger the better, so we set the symbol of T_{max} to minus.

The process of optimizing loading path can be divided into 3 steps.

Step 1: initialize the parameters of particle swarm optimization; make the input parameters mapping a group of particles and initialize speed and position of particle.

Step 2: calculate the output parameters by trained BP neural network, and then calculate value of the objective function.

Step 3: judge whether the value are function's the optimal solution or whether the number of iterations have attained to the maximum. If meet the condition, the value is the expecting result. If don't meet the condition, renew the speed and position of particle and go to the step 2.

Optimization results are shown as Figure 5. In the figure, the blue points indicate all particles' position in the internal high pressure forming process, the red points indicate each generations of particles' optimal position. When the function get the optimal result, P_1=30.8599, P_2=74.959, P_3=12.1239, Δl=59.5066. We make the required value of the parameters into the DYNAFORM to simulate, and the result is shown as Figure 6. From the Figure 6 we can see that the branch tube's height ΔH is 19.95 mm and the minimum thickness of the tube T_{min} is 1.167 mm and the maximum thickness of the tube T_{max} is 3.079 mm, all results entirely reached the expected target.

5. Conclusions

(1) Using BP neural network to simulate tube's internal high pressure forming process can solve that tube's internal high pressure forming process has so many factors that it can't be described by mathematical model.

(2) Particle swarm optimization make the work of searching optimal loading path more simple and convenient and freed the craft workers from complex work which need to be calculate and simulate. Obtained many optimal solution can offers several options to craft workers.

(3) In the test, we need experience to select weighting factor of the objective function's parameters and set parameters of particle swarm optimization. Different selecting method has different effect on the result of the test. How to select all parameters scientifically is the key to further research.

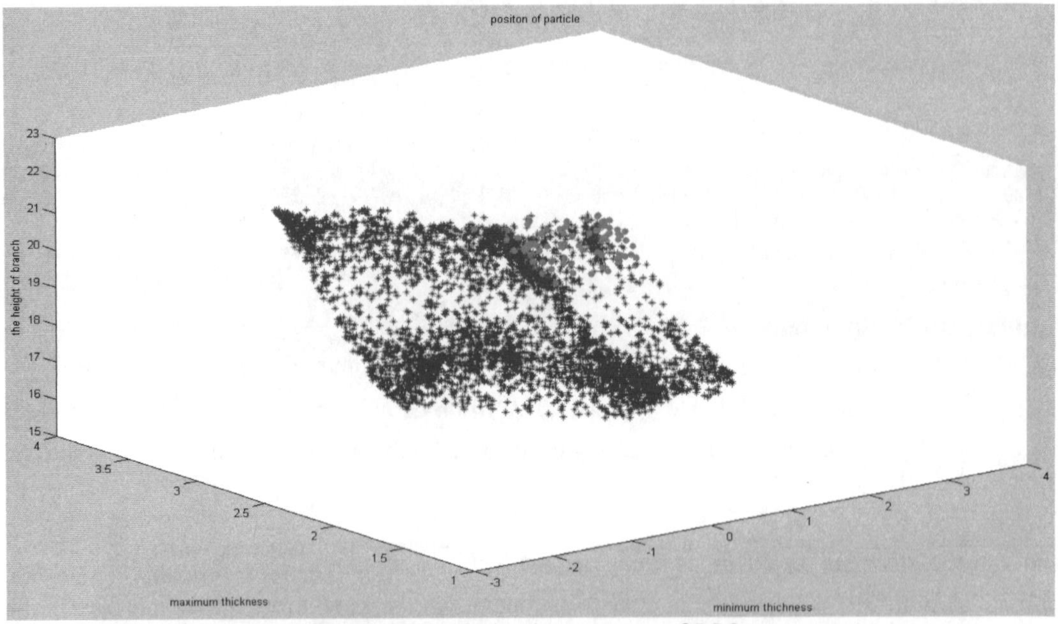

Figure 5. Optimization process of PSO

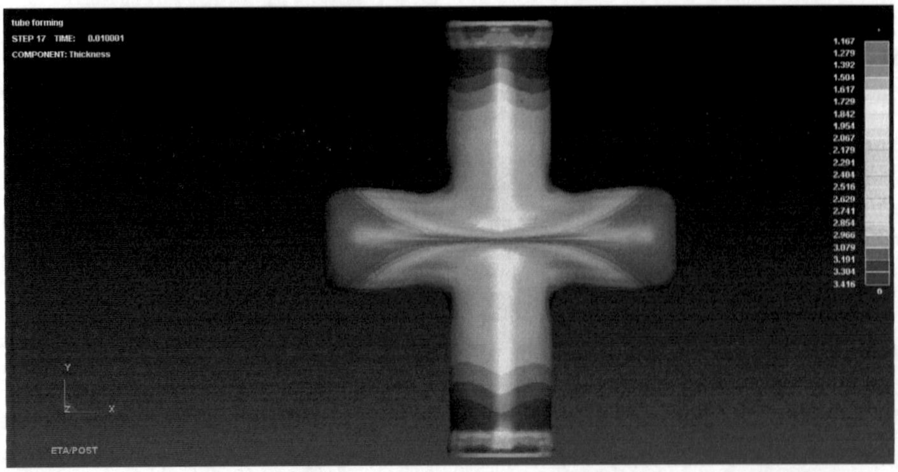

Figure 6. Simulating result

Acknowledgment
This work was supported by Qiqihar City Science and Technology Project (Grant No. GYGG-201101 and GYGG-201420) and Qiqihar University Scientific Research Project (Grant No. 135109217).

References

[1] M Imaninejad, G Subhash, A Loukus 2005 Loading path optimization of tube hydroforming of tube hydroforming process *International Journal of Machine Tools & Manufacture* **45** 1504-151.

[2] Koc M 1999 Development and design guidelines for part, tooling and process in the tube

hydroforming technoilogy *USA: The Ohio State University*.

[3] Sule Feng, Biyan Zhao, Yimin Luo 2014 Study on the application of hydraulic forming in the booster conveying system of the launch Vehicle *Manufacturing Technology* (**5**) 38-42.

[4] Altan T, Jirathearanat S, Strano M 2001 Adaptive FEM simulation for hydroforming tubes *Proceedings of 2nd International Conference on Hydroforming, Fellbach, Germany* 363-384.

[5] Bing Du, Changcai Zhao, Yijiang Liu 2014 Pipe hydroforming model *Chinese Journal of Mechanical Engineering* (**16**) 126-134.

[6] He Yang, Luoxing Li, Qu Wang 2010 Research on the development of science and technology in the field of light alloy forming *Journal of Mechanical Engineering* (**12**) 31-42.

[7] Choueiki M H, Mount-Campbell C A 1999 Training data development with the doptimality criterion *IEEE Transactions on Neural Networks* **10** 56-63.

[8] Shigang Wang, Dan Wang, Fusheng Gao 2014 Analysis of tee pipe hydroforming process parameters *Applied Mechanics and Materials* **651-653** 643-646.

[9] ZHOU Yi, XU Bai-ling 2001 Orthogonal method for training neural networks *Journal Nanjing University* **37** 72-78.

[10] Ziari I, Jalilian A 2012 Optimal allocation and sizing of active power line conditioners using a new particle swarm optimization-based approach *Electric Power Components and Systems* **40** 273-291.

Derivation of reliability formula design for cylinder- elliptical head structure

Xinchen Wei and Chenghong Duan*

Beijing University of Chemical Technology, Beijing 100029, China

duanchenghong@163.com

Abstract: Based on the JC method (reliability calculation method) and the fourth strength theory proposed by Lacvitz et al.[1] and recommended by the Joint Committee on International Security (JCSS), the reliability design formula for pressure vessel elliptical head and thin-walled circular cylinder is deduced. Monte Carlo reliability calculation method of ANSYS is used to verify the correctness of the formula, and the factors that have the greatest impact on the structural reliability are analyzed, which can provide some reference for the design of pressure vessel reliability.

1. Introduction

With the trend toward large-scale petrochemical industry development, the pressure vessel design requirements are getting higher and higher. Under the premise of ensuring the safety of the pressure vessel, taking into account its economy has become increasingly important [2]. In the design of the pressure vessel, the design pressure, material strength and structure size are all random, while existing standards treat these parameters according to the established amount. Using a larger safety factor to ensure the safety of the structure, obviously can't take into account the economy of the pressure vessel [3, 4]. In this paper, the reliability formula of the pressure vessel elliptical head and circular cylinder is deduced by JC method (reliability calculation method) and fourth strength theory, and applied to the design of the cylinder-elliptical head structure. Monte Carlo reliability calculation method [5] in ANSYS is used to verify the correctness of the formula. The factors that have the greatest influence on the structural reliability are analyzed, which can provide some reference for the design of pressure vessel reliability.

2. Calculation of reliability theory

2.1. Calculation of cylindrical reliability

Radial stress of thin-walled cylinder is much smaller than the axial stress and circumferential stress. According to the elastic mechanics, the three-way stress state of the cylinder can be obtained as follows:

$$\sigma_1 = \frac{PD}{2t} \tag{2.1}$$

$$\sigma_2 = \frac{PD}{4t} \tag{2.2}$$

$$\sigma_3 = 0 \tag{2.3}$$

Content from this work may be used under the terms of the Creative Commons Attribution 3.0 licence. Any further distribution of this work must maintain attribution to the author(s) and the title of the work, journal citation and DOI.
Published under licence by IOP Publishing Ltd

In the above formulas, P denotes the design pressure, D denotes cylinder diameter, t denotes the wall thickness of cylinder, and $\sigma_{i(i=1,2,3)}$ denotes the i-th principal stress.

According to the cylinder stress calculation formula and the fourth strength theory, it can be seen that the equivalent stress S (the stress strength of the cylinder) is:

$$S=\sqrt{\sigma_1^2-\sigma_1\sigma_2+\sigma_2^2}=\frac{\sqrt{3}PD}{4t} \tag{2.4}$$

P, D, and t in equation (2.4) are treated as random variables, it can be assumed that these random variables conform to the normal distribution according to statistics analysis and design experience. The Taylor formula expansion of S is carried out at the mean of the variable, and more than one level of expand items are neglected. Then according to the functional function of the cylinder (in equation (2.5)), the reliability formula of the cylinder is calculated by JC method as follows:

$$Z=R-S \tag{2.5}$$

$$\beta=\frac{\mu_R-\mu_S}{\sqrt{\sigma_R^2+\sigma_S^2}}=\frac{\mu_R-\dfrac{\sqrt{3}\mu_P\mu_D}{4\mu_t}}{\sqrt{\sigma_R^2+\left(\dfrac{\sqrt{3}\mu_D}{4\mu_t}\right)^2\sigma_P^2+\left(\dfrac{\sqrt{3}\mu_P}{4\mu_t}\right)^2\sigma_D^2+\left(\dfrac{\sqrt{3}\mu_P\mu_D}{4\mu_t^2}\right)^2\sigma_t^2}} \tag{2.6}$$

The intensity R in equation (2.5) is the yield strength of the material at the design temperature, taking 1.5 times the allowable stress at the design temperature as its value. In equation (2.6), $\mu_{i(i=R,S,P,D,t)}$ is the mean value of the variable, $\sigma^2_{i(i=R,S,P,D,t)}$ is the variable variance, β is the variable value of the standard normal distribution function.

2.2. Calculation of elliptical head reliability

The stress composition of the elliptical head is similar to that of the spherical head. In addition to the membrane stress caused by internal pressure, there are also discontinuous stresses of the junction. According to a large amount of engineering practice experience and theoretical research, the stress condition of the elliptical head is related to the ratio of the length of the ellipse as a/b. When a/b = 1.0 ~ 2.6, the following formula is used to calculate the stress enhancement factor K (also known as the shape factor) of elliptical head [6] in the engineering.

$$K=\frac{1}{6}[2+\left(\frac{D}{2h}\right)^2] \tag{2.7}$$

Equation (2.7) indicates that 2K equals head total stress divided by circumferential membrane stress. When a/b = 1.0 ~ 2.6, the total stress of the elliptical head is K times the membrane stress of the spherical head at the same radius, while the radius of spherical head equals to the long axis of elliptical head. Therefore, the functional function of the elliptical head is:

$$Z=R-S=R-K\frac{PD}{2t} \tag{2.8}$$

Reliability index of the elliptical head can be obtained after the corresponding mean and variance obtained by JC method.

$$\beta=\frac{\mu_R-\mu_S}{\sqrt{\sigma_R^2+\sigma_S^2}}=\frac{\mu_R-\dfrac{K\mu_P\mu_D}{2\mu_t}}{\sqrt{\sigma_R^2+\left(\dfrac{K\mu_D}{2\mu_t}\right)^2\sigma_P^2+\left(\dfrac{K\mu_P}{2\mu_t}\right)^2\sigma_D^2+\left(\dfrac{K\mu_P\mu_D}{2\mu_t^2}\right)^2\sigma_t^2}} \tag{2.9}$$

2.3. Reliability of cylinder-elliptical head structure

The reliability of the structure is:

$$P_r = \phi(\beta) \tag{2.10}$$

The failure probability is:

$$P_f = 1 - P_r = \phi(-\beta) \tag{2.11}$$

In the equations (2.10) and (2.11), $\phi(x)$ is the probability cumulative distribution function of the standard normal distribution.

3. Calculation of a case

3.1. Problem Description

There is a case of cylinder-elliptical head structure, the parameters and design conditions of a pressure vessel are shown in Table 1.

Table 1. Structure parameters of cylinder - elliptical head.

Parameters	Mean value μ	Standard deviation σ
Design pressure P/MPa	5	0.2
Design temperature T/℃	100	——[a]
Inner diameter of cylinder D/mm	1500	3
yield strength of material (16Mn) S/MPa	282	19.74
Long axis of head Rt/mm	1500	3
Length of cylinder L/mm	2000	——[a]

[a] The variable here is not a random variable.

According to the experience of pressure vessel design, we can see the reliability of cylinder is 99.999%. Find that $\beta = 4.27$ from standard normal function table, and then calculate the wall thickness of the cylinder and the head shown in Table 2 using the iterative method according to the formula (2.6) and formula (2.9).

Table 2. Calculation results of structure reliability method for cylinder - elliptical head.

Parts	Mean value	Standard deviation
Cylinder thickness t1/mm	20	0.134
Elliptical head wall thickness t2/mm	20	0.134

3.2. Reliability analysis

Take the above-mentioned cylinder-elliptical head structure calculated by theoretical formula as the object of study. The finite element model shown in Figure.1 is established by ANSYS. The reliability analysis of the cylinder - elliptical head structure is carried out. The inner diameter (D), head long axis (Rt), cylinder wall thickness (t1), head wall thickness (t2), design pressure (P) and yield strength (yies) of the cylinder are taken as random variables that follow the normal distribution during the analysis process.

$$Z = yies - maxstr \tag{3.1}$$

In the formula (3.1), maxstr represents the maximum von Mises stress in the structure. The failure probability P_f of the structure can be obtained through Monte Carlo Method in ANSYS, which is the probability of $Z \le 0$.

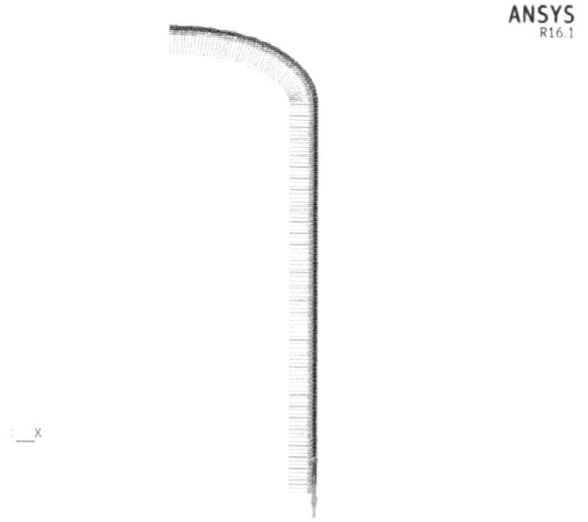

Figure 1. Finite element model of cylinder - elliptical head structure.

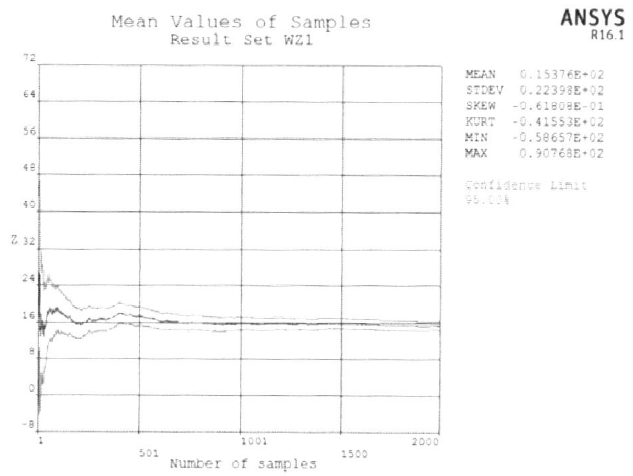

Figure 2. Sampling times of functional function Z.

It can be seen from ANSYS reliability analysis results that:

1) The functional function Z sampling process (see Figure.2) is within 95% confidence level, and indicates that the number of simulations is sufficient.

2) The sensitivity analysis of the output variable Z is shown in Figure 3, and it is known from the sensitivity analysis results that yies, P and t1 have a great effect on functional functions, and the influence of other parameters is small. When yies increases, the carrying capacity of the structure increases, and the reliability increases.

3) The value of yies is shown in Figure 4. There is no overshoot and step in the histogram and indicates that the number of sampling is sufficient and the simulation results are credible.

4) In the case of 95% confidence, the probability of Z<0 is 0%. That is, the failure probability of the container is 0%, so that the reliability of the structure is 100%.

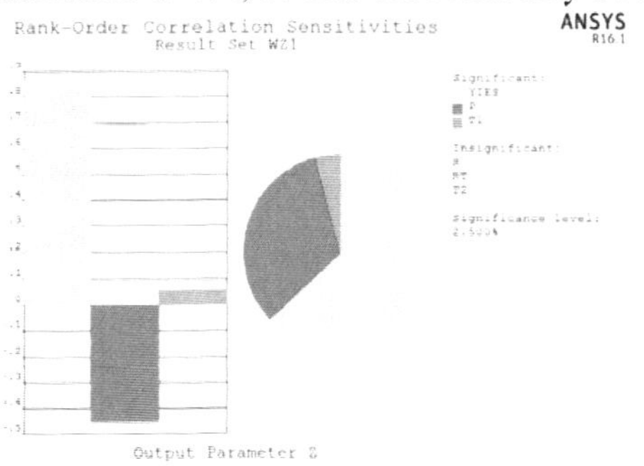

Figure 3. Sensitivity analysis of functional function Z.

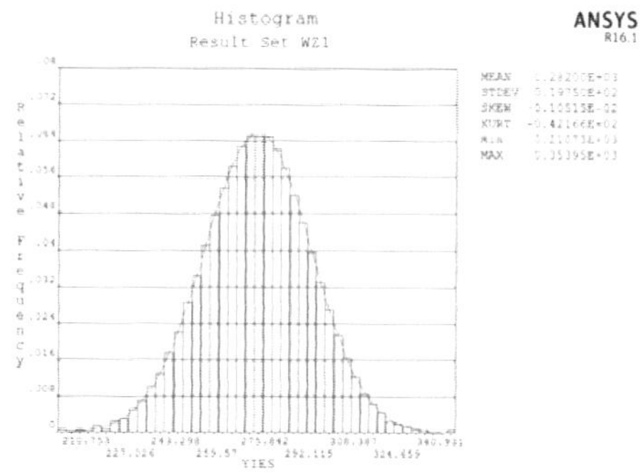

Figure 4. Histogram of material yield strength (yies) Value.

4. Conclusion

In pressure vessel design process, design pressure, diameter of cylinder, yield strength of material and other factors are not independent of each other. There is a certain correlation between each other, so it is necessary to carry out reliability analysis of the pressure vessel. In this paper, the formula for reliability calculation of cylinder-elliptical head is deduced by JC method. And Monte Carlo reliability calculation method is used in ANSYS. The results show that the reliability formula of this paper can be used to design the reliability of cylinder-elliptical head.

IOP Conf. Series: Materials Science and Engineering **242** (2017) 012053 doi:10.1088/1757-899X/242/1/012053

References

[1] Rackwitz R and B Flessler 1978 Structure reliability under combined random load sequences. *Computers & Structure* 9(5) pp 489-494
[2] Zhiming Lu 2009 *Reliability design of pressure vessel and its application on solid rocket motor vessel.* Beijing Jiaotong University p 121
[3] Yao and LRFD 2013 *Calibration of bridge foundations subjected to scour and risk analysis.* Dissertations & Theses – Gradworks
[4] Marquez B and J. 2015 *A simplified method to develop LRFD preliminary design charts for prestressed concrete bridges.* Dissertations & Theses – Gradworks
[5] Bo Sun 2014 *Stress analysis and reliability analysis of pressure vessel in sintering furnace. Shenyang University of Technology p 77*
[6] Cai B et al. 2013 *Exploratory study on load and resistance factor design of pressure vessel for subsea blowout preventers.* Engineering Failure Analysis 27(1) pp 119–129

Effect on property of HIC-Resistance of vessel steel of PWHT

Xinyu Zhao[1,2] **, Yang Zou**[1,2]**, Liye Qin**[1,2] **, Yanchun Lv**[1,2]

[1] Plate Technology Department, Shougang Research Institute of Technology, Beijing 100043

[2] Beijing Engineering Research Center of Energy Steel, Beijing 100043

xyz1076@sina.com

Abstract. Post-Welding Heat Treatment (PWHT) is usually taken after welding and joining of vessel steel, which effects the mechanical and Hydrogen Induced Cracking (HIC) resistance property of vessel plates. Simulating PWHT experiment was taken to research on the effect on mechanical property and HIC-resistance of PWHT of vessel plates. Some conclusions can be summarized as following. Comparing with the normalizing samples, the tensile strength of the samples after PWHT with holding time of 2h, 6h and 20h decreases by 27MPa, 44MPa and 47MPa. Ductile Brittle Transition Temperature (DBTT) of the normalizing samples was almost close to those of the samples after PWHT. But the impact energy of samples at 0℃ increased with rise of PWHT holding time. And the hardness of samples decreases with rise of PWHT holding time. As shown in morphology structure, the precipitation of carbonide increases with the rise of holding time of PWHT, which decreases the strength and hardness of samples and raises the impact energy. But PWHT has a little effect on HIC-Resistance, which means PWHT don't deteriorate property of HIC-Resistance of vessel plates severely.

1. Introduction

PWHT is a heat treatment method in order to improve the mechanical property and structure and relief the residual stress of welding joint, which is kind of stress relief annealing. A part or component is heated to a special temperature with a special heating rate and maintained to be a fixed time and then cooled in furnace with very low cooling rate.

An amount of reports had been published about the research on effect on mechanical property of PWHT through simulating method in Lab [1]-[6]. Also the effect of the PWHT on the HIC-Resistance property had not been investigated sufficiently. The simulating PWHT in laboratory was taken to research on the effect of PWHT on HIC-Resistance of vessel plates in this paper.

2. Experimental procedure

The samples with a chemical composition, given in table 1, were prepared. And the samples conform to the requirements provided by criterion that Yield Stress (YS) ≥220MPa, Tensile Stress (TS) ≥ 415MPa.

Table1 Chemical composition

C	Si≤	Mn≤	P≤	S≤	Alt≤	Nb+V+Ti≤	Mo≤	Cu≤
0.12%-0.18%	0.45%	1.50%	0.01%	0.003%	0.05%	0.1%	0.1%	0.1%

The industrial manufacturing procedure of samples is BOF-LF-RH-CC slab-TMCP. Then 4 samples were cut to normalize in labarotory. After that 3 samples of these were heat treated by PWHT.

Content from this work may be used under the terms of the Creative Commons Attribution 3.0 licence. Any further distribution of this work must maintain attribution to the author(s) and the title of the work, journal citation and DOI.
Published under licence by IOP Publishing Ltd

The samples, with a thickness of 37.7mm, were heat treated in table 1. The sample code, given in table, is the abbreviation of samples with different heat treatment and different holding time.

Table 2 Experimental scheme

	Sample code	Sample thickness	Heat treatment	Temperature	Holding time
Sample 1	880N	37.7mm	Normalization	880℃	1h
Sample 2	635-2h	37.7mm	PWHT	635℃	2h
Sample 3	635-6h	37.7mm	PWHT	635℃	6h
Sample 4	635-20h	37.7mm	PWHT	635℃	20h

The thermal histories of normalization and PWHT are given in Fig. 1. The normalizing temperature is 880℃ and the heating and cooling rate ranged from 50℃/h to 1000℃/h. Then the simulating PWHT process in labrotory is that a sample was thrown into heat furnace with temperature 25℃ and heated to a uniform temperature, 635℃ in this experiment. Also the sample was held with a special holding time at this temperature for a period of time and then cooled in the heat furnace. The heating rate is not larger than 100℃/h. Then the holding time are 2h, 6h and 20h respectively. After that the cooling rate is not larger than 100℃/h.

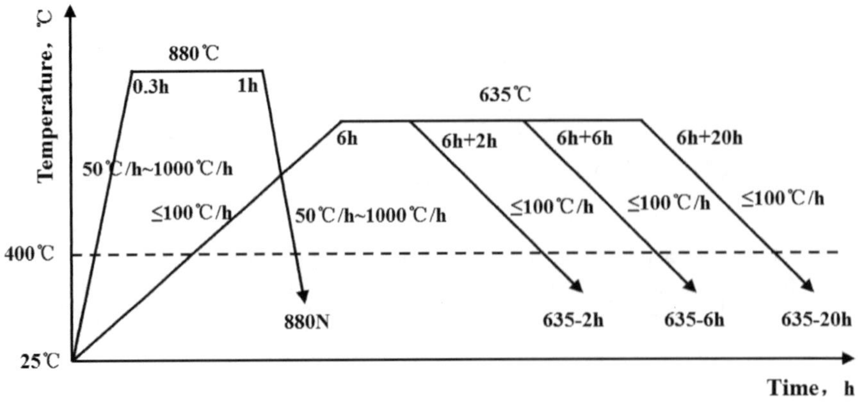

Fig. 1 Thermal histories of heat treatment

The HIC test experiment is accordance with criterion of NACE 0284. And the solution A provided by the criterion is taken in the experiment. Finally three ratios of Crack Length Ratio(CLR), Crack Thickness Ratio(CTR) and Crack Senstivity Ratio (CSR) were got[7]. A HIC-Resistance Index (HRI) was proposed to estimate the property of HIC-Resistance. And

HRI=max((CLR)/15, (CTR)/5, (CSR)/2)

Then in case of HRI＞1 the property of HIC-Resistance of plate is not standards-compliant. Otherwise, in case of HRI≤1 the property is standads-compliant.

3. Results and discussion

The effect on strength of PWHT is given in Fig.2. As shown in figure, the strength of the sample after simulating PWHT, comparing with normalizing sample, decreases with rise of holding time.The YS of normalizing sample is 320MPa. But the YS of the samples after simulating PWHT, comparing with normalizing sample,decreases by about 20MPa. The YS of the samples after simulating PWHT with different holding time ranges from 297MPa to 307MPa. At the same time TS of normalizing sample is 437MPa. But TS after simulating PWHT samples significantly decreases with rise of holding time. TS decreases by 27MPa 44MPa and 47MPa, corresponds with the holding time of 2h, 6h and 20h respectively.

The charpy impact tests were taken at the position of 1/4 thinckness direction. And tests were conducted at 0℃、-20℃、-40℃、-60℃and-80℃. The effect on DBTT of PWHT is given in Fig.3.

As shown in figure, DBTTs of the normalizing and PWHT samples are all approximately 40℃, whatever the holding time is 2h, 6h or 20h. But charpy impact energy of 0℃ increases with a rise in the holding time.

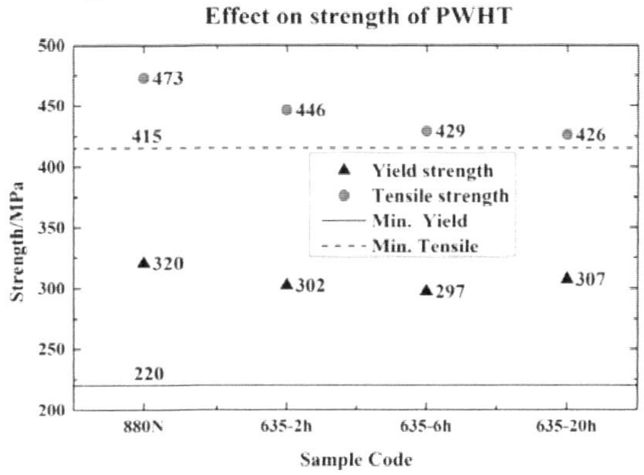

Fig.2 Effect on strength of PWHT

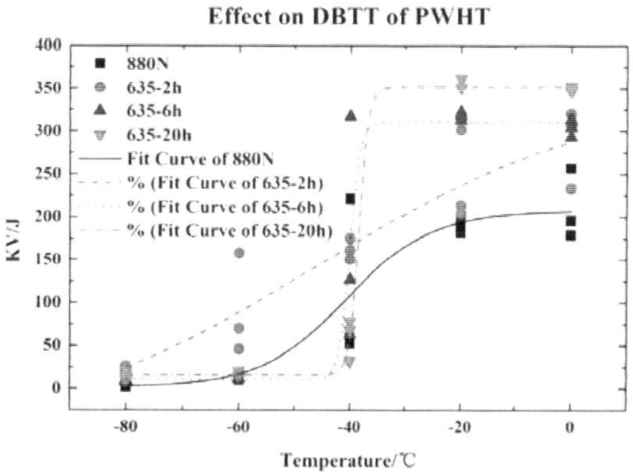

Fig.3 Effect on DBTT of PWHT

The effect on HRI of PWHT is given in Fig. 4. As shown in figure, the HRI has a litte increase with a rise in the holding time. But all HRIs are under the Max HRI, which means PWHT don't deteriorate the propety of HIC-Resistance of vessel steel severely.

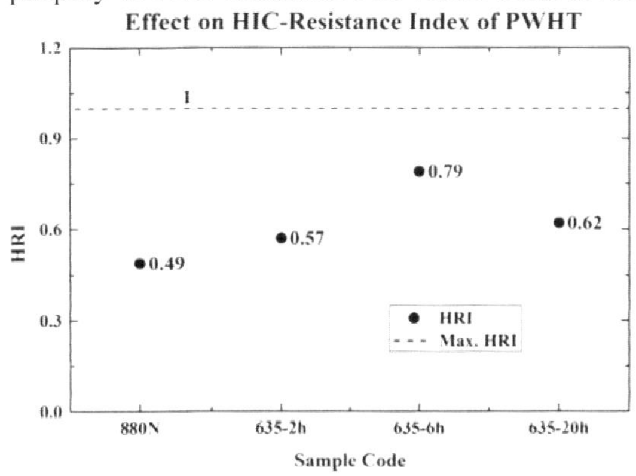

Fig.4 Effect on HRI of PWHT

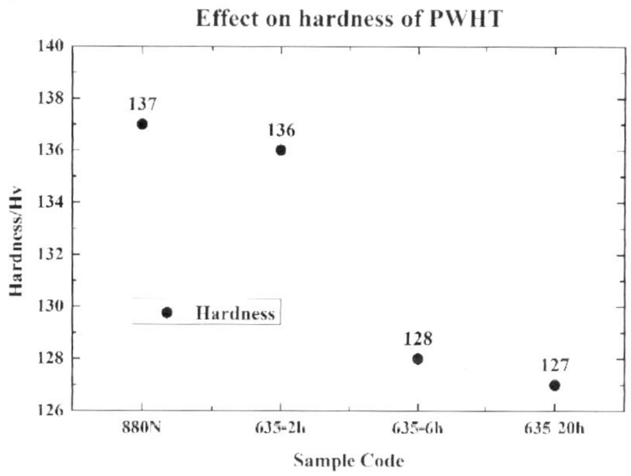

Fig. 5 Effect on hardness of PWHT

The effect of hardness of PWHT is shown in Fig. 5. As shown in figure, hardness of the samples after simulating PWHT,comparing with normalizing samples, decreases apparently, especially with a rise in holding time.

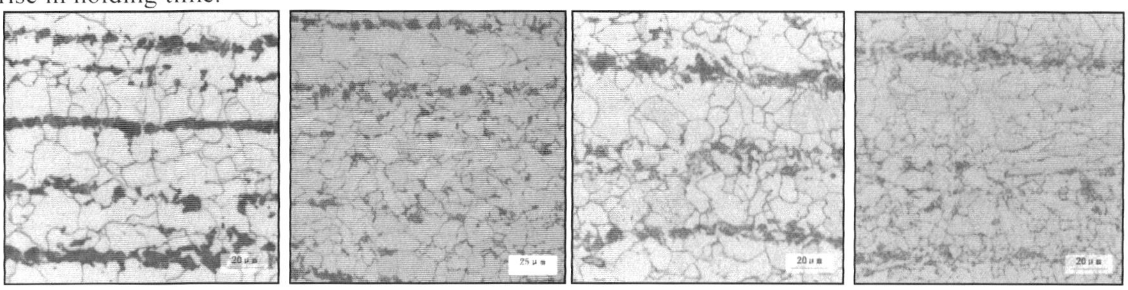

880N 635-2h 635-6h 635-20h

Fig.6 OM images of the effect on morphology structure of PWHT

The OM images of the effect on morphology structure of PWHT are given in fig.6. As shown in figure, the microstructures of 4 samples were all composed of ferrite and pearlite. Moreover more dot-like carbonides could be found on pearlite of simulating PWHT samples comparing with normalizing samples. Then the content of pearlite of the samples after simulating PWHT decreases because of precipitation of carbonides, which decreases the strength and hardness but increases the toughness of the samples after PWHT. However the carbonides don't deteriorate the HIC-Resistance of the samples significantly.

4. Conclusions

PWHT is usually taken after welding and joining of vessel steel, which effects the mechanical and HIC resistance property of vessel plates. Simulating PWHT experimental process was taken to research on the effect of PWHT on mechanical property and HIC resistance of vessel plates. Some conclusions can be summarized as following.

1) The strength of the samples after PWHT, comparing with normalizing samples, decreases with a rise in holding time. And TS of samples after PWHT with holding time of 2h, 6h and 20h, comparing with TS of normalizing samples, decreases by 27MPa, 44MPa and 47MPa.

2) The DBTTs of normalization and PWHT samples are approximately 40℃, which means PWHT don't change the DBTT no matter how long holding time is. But the impact energy of 0℃ increases with a rise in holding time of PWHT.

3) The hardness of PWHT samples decreases with a rise in holding time.

4) More dot-like carbonides could be found on pearlite of simulating PWHT samples comparing with normalizing samples. Then the content of pearlite of the samples after simulating PWHT decreases because of precipitation of carbonide, which decreases the strength and hardness but increases the toughness of the samples after PWHT. But PWHT has little effect on HRI, which means PWHT don't deteriorate property of HIC-Resistance of vessel plates severely.

References

[1] WANG J, LU H, MURAKAWA H. Mechanical Behavior in Local Post Weld Heat Treatment (Report I): Visco-Elastic-Plastic FEM Analysis of Local PWHT (Mechanics, Strength & Structure Design)[J]. Transactions of JWRI, 1998, 27(1): 83-88.

[2] Guozheng Zhang, Feng Zhao. Research on Post weld Heat Treatment for Steel Pressure Vessel [J]. Industrial heating, 2013,01:66-68.

[3] Ravi S, Balasubramanian V, Nasser S N. Influences of post weld heat treatment on fatigue life prediction of strength mis-matched HSLA steel welds[J]. International journal of fatigue, 2005, 27(5): 547-553.

[4] Mitchell D R G, Moss C J, Griffiths R R. Optimisation of post-weld heat treatment of a 1.25 Cr–0.5 Mo pressure vessel for high temperature hydrogen service[J]. International journal of pressure vessels and piping, 1999, 76(4): 259-266.

[5] WANG Z, CHEN W. Local Post Weld Heat Treatment on Site for Pressure Vessel of Large-diameter and Thicker Wall [J][J]. Petro-Chemical Equipment, 2010, 1: 015.

[6] Stout R D. Postweld heat treatment of pressure vessel steels[J]. Weld. Res. Counc. Bull., 1985 (302): 1-14.

[7] NACE TM 0284. Standard Test Method—Evaluation of Pipeline and Pressure Vessel Steels for Resistance to Hydrogen Induced Cracking, 2003.

IOP Conf. Series: Materials Science and Engineering **242** (2017) 012055 doi:10.1088/1757-899X/242/1/012055

Dimensional optimization of the crossbeam of the spectacle frame CNC polishing machine based on sensitivity analysis

Kang Chen[1,2,3], Kai He[1,2,*], Ming Ge[4] and Ruxu Du[5]

[1] Shenzhen Institutes of Advanced Technology Chinese Academy of Science, Shenzhen, Guangdong, China
[2] Shenzhen Key Laboratory of Precision Engineering, Shenzhen, Guangdong, China
[3] School of mechanical and Automotive Engineering, South China University of Technology, Guangzhou, Guangdong, China
[4] The Hong Kong Productivity Council, Kowloon, Hong Kong SAR, China
[5] Institute of Precision Engineering, The Chinese University of Hong Kong Hong Kong SAR, China

***Email:** kai.he@siat.ac.cn

Abstract. Based on finite element analysis, we obtained the first six natural frequencies and corresponding vibration modes of the crossbeam of the spectacle frame CNC polishing machine, then we proposed some improvement according to the FEM analysis results. In order to determine the reasonable size parameters, we conducted the sensitivity analysis to the mass, total deformation and first frequency of the crossbeam. The results showed that three key parameters had greater impact to the structural characteristics of the crossbeam, while the other parameters had almost no effect. An optimization problem had been established to determine the three parameters (L, H and B), and the mass, total deformation and first frequency of the crossbeam were set as optimization target. With the help of the ANSYS optimization tool, we got three groups of candidate points, and later we used an evaluation function to determine the suitable parameters.

1. Introduction

The crossbeam of numerical control grinding machine is an important part to connect the equipment and polishing tool structure of machine tools. When the machine is working properly, the mass of the crossbeam will affect the rapidity and sensitivity of the response of the control system. The machine is equipped with motor and transmission parts inside the grinding machine, the vibration of the motor can be transmitted to the tool by means of the mounting plate, and it would have a bad effect on the process. Therefore, it is important to improve the machining accuracy and stability of the machine tool.

Structure optimization is an effective measure to enhance the dynamic characteristics of the machine tool. The research of structural optimization is mainly carried out in two aspects[1], one is structural topology optimization, which is based on finite element analysis, the other is to optimize the shape feature with the help of the optimization algorithm. Dimensional optimization is a mature method of detailed design, the advantage of dimensional optimization is that the performance of the structure is taken into account in the design process, so we could obtain the reasonable structure parameters that meet the requirement.

The sensitivity analysis method is a kind of optimization analysis method, which is based on the establishment of an accurate finite element model and a reasonable mathematical model[2]~[5].The

Content from this work may be used under the terms of the Creative Commons Attribution 3.0 licence. Any further distribution of this work must maintain attribution to the author(s) and the title of the work, journal citation and DOI.
Published under licence by IOP Publishing Ltd

shape and size parameters of different parts have different effects on the structure performance, that is to say, the sensitivity is different[6]. By analyzing the sensitivity of the structural parameters, the most sensitive parameters are found out and the optimization design is carried out.

In this paper, the finite element model of the crossbeam of the frame grinding machine was established, later we obtained the first six natural frequencies and the corresponding vibration modes of the beam based on the basic theory of modal analysis. According to the simulation results, the improvement measures were put forward, then we carried out sensitivity analysis to the original dimension of the crossbeam. Based on the sensitivity analysis, we took 3 key parameters to conduct the optimization, and the mass, total deformation and frequency were set as the optimization target. We used the direct optimization tool in ANSYS to get 3 groups of candidate points, and later we chose the best one by evaluation function.

2. The introduction of the polishing machine structure

2.1. Introduction of the polishing machine

The spectacle frame CNC polishing machine was used to burnish plastic resin frames of the spectacles. To make the polishing process more smooth, we took a four-axis scheme, in which three sliders was fixed on the holder and the sliders were orthogonal to each other. As shown in Figure 1, the three sliders could move in straight-line motion independently. The crossbeam was fixed on the slider of Z-direction, and it could move in compound space motion by following the sliders. The turntable was the fourth axis which could turn around the Y-axis of the machine tool coordinate system, the fixtures and workpiece were fixed on the turntable. Under the coordination of the four shafts, the machine tool can finish the machining of the complex surface of the workpiece.

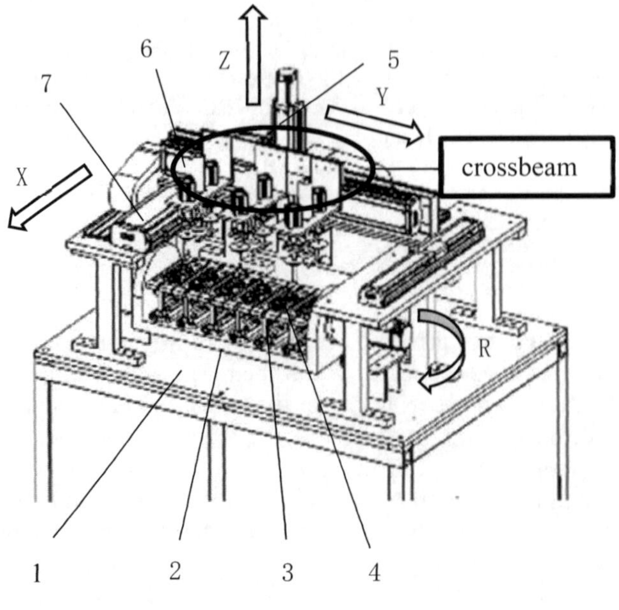

1. Holder 2. Turntable 3. Fixture 4. Workpiece 5. Z-axis slider
6. Y-axis slider 7. X-axis slider

Figure 1. The structure of the polishing machine

2.2. Introduction of the crossbeam

The crossbeam plays a role of coupling and transmission, and the tool driven by the crossbeam could be able to reach any point in space. There are three groups of tool fixed on the crossbeam, so the stability and the static rigidity of the crossbeam could affect the working performance of the machine tool. What's more, the weight of the crossbeam reflected the rapidity and sensitivity of the sliders.

As shown in Figure 2, the crossbeam was made of alloy steel, the dimensional parameters were shown in Table 1, and we listed the interval of these parameters, we would conduct modal analysis by finite element method to get the crossbeam.

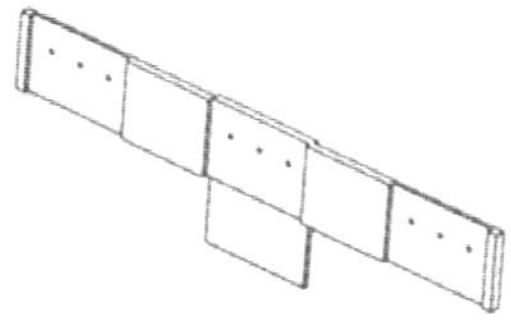

Figure 2. Crossbeam Model

Table 1. Structural parameters of the crossbeam

Item	Initial dimension (mm)	Variation (mm)
Lateral span: L	580	560~600
Beam height: H	80	75~95
Beam thickness: B	10	5~15
Groove width: W	120	120
Groove center distance: S	220	220
Thickness of connecting plate: D	5	5~10
Groove depth: A	5	5
Height of connecting plate: T	150	150~200
Board width: K	120	120

3. FEM modal analysis of the crossbeam

3.1. Modal analysis

The crossbeam model was established in ANSYS Design Module, we imported the built model into ANSYS Workbench, and we set the parameters of the simulation environment, as shown in Table 2.

Table 2. Properties and parameters of the FEM model

Properties	Parameters
Material of frame	Alloy steel
Density	7.9*103
Young's modulus	0.3
Grid quantity	2187

After we set the primary parameters, we adopted the hexahedral element to generate the mesh. To avoid stress concentration, we made the boundary dense and the transition was gentle, in the non key area, we used larger mesh, so the computing time was reduced. According to the actual installation of the crossbeam, the back connecting plate of the cross beam was provided with a fixed restraint. After we set the solution condition, the first six order natural frequencies and vibration modes of the beam were calculated, as shown in Table 3 and Figure 3~ Figure 8.

Table 3. The first six frequencies and modes of the crossbeam

Order	Frequencies(Hz)	Description of modal shape
First order	147.06	The left wing plate swung back and forth
Second order	147.72	The offside plate swung back and forth
Third order	364.74	The offside plate twisted around the horizontal center
Fourth order	364.79	The left side plate twisted around the horizontal center
Fifth order	576.85	The left side plate swung back and forth in a wave motion
Sixth order	578.31	The offside plate swung back and forth in a wave motion

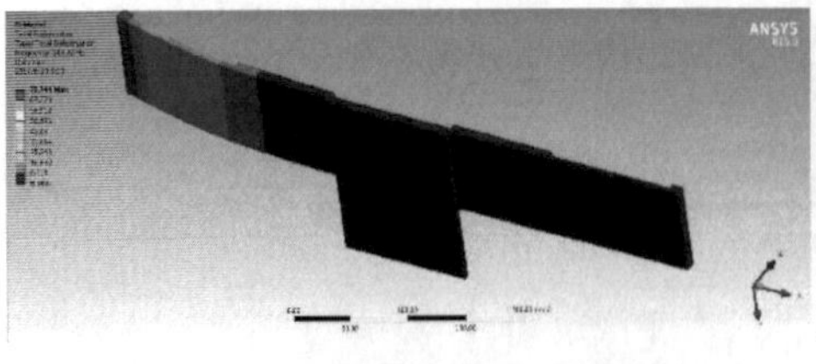

Figure 3. The first order vibration mode

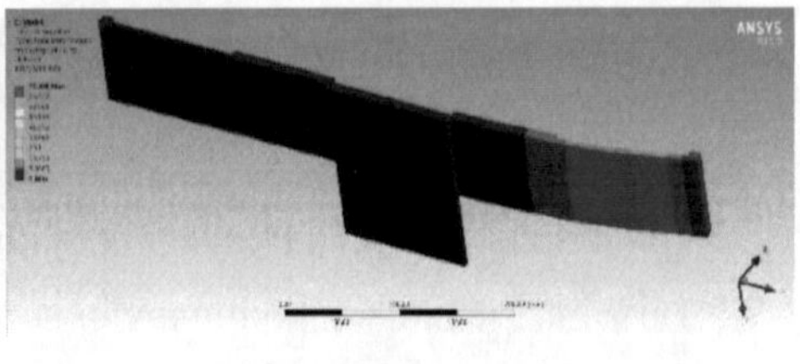

Figure 4. The second order vibration mode

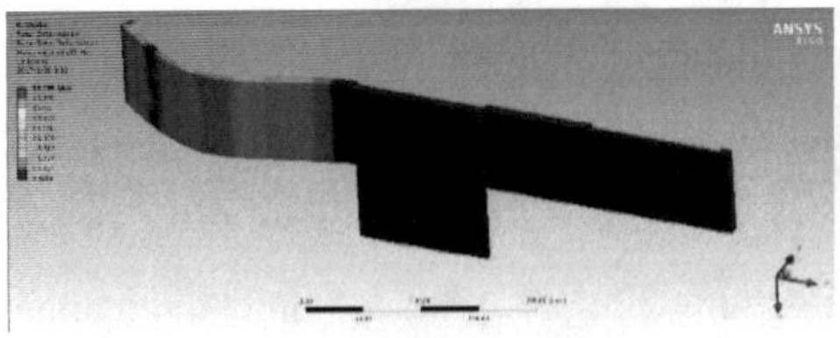

Figure 5. The third order vibration mode

Figure 6. The forth order vibration mode

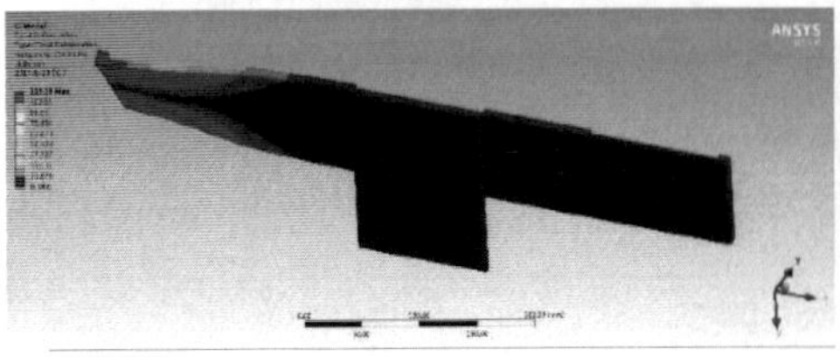

Figure 7. The fifth order vibration mode

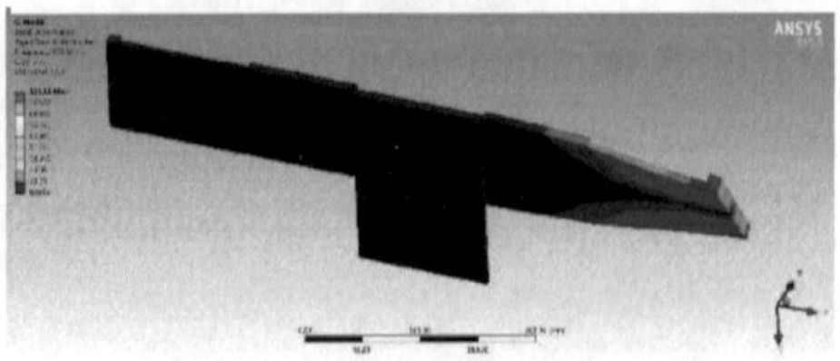

Figure 8. The sixth order vibration mode

3.2. Analysis of the results

From the Figure 3~8, we can see the first six frequencies of the beam were lower, and among them, the first two frequencies were close to the free rotation frequency of the motor, which was around 100 Hz. The vibration of the crossbeam would increase with the increasing of the motor rotating speed, so we need to improve the natural frequencies of the crossbeam. If the tool is subjected to greater damping after the motor starts, the speed may decrease slightly. The operating frequency is close to the low order natural frequency of the crossbeam, in this condition , the resonance occurs.

We can see from the modal shape that the wing of the crossbeam swung on either side, so we can add some stiffened plate or apply prestressing to increase the stiffness in that direction. Later we would use optimization method to change the characteristic of the crossbeam.

4. Sensitivity of the crossbeam

4.1. Dimensional sensitivity analysis

The sensitivity analysis method is a kind of optimization analysis method based on the establishment of an accurate finite element model and a reasonable mathematical model. In the process of structural

IOP Conf. Series: Materials Science and Engineering **242** (2017) 012055 doi:10.1088/1757-899X/242/1/012055

optimization, sensitivity can be used to indicate the extent to which the performance parameters of a system are affected by the size parameter. Usually the size parameter can be height, thickness, performance parameters can be the quality of the components and the natural frequency, dimensional sensitivity analysis can optimize the design parameters of the most sensitive component.

Assume a set of variables are: X_1, X_2 ... X_n , function S is the partial derivative of function $f(X_1,X_2...X_n)$, S_i is the sensitivity of a function to variable X_i, we call the sensitivity function S,

$$S_i = \frac{\partial f\left(X_1, X_2, \cdots X_n\right)}{\partial X_i} \quad (1)$$

According to function S, we could analysis the sensitivity of different dimension to the quality and the nature frequency of the crossbeam.

4.2. Sensitivity of dimensional parameters

In this sensitivity analysis, the structure parameters of the crossbeam were taken as the design variables. We used finite element analysis software ANSYS Workbench to get the sensitivity of mass, total deformation and frequency, as shown in Figure 9. In the left part of Figure 9, we could see that the first three variables (L, H and B) had significant influence on the mass of the crossbeam, while the other variables almost has no effect, the thickness of the crossbeam affected the most especially. In the middle part of Figure 9, we can see that only the thickness of the crossbeam affected the total deformation, and the other variables almost had no effect. In the left part of Figure 9, the length was the only influence factor of the first frequency.

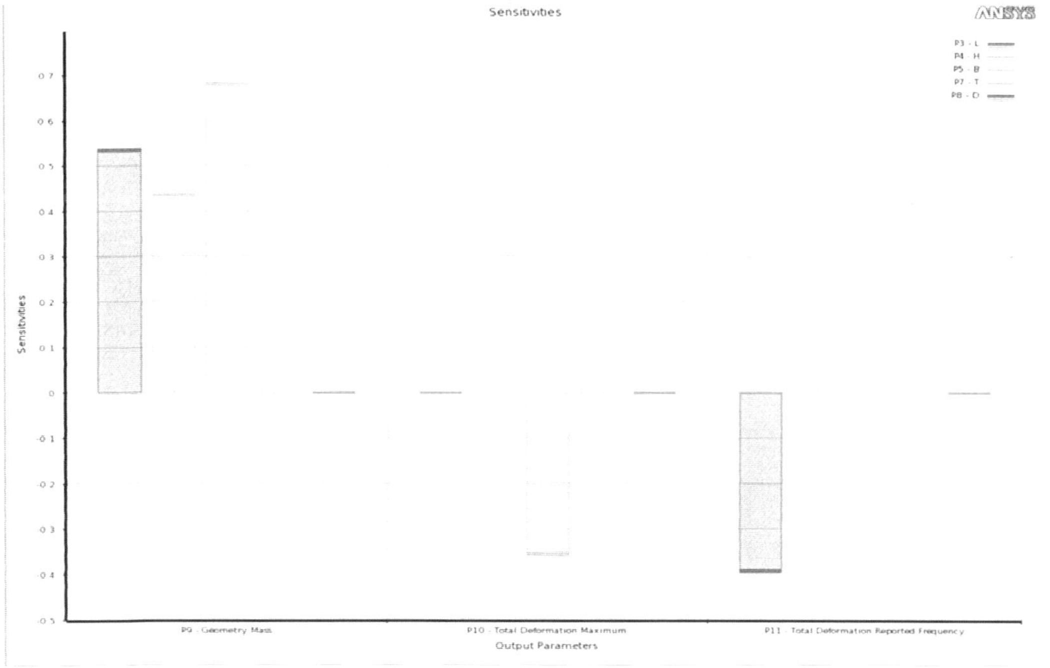

Figure 9. Sensitivities of the variables

In the subsequent optimization process, to simplify the optimization model, we focus on the first three variables (L, H and B) and ignored the other variables.

5. Dimensional optimization and analysis

5.1. Establish the optimization problem

The main content of the optimization design is to guarantee the constraint condition of the model, and to achieve small deformation, light weight, small size, reasonable shape, the lowest cost, and the best mechanical properties as far as possible. We set the range of the three parameters and defined optimization goal, as shown in Table 4.

Table 4. The optimization parameters

Parameter	Type	Lower bound	Upper bound	Requirement
L	Variable	560 mm	600 mm	none
B	Variable	5 mm	15 mm	none
H	Variable	75 mm	95 mm	none
Mass	Target	none	none	Minimum
Total deformation	Target	none	none	Minimum
First frequency	Target	none	none	Maximum

5.2. Solve the optimization problem

We used direct optimization tool in ANSYS to solve the optimization problem, 100 groups of points were generated by the suitable algorithm, after the computer finished computing, three groups of candidate points were selected, the candidate points and corresponding results were shown in Table 5.

Table 5. The candidate points and corresponding results

Parameter	Candidate Point 1	Candidate Point 2	Candidate Point 3
L (mm)	596.2	592.6	599.8
H (mm)	82.13	85.88	90.56
B (mm)	5.46	5.09	5.83
Mass (Kg)	1.65	1.52	1.91
Total deformation maximum (mm)	3701.6	4138.4	3253.3
First frequency (Hz)	54.93	50.62	52.54

5.3. Determine the three parameters

From the Table 5, we could see that the three candidate points behaved differently on different objective functions, the mass of candidate point 2 was the minimum of the three points, and the total deformation is the maximum of them. The first frequencies of the three candidate points were very close, we needed to take into account of both the mass and the total deformation maximum.

We established an evaluation function to determine the best value of the three candidate points:

$$\min\left\{E=0.7\times M+2\times D/10000-F/100\right\} \tag{2}$$

Where M is the mass, D is the total deformation maximum, F is the first frequency. We gave different weights to the two variables according to the influence degree of the performance of the polishing machine tool, the mass of the crossbeam was more important so we set the weight of the mass to 0.7, and the weight of total deformation maximum was set to 2/10000, the weight of the first frequency was set to 1/100. We calculated the value of the evaluation function, and chose the minimum. The three evaluation function values respectively were 1.35, 1.39 and 1.46, we chose the candidate point 1 as the design value according to the evaluation function, after rounding, we took the L as 596 mm, H as 82 mm and B as 5.5 mm.

6. Conclusion

In this paper, we obtained the vibration mode and natural frequency of the first 6 steps of the spectacle frame CNC polishing machine tool through the finite element analysis, and we found the weak link of the bed, which provided the basis for the structural optimization design. Later we conducted sensitivity

IOP Conf. Series: Materials Science and Engineering **242** (2017) 012055 doi:10.1088/1757-899X/242/1/012055

analysis of the crossbeam, and we found the three key parameters which had greater impact to the crossbeam. Focused on the mass, total deformation and first frequency, we set an optimization problem and solved it with the help of the ANSYS software. Eventually, we got the best group of parameters of the crossbeam.

Acknowledgments
This paper is supported by the Shenzhen-Hong Kong Innovative Circle Project (SGLH20150213143207909). The authors would like to thank OKIA Optical limited company for their support.

References
[1] FU Shi-xin QIN Gang WANG Qiang LIU Jun CHEN Yun-y. Size Optimization Design of the Crane Arm Structure [J]. *Ship & Ocean Engineering*, 2017(1):**106-108.**
[2] ZHENG Jiahuan JIN Weiya SONG Yaling. Research Progress of Structural Sensitivity Analysis [J] . *Light Industry Machinery,* 2016, 34(1):**93-96.**
[3] Bolognani S, Peretti L, Zigliotto M. Parameter Sensitivity Analysis of an ImprovedOpen-Loop Speed Estimate forInduction Motor Drives[J]. *IEEE Transactions on Power Electronics,* 2008, 23(4):**2127-2135.**
[4] Zhu J, Cai X, Pan P, et al. Parameter sensitivity analysis and optimization for structure of wind turbine blades[J]. *Journal of Hohai University*, 2015, 43(2):**156-162.**
[5] Chanal H, Paccot F, Duc E. Sensitivity Analysis of an Overconstrained Parallel Structure Ma chine Tool, the Tripteor X7[J]. *Applied Mechanics & Materials*, 2012, 162(4):**394-402.**
[6] Ren Y T, Qiu X M, Yu T X. The sensitivity analysis of a geometrically unstable structure under various pulse loading[J]. *International Journal of Impact Engineering,* 2014, 70(4):**62-72.**

IOP Conf. Series: Materials Science and Engineering **242** (2017) 012056 doi:10.1088/1757-899X/242/1/012056

Design of a new separable rotary transformer

X F Gong[1], L Zhang[1] and E J Feng[1]

[1] School of mechanical engineering, University of Science & Technology Beijing, Beijing, China

bkjxzhangli@163.com

Abstract. A new-type separable rotary transformer which can be used in rotary steerable drilling is designed to deliver power efficiently from a stationary primary source to a rotary secondary load over a relatively large air gap via magnetic coupling. In this paper, E-type magnetic cores are reasonably distributed so that rotation of the rotary secondary has the least influence on reluctance of magnetic coupling. The influence of different winding layouts and connection modes on self-inductance and coupling coefficient is studied. By analysing the influence of the different geometrical shapes of cores on magnetic path, a design principle is proposed.

1. Introduction

Rotary steerable drilling and Logging While Drilling(LWD) system represent the highest level of drilling technology in today's world. Hydraulic driving mechanism of rotary steerable system is placed in the straightening looper, and the required electric energy is supplied by a mud generator on the drill collar, while there is relative rotation between straightening looper and drill collar [1-6]. Traditionally, slip ring is used to realize energy transfer in the rotating and non-rotating part. However, underground operating environment is so harsh that the slip ring frays easily with the strong shock and vibration, and its life and reliability will decrease [7].

In this paper, a new-type separable rotary transformer is designed. E-type magnetic cores are arranged in the outer surface of drill collar spindle and inner surface of straightening looper, forming a closed magnetic path between the outer surface core and inner surface core. In this design, coupling coefficient remains the same under relative rotation, which ensures the stability of the transmission efficiency.

2. Separable transformer design

The new mechanical structure of separable transformer is shown in Figure 1. On the inner cylindrical surface of stator and outer surface of rotor, respectively, with machining methods, a number of evenly spaced rectangular grooves are created around the axis of the circular surface, which are filled with E-type ferrite cores. Stator and rotor cores together constitute a transformer flux loop.

2.1. Core design

Since the magnetic core of the transformer are not continuously distributed, magnetic resistance may change when the rotor rotates, which will affect the system transmission efficiency. Therefore, number and width of cores are reasonably designed to ensure the stability of magnetic resistance while rotating. Core's cross-sectional area and window area will influence the output power.

Content from this work may be used under the terms of the Creative Commons Attribution 3.0 licence. Any further distribution of this work must maintain attribution to the author(s) and the title of the work, journal citation and DOI.
Published under licence by IOP Publishing Ltd

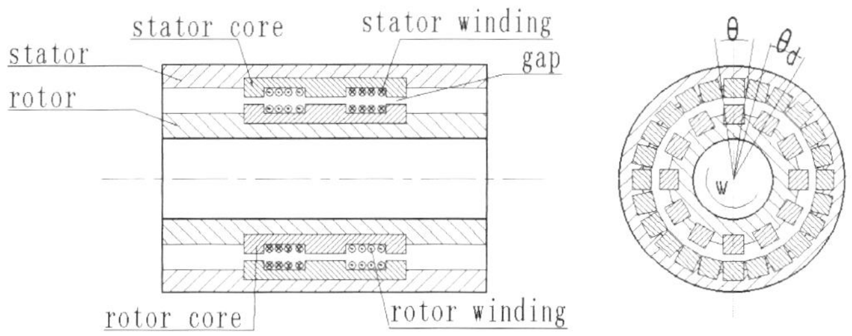

Figure 1. Separable rotary transformer.

2.1.1. Core width angle design. The core width angle and core number must meet the formula (1) to ensure that magnetic resistance shouldn't be affected by rotation of rotor.

$$\begin{cases} 0 < \theta < \dfrac{360}{m} \\ 0 < \theta < \dfrac{360}{n} \\ \theta = k\dfrac{360}{p}, k = 1,2,3... \end{cases} \quad (1)$$

m, n are respectively the number of rotor and stator cores; p is the least common multiple of m, n; θ is the angle of a single core width to the axis (core width angle).

Change of magnetic resistance can be measured by the angle of the coincident part of primary and secondary core to axis (θ_e, core coincidence angle). Figure 2 shows the effect of rotor rotation angle θ_d on core coincidence angle θ_e when $m = 20, n = 24$. When the core width angle doesn't meet the formula (1), such as $\theta = 7.5°, 10.5°$, the change of θ_e caused by the rotation is up to $9°$; When the core width angle meets the formula (1), such as $\theta = 9.0°, 12.0°$, θ_e is substantially unaffected by rotation of the rotor, which ensures effective cross-sectional area and magnetic reluctance constant.

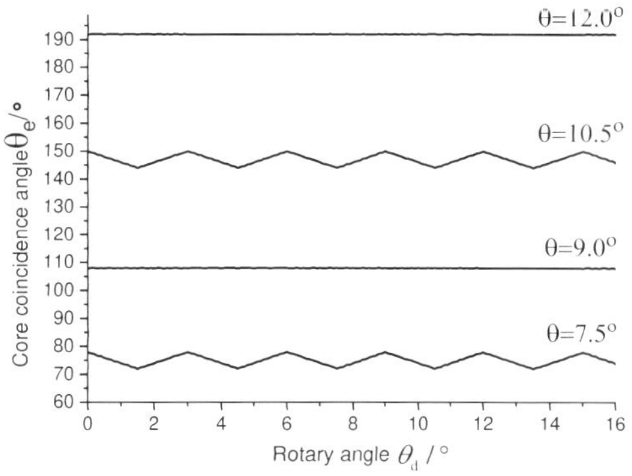

Figure 2. The relationship between coincidence angle, rotary angle and core width angle.

Increasing core utilization η is an important way to improve the transmission efficiency of the transformer. Core utilization can be described as following:

$$\eta = \frac{\theta_e}{m\theta} \quad (2)$$

Table 1 shows the relationship between core utilization and core width angle when $m = 20, n = 24$. It can be seen that the utilization increases along with the core width angle, but the rotor's mechanical

IOP Conf. Series: Materials Science and Engineering **242** (2017) 012056 doi:10.1088/1757-899X/242/1/012056

structure strength decreases simultaneously. Therefore, core width angle should not be too large, $\theta = 9^{\circ}$ is recommended in this paper.

Table 1. The relationship between core utilization and core width angle.

θ / °	3	6	9	12	15
θ_e / °	12	48	108	192	300
η / %	20	40	60	80	100

2.1.2. Core cross-sectional area and window area design. Assuming that the input voltage of the system is sinusoidal, the sectional area of the magnetic core can be described as:

$$A_m = \frac{V_{wp}}{4.44\eta mfNB_m} \quad (3)$$

Where N is winding turn; B_m is saturation magnetization.

The formula (3) shows A_m relates to the peak input voltage V_{wp} and operating frequency f. When V_{wp} is constant, A_m can be adjusted by changing f to adapt to different working space.

Core window area W_a is as following:

$$W_a = \frac{V_{wp}I_P + V_{ws}I_S}{4.44\eta mK_u fB_m JA_m} \quad (4)$$

Where K_u is the core window utilization; J is the maximum current density; I_P , I_S are respectively the primary and secondary current amplitude [8].

The formula (4) shows the core size is determined by the primary and secondary apparent power.

2.2. Winding design

Separable transformer belongs to loosely coupled transformer, which has large leakage inductance, low coupling coefficient and low transmission efficiency. Winding layout and connection will affect electrical parameters of the transformer, so appropriate winding layout and connection can improve transmission efficiency.

2.2.1. Winding layout selection. There are two winding layout schemes. One is that the primary and secondary windings are radially separated as shown in Figure 3(a); the other is that windings are radially staggered as shown in Figure 3(b) [9].

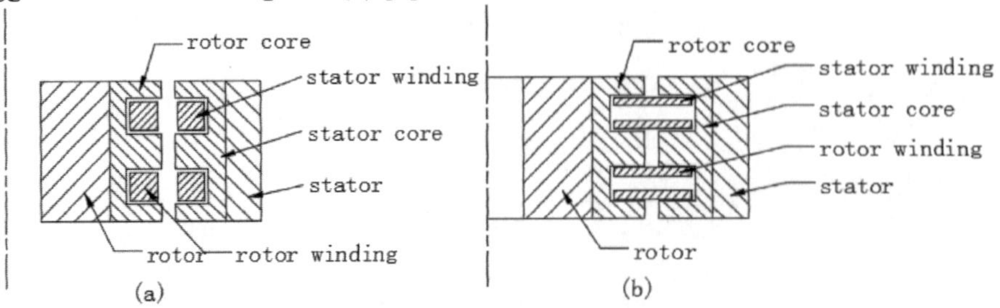

Figure 3. Winding layout.

(1) Separated layout. According to the winding position in the core window, separated layout can be subdivided into axially spaced and radially spaced layout as shown in Figure 4(a) and 4(b).

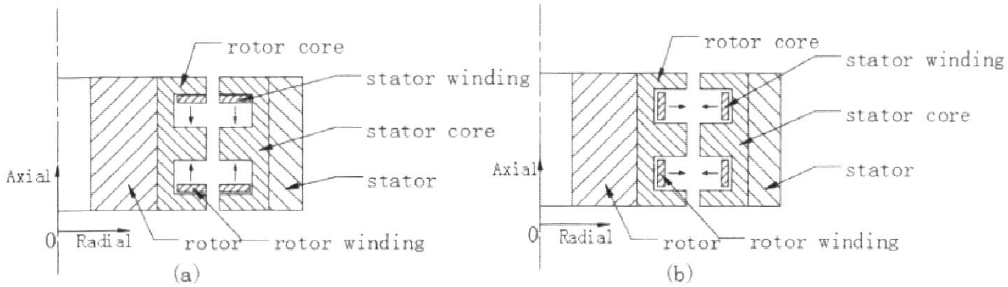

Figure 4. Separated layout.

Simulation results of ANSYS Maxwell are shown in Figure 5 and Figure 6. Winding axial movement distance has little effect on self-inductance, mutual inductance and coupling coefficient as shown in Figure 5; as shown in Figure 6, the larger the winding radial distance is, the smaller the gap becomes, and the larger the coupling coefficient is. Thus, coupling coefficient can be increased by reducing winding radial distance.

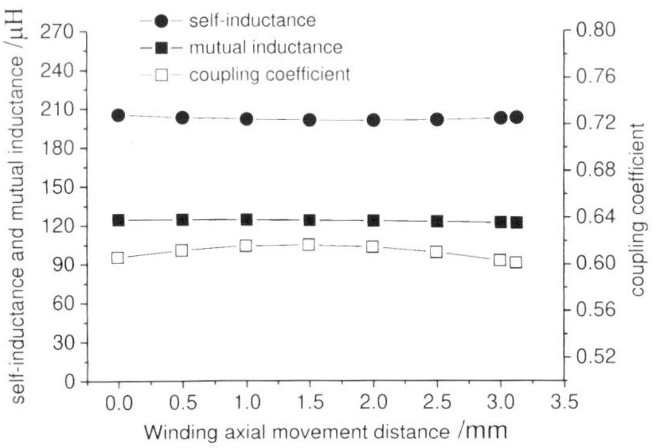

Figure 5. The effect of winding axial position on coupling coefficient.

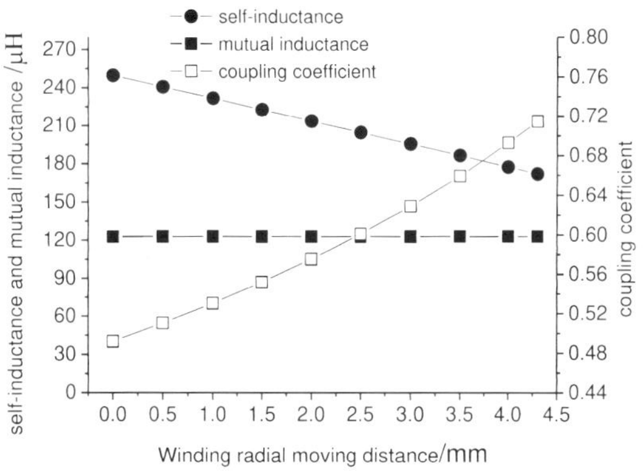

Figure 6. The effect of winding radial position on coupling coefficient.

(2) Staggered layout. As shown in Table 2, compared with the radial separated layout, rotor and stator windings of staggered layout have smaller interval, lower leakage inductance and higher coupling coefficient, but the staggered layout has complex winding process, low reliability and is easy to wear. Therefore, the radial separated layout is recommended with minimizing the winding radial distance.

Table 2. Comparison between staggered and separated layout.

	self-inductance L_r / μH	mutual inductance M / μH	coupling coefficient
Staggered layout	158.58	149.49	0.94
Radial separated layout	172.18	123.06	0.71

2.2.2. Winding Connection Options. E-cores have four wire connection ways in the two windows as shown in Figure 7: Series with same direction, series with opposite direction, parallel with same direction, parallel with opposite direction. When wire connects in same direction, the magnetic flux is along upper, lower and sides of the magnetic circuit as shown in Figure 7a, 7c; When connecting in opposite direction, the flux is along upper and lower windows of the magnetic circuit as shown in Figure 7b, 7d.

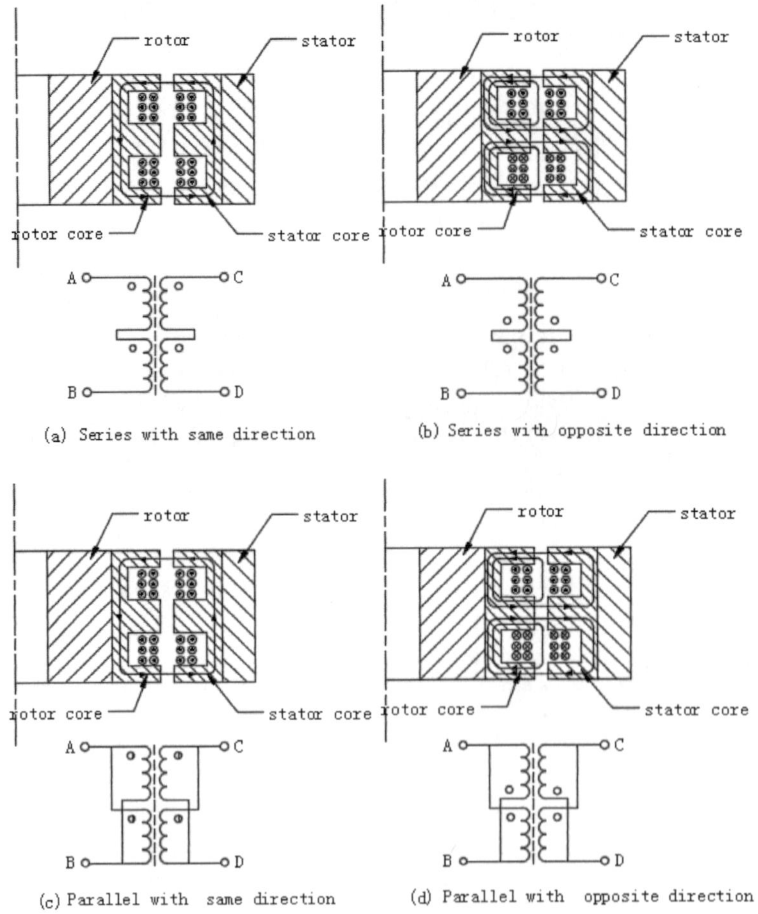

(a) Series with same direction (b) Series with opposite direction

(c) Parallel with same direction (d) Parallel with opposite direction

Figure 7. The ways of winding connection

Table 3 shows the measured inductance values of four kinds of connection ways when the air gap is 3 mm. As can be seen from the table: series with same direction connection has maximum value of inductance; parallel with opposite direction has the minimum; inductance of connecting with same direction is higher than with opposite direction.

Table 3. Inductance comparison of four wire connection ways.

	self-inductance $L_r / \mu H$	mutual inductance $M / \mu H$	coupling coefficient
Series with same direction	1310.3	953.9	0.75
Series with opposite direction	674.7	327.7	0.51
Parallel with same direction	324.6	234.8	0.74
Parallel with opposite direction	169.6	80.9	0.50

To improve the power transfer efficiency, a connection way with high inductance and low leakage inductance is feasible, such as series with same direction and series with opposite direction. Material of drill collar has a certain permeability and conductivity, series with opposite direction mode generates equal and opposite vortex in two windows, which can cancel each other out and improve the transmission efficiency. Therefore, the connection way of series with opposite direction is recommended in this paper.

3. Core size optimization

Reducing the leakage inductance of the transformer and improving inductance and coupling coefficient can effectively improve the transmission efficiency. Theoretical derivation and finite element analysis method are used to optimize the core geometry here.

The magnetic circuit of series with opposite direction mode is shown in Figure 8, since the magnetic circuit A, B of the upper and lower windows are symmetrical, it's feasible only to study magnetic circuit A [10].

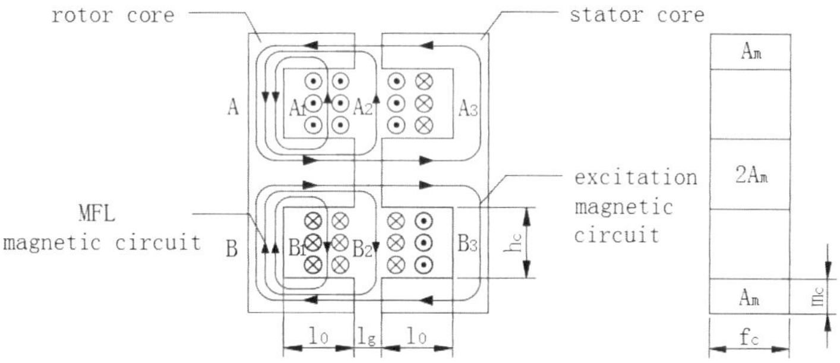

Figure 8. Magnetic circuit of series with opposite direction mode.

3.1. Magnetizing inductance

To simplify the model, by ignoring the flux of air gap edge, the magnetizing inductance of circuit A is as following:

$$L_{mA} = \frac{N^2}{\frac{1}{\mu_0 A_m}\left(\frac{l_m}{\mu_r} + 2l_g\right)} \tag{5}$$

Where N is primary winding turns; μ_0 is air permeability; μ_r is relative permeability of Mn-Zn ferrite; l_m is length of excitation magnetic path; l_g is length of the air gap.

Magnetizing inductance of the transformer is twice of L_{mA}. As relative permeability is generally up to 2500～3000, $l_m / \mu_r << 2l_g$, magnetizing inductance is mainly decided by air gap length [11].

3.2. Leakage inductance

As shown in Figure 8, leakage flux mainly distributes in the rotor window (as A_1) and the air gap (as A_2). Leakage inductance can be calculated as following:

$$L_{lk} = \frac{2}{3}\mu_0 \eta m f_c N^2 \frac{l_0 + l_g}{h_c} \tag{6}$$

Where η is the core utilization; f_c is the core width; h_c is the core window height; l_0 is the width of the core window; l_g is the air gap length.

As $l_0 << l_g$, flux leakage is mainly decided by h_c and l_0.

3.3. Size Optimization

Transformer coupling coefficient can be calculated as following:

$$k = \frac{L_m}{L_m + L_{lk}} = \frac{1}{1 + \frac{2}{3}\eta m f_c l_g \frac{l_0 + l_g}{A_m h_c}} \tag{7}$$

Formula (7) shows that increasing magnetizing inductance and reducing leakage inductance can effectively improve the coupling coefficient.

From formula (5), the magnetic energy of the excitation circuit mainly distributes in the core gap, reducing air gap can reduce the reluctance of the excitation magnetic circuit and increase magnetizing inductance. From the formula (6), the magnetic field energy of leakage magnetic circuit is mainly in the window, increasing the core window height and reducing the core width can improve the reluctance of magnetic flux leakage and reduce leakage magnetic flux.

To further analyse the influence of the air gap and height of the window on the transformer inductance, mutual inductance and coupling coefficient, setting $\Delta = h_c / l_g$, ANSYS Maxwell is used to simulate how parameters of transformer change when only Δ changes.

As shown in Figure 9, inductance and coupling coefficient are more sensitive when $\Delta < 2$; when $\Delta = 2$, the length of the air gap of excitation magnetic path and the leakage path are approximately equal, $L_m \approx L_{lk}$, k is nearly 50%; When $\Delta > 2$, the change rate of inductance and coupling coefficient gradually reduce, and with the increase of Δ, core's volume gradually increases. Therefore, the height of E-type core window must be more than twice of air gap width.

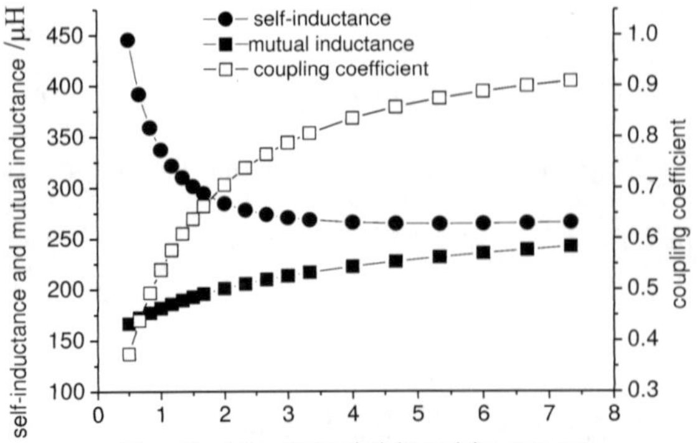

Figure 9. The relationship between the inductance of transformer and Δ.

4. Epilogue

The theoretical analysis, simulation and experimental results show that the inductance of transformer can avoid the interference of rotor' rotation; the winding's radial separated layout and series with opposite direction connection mode can improve the coupling coefficient and reduce the eddy current loss. The core cross-sectional area and the window area should be designed according to formula (3) and (4), minimizing the air gap and complying with the principle that the height of E-type core window must be more than twice of the air gap width.

References

[1] Dai Yan, Flourish, Su Yugang, Tangchun Sen, et al. Study [J] non-contact power two-way push mode. China CSEE, 2010, 18: 55-61.

[2] Zhang Zongming non-research [D] contactless power transmission system electromagnetic mechanism. Chongqing University, 2007.

[3] Huang Xueliang, Tan Linlin, Chen, and other wireless power transmission technology research and application review [J] Electrical Technology, 2013, 10: 1-11.

[4] The full-Ru Dai, Xie light Yong, Wang Lin, a non-contact power transmission system based on electromagnetic coupling mechanism [J] college physics experiment, 2009, 04: 1-3.
[5] Cheng Shijie Chen Xiaoliang, Wang Junhua, and other key wireless transmission technology and its application [J] Electrical Technology, 2015, 19: 68-84.
[6] Qin Haihong, PROCEEDINGS, Yan Yangon non-contact analysis and design principles of loosely coupled inductive power transfer system [J] Power Technology, 2004, 05: 257-262.
[7] Zhou Wentao. Contact mechanics characteristics and wear life slip ring analysis [D]. Xiangtan University, 2014.
[8] YH Kim, KH Jin.Design and Implementation of a Rectangular-Type Contactless Transformer[J]. Industrial Electronics IEEE Transactions on, 2011, 58(12): 5380- 5384.
[9] Phaengkieo D, Somlak W, Ruangsinchaiwanich S. Transformer Design by Finite Element Method with DOE Algorithm[C] Electrical Machines and Systems (ICEMS), 2013 International Conference on. IEEE, 2013: 2219-2224.
[10] Lin Ning, Yao Ying Ying, Li Yuling, etc. inductively coupled power transfer system design [J] Zhejiang University: Engineering Science, 2012, 46 (2): 199-205.TA Stuart，RJ King，H Shamseddin.Rotary transformer design with fixed magnetizing and/or leakage inductances[J]. IEEE Power Electronics Specialists Conference, 1985, 22(5): 480-487.
[11] EE Landsman.Rotary transformer design[J].IEEE Power Electronics Specialists Conference, 1970: 139-152.

IOP Conf. Series: Materials Science and Engineering **242** (2017) 012057 doi:10.1088/1757-899X/242/1/012057

Experimental research and manufacturing technology of heavy gauge WPHY-80 pipe fitting plate with uniform microstructure

Shao-Po Li[1,2,3], Wen-Hua Ding[2,3] and Hai Zhang[2]

[1]University of Science & Technology Beijing, Beijing, 100083, China;
[2]Shougang Research Institute of Technology; Beijing, 100043, China
[3]Beijing Engineering Research Center of Energy Steel, Beijing, 100043, China;

E-mail: ustbmp@126.com, dingwh31@163.com, zhanghai1861@163.com

Abstract: In order to develop heavy gauge WPHY-80 pipe fitting steel with high strength and low temperature toughness, the experimental research and manufacturing technology have been studied in this paper. The continuous cooling phase transformation behavior of the material under deforming condition was investigated. The microstructure of the steel at different cooling rates was observed by microscope. The temperature gradient rolling process in rough rolling stage was developed, which prompt the deformation penetrated from the surface to the center. It could refine the grain size in the center of the thick plate to enhance uniformity of the microstructure along the thickness direction. By utilizing this process route, the heavy gauge (54mm) WPHY-80 pipe fitting steel with excellent mechanical properties was developed in Shougang steel (SG). The WPHY-80 plates obtained high strength and stable low temperature impact toughness.

1. Introduction

As well known, natural gas is very important energy resource to maintain the development of national economy, in the environment of the global financial crisis, some countries have started to preempt energy resources, strengthen the exploitation of oil and gas fields. In oil and gas pipeline projects, there is a series of steel pipe fittings, such as tee, cross, reducer, cap and elbow.

The development of pipeline steel trends toward higher strength, heavier wall thickness and larger diameters[1,2,3]. For example, the pipe fitting used for West to East Ⅱ pipeline project is WPHY-80 grade, the thickness of the pipe fitting reached to 54mm. Deformation of thick steel plate along the thickness direction was non-uniform during the traditional hot rolling, with which was a small compression ratio in center. The deformation cannot penetrate from the face layer to the center made the defects unbinding. This is the driving force that temperature gradient rolling process was developed by Shougang steel (SG). Stable high strength WPHY-80, heavy gauge (54mm) mechanical properties in the steel plates can be achieved by utilizing this process route.

2. Experimental research of WPHY-80 Steel

Table 1 shows the chemical compositions of 54mm WPHY-80 pipe fitting steel. Low carbon, high manganese, Nb, V and Ti additions were increased, and added some Ni, Cu, Mo etc elements for achieving bainite microstructure.

Content from this work may be used under the terms of the Creative Commons Attribution 3.0 licence. Any further distribution of this work must maintain attribution to the author(s) and the title of the work, journal citation and DOI.
Published under licence by IOP Publishing Ltd

Table 1. Chemical composition of 54mm WPHY-80 pipe fitting steel (wt%)

C	Si	Mn	P	S	Nb+V+Ti	Ni+Mo+Cu+Cr
0.075	0.25	1.70	0.008	0.0015	≤0.12	≤0.80

Table 1 shows chemical composition of the heavy gauge WPHY-80 pipe fitting steels. Low carbon levels (0.075%) contribute to a number of performance improvements, such as low temperature toughness, the weldability. Certain amount of niobium/vanadium solution and diffuse small (Nb,V) (C, N) precipitates can significantly refine the grain. And the molybdenum element can affect CCT curves [4], to avoid uneven cooling inside and outside the heavy plate, the bainite ferrite can be obtained in a wide range of cooling rate.

The sample was reheated to 1180℃ and holding 5 minutes on the Gleeble-2000 thermal simulation machine, then cooled at 15℃/sec to 1050℃, for 20% of compression deformation, the deformation rate was 20S-1, and then cooled to 830℃ by 15℃/sec, 30 % deformation , then cooled to room temperature, the cooling speed respectively 26℃/s, 22℃/s, 18℃/s, 15℃/s, 10℃/s, 5℃/s, 1℃/s, 0.5℃ /s, 0.1℃/s. Figure 1 shows the curve of thermal simulation test.

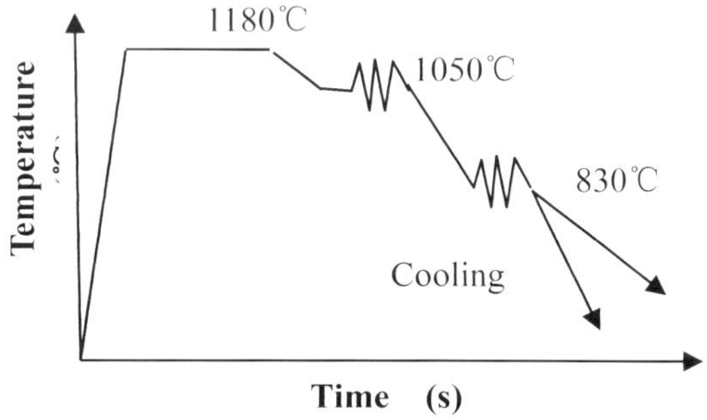

Fig 1. The curve of thermal simulation test

The microstructure of at different cooling rates was observed by microscope. From Figure 2, we can conclude that the ferrite(F) and pearlite(P) were obtained at a lower cooling speed range for 0.1~0.5℃/s. Granular bainite (GB) was obtained when the cooling speed increase to the range of 1~5℃/s. With the increase of the cooling speed, the transformation phase were mainly composed of lath bainite (LB) and martensite (M). With the increase of the cooling rate, the bainite microstructure was gradually refined. Figure 3 shows the curve of continuous cooling transformation (CCT). It was shown that the bainite microstructure was obtained in a wider cooling speed range.

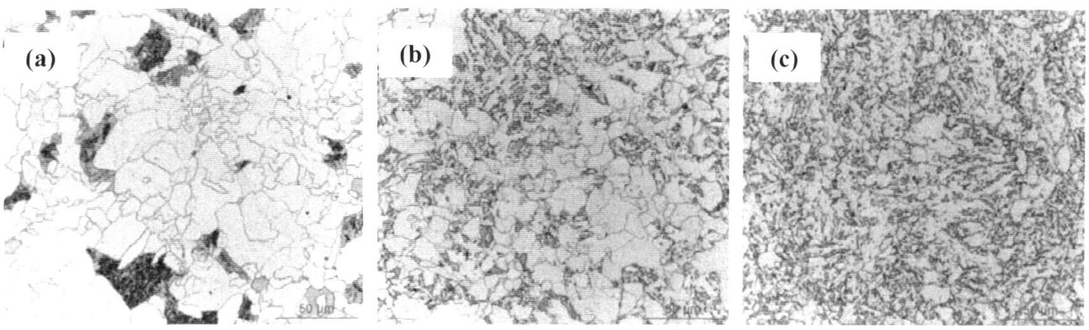

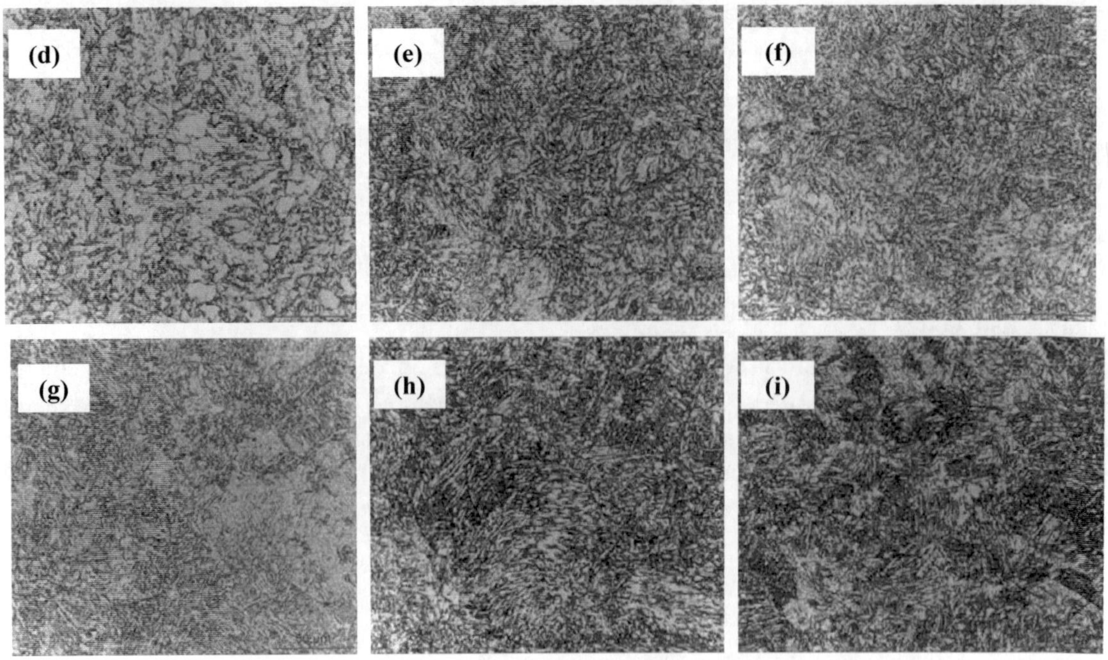

Fig 2. The microstructure of WPHY-80 with different cooling rate

(a)0.1℃/s; (b) 0.5℃/s; (c) 1℃/s; (d) 5℃/s; (e)10℃/s; (f) 15℃/s; (g) 18℃/s; (h) 22℃/s; (i) 26℃/s

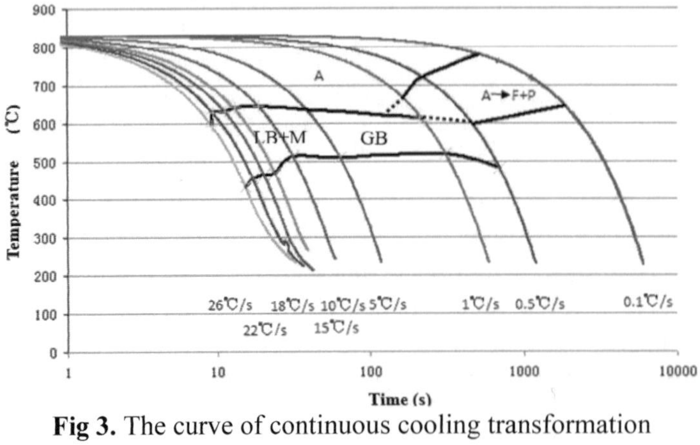

Fig 3. The curve of continuous cooling transformation

3. Manufacturing technology of WPHY-80 Steel

The manufacturing technology starts from the slab reheating which needs to address both dissolution of the main microalloy elements and controlling of the austenite grain coarsening[5,6]. In the rough rolling stage, temperature gradient rolling method was developed, which was a process that rapid cooling while rolling, and the center of thick plate could not cool down, so that a non-uniform temperature distribution was forming that low temperature on surface and high temperature in center. With this rolling method, the harden layer on the surface could prompt the deformation penetrated from the surface to the center, which could remove the defects in the center of the thick plate to enhance uniformity of the deformation along the thickness direction. Post rolling cooling was controlled to achieve the appropriate cooling stop temperature and cooling rate to obtain the ideal microstructure and plate flatness[7].

According to the cooling conditions while rolling, the symmetry model was established. The temperature distribution nephogram of the plate during rough rolling stage was simulated under the temperature gradient rolling process and the process without cooling while rolling, and the temperature contrast between the two processes was shown in Figure 4. Also, the rolling deformation contrast from surface to center of the plate under the two processes was shown in Figure 5.

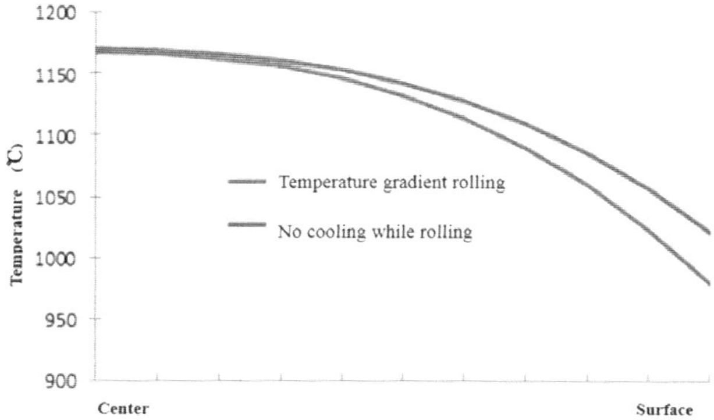

Fig 4. The temperature contrast form surface to center under the two processes

It can be seen from Figure 4 that the surface temperature under the temperature gradient rolling process is about 42℃ lower than that of the process without cooling while rolling, and the center temperature of the plate is basically the same under the two processes.

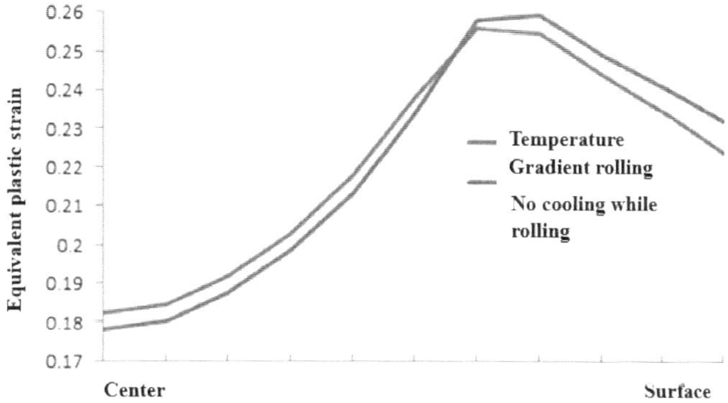

Fig 5. The rolling deformation contrast form surface to center under the two processes

It can be seen from Figure 5 that the surface deformation of the plate under the temperature gradient rolling process is smaller than that of the process without cooling while rolling, and the center deformation of the plate under the temperature gradient rolling process is greater than that of the process without cooling while rolling. With this rolling method, the deformation penetrated to the center of the plate, which could remove the defects in the center of the thick plate to enhance uniformity of the microstructure along the thickness direction.

The microstructure on the center of 54mm WPHY-80 pipe fitting plate under the two different processes was observed by optical microscope, as shown in Figure 6.

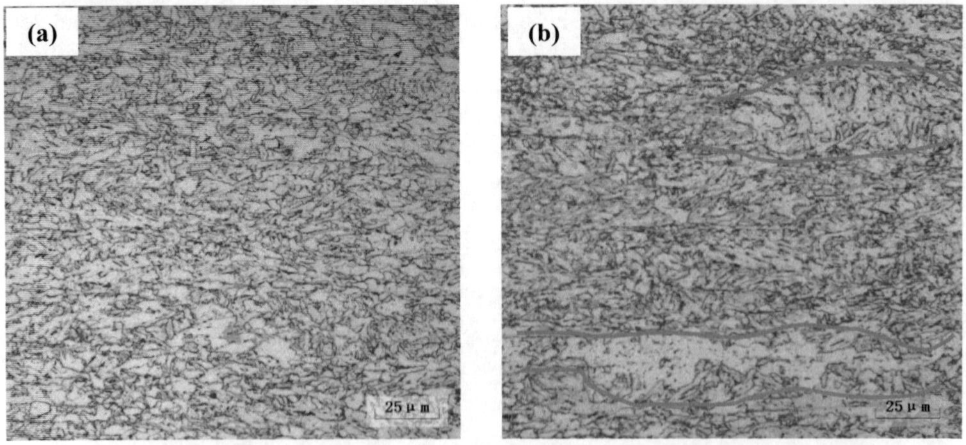

Fig 6. The microstructure on the center of the plate under the two processes
(a) Temperature gradient rolling process; (b) Process without cooling while rolling

It can be seen from Figure 6 that the center microstructure of the plate under the process without cooling while rolling is uneven, and there exists some large size grains, while the center microstructure of the plate under temperature gradient rolling process is uniform and small.

4. Mechanical properties of WPHY-80 Steel

According to the continuous cooling phase transformation characteristics and the manufacturing technology, the heavy gauge WPHY-80 pipe fitting steel was produced in Shougang steel (SG).

Table 2 shows the tensile properties on the center of the 54mm WPHY-80 plate under the two processes, Table 3 shows the impact toughness on the center of the 54mm WPHY-80 plate under the two processes.

Table 2. Tensile properties on the center of the plate under the two processes

Process	Tensile properties on the center			
	Rt0.5/MPa	Rm/MPa	Rt0.5/Rm	A/%
Process without cooling while rolling	540	645	0.84	39
Temperature gradient rolling process	575	695	0.83	40

Table 3. Impact toughness on the center of the plate under the two processes

Process	Test temperature /℃	Impact toughness on the center /J			
		1	2	3	Avg
Process without cooling while rolling	-30	246	260	178	228
Temperature gradient rolling process		313	332	317	321
Process without cooling while rolling	-40	192	146	178	172
Temperature gradient rolling process		287	254	269	270
Process without cooling while rolling	-60	113	86	127	109
Temperature gradient rolling process		215	203	196	205

From the results of Table 2 and 3, we can see that the plate under the temperature gradient rolling process obtained more stable and excellent performance.

5. Conclusions

According to the experimental research and production practice on 54 mm WPHY-80 pipe fitting plates, the following conclusions can be drawn:

1. For the WPHY-80 pipe fitting steel, bainite microstructure can be obtained in a wider cooling

speed range, and with the increase of the cooling rate, the bainite microstructure was gradually refined.

2. The temperature gradient rolling process was developed and applied to produce low temperature heavy gauge WPHY-80 pipe fitting plates. The plates obtained high strength, low Y/T ratio, and good low temperature toughness.

3. By using the temperature gradient rolling process, a stable control of uniform microstructure can be realized and further more result in the consistent and stable control of mechanical properties of plates.

References

[1] F. Siciliano, D. Stalheim, J. Gray, "Modern High Strength Steels for Oil and Gas Transmission Pipelines", Proceedings of 7th International Pipeline Conference 2008, IPC2008-64294.
[2] W. Li, C. Huo, Q. Ma, et al., "The Development of Large Diameter & Thickness X80 HASW Linepipe", Proceedings of 7th International Pipeline Conference 2008, IPC2008-64500.
[3] D. Stalheim, S. Li, W. Ding, Z. Jiang, et al., "Research and Development Into Low Temperature Toughness of Large Diameter Heavy Wall X80 Pipeline Steel at Shougang Steel", Proceedings of 9th International Pipeline Conference 2012, IPC2012-90326.
[4] Shouhua Zhang. Alloyed steel[M]. Beijing: Metallurgy Industry Press. 1988, 25-30
[5] Z. Zhang, Q. Zhang, "Effect of Heating Temperature on Prior Austenite Size and Solution of Nb in Pipeline Steel", Journal of Iron and Steel Research, 2008(10): 36~39.
[6] D. Stalheim, Y. Yin, et al., " Development and Capability of High Temperature Processing (HTP) Pipeline Steel at Nanjing Iron and Steel Company", Proceedings Pipeline Technology Conference 2009, Ostend Belgium, 2009.
[7] S. Li, Q. Ma et al., "Research of Basic Parameters of Nb-Ni series X70 Pipeline Steel", [J]. Steel Rolling, 2009, 26(5): 5-8.

Investigation of deformation in stretch forming based on distributed displacement loading

Qingmin Chen, Lirong Sun* and Xueping Yan

Roll Forging Research Institute, Jilin University, Nanling Campus, 5988 Renmin Street, Changchun 130025, People's Republic of China

*Corresponding author: 18843011405@163.com

Abstract. Stretch forming based on distributed displacement loading is a new stretch forming process, in which distributed displacements are applied at a series of discrete points, therefore, the deformation in different positions of sheet metal can be individually controlled. To investigate the deformation of sheet metal in stretch forming, the rational loading trajectory is designed, and it is determined by the respective length of longitudinal cross-section curve. The numerical simulation results show that, the loading trajectory is valid to form three-dimensional surface parts, and the relatively small shape errors between the simulated results and target results make it possible to form qualified parts. Meanwhile, the longitudinal strains are uniformly distributed along the longitudinal material lines, and longitudinal strains increase evenly with the process of stretch forming. Finally, experimental tests proved the effectiveness and feasibility of the stretch forming based on distributed displacement loading.

1. Introduction

Stretch forming, as a vital process for manufacturing three-dimensional surface parts, has been widely applied in diverse industries, especially to the external bodies of aircraft, automobile and high-speed train [1]. Conventional stretch forming requires rigid gripping jaws, obtaining equal displacement at the end of sheet metal clamped by rigid clamps [2]. Since sheet metal in conventional stretch forming can't deform automatically with the die surface in the transverse direction (perpendicular to the tensile direction), forming defects such as crack, wrinkle and severely local deformation may customarily occur in large-curvature surface parts.

Multi-point stretch forming introduces the idea of discretization, and adopts flexibly discrete die and clamps, which can form three-dimensional surface parts of diverse shapes [3-6]. Multi-point stretch forming is controlled by using force loading on the discrete clamps, including horizontal force, vertical force and tilting force, so curved surface parts with large curvatures are susceptible to springback after unloading tensile force, which could lead to larger shape errors compared with acquired ideal forming result.

The stretch forming based on distributed displacement loading (SF-DDL), which is a new forming method for three-dimensional surface parts, applies different displacements loading on a series of discrete points on the two ends of sheet metal, and each individual material fibre is locally prescribed to deform along a respective deformation path [7]. The reasonable loading trajectory is the key to forming qualified curved surface parts, and the sheet metal can deform automatically with the die surface in the transverse and longitudinal directions, which can achieves to manufacture qualified large-curvature surface parts by deformation as small as possible [8].

Content from this work may be used under the terms of the Creative Commons Attribution 3.0 licence. Any further distribution of this work must maintain attribution to the author(s) and the title of the work, journal citation and DOI.
Published under licence by IOP Publishing Ltd

2. Principle of stretch forming and design of loading trajectory

In the forming process, the corresponding displacements are applied at the reference points of the discrete clamps so that the sheet metal can be stretched and wrapped around the die surface. Each loading unit moves along a combined horizontal and vertical path, furthermore, it rotates a certain angle in order to maintain tangent relation between the die surface and the sheet metal.

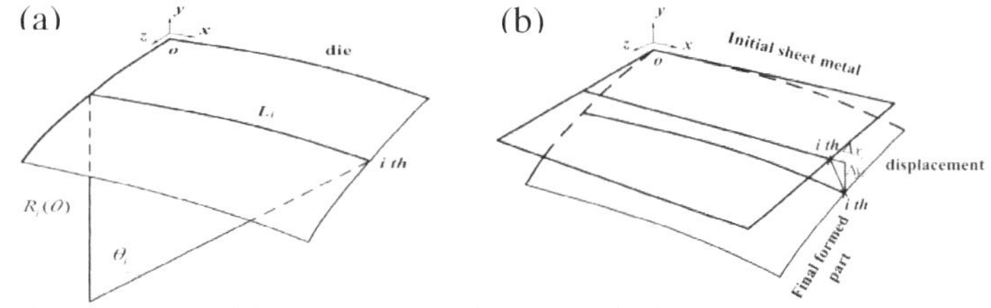

Figure 1. Diagram of SF-DDL: (a) diagram of die (b) sheet metal in forming.

It is assumed that the strains on the longitudinal material lines are uniformly distributed along the longitudinal direction (the tensile direction), and longitudinal strains satisfy $\varepsilon_{t_{ij}} = \xi(t_j)\varepsilon_{if}$, where ε_{if} is the final strain of longitudinal liber at the ith loading point, $\varepsilon_{t_{ij}}$ is the longitudinal strain at t_j time, and $\xi(t_j)$ is the function of the loading time. [7].

At time $t = t_j$, the length of longitudinal fibre of the deformed sheet metal can be calculated by

$$l(t_j) = \int_0^{l_0} e^{\xi(t_j)\varepsilon_{if}} dx \approx l_0[1 + \xi(t_j)\varepsilon_{if}] \tag{1}$$

The loading trajectory of the ith loading point can be described by the displacement of the ith clamp Δx_{ij} in x direction and Δy_{ij} in y direction, as shown in figure 1. Then, the displacements of the ith loading point at time $t = t_j$ should be

$$\begin{cases} \Delta x_{ij} = \{l_0[1+\xi(t_j)\varepsilon_{if}] - \int_0^{\theta_{ij}} R_i(\theta)d\theta\}\cos\theta_{ij} + R_i(\theta)\sin\theta_{ij} - l_0 \\ \Delta y_{ij} = R_i(\theta)\cos\theta_{ij} - \{l_0[1+\xi(t_j)\varepsilon_{if}] - \int_0^{\theta_{ij}} R_i(\theta)d\theta\}\sin\theta_{ij} - R_i(\theta) \end{cases} \tag{2}$$

Where l_0 is the initial length of sheet metal, $R_i(\theta)$ is the curvature radius of longitudinal section curve of the die surface, and θ_{ij} is the normal angle of the ith loading point at time $t = t_j$.

3. Numerical simulations of stretch forming based on distributed displacement loading

3.1 deformation of sheet metal under the different loading trajectories

Numerical simulations of SF-DDL are performed under ABAQUS/Explicit environment, and 1/4 finite element model (as shown in figure 2a) is applied to forming three-dimensional surface parts on the basis of the symmetrical relation of the formed parts. The aluminum alloy of 2024-O is used, and the relevant mechanical properties are the following: the density $\rho = 2.78 g/cm^3$, the Young modulus $E = 40.6GPa$, the yield stress $\sigma_s = 77.5MPa$ and the Poisson's ratio $\nu = 0.33$.

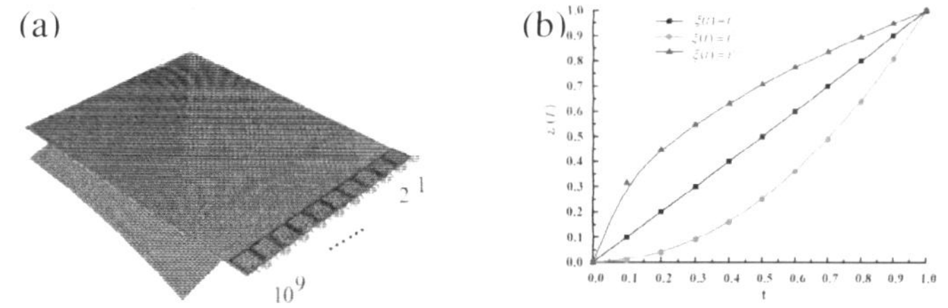

Figure 2. Diagram of numerical simulation: (a) a quarter FE model (b) loading curve

The spherical parts of $R400$ as target radius are simulated in SF-DDL, and the dimension of the initial blank is 400mm × 300mm with the thickness 1mm.The elongation of sheet metal at ith loading point is determined by equation (1), and $\xi(t)$ respectively adopts $\xi(t)=t^{1/2}$, $\xi(t)=t$, $\xi(t)=t^2$, (as shown in figure 2b)at different times, so three different loading trajectories are obtained. In the three trajectories, the elongation of longitudinal fibre is different at diverse loading steps, and the displacement can be respectively acquired by equation (2).The longitudinal strain distributions of three trajectories are shown in figure 3, and it can be observed that, longitudinal strains are more uniform at the effective forming area when adopt $\xi(t)=t$. So it is possible that the deformation of the sheet metal is more stable when the elongation is uniformly increased.

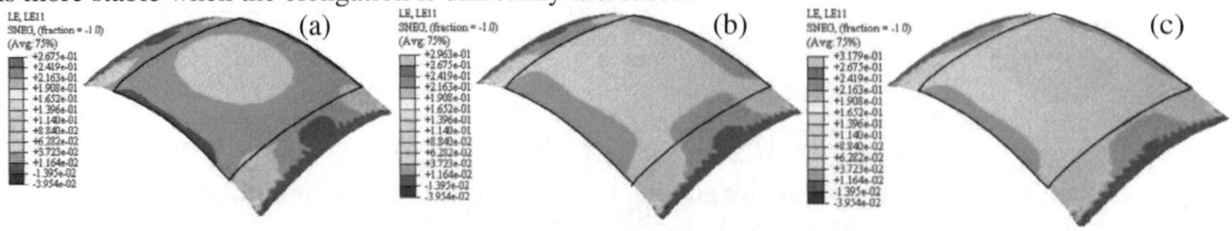

Figure 3. Longitudinal strain distributions of different trajectories:
(a) $\xi(t)=t^{1/2}$ (b) $\xi(t)=t$ (c) $\xi(t)=t^2$

3.2 Longitudinal strain distributions and the shape error

The spherical and saddle parts of *R*500 as target radius are simulated in SF-DDL, and the elongation of sheet metal increase evenly in the forming process. The longitudinal strain distributions on spherical parts and saddle parts are shown in figure 4a and 4b respectively, and formed parts embrace high quality without defects such as crack and wrinkle, which can demonstrate that the loading trajectory of distributed displacement loading is valid to form three-dimensional surface parts.

Then, extract strains of cross section of formed parts at different steps, as shown in figure 4c and 4d, in which the longitudinal strains basically reveal horizontal lines at different times. It indicates that the longitudinal strain distributions are relatively uniform along longitudinal direction, and longitudinal strains increase evenly with the stretch forming process, which verify the assumption that the strains on the longitudinal material line are uniformly distributed and satisfy $\varepsilon_{t_{ij}} = \xi(t_j)\varepsilon_{if}$ as mentioned above.

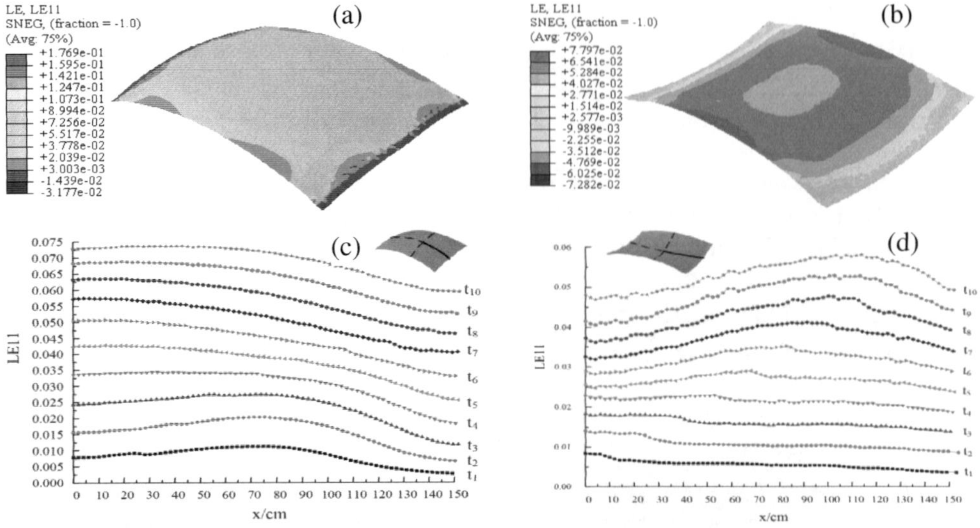

Figure 4. Longitudinal strain distributions: (a) spherical part (b) saddle part (c) strains of spherical part at different times (d) strains of saddle part at different times.

To verify the validity of the loading trajectory, the three-dimensional point-cloud data of simulated parts are compared with the target parts, then, shape error distributions of the spherical parts and saddle parts are obtained, as shown in figure 5. The shape error distributions of spherical part are mainly in 0.5mm to -0.5mm, which is quite small and is allowed in mechanical manufacturing. While shape error distributions of saddle part change from 0.79mm to -0.79mm, which are larger slightly than the ones of spherical part, maybe, the saddle part surface is under the deformation in both positive and negative directions simultaneously. In brief, the relatively small shape errors make it possible to form qualified parts in accordance with the trajectory of distributed displacement loading.

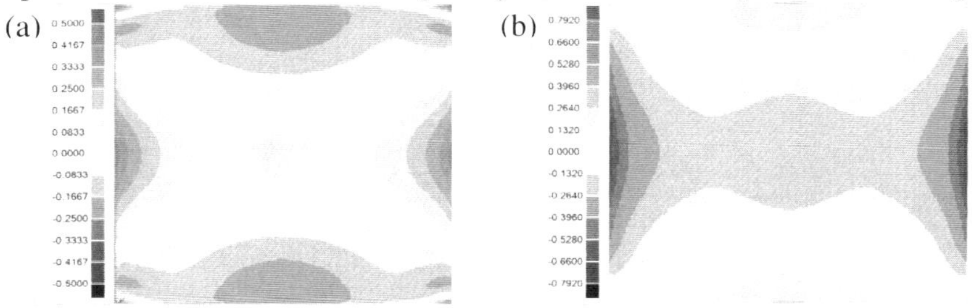

Figure 5. Shape error distributions: (a) spherical part (b) saddle part.

4. Numerical simulation results and experimental validation

In SF-DDL process of spherical part, the sheet metal is susceptible to no contact with the surface of die on the edge. However, in order to manufacture three-dimensional surface parts of high quality, the effective forming area of the sheet metal should be full contact with the die surface. In numerical simulation, the dimension of the initial blank is 1000mm×800mm with the thickness 1mm, and the target radius of spherical part is 1500mm.

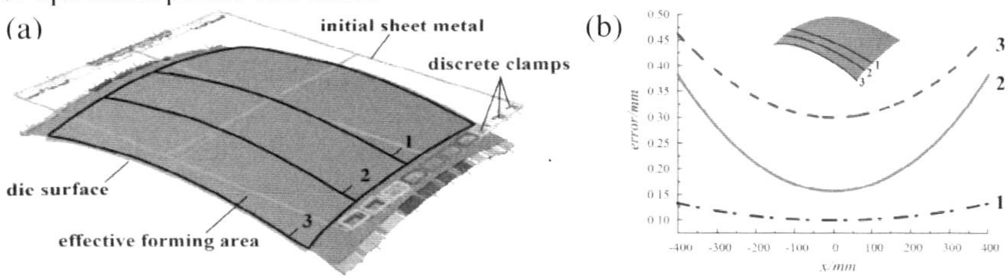

Figure 6. Numerical simulation results: (a) deformation process (b) shape error

Numerical simulation results of spherical part are shown in figure 6, it can be seen that the effective forming area can full contact with the surface of die. Extracting the cross-section line in different locations, shape errors between the simulated part and the target part are shown in figure 6b. And it can be observed that shape errors are minimum in the middle area and change from 0.10mm to 0.15mm, however, shape errors are greater on the edge region and fluctuate in the range of 0.33mm to 0.47mm, which have illustrated that the shape error is extremely small and ideal shape of the curve surface part can be obtained in SF-DDL. By fitting the curvature radii of extracting contours, the radii of the curves 1, 2 and 3 (as shown in table 1) are respectively 1532mm, 1514mm and 1483mm, which are closed to the target radius ($R = 1500mm$).

Table 1 curvature radius of spherical part in different positions

Location	1	2	3
Radius(mm)	1532	1514	1483

To verify the feasibility of the loading trajectory and the accuracy of the numerical simulation, the experiments of stretch forming spherical parts have been carried out on a home-made apparatus, and the experimental spherical part is obtained, as shown in figure 7. In experiment, the aluminum 2024-O material is put into use, and the target radius of the spherical part is 1500mm.

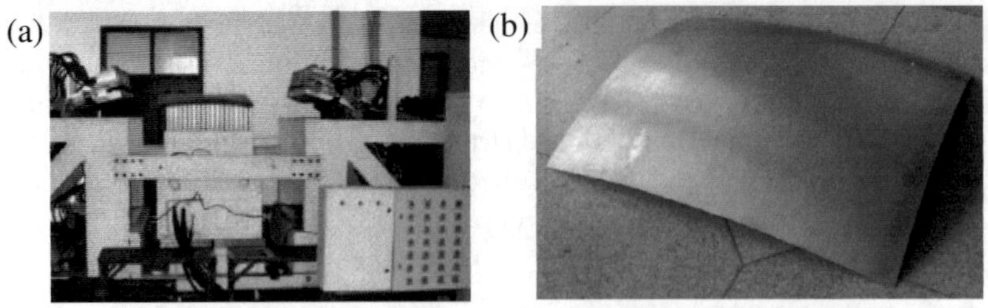

Figure 7. Diagram of experimental equipment and spherical part:
(a) experimental equipment (b) spherical part.

Different displacements are applied at a series of discrete points on the two ends of sheet metal, and the deformation of sheet metal can be individually controlled at the discrete points. By the loading method, we have been realized stretch forming based on distributed displacement loading. It can be observed that the spherical part is with good surface quality and without any defects such as crack and wrinkles. The experiment results demonstrate that the qualified three-dimensional surface parts can be acquired and used in engineering manufacture.

5. Conclusions

To study the deformation in SF-DDL, the optimal loading trajectory was designed, meanwhile, loading trajectory validation and longitudinal strains variation were investigated, and some experiments were carried out to verify the results of numerical simulation. Finally, the following conclusions can be acquired:

(1) The loading trajectory is determined by the length of longitudinal cross-section curves in different clamps. And the numerical simulations of spherical parts and saddle parts demonstrate that the loading trajectory is valid to form three-dimensional surface parts.

(2) The strains on the longitudinal material lines are uniformly distributed along the longitudinal direction, and longitudinal strains increase evenly with the process of stretch forming, which can prove that it is reasonable to adopt the mode of distributed displacement loading to forming three-dimensional surface parts.

(3) In experiment, the effective forming area of the sheet metal can be full contact with the surface of die and obtain high-precision curved surface parts. And the experimental results show that the feasibility of the loading trajectory and the accuracy of the numerical simulation.

Acknowledgement

The authors would like to acknowledge the financial support provided by the National Science Foundation of China (Grant No. 51575231).

References

[1] Shim D S, Baek G Y, Shin G Y, et al. 2016 Investigation of tension force in stretch forming of doubly curved aluminum alloy (Al5083) *sheet International Journal of Precision Engineering and Manufacturing* **17** 433-444

[2] Hardt E D, Norfleet W A, Valentin V M, Parris A 2000 In process control of strain in a stretch forming process *Journal of Engineering Materials and Technology* **123** 496–503

[3] Cai Z Y, Wang S H, Xu X D, Li M Z 2009 Numerical simulation for the multi-point stretch forming process of sheet metal *Journal of materials processing technology* **209** 396-407

[4] Wang S H, Cai Z Y, Li M Z 2010 Numerical investigation of the influence of punch element in multi-point stretch forming process *International Journal of Advanced Manufacturing Technology* **49** 475-483

[5] Cai Z Y, Wang S H Li MZ 2008 Numerical investigation of multi-point forming process for sheet metal: wrinkling, dimpling and springback *International Journal of Advanced Manufacturing Technology* **37** 927-936

[6] Wang S H, Cai Z Y, Li M Z, et al. 2012 Numerical simulation on the local stress and local

deformation in multi-point stretch forming process *International Journal of Advanced Manufacturing Technology* **60** 901-911

[7] Cai Z Y, Yang Z, Che C J, Li M Z 2016 Minimum deformation path sheet metal stretch- forming based on loading at discrete points *International Journal of Advanced Manufacturing Technology* **87** 2683-2692

[8] Yang Z, Cai Z Y, Che C J, Li M Z 2016 Numerical simulation research on the loading trajectory in stretch forming process based on distributed displacement loading *International Journal of Advanced Manufacturing Technology* **82** 1353-1362

Research and development of weathering resistant bridge steel of Shougang

Yongda Yang [1,2], Yanfeng Wang [1,2], Leqing Huang [1,2], Guobiao Di [1,2], Changwen Ma [1,2], Qingshen Ma [1,2]

1. Shougang Research Institute of Technology, Beijing 100043;
2. Beijing key Laboratory of Green Recyclable Process for Iron & steel Production Technology, Beijing 100043

wangyf@shougang.com.cn

Abstract. To introduce the composition design and mechanical properties and microstructure of the weathering bridge steel which would be used for bridge of Guanting reservoir. We adopt cyclic immersion corrosion test to study corrosion resistance difference of weathering bridge steel and common bridge steel. At the same corrosion time, the weight loss and corrosion rate of weathering bridge steel are lower than the common bridge steel's. Testing phase composition of rust layer by X-ray diffraction, two kinds of test steel's rust layer is mainly composed of Goethite and Fe_3O_4 and Fe_2O_3. At the same corrosion time, the percentage composition of goethite in rust layer of weathering bridge steel are significantly higher than common bridge steel's, the higher goethite content is, the compacter rust layer structure is. The compact rust layer would prevent the water and air passing the rust layer, and then preventing the further corrosion reaction, improving the corrosion resistance performance of weathering bridge steel.

1. Introduction

Since the beginning of the 20th century, the United States, Germany, Britain and Japan have been researched weathering resistant steel deeply[1-2]. Corten steel which has excellent properties of high-strength and weathering resistant was developed by The United States Steel Corporation in 1930 s. Weathering resistant steel has advantages in atmospheric corrosion compared to ordinary steel and economical efficiency compared to stainless steel. So the United States first time to apply weathering resistant steel to the bridge of the New Jersey highway, after that, Japan, Britain, Germany began constructions of weathering resistant steel bridge.

The reason of China's weathering resistant bridge development slowness is design and costs. In 2016, Guanting Reservoir Bridge is designed to use weathering resistant steel to build the first large-scale highway avoid coating steel bridge project in China, The bridge is crossed the Guanting Reservoir which is the important fresh water resource for city of Beijing. The construction of the bridge in water resource pays attention to environmental protection, so that the bridge is strictly requested in designing, materials and manufacturing. Shougang supply weathering resistant steel for its scientific research on bridge processing. In this paper, we focus on research procedure, microstructure, mechanical prosperities and weather resistant prosperity of Q345qENH produced by Shougang.

Content from this work may be used under the terms of the Creative Commons Attribution 3.0 licence. Any further distribution of this work must maintain attribution to the author(s) and the title of the work, journal citation and DOI.
Published under licence by IOP Publishing Ltd

2. Technical requirements and research procedure

2.1. Technical requirements

The chemical composition (mass fraction,%) of the weathering resistant bridge steel according to GB / T714-2015 is listed in table 1. Weather resistance index I is calculated by chemical composition according to ASTM G101, I (weather resistance index) of weathering resistant bridge steel require≥6.0. In terms of welding, Pcm(crack sensitivity coefficient)is decided to designers and manufactures of Guanting Reservoir Bridge. Table 2 is mechanical performance requirements.

Table 1 Chemical composition requirements of Q345qENH.

Element	C	Si	Mn	Nb	V	Ti	Cu	Ni	Cr
GB714 -2015	≤ 0.11	0.15 ~ 0.50	1.10 ~ 1.50	0.01 ~ 0.10	0.01 ~ 0.10	0.006 ~ 0.030	0.25 ~ 0.50	0.30 ~ 0.40	0.40 ~ 0.70

Note 1: I =26.01(%Cu)+3.88(%Ni)+1.20(%Cr)+1.49(%Si)+17.28(%P)-7.29(%Cu)(%Ni)-9.10(%Ni)(%P)-33.39(%Cu)^2;
2: Pcm = C + Si/30 + Mn/20 + Cu/20 + Ni/60 + Cr/20 + Mo/15 + V /10 + 5B;

Table 2 Mechanical performance requirements of Q345qENH

Rel /MPa	Rm /MPa	Y/T ratio	A%	-40℃impact energy /J
≥345	≥490	≤0.85	≥20	≥120

2.2. Research procedure

In terms of composition, according to the formula of weather resistance index I, it is known that the improvement of weather resistance index mainly needs to ensure the content of Cu, Ni, Cr, Si and P, however, these elements will deteriorate the welding performance as well. In this way, we design composition on the foundation of weathering resident element content, at the same time reduce the carbon and manganese content to ensure the welding performance.

In terms of microstructure, In order to ensure the steel has high strength, low yield ratio and high toughness, microstructure is designed as AF (acicular ferrite) +PF (polygonal ferrite).AF is the assurance of strength and low temperature toughness, PF control yield strength, so as to ensure low yield ratio.

3. Mechanical prosperities and microstructure

3.1Mechanical prosperities

Industrial production of 8-60mmQ345qENH weather resistant bridge steel is showed as table2. 375 MPa ≤ yield strength ≤ 455MPa, 547MPa ≤ tensile strength ≤ 606MPa, 0.67≤Y/Tratio≤0.82, 24% ≤elongation≤29%,- 40 ℃ impact energy ≥ 182J.These prosperities satisfy performance requirement of Guanting Reservoir Bridge project and national standard of GB / T714-2015.At the same time, Charpy V-Notch impact test were conducted at -20℃、-40℃、-60℃、-80℃、-100℃ to determine the ETT_{50}(ductile-brittle transition temperature). ETT_{50} of 16mm and 60mm steel is around -60 ℃ as shown in Fig. 1.

Table3. Mechanical prosperities of Q345qENH produced by Shougang.

Steel	Thickness/mm	Rel/MPa	Rm/MPa	Y/T ratio	A%	-40℃impact energy /J		
Q345qENH	8	409	606	0.67	26.5	182	196	188
Q345qENH	16	434	547	0.79	24.0	247	257	202
Q345qENH	24	407	564	0.72	29	294	383	306

Q345qENH	44	455	553	0.82	25.0	290	290	281
Q345qENH	60	375	561	0.67	25.0	197	234	197

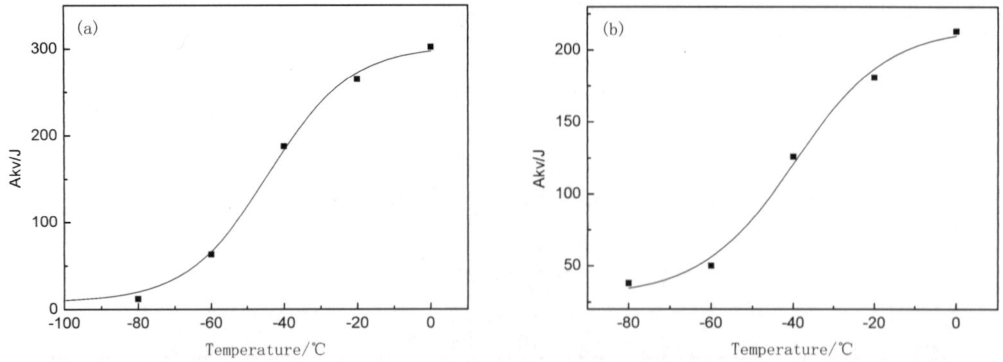

Fig.1 Ductile-brittle transition temperature curve of Q345QENH.(a)16mm,(b)60mm.

3.2 Microstructure

Microstructure examination of 16mm and 60mm steel is shown in Fig. 2, the examination position is 1/4 thickness of the sample. Fig. 2a shows PF+AF and Fig. 2b shows PF+AF+ a small amount P. Acicular ferrite is conducive to improving the strength of steel and low temperature toughness.

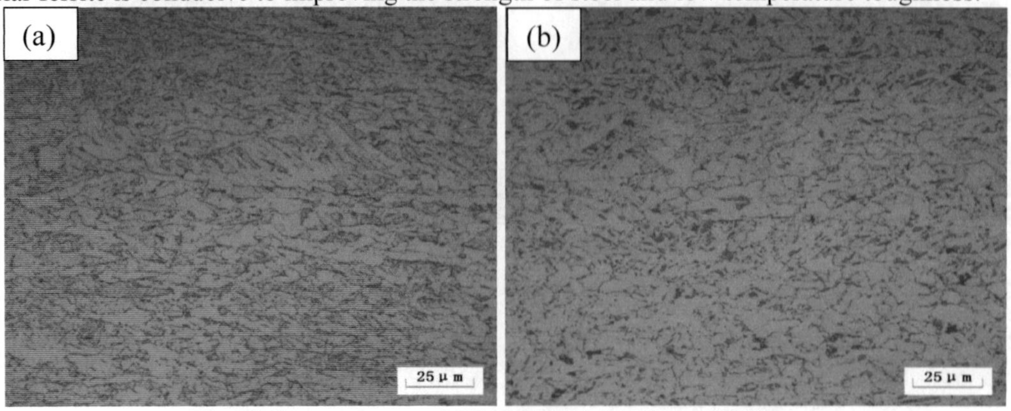

Fig.2 Microstructure of Q345QENH.(a)16mm,(b)60mm.

4. Weather resistant prosperity

In this chapter, we contrast weathering resistant steel Q345qENH with ordinary bridge steel Q345qE. To evaluate the weather resistant prosperities, simple size (60mm×40mm×3mm) was employed according to TB2375-1993. Corrosion solution selected 0.01mol / L $NaHSO_3$, testing temperature was 45±2℃, humidity choose 70 ± 5% RH, each cycle was 60 ± 3min, and infiltration time is 12 ± 1.5 min.The test cycle were 1 day, 2 days, 4 days, 8 days, weight loss and corrosion rate of different test cycles were measured, the composition and structure of rust layer were analyzed and observed.

4.1 Discussion of weight loss

Test cycles were 1,2,4,8 days of two kinds of steel, weight loss and corrosion rate are shown as Figure3. At the same cycle, weight loss and corrosion rate of weathering resistant bridge steel Q345qENH is significantly lower than Q345qE, the difference of weight loss will be further expanded with the increase of experimental time, and the corrosion rate of Q345qENH is 51.5% of Q345qE at 8 days.

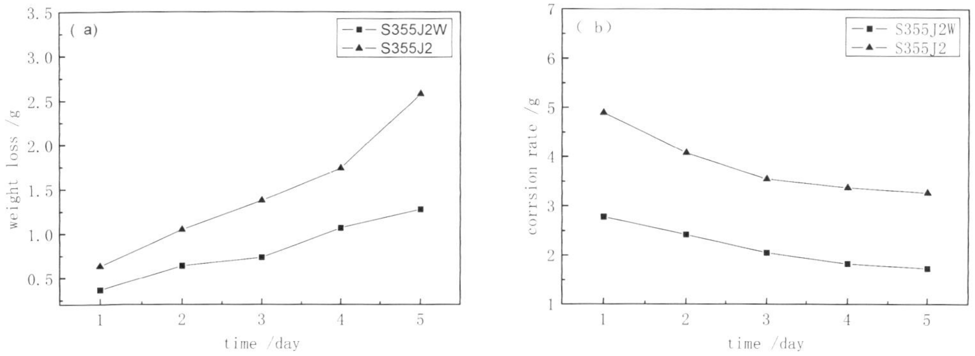

Fig.3 Curves of weight loss and corrosion rate.(a) weight loss,(b) corrosion rate.

4.2 Discussion of rust layer

As shown in table 4, component of rust layer at 8 days are composed of α-FeO (OH), Fe_3O_4 and Fe_2O_3, α-FeO (OH) has a compact structure, so the more α-FeO (OH), the better corrosion resistance performance [4-5]. The content of α-FeO (OH) in Q345qENH rust layer is obviously higher than Q345qE,we can conclude weathering resistant steel has a better performance at corrosion resistance. As shown in Fig. 4, rust layer appearance of Q345qENH and Q345qE at 8 days under SEM, Fig. 4a of Q345qENH possesses a more compact structure than Fig. 4b of Q345qE, which is exactly consistent with component analysis. The dense rust layer would prevent oxygen and water in the air from traversing protecting coat, and then protect steel substrate from corrosion.

Table 4 Component of rust layer at 8 days.

Steel	Test cycle/day	α-FeO(OH) /%	Fe_3O_4/%	Fe_2O_3/%
Q345qE	8	42.4	26.4	31.2
Q345qENH	8	51.8	13.0	35.2

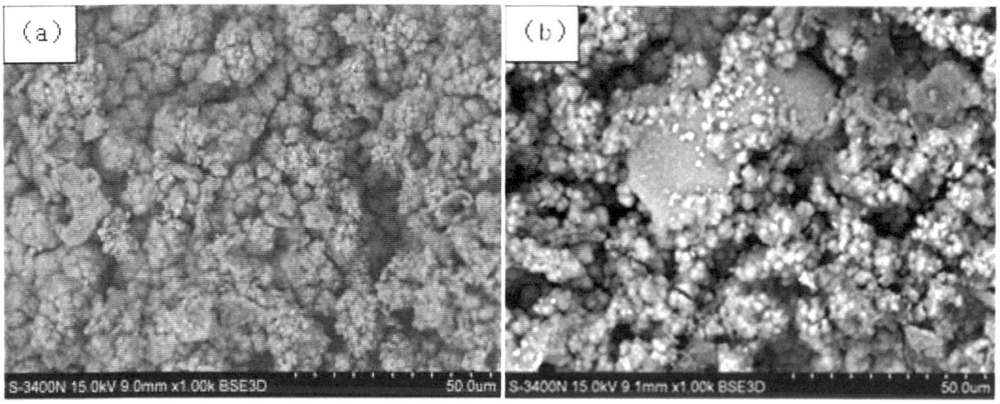

Fig.4 Rust layer appearance at 8 days.(a) Q345qENH,(b) Q345qE

5. Conclusion

1. Q345qENH weathering resistant bridge steel produced by Shougang, weather resistance index> 6.0,strength , plastic and toughness can meet the national standards of ASTM G101 and GB714-2015. The ductile-brittle transition temperature of the steel is around -60 ℃, which provides higher security for bridge.

IOP Conf. Series: Materials Science and Engineering **242** (2017) 012059 doi:10.1088/1757-899X/242/1/012059

2. Q345qENH weathering resistant bridge steel produced by Shougang has a better corrosion performance than Q345qE.The results of alternate immersion test show that weight loss and corrosion rate of Q345qENH are obviously lower than t Q345qE bridge steel. Especially, the corrosion rate of Q345qENH is 51.5% of Q345qE at 8 days.

3. α-FeO (OH) content of Q345qENH is higher than Q345qE, the dense rust layer would prevent oxygen and water in the air from traversing protecting coat so that protect steel substrate from corrosion.

References

[1] WANG Dongcai,GUO Liang. Current status and development of weathering resistant steel[J]. Science & Technology of Baotou Steel, 1993,3:59-73.

[2] WANG Lei,QIU Ruipang.Current status and development of weathering resistant steel[J].Journal of Highway and Transportation Research and Development , 2014,5:257-260.

[3] YU Qian. Current status and development of weathering resistant steel[J]. Journal of Iron and Steel Research,2007.19(11):1-4.

[4] CHEN Xiaoping, WANG Xiangdong, etc. Anti-corrision mechanism of rust layer with atmospheric corrosion resitiance [J]. corrision and protection,2009,30（4）:241-243.

[5] Keiser J T，Brown C W. Characterization of the passive film formed on weathering steels δ-FeOOH [J]. Corrosion Science,1983,（23）:251-253.

IOP Conf. Series: Materials Science and Engineering **242** (2017) 012060 doi:10.1088/1757-899X/242/1/012060

The analysis of critical cooling rate for high-rise building steel S460

Shiping Lu [1,2], Xia Chen [1,2], Qun Li [3], Haibao Wang [1,2], Linhao Gu [1,2]

[1] Shougang Research&Design Institute of Technology;
[2] Laboratory of Green Recyclable Process for Iron & steel Production Technology;
[3] Shouqin metal materials CO.,LTD

lushiping@shougang.com.cn

Abstract High-rise building steel S460 is an important structure steel.The product process of the steel is Quenching&Tempering. The critical cooling rate of steel is very important in heavy plate quenching process, and it is also the basis of the cooling process[1].The critical cooling rate of HSLA steel S460 is obtained from the Thermal simulation method,and the differences about the microstructure and properties of different cooling rate is also analyzed.In this article, the angle of the grain boundary and the average grain size are analyzed by EBSD under different cooling rate. The relationship between grain boundary angle and grain size with the cooling rate is obtained. According to the experiment,it provides the basis for the formulation of the quenching process of the industrial production.

1. Backgroud

High strength structural steel has significant advantages in structure performance,construction use function social economic benefits and environmental benefits[2].High strength structural steel can improve the safety and reliability of structure, and can create greater use of space,and the flexible architecture,at the same time can save construction cost,belonging to the environmental protection building materials.The using condition of the high strength structural steel S460 is QT.The critical cooling rate should be cleared by experiment to determine reasonable processing technology.

2. The purpose of the test

High strength structural steel is a kind of HSLA steel, the steel grade of the critical cooling rate vary according to the composition, usually with the increasing of carbon equivalent the critical cooling rate has the tendency to decline. The critical cooling rate can be obtained by thermal simulation method.The composition of S460 as below.

Table 1. Chemical composition of S460.

	C	Si	Mn	P	S	Alt	CEQ
S460	0.09	0.28	1.4	0.01	0.0024	0.038	0.33

Quenching is to heat the steel to the temperature above Ac3 or Ac1,heat preservation for some time,cooling rate greater than the critical cooling speed,finally get the microstructure of martensite or

Content from this work may be used under the terms of the Creative Commons Attribution 3.0 licence. Any further distribution of this work must maintain attribution to the author(s) and the title of the work, journal citation and DOI.
Published under licence by IOP Publishing Ltd

low bainite[3].The aim of this experiment is to grope for the critical cooling speed,and also for the microstructure difference under different quenching cooling speed.

The thin steel plate is easier to hardening, after quenching the strip shape is difficult to control.Based on this,the cooling rate has to be moderate[4].

3. Quenching process

In this experiment,7 round rods is selected,The size is Φ 6*80 mm,steel grade is high strength structural steel S460.

The thermal simulation testing machine GLEEBLE 2000D is used.The samples heating to Ac3 temperature (S460 is above 900 °C) from room temperature at heating rate of 10°C/S,keep 240 s, rapidly cooling to room temperature.The cooling rate is 5℃/S(1#),10℃/S (2#),20℃/S(3#),30℃/S(4#),40℃/S(5#),50℃/S(6#),60℃/S(7#) respectively.Schematic diagram as below.

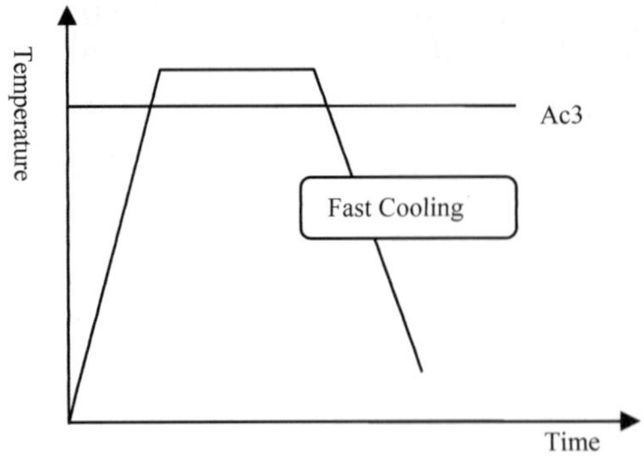

Figure1. Schematic diagram

4. Experimental analysis of quenching

Making metallographic and electron microscope observation for quenching samples,and compared the differences of the microstructure and performance at different cooling speed,to determine the critical cooling rate of S460.

4.1 Microstructure and microhardness analysis

Figure2-Figure8 is the microstructure of samples.

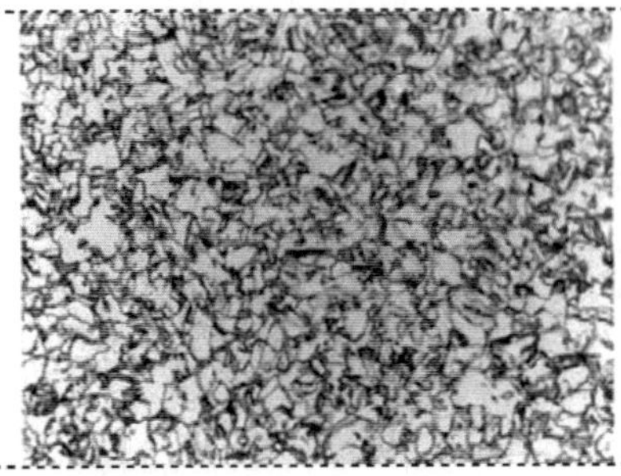

Figure 2. Microstructure of sample 1

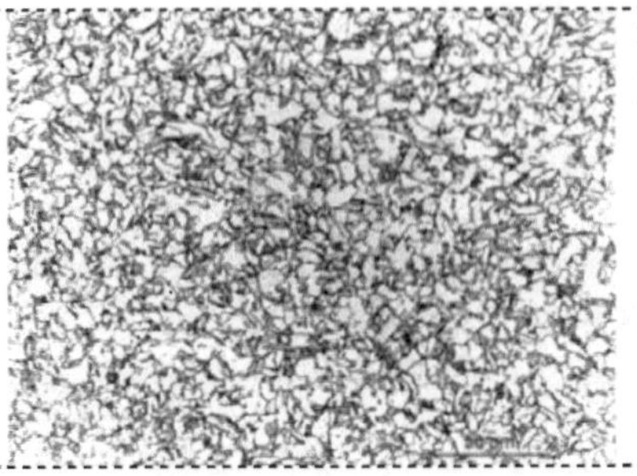

Figure 3. Microstructure of sample 2

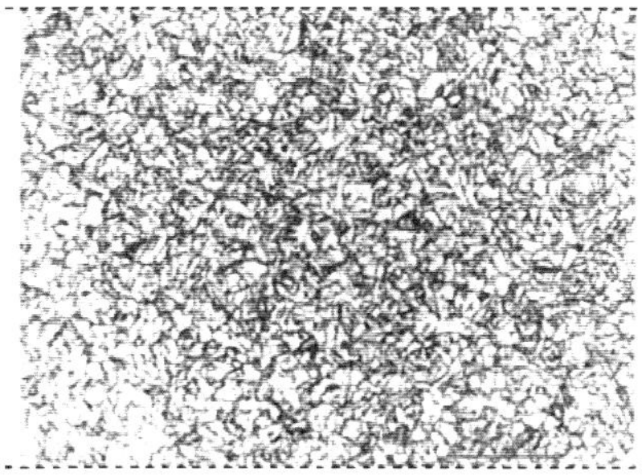

Figure 4. Microstructure of sample 3

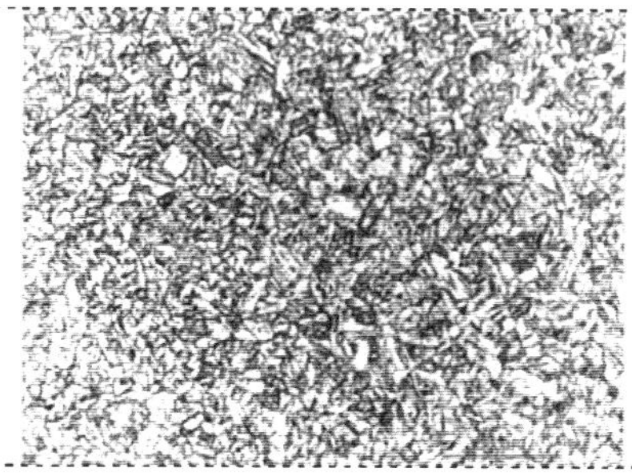

Figure 5. Microstructure of sample 4

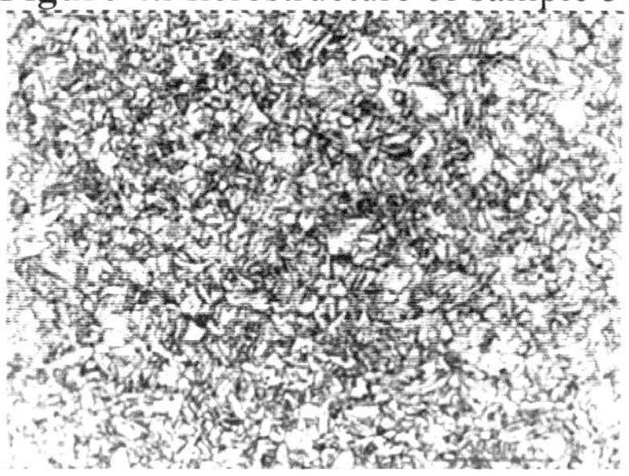

Figure 6. Microstructure of sample 5

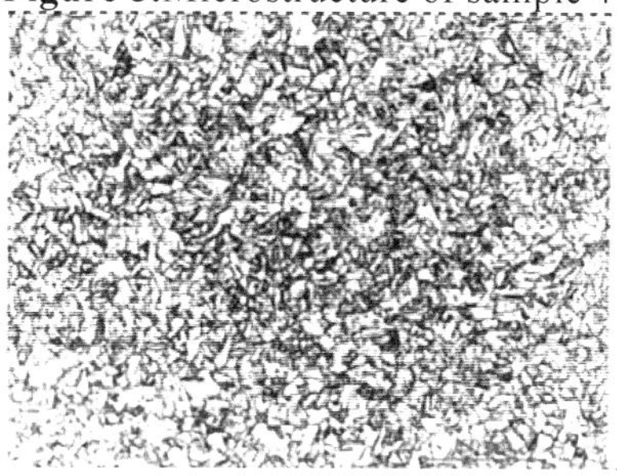

Figure 7. Microstructure of sample 6

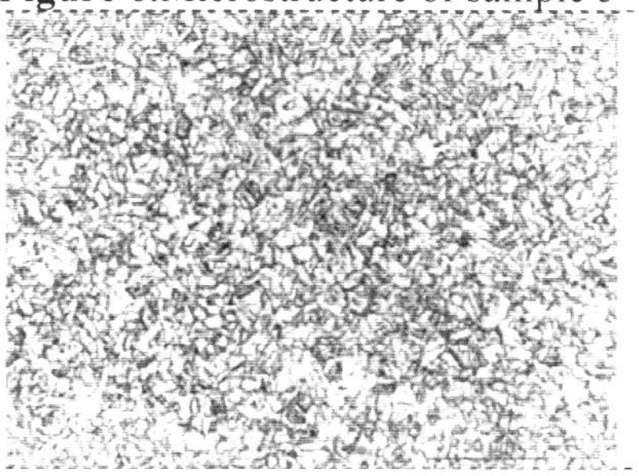

Figure 8. Microstructure of sample 7

Microstructure microhardness and MA rate as Table 2.

Table2. Microstructure result

Sample	Microstructure	Cooling Rate	Microhardness	Mean value of Microhardness	MA Rate (%)
1	Ferrite+Bainite+ Pearlite	5℃/S	189.84 197.54 190.88	192	6.9
2	Ferrite+Bainite+ Pearlite	10℃/S	205.71 208.13 198.64	204	6.6
3	Bainite+ Ferrite	20℃/S	215.72 233.61 229.28	226	4.8
4	Bainite+ Ferrite	30℃/S	236.54 236.54 230.82	234	-
5	Bainite+ Ferrite	40℃/S	214.48	230	-

6	Bainite+ Ferrite	50℃/S	235.03 239.53 232.13 210.67 227.84	223	-
7	Bainite+ Ferrite	60℃/S	225.07 229.28 238.07	230	-

From the microstructure observation result.The cooling rate of sample 1# 2# is lower.The final microstructure is Ferrite+Bainite+Pearlite.The microstructure of sample 3#(cooling rate 20℃/S)-7#(cooling rate 60℃/S)is Bainite+ Ferrite.With the increasing of the cooling rate,the Bainite is more and the Ferrite is less.And sample 1# 2#also contained some Pearilite.

Microhardness value also correspond to microstructure.The microhardness of sample 1# 2# is lower because of the contained pearilite .The value is between 190~205HV.The value of sample 3#-7# is between 225~235HV.Depending on the microstructure and microhardness result,the critical cooling rate of S460 is 20℃/S.

4.2 Scanning electron microscopy (SEM) analysis

In order to further analysis the microstructure under different cooling speed, main contrast of bainite, the electron microscope analysis is used.The cooling rate from 20℃/S~60℃/S.The microstructure is Bainite+Ferrite.The microstructure of SEM under different cooling rate as below.The difference of B+F from different samples is observed.

Figure 9. Sample 4#(1000X)

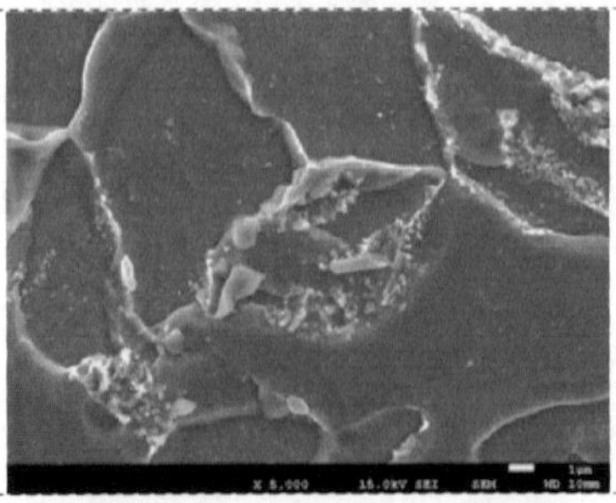

Figure10. Sample4#(5000X)

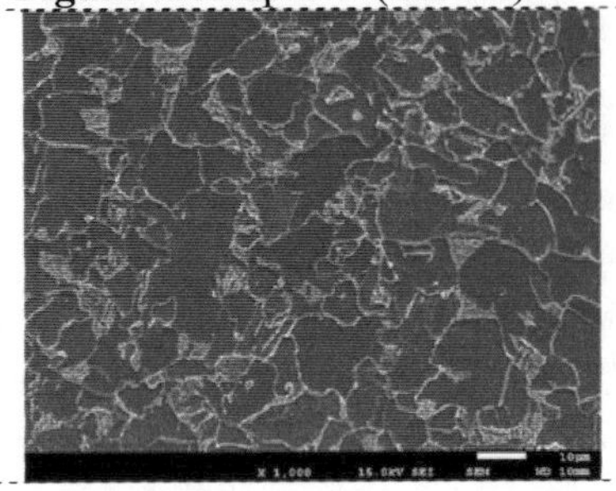

Figure 11. Sample5#(1000X)

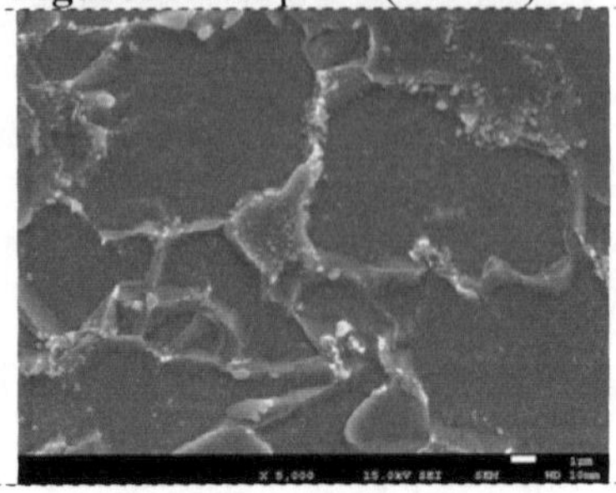

Figure 12. Sample5(5000X)

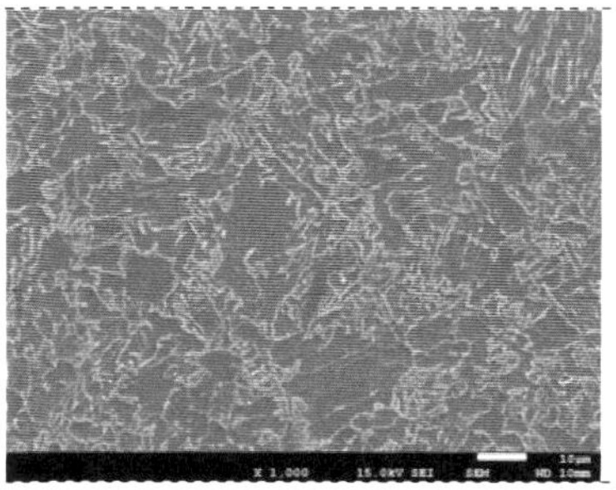

Figure 13. Sample 6#(1000X)

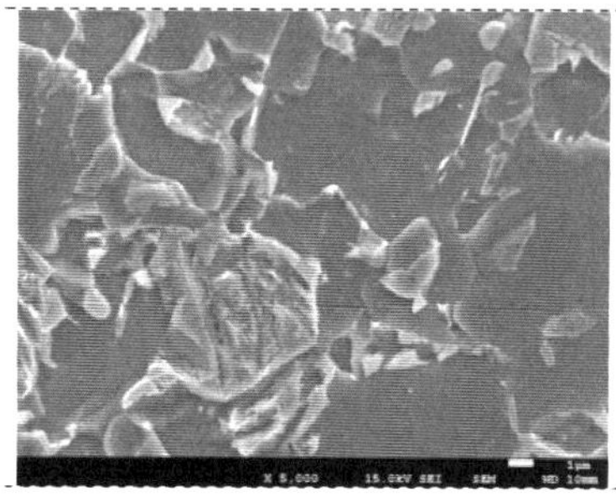

Figure 14. Sample6#(5000X)

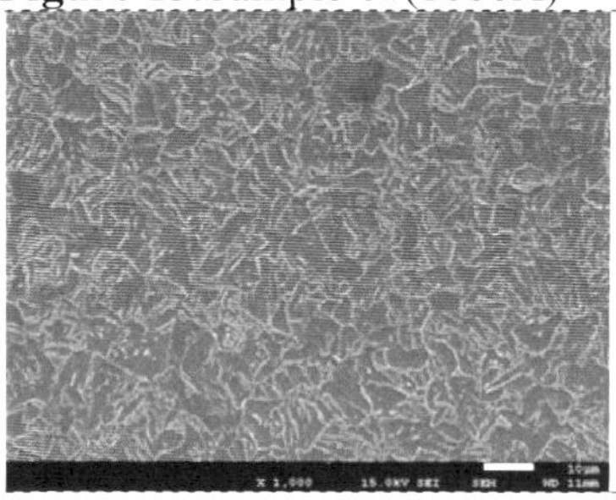

Figure 15. Sample 7#(1000X)

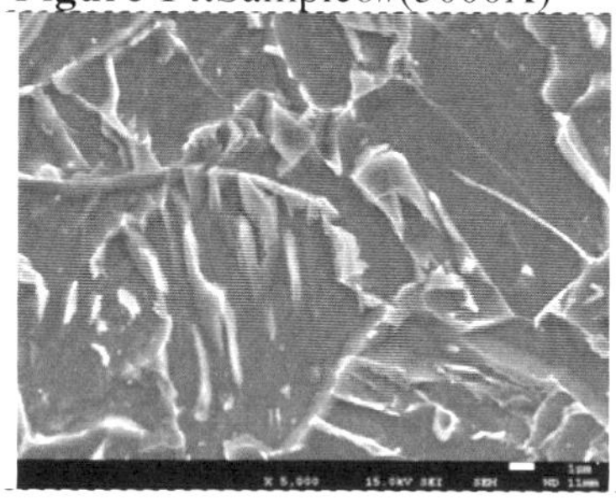

Figure 16. Sample7#(1000X)

The differences of B+F from 3#-7# is observed by SEM.The conclusion as below.Some pearlite phase and some dot carbides, M - A island of bainite phase is observed when the phase of bainite area of sample 4# is zoom to 5000X.Little pearlite phase and some dot carbides,M - A island of bainite phase is observed when the phase of bainite area of sample 5# is zoom to 5000X.There is no pearlite phase in sample 6# 7#,the Dot bainite transform to lath bainite.it can be concluded that above the critical cooling rate,with the increase of cooling speed ,bainite content increased gradually.

4.3 Electron Back-Scattered Diffraction (EBSD) analysis

Average grain size and grain boundary angle measurement analysis is carried out by EBSD on the sample.The samples are 4#-7#,Figure3 is grain boundary angle measurement analysis.Table2 is grain size and proportion of the large Angle grain boundary for each sample.

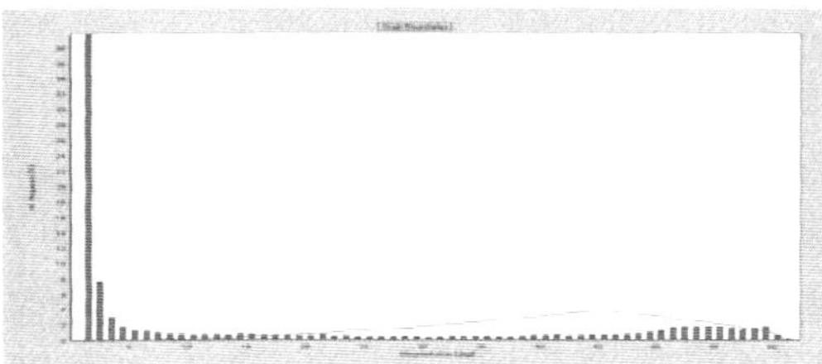

Figure 17. Grain boundary angle of 4#.

Figure 18. Grain boundary angle of 5#.

Figure 19. Grain boundary angle of 6#.

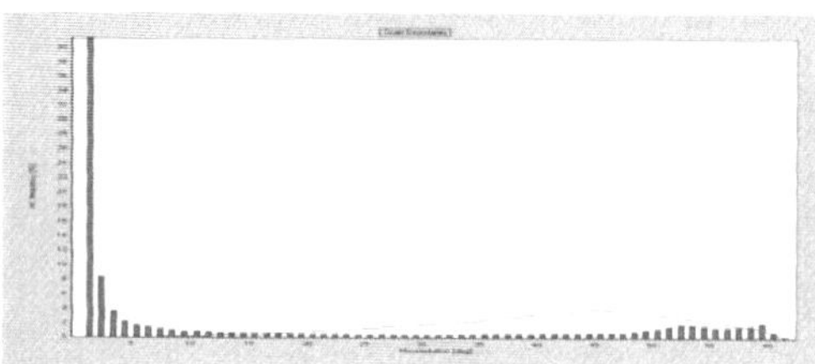

Figure 20. Grain boundary angle of 7#.

Table 3. Grain size and grain boundary angle

Sample	Grain boundary angle(>15°)	Grain size	Average grain size (μm)
4	38.25%	1.13~34.97	6.63
5	37.53%	1.13~34.02	5.66
6	31.18%	1.13~26.41	4.25
7	34.57%	1.13~18.57	4.21

It can be concluded from Figure 3 that the grain boundary angle distribution is similar from sample 4#-7# . The proportion is higher in less than 5 ° and greater than 50 °, and the proportion of 1.5 ° angle grain boundary is as high as 40%.

It can be concluded from Table 3 that the proportion of high-angle boundary(>15°)is 31-38% in sample 4#-7#. The average grain size reduced gradually from 6.63 um ~ 4.21 um.

5. Conclusion

- From thermal simulation experiment results and microstructure microhardness analysis of the 1#-7# samples, preliminary determine critical cooling rate of S460 is 20℃/S.
- The SEM indicate the result of metallographic observation.And low cooling rate samples as 1# 2# appears widmanstatten structure.From sample 4#, bainite structure content gradually increased, with the higher cooling rate.
- EBSD indicate that greater rate than the critical speed range, with the improvement of cooling speed, the average grain size decreases,the grain boundary Angle proportion is higher in less than 5 ° and greater than 50 °.

References

[1] Cui Zhongqi 1998 *Principle of metallography and heat treatment* (Haerbin: Harbin industrial university press) p302

[2] Wang Youming 2009 *The controlled rolling and controlled cooling of steel*(Beijing: Metallurgical industry press) p45

[3] Yu Yongning 2006 *Material science basis*(Beijing:Higher education press) p433

[4] Gao Weiguo 2003 *The microstructure and mechanical properties of steel widmanstatten structure vol 28* (Beijing:Journal of metal heat treatment) p39

IOP Conf. Series: Materials Science and Engineering **242** (2017) 012061 doi:10.1088/1757-899X/242/1/012061

Effect of solution treatment on microstructure and properties of duplex stainless steel

X Y Wang, J M Luo, L Q Huang, H B Wang and C W Ma

Shougang Research Institute of Technology, Beijing, 100043, China

Corresponding author's e-mail: wwxyong2008@163.com

Abstract. The influence of solution treatment on microstructure and properties of 2205 duplex stainless steel (DSS) was studied. The microstructure, precipitates and corrosion resisting property were observed and analyzed by means of optical microscopy (OM), scanning electron microscopy (SEM) and electrochemical methods. The results showed that a large number of brittle σ-phase precipitates, which deteriorate the plasticity and corrosion resistance of the material, were easy to produce in the duplex stainless steel under the low temperature. The precipitation of σ-phase can be decreased and the plasticity and corrosion resistance can be improved by increasing solution temperature. In addition, the ferrite content increases with the increase of solution temperature, while less affected by cooling rate.

1. Introduction

Duplex stainless steel consists of austenite(γ) and ferrite(δ) phases, which is produced by increasing Cr content or other ferrite formation elements and decreasing Ni content on the basis of austenite stainless steel [1-5]. Compared to traditional austenitic stainless steel and ferritic stainless steel, duplex stainless steel possesses good strength, high toughness, formability, anticorrosion, biocompatibility and weldability. This grade of steel has found extensive application in heat exchangers, water heaters, pressure vessels, storage tanks, rotors, impellers and shafts, digesters and other equipment in pulp and paper production, cargo tanks in chemical tankers, desalination plants, waste gas purifiers and sea water systems etc[6-12].

UNS S2205 is the typical representative of duplex stainless steel, the two phases each account for about 50% in normal state (solution treatment), to ensure its high strength, high corrosion resistance and good welding properties. Some studies have shown that the performance of the DSS is closely connected with the proportion of the two phases and other precipitated phase [13-23]. However, the proportion and other precipitation are depend on the chemical composition, heat treatment temperature and other factors. An improper heat treatment can change the proportion and produce a large amount of brittle phase (such as σ-phase), which is bad for the plasticity, toughness and corrosion resistance of the plate. Especially it is very important and difficult for both materials to achieve good performance with the same heat treatment when DSS is used for clad plates or devices welded with carbon steel. Therefore, the influence of different solution conditions (especially at 900℃) on the microstructure and precipitates of DSS is studied in present study, in order to provide a reasonable process reference for the subsequent processing of using of DSS in clad plates or other structure.

2. Experimental materials and procedures

2.1 Experimental materials

Content from this work may be used under the terms of the Creative Commons Attribution 3.0 licence. Any further distribution of this work must maintain attribution to the author(s) and the title of the work, journal citation and DOI.
Published under licence by IOP Publishing Ltd

The material used was a commercial 2205 steel plate (16 mm thickness) produced by rolled at 1160–1180°C and followed by water quenching. The chemical composition of the steel is listed in Table 1.

Table 1. Composition of the tested steel（wt. %）.

Element	C	Si	Mn	P	S	Ni	Cr	Mo	N	Fe
2205 steel	0.02	0.50	1.30	0.02	0.008	5.2	22.5	3.1	0.16	Bal.

Note: Its yield strength is 591MPa, ultimate tensile strength is715MPa, and total elongation is 25%.

2.2 Experimental procedures

All the solution treatments were carried out in laboratory, and the sample size is 200mm×300mm. The specimens were heated at the rate of 10°C $s^{-1}$from room temperature, then kept at the temperatures of 900,1000 and 1050°C, respectively, for 40 min, and continuously cooled to the ambient temperature by air cooling (AC) or water quench (WQ).

After solution treatments, the tensile test is according to GB/T228.1-2010 and the samples for optical microscope (OM), scanning electron microscopy (SEM & BSE) observation, were etched in copper sulfate and hydrochloric acid aqueous solution (1:5:5). The intergranular corrosion testing carried out by the E method of ASTM A262.In addition, the electrochemical polarization curve of the samples with different solution treatments were determined by the Parstat2273 electrochemical test system, and the pitting broken potential of the samples were characterized also.

3. Results

The mechanical properties of samples with different solution process are shown in Figure1.The strength of the 2205 duplex stainless steel is increased and elongation decrease after solution treatment at 900 ℃, relative to the original state (as rolled). While the solution temperature is higher to 1000 ℃, the plasticity is improved, but the strength is little reduced. Further raise the temperature (1050 ℃), the plasticity is further improved, and the strength of samples by air cooling are not different from that of water cooling. Intergranular corrosion test results showed that there is a small amount of tiny cracks in the samples after 900 ℃ quenching. However, the other samples have no cracking and show a good intergranular corrosion resistance.

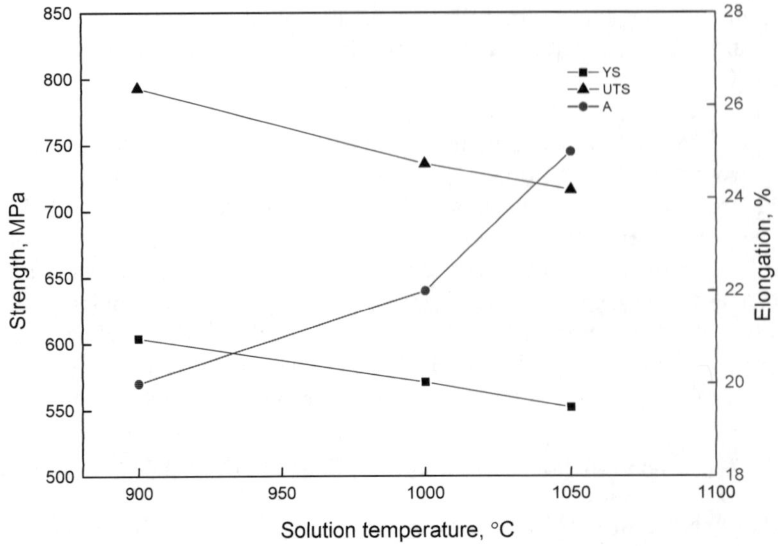

Figure1. Effect of solution temperature on the properties of DSS.

The microstructures of DSS after different solution treatment were observed, and the proportions of ferrite were calculated also. The results were showed in figure 2 and 3.There are a large number of precipitates at the flexural grain boundaries/phase boundaries in the samples after solution treatment at 900 ℃（Figure2a）. With the temperature is higher (1000 ℃), the precipitates at the grain boundary are decreased, and grain boundary is also more rules(Figure2b). Further raise solution temperature to 1050 ℃, the grain boundary is more clear and there is no more precipitation (Figure2c and d). The ferrite content of DSS with rolling or solution at 900 ℃ are both 46% (Figure 3). The ferrite content is increased with the solution temperature increasing. After solution at 1050 ℃, the ferrite content is as much as 54%.

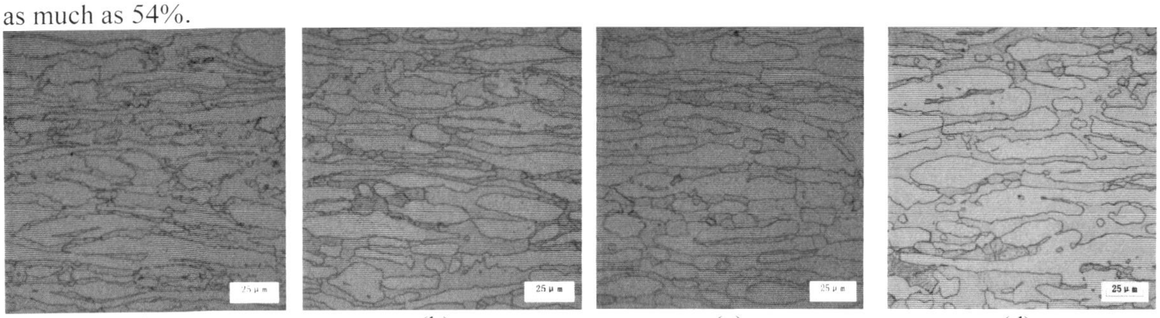

(a) (b) (c) (d)

Figure 2. The microstructure of the DSS with different solution temperatures (*a*) 900-WQ, (*b*) 1000-WQ, (*c*) 1050-WQ and (*d*) 1050-AC.

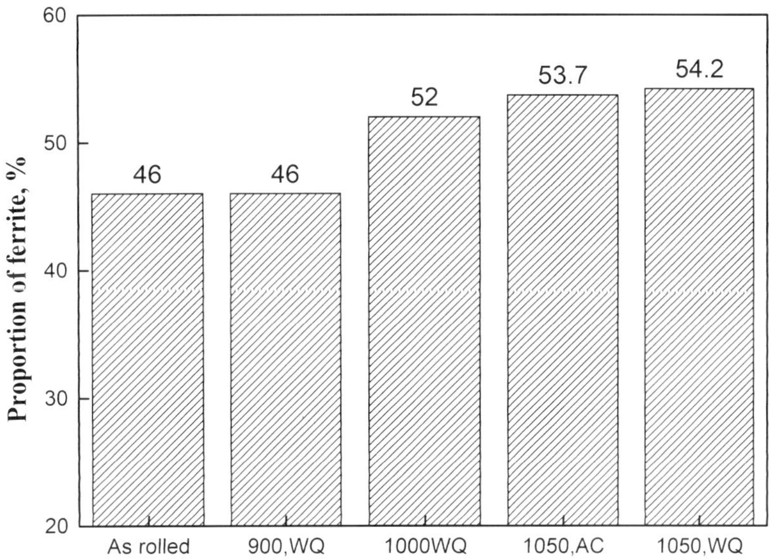

Figure 3. The ferrite proportion of the DSS with different solution process.

4. Analysis and discussion

4.1 Effect of solution treatment on precipitates

According to the characteristics of the DSS, the brittle precipitate such as σ-phase, chi-phase, pi-phase and alpha is easily appeared in 2205 with different process, in which the σ-phase is the most significant to the properties of DSS. The results of scanning electron microscopy of samples after different solution treatment are shown in figure 4.There are a lot of precipitates, which are clear and distinguishable with backscattered electron (BSE) mode in the ferrite along the interface of austenite and ferrite, for a long time under the 900 ℃. The morphology is white block or small dot and the energy spectrum shows that the precipitate is rich of Cr and Mo (it's σ-phase, table 2).Its volume is around 8 ~ 10%, that is consistent with other literature research results that σ-phase occur at a specific temperature range (commonly in 650 ~ 950℃), the peak temperature is around 900 ℃. And some studies have also found that the rolling deformation can promote the formation of σ- phase, but does not affect the precipitation peak temperature. Thermodynamic calculation results also show that σ-phase precipitate begin at about 950 ℃, and the amount increase with the temperature decrease, intense precipitate around 900 ℃ (Figure 5).The existence of such Cr -and Mo- rich brittle phase, that deteriorating the plasticity of DSS, and leads to the poor chromium near the grain boundary with reducing its corrosion resistance [13-18].

The Cr -and Mo- rich σ-phase is decreased significantly with the solution temperature increasing, which only exist in local area with a small amount at 1000 ℃(Figure 4b). Its morphology is clusters form with the surface is not smooth different from the rules smooth block under 900 ℃. Energy spectrum show that Mo content is also reduced, suggests that it's in dissolve state. The σ-phase dissolve more fully when solution temperature further increase to 1050 ℃. There is sporadic precipitation by air cooling but nothing in the case of water cooling (Figure 4c and d). According to the change rule of the precipitation, it is not advisable for this type of DSS to be used at lower temperature solution, for the precipitation of σ-phase. In order to avoid the precipitation of σ-phase and have a good comprehensive performance, the solution temperature should not be below 1000 ℃.

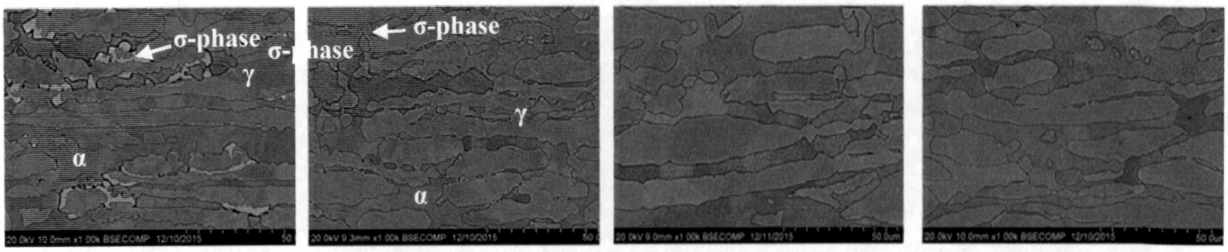

(a) (b) (c) (d)

Figure 4. The backscatter electron image of precipitates in the DSS with solution temperature at (*a*)900℃-WQ,(*b*) 1000℃-WQ, (*c*)1050℃-WQ and (*d*)1050℃-AC.

Table 2. Average chemical composition of different phase present in DSS 2205 at 900℃.

Phase	Cr	Fe	Ni	Mo
Ferrite	26.14	66.14	3.41	4.30
Austenite	23.59	69.17	5.03	2.20
Sigma(σ-phase)	29.44	56.30	3.18	10.00

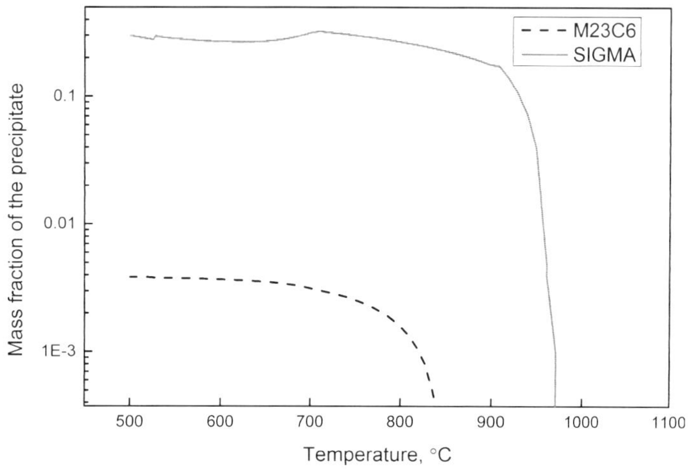

Figure 5. Thermodynamic calculation of sigma phase in duplex stainless steel.

4.2 Effect of solution treatment on corrosion resisting property

The electrochemical analysis of the samples by the different solution treatments were carried out. Results of the corrosion resistance of each process are shown in figure 6and 7.The obtuse potential, over-blunt potential and pitting broken potential of the samples with different solution process were compared. It is visible that pitting broken potential and blunt potential are low while the passivation range is very narrow with the solution temperature is 900 ℃(Figure 7).When the solution temperature rise to 1000 and 1050 ℃, the pitting broken potential and blunt potential rise obviously, and the passivation range is wider. And the pitting broken potential of sample by water cooling is slightly higher and the passivation zone is wider also. This shows that the solution temperature had significant effect on the corrosion resistance of DSS. There is a risk to reduce the corrosion resistant ability while the solution temperature is too low. In order to get a good corrosion resistance, the solution temperature should not be below 1000 ℃, This is consistent with the precipitation rule of the σ-phase in DSS. In addition, moderate fast cooling is beneficial to the corrosion resistance of DSS after solution treatment. It is consistent with the relevant research results also [7, 23].

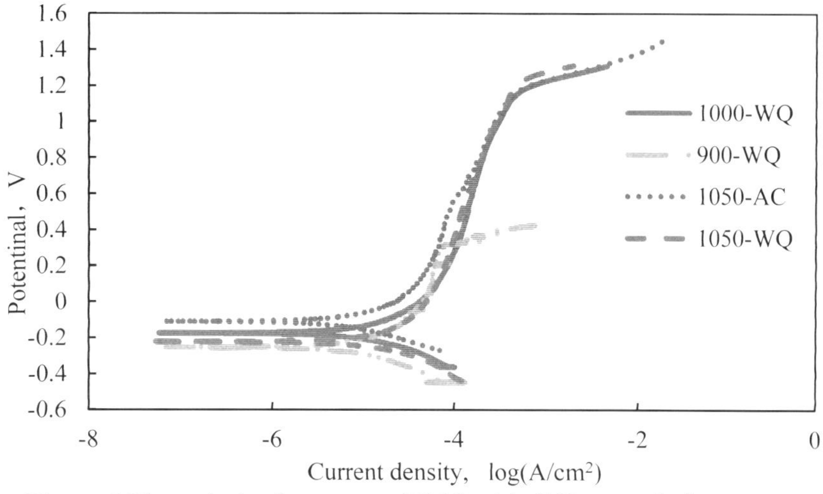

Figure 6. The polarization curve of DSS with different solution process.

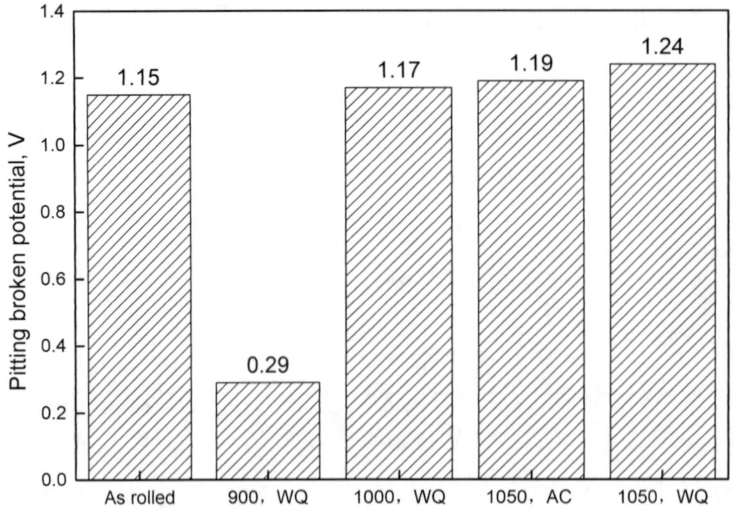

Figure 7. The pitting broken potential of DSS with different solution process in 6.0%NaCl solution at 70℃.

5. Conclusion

The main conclusions of this research work are the following.

(1) The physical and chemical properties of duplex stainless steel are affected by the solution temperature significantly. It has poor plasticity and corrosion resistance at lower temperature (900~1000 ℃) for the large amount of σ-phase.

(2) Precipitation of Cr- and Mo -rich σ-phase, is closely related to the process of solution treatment. It would result in deterioration of plastic and corrosion resistant properties when there are a large amount of precipitation at low temperature (900 ℃). With the increase of temperature, the σ-phase is gradually dissolved, and the precipitation volume decreases. After the temperature is up to 1050 ℃, the plasticity and corrosion resistant properties of the duplex stainless steel are improved significantly for the dissolution of σ-phase.

(3) The content of ferrite in duplex stainless steel increases with the increase of solution temperature, while the proportion of ferrite and austenite is less affected by the cooling rate.

References

[1] Chen T H and Yang J R 2002 *Mater. Sci. Eng., A.* **A338** 166.
[2] Wu J 1996 *Petrochemical Corrosion and Protection* **13** 6.
[3] Muthupandi V, Bala Srinivasan P, Seshadri S K, and Sundaresan S 2003 *Mater. Sci. Eng., A.* **A358** 9.
[4] GaoW, Luo J M and Yang J J 2005 *Ordnance Material Science and Engineering* **28** 61.
[5] Charles J and Chemelle P 2011*World Iron & Steel* **11** 1.
[6] Nelson D E, Baeslack W A and Lippold J C 1997 *Mater. Charact.* **39** 467.
[7] Chen T H and Yang J R 2001 *Mater. Sci. Eng., A.***A311** 28.
[8] Klyk-Spyra K and Sozanska M 2006 *Mater. Charact.* **56** 384.
[9] Atamert S and King J E 1991 *Acta Metall. Mater.* **39** 273.
[10] Ramirez A J, Lippold J C and Brandi S D 2003 *Metall. Mater.Trans. A* **34A** 1575.
[11] Weng K L, Chen T H and Yang J R 2003 *Bull. Coll. Eng., N.T.U.* **39** 45.
[12] Ahn Y S and Kang J P 2000 *Mater. Sci. Technol.***16** 382.
[13] Iacoviello F, Casari F and Gialanella S 2005 *Corros. Sci.* **47** 909.
[14] Tseng C M and Tsai W T 2004 *Mater. Chem. Phys.* **84** 162.
[15] Kobayashi D Y and Wolynec S 1999 *Mater. Res.* **2** 239.

[16] Cvijovic Z and Radenkovic G 2006 *Corros. Sci.* **48** 3887.
[17] Tsai W T and Chen M S 2000 *Corros. Sci.* **42** 545.
[18] Rondelli G, Vicentini B and Sivieri E 1997 *Corros. Sci.* **39** 1037.
[19] Feng H, Zhou X Y, Liu H and Song Z G 2015 *Journal of Iron and Steel Research* **27** 1.
[20] Wei B, Bai Z Q, Yi C X and Li W G 2009 *Transactions of Materials and Heat Treatment* **30** 73.
[21] Liu X, He Y L, Ding X P, Song H M, Hu J C, Chen X and Li L 2010 *Transactions of Materials and Heat Treatment* **31** 48.
[22] Bin Y H, Li P F, Li Z Z and Fan Z Y 2013*Transactions of Materials and Heat Treatment* **34** 136.
[23] Wang Y X, Chen Y M, Liu Z Y, Fang Y T and Wang G D 2009 *Journal of Iron and Steel Research* **21** 51.

IOP Conf. Series: Materials Science and Engineering **242** (2017) 012062 doi:10.1088/1757-899X/242/1/012062

Development and application of super heavy gauge high-strength structural steel for high-rise buildings

Linhao Gu [1,2,3,a], Shiping Lu [1], Chunming Liu [1], Jingang Liu [1], Suyuan Zhang [1], Rensheng Chu [1], Changwen Ma [1]

[1] Shougang Research Institute of Technology, No.69, Yangzhuang Street, Shijingshan District, Beijing China
[2] Beijing key Laboratory of Green Recyclable Process for Iron & steel Production Technology, No.69, Yangzhuang Street, Shijingshan District, Beijing China
[3] Beijing Engineering Research Center of Energy Steel, No.69, Yangzhuang Street, Shijingshan District, Beijing China

[a] email:464757839@qq.com

Abstract. This paper presents development of 130mm S460G1-Z35 by using low carbon Nb-Ni-Mo-V-Ti micro-alloying design and two-stage rolling, quenching and tempering process. For the super heavy gauge high-strength structural steel, the yield strength is higher than 450MPa, the tensile strength is higher than 550MPa, the elongation is greater than 20%, the low temperature(-40) impact energy value is not less than 250J, the z-direction section shrinkage is more than 65%, and the welding performance is good. The plate are successfully applied to the engineering construction of the city of dreams in Macau.

1. Introduction

High strength structural steel with high strength, high toughness, low yield to tensile strength (Y/T) ratio and good weldability is widely applied in welded steel structure of high-rise buildings, large-span stadiums, conference and exhibition center, airport terminals, large Bridges, large offshore oil platform, large power stations, large engineering machinery, large hoisting equipment, etc [1-2]. In recent years, with the booming of the construction of high-rise building foundation, the quality demand of super heavy gauge high-strength structural steel is getting bigger and bigger. In the meanwhile, the various performance requirements are becoming higher and higher, such as high strength, high toughness, low Y/T ratio, low temperature resistance, lamellar tearing resistance, easy welding and so on. In order to meet the demand of steel for the large-scale high-rise buildings, high-strength structure steel of 130mm thick S460G1-Z35 for high-rise buildings has been developed, which is manufactured through resonable composition design, steelmaking, continuous casting, rolling and heat treatment. And the plate is successfully applied in the construction of new landmark building - the city of dreams in Macau.

2. Chemical Composition Design

In order to ensure the welding property of the structure steel for the building, the steel has low carbon content and carbon equivalent. Besides, Nb-V-Ti microalloying elements are adopted to increase strength, toughness and uniform microstructure. Ni is added in the steel to improve low temperature impact toughness and Mo to improve the hardenability of the plate.

Content from this work may be used under the terms of the Creative Commons Attribution 3.0 licence. Any further distribution of this work must maintain attribution to the author(s) and the title of the work, journal citation and DOI.
Published under licence by IOP Publishing Ltd

C element is the main factor that affects the sensitivity of welding cracks in steel. With the same Pcm, low carbon can reduce the hardness and hardening tendency of the heat affected zone (HAZ). When the carbon level is below 0.10%, carbon equivalent is not significant influence of cold cracking sensitivity, even though a great amount of alloy in order to ensure the strength, carbon equivalent is not significant influence of the cold cracking sensitivity [3].

Nb is added in the structural steel as a typical microalloying element. Nb has a wide influence on each rolling stage microstructure change and the final mechanical properties during hot working process of steel plate. Nb is also used in off-line heat treatment (RQ-T) of high-strength steel, the combination of Nb with other alloys can improve hardenability of steel, refine microstructure and improve toughness [4].

Ni can improve the strength of steel, greatly improve the low temperature toughness without reducing ductility, and also reduce the tendency to produce cracks during slab reheating and hot rolling.

The chemical composition of S460G1-Z35 steel plate is designed according to the influence of different elements on the properties of steel plate and the customer's technical requirements. The basic alloying concept is given in table 1.

Table 1. Basic chemical composition (Wt.%)

C	Si	Mn	Nb	Ni	Others	Ceq[a]	Pcm[b]
0.08	0.25	1.25	0.025	0.55	Mo,V,Ti	0.38	0.18

[a] Ceq=C+Mn/6+(Mo+V+Cr)/5+(Ni+Cu)/15
[b] Pcm=C+Si/30+Mn/20+Ni/60+(Cu+Cr)/20+Mo/15+V/10+5B

3. Production process of steelmaking, rolling and heat treatment

3.1. Steelmaking
During the steelmaking process, the contents of H, S, P, O, N and the amount of inclusions are strictly controlled to reduce the anisotropy of steel. Through the optimization of steelmaking and continuous casting process, the reduction of center segregation of casting slabs could, so as to guarantee good Z-performance.

3.2. Rolling
Two-stage rolling process is used in the hot rolling of steel, rolling speed, deformation temperature and deformation amount are strictly controlled in the process. In the high-temperature austenite recrystallization region, high-temperature low-speed and high pressure process is used, so that the deformation in the thickness direction fully penetrate into the core and austenite grain gets very fine through the repeated deformation.[5]. In finish rolling stage, steel is deformed below the recrystallization temperature in order to produce more ferrite nucleation sites in the deformed austenite, causing uniform and fine ferrite, pearlite and bainite after phase transformation [6].

3.3. Heat Treatment
The quenching and tempering process is the important production technology of the steel plate, which has the advantages of good shape, stable performance and strong adaptability of thickness. The quenching temperature is Ac3 + (30 ~ 50) ℃, according to the actual chemical composition, the temperature of Ac3 is 863℃, the austenitizing temperature is 900 ℃. Considering the thickness is very heavy, high quenching temperature will result in coarse grains, so combining with CCT curve to develop a reasonable quenching and tempering process. The quenching temperature is 900 ~ 910 ℃, the tempering temperature is 530 ~ 560 ℃.

4. Results and Discussion

4.1. Mechanical properties

The mechanical properties of S460G1-Z35 are given in Table 2. The high-strength structural steel plate of industrial production has excellent mechanical properties. The yield strength is higher than 450MPa, the tensile strength is more than 550MPa, the elongation is greater than 20%, the average impact energy at -40℃ exceeds 250J, and the z-direction section shrinkage is more than 65%.

Table 2. Mechanical properties of S460G1-Z35

Steel grade	Thickness mm	YS MPa	TS MPa	A %	Akv at -40℃ J			Z %		
Technical requirements	130	≥380	500-660	≥17	≥60			≥35		
S460G1-Z35	130	469	573	26	276	270	298	70	69	72

In Figure 1 the results of the Charpy impact tests at different temperatures are presented. The testing temperature is from -120 ℃ to 20 ℃ .It can be seen from this figure that the ductile-brittle transition temperature (DBTT) at the quarter of the thickness is about -80 ℃, the DBTT at the center of thickness is about -65 ℃. Therefore, the steel plate has excellent low temperature impact toughness.

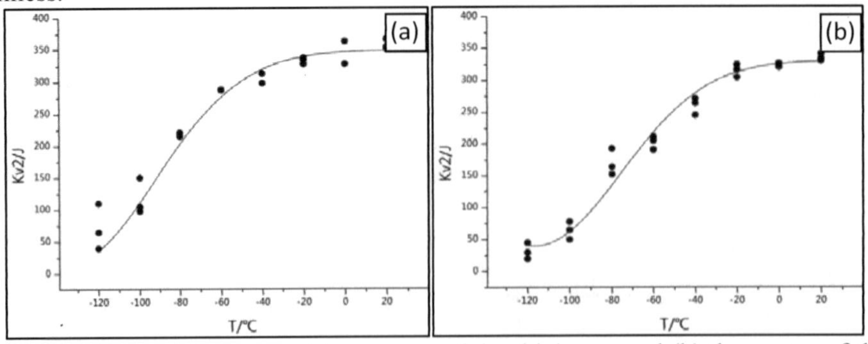

Figure 1. Charpy impact energy at (a) a quarter of the thickness and (b) the center of thickness

4.2. UT

The ultrasonic testing of the steel plate is carried out, and the defect of flaw detection is not found across the whole thickness of the steel plate, which meet the customer requirement of EN1016:S1/E3.

4.3. Microstructure

The micrograph of microstructure at different positions in the thickness direction is shown in Figure 2. During the quenching process, as the cooling rate of the surface location is higher, the microstructure is mainly bainite. From surface to the center of steel plate, the cooling rate decreases gradually, the microstructure turned to be mainly ferrite containing small amounts of bainite at the quarter of thickness, and the ASTM grain size number is about 10. The cooling rate at the center is the lowest, so the microstructure is mainly ferrite and pearlite, and the ASTM grain size umber is about 9. The mechanical properties of steel plates are excellent due to the uniform microstructure and small grain size at a quarter and center of the thickness.

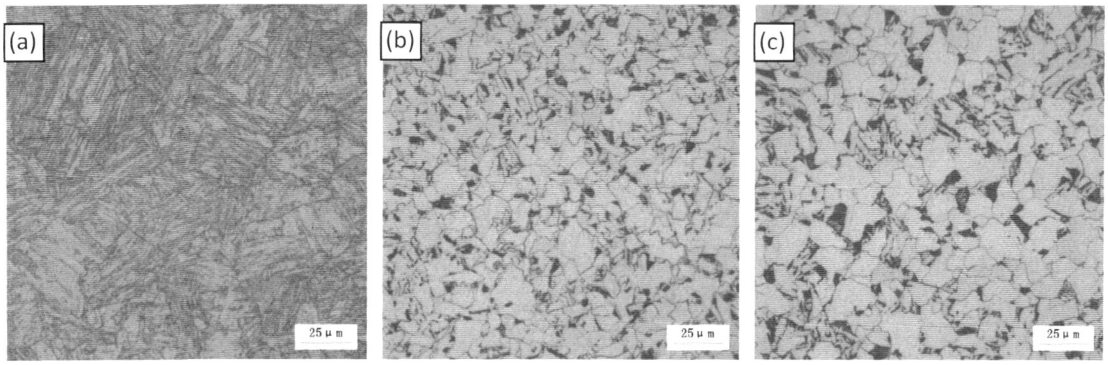

Figure 2. Typical microstructure of 130mmS460G1-Z35 at (a) surface (b) a quarter (c) center

4.4 Welding
According to the requirements of super heavy gauge high-strength structure steel for buildings, the welding performance of industrial production of 130mm S460G1-Z35 is tested and evaluated. The maximum hardness test of the HAZ, the cold crack test of inclined Y-groove and the welding test of the butt joint are respectively carried out.

4.4.1 maximum hardness test. The different wilding process of manual arc welding(MAW) and CO2 gas shielded welding(GSW) are used to the maximum hardness test in the HAZ under the situation of no-preheating, 100 ℃ preheating and 150 ℃ preheating. The results show that the maximum hardness of the HAZ is HV362 and HV390 without preheating, and the maximum hardness is HV 313 and HV 312 with a preheating temperature of 100 ℃. When the preheating temperature is 150 ℃, the maximum hardness HV 287 and HV 291. Using two kinds of wielding process, the maximum hardness of the HAZ is below HV 350 when the preheating temperature is no less than 100 ℃, which is less than the critical value HV 350 of the cold welding sensitive specified by the International Welding Association.

4.4.2 Y-groove cracking test. The crack sensitivity test of oblique Y-groove is a kind of cold crack test method which is more restrictive, which is mainly used to evaluate the cold crack sensitivity in the HAZ of carbon steel and low alloy high strength steel. It can be seen from the test results in Table 3, when the preheating temperature is not less than 100 ℃, there are no any cold cracks in the HAZ by using two kinds of welding process.

Table 3. The results of Y-groove cracking test

No.	Environment temperature(℃)	Environment humidity(%)	Preheating temperature(℃)	Cracking rate(%)	
				Surface	Section
1	8.5	70	no	0	21
2	20	90	50	0	1.5
3	24	76	100	0	0
4	32	70	150	0	0

4.4.3 butt joint wielding test. Due to 130mmS460G1-Z35 steel has heavier thickness, using layered wielding process for the butt joint wielding test. At preheating temperature of 100 ℃, welding with MAW at the bottom of the region, all of the CO2 GSW and MAW and submerged arc welding(SAW) are used at the middle of the filling bead and the surface of cover welds.

IOP Conf. Series: Materials Science and Engineering **242** (2017) 012062 doi:10.1088/1757-899X/242/1/012062

Table 4 and Table 5 describe the results of the mechanical properties of butt welding joints. It can be seen here that the properties of the tensile and cold bending and impact of welding joints are better meeting the technical requirements.

Table 4. The mechanical properties of butt welds

Wilding method	No.	YS(MPa)	TS(MPa)	Fracture position	Cold bending(D=3a,180°)
MAW+ GSW	1-1	475	545	parent metal	Ok
MAW+ GSW	1-2	455	530	parent metal	OK
MAW	2-1	470	555	parent metal	OK
MAW	2-2	460	550	parent metal	OK
MAW+ SAW	3-1	473	545	parent metal	OK
MAW+ SAW	3-2	470	546	parent metal	OK

Table 5. The impact properties of butt welds

Wilding method	No.	Kv2 at -40℃(J)	
		Wilding line	HAZ
MAW+ GSW	1-1	98，118，95	186，198，127
MAW	2-1	68，63，77	118，165，139
MAW+ SAW	3-1	84，95，67	74，64，138

During the production steel, a low carbon and Nb-Ni-Mo-V-Ti micro-alloying are designed, the elements that influences the carbon equivalent are controlled and the contents of P, S and gases (e.g. N, O, H) are also strictly controlled. Therefore, the steel has a low sensibility of welding cracking. At the same time the super heavy gauge steel has a good weldability.

5. Application

Figure 3 shows that the city of dreams in Macau is designed by Zaha Hadid who is renowned designer. It is 40 stories high and becomes the new landmark of Macau. The super heavy gauge high-strength structure steel of 130mm S460G1-Z35 is successfully applied in the construction of the project.

Figure 3. The city of dreams

6. Conclusion

In this paper with the use of low carbon equivalent design and reasonable process, the 130mm high-strength structural steel plate is successfully developed. In the production of super heavy gauge high-strength steel plate it is good to meet the requirements of steel for high-rise buildings. The welding performance of 130mm S460G1-Z35 is good, which meets the customer's requirements of construction

IOP Conf. Series: Materials Science and Engineering **242** (2017) 012062 doi:10.1088/1757-899X/242/1/012062

welding. The high-strength structural steel of 130mm Q460G1-Z35 is successfully applied in the construction of the city of dreams in Macau.

References

[1] Shi Gang, Ban Huiyong, Shi Yongjiu, et al. Engineering application and recent research progress on high strength steel structures, *Industrial Construction*,2012(1):p.1-7.

[2] Qiu Linbo, Liu Yi,Hou Zhaoxin, et al. State Application Research of High Strength Steel in Steel Strustures,*Industrial Construction*,2014(3):p.1-5.

[3] He Xinlai, Shang Chengjia, Yang Shanwu, et al. High performance low-carbon bainite steel -composition, process, microstructure, performance and application. *Beijing:Metallurgy Industry Press*, 2008: p.3-6.

[4] Gu Linhao, Wang Zhenqiang, Zhang Yongqing, et al. Development and Application of the High-strength Wear-resistance Steel Used for the Lightweight Heavy-duty Dump Mining Truck. *HSLA Steels 2015 & Microalloying 2015 & Offshore Engineering Steels 2015*, 2015: p. 549-556.

[5] Li Jing, Zhao Dewen, Liu Xianghua, et al. Development of Q345E-Z35 High Strength Ultra-heavy Plate. *Steel Rolling*,2009(1):p.14-16.

[6] Liu Lixiang, Wang Jianguo, Luo Yingming, et al. Research and Development of Large Thickness Steel Plate S460G2+QT for Stationary Offshore Structure. *Wide and Heavy Plate*, 2016(1):p.1-4.

Transient vibration of wind turbine blades

Yuanzhe Li, Minghai Li and Feng Jiang

Internal Combustion Engine Research Institute of Dalian Jiaotong University, Dalian 116028, China

zim_018@163.com

Abstract. This article aims to the transient vibration of wind turbine blades. We firstly introduce transient vibration and previous studies in this area. The report then shows the fundamental equations and derivation of Euler Equation. A 3-D beam are created to compare the analytical and numerical result. In addition we operate the existing result and Patran result of a truncation wedge beam, especially the frequencies of free vibration and transient vibration. Transient vibration cannot be vanished but in some case it can be reduced.

1. Introduction

1.1. Transient vibration

Transient vibration is the vibration which is defined as a temporarily sustained one in a mechanical system, consisting of one of forced or free vibration, or both of them. Transient loading, also known as impact or mechanical shock, is a non-periodic excitation, which is characterized by a sudden and severe application.

In many cases, transient vibration can be ignored in steady state vibration. However, large margins of safety, in which ignoring of transient conditions can destroy, are not always possible. Today the study about transient phenomena is one of the largest areas of concern in mechanical vibration. The analysis of transient motion requires extensive mathematical treatment and modern computational methods.

1.2. Study of Method

In our article twisting structure is ignored. Blade is considered as a wedge cantilever beam for simplicity. Studying dynamic characteristics is a significant part of the design and control processes of a wind turbine blade. It provides necessary modal information to do aeroelastic dynamics and stability analysis [1].

A wind turbine rotor system containing blades has its own characteristics. The vibration characteristics of blades with special airfoil can influence the vibration of the rotor system [2]. The Finite Element Method (FEM) can be applied to establish the rotor model. Meanwhile we choose a finite element software named MSC Patran to create model and calculate.

2. Fundamental Equations

As considering the situation of a uniform beam, it is assumed that the density, area, Young's modulus and second moment of area are all constant. The free vibration of the beam is defined as well. As a result, rotary inertia and shear deformation can be ignored.

Consider an arbitrary element of a beam as shown in figure 1 below:

Content from this work may be used under the terms of the Creative Commons Attribution 3.0 licence. Any further distribution of this work must maintain attribution to the author(s) and the title of the work, journal citation and DOI.
Published under licence by IOP Publishing Ltd

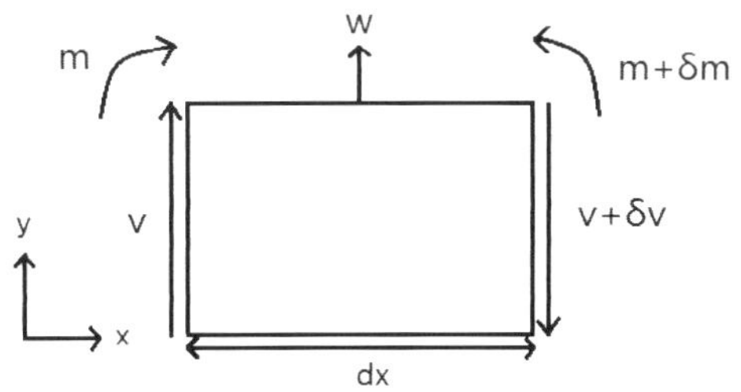

Figure 1. Element beam.

Where: v = shear force; m = bending moment; w = displacement of a beam; the mass of the element is given as: ρAδx. Since shear force is applied:

$$v - (v + \delta v) = \rho A \delta x \frac{\partial^2 w}{\partial t^2} \tag{1}$$

Simplifying equation above gives:

$$-\frac{\partial v}{\partial x} = \rho A \frac{\partial^2 w}{\partial t^2} \tag{2}$$

It is now necessary to sum the moments from the right hand side of the element of the beam:

$$m + \delta m = m + v \delta x \tag{3}$$

Therefore:

$$\frac{\partial v}{\partial x} = \frac{\partial^2 m}{\partial x^2} \tag{4}$$

With simple bending theory:

$$m = EI \frac{\partial^2 w}{\partial x^2} \tag{5}$$

Substitute and rearrange equations above, we have:

$$EI \frac{\partial^4 w}{\partial x^1} + \rho A \frac{\partial^2 w}{\partial t^2} = 0 \tag{6}$$

The equation above is the Bernoulli-Euler Equation.

3. Early Work for free Vibration

To start with the software, a 3-D beam is created. The original choice was 2-D beams while the software require a thickness, so we use 3-D element instead. And also because of the beginning of the study, we consider that the beam is flat and assumed no twist.

3.1. Analytical Results

For a cantilever beam is subjected to free vibration, and the system is considered as continuous system in which the beam mass is considered as distributed along with the stiffness of the shaft, the equation of motion can be written as:

$$\frac{d^2}{dx^2}\left\{EI(x)\frac{d^2Y(x)}{dx^2}\right\} = \omega^2 m(x)Y(x) \tag{7}$$

Where E is the modulus of rigidity of beam material, I is the moment of inertia of the beam cross-section, $Y(x)$ is displacement in y direction at distance x from fixed end, ω is the circular natural frequency, m is the mass per unit length, $m = \rho A(x)$, ρ is the material density, x is the distance measured from the fixed end.

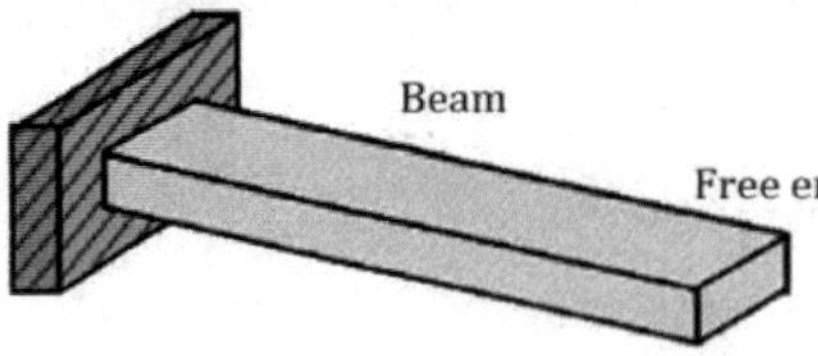

Figure 2. A cantilever beam.

For a uniform beam under free vibration as in equation (7), we have:

$$\frac{d^4Y(x)}{dx^4} - \beta^4 Y(x) = 0 \qquad \beta^4 = \frac{\omega^2 m}{EI} \tag{8}$$

The mode shape for a continuous cantilever beam is given as:

$$f(x) = A_n\{(\sin\beta_n L - \sin h\beta_n L)(\sin\beta_n x - \sin h\beta_n x) + (\cos\beta_n L - \cos h\beta_n L)(\cos\beta_n x - \cos h\beta_n x)\} \qquad (n = 1,2,3\cdots) \tag{9}$$

A closed form of the circular natural frequency ω_{nf} from above equation of motion and boundary conditions can be written as:

$$\omega_{nf} = \alpha_n^2\left(\frac{EI}{mL^4}\right)^{\frac{1}{2}} \tag{10}$$

We chose steel to calculate. Young's Modulus is $E = 2.1\times10^{11}$ N/m^2, the density is $\rho = 7850$ kg/m^3, the geometries are defined that the undamped natural frequency of a steel beam with L = 0.45 m, d = 0.003 m, and b = 0.02 m.

First and second natural frequency:

$$\omega_{nf1} = 1.875^2\left[\frac{E(bh^3/12)}{\rho(bh)L^4}\right]^{\frac{1}{2}} = 1.875^2\left[\frac{2.1\times10^{11}\times0.003^2}{12\times7850\times0.45^4}\right]^{\frac{1}{2}} = 77.76\,rad/\sec \qquad f_{m1} = 12.37\,Hz \tag{11}$$

$$\omega_{nf2} = 4.694^2\left[\frac{E(bh^3/12)}{\rho(bh)L^4}\right]^{\frac{1}{2}} = 4.694^2\left[\frac{2.1\times10^{11}\times0.003^2}{12\times7850\times0.45^4}\right]^{\frac{1}{2}} = 487.34\,rad/\sec \qquad f_{m2} = 77.53\,Hz \tag{12}$$

The frequencies above have to be modified since there is a mass in the form of an accelerometer at the free end of the continuous beam. By continuous approach the solution is difficult since with tip mass the boundary condition at free end is now time dependent. Now we can explain a simpler procedure by which corrections to the natural frequency could be made so as to get closer to the measured natural frequency. In case of non-contacting sensors, there will not be any correction required since there will not be any additional tip mass on the beam.

3.2. Numerical Result

We type the data of steel into Patran, in which we have built a FEM model of 3-D beam with specific shape and dimensions

After operating we can work out the data:

Table 1. Patran result of 3-D beam.

	First	Second
ω (rad/s)	77.8	475
f (Hz)	12.4	75.5

Comparing with equation (11) and equation (12), we shall confirm that the numerical and analytical results are quite similar.

4. Truncated Beam for Wedge

4.1. Existing Results
However it is discovered that very little researches have been published with regards to finite element modelling in this area.

The need for accurate numerical results in this area is discussed in Downs' paper [3]. Therefore we put forward a method for determining these frequencies and produced results for reference. In order to allow simple comparisons between beams of different dimensions, Downs has produced results in a dimensionless form. These dimensionless Euler frequencies are produced using the following equation:

$$\lambda = \omega(\frac{\rho A_0 l^4}{E I_0})^{\frac{1}{2}} \tag{13}$$

In the equation above the subscript zero denotes the conditions at the fixed end of the beam. The equation for calculating of natural frequencies of the truncated wedged cantilever is as follows:

$$\begin{vmatrix} J_2(B) & Y_2(B) & I_2(B) & K_2(B) \\ J_3(B) & Y_3(B) & -I_3(B) & K_3(B) \\ J_4(A) & Y_4(A) & I_4(A) & -K_4(A) \\ J_5(A) & Y_5(A) & I_5(A) & K_5(A) \end{vmatrix} = 0 \tag{14}$$

Where: $= \frac{2\lambda^{\frac{1}{2}} r^{\frac{1}{2}}}{1-r}$; $B = \frac{2\lambda^{\frac{1}{2}}}{1-r}$; r = (Beam diameter at tip/Beam diameter at root).

4.2. Project Results
From equation (13) it can be derivate that:

$$\omega = \lambda(\frac{E I_0}{\rho A_0 l^4})^2 \tag{15}$$

Where: ρ = density, A = area at abutment, l = length of beam, E = Young's Modulus, I = moment of inertia.

The only condition we work about is truncated wedged cantilever, because it is the most similar shape of model with real wind turbine blade. The difference is the wind turbine blade is twisted but the truncated wedge is not.

As we can figure out a real frequency with assuming the real property, it is assumed that: l = 40 m, b = 4 m, d = 0.5 m. The material is aluminium so the density is 2700 kg/m^3, E = 0.7×10^{11} N/m^2. Determine the truncation ratio is 0.5, then the frequency of vibration of the beam, ω, can be calculated. The result as in table 2:

Table 2. Frequency of truncated beam.

	The truncation ratio, r = 0.5				
λ	3.82381	18.3172	47.2648	90.4505	148.002
ω	0.01756	0.08413	0.2171	0.41547	0.67982

4.3. Patran Results
After creating the wedged beam in Patran using the value of real aluminum and the dimension assumed before, figure 3 and figure 4 are worked out to show the mode shapes of the beam. The mode shapes belong to natural frequency of free vibration.

IOP Conf. Series: Materials Science and Engineering **242** (2017) 012063 doi:10.1088/1757-899X/242/1/012063

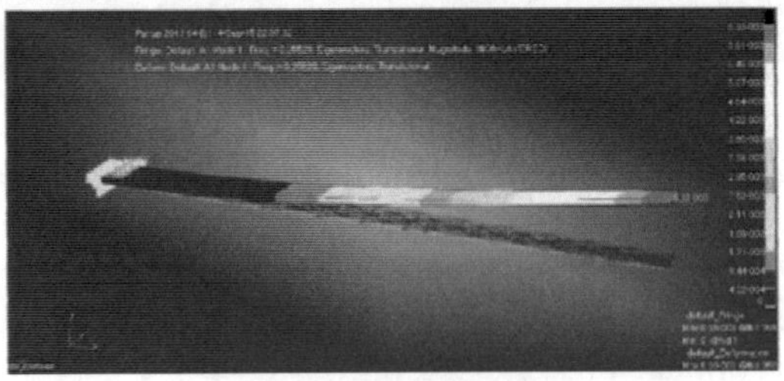

Figure 3. Mode 1.

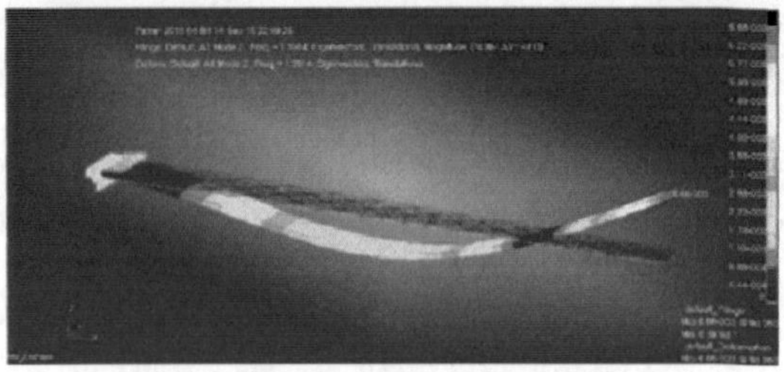

Figure 4. Mode 2.

The mode shapes of beam's vibration take the form of sine waves but in reality they are not exact sine waves. The beam is not free to rotate around the fixed end due to the clamped cantilever. So the beam cannot form into an exact sine wave. Moreover, there is a small amount of elements that do not deflect. As shown in figure 4. The elements on the left hand side are not deflecting.

5. Transient Vibration of 3D Wedge

A wedged beam is created in Patran in order to determine the transient frequency for truncated wedged cantilevers. As we use the seam beam for free vibration of truncated wedged beam because of all the properties and boundary conditions are the same. The only thing we need is to add pressure on the surface of the beam.

Assuming the pressure of the load case is 100 N/m^2. Figure 5 is worked out to show four nodes on the surface of the beam which are assumed to endure the pressure from wind. The line of four nodes are parallel to the cantilevers.

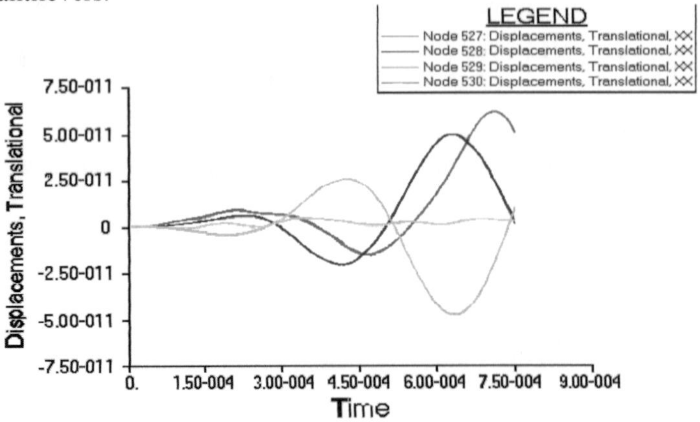

Figure 5. Transient displacement of wedged beam.

6. Discussion

In part 4 we focus on the non-dimension result of truncated wedge especially with the truncation ratio 0.5. As the result of the limitation of computing power in the ages of Downs, he was not able to determine some of the results for beams with part of truncation ratio. However it is now possible to calculate all of the ratio and to go through a good result omitted from Downs' study. The results coming from Matlab and Patran agree with Downs' theory. Also as assuming a real condition, the natural frequencies can be figured out in ratio of 0.5.

The first natural frequency of the wedge is a kind of banding in Mode 1. While in Mode 2 it is also bended but there is one cross point to the initial location. The data of truncated cantilever also show pairs of results caused by the deflections in the y and z axis, which are complicated types of frequency. As decided before, the condition of twist would not be considered in the analysis of transient results.

In the study of the transient vibration of truncated wedge, we load the pressure with 100 N/m^2. And then chose four nodes on the surface with loads. Figure 5 shows that as the time goes by, the displacements of the nodes change irregularly and there is no relevance between these four nodes.

Furthermore, another two nodes, 117 and 258, which distributing along the length of beam a chosen.

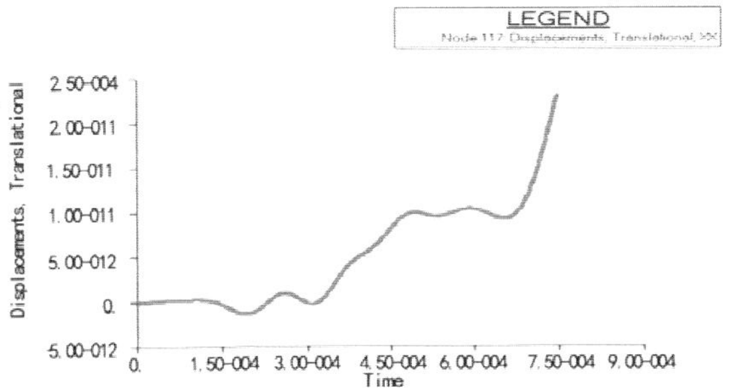

Figure 6. Node 117.

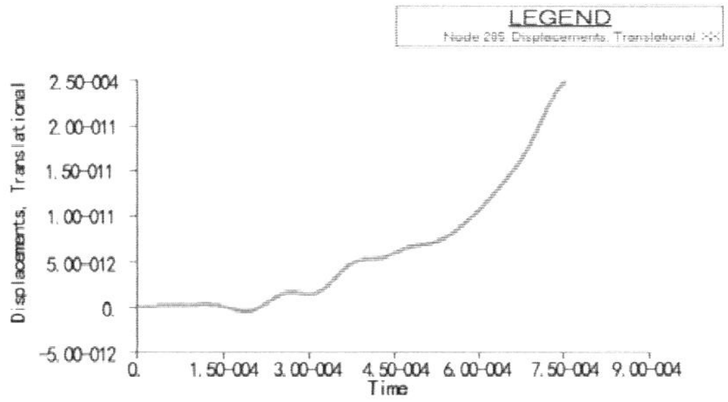

Figure 7. Node 285.

Node 285 is further from the cantilever fixed end than Node 117. A result can be come out that the transient vibration of the node or element further from the cantilever fixed end is more rapid (figure 6 and figure 7). Anyway, transient vibration is determined as a vibration with no constant period or amplitude. We can only analyze the situation and the frequency in a specific time period with specific load condition.

7. Conclusion

Free vibration can be regarded as the basic version of transient vibration, as they may have same boundary conditions. Transient vibration happens in random directions and goes irregularly. However transient motion cannot be ignored in a developing study area such as wind turbine. Most of time transient is harmful. In general, any change of transient motion accompanies forms of energy storing in the vibrating system. Too much transient vibration may damage the internal structure of elements of a subject. When it happens to a mechanical structure, it will be much more difficult to control the condition of the structure. Transient vibration cannot be vanished but in some case it can be reduced. It can be developed in future works.

However, the wind turbine blades are still twisted and much more complex. A future step with adding twisting, more reality load and boundary conditions should be expected.

References

[1] Wei L, Yin J C, Pu C, *et al.* 2011 Dynamic analysis and aeroelastic stability analysis of large composite wind turbine blades *Acta Aerodynamica Sinica*. **29** 391–395

[2] Liu C, Jiang D, Chen J 2010 Vibration characteristics on a wind turbine rotor using modal and harmonic analysis of FEM *World Non-Grid-Connected Wind Power and Energy Conf.* (Nanjing: IEEE) pp 1–5

[3] Downs, B 1978 Reference frequencies for the validation of numerical solutions of transverse vibrations of non-uniform beams *Journal of Sound & Vibration* **61** 71–78

[4] Hansen M O L 2000 Aerodynamics of wind turbines *Rotors Loads & Structure James* **5** 141–167

[5] Rahimi M, Parniani M 2009 Dynamic behavior and transient stability analysis of fixed speed wind turbines *Renewable Energy* **34** 2613–2624

[6] Staino A, Basu B 2013 Dynamics and control of vibrations in wind turbines with variable rotor speed *Engineering Structures* **56** 58–67

[7] Kaczmarczyk S, Ostachowicz W 2003 Transient vibration phenomena in deep mine hoisting cables Part 1: Mathematical model *Journal of Sound & Vibration* **262** 219–244

IOP Conf. Series: Materials Science and Engineering **242** (2017) 012064 doi:10.1088/1757-899X/242/1/012064

Study of vehicle-terrain interaction based on ANCF

G B Li [1, 2], H S Wang [2]

[1] Department of Mechanical Engineering, Shanghai Normal University, 100 Guilin Road, Shanghai, 200234, China
[2] Department of Mechanical Engineering, Shanghai Lida Polytechnic Institute, 1788 Cheting Road, Shanghai, 201609, China

liguangbu@163.com

Abstract. A brief review of the empirical and analytical models are presented. The pressure-sinkage relationship proposed by Bekker and the tractive performance (drawbar pull-slip) of vehicle is set up. A figure of drawbar pull vs. slip at different time and slip is illustrated. The modified Bekker soil models are integrated finite element (FE) Absolute Nodal Coordinate Formulation (ANCF) interpolations to determine the generalized forces. The paper discusses important fundamental issues that must be addressed when implementing terramechanics models into MBS algorithms in order to model complex tracked vehicle-soil interactions.

1.Introduction

Terramechanics is the study of the relationships between a vehicle and its environment. Some of the principal concerns in terramechanics are developing functional relationships between the design parameters of a vehicle and its performance with respect to its environment, establishing appropriate soil parameters, and promoting rational principles which can be used in the design, and evaluation of vehicles (Wong, 2010). Terramechanics models can be categorized into empirical and analytical models.

Empirical terramechanics models use established experimental measurements of appropriate parameters, properties, and behaviors of soil; these experimental results are then used to establish empirical relationships that could be used to predict at least qualitatively the response of soils under various conditions (Bekker, 1969). Parametric models, which are based on experimental work and have been widely used, offer practical means by which an engineer can qualitatively evaluate tracked vehicle performance and design.

Research in the field of terramechanics has continued over many decades. Since the tractive performance of a vehicle depends significantly on the normal and shear stress distribution at the soil-vehicle interface, many attempts have been made to predict such distributions beneath tracks and wheels (Schmid, 1995).

Bekker (1956, 1960) first proposed a theoretical study regarding the pressure-sinkage relationship of terrain, as well as the response to repeated loading and unloading. Janosi and Hanamoto (1961) proposed a model for predicting the ground pressure distribution beneath a rigid track. They assumed the normal stress distribution to be higher at the rear end of the track. The predicted ground pressure distribution was trapezoidal with higher pressure at the rear end of the track. Reece (1965) improved Bekker's model by making the parameters dimensionless. This single equation could then account for different plate shapes.

Content from this work may be used under the terms of the Creative Commons Attribution 3.0 licence. Any further distribution of this work must maintain attribution to the author(s) and the title of the work, journal citation and DOI.
Published under licence by IOP Publishing Ltd

Garber and Wong (1981) developed analytical methods for predicting the pressure distribution beneath tracks taking into account all major design parameters of the vehicle and the pressure-sinkage characteristics of the soil. Wong et al. (1984) continued this work and established a model for the prediction of the distribution of ground pressure and the tractive performance of a tracked vehicle, including the effects of repetitive loading. Wong and Huang (2006) evaluated the tractive performance of wheeled and tracked vehicles assuming that the tracks on a tracked vehicle have an essentially flat and rectangular contact area with uniform normal pressure and the same contact length, as well as assuming the vehicle weight is uniformly distributed among the tracks.

2.Terramechanics based soil models

In study of vehicle-terrain interaction, terramechanics is one of the most important methods. Many attempts have been made to predict the normal and shear stress distributions beneath the tracks of moving vehicles ranging from empirical, analytical, to finite element models. Besides the tractive performance of a vehicle depends significantly upon its normal and shear stress distributions in the context of the soil-vehicle interface, and these stress distributions rely on the mechanical interactions inherent to the interface itself.

2.1 Pressure-sinkage relationship

Through above analysis, three common formulations on pressure–sinkage relationship are widely used in vehicle-terrain interaction investigation.

The first common pressure–sinkage relationship for the mineral terrain proposed by Bekker :

$$p = (k_c/b + k_\phi)z^n = k_{eq}z^n \quad . \tag{1}$$

p is pressure; b is the radius of a circular plate or the smaller dimension of a rectangular plate; n, k_c and k_ϕ are pressure-sinkage parameters related to the cohension and the angle of shearing of the material; $k_{eq} = (k_c / b + k_\phi)$, and z is sinkage.

Besides, equation (1) is now widely used in track-terrain interaction, tire-terrain interaction and wheel-terrain interaction.

Another common formulation of the pressure–sinkage relationship was proposed by Reece:

$$p = (ck_c^{'} + \gamma bk_\phi^{'})\left(\frac{z}{b}\right)^n \tag{2}$$

$k_c^{'}$ and $k_\varphi^{'}$ are dimensionless soil parameters which replace Bekker's k_c and k_φ parameters.

Equation (2) can be used on wheel-terrain interaction, and tire-terrain interaction.The last common formulation of the pressure-sinkage relationship was proposed by Reece

$$\sigma_1 = (k_1 + k_2 b)\left(\frac{r}{b}\right)^n (\cos\theta - \cos\theta_1)^n \tag{3}$$

$$\sigma_2 = (k_1 + k_2 b)\left(\frac{r}{b}\right)^n \left[\cos\left(\theta_1 - \frac{\theta}{\theta_m}(\theta_1 - \theta_m)\right) - \cos\theta_1\right]^n \tag{4}$$

n is the sinkage exponent, k_1 and k_2 are pressure sinkage moduli, b is the wheel width, r is the wheel radius, θ_1, θ_m is entry angle and the angular location of the maximum normal stress.

Equation (3) and (4) can be used on wheel-terrain interaction. For the flexible tire acting on the deformable terrain, Equation (3) and (4) also can be used to tire-terrain interaction if radius r is effective radius.

From above study, the pressure-sinkage relationship proposed by Bekker is the best choice for the vehicle-terrain interaction analysis.

2.2 The Tractive Performance Study of vehicle

Besides the normal pressure, terrain shear stress is another parameter to determine vehicle tractive force.

For 'plastic' terrain which do not exhibit a 'hump' of maximum shear stress, a modified version of Bekker's equation containing only one constant was proposed by Janosi and Hanamoto and is widely used in practice:

$$s / s_{\max} = 1 - e^{-j/K} \quad (5)$$

K is usually referred to as the shear deformation parameter and is a measure of the magnitude of the shear displacement required for the development of the maximum shear stress.

Based on the shear stress-shear displacement curves measured under different normal pressures, the relation between the maximum shear stress and applied normal pressure can be derived. It has been found that the Mohr-Coulomb equation describes this relation adequately in many cases:

$$s_{\max} = c + p \tan\phi \quad (6)$$

In this equation, $s_{\max}$ is the maximum shear stress, p is the normal stress, and c and ϕ are the cohesion (or adhesion) and angle of shearing resistance, respectively. In deriving the values of c and ϕ from measured shear data, a procedure proposed by Reece in which the effect of grouser height of the shear ring is taken into consideration should be followed.

When the normal pressure and shear stress distributions under a vehicle at a given slip have been determined, the tractive performance of the vehicle can then be predicted. The tractive performance of a vehicle is usually characterized by its motion resistance, tractive effort, and drawbar pull (the difference between tractive effort and motion resistance) as functions of slip.

If it is assumed that a track link has an essentially flat and rectangular contact area with uniform normal pressure and the same contact length l, and the link and terrain is horizontal, then under steady-state operating conditions, the link tractive effort F and drawbar pull F_d at a given slip i can be expressed as following:

$$F = b\int_0^l (c + p\tan\varphi)(1 - e^{-ix/K})dx = bl(c + p\tan\varphi)\left[1 - \frac{K}{il}(1 - e^{-il/K})\right]. \quad (7)$$

$$F_d = F - R_v. \quad (8)$$

b is the contact width of the track link, l is the length of track link in contact with the terrain, p is normal pressure, and R_v is the motion resistance of track link. Figure 1 illustrates a drawbar pull vs. slip at different time and slip.

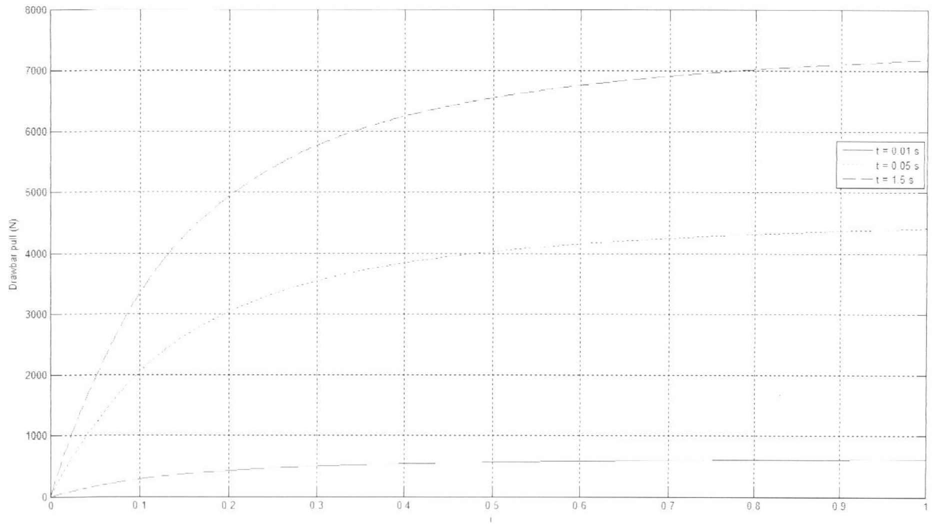

Figure 1. Drawbar pull vs. slip

3. ANCF implementation

The FE implementation of the soil mechanics plasticity equations requires the use of an approach that allows employing general constitutive models. The vehicle/soil interaction can lead to a significant change in geometry that cannot be captured using finite elements that employ only translational displacement coordinates without significant refinement. In some soil applications, such a significant change in geometry may require the use of elements that employ gradients and accurately capture curvature changes. This requirement can be met using the FE absolute nodal coordinate formulation (ANCF).

ANCF finite elements do not employ infinitesimal or finite rotations as nodal coordinates; instead, absolute slopes and displacements at the nodal points are used as the element nodal coordinates. The position vector $\mathbf{r}^j$ of an arbitrary point on element j can be defined in a global coordinate system XYZ as $\mathbf{r}^j = \mathbf{S}^j\left(x^j, y^j, z^j\right)\mathbf{e}^j(t)$. In this equation, x^j, y^j, and z^j are the element spatial coordinates, $\mathbf{S}^j$ is the shape function matrix, $\mathbf{e}^j$ is the vector of element nodal coordinates, and t is time. The nodal coordinate vector $\mathbf{e}^{jk}$ at node k can be defined as follows (Shabana, 2005)

$$\mathbf{e}^{jk} = \left[\mathbf{r}^{jkT} \quad \left(\frac{\partial \mathbf{r}^{jk}}{\partial x^j}\right)^T \quad \left(\frac{\partial \mathbf{r}^{jk}}{\partial y^j}\right)^T \quad \left(\frac{\partial \mathbf{r}^{jk}}{\partial z^j}\right)^T\right]^{\mathrm{T}} \tag{9}$$

Fully parameterized ANCF finite elements allow using a general continuum mechanics approach to define the Green-Lagrange strain tensor $\varepsilon = \left(\mathbf{J}^{\mathrm{T}}\mathbf{J} - \mathbf{I}\right)/2$, where $\mathbf{J}$ is the matrix of position vector gradients. In dynamic soil problems, ANCF leads to a constant inertia matrix and to zero Coriolis and centrifugal forces. The mass matrix obtained using ANCF finite elements can always be written a $\mathbf{M}^{\mathrm{j}} = \int \rho^j \mathbf{S}^{j^T} \mathbf{S}^j dV^j$, where ρ^j and V^j are, respectively, the initial mass density and initial volume of the finite element. ANCF finite elements allow for straight forward implementation of general constitutive models including the continuum mechanics-based soil model discussed in this paper.

For a finite element or deformable body, the principle of virtual work can be written using the reference configuration as

$$\int_V \rho \ddot{\mathbf{r}}^T \delta \mathbf{r} dV + \int_V \boldsymbol{\sigma}_{P2} : \delta \boldsymbol{\varepsilon} dV - \int_V \mathbf{f}_b^T \delta \mathbf{r} dV = 0 \tag{10}$$

In this equation, V is the initial or reference volume, ρ is the initial mass density, $\mathbf{r}$ is the global position vector of an arbitrary point, $\boldsymbol{\sigma}_{\mathbf{p2}}$ is the second Piola Kirchhoff stress tensor, $\boldsymbol{\varepsilon}$ is the Green-Lagrange strain tensor, and $\mathbf{f}_b$ is the vector of body forces. The second term in the preceding equation can be recognized as the virtual work of the elastic forces, it can be rewritten to define the generalized elastic forces, that is

$$\delta W_s = \int_V \boldsymbol{\sigma}_{P2} : \delta \boldsymbol{\varepsilon} dV = \mathbf{Q}_s^T \delta \mathbf{e} \tag{11}$$

$\delta \mathbf{e}$ is the virtual change in the nodal coordinates associated with a particular ANCF finite element or a body, and $\mathbf{Q}_s$ is the vector of the generalized elastic forces. The vector of elastic forces often takes a fairly complicated form, especially in the case of plasticity formulations, and is obtained using numerical integration methods. The principle of virtual work leads to the following equations of motion:

$$\mathbf{M}\ddot{\mathbf{e}} + \mathbf{Q}_s - \mathbf{Q}_e = \mathbf{0} \tag{12}$$

$\mathbf{M}$ is the symmetric mass matrix, and $\mathbf{Q}_e$ is the vector of body applied nodal forces.

Knowing the strains, the soil properties, yield function, and the flow rule; the state of soil deformation (elastic or plastic) can be determined. Knowing the state of deformation, the constitutive model appropriate for this state can be used to determine the elastic force vector $\mathbf{Q}_s$.

4. Integration with MBS Algorithms

The FE implementation of the soil mechanics plasticity equations requires the use of an approach that allows employing general constitutive models. The vehicle/soil interaction can lead to a significant change in geometry that cannot be captured using finite elements that employ only translational displacement coordinates without significant refinement. In some soil applications, such a significant change in geometry may require the use of elements that employ gradients and accurately capture curvature changes. This requirement can be met using the FE absolute nodal coordinate formulation (ANCF). The objective of this paper is to present a review of soil mechanics formulations that can be integrated with computational MBS algorithms used for the virtual prototyping of vehicle systems. These algorithms allow for modeling rigid, flexible, and very flexible bodies. The small deformation of flexible bodies in vehicle systems are often examined using the floating frame of reference (FFR) formulation. Therefore, efficient modeling of complex vehicle system dynamics requires the implementation of different formulations that can be used for rigid body, small deformation, and large and plastic deformation analyses. A Newton-Euler or Lagrangian formulation can be used to model rigid bodies, the FFR formulation that employs two sets of coordinates (reference and elastic) can be used to model small deformations, and ANCF finite elements can be used to model large and plastic deformations including soil deformations.

MBS algorithms are designed to exploit the sparse matrix structure of the resulting dynamic equations. Because ANCF finite elements lead to a constant inertia matrix, Cholesky coordinates can be used to obtain an identity generalized mass matrix, leading to an optimum sparse matrix structure. Computational MBS algorithms are also designed to solve a system of differential and algebraic equations. The differential equations define the system equations of motion, while the algebraic equations define the joint constraints and specified motion trajectories. Using the constraint equations and the equations of motion, the augmented form of the equations of motion can be got as (Shabana, 2005).

5. Summary
In this paper, several simple models including Bekker's model are discussed. Bekker's model as well as other parametric and analytical terramechanics models have been used in the study of track/soil interaction and can be implemented in MBS algorithms using simple discrete force elements. The absolute nodal coordinate formulation (ANCF) which allows for the study of vehicle/soil interaction is also discussed.

References
[1] Bekker, M.G., 1969, *Introduction to Terrain-Vehicle Systems*, The University of Michigan Press, Ann Arbor.
[2] Chen, W.F., and Baladi, G.Y., 1985, "Soil Plasticity: Theory and Implementation", Elsevier, Amsterdam.
[3] Drucker, D.C., Gibson, R.E., and Henkel, D.J., 1957, "Soil Mechanics and Work Hardening Theories of Plasticity", *Trans., ASCE*, 122: pp. 338-346.
[4] Garber, M., and Wong, J.Y., 1981, "Prediction of ground pressure distribution under tracked vehicles – Part I. An analytical method for predicting ground pressure distribution", *Journal of Terramechanics*, 18(1): pp.1–23.
[5] Goldscheider, M., 1984, "True Triaxial Tests on Dense Sands", *Constitutive relations for soils*, Balkema, Rotterdam.
[6] Harnisch, C., Jakobs, R., Troulis, M., and Nehls, O., 2005, "A new tyre–soil interaction model for vehicle simulation on deformable ground", *Vehicle System Dynamics*, 43; pp. 384-394.
[7] Irani, R.A., Bauer, R.J., and Warkentin, A., 2011, "A dynamic terramechanic model for small lightweight vehicles with rigid wheels and grousers operating in sandy soil", *Journal of Terramechanics*, 48; pp 307-318.
[8] Ishigami, G., Miwa, A., Nagatani, K., and Yoshida, K., 2007, "Terramechanics-Based Model for Steering Maneuver of Planetary Exploration Rovers on Loose Soil", *Journal of Field Robotics, Special Issue: Special Issue on Space Robotics,* Part I 24(3); pp. 233-250.
[9] Li, L., and Sandu, C., 2007, "On the impact of cargo weight, vehicle parameters, and terrain characteristics on the prediction of traction for off-road vehicles", *Journal of Terramechanics*, 44; pp. 221-238.
[10] Shabana, A.A., 2005, *Dynamics of Multibody Systems*, Cambridge University Press, New York.

IOP Conf. Series: Materials Science and Engineering **242** (2017) 012065 doi:10.1088/1757-899X/242/1/012065

The influence of high heat input and inclusions control for rare earth on welding in low alloy high strength steel

Rensheng Chu [a], Shukun Mu [a], Jingang Liu [a], Zhanjun Li [a]

[a] Shougang Research Institute of Technology, Beijing 100043, China

churensheng@163.com

Abstract. In the current paper, it is analyzed for the influence of high heat input and inclusions control for rare earth on welding in low alloy high strength steel. It is observed for the structure for different heat input of the coarse-grained area. It is finest for the coarse grain with the high heat input of 200 kJ / cm and the coarse grain area with 400 kJ / cm is the largest. The performance with the heat input of 200 kJ / cm for -20 ℃ V-shaped notch oscillatory power is better than the heat input of 400 kJ / cm. The grain structure is the ferrite and bainite for different holding time. The grain structure for 5s holding time has a grain size of 82.9 μm with heat input of 200 kJ/cm and grain size of 97.9 μm for 10s holding time. For the inclusions for HSLA steel with adding rare earth, they are Al_2O_3-CaS inclusions in the Al_2O_3-CaS-CaO ternary phase diagram. At the same time, it can not be found for low melting calcium aluminate inclusions compared to the inclusions for the HSLA steel without rare earth. Most of the size for the inclusions is between 1 ~ 10μm. The overall grain structure is smaller and the welding performance is more excellent for adding rare earth.

1. Introduction

At present, high heat input welding has become a trend. But the high heat input welding makes toughness and strength reduced for the weld joint. The Rare Earth (Re) can refine the grain to improve strength and toughness and also can change the shape and distribution of the inclusions in steel. There are also some studies on the effect of Re in high heat input for welding and the induction of acicular ferrite behaviour for laboratory studies[1~5]. But there is little research for the base metal especially the control for the Re to improve the toughness in the industrial production in low alloy high strength (HSLA) steel. Re elements play an important role in the refinement of grains and the formation of intragranular ferrite. Also it can enhance the impact toughness at low temperature and welding performance for HSLA steel for refinement of heat affected zone grain size for high heat input welding. Therefore, it is important for the influence of inclusions for rare earth on welding in HSLA steel.

2. Experiment Procedure

2.1 Steelmaking process for HSLA steel

In this paper, testing and analysis in steelmaking process were tracked and done in some manufacturers for HSLA steel plate production. The steelmaking process for HSLA steel is Pretreatment of hot metal → LD → LF refining → RH refining → Calcium treatment → CC as shown in Fig. 1. The Re content in alloy is more than 95% which the ratio of cerium and lanthanum is 2 for HSLA steel.

Content from this work may be used under the terms of the Creative Commons Attribution 3.0 licence. Any further distribution of this work must maintain attribution to the author(s) and the title of the work, journal citation and DOI.
Published under licence by IOP Publishing Ltd

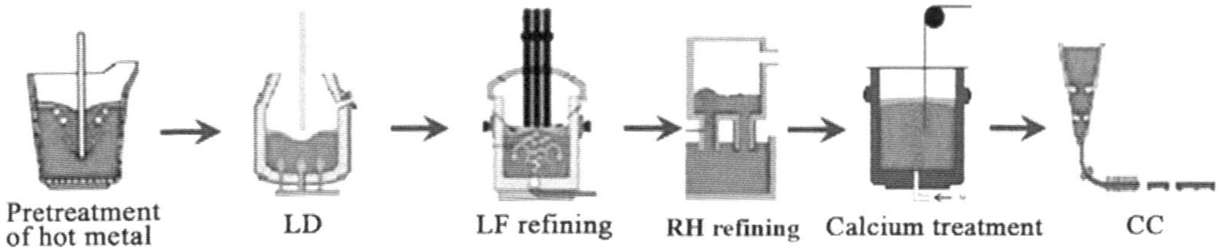

Fig. 1 Steelmaking process for HSLA steel

3. Results and Discussion

3.1 The thermal simulation test for high heat input

The structure and properties of 200kJ / cm heat input for welded coarse grain regions were simulated by Gleeble 2000D thermal simulation test machine. The peak heating temperature is 1350 ℃ and the holding time is 3S. The corresponding welding characteristic parameter $t_{8/5}$ is 120s, and the thermal cycling curve is shown in Fig. 2.

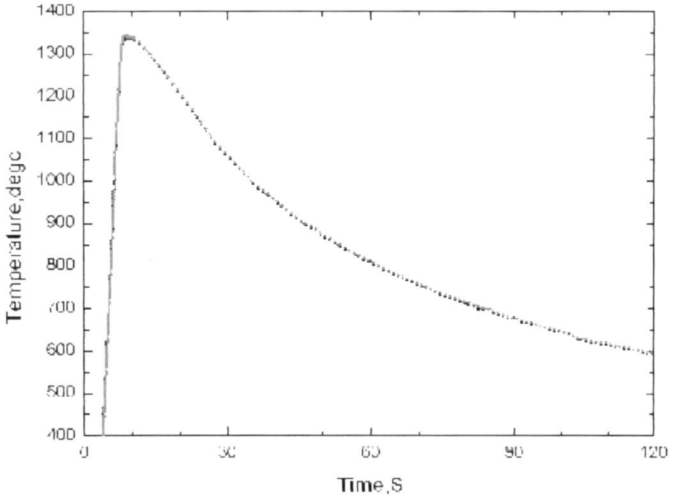

Fig. 2 Peak temperature 1350 ℃ for 200KJ / cm welding thermal cycle curve

3.2 The structure for different high heat input

It is observed for the structure for different heat input of the coarse-grained area. We can draw a conclusion from Fig. 3 that it is finest for the coarse grain with the high heat input of 200 kJ / cm and the coarse grain area with 400 kJ / cm is the largest. The performance with the heat input of 200 kJ / cm for -20 ℃ V-shaped notch oscillatory power is better than the heat input of 400 kJ / cm.

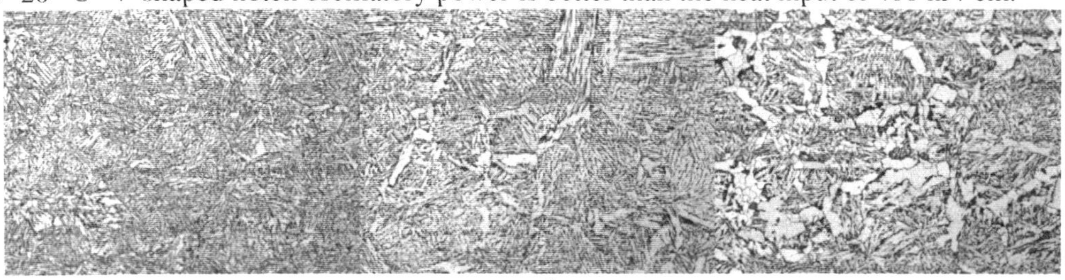

a b c

Fig. 3 Grain structure for different heat input welding (a: 200 kJ/cm; b: 300 kJ/cm; c: 400 kJ/cm)

3.3 The structure for different holding time

It is analyzed for the structure for different holding time of the coarse-grained area. We can see from Fig. 4 that it is finest for the coarse grain for 5s holding time with the high heat input of 200 kJ / cm and the coarse grain area with 10s holding time is the largest. From Fig. 4 we can also see that the grain structure is the ferrite and bainite. The grain structure for 5s holding time has a grain size of 82.9 μm with heat input of 200 kJ/cm and grain size of 97.9 μm for 10s holding time.

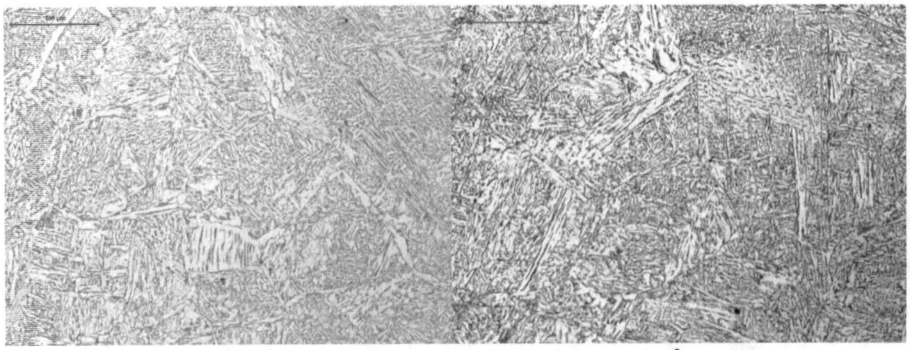

a b

Fig. 4 Grain structure for different holding time with heat input of 200 J/cm (a: 5s for holding time; b: 10s for holding time)

3.4 The control of inclusions

Re plays a role on the inclusions for a certain degree of denaturation for the inclusions spheroidization to improve the strength of steel. The study was carried out by means of ASPEX to automatically scan the sample inclusions by the effect of the addition of Re on the denaturation behavior of the inclusions. The control of inclusions has become a key to the HSLA steel with Re. Size and morphology of the inclusions for the HSLA steel with Re in some factory are shown in Fig. 5. For the inclusions for HSLA steel with adding Re, they are Al_2O_3-CaS inclusions in the Al_2O_3-CaS-CaO ternary phase diagram. At the same time, it can not be found for low melting calcium aluminate inclusions compared to the inclusions for the HSLA steel without Re. Most of the size for the inclusions is between 1 ~ 10μm. We can see from Fig. 6 ~ Fig. 8 that the inclusions contain La and Ce for Res and they really play an important effect on the control of inclusions and the performance of plates.

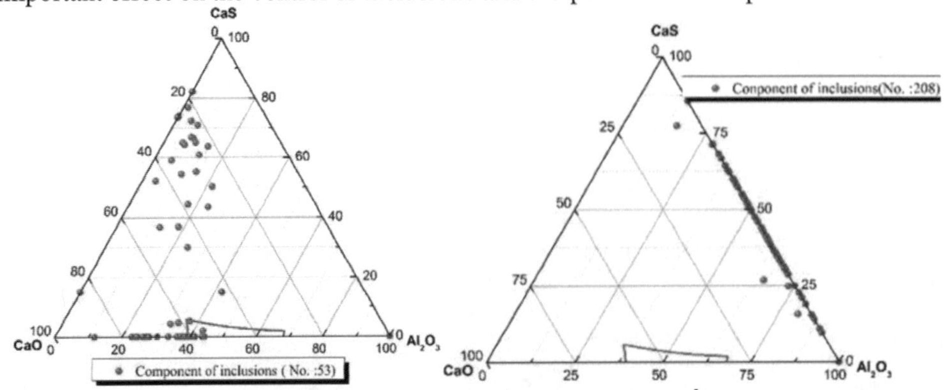

a b

Fig. 5 Component distribution of inclusions for the HSLA steel (a: inclusions with Re; b: without Re)

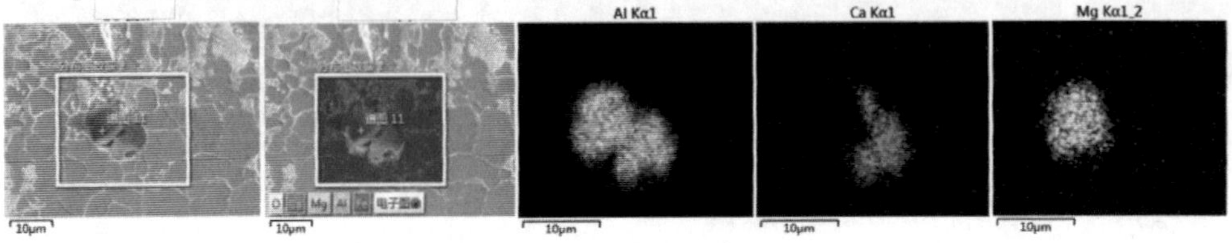

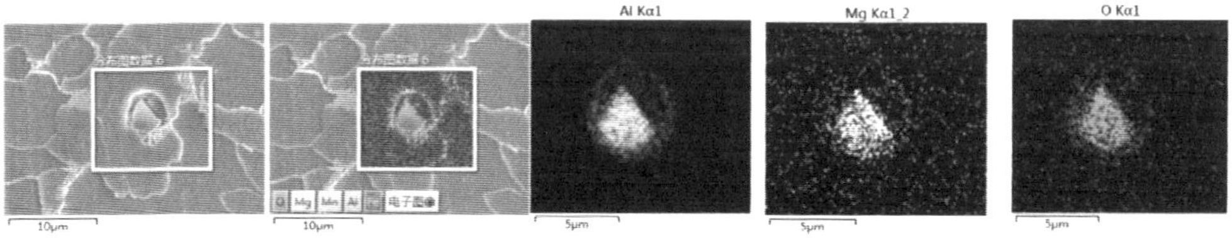

Fig. 6 Inclusion surface scan results for HSLA steel without Re

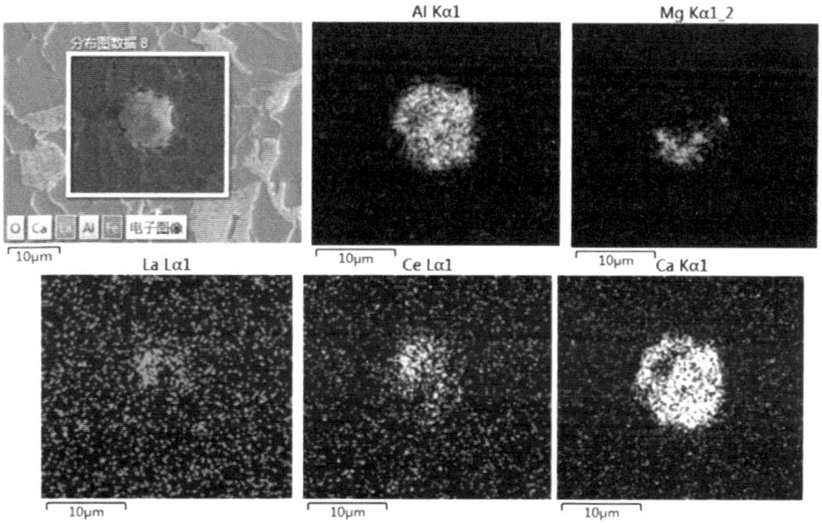

Fig. 7 Inclusion surface scan results for HSLA steel with Re

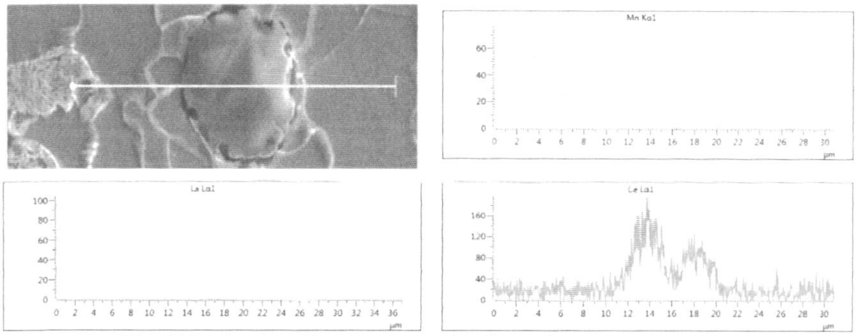

Fig. 8 Inclusion line scan results for HSLA steel with Re

3.5 The grain structure for plate adding Re and without Re

It is observed for the structure of the welded joint with 200 kJ / cm heat input in the coarse-grained area. We can draw a conclusion from Fig. 9 that it is finer for the coarse grain with the high heat input of 200 kJ / cm for adding Re. It is a small amount of acicular ferrite and a large number of massive ferrite for the grain structure for plate without Re of the weld metal. We can also draw a conclusion that the deposited metal structure contains a large amount of acicular ferrite. The overall grain structure is smaller and the welding performance is more excellent.

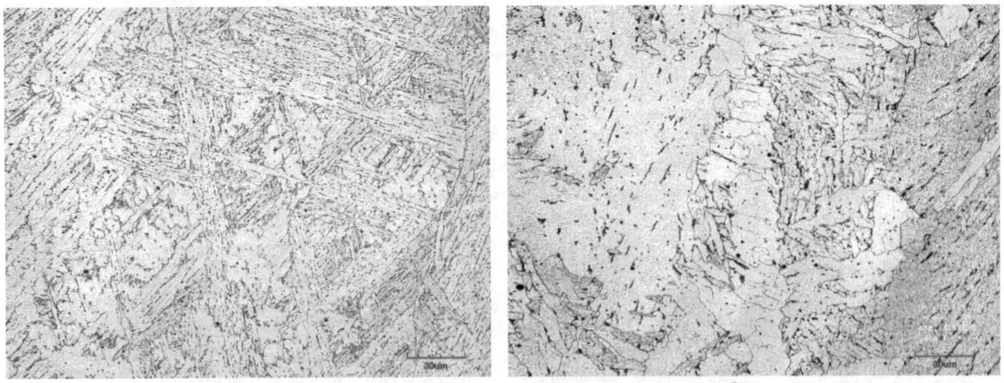

a b

Fig. 9 The grain structure for plate adding Re and without Re(a: inclusions with Re; b: without Re)

4. Conclusions

(1) It is observed for the structure for different heat input of the coarse-grained area. It is finest for the coarse grain with the high heat input of 200 kJ / cm and the coarse grain area with 400 kJ / cm is the largest. The performance with the heat input of 200 kJ / cm for -20 ℃ V-shaped notch oscillatory power is better than the heat input of 400 kJ / cm.

(2) The grain structure is the ferrite and bainite for different holding time. The grain structure for 5s holding time has a grain size of 82.9 μm with heat input of 200 kJ/cm and grain size of 97.9 μm for 10s holding time.

(3) For the inclusions for HSLA steel with adding Re, they are Al_2O_3-CaS inclusions in the Al_2O_3-CaS-CaO ternary phase diagram. At the same time, it can not be found for low melting calcium aluminate inclusions compared to the inclusions for the HSLA steel without Re. Most of the size for the inclusions is between 1 ~ 10μm

(4) The overall grain structure is smaller and the welding performance is more excellent for adding Re.

References:

[1] Y.Ito, S.Nakanishi. Yosetsu Kakaishi. The Relation between Charpy Impact Properties and Control of the Microstructure of Wled Metals[J]. Journal of The Japan Welding Society, 1975, 44(10) :815-821.

[2] C. van der Eijk, O. Grong and J. Hjelen. Quantification of inclusion-stimulated ferrite nucleation in wrought steel using the SEM-EBSD technique， Proc. of Int. Conf. on Solid-Solid Phase Transformations[C], JIM, Sendai, 1999, 1.

[3] J. L. Lee , Y. T. Pan. The Formation of Intragranular Acicular Ferrite in Simulated Heat-affected Zone[J] ， ISIJ International, 1995,35 (8): 1027-1033.

[4] Z. Yang, F. Wang, S. Wang, et al. Intragranular ferrite formation mechanism and mechanical properties of non-quenched-and-tempered medium carbon steels[J]. Steel Res. Int., 2008, 79 (5): 390-395.

[5] T. Yamada, H. Terasaki , Y. Komizo. Microscopic observation of inclusions contributing to formation of acicular ferrite in steel weld metal[J], Sci. Technol. Weld. Joining,2008,13 (3): 118-126.

[6] R.A.Ricks, P.R.Howell , G.S.Barritte. The Nature of Acicular Ferrite in HSLA Steel Weld Metals[J]. J.Mat.Sci.,1982, 17 (2): 732-740.

Modelling of micro vibration energy harvester considering size effect

Chuangye Li[1,2], Rui Huo[1,2] and Weike Wang [1,2]

[1] School of Mechanical Engineering, Shandong University, Ji'nan 250061, China;
[2] Key Lab. of High Efficiency and Clean Mechanical Manufacture (Shandong University) Ministry of Education, Ji'nan 250061, China

lcyzzu@126.com

Abstract. Considering increase of stiffness caused by size effect, equivalent Young's modulus was introduced for futher analysis. Experimental platform was established to test vibration characteristics. Dynamic equation for micro piezoelectric cantilever beam considering size effect was studied with finite element analysis and experiment. Results shows it is accurate. Based on that, dynamic model for micro vibration energy harvester was improved, a T-type micro vibration energy harvester was designed and fabricated. Resonant frequency, tip displacement and output voltage of the harvester were obtained. Comparing with macroscopic model for vibration harvester, improved one reduces errors by 13%, 35% and 22%.

1. Introduction

"Internet of Things"[1-2] includes a huge amount of sensors, traditional power supply can not meet its demands as it is nondurable, storage limited, environmental polluting[3]. Research of new power supply technology is important. Mechanical energy can be transformed to electrical one by piezoelectric energy harvester which is efficient, simple, durable, environmental friendly, and compatible with micro-nano manufacture[4]. Due to size effect, microstructure performance is inconsistent with macroscopic model[5-7]. Micro piezoelectric cantilever beam is a basic structure of micro vibration energy harvester[3], Dynamic model for micro-beam was verified, and a model for harvester was refined. Improved model for harvester reduces error dramaticly.

2. Improved dynamic model for micro piezoelectric cantilever beam and verification

Not considering size effect, dynamic model of micro-beam can be expressed as:

$$EIw^{(4)} + \rho A\ddot{w} = 0 \tag{1}$$

In [8], intrinsic length l that reflects size effect was introduced. Dynamic model for micro-beam was improved by couple stresses theory:

$$\left(EI + \mu A l^2\right) w^{(4)} + \rho A\ddot{w} = 0 \quad (h/l = 3.5) \tag{2}$$

Using mode superposition method, resonant frequency equation is obtained:

$$\omega_r = (\beta_r)^2 \left(\frac{EI + \mu A l^2}{\rho A L^4}\right)^{1/2} \quad \left(\beta^4 = \frac{\omega^2 \rho A}{EI + \mu A l}; r = 1, 2, 3, \ldots\right) \tag{3}$$

2.1. Finite element analysis of micro piezoelectric cantilever beam

Content from this work may be used under the terms of the Creative Commons Attribution 3.0 licence. Any further distribution of this work must maintain attribution to the author(s) and the title of the work, journal citation and DOI.
Published under licence by IOP Publishing Ltd

Considering formula (2), μAl^2 is main reason for increasing of stiffness under size effect[9-12]. Equivalent Young 's modulus E_l was introduced:

$$E_1 = E + \frac{\mu A l^2}{I} = E\left[1 + \frac{6}{1+v}\left(\frac{l}{h}\right)^2\right] \tag{4}$$

Micro-cantilever can be simplified to two-layer whose size are shown in table 1.

Table 1. Size of silicon beam and PZT-5H

	Length(μm)	Width(μm)	Thickness(μm)
Si-Beam	3150	1000	13
PZT-5H	3150	1000	1.6

Reference to parameters in [13] and formula (4), equivalent Young's modulus is shown in Table 2.

Table 2. Equivalent Young's modulus of matirials

	Si	**PZT-5H**
Equivalent Young's Modulus (GPa)	178.98	$\begin{bmatrix} 18.1733 & 7.7750 & 10.0504 & 0 & 0 & 0 \\ & 18.1733 & 10.0504 & 0 & 0 & 0 \\ & & 15.8328 & 0 & 0 & 0 \\ & & & 4.1303 & 0 & 0 \\ & & & & 3.5796 & 0 \\ & & & & & 3.5796 \end{bmatrix}$

Infinite element analysis for micro-beam was done. Figure 1 and figure 2 show that first-order vibration modes of diffrent models are 1536Hz and 1803Hz, the difference is 267Hz.

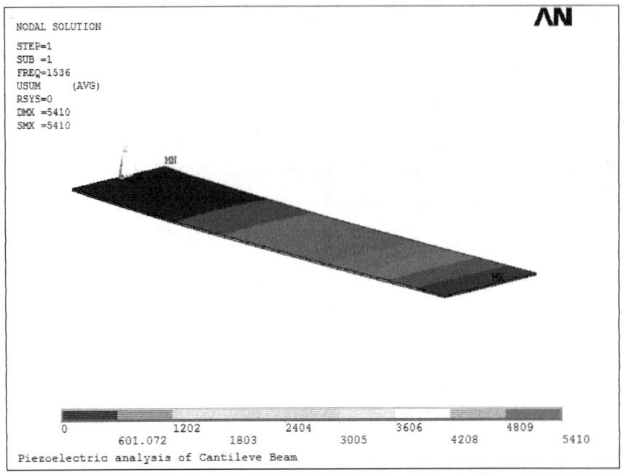

Figure 1. Results with macroscopic model

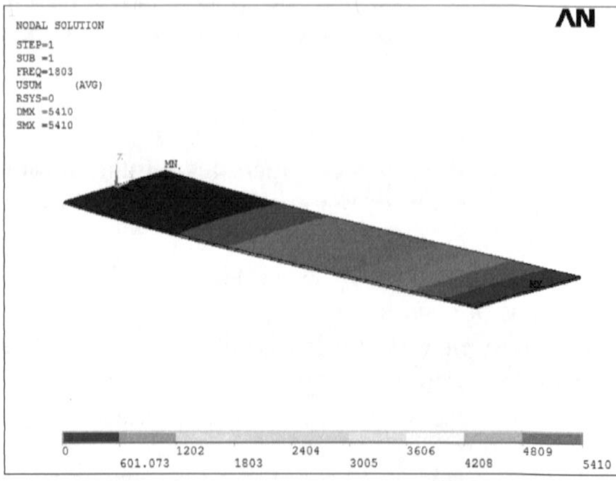

Figure 2. Results with improved model

2.2 Fabrication and experiment test of piezoelectric cantilever beams

Photolithography, sputtering, sol gel, dry etching process was used in a 4-inch SOI substrate to complete fabrication and packaging, the SEM is shown in figure 3 and figure 4.

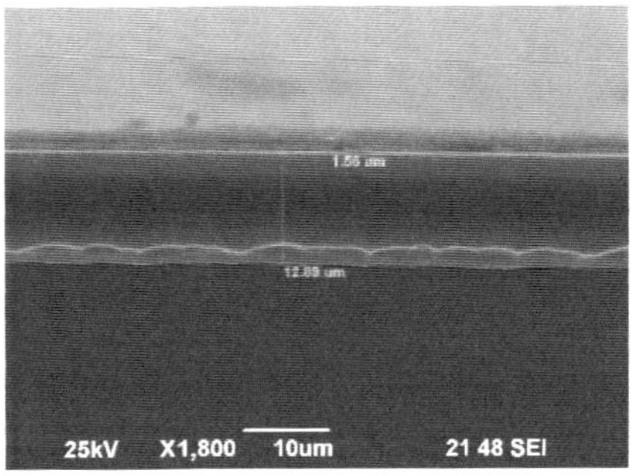

Figure 3. Top view of SEM image

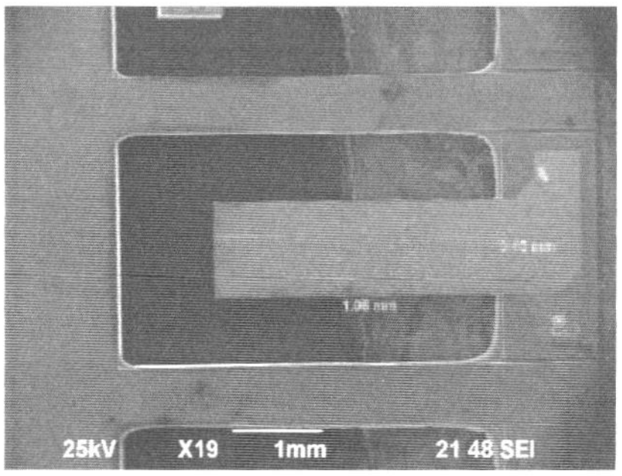

Figure 4. Thickness enlargement of SEM image

Use signal generator (YE1311), power amplifier (E5871A), vibration table (E-JZK-5), laser vibration meter (OFV505/5000), data acquisition card (INTEST Main Unit INJ9008U-I), acceleration sensor (CA-YD-1182) and other divices to build a vibration experimental platform for cantilever beam, as shown in figure 5.

Figure 5. Experimental platform

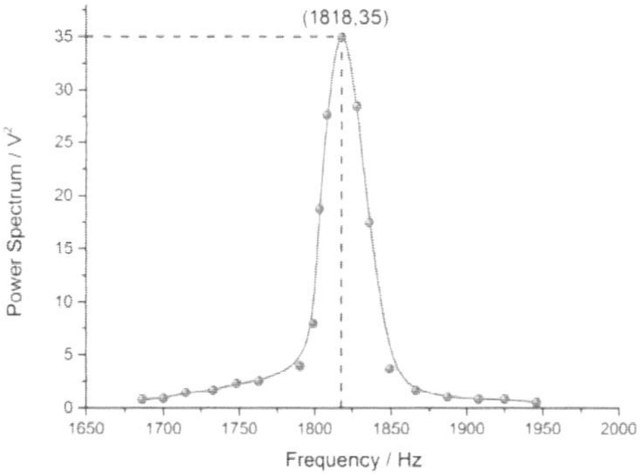

Figure 6. Power spectrum-frequency curve

Set acceleration excitation to 6g (1g=9.8m/s2), frequency of acceleration was sequentially changed in a range of 1600Hz~2000Hz, tip displacement was recorded. Using power spectrum to deal with it, power spectrum and vibration frequency curve were generated, as shown in figure 6. At 1818Hz, the power spectrum value to the maximum, that is the resonant frequency.

As shown in table 3, iproved model reduces errors efficiently.

Table 3. Frequency comparison between experiment and theoretical analysis

	Resonant Frequency(Hz)	Error
Experimental Test	1818	0%
Improved Model	1803	0.8%
Macroscopic Model	1536	15.5%

3. Experimental analysis of micro vibration energy acquisition

Micro-beam resonant frequency is generally above 1000Hz, many scholars[14,15,16] try to reduce resonant frequency by adding a Si mass or a Ni one to the free end of cantilever beam, so that it can perfomance well in low-frequency vibration environment. In this paper, a Si mass was etched onto the T-type cantilever beam, and a "T-type" harvester was made. The "T-type" vibration energy harvester consists of five layers, which can be simplified to three layers that consists of PZT-5H, Si-beam and mass blocks. The structure is shown in figure 7.

IOP Conf. Series: Materials Science and Engineering **242** (2017) 012066 doi:10.1088/1757-899X/242/1/012066

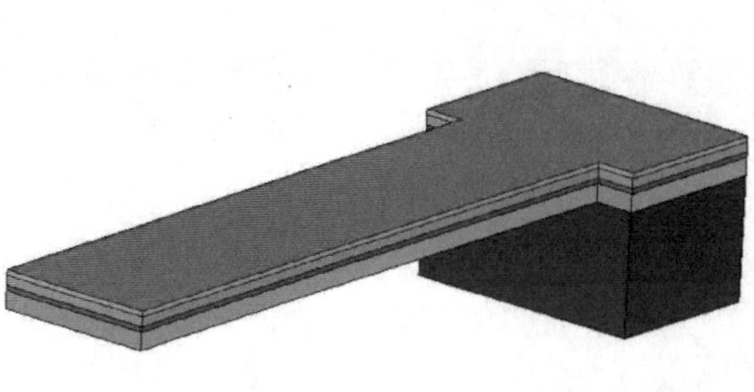

Figure 7. T-type harvester structure

3.1. Micro vibration energy harvester dynamic analysis

Considering size effect and improved model of micro-beam, the dynamic model of the harvester for free vibration can be expressed as:

$$\left(EI+\mu Al^2\right)\frac{\partial^4 w(x,t)}{\partial x^4}+\rho A\frac{\partial^2 w(x,t)}{\partial t^2}=0 \tag{5}$$

When μAl^2 was removed, formula (5) turns to be a macroscopic model. Charge Q produced by the harvester deformation gathers on the surfaces of PZT-5H to form a capacitor, output voltage of the micro harvester was obtained:

$$V_s=\frac{Q}{C_s}=\frac{2k\pi w_b d_{31}E_p\left(t_{Si}+t_{PZT-5H}\right)W'(L,t)}{\varepsilon_r S_s} \tag{6}$$

3.2. Finite element analysis of harvester

Finite element analysis can reflect the actual motion state of harvester in a form of cloud map. According to the equivalent elastic modulus in Table 1 and relevant parameters in [13], finite element model of harvester was established. DOF of fixed end of harvester was set to zero and the acceleration was 1g. Output voltage and tip displacement reach maximum at the resonant frequency 300Hz, which were 0.17V and 10.7μm.

3.3. Micro vibration energy harvester performance test

Acceleration excitation was set to 1g, vibration table was in the range of 50Hz~500Hz by specific frequency. Figure 8 and figure 9 show the tip displacement frequency, the output voltage frequency of micro harvester which is obtained from macroscopic model, improved model and experiment. Main information and error compared with experiment are shown in table 4.

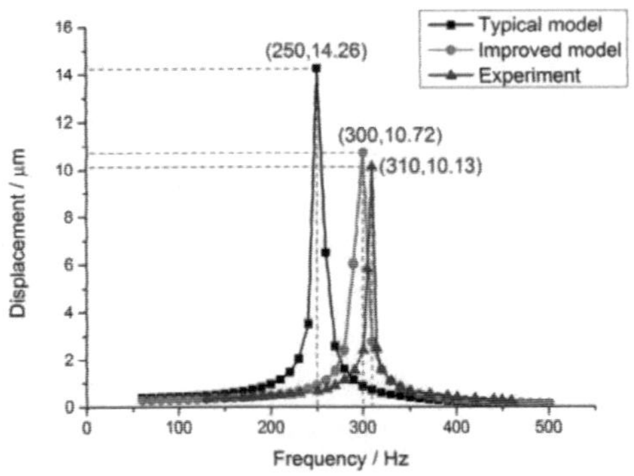

Figure 8. Tip displacement of energy harvester

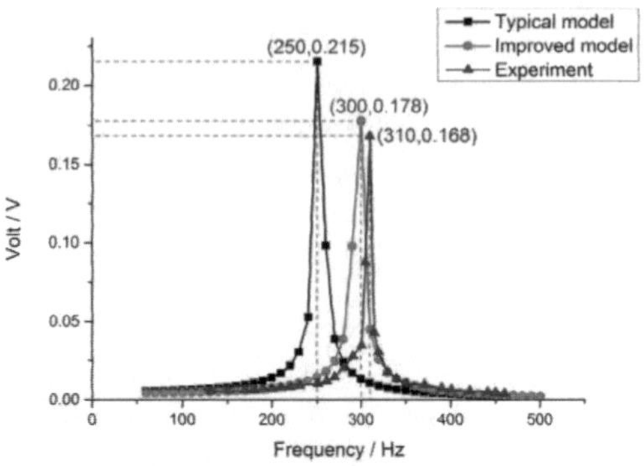

Figure 9. Output voltage of energy harvester

Comparing macroscopic model to experiment test, error of resonant frequency was large. The macroscopic model failed to describe dynamic performance of micro energy harvester. When improved model was compared to experiment test, error turned out to be relatively mall.

Table 4. Main information of resonance point and the contrast error

		Macroscopic model	Improved model	Experiment test
Resonant frequency	Value (Hz)	250	300	310
	error	19.4%	3.2%	0%
Tip displacement	Value (μm)	14.26	10.72	10.13
	error	40.8%	5.8%	0%
Output voltage	value (V)	0.215	0.178	0.168
	error	28.0%	6.0%	0%

4. Conclusion

Mechanical properties of microstructure considering size effect was studied using micro piezoelectric cantilever beam. Intrinsic length of material was introduced, and dynamic model was improved by couple stresses theory. Mechanical properties of micro-beam were studied, the improved model could describe mechanical properties of microstructures more accurately. Based on that research, dynamic model of micro vibration energy harvester was refined. Output performance was analyzed with finite element and vibration test. Results show that the improved model of harvester can provide theoretical support for design of microstructure.

References

[1] Lee I, Lee K 2015 *Business Horizons* 58 431-40
[2] Turkanović M, Brumen B, Hölbl M 2014 *Ad Hoc Networks* 20 96-112
[3] Tang G, Liu J, Ma H, Lu H 2010 *Machine Design and Research* 26 61-4
[4] Zhu L, Chen R , Lei X 2011 *China Mechanical Engineering* 22 3016-22
[5] Yu Shouwen 2001 Journal of Mechanical Strength 23 380-4
[6] Yu Shouwen 1995 *Advances in Mechanics* 25 249-59
[7] Nie Z, Zhou S, Han R, Xiao L, Wang K 2012 *Engineering Mechanics* 29 38-46.
[8] Wang Z, Huo R, Zhang D. 2016 *Piezoelectrics and Acoustooptics* 38 320-3
[9] Mcfarland A W, Colton J S 2005 Journal of Micromechanics and Microengineering 15 1060
[10] Lam D C C, Yang F, Chong A C M ,Wang J, Tong P 2003 *Journal of the Mechanics and Physics of Solids* 51 1477-508
[11] Fleck N A, Muller G M, Ashby M F, Hutchinson J W 1994 *Acta Metallurgica et Materialia* 42 475-87
[12] Stelmashenko N A, Walls M G, Brown L M 1993 Milman Yu V *Acta Metallurgica et Materialia* 41 2855-65
[13] Fang H 2007 Shanghai Jiao Tong University
[14] Defosseux M, Allain M, Defay E, Basrour S 2012 *Sensors and Actuators* A: Physical 188 489-94
[15] Kudimi J M R, Mohd-Yasin F, Dimitrijev S 2012 *Procedia Engineering* 47 1165-72
[16] Tang G, Yang B, Liu J, Xu B, Zhu H, Yang C 2014 *Sensors and Actuators* A: Physical 205 150-5

IOP Conf. Series: Materials Science and Engineering **242** (2017) 012067 doi:10.1088/1757-899X/242/1/012067

Research on the application of vehicle network in optimization of automobile supply supply chain

Xuelei Jing, Baoxian Jia

Liaocheng University, Liaocheng, China

Abstract: The four key areas of the development of Internet-connected (intelligent transportation) with great potential for development,environmental monitoring, goods tracking, and the development of smart grid are the core supporting technologies of many applications. In order to improve the adaptability of data distribution, so that it can be used in urban, rural or highway and other different car networking scenarios, the study test and hypothetical test of the technical means to accurately estimate the different car network scene parameters indicators, and then different scenarios take different distribution strategies. Taking into account the limited nature of the data distribution of the Internet network data, the paper uses the idea of a customer to optimize the simulation

1 Introduction

In 2015, the State Council on the issuance of "China made 2025" notice, which was classified as one of the top ten "vigorously push the focus area breakthrough development". Then the Ministry of Industry released a article to promote energy and new energy Car development.The article talked about the four perspectives of the automotive industry, which is the automotive industry, the next decade development of the car manufacturers, the consumers driving the development of the car, we look at what the future development of the car.

In the "Made in China 2025" notice, energy conservation and new energy vehicles have been given full attention.It is worth noting that this includes not only the new energy vehicles, but also energy-efficient vehicle. In the technical details of the description, include the "electric vehicle, fuel cell vehicle development, master car low carbon, information technology, intelligent core technology, enhance the power battery, drive motor" , and other energy-efficient vehicles include "efficient internal combustion engine, advanced transmission, lightweight materials, Control and other core technology of engineering and industrialization capabilities. "

2 Supply chain visualization

Internet of Things is a huge network of information sensing devices, including radio frequency identification devices, infrared sensors, global positioning systems, laser scanners and network interconnection. Its purpose is to facilitate the identification and management. Through the device in the various types of objects on the electronic tags (RFID), sensors, two-dimensional code through the interface and wireless network connected to the object to give intelligence, can achieve human and object communication and dialogue, but also can achieve objects and objects communication and dialogue. Internet use, environmental protection, government work, public safety, safe home, intelligent fire, industrial monitoring, elderly care, personal health and other fields. In the concept of Internet of Things, RFID tags contain standardized and interoperable information, through the wireless data communication network to automatically collect them to the central information system, to achieve the identification of items, and then through the open Computing network to achieve

Content from this work may be used under the terms of the Creative Commons Attribution 3.0 licence. Any further distribution of this work must maintain attribution to the author(s) and the title of the work, journal citation and DOI.
Published under licence by IOP Publishing Ltd

information exchange and sharing, to achieve the transparent management of goods, RFID has become the most critical technology applications.

It is particularly important to realize perception. Numerous sensors with communication and computing capabilities are connected wirelessly, interacting with each other, interacting with the physical world, and accomplishing specific application tasks, called the Sensor Network. It integrates the research results of sensor technology, embedded computing technology, distributed information processing technology and wireless communication technology. In the sensor network, a multi-hop self-organizing system is constructed through wireless communication. And deal with the network coverage area in the perception of information, flexible monitoring and processing of these information, while the detailed information transmitted to the user, Internet of Things to all items through the RFID and other devices connected with the Internet to achieve intelligent identification and management.

3 RFID applications to promote the development of car network

Auto Internet: through information sensor equipment, in accordance with the owner's idea, will be closely linked with their own cars, anytime, anywhere information exchange and communication, in order to achieve the car and its surrounding environment, intelligent identification, positioning, tracking, monitoring and Management of a network. In recent years, the rapid development of sensor network based on RFID tag technology has been developed rapidly. By means of information identification equipment such as radio frequency identification, infrared sensor, global positioning system and laser scanner, any item can be connected with the communication network,information exchange,to achieve intelligent identification, positioning, tracking, monitoring and management. This solves the problem of how the car joins the public network. Vehicle self-organizing network is the main component of vehicle wireless communication network, is a service in a variety of vehicles (vehicle terminal) and road traffic management information network.The research and development in this field has been widely concerned, and the research and application of car networking has gradually become a research hotspot. US and Japanese car network has developed into a more mature application, the Chinese car network has just started. The United States to promote the wisdom of the Earth, has developed into a $ 200 billion market, Japan's VICS system has been the formation of advertising, value-added business model.In all aspects yet to be perfect, but the future will maintain a high growth rate for a long time. Compared with the traditional wireless network, the vehicle network has its unique network characteristics [4]: high dynamic topology, intermittent network connection, the network link is highly unreliable, node movement has a certain regularity, Wireless signal transmission model has an impact, energy and space are not limited, the network size is relatively large, some applications on the packet submission delay and reliability is strictly limited, the global satellite positioning system GPS and other equipment support. Car network needs a new network architecture support, which is a major aspect of car networking research. At the same time, the key technologies of each layer under the new network architecture, including physical layer technology, media access control protocol, routing layer technology and security technology, constitute the main content of the second aspect of car networking. In addition, some common technologies closely related to the application are also an important part of the basic research of car network, including positioning system, coverage and connectivity.

Based on the technical means of measurement and hypothesis test, the parameters of different vehicle network are estimated, and the parameters of several parameters are analyzed, and the change types of several performance indicators in different application scenarios are analyzed and the description method is established. Describe the car network performance parameters, and adapt to different scenarios for different scenarios. In order to reduce the occupancy of the limited data distribution resources of the vehicle network environment, a new adaptive message loading mechanism is designed to reduce the need for the network bandwidth to occupy the necessary information for the vehicle nodes to obtain the data distribution. Based on the theoretical knowledge of the experimental data collection and the theoretical knowledge of the stochastic process, the statistical characteristics of the nodes in the vehicle network are revealed. The distribution rules of the nodes in the network are analyzed, the motion of the nodes and the influence of the node motion on the distribution of nodes. Link stability, the relationship between the statistical characteristics of the above, the establishment of the statistical characteristics of qualitative or quantitative expression. This paper explores the new data distribution decision mechanism from the macroscopic level and the micro level according to the mainstream and effective methods used in the current vehicle network data distribution mechanism and algorithm.

4 Application of a Maker

The use of graphical operating system, 4-year-old children can also use 3d printer. Wearing a miniature electronic wrist device, your pulse, heart rate, blood pressure, blood sugar and other data will be uploaded to the mobile phone and cloud platform, It can predict Cardiovascular possible lesions, and early warning of health emergencies.

In the 2016 China International Electronic Information Starter Competition and cloud on Guizhou big data business model contest,from India, South Korea and more than 20 provinces in the 13,000 teams participating in the project type covers intelligent manufacturing , Large data services, large data finance government governance and other fields.

College entrance examination near, how can accurate registration volunteer, do not waste every 1 minute? "If you do not know their scores for the school which school, may wish to let big data to help you." "Tree elite data selection platform" project leader Yu Yueqing said.Yu Yueqing introduction, "tree elite" the use of massive data resources and years of scientific analysis of the results of conduct for the user to accurately find suitable and feasible educational opportunities.

Big data recorders not only tap the college entrance examination scores, but also "eyeing" people's clothes. When you see someone wearing a nice dress, just need to take pictures to upload, you can search for this piece of clothing related information. Won the cloud on Guizhou data contest first prize is Shenzhen Code Long Technology Co., Ltd. developed "artificial intelligence based on the visual decision engine" project.

5 Concluding remarks

Although the research has been carried out in the recent time, some basic data distribution mechanisms based on anchor routing, vehicle traffic, road topology and so on are proposed. Based on their implementation algorithm ,but there are some shortcomings and deficiencies, including the application of a single scene, the network load is high, poor scalability, assumptions can not be met in the project to achieve, so the current research is still in the early of the exploratory stage, to the final formation of mature application technology, may also need more innovative research, or a lot of detailed and perfect work. It is necessary and urgent to start some research topics such as data distribution mechanism and algorithm, such as data, in the context of increasing emphasis on car networking and its application in China.

References

[1] Mittag J, Papanastasiou S, Hartenstein H, et al. Enabling Accurate Cross-Layer PHY / MAC / NET Simulation Studies of Vehicular Communication Networks [J]. Proceedings of the IEEE. 2011, 99 (7): 1311-1326 The

[2] Annese S, Casetti C, Chiasserini C F, et al. Seamless Connectivity and Routing in Vehicular Networks with Infrastructure [J]. IEEE J. Select. Areas Commun. 2011, 29 (3): 501-514.

[3] Camp J, Knightly E. Modulation Rate Adaptation in Urban and Vehicular Environments: Cross-Layer Implementation and Experimental Evaluation [J]. IEEE / ACM Trans. Netw. 2010, 18 (6): 1949-1962.

[4] Niyato D, Hossain E, Wang P. Optimal Channel Access Management with QoS Support for Cognitive Vehicular Networks [J]. IEEE Trans. Mobile Comput. 2011, 10 (4): 573-591.

[5] I. F. A kyildiz, W.Su, Y.Sankarasubramaniam and E. Cayirci. Wireless Sensor Networks: A Survey [J] .Computer Network s, 2002, 38 (3): 393- 422.
[6] Bottani Eleonora, Montanari Roberto, Volpi Andrea. The impact of RFID and EPC network on the bullwhip effect in the Italian FMCC supply chain [J]. International Journal of Production Economics. 2009,124 (2): 4 26-432

Research on NC laser combined cutting optimization model of sheet metal parts

Z Y Wu[1], Y L Zhang[1], L Li[2], L H Wu[1] and N B Liu[1]

[1] School of Mechanical & Electrical Engineering, Henan University of Technology, Zhengzhou, China
[2] Automobile Engineering Department, Henan Polytechnic, Zhengzhou, China

Corresponding author's e-mail: wzhaoyun@163.com

Abstract. The optimization problem for NC laser combined cutting of sheet metal parts was taken as the research object in this paper. The problem included two contents: combined packing optimization and combined cutting path optimization. In the problem of combined packing optimization, the method of "genetic algorithm + gravity center NFP + geometric transformation" was used to optimize the packing of sheet metal parts. In the problem of combined cutting path optimization, the mathematical model of cutting path optimization was established based on the parts cutting constraint rules of internal contour priority and cross cutting. The model played an important role in the optimization calculation of NC laser combined cutting.

1. Introduction

NC laser combined processing technology combining computer numerical control technology and laser cutting technology, has the advantages of reliable quality and high cutting precision etc., and has now become the important processing method in the process of plate cutting. The laser cutting path generated by CAM (Computer Aided Manufacturing) NC program, has a great influence on cutting process and machining efficiency. Optimizing the cutting path can significantly shorten the cutting time and improve the quality of processing.

Han etc.[1] used the simulated annealing algorithm to solve the cutting path optimization problem of irregular sheet metal parts; Pan, etc. [2] set up a "punching and cutting" integrated planning CAD/CAM system based on the technology of knowledge base; The three-step algorithm based on hierarchical planning idea, proposed by Hui-xia Liu [3], used to optimize the cutting path problem; according to graph theory, Sa Chen [4] proposed an effective algorithm to optimize cutting path based on Prim algorithm; Jian-tao Li [5] carried on the global and local optimization of two-dimensional rectangular parts cutting path.

The problem of cutting path optimization was researched at many different angles in the above documents, but most of them researched cutting metal parts inside a rectangular plate, which is not fit for assembled cutting conditions based on several plates. On the other hand, it is inevitable that there will be remnants during long-term production process in large sheet metal parts manufacturing enterprises, in order to save material and reduce cost, the excess material is preferentially used. Currently, the related research of excess stock layout and cutting optimization method is relatively small.

Content from this work may be used under the terms of the Creative Commons Attribution 3.0 licence. Any further distribution of this work must maintain attribution to the author(s) and the title of the work, journal citation and DOI.
Published under licence by IOP Publishing Ltd

Taking the combined cutting optimization problem inside several plates as the research object, the NC laser combined packing and cutting path optimization inside irregular remnants was mainly discussed in the paper to solve some key problems in excess material recycling.

2. Description of combined cutting problem of excess material

Assumed to be the location of the cutting parts inside excess material has been determined by relevant layout algorithm. Due to remnants are relatively small and irregular, in the process of cutting, machine need to restart and plate need to load and unload constantly, which is tedious and time consuming, and seriously affects the processing efficiency. Therefore, currently, the combined cutting method of excess material is adopted, that is, let several plates place in the cutting area of the cutting machine, then one-time cutting is finished by laser head and the next batch of excess material is loaded. As shown in fig.1, the remnant S1 has three parts with hole P_{11}, P_{12} and P_{13} to be cut, similarly, S2, S3 and S4 have P_{21}, P_{31}, P_{41} and P_{42} respectively, let S1, S2, S3 and S4 in the rectangular cutting area of the cutting machine without overlapping, then 7 parts (P_{11}, P_{12}, P_{13}, P_{21}, P_{31}, P_{41} and P_{42}) were cut by laser head at a time, the processing method is called excess material combined cutting. The method can effectively save the processing time and improve processing efficiency.

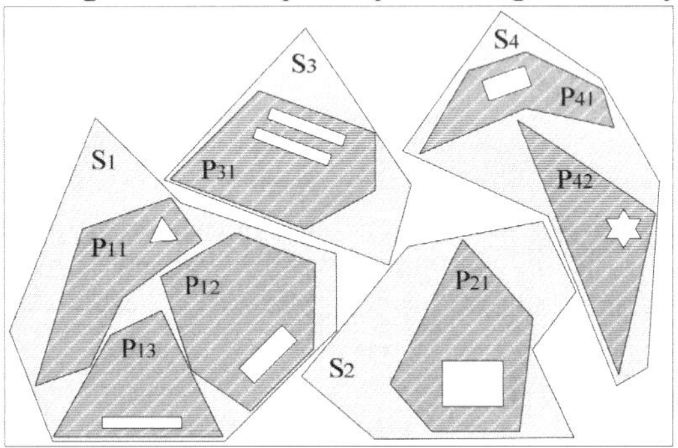

Figure 1. Example of combined cutting problem inside excess material

The combined cutting problem of excess material includes two contents: 1) Combined packing optimization. Through packing optimization, more remnants can be packed with the best location in the cutting area to improve the single cutting production; 2) Combined cutting path optimization. Through laser cutting path optimization during processing, the idle travel of laser head can be effectively reduced so as to shorten the cutting time and improve efficiency.

3. Combined packing optimization of excess material

3.1. *The method of combined packing optimization of excess material*

Combined packing optimization of excess material is a typical two-dimensional irregular part packing optimization. So the developed method can be used for the solution, such as the packing method of rectangular fitting, NFP (No Fit Polygon, critical polygons) algorithm, intelligent search packing algorithm, etc. The geometry contour of remnants is equivalent to the external contour of unpacked part in conventional two-dimensional irregular packing, but the internal and external contour of unpacked parts inside excess material is equivalent to the internal contour of unpacked part in conventional two-dimensional irregular packing. In order to improve the computation efficiency, first the geometry contour of remnants is calculated, and then the geometry coordinate of graphics inside excess material is updated by the method of geometric transformation. The method reduces the packing calculation effectively, and greatly increases packing speed. Based on the current literature, "genetic algorithm and gravity center NFP" are adopted to solve packing problem in the paper. Algorithm steps are as follows:

Step1. Initialization. First sort all plates by non-increasing ordering of area and obtain initial ordering, then generate the first generation by variation; let the current evolutionary generation gn = 0; Initialize the biggest evolutionary generation GN.

Step2. Population evolution.(1) choice: randomly select two individuals in the population as the parent individual Parent1, Parent2;(2) cross: if the crossover probability is not less than Random [0,1], new child Child1and Child2 are produced through parent individual genetic crossover, otherwise the genes is passed on directly;(3) variation: if the mutation probability is not less than Random [0,1], the gene of the individual Child1, Child2 gets mutation;(4) calculate the fitness of Child1 and Child2;(5) generate a new population: join the individual Child1, Child2 to the new group, if the number of individuals in the new group is less than the number of individuals in original group, go to step (1); Otherwise, go to step 3.

Step3. Calculate the fitness of individual. Update the current best individual and the global optimal individual.

Step4. Determine whether terminate evolution. If gn≥GN, the population evolution terminates, turn to Step5; Otherwise, go to Step2.

Step5. According to the gene coding of the global optimal individual, the packing sequence of plates is acquired.

Step6. According to the packing sequence of plates and the location selection rules based on gravity center NFP, packing optimization is done for all the remnants to obtain the packing result.

3.2. *The geometric transformation of graphics inside excess material*

After combined packing of excess material, the geometry coordinates of the new plates outline are acquired, and the geometry coordinates of graphics inside excess material need to be updated. It is known that the shape and size of plates remains the same, just their locations change by analysis. Thus, the process of packing is essentially the geometric transformation process of graphics. The graphics before and after packing can be called old graphics and new graphics respectively, according to the geometric transformation theory of two-dimensional graphics, the process of graphics transform should satisfy the following formula:

$$A \times T = B \tag{1}$$

In the formula, A denotes the point set matrix of old graphics, T represents transformation matrix, and B denotes the point set matrix of new graphics. For the convenience of calculation, select three vertices from all vertices of remnant S_i as reference points, the original coordinate matrix before packing is:

$$A_i = \begin{bmatrix} x_1 & y_1 & 1 \\ x_2 & y_2 & 1 \\ x_3 & y_3 & 1 \end{bmatrix} \tag{2}$$

The new coordinate matrix after packing is:

$$B_i = \begin{bmatrix} x_1' & y_1' & 1 \\ x_2' & y_2' & 1 \\ x_3' & y_3' & 1 \end{bmatrix} \tag{3}$$

For there are only translation and rotation in the process of packing, the transformation matrix is:

$$T_i = \begin{bmatrix} a & b & 0 \\ c & d & 0 \\ l & m & 1 \end{bmatrix} \tag{4}$$

According to the formula (1), the formula (5) was obtained:

$$\begin{bmatrix} x_1 & y_1 & 1 \\ x_2 & y_2 & 1 \\ x_3 & y_3 & 1 \end{bmatrix} \times \begin{bmatrix} a & b & 0 \\ c & d & 0 \\ l & m & 1 \end{bmatrix} = \begin{bmatrix} x_1' & y_1' & 1 \\ x_2' & y_2' & 1 \\ x_3' & y_3' & 1 \end{bmatrix} \tag{5}$$

The formula (5) is a standard linear equation with 6 unknown numbers, the unknown numbers a , b , c , d , l and m are solved easily (the solving process is abbreviated). Thus, the transformation matrix T_i can be obtained. In the process of packing, the relative location of graphics inside remnant S_i keeps invariant, so the geometry coordinates are also change according to the transformation matrix T_i. In the correlation model of graphics data, denote the original coordinate's matrix of outside contour vertices of parts inside remnant S_i as A_{io} , the new coordinate's matrix after packing is denoted as B_{io} ; accordingly, denote the matrix of inside contour vertices as A_{ii} , the new coordinate's matrix after packing is denoted as B_{ii} . According to the formula (1), there are:

$$\mathrm{B}_{io} = A_{io} \times T_i \tag{6}$$

$$\mathrm{B}_{ii} = A_{ii} \times T_i \tag{7}$$

The transformation matrix of each plate can be calculated according to the formula (5), and then according to the formula (6) and (7), the geometric transformation of graphics inside all remnants is carried on, finally, the relevant geometry coordinates in the correlation model of graphics data are updated according to the calculation results.

4. Combined cutting path optimization inside excess material

4.1. *Parts cutting constraint rules*

According to the requirements of cutting process, in cutting process, the following two constraint rules should meet:

(1) The principle of from internal to external

In order to reduce the deformation of parts in the process of cutting, usually adopt "first cutting internal contour cutting, then external contour" method, the principle has been approved in general. The three layer correlation model of graphic data is introduced to describe the geometry information of graphics in section 3.1, in the model, the geometry data layers of internal and external contours of parts are clear and interconnected, while cutting path optimization calculation, the data of part's internal and external contours could be quickly identified.

(2) The principle of cross cutting

At present, under the influence of "equivalent TSP method", the serial cutting principle of parts is adopted in most of the literature, that is, another part can be cut only after the previous part is cut by the laser head. The serial cutting principle has certain influence on the cutting path optimization result, sometimes, it may lead optimization algorithm not to find the global optimal solution. For solving the problem, the cross cutting principle of parts is put forward, that is, in the process of cutting, let the laser head removes among different parts and does cross cutting. The principle leads optimization algorithm to find a better optimal solution.

4.2. *The optimization mathematical model of cutting path*

In the process of machining, when the laser head cuts parts, it is in working state; when it removes among the contours of different parts, it is idle, known as idle travel. The goal of cutting path optimization is to make the distance of idle travel the shortest in order to improve machining efficiency. For adopting the cross cutting principle, all internal and external contours of parts to be cut must be discretized into line segment groups, so the cutting path optimization problem is equivalent into cutting sequence planning problem of each contour line segment. By the two examples shown in table 1, it is known that the distance of transferring from endpoints of various line segments is bigger than that of transferring with the shortest distance between various line segments, so in order to shorten the idle travel, the principle of “transferring with the shortest distance between various line segments" should be followed. The shortest distance formula between line segment and is:

$$d_{\min} = \min(d_1, d_2, d_3, d_4) \tag{8}$$

In the formula, d_1 denotes the shortest distance from point A to line segment CD, d_2 represents the shortest distance from point B to line segment CD, d_3 signifies the shortest distance from point C to line segment AB, and d_4 denotes the shortest distance from point D to line segment AB.

Table 1. Example of laser head transfer path

Name	Transfer from endpoints	Transfer with the shortest distance	Transfer path
Example 1	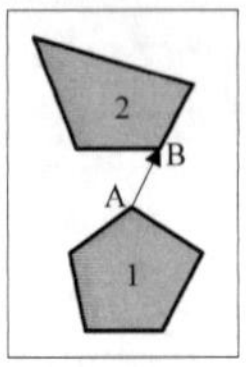	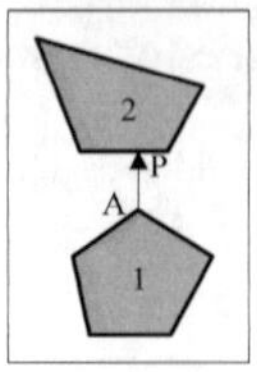	AB>AP
Example 2	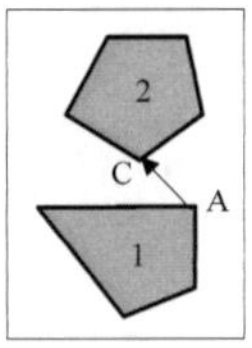	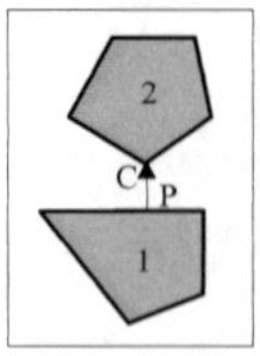	AC>PC

Assuming that the cutting sequence of contour lines is shown in table 2. Z denotes the number of all contour lines, the laser head start from origin point O (0, 0). The cutting path of laser head is: O→ [LP_{11}→LP_{12}]→[LP_{21}→LP_{22}]→……→[LP_{Z1}→LP_{Z2}]→O。 Idle travel

$$S_e = \overline{OLP_{11}} + \overline{LP_{12}LP_{21}} + \overline{LP_{22}LP_{31}} + \cdots + \overline{LP_{Z-1,2}LP_{Z1}} + \overline{LP_{Z2}O} \tag{9}$$

Therefore, the mathematical model of cutting path optimization problem is as follows:

$$\min S_e = \sqrt{x_{11}^2 + y_{11}^2} + \sum_{i=1}^{Z-1}\sqrt{(x_{i+1,1} - x_{i,2})^2 + (y_{i+1,1} - y_{i,2})^2} + \sqrt{x_{Z2}^2 + y_{Z2}^2} \tag{10}$$

s.t. $x_{ij} \geq 0 \ (i = 1,2,\cdots,Z;\ j = 1,2)$

$y_{ij} \geq 0 \ (i = 1,2,\cdots,Z;\ j = 1,2)$

Table 2. Cutting sequence of parts contour line

No.	Line segment	Endpoint 1 and its coordinate	Endpoint 2 and its coordinate
1	L_1	$LP_{11}(x_{11}, y_{11})$	$LP_{12}(x_{12}, y_{12})$
2	L_2	$LP_{21}(x_{21}, y_{21})$	$LP_{22}(x_{22}, y_{22})$
3	L_3	$LP_{31}(x_{31}, y_{31})$	$LP_{32}(x_{32}, y_{32})$
4	L_4	$LP_{41}(x_{41}, y_{41})$	$LP_{42}(x_{42}, y_{42})$
……	……	……	……
Z	L_Z	$LP_{Z1}(x_{Z1}, y_{Z1})$	$LP_{Z2}(x_{Z2}, y_{Z2})$

5. Conclusion

In the process of cutting parts inside irregular excess material, it is very necessary to solve the optimal combined packing scheme and the shortest cutting path; the optimization solution can effectively improve the cutting efficiency and shorten the processing time. The method proposed in the paper to solve NC laser combined cutting optimization problem inside irregular excess material is also has certain reference value to the similar problem. However, it is a new subject, which also has a lot of unsolved problems, and still need to be further in-depth studied.

Acknowledgments

This work was supported by the National Nature Science Foundation of China (Grant No. 51405133), the Province Science and Technology Research Projects of Henan (Grant No. 172102210213), and the Fundamental Research Funds for the Henan Provincial Colleges and Universities in Henan University of Technology (Grant No. 2016RCJH02).

References

[1] Gukchan H and Suckjoo N 1999 *Journal of Manufacturing Processes* **1** 62-70
[2] Pan M and Rao Y 2009 *Knowledge-Based Systems* **22** 368-75
[3] Huixia L, Xiao W and Lan C 2004 *Journal of CAD & CG* **16** 660-5
[4] Sa C and Zhan S 2003 *Journal of University of Shanghai for Science and Technology* **25** 376-8
[5] Jiantao L, Xingmei H and Zhihua Z 2005 *Machinery Design & Manufacture* **21** 86-7

IOP Conf. Series: Materials Science and Engineering **242** (2017) 012069 doi:10.1088/1757-899X/242/1/012069

Analysis on spectra of hydroacoustic field in sonar cavity of the sandwich elastic wall structure

Xuetao W [1], Rui H[2] and Weike W[2]

[1] Key Laboratory of Shipping Noise and Vibration, Naval University of engineering, Wu'han 430033, China

[2] School of Mechanical Engineering, Shandong University, Ji'nan250061, China;

449200779@qq.com

Abstract. In this paper, the characteristics of the mechanical self - noise in sonar array cavity are studied by using the elastic flatbed - filled rectangular cavity parameterization model. Firstly, the analytic derivation of the vibration differential equation of the single layer, sandwich elastic wall plate structure and internal fluid coupling is carried out, and the modal method is used to solve it. Finally, the spectral characteristics of the acoustic field of rectangular cavity of different elastic wallboard materials are simulated and analyzed, which provides a theoretical reference for the prediction and control of sonar mechanical self-noise. In this paper, the sandwich board as control inside the dome background noise of a potential means were discussed, the dome background noise of qualitative prediction analysis and control has important theoretical significance.

1. Introduction

The mechanical self - noise system in sonar array is transmitted to the sonar position by the vibration of the ship's vibration source device, it is particularly important for the low-speed submarine underwater navigation. According to some test and analysis results, the typical mechanical noise sources such as main motor slot frequency and high order harmonic, podium shell around the low-frequency pattern, they transmit distance and stability, which is an important factor in the low frequency detection ability of sonar[1].

Dowell[2] has made theoretical analysis on the model of the coupling system of the plate and closed space, and the general form of the vibration coupling problem of the acoustic field and solid structure is given by the green function method[3]. Pan[4] shows that the response of the coupled system can be regarded as the result of the coupling of plate mode and cavity mode. Because the sandwich panel has the advantages of high strength, good sound insulation performance, strong design sex. As early as 1959, was proposed to improve the sound insulation performance of plate[5], used in the field of aerospace and Marine industry more and more widely. Zhenyu Feng[6] studied free vibration problems of several main viscoelastic model sheet. Cupial[7] calculated the free vibration problem of core material based on the constant complex model of constraint damping plate, studied the free vibration characteristic of the sandwich board. Narayanan[8] calculated the sound field of noise through the sandwich plate into the cavity.

2. Theoretical model

Content from this work may be used under the terms of the Creative Commons Attribution 3.0 licence. Any further distribution of this work must maintain attribution to the author(s) and the title of the work, journal citation and DOI.
Published under licence by IOP Publishing Ltd

2.1. The closed-cavity indoor underwater acoustic field with a fluid-elastic structure coupled boundary

Figure 1 sandwich board - rectangular acoustic field coupling model, the upper is subjected to external excitation pe has elastic and viscoelastic - elastic three layers structure of sandwich board D_B, the rest of the wall are rigid (D_S). Underwater acoustic field of water filling airtight chamber interior ministry satisfies wave equation.

$$\nabla^2 p(\sigma,t) - \frac{1}{c_0^2}\frac{\partial^2 p(\sigma,t)}{\partial t^2} = 0 \tag{1}$$

In this equation, p is the sound pressure, and c_0 is the sound velocity.

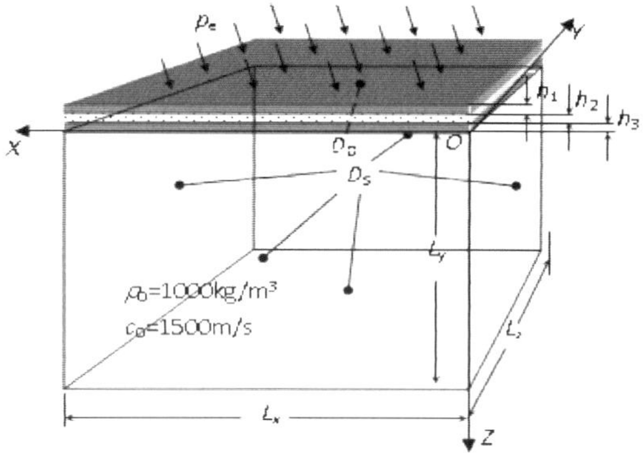

Figure 1. Coupling model of sandwich panel - rectangular underwater acoustic field

On the elastic sandwich panel($\sigma \in D_B$), Eq. (1) has boundary conditions

$$\partial p(\sigma,t)/\partial n = -\rho_0 a_n(\sigma,t) \tag{2}$$

In the formula, n is the outer normal direction of the boundary wall plate, ρ_0 is the medium (water)density, and an is the normal acceleration of the vibration of the sandwich plate.

On the rigid wall boundary($\sigma \in D_S$), Eq. (1) has boundary conditions

$$\partial p(\sigma,t)/\partial n = 0 \tag{3}$$

If all the boundary are rigid, Eq.(1) has the general solution of the following form

$$p = \rho_0 c_0^2 V \sum_r \frac{F_r^A(\sigma) e^{j\omega_r^A t}}{(\omega_r^A)^2 M_r^A} \tag{4}$$

For rectangular chamber

$$\omega_r^A = c_0 \sqrt{\left(\frac{m_r^A \pi}{L_x}\right)^2 + \left(\frac{n_r^A \pi}{L_y}\right)^2 + \left(\frac{l_r^A \pi}{L_z}\right)^2} \tag{5}$$

$$F_r^A(\sigma) = \cos\left(\frac{m_r^A \pi x}{L_x}\right)\cos\left(\frac{n_r^A \pi y}{L_y}\right)\cos\left(\frac{l_r^A \pi z}{L_z}\right) \tag{6}$$

$$\rho_0 c_0^2 (\omega_r^A)^2 M_r^A = \iiint_V [F_r^A(\sigma)]^2 d\sigma \tag{7}$$

When there is a non rigid boundary, referencing Gauss formula can be obtained

$$\ddot{P}_r + (\omega_r^A)^2 P_r = -\frac{A_r}{V} \quad (r = 0,1,2,\ldots) \tag{8}$$

2.2. Vibration response of sandwich elastic wall plate

The motion differential equation of viscoelastic sandwich board under transverse excitation is established under the following assumptions by Mead: (1) In the sandwich panel, there is no significant positive strain in the direction perpendicular to the plate, so the three layers have uniform lateral vibration displacement. (2) The bending stress within the sandwich layer is negligible compared with the normal stress in the panel. (3) The shear strain of the sandwich layer in the direction of the panel is ignored. (4) The contact surface of the panel and sandwich layer has the continuity of displacement, without the continuity of stress.

$$D_t\nabla^4 w-\frac{G_c(h_1+h_2)}{h_2}\left[(h_1+h_2)\nabla^2 w-2\left(\frac{\partial u}{\partial x}+\frac{\partial v}{\partial y}\right)\right]+(2\rho_p h_1+\rho_c h_2)\frac{\partial^2 w}{\partial t^2}=q(x,y,t) \tag{9}$$

$$\frac{E}{1-\mu^2}\frac{\partial^2 u}{\partial x^2}+\frac{E}{2(1+\mu)}\frac{\partial^2 u}{\partial y^2}-\frac{2G_c}{h_1h_2}u+\frac{E}{2(1-\mu)}\frac{\partial^2 v}{\partial x\partial y}+\frac{G_c(h_1+h_2)}{h_1h_2}\frac{\partial w}{\partial x}=0 \tag{10}$$

$$\frac{E}{1-\mu^2}\frac{\partial^2 v}{\partial y^2}+\frac{E}{2(1+\mu)}\frac{\partial^2 v}{\partial x^2}-\frac{2G_c}{h_1h_2}v+\frac{E}{2(1-\mu)}\frac{\partial^2 u}{\partial x\partial y}+\frac{G_c(h_1+h_2)}{h_1h_2}\frac{\partial w}{\partial y}=0 \tag{11}$$

The differential equations of lateral displacement w can be obtained by algebraic manipulation of equation (9) ~ (11)

$$D_t\nabla^6 w-D_t g(1+Y)\nabla^4 w+(\rho h)\partial^2(\nabla^2 w)/\partial t^2-g(\rho h)\partial^2 w/\partial t^2=(1-g)q(x,y,t) \tag{12}$$

Applying the simple damping mode, the solution of equation (12) has the superposition of vibration mode

$$w(x,y,t)=\sum_j W_j^{\mathrm{B}}(x,y)B_j(t) \tag{13}$$

$$\nabla^6 W-g(1+Y)\nabla^4 W-\Omega^2\nabla^2 W+g\Omega^2 W=0 \tag{14}$$

In order to simplify the calculation, consider the situation of the boundary conditions of the four sides

$$W_j^{\mathrm{B}}(x,y)=\cos\left[m_j^{\mathrm{B}}\pi\left(\frac{x}{L_x}-\frac{1}{2}\right)\right]\sin\left(\frac{n_j^{\mathrm{B}}\pi y}{L_y}\right) \tag{15}$$

$$\omega_j^{\mathrm{B}}=\bar{\omega}_j^{\mathrm{B}}\sqrt{1+\mathrm{j}\eta_j^{\mathrm{B}}}=\pi^2\gamma_j^{\mathrm{B}}\sqrt{\frac{D_t}{(\rho h)}\left[1+\frac{gYL_x^2}{\gamma_j^{\mathrm{B}}+gL_x^2}\right]} \tag{16}$$

The modal coordinate equation can be obtained

$$M_j^{\mathrm{B}}\left[\ddot{B}_j+(\omega_j^{\mathrm{B}})^2 B\right]=Q_j \quad (j=1,2,3,\ldots) \tag{17}$$

2.3. Coupling vibration response of underwater acoustic field - sandwich plate

The underwater acoustic field, On the rigid wall D_S, a_n=0, On the fluid-structure coupling boundary, $a_n=\partial^2 w/\partial t^2$.

$$A_r(t)=\iint_{D_B} F_r^{\mathrm{A}}\big|_{z=0}\sum_j W_j^{\mathrm{B}}\ddot{B}_j\mathrm{d}A=\sum_j L_{rj}\ddot{B}_j \tag{18}$$

The sandwich board, making the generalized force $Q_j= Q_j^{\mathrm{A}}+Q_j^{B}$, Q_j^{A} is the generalized force corresponding to the internal fluid excitation of the cavity. Q_j^{B} is the generalized force corresponding to the external excitation p_e in the direction of the direction of the panel.

$$Q_j^{\mathrm{A}}=\iint_{D_B} p\big|_{z=0}W_j^{\mathrm{B}}\mathrm{d}A=\iint_{D_B} V\sum_r\frac{F_r^{\mathrm{A}}\big|_{z=0}P_r}{(\omega_r^{\mathrm{A}})^2 M_r^{\mathrm{A}}}W_j^{\mathrm{B}}\mathrm{d}A=V\sum_r\frac{L_{rj}P_r}{(\omega_r^{\mathrm{A}})^2 M_r^{\mathrm{A}}} \tag{19}$$

The fluid - structure coupled vibration equation

$$V\left[\ddot{P}_r + \left(\omega_r^{\mathrm{A}}\right)^2 P_r\right] = -\sum_j L_{rj}\ddot{B}_j \;,\; M_j^{\mathrm{B}}\left[\ddot{B}_j + \left(\omega_j^{\mathrm{B}}\right)^2 B\right] = \sum_r \frac{VL_{rj}P_r}{\left(\omega_r^{\mathrm{A}}\right)^2 M_r^{\mathrm{A}}} - Q_j^{\mathrm{B}} \tag{20}$$

Taking the first $R+1$ order of hydroacoustic field mode, the first N order of the sandwich plate mode. $\boldsymbol{P}=[P_0, P_1, P_2, \ldots, P_R]^{\mathrm{T}}$, $\boldsymbol{B}=[B_1, B_2, \ldots, B_N]^{\mathrm{T}}$, matrix equation form of equation (20)

$$V\left(\ddot{\boldsymbol{P}} + \boldsymbol{\Omega}_{\mathrm{A}}^2 \boldsymbol{P}\right) = -\boldsymbol{L}\ddot{\boldsymbol{B}} \tag{21}$$

$$\boldsymbol{M}_{\mathrm{B}}\left(\ddot{\boldsymbol{B}} + \boldsymbol{\Omega}_{\mathrm{B}}^2 \boldsymbol{B}\right) = V\boldsymbol{L}^{\mathrm{T}}\boldsymbol{\Omega}_{\mathrm{A}}^{-2}\boldsymbol{M}_{\mathrm{A}}^{-1}\boldsymbol{P} + \boldsymbol{Q}_{\mathrm{B}} \tag{22}$$

Making $\boldsymbol{Q}_{\mathrm{B}}=0$, and assuming that it has vibration solution

$$\boldsymbol{P} = \frac{\boldsymbol{\Omega}_{\mathrm{A}}\boldsymbol{M}_{\mathrm{A}}^{1/2}}{\sqrt{V}}\chi_{\mathrm{A}}\mathrm{e}^{\mathrm{j}\omega t} \;,\; \boldsymbol{B} = \sqrt{V}\left(\boldsymbol{\Omega}_{\mathrm{B}}\boldsymbol{M}_{\mathrm{B}}^{1/2}\right)^{-1}\chi_{\mathrm{B}}\mathrm{e}^{\mathrm{j}\omega t} \tag{23}$$

Substituting the upper equation into the equation (21) and (22), the eigenvalue problem

$$\boldsymbol{A}\chi = \begin{bmatrix} \boldsymbol{A}_{11} & \boldsymbol{A}_{12} \\ \boldsymbol{A}_{21} & \boldsymbol{A}_{22} \end{bmatrix}\begin{bmatrix} \chi_{\mathrm{A}} \\ \chi_{\mathrm{B}} \end{bmatrix} = \omega^2\begin{bmatrix} \chi_{\mathrm{A}} \\ \chi_{\mathrm{B}} \end{bmatrix} = \omega^2\chi \tag{24}$$

$\boldsymbol{A}_{11}=\boldsymbol{\Omega}_{\mathrm{A}}^{2}+(\boldsymbol{\Omega}_{\mathrm{A}}\boldsymbol{M}_{\mathrm{A}}^{1/2})^{-1}\boldsymbol{L}\boldsymbol{M}_{\mathrm{B}}^{-1}\boldsymbol{L}^{\mathrm{T}}(\boldsymbol{\Omega}_{\mathrm{A}}\boldsymbol{M}_{\mathrm{A}}^{1/2})^{-1}$, $\boldsymbol{A}_{12}=\boldsymbol{A}_{21}^{\mathrm{T}}=-(\boldsymbol{\Omega}_{\mathrm{A}}\boldsymbol{M}_{\mathrm{A}}^{1/2})^{-1}\boldsymbol{L}(\boldsymbol{\Omega}_{\mathrm{B}}\boldsymbol{M}_{\mathrm{B}}^{1/2})^{-1}$, $\boldsymbol{A}_{22}=\boldsymbol{\Omega}_{\mathrm{B}}^{2}$

The coupling mode function of cavity and sandwich plate is respectively

$$F_k^{\mathrm{C}} = \sqrt{V}\boldsymbol{F}_{\mathrm{A}}^{\mathrm{T}}\left(\boldsymbol{\Omega}_{\mathrm{A}}\boldsymbol{M}_{\mathrm{A}}^{1/2}\right)^{-1}\chi_{\mathrm{A}}^{(k)} = \boldsymbol{F}_{\mathrm{A}}^{\mathrm{T}}\boldsymbol{\kappa}_{\mathrm{A}}\chi_{\mathrm{A}}^{(k)} \tag{25}$$

$$W_k^{\mathrm{C}} = \sqrt{V}\boldsymbol{W}_{\mathrm{B}}^{\mathrm{T}}\left(\boldsymbol{\Omega}_{\mathrm{B}}\boldsymbol{M}_{\mathrm{B}}^{1/2}\right)^{-1}\chi_{\mathrm{B}}^{(k)} = \boldsymbol{W}_{\mathrm{B}}^{\mathrm{T}}\boldsymbol{\kappa}_{\mathrm{B}}\chi_{\mathrm{B}}^{(k)} \tag{26}$$

Fluid-structure coupling vibration response (initial conditions of 0)

$$p(\sigma,t) = -\boldsymbol{F}_{\mathrm{A}}^{\mathrm{T}}\boldsymbol{\kappa}_{\mathrm{A}}\boldsymbol{X}_{\mathrm{A}}\boldsymbol{\alpha}(t) \tag{27}$$

$$w(x,y,t) = \boldsymbol{W}_{\mathrm{B}}^{\mathrm{T}}\boldsymbol{\kappa}_{\mathrm{B}}\boldsymbol{X}_{\mathrm{B}}\boldsymbol{\alpha}(t) \tag{28}$$

3. Numerical simulation and analysis

3.1. Modal frequency and influencing factors of fluid-solid coupled vibration

Table 1 is the modal frequency of the rectangular plates with four edges and the rectangular plates obtained by applying formula (16), and compared with the single layer board. Sandwich board of the upper and lower panels are 45 # steel, sandwich layer is viscoelastic damping materials (taking its density ratio 0.2 with panel material, the damping loss factor β=0.3, the dimensionless shear modulus g=1+jβ).

Table 1. The modal frequency of the rectangular plate with four edges and the comparison with the single layer plate

	Single sheet elastic sheet L_x=0.4m,L_y=0.6m,h=0.005m		Sandwich board 1 L_x=0.4m,L_y=0.6m h_1=0.0015m,$_2$=0.002m		Sandwich board 2 L_x=0.4m,L_y=0.6m,h_1=h_2=0.005m	
j	$(m_j^{(1)}, n_j^{(1)})$	$f_j^{(1)}$(Hz)	$(m_j^{(1)}, n_j^{(1)})$	$f_j^{(1)}$(Hz)	$(m_j^{(1)}, n_j^{(1)})$	$f_j^{(1)}$(Hz)
1	1,1	107.963	1,1	43.9552	1,1	138.018
2	1,2	207.622	1,2	73.5652	1,2	236.412
3	2,1	332.195	2,1	109.445	2,1	356.831
4	1,3	373.719	1,3	121.303	1,3	396.749
5	2,2	431.854	2,2	137.859	2,2	452.538

3.2. The spectral characteristics of sound pressure of underwater acoustic field

Assuming that the sandwich panel is used to load the uniform-changing mechanical load $p_E(\sigma, t)=P_E e^{j\omega t}, P_E=1$.Taking the excitation frequency of continuous change and the acoustic pressure of the cavity is drawn, water sound pressure spectrum are shown in figure 2 ~ 7.

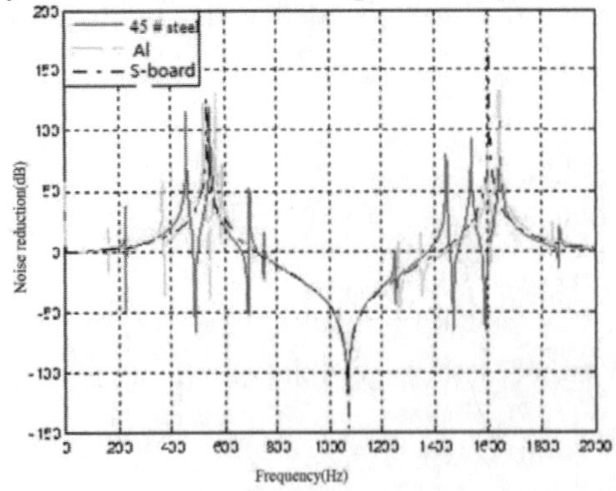

Figure 2. Comparison of internal water pressure in rectangular cavity of elastic wall of different materials

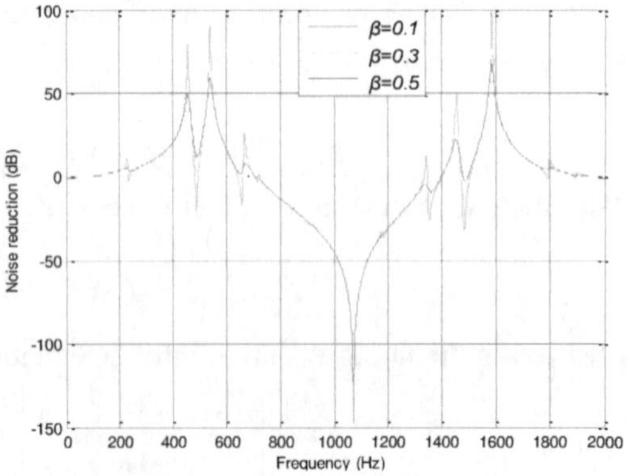

Figure 3. The influence of sandwich damping on the acoustic pressure of the internal water

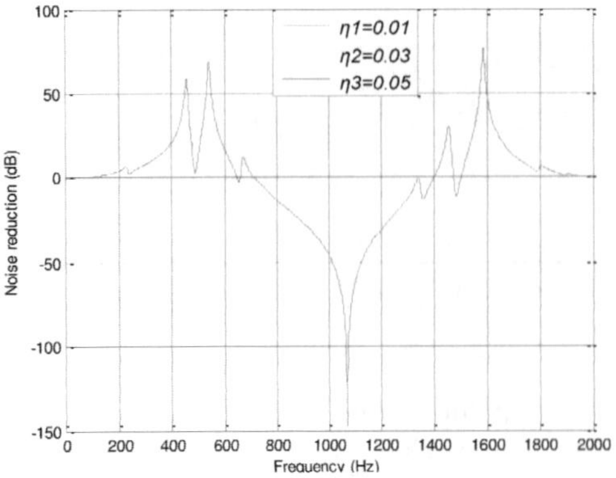

Figure 4. The effect of panel damping on noise reduction in cavity is studied

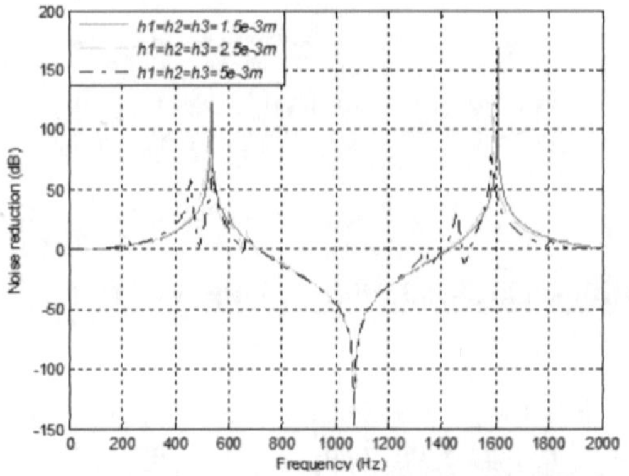

Figure 5. The influence of different plate thickness on noise reduction of cavity

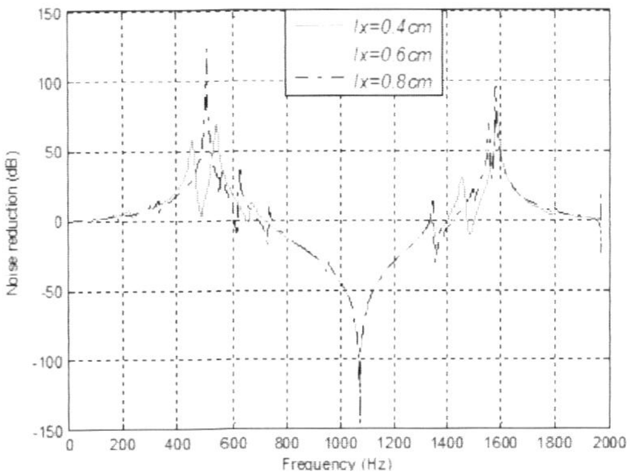

Figure 6. The influence of different *Lx* sides on the acoustic field of the cavity

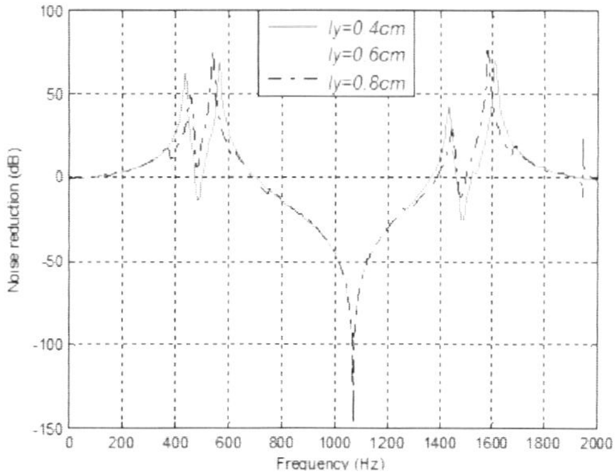

Figure 7. The influence of different *Ly* sides on the acoustic field of the cavity

Figure 2 compares the acoustic pressure spectrum of the rectangular cavity of three different elastic wall plates, sandwich plate, 45 # steel and aluminum alloy. The structure of sandwich plate greatly reduces the number of resonant peaks.In addition, the rectangular cavity of aluminum alloy wall plates has lower resonance frequency than 45 # steel wall rectangular cavity. Figure 3 and 4 show that the inner damping of the sandwich and panel is used only to weaken the resonance peak. Figure 5 shows that the thickness of sandwich plate is not affected by the thickness of the laminates. In detail, the resonant peaks of the thicker plates are more known, but the peaks of each resonance are reduced. Figure 6 changes the length of the sandwich plate when L_y=0.6m, L_z=0.7m. Figure 7 When L_x=0.4m, L_z=0.7m, the length of the sandwich panel is changed. The size change of sandwich plate will change the mode of vibration mode, and the coupling resonant frequency will change. The overall change trend is that the size of the plate decreases and the resonant frequency increases, which is conducive to the control of low frequency noise. In addition, it is worth pointing out that the number of resonant peaks that are stimulated in all square plates is significantly less than that of rectangular plates.

4. Conclusion

This paper presents a rectangular cavity structure composed of a sandwich panel by establishing a plate-filled rectangular cavity fluid-solid coupled vibration model and solving the cavity of the cavity. The coupling vibration response of the underwater acoustic field is derived. The results of MATLAB simulation show that the clamping plate of the same thickness is better than that of the rectangular laminates with the same thickness, and the noising effect of the rectangular sandwich panel of the square is better than that of the rectangle. When the plate is rectangular, as the size of the plate increases, its noise reduction effect is better. The damping of the sandwich plate and the loss factor of the sandwich layer also have an effect on the noise reduction of the plate. The damping and increasing loss factors can reduce the mutation of the internal noise in the vicinity of the resonance point.

References

[1] Huaiying W, Guojian J. *Acoustical and electronic engineering.* **1997 (02)** : 8-13.
[2] A Dowell E H, Voss H M. *AIAA journal.* **1963, 1(2)**: 476-477.
[3] Dowell E H, Gorman G F, Smith D A. *Journal of Sound and vibration.* **1977, 52(4)**: 519-542.More references
[4] Pan J, Hansen C H, Bies D A. *The Journal of the Acoustical Society of America.* **1990, 87(5)**: 2098-2108.
[5] Kurtze G, Watters B G. *The Journal of the Acoustical Society of America.* **1959, 31(6)**: 739-748.
[6] Zhenyu F. *Journal of xi 'an highway traffic university.* **1998, 18 (3)** : 59-63.
[7] Cupiał P, Nizioł J. *Journal of Sound and Vibration.* **1995, 183(1)**: 99-114.
[8] Narayanan S, Shanbhag R L. *Journal of Sound and Vibration.* **1981, 78(4)**: 453-473.

Development of engineering drawing ability for emerging engineering education

Jian-Wen Guo*, Xiao-Chang Cao, Li Xie, Jian-Jun Jin, Chu-Diao Wang

Dongguan University of Technology, Dongguan, GuangDong, 523982, People's Republic of China

*E-mail:given_gjw@163.com

Abstract: Students majoring in engineering is required by the emerging engineering education (3E) in the aspect of their ability of engineering drawing. This paper puts forward training mode of engineering drawing ability for 3E. This mode consists of three kinds of training including training in courses, training in competitions and training in actual demand. We also design the feasible implementation plan and supplies viable references to carry out the mode.

1. Introduction

The new round technological revolution and industrial transformation are accelerating and the competition in overall national strength is fiercer. The new booming economy which is characteristic of new technology, new condition and new industry asks engineering talents for higher ability of innovation and entrepreneurship and of transboundary integration to boost economic transformation and upgrading. Now, many universities are urgent to carry out emerging engineering education (3E) [1]. On the one hand, they need to establish and develop several new engineering majors. On the other hand, they must promote the existing engineering majors to reform and innovate and explore the application of more modern curriculum as well as training engineering talents with innovative ability.

Engineering drawing is a required skill in engineering majors, such as Mechanical Manufacturing, Industrial Design, Industrial Engineering, Environmental Engineering, Civil Engineering, Process Equipment, etc.. The rapid updating of manufacturing technology and the unending changes of manufacturing market demand put forward new requirements for students' drawing ability. In the new background of new engineering, the training of engineering drawing ability needs to be based on the needs of society, closely combined with the development of the market, cultivating the training of innovative engineering talents. Combined with the development needs of new engineering majors, we establish engineering drawing ability training mode. This mode creates an environment to study, research and innovate for students and provides a referenced scheme to train engineering drawing ability of engineering majors for the requirements of 3E.

2. Training mode of engineering drawing ability

Development of engineering drawing ability for 3E is a kind of teaching mode which relies on engineering drawing courses and uses innovative teaching measures to enable students to master the skills of engineering drawing and to improve the professional ability. This training mode is shown in figure 1.

Content from this work may be used under the terms of the Creative Commons Attribution 3.0 licence. Any further distribution of this work must maintain attribution to the author(s) and the title of the work, journal citation and DOI.
Published under licence by IOP Publishing Ltd

(1) Training in courses. When students have their courses Mechanical Drawing in their first year, we are going to improve the quality of course teaching through establishing the teaching system-Virtual Reality based on Mechanical Drawing system.

(2) Training in competitions. When students have the course Three-Dimensional Drawing Skills Training in their second year and third year, we are about to integrate competitions at home and abroad and professional skills certification requirements as well as theories into the creative education in engineering drawing. What's more, in order to promote the teaching effect, we will mobilize students to compete and obtain certificates.

(3) Training in actual demand. At the time of learning Three-Dimensional Software Integrated Training in students' fourth year, we will make full use of students' spare time to organize them to have practical training using the mode of Witkey. Therefore, students' ability of engineering drawing will develop under the motivation of daily and common tasks.

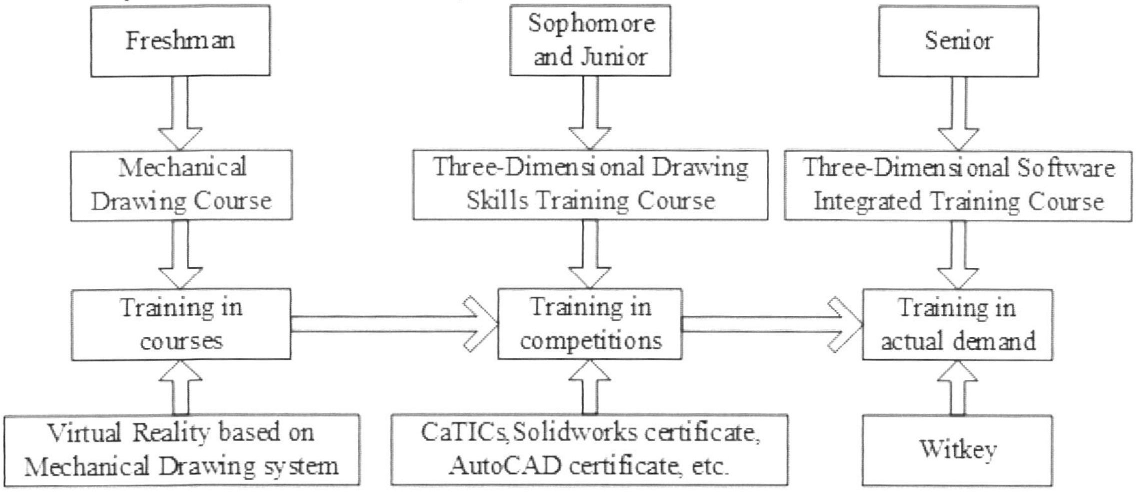

Figure 1. Training mode of engineering drawing ability

3. Implementation

3.1. Engineering drawing ability training in courses

The core of this kind of training is to enhance students' spatial imaginative ability and drawing capacity. The training adopts engineering drawing teaching system-Virtual Reality to assist students to study. The Virtual Reality based on mechanical drawing system is explained in our previous research [3]. The system is realized by VRML and Java. According to the teaching process of mechanical drawing course, this system provides Virtual Reality to support students to learn by themselves. Through this system, students can have interaction with relevant virtual object or event in the virtual environment and realize the perception of 3D Model. Figure 2 is the system interface.

3.2. Engineering drawing ability training in competition

This kind of training regards engineering drawing skills competitions as its carriers and encourages students to take part in various kinds of engineering drawing competitions and to study for their certificates. It combines knowledge of different subjects and skills to arouse students' interests and to offer a platform to them to study and improve themselves. In doing so, not only can students perfect their quality in competitions and practical activities, but also form a mechanism which depends on drawing skills competitions and obtaining certificates and other creative activities to advance students' creative skills.

In the course of Three-Dimensional Drawing Skills Cultivation, the training attaches importance to drawing standards and three-dimensional drawing skills training. In the extracurricular practice, teachers divide students into several after-school clubs so as to further foster students' ability by the form of training and lectures. We mobilize students to join in professional competitions as well as certificates. Figure 3 is students' training for competition.

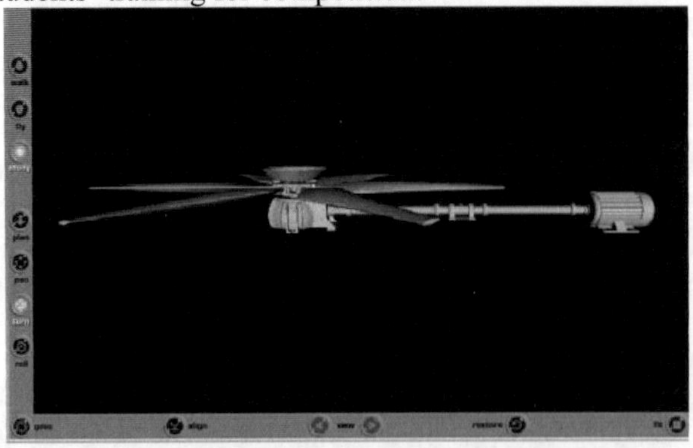

Figure2. Virtual Reality based on Mechanical Drawing teaching system

Figure 3. Students' training for competition

3.3. Engineering drawing ability training in actual demand

Engineering drawing ability training in actual demand will be conducted by the form of Witkey. Witkey is the person transforming knowledge and ability into actual income through internet [3]. This mode will be carried out in the course of Three-Dimensional Software Integrated Training in the fourth year. The Witkey platform is referred specifically in our previous research. [4] It is more similar to operation pattern of great enterprises. Students will establish virtual studio and participate in corresponding Witkey tasks according to their own hobbies, interests and professional expertise.

4. Conclusion

Under the background of new technology, students' ability of engineering drawing is required and therefore this article puts forward training mode of engineering drawing ability. This mode can provide feasible reference for implementation of engineering drawing training.

Acknowledgments

The study was supported by the Project of Department of Education of Guangdong Province (Project Name: Research of Witkey-Based Teaching Pattern of Innovative and Entrepreneurial Practice for Mechanical Specialty), the Project of Education of Dongguan University of Technology (Project Name: Mechanical Drawing Teaching Team), the Project of Education of Dongguan University of Technology (Project Name: Reform and Improvement of Teaching Quality of Mechanical Basic Courses Based on the Cultivation of Innovative Talents), the Project of Education of Dongguan University of Technology (Project Name: Design Thinking and Expression), the Dongguan Social

Science and Technology Development Project (NO.2013108101011, NO.2017507140058, NO.2017507140059), and Dongguan Industrial Science and Technology Development Project (NO.2015222119).

References

[1] Saskia H and Henk S 2012 *IEEE Engineering Management Review* **40** 164-74
[2] Guo J W, Xu C, Jing J J, Chen H B and Wang S 2014 *WIT Transactions on Engineering Sciences* **97** 1461-6
[3] Zhang L 2015 *5th Int. Conf. on Humanities and Social Science Research(Rome)* p178
[4] Guo J W, Cao X C, Huang H Y, Xie L, Sun Z Z and Wu G H, 2017 *2nd Int. Conf. On Electrical And Electronics: Techniques and Applications (Beijing)* p237

IOP Conf. Series: Materials Science and Engineering **242** (2017) 012071 doi:10.1088/1757-899X/242/1/012071

Application of dynamic milling in stainless steel processing

Wenju Shan[1]

[1] Zhuhai College of Jilin University, Zhuhai, Guangdong Province, China

E-mail: 36486846@qq.com

Abstract. This paper mainly introduces the method of parameter setting for NC programming of stainless steel parts by dynamic milling. Stainless steel is of high plasticity and toughness, serious hard working, large cutting force, high temperature in cutting area and easy wear of tool. It is difficult to process material. Dynamic motion technology is the newest NC programming technology of Mastercam software. It is an advanced machining idea. The tool path generated by the dynamic motion technology is more smooth, more efficient and more stable in the machining process. Dynamic motion technology is very suitable for cutting hard machining materials.

1. Cutting characteristics of stainless steel

Stainless steel generally refers to steel that resists atmospheric corrosion and resists corrosion in chemical mediator. Stainless steel is divided into martensitic stainless steel, ferritic stainless steel and austenitic stainless steel. Compared with medium carbon steel, stainless steel is hard-to-cut material, and its relative machinability is about 0.3 - 0.5. [1]

2. About dynamic motion technology

Mastercam is a commonly used CAD/CAM integration software. Consecutive years ranked first in the global number of similar software installed. dynamic motion technology is the newest NC programming technology of Mastercam software. It is different from traditional processing, and it is a breakthrough advanced processing idea.

A smooth tool path can improve machining speed more than any machine tool. With less retract and better tool contact control more billets can be safely removed in less time. Dynamic motion technology provides accurate cutting patterns and enables accurate machining of workpieces. Almost all machine tools can experience the acceleration of the speed and efficiency of the dynamic toolpath.

Dynamic milling uses cutting tool edges to cut. Instead of simply calculating the moving path of the tool, a series of algorithms are used to analyze the process of tool cutting and material removal. Adjust the cutting according to the movement of the cutter in the process, with stable material removal rates for rapid, smooth cutting. [2]

Dynamic milling with smooth tool path can uses the entire side edge of the milling cutter to cut, not only shorten the processing time of the workpiece, but also avoid unnecessary sudden swerve in the movement of the spindle of the machine tool. The cutting method can ensure the uniform fracture and stable discharge of the chip. A smooth chip removal process can quickly remove heat from the process and prevent heat from accumulating in the machined area. The surface hardening process of the material is suppressed and the tool wear during processing is reduced, Therefore, reduce the difficulty of processing hard cutting materials.

Content from this work may be used under the terms of the Creative Commons Attribution 3.0 licence. Any further distribution of this work must maintain attribution to the author(s) and the title of the work, journal citation and DOI.
Published under licence by IOP Publishing Ltd

IOP Conf. Series: Materials Science and Engineering **242** (2017) 012071 doi:10.1088/1757-899X/242/1/012071

3. 3D dynamic milling parameter setting

CNC programming of the workpiece using Mastercam2017, The Part drawing is shown in Figure 1

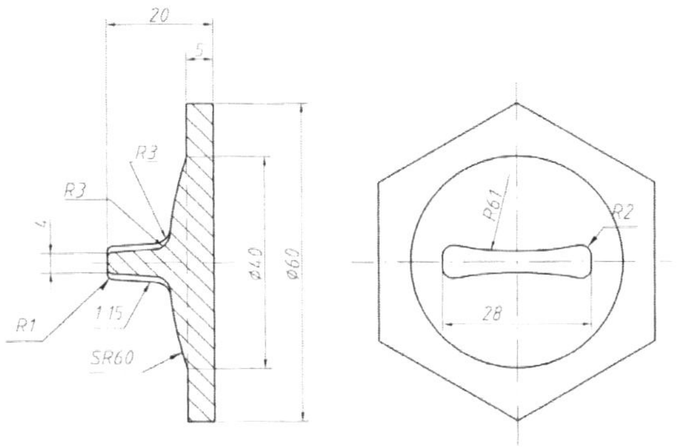

Figure 1 The part drawing

The workpiece is made of stainless steel, which is difficult to process. it has the advantages of simple structure, low surface roughness and low tolerance grade, but its material removal rate is relatively large. Milling efficiency can be improved by using 3D dynamic milling. The following is a brief introduction to the parameter setting process of dynamic milling.

3.1 Workpiece blank setting

The blank is the state before the workpiece is machined. It should be set according to the actual processing. The correct blank setting can guarantee the correctness of the actual simulation before the actual processing. After drawing the solid model according to the drawing, ensure the center of the upper surface part of the workpiece is located at the origin of the system by drawing the boundary box. Then enter the milling Modular, Sets the blank size in the machine property group.

3.2 Tool parameter setting

Because the stainless steel material belongs to the difficult to processing material, therefore selects the good cutting performance carbide cutting tools to carry on the processing.The process of the workpiece is subjected to rough machining and finishing operation. At the beginning of the rough, you can use a larger diameter tool to improve the material removal rate. A flat end mill with a diameter of 16mm is used here. In finishing, in order to guarantee the processing of R3mm fillet, the ball end mill with 6mm diameter is selected. [3]

After the tool size is set, the machining parameters of the tool must be set. The main machining parameter settings include the choice of cutting speed and feedrate per blade. Then the spindle speed and feed speed are calculated automatically.

Reference NC machining manual, The cutting speed range of carbide milling cutters for milling of stainless steel is.90-100m/min, The range of feedrate per blade is0.007~0.009*dmm/z. In accordance with the above range, select the cutting parameters and input to the tool parameter settings dialog box, automatically calculate the spindle speed and feed speed. [4]

3.3 Dynamic optirough parameter setting

The goal of roughing is to remove as much material as possible from the workpiece in the shortest possible time. Select the dynamic optirough command icon, select the whole entity as the drive surfaces, and draw a rectangle of 61*61mm size as the containment boundary. Select the 16mm flat end mill as the tool used for roughing.Then set the cut parameter. The cut parameter setting is shown in figure 2.

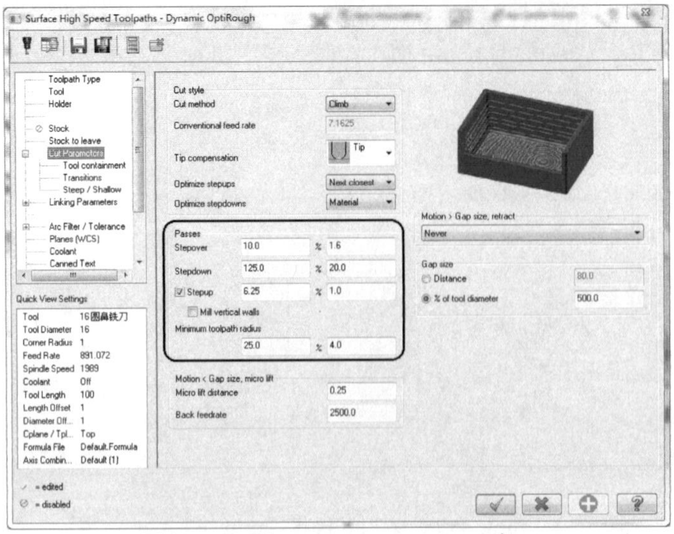

Figure 2 The cut parameter setting

The key of parameter setting is the reasonable selection of step size and the setting of minimum toolpath radius. There are many factors that affect the choice of step size, and it is difficult to choose with a uniform standard, It can only be selected based on some principles or recommendations, And then revise it according to the specific situation and experience.

The step size includes the settings of cutting depth and the side cutting depth. It determines the processing efficiency, However, excessive step size will lead to an increase in cutting resistance. The fracture of the cutting edge may occur during machining. [5] Generally choose the larger cutting, depth, and then select the corresponding side cutting depth according to different materials. For the flat end milling cutter, the cutting depth generally should not exceed 1.5 times of the diameter of the tool. This workpiece most deep processing position is 20mm, so the cutting depth can be set to 20. Input 20 to the stepdown dialog box Reference NC machining manual, The side cutting depth of stainless steel materials can be set to 10%-15% of the tool diameter. input 10% to the stepover dialog box. In dynamic optirough, you also need to set up an important parameter, "Stepup." This parameter is mainly to set the Z layered of milling. This value determines the surface quality of the workpiece after roughing. The smaller of the value setting, the more uniform the rough machining allowance. input 0.5 to the stepup dialog box.

minimum toolpath radius determines the degree of smoothness of the toopath. The bigger of the radius setting , the smoother of the tool path , the more stable of the machining process. However, excessive minimum tool radius may result in increased machining allowance at corners. input 25% to the minimum toolpath radius dialog box. Then set stock to leave, input 0.3 to the leave stock on drive walls dialog box, input 0.5to the leave stock on the drive floors dialog box.

Then set the linking parameters. In order to reduce the time to raise the tool, set the lifting operation to"Minimum Vertical Retract". In the rough machining operation, the total tolerance value can be set to a larger value to improve the software operation speed. it can be set to 20-30% of the allowance. Set the total tolerance to 0.06 on the Arc Filter/Tolerance tab, Turn on the "Line/Arc Filtering Setting" option. Create toolpath after setting is complete. Rough milling theory processing time is about 6 minutes.

3.4 Dynamic finishing parameter setting

The aim of finishing is to guarantee the geometrical dimension and surface precision of the parts. Sometimes, due to the larger diameter of roughing tools or rough machining parameters, the allowance values in some areas are obvious larger than the parameter settings. In order to ensure finish machining smoothly, the machining allowance must be uniform. Remaining operations are usually required. Copy and paste previous rough machining operations. Select the ball end mill with 6mm diameter. Check the "Rest material" option. Set the "Compute remaining stock option from" to "Toolpath group only". input 20% to the "stepover" dialog box. input 10% to the "stepup" dialog box. Limit the processing depth from -9 to -14.7, Remaining is applied to the flat region of the workpiece. Remaining theory processing time is about 2 minutes. [6]

In the surface finishing setting, the machining parameters of the curved surface in the flat region and the steep region need to be set separately. The necessary clearance operations are required in the small fillet area. Select "hybrid" command icon. then select all surfaces except the plane region as the machining surface, and select the other surfaces as interference surfaces. input 0 to the leave stock on drive walls dialog box, input 0 to the leave stock on the drive floors dialog box. input 0.1 to the check walls and fioors dialog box.Then input 0.1 to the "3D stepover" dialog box. Other parameters follow default, Generate finishing operations toolpath. Finally, complete the programming of the horizontal and vertical planes with flat end mill. Finish milling theory processing time is about 8 minutes. The machining simulation results are shown in Figure 3

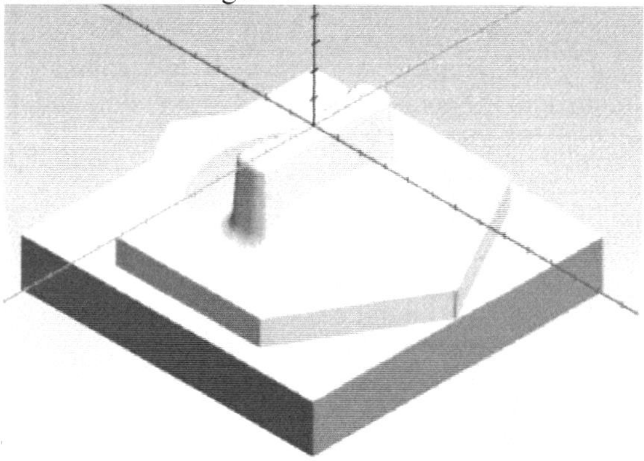

Figure 3 The machining simulation results

4. Conclusion

Through the examples above, we can get a conclusion, that is, the dynamic milling motion of Mastercam is a safe, efficient programming. The dynamic milling programming is a more reliable method compared to the traditional milling methods, especially suit the stainless steel NC milling.

Acknowledgement

Premier-Discipline Enhancement Scheme Supported by Zhuhai Government Fund. (Fund code: 2015YXXK14)

References

[1] DH.Zhang, NC machining Handbook, Chemical industry press,2013, vol. 2, pp. 364–365.
[2] Q.Guo, Research on the methods and technologies for high performance peripheral machining of complex surface, Dalian University of Technology 2013, pp. 1–5.
[3] JX.Pan,Application of Mastercam software in high speed milling,NC Technology ,2006,vol. 1, pp. 43–46.
[4] Wg.Chen, Hao.Chen, NC machining tool, material structure and selection Handbook,China machine press, 2016, pp. 230–232.
[5] LD. Kang, FZ.Yang,CM.Tong, Research status and analysis of key technologies for high speed machining of aluminum alloy, Modern Manufacturing Engineering. 1(2014)83-86.
[6] YB.Wang,A side edge milling method based on Mastercam programming,Die manufacturing technology vol. 7, pp. 74–76,2015.

IOP Conf. Series: Materials Science and Engineering **242** (2017) 012072 doi:10.1088/1757-899X/242/1/012072

Creep behavior of Grade 91 steel under uniaxial and multiaxial state of stress

Facai Ren [1]*, Xiaoying Tang [1]

[1] Shanghai Institute of Special Equipment Inspection and Technical Research, No. 915 Jinshajiang Road, Shanghai, 200062, PR China

corresponding author : caifaren@163.com

Abstract. Creep rupture behavior of Grade 91 heat-resistant steel used for steam cooler under uniaxial and multiaxial state of stress was investigated. Creep tests were conducted at the temperature of 923K under the stress 125MPa. The notch root radii (r) of doubled circumferentially U-notched specimens were 0.6 and 6 mm. The creep rupture life of Grade 91 steel was found to increase with the increasing of notch acuity ratio. The creep rupture mechanism was investigated based on the SEM fractography analysis.

1. Introduction

In comparison with creep resistant austenitic steels, Grade 91 steels have been considered as superior material due to lower coefficient of thermal expansion and high thermal conductivity [1]. Grade 91 steel contains alloying elements like Nb and V, forming fine and stable carbides and/or carbonitrides, such as $M_{23}C_6$ and MX [2]. It is used for the steam and gas turbines or boiler at elevated temperatures due to the good creep strength.

In the past years, the creep rupture behavior under uniaxial state of stress of Grade 91 steel has been studied by many researchers [3-6]. In practical application, the components may be subjected to multiaxial state of stress due to mode of loading or from sharp changing in geometry which causes local stress concentration. So it's necessary to investigate the creep deformation behavior of Grade 91 steel under multiaxial state of stress in order to accurately predict the life of such components. Goyal et al. [7] investigated the effect of multiaxial state of stress on creep rupture behavior of 2.25Cr-1Mo steel using two different circumferential U-notches in the uniaxial creep specimen. The results show that the creep rupture life of the material increases in presence of the notch with the consequent decrease in ductility. Goyal et al. [8] investigated the creep rupture life of 9Cr-1Mo steel under multiaxial state of stress at 873K over the net applied stresses in the range of 110-210MPa. The creep rupture life was found to be higher in the presence of notch than that of smooth specimen, which indicates notch strengthening behavior of this steel. Huang et al. [9] studied the effect of multiaxial stress state on creep behavior and stress rupture life of a Ni-based directionally solidified superalloy. Chang et al. [10] studied the notch strengthening effect of P92 steel using both plain and notched specimens at 650°C over a stress range of 120-185MPa. The influence of the multiaxial stress states on creep rupture behavior of P92 steel was investigated by the relationship between the fracture ductility and multiaxiality.

The effects of notch on creep behavior of Grade 91 steel used for steam cooler were investigated in this paper. The multiaxial state of stress at 923K and 125MPa has been studied by the introduction of different notch acuities.

Content from this work may be used under the terms of the Creative Commons Attribution 3.0 licence. Any further distribution of this work must maintain attribution to the author(s) and the title of the work, journal citation and DOI.
Published under licence by IOP Publishing Ltd

2. Material and experimental

The studied hot rolled Grade 91 steel was obtained in normalized and tempered condition. The chemical composition (wt.%) of the Grade 91 steel is given in Table 1. As shown in figure 1, the creep tests specimens were machined from the Grade 91 plate with a diameter of 10mm and gauge length of 100mm. Uniaxial and multaxial creep tests were conducted at 923K and 125MPa by GWT2504 high temperature creep and stress-rupture testing machines. The doubled circumferentially U-notched specimens having notch root radii (r) of 0.6 and 6 mm and notch throat diameter (d) of 6mm as that of smooth specimens. These notch geometries resulted in notch acuity ratio (d/r) of 1 and 10. The creep experiments refer to GB/T2039-2002 Metallic material-creep and stress-rupture test in tension.

Table 1 Chemical composition of Grade 91 steel (wt.%).

Element	C	Mn	Si	Ni	Cr	Mo	Cu	Al	S	P
Amount (wt.%)	0.10	0.40	0.23	0.13	8.34	0.98	0.06	0.009	0.002	0.0101
Element	Sn	V	Nb	N	Ti	Zr	N/Al			
Amount (wt.%)	0.005	0.229	0.079	0.044	0.002	0.001	4.9			

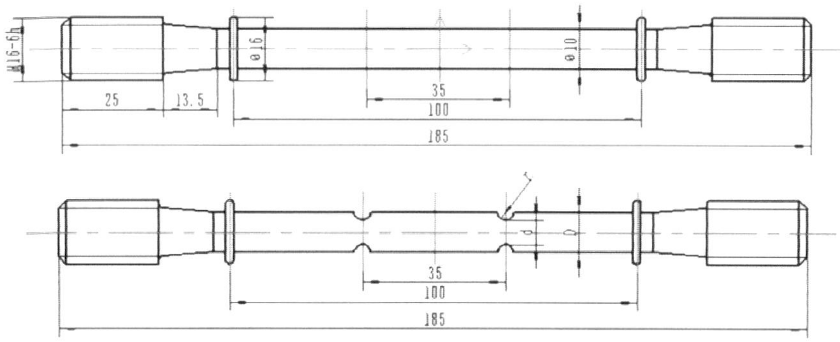

Figure 1. Diagram of uniaxial and multiaxial creep test specimen.

3. Results and Discussions

Uniaxial and multiaxial creep strain-time curves of Grade 91 steel at 923K and 125MPa are shown in figure 2. It can be seen that the curves of plain specimen and the notched specimen (r=6mm) exhibit typical primary, steady-state and tertiary creep stages. The notched specimen (r=0.6mm) has no obvious tertiary creep stage, which means that the creep rupture mechanism is different from the others. It also can be seen that the creep rupture life was found to increase with the increasing of notch acuity ratio. It is generally known that high ductile materials show notch strengthening behavior when notches are present in the components.

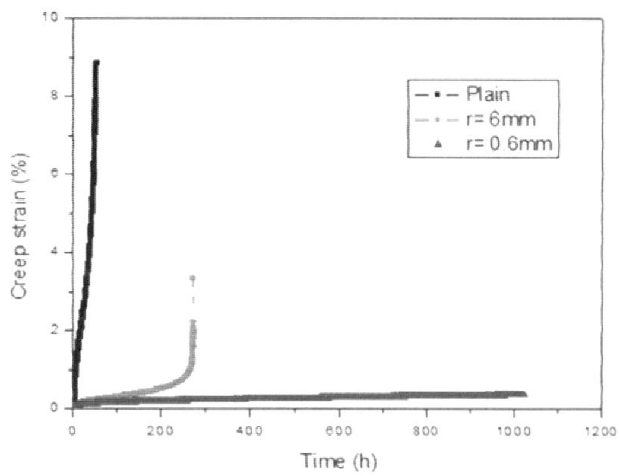

Figure 2. Creep strain-time plots for Grade 91 steel.

The global fractography of plain and notched specimens (d/r=10) conducted at 923K and 125MPa is shown in figure 3. The typical cup and cone type appearance of transgranular ductile fracture of plain specimen is observed as shown in figure 3(a), which means that Grade 91 steel possesses fine plasticity under this test condition. It can be seen from figure 3(b) that the reduction of area decreases dramatically when the notch acuity ratio (d/r) is 10. It implies that the ductility of Grade 91 steel gets worse due to the multiaxial state of stress. Fracture appearance of the creep rupture specimens was found to be dependent on the notch geometry.

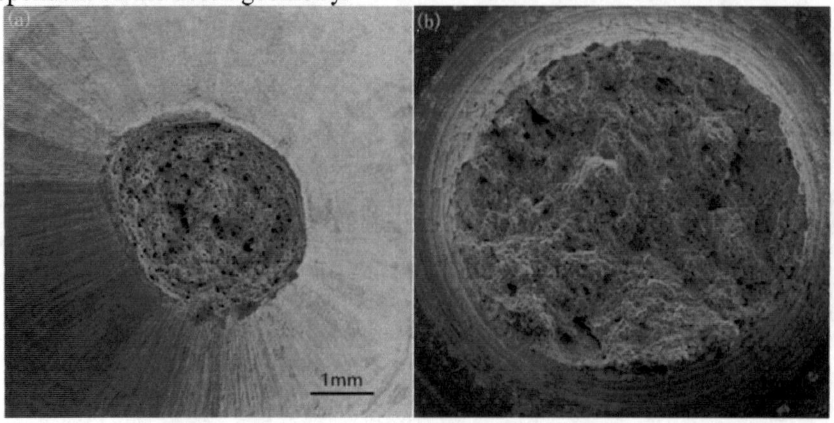

Figure 3. Global fracture appearance (a) plain specimen, (b) notch acuity ratio (d/r) =10.

The detailed SEM micrographs of center and root area of notched specimen (d/r=10) are shown in figure 4. It can be seen that more shallow dimples exist at the center (figure 4(a)), while rarely dimples are found at the root of the notched specimens (d/r=10) (figure 4(b)). It implies that the ductile fracture and intergranular brittle fracture co-exist under this test condition during creep. Comparing with the plain specimen, the ductility of Grade 91 steel declines due to the presence of notches.

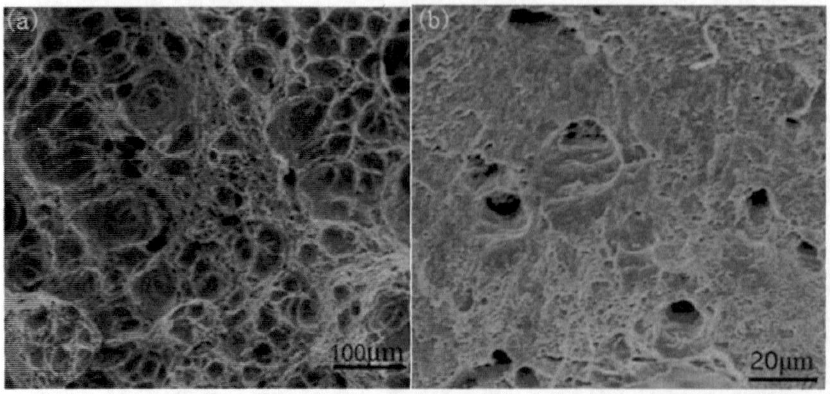

Figure 4. SEM fractography of notched specimen (d/r=10): (a) center area, (b) root area.

4. Conclusions

Creep rupture behavior of Grade 91 steel under uniaxial and multiaxial stresses at 923K and 125MPa was investigated. Grade 91 steel showed notch strengthening at both notch acuities. Creep rupture life was found to increase with the increasing of notch acuity ratio.

Acknowledgement

The authors gratefully acknowledge the financial supports by Shanghai Municipal Bureau of Quality and Technical Supervision (No. 2016-40).

IOP Conf. Series: Materials Science and Engineering **242** (2017) 012072 doi:10.1088/1757-899X/242/1/012072

References

[1] P.J. Ennis, A. Czyrska-Filemonowicz, Recent advances in creep-resistant steels for power plant applications, Sadhana 28(2003)709-730.

[2] Fujio Abe, Precipitate design for creep strengthening of 9% Cr tempered martensitic steel for ultra-supercritical power plants, Sci. Technol. Adv. Mater. 9(2008)1-15.

[3] P. Anderson, T. Bellgardt, F. L. Jones, Creep deformation in a modified 9Cr-1Mo steel, Mater. Sci. Technol. 19(2003)207-213.

[4] H. Ghassemo-Armaki, R.P. Chen, K. Mauyama, M. Igarashi, Contribution of recovery mechanisms of microstructure during long-term creep of Gr.91 steels, J. Nucl. Mater. 433(2013)23-29.

[5] R.P. Chen, H. Ghassemi Armaki, K. Maruyama, M. Igarashi, Long-term microstructural degradation and creep strength in Gr.91 steel, Mater. Sci. Eng. A. 528(2011)4390-4694.

[6] T. Shrestha, M. Basirat, I. Charit, G.P. Potirniche, K.K. Rink, U. Shaym, Creep deformation mechanisms in modified 9Cr-1Mo steel, J. Nucl. Mater. 423(2012)110-119.

[7] Sunil Goyal, K. Laha, C.R. Das, S. Panneer Selvi, M.D. Mathew, Finite element analysis of uniaxial and multiaxial state of stress on creep rupture behavior of 2.25Cr-1Mo steel, Mater. Sci. Eng. A. 563(2013)68-77.

[8] Sunil Goyal, K. Laha, Creep life prediction of 9Cr-1Mo steel under multiaxial state of stress, Mater. Sci. Eng. A. 615(2014)348-360.

[9] J. Huang, D.Q. Shi, X.G. Yang, X.A. Hu, Effect of multi-axial stress state on creep behavior and stress rupture life of a Ni-based DS superalloy, Comp. Mater. Sci. 85(2014)20-31.

[10] Y. Chang, H. Xu, Y.Z. Ni, X. Lan, H.Y. Li, Research on representative stress and fracture ductility of P92 steel under multiaxial creep, Eng. Fail. Anal. 59(2016)140-150.

IOP Conf. Series: Materials Science and Engineering **242** (2017) 012073 doi:10.1088/1757-899X/242/1/012073

Design of 120t converter furnace body

Xiang Li and Biao Tang*

School of Materials Science and Engineering, Jiamusi University, Jiamusi 154007

Corresponding author: tanglaobiao@163.com

Abstract. At present, the oxygen converter steelmaking equipment of large-scale, production of continuous and high-speed, which has achieved a high productivity, and this requires matching equipment to complete together. The process about the layout of these equipment and the workshop of various materials must be reasonable in order to enable the production to proceed smoothly. The furnace body for 120t converter is designed in the paper. The results are as follows. The converter raw data has been used, the experimental step of the design has been listed to determine the data required for the furnace size design, including the steel yield, scrap ratio, furnace type, etc. The size of the data has been obtained by the calculation, and, check the results to determine the design of the furnace for the specific data according to the empirical formula of the size of the calibration, the data can be used for practical production. Through the rigorous calculation of the design, the converter furnace capacity ratio, aspect ratio, pool diameter and depth of experimental data have been determine. The results show that the design is reasonable. The design data of the converter and the corresponding furnace can meet the equipment of the effective volume of 120t converter design. That can ensure the smooth production process of the converter, while achieve the requirements of lining longevity.

1. Introduction

Starting from the industrial age, the iron and steel has been the main application of structural materials, which is also the largest and the most common functional materials production in human society, and it plays an important role in the economic development, known as "the backbone of industry". Although in recent years the iron and steel is faced with ceramic materials, polymer materials, non- ferrous metal materials (such as aluminum) in the competition, but it has some significant advantages, such as ore reserves, the cost of production, recycling utilization, and comprehensive performance. In the foreseeable future, compared with other kinds of materials iron and steel in the industrial production has the superiority and important position still won't change [1-5].

Bessemer converter steelmaking process with a bottom blowing on appeared Britain in 1856. French Martin used heat storage principle, and founded that pure oxygen top-blown converter in Austria in 1952. In the mid and late 1980s, Western Europe, Japan and the United States developed the process of remaking and remaking of oxygen converter. The advantage is good slag and the amount of steel used in converter steelmaking in the bottom and back converter steelmaking, at the same time, it also has the advantage of a mixture of oxygen bottom blown converter steelmaking melting pool, iron and manganese oxidation loss, and less metal spraying, etc. Therefore, most of the world's large-capacity converters are used in the process of steelmaking [6-15].

Content from this work may be used under the terms of the Creative Commons Attribution 3.0 licence. Any further distribution of this work must maintain attribution to the author(s) and the title of the work, journal citation and DOI.
Published under licence by IOP Publishing Ltd

IOP Conf. Series: Materials Science and Engineering **242** (2017) 012073 doi:10.1088/1757-899X/242/1/012073

2. Parameters and dimensions of converter

2.1 Determine the original parameters

(1) The nominal capacity (T) is the name of the size of the converter, the tonnage of converter. It is the main symbol of the converter production ability and the important basis of furnace design.

(2) The average amount of steel in the furnace is 120t, the rate of steel is 90%, the maximum scrap is 10%, and the quantity of oxygen is 51.65 (m^3/t), which is cooled by scrap steel.

(3) Volume ratio (V/T): the ratio of the effective volume (V) to the nominal capacity (T) of the furnace in the new furnace（m^3/T）. In domestic 100-200t converter capacity ratio between 0.90-1.0. At the same time, according to the stove, the greater the furnace capacity and the smaller the experience rule, in the design of the furnace capacity of 0.92m^3/t is used

(4) Cap Angle (θ) referred to the angle between the furnace cap and the furnace body, the furnace cap and the horizontal line of the furnace. The recommended value of Θ = 60-68°, the big stove takes the lower limit, the small furnace takes the upper limit. According to domestic production experience, the number is 64°。

2.2 The size of the converter is determined

(1) Molten pool diameter D

The most common formula is recommended by the Beijing steel design institute：

$$D = K\sqrt{\frac{G}{t}}$$

Type: D-melt diameter, m; G-new furnace metal load, T; t-oxygen blowing time, min; K-proportional coefficient, K of 120t converter is 1.79.

According to the formula：

$$G = \eta 2T / (2+B) = 121\text{t}$$

Oxygen blowing time (t) =2.83m^3/ (t·min)

So D_{molten} =4.64m

(2) The depth of molten pool h

The calculation formula of depth h of molten pool is:

$$V_{\text{molten}}=0.79D^2\text{h}-0.046D^3$$

Determine the diameter of the molten pool is 4.64 m, h=1.32m

Check: the depth diameter ratio of molten pool (h/D) is 0.23-0.54. This is h/D =1.32/4.64=0.284,which meets the requirements.

Determination of furnace cap size

Glory hole diameter d_0=0.51D=0.51×5.393=2.37m;

Furnace cap height H_{cap}： $H_{\text{cap}}=H_{\text{mouth}}+H_{\text{cone}}$;

The design of the furnace mouth is 400mm.

$$H_{\text{core}} = 1/2(D_0 - d_0)\tan\theta = 1/2(4.64-2.37)\tan 64 = 2.33\text{m};$$

$$H_{\text{cap}} = H_{\text{mouth}} + H_{\text{core}} = 2.33 + 0.4 = 2.73$$

The furnace volume cap V_{cap}:

$$V_{\text{cap}}=\pi/12H_{\text{core}}(D^2+Dd_{02}+\pi/4d_{02})+d_{02}H_{mouth} = \pi/12*2.33(4.642+4.64*2.37+2.372)+\pi/4*2.372*0.4=25.03\text{m}^3$$

Determination of body size

According to the selected furnace ratio of 0.92, the total capacity of the furnace has been calculated.

$$V_{\text{total}}=0.92\times120=110.4\text{m}^3$$

As V_{total} was known, the body product of the converter was calculated, thus the height of the furnace has been obtained.

$$H_{\text{body}} = 4V_{\text{body}} / \pi D^2 = 4\text{m}$$

3. Furnace lining and furnace shell

3.1 Furnace lining

The design of 120t converter furnace was completed by referring to relevant literature on the design of converter furnace. This design aims at the design of 120t converter furnace body, and follows the design idea of modern furnace with practical, advanced, high quality, low consumption, long life and environmental protection. Through this design, the conclusion is as follows:

(1) The development status and furnace structure of the converter are studied. The original conditions include: the nominal capacity *T* takes 120t, the rate of steel is 90%, the maximum scrap ratio is 10%,and the oxygen time is 18min.

(2) According to the previous experience and the original conditions of the person, of the tube ball type furnace type was selected, and the ratio of the furnace volume of 0.92 m^3/t was determined, and the high ratio of the check height of 1.59m. The diameter of the molten pool D = 4.64 m was determined, the depth of the molten pool is 1.32 m, the furnace cap h_{cap} =2.73 m, and the furnace body h_{body} = 4m.

(3) The high quality furnace lining and furnace shell is selected to ensure the normal working of the converter, and maintain the service life of the converter and the safety of the staff regularly.

The design determines the parameters of the 120t converter furnace, and the corresponding furnace structure is obtained. The design can meet the design requirements of the effective volume for the operation of the 120t converter.

Converter lining life reflects an enterprise's management and technical level, and directly affects the productivity of the converter, which can not be ignored in the design of furnace lining life, and it is also a type of a problem. To improve the life of lining, from the technical aspect, it is to insist on reasonable operating system, from the design point of view:

(1) The high quality refractory for furnace lining should be chose;

(2) The best lining thickness should be selected.

3.2 The furnace shell

The furnace shell: the converter shell is similar to a container bearing high temperature and high pressure. It has the following forces in its use:

(1) Stress caused by static load;

(2) The stress generated by the dynamic load;

(3) The expansion stress of the lining;

(4) The thermal stress.

According to the use characteristics of the shell, the material of the furnace shell is required to withstand the time of high temperature resistance, creep resistance and good forming and welding property. Generally small converter with carbon structural steel, large converter with low structural steel, such as Q235, 16Mn, 15MnTi, 14MnNb, etc.

4. Conclusion

The design of 120t converter furnace was completed by referring to relevant literature on the design of converter furnace. This design aims at the design of 120t converter furnace body, and follows the design idea of modern furnace with practical, advanced, high quality, low consumption, long life and environmental protection. Through this design, the conclusion is as follows.

(1) The development status and furnace structure of the converter are studied. The original conditions include: the nominal capacity takes 120t, the rate of steel is 90%, the maximum scrap ratio is 10%, and the oxygen time is 18min.

(2) According to the previous experience and the original conditions of the person, the selection of the tube ball type furnace type, determined the ratio of the furnace volume to 0.92 m^3/t, and the high ratio of the check height to 1.59m. It is calculated to determine the diameter of the molten pool D = 4.64 m, the depth of the molten pool h = 1.32 m, the furnace h_{cap} = 2.73 m, and the furnace body h_{body}

= 4m.

(3) Select the high quality furnace lining and furnace shell to ensure the normal working of the converter, and maintain the service life of the converter and the safety of the staff regularly.

The design determines the parameters of the 120t converter furnace, and the corresponding furnace structure is obtained. The design can meet the design requirements of the effective volume for the operation of the 120t converter.

Acknowledgments
Authors wishing to acknowledge financial support from Science and Technology Innovation Projects of College Students of Jiamusi University (xslm2016-05).

References

[1]Omran M, Fabritius T. Effect of steelmaking dust characteristics on suitable recycling process determining: Ferrochrome converter (CRC) and electric arc furnace (EAF) dusts [J]. Powder Technology, 2016, 308 (2017): 47-60.
[2]Bodor M, Santos R M, Cristea G, et al. Laboratory investigation of carbonated BOF slag used as partial replacement of natural aggregate in cement mortars[J]. Cement & Concrete Composites, 2016, 65:55-66.
[3]Brezny B. The Microstructures and Properties of Magnesia-Carbon Refractories [J]. Key Engineering Materials, 1993, 88: 21-40.
[4]Cunha A P, Pacianotto T A, Fileti A M F. Steelmaking process: Neural models improve end-point predictions [J].Computer Aided Chemical Engineering, 2004, 18(04):631-636.
[5]Omran M,Fabritius T.Effect of steelmaking dust characteristics on suitable recycling process determining: Ferrochrome converter (CRC) and electric arc furnace (EAF) dusts[J].Powder Technology,2016,308 (2017):47-60.
[6]Bodor M, Santos R M, Cristea G, et al. Laboratory investigation of carbonated BOF slag used as partial replacement of natural aggregate in cement mortars[J].Cement & Concrete Composites,2016,65:55-66.
[7]Brezny B. The Microstructures and Properties of Magnesia-Carbon Refractories [J]. Key Engineering Materials, 1993, 88: 21-40.
[8]Cunha A P, Pacianotto T A, Fileti A M F. Steelmaking process: Neural models improve end-point predictions[J]. Computer Aided Chemical Engineering, 2004, 18(04):631-636.
[9]Almeida S M,Lage J, Fernández B, et al. Chemical characterization of atmospheric particles and source apportionment in the vicinity of a steelmaking industry[J]. Science of the Total Environment,2015, 521–522:411-420.
[10] Hleis D, Fernándezolmo I, Ledoux F,et al. Chemical Profile Identification of Fugitive and Confined Particle Emissions from an Integrated Iron and Steelmaking Plant[J]. Journal of Hazardous Materials, 2013, 250–251(2):246-255.
[11] Cullen J M, Allwood J M, Bambach M D. Mapping the Global Flow of Steel: from Steelmaking to end-use goods[J]. Environmental Science & Technology, 2012, 46(24): 48-55.
[12] Singh R K, Murty H R, Gupta S K, et al. Development of Composite Sustainability Performance Index for Steel Industry[J]. Ecological Indicators, 2007, 7(3): 565-588.
[13] Draney C T. Converter Furnace,US2288272[P]. 1942. US2288272.
[14] Jiang H. Design and Practice of Converter Lining Protection Technology by Slag Splashing [J].Steelmaking, 2000.
[15] Wang L. New Technology on a 300t-Converter Shell Replacement at Baosteel[J].Metallurgical Equipment, 2002.

IOP Conf. Series: Materials Science and Engineering **242** (2017) 012074 doi:10.1088/1757-899X/242/1/012074

Study on electrochemical corrosion mechanism of steel foot of insulators for HVDC lines

Weihua Zheng[1,a)],Xiaoyu Sun[1,b)] and Youping Fan[1,c)]

[1]Department of Electrical Engineering, WuHan University, Wuhan, Hubei 430072 China.

E-mail: [a)]Corresponding author: 895298286@qq.com, [b)] sunxiaoyu@whu.edu.cn, [c)] ypfan@whu.edu.cn

Abstract. The main content of this paper is the mechanism of electrochemical corrosion of insulator steel foot in HVDC transmission line, and summarizes five commonly used artificial electrochemical corrosion accelerated test methods in the world. Various methods are analyzed and compared, and the simulation test of electrochemical corrosion of insulator steel feet is carried out by water jet method. The experimental results show that the experimental environment simulated by water jet method is close to the real environment. And the three suspension modes of insulators in the actual operation, the most serious corrosion of the V type suspension hardware, followed by the tension string suspension, and the linear string corrosion rate is the slowest.

1. Introduction

Our land is vast. The reserves of natural resources and energy consumption in the eastern and western regions are uneven. With the advantages of long distance transmission, large transmission capacity, low transmission environment and low comprehensive cost, UHVDC has effectively solved this contradiction in our country. However, part of the DC transmission line in the running section of the natural environment is relatively humid, large rainfall. In recent years China's haze increasing, insulator fittings along with easy water, dirt form electrolyte, in the action of an external DC voltage, forming a conductive loop, local corrosion phenomenon.

The corrosion of insulators has occurred in most of the transmission lines, and their harm can cause the mechanical and electrical performance of insulators to decline in different degrees. The decline of mechanical and electrical performance is likely to cause serious transmission accidents such as flashover or broken string, thus threatening the safe and stable operation of power system.

Therefore, the electrochemical corrosion mechanism of the steel foot of insulator is studied, and the main factors affecting the electrochemical corrosion of insulator are obtained based on the experimental data. It is very important to study the restraining measures of insulator corrosion and prolong the service life of insulators. In view of the electrochemical corrosion of insulators, a sacrificial anode cathodic protection method is adopted to slow down the corrosion of steel materials. The sacrificial electrode is made of pure zinc sheath. The scheme can inhibit the corrosion caused by the electrochemical principle, but the volume expansion of the corrosion product will cause the porcelain parts of insulators to crack. American scholars put forward a kind of anticorrosive cover improvement measures. This method can effectively reduce the expansion of the insulator caused by the expansion of the corrosion products, but the project is difficult to implement and is not conducive to promotion. Soviet scholars proposed a method for improving the structure of steel feet. The steel foot at the joint of the cement and air is

Content from this work may be used under the terms of the Creative Commons Attribution 3.0 licence. Any further distribution of this work must maintain attribution to the author(s) and the title of the work, journal citation and DOI.
Published under licence by IOP Publishing Ltd

improved into the shape of the boss. The relevant tests conducted by the domestic researchers show that the convex platform of the Soviet scholars is replaced by pure zinc sheath, and this combination structure has a good inhibitory effect on the electrochemical corrosion of insulators in the current running insulators in china. At present, the technical requirements for insulator protection, zinc ring and zinc sleeve are also different at home and abroad. In this paper, based on the electrochemical corrosion of the steel foot of insulator, the simulation experiment scheme is summarized and the main factors affecting the electrochemical corrosion of steel foot are put forward.

2. Electrochemical corrosion mechanism of steel foot

The mechanism of DC electrochemical corrosion is shown in figure 1.1. The applied DC power supply, the positive and negative metal electrode and the electrolytic solution together form the conductive circuit of the three. The anode connected to the applied power source is an anode. Under the action of the DC electromotive force, the anode is oxidized and loses electrons, and the cation is formed from the surface of the metal material, which is anodic corrosion.

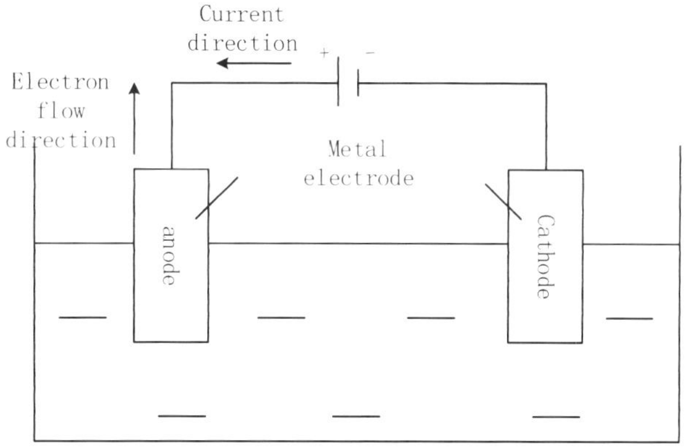

Figure 1. Schematic Diagram of Direct Current Electrolysis

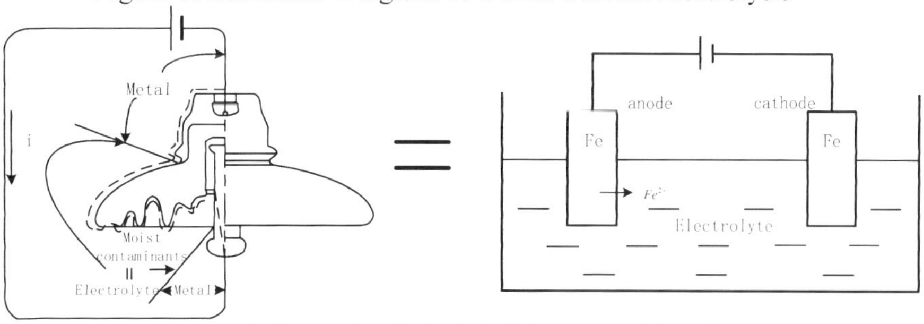

Figure 2. Corrosion Principle Diagram of Steel Foot

An analogy to the process of electrochemical corrosion of insulators is shown in figure 1.2. When the porcelain surface is damp and the contaminants dissolve in the water and form the flow path of the leakage current, the metal at the anode will undergo an oxidation reaction, losing electrons and forming a positive valence cation.

$$Fe \rightarrow Fe^{2+} + 2e^{-} \quad (1)$$

While the cathode metal will get electrons, the reduction reaction, the formation of anions.

$$O_2 + 2H_2O + 4e^{-} \rightarrow 4OH^{-} \quad (2)$$

Anions and cations combine in electrolyte solutions to produce hydrogen and oxygen compounds that are not soluble in water.

$$Fe^{2+} + 2OH^{-} \rightarrow Fe(OH)_2 \downarrow \quad (3)$$

And the hydrogen and oxygen compounds will be further oxidized under the action of water and oxygen.

$$4Fe(OH)_2 + O_2 + 2H_2O \rightarrow 4Fe(OH)_3 \quad (4)$$

If at different temperatures, different humidity, different PH conditions, as well as different oxygen content and other conditions, you can get different elements of the proportion of hydrogen oxidation and substance, so you can rust composition as follows:

$$mFeO + nFe_2O_3 + pH_2O \quad (5)$$

The m, n, and p here represent the proportions of the elements in the compound, and their values vary with the conditions[1].

In the DC transmission line, the insulator on the positive side is prone to corrosion of the steel foot. The utility model is mainly provided with an insulator positioned on the positive side, and a transmission line is connected on one side of the steel foot. Therefore, the steel foot side belongs to the high potential side relative to the cap. When the rain or moisture insulator, insulator surface contamination in water soluble form electrolyte environment, and steel feet and were served as iron anode and cathode, together constitute the conditions of electrolytic cell. As an anode, the steel foot oxidizes, loses electrons, forms two valent iron ions, dissolves in the electrolyte, and combines with anions to form hydrogen and oxygen compounds. At the same time, the electrolyte on the metal surface flows through the leakage current, the corrosion rate and the leakage current density in the anode region (i.e., the current on the metal unit area I). The current flowing through the insulator surface can cause electrochemical corrosion of the metal parts. The main influence is as follows: the steel foot loses the electron, forms the ion, causes the steel foot to dissolve unceasingly, and the mechanical strength of the insulator reduces. In the vicinity of the steel feet inside the porcelain, if the accumulation of hydrogen and oxygen compounds, it will lead to volume expansion, and will cause the insulator porcelain to crack when it is serious[2].

3. Simulation test method for electrolytic corrosion of insulators

The simulation test methods of electrolytic corrosion of insulators are mainly divided into two major categories. One is the long time field simulation method. The other is to build the accelerated simulation method of test platform, including salt spray method, electrolytic tank method, solid soil layer method, spray method, etc..

3.1 Field Simulation Method

Field simulation is a new insulator placed in the outdoor environment, exposed to outdoor insulators, long-term exposure to heavy industrial waste gases and coal, gas and other corrosive gases erosion. The voltage applied to the insulator string is 44kv, and the corrosion data of the fittings are recorded regularly by the experimental personnel. The field test proves the effect of the polarity of the voltage on the corrosion of the steel foot.

3.2 Salt Fog Method

Salt mist method is a certain concentration of Nacl solution through high-pressure electric spray gun form fog particles, sprayed on the clean surface of the insulator. As a result of high pressure, the fog particles will form a conductive film directly on the surface of the insulator to produce a certain leakage current. The test will clean insulators hanging in the fog chamber, gossan ground, steel pin is connected with a power supply cathode. The researchers used high-pressure electric spray guns, pre spraying salt fog of about 5 to 6min. When the visibility of fog chamber is less than 0.3m, 7 ~ 10kV DC voltage is applied to the insulator. The leakage current through the insulator surface is about 60 ~ 80mA. The conductivity of the salt fog solution is 17.6mS/cm, and the current fluctuates greatly. Salt spray test can be carried out continuously, and no other follow-up operation is needed during the test. When salt spray test is carried out abroad, the insulator is placed in the airtight box with relative humidity of more than 95%, and the sulfide gas is used to corrode, so as to simulate the corrosion of fossil fuel burning area.

3.3 Electrolytic Bath Method

The electrolytic bath is the process of filling the electrolyte into the electrolyte, which is a common solution, such as Nacl. Then, the metal parts which need to be tested are placed in the electrobath, and the test DC power supply is tested. In order to ensure that electrolytic corrosion occurs only at the corrosion site of the hardware, the corrosion free parts of the fittings shall be sealed with waterproof materials such as sealant. For example, in the simulation test of steel foot corrosion, the wire and the steel foot should be sealed to ensure that the corrosion shape is similar to the corrosion situation at the site, and only the annular area at the junction of the steel foot and the cement can be exposed. In order to achieve the purpose of quantitative corrosion simulation experiment, the electrolysis current should be monitored in real time, and the accumulated amount of corrosion charge in the electrolytic circuit is calculated. The sample preparation method of electrolysis process is complicated, especially the gossan corrosion, sealing area is large, the sample is ready.

3.4 Solid waste layer method

Solid waste layer method is one of the most popular test methods for electrolytic corrosion of steel feet. Similar to the pollution flashover test, a layer of salt and insoluble material is coated on the surface of the cleaned insulator. When it is dry, it is damp in the fog chamber. And test the DC voltage to do the electrolytic corrosion test. The fog chamber and power supply condition are the same as the salt fog method. The suspension composition is as follows: 250g diatomite, 1000g tap water, a certain amount of Nacl. Surface average contaminant density 3mg/cm^2. The test is carried out in a 220mm×220mm×260mm container. The filter paper is placed in the container, the bottom is water, the filter paper is kept moist, and the humidity inside the container is close to 100%. Dalian electric porcelain factory and Chinese EPRI experts in the test by solid pollution method, the leakage current is maintained at 80mA, and to mark steel foot protection zinc sleeve exposed part is etched off half the zinc sleeve end of life. Some researchers on the adhesion of different components of pollution layer were studied, pointed out that the first grade product with hot tap water plus bone paste lovastatin is heated to 90°C with a glass rod to stir, then add the right amount of formaldehyde additives (40% formaldehyde solution) attachment delay solution prepared has strong adhesion and long the (in insulators attached to the surface of 8 ~ 10 hours). The thickness of the sewage layer is 40~60 microns. If the porcelain surface is short with aluminum foil near the steel foot, it can solve the problem that the corrosion rate decreases due to the frequent discharge of dry zone in the area. At the same time, the adhesion time can be extended to 20~24 hours. During the test, the applied DC voltage is 10kV, and the current remains at 60±2mA[3].

3.5 Water Spraying Method

The method of spraying water is to hang the insulator according to the method of hanging in the field, and spray a certain concentration of Nacl solution on the corrosion of the steel foot, and form a conductive channel with the additional DC voltage. Anode corrosion occurs when the steel pin is connected to the positive electrode of the power supply. The simulation formed by the method of accelerated corrosion tests in which the electrical conductivity of solution flow rate and conductivity of the solution has a great influence on the corrosion rate, flow rate can be adjusted by the water speed, the conductivity can be calculated by the concentration of Nacl solution[4].

4. Comparison of accelerated simulation experiments for electrochemical corrosion of insulators

According to the existing research results, compare the advantages and disadvantages of the above five experimental methods:

1) The scene simulation method is a method of simulation experiment results in all the most close to the actual situation, but the duration is too long, unable to obtain experimental data in a short period of time, so it is not suitable to be used as artificial accelerated corrosion test methods.
2) Salt fog has the advantages of large corrosion current, long experimental duration and small amount of labor. It is a practical and convenient simulation experiment method. It is also suitable for the artificial accelerated corrosion of the steel foot of insulators, especially for the simulation experiment of the DC suspension steel foot corrosion. However, in the experiment, we should pay

attention to limit the flashover of insulators, and prevent the salt fog corrosion of the test equipment and flashover of the wall through the fog chamber. In addition, salt spray test should pay attention to salt mist particles for high pressure nozzle corrosion and blockage.

3) The electrolytic cell method is a simple and easy to operate accelerated corrosion simulation method, which is suitable for studying the corrosion of steel feet separately. But for other complex suspension insulator, its preparation process is complicated, especially the gossan corrosion, sealing area is large, the sample is ready. For the steel foot corrosion test, the corrosion area is mainly at the exposed part of the steel foot and cement, and the corrosion shape is close to a circle shape. There is a certain difference between the simulation result and the actual corrosion condition of the insulators, so it is not suitable for use.
4) The method of solid waste layer can be applied to the electrochemical corrosion test of steel foot of insulator. However, if a stable leakage current is to be obtained, the coating material and coating process must be highly technically required. Coated and dirt layer workload, must enter into the spray chamber repeatedly remove repeated smear samples. Not only the work is tedious, but also the repeated operation is also easy to have the error influence to the experiment, therefore also does not apply to the insulator steel foot corrosion simulation experiment.
5) Compared with the previous four simulation methods, the water jet test has lower requirements for the test equipment, and the test platform is easy to build. The test is easy, simple and easy to prepare. Moreover, the simulated test results are in good agreement with the field operation, so the water jet method is suitable for studying the accelerated corrosion test of the electrochemical corrosion of the steel foot of insulator.

5. The accelerated simulation experiment of electrochemical corrosion of insulators is carried out by water jet method

Set up the following experimental platform for simulation experiments:

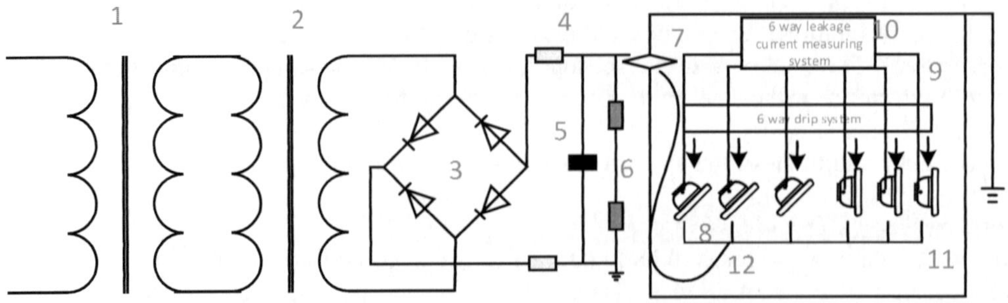

1-regulator; 2- test transformer; 3- full bridge rectifier system 4- protection; resistance; 5- filter; 6- DC voltage divider wall; 7- cable; 8- insulator suspension system; 9-6 drip system; 10-6 leakage current measurement system; 11- drainage system; 12- accelerated electrical corrosion test room[5].

Figure 3. Experimental Platform of Spray Irrigation

In order to compare the effect of three kinds of suspension methods on the electrochemical corrosion of steel foot in practical operation, 3 new ceramic insulators were selected in the experimental group. Wherein, the plane and the horizontal plane of the 3 insulators form parallel, 45 and 90 angles respectively, and simulate the running conditions of the linear string, the V string and the strain string, and the numbers are 1#, 2# and 3# respectively. The leakage charge of each of the three insulators was recorded at intervals of one hour. The three insulators were removed at 20 hours, and the data were plotted as follows:

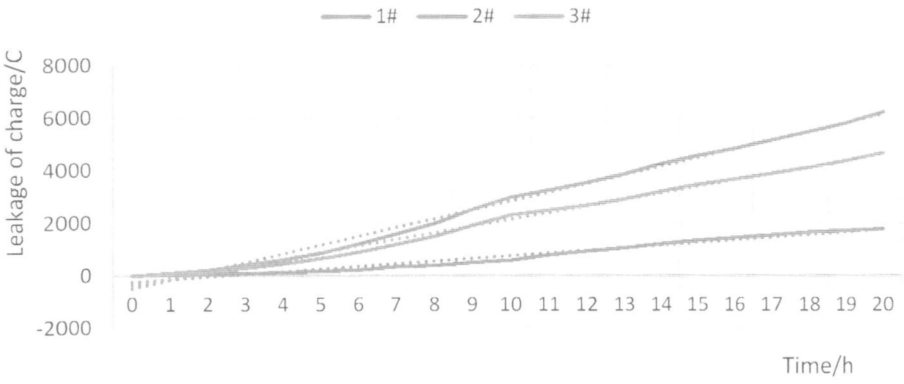

Figure 4. Leakage Charge of Insulator.

As a result, the corrosion rate of the steel foot is the fastest in the V shape suspension mode, followed by the stress absorption string suspension mode, and the linear string corrosion rate is the slowest in the actual operation.

Select one of the V insulators suspended from the actual running line 4#. The service life of the insulator is 5.5 years, and the running time of the insulator 2# is equivalent to that of the leakage charge amount in this simulation test, and the corrosion situation of the two is contrasted.

Figure 5. Corrosion Samples of 2# (left) and 4# (right) Insulators.

Through preliminary observation, we can get the corrosion test sample under the accelerated simulation of water jet method. The corrosion state is more serious than that in the field. This and we select the worst case average leakage charge as the benchmark consistent, corrosion area and similar shape, showed that spraying under laboratory conditions accelerated corrosion test and field operation of corrosion equivalent better.

6. Conclusion

From the experimental results and analysis, we can draw the following conclusions: the main reason for the electrochemical corrosion of the steel foot is that the insulator surface is polluted, and the electrolyte environment is formed in the humid environment. And steel foot and gossan exists between the potential difference between the steel pin connected to the positive potential of oxidation reaction. Compared with several simulation experiments, the simulation experiment with water jet method is most feasible, and the simulated experimental environment is in line with the actual environment. In the actual operation process, the three kinds of suspension modes of insulators, V type suspension of the most serious corrosion of metal fittings, tensile string suspension mode followed, linear string corrosion rate is the slowest.

Acknowledgments

Thank you for your great support from the school and the college! Thanks to Mr. Fan for my help, we have put forward a lot of valuable suggestions for our thesis, so that our papers can be more rigorous and scientific! Thank you Xiaoyu Sun for helping us with our thesis and help us solve it when I have any professional problems! Thank you for your monographs!

References

[1] Ling Luo, Fuzeng Zhang, Liming Wang, Hongwei Mei, Zhicheng Guan, Liying Li. The protection of zinc inhibited HVDC porcelain and glass insulator cap electrolytic corrosion ring design [N]. China ofthecsee, 2015, 35 (17).

[2] Xuan Wang, Xiaolong Cao, man Xu. HVDC insulators electrochemical corrosion test of [A]. high voltage engineering, 2006, 32 (3): 31-33.

[3] Junfeng Zhang, Haiping Miao. On HVDC suspension insulator pin of the artificial accelerated corrosion test of [J]. electromagnetic arrester, 1989, 6 (4): 19-23.

[4] Ling Luo, Fuzeng Zhang, Liming Wang, Hongwei Mei, Zhicheng Guan, Liying Li. Electrolytic corrosion test methods for metal fittings of high voltage direct current porcelain and glass insulators [A]. transactions of the Electrotechnical Society, 2015, 30 (9): 103-111.

[5] Fuzeng Zhang, Yifan Liao, Ling Luo, Ruihai Li, Guoli Wang, Yining Wang. UHVDC porcelain insulator cap accelerated electrical corrosion test of [A]. power system technology, 2014,38 (2): 311-315.

IOP Conf. Series: Materials Science and Engineering **242** (2017) 012075 doi:10.1088/1757-899X/242/1/012075

The research of full automatic oil filtering control technology of high voltage insulating oil

Gangjun Gong[1, a)], Tong Zhang[1, b)], Guozeng Yan[2, c)], Han Zhang[1, d)], Zhimin Chen[1, e)], Chang Su[1, f)]

[1] Beijing Engineering Research Center of Energy Electric Power Information Security, North China Electric Power University, Beijing 102206, China.

[2] AC Engineering Construction Company of State Grid Corporation of China, 100052, China.

E-mail: [a)] gonggangjun@126.com, [b)] Corresponding author: Jocanya@126.com, [c)] 2686952245@qq.com, [d)] 280170073@qq.com, [e)] 18010052640@163.com, [f)] suchang@ncepu.edu.cn

Abstract. In this paper, the design scheme of automatic oil filter control system for transformer oil in UHV substation is summarized. The scheme specifically includes the typical double tank filter connection control method of the transformer oil of the UHV substation, which distinguishes the single port and the double port connection structure of the oil tank. Finally, the design scheme of the temperature sensor and respirator is given in detail, and the detailed evaluation and application scenarios are given for reference.

1. Introduction

If the quality of the transformer oil is not up to standard, it will cause partial discharge of insulating oil, and will accelerate the aging process of the transformer insulation, and also directly affect the safe operation of the whole power grid. At present, the transformer oil filter using manual operation, there is a large workload, low efficiency, easy to misuse and other deficiencies. With the development of UHV project, it is urgent to study a fully automatic oil filter control system for transformer oil to meet the requirement of UHV engineering.

2. Overall Design of Double Tank Oil Filtering System

The double tank oil filtering system consists of oil tank, empty tank, respirator, two-position three-way solenoid valve, flow sensor, temperature sensor, oil filter, sub controller, and condition monitoring and control module. The instantaneous flow monitoring of the transformer oil in the process, and the instantaneous flow as the switching criterion, while monitoring the total flow and the total flow as protection criterion, the formation of double control criterion, the double protection function, improve the stability and safety of the system. According to the different connection schemes of oil tank, the system is divided into single interface connection and dual interface connection scheme.

2.1 Single Interface Connection Scheme

The single interface connection scheme is shown in Fig.1. Only through the bottom of an oil tank is connected with the interface of oil pipeline, and add the 'T' type of mouth fittings, which are respectively connected with the two two-position three-way solenoid valves, two-position three-way

Content from this work may be used under the terms of the Creative Commons Attribution 3.0 licence. Any further distribution of this work must maintain attribution to the author(s) and the title of the work, journal citation and DOI.
Published under licence by IOP Publishing Ltd

solenoid valve is connected with the oil filter, through state monitoring and control module of flow, temperature and other parameters of the real-time monitoring, and as a criterion control valve position switch, complete the work cycle of oil filter. Through the bottom interface as the oil outlet and the oil inlet, the process of oil filling and oil extraction is completed. A single interface reduces the oil tank and the external environment contact interface, while the installation of respirator in the tank, filter the air of impurities such as water, is conducive to maintaining tank sealing, enhanced tank sealing effect, ensure the quality of transformer oil.

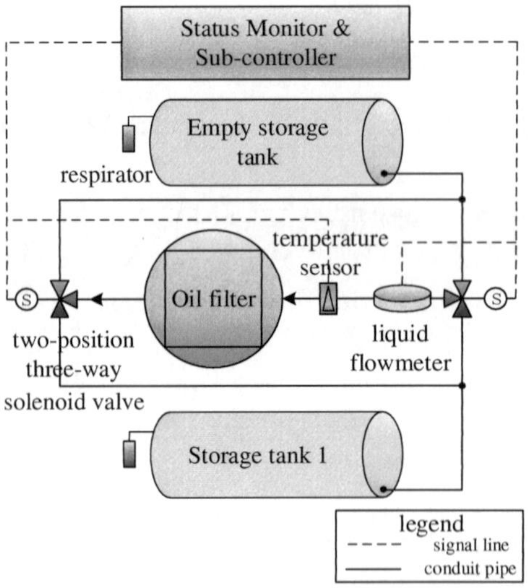

Figure 1. Double tank type oil filter system single interface connection structure

During refueling, when the oil tank is filled to a certain amount, the oil gravity creates a reverse pressure on the oil in the oil inlet pipe, as shown in Fig. 2. Therefore, the power requirement of the oil filter is higher, the overall adaptability of the system is reduced, and the energy consumption is larger. For the high power oil filter in the field, the automatic oil filtration system with single interface connection can be considered.

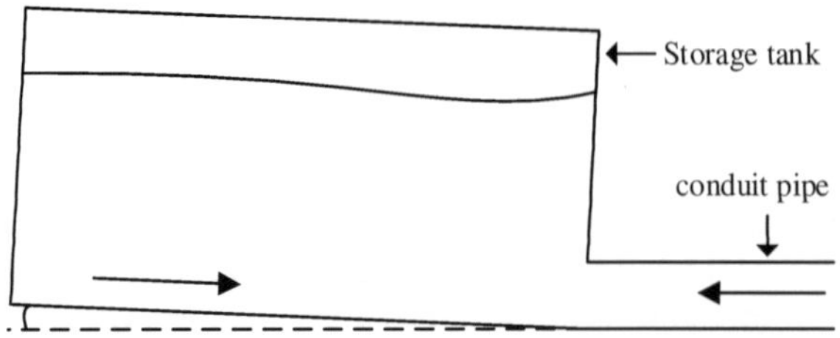

Figure 2. Single interface refueling schematic diagram

2.2 Dual Interface Connection Scheme

In order to enhance the adaptability of the system, the project proposed a double interface connection, as shown in Fig. 3. Make the bottom interface of tank as an oil outlet, and the top interface of tank as an oil inlet, connecting pipeline respectively and connected with the two and two-position three-way solenoid valve, and then the two position three way solenoid valve and oil filter is connected through the switch valve, complete cycle oil filter work. Thus, the problem that the power requirement of the oil filter machine is higher in the single pipe connection scheme is solved. Application of double interface connection, can use the function of transformer oil gravity, transformer oil tank in the auxiliary injection and extraction, the liberation of the requirements on power oil filter system to reduce energy consumption, improve the overall adaptability of automatic oil filtering system, is conducive to the promotion of the system.

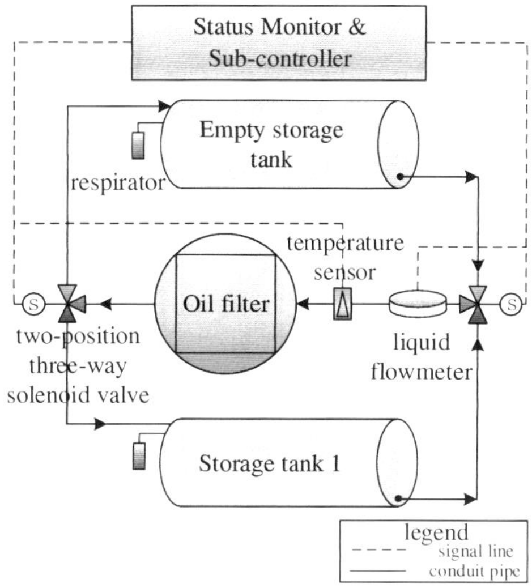

Figure 3. Double tank type oil filter system dual interface connection structure

3. Scheme Design of Cluster Oil Filtering System

The design scheme of cluster filter oil is promoted by double tank scheme. It also consists of oil tank, empty tank, respirator, two-position three-way solenoid valve, flow sensor, temperature sensor, oil filter, sub controller, and condition monitoring and control module. The instantaneous flow monitoring of the transformer oil in the process, and the instantaneous flow as the switching criterion, while monitoring the total flow and total flow as a standby switch optional criterion, the formation of double control criterion, the double protection function, improve the stability and safety of the system. According to different oil tank connection schemes, the cluster oil filter design scheme is divided into single interface connection and double interface connection scheme.

3.1 Single Interface Connection Scheme

Cluster type oil filter design scheme of single interface connection scheme is the addition of one-way solenoid valve for each tank except No. 0 empty tank based on double tank scheme, solenoid valve opening and closing state controlled by the module control, each only allowed a single solenoid valve is open. As shown in Fig. 4. The principle, advantages and disadvantages are same as that of double tank, so no longer details.

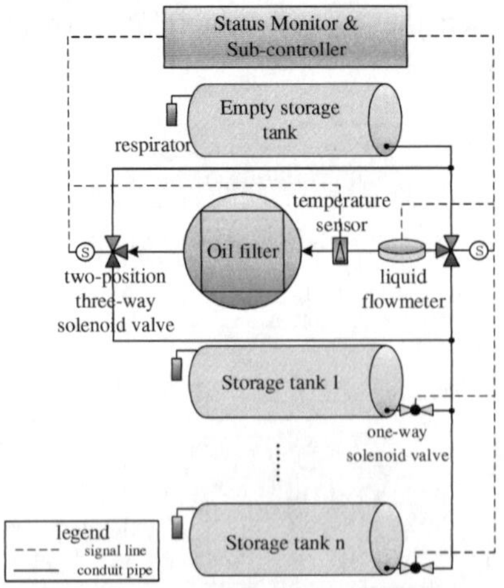

Figure 4. Cluster type oil filter system single interface connection structure

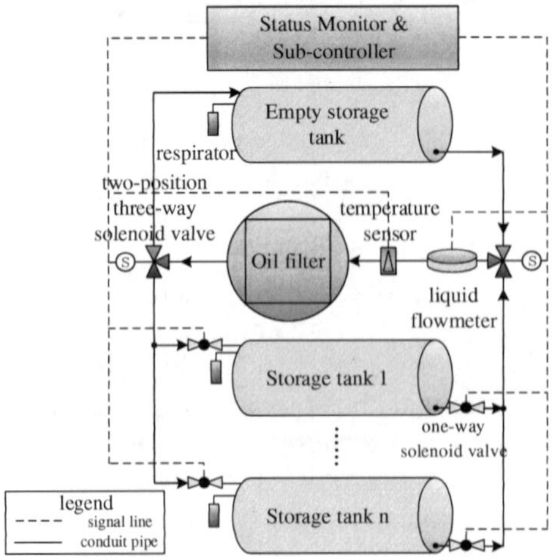

Figure 5. Cluster type oil filter system dual interface connection structure

3.2 Dual Interface Connection Scheme

Cluster type oil filter design scheme of dual interface connection scheme is the addition of one-way solenoid valve for each tank except No. 0 empty tank based on double tank scheme, solenoid valve opening and closing state controlled by the module control, each only allowed two one-way solenoid valve belongs to the same tank a single solenoid valve is open. As shown in Fig. 5. The principle, advantages and disadvantages are same as that of double tank, please refer to double tank as they are similar.

4. Design Scheme of Temperature Sensor

In order to improve the applicability of the system, the temperature sensor is converted into a stud, and a set of special hardware fittings are designed for the temperature sensor, as shown in Fig. 6. The sensor is fixed on the upper fixed with bolts and fittings, and other means to keep the whole sealing fittings, probe into the internal pipeline, to monitor the temperature of the liquid flowing through the flange fittings, access filter oil pipeline. This design avoids the installation of the temperature sensor in field drilling package, saving construction time, and avoid the potential fire risks and other issues caused by uncontrollable error of site construction, drilling result. If the probe fails, it can be replaced with the whole metal fittings, and the probe can be replaced independently. The flexible workpiece is flexible and convenient to use and has good applicability.

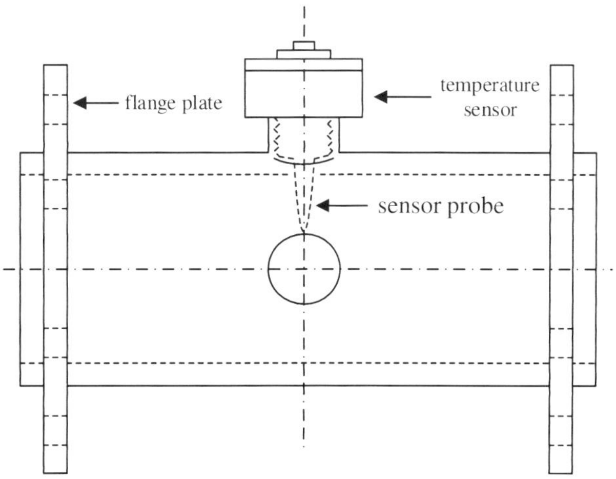

Figure 6. Schematic diagram of temperature sensor and fittings

5. Design Scheme of Respirator

There are two kinds of breathing apparatus design in this system: centralized scheme and split scheme. Centralized breathing apparatus is designed to be relatively cost effective, and if the respirator is damaged, all tanks will be affected, so the risk is more concentrated. But because of the low number of respirators used in centralized design, the failure rate is relatively low. As shown in Fig. 7.

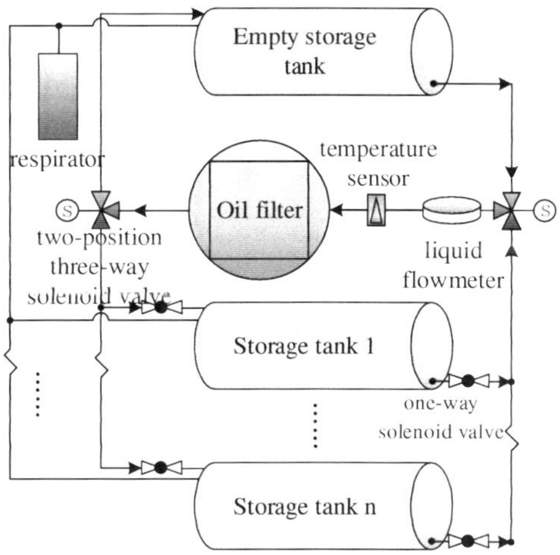

Figure 7. Design scheme of centralized respirator

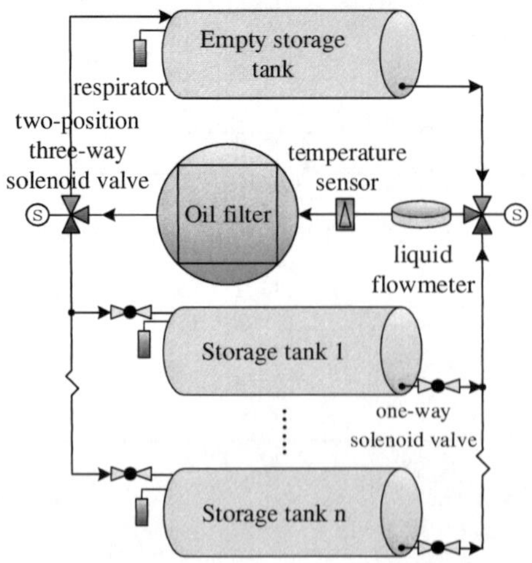

Figure 8. Design scheme of split respirator

The design of split breathing apparatus is more expensive, and each oil tank respirator is exclusive. It can maximize the risk caused by respiratory failure and enhance the reliability of the system. However, due to the larger number of ventilators used, the rate of failure is higher in theory than in the centralized scheme. As shown in Fig. 8.

6. Conclusion

In this paper, we summarized the related control technology of transformer oil automatic oil filtration system and the connection way of different requirements. The application scenarios and advantages and disadvantages of various schemes are analyzed in detail. It is important to improve the transformer oil automatic oil filter system.

References

[1] Lou F. The Research of Wire Grinding Machine and Wire Stripping Machine Control System Based On PLC [J]. International Journal of Control & Automation, 2015, 8.
[2] Chai S, Chen L, Cao Y. Application of inverter fault diagnosis technology based on control system in actual production [J]. Automation & Instrumentation, 2015.
[3] Tatzel D W, Jacobson A A, Cipollo N J, et al. AUTOMATIC CONFIGURATION OF CONTROL DEVICE USER INTERFACE IN A HOME AUTOMATION SYSTEM:, US20150341184[P]. 2015.
[4] Tan C. Design of wireless remote monitoring and control system based on radio modem [J]. Industry & Mine Automation, 2015.
[5] Feng C, Yang X, Wang Y, et al. Study on intelligent optimal control of fuzzy-PI applying in buck circuit [J]. Electrical Measurement & Instrumentation, 2015.
[6] Kim J S. Development of the Visualization Tool for the Automatic Variable Message Generator of VMS Emulator System [J]. International Journal of Control & Automation, 2015, 8
[7] Zhao J, Wang L, Ren S, et al. Automatic control system of spray and dust reduction in fully-mechanized working face of coal mine [J]. Industrial Instrumentation & Automation, 2015
[8] Reichard D J, Reid D W, Plache K S, et al. Automation control and monitoring system and method[J]. 2015.

Applications of polynomial optimization in financial risk investment

Meilan Zeng [1, a)] **and Hongwei Fu**[1, b)]

[1] College of Mathematics and Statistics, Hubei Engineering University, Xiaogan, Hubei, 432000, China.

E-mail: [a)] zml571@ 126.com, [b)] Corresponding author: gcxyfhw@ 163.com

Abstract. Recently, polynomial optimization has many important applications in optimization, financial economics and eigenvalues of tensor, etc. This paper studies the applications of polynomial optimization in financial risk investment. We consider the standard mean-variance risk measurement model and the mean-variance risk measurement model with transaction costs. We use Lasserre's hierarchy of semidefinite programming (SDP) relaxations to solve the specific cases. The results show that polynomial optimization is effective for some financial optimization problems.

1. Introduction

Financial risk refers to the uncertainty of investment results that is affected by one or more random factors, including operational risk, credit risk, exchange rate risk and market risk. The greater the risk is, the greater the return or loss is. To remain invincible in the face of risk, it is necessary to implement the risk management. The financial risk management has its strategic significance, especially in market deregulation, globalization and cross business today. Quantification and control of risk has become the core of all the activities of investors and financial regulatory authorities. To manage risk, we need to quantify or measure the size of risk. After more than half a century of development, the financial risk measurement methods have made the great progress. Especially with the successful use of complex mathematical methods, there have been a lot of good measurement models in recent years, so the research on the risk measurement has become a hot research field of Applied Mathematics.

In 1952, Markowitz first gave a quantitative method of selecting the optimal portfolio, which is known as the mean-variance model. In this model, he proposed to use the mathematical expectation $E[X]$ of portfolio random income X to represent its expected return, and use the variance $Var(X)=E[(X-E[X])^2]$ to measure the risk of portfolio. The Markowitz method is easy to achieve, because $Var(X)$ can be calculated through the covariance, which is based on the joint distribution of the return of each asset. This model and the inherent risk decentralization thought become the foundation of modern investment theory.

Markowitz portfolio theory has become an important tool to help investors how to invest capitals in a certain proportion, which makes the investment income big and small risk. $Var(X)$ is generally a nonconvex function, and there are many local extremums, so the optimal portfolio choice problems are very difficult to use the usual optimization algorithm [10] for control and optimization, and results may not be stable. It is precisely this reason that only a few literatures [1, 3, 6] really use Var as a risk measure for the actual large-scale investment portfolio optimization problems.

Content from this work may be used under the terms of the Creative Commons Attribution 3.0 licence. Any further distribution of this work must maintain attribution to the author(s) and the title of the work, journal citation and DOI.
Published under licence by IOP Publishing Ltd

In recent years, polynomial optimization theory has made great progress (see [5, 7-9]), which can find the global optimal point from some local extremums for the nonconvex polynomial functions. In this paper, we construct the standard mean-variance risk measurement model and the mean-variance risk measurement model with transaction costs. We use Lasserre's hierarchy of SDP relaxations to solve two cases in financial risk investment, and the results show that polynomial optimization can effectively solve some actual financial optimization problems.

2. Polynomial optimization theory

In this section, we introduce the preliminary knowledge of polynomial optimization (see [5, 7, 8]).

Let $\mathbb{N},\mathbb{R}$ be the sets of nonnegative integer numbers and real numbers, respectively. The symbol $\mathbb{R}[x]:=\mathbb{R}[x_1,x_2,\ldots,x_n]$ denotes the polynomial ring in $x:=[x_1,x_2,\ldots,x_n]$ with real coefficients. For a degree d, $\mathbb{R}[x]_d$ denotes the set of all polynomials in $\mathbb{R}[x]$ whose degrees are at most d. Let $\mathbb{N}_d^n:=\{\alpha\in\mathbb{N}^n \mid \alpha_1+\alpha_2+\cdots+\alpha_n\le d\}$. The symbol deg(p) denotes the degree of polynomial p. For a positive integer k, $[k]=\{1,2,\ldots,k\}$.

An ideal I in $\mathbb{R}[x]$ is a subset of $\mathbb{R}[x]$ such that $I\cdot\mathbb{R}[x]\subseteq I, I+I\subseteq I$. For a tuple $h=(h_1,\ldots,h_m)$ in $\mathbb{R}[x]$, $\langle h\rangle$ denotes the smallest ideal containing all h_i, i.e. $\langle h\rangle=h_1\cdot\mathbb{R}[x]+\cdots+h_m\cdot\mathbb{R}[x]$. The k-th truncation of the ideal $\langle h\rangle$ is denoted by $\langle h\rangle_k$, which is the set

$$\langle h\rangle_k=h_1\cdot\mathbb{R}[x]_{k-deg(h_1)}+\cdots+h_m\cdot\mathbb{R}[x]_{k-deg(h_m)}.$$

A polynomial σ is called a sum of squares (SOS) if $\sigma=p_1^2+\cdots+p_k^2$ for some $p_1,\cdots,p_k\in\mathbb{R}[x]$. $\Sigma[x]$ denotes the set of all SOS polynomials in x. For a degree m, $\Sigma[x]_m$ denotes the truncation $\Sigma[x]\bigcap R[x]_m$. For a tuple $g=(g_1,\ldots,g_t)$, its quadratic module is the set

$$Q(g)=\Sigma[x]+g_1\cdot\Sigma[x]+\cdots+g_t\cdot\Sigma[x].$$

The k-th truncation of $\mathrm{Q}(g)$ is the set

$$Q_k(g)=\Sigma[x]_{2k}+g_1\cdot\Sigma[x]_{d_1}+\cdots+g_t\cdot\Sigma[x]_{d_t},$$

Where $d_i=2k-deg(g_i), i=1,2,\cdots,t$. For polynomial tuples h and g, the sum $\langle h\rangle+Q(g)$ is called archimedean if there exists $p\in\langle h\rangle+Q(g)$ such that $p(x)\ge 0$ defines a compact set $\mathbb{R}[x]$. Suppose that $p\in\langle h\rangle+Q(g)$ is archimedean, if a polynomial f is positive on the set $\{h(x)=0, g(x)\ge 0\}$, then $f\in\langle h\rangle+Q(g)$ (see Putinar [11]).

Let $\mathbb{R}^{\mathbb{N}_d^n}$ represent the space of real sequences indexed by $\alpha\in\mathbb{N}_d^n$. A vector y in $\mathbb{R}^{\mathbb{N}_d^n}$ is called a truncated moment sequences (tms) of degree d, i.e.

$$y=(y_\alpha)_{\alpha\in\mathbb{N}_d^n}.$$

A tms $y\in\mathbb{R}^{\mathbb{N}_d^n}$ defines a Riesz function L on $\mathbb{R}[x]_d$ as

$$L_y\left(\sum\nolimits_{\alpha\in\mathbb{N}_d^n} p_\alpha x_1^{\alpha_1}\cdots x_n^{\alpha_n}\right):=\sum\nolimits_{\alpha\in\mathbb{N}_d^n} p_\alpha y_\alpha.$$

For convenience, let $x^\alpha=x_1^{\alpha_1}\cdots x_n^{\alpha_n}$ and $\langle p,y\rangle:=L_y(p)$.

Let $q\in\mathbb{R}[x]$. For every $y\in\mathbb{R}^{\mathbb{N}_{2k}^n}$, the function $L(qp^2)$ defines as

$$L(qp^2)=vec(p)^T(L_q^k(y))vec(p),$$

Where $\mathrm{vec}(\mathrm{p})$ denotes the coefficient vector of polynomial p with $deg(qp^2)\le 2k$, the symmetric matrix $L_q^k(y)$ is the k-th localizing matrix of q, generated by y, which is linear about y. For example, $n=2,k=2,q=1-x_1^2-x_2^2$, it follows

$$L^{(2)}_{1-x_1^2-x_2^2}(y)=\begin{pmatrix} 1-y_{20}-y_{02} & y_{10}-y_{30}-y_{12} & y_{01}-y_{21}-y_{03} \\ y_{10}-y_{30}-y_{12} & y_{20}-y_{40}-y_{22} & y_{11}-y_{31}-y_{13} \\ y_{01}-y_{21}-y_{03} & y_{11}-y_{31}-y_{13} & y_{02}-y_{22}-y_{04} \end{pmatrix}.$$

When $q=1, L_1^{(k)}(y)$ is called the k-th moment matrix generated by y, denoted by $M_k(\mathrm{y})$. For instance, $\mathrm{n}=2,\ \mathrm{k}=2$,

$$M_2(y)=\begin{pmatrix} y_{00} & y_{10} & y_{01} & y_{20} & y_{11} & y_{02} \\ y_{10} & y_{20} & y_{11} & y_{30} & y_{21} & y_{12} \\ y_{01} & y_{11} & y_{02} & y_{21} & y_{12} & y_{03} \\ y_{20} & y_{30} & y_{21} & y_{40} & y_{31} & y_{22} \\ y_{11} & y_{21} & y_{12} & y_{31} & y_{22} & y_{13} \\ y_{02} & y_{12} & y_{03} & y_{22} & y_{13} & y_{04} \end{pmatrix}.$$

For the polynomial optimization problem

$$\begin{aligned} &\min_{x\in\mathbb{R}^n} \quad f(x) \\ &\text{s.t.} \quad h_1(x)=\cdots=h_{m_1}(x)=0, \\ &\qquad g_1(x)\ge 0,\ldots,g_{m_2}(x)\ge 0, \end{aligned} \tag{2.1}$$

Where $f(x),h_i(x),g_j(x)$ are polynomial functions. Let u be a local minimizer of (2.1), $J(u)=\{j_1,\dots,j_r\}$ be the active set at u. We can construct new polynomials $\varphi_1,\dots,\varphi_r$, such that (2.1) is equivalent to the following problem

$$\begin{aligned} &\min_{x\in\mathbb{R}^n} \quad f(x) \\ &\text{s.t.} \quad h_i(x)=\varphi_j(x)=0, i\in[m_1], j\in[r] \\ &\qquad g_1(x)\ge 0,\ldots,g_{m_2}(x)\ge 0. \end{aligned} \tag{2.2}$$

For solving (2.2), the hierarchy of semidefinite programming (SDP) relaxations (see[5]) is

$$\begin{cases} \rho_k = \min \quad \langle f, y\rangle \\ \text{s.t.} \quad L_{h_i}^{(k)}(y)=0, L_{\varphi_j}^{(k)}(y)=0, i\in[m_1], j\in[r] \\ \qquad M_k(y)\succeq 0, L_{g_i}^{(k)}(y)\succeq 0, i=1,2,\ldots,m_2 \end{cases} \tag{2.3}$$

The dual problem of (2.2) is

$$\eta_k := \max \gamma \quad \text{s.t.} \quad f-\gamma\in\langle h\rangle_{2k}+Q_k(g). \tag{2.4}$$

Denote by

$$f_{\min} := \min f(x), \text{s.t.} \quad h(x)=0, g(x)\ge 0,$$

Where $h=(h_1,\cdots,h_{m_1}), g=(g_1,\cdots,g_{m_2})$. For $k=1,2,\cdots$, solving the SDP (2.4), we get a sequence with lower bounds

$$\eta_1 \le \eta_2 \le \eta_3 \le \cdots \le f_{\min}.$$

Theorem 1 ([5]) If $\langle h\rangle + Q(g)$ is archimedean, then

$$\lim_{k\to\infty} \eta_k = f_{\min}.$$

In fact, when k is big enough, Lasserre's hierarchy generally has finite convergence (see Theorem 1.1 in [8]). The software Gloptiploy 3 [4] can solve (2.3) and (2.4).

3. Optimization models in financial risk investment

According to the demands of different style investors, we can construct different optimization models with the Markowitz theory.

3.1 Standard mean-variance risk measurement model

With n kinds of securities available for investment options, note that the yield of the i-th security during the j-th period is $r_{ij}, i=1,2,\cdots,n, j=1,2,\cdots,m$, the proportion of the total assets of the i-th security is $x_i, i=1,2,\cdots,n$, the average yield and it variance of the m-th period of the i-th security are

$$\overline{r_i}=\frac{1}{m}\sum_{j=1}^{m} r_{ij}, \sigma_i^2=\frac{1}{m}\sum_{j=1}^{m}(r_{ij}-\overline{r_i})^2,$$

the covariance of the i-th and j-th securities' yield is

$$\sigma_{ik}=\frac{1}{m}\sum_{j=1}^{m}(r_{ij}-\overline{r_i})(r_{ij}-\overline{r_k}).$$

The investment risk can be defined as

$$V=\sum_{i=1}^{n}\sigma_i^2 x_i^2+2\sum_{i=1}^{n-1}\sum_{j=i+1}^{n}\sigma_{ij}x_i x_j. \tag{3.1}$$

Denote $x=(x_1,x_2,\cdots,x_n)^T$ as the portfolio vector, then the above investment risk V can be expressed as $V=x^T Qx$, where

$$Q=\begin{pmatrix} \sigma_{11} & \sigma_{11} & \cdots & \sigma_{1n} \\ \sigma_{21} & \sigma_{22} & \cdots & \sigma_{2n} \\ \cdots & \cdots & \cdots & \cdots \\ \sigma_{n1} & \sigma_{n2} & \cdots & \sigma_{nn} \end{pmatrix}, \sigma_{ij}=\sigma_i^2, \sigma_{ik}=\sigma_{ki}, i=1,2,\cdots,n, j=1,2,\cdots,n,$$

is a symmetric and semidefinite matrix.

Denote the yield vector as $\overline{r}=(\overline{r_1},\overline{r_2},\cdots,\overline{r_n})^T$, then the investment yield can be expressed as

$$R=\sum_{i=1}^{n}\overline{r_i}x_i=x^T\overline{r}.$$

In the case of allowing the short-term trading, if the decision is to minimize risk, then the risk model is obtained as follows

$$\min \quad V=x^TQx, \quad \text{s.t.} \quad \sum_{i=1}^{n}x_i=1.$$

The above model is only aware of the risk, not considering the return. In the optimization problems of financial portfolio, if different conditions are considered, the optimization model can be different.

For instance, considering how to purchase a number of assets from n kinds of risk assets, the variance risk is as small as possible under the premise of the yield to a certain level. Based on the above conditions, we can construct the following model [2]

$$\begin{cases} \min \quad V = x^T Q x, \\ \text{s.t.} \quad \sum_{i=1}^{n} x_i = 1, \\ \sum_{i=1}^{n} \overline{r_i} x_i = r_p, \\ x_i \geq 0, i = 1, 2, \cdots, n, \end{cases} \tag{3.2}$$

Where r_p is the expected return of investment portfolio.

3.2 Mean-variance risk measurement model with transaction costs

Transaction cost generally refers to the amount of money that is paid for the purchase of an asset more than its present value, which is an important factor that investors must consider when choosing a portfolio.

With n kinds of risk assets available for selection, note that the mean value of the return of the i-th asset is $\overline{r_i}, i = 1, 2, \cdots, n$, the covariance matrix is $Q = (\sigma_{ij})$, where σ_{ij} is the covariance of the returns of the i-th and j-th assets. The amount of the purchase of the i-th asset is x_i, the transaction cost is denoted by $c_i(x_i)$. How to purchase a number of assets from the n kinds of assets, the variance risk is as small as possible under the premise of the yield to a certain level. Based on the above conditions, we also can construct the following model [12]

$$\begin{cases} \min \quad V = x^T Q x, \\ \text{s.t.} \quad \sum_{i=1}^{n} x_i = 1, \\ \sum_{i=1}^{n} (\overline{r_i} x_i - c_i(x_i)) = r_p, \\ x_i \geq 0, i = 1, 2, \cdots, n, \end{cases} \tag{3.3}$$

Where r_p is the expected return of investment portfolio. The cost function $c_i(x)$ is not given, and is generally a nonlinear function. $c_i(x)$ is generally obtained by using least squares fitting based on the actual data.

4. Applications

In this section, we use polynomial optimization to solve the financial risk investment cases.

Case 1. An investor intends to choose three stocks A, B, C with outstanding performance to carry out the long-term portfolio investment. Through the market analysis and statistical forecast of three stocks, we get the relevant data shown in Table 1. Please give the investment proportion of three stocks from two aspects.

(1) Such that the variance of stock returns in the portfolio is reduced to a minimum, to reduce the risk of investment, and hope that the expected yield after five years is less than 65%.

(2) Hope to get the maximum benefit, in the case that the standard deviation is not more than 12%.

Table 1. Related data of stocks.

Name of stocks	Expected yields in five years%	Covariance in five years %		
		A	**B**	**C**
A	92	180	36	110
B	64	36	120	-30
C	41	110	-30	140

Let x_1, x_2, x_3 respectively represent the investment ratio of three shares A, B, C. Denote the expected yield in five years as r_1, r_2, r_3, which are random variables. The total return of the portfolio in five years is

$$R = x_1 r_1 + x_2 r_2 + x_3 r_3,$$

the variance of the portfolio is

$$var(R) = x_1^2 var(r_1) + x_2^2 var(r_2) + x_3^2 var(r_3) + 2x_1x_2 cov(r_1, r_2) + 2x_1x_3 cov(r_1, r_3) + 2x_2x_3 cov(r_2, r_3).$$

Based on the data of Table 1, the portfolio risk is obtained

$$V = var(R) = 180x_1^2 + 120x_2^2 + 140x_3^2 + 72x_1x_2 + 220x_1x_3 - 60x_2x_3.$$

Therefore, the standard deviation of the portfolio is

$$D = [180x_1^2 + 120x_2^2 + 140x_3^2 + 72x_1x_2 + 220x_1x_3 - 60x_2x_3]^{\frac{1}{2}}.$$

According to the conditions of item (1), we construct the mathematical model of the problem

$$\begin{cases} \min V = 180x_1^2 + 120x_2^2 + 140x_3^2 + 72x_1x_2 + 220x_1x_3 - 60x_2x_3, \\ \text{s.t.} \quad x_1 + x_2 + x_3 = 1, \\ \qquad 0.92x_1 + 0.64x_2 + 0.41x_3 \ge 0.65, \\ \qquad x_1, x_2, x_3 \ge 0. \end{cases} \tag{4.1}$$

Using Lasserre's hierarchy of SDP (2.3) and (2.4) to solve, we get

$$x_1 = 0.2351,\ x_2 = 0.5222,\ x_3 = 0.2427,$$

the optimal solution of the objective function is $V = 64.7054$. In other words, under the conditions of ensuring the minimum risk and the total yield in five years being 65%, the portfolios of three stocks A, B, C are 23.51%, 52.22%, 24.27%, the minimum variance is 0.6471%.

According to the conditions of item (2), we construct the mathematical model of the problem

$$\begin{cases} \max \quad R = 0.92x_1 + 0.64x_2 + 0.41x_3, \\ \text{s.t.}\ x_1 + x_2 + x_3 = 1, \\ \quad 180x_1^2 + 120x_2^2 + 140x_3^2 + 72x_1x_2 + 220x_1x_3 - 60x_2x_3 \le 144, \\ \quad x_1, x_2, x_3 \ge 0. \end{cases} \tag{4.2}$$

Using Lasserre's hierarchy of SDP (2.3) and (2.4) to solve, it follows

$$x_1 = 0.8593,\ x_2 = 0.1407,\ x_3 = 0.0000,$$

the optimal solution of the objective function is R=0.8806. In other words, under the conditions that the standard deviation is not more than 12% and the maximum income is obtained in five years, the portfolios of three stocks A, B, C are 85.93%, 14.07%, 0%, the highest yield can reach 88.06%.

Case 2. An investor intends to choose five securities A, B, C, D, E to carry out the portfolio investment. Through the market analysis and statistical forecast of five securities, we obtain the relevant data shown in Table 2. The functions of transaction costs are defined as

$c_i(x_i) = 0.004x_i - 0.001x_i^2, i = 1,2,\ldots,5$. Please give the investment proportion of five securities, such that the expected yield is 10% and the risk of investment is the minimum.

Table 2. Related data of securities.

Name of securities	The average yields	Covariance of yields				
		A	**B**	**C**	**D**	**E**
A	0.16	0.56	0.11	0.09	0.08	0.13
B	0.11	0.11	0.32	0.20	0.14	0.11
C	0.08	0.09	0.20	0.22	0.13	0.12
D	0.13	0.08	0.14	0.13	0.48	0.11
E	-0.02	0.13	0.11	0.12	0.11	0.24

Let x_1, x_2, x_3, x_4, x_5 respectively represent the investment ratio of five securities A, B, C, D, E. This problem is an investment portfolio with transaction costs. According to the optimization model (3.3), we construct the mathematical model of the problem

$$\begin{cases} \min \quad V = 0.56x_1^2 + 0.32x_2^2 + 0.22x_3^2 + 0.48x_4^2 + 0.24x_5^2 + 0.22x_1x_2 + 0.18x_1x_3 + 0.16x_1x_4 \\ \quad + 0.26x_1x_5 + 0.4x_2x_3 + 0.28x_2x_4 + 0.22x_2x_5 + 0.26x_3x_4 + 0.24x_3x_5 + 0.22x_4x_5, \\ \text{s.t.} \quad \sum_{i=1}^{5} x_i = 1, \\ 0.16x_1 + 0.11x_2 + 0.08x_3 + 0.13x_4 - 0.02x_5 - 0.004\sum_{i=1}^{5} x_i + 0.001\sum_{i=1}^{5} x_i^2 = 0.1, \\ x_i \geq 0, i = 1,2,\cdots,5, \end{cases}$$

Using Lasserre's hierarchy of SDP relaxations (2.3) and (2.4) to solve, we obtain

$$x_1 = 0.2191,\ x_2 = 0.1156,\ x_3 = 0.4110,\ x_4 = 0.1878,\ x_5 = 0.0666.$$

The optimal solution of the objective function is V=0.1746. Under the condition that the expected yield is 10%, the portfolios of five securities are 21.91%, 11.56%, 41.10%, 18.78%, 6.66%, the minimum risk is 17.46%.

The above cases show that polynomial optimization can effectively solve some actual financial optimization problems.

Acknowledgments
This work is supported by the Science and Technology Foundation of the Department of Education of Hubei Province (D20152701) and National Natural Science Foundation of China (11601138).

References

[1]S. Basak, A.Shapiro, "Value-at-risk-based risk management: optimal policies and asset prices," The Review of Financial Studies, 2001, 14(2), 371-405.
[2]M. B. Biggs, "Nonlinear optimization with financial applications,"*Netherlands: Kluwer Academic Press*, 2005.
[3]R. Campbell, R. Huisman, K. Koedijk, "Optimal portfolio selection in a Value-at-Risk framework," Journal of Banking & Finance, 2001, 25(9):1789-1804.
[4]D. Henrion, J. B. Lasserre, J. Loefberg, "GloptiPoly 3: moments, optimization and semidefinite programming," Optim. Methods Software, 2009, 24(4-5): 761-779.
[5]J. B. Lasserre, "Global optimization with polynomials and the problem of moments," SIAM Journal on Optimization, 2001, 11(3): 796-817.

[6]A. Lucas, P. Klaassen, "Extreme returns, downside risk, and optimal asset allocation," Journal of Portfolio Management, 1998, 25(1): 71-79.
[7]J. Nie, "An exact Jacobian SDP relaxation for polynomial optimization," Mathematical Programming, 2013, 137(1-2): 225-255.
[8]J. Nie, "Optimality conditions and finite convergence of Lasserre's hierarchy," Mathematical Programming, 2014, 146(1-2): 97-121.
[9]J. Nie, "The hierarchy of local minimums in polynomial optimization," Mathematical Programming, 2015, 151(2): 555-583.
[10] J. Nocedal, S. Wright, "Numerical optimization," *Springer Science & Business Media*, 2006.
[11] M. Putinar, "Positive polynomials on compact semi-algebraic sets," Indiana University Mathematics Journal, 1993, 42(3): 969-984.
[12] Z. Zhang, "Two programming-nonlinear programming and portfolio algorithm," *Wuhan: Wuhan University press*, 2006.

IOP Conf. Series: Materials Science and Engineering **242** (2017) 012077 doi:10.1088/1757-899X/242/1/012077

Research on the construction of three level customer service knowledge graph

Shi Cheng[1,2, a)], Jiajie Shen[1, b)] ,Quan Shi[1,2, c)] and Xianyi Cheng[1,2, d)]

[1] School of Computer Science and technology, Nantong University, JiangSu Nantong 226019, China.

[2] Nantong Research Institute for Advanced Communication Technologies, JiangSu Nantong 226019,china.

E-mail: [a)] chenshi@ntu.edu.cn, [b)] 598864151@qq.com, [c)]sq@ntu.edu.cn, [d)] Corresponding author: xycheng@ntu.edu.cn

Abstract. With the explosion of knowledge and information of the enterprise and the growing demand for intelligent knowledge management and application and improve business performance the knowledge expression and processing of the enterprise has become a hot topic. Aim at the problems of the electric marketing customer service knowledge map (customer service knowledge map) in building theory and method, electric marketing knowledge map of three levels of customer service was discussed, and realizing knowledge reasoning based on Neo4j, achieve good results in practical application.

1. Introduction

Power marketing refers to the power production, transportation and sales, and meets the power customer's economic and reasonable, safe and reliable use of electric products, and constantly improves the floorboard of a series of economic activities of the economic benefit of power enterprise. The power enterprise through appropriate and feasible marketing strategy, including quality service strategy, price strategy, brand strategy, promotion strategy and other means to continuously improve the electricity market share, in order to meet the need of customers, and to achieve the desired target for power Enterprises [1]. In current society, the role and status of power marketing is increasingly evident. Power supply electricity knowledge graph is the basis of power customers' behavior modeling, we did not find the relevant knowledge and power marketing knowledge graph from the open literature [2].Knowledge modeling and application technology based on knowledge graph theory have obtained and achieved good results the considerable development for many years at home and abroad in many industries, but in the power industry of our country, the research and application remain to be promoted. The knowledge graph can be used as the basis to explore the implicit relations among the concept terms through logical reasoning, to make the tacit knowledge explicit and to realize the sharing and reuse of domain knowledge.

Knowledge graph is an indispensable basic knowledge database for computer Natural Language Processing, semantic search, computer reasoning and artificial intelligence. But there is no public Chinese knowledge graph database, which has caused great difficulties to the relevant research. Natural Language Processing, semantic search is quite developed in foreign countries, one is easier to handle English than Chinese, and also English has many public knowledge graph database (http://www.daml.org/onto- logiest/keyword.html [3]).

Content from this work may be used under the terms of the Creative Commons Attribution 3.0 licence. Any further distribution of this work must maintain attribution to the author(s) and the title of the work, journal citation and DOI.
Published under licence by IOP Publishing Ltd

With the explosion increasing of enterprise knowledge and information, and the enterprise for knowledge intelligent management and application is increasing, in order to facilitate the discovery and application of knowledge and to improve business performance, the expression and processing of enterprise knowledge have become the focus [4].

2. Related work

The concept of knowledge graph is proposed by Google in May 2012, and announced the basis for the construction of the next generation of intelligent search engine. Although the concept of knowledge graph is new, it is not a new field of study. In 2006 Berners-Lee proposed the idea of data link, thus improved the relevant technical standards, such as URL (Uniform Resource Identified), RDF (Resource Description Framework), WL (Web Ontology Language).Knowledge graph is based on the relevant research results, and it is a revolution of the existing semantic network technology [5].The knowledge graph that has been published and applied is shown in TABLE 1.

Table 1. The knowledge graph applied

	Content
Baidu knowledge graph [6]	When Baidu search some key words of public figures, there will be the relevant information of the character, the search results in the "Encyclopedia" style are displayed. But now not only search popular characters, when users search for names, subject or popular "facts", Baidu will give regular search results on the left side of the search results, and search results on the right show Baidu encyclopedia content with relevant keywords, and related search links.
Sogou cube [7]	Sogou said in its official micro-blog: in order to allow users to obtain information more easily, Sogou search released a new knowledge base search engine --"known cube". This is the first search engine in the domestic search engine industry.
Fudan GDM Chinese knowledge graph [8]	Data mining analysis of micro-blog with knowledge graph to listen to public opinion and improve people's livelihood research results, which are published in the Liberation Daily, Xinmin Evening News and other newspapers and reproduced by a number of network media.
Google knowledge graph [9]	In order to allow users faster and easier to find new information and knowledge, Google search will release "knowledge graph" to search results for the knowledge of the system, and through any key words can get a complete knowledge system.

The relationship between the related concepts is given in figure.

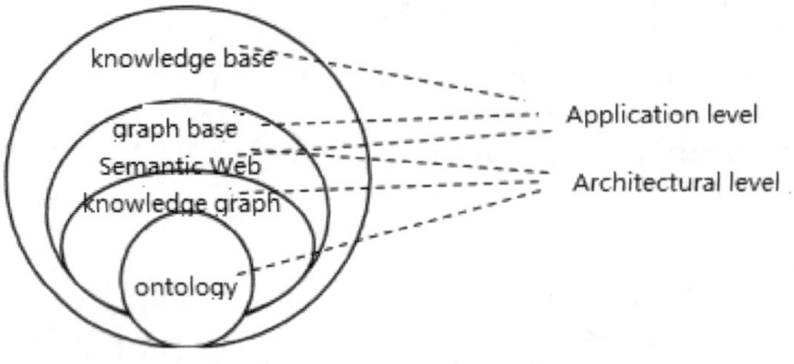

Figure 1. The relationship between related concepts

As can be seen from Figure 1, ontology and knowledge graph is the basis of the construction of knowledge system. Semantic web serves as a bridge between modeling and application. The advantages and disadvantages of ontology knowledge graph have great influence on the quality of graph database and knowledge base.

3. Three level architecture of power marketing customer service knowledge graph

3.1 Self-service architecture
Three levels knowledge graph used of power marketing customer service, providing customers with a self-service customer service system (shown in Figure 2).Realize implementation of data extraction from analysis to knowledge application.

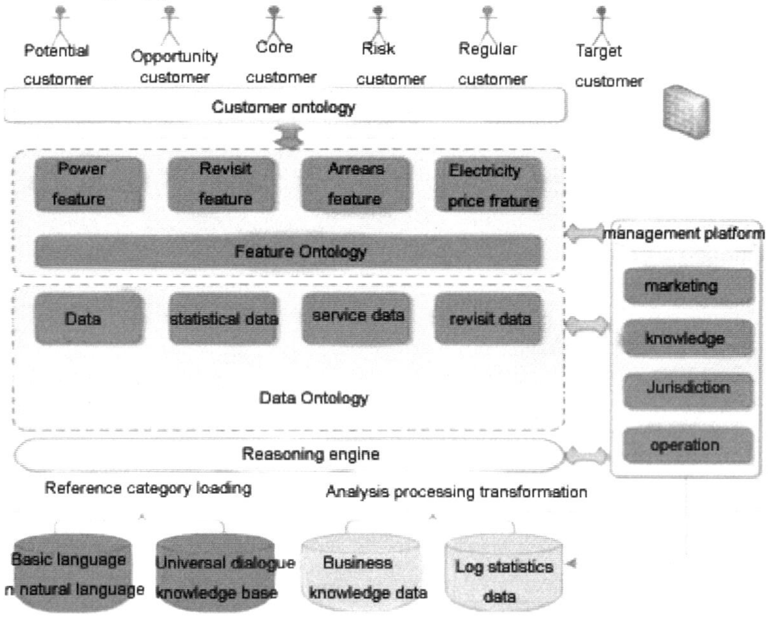

Figure 2. Self-customer service based on customer service knowledge graph

Figure 7 shows the self-service customer service system architecture can be targeted for the user's search information for context correlation, the relevant issues recommended the fuzzy intention of intelligent guidance, flexible push key business, error correction and Pinyin analysis. The main functions of the scene include business management, multi candidate output, sensitive word filtering, intelligent input, voice recognition, context, customer care etc. Self-service customer service system is low cost, 24h response throughout the day,the cost of a single service is 1/50 of an artificial hot line, the effective diversion of 10%~20% artificial seat business, and the first time diversion of the business can be maintained at more than 85%.

3.2 Application case analysis
This section mainly discusses the application of knowledge graph in power marketing customer service. Of course, many application scenarios and ideas can be extended to other industries. The application scene mentioned here is just the tip of the iceberg, and in many other applications, the knowledge graph can still play its potential value.

3.2.1 Customer value analysis. Customer value analysis is a very important link in customer service. The difficulty of analysis based on big data is how different sources of data (structured and unstructured) to be integrated together. And which build the value analysis engine, so as to effectively identify valuable customers.

A valuable customer is likely to be the beginning of a brand. A customer who has important value is the key to a brand. The reputation of customers with important value directly affects and promotes company's market influence and even market open, development. Having large customers can improve the company's market value. An account for 20% of the total number of customers brings 80% of the company's profit, which is the most important value of the customer to the company's most direct impact on the value and benefits. In addition to the growth of corporate profits, the value of the customer has a profound impact on the company.

And the important value of the customer will involve complex network, which also brings new challenges to the value analysis. The knowledge graph, as a direct representation of the relationship, can well solve these two problems. First, the knowledge graph provides a very convenient way to add new data sources, as mentioned earlier. Secondly, the knowledge map itself is used to represent the relationship, which can help us to more effectively analyze the specific potential value of complex relationships.

Value analysis is a tool to identify the competitive advantage of enterprises. Value analysis pays more attention to the value of customer service activities. Through the analysis of customer service, the customer service activities are classified and analyzed, and the key link of customer value chain is determined, so as to establish the source of competitive advantage.

The core of value analysis is customers, first need to put all the data source associated with a particular customer to get through, and build the knowledge graph contains multiple data sources, so as to integrate into a machine understandable structured knowledge. Here, we should not only integrate the basic information (such as customer information when completing the application),can also put the customer's demand information and behavior information, value information to integrate the whole knowledge graph, thus reasoning.

3.2.2 New attribute inference.

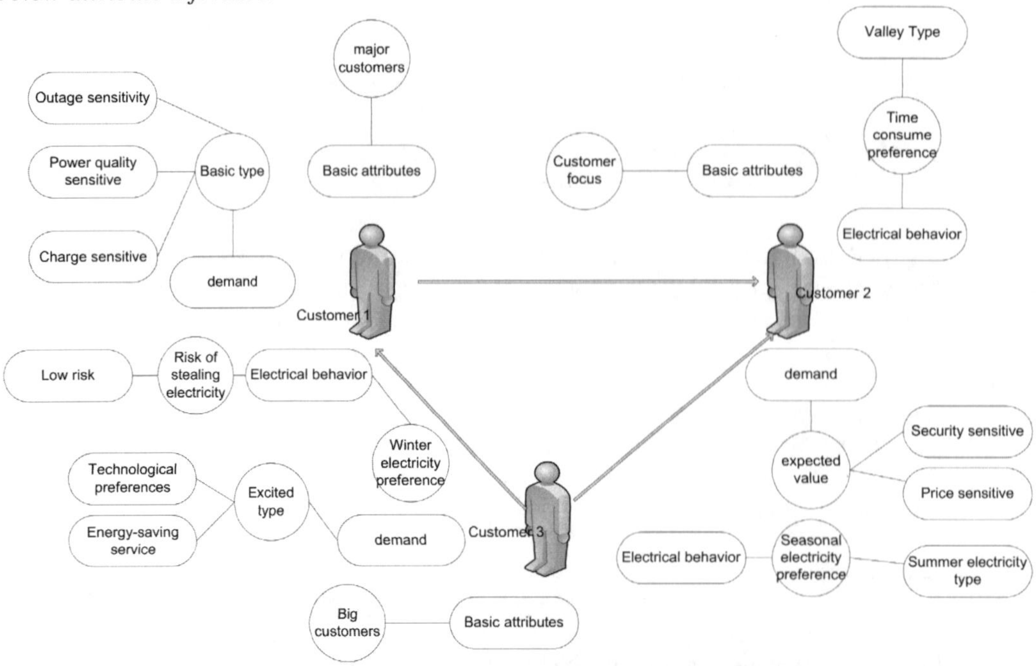

Figure 3. Power marketing knowledge graph reasoning

In Figure 3, the demand of customer 1 is “energy saving” and the demand of customer 2 is “billing sensitive”, then we can the demand of customer 2 is also "energy saving”. Customer 3’s power preference is “valley power type”, can also be inferred that the demand of customer 3 including "energy saving”. According to the customer 2’s "steal electricity risk" low and customer 3 "security sensitive”, can infer that the customer 3 is a potential customers.

The self-help service system based on knowledge graph can propose current knowledge graph for user, which cannot answer content records and preliminary classification, to facilitate later artificial intervention, the hot spot cared by new users added to the existing knowledge graph. With the continuous improvement of user information in the customer service knowledge graph, according to the operation characteristics and concerns of the user can intelligently remind the user whether needs to query often concerned content, even users can regularly push attention information, gradually play a service active perception and active service role in power marketing customer.

4. Conclusions

The knowledge graph is based on the new knowledge representation field of big data, is also the basis of constructing knowledge base. There are many challenges in the exploration and practice of power marketing customer service knowledge graph, this paper discusses three levels customers service knowledge graph framework of power marketing.

The application of knowledge graph has a certain randomness, and the interpretation of knowledge graph relies heavily on qualitative description and judgment, and cannot solve many practical problems; The mapping of knowledge graph has certain technical barriers, and it is difficult for non-experts to use it, It is a constraint factor that cannot have a significant impact on other disciplines and social fields. Although some research makes a contrastive analysis of different data, methods and tools of knowledge graph, but it did not form a specific application research standard, and the existing research results are different or even contradictory, standardized data samples set for comparative study is also a lack of.

In conclusion, the role of knowledge graph is not fully realized, and its conclusion is only auxiliary verification, and the application needs to be further strengthened; At the same time, developing in breadth, not only in the field of discipline (domain) structure visualization, but also in science and technology management and scientific decision-making, enterprise innovation and competitive intelligence play an important role. In addition, the application of knowledge graph is constrained by the improvement of its theory and software tools; It is closely related to the development of knowledge mining and artificial intelligence, and also needs to be combined with other semantic web technologies to visualize applications; Much depends on a better unde-rstanding of the human brain. In a word, go through the phenomenon of surface, it still has a long way to discover the trends and laws of subject knowledge, and to visualize visually.

Acknowledgments

This paper is supported by the National Natural Science Foundation of China (No. 61340037), Nantong Research Institute for Advanced Communication Technologies

References

[1]Junzhang Kang. Research on fine power marketing service management measures (Management Science, China, 2017.09),pp.24-25.

[2]Guibin Wang. *The Internet _ power marketing intelligent interactive service construction* (China power enterprise management, China, 2017.01),pp.48-49.

[3]Xiao Tian, Yongchao Liu,Jing Wang, et al. *Application value of customer service knowledge graph construction of Power Grid Corp*(Shandong electric power technolo,China,2015,42(12)),pp.65-69.

[4] Changle Xu. *Analysis of power network marketing mode under smart grid* (communication world, China, 2017, 03),pp.219-220.

[5]Jiao Liu, Yang Li,Hong Duan, et al. *Overview of knowledge graph technology*(computer research and development,China,2016,53(3)),PP.582-600.

[6]*Baidu next generation search engine prototype exposure application knowledge mapping technology* (Computer programming skills and maintenance, China, 2013,19),pp.4-4.

[7]Sogou cube.http://baike.sogou.com/v66616234.htm.In Chinese.

[8]Wei Li,Yanghua Xiao,Wei Wang. *Character entity recognition based on Chinese knowledge graph* (computer engineering, China, 2017,43(3)),pp.226-234.

[9]Ochs C, Tian T, Geller J,et al. *Google knws who is famous taday- building an ontology trom search engine knowledge and DBpedia*[C]//Proc of the 5th IEEE Int Conf on Semantic Computing.Piscataway, NJ:IEEE, 2011:320—327.

IOP Conf. Series: Materials Science and Engineering **242** (2017) 012077 doi:10.1088/1757-899X/242/1/012077

[10] Yisong Ma,Zhigang Wu.*Modeling and analysis of large power data based on Neo4j*(new technology of electrical and electrical energy,China,2016,35(2)),pp.24-31.

IOP Conf. Series: Materials Science and Engineering **242** (2017) 012078 doi:10.1088/1757-899X/242/1/012078

Research on intelligent recommendation algorithm of e-commerce based on association rules

Jiajie Shen[1,2,a)] and Xianyi Cheng[1,2,b)]

[1]School of Computer Science and technology, Nantong University, JiangSu Nantong 226019, China

[2] Nantong Research Institute for Advanced Communication Technologies,JiangSu Nantong 226019,china.

E-mail: [a)]598864151@qq.com, [b)]Corresponding author:xycheng@ntu.com

Abstract. As the commodities of e-commerce are more and more rich, more and more consumers are willing to choose online shopping, because of these rich varieties of commodity information, customers will often appear aesthetic fatigue. Therefore, we need a recommendation algorithm according to the recent behavior of customers including browsing and consuming to predicate and intelligently recommend goods which the customers need, thus to improve the satisfaction of customers and to increase the profit of e-commerce. This paper first discusses recommendation algorithm, then improves Apriori. Finally, using R language realizes a recommendation algorithm of commodities. The result shows that this algorithm provides a certain decision-making role for customers to buy commodities.

1. Introduction

With the rapid development of Internet and electronic commerce, more and more people are inclined to online shopping. But customers often lose in a large amount of information and become aesthetic fatigue, which lead to not find the goods they need[1]. If businesses can recommend commodities which meet users' demand from mass of commodities to the users, it is expected to transform the potential demand into actual demand. So according to users' characteristics the recommendation system come into being[2].

The recommendation system has important application value, but it is also faced with many difficult problems in theory and application, among which sparse data, over fitting, extendibility and multimedia information feature extraction are the main problems[3]. And now, although R language can use Apriori algorithm to explore mining, but it can only find the association rules between goods, and cannot achieve the recommended effect. Therefore, this paper presents an intelligent e-commerce recommendation algorithm based on association rules. Through the recent customer behavior analysis, model by Apriori algorithm and mine the relationship between the commodities, and then according to the recent behavior of customers, intelligently recommend products for customers, to play an auxiliary decision-making effect on helping customers to buy goods[4].

2. Association rules and improvement of apriori algorithm

The association rules analysis is one of the most active research methods in data mining, the purpose is to find the association relationships between items in a data set that is not directly represented in the data[5,6]. Some definitions of association rules: The support of association rule is the ratio of the number of transactions containing X and Y to all transactions at the same time:

Content from this work may be used under the terms of the Creative Commons Attribution 3.0 licence. Any further distribution of this work must maintain attribution to the author(s) and the title of the work, journal citation and DOI.
Published under licence by IOP Publishing Ltd

$$\text{support}(X \Rightarrow Y) = \frac{\text{count}(X \cup Y)}{|D|} \quad 1)$$

The confidence of association rules is the ratio of the number of transactions containing X and Y to the number of transactions containing X:

$$\text{confidence}(X \Rightarrow Y) = \frac{\text{support}(X \cup Y)}{\text{support}(X)} \quad 2)$$

But the confidence measure ignores the support of the item sets in the rule post, the rules of high confidence may be misleading, so one of the ways to solve this problem is to use the lift degree. The lift degree of association rules represents the ratio of the probability that contains Y after contains X to the probability that contains Y under the condition without X:

$$\text{lift}(X \Rightarrow Y) = \frac{\text{confidence}(X \Rightarrow Y)}{\text{support}(Y)} \quad 3)$$

If the value =1, indicating that two conditions without any correlation, if the value <1, X and Y are mutually exclusive, generally in data mining when the value is greater than 3, the rule is valuable.

If the support of an item is greater than the min_sup specified by the user, it is called frequent. The task of association rule mining is to find the rule A⇒B which the support and confidence in transaction database D satisfy the min_sup and the min_con specified by the users, that is strong association rules. Association rules mining is a two step process: (1) finding all frequent item sets; (2) generating strong association rules by frequent item sets[7].

Apriori algorithm is a classic frequent item sets mining algorithm. Discovery process of Apriori algorithm is divided into two steps: the first step by iteration, retrieve all the frequent item sets in a transaction database, which supports not less than the threshold value set by users; the second step is the use of frequent item sets to construct rules meet the minimum confidence set by users.

3. Intelligent recommendation algorithm for goods

The implementation steps of improved intelligent e-commerce recommendation algorithm based on association rules as shown in Fig.1, the original customer transaction records as the data source for the recommendation system, through the data pretreatment, will be converted into standard customer transaction data. Based on a large number of user transaction data, association rule model for the above research as the basic algorithm, looks for user transactions including the purchase mode and buying interest, in the form of association rules stored in the rule base. After obtaining the user inquiry or purchase information, according to the association rule, the recommendation is produced, and the commodity which may be of interest to the users is recommended to the users in real time.

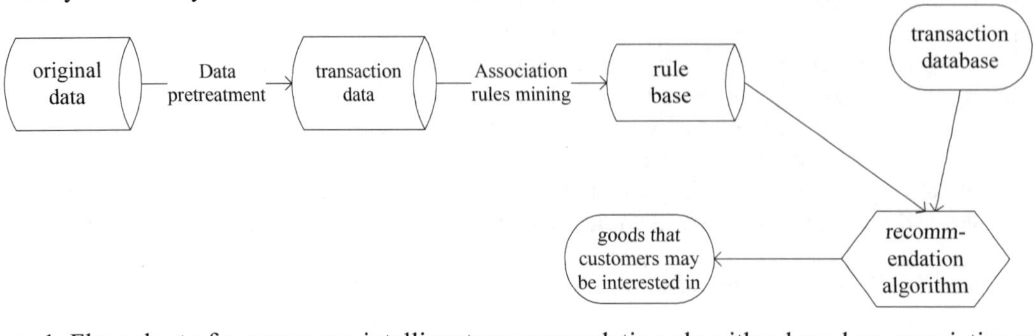

Figure 1. Flow chart of e-commerce intelligent recommendation algorithm based on association rules

3.1 Data pretreatment

According to three months of customer information provided by an electricity supplier website, the name of each field is explained in Table 1.

Table 1. Description of customer behavior information

user_id	User number	desensitization
sku_id	Commodity number	desensitization
time	Behavior time	
type	1. browse (refer to browse details page); 2 plus; 3 shopping cart; 4 deletion; 5 orders; 6 Click;	

Because the original data has a large number of missing, inconsistency and abnormal data, if the data is not removed, it will lead to deviation of mining results, so data pretreatment is very important.

```
> dim(act)        [1] 13199934        4
> act<-act[!duplicated(act)]
> dim(act)        [1] 9527224         4
```

By comparing the two dimensions, it is very clear that there is 3672710 duplicate data in the original data. If it is not removed before mining, it will seriously affect the result of the mining.

3.2 Implementation of recommendation algorithm based on apriori

Then the transaction record of each user is constructed. First, removing the other columns, only leaving the user ID and the commodity ID, because only two columns are needed to build the transaction data. Here's the transaction data for type=1 and type=4, because it's better to understand what you'll buy after you've browsed that product, or what you'll browse after you buy it. Then use Apriori algorithm to mine the transaction database constructed to find association rules, create a set of association rules (setting the minimum support and minimum confidence, and in accordance with the lift value from high to low ranking). Association rules are shown in Fig.2. Because general default rule which lift value is greater than 3 is valuable, select the top 4 rules in the rule base.

```
lhs                  rhs     support    confidence lift
{79636}           => {32465} 0.04315787 0.6375895  4.160353
{5505,63006}      => {12564} 0.04292017 0.7860003  3.190353
{12564,164258}    => {63006} 0.05727760 0.7646611  3.091805
{39425}           => {63006} 0.04842544 0.7475415  3.022584
{164258,63006}    => {12564} 0.05727760 0.7324012  2.972795
{12564,18412}     => {63006} 0.04967102 0.7297109  2.950488
```

Figure 2. Association rules

After having the association rules base and the customer transaction database, first create a recommendation set for each user and initialize it, the lhs in rules is matched with the items in the transaction data through the loop by the defined function. Once the match is successful, the rhs in the rules will be recommended to the user, which is exactly what the Apriori algorithm lacks in the R.

4. Conclusions

In this paper, the related knowledge of recommendation algorithm based on association rules is discussed. Based on the research of traditional Apriori algorithm, the recommendation algorithm is studied. By selecting the rules of high lift value, we can effectively avoid the misleading of high confidence, and then find the valuable association rules, and then recommend the customer's interested goods to the customer through the recommendation algorithm. The experiment proves that the algorithm is effective and feasible, and it can provide data support for the management decision of the merchant, and it is helpful to improve the service level and quality, and also provides a basis for the customer to buy the goods.

Acknowledgments
This paper is supported by the National Natural Science Foundation of China (No. 61340037).

References
[1] Yongjie Yang, Pengfei Yang, Qun Duan, Lina Han. Intelligent recommendation algorithm of goods based on association rules [J]. Modern computer, 2016, 10 (04), pp. 25-27.
[2] Xuan Li, Xiaoyan Wang, Zhenquan Zhuang, Personalized intelligent recommendation service based on Association Rule Mining [J]. Computer Science, 2002, 38 (7), pp. 200-204.
[3] Bo Yang, Peng-fei Zhao, Overview of recommendation algorithms [J]. Journal of shanxi university, 2011, 34 (3), pp. 337-350.
[4] Yu-lin Liu, Research on hot spot of combination sale of e-commerce based on association rule [J]. Journal of Zhejiang Sci-Tech University, 2016, 6(36), pp. 528-541.
[5] Xue Ding, Research on Intelligent Recommendation System of books based on association rules [J]. Library Information Forum, 2010, 4, pp.48-52.
[6] Carlos Ordonez, Norberto Ezquerra, Cesar A.Santana, Constraining and summarizing association rules in medical data [J]. knowledge and Information Systems,2006,3(9),pp.259-283.
[7] Jan Rauch, Logic of Association Rules [J]. Applied Intelligence, 2005, 22 (1), pp. 9-28.

IOP Conf. Series: Materials Science and Engineering **242** (2017) 012079 doi:10.1088/1757-899X/242/1/012079

Modeling and simulation for micro DC motor based on simulink

Hanxin Shen[1, a], Qiao Lei[1,b], Wenxiang Chen[1,2,c]

[1]. School of Optoelectronic and Communication Engineering, Xiamen University of Technology, Fujian Xiamen, 361005, China

[2]. Xiamen University School of Aerospace Engineering, Fujian Xiamen, 361005, China

E-mail: [a)] hxshen@xmut.edu.cn, [b)] lei929193@163.com, [c)] Corresponding author: wxchen@163.com

Abstract. The micro DC motor has a large market demand but there is a lack of theoretical research for it. Through detailed analysis of the commutation process of micro DC motor commutator, based on micro DC motor electromagnetic torque equation and mechanical torque equation, with the help of Simulink toolkit, a triangle connection micro DC motor simulation model is established. By using the model, a sample micro DC motor are simulated, and an experimental measurements has been carried on the sample micro DC motor. It is found that the simulation results are consistent with theoretical analysis and experimental results.

1. Introduction

Micro DC motor has the advantages of small volume, high efficiency, high torque, convenient control and low cost. It has been widely used in the fields of consumption, office, computer, home appliance, medical treatment and industrial control [1]. Although China's micro-motor production started late, the basic micro-motor manufacturing technology has been mastered, and there are more than a thousand micro-motor production enterprises in Zhejiang, Fujian, Guangdong and other coastal areas [2]. In order to regulate the quality of micro-motor products, the National Micro-Motor Standardization Technical Committee has revised a series of relevant national standards for the quality of micro-motor [3]. Likewise, in order to ensure the quality of manufactured products, enterprises have to conduct 100% testing of the products in the production line. However, due to the lack of effective automated detection methods for testing, the current micro-motor product testing is through artificial methods. The manual detection method cannot be guaranteed because of the experience, emotion and state of the testing operator. So the effectiveness of the testing cannot be guaranteed. Therefore, it is necessary to find an on-line automatic detection method of micro DC motor for the assurance of product quality. There are lots of research in Motor health diagnosis and fault detection at home and abroad, but the research mainly focuses on large motors, AC induction motor and brushless DC motor. The literature of the fault detection and diagnosis of Micro-DC motor is less common [4]. In order to detect the quality of micro-DC motor, paper [5] uses two motors with the same type, to do the same motor, a generator composed of electric power drag system, given a different drive voltage conditions, measuring speed, power voltage, armature current parameters, to determine the five indicators reflect the motor consistency model for motor performance testing. The paper [4] according to the time-frequency characteristics of the armature current, a fault diagnosis model of micro-DC motor

Content from this work may be used under the terms of the Creative Commons Attribution 3.0 licence. Any further distribution of this work must maintain attribution to the author(s) and the title of the work, journal citation and DOI.
Published under licence by IOP Publishing Ltd

based on fault multi-feature quantity is proposed, which is used to diagnose the fault of micro DC motor. The above method can realize the diagnosis of typical faults under certain conditions, but because the model used for the diagnosis cannot fully reflect the electromechanical characteristics of the micro DC motor, the diagnosis cannot be of universal significance. With the modern signal processing technology widely used in fault diagnosis, it should also be fully applied to micro-DC motor fault diagnosis, but because the micro-DC motor mathematical model cannot be resolved, most scholars can only numerical methods and computer simulation methods, such as the establishment of micro-DC motor model and its parameters are estimated, the use of model parameters and micro-DC motor electromechanical characteristics of the corresponding relationship between the micro-DC motor characteristics (armature inductance, coil resistance, induced electromotive force, mechanical rotation Inertia and friction coefficient, etc.)[6,7,8,9], according to the characteristics of the characteristics of the trend can be identified on the fault and then its fault diagnosis and diagnosis, but the motor modeling of the literature mostly for large motors or brushless motors, micro-DC motor modeling research rarely reported. The paper [8-12], modeling of DC motor is studied based on Simulink, but the simulation only simulates the no-load situation and lacks versatility. In this paper, the state equation of micro DC motor is established according to the electrical and mechanical energy transfer relation. By introducing the electromagnetic torque and mechanical torque equation, the simulation problem of DC micro-motor under load condition is solved, and the simulation model based on MATLAB/ Simulink is established. Through this model can be a good motor and a variety of fault motor simulation analysis.

2. Dynamic model of micro DC motor

Fig.1 for the micro-DC motor electrical, mechanical model, In Fig.1, u 、 r 、 i 、 L_m 、 e respectively represent drive voltage, armature coil resistance, armature current, coil inductance and back electromotive force, Φ represent the magnetic flux generated by the armature current, T 、 T_0 、 T_1 respectively represent electromagnetic torque, no-load torque and load torque; J 、 ω respectively represent the moment of inertia and angular velocity.

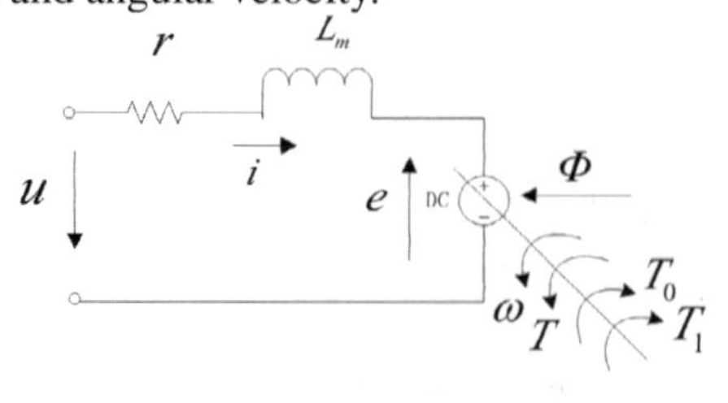

Figure 1. Electricity, machine characteristics model of micro DC motor

According to the principle of motor mechanics, the electromagnetic torque generated by the magnetic flux should be equal to the mechanical torque of the motor drive shaft. Therefore, the electrical and mechanical characteristics of Fig. 1 can be expressed by the dynamic equation [9],

$$\begin{cases} u = ri + \dfrac{di}{dt} + e \\ T = T_0 + T_1 + J\dfrac{d\omega}{dt} \end{cases} \tag{1}$$

Where $e = C_e \Phi n$ is the back electromotive force, $T = C_T \Phi i$ is the output torque, n is the motor speed, C_T is the back electromotive force coefficient, According to (1), in the motor characteristics, the electromagnetic equation and the mechanical equation are coupled with each other and have strong

IOP Conf. Series: Materials Science and Engineering **242** (2017) 012079 doi:10.1088/1757-899X/242/1/012079

nonlinear. It is difficult to obtain the analytical solution. In most cases, the iterative method is used to obtain the numerical solution or the computer simulation.

3. Simualation model

3.1. Equivalent circuit model of micro DC moter

The motor commutation of the state can be described as:

(1) The three connecting ends of the triangular connecting coil windings correspond to three brush commutators K_1, $K2$, K_3;

(2) The three commutators are connected to the positive or negative terminals of the drive power at a certain timing during operation. Can be used to switch the six switches to simulate the three commutators need to achieve positive and negative two working conditions, each commutator corresponds to two switches, a switch will be connected to the drive power supply positive pole, a switch will change Device connected to the power supply negative, according to this idea can be micro-DC motor equivalent circuit model shown in Fig.2:

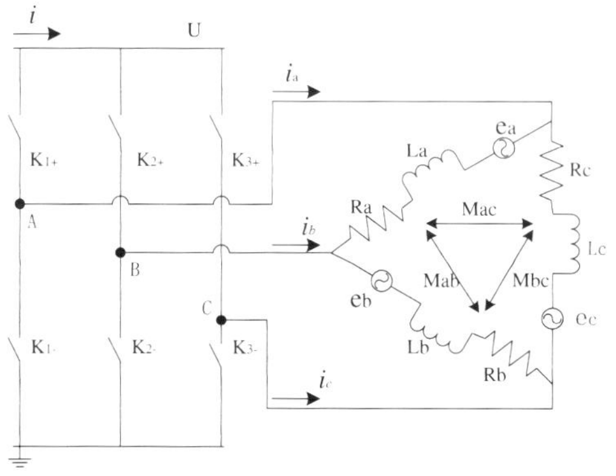

Figure 2. Micro DC motor equivalent circuit diagram

K_{1+} conduction means that the brush K_1 in Fig.2 is connected to the positive pole of the driving power source, K1- conduction means that the brush K_1 in Fig. 2 is connected to the negative pole of the driving power source, and the other switches in the figure are so pushed. 4; R_a is the equivalent resistance of the a-phase winding, L_a is the self-inductance of the a-phase winding, M_{ba} is the mutual inductance of the b-phase winding to the a-phase winding, M_{ca} is the mutual inductance of the c-phase winding to the a-phase winding, and so on.

In Fig. 2, A phase winding voltage $U_a=U_A-U_B$, that is, the voltage between point a and point b; B phase winding voltage $U_b=U_B-U_C$, that is, between b and c point voltage; C phase winding end voltage $U_c=U_C-U_A$, that is, point c and a point of the Between the voltage. The KVL equation is established for the three windings of the armature.

$$\begin{bmatrix} u_a \\ u_b \\ u_c \end{bmatrix} = \begin{bmatrix} R_a & 0 & 0 \\ 0 & R_b & 0 \\ 0 & 0 & R_c \end{bmatrix} \begin{bmatrix} i_a \\ i_b \\ i_c \end{bmatrix} + \frac{d}{dt} \begin{bmatrix} L_a & M_{ba} & M_{ca} \\ M_{ab} & L_b & M_{cb} \\ M_{ac} & M_{bc} & L_c \end{bmatrix} \begin{bmatrix} i_a \\ i_b \\ i_c \end{bmatrix} + \begin{bmatrix} e_a \\ e_b \\ e_c \end{bmatrix} \quad (2)$$

The timing sequence of u_a, u_b and u_c is corresponds to K_1, K_2 and K_3; i for the armature current, that is, through the brush current, ia, ib and ic is the corresponding phase winding current, armature current and three-phase current between (2);ea, eb and ec for the winding back electromotive force.

3.2 Simulation model based on simulink

IOP Conf. Series: Materials Science and Engineering **242** (2017) 012079 doi:10.1088/1757-899X/242/1/012079

Combined with Fig. 2 and the formula (2), the application of the torque sub-module and the induced electromotive force and mutual sub-module can be established micro-DC motor simulation model shown in Fig. 3. In Fig.2, six switches K_{1+}~K_{3-} is corresponding to the corresponding MOSFET in Figure 3 Tube K_{1+}~K_{3-} replacement; *(R-L)a, (R-L)b, (R-L)c* represent the self-inductance and equivalent resistance of the armature *a*、*b*、*c* respectively. The Te-model and the (M + e) -model modules are the torque sub-module and the induced electromotive force and the reciprocal sub-module respectively.

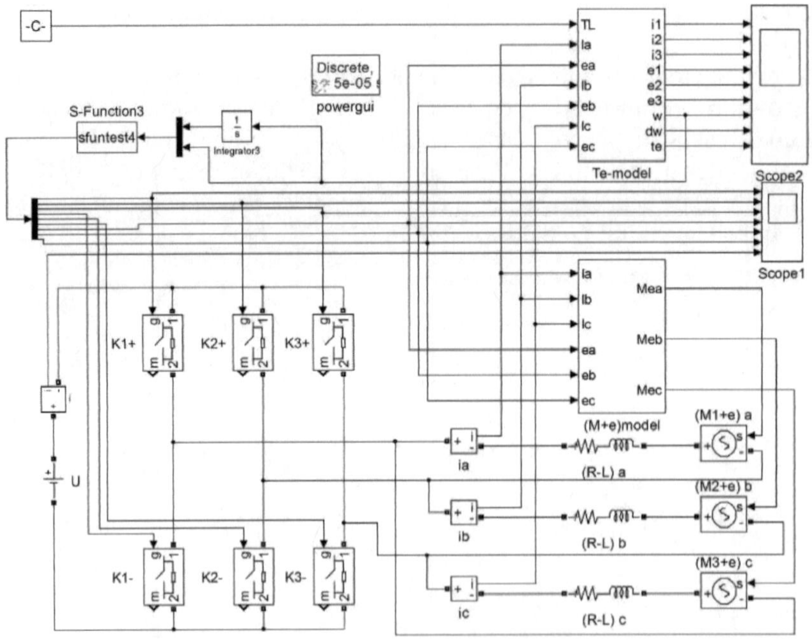

Figure 3. Micro DC motor simulation model

4. Simulation and experiment results

The motor is widely used in the fields of optical drive, printer, home appliance and so on. The social demand is very large, and it is of great theoretical and engineering significance to study the general micro DC motor produced by a motor factory. Its electrical parameters as shown in Tab. 1:

Table 1. Parameters of experiment micro DC motor

Parameter name	Value
Rated voltage	5.0 V DC
Working voltage range	1.5 V ~ 9.0 V DC
Rated torque	0.0015Nm
No-load speed (rated voltage)	6050±650 r/min
Fixed load speed	2800±300 r/min
No-load current	50mA max
Starting current	500mA max

According to the data provided by the manufacturer, the parameters in the simulation model are as follows: self-inductance 4.56mH, mutual inductance 120uH, winding resistance 16Ω.

IOP Conf. Series: Materials Science and Engineering **242** (2017) 012079 doi:10.1088/1757-899X/242/1/012079

In the no-load and full load (0.0015Nm) under the conditions of simulation, the results shown in Fig. 4,

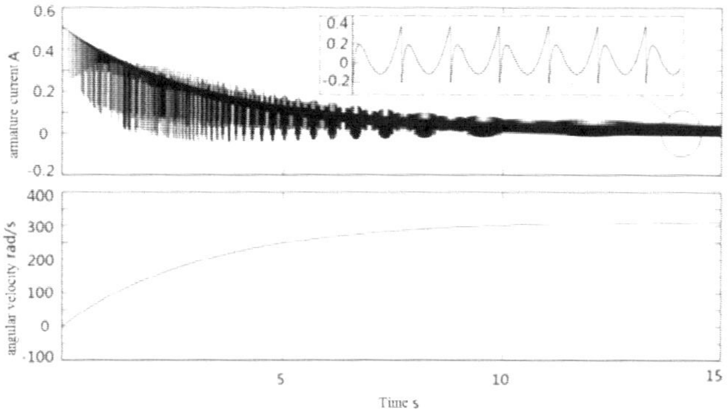

a No load;

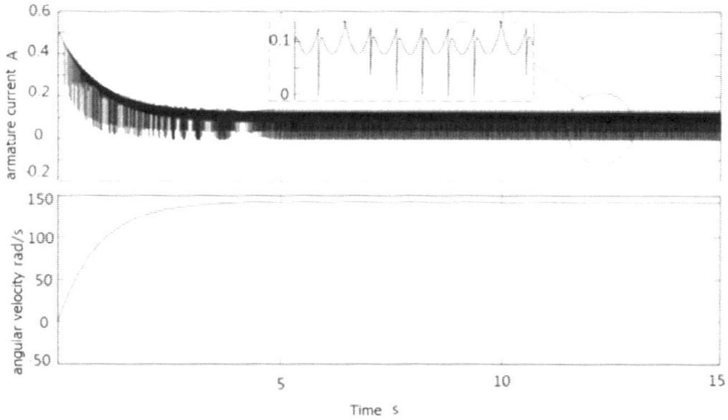

b rated load

Figure 4. The armature current, rotation velocity simulation results

In Fig.4, *a* is the case of no-load simulation results, to be able to see the current waveform, the figure of the circle within the current waveform expansion in the virtual box (later analyzed the same), you can see the armature starting current 500mA and decreases with time index, the instantaneous output torque is 0.008Nm and decreases exponentially. The speed rises from 0 to exponential. After stabilization, the armature current drops to 5mA, the torque drops to 0, the speed is stable at 6000 rpm (Angular velocity of 310 rad/s). B is the simulation results in the case of rated output torque, the armature starting current is 500mA and decreases with the time index, the instantaneous output torque is 0.015Nm and exponentially decreases, the speed starts from 0 and the armature current drops to 15mA, the output torque is 0.0015 rated torque, the speed down to 2030 r / min (angular velocity 280rad/s). The above data show that the simulation results of the simulation model are consistent with the motor data given in Tab. 1.

Take an actual motor, the actual test under rated torque conditions, Fig.5 for the test results. From the two oscilloscope screenshots can be seen, the actual test results and simulation results consistent.

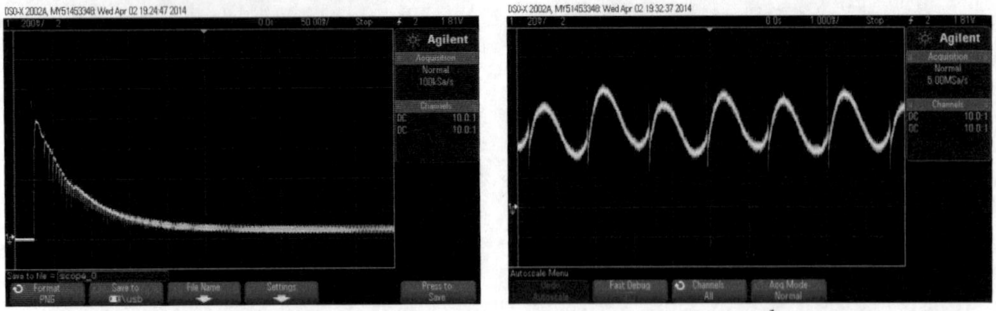

a b

Figure 5. Actual sample motor armature current waveform

For further analysis, Fig. 6 shows the rated load case of a phase winding back electromotive force and armature current in the case of rated load. The winding back electromotive force is sinusoidal. The armature current has six changes in a 360°electrical angle. Each process corresponding to a commutation process, which is consistent with the theoretical analysis of commutation process; in each commutation process armature current for the oscillation changes, the waveform has a second-order system characteristics, and Fig. 2, Fig.3 and the theoretical analysis of formula 2 Consistent.

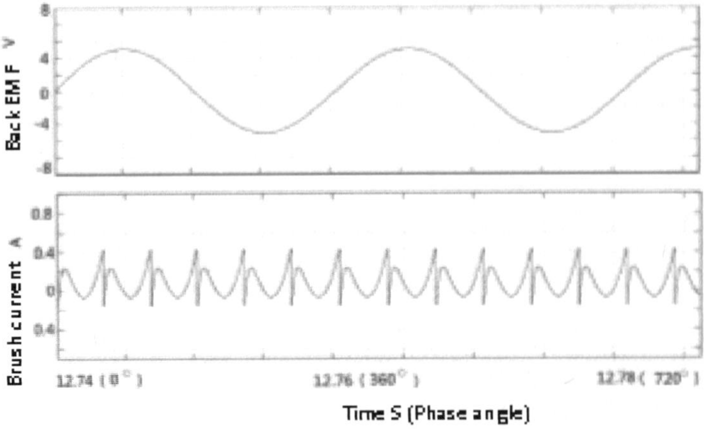

Figure 6. Simulation results of armature current and electromotive force

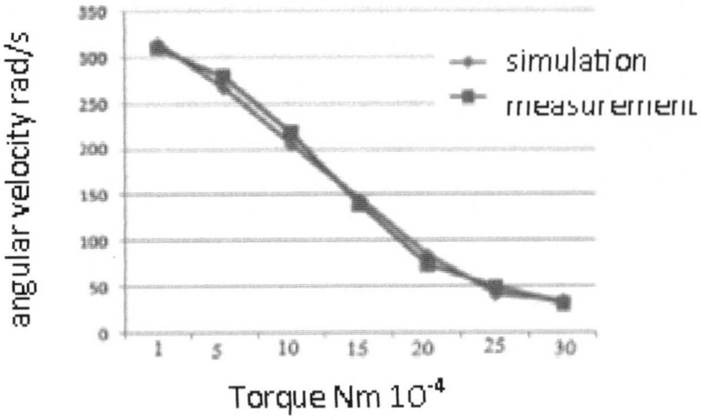

Figure 7. Simulation and measured results of mechanical properties

Fig.7 for the simulation and the actual measurement of the mechanical properties, mechanical properties of the actual measurement is the use of paper [1]. The simulation results are in good agreement with the measured results.

5. Conclusion

Based on the analysis of the commutation process, the micro-DC motor simulation model is established in the Simulink environment based on the electromagnetic torque and the mechanical torque equation. The model is modular and has the generality. The simulation results of the model are in accordance with the theoretical analysis and the actual test results. The work done in this paper can provide a better experimental simulation platform for micro DC motor.

Acknowledgments

This work was financially supported by the Xiamen Science and Technology Project of Fujian Province of China (3502Z20173034) and the Collaborative Innovation Center for R&D of Coach and Special Vehicle Project of Fujian Province of China (2016BJC004).

References

[1] Zhengwen Die. Micro-motor application in electronic equipment, household appliances and cars. Micro-motor. 2005, 38 (1): 76-79.
[2] Lindi Huang, Daxu Huang. Micro-motor market overview at home and abroad. Micro-motor.2007, 40 (7): 75-78.
[3] Qiaobin Guo. The development of micro motor industry in 2007.Micro-motor. 2008, 41 (8): 60-62.
[4] Zhiping Xie, Wei Wu, Jianfeng Hong. DC micro-motor armature current features and fault characteristics study. Electric Machines and Control. 2015, 19 (8): 107-114.
[5] Lianghe Ma. A New Method for Measuring and Performance of DC Micro - motor. Development & Innovation of Machinery & Electrical Products., 2014, 27 (1): 65-67.
[6] MOSELER O, JURICIC D, RAKAR A, et al. Model-based fault diagnosis of an actuator system driven by the brushless DC motor [C]. // proceedings of the Proceedings of the 1999 American Control Conference (99ACC). San Diego, USA: IEEE. 1999, 6: 3779-3783.
[7] MOSELER O, ISERMANN R. Application of model-based fault detection to a brushless D C motor [J]. IEEE Transactions on Industrial Electronics, 2000, 47(5): 1015-1020.
[8] Zhiping Xie. Research on DC micro - motor modeling and fault diagnosis. Xiamen: Xiamen University, 2015
[9] Manlan Liu, Haifang Yu, Shumei Cui. A fault diagnosis method of permanent magnet DC motor. Transactions of China Electrotechnical Society. 2007, 22 (10): 42-48.
[10]Yanfeng Xu, Hanxu Sun, Qingxuan Jia, Wei Xi. Simulation model of fault detection and diagnosis for brushless DC motor.Journal of Vibration, Measurement & Diagnosis. 2015, 25 (3): 190-192.
[11]Yunhua Yin, Bin Zheng, Haoxin Zheng. Modeling and simulation of brushless DC motor control system based on MATLAB. Journal of System Simulation. 20 (2): 293-298.
[12]Dongliang Liu, Yanfei Cui, Meibin Chen. Method of back EMF estimation for brushless DC motor.Transactions of China Electrotechnical Society. 2013, 28 (6): 52-58.

IOP Conf. Series: Materials Science and Engineering **242** (2017) 012080 doi:10.1088/1757-899X/242/1/012080

A new data collaboration service based on cloud computing security

Ren Ying[1], Li Hua-Wei[2], Wang Li_na[1]

[1]Naval Aviation University, Shandong, China

[2]Shan dong Business Institute, Shan dong Yantai, 264001, China

Abstract: With the rapid development of cloud computing, the storage and usage of data have undergone revolutionary changes. Data owners can store data in the cloud. While bringing convenience, it also brings many new challenges to cloud data security. A key issue is how to support a secure data collaboration service that supports access and updates to cloud data. This paper proposes a secure, efficient and extensible data collaboration service, which prevents data leaks in cloud storage, supports one to many encryption mechanisms, and also enables cloud data writing and fine-grained access control.

1. Introduction

With the rapid development of cloud computing [1], the storage and usage of data have undergone revolutionary changes. Data owners can store data in the cloud and outsource data to the cloud. Users of cloud computing will inevitably worry about privacy and privacy protection for private data. Therefore, maintaining the availability and confidentiality of data becomes critical in high-quality data services provided by cloud service providers (CSPs).How to ensure secure data storage has become an important issue[2][3][4].At present, some people have done some work to solve the secure cloud storage and data access problems, such as the introduction of attribute based encryption (ABE) to achieve fine-grained access control. Programs based ABE are data read-only shared services, a one to one encryption mechanism, that is, encrypted data can only be decrypted by a particular recipient. Existing work does not consider multiple user collaboration operations (readable, writable) encrypted data, that is, data collaboration services. This paper proposes a scalable data collaboration service called -SECO.SECO enables secure data collaboration services and supports dynamic user/data operations[5][6].Using a multitier identity based encryption architecture (HIBE), the architecture includes a root private key generator (R-PKG),a series of layer key generators (L-PKG), and an independent working domain. R-PKG just generates the private key for L-PKG, and L-PKG generates the key for the next layer in turn. A domain consists of a domain key generator (D-PKG) and a number of users who collaborate to complete a project.

2. Problem description

Figure 1 depicts the system model of SECO, a multilayer HIBE architecture. It consists of a R-PKG, a series of L-PKG and D-PKG, and individual users. In this hierarchy, the R-PKG generates parameters for the system's all entities and generates the keys for the next layer's L-PKG. Then L-PKG generates the key for the next layer of entity. L-PKG can share the key generation and identity authentication loads for R-PKG. Therefore, the transmission of keys and authentication can be implemented locally. A domain consists of a D-PKG and some users who collaborate to complete a project。In each domain, D-PKG stores a user list UL_{dom} that records the public key of all legitimate users in the domain.

Content from this work may be used under the terms of the Creative Commons Attribution 3.0 licence. Any further distribution of this work must maintain attribution to the author(s) and the title of the work, journal citation and DOI.
Published under licence by IOP Publishing Ltd

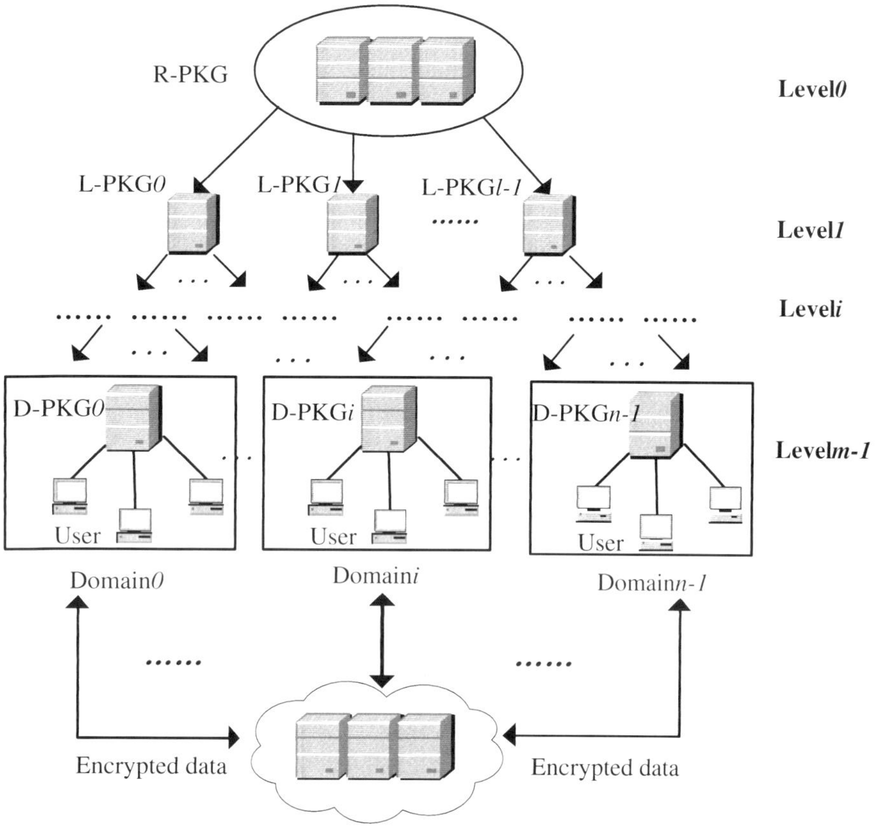

Figure 1 system model

D-PKG sends the latest list of users to all legitimate users in the domain. All entities in the domain store their data on the server side of the cloud. Users use their keys to decrypt data stored on the cloud server. All entities in the domain can dynamically access (read, write, update, etc.) data stored on the server side of the cloud.

In order to implement secure data collaboration services in cloud computing, this paper mainly aims to ensure domain data cannot be obtained by cloud server or an attacker. All attacks can either be active or passive. In order to implement secure data collaboration services, we implement the following security requirements:

1.Fine grained access control: Each user can access only own authorized data and cannot access unauthorized data.

2. Collusion attack: Users cannot conspire to access unauthorized data with other users or cloud servers by sharing their keys

3.Backward secrecy: The data access control policy must ensure that the cancelled user cannot access the data in the cloud server

3. Secure Cloud Data Collaboration Scheme

In order to implement secure cloud data collaboration services, this paper proposes a three layer SECO scheme.

3.1 System model

The system model of the three layer SECO scheme includes the following four entities:

1. Cloud Servers: Manage a large number of servers, have considerable storage space and computing power, and use these servers to provide high quality cloud computing services for the outside.

2.Root key generator (R-PKG): Have a master key and generate the corresponding key for D-PKG.

3.Domain key generator (D-PKG): Requests the key from the R-PKG and generates the key for all the entities in that domain.
4.User: Work together to complete a project, receive their private keys from the D-PKG, and rely on the cloud server to store their data.
Figure 2 depicts a three tier SECO system model, which is a three layers HIBE architecture. From the system model, we can see that it consists of a R-PKG and a range of work domains. A domain consists of a D-PKG, and some users who collaborate to complete a project. In practice, R-PKG is a trusted third party, and D-PKG is a team leader to manage all domain users.

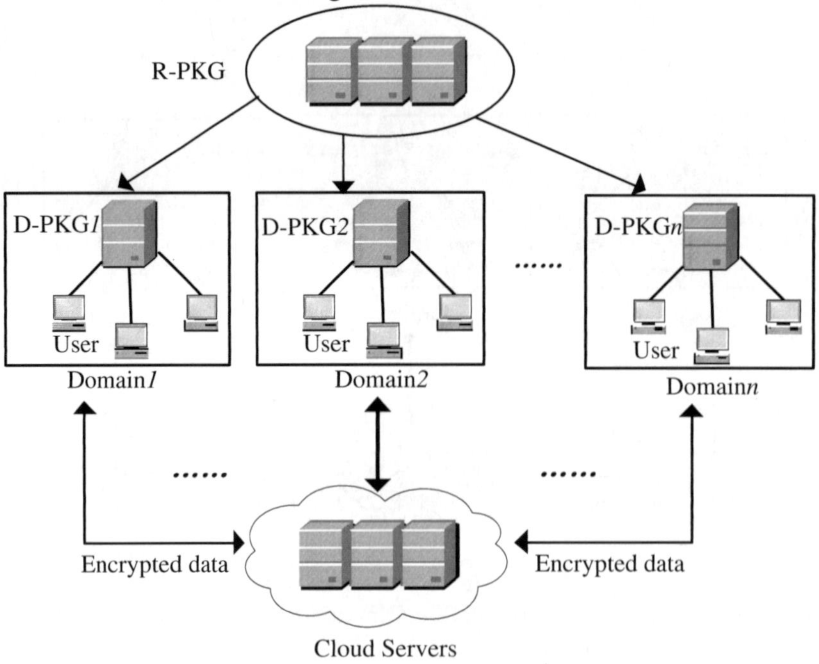

Figure 2 three tier SECO system model

3.2 Design and Implementation
In the three layers of the SECO scheme, *Level*0 = {R-PKG} and *Level*1 = {D-PKGs}.RPKG generates keys for D-PKGs, and then D-PKG generates keys for domain users. D-PKG has two keys: one private key and one master key. D-PKG uses these two keys to generate the private key for all users in its domain. Each user selects a random seed as their master key. In a domain, D-PKG has a unique ID with each user, and ID is an arbitrary string that can be the user's identity card number or mailbox. The user's public key is a ID group that consists of the D-PKG of the domain and the user's own ID, such as (D-PKG's ID, User's ID)[7]. In addition, R-PKG will also disclose some system parameters for encrypting and decrypting cloud data.
3.2.1System initialization. Set K as the security parameter used by the BDH parameter generator IG .
BDH(Bilinear Diffie-Hellman)Parameter generator[8][9]:if the random algorithm IG in polynomial time, by entering a security parameter $K > 0$, the output of two orders of prime q Multiplication cycle group $\mathcal{G}_1$ *and* $\mathcal{G}_2$ and a bilinear mapping: $\hat{e}:\mathbb{G}_1 \times \mathbb{G}_1 \rightarrow \mathbb{G}_2$.Then the random algorithm is the BDH parameter generator. The public key of the user E_i is (ID_{dom},ID_i), where ID_{dom} is ID of D-PKG, andID_i is the ID. of the userE_i.
The R-PKG runs the BDH parameter generator $I\mathcal{G}$, takes the security parameter K as input, and outputs params (system parameters) and a root master key s_0.System parameters include clear text space M and cipher text space C description, and other parameters. The system parameter is open, but the root master key is only R-PKG aware.

R-PKG first executes BDH parameter generator $I\mathcal{G}$ to generate two order of prime q multiplication cyclic group $\mathcal{G}_1$and $\mathcal{G}_2$,and a bilinear mapping $\hat{e}:\mathbb{G}_1\times\mathbb{G}_1\rightarrow\mathbb{G}_2$.

The mapping has the properties of bilinear, computability and non degeneracy.

R-PKG then chooses an arbitrary generators $P_0\epsilon\mathbb{G}_1$ and a random seed $s_0\in\mathbb{Z}_q$,and make $Q_0 = s_0P_0$.Among them,$\mathbb{Z}_q=\{0,1,2,\ldots\ldots q-1\}$.Finally,R-PKG defines four hash functions:$H_1:\{0,1\}^*\rightarrow\mathbb{G}_1$, $H_2:\mathbb{G}_2\rightarrow\{0,1\}^n$, $H_3:\{0,1\}^n\times\{0,1\}^n\rightarrow\mathbb{Z}_q$ and $H_4:\{0,1\}^n\rightarrow\{0,1\}^n$, The four hash functions will be treated as random oracles.

Clear text space is $M=\{0,1\}^n$,Cipher text space is $\mathbb{C}=\mathbb{G}_1^t\times\{0,1\}^n$,Where t is the number of receivers in the cipher text.System parameter is: $params=\langle\mathbb{G}_1,\mathbb{G}_2,\hat{e},P_0,H_1,H_2,H_3,H_4\rangle$ R-PKG's root master key is s_0.

3.2.2 Key generation algorithm. Each D-PKG receives the system parameter (params) from the R-PKG.Every D-PKG randomly selected a seed $s_{dom}\in Z_q$ as the main key. s_{dom} will be used to generate the private key of the domain user. In addition to s_{dom} ,each D-PKG does not allow to generate any other params. R-PKG uses his master key to generate the private key for the next layer of D-PKGs, while D-PKG uses the system params and its own private key to generate the private key for all domain users. Make s_0. the unit element for group $\mathbb{G}_1$.

For each D-PKG *Edom* $\in$ *Level*1, he first selects a random seed $s_{dom}\in Z_q$ as his master key.Given the public key ID_{dom} of E_{dom} ,R-PKG generates the private key SK_{dom} for E_{dom}. R-PKG at first calculated $P_{dom}=H_1(ID_{dom})\in G_1$ and R-PKG then calculate the private key for D-PKG as follows:

$$SK_{dom}=S_0+s_0P_{dom} \qquad (3.1)$$

After the private key is calculated, R-PKG sends the value $Q_{dom}=s_{dom}P_0$ to the E_{dom}.For each D-PKG, he has two keys: a master key s_{dom} and a private key SK_{dom}.The private key SK_{dom} is used to decrypt all data stored in the cloud. Each D-PKG uses its master key and private key to generate the private key for all users in the domain.For each user E_i which is E_{dom} in the domain D-PKG , its ID group is(ID_{dom}, ID_i). E_i randomly selects an element $s_i\in Z_q$ as the primary key.To generate a private key SK_i for E_i.

For each E_i ,D-PKG E_{dom} first calculated $P_i=H_1(ID_{dom},ID_i)\in\mathbb{G}_1$,D-PKG then calculate the private key for E_i as follows:

$$SK_i=SK_{dom}+s_{dom}P_i \qquad (3.2)$$

After the private key is calculated, D-PKG sends the values Q_{dom} and Q_i to the E_i ,where $Q_i=s_iP_0\cdot E_i$ also has two keys: a master key s_i and a private key SK_i. E_i uses s_i and SK_i to decrypt authorized data in the cloud.

3.2.3 Encryption algorithm. In the process of encryption, user input system parameters params,clear text $M\in M$ and ID express group of potential recipients, and then calculate the ciphertext $C\in\mathcal{C}$. After modifying the data of *D*,users use ID(public key)(ID_{dom}, ID_i)group in the domain t receiversrs,and $1\le i\le t$ to encrypt D. The user first for all $1\le i\le t$ calculation $Pi=H1(IDdom, IDi)$ and $P_{dom}=H_1(ID_{dom})$, then, the user selects a random number $\sigma\in\{0,1\}$ and $r=H_3(\sigma,M)$ .Therefore, ciphertext is calculated as follows:

$$C=[rP_0,rP_1,\ldots,rP_t,\sigma\oplus H_2(g^r),M\oplus H_4(\sigma)] \qquad (3.3)$$

IOP Conf. Series: Materials Science and Engineering **242** (2017) 012080 doi:10.1088/1757-899X/242/1/012080

among $g = \hat{e}(Q_0, P_{dom}) \in \mathbb{G}_2$。 From the formula 3-3, you can see that the user encrypts the data D with the public key of the t receiver in the same domain, and sends the ciphertext to the cloud server. It should be noted that since D-PKG manages all domain users, users can obtain the public key of the receiver from D-PKG or other users.

3.2.4 Decryption algorithm. In the process of decryption, D-PKG or user input system parameters params, cipher text $C \in \mathcal{C}$ and their encrypted private key *SK*, and then decrypt the clear text data $D \in \mathcal{M}$.D-PKG can decrypt all the cipher text data in the domain, but the user in the domain can only decrypt the authorized data.

4. Experimental result

We used a three layer SECO scheme to evaluate performance, where $Level_0 = \{RootPKG\}$ and $Level_1 = \{D - PKGs\}$ where all domain users are in *Level*2.We calculate the time cost of each algorithm, and all the experimental results are the average of the 100 experiments. We compare SECO and ABE based schemes, overhead based on the BE scheme, and then evaluate the scalability of the SECO scheme

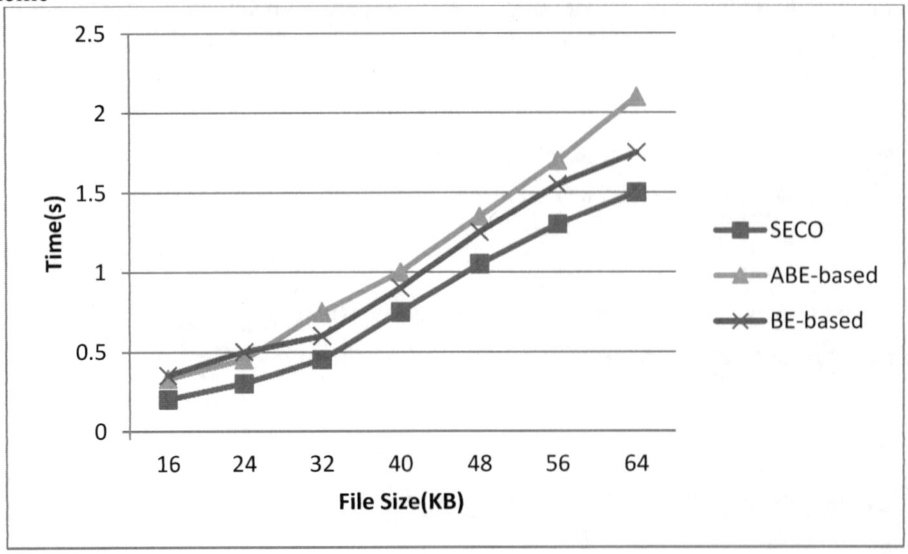

Figures 3 the time overhead of the encryption algorithm in SECO and the scheme based ABE, the scheme based on BE

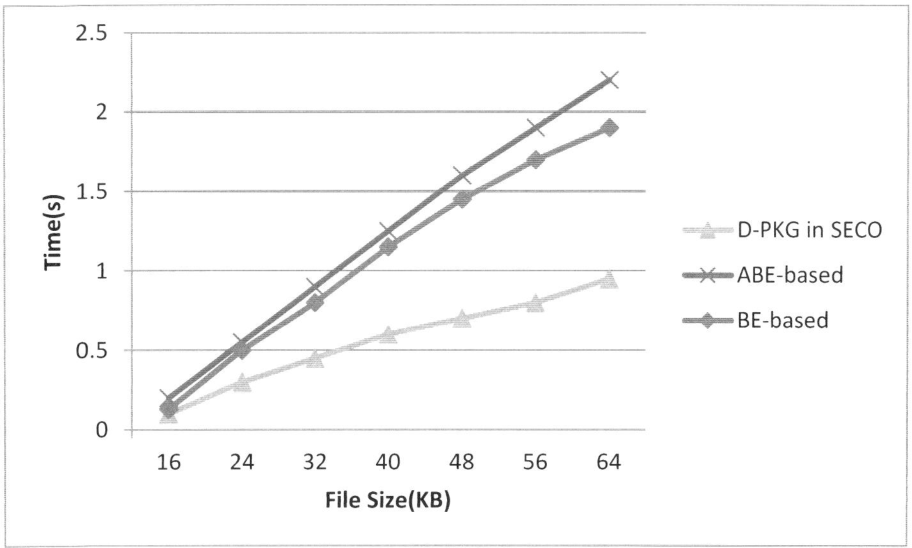

Figure 4 the time overhead of decryption algorithms SECO and scheme based on ABE, scheme based on BE

We evaluated the SECO scheme, the overhead based on the ABE scheme[10][11],and the encryption and decryption algorithms based on the BE scheme. Figures 3 and 4 illustrate the overhead of SECO and ABE based scenarios, based on the BE scheme, as the data size increases. Figure 3 shows the time overhead of the encryption algorithm in SECO and the scheme based ABE, the scheme based on BE.As can be seen from Figure 3,the time overhead of the encryption algorithm increases linearly with the size of the data in all three scenarios. With the increase of data size, SECO spend less time on Expenses scheme based on ABE, scheme based on BE. Figure 4 shows the time overhead of decryption algorithms SECO and scheme based on ABE, scheme based on BE. From Figure 4 we find that the time overhead for all three scenarios varies with the size of the data

5. Summary

This paper studies one to many encryption mechanisms, data write operations and fine-grained access control issues, and proposes a secure Cloud Data Collaboration Scheme- SECO. SECO takes advantage of a multilayer HIBE scheme to ensure data security in the cloud. SECO also implements one to many encryption mechanisms and data write operations, and then implements secure data collaboration services in cloud computing. Security analysis shows that SECO is safe for IND-ID-CCA under BDH assumption, and fine grained access control, anti collusion attack and backward secrecy are also implemented. In addition, we evaluate the performance of SECO. The experimental results show that the SECO scheme is very inexpensive and efficient

Reference

[1] Armbrust M, Fox A, Griffith R,et al. Above the clouds: A Berkeley view of cloud computing[R].[S.l.]: EECS Department, University of California, Berkeley, Tech. Rep. UCB/EECS-2009-28, 2009.

[2] Vimercati S D C D,Foresti S,et al.Encryption policies for regulating access to outsourced data[J]. ACM Transactions on Database Systems (TODS), 2010, 35(2):12:1–12:46.

[3] Samarati P,di Vimercati S D C.Data protection in outsourcing scenarios: issues and directions[C]//Proceedings of the 5th ACM Symposium on Information, Computer and Communications Security (ASIACCS). .[S.l.]: [s.n.], 2010:1–14.

[4] Arora R, Parashar A, Transforming C C I.Secure User Data in Cloud Computing Using Encryption Algorithms [J]. International Journal of Engineering Research and Applications (IJERA), 2013, 3(4):1922–1926.

[5]Vimercat1 S D C D,Foresti S,Jajodia S,et al.Encryption policies for regulating access to outsourced data[J].ACM Transactions on Database Systems (TODS), 2010, 35(2):12:1–12:46.
[6]Samarati P,di Vimercati S D C.Data protection in outsourcing scenarios: issues and directions[C]//Proceedings of the 5th ACM Symposium on Information, Computer and Communications Security (ASIACCS).[S.l.]: [s.n.] ,2010:1–14.
[7]Horwitz J,Lynn B.Toward hierarchical identity-based encryption[C]//Advances in Cryptology–EUROCRYPT. .[S.l.]: [s.n.] , 2002:466–481.
[8]Gentry C,Silverberg A.Hierarchical ID-based cryptography[C]//Advances in Cryptology–ASIACRYPT. .[S.l.]: [s.n.] , 2002:149–155.
[9]Boneh D,Franklin M.Identity-based encryption from the Weil pairing[C]//Advances in Cryptology–CRYPTO..[S.l.]: [s.n.],2001:213–229.
[10]Delerablee C.Identity-based broadcast encryption with constant size ciphertexts and private keys [M]//Advances in Cryptology–ASIACRYPT.[S.l.]: Springer, 2007:200–215.
[11]Gentry C,Waters B.Adaptive security in broadcast encryption systems (with short ciphertexts)[M]//Advances in Cryptology–EUROCRYPT.[S.l.]: Springer, 2009:171–188.

IOP Conf. Series: Materials Science and Engineering **242** (2017) 012081 doi:10.1088/1757-899X/242/1/012081

A livability rating system of a high-rise housing and its computer simulation

Zhengrui Liu

School of Computer Science, Beijing University of Posts and Telecommunications, Beijing 100876, China.

Corresponding author: lbwbj@163.com

Abstract. Aiming at the problems in housing choosing and purchasing in the high-rise residential buildings, this paper considers the factors that affect various livable degrees and analyzes the important degrees of the factors through the Analytic Hierarchy Process (AHP), quantifying the various factors by 10 point scoring system. Accordingly, this paper puts forward a habitable housing index, validating the correctness of the indicators by simulating the process of housing choosing through computer program.

1. Introduction

Due to the conflict between population and land, the construction of high-rise residential buildings in modern cities becomes very common and the housing quality is largely affected by the floor, direction, type of apartment and even apartment number, etc. The common problem for the buyer is how to choose the most desirable and livable apartment in certain unsold high-rise residential neighborhood, but there is currently no a general evaluation standard on the housing habitable degree. Based on the Analytic Hierarchy Process (AHP), the important degree of various factors on housing choice quantitative relation is calculated, thus a 10-point evaluation mechanism is put forward. Meanwhile, the process of 200 buyers selecting apartments is simulated by the computer program, verifying the agreement and integrating degree between the standard and the actual situation of housing choice, and confirming the correctness of this kind of evaluation method.

2. The establishment of the scoring rules

2.1 The application scenario for the scoring rule: the object function

This paper aims to investigate and study the livable degree of an apartment building itself, and takes into consideration the evaluation criteria by which the buyer evaluates the apartment situation in the known residential buildings, defaulting the unrelated factors such as the buyer having selected the location. Based on concerns of buyers about the apartment building itself, six factors are given, containing visual field, safety coefficient, air quality, noise, ventilation and lighting. In the study, the satisfaction degree of the housing buyer on the six factors is confirmed as that on the whole apartment. A 50-storey building with four apartments and two types of model each level is used in the validation process in this paper.

The score is given in the form of a function:

Content from this work may be used under the terms of the Creative Commons Attribution 3.0 licence. Any further distribution of this work must maintain attribution to the author(s) and the title of the work, journal citation and DOI.
Published under licence by IOP Publishing Ltd

$$S = \sum_{i=1}^{6} p_i x_i \quad (1)$$

Pi is an important factor in the impact factor Xi; S means that a room has a maximum of 10 points.

2.2 The importance of influencing factors is modeled using hierarchical analysis

2.2.1. Constructing the judgment matrix. Hierarchical structure reflects the relationship between various factors, and their importance in the eyes of buyers is not the same, so the judgment matrix needs to be constructed to evaluate these factors. By the Analytic Hierarchy Process, matrices for n factors need to be set up (the two criteria layers in this paper are: n1=2, n2=6). Each time two factors of x_i and x_j are taken, and a_{ij} is used to represent the comparison of the relative importance of the two factors, thus the matrix can be established: $A=(a_{ij})_{n\times n}$..

Based on the questionnaire data, according to the ventilation, lighting, air quality, noise, safety coefficient, the order of the vision, the floors and the order of the apartments type, the three-scale comparison matrix is set up, and statistics are as following:

$$C_1=\begin{bmatrix}1&1&2&2&2&2\\1&1&2&2&2&2\\0&0&1&2&1&1\\0&0&0&1&0&0\\0&0&1&2&1&1\\0&0&1&2&1&1\end{bmatrix} \quad (i,\ j=1,2,\ \ldots 6) \quad (2)$$

2.2.2 Single hierarchy arrangement and consistency checking. Consistency indicators are defined as: $CI=\frac{\lambda-n}{n-1}$

The eigenvectors are normalized: {0.345,0.345,0.092,0.034,0.092,0.092}

The vector is the weight vector of the six requested factors. According to the example, it can be judged that the importance of the floor and the housing type are the same, and the weight vector is the weight of the six factors for the target. The substitution formula (1) is:

$$S=0.0920X_1+0.0340X_2+0.3450X_3+0.3450X_4+0.0920X_5+0.0920X_6$$

2.3 The influence factors are quantified by 10-point system

2.3.1 The modeling of the influence factor x1- vision. The vision increases gradually as the floor increases. Questionnaire survey shows most people believe that view field on the floors 1-5 is poor, broad on 6th to 7th floor, and broader above the eighth floor, but the change is small, as shown in the figure below:

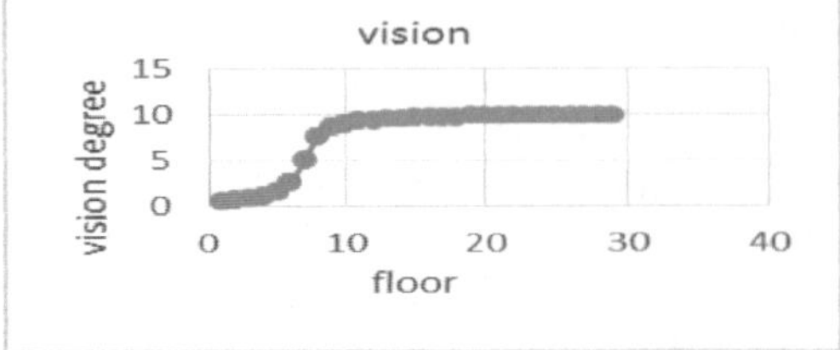

Based on this, the effect of floor f on the vision x1 is:

$$x1 = \frac{10}{PI}\arctan(f-7)+5 \quad \text{(PI is 3.14)}$$

2.3.2 The modeling of the influence factor x2-safety. The results of the survey show that it is easier for residents on the first floor to escape in disaster such as fire or earthquake, but easier to be threatened by theft and other security threats. It has the advantage of easy escape on the 2-4 floors, while the threat of theft is reduced to the highest level of safety. It is easier for residents on the top floors near the roof to escape to the roof of the building in fire, so the top floor is safer than the middle floor.

A function of the ratio of the floor to the maximum floor for safety is shown below:

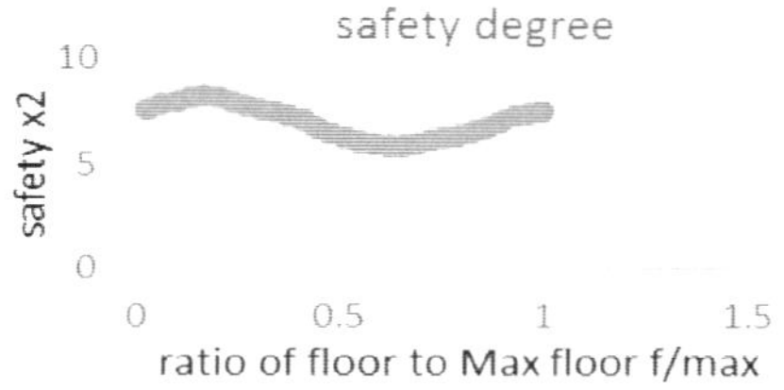

Using the excel fitting function, the effect of the floor f on the view x1 is simulated as:

$$X2=3.925x^6-66.521x^5+112.12x^4-29.842x^3-30.505x^2+11.04x+7.1675 \quad (X=f/max)$$

2.3.3 The modeling of the influence factor x3-daylight. According to the requirement of constructing a residential building, assuming that two adjacent buildings are equal in height and the minimum of the distance between the buildings is required to make the first floor get daylight, which is to satisfy the following relationship: $\tan 26.5° = \frac{H}{d}$

so, $d = \frac{H}{\tan 26.5°} \approx 2H$

Let's say t is the first time the sun shines on a certain floor on the winter solstice, and T is the time of the floor with daylight during the whole day, so it is: T= (12-t) ×2. The (H-h)/d is the sine value of the height of the sun at the time t, and h is the height of the floor above the ground.

$$\text{so: } t = \frac{\arctan(\frac{H-h}{d})}{V} = \frac{\arctan(\frac{H-h}{d}) \times 12}{26.5°} \tag{3}$$

$$\text{then: } x_3 = 10 \times \frac{T}{24} = 10 \times \left(1 - \frac{\arctan(\frac{H-h}{2H})}{26.5°}\right) = 10 \times \left(1 - \frac{\arctan(\frac{\max - f}{2\max})}{26.5°}\right); \tag{4}$$

2.3.4 The modeling of the influence factor x4 ventilation. In the modeling of the ventilation evaluation criteria, it is in consideration that the ventilation quality is not only related to the height of the building, but also related to the ratio (r) of the natural ventilation open area to floor area. The national regulations stipulate that this ratio mentioned above should not be less than 5%, and the room ventilation area of a 20-square-meter house should be between one and two square meters. Therefore, we find that the ratio is 5% to 10%. It is widely believed that housing type has a greater effect on ventilation, and the survey results show that the value is about 2/3 and 1/3. The Evaluation score is:

$$=10 \times (\frac{1}{3} \times \text{height} + \frac{2}{3} \times \text{room type})$$

Of: $V = h^{0.2375} + V_0$ (V_0 is wind speed on the ground)

$$R1 = \frac{V_h}{V_{max}} = \frac{h^{0.2375} + V_0}{H^{0.2375} + V_0}$$

So the final score is:

$$X_4 = 10 \times (\frac{1}{3} \times \frac{h^{0.2375} + V_0}{H^{0.2375} + V_0} + \frac{2}{3} R_2) = 10 \times (\frac{1}{3} \times \frac{(f*p)^{0.2375} + V_0}{(\max * p)^{0.2375} + V_0} + \frac{2}{3} R_2)$$

(The constant p is the height of each floor)

2.3.5 The modeling of the influence factor x5 –noise. The function of the linear sound source attenuation is: $\Delta L=10*\lg\frac{1}{2\pi r}$

Of which r is the line distance from the linear source; l is the length of the line source. In particular, when r/l < 0.1 (such as highways, etc.), it can be regarded as an infinite long line source. At this point, the attenuation value of the sound source at r1 to r2 is: $\Delta L=10*\lg\frac{r_1}{r_2}$

Corresponding to the current modeling situation, $r_1=\sqrt{d^2+h^2}$, of which the constant d is the horizontal distance between the highway and the residential building; h for one certain floor height in the residential building. Because the sound source strength is obtained directly from the calculation, thereby $r_2=1$, the final attenuation relation is obtained i from the above analysis: $L=\lambda-10*\lg\sqrt{d^2+h^2}$

（λ=70dB） then: $L=\frac{\int_{(f-1)*p}^{f*p}70-10*\lg\sqrt{d^2+h^2}dh}{p}$

With constant p for the height of each floor, and constant d for the horizontal distance of highway to the residential building, according to the calculated noise degree, the habitable evaluation score x5 with noise factor in the corresponding floor can be obtained in the corresponding score table.

$$X_5=-0.00002L_3-0.0004L_2+0.0216L+9.8611$$

2.3.6 The modeling of the influence factor x6-air quality. Normally, regardless of the pollution, with the increase of floor height, air quality is gradually getting better, but according to the survey result, air quality above 40 meters high may decline:

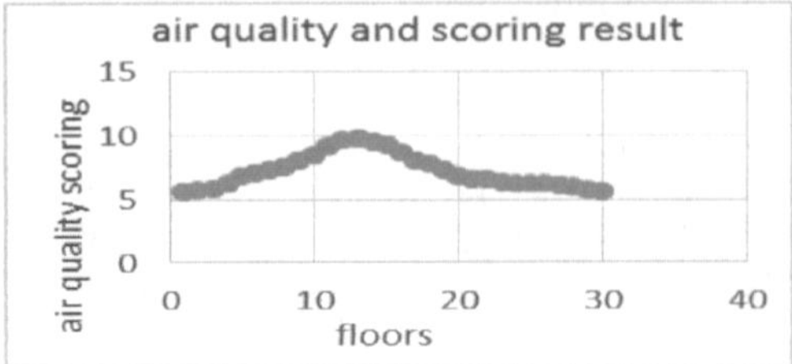

The polynomial fitting is: $X_6=-0.00007f^6+0.00006f^5-0.0015f^4+0.0122f^3+0.0057f^2-0.0002f+5.5627$

On this basis, we have a function of the livable fraction of the house:

$$S=0.0920X_1+0.0340X_2+0.3450X_3+0.3450X_4+0.0920X_5+0.0920X_6$$

3. Computer simulation of the scoring rules

Computer program is taken to test the correctness of the model and to find the fitting degree between the satisfaction of the buyer and the result of the rating. The purpose of the program is to simulate a process of apartment selection. Assuming a residential building has 50 floors; each floor consists of 4 apartments of ABCD, of which the R2 value of AB is 0.4, that of CD is 0.8; the daylight condition of the 4 apartments on the same level are the same (nothing to do with the apartment type); each floor is 3 meters in height, the annual average wind speed is 2 m/s on the ground, and the distance between the building and the nearest highway is 30 meters. Now assume that there are 200 buyers entering to choose apartments in order, and each buyer will choose an apartment. So in all customers' minds, the corresponding weight distribution to the six factors is the normal distribution with the above obtained standard as the average value.

The input of the program is 200 six-tuples, corresponding to the weight in each buyer's mind. The 200 six-tuples are randomly generated by the MATLAB (200 samples to a single element are normal distribution with the average given weight). The output is the results of the 200 buyers choosing flats (200 triples (floor f, room type K (K ∈ {A, B, C, D}), unsatisfying indicators (integer, the bigger the unsatisfying))).

We analyze the similarity of the 200 people's process of choosing apartments and the actual apartments purchase in the reality list. The assumptions list of the program simulation is the following:

max	Total floors	50th floor
P	Height of each floor	3m
R_2	Ventilation	0.4/0.8(AB/CD)
V_0	Average wind speed	2m/s
D	Distance of buildings and the nearest freeway	30m

First, according to the assumption, put the corresponding data into object function S to get new S, a normal distribution of 200 six-tuples is given by Matlab. Based on each six-tuple, with Mathematica, we calculated each buyer's evaluation on each apartment. There is a total of 40000 grades. The unsatisfactory indicator is the ranking of the flat ratings. And the simulation itself used the C language program to simulate the choice of flats for 200 people, and the core code is as follows:

```
for(i=0;i<200;i++){//housing choosing one by one
  for(j=0;j<200;j++){//choosing one by one from the most favored one
    for(k=0;k<i;k++){//compared to the chosen one
      if(in[i][j].floor==out[k].floor&&in[i][j].type==out[k].type){//chosen
      out[i].sadrate++; //unsatisfying indicator +1
      break;//considering minor favored house
      }
      else{//not chosen
      }
      }
    if(k==i){
    out[i].floor=in[i][j].floor;//add to chosen
     out[i].type=in[i][j].type;
    }
  }
}
```

The run program gets 200 three-tuples that reflect the process of the flat buyers' choices and the distribution of dissatisfaction. The analysis concludes:

(1) The objective function s of evaluating livability of is credible, because the simulation is the same as the actual scene of buying an apartment in reality. The buyers can avoid the tallest flat and prefer to buy a higher the middle one. The flats with better ventilation are more popular (C and D).

(2) According to the analysis of dissatisfaction, the later the buyers enter the scenery, the less likely they are to buy a satisfying flat.

4. Conclusion

Decision making of choosing a flat is crucial for every buyer. There are various factors that influence the decision making of the selected flat. This article enumerates the six factors that are selected by the questionnaire to analyze the decision-making process. An apartment scoring system was developed based on the Analytic Hierarchy Process (AHP), and described the composition of the score function and its concrete relationship with the six factors. The score method fully reflects the importance of each influence factor, which could give the buyers intuitive and referable grades, reducing the degree of difficulty in the decision-making. This paper also uses a computer program to simulate the apartment purchase process, to simulate a normal distribution of 200 apartment buyers, and compares with the process of choosing apartments in the early surveyed scene to testify the reasonability of the scoring system. The goal of the future research is to put forward a new evaluation system in combination with the location factor of the building, further reducing the buyers' difficulty of selecting houses.

IOP Conf. Series: Materials Science and Engineering **242** (2017) 012081 doi:10.1088/1757-899X/242/1/012081

References

[1] Zuo, Jun, The Indirect Method of Judging Matrices in Analytic Hierarchy Process. [J], System Engineering. 1988(11): pp35-38
[2] Yang, Yi: "Analytic Hierarchy Process (AHP) to Generate Judgment Matrix Simple Algorithm and Its Application". [N]. *Journal of Hengyang Normal College.* 2014(6): pp53-56
[3] "Housing Design Specification" (gb50096-2011).
[4] "GB 3096-2008" and "Ambient Quality Standards".
[5] Guo, Jinyu, zhang Zhongbin, Sun Qingyun: "Research and Application of Analytic Hierarchy Process" [N]. *Journal of Chinese Safety Science.* 2015(8): pp26-29
[6] "The Comparison of Several Scales in Analytic Hierarchy Process". [J], *Systems Engineering Theory and Practice*, 2004(12): pp39-42
[7] S. Blake et. al. An Architecture for Differentiated Services. RFC 2475, December 1998.
[8] Wieczorek M. A. "The Gravity and Topography of the Terrestrial Planets." *Treatise on Geophysics.* 2012 （02）： 47-51

IOP Conf. Series: Materials Science and Engineering **242** (2017) 012082 doi:10.1088/1757-899X/242/1/012082

Scratch detection technology for product surface based on improved contourlet transform

Kun Hu[1,2,a)], **Shuyou Zhang**[1,b)] **and Jinghua Xu**[1,c)]

[1]Department of Mechanical Engineering,Zhejiang University, Hangzhou 310027,China.

[2]Department of information technology,PLA University of Science and Engineering, Nanjing 210000, China.

[a)]Corresponding author: nanjing_hukun@sina.com, [b)]zsy@zju.edu.cn, [c)]xujh@zju.edu.cn

Abstract. A method of surface scratch detection is proposed in order to realize the automatic detection of surface scratches of molded products. This method uses contourlet transform to decompose the image, extracts the mean and variance of the different sub-bands and different directions for the matrix as the eigenvectors, and uses the distance between the calculation point and the scoring center to identify scratches, to construct a product scratch detection system based on contourlet transform. Experimental results show that compared with traditional methods, the method has higher retrieval accuracy and retrieval speed.

1. Introduction

In the molding process, scratches that differ in the length, direction and depth often exist in product surface. Defective products need to be discarded through scratch detection, which can be realized by real-time monitoring of molded products using machine vision technology. The product surface image collected by the camera isprocessed to decide whether the productis scratched and forwarded to the controller to be dealt with.

The surface scratches of the product are caused by the collision friction of the mold during the manufacturing process. The main features of the collected images are as follows:

1) Due to the differenceinillumination and depth, the surface showed irregular grayscale changes.

2) Due to frequent mechanical collision, fixed image acquisition device cannot guarantee zero geometric deviation between the images collected at different times, and the high frequency signal at the edge of the image has an influence on the extraction.

3) A scratch is usually a multi-directional irregular line of 3 ~ 10pixels width.

In view of the above problems, research has been done from different aspects. Standard detection techniques[1,2,3]deal with regions of interest based on different space-time averages and median filtering. These low-computational-cost methods work perfectly on still images, but are less effective on sequence images, are prone to blurring, and are more severe due to the fixed position of vertical scratches. The Kokaram model mentioned in literature[4,5] is one of the most common airspace models, which proposes cosine distribution for the brightness attenuation of the scratch, and uses the median filter and the Hough transform to perform the initial screening, and then uses Gibbs sampling to get the edge distribution of the center brightness of the scratch to decide whether it is scratch or false alarm. This method tends to missscratches that are shorter than half the length of the image. In

Content from this work may be used under the terms of the Creative Commons Attribution 3.0 licence. Any further distribution of this work must maintain attribution to the author(s) and the title of the work, journal citation and DOI.
Published under licence by IOP Publishing Ltd

literature [6], a vertical scribe detection method based on wavelet analysis is proposed. The algorithm first carries out two-layer wavelet decomposition of the image, and then marks all possible scratches according to the relevant characteristics of the wavelet coefficients in the horizontal and vertical directions. But this method is applied to film scratches, which is different from mold surface scratches.

Based on the above literature, this paper adopts the Contourlet transform in multi-scale analysis, and proposes a product scratches detection algorithm using non-subsampled Contourlet transform. The mean and variance of the transform coefficient matrix are extracted as scratch features by non-subsampled Contourlet transform, to achieve automatic detection of the product surface scratches.

2. Non-subsampled contourlet transform

The Contourlet transform is proposed by M.N.Do and Martin Vetterli based on the idea of clustering base functions near the same scale with a local direction transformation to obtain a sparse representation of an image, also called the Pyramide Directional Filter Bank (PDFB). In the Contourlet transform, multi-scale analysis and direction analysis are separate processes.Laplacian pyramid (LP) proposed by Burt and Adelson [7] is used to obtain multi-scale decomposition, allowing different numbers of decomposition directions for each scale.The base structure of contour segmentapproximates the original image, the support interval of the base is a "long strip" structurewith changing aspect ratio, with the directional anisotropy, so the coefficients representing the edge of image have more energy compaction.

However, LP uses two-bit separable biorthogonal filter group for decomposition and reconstruction, interlaced sampling of the filtered image will produce spectral aliasing. In order to eliminate the aliasing of spectrum and to enhance the invariance of the direction translation, a non-subsampled Contourlet transform (NSCT) using àtrous[8][9]algorithm is proposed. NSCT is composed of non-subsampled pyramid (NSP) and nonsubsampled directional filter banks (NSDFB), where the NSP is used to provide a multi-scale analysis feature, and the NSDFB is used to provide directional features. The structure and frequency domain decomposition of the two layers of NSCT are shown in Fig.1.

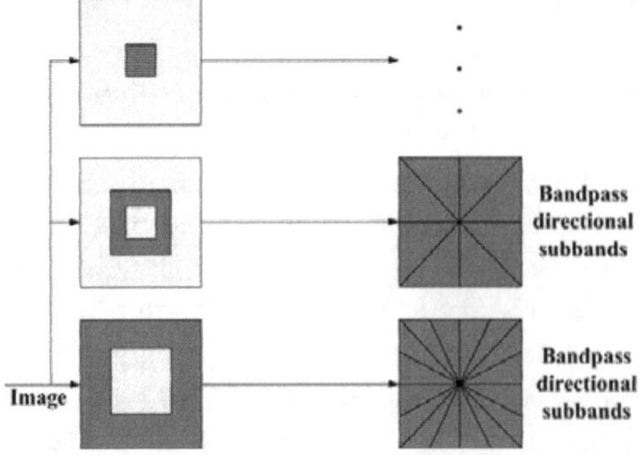

Figure 1.Structure and frequency domain decomposition diagram of NSCT.

On the basis of the Contourlet transform, the required non-downsampling operations in the NSCT construct are implemented by the àtrous algorithm. The multi-resolution decomposition in NSCT is achieved by a translation invariant filter bank that satisfies the Bozout identity (fully reconstructed) condition, instead of LP decomposition. NSP and NSDFB can ensure that the signal is fully reconstructed by the condition that the filter is as follows:

$$H_0(z)G_0(z) + H_1(z)G_1(z) = 1 \tag{1}$$

$H_0(z)$ and $H_1(z)$represents the decomposition filter, $G_0(z)$ and $G_1(z)$ represents the reconstruction filter.

The structure of NSP is similar to that of the non-sampling wavelet transform designed with àtrousalgorithm,the output of the up-samplingis used as the next filter, to avoid the displacement between the pixels due to sampling and to satisfy translation invariance. NSDFB decomposesthe image spectrum into wedge-shaped regions, but only upsamples individual filters, thus obtaining a non-down-sampling filter.

It can be seen from above that the NSCT transform has the advantages of translational invariance and high accuracy in retrieving geometric structural information of the image, in addition to the advantages of Contourlet transform.

3. Surface scratch detection algorithm

3.1 Contourlet Decomposition of Scratch Image

In the Contourlet-based scratches detection algorithm, it is assumed that the noise variance is known and the noise variance is estimated using the Monte Carlo method in the Contourlet field. But it is not realistic to know the noise variance in advance, and the Monte Carlo method is too costly. So we need to estimate the noise variance from the observable image, and to find a simple way to estimate the noise variance in the Contourlet field.

Since the Contourlet transform is not an orthogonal transformation, the noise coefficient $\varepsilon_{j,k}(m,n)$ in the contour wave domain is not equivalent to the mean 0 variance for the σ_ε^2 Gaussian random variable, it is difficult to estimate the noise variance in the Contourlet field. In this paper, we propose a two-step method to estimate the noise variance in the Contourlet field. First, we use the robust median absolute deviation (MAD) to estimate the noise variance in the Contourlet field. Due to the orthogonal nature of the wavelet transform, the estimated noise variance is equivalent to the noise variance at the pixel level. Then, the noise variance of each contour decomposition sub-band can be calculated by the corresponding equivalent Contourlet filter. This process is summarized as follows:

1) First-order wavelet transform of the image, the standard deviation of noise is firstly estimated as:

$$\tilde{\sigma}_\varepsilon = \frac{media(|y_{HH}(m,n)|)}{0.6745}, (m,n) \in y_{HH} \quad (2)$$

In which $y_{HH}(m,n)$is the wavelet coefficients of the first order wavelet decomposition diagonal sub-band.

2) {h,g}represent for low-pass analysis and Laplacian pyramid synthesis filter, respectively. M is a sampling matrix. $d_{j,k}, j=1,2,...,2^l-1$represents the equivalent filter used in the j-layer k-direction sub-band in the directional filter used, while the equivalent Contourlet filter $h^\varepsilon_{j,k}(m,n)$ j = 1,2,...,J, k = 0,1,...,2^l-1 pressed in the z transform field as:

$$H^\varepsilon_{j,k}(Z_1,Z_2) = (\prod_{i=1}^{j} H(Z^{M^{i-1}})(I - H(Z^{M^i})G(Z^{M^i}))D_{j,k}(Z_1,Z_2)) \downarrow M^j \quad (3)$$

In which $D_{j,k}(Z_1,Z_2)$ and $\downarrow M^j$ represent z-transform domain and two-dimensional reduction in $d_{j,k}(m,n)$. Note that the contours of each sub-band are calculated, $\sigma_{\varepsilon_{j,k}}(m,n) = \sigma_{\varepsilon_{j,k}}$ to all (m,n). Then, the J-level and k-sub-bands of the noise variance can be calculated using the following equation:

$$\sigma^2_{\varepsilon_{j,k}} = \sum_{m,n} \left(h^2_{j,k}(m,n)\right)^2 \cdot \tilde{\sigma}^2_\varepsilon \quad (4)$$

The scratch images decomposed by Contourlet transform are as follows:

IOP Conf. Series: Materials Science and Engineering **242** (2017) 012082 doi:10.1088/1757-899X/242/1/012082

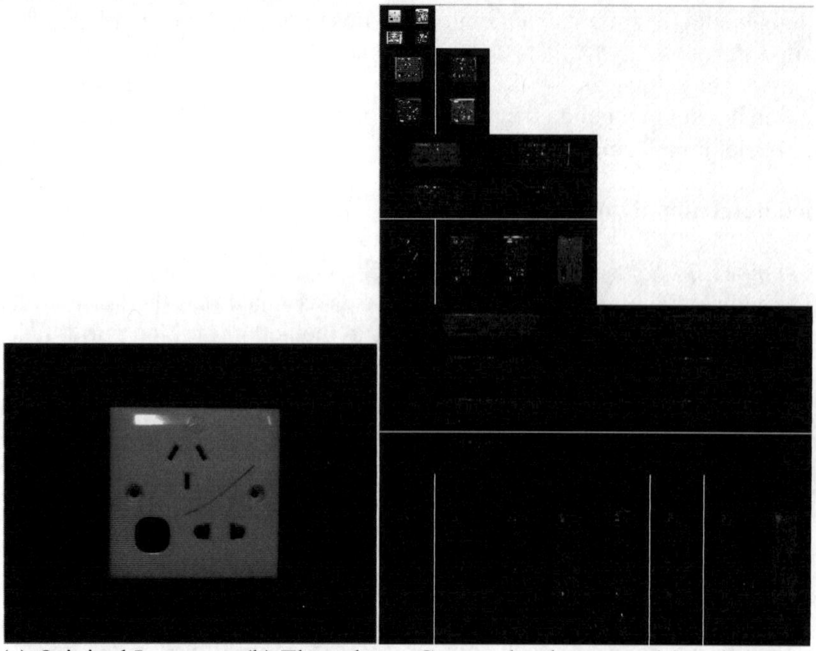

(a) Original Image (b) Three-layer Contourlet decomposition diagram

(c)Reconstructed image

Figure 2.The scratch images decomposed by Contourlet transform

In Fig. 2, (a) shows the product image with scratches, and (b) is the exploded view of each sub-band obtained by three-layer Contourlet transform. (c) is the image obtained after decomposition with main coefficients of each sub-band reconstructed

3.2 Detection of High Frequency Sub-bandScratches

3.2.1 Modulus Maxima Algorithm. Only a few wavelet-based functions with vanishing moments near the singularity can "feel" the singularity of these points and produce important coefficients. Based on the wavelet maximum modulus detection method, it is necessary to determine the direction of the edge curve gradient by the phase angle of the model. The detected modulus maxima also require the connection operation along the edge of the modulus maxima curve. While the filter bank of the contourlet transform is inseparable, the coefficient matrix of each direction sub-band already contains enough direction information, that is, the specific sub-band contains the modulus maxima in the specific direction.

The direction filter decomposes the image in eight directions, and each point $Coeffs^{i}_{j,k}(n)$ in the sub-band has eight directions that can be the equivalent gradient direction of the edge of the $ArgCoeffs^{i}_{j,k}(n)$direction. Compare mod[$Coeffs^{i}_{j,k}(n)$]and the model values of two adjacent elements of r in the Arg($gradCoeffs^{i}_{j,k}(n)$) which is an equivalent gradient direction of Arg$Coeffs^{i}_{j,k}(n)$(i.e. the direction perpendicular to Arg$Coeffs^{i}_{j,k}(n)$), andit can be determined whether the point modulus is local maximum.

$$\begin{cases} 0 & \mathrm{mod}\left[coeffs^{i}_{j,k}(n_1,n_2)\right] < \mathrm{mod}\left[coeffs^{i}_{j,k}(n_1 - r_1(n_1), n_2 - r_1(n_2))\right] \\ 0 & \mathrm{mod}\left[coeffs^{i}_{j,k}(n_1,n_2)\right] < \mathrm{mod}\left[coeffs^{i}_{j,k}(n_1 - r_2(n_1), n_2 - r_2(n_2))\right] \\ \mathrm{mod}\left[coeffs^{i}_{j,k}(n_1,n_2)\right] & other \end{cases} \tag{5}$$

In the formula,$r(n_1)$ and $r(n_2)$ denote the horizontal and vertical coordinate deviations from the $Coeffs^{i}_{j,k}(n)$in the direction of Arg($Coeffs^{i}_{j,k}(n)$).

3.2.2 Detection of High Frequency Edge using Double Threshold Method. The two-threshold edge arrays T1 and T2 are obtained by using the double threshold$\tau 1$ and $\tau 2$ ($\tau 2 = \alpha \times \tau 1$, $\alpha$is the scale factor) respectively. T2 is obtained with a high threshold thus contains fewer false edges, but also some useful edge information are missing. Edge array T1obtained with a low threshold thus retains more information. Therefore, based on the edge array T2, the edge array T1 is used to complement, and finally the edge image is obtained.

The specific steps for connecting the edges are as follows.

Step1 Scan in array T2. When a nonzero pixel P is encountered, track the contour line with P as the starting point until the end point Q of the contour.

Step 2 Takes Q as the reference point, scan in the 8 areas at the Q point ofarray T1, if the non-zero pixel exists, it is a pixel in the outline of array T1.Include it to the array T2as Rpoint. Starting from R, repeat Step 1 until it cannot continue in arrays T1 andT2.

Step 3 After completing the connection of a contour line containing P in the two matrices, mark this contour line as "viewed" and return to Step 1 to find the next contour line. Repeat Steps 1 through 3 until no new contours are found in the array T2.

3.2.3 Description of the Algorithm. In this paper, the algorithm for Contourlet transform of the detected image is described as follows:

1) Contourlet transform the image into two parts: low frequency image and high frequency image.

2) High frequency image edge detection. First set the low frequency coefficient to zero, keep the coefficient unchanged. The sub-bands of each direction on each scale are tested for modulus maxima. The edge matrix is detected and compensated by the double threshold. Finally, the Contourlet inverse transform is carried out to obtain the high frequency edge image.

3) Low frequency image edge detection. The low frequency edge matrix is detected by the canny operator to obtain the low frequency edge image.

4) Fusethe high frequency coefficient and low frequency coefficient to obtain the reconstructed image.

4. Experimental methods and results analysis

In this paper, industrial CCD camera is used. The relevant parameters are as follows: resolution 800 × 600, frame rate 14Hz, pixel size 7.4 × 7.4μm, exposure time 3μs-2s. The background computer configuration is as follows: Inter P4 1.6G, 2G memory.

Figure 3 shows the results of different experimental images:

(a) (b) (c) (d) (e) (f)

Figure 3.Processing results of scratch image

In this paper, the peak signal-noise ratio (PSNR) and the subjective quality evaluation are used as the standard to measure the effect of the detection. The scraches are compared in the wavelet domain and the Contourlet field respectively. The subjective quality is divided into five levels: excellent, good, average, poor and very poor. Table 1 and Table 2 show the PSNR and subjective evaluation for different image restorations in the wavelet domain and Contourlet field.

Table 1. Detection results in wavelet domain

Evaluation	**Wavelet domain detection method**				
	a	b	c	d	e
PSNR(db)	37.020	37.023	36.770	35.560	36.021
Subjective quality	average	good	average	good	excellent

Table 2.Detection results in Contourlet field

Evaluation	**Contourlet domain detection method**				
	a	b	c	d	e
PSNR(db)	43.121	41.820	41.410	41.301	47.273
Subjective quality	good	good	good	good	poor

It can be seen from Table 1 and Table 2 that the detection in the Contourlet domain obtains a larger PSNR compared with that in the wavelet domain. From a subjective point of view, the boundaries obtained by the wavelet domain method are more likely to be blurred, whereasonly minor scratches can be seen on the partial imagein the Contourlet field.

5. Conclusion

To realize automatic detection of scratched molded products with image monitoring system, we propose a surface scratch detection technology based on Contourlet transform. The main innovations of this paper are as follows:

(1) A non-subsampled Contourlet transform image decomposition method for scratched images is proposed. The processed images are not easily affected by factors such as light illumination and scratch depth, and have strong stability.

(2) Modulus maxima and double threshold method are used to detect high frequency edges. This method can sense the category of the corresponding pixels of scratch points in the original image, and accurately detect the scratches in the current image, and has high precision and low false positive rate.

(3) An automatic detection system of the product surface scratchesis designed and implemented. The system overcomes the adverse effects of the production environment on the image of the molded product, and further combines the Contourlet transform image processing technology, which guarantees the robustness of the light and the robustnessof geometric deviation.

Experiments show that the algorithmnot onlysuccessfully detectsand removes the scratches without affecting the rest of the image, but also is simple and effective. This method can be widely used in motion target detection, surface defect detection and environmental change detection, applicable to many fields such as military service, medicine, electronics, packaging, printing, security, intelligent transportation and robot engineering.

Reference

[1] Abreu E.Mitra S K."A simple algorithm for restoration of images corrupted by streaks"[A].In:Proceedings of IEEE International Conference on Circuits and Systems[C], Atlanta,GA,USA,1996,2:730~733.

[2] Venetianer P L,Werblin F,RoskaT,et al."Analogic CNN algorithm for some image compression and restoration tasks"[J].IEEE Transactions on Circuits and Systems,1995,42(5):278~284

[3] Kao O,Engehausen J."Scratch removal in digitized film sequences" [A].In:Proceedings of the International Conference on Imaging Science, Systemsand Technology[C],Las Vegas ,NV,USA,2000,1:171~179

[4] KokaramAC."Motion picture restoration digital algorithms for artifact suppression in degraded motion picture film and video".Berlin:Springer-Verlag,1998.Chapter 2

[5] KokaramAC."Detection and removal of line scratches in degraded motion picture sequences".In:Pro Signal Processing VIII,Vol 1.Trieste,Italy,1996.9:5~8

[6] LaccettiG,Maddalena L,Petrosino A.P-LSR:"A parallel algorithm for line scratch restoration"[A].In:Proceeding of the Seventh International Workshop on Computer Architecture for Machine Perception[C],Palermo,Italy,2005.

[7] P.J.burt, E.H.Adelson. "The Laplacian pyramid as a compact image code". IEEE Trans. Commun, April1983, 31(4): 532-540.

[8] M.J.Shensa. "The discrete wavelet transform: wedding the àtrous and mallat algorithms". IEEE Trans. on Signal Processing, 1992, 40(10): 2464-2482.

[9] S.Mallat. *A wavelet tour of signal processing*.2nd edition.Boston: Academic Press, 1998.

IOP Conf. Series: Materials Science and Engineering **242** (2017) 012083 doi:10.1088/1757-899X/242/1/012083

Review of the research on "structural bionic" method of large sculpture

Jiang Yin and Wenchang Yang

School of Civil Engineering ,Central South University of Forestry and Technology,Changsha,Hunan,China

1425147124@qq.com

Abstract. This paper presented the basic concept of bionic sculpture and summarized the application status of "structural bionic"theory in large bionic sculpture field. Introduced the development trend and challenges of large bionic sculpture and pointed out that the sculpture's "structural bionic" can bring higher mechanical performance of the new structure and system, The evaluation method and structure design for large bionic sculpture are urgently needed.Finally prospected the market of the large bionic sculpture .

1. Introduction

Bionic architecture is based on the constitution of some biological function-organization and image, to explore the construction law of scientific and reasonable in nature. Through the application of the research means to enrich and perfect the building, this research can also promote the architecture structure and construction function and reasonable composition of efficient design layout. Architectural bionics believes that the nature already has a corresponding solution for the problems that encountered in the construction of human. It is the main research subject to find and make use of the constitution rules of the biology (including human being) and apply it to the practice of architecture.

As a special kind of bionic architecture, large bionic sculpture belongs to bionic architecture broadly, which mainly focuses on two aspects: "bionic configuration" and "bionic structure". At present, the "bionic configuration" sculpture mainly focus on the ornamental features and seldom consider the structural stress. The "bionic structure" sculpture focuses on the imitation of the internal constitution of the organism, which promotes the development and utilization of the new and efficient structure form in the field of sculpture.

2. Research status at home and abroad

The "bionic structure" based on engineering mechanics principle, Getting inspiration from the research of organism on morphology and properties that including different structural levels (micro, meso and macro) , and then simulating the material, structure, system which can achieve the function of the traditional structure cannot be achieved .Finally providing a new design thought to improve the utilization efficiency of engineering structure[1] . Both the sculpture and the generalized architecture must rely on the structure, and the structural bionics has become a very rapid branch in the development of bionics.

The development and application in the field of "bionic structure" began at the beginning of the last century. The structural engineers created a series of new bionic structural systems by using modern

Content from this work may be used under the terms of the Creative Commons Attribution 3.0 licence. Any further distribution of this work must maintain attribution to the author(s) and the title of the work, journal citation and DOI.
Published under licence by IOP Publishing Ltd

technology. At present, the research achievements in the field of "bionic structure" have gradually moved towards engineering practice:

(1)Tree structure [2-3]: the branches of the tree are graded to form more branches and the mechanical load is very reasonable. Meinhard v.Gerkan and Karsten Brauer[4] according to the principle of tree structure designed of the Stuttgart airport lounge in 1980 , see figure 1. The tree structure can save the material with beautiful shape and easy to get the larger space by the smaller rods. Nowadays the utility model is widely applied to the supporting system of the exhibition hall, the venue and the large span space in foreign countries.

(2) Membrane structure [5]: there are many bubbles in nature, such as soap bubbles. In 1967, Freiotto [6] according to the working principle of the bubble structure designed the West German Pavilion in Montreal, Canada, with light and transparent organic film as the top. Pioneered the commercial use of membrane structure. The dome stadium in Tokyo, Japan is also a typical case of membrane structure design, see figure 2. In recent years, many scholars study on the structure of this new form. such as: Wang[7] Combined with the modal analysis and fold analysis to study the design elements that the flexible membrane space structure, asymmetric tension distribution, vibration characteristics and the distribution of fold detailed.

(3) Suspended-cable structure: the spider web is soft and flexible, so it can make full use of the tensile properties of the fiber structure. Inspired by the spider web that people designed a very economical suspended-cable structure [8]. The suspended-cable structure is characterized by the fact that all the "components" are drawn to ensure maximum material properties, while the structure has a smaller structural weight and a greater economic span. The appearance of suspended-cable structure promotes the development of suspension bridge. Japan in 1998 built a span of 1991 meters of the Akashi kaikyo bridge[9] , which have two main cables each about 4000m in length and 1.12m in diameter, consisting of 290 thin strands and weighing about 50 thousand tons. So far this bridge is still the world's first suspension bridge, see figure 3.

(4) Shell structure: the shell of the scallop can bear the enormous pressure of the sea water, the wavy of the geometric surface make the shell obviously improve the compressive strength under the same thickness. Dong Shilin [10] from the structure of deep Scallop's shaped shell, proposed a new type of shell structure -Scallop shell structure, the surface of this structure is wavy, which have graceful shape, material saving and reasonable structure stress.

(5) Grid structure [11-12]: The vein of Wang Lian's leave interwoven looks like reticulation, which bear the organization overall load, this structure stress is very reasonable. Philip Cox and Richardson Taylor [13] according to the structure of the imitation grid designed the court Museum of Sydney Olympic Games--"Stadium Australia".

(6) Imitated bamboo cylinder structure: hollow bamboo body to ensure the overall stability of compression, the bamboo joint ensure local compression and stability of bamboo and increased the level of lateral stiffness and bending stiffness[14]. Li Zuyuan [15] follow the characteristics of bamboo structure designed the Taipei 101 building, the height of Taipei 101 building is 509 meters, which have 101 floors on the ground and 5 floors underground, from the beginning of the twenty-seventh layer every 8 layers with transverse truss around the column, similar to the role of bamboo, which greatly improves the rigidity and stability and seismic performance of the building,see Figure 4.

Fig.1 Tree structure

Fig.2 Membrane structure

Fig.3 Cable structure

Fig.4 Imitated bamboo cylinder structure

In China, the research on the bionic structure of large sculpture started late. In recent years, the scholars mainly focused on the study of the mechanical properties, design theory and methods of bionic building (or sculpture). Zhou Qingqing [16] analyzed the swallow shaped space truss structure static performance, stability and dynamic performance and evaluated the safety of the structure, she explore the impact of support form and different support conditions on the performance of structure and get the optimization design of support and support for building . He Yong Jun [17] based on bamboo's excellent mechanical properties validated the rationality of hoop layer arrangement in imitation bamboo type parking structure. Zhang long [18] studied the application of structural bionics in the design of large span buildings. It is pointed out that the principle of systematic and scientific design should be followed when Structural bionics in the design of large span buildings. Zhang Qian [19] presents a new system of forest structure with the ability of structural synergy based on the method for finding the shape that based on continuous broken line cable element with 3D tree structure.Wang Ke [20] have done topological optimization for crane boom based on Wang Lian as bionic prototype and analyzed the optimization model's statics and modal characteristics. It is proved that the bionic design method is effective in the optimization of the truss cantilever structure. Based on the theory of bionic design, Wang Qunwang [21] proposed the imitation bamboo tower structure and analyzed its static and dynamic characteristics and stability based on the bionic design theory. It is proved that the bionic design theory applied to the design of tower is scientific.

To sum up, the "bionic structure" focuses on imitating the structure or structure of various biological prototypes. The natural environment determines the diversity of organisms in order to adapt to the external environment and structure evolve also vary, so the "bionic structure" is different from the general structure design, it has a strong personality and pertinence, it is difficult and not have to follow the rules of the existing structure, in fact, there may break the traditional concept of bondage and achieve a major breakthrough in the form of force structure when do personalized structure bionic design according to the different biological prototype the bionic structure design , in the history it appeared similar precedent such as tree structure and shell structure etc.

3 Development trends and challenges

In recent years, with the development of large bionic sculpture market, the prospect of sculpture industry is very considerable. At present, large bionic sculpture is developing towards modernization and complexity. However, the stress analysis, fabrication and installation, design method, evaluation and acceptance for large sculpture still rely on traditional experience for a long time, so it is difficult to adapt to the development trend of large-scale and complicated bionic sculpture. Specific performance:

(1) Bionic sculpture often designed by the artist dominantly. They only focus on the external shape of the bionic sculpture and ignoring the bionic structure, which often lead to the existence of security risks or construction materials unreasonable use of bionic sculpture.

(2) The organism's perfect external form is not only an art, but also a result of its adaptation to the environment. This form often with a reasonable degree of mechanical efficiency. In the future the "bionic configuration" which belong to the field of large bionic sculpture in addition to emphasizing the ornamental value, should also analyze the macroscopic mechanical properties of sculpture shape in order to get highly ornamental value, at the same time try to retain the mechanical logic its biological prototype.

(3) The large bionic sculpture is a special kind of building, there is no uniform design and evaluation methods and implementation steps can be followed, it cannot be directly applied to various types of design specifications. There is an urgent need for large bionic sculpture to develop a "performance goal" control system (such as seismic index, displacement control index) and establish a design method and general steps for large bionic sculpture to guide similar engineering practice.

To sum up, there are still a series of important problems in the field of large-scale bionic sculpture. It requires based on structure bionics theory and using advanced methods and means of modern mechanical engineering field to develop the large-scale bionic sculpture's design and evaluation method in system, which can guide the similar engineering practice and stimulate the healthy and rapid development of sculpture industry chain.

4 Conclusion

At present, the research on the structure of the large bionic sculpture field in our country started soon, although the "bionic sculpture" belongs to the category of building broadly, the exterior and interior structures are significantly different from the traditional buildings (such as buildings, bridges, etc.). Bionic sculpture is a kind of special building [22]. Compared with the traditional building the bionic sculpture can mimic a variety of biological prototype that almost covering all the life body in the nature. This is a very promising field of innovation and development direction in structural engineering and bionics. Personalized bionic structure design process of large sculpture, most likely getting the new structure of higher mechanical efficiency, the general design and evaluation methods should be improved, the resulting of social and economic benefit is very considerable and also can promote the transformation and development of traditional large sculpture industry.

References

[1] Lv Fu Xun.Towards 21 Century: the past and future of architectural bionics[J].*Architectural Journal*,1995,06:14-17.

[2] Wu Yue, Zhang Jian Liang ,Cao Zheng Gang.Form-finding analysis and engineering application of branching structures[J].*Journal of Building Structures*,2011, 32(11):162-168.

[3] Zhu Zheng Hao. The bionic optimization and analysis and calculation of the cast-steel joint with three branches[J]. *Applied Mechanics and Materials*. 2014, Vol(548-549):548-549.

[4] Zeng Xiao Hong. Stuttgart Airport[J]. *World Architecture Review*,2000, 09: 116-121.

[5] Noor, M.S.M. HooiMin Yee, Kok Keong Choong, et al. Tensioned Mem brane Structures in the Form of Egg Shape[J]. *Applied Mechanics and Materials*, 2013,Vol(405-408), 989-92.

[6] Zhou Wei Hua.Frei Otto and Light-qualitied load-bearing Structure[J]. *World Architecture,*1998,08:70-73.
[7] Wang Changguo, Li Yunliang, Du Xingwen, He,etal. Simulation analysis of vibration characteristics of wrinkled membrane space structure[J]. *International Journal of Space Structures,*2007, 22(4):239-246.
[8] Ye Qing hui，Tao Jian,Yu Dong Liang，Li Zhong Xue.A Survey of bionic design in spatial structures[J].*Structural Engineers，*2010, 26(3): 13-18.
[9] Jin ZengHong. Brief introduction of Akashi [J]. *Foreign bridge,*2001,02:13-18.
[10] Zhuo Xin,Dong Shi Lin.Research on structual from bionics architecture of scallop form[J].*Spatial Structures,*2004,04:19-22.
[11] Park Young-Min. Semi-active vibration control of space truss structures by friction damper form aximization of modal damping ratio [J].*Journal of Sound and Vibration,* 2013, 332(20): 4817-4828.
[12] Sofla A.Y.N.A rotational joint for shape morphing space truss structures[J]. *Smart Materials and Structures,* 2007, 16(4)：1277-1284
[13] Wang Da Sui，Lu Dao Yuan，Huan，et al.Structure design of jinta in Tianjin[J].*Journal of Building Structures,*2009,3 (S1)：1-7.
[14] Richard Green B.E.Sydney Homebush Multi Use Arena,Proceedings of the LSA98 Conference，Australia，2000:11-19.
[15] Li Zu Yuan.Taipei 101 building[J].Architectural Journal，2005,05：32-36.
[16] Zhou Qing Qing.Analysis on the mechanical properties of Swallow-like Space Truss Structure[D].Tianjin：Tianjin University，2012。
[17] He Yong Jun,Liu Xiao Hua，Zhou Xu Hong.Research on Arrangement of hoop-layers of imitating bamboo drum type tridimensional parking structure[J].*Journal of Hunan University,*2013,40（4）:1-7.
[18] Zhang Long。Bionic In Large Span Architectural Design Applied Research[D].Tianjin：Hebei University of Technology，2013.
[19] Zhang Qian.Form-finding Optimization Analysis and Structure Performance Research of Tree Structure[D].Tianjin：Tianjin University，2013.
[20] Wang Ke.Optimization Design of Tower Crane Boom Based on Structural Bionic Method[D].Taiyuan：North University of China,2015.
[21] Wang Qun Wang.Bionics design and mechanical properties analysis of large-scale wind tubine Tower[D].Lanzhou: Lanzhou University of Technology,2016.
[22] Zhu Yu Long,Jiang Huan Jun.Applications and future outlook of bionic architecture of steel structures in china[J].*Structural Engineers,*2014,30（2）:164-169.

IOP Conf. Series: Materials Science and Engineering **242** (2017) 012084 doi:10.1088/1757-899X/242/1/012084

The numerical simulation study of hemodynamics of the new dense-mesh stent

Jiali Ma, Zhishan Yuan, Xuebao Yu, Zhaowei Feng, Weidong Miao, Xueli Xu, Juntao Li

GRIMED Co., Ltd,, Beijing 102200,P. R. China

E-mail: majialidd@126.com

Abstract. The treatment of aortic aneurysm in new dense mesh stent is based on the principle of hemodynamic changes. But the mechanism is not yet very clear. This paper analyzed and calculated the hemodynamic situation before and after the new dense mesh stent implanting by the method of numerical simulation. The results show the dense mesh stent changed and impacted the blood flow in the aortic aneurysm. The changes include significant decrement of blood velocity, pressure and shear forces, while ensuring blood can supply branches, which means the new dense mesh stent's hemodynamic mechanism in the treatment of aortic aneurysm is clearer. It has very important significance in developing new dense mesh stent in order to cure aortic aneurysm.

1. Introduction

With the improvement of the living standards of people, aortic disease, a great vessel dilated disease which has higher morbidity and mortality, increases rapidly [1]. Intracavitary minimally invasive interventional prosthesis has become a new clinical treatment method and is widely used in the treatment of all kinds of great vessels aneurysm disease based on the rapid development of interventional radiology and the research of intravascular stent [2].The development and clinical application of the new type of stent which has greatly increased the success rate of surgery and simplify the clinical operation is a kind of mainstream treatment.

This research develops a new dense mesh stent mainly based on the current difficulties of the process of interventional therapy in which commonly the coated stent is used in the treatment of aortic aneurysm and the blood-supply of branch blood vessels is blocked at the same time. The new stent made of nickel titanium shape memory alloy mainly uses excellent phase change hyperelasticity, shape memory properties, good biocompatibility and corrosion resistance of the nickel titanium shape memory alloy and is a very excellent biomedical material. Based on the theory of reducing the mesh of stent making the hemodynamic change obviously [3], multi-layer dense mesh stent will be produced and it will change the smooth of blood flow in aortic aneurysm, then it will allow blood to supply to the branch blood vessels and reduce the blood pressure and shear force of tumors so as to the goal of treatment will be achieved.

2. Materials and Methods

2.1 The Simulation of the Abdominal Aortic Aneurysm

For the convenience of the numerical simulation calculation and to ensure that the approximation of results of calculation and clinical actual state, the model size and the actual size is not completely

Content from this work may be used under the terms of the Creative Commons Attribution 3.0 licence. Any further distribution of this work must maintain attribution to the author(s) and the title of the work, journal citation and DOI.
Published under licence by IOP Publishing Ltd

consistent, and the size is chosen to ensure that the Reynolds number approximation. Abdominal aortic aneurysm numerical model for geometric parameter is set as follows: abdominal aorta's diameter D = 12 mm, aneurysm's diameter d1 = 30 mm, branch blood vessels' diameter d2 = 6 mm, artery' length L = 180 mm, distance between the entrance and aneurysm L1= 60 mm, the length of branch blood vessels L2 = 40 mm. Artery blood vessels and branch length are comparatively long to ensure the free flow of blood, which is closer to the clinical condition.

2.2 Physiological Condition Assumptions
Due to the special nature of the blood vessels and blood, some physical assumptions and reasonable simplification needed to be done before the simulation of stent implanting blood vessels and they can be as close to the ideal type of hermal wall and blood flow as possible 1) due to the numerical simulation results of the rigid hemal wall and flexible hemal wall are similar and the rigid hermal wall is easier to be handled, so the blood vessel and tumor walls are assumed rigid hermal wall; 2) the blood flow is named as Newtonian fluid [4], which is incompressible. The blood flow form in the aneurysm [5] is set for laminar flow. Pulsed flow is used as blood flow. Heart rate is adopted as 72 times/min and the constant is adopted as the coefficient of blood viscosity; 3) the form of aortic blood flow is laminar flow generally. The blood density: ρ= 1.05 x 103 kg/m3, the coefficient of kinematic viscosity = 0.0035 Pa/s, V = 0.3 m/s, D = 0.12 m, Re = 1080, which accords with the physiological condition of human body.

2.3 The Setting of Meshing and Boundary Conditions
Gambit software is used for meshing and the quadrilateral mesh grid is selected to divide and verify the model mess of abdominal aortic aneurysm. In addition, the grid number is 11280, node number is 11789. Setting the inlet, outlet and boundary conditions: the inlet condition is set as speed inlet, the average speed v = 0.3 m/s, the pressure of outlet including aortic outlet1 and branch vascular outlet2 takes the average pressure in artery blood vessel. In this paper, outlet pressure is set as P = 100 mmHg (about 13.3 KPa) and abdominal aortic aneurysm boundary is set as solid wall boundary.

2.4 Calculation
With the fact of hemodynamics numerical simulation about abdominal aortic aneurysm in the condition of having and being without stents, the speed, shear force, pressure, streamline contours and the vector diagram are o calculated under the above two cases by using Fluent, the commercial software of fluid mechanics. And through the map, the condition of different numerical changes and flow regime changes can be directly observed.

3. The Result Analysis

3.1 The Streamline Characteristics of Dense Mesh Stent Before and After Implanting
As the flow chart (Fig. 3.1) of dense mesh stent before and after implanting shows, a part of blood in the artery without implantation of stent flows into the tumor cavity and forms vortex, then a small amount of blood flows into the branch blood vessels after the vortex have twisted in tumor cavity, because most of blood flow in the tumor cavity don't be in the process of blood-supply for the branching vascular but twists in tumor cavity; When the dense mesh stent is implanted in the abdominal aortic aneurysm, a small amount of blood flow in arteries flows into the tumor cavity, and all flows into the branching vessels after slight twist. This suggests that when the stent is placed, it does not affect the blood supply of branching vessels but limits the small amount of blood flow that originally twists in the tumor cavity to flow into the tumor cavity at the same time, which radically reduced blood flow pressure and wall shear force in the tumor cavity.

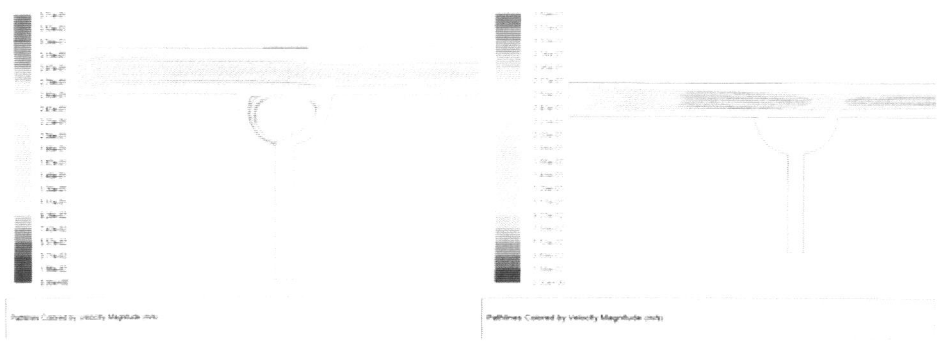

Figure 3.1 flow chart (a. no stent implanted b. with stent implanted)

3.2 The Velocity Field

The velocity distribution within the abdominal aortic aneurysm that having no stent implanted is listed as flows: Blood flow velocity in tumor cavity is lower than aortic's; Tumor cavity shows flow distribution in large eddy; Fringes of tumor cavity appear higher velocity.

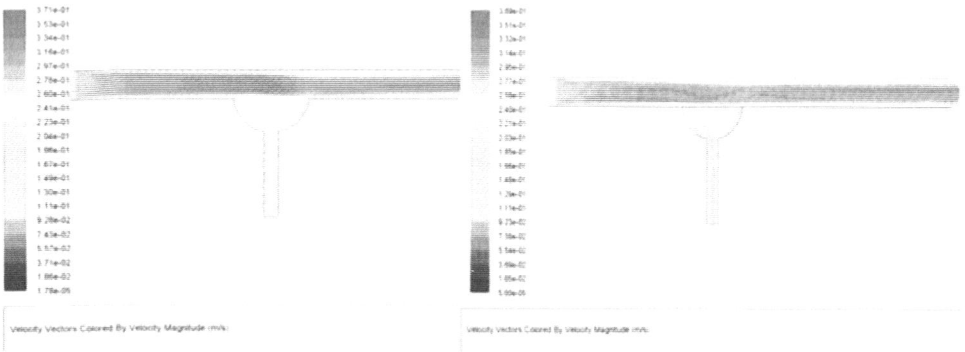

Figure 3.2 velocity chart (a. no stent implanted b. with stent implanted)

As the figure 3.2 shows: 1) The tumor cavity velocity with stent implanted was lower than that without stent implantation, and a large area of low speed flow area is showed as a whole; 2) The disturbance flow field in tumor cavity arises after the dense mesh implanted, and both the decline of velocity and flow distribution of disturbance flow can reduce the impact of the blood flow to tumors and reduce the risk of tumors' fracture. After calculating, it turns out that the flow rate in tumors with the dense mesh stent implanted is reduced by about 50.74%, and the flow rate in the neck of branch tumor is reduced by about 62.44%. This shows that the dense mesh stent implantation significantly reduces the blood flow velocity of the neck of tumors and tumors, and the risk of tumors' fracture is greatly reduced.

3.3 Pressure Field

From the figure 3.3, it can be directly observed that the color of tumor cavity with stent implanted is darker, and the disturbance state arises in the pressure chart. After numerical simulation calculating, it turns out that the pressure of tumors, tumor necks, branch vessels, and so on, with stent implanted is dropped respectively by 46.38%, 43.2% and 59.44%. This suggests that: 1) the blood pressure of tumor cavity, tumor cavity wall, tumor's neck and branch vessel is decreased, and the pressure distribution area is relatively scattered. So this can reduce the impact of the blood flow to tumors and the risk of tumors' fracture is also reduced; 2) the way of disturbance of flow state that in a model with stent implanted suggests that the decrease of the static pressure makes the flow state in the blood flow change, which will help for the thrombosis and create favorable conditions for the tumor wall reinforcement.

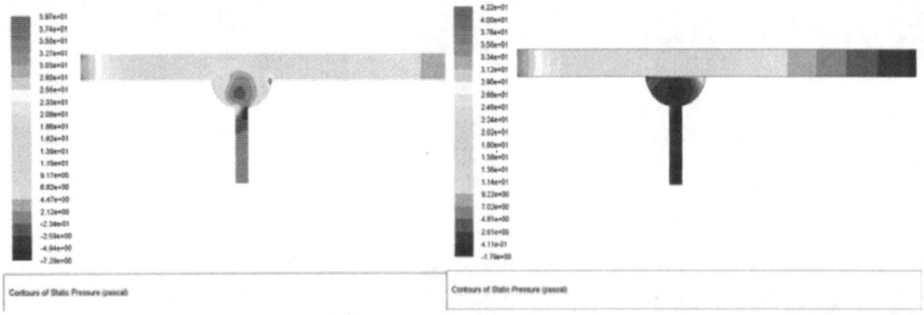

Figure 3.3 pressure field (a. no stent implanted b. with stent implanted)

3.4 The Wall Shear Force

As the figure 3.4 shows that the wall shear force in tumors with stent implanted is greatly reduced, especially in the place where tumors are the most easily broken[6] ---- the wall shear force in the places where tumor necks of the connection between tumors and the arteries, and tumors and the branch vessel reduced a lot.

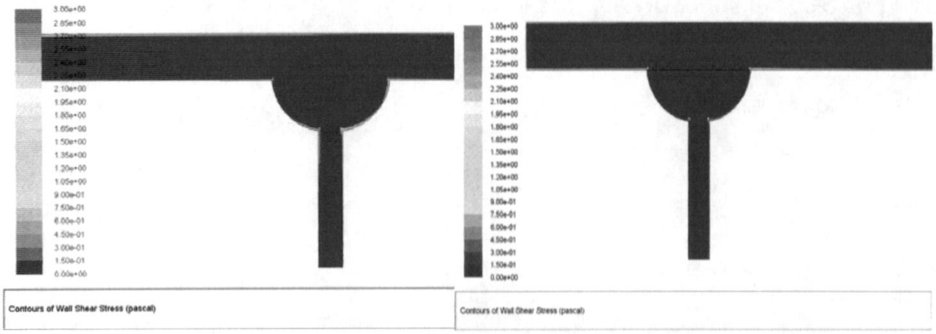

Figure 3.4 wall shear force (a. no stent implanted b. with stent implanted)

According to the Fluent software numerical simulation, it suggests that the wall shear force of aortic aneurysm neck, aneurysm neck of branch vessels, and tumors walls is dropped significantly, and the reduction rate is about 41.3% on average. This trend possesses extremely positive significance in the treatment of aortic aneurysm because the decrease of wall shear force in aortic aneurysm can directly reduce the possibility of tumors' fracture, and lipid deposition and the growth of the wall thickness also arise near the tumor wall, which will more directly play a role of thickening the wall and treating aortic aneurysm.

4. Conclusion

With the gradual understanding of etiology, treatment mechanism of aneurysm, hemodynamic factor is thought to have very important influence on the process of formation, development and rupture of aneurysm [7]. More and more domestic and foreign researchers have a study on hemodynamics of aneurysm[8], and with the development of computer operation speed and software, the numerical simulation about hemodynamic factor in the aneurysm has become a very important technological research method[9,10].

In this paper, the author aims to conduct hemodynamic numerical simulations of abdominal aortic aneurysm and abdominal aortic aneurysm with dense stent implanted. Under these two conditions, this paper respectively analyzes blood velocity contour, the blood flow velocity vector, pressure contour, wall shear force contour and streamline chart. The conclusions are listed as follows:

When the dense mesh stent is implanted in abdominal aortic aneurysm, it can reduce blood flow speed and pressure in the tumor cavity, and the wall shear force in aneurysm neck. In addition, it can form the spoiler to reduce the impact force of the fluid. Therefore, the implantation of a multi-layer dense mesh stent can reduce blood flow speed in tumor and the impact force, promote the lipids and cell in blood to deposit, accelerate thrombosis, and strengthen aneurysm wall, so as to it can ensure the blood-supply of branch vessels, and promote thrombosis of that can lower risk of aneurysm rupture, and achieve the effect of the treatment of abdominal aortic aneurysm.

In this dissertation, the author has carried on the preliminary exploration of the hemodynamic mechanism of the dense mesh stent, but further simulation studies and the evaluation and analysis of vitro fluid test and animal test needed to be done if we want to thoroughly know the influence relationship between the structure of dense mesh stent and hemodynamic.

References

[1] Liu Rengui & Zhao Jichun. The pathogenesis and treatment development of abdominal aortic aneurysm [J]. Chinese journal of general basic and clinical, 2008, 6 (15) : 422-425

[2] Xu Qiang. The development of high performance of shape memory alloy stent [D]. Hunan: zhong shan university, 2007

[3] Zhang Xing. Three-dimensional numerical simulation about stent to the hemodynamics of brain aneurysm [D]. Shanghai: the second military medical university, 2008

[4] Huang qing, Li Tielin & Ling Feng. The hemodynamic of intracranial aneurysm [J]. Foreign medical (archies of cerebrovascular disease), 2004, 10 (12) : 768-770

[5] Cheng Jiyong. A study on hemodynamic numerical simulation of pure dense mesh stent in the treatment of rabbit artery aneurysms [D]. Shanghai: the second military medical university, 2010

[6] Zhao Junwei, Yin Wenyi, Ding Guanghong et al. The hemodynamic numerical simulation and analysis of the two-dimensional elastic aneurysm model [J]. Journal of biomedical engineering in China, 2007, 26 (5) : 730-738

[7] Meng Qingxin, Yang Bin, Ding Weiwei et al. The abdominal aortic bottom stenosis have influence on adjacent upstream hemodynamics [J]. Chinese journal of ultrasound imaging, 2009, 18 (1) : 67-70

[8] Qiao Aike & liu Youjun. Blood dynamics numerical simulation for medical applications (I) [J]. Journal of Beijing university of technology, 2008, 2 (34) : 189-194

[9] Han Zhanzhong, Wang Jing, & Lan Xiaoping. FLUENT fluid simulation calculating examples and engineering application [M]. Beijing: Beijing university of science and technology press, 2008:34-38

[10] Zhao Junwei, Yin Wenyi, Ding Guanghong et al. The hemodynamic numerical simulation and analysis of the two-dimensional elastic aneurysm model [J]. Journal of biomedical engineering in China, 2007, 26 (5) : 730-738

IOP Conf. Series: Materials Science and Engineering **242** (2017) 012085 doi:10.1088/1757-899X/242/1/012085

The algorithm of central axis in surface reconstruction

Bao Ping Zhao[1], Zheng Mei Zhang[1], Ji Cai Li[1], Da Ming Sun[1], Hui Ying Cao[1] and Bao Liang Xing[1]

[1] Shandong Labor Vocational and Technology College, Jinan 250022, China

zmzhang907@163.com

Abstract. Reverse engineering is an important technique means of product imitation and new product development. Its core technology -- surface reconstruction is the current research for scholars. In the various algorithms of surface reconstruction, using axis reconstruction is a kind of important method. For the various reconstruction, using medial axis algorithm was summarized, pointed out the problems existed in various methods, as well as the place needs to be improved. Also discussed the later surface reconstruction and development of axial direction.

1. Introduction

Reverse Engineering is a new emerging discipline with the development of computer development. At the same time, maturity of data measurement technique further promote the development of Reverse Engineering, making this technology a new way for imitation and new product development [1]. This technology has been widely used in the field of machinery, aviation, automobile and shipbuilding. Surface reconstruction is one of the most important and difficult problems, which solves how to construct surface by scattered points. Lots of domestic and overseas scholars pay more attention to surface reconstruction, and rise numerous algorithms and methods, many algorithms revolve around central axis. Central axis is an important tool of describing sampling density and solid shape, which plays an important way in surface reconstruction algorithm. Central axis has been used in many fields, such as picture processing, computer visualization, solid modeling, grid production, kinematics.

2. Implication of central axis

There is no single argument about central axis in the sphere of computer graphics and computation geometry. Reaching the definition by consulting related articles [2]. First one is the definition of astrosphere. For one smooth surface named S, if one ball is tangent to S and has at least two pointcuts, the ball is called astrosphere. Central axis is the set of center of central axis, as shown in figure 1 in two dimensional state. It is observed that it can capture all characteristic of S if have enough sampling points, then central axis of S can be approached by set of sampling points named P. Central axis is a curve on the plane, the branch of it is different on the basis of different shape of entity.

The density of sampling points is request in the counting process of central axis and in the reconstruction of surface, so sampling should has enough density to contain all characteristic of S. Here is a definition of local characteristic function $f(x), f: S \rightarrow R, f(x)$ represents distance between point x and $f(x)$. For point x in the set of sampling point, point y can be fined, it makes the distance between x and y less than $\varepsilon f(x)$, the smaller ε is, the more closely spaced is the sampling, on that way, it is better for the calculation of central axis. Generally, $\varepsilon < 0.25$, small value of ε is used under certain circumstances.

Content from this work may be used under the terms of the Creative Commons Attribution 3.0 licence. Any further distribution of this work must maintain attribution to the author(s) and the title of the work, journal citation and DOI.
Published under licence by IOP Publishing Ltd

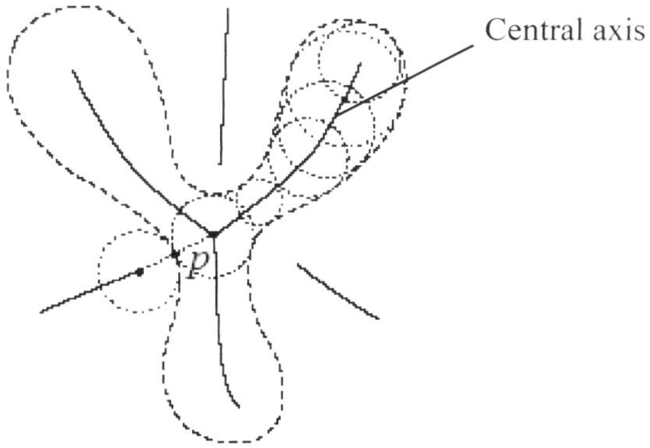

Figure 1. Central axis in two dimensional state

3. The method of approaching to the central axis

3.1. The rule of surface reconstruction based on Voronoi filtering

Although the central axis plays an important role in the surface reconstruction, but calculating the central axis directly is difficult, it is the key point to find suitable algorithmto approaching the central axis. The study find that all approaches are based on Voronoi and Delaunay triangulation. Voronoi vertices in two dimensions are able to approximate the central axis, which is already achieved, and when the sampling density tends to infinity, the Voronoi vertices converge to the central axis. But this property is broken by the Sliverin three dimensions, because the split in three dimensions, regardless of the density of the sampling, is able to close to the surface. In order to mitigate the effects of the whole surface reconstruction, Amenta and Bern, put forward the concept of pole [3], that is, each sample point corresponds to the Voronoi polygon and two points of Voroni polygon farthest from the sample point are respectively referred to positive pole and negative pole. It shows that the pole is close to the central axis and converges to it for a continuous smooth entity, when sampling density tends to zero, the sample point tends to infinity.

The discovery of the pole is a huge progress in 3D reconstruction. The output of most surface is discrete by pure approximation with pole axis, while many continuous approximation are needed in practical application. A series of algorithms arise at the historic moment. Amenta et al, on the basis of the pole and the pole of Delaunay ball, view the radius of Delaunay ball as weight, then weight to the pole of Delaunay triangulation, finally calculate the central axis by the boundary of the balls [4]. It is the first time that the approach method is of continuity for central axis in three dimensions. However, this algorithm requires a second Voronoi diagram to achieve the calculation of the central axis, and the central axis is noisy in some cases.

It is well known that noise has a large effect on surface reconstruction, and it will invalidate the reconstruction algorithm in many cases. The above algorithm is not available because of noise. Tamal k. Dey et al., on the basis of the previous algorithm, propose two independent conditions of the ratio and density, and the Voronoi plane is selected from the Voronoi diagram to approach the central axis. The main idea of the algorithm is to filter the Delaunay edge from the Delaunay triangle in the sample point set and output its dual Voronoi surface as approach of center axis. The Angle condition named θ is used to filter the Delaunay. θ is the maximum Angle, which the intersection point of the center ball B and plane S is intersected to the sectional polygon (sectional polygon refers to the crossing plane by plane passing through ponit P with normal vector and Voronoi diagram of ponit P), as shown in figure 2. Angle θ needs to increase with the decrease of appropriate sampling density. So we can not find a continuous θ to adapt to any model. Proportional conditions are needed. The second is the proportional conditions ρ. When the side of the Delaunay triangle is basically parallel to the cut polygon, the condition fails and the proportional condition is required to constrain it. The defect of proportional condition is that one ρ can not capture the central axis of the whole entity if sampling density is irregular on the entire surface, So, the Angle and proportional conditions are cross-over many times.

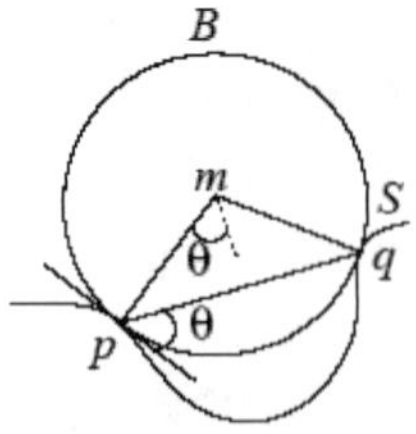

Figure 2. Angle condition θ

Specifically, it may lead to empty of reconstructing surface when the scope of θ is too large, while it may appear for something looks like nail when the scope of θ is too small. The experiment has proved that θ for $\pi/8$ is the most appropriate. ρ also has a limitation, it is found that eight is the best.

So, in summary, this is the first time that a subset of Voronoi is used to approximate the central axis and the convergence is a guaranteed algorithm. But this algorithm does not prove that the output is consistent with the original topology. Although it is very difficult to prove that the axis of the output is identical to the original axis, it is not impossible. So future studies should focus on the problem of homeomorphism.

3.2 Algorithm of surface reconstruction based on Crust

In the algorithm of approaching the medial axis, there is also a classic approach based on the Crust algorithm. The algorithm idea uses a subset of the vertices of the Voronoi to remove the redundant triangles from Delaunay triangulation, and it is proved that the sampling density depends on the local feature surface, the output of this algorithm can also be consistent with the original surface topology and converges [5,6]. Its uniqueness is that after the initial crust, there are two steps of subsequent processing. One is to remove the extra triangles according to the surface normal vector. The specific approach is that T to be a triangle, S is its maximum angle vertex. If the angle between the normal vector of T and the normal vector of S and p is too large, more than the input parameters θ, then remove the triangle T. The second is to adjust all the triangles, mainly for Sliver. Starting at any point S on the convex shell S, find the positive and negative poles of S, and the triangle T near S, define its outward side to be visible. Then adjust the other vertices of T also follow this rule, and each triangle having the same vertex with T follows this rule, and goes on until the end.

4. Conclusion

Summarizing the algorithms for approaching the medial axis, most of them are based on Voronoi diagrams, and the effect of most refactoring is better. But in all kinds of algorithms, there are still many problems in 3D reconstruction. For example, the algorithm mentioned above, in some parts of the machine parts of the area of creases or corners, the approximation of the axis will often appear error, resulting in the result error; the other is the presence of lobes (Sliver) in three dimensions. For the characteristic of being tangent and close to the surface makes many algorithms failure, so most of the algorithms only mitigate the impact of the lobes on the reconstruction results to some extent, and not completely eliminated, these are the problems that need to be solved later.

Acknowledgments

The research was financially supported by the Science and Technology Foundation of Shandong Labor Vocational and Technology College (Grant No. 2015KJZ003).

References

[1] Guohui Wang and Jin Chen 2016 Three-dimensional reconstruction of hybrid surfaces using perspective shape from shading *Optik.* **127** 7740-7751

[2] Tamal K. Dey and Wulue. Zhao 2004 Approximate medial axis as a Voronoi subcomplex

Computer-Aided Design **36** 195-202.
[3] N. Amenta, M. Bern and M. Kamvysselis 1998 A new voronoi-based surface reconstruction algorithm *In Proceedings of ACM SIGGRAPH 1998*. **22** 415-421.
[4] Nina Amenta, Sunghee Choi and Ravi Krishna Kolluri 2001 The power crust union of balls, and the medial axis transform *Computational Geometry: Theory and Applications*. **19** 127-153.
[5] N. Amenta and T. K. Dey 2002 A simple algorithm for homeomorphic surface reconstruction *Internat. Comput, Geom. Applications*. **12** 125-141.
[6] M. Fanfoni, A. Filabozzi, E. Placidi, F. Patella, A. Balzarotti and F. Arciprete 2016 2D Voronoi tessellation generated by lines and belts of dots *Physics Letters A*. **380** 516-519.

IOP Conf. Series: Materials Science and Engineering **242** (2017) 012086 doi:10.1088/1757-899X/242/1/012086

Study on an active correction algorithm of primary mirror

Xiaolin Dai[1,2,3], Hao Xian[1,2], Xuejun Zhang[1,2] and Yudong Zhang[1,2]

[1]The key laboratory of adaptive optics, Chinese Academy of Sciences, 1 Xiahanggang Ave. Chengdu, China

[2]Institute of Optics and Electronics, Chinese Academy of Sciences, 1 Xihanggang Ave. Chengdu, China

[3]University of Chinese Academy of Sciences, Beijing, China

Abstract. The effects of active correction for the telescopes' primary mirrors are often degraded because the axial fixed points of the primary mirrors usually aren't able to perform active correction. This paper analyzes this problem and proposes a modified active correction algorithm to eliminate the adverse influence of the axial fixed points. We use a 1.2m thin mirror to simulate the modified algorithm, it shows the modified algorithm can significantly eliminate the adverse influence of the fixed points and improve the mirror's active correction effect: the fitting error of the 1.2m thin mirror on the 5th Zernike aberration reduces from 13nm to 1.6nm, the fitting errors for other Zernike aberrations also significantly reduce.

1. Introduction

Active optics is one of the key technologies to build large modern telescopes. By detecting the wavefront and correcting the mirror surface correspondingly, active optics can reduce the influence of the mirror's deformations on the beam quality and improve the observation capability of the large telescopes[1,2].

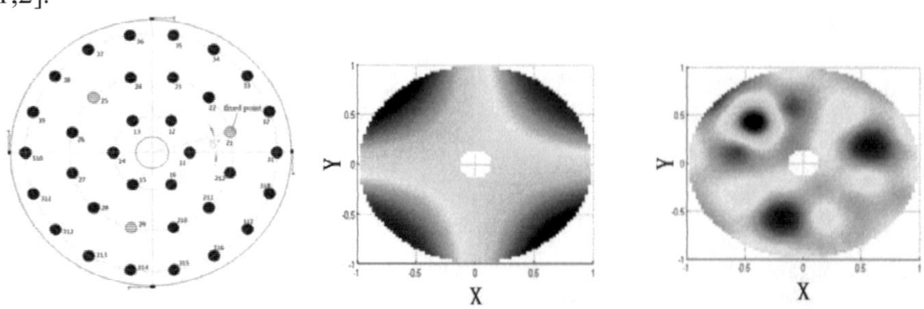

(A) (B) (C)

Figure. 1: (A): configuration of axial supports for an 1.2m thin mirror;

(B): 1.2 mirror's surface before active correction, a 5th Zernike aberration is taken for example;
(C): 1.2m mirror's surface after active correcting the deformations in (B).

Thin mirrors are widely used in large modern telescopes. Figure. 1(A) shows a typical axial supports' configuration of a thin mirror, there are 3 axial fixed points on the back of the mirror which are used to locate the primary mirror. These points can't generate axial motion or apply active correction forces, which causes an issue: mirror surface around these points can't be actively corrected thus degrading the active correction effect of the mirror(Figure. 1(B) and Figure. 1(C)). Floatation support is the most frequently adopted method to solve this problem[3-8], it consists of two steps:

Content from this work may be used under the terms of the Creative Commons Attribution 3.0 licence. Any further distribution of this work must maintain attribution to the author(s) and the title of the work, journal citation and DOI.
Published under licence by IOP Publishing Ltd

firstly, measures the deformations of the mirror, calculates the correction forces(CF) and corrects the surface of the mirror; secondly, after the correction in the first step, measures the forces supported by the fixed points and modifies CF in term of eliminating the unbalanced moment on the mirror caused by the fixed points' forces. The second step is called the second step correction. But this method also has some issues: on one hand, a primary mirror which adopts floatation support has to apply the forces to the actuators for two times during each active correction process, which increases the time cost and leads to the decrease of correction frequency; on the other hand, the range of the correction forces in the first step is usually larger than the one after the second step(which is the range of the final correction forces), this increases the motion requirement of the actuators.

This paper firstly introduces the active correction's process and analyzes the adverse influence of the axial fixed points on the mirror surface's active correction in section 2. In section 3 we proposes a modified algorithm, this new algorithm can eliminate the axial fixed points' influence without introducing the second step correction. In section 4, we verifies the new algorithm via a simulation. The result shows the new algorithm is significantly effective: compared with the unmodified algorithm, the residual mirror surface RMS after the modified algorithm is greatly reduced.

2. Principle of the Active Correction

2.1 Principle of the Active Correction

Figure. 2 shows the principle of the active correction, it consists of 3 steps:

(1) Calibrate the actuators' interacting functions to the primary mirror surface. Firstly, measure W_0 by shooting the reference wave front W_{ref} to the mirror. Secondly, apply unit force to each actuator to make the mirror deform from M_1(ideal parabolic surface) to $M_{1(1)}$, $M_{1(2)}$, $M_{1(3)}$, ... $M_{1(m)}$, then measure the corresponding wave fronts W_i(i=1,2,3...m). Lastly, because of the beam reflection at the mirror surface, the interacting function of each actuator to the mirror surface is $P_i = (W_i - W_0)/2$, (i=1,2,3...m).

(2) Measure the deformations of the primary mirror. Firstly, measure and reconstruct the wavefront W_{origin} at the exit pupil of the telescope, as Zernike modes are the most widely used bases in active correction, W_{origin} can be decomposed into Zernike modes here. Secondly, eliminate the first three Zernike modes from W_{origin} to get W. Thus the mirror's surface before active correction can be calculated as $W_{mirror} = W/2 = \sum_{i=4}^{n} a_i Z_i$, where Z_i is Zernike mode and a_i is the coefficient.

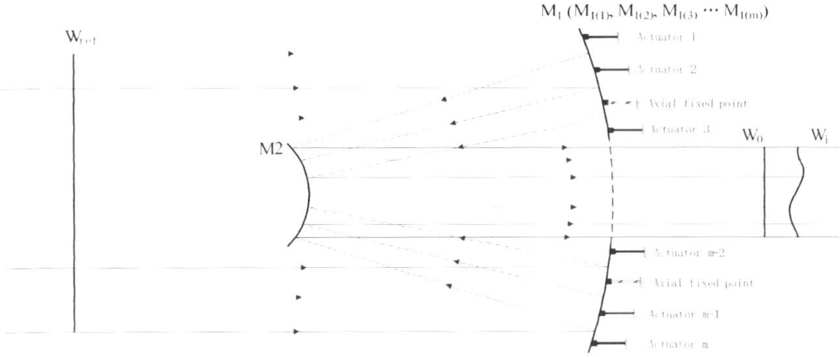

Figure. 2: Schematic of the Active Correction Algorithm.

(3) Calculate the correction force matrix of the actuators via the least square estimation. Assume the interacting function of each actuator measured in step (1) is P_i(i=1,2,3...m), the interacting function matrix will be $P=[P_1, P_2, \dots P_m]$, assume the correction force matrix of the actuators is $F=[F_1, F_2, \dots F_m]^T$, we have:

$$-W_{mirror} = -\sum_{i=4}^{n} a_i Z_i = P \cdot F \tag{1}$$

F can be calculated via the least square estimation as follows:

$$F=(P^T\cdot P)^{-1}\cdot P^T\cdot -W_{mirror}=(P^T\cdot P)^{-1}\cdot P^T\cdot -\sum_{i=4}^{n}a_iZ_i \tag{2}$$

By applying F to the actuators, the mirror deformations and the wavefront distortion at the exit pupil of the telescope could be corrected, which also means the surface of the primary mirror is corrected to the ideal parabolic M_1. Circulate steps (2) and (3) during the operation of the telescope and the mirror surface can be corrected in real time.

2.2 The Defection caused by the fixed points

During the implementation of the active correction, when the mirror deformations W_{mirror} is obtained, the algorithm usually will go to step (3) to calculate the correction force matrix F, but this leads to a problem. Shown in Figure. 3, curve 1 is the ideal parabolic mirror surface M_1 while curve 2 is W_{mirror}, if now we use the actuators to directly correct W_{mirror} to M_1, because the fixed points can't generate active motion, the final surface after active correction will be curve 3.

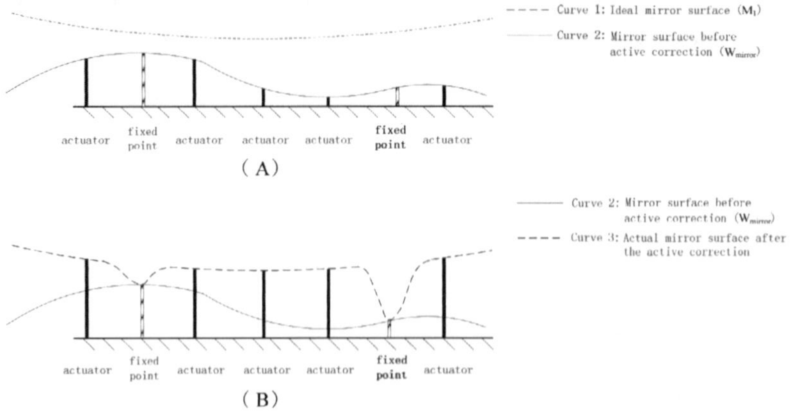

Figure. 3: the effect when using the actuators to directly correct W_{mirror} (profile map).

Figure. 3 shows that the mirror's active correction capability at the fixed points is limited, as a result the mirror surface after active correction is not an ideal parabolic but a curve with some "dents" at the fixed points, which degrades the effect of the active correction.

3. Modified Active Correction Algorithm

To reduce the adverse influence of the fixed points, a modified algorithm is proposed. This new algorithm firstly modifies the ideal parabolic surface M_1 to M_1', then calculates the correction force matrix according to W_{mirror} and M_1'. The core idea of the modification is: by adding M_1 a piston and one or two tilts to get M_1' which overlaps with W_{mirror} at the fixed points. Below is the detailed instruction of this algorithm.

Figure. 4(A) shows the modification of a 2-dimension M_1, assume the coordinates of the two fixed points are x_1,x_2 and the deviations of W_{mirror} from M_1 are s_1,s_2 (Figure. 4(A) is a profile map of a primary mirror, where M_1 and W_{mirror} are 2-dimension curves). Assume a modification curve $M_{correct}$ whose function is $y=a+lx$, $M_{correct}$ means a piston and a tilt, we have:

$$a+l\begin{bmatrix}x_1\\x_2\end{bmatrix}=-\begin{bmatrix}s_1\\s_2\end{bmatrix} \tag{3}$$

Equation (3) means: by adding a modification curve $M_{correct}$, the modified curve M_1'(=M_1+$M_{correct}$) will overlap with W_{mirror} at the fixed points(shown in Figure. 4, curve 3).

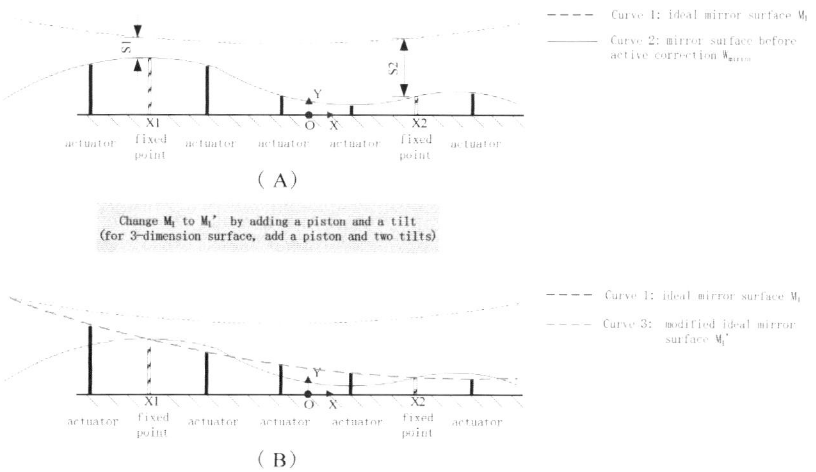

Figure. 4: Process of the modification of M_1.

After obtaining a and l via Equation. (3), we have:

$$M_1' = M_{correct} + M_1 \tag{4}$$

Figure. 4(B) shows the modified ideal mirror surface M_1' satisfies $M_1'(x=x_1) = W_{mirror}(x=x_1)$, $M_1'(x=x_2) = W_{mirror}(x=x_2)$, which means M_1' overlaps with W_{mirror} at the fixed points. To correct W_{mirror} to M_1', because the deviations of W_{mirror} from M_1' at $x=x_1$ and $x=x_2$ are 0 now, the fixed points actually don't need to generate any shift or motion during the correction, thus the adverse influence of the fixed points is eliminated. The mirror surface after active correction will be precisely M_1'.

Notice that M_1' is not the ideal surface M_1, the differences between them are a piston(a) and a tilt(l), these are actually some new distortions introduced into the telescope system, but they can be corrected by the telescope(for example by the secondary mirror) easily at the same time with the primary mirror's active correction process, so at last, we can virtually obtain the ideal parabolic surface M_1.

The reason why we don't adopt the second and third Zernike modes which are eliminated in step (2), section 2.1 as the tilts l_x and l_y which we need in the modification is that these two Zernike distortions can be caused by both the primary mirror and the second mirror, and the two parts can't be separated, thus they are different from l_x, l_y and can't be utilized directly here.

4. Simulation of the Modified Active Correction Algorithm

A 1.2m thin primary mirror is used to simulate the modified algorithm. Figure. 1(A) shows the configuration of the axial supports of the mirror, the diameters of the mirror and its central hole are 1.2m and 0.14m, the thickness is 50mm, radius-thickness ratio is 24. The material of the mirror is glass ceramics. There are 36 axial supports on the back of the mirror, the 21th, 25th and 29th supports are axial fixed supports or fixed points, which uniformly distribute on the 2th support ring, the rest 33 supports are active supports(actuators) which can apply active forces to the mirror. On the lateral side of the mirror there are 12 uniformly distributed lateral supports which can apply tangential support forces. In addition, there are 4 lateral fixed supports on the top, bottom, left and right of the mirror, which limit the 6 degrees of freedom of the mirror together with the 3 axial fixed points.

The active correction of the 5th Zernike aberration(astigmatism) via the 1.2m mirror is taken as an example to verify the effect of the modified algorithm. The simulation is implemented as follows:

(1) Calibrate the 33 actuators' interacting functions to the mirror surface and record them as P_i(i=11,12,13…22,23…317,318), part of the interacting functions are shown in Figure. 5.

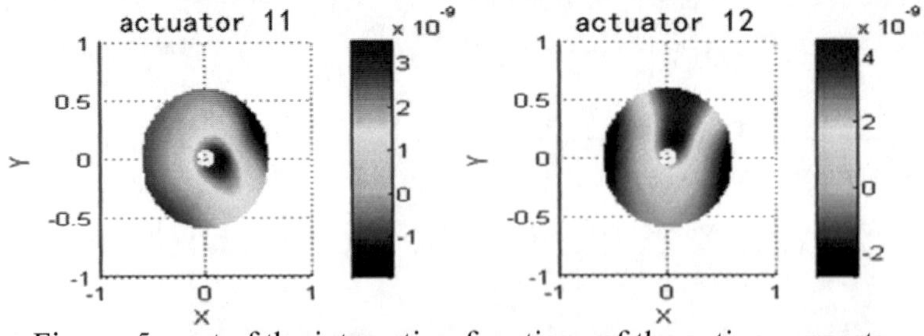

Figure. 5: part of the interacting functions of the active supports.

(2) Measure the deformations of the mirror. The amplitude of M_l on Z axis is assumed to be 0. The mirror surface before active correction is shown in Figure. 6(left) as W_{mirror} (the radius has been normalized) whose RMS is about 100nm. Figure. 6(left) shows that the deviations of W_{mirror} from M_l at the 3 fixed points are not zero, which means M_l doesn't overlap with W_{mirror} at the fixed points. If now we correct W_{mirror} directly without any modification, the surface after the correction is shown in Figure. 6(right), the residual deviation has obvious peaks and valleys at the axial fixed points, the RMS of the residual deviation is about 13nm. Meanwhile, the correction forces of the actuators are shown in Figure. 7(blue line), the range of the forces is about -27N~40N.

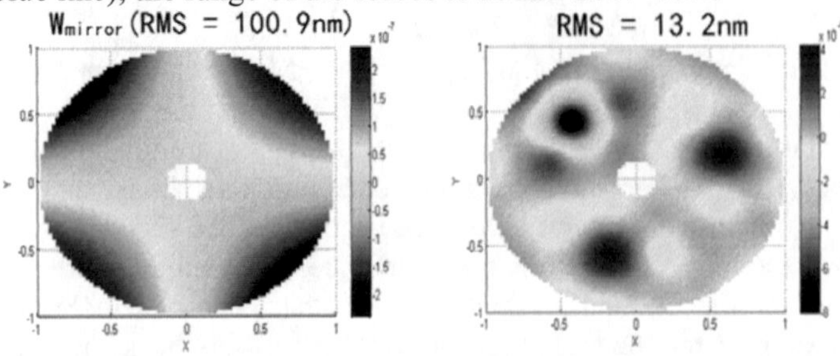

Figure. 6: left: mirror surface before active correction;
right: mirror surface after directly correction of W_{mirror}.

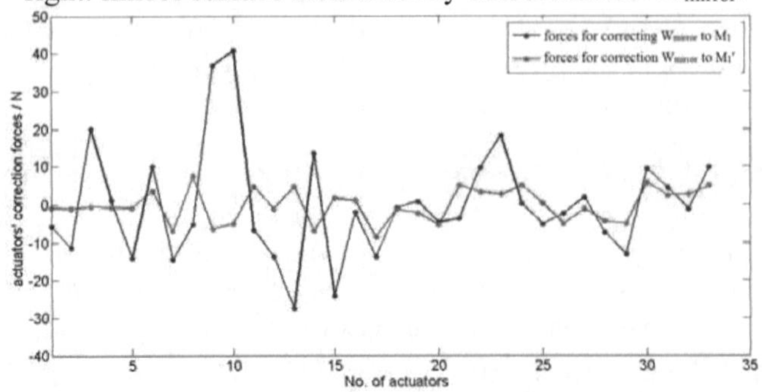

Figure. 7: correction forces of the 1.2m mirror's actuators for correcting W_{mirror} to M_l(blue line) and W_{mirror} to M_l' (red line). The 33 actuators on X axis corresponds sequentially to the actuators 11,12,13…16,22,23…212,31,32…318 in Figure. 1(A).

(3) Modify M_l. Firstly, the coordinates of the 3 fixed points on X-Y plain are calculated out: (0.5586m,0.1497m), (-0.4089m,0.4089m), (-0.1497m,-0.5586m), the deviations from W_{mirror} to M_l at these 3 points are 33.7nm, -83.3nm, 42.3nm. Secondly, assume a modification curve $M_{correct}$ whose function is $z=a+l_x x+l_y y$ ($M_{correct}$ means a piston aberration and two tilt aberrations, where z axis overlaps with the optical axis of the primary mirror), we have the following equations:

$$a+\begin{bmatrix} 0.5586 & 0.1497 \\ -0.4089 & 0.4089 \\ -0.1497 & -0.5586 \end{bmatrix}\begin{bmatrix} l_x \\ l_y \end{bmatrix}=-\begin{bmatrix} 3.37\times10^{-8} \\ -8.33\times10^{-8} \\ 4.23\times10^{-8} \end{bmatrix} \tag{5}$$

Solve the equations above and we can obtain $M_{correct}$:

$$a = 2.45e-9 \quad l_x = -1.05e-7 \quad l_y = 9.28e-8 \tag{6}$$

Thus M_1' can be obtained via Equation (4). After the modification, the deviations of W_{mirror} from M_1' at the 3 fixed points are nearly zero.

(4) Calculate the correction forces for correcting W_{mirror} to M_1' via the least square estimation. The interacting function matrix of the actuator alignment is P=[P_{11} P_{12} … P_{317} P_{318}], assume the correction forces matrix to be F, then F could be computed via Eq. (7). At last, apply F to the actuators and at the same time use the second mirror to correct the extra piston a and tilts(l_x, l_y) to accomplish the active correction of the primary mirror.

$$F = (P^T \cdot P)^{-1} \cdot P^T \cdot (M_1 ' - W_{mirror}) \tag{7}$$

Figure. 7(red line) shows the values of F, the range of the forces is about -8.5N~7.3N, which significantly decreases compared with the range of the blue line.

The residual deviation of the 1.2m mirror surface after the correction via the new algorithm is shown in Figure. 8, contrasted with Figure. 6(right), the residual deviation of the mirror surface is significantly reduced, the peaks and valleys at the 3 fixed points basically disappear, the RMS of the residual deviation is reduced from 13nm to 1.6nm.

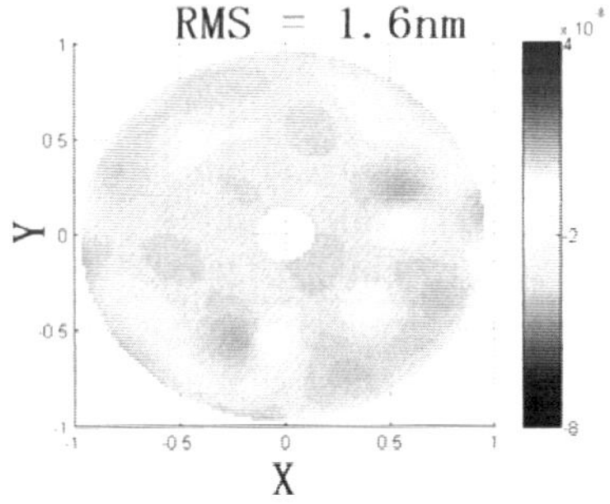

Figure. 8: the residual deformations of the 1.2m mirror surface after the correction with modified algorithm.

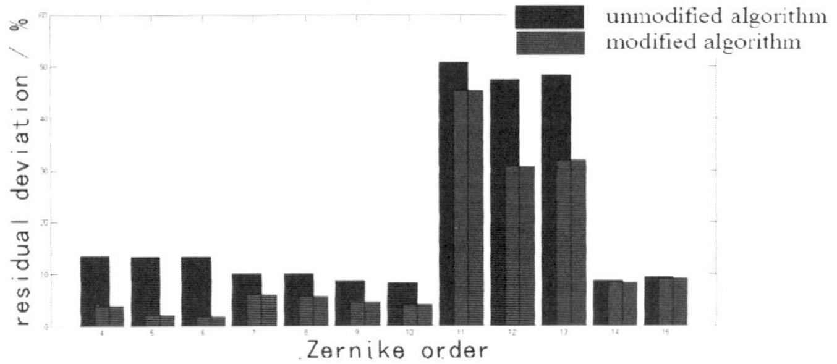

Figure. 9: the effects of the correction of the 1.2m thin mirror on different Zernike aberrations with unmodified and modified algorithm.

The effect of the modified correction algorithm on correcting the 4th~15th Zernike aberration via the 1.2m mirror is also simulated. Set the original mirror deformation to about 100nm, the effects of the active correction of the mirror with unmodified and modified algorithm are shown in Figure. 9, where the residual deviation is the ratio of mirror surface RMS after and before active correction. Figure. 9 shows that contrasted with the unmodified algorithm, the residual deviation of most Zernike aberrations after the correction improve significantly when adopting the modified algorithm. In addition, the range of the correction forces also decreases.

5. Conclusion

This paper firstly introduces the principle of active correction and analyzes one of its problem - the axial fixed points may degrade the primary mirror surface's active correction effect because they can't perform active axial motion; Secondly proposes a modified active correction algorithm which can eliminate the adverse influence of the fixed points; Lastly applies the algorithm to a 1.2m thin mirror to verify the new algorithm. The simulation shows that this new algorithm can effectively eliminate the influence of the fixed points, the RMS of the residual deviations of the mirror surface decreases significantly.

References

[1] L. Noethe. 2002. Active Optics in Modern, Large Optical Telescopes. *Progress in Optics*, vol. **43**, 3-69.

[2] Cheng Jingquan. 2003. Principles of Astronomical Telescope Design. Chinese Science & Technology Press, Beijing.

[3] B. C. H. M. Martin, L. R. Dettmann. 2004. Active optics and force optimization for the first 8.4 m LBT mirror. *Proc. SPIE* **5489**, 826-837.

[4] S. P. C. H. M. Martina, B. Cuerdena. 1998. Active supports and force optimization for the MMT primary mirror. *Proc. SPIE* **3352**, 412-423.

[5] D. G. James E. Kimbrell. 1998. AEOS 3.67 m Telescope Primary Mirror Active Control System. *Proc. SPIE* **3352**, 400-411.

[6] X. C. M. Schneerrnann, D. Enard. 1990. ESO VLT III : the support system of the primary mirror. *Proc. SPIE* **1236**, 920-928.

[7] E. D. Thomas A. Sebring, Robert L. Millis. 2004. The Discovery Channel Telescope A Wide Field Telescope in Northern Arizona. *Proc. SPIE* **5489**, 658-666.

[8] Li Hongzhuang, Zhang Zhenduo, Wang Jianli. 2013. Active Surface-Profile Correction of 620mm Thin-Mirror Based on Flotation Support. *Acta optica sinica*, vol. **33**, 0511001.

IOP Conf. Series: Materials Science and Engineering **242** (2017) 012087 doi:10.1088/1757-899X/242/1/012087

A feasible DY conjugate gradient method for linear equality constraints

Can LI

College of Mathematics, Honghe University, Mengzi 661199, China

E-mail: Lican841014@163.com

Abstract. In this paper, we propose a feasible conjugate gradient method for solving linear equality constrained optimization problem. The method is an extension of the Dai-Yuan conjugate gradient method proposed by Dai and Yuan to linear equality constrained optimization problem. It can be applied to solve large linear equality constrained problem due to lower storage requirement. An attractive property of the method is that the generated direction is always feasible and descent direction. Under mild conditions, the global convergence of the proposed method with exact line search is established. Numerical experiments are also given which show the efficiency of the method.

1. Introduction

Consider the unconstrained optimization problem

$$\min f(x),\ x \in R^n, \tag{1}$$

where $f: R^n \to R$ is a continuously differentiable function. The conjugate gradient method for solving (1) is given by

$$x_{k+1} = x_k + \alpha_k d_k, \tag{2}$$

where α_k is the step length, and d_k is search direction defined by

$$d_k = \begin{cases} -\nabla f(x_k) & k = 0, \\ -\nabla f(x_k) + \beta_k d_{k-1} & k \geq 1, \end{cases} \tag{3}$$

where β_k is a parameter. The conjugate gradient method is available for large-scale unconstrained optimization because its storage is relatively small. Numerical results [1] showed that, if f is easy to be computed and if its dimension n is vary large, the conjugate gradient method is still the best choice for solving unconstrained optimization problem (1).

If the objective function $f(x)$ is a strictly convex quadratic function, and if α_k is computed by the exact line search, namely,

$$f(x_k + \alpha_k d_k) = \min_{\alpha > 0} f(x_k + \alpha d_k). \tag{4}$$

then the method (2) and (3) is called the linear conjugate gradient method. The linear conjugate gradient method was originally propose by Hestenes and Stiefel [2] for solving the linear system of equations $Ax = b$ and several formulas of β_k were considered, which are equivalent for strictly convex quadratic objective function.

On the other hand, (2) and (3) is called the nonlinear conjugate gradient method for general

Content from this work may be used under the terms of the Creative Commons Attribution 3.0 licence. Any further distribution of this work must maintain attribution to the author(s) and the title of the work, journal citation and DOI.
Published under licence by IOP Publishing Ltd

unconstrained optimization problem. The nonlinear conjugate gradient method was first proposed by Fletcher and Reeves [3]. Some well-known conjugate gradient methods include the Hestenses-Stiefel method , Fletcher-Reeves method, Polak-Ribiere-Polyak method, Conjugate-Descent method, Liu-Storey method, and Dai-Yuan method . In this paper, we are interested in the DY method[4], in which β_k is defined by

$$\beta_k^{DY} = \frac{\| \nabla f(x_k) \|^2}{d_{k-1}^{\mathrm{T}} y_{k-1}} \tag{5}$$

where $y_{k-1} = \nabla f(x_k) - \nabla f(x_{k-1})$, and $\| \cdot \|$ stands for the Euclidean norm. It was shown in [4] that such a method can guarantee the descent property of each direction provided the steplength satisfies the Wolfe conditions. In this case, the global convergence of the method was also proved in [4] under some mild assumptions on the objective function.

Recently, in view of the advantage of the conjugate gradient method for solving unconstrained optimization problem, Li and Li[5] extend it to the following linear equality constrained optimization problem

$$\begin{aligned} &\min f(x) \\ &\text{s.t. } Ax = b, \end{aligned} \tag{6}$$

where $f : R^n \to R$ is continuously differentiable function, $A \in R^{m \times n}$ with $m < n$ is of full rank, and $b \in R^m$. The direction generated by the method is always feasible and descent direction. Under suitable conditions, the feasible Fletcher-Reeves method with exact line search is globally convergent.

In this paper, we further study the feasible conjugate gradient method for linear equality constrained optimization problem. We focus our attention to the DY conjugate gradient method. We will extend the DY method for solving unconstrained optimization problem to the linear equality constrained optimization problem.

2. Algorithm

In this section, we extend the DY method to linear equality constrained optimization problem (6). First we simply recall the feasible Fletcher-Reeves method in [5] for solving equality constrained optimization problem. For the problem (6), we separate the coefficient matrix A into $A = (B, N)$, where $B \in R^{m \times m}$ is nonsingular. For any $x \in D$, let $x = \begin{pmatrix} x^B \\ x^N \end{pmatrix}$, we get

$$x^B = B^{-1}b - B^{-1}Nx^N. \tag{7}$$

The problem (6) is equivalent to the following unconstrained optimization problem

$$\min F(x^N) \overset{\Delta}{=} f(x^B, x^N) = f(B^{-1}b - B^{-1}Nx^N, x^N).$$

In order to get feasible direction, the following lemma is given by Li and Li in [5].

Lemma 1. Let x_k be the current iterate, $\bar{d}_k = \begin{pmatrix} \bar{d}_k^B \\ \bar{d}_k^N \end{pmatrix}$ is answer of the following linear programming

$$\begin{aligned} \min \quad & \nabla f(x_k)^{\mathrm{T}} d \\ s.t \quad & Ad = 0, \quad \| d^N \| \le 1. \end{aligned}$$

Then the answer $\bar{d}_k = \begin{pmatrix} \bar{d}_k^B \\ \bar{d}_k^N \end{pmatrix}$ is given by

$$\bar{d}_k^N = -\frac{\nabla F(x_k^N)}{\| \nabla F(x_k^N) \|}, \quad \bar{d}_k^B = -B^{-1}N\bar{d}_k^N. \tag{8}$$

Let $z_k = \nabla f(x_k)^{\mathrm{T}} \bar{d}_k = \nabla F(x_k^N)^{\mathrm{T}} \bar{d}_k^N = -\| \nabla F(x_k^N) \|$, and $-g_k = -z_k \bar{d}_k$. Similar to the DY conjugate method in [4] for solving unconstrained optimization problem, we propose the following DY- type conjugate gradient method for solving the linear equality constrained optimization problem (8) which we call FDY method, that is, the direction d_k is defined by

$$d_k = \begin{cases} -g_k & k=0, \\ -g_k + \beta_k^{FDY} d_{k-1} & k \geq 1, \end{cases} \tag{9}$$

where $\beta_k^{FDY} = \dfrac{z_k^2}{d_{k-1}^{\mathrm{T}} y_{k-1}}$, $y_{k-1} = \nabla f(x_k) - \nabla f(x_{k-1})$. It is easy to see from (8) and (9) that for any $k \geq 0$, $Ad_k = 0$.This implies d_k that provides a feasible direction of f at x_k.

Based on the above discussion, we propose a feasible DY method which we call FDY method as follows.

Algorithm 1(FDY method).

Step1: Given constant $\varepsilon > 0$, choose an initial point $x_0 \in D$, let $k := 0$.

Step2: If $| z_k | \leq \varepsilon$, stop, get solution x_k . Otherwise go to step 3.

Step3: Compute d_k by (9).

Step4: Determine step size α_k by some line search.

Step5: Let $x_{k+1} = x_k + \alpha_k d_k$ and $k := k+1$ .Go to step2.

3. Global convergence

In this section, we prove the global convergence of Algorithm 1 with exact line search and with Wolfe line search respectively. To establish the global convergence theorem of Algorithm 1, we assume that the objective function satisfies the following condition that the level set $\Omega = \{x \in R^n \mid f(x) \leq f(x_0)\}$ is bounded.Next, we prove the global convergence of Algorithm 1 with exact line search (4).

Theorem 2. Let the sequence $\{x_k\}$ be generated by Algorithm 1 with exact line search, then

$$\liminf_{k \to \infty} \| \nabla F(x_k^N) \| = 0. \tag{10}$$

Proof. For the sake of contradiction, we suppose that the conclusion is not true. Then there exists a constant $\varepsilon > 0$ such that

$$\| \nabla F(x_k^N) \| \geq \varepsilon, \ \forall k. \tag{11}$$

From the definition of d_k, we have

$$\nabla f(x_k)^{\mathrm{T}} d_k = -\nabla f(x_k)^{\mathrm{T}} g_k + \beta_k^{FDY} \nabla f(x_k)^{\mathrm{T}} d_{k-1}.$$

Since the steplength α_k is determined by exact line search, it holds that

$$\nabla f(x_k)^{\mathrm{T}} d_k = -\nabla f(x_k)^{\mathrm{T}} g_k.$$

Recalling the definition of g_k, we obtain from (17) that

$$\nabla f(x_k)^{\mathrm{T}} d_k = -\| \nabla F(x_k^N) \|^2 < 0. \tag{12}$$

Since $\{f(x_k)\}$ is decreasing, it is clear that the sequence $\{x_k\}$ generated by Algorithm 1 is contained in Ω. This implies that $\{x_k\}$ is bound. That is, there is an infinite index set K_1 such that

$$f^* = \lim_{k(\in K_1) \to \infty} f(x_k) = f(\lim_{k(\in K_1) \to \infty} x_k) = f(x^*).$$

At the same time, the sequence $\{x_{k+1} | k \in K_1\}$ is bound. Similarly, there exists an infinite index set $K_2 \subset K_1$ such that

$$f^* = \lim_{k(\in K_2)\to\infty} f(x_{k+1}) = f(\lim_{k(\in K_2)\to\infty} x_{k+1}) = f(\overline{x}^*).$$

That is

$$f(x^*) = f(\overline{x}^*) = f^*. \quad (13)$$

Notice that the step length α_k is determined by exact line search, that is

$$f(x_{k+1}) = f(x_k + \alpha_k d_k) \le f(x_k + \alpha d_k), \quad \forall \alpha > 0.$$

Let $k \to \infty$, we have that for all $k \in K_2 \subset K_1$

$$f(\overline{x}^*) \le f(x^* + \alpha d^*), \quad \forall \alpha > 0. \quad (14)$$

By Taylor's expansion, there is

$$f(x^* + \alpha d^*) = f(x^*) + \alpha \nabla f(x^*)^{\mathrm{T}} d^* + o(\alpha). \quad (15)$$

It is follows from (15) and (12) that when $\alpha > 0$ is sufficient small $f(x^* + \alpha d^*) < f(x^*)$. With (14), we have $f(\overline{x}^*) \le f(x^* + \alpha d^*) < f(x^*)$. This contradicts (13). This contradiction shows that (10) is true.

4. Numerical experiment

In this section, we report some numerical experiments about Algorithm 1 with exact line search, and compare it with the feasible FR method proposed in [5]. We stop the iteration if the iteration number exceeds 5000 or the function evaluation number exceeds 7000 on the following inequality is satisfied $| z_k | \le 10^{-5}$.

In this paper, all codes are written in Matlab and run on PC with 2.66 GHZ CPU processor and 1GB RAM memory and Windows XP operation system. The results are listed in Table1. In Table1, Name denotes the name of the test problem, Dim denotes the dimension of the test problems, NF and NI denote the number of function evaluations and the iteration number respectively, Time denotes CPU time in second.

In table 1, we test the Algorithm 1 by eleven test problems. We can see from table 1 that the method terminates at the solution of all test problems. The results of numerical experiments indicate that the proposed method in this paper is efficiency method for some test problems.

Table 1 Numerical Results

Name	Dim	FFR method			FDY method		
		NI	NF	Time	NI	NF	Time
MAD1	2	2	3	0.0160	1	2	0.0321
MAD2	2	1	2	0.0150	4	5	0.0160
MAD4	2	46	47	0.1410	13	14	0.0160
Hs028	3	14	15	0.0160	58	59	0.0150
Hs048	5	3	4	0.0160	3	4	0.0160
Hs049	5	41	42	0.0150	37	38	0.0160
Hs050	5	10	11	0.0150	11	12	0.0310
Hs051	5	2	3	0.0160	2	3	0.0160
Hs052	5	2	3	0.0160	2	3	0.0150
Wong2	10	130	131	0.2030	106	107	0.0940
Wong3	20	817	818	1.6400	1266	1267	1.6400

Acknowledgment
This work is supported by Yunnan Natural Science Foundation (2014FD053)

References

[1] Liu D C, Nocedal J. On the limited memory method for large scale optimization, Math. Prog., 1989,45:503-528.
[2]Hestenes M R, Stiefel E L. Method of conjugate gradient for solving linear systems. J.Res.Nat.Bur.Stand., 1952, 49:409-432.
[3]Fletcher R, Reeves, C. Function minimization by conjugate gradient method. J.Comput., 1964, 7:149-154.
[4]Dai Y H, Yuan Y. A nonlinear conjugate gradient method with a strong global convergence property. SIAM J.Optim., 1999, 10:177-182.
[5] Li C, Li D H. An extension of conjugate gradient method to linear equality constrained optimization problem, Applied Mathematics and Computation. 2013, 219: 10909-10914.

IOP Conf. Series: Materials Science and Engineering **242** (2017) 012088 doi:10.1088/1757-899X/242/1/012088

Modeling delamination of FRP laminates under low velocity impact

Z. Jiang[1, a], H. M. Wen[1], S.L. Ren[1]

1.CAS Key Laboratory for Mechanical Behaviour and Design of Materials, University of Science and Technology of China, Hefei 230026,Anhuiprovince,China

E-mail: [a]zj0510@mail.ustc.edu.cn

Abstract. Fiber reinforced plastic laminates (FRP) have been increasingly used in various engineering such as aeronautics, astronautics, transportation, naval architecture and their impact response and failure are a major concern in academic community. A new numerical model is suggested for fiber reinforced plastic composites. The model considers that FRP laminates has been constituted by unidirectional laminated plates with adhesive layers. A modified adhesive layer damage model that considering strain rate effects is incorporated into the ABAQUS / EXPLICIT finite element program by the user-defined material subroutine VUMAT. It transpires that the present model predicted delamination is in good agreement with the experimental results for low velocity impact.

1. Introduction

Fiber reinforced plastic (FRP) laminates have been widely used in engineering such as aeronautics, astronautics, transportation, naval architecture due to their high specific strength, high specific stiffness, design flexibility and high durability. However, during manufacturing or in service FRP laminates may be subjected to impact by dropped objects such as tools or fragments generated from a nearby accidental explosion or even by a military projectile. Hence, the response and failure of FRP laminates under impact loading are a major concern both for academic community and industry.

Delamination is a major damage pattern of FRP laminates subjected to impact loadings and it is still difficult to simulate this damage pattern so far. Olsson [1] proposed an analytical method to predict the delamination initiation and growth of composite laminates struck transversely by large masses at low velocities. Espinosaet.al[2]examined the dynamic delamination in GFRP laminates using a 3D finite deformation anisotropic viscoplastic model in conjunction with contact/cohesive laws whilst Johnson et.al [3] studied the dynamic response of a woven GFRP composite structure utilizing a simple 2Ddamage model in conjunction with cohesive elements. Singh and Mahajan[4]proposed a 3D elasto-plastic damage model combining the surface based cohesive interaction to simulate the delamination behavior of graphite/epoxy laminates under low velocity impact and reasonable agreement was obtained between the numerically predicted delamination patterns and the experimental observations [5]. Aymerichet.al [6] investigated the usefulness of cohesive interface elements for damage prediction in cross-ply composite laminates subjected to low velocity impact. Francesconi and Aymerich [7]also conducted numerical simulation of the effect of stitching on the delamination resistance of laminated composites subjected to low velocity impact.

The objective of this paper is to formulate an FE model which can be used to predict the delamination of fiber reinforced plastic (FRP) laminates subjected to impact loadings.

Content from this work may be used under the terms of the Creative Commons Attribution 3.0 licence. Any further distribution of this work must maintain attribution to the author(s) and the title of the work, journal citation and DOI.
Published under licence by IOP Publishing Ltd

1.1 Formulation of an FE model

An FE model will be constructed in the following. A recently developed progressive damage model for fiber reinforced plastic (FRP) laminates subjected to impact loadings is first briefly described and then a modified adhesive layer damage model in some more details.

1.1.1 A progressive damage model. A progressive damage model for uniderictional laminated plates under impact loadings was developed recently by Xin and Wen[8]. The model includes three parts, namely damage initiation, damage evolution and strain rate effects.

Quadratic stress based damage initiation criteria were given to predict the onset of various damage modes. When one of the failure mechanisms takes place, the properties of the material may reduce gradually until to complete failure. This new CDM model introduced a concept of stiffness degradation which is controlled by damage variables and the values of damage variables suggested to be between zero (undamaged state) and one (fully damaged state). The fracture energy approach in terms of stress-displacement relationship was adopted to compute the damage variables under different failure modes.

In the meantime, the strain rate effects on the strengths as well as moduli were considered in the model by introducing the dynamic increase factor (DIF), which is defined as the ratio of dynamic to static values.

Due to space limitation the detailed information about damage evolution and strain rate effects will not be given here. Interested readers are referred to Ref.[8] for more details.

1.1.2 Adhesive layer damage model. To simulate the delamination of FRP laminates, one of the common and successful methods is cohesive zone method (CZM) which was suggested by Barenblatt[9]. This method first uses failure criterion that combine aspects of strength based analysis to predict the onset of the damage of the interface and then fracture mechanics to describe the delamination propagation. However, the cohesive zone method (CZM) and nearly all its subsequent developments [1]-[7]took no consideration of strain rate effects on the strengths as well as moduli of FRP laminates. In the following a modified CZM method that considers strain rate effects is proposed.

1.1.3 Traction-separation law. Adhesive layers damage model are employed to simulate the delamination between layers of FRP laminates. The basis of the method is the cohesive behavior interaction of two adjacent surfaces. The traction stress and separation displacement of the nodes on the surfaces are governed by traction-separation law as shown schematically in Figure 1 which consists of initial elastic response, delamination initiation and delamination propagation.

The elastic part of the traction-separation law can be written by the following expression, viz.

$$\mathbf{t}=\begin{Bmatrix} t_n \\ t_s \\ t_t \end{Bmatrix}=\begin{pmatrix} K_n & 0 & 0 \\ 0 & K_s & 0 \\ 0 & 0 & K_t \end{pmatrix}\begin{Bmatrix} \delta_n \\ \delta_s \\ \delta_t \end{Bmatrix}=\mathbf{K\delta} \quad (1)$$

where t is nominal traction stress vector, t_n is the out-of-plane normal stress, t_s and t_t are the transverse shear stresses; K is the penalty stiffness vector, K_n、K_s、K_t are the normal, shear and tearing stiffnesses, respectively; $\boldsymbol{\delta}$ is relative displacement vector. δ_n、δ_s、δ_t are the normal, shear and tearing displacements, respectively.

The delamination initiation of the traction-separation law can be expressed by the quadratic failure criterion, namely

$$\left\{\frac{\langle t_n\rangle}{t_n^0}\right\}^2+\left\{\frac{t_s}{t_s^0}\right\}^2+\left\{\frac{t_t}{t_t^0}\right\}^2\geq 1 \quad (2)$$

Where t_n^0, t_s^0, t_t^0 are peak values of the tractions, respectively. It is noted that the compressive normal tractions do not cause delamination to occur and the effective separation displacement corresponding to the peak stress is used in this paper to predict the initiation of delamination. Thus, the effective displacement corresponding to the onset of the delamination (δ_m^0) can be expressed by the following equation [10]

$$\delta_m^0 = \begin{cases} \delta_n^0 \delta_s^0 \sqrt{\dfrac{1+\beta^2}{\left(\delta_s^0\right)^2 + \left(\beta\delta_n^0\right)^2}} & (\delta_n > 0) \\ \delta_s^0 & (\delta_n \le 0) \end{cases} \quad (3)$$

in which $\beta = \dfrac{\delta_{shear}}{\delta_n}$ with δ_{shear} being defined as $\delta_{shear} = \sqrt{\delta_s^2 + \delta_t^2}$ and δ_n^0, δ_s^0, δ_t^0 are the separation displacements corresponding to the peak values of the tractions t_n^0, t_s^0, t_t^0, which are defined as follows:

$$\delta_n^0 = \frac{t_n^0}{K_n}, \delta_s^0 = \frac{t_s^0}{K_s}, \delta_t^0 = \frac{t_t^0}{K_t} \quad (4)$$

It can be seen clearly from equation (12) that when $\beta = 0$ (i.e., $\delta_{shear} = 0$) $\delta_m^0 = \delta_n^0$ implying delamination is caused by crack opening mode (mode I) only and that when $\beta \to \infty$ (i.e., $\delta_n = 0$) $\delta_m^0 = \delta_s^0$ implying delamination by either shear or tearing mode.

On the basis of assumption that the fracture toughness for tearing mode (Mode III, G_{IIIC}) is equal to the fracture toughness for shearing mode (Mode II, G_{IIC}), namely $G_{IIC} = G_{IIIC}$, Benzeggagh and Kenane [11] proposed a criterion for delamination of FRP laminates which can be written as

$$G_{IC} + \left(G_{IIC} - G_{IC}\right)\left(\frac{G_{shear}}{G_T}\right)^{\eta} = G_C \quad (5)$$

where G_{IC} is the fracture toughness for crack opening mode (Mode I) and η is an empirical constant dependent upon material system and in the present paper it is taken to be unity; $G_{shear} = G_{II} + G_{III}$ and $G_T = G_I + G_{shear} = G_I + G_{II} + G_{III}$ with G_I, G_{II}, G_{III} being defined by the following equations

$$G_I = \int_0^{\delta_n^f} t_n d\delta_n, G_{II} = \int_0^{\delta_s^f} t_s d\delta_s, G_{III} = \int_0^{\delta_t^f} t_t d\delta_t \quad (6)$$

G_C in equation (14) is the critical energy release rate which is determined by

$$G_C = \frac{1}{2} K \delta_m^0 \delta_m^f \quad (7)$$

where K is the effective stiffness. Substituting equation (16) into equation (14) and rearranging yields[10]

$$\delta_m^f = \begin{cases} \dfrac{2}{K\delta_m^0}\left[G_{IC} + \left(G_{IIC} - G_{IC}\right)\left(\dfrac{\beta^2}{1+\beta^2}\right)^{\eta}\right] & (\delta_n > 0) \\ \dfrac{2G_{IIC}}{K\delta_m^0} & (\delta_n \le 0) \end{cases} \quad (8)$$

after using expressions for G_{shear} and G_T and equations (15) and (10), together with the assumption that $K_n = K_s = K_t$. δ_m^0 is estimated by equation (12) and δ_m^f in equation (17) is the critical effective separation displacement at which two adjacent surfaces separates from each other or delamination occurs.

The critical effective separation displacement (δ_m^f) can be utilized to calculate the damage variable (D) which can be expressed as

$$D = \frac{\delta_m^f(\delta_m^{\max} - \delta_m^0)}{\delta_m^{\max}(\delta_m^f - \delta_m^0)} \quad (9)$$

where $\delta_m^{\max}$ is a state variable which is defined as

$$\delta_m^{\max}(\mathrm{t}) = \max\left\{\delta_m(\mathrm{t}), \delta_m^{\max}(\mathrm{t} - \Delta\mathrm{t})\right\} \quad (10)$$

in which $\delta_m^{\max}(\mathrm{t})$ is the value of effective separation displacement (δ_m) at current time step while $\delta_m^{\max}(\mathrm{t}-\Delta\mathrm{t})$ is the value of previous time step. Thus, by using this state variable, the irreversibility of damage is taken into account.

The traction-separation law, as shown in Figure 1, can be described by the following equation

$$\mathbf{t} = \begin{cases} \mathbf{K\delta} & (\delta_m < \delta_m^0) \\ (1-D)\mathbf{K\delta} & (\delta_m^0 \le \delta_m \le \delta_m^f) \\ \mathbf{0} & (\delta_m > \delta_m^f) \end{cases} \quad (11)$$

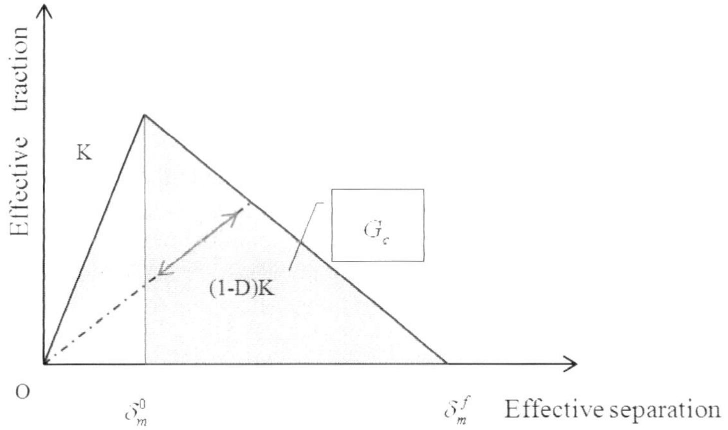

Figure 1 Schematic diagram of the traction-separation law.

1.1.4 Strain rate effects. The rate dependency of strength as well as modulus can be described within a unified framework by introducing dynamic increase factor (DIF)[8], namely

$$t_i^0 = t_i^s \times DIF,\ K_i = K_i^s \times DIF,\ (i = n, s, t) \quad (12)$$

Where t_i^0, K_i are the dynamic strength and modulus respectively and t_i^s, K_i^s are the corresponding static values. The dynamic increase factor (DIF) can be expressed in the following form [8]

$$DIF = \left\{\left[\tanh\left(\left(\log\left(\frac{\dot{\varepsilon}}{\dot{\varepsilon}_0}\right) - A\right) \times B\right)\right] \times \left[\frac{C}{(C+1)/2} - 1\right] + 1\right\} \times \frac{C+1}{2} \quad (13)$$

in which A, B and C are empirical constants to describe the rate sensitive behavior of FRP laminates.

2. Results and discussion

The progressive damage model for uniderictional laminated plates and the adhesive layer damage model described above are implemented in ABAQUS finite element code as user defined materials (VUMAT). In the following a low velocity impact is given to demonstrate the validity and accuracy of the FE model developed in section above.

Table 1 cohesive interface properties used in adhesive layer damage model.

Parameter	Data	Parameter	Data	Parameter	Data
t_n^s	11(Mpa)	t_s^s	17(Mpa)	t_t^s	17(Mpa)
G_{IC}	0.3(N/ mm[4])	G_{IIC}	0.8(N/ mm[4])	G_{IIIC}	0.8(N/ mm[4])
K_n^s	850(MPa[12])	K_s^s	850(MPa[12])	K_t^s	850(MPa[12])
A_s	2.5[8]	B_s	0.9[8]	C_s	3.7[8]
A_m	1.85[8]	B_m	0.5[8]	C_m	1.3[8]

Table 2 Parameters for graphite/epoxy laminates[5] used in the progressive damage model .

Parameter	Data	Parameter	Data	Parameter	Data
E_{11}	143.4(GPa[4])	Y_t	54(MPa[4])	v_{23}	0.52[4]
E_{22}	9.27(GPa[4])	X_c	1650(MPa[4])	v_{13}, v_{12}	0.31[4]
E_{33}	9.27(GPa[4])	Y_c	240(MPa[4])	δ_1^f, δ_2^f	$5.7\text{x}10^{-2}$(mm[8])
G_{12}	3.8(GPa[4])	Z_t	54(MPa[4])	δ_7^f,	$2.7\text{x}10^{-2}$ (mm[8])
G_{23}	3.2(GPa[4])	Z_c	240(MPa[4])	δ_3^f, δ_4^f	$7.65\text{x}10^{-2}$ (mm[8])
G_{31}	3.8(GPa[4])	S_{12}	100(MPa[4])	δ_5^f	$5\text{x}10^{-3}$ (mm[8])
X_t	2945(MPa[4])	S_{31}, S_{23}	100(MPa[4])	δ_6^f	$5\text{x}10^{-2}$ (mm[8])

Figure 2 shows an FE model for low velocity impact tests which were reported by Aymerich et al. [5]. The specimen was made of Graphite/epoxy laminates (stacking sequence $[0_3/90_3]_s$) which had a size of 87.5 × 65mm ×2mm and it was supported in a steel frame and clamped at four corners using G-clamps. The open window (aperture) area was 67.5mm ×45mm. C3D8R solid elements are used to model both the uniderictional laminated plates and the impactor which had a 12.5mm diameter hemispherical end and a mass of 2.3kg. The adhesive layers are created by COH3D8 cohesive elements . The elements in the open window (67.5mm×45mm) are refined with in-plane size of about 1mm × 1mm ,the total number of the elements through the thickness of the sample is four (one element for three plies which aligned in the same direction). COH3D8 cohesive elements are inserted at the interface between the layers with different fiber orientations (0°/90° and 90°/0°) to model delamination damage. The impactor is assumed to remain rigid during impact. A general contact algorithm is employed to model the contact between the impactor and the CFRP laminates. A friction coefficient value u=0.2 is included in the contact formulation.

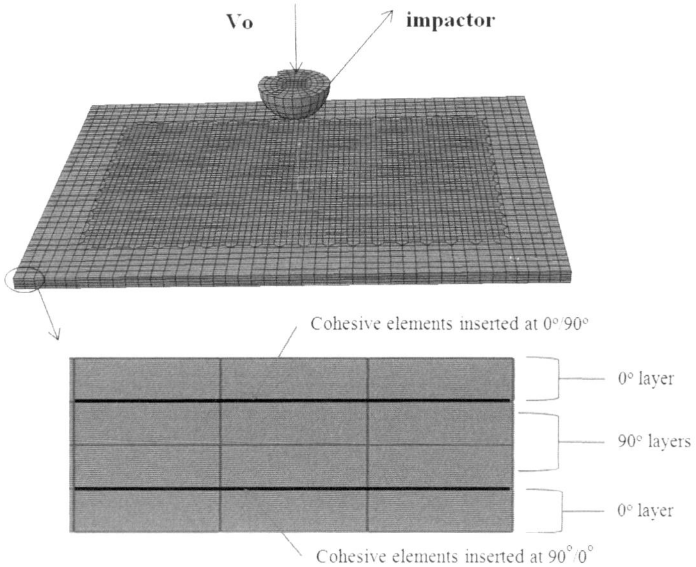

Figure 2 Finite element model employed in low velocity impact

Figure 3 shows comparison of the numerically predicted delamination shape and area with those observed for the 2mm thick CFRP laminates struck transversely by a 2.3kg, 12.5mm diameter hemispherical-ended impactor [5]at three different impact energy of 2.1 J (1.35 m/s),4.9 J (2.06 m/s) and 9.3 J (2.84 m/s). Here, a state variable is defined in the VUMAT subroutine which represents the stiffness degradation of cohesive elements. The left picture shows the stiffness degradation contour of cohesive elements calculated by the present model at lower 90°/0° interface. When cohesive element's stiffness degradation is 0 it means no damage at all whilst the stiffness degradation is 1 it means full damage. The fully damaged cohesive elements represent delamination between the adjacent laminate layers. It is clear from Figure 3 that the present model predictions are in good agreement with the experimental observations [5].

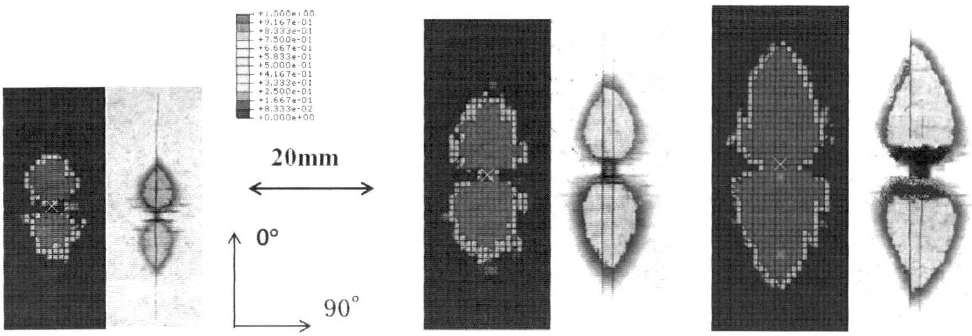

(a) 2.1J (b) 4.9J (c) 9.3J

Figure 3 Comparison of the numerically predicted delamination area with the experimental observation[5] for the 2mm thick CFRP laminates struck transversely by a 2.3kg, 12.5mm diameter hemispherical-ended impactor.

FRP laminates are prone to delamination under impact loading due to their manufacturing process. They are formed by superposition of anisotropic monolayers through cementing and curing. Under impact loadings, the deformation of the unidirectional layers with different laying direction is not coordinated, which more likely leads to the damage and destruction of the adhesive layer. This model is capable to reflect the non-coordinated deformation of the adjacent unidirectional layers by introducing adhesive layer.

IOP Conf. Series: Materials Science and Engineering **242** (2017) 012088 doi:10.1088/1757-899X/242/1/012088

3. Conclusions

A numerical study has been conducted to predict the delamination of fiber reinforced plastic (FRP) laminates subjected to impact loadings. A new FE model is suggested for fiber reinforced plastic composites which considers that FRP laminates has been constituted by unidirectional laminated plates with adhesive layers. A modified adhesive layer damage model that considering strain rate effects is incorporated into the ABAQUS / EXPLICIT finite element program by the user-defined material subroutine VUMAT. It transpires that the present model predicted delamination is in good agreement with the experimental results for low velocity impact.

References

[1] Olsson R. Analytical prediction of large mass impact damage in composite laminates. Composites Part A: Applied Science and Manufacturing, 2001, 32(9): 1207-1215.

[2] Espinosa H D, Dwivedi S, Lu H C. Modeling impact induced delamination of woven fiber reinforced composites with contact/cohesive laws. Computer Methods in Applied Mechanics and Engineering, 2000, 183(3): 259-290.

[3] Johnson H E, Louca L A, Mouring S, et al. Modeling impact damage in marine composite panels. International Journal of Impact Engineering, 2009, 36(1): 25-39.

[4] Singh H, Mahajan P. Modeling damage induced plasticity for low velocity impact simulation of three dimensional fiber reinforced composite. Composite Structures, 2015, 131: 290-303.

[5] Aymerich F, Pani C, Priolo P.Damage response of stitched cross-ply laminates under impact loadings. Engineering Fracture Mechanics, 2007, 74(4):500-14.

[6] Aymerich F, Dore F, Priolo P. Prediction of impact-induced delamination in cross-ply composite laminates using cohesive interface elements. Composites science and technology, 2008, 68(12): 2383-2390.

[7] Francesconi L, Aymerich F. Numerical simulation of the effect of stitching on the delamination resistance of laminated composites subjected to low-velocity impact. Composite Structures, 2017, 159: 110-120.

[8] Xin S H, Wen H M. A progress damage model for fiber reinforced plastic composites subjected to impact loading. International Journal of Impact Engineering, 2015, 75: 40-52.

[9] Barenblatt G I. The mathematical theory of equilibrium cracks in brittle fracture. Advances in applied mechanics, 1962, 7: 55-129.

[10] Camanho P P, Davila C G, De Moura M F. Numerical simulation of mixed-mode progressive delamination in composite materials. Journal of composite materials, 2003, 37(16): 1415-1438.

[11] Benzeggagh M L, Kenane M. Measurement of mixed-mode delamination fracture toughness of unidirectional glass/epoxy composites with mixed-mode bending apparatus. Composites science and technology, 1996, 56(4): 439-449.

[12] Long S, Yao X, Zhang X. Delamination prediction in composite laminates under low-velocity impact. Composite Structures, 2015, 132: 290-298.

IOP Conf. Series: Materials Science and Engineering **242** (2017) 012089 doi:10.1088/1757-899X/242/1/012089

Discussion on the Technology and Method of Computer Network Security Management

Jianlei Zhou

Baoding Technical College of Electric Power. 071051

Abstract: With the rapid development of information technology, the application of computer network technology has penetrated all aspects of society, changed people's way of life work to a certain extent, brought great convenience to people. But computer network technology is not a panacea, it can promote the function of social development, but also can cause damage to the community and the country. Due to computer network' openness, easiness of sharing and other characteristics, it had a very negative impact on the computer network security, especially the loopholes in the technical aspects can cause damage on the network information. Based on this, this paper will do a brief analysis on the computer network security management problems and security measures.

1. The analysis of the computer network security problems and causes

With the development of information technology, the state and the community are more a concerned about the computer network security and network information security, the citizen's network information protection attracts more and more attention, and has adopted the relevant protective measures, but because the development time of the network in China is not long, the technical means of network security management is still flawed, the complexity of all aspects of citizen information rights, have caused many problems in China's current computer network security management. It can be summarized as the following five aspects.

1.1 Vulnerability in computer hardware and software

The most important problem of computer network security problem is network information security, the most direct reason caused the damage to information security due to computer hardware and software shortcomings. First of all, there are hidden dangers on the computer hardware, once the computer network hardware equipment is damaged, viruses and Trojans can entered through the hardware into the computer system, the computer network file data and other things will damage, and the virus' spread speed is fast, it is easy to cause the computer system paralyze, resulting in the loss of documents and information resources. As the Trojans and viruses are highly concealed, and the technology is also more and more advanced, so the current protection technology is difficult to achieve foolproof. Second, there are loopholes in computer software, which is due to the negligence caused by the procedural staff in a certain part [1]. Once the software program has problem, hackers and criminals will be invaded into the computer system, attack the computer software programs, programs and systems, cause the computer network security is threatened, making the computer network information get theft and destruction.

1.2 Laws and regulations are not perfect

In fact, China has long been concerned about the computer network information security issues, but also promulgated a number of related laws and regulations, such as 2012 issued "decision on the strengthening of network information protection " has the initial provisions on the protection of citizen

Content from this work may be used under the terms of the Creative Commons Attribution 3.0 licence. Any further distribution of this work must maintain attribution to the author(s) and the title of the work, journal citation and DOI.
Published under licence by IOP Publishing Ltd

network information , but these similar laws and regulations are only for electronic data which is related to citizenship or citizen privacy, but they are not yet fully related to all personal information, including structured information and unstructured information which is related to citizens. In addition, even in some other laws have the content of citizens on the protection of the right to information, such as administrative law, civil law, but they are only as an additional provision, not in specific measures to implement these laws and regulations, it doesn't define and divide the scope of protection of personal privacy and citizens information rights .

1.3 Industry lacks self-discipline
Network New media era gave the birth to many new media companies, these enterprises are more chaotic in their own management, lack of sense of responsibility. Such as the financial sector industry, some financial enterprises in order to expand their business or for the private interests of some side they do the excessive collection even modify the citizen information, or directly sale the public information. The industry lacks self-discipline, arbitrarily destroys and leaks the information on citizens, so that the right of citizens information is not guaranteed, not only bring risks of the users' property and life who they services to , in fact, it is also harmful to its own long-term development of the enterprise.

1.4 The user's safety awareness and protection measures are not in place
In addition to the objective computer hardware and software vulnerabilities, laws and regulations and industry self-discipline issues, the other one makes the network information security threat is because the user's safety awareness is not in place, they don't take the correct protective measures. Such as login account in the public online forget to hide the password or choose to remember the password and other actions will make the virus attack personal account easily; for example, fill in important data and didn't use encryption technology to deal with the data, didn't backup the data either, causing the data get lost. The existence of these security risks is because the user does not have the awareness of security, do not pay attention to the protection of personal privacy, fail to grasp the main point of the network technology [2].

1.5 Law enforcement agencies to the crackdown behavior of breaking the network security is not enough
In addition to the objective of the technical reasons which cause the computer network information security risks, the personal subjective security awareness is not strong, there is an another important reason that is in the fight against saboteurs law enforcement agencies did not take effective measures. On the one hand it is due to that the extensive distribution of destructive elements, mobility and concealment are very strong, coupled with their own very high technical means, it enhance the law enforcement departments hunting difficulty On the other hand, because the relevant departments are not in accordance with the national policy of effective implementation, in the fight against sabotage elements, some law enforcement departments don't have strong sense of responsibility, their security awareness is weak, coupled with the need to improve the fighting and other reasons, punishment is still not strong enough. In the supervision of the industry, the relevant regulatory authorities did not strengthen the link between the rows of their own, the responsibility of each other repeatedly occur, the improper management and disorder cause the information of the citizens is leaked, the right to information of citizens cannot be guaranteed.

2. The protecting countermeasures of the computer network information security
In view of the above problems of the computer network security management, I believe that we they should do security management mainly from technical aspects, and then strengthen management in other areas, take reasonable measures to solve the corresponding problems. In the technology mainly use computer security technology to protect and manage the computer network security, in other aspects, strengthen law enforcement and government regulation including legislation, industry self-discipline, individuals should pay attention to protect the network security and personal information security.

2.1 Use computer technology to protect information

The most direct means to protect the network information security is to improve the computer hardware and software, improve the professional level and overall quality of the program staff, technical staff. But any advanced hardware and software will be able to attack the vulnerabilities and have hidden dangers, so in addition to improve the level of hardware and software constantly, but also we need to use computer technology to reduce the risk of attacking network information. The computer security technology mainly in the following (see Table 1). First, use firewall technology in the network of the external and internal barriers. Firewall on the one hand can control the external network data onto the internal network system, make the external data source audit control , on the other hand it can protect the internal data structure, making the internal data is not easily seen by the outside. The control of external data entry and internal data out has played a very good protective effect in order to protect the data security. Second, use the access authorization technology to protect the network information. Access authorization is popular to allow legitimate users to enter the network, the illegal users are refused to go outside the network. To determine whether the user is legitimate mainly through the access, access to passwords, the user identity. Access refers to do the security restriction when the users to be accessed, the user can access only through the inspection, the password visiting is equivalent to the access key, only the password that matches the original password can be recognized by the system and then entered into the internal network. The password must be set to strengthen the settings, otherwise it's easy to be broken. Third, use the intrusion detection technology, that is, all-weather monitoring of intrusion, intrusion detection technology with monitoring, early warning and emergency response and other functions. Fourth, use the encryption technology to protect the user information. Each user needs to learn to use the encryption technology, we also need to increase the difficulty of the secret key, making the outside world will not easily break the protection line. These kinds of computer technology can effectively guarantee the network information security [3].

Table 1 Computer Security Technology

Computer technology	The role of technology	Technical security defense measures
Firewall technology	Control external data into the network; protect the internal data structure	Set up obstacles inside and outside the network
Access authorization technology	Allowing legitimate users to enter the network, the illegal users outside the network	Set access, access password, and user identity
Intrusion detection technology	Monitoring, early warning and emergency response	Monitor intrusion every day
Encryption technology	Protect the security line isn't break by the outside	Set the cipher key

2.2 Strengthen and perfect laws and regulations

There is a need to strengthen legislation at the national legal level. In the legislation, we must first clear the relevant legislative principles, in addition to ensuring the safety of computer networks, but also need to develop appropriate laws and regulations in the network security information management, such as computer network security management needs to protect the privacy of citizens, besides the privacy information other information which can be circulated in accordance with laws and regulations, other citizens information must be protected by laws and regulations. Second, we must determine the scope of the protection for civic information and specific protection measures. Because we are going to protect the information on citizens, so which areas and the specific protection measures needs to be protected should be stipulated in the law clearly. In addition, it should establish a clear and complete personal information protection law, so that the information rights of citizens are guaranteed.

2.3 To improve the industry's self-discipline

The above mentioned in the computer network security, some enterprises or individuals due to their personal interests make damage to other citizens' interests, free disclosure of citizen information, endanger the computer network security behavior. To improve computer network security, in addition to technical and legal aspects of the method, we also need the joint efforts of enterprises and individuals. First of all, require units and individuals to enhance their own quality, the unit cannot lay out the user's information for the sake of the interests, individuals cannot reduce the protection of user information because lacking responsibility. Second, once the unit organization found that employees have leaked information behavior, they must deal with these kind of behavior severely; individuals need to strengthen self-cultivation ability, improve cultural quality, not be tempted by outside and don't does the act of breaking a law.

2.4 Strengthen government supervision

In addition to set up a clear supervision and management of all walks of life outside, the government also needs to strengthen the guidance of these regulatory management. For example, the CBRC, while monitoring the financial institutions, also should contact with other regulators and strengthen the cooperative relationship, so the government should unified the management, divide the scope of the role to form a joint force. Regulatory authorities on the supervision and disciplinary mechanisms needs to determine which important information of the citizen information and other computer networks cannot be transmitted. Second, it is necessary to strengthen the supervision of new media organizations, so that these enterprises needs to improve the citizen information and other computer network information protection measures. In addition we should also regulate the personal information behavior of the citizens, when necessary, regulate and punish the behavior of citizens.

2.5 Strengthen the user's safety awareness and protection awareness

First of all, users must pay attention to the computer network security and computer network information protection, have a clear understanding of the existence of computer network defects, protect their privacy in a timely manner to prevent the disclosure of personal information, enhance personal safety awareness and protection awareness. For example, don't login personal account in the public network, outside the Internet at random, encrypted to their own information, do not tell the password to others and so on. Only the security awareness is rising, they will take favorable measures to protect the computer network information. Second, the relevant departments should also strengthen the computer network security publicity and education, and guide people use the Internet legally, do not believe that online information and links arbitrarily and have a sense of protecting rights, once found their information has been tampered or leaked, do timely alarm processing to these illegal behavior.

2.6 Strengthen law enforcement efforts of the law enforcement departments, punish disciplinary units and individuals severely

In addition to legislation to improve the computer network security laws and regulations, in law enforcement should also strengthen the intensity of disciplinary units or individuals to severely punish. On the one hand we mentioned in the previous article, some units or individuals do some damage to the computer network security for their own interests, such as leakage of personal information and so on. For this deliberate damage to computer network security behavior, the unit itself needs to recognize the importance of protecting the computer network security, while enhancing self-awareness at the same time, but also the law enforcement departments need to regulate the law. For example, once the unit or the individual in unit makes a similar violation of law, law enforcement agencies required to investigate the unit or individual in accordance with laws and regulations first, and then do the corresponding punishment according to the damage degree. On the other hand, law enforcement agencies must adhere to the principle of fairness and impartiality and openness in law enforcement. Treat any unit and individual in accordance with the laws and regulations of the objective requirements, do not cover and condone, in order to ensure the effectiveness of law enforcement. Such as serious acts of wrongful acts and lawless elements, developed the clear and specific punitive measures to avoid reduce the punishment of illegal behavior which lack of operational law enforcement measure.

3. Conclusion

With the rapid development of information technology and the network, individuals, society, countries and even over the world all need to pay attention to computer network security, because the security of computer network information is related to personal interests, stable and harmonious development of society, development of economic, technology, culture military and other aspects,as well as the safety of the country. Because once the computer network is destroyed, important network information will be leaked, which will cause extremely adverse political and military impact to the country. Therefore, for the existing computer network security problems, countries, society and individuals can take different means to do computer network information security protection.

References

[1] Li Linteng, Zhao Yizhang. Computer network information security and protective measures [J]. China's strategic emerging industries, 2017 (02).
[2] Chen Dongmei.Computer network information security and its protection strategy based on computer network technology [J]. Electronic Testing, 2017 (03): 131-132.
[3] Cao Cheng, Zhou Jian, Zhou Hong, Wang Mingfu. Network security and countermeasures [J]. Journal of Hefei University of Technology (Natural Science Edition), 2007 (09): 14-16.
[4] Zhai Sujuan, Sun Yuxiu Computer network security management related issues [J]. Coal Technology 2011
[5] Lan Hao. Talk about the computer network security management measures [J] computer CD-ROM software and application 2011

On Study of Application of Power Electronical Technology in Field of Energy Conservation

Ke Song

Sichuan Aerospace Vocational college, 618300, China

Abstract: aiming at achieving the sustainable development in economy and improving living standard of the residents, the government departments in China have strengthened the promotion of green, low-carbon and sustainable development concepts. Under this context, the energy conservation concept is implemented during the implementation of social production and life activities. Currently, in order to ensure the improvement in social energy conservation benefits, the advanced technologies are applied by China's enterprises and residents. This paper, based on this, discusses the development situations of energy conservation in China, and analyzes the application of power electronical technology in the field of energy conservation, hoping to realize the sustainable development of energy conservation cause in China and to gain high yield.

1. Introduction

Serving as a technology type integrating control, power and electronical technology, the power electronical technology tends to effectively transform energy output by virtue of high power electronic devices in the process of promotion, which saves various energies at maximization and realizes the social sustainable development in China. The recent years see the implementation of low-carbon and green development concepts in China, and the application of power electronical technology is strengthened during the process of social production and life activities implementation, facilitating benefits in various aspects.

2. Introduction of Power Electronical Technology

2. 1 Definition of Power Electronical Technology

The so-called power electronical technology refers a science which solves practical problems through designing and applying various electronic components according to principle of electronics. Generally speaking, the research of power electronical technology has gone through three eras (rectifier era, inverter era and inverter era), and is widely applied in various fields in social production and life.

Actually, with the wide application of power electronical technology, China's social production and life will gradually develop towards the digital, green, intelligent, humanity and network orientation, and promote higher benefits.

2.2 Reasons for Application of Power Electronical Technology in Energy Conservation Field

With the continuous improvement of urbanization and industrialization in China, the scale of cities in China is expanding day by day, and an irrational situation occurs in industrial structure. Under such a background, China faces severe environmental pollution gradually, and our ecological environment is further damaged by garbage, sewage and gas produced by industrial production.

Content from this work may be used under the terms of the Creative Commons Attribution 3.0 licence. Any further distribution of this work must maintain attribution to the author(s) and the title of the work, journal citation and DOI.
Published under licence by IOP Publishing Ltd

Considering this situation, relevant departments in China strengthen call on green and low-carbon economy, and consolidate promotion of energy-saving work. In this process, in order to avoid the waste of power resources as well as pollution caused by electricity resources production, the relevant departments strengthen the application of power electronical technology in order to improve energy efficiency of power systems, avoiding power resources waste and facilitating the environmental improvement.

3. Key Technologies of Power Electronical Energy Conservation

In order to make the further improvement in energy efficiency in China's power system, the relevant units and persons enhance application of power electronical energy-saving technology. In terms of commonly used power electronical energy conservation technologies, the author draws relevant conclusions as follows.

3.1 Switching Frequency Conversion Technology

The so-called switching frequency conversion technology refers to the energy conservation technology which adjusts fixed frequencies. Currently, China's power equipment faces low efficiency and low power factors in the process of running, further reducing power consumption. It is reported that about 2/3 of annual total energy consumption is wasted caused by power equipments in the non-optimal efficiency.

Actually, the promotion of frequency conversion technology guarantees that the power equipment is in the best operating efficiency state to a large extent, improves electricity efficiency, avoids line loss and reduces pollutant emissions. Here is an example about testing frequency conversion and speed regulation of three 300kW water pumps in one 400kVA transformer frequency conversion device, and we have to ensure that three pumps mentioned above will save 600kW electricity or so during running in the process of test implementation.

Relevant research shows that the traditional switching power supply only utilizes 65% to 70% of energy in the design and application, however, with the promotion and application of frequency conversion technology, the utilization of power resources reaches as high as 85 %. The changes of these data show that: frequency switching technology is characterized by high efficiency, good performance and small size in the running, and is able to reduce energy consumption in the process of promotion and application.

3.2 Soft Switching Technology

Actually, in the application of switch, energy consumption will occur during the process of opening and closing, which will result in low energy conservation efficiency of power systems. In order to avoid this problem, the technical persons strengthen development, promotion and application of soft switching technology. It is said that soft-switching technology is capable of reducing switch energy consumption and noise in the process of running, and of upgrading frequency of switches.

The so-called soft-switching technology is a relative speaking to hard switching technology. Generally speaking, the hard switching technology refers to one that, in the switching process of switch operation, will avoid no-voltage and zero current switch which will lead to overlap various indexes. In general, the soft-switching technology will cause noise caused by rapid changes in voltage and current in the implementation process. However this technology means that inductance, capacitance and other resonant components should be added in hard switching technology circuits in order to avoid voltage and current overlap situations.

Generally speaking, the promotion of such technology is able to ensure components to switch under no-voltage and zero current state and to greatly reduce energy consumption, upgrading utilization of power efficiency and realizing energy conservation.

3.3 High Frequency Technology

High-frequency technology develops on the basis of frequency conversion technology. It aims at

IOP Conf. Series: Materials Science and Engineering **242** (2017) 012090 doi:10.1088/1757-899X/242/1/012090

improving frequency on the basis of frequency conversion, and making full use of energy and materials. In fact, the technical persons find that there is an inverse ration between volume and weight of transformers, inductors and capacitors of product with the square root of power supply frequency when analyzing relevant experiments

Guided by such study, the technical persons raise frequency of electrical equipment from 80Hz to 20kHz, and the volume and weight of power equipment will be lightened to a large extent, besides, the decline ranges are between 5% and 10%. Actually, according to this principle, the technical members are able to transform power supply in electrical processing industry, which shall save electricity with the minimum rate of 30%. A conclusion can be drawn from such result that the promotion application of high-frequency technology is capable of improving the maximum energy utilization and its operating result is better than that of frequency conversion technology.

3.4 Active Power Factor Correction Technology

The so-called power factor refers to the cosine produced by voltage at two ends of two-port network and the current. In general, the power factors are significant for researching utilization rate of power equipments.

There is a certain rated voltage and current generally in the operation of power equipments, and once they exceed a rated value in operation, safety loopholes tend to occur in working state of equipments. Taking this into consideration, operators need to improve power factors in order to realize a perfect energy conservation property of power equipments, to strengthen output of active power, to reduce the reactive power and to fully make use of electricity.

3.5 Other Energy Utilization Technologies

Actually, with the continuous development and wide application of science and technology, many ways for energy conservation are adopted in China's power electronical technology in the promotion and application. At present, there are other energy utilization technologies like new magnetic components, new transformers and smart power. In fact, these technologies are able to greatly utilize energies and further improve energy conservation in the extensive promotion process.

4. Energy Conservation Results of Power Electronical Technology

In order to further verify outstanding contribution of power electronical technology in energy conservation aspect, the author summarizes achievements of various power electronical technologies in the process of promotion and application. The specific content about energy conservation results of power electronical technologies is as follows.

Currently, with the promotion and application of power electronical technology as well as the implementation of transformation of traditional electric power equipments, the electricity-saving rate of relevant equipments has reached about 20%. If the data are introduced into the field of national electricity, about 50 billion kwh of electricity will be saved annually.

In this process, the power technical persons strengthen the promotion and application of AC highly efficient speed regulation device based on power integrated devices such as the giant transistor, and such equipment is able to regulate speed of fan and pump equipment as well as save energy during promotion, besides, it has to ensure that its operating power consumption must save energy as 30% as that in traditional ways. Currently, over 20 million units of fans and pump equipments are adopted in first-tier and about 15 billion kwh of electricity will be saved after such energy-saving device is adopted.

Besides, the technical persons also develop and promote the carrier wave device with DC high-efficiency regulation way through grid thyristor. It is said that the equipment tends to replace resistors in the process of application and the energy efficiency shall up to 20% when it is used in fields like urban trams, industrial and mining electric locomotives. Here is an example about Shenyang, about 500 trolleybuses are modified to cater to the call of developing green transport, besides, the carrier wave device with DC high-efficiency speed regulation way. In such a circumstance, 4 million kwh of electricity will be saved by vehicles above and at least 103 billion kwh of electricity will be saved if such technology is introduced nationally.

Besides, the application of electrostatic induction thyristors and field effect transistors tends to replace power frequency inductance ballast and about 20% electricity will be saved in its operation. Moreover, the promotion and application of rare earth three-band fluorescent lamp and inductive ballast will save over a half electricity in power system. Currently, 8% electricity in China power generation is used for lighting. In this context, if technical persons transform 2/3 lighting systems with such technology, about 13 billion kwh electricity will be saved in the running of power system.

Besides, the technical persons strengthen the application of asymmetric thyristor in the promotion of power electronical technology, realizing high frequency processing to intermediate frequency power. The measure not only improves efficiency of electric-thermal conversion to a large extent, but also guarantees the application of such technology. Currently, there are 12,000 standard high-frequency electric furnaces (with 100 kilowatts as the standard) operated by China's production system, but these devices still adopt tube-shaped high-frequency oscillator for operation, so the operating efficiency is only 50% or so. However, if the electric power technical persons transform equipments by adopting electrostatic induction transistor, the operating efficiency will be maintained as high as 80% of equipments and about 1 billion kwh electricity will be saved at the macro level.

5. Prospect of Application of power Electronical Technology

It shows from previous analysis that: with the promotion of power electronical technology, China's power system equipment is capable of improving energy efficiency benefits in operation. Actually, with the implementation of China's scientific outlook on development, the power electronical technology will experience a permanent development and usher in a promising prospect in the future. Besides, with the continuous development of science and technology in China, the improvement of power electronics theories as well as development and application of new power materials, the technical persons in China are capable of developing and applying high-performance circuit topologies so as to save more energy resources, and obtain economic and social benefits widely, realizing the sustainable development of national economy.

6. Conclusions

Aiming at achieving steady improvement in energy conservation efficiency in China, our technical persons strengthen development and application of power electronical technology, obtaining benefits in various aspects. This paper, based on such situation, analyzes the connotation of power electronical technology (definition of power electronical technology and application reasons of it in energy conservation field), and discusses key technologies of power electronical energy conservation (switching frequency conversion technology, soft switching technology, high frequency technology, active power factor correction technology and other energy utilization technologies); finally, the energy conservation effects and application prospect of power electronical technology are mentioned. The author believes that with the implementation of relevant measures in place and development of technology, permanent and steady progress will be made in our energy conservation cause, which shall bring economic profits and social benefits, and is conducive to the implementation of sustainable development.

Reference

[1] Zhang Wenliang, Tang Guangfu and Cha Kunpeng. *On Study of Application of Advanced Power Electronical Technology in Smart Grid*[J]. Proceedings of the CSEE, 2014, (4):1-7

[2] Guo Shiming. *On Study of Application of Power Electronics Technology in Energy Conservation and Strategic Emerging Industry*[J]. Electronic Component and Device Applications, 2012, (6): 1-2 + 11.

[3] Wei Lin, Liao Huixin and Yi Ganhong. *On Study of Application of Power Electronical Technology in Power System*[J]. Digital Technology & Applications, 2012, (10): 97-98.

[4] Huang Zongjian, Wei Hongfei and Wu Huimin. *On Study of Application of Power Electronical Technology in Power System*[J]. Electronic Test, 2014, (13): 91-92 + 134.

[5] He Xiangning, Wu Jiande and Li Wuhua. *A Study on Interconnection and Network Technology of Power Electronical Equipment in Distribution Network*[J]. Proceedings of the CSEE, 2014, (29): 5162-5170
[6] Li Daxing and Luo Wei. *On Study of Application of New Energy Saver Based on Advanced Power Electronical Technology in Urban Green Lighting Project*[J]. Electrotechnical Application, 2017, (2): 20-24 + 65.
[7] Zheng Jinbiao. *On Study of Application and Research of Power Electronical Technology in Power System*[J]. Heilongjiang Science and Technology Information, 2017, (5): 3.
[8] Cao Chengjie and Zhu Jun. *On Study of Application and Development of Power Electronical Technology in Ship Power System*[J]. Marine Electric, 2017, (4): 216-220.
[9] Bo Baozhong, Su Yanmin and Hou Lijun. *On Study of Application Prospect of Power Electronical Technology in Pumping Unit Motor Energy Saving*[J]. Energy Conservation, 2014, (3): 14-16 + 2.

IOP Conf. Series: Materials Science and Engineering **242** (2017) 012091 doi:10.1088/1757-899X/242/1/012091

On Study of Construction of New Generation Intelligent Communication Network for Distribution System

Dan Wu[1], Xueyan Zhang[2]

[1]Sichuan Aerospace Vocational college, 610100, China

[2]Chengdu Aeronautic Polytechnic, 610100, China

Abstract: many new technological means are integrated in the new network of smart electricity distribution including electric power technology, control technology and information technology, and in this area, people carry out profound research to the construction of communication network and power distribution network. This paper analyzes specific structures of the communication network for distribution system, discusses the development trend of it, explores its technical system in depth, and finally proposes some concrete constructive strategies, hoping to be a valuable reference for related research persons.

1. Introduction

The smart power grid integrates various technological means including information technology, power technology and automation technology, and especially the communication network for intelligent distribution system is a core in this project, besides, the contradiction problem between technical needs and costs must be handled properly in the construction of a new generation intelligent communication network for distribution system. In order to solve such problem properly, we need to carry out a profound research in industrial demands and development trend of the intelligent communication network for distribution system, and carefully analyze strategies of constructing intelligent communication network for distribution system so as to assist the construction of new intelligent communication network for distribution system.

2. Analysis of Network Structures

The structures of communication network for distribution system and grid topology are roughly same as shown in Figure 1 below. They mainly construct a tree or ring network structure through low-voltage communication network and medium voltage communication network, and view many remote terminals like distribution automation and electricity information collection a downlink terminal points or view backbone communication network node of 35kV or 110kV transformer substation as uplink terminal points in order to offer access channel to information system[1]. There are still severe challenges in constructing communication network for distribution system in 10kV section in this regard, therefore this paper emphasizes on analyzing problems in this network segment.

Content from this work may be used under the terms of the Creative Commons Attribution 3.0 licence. Any further distribution of this work must maintain attribution to the author(s) and the title of the work, journal citation and DOI.
Published under licence by IOP Publishing Ltd

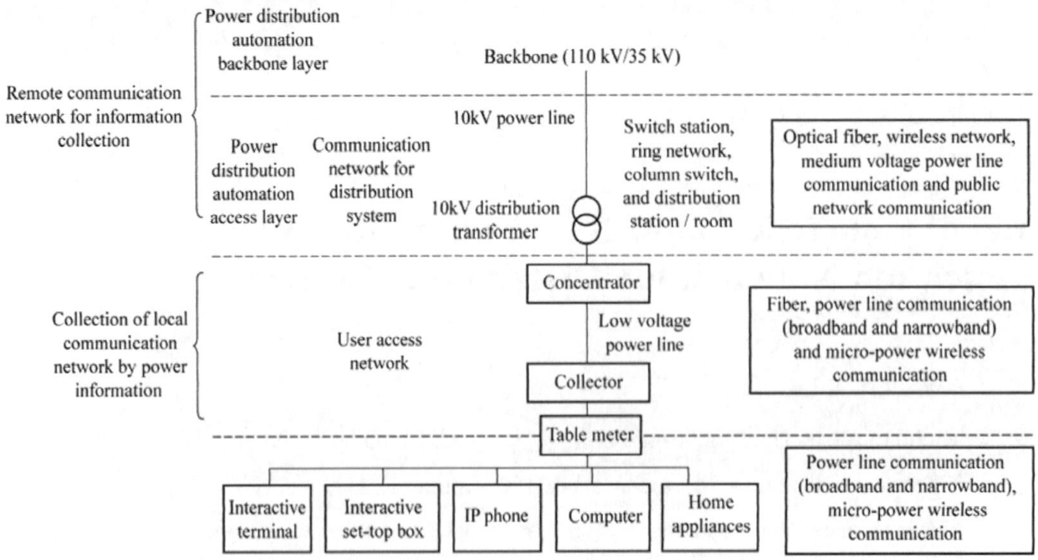

Figure 1: Structures of Communication Network for Distribution System

3. Development Trend of Communication Network for Distribution System

3.1 Business Orientation

The business of smart electricity distribution can be divided into three types according to the level of information application: (1) spot voice, video, data and other ancillary business of communication network for distribution system; (2) remote monitoring and control to distribution network and equipment including distribution transformer control and distribution automation; (3) the interactive operation between electric power enterprise and end users including electricity information collection and expense control, the electric power load supervisory control and power supply access control.

3.2 Communication Network Platform

After realizing platform, the communication network is capable of providing bearer services with high efficiency for integrated business and of improving the overall application rate of network comprehensively. Communication and information platform can be divided into three major levels, namely: (1) application layer, including I, II and III master station systems; (2) platform information layer. It is in the internal partition for centralizing storage; (3) platform communication layer. It is also in the internal partition and responsible for centralized transmission. In terms of specific application, it is mainly responsible for handling information through facing the application layer in business departments directly, which requires that the communication network for distribution system should not only consider the overall carrying capacity of information to distribution business but also realize a unified access, management, backup and expansion in terms of each device interface in the communication layer.

3.3 High Precision Transmission Requirement

The current distribution business is still in the process of integration stage, and communication information in 10kV section also involves in power distribution automation and marketing business processes. The amount of single-point data in monitoring terminal, feeder terminal units, concentrators, collectors and other intelligent terminal systems in distribution transformer is larger than that in 10Mbit/s, and data information collected by transformer substation in sub-granular zone is basically 10Mbit/s, therefore, compared with the public network media data volume, the industrial control information is relatively small in amount which focuses more on real-time and stability requirements. In the whole business process, the communication part in rigid indicator can be selected according to a higher order of magnitude, which means that communication delay should be less than 100ms and the accuracy rate must be over 99.999%.

3.4 High Security Requirement
Currently, in the whole distribution network system, the distribution business is still in the edge position, and we have to conduct encryption processing to business content involved and user identity information. Besides, we have to ensure the safety of transmission approaches and remote device of communication network, which means that we need to guarantee integrity, availability and confidentiality of data packet and the device itself [2]. Specific measures below can be taken to strengthen the security of communication network for distribution system: (1) encryption processing algorithm can be adopted like Ipsec, WPA, SSL and other transmission pates to avoid the disclosure of communication packets; (2) measures including security authentication, user authentication, packet filtering and others should be taken to avoid illegal communication equipment and data packets to enter into the network environment when the required remote device must be accessed; (3) the exclusive transmission media or logical physical channel can be adopted to effective avoid physical isolation effect in case of encrypted information leak in physical media; (4) measures like network address translation technology and isolation areas as well as others should be taken to hide locations of important information equipments, greatly reducing the chance of direct attacks.

4. Study on Technical System

4.1 Standards of Intelligent Communication Network for Distribution System
When constructing intelligent communication network for distribution system, we must learn to apply many mixed communication technologies and adopt standard system specifications of network control which can be applied in a wider range. The local area networks with relatively small size are mostly from the IEEE802 protocol cluster which is relatively more convenient compared with smaller local area network implementation, but its size and distance will be greatly limited. We have to realize the adoption of telecommunications standards including 3GPP, ITU in order to carry out a relatively large scale network construction with the purpose of properly solving bandwidth loss, network management and business control issues in large-scale terminal equipments and a variety of access network technologies. In terms of Ethernet, the adoption of external passive optical network, global microwave interconnection and multi-carrier wireless information local loop and other information and communication technologies is mainly applied in construction of some relatively small scale networks by virtue of improvement of packet wireless services and network information transmission speed, and this type of carrier-class standard is a main reference basis of large-scale networking technology during the long process of development and evolution. [3]. Figure 2 below refers technology application routes of communication network for distribution system in the wide area of industrial control communications network. Meanwhile, considering the amount of capital investment, operation and maintenance scale and industrial chain development of network access, we think that relatively reasonable technical routes should be based on the current telecommunications wide area network technology system, and industrial customization can be carried out through the real-time characteristic of industrial control communication technology.

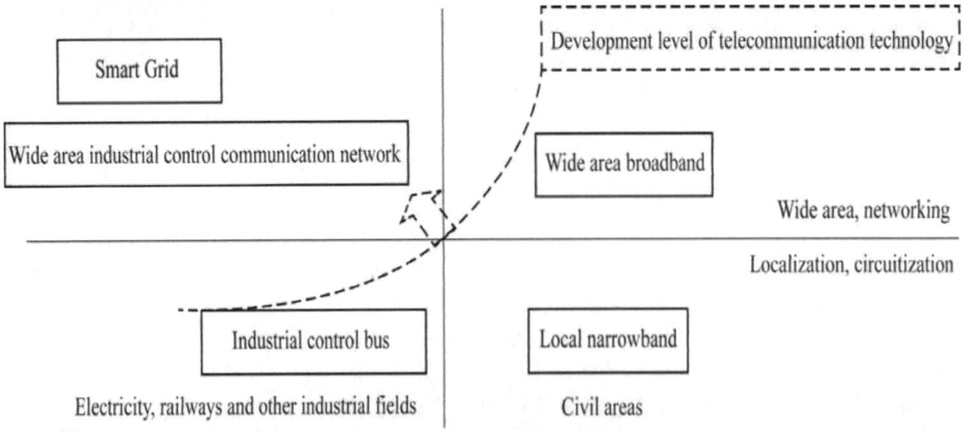

Figure 2: Wide Area Industrial Control Communication Network Technical Route

4.2 Design of Information Transmission Channel

When designing information transmission channels, we have to take bandwidth efficiency, stability, isolation performance and other aspects into consideration and principles of information transmission channel design are summarized as the following:

(1) As for conventional TCP / IP protocol, it can be established on the basic level of statistical multiplexing; as for the transmission of control information, the highly efficient communication resource management technology can be selected to establish a safe and stable circuit channel.

(2) As for various security levels in electricity business, the communication network platform should be adopted to effectively improve practical application of network guided by a design principle that horizontal isolation and longitudinal control. [4].

(3) The distribution electricity network information boasts of a large amount and a single node data volume is relatively small, therefore we had better to avoid over-encapsulation to data packet so as to prevent the communication bandwidth from reducing efficiency.

4.3 Operational Control of Communication Network

The communication network itself is characterized by a certain hierarchy and regionalization and requirements in operation must be taken into consideration by communication network for distribution system.

(1) The scale of network itself also determines that the actual communication network will be divided into multiple levels and a number of logical sub-network systems. When the networks are divided, various problems will occur about stability, application, security and network' bandwidth, therefore, the network division must be implemented according to the actual situations of the entire information in electricity business information.

(2) The communication network for distribution system must take business network address into consideration when passing through the communication network. Address itself can be divided into physical and logical addresses, and IP and MAC may be other network physical addresses.

(3) In order to ensure a safe operation of distribution network, we have to strengthen control to network equipment and business, and prepare network testing, authorization and data statistics servers [5].

5. Constructive Strategies

In the system of communication network for distribution system exists a high integration of technologies and asset attributes, and we should comprehensively consider network structures and industrial control details features when carrying out engineering projects; the following constructive strategies and methods are proposed in terms of construction of a new generation intelligent communication network for distribution system:

5.1 Mixed Public and Private Network

Viewing of the investment scale and industry division existed naturally, the smart electricity distribution network is bound to adopting mixing application of leased telecommunications public network and self-built dedicated network. While the latter is mainly responsible for production control and other major core businesses, and as for some ancillary businesses, the leased public network can be adopted to make compensation so as to achieve excellent cost performance. When carrying out self-construction of special network, we must meet the practical application of objectives and requirements, and ensure the safety to compliance with the standard specifications. Meanwhile, we have to realize that there is still unreliability for self-built special network in technical maturity, scalability, adaptability, operation and maintenance supports and other aspects, and the technical risks and costs of equipment itself are relatively high, therefore we are able to meet the actual development demands in power industry by virtue of public network communications industry customization with relatively mature technologies. Demands about division of labor and cost will occur when the power industry rents public network, but the management is still weak in this field. Therefore, it requires telecommunications enterprises to fully take communications services demanded by industrial environment development into consideration and turn the target of "private network operators" into one which provides industrial control style industry customized services for many industrial sectors including power sector.

5.2 Separation and Coupling of Communication Network Functions

The continuously development and evolution process of intelligent communication for distribution system should realize the modularity of function itself at the same time so as to avoid serious impact on systems due to improvement of the single functions. For instance the new generation network is divided into network management control layer, information application layer, core transport layer and access layer, and the definition is re-defined to information service interfaces and communication access network interfaces [6].

The telecom network has a large scale and carries out a fixed network individual construction and mobile network construction, effective reducing difficulty in operation and maintenance. However, this also leads to multiple access and acquisition of communication access network management resources, control and management equipments. In terms of power communication network with a relatively small scale, it is capable of avoiding equipment and line resource consumption in the construction process by virtue of coupling of equipment itself on the basis of effectively ensuring that the functions can be separated. For instance, the application of wired or wireless network coupling can improve the overall stability of communication network, greatly reduce cost investment of the entire project or coordinate and apply stability of cable channel and the flexibility of wireless channel, allowing them to display functions respectively.

5.3 Evolution Strategies of Communication Network Engineering of Distribution System

One evolution strategy for intelligent communication network for distribution system refers gradual evolution which highly utilizes the existing network infrastructure in order to fully protect existing assets, to comprehensively upgrade network and to expand capacity. The intelligent communication network for distribution system is characterized by a significant "industrial control" and "wide area broadband"[7], and particularly, a core principle should be fully implemented in the construction process is "private network technology and public network management network. The guarantee layer, a core business in electricity distribution grid should take it into consideration and determine principles with application of private network as the main body and that of the public network as the assistance; consideration should be given to equipment technologies and supplement means include fiber technology and wireless technology. However, you still can select a route transition mode with private network covering and public network compensating covering, and other regions should choose starting and ending points suitable for construction according to actual conditions.

IOP Conf. Series: Materials Science and Engineering **242** (2017) 012091 doi:10.1088/1757-899X/242/1/012091

6. Conclusions

In a word, with the rapid development of current information technology, the power grid system develops toward automation and intelligent orientation, and in order to promote the construction of intelligent power grid, we must build a new generation of intelligent communication network for distribution system, which is also an essential way tackling with flaws in construction of traditional intelligent communication network for distribution system. The construction of intelligent communication network for distribution system can not only effectively satisfy control requirements of industrial communication network in wider regions, but also indicate directions and propose goals for the development in power and communication industries.

Reference

[1] Wang Qiang, Zhu Yanzhang and Ge Guangsheng. Application Discussion of Packet Transfer Technology for Smart Power Distribution and Utilization Networks[J]. Power System Protection and Control, 2014, (9).

[2] Zhen Chong, Huang Hui And Zheng Bao. Study on Intelligent Communication Network[J]. Electrical Measurement & Instrumentation, 2017, 54 (4).

[3] Tao Yongjian. Prediction Analysis of Power Communication Network in Distribution Network Construction[J]. Commodities and Quality, 2016, (30).

[4] Hu Cheng and Shen Hanyang. Discussion on Channel Coding Technology in Intelligent Power Distribution Communication Network[J]. Information & Communications, 2015, (5).

[5] He Shisi. Research on Communication Network Construction in Intelligent Distribution Network[J]. China New Telecommunications, 2015, (5).

[6] Dai Yuping. Discussion on Key Problems of Distribution Communication Network in Smart Grid[J]. Architectural Engineering Technology and Design, 2016, (17).

[7] Zhang Yu, Gang Jianxun and Du Qian. On Study of Measures for Constructing a New Generation Intelligent Communication Network for Distribution System[J]. Digital Communication World, 2015, (6).

Design and Analysis of Secure Routing Protocol for Wireless Sensor Networks

Jiong Wang, Hua Zhang

Fujian Provincial Key Laboratory of Information Processing and Intelligent Control(MinJiang University),Fuzhou 350121 China

Abstract: In recent years, with the development of science and technology and the progress of the times, China's wireless network technology has become increasingly prosperous and it plays an important role in social production and life. In this context, in order to further to enhance the stability of wireless network data transmission and security enhancements, the staff need to focus on routing security and carry out related work. Based on this, this paper analyzes the design of wireless sensor based on secure routing protocol.

1. Introduction

With the development of the times and the progress of science and technology, China's wireless network technology has developed by leaps and bounds. In this context, the role of the wireless sensor network in the social production and life is increasingly prominent. In fact, as a combination of sensors, embedded computing, distributed processing and other high-tech networks systems, in the process of running wireless sensor networks often encounter different levels of security attacks, which led to the user's network security to further reduced. Therefore, the technical staff need to design a secure routing protocol, to promote the acquisition and promotion of the benefits.

2. The overview of wireless sensor network

2.1wireless sensor network connotation

The so-called wireless sensor network, refers to the use of various types of micro-sensor nodes combined network system. In the process of running this type of network collect, analyze, organize information on coverage area, thus driving the smooth development of the operation.

2.2 the characteristics of wireless sensor networks

As one of the three high-tech industries of the future society, in the process of running the wireless sensor network system has many characteristics. The author gives the relevant summary, the specific content is as follows.

2.2.1 large-scale network. In order to improve the accuracy of information acquisition, the wireless sensor network system often needs to set a large number of sensor nodes in the monitoring area, and the number of nodes in this category is often more. At present, the large-scale sensor network system is mainly reflected on two aspects: First, the sensor node distribution area is larger; Second, the sensor is deployed more intensive in the node.

2.2.2 self-organizing network . In the process of building a sensor network, the technician often needs to install it in an area which lacks of infrastructure. Under such background, the neighbor relationship of nodes is often unknown, but the self-organization ability of sensor nodes is strong. So it can realize the

Content from this work may be used under the terms of the Creative Commons Attribution 3.0 licence. Any further distribution of this work must maintain attribution to the author(s) and the title of the work, journal citation and DOI.
Published under licence by IOP Publishing Ltd

autonomous configuration and management. Based on this, through the topology control mechanism and network protocols the technical staff can promote the construction of network systems. In addition, in the use of domestic production the network system there are often some nodes fail, but some nodes of the monitoring accuracy is improved. In this context, the number of nodes in the sensor network system is in a dynamic change, and this leads to the stability of the network system.

2.2.3 Dynamic network. In addition, the network system also has a dynamic feature. This feature is mainly manifested in two aspects: First, the network node is out of the dynamic change of the situation, thus ensuring the safety of the entire system and stability of the upgrade; Second, the three factor ： the wireless sensor network sensor, the perception of the object and the observer is also in motion. In fact, in such a system, the technician needs to ensure that the sensor network system is reconfigurable.

2.2.4 Reliable network. In general, the wireless sensor networks system in the process of running to adapt to all kinds of natural environment, and can not be designed in the human region to play an important role.

In this context, in order to ensure the stability and improve the security of the sensor network system, the staff need to ensure that the sensor node quality to meet the design requirements, with the characteristics of strong, difficult to damage and adapt to a variety of harsh environmental conditions. Therefore, in the process of running China's common wireless sensor network system has a strong reliability, thus, it able to adapt to various environments, access to a wide range of promotion and use.

3. The design principles of wireless sensor network routing protocol

In the process of design wireless sensor network routing protocol, technical personnel often needs to strengthen the energy efficiency, the accuracy of data transmission and other factors, and follow various principles to carry out design operations. The author made a summary on the wireless sensor network routing protocol designs principles, the specific content is as follows.

3.1 the principle of energy utilization priority

In the process of wireless sensor network system designs, the WSN nodes are used in the use of a class of energy such as batteries. Under such a background, the energy consumption of WSN nodes often has a more direct impact on the efficiency of system operation. Based on this, designers need to follow the energy utilization priorities to ensure that WSN survival time is extended. In general, the technical staff need to use data fusion, localization algorithms and other means to reduce the communication consumption, to ensure that the node battery consumption is fair and average, to promote the improvement of the benefits.

3.2 the principle of data center

At present, in the application process China's wireless sensor network system, need to collect and analyze various types data of the task area, so as to provide users data support with all kinds of work. Based on this, in the relevant network in the process of the system construction and design operations the staff need to strengthen the principle of data center compliance, and thus ensure that the sensor node can be based on the naming mechanism to describe the various types of data, and collect the interest of a named data by sending to all nodes data.

In fact, in the actual operation of the process the data-centric routing protocol is often able to achieve the entire system of node data concerns, thus avoiding a single node failure for system damage, help the promotion of the entire network system security and reliability.

3.3 the principle of data aggregation

Because in the process of running the sensors in the WSN can use different sense of precision to carry out the perception of the target environment, so the technical staff often collect and transmit a large number of duplicates and redundant data in the process of constructing data-centric routing protocol, that leads to further reduction of system operation efficiency, which is not conducive to the acquisition of relevant benefits.

Based on this, in order to further achieve the processing of various types of data, the technical staff need to strengthen the use of data aggregation technology, which as a basis for the realization of various types of duplication of data to eliminate and reduce, and by simplifying the scale of data transmission, achieve the energy conservation purposes.

In fact, the principle and the use of technology can not only simplify the node data, but also can be a number of sensor nodes which can combine the data onto meaningful information according to the characteristics of the data, thereby enhancing the accuracy of the information, while enhancing the robustness of the system.

3.4 node positioning, target tracking principle

The so-called node location, refers to in the process of construction and operation of the sensor nodes the system can be marked on the system location, so the wireless sensor network routing protocol decision-making work can provide a variety of information reference. In addition, through the smooth positioning of the node to carry out the work, the operating staff can be dedicated to the intelligent selection of various types of node tasks. In general, the application of this method not only can reduce the energy consumption of the system, but also can further to improve the system's life expectancy, to ensure that the system in the process of running to obtain higher economic and social benefits.

4. Security threats and coping strategies

4.1 routing attacks

In the transmission and use of the process routing information often encounter a variety of forms of attack damage, the current attack are divided into two categories, namely: passive attack and active attack. Passive attack refers to the attacker through eavesdropping routing information, and the direction of data transmission and content changes, resulting in damage, and active attack is through false routing information, selective forwarding, sinkhole attacks and other means to attack, and thus lead to wireless the sensor network system is faulty and can not run efficiently.

4.2 the connotation of secure routing protocol

In order to ensure the security and stability of the transmission of the network system, designers need to ensure that the routing process can isolate the non-authorized nodes during the process of protocol design and ensure the formal nature of the route discovery and ensure Node topology confidentiality, to avoid all kinds of problems which can caused system failure. In this context, the designers are often able to take measures to solve the above problems of the node authentication, key management, message source certification and other aspects.

4.3 commonly used security technology

4.3.1 secret sharing technology. The so-called secret sharing technology, refers to in the process of running the network system sent the network operation password and other confidential information on many participants, in order to achieve the confidential file shunt, only the participants of the sub-combination of authorized can restore the secret file information. In the process of using the technology for system design and work, the technical staff often need to use the secret distribution, reconstruction algorithm for specific operations. In the process of doing a job of a secret distribution algorithm, the designer needs to divide the secret document data onto several parts and distribute it among the participants to ensure that each participant gets a secret share. The secret reconstruction algorithm is often able to ensure that the participant union with the qualified subset can recover the secret correctly, while the non-qualified subset can not participate in the relevant job.

4.3.2 ID-based authentication and key negotiation. WSN systems are often built on the exchange of broadcasts (unicast messages). Under such a background, the sensor nodes are able to carry out periodic broadcasts and to the neighbor nodes. In this process sensor can use ID authentication and key agreement technology to ensure the safe operation of the system. In this process, when the A node is added to the network system, it will often broadcast to the neighboring nodes Hello message, and the surrounding node will be initiated into A node authentication, key establishment process. Under this background, saving the interaction between nodes, and promote the system to achieve a substantial increase in efficiency. In fact, with the node ID authentication, it can further negotiate the unicast key and obtain the other party's broadcast key.

4.3.3 the design of secure routing protocols. In the process of building a secure routing protocol, there is an offline trusted center of the designer erection protocol, and it can distribute the ID private key to each node during the running process and it can realize the data onto multiple node data by means of secret technology sharing. These shared informations about multiple nodes that cooperate to update the key for the node and distribute the private key to the newly added node. On the routing protocol design flow chart, the author made a summary,the specific content is shown in Figure 1.

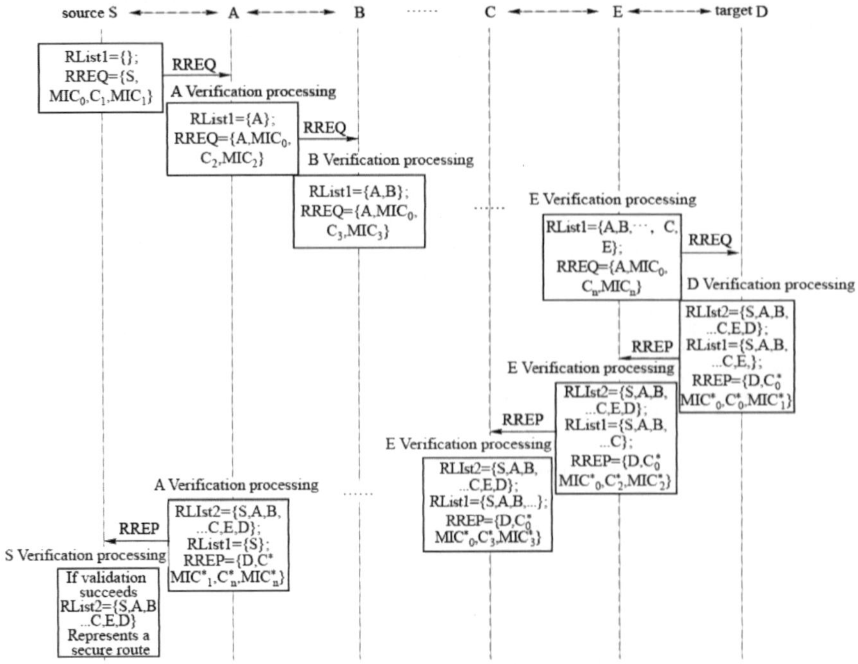

Figure 1 routing protocol flow chart

In this process, the protocol uses the D-identity information to carry out the calculation of the shared key ksD, and the unicast encryption key between the S and D and the unicast integrity key is exoprted, and then the route requests message to construct RREQ. Subsequently, the intermediate node A will accept the request information and use the broadcast integrity key of the RS to calculate the MICi and compare the value with the design value of ensure that the same information is transmitted to the neighboring node for broadcast. And in order to find the target node of the Keei warehouse, to ensure that the source node S receives the routing response messages RREP, and take measures to achieve the entire routing system security and stability of the upgrade. Through the design and development of the protocol, the technical staff also summarizes the processing methods of various types of routing attacks, and draws the relevant forms, the specific contents are as follows.

Table 1 scenario running table

Attack category	Attack processing	effect	case analysis
case analysis	Using identity authentication, message source authentication, news freshness and other technologies	Great	the authentication and key program communication and computing performance which based on IDand bilinear pairs is higher
Unauthorized access	Using identity authentication technology	Great	Can be used to prevent Sybile technology
Retry attack	Using a unique serial number method	Great	The serial number needs to be transmitted, increase the communication burden, but can be used to prevent Hello attacks
Transmission analysis	Encrypt protection of routing information	Great	Able to prevent passive eavesdropping attacks
Black hole attack	Avoid the intermediate point to provide a response message to prevent the intermediate point from attracting the network data stream itself	Great	The routing efficiency is lost, but the reliability will be greatly increased is conducive to the prevention of Sinkhole attacks attacks and selective forwarding attacks
Tunnel attack	Using the probability of combining the number of hops and the time extension	general	This method can make the probability of tunnel attack greatly reduced, so that the tunnel attack failed. Reasonable use of this method, can effectively prevent Wormholes attacks
Critical point attack	Secret sharing technology	better	Through the secret sharing technology in a number of nodes to share, which makes the key point of failure, but the system performance will be affected
Isolate malicious attacks	Secret sharing and secret update technology	general	When a certain number of broadcasts of a node's malicious attacks reaches a certain number, the network node key can be updated by secret sharing technology to exclude the malicious node from the network but with poor performance.

Through the analysis of the program we can learn: in the process of running the program can be based on various types of routing security risks and take positive measures, and thus achieve the entire system operating efficiency and quality improvement, to ensure that the user network information security, to further enhance the realization of the benefits of the acquisition.

5. Intermodulation Interference Analysis Model

In the process of wireless router network security routing protocol design, the technical staff also need to strengthen the intermodulation interference analysis model for the construction and use. At present, the relationship between the output current ic of the nonlinear device and the input voltage can be written as: ic = a0 + a1u + a2u1 + a3u2 + ... In this formula, ai refers to the characteristic coefficient of the nonlinear device. According to the above analysis, it is assumed that two signals simultaneously act on the nonlinear device, that is, u = AcosωAI + BsinωBI. In general, the distortion term of this relation can be expressed as

$$(\cos\omega AI + B\sin\omega BI)\ n.$$

Through the expansion of the above formula, the analysis can be learned: 2ωA-ωB, 2ωB-ωA two frequency interference on the receiver of the greater harm, and by the combination of these two frequency interference is often referred to as the third order Intermodulation interference.

Concluding remarks:

Based on these above mentioned information, this paper mainly discusses the connotation and characteristics of wireless sensor networks, and discusses the design principles of routing protocols and the security threats and coping strategies of wireless sensor networks. Finally, the design of secure routing protocol is analyzed. I believe that with the implementation of relevant measures in place, China's wireless sensor network security routing protocol will be developed by leaps and bounds, and thus meet the needs of social needs of the community, to promote the realization of the benefits and enhance the realization of China's network security. As well as the improvement on reliability, and ultimately lead the development of the network.

Acknowledgment

1. Research on Security Access Technology of Mobile Ad hoc Network, JAT160385，2016 Young and Middle-Aged Project Funded by Fujian Provincial Education Bureau
2. Research and Implementation of Campus Network IVI Transition Technology Based on IPv6, JB12159，2012 Class B Project Funded by Fujian Provincial Education Bureau
3. Research on Aodv Routing Protocol Simulation and Multipath Optimization. YKY12006, 2012 Science and Technology Incubator Project of Minjiang University
4. Research on IT Management and Service Model of Library Based on Cloud Computing. YSY12015, 2012 Social Science Project of Minjiang University

References

[1] Ren Xiao-gang.Study on Secure Routing Protocol Technology of Wireless Sensor Networks [J]. Computer Programming Skills & Maintenance, 2013, (24): 111-112 + 114.
[2] Zhao Qi, Wang Ru Chuan. Wireless sensor network routing protocol security problem analysis [J]. Journal of Nanjing University of Posts and Telecommunications (Natural Science Edition), 2016, (3): 83-87.
[3] Tan Bo-ping, Zhou Xian-wei, Yang Jun, LI Xiao-qin.Study on security mechanism based on routing protocol of wireless sensor networks [J] .Technical Journal of Sensors and Actuators, 2016 (4): 1276-1278 + 1283.
[4] Long Zhao-hua, Gong Jun, Wang Bo, Qin Xiao-huan, Liu Da-ming.Study on Energy Efficiency of Clustered Secure Routing Protocol in Wireless Sensor Networks [J] .Journal of Electronics & Information Technology, 2015, (8): 2000-2006.
[5] Song Zhigao, Chen Fei, Chen Kefei, Li Hui. Wireless Sensor Network Routing Protocol Security Analysis and Research [J]. Computer Simulation, 2015, (5): 134-136 + 140.
[6] Yu Xiehua, Zhang Xiaofang, Huang Zexin. Wireless sensor network security routing protocol research [J]. Journal of Yangtze University (Natural Science Edition), 2011, (8): 98-101 + 279.
[7] He Liyuan, Li Yongming, Wang Quandi. Analysis of cluster routing protocols for wireless sensor networks [J]. Journal of Chongqing University (Natural Science Edition), 2017 (1): 50-53.
[8] Jiang Yi, Zhang Ruonan, Shi Haoshan. A geo-based wireless sensor network security routing protocol [J]. Journal of Northwestern Polytechnical University, 2012, (1): 11-16.
[9] Wang Wei, Tang Minghao, Xu Yunmin.Design of multi-path routing protocol for wireless sensor networks [J]. Computer Engineering and Applications, 2014, (12): 136-138

IOP Conf. Series: Materials Science and Engineering **242** (2017) 012093 doi:10.1088/1757-899X/242/1/012093

Detection and analysis of the stored grain insect creeping sound

Senlin Geng [1,a)], Xiuqin Zhang[2,b)]and Wei Zhao [1,c)]

[1] College of Mathematics and Physics, Weinan Normal University, Weinan 714000, P.R. China

[2] College of Chemistry and Material, Weinan Normal University, Weinan 714000, P.R. China

E-mail: [a)]senlingeng@163.com, [b)]1359444636@qq.com, [c)]senlingeng@sina.com

Abstract: With the random acoustic source model as the theory model of the stored grain insects creeping, the sounds, of 20 Alphitobius diaperinus Panzer in wheat and 20 Tribolium castaneum Herbst adults in corn, are detected, respectively. By using Matlab, the original sound signals are reproduced and the de-noised signals are obtained. The power spectrums characteristics analysis are made. It is shown that the random acoustic source model is effective for the stored grain insect creeping sound detection, and their power spectrums are all discrete, where the highest frequency is 1600 Hz, the main frequency 205 Hz in the former, and the highest frequency is 800 Hz, the main frequency 350 Hz the latter, which may be used to distinguish different types insects in grain.

1. Introduction

To detect insects by using their action sound is a quick and practical method to detect insect in agricultural post-harvest commodities to be newly developed in recent years [1-3]. The numbers of insect activity sound have been used to measuring insects numbers in grain [4-6]. To use insects activity sound to identify them is an important in detecting insects. Coggins etc have used artificial neural network to study the sound of insects in corn and classificatie insects by the time domain features of insect acoustical signal [7]. It can be rarely reported to study the spectrum of activity sound in grain to identify insects. Guo Min etc study the creeping sound of insects on film and find that the basic frequency of acoustical signal has to do with the insect species [1,8]. In the paper, the power spectrum of activity sound of insects in stored grain will be studied, which has a role in promoting to implement the acoustic differentiation technique of insects in grain.

2. The random acoustic source model

The creeping sound of single insect in grain is generally not easy to be detected for it being weak , additionally absorbed and scattered by grain, but if we regard the creeping sound as random acoustic source, we can achieve detecting it.

For 2 random acoustic sources with the phase position varying according to certainly statistical law, the sound pressures are written as, $p_1 = p_{1a}\cos(\omega_1 t + \varphi_1)$ and $p_2 = p_2\cos(\omega_2 t + \varphi_2)$, where φ_1 and φ_2 are the initial phase of p_1 and p_2, respectively, and p_{1a} and p_{2a} are the amplitude of p_1 and p_2, respectively. With φ_1 and φ_2 taking any value between 0 and 2π, the total sound pressure

Content from this work may be used under the terms of the Creative Commons Attribution 3.0 licence. Any further distribution of this work must maintain attribution to the author(s) and the title of the work, journal citation and DOI.
Published under licence by IOP Publishing Ltd

is $p = p_1 + p_2 = p_a \cos(\omega t - \varphi)$, where p_a is the amplitude of p, $p_a^2 = p_{1a}^2 + p_{2a}^2 + 2p_{1a}p_{2a}\cos(\varphi_1 - \varphi_2)$, and φ is the initial phase of p, $\varphi = \arctan((p_{1a}\sin\varphi_1 + p_{2a}\sin\varphi_2)/(p_{1a}\cos\varphi_1 + p_{2a}\cos\varphi_2))$,where φ_1, φ_2 and φ randomly vary with time. The total average acoustic energy density [9] $\bar{\varepsilon} = p_a^2 / 2\rho_0 c_0^2 = \bar{\varepsilon}_1 + \bar{\varepsilon}_2 + \overline{\cos(\varphi_2 - \varphi_1)} p_{1a} p_{2a} / \rho_0 c_0$, where $\bar{\varepsilon}_1$ and $\bar{\varepsilon}_2$ are the average acoustic energy density of p_1 and p_2,respectively, and ρ_0 and c_0 are medium density and acoustic propagation velocity, respectively. With a sufficient time, we thick $\varphi_1 - \varphi_2$ take any value between 0 and 2π at an equal probability, $\overline{\cos(\varphi_1 - \varphi_2)} = 0$, or $\bar{\varepsilon} = \bar{\varepsilon}_1 + \bar{\varepsilon}_2$ [9]. The same conclusion can apply to two or more random acoustic sources. We consider that for the same species insects being in the same growth phase, the creeping frequencies are the same, only the initial phase take random value. Except the energy is different, the creeping sound requency of insects in grain is the same as one of single insect. The creeping sound of insects in grain may be regarded as random acoustic source, therefore we may overcome the shortcomings of single insect creeping sound too weak to be detected. In experiment, we find that insect in grain has the characteristic of apparent death and the all insects seldom act together. Only when most of insects creep in grain all at once, the obvious creeping sound signal appear.

3. The sound signal collection

3.1 The Main Experimental Apparatuses and Samples

The Microphone Shuer BG4 model with frequency response from 40 Hz to 18000 Hz and output impedance 600 ohms; Professional sound card MAYA44 with frequency response from 20 Hz to 20000 Hz, the highest sampling rate being 48 k Hz and analog-to-digital conversion being 18 bits; The ordinary earphone; PC with Celeron(R) CPU2.40GHz, memory 128MB,WindowsXP; The sound-proof chamber with about 22dB average acoustic insulation mass[10]; The barrel, which is cylindrical glass barrel with 15 cm tall and 8 cm long diameter; Grain samples with wheat and corn; Insect samples with *Tribolium castaneum Herbst* adults and for *Alphitobius diaperinus Panzer* adults.

3.2 The Experiment

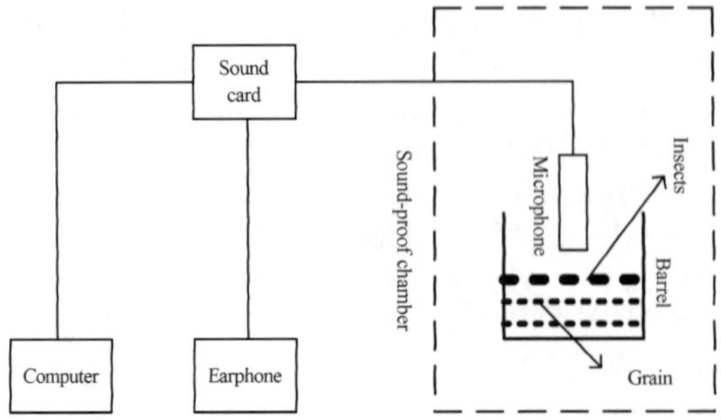

FIGURE 1. The scheme of sound detection setup

The scheme of sound collecting setup is shown in Fig.1. The barrel and microphone are placed in the sound-proof chamber. Put insects in the barrel with grain. The microphone is placed about 4 cm opposite from the grain surface. When insects move in grain, sound signals are sampled by sound card, which is saves in PC with format wav. Ear phone is used to listen to sound, so that the best time and duration may be determined. The 30 sound sample to each kind of sound are sampled so as to do the statistical analysis. When detecting, select sampling frequency 22050 Hz, adjust temperature 28°C and relative humidity 79%, and get 20 insects and sample 30 acoustical specimens with 5 minutes duration every time. The signals collected are sent to the computer so as to be further processed in off-line.

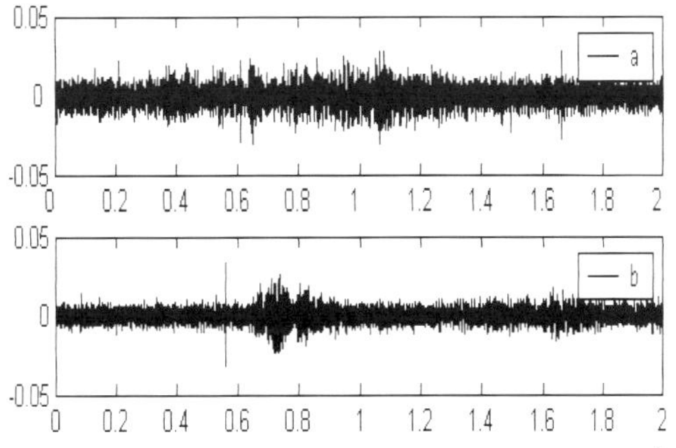

FIGURE 2. The creeping sound signal .
a *Alphitobius diaperinus Panzer* adults in wheat ,
b *Tribolium castaneum Herbst* adults in corn.

For *Alphitobius diaperinus Panzer* adults in wheat, and for *Tribolium castaneum Herbst* adults in corn, the creeping sound signals are detected, respectively. By use of MTLAB, selecting the clearest of acoustical specimens, the creeping sound signals are reproduced shown in (a) and (b) in Fig.2, respectively, where the x-axis is sampling point, y-axis is relative amplitude. From Figures 2, it can be seen that for the acoustical specimen studied, the creeping sound of *Alphitobius diaperinus Panzer* adults in wheat happens to about point 7000 and point 11000, about point 8000 for *Tribolium castaneum Herbst* adults in corn, respectively. In addition, some sharp pulses appear in the individual parts, and the background noise is amplified because of preamplifition, and the creeping sound signal is almost drowned.

4. The sound signal processing and power spectrum analysis

A one-dimensional sound signal containing noise may be represented as a function: $x(i) = f(i) + \sigma \times e(i)$, $i = 1, 2, 3, \cdots, n-1$. where $f(i)$ is real signal, $\sigma$is the coefficient, $e(i)$ is the noise signal depending on the noise type, and $x(i)$ is the signal containing noise. In general, how to process noise signal depends on the noise type. In the experiment, there are two type noise in the sound signal, which are the background noise and the sudden noise, where the former is from thermal motion continuous interference in the circuit, presenting uniform amplitude distribution and wide frequency range, and the latter from the insect collision with the environment, presenting bigger amplitude and shorter duration, frequency distribution in bigger frequencies. The insects creeping sound is distribution in shorter frequencies [11].

4.1 The Filter De-Noising

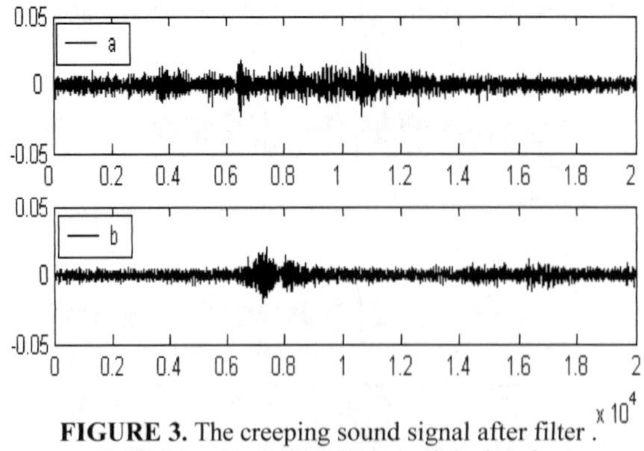

FIGURE 3. The creeping sound signal after filter .
a *Alphitobius diaperinus Panzer* adults in wheat ,
b *Tribolium castaneum Herbst* adults in corn.

According to the characteristics of the insects creeping sound, the filter de-noising may be used. The digital filter contains two type filters of IIR and FIR, where the former has bitter function in pass band and stop band, and in the same performance , the order is higher and the delay time is longer than the latter's. In the sound signal processing, the bitter function in pass band and stop band meet the requirements, and so IIR filter may be used. Considering the sound frequency being from 50 Hz to 1500 Hz, IIR Elliptic low pass digital filter is designed by using minimum order time estimation, where pass band frequency is 2000 Hz and the sampling frequency is 22050 Hz, and it is to be sure that the higher sampling frequency do not reduce high frequency components but only the computer memory. The filtered signals are as shown in Fig.3. It is can be seen that the spike pulse noise signals are filtered, the background noises are a certain degree of reduced and the signal to noise ratio is greatly improved, but the background noises show that the signal frequency and the noise frequency overlap each other.

4.2 The Wavelet De–Noising

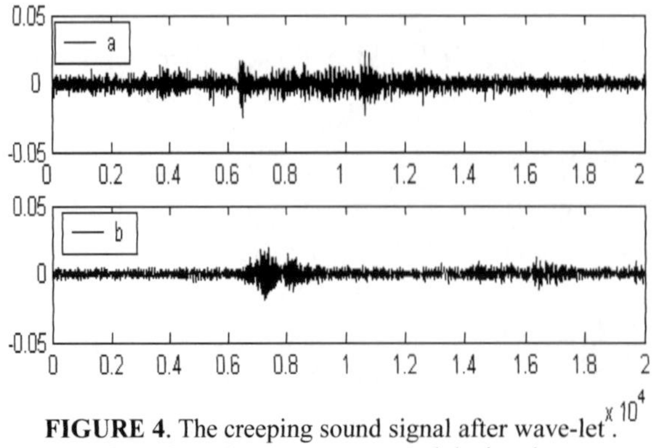

FIGURE 4. The creeping sound signal after wave-let .
a *Alphitobius diaperinus Panzer* adults in wheat ,
b *Tribolium castaneum Herbst* adults in corn.

The wavelet de-noising to processing signal is through decomposing signal into layers, then selecting threshold and quantifying for each layer. The key is the selection of threshold value. There are four principles in threshold selecting which are unbiased likelihood estimation, fixed, heuristic and mini max principle. There are 3 methods in the wavelet de-noising: ① Forced de-noising, it is complete through setting all of high frequency coefficients to zero in the wavelet decomposition structure, or filtering all of high frequencies, then reconstructing the signal, which is simple and reconstructed signal curve is smooth, but only the useful signal compose is easily loss; ② The default threshold de-noising, it is complete by calculating the threshold value, which is useful to the signal with higher signal to noise ratio and signal frequency to noise frequency ratio; ③ Given the soft (hard) threshold, it is complete by the empirical threshold, which has better credibility than the default threshold. In the actual signal de-noising, the threshold generally is empirically determined, where if T ($T > 0$) is for the threshold, the signal becomes $y = x \cdot (|x| - T)_+$, or the points which are equal to or less than the threshold are set to zero, and the points which are more than threshold are set to the difference between the point and the threshold. The signals after the hard threshold generally are represented as $y = x \cdot 1_{(|x|>T)}$, or, comparing signal value with threshold, the points which are equal to or less than the threshold are set to zero, and the points which are more than threshold remain the same. Generally, the hard threshold signals are rough than the soft threshold signals. In order to get more high frequency signal, the unbiased likelihood estimation adaptive threshold is selected in the experiment. Through calculating, the creeping sound signal soft threshold for *Alphitobius diaperinus Panzer* is 0.0014 and 0.0197 for *Tribolium castaneum Herbst.* After the wavelet function selected is "dB5", the decomposition layer number is 5, and the global variable is used and the low frequency coefficient keeps immutable, the filtered signals are processed shown as in Fig.4. It can be seen that after given the soft (hard) threshold processing, the background noise in the signal is reduced a lot and signal curve is clearer. The duration is about 0.2 s for *Alphitobius diaperinus Panzer,* and about 0.4 s for *Tribolium castaneum Herbst.* Repeating the experiment shows that changing parameters to deduce the background noise must be at the expense of damaging useful signal.

4.3 The Power Spectrum

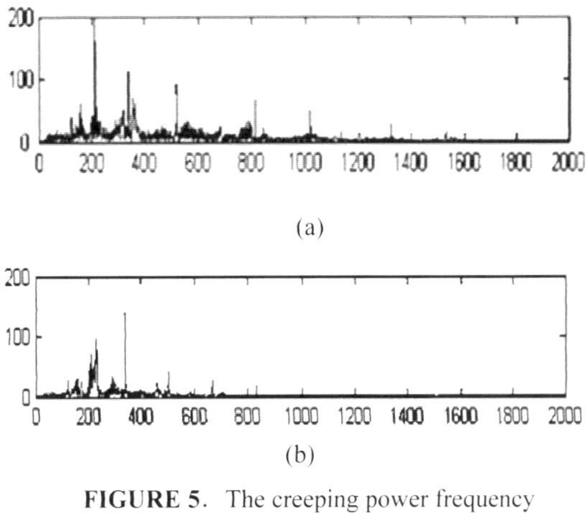

FIGURE 5. The creeping power frequency
(a) *Alphitobius diaperinus Panzer* adults in wheat ,
(b) *Tribolium castaneum Herbst* adults in corn.

In order to estimate the power spectrum, the data is first captured. Generally, only the data length is minimally the eight power of 2, or 256, can all the information be contained, and there is no limit to the upper limit of the data, only the processing signal runs fast with integer power of 2. Selecting 2048 for the signal data, the power spectrum curve of the signal is shown as in Fig.5, where the x-axis represents frequency (Unit: Hz), and y-axis relative amplitude (the same below). It can be seen that the power spectrum presents the discrete. For *Alphitobius diaperinus Panzer* adults in wheat, the highest frequency is 1600 Hz and the main peak frequency is 205 Hz. for *Tribolium castaneum Herbst* in corn, the highest frequency is 800 Hz and the mine peak frequency is 350 Hz. From the signal waveform the duration of the former is smaller, which is caused not only insect species but also grain species.

5. The conclusion

The sounds of the same kinds of insects creeping in the same grain to the same growth stage , because of the small differences in size、physical and structure, have same frequency. While the many pests creep in the same time, the sound can be regarded as the random acoustic source model. For *Alphitobius diaperinus Panzer* adults in wheat and *Tribolium castaneum Herbst* adults in corn, respectively, the creeping sound signals are detected. According to the noise characteristics, filter method and wavelet method are used to reduce the noise to get clear noise signals. The power spectrums of two kinds are all discrete spectrum.

Acknowledgements

This work was supported by the Agricultural Science and Technology Project of Shaanxi Province (No. 2016NY-198) and The Mathematics Characteristic Discipline Construction of Weinan Normal University.

References

[1] Min Guo, Zhiyuan Shang.A new method detection pests in agricultural products by acoustical signals detection. Journal of Shaanxi Normal University (Natural Science Edition),2001,29(1):107-112. (In Chinese with English abstract)
[2] Senlin Geng, Shang Zhiyuan. Research progress and prospects of stored grain insect sound detection technology [J]. Transactions of the Chinese Society of Agricultural Engineering, 2006,22(4):204-207.(In Chinese with English abstract)
[3] Hagstrum, D.W., Flinn, P.W., Comparision of acoustical detection of several species of stored-grain beetles (Coleoptera:Curcul-ionidae, Tenebrionidae, Bostrichidae, Cucujidae) over a range of temperatures, J. Econ. Entomol., 1993, 86(4): 1271-1278
[4] Shuman D, Coffelt J A, Vick K W, et al. Quantitative acoustical detection of larvae feed inside kernels of grain [J]. J. Econ. Entomol, 1993, 86(3): 933-938
[5] Hagstrum D W, Flinn P W, Shuman D. Automated monitoring using acoustical sensors for insects in farmstored wheat[J]. J. Econ. Entomol, 1996, 89 (1):211-217
[6] Shuman D, Weaver D K, Mankin R W. Quantifying larval infestation with an acoustical sensor array and cluster analysis of cross-correlation outputs [J]. Appl. Acoust, 1997, 50 (4):279-296
[7] Coggins K M, Principe J. Detection and classification of insect sounds in a grain silo using a neural network [J]. IEEE Int Conf Neural Networks Conf Proc, IEEE Piscataway N J(USA),1998,3:1760-1765
[8] Min Guo. Propagation of Sound Signals in Quasi-porous Media and Analyze the Sound Properties of Pests [D]. Shaanxi: Shaanxi Normal University, 2003.4
[9] Du Gonghuan, Zhu Zhemin, Gong Xiufen. Fundamentals of Acoustics, Shanghai, Shanghai scientific technology Press, 1981, 194-196
[10] Senlin Geng, Fang-ju Li. Design of isolation booth for stored grain insect sound detection [J]. Applied Mechanics and Materials Vols. 220-223(2012)pp 1598-1601, 2012
[11] Junli Zheng, Weili Yang, Ying Qiyan. Signals and Systems [M], Beijing: Higher education Press, 1988, 294-251. (In Chinese)

IOP Conf. Series: Materials Science and Engineering **242** (2017) 012094 doi:10.1088/1757-899X/242/1/012094

Study of Agricultural Product Options Pricing

Qiu HONG

School of Economy and Management, Tianjin University of Science & Technology, Tianjin, China

Corresponding author:qiuhong_1991@126.com

Abstract: China is a large agricultural country, and the healthy development of agriculture is related to the stability of the whole society. The agricultural production and management of agricultural products are confronted with many risks, especially the market risks. Option contract is the object of option market transaction, so it is very important to study the option contract of agricultural products. Option trading separates the risk and profit, so that the trader can avoid the risk while retaining the opportunity to obtain income. The option has the characteristics of low transaction cost, simple and efficient, so it is suitable for small and medium investors.

1 Introduction

At the end of 2016, China Securities Regulatory Commission formally approved the Dalian Commodity Exchange soybean meal futures options and Zhengzhou Commodity Exchange sugar futures options. The Chinese derivatives market will officially open the option era. To soybean meal, sugar as the representative of the agricultural options listed for China has an unusual significance. It brings new tools and ways for China's agriculture related subject risk management and service of the real economy [1,2]. At the same time, this is also an indispensable step for the construction of commodity pricing in China.

The agricultural product options are kinds of commodity options. They are financial derivative tools with agricultural products. They have the characteristics of low cost, low risk and high profit. The low cost and high income of agricultural products have attracted the attention of the hedge, the speculative and the arbitrage[3,4]. They play an important role in stabilizing the price of agricultural products and avoiding the market risk.

The so-called option, also known as the right to choose, is the right of the buyer within the agreed period, according to a predetermined price, buy or sell a certain number of specific goods or financial instruments[5,6].Options are the basic risk management tools of the international derivatives market, and constitute a complete risk management tool system for futures markets together with futures, forwards and swaps.

2 The importance of agricultural product options

In the actual agricultural products trading, the agricultural market price fluctuations are relatively large, seasonal strong. Farmers worry that the price of agricultural products fell, the rise in agricultural prices means that their income will increase, but if the price falls, will give them economic losses. Therefore, in the options market, farmers often buy the put options to lock in prices caused by the risk and gain the benefits of rising prices of agricultural products, and the price of a call option to pay a certain fee options.

In the agricultural options, the farmer as the option buyer, the agricultural market is higher than the agreement price, the farmers giving up the implementation of the option, engaged in the spot market product transactions, to obtain more benefits. The price is only paid to the enterprise. The final stop is limited to the option fee.

Content from this work may be used under the terms of the Creative Commons Attribution 3.0 licence. Any further distribution of this work must maintain attribution to the author(s) and the title of the work, journal citation and DOI.
Published under licence by IOP Publishing Ltd

IOP Conf. Series: Materials Science and Engineering **242** (2017) 012094 doi:10.1088/1757-899X/242/1/012094

The farmer chooses to execute the option contract and transfers some of the risk of falling prices to the firm through the contract when the market sentiment is lower than the agreed price, which ensures that the proceeds are higher than those of the non-taking option contracts. As an option contract, the seller's enterprise, whether farmers choose to execute the option, can receive an option fee, as a profit guarantee, to a certain extent, avoiding the operational risk. The use of agricultural option model, on the one hand, can ensure that farmers do not lose access to higher income, on the other hand, compensate for the enterprise price fluctuations caused by the cost risk.

3 The emergence of agricultural product options market

Option market is a kind of advanced market form, which has the functions of price discovery, risk transfer and risk redistribution. The beginning of the world's options market can be traced back to the tulip trade in Holland in seventeenth Century.

The construction of agricultural product options market can not only improve the market system of agricultural products in China, but also guide the production, trading and consumption of agricultural products spot market, becoming a regulator of stable spot price of agricultural product and long-term supply and demand.

The agricultural product option market is the normative and the flexibility. Its contract price has the characteristics of advanced nature, continuity, openness, competition and high risk management performance.

The emergence of agricultural options is later than financial options, although agricultural futures are early. In April 1973, the Chicago Board Options Exchange launched a stock option contract. The emergence of agricultural options was late for ten years. By 1983, the Chicago Board of Trade for the first time launched a corn option deal, and then in 1984 launched a soybean option deal. Since then, agricultural options trading has gradually developed, a number of futures exchanges have opened with its listing futures related to the options business. Brazil, Argentina and many other developing countries, regional futures exchanges, have also carried out agricultural options trading. The option contract has basically become a related contract for the futures contract, with a certain size of the futures contract has a corresponding option contract. At present, there are more than 60 varieties of agricultural products in the world, such as corn, wheat, soybean, soybean oil, rapeseed, cotton, sugar and orange juice.

The subject of an option transaction is a right. The option buyers pay the option fee and thus enjoy the right to buy or sell an asset at a fixed price in the future. Different options and futures, the option buyer has no obligation to the seller only option right. The seller of the option is only the obligation and no right futures. They must be carried out in accordance with the provisions of the liquidation or delivery of a futures contract. The rights of both sides is also an obligation. The rights and obligations enjoy by both parties are reciprocal.

Option trading is of great significance to form a perfect futures market system and risk management system. Option trading, as a risk management tool for futures trading, will not produce new risks outside the futures transaction. The biggest risk to the option seller is the risk of futures trading and no risk of amplification. At present, the international commodity futures market option contract is very common. The option transaction shows a growing trend, fully embodies the strong demand in the market.

Since the 80s of last century, options trading has developed at an alarming rate. According to the statistics of the American futures industry association, the global turnover contract reached 8 billion 550 million hands in 2016, nearly 60% of turnover in the same year contract, 45 times in 1985, and an average annual growth rate of 13.5%. In the past a long period of time, the agricultural options trading volume has been stable in the fourth place trading volume. In 2016, the agricultural product options trading volume was 78 million, an increase of 2.8%. CBOT is the world's largest agricultural futures exchange, a large number of spot traders, farmers and enterprises to participate in options trading, hedging.

4 Empirical study

The basket price of agricultural products at maturity date is $\left(S(T)-K\right)^+$. The price of the basket option of the agricultural products is as follows

$$C[K,T]=e^{-rT}E\left[S(T)-K\right]^+$$

We consider corn and wheat as 2 basket options for agricultural products with a storage ratio of 0.5% and 0.6%, where $r=6\%$, $v=0.9$. The model parameters and weights are shown in the table. The expiration date of the option is 6 months. The execution price K is 4600. The initial price $S_i(0)$ is the unit (per ton price).

Table1 2 underlying assets of agricultural product options

	1	2
$S_i(0)$	3000	3500
μ_i	0.006	0.009
σ_i	0.2	0.09
ω_i	$\frac{1}{2}$	$\frac{1}{2}$

Using the Monte Carlo method to simulate 10^5 times, the agricultural option price are 34.4698 and 39.6895 respectively by Black-Scholes model and Variance Gamma model. The value of the Variance Gamma model is higher.

The Black-Scholes model and the Variance Gamma model are compared by a set of execution prices, where $K\in[4550,4650]$. The results of the simulation are shown in the Table2.

Table 2. Call price of agricultural products

T	K	$BLSC[K,T]$	$VGL[K,T]$
Six months	4550	84.3512	88.6987
	4555	72.5412	76.3258
	4560	68.2541	72.2415
	4565	62.1354	66.3548
	4570	57.2154	61.2543
	4575	52.3147	57.3591
	4580	47.3564	52.2367
	4585	42.2574	47.6584
	4590	37.2698	41.2368
	4595	32.6121	37.2514
	4600	27.6354	31.2635
	4605	22.0354	26.3214
	4610	17.3621	21.8584
	4615	11.9845	15.3298
	4620	6.9854	1.9654
	4625	1.2654	5.6478
	4630	0	1.3657
	4635	0	0
	4640	0	0
	4645	0	0
	4650	0	0

The values obtained by the A model are higher than the values of the B model. The results are consistent with the reality.

5 Suggestions for developing agricultural product options in China

5.1. Establish a sound risk control and monitoring system

The government should introduce policies and regulations to prevent and control the risks of options trading. From the US futures trading risk management and control strategy point of view, the systems generally include margin system, price limit system and market maker system, etc. These systems, after decades of practice in the United States, are enough to be used as reference in the establishment of our country's risk control system. For example, the price limits set, helps to set the price fluctuations, can effectively control the market risk. Raise the option market access principle, monitor the market position risk of the market maker, assess the performance of the market maker's obligations, and establish a complete financial monitoring system. In an open, fair and impartial manner, efficient operation of the long hours, the development of risk response strategies, standardization of risk, market transactions monitoring.

5.2. Government support and guide farmers to use agricultural product options

Encourage other commercial enterprises, banks, trust companies, insurance companies and other competent and willing to take the risk of institutional entities into the options market speculation, increase the number of agricultural options market players, to improve the liquidity of the options market. To guide farmers to buy the lowest target price put options, and the government can pay a certain amount of options for farmers to pay the right to give farmers a certain financial subsidies to reduce the government's financial burden and the agricultural market price signal fully reflects the actual relationship between supply and demand, Improve the farmers in the agricultural market environment in the survival and competitiveness. Speed up the introduction of the laws and regulations on the management of agricultural options market, the options market trading activities under the legal framework to run, improve the transparency of agricultural options market management, to ensure that the agricultural options trading policy stability and continuity.

5.3. Increase the personnel training

Institutions of higher learning should actively introduce options trading courses, understand the rules of the options trading process, and participate in options trading through practice. Futures exchanges and Futures Company should strengthen the training of options for the delivery of more advanced talents for society. The government education department will increase the advocacy of the option, and can draw on the experience of the United States and use part of the agricultural subsidies in the main agricultural production areas for the pilot trading of options. At the same time, we should pay more attention to strengthening investor education, increase the awareness of the investors, especially the farmers, and make rational use of the option to give full play to the role of option trading in agricultural products.

6 Conclusion

The development of rural economy in China needs to rely on agriculture to pull, to emerging financial derivatives options, for example, confirmed the current rural financial product innovation needs. The rapid development of options since the emergence of the main reason is that the option itself has a unique advantage, such as small capital costs, simple and efficient. The option divides the risk and the return so that the trader retains the opportunity to obtain the benefit while avoiding the risk.

In 2017, the listing of two options for sugar and soybean meal in the field ended the history of China's commodity options blank, and provided new ideas for the implementation of China's agricultural support policy. It is recommended to draw lessons from foreign experience and combine with China's national conditions to study the feasibility and specific implementation path of agricultural subsidies to explore the use of agricultural products. The agricultural product options help farmers lock in the lowest sales income and enrich the agricultural support policy tools.

The agricultural product options market has just been established in China, and the option products should be tested by the market and investors. The exchange should improve the contracts and keep up with the requirements of the market development.

To prepare and timely introduction of new varieties of agricultural options, such as corn options, wheat options, etc., not only can enrich the options system, but also better serve the important grain varieties of the purchasing and storage system reform and agricultural supply side structural reform.

References

[1] Whaley R E. Does Net Buying Pressure Affect the Shape of Implied Volatility Functions?[J]. Social Science Electronic Publishing, 2004, 59(2):711-753.
[2] Garcia P, Sherrick B J. The Term Structure of Implied Forward Volatility: Recovery and Informational Content in the Corn Options Market[J]. American Journal of Agricultural Economics, 2007, 89(1):1-11.
[3] Urcola H A, Irwin S H. Hog options: contract redesign and market efficiency.[J]. Journal of Agricultural & Applied Economics, 2010, 42(4):773-790.
[4] Constantinides G M, Jackwerth J C, Perrakis S. Mispricing of S&P 500 index options[J]. The Review of Financial Studies, 2008, 22(3): 1247-1277.
[5] Tomek W G, Kaiser H M. Agricultural product prices[M]. Cornell University Press, 2014.
Understanding options for agricultural production[M]. Springer Science & Business Media, 2013.

IOP Conf. Series: Materials Science and Engineering **242** (2017) 012095 doi:10.1088/1757-899X/242/1/012095

Analysis of self-oscillating behaviors aimed at the development of a molecular robot with organic acids as fuel

Tomoka Nakazumi[1] and Yusuke Hara[1,*]

[1]Research Institute for Sustainable Chemistry, ISC, National Institute of Advanced Industrial Science and Technology, AIST, Central 5-2, 1-1-1 Higashi, Tsukuba 305-8565, Japan; E-Mail: nakazumi.t@aist.go.jp

*Correspondence: y-hara@aist.go.jp; Tel.: +81-29-81-6236

Abstract. We studied the transmittance self-oscillation of a polymer chain driven by an organic acid as the fuel. The self-oscillating polymer chain consists of 4-acryloylmorpholine (ACMO) and the Ru catalyst ($Ru(bpy)_3$) of the Belousov–Zhabotinsky (BZ) reaction. The transmittance self-oscillating behavior was affected significantly by the temperature. As the amplitude of the transmittance self-oscillation, which is reflected by the aggregation state, decreased with time, the oscillation period also decreased. This trend indicates that the polymer aggregation affects the rate of the BZ reaction significantly. The activation energy of the self-oscillating value was almost the same in the normal BZ reaction, which does not include $Ru(bpy)_3$ complexes in the polymer chains. In addition, we demonstrated the effect of one BZ substrate (sodium bromate or malonic acid) on the transmittance self-oscillation period.

1. Introduction

Mechanical robots have actuators, sensors, and intelligence, and these three essential components are independent yet connected to each other. On the other hand, living organisms also have actuators, sensors, and intelligence, but it is impossible to clearly separate three essential components. Moreover, a living organism has specific functions, such as a self-motion, self-repairing, self-reproduction, and self-organization [1], and it is difficult to imitate these four functions using a mechanical framework. In recent years, molecular robots consisting of organic substances have been investigated with the aim of building a robot like a living organism. These molecular robots have been developed by using DNA, proteins, gels, and polymers [2–12]. Molecular robots assembled using DNA are of particular interest because it has become possible to synthesize various DNA structures [13–20]. However, molecular robots made of DNA nanostructures have several technical problems; for example, (1) it is difficult to drive these robots dynamically, such as polymer robots, and, (2) from the viewpoint of the cost, large DNA nanostructures are not realistic. On the other hand, by careful choice of polymeric material, large molecular robots can be developed [21, 22].

In many previous investigations, polymer based molecular robots have been fabricated by using stimuli-responsive polymers [23–25]. To date, stimuli-responsive polymers driven in response to various external stimuli, such as light, temperature, magnetic fields, and electric fields, have been developed [26–31]. In particular, temperature responsive polymers have been heavily studied because temperature control is convenient for chemists. To drive and control a robot made of stimuli-responsive polymers, external devices to control the external stimuli are necessary. That is, stimuli-responsive polymer-based robots are dependent on external control devices, as well as batteries as the energy source. This means that we cannot make small, self-sustaining ant-like robots.

Content from this work may be used under the terms of the Creative Commons Attribution 3.0 licence. Any further distribution of this work must maintain attribution to the author(s) and the title of the work, journal citation and DOI.
Published under licence by IOP Publishing Ltd

In this study, to develop a molecular robot with the autonomous motion like a living organism, we synthesized a self-oscillating polymer chain [32–34] that is driven by the consumption of an organic acid (malonic acid (MA)) induced by the Belousov–Zhabotinsky (BZ) reaction [35–40]. In the BZ reaction, an organic substrate, such as malonic acid, is oxidized by an oxidizing agent (sodium bromate) in strongly acidic conditions in the presence of a catalyst. In this reaction process, a spontaneous redox oscillation of the metal catalyst (ruthenium tris(2,2′-bipyridine)) is generated. The solubility of the Ru catalyst that is covalently bonded to the polymer chains periodically changes, induced by the BZ reaction. The change in the solubility of the Ru catalyst moiety results in the periodic aggregation and disaggregation of the polymer chain. This behavior can be observed as the transmittance self-oscillations of the polymer chains [41–43]. In other words, the self-oscillating polymer chain autonomously converts chemical energy into mechanical energy.

Hara et al. studied self-oscillating polymer chains and gels to fabricate molecular robots. In particular, many studies have been published concerning self-oscillating polymer systems for operation under biological conditions. To drive the polymer under the biological conditions, Hara et al. synthesized a temperature-responsive N-isopropyl acrylamide (NIPPAm)-based polymer chain that was incorporated into a pH control site (acrylamide-2-methylpropane sulfonic acid (AMPS)) or a cationic capture site (methacrylamidopropyltrimethylammonium chloride (MAPTAC)) for an anionic oxidizing agent (bromate ions). As a result, these polymer systems underwent an aggregation–disaggregation self-oscillation under acid-free or oxidant-free conditions [44, 45]. Moreover, Hara et al. succeeded in the development of a novel self-oscillating polymer chain that underwent the self-oscillation in the presence of a BZ substrate (malonic acid) [46]. Subsequently, by improving the polymer structure, Hara et al. succeeded in the development of a novel self-oscillating polymer gel that undergoes self-oscillation under acid-free conditions [47]. In the process of these investigations, Hara et al. clarified that the NIPAAm-based AMPS-containing self-oscillating polymer chain has a high potential that can cause many types of self-oscillating behavior. For example, AMPS-containing self-oscillating polymer chains can switch the transmittance self-oscillation by external temperature control [48]. Moreover, NIPAAm-based AMPS-containing polymer gels undergo large displacements. By using the large displacement of the gel, a self-walking gel and a matter transport system were realized [49–51]. In addition, to improve the frequency of the self-oscillation, Hara et al. developed nonthermoresponsive self-oscillating polymer systems, such as the NIPAAm-based polymer chains [52–54]. In these polymer systems, the transmittance self-oscillation occurs at elevated temperatures. Furthermore, to reduce the cost of the self-oscillating polymer chain, a self-oscillating polymer chain with an Fe catalyst moiety, instead of the expensive and scarce $Ru(bpy)_3$ catalyst, was developed [55–57].

In this study, for the purpose of the development of an autonomous molecular robot, we synthesized a 4-acryloylmorpholine (ACMO)-based self-oscillating polymer chain, and analyzed in detail the self-oscillating behavior. The ACMO-based self-oscillating polymer gel has already been utilized as the main-chain of a self-oscillating polymer gel [57], but the detailed self-oscillating behavior has not been analyzed. To demonstrate the influence of the temperature and the concentration of the BZ substrates on the self-oscillating behavior, we measured the transmittance self-oscillation of the ACMO-based polymer chain by changing the temperature and the concentration of the BZ substrates. Moreover, the activation energy of the self-oscillation was calculated by utilizing the period of the transmittance self-oscillation at four different measuring temperatures. We believe that detailed analysis of the self-oscillating behavior is very important for the development and control of an autonomous molecular robot.

2. Method

2.1. Measurement of the isosbestic point of poly(ACMO-co-Ru(bpy)$_3$)

The isosbestic point of the poly(ACMO-*co*-Ru(bpy)$_3$) solutions was determined by measuring the absorbance in the oxidized and reduced states. By using sodium chloride, the ionic strength of the polymer solutions was adjusted to be the same in both states. For the oxidized state measurements, a 3.0 wt.% polymer solution including 0.1 M NaCl, 0.3 M sodium bromate ($NaBrO_3$), and 0.4 M nitric acid (HNO_3) was prepared. For the reduced state, the polymer solution (3.0 wt.%) contained 0.3 M NaCl, 0.1 M MA, and 0.4 M HNO_3. As a result of the measurements of the isosbestic point of the poly(ACMO-*co*-Ru(bpy)$_3$) solutions in the reduced and oxidized states (see Figure 1), we adopted a wavelength of 548 nm for all other measurements in this study.

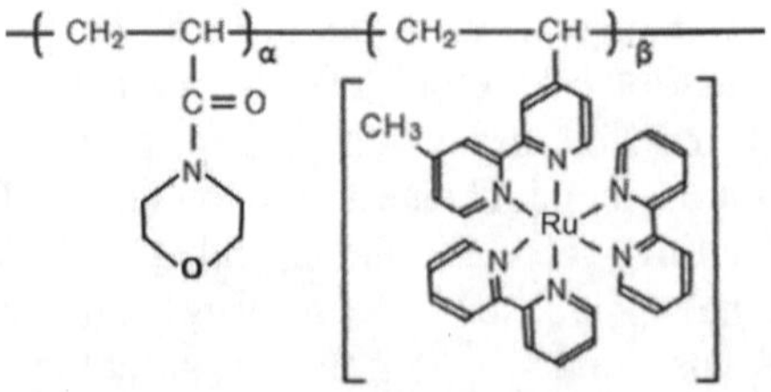

Figure 1. Chemical structure of poly(ACMO-*co*-Ru(bpy)$_3$).

2.2. Measurement of the lower critical solution temperature (LCST) of poly(ACMO-co-Ru(bpy)$_3$)
The lower critical solution temperatures (LCST) for the poly(ACMO-*co*-Ru(bpy)$_3$) solutions in the reduced and oxidized states were measured by UV-vis spectrophotometry at the isosbestic point (wavelength: 548 nm) (Figure 2) at temperatures from 15 to 60 °C. To equalize the total ionic strength of the polymer solutions in the reduced and oxidized states, NaCl was dissolved in the polymer solutions. For the oxidized state measurement, the 3.0 wt.% polymer solution contained 0.1 M NaCl, 0.3 M $NaBrO_3$, and 0.4 M HNO_3. On the other hand, for the reduced state, a polymer solution (3.0 wt.%) including 0.3 M NaCl, 0.1 M MA, and 0.4 M HNO_3 was prepared.

2.3. Measurement of the transmittance of the self-oscillations
To measure the transmittance self-oscillations, we prepared a 3.0 wt.% poly(ACMO-*co*-Ru(bpy)$_3$) solution containing the three BZ substrates, $NaBrO_3$, MA, and HNO_3. The transmittance measurement was monitored by UV-vis spectrophotometry at the isosbestic point (wavelength: 548 nm) under constant temperature conditions to measure the autonomous aggregation–disaggregation changes of the polymer solutions.

3. Results and discussion
Figure 2 shows the relationship between the absorbance and wavelength of the poly(ACMO-*co*-Ru(bpy)$_3$) solutions in the reduced and oxidized states. The absorbance of the polymer solution in the reduced and oxidized states had two intersections at 399 and 548 nm in the range from 300 to 700 nm; that is, the isosbestic point of the polymer solutions in the reduced and oxidized state has two wavelengths. In this study, we adopted a wavelength of 548 nm in all measurements. For the transmittance self-oscillation of the polymer solutions detected at the isosbestic point, the self-oscillating behavior did not result in a color change of the Ru(bpy)$_3$ moiety but a change in the turbidity of the polymer solutions. That is, the transmittance self-oscillation reflects the changes in the solubility of the reduced and oxidized states. Therefore, by measuring the self-oscillating behavior, it is possible to discuss the dissolution state of the poly(ACMO-co-Ru(bpy)$_3$) chains induced by the BZ reaction. The turbidity of the polymer solution is attributed to the size of the polymer aggregates. As the solubility of the polymer chain decreased, the size of the polymer aggregates increased. Therefore, the transmittance self-oscillation is attributed to the aggregation-disaggregation self-oscillation arising from the different solubilities of the polymer chains in the reduced and oxidized states.

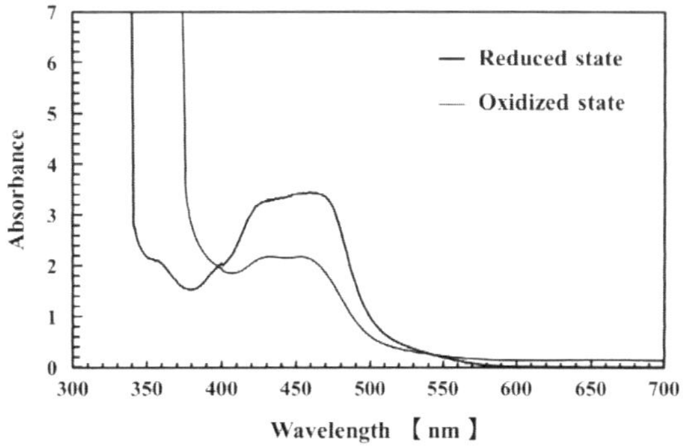

Figure 2. The relationship between absorbance and wavelength of the 3.0 wt.% poly(ACMO-*co*-Ru(bpy)$_3$) solutions in the reduced and oxidized states. MA and $NaBrO_3$ were used as the reducing and oxidizing agents, respectively. The ionic strength was adjusted to the same level in the reduced and oxidized state using NaCl.

Figure 3 shows the results of the transmittance measurements for poly(ACMO-*co*-Ru(bpy)$_3$) in the reduced and oxidized states from 15 to 60 °C. As shown in Figure 3(A), the LCST of the polymer solution in the oxidized state was about 42 °C. In contrast, there was no LCST of the polymer solution in the reduced state because the transmittance value in the reduced state was stable over the measured temperature range. This result indicates that the LCST of poly(ACMO-*co*-Ru(bpy)$_3$) is present only in the oxidized state. As shown in Figure 3, from 15 to 33 °C, the transmittance value in the oxidized state was higher than that in the reduced state. This different transmittance value between the reduced state and in the oxidized state is the driving force for the self-oscillation of the poly(ACMO-*co*-Ru(bpy)$_3$) solutions. For this reason, in this study, the transmittance self-oscillations of the 3.0 wt.% poly(ACMO-*co*-Ru(bpy)$_3$) solution were measured below 25 °C to detect the stable self-oscillating behaviors.

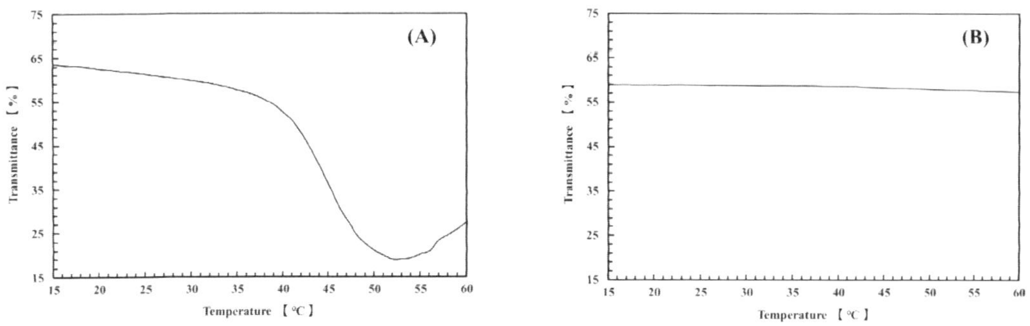

Figure 3. Relationship between the transmittance and temperature for the 3.0 wt.% poly(ACMO-*co*-Ru(bpy)$_3$) solutions with: (**A**) [MA] = 0.1 M, [HNO_3] = 0.4 M, and [NaCl] = 0.3 M; (**B**) [$NaBrO_3$] = 0.3 M, [HNO_3] = 0.4 M, and [NaCl] = 0.1 M.

Figure 4 shows the transmittance self-oscillation of the poly(ACMO-*co*-Ru(bpy)$_3$) solutions with a fixed concentration of the three BZ substrates (0.3 M $NaBrO_3$, 0.1 M MA, and 0.4 M HNO_3) at four constant temperature conditions (13, 19, 23, and 25 °C). Figure 5 shows the results of the amplitude and the period for the self-oscillation of the poly(ACMO-*co*-Ru(bpy)$_3$) solutions at each temperature. As shown in Figure 4, the amplitude of the self-oscillation decreased with time. In addition, as the temperature increased, the damping arose earlier. At 13 °C, the amplitude reduction was much slower than that above 19 °C. In contrast, in the case of 19 and 23 °C, the amplitude value rapidly decreased with time (see Figure 5(C-1) and 5(D-1)). At 25 °C, the damping behavior occurred about 600 s from the start of the self-oscillation; subsequently, the self-oscillation stopped quickly. This phenomenon was attributed to the decreasing solubility of the polymer chain on increasing the measurement temperature. As shown in Figure 3, in the oxidized state, the transmittance value gradually decreased with increasing temperature. That is, when the aggregation size of the polymer chains increased, the polymer aggregation causes light scattering [53]. Therefore, the amplitude of the transmittance self-oscillation of the poly(ACMO-*co*-Ru(bpy)$_3$) solutions decreased more rapidly as the measuring temperature increased.

Figure 5 shows the time dependence of the period for the transmittance self-oscillation of the poly(ACMO-*co*-Ru(bpy)$_3$) solutions at each temperature. As shown in Figures 5(A-2) and 5(B-2) (measuring conditions: 13 and 19 °C), the period of the self-oscillation gradually decreased with increasing self-oscillation number, while the period remained almost the same, except for the initial 10 cycles. Except for at 25 °C, the periods of the first 10 cycles were longer until the period stabilized. When the temperature rose, the stabilization time was shorter than at lower temperatures. At 23 °C (see Figure 5(C-2)), as the period decreased, the amplitude also decreased. Moreover, at 25 °C, the period sharply decreased in synchrony with the dramatic decrease in the amplitude. This result demonstrates that the period of the transmittance self-oscillation is affected by the amplitude of the transmittance self-oscillation. That is, as the size of the polymer aggregates increased in the self-oscillations, the amplitude decreased. As the same time, the density of the Ru(bpy)$_3$ moiety in the polymer aggregation increased. The density of the Ru(bpy)$_3$ moiety affected the reaction rate because the density of the substrates of the BZ reaction influences the reaction rates originating in the increasing collision rate among the reaction substrates. Therefore, the period of the self-oscillation was significantly influenced by the measuring temperature and the amplitude of the transmittance self-oscillation.

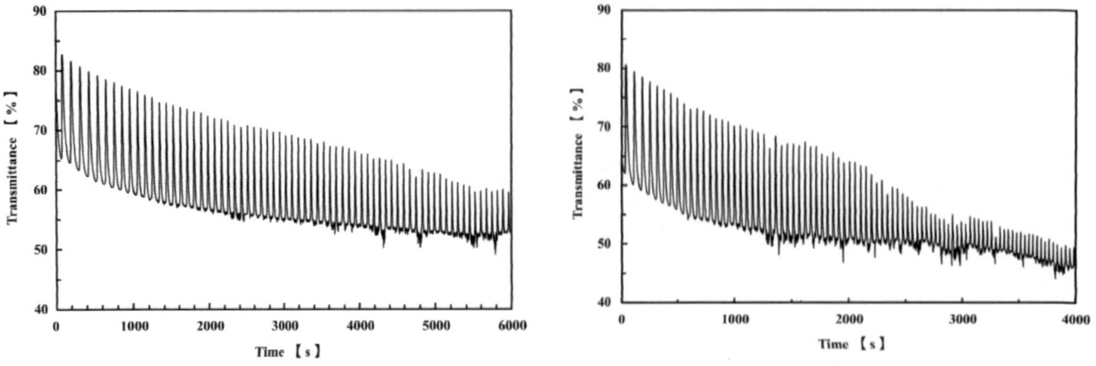

(A) (B)

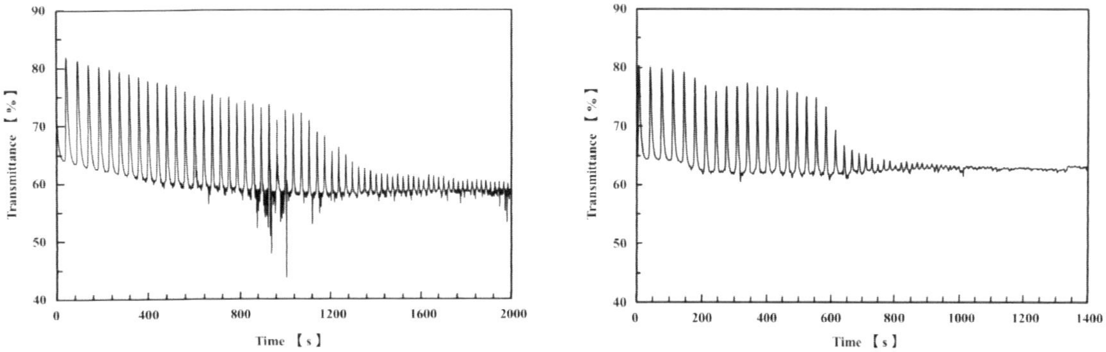

(**C**) (**D**)

Figure 4. Transmittance self-oscillating profiles for 3.0 wt.% poly(ACMO-*co*-Ru(bpy)$_3$) solutions with fixed concentrations of HNO_3, $NaBrO_3$, and MA ($[HNO_3] = 0.4$ M, $[NaBrO_3] = 0.3$ M, and $[MA] = 0.1$ M) at four different temperatures: (**A**) 13 °C, (**B**) 19 °C, (**C**) 23 °C, and (**D**) 25 °C.

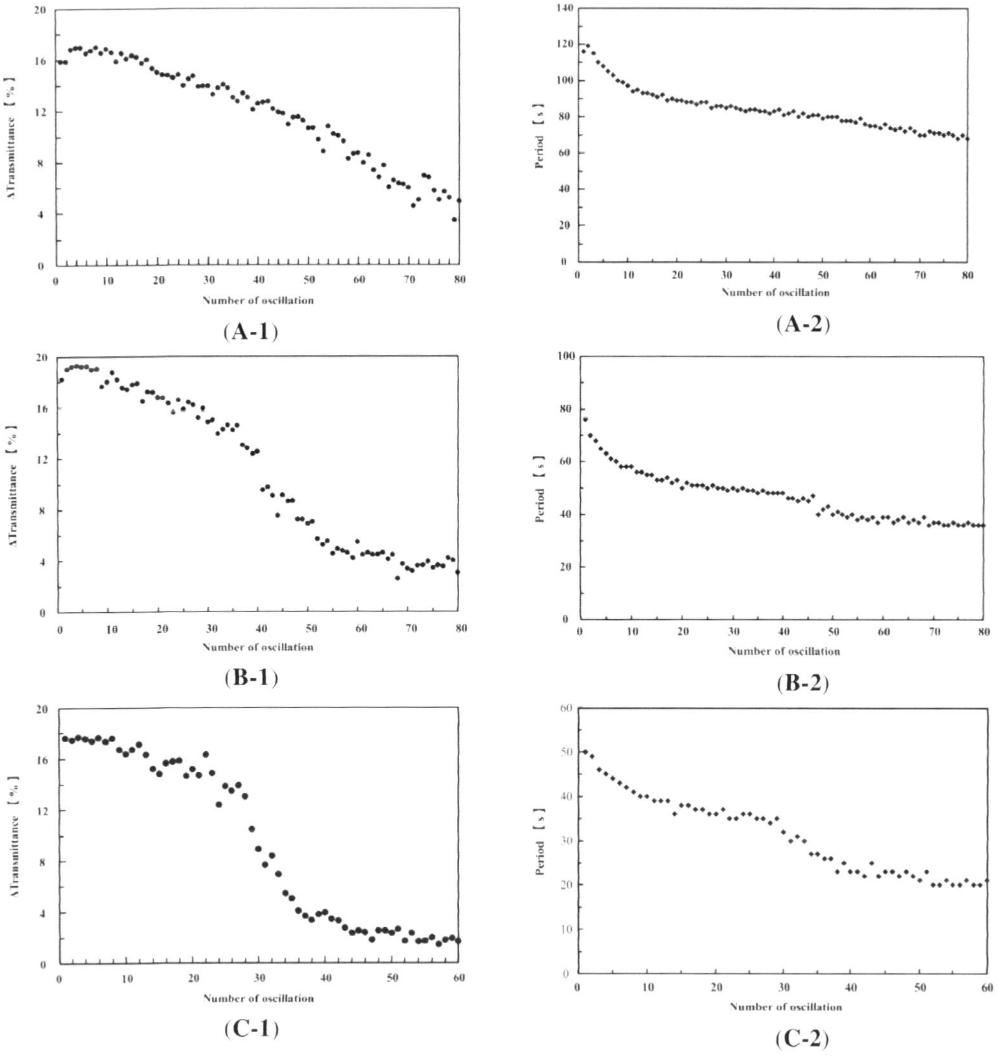

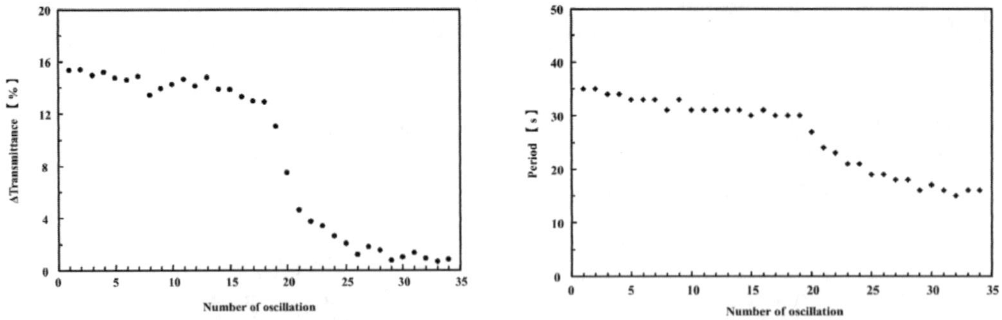

(D-1) (D-2)

Figure 5. Amplitude (Δtransmittance) and period of transmittance self-oscillation for 3.0 wt.% poly(ACMO-*co*-Ru(bpy)$_3$) solutions with fixed concentrations of HNO_3, $NaBrO_3$, and MA ([HNO_3] = 0.4 M, [$NaBrO_3$] = 0.3 M and [MA] = 0.1 M at (**A**) 13 °C, (**B**) 19 °C, (**C**) 23 °C, and (**D**) 25 °C).

Figures 6 shows the Arrhenius dependence on the measuring temperature at a fixed concentration of two of the BZ reaction substrates. In Figure 6, the average periods were adopted by choosing the stable period region from Figure 5. As shown in Figure 6, the periods of the transmittance self-oscillation have a linear relationship; that is, the ACMO-based self-oscillating polymer chain follows the Arrhenius equation. The activation energy of the transmittance self-oscillation was about 67.0 kcal. This activation energy is almost the same value as that of the normal BZ reaction, which consists of a Ru(bpy)$_3$ catalyst that is not bonded to the polymer chain [58]. This result indicates that the Ru(bpy)$_3$ moiety that is bonded to the ACMO-main chain does not inhibit the BZ reaction. Furthermore, the activation energy of the poly(ACMO-*co*-Ru(bpy)$_3$) solutions is the almost same value as that of an NIPAAm-based [59] and a vinylpyrrolidone (VP)-based polymer chain [52], which have a nonthermoresponsive nature. In the previous investigation, a nonthermoresponsive VP main-chain was used to undergo the self-oscillation in high-temperature condition compared to the NIPAAm-based self-oscillating polymer system.

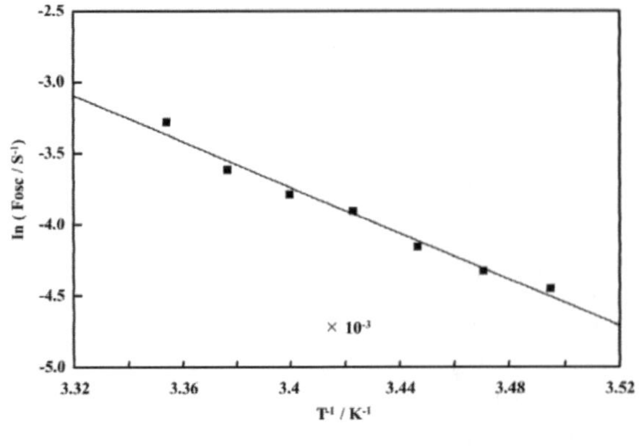

Figure 6. Arrhenius dependence on the temperature of self-oscillation for poly(ACMO-*co*-Ru(bpy)$_3$), oscillating frequency (*Fosc*), with a fixed concentration of the other three BZ substrates: [NaBrO3] = 0.3 M, [MA] = 0.1 M, and [HNO_3] = 0.4 M.

Figure 7 shows the effect of one BZ substrate ([$NaBrO_3$] or [MA]) in the polymer solution on the period of the transmittance self-oscillation. In these measurements, the concentrations of two of the BZ substrates were fixed, while that of the remaining BZ substrate ($NaBrO_3$ (Figure 7(A)) and MA (Figure 7(B)) was changed at a constant temperature (T = 13 °C). As shown in Figure 7, the graphs have good linear relationships. Therefore, we expressed this relationship as $T = a[\text{substrate}]^b$, where the T indicates the period and the brackets show the initial concentration, and a and b denotes the experimental constants. The values of a and b of the poly(ACMO-*co*-Ru(bpy)$_3$) solutions were almost the same as those of poly(NIPAAm-*co*-Ru(bpy)$_3$) and poly(VP-*co*-Ru(bpy)$_3$) solutions [52, 59]. The gradient of the graph in Figure 7(A) is steeper than that in Figure 7(B). This trend indicates that the effect of the concentration of $NaBrO_3$ for the oscillating period was larger than that of MA. Moreover, these results demonstrate that the period of the transmittance self-oscillation can be controlled by changing the BZ substrate conditions, as well as controlling the measuring temperature.

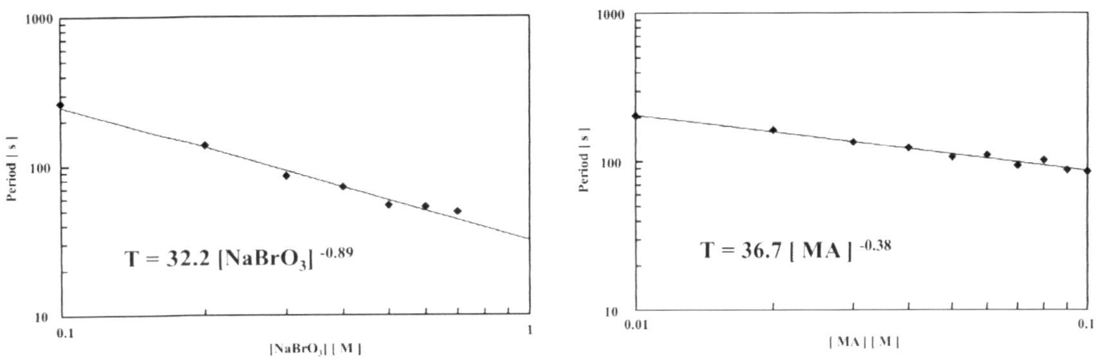

(A) (B)

Figure 7. Relationship between period T (s) and initial molar concentration of one BZ substrate at constant temperature (T = 13 °C) with fixed concentrations of the other two BZ substrates: (A) [MA] = 0.1 M and [HNO_3] = 0.4 M; and (B) [$NaBrO_3$] = 0.3 M and [HNO_3] = 0.4 M.

4. Conclusions

In this study, we measured the transmittance self-oscillation of the poly(ACMO-*co*-Ru(bpy)$_3$) solutions as the function of the temperature and concentration of the BZ substrates. The amplitude of the transmittance self-oscillation was affected significantly by the measuring temperature. The amplitude of the self-oscillation decreased with time because the polymer chain aggregated in the self-oscillating behavior. As the size of the polymer aggregation increased with time, the period of the self-oscillation decreased. That is, the change in the density of the Ru(bpy)$_3$ moiety in the polymer aggregation increased the collision rate among the BZ substrates. Therefore, as apparent from the decreasing period, the reaction rate of the BZ reaction increased because of the increasing collision rate. The activation energy of the transmittance self-oscillation was almost the same value as the normal BZ reaction and the NIPAAm-based and VP-based self-oscillating polymer chains. Moreover, in this study, we have demonstrated that the period of the self-oscillation for the poly(ACMO-*co*-Ru(bpy)$_3$) solutions can be controlled by changing the concentration of the BZ substrates, as well as by changing the temperature.

Acknowledgments

This work was supported by Grants-in-Aid (KAKENHI) for Young Scientists (B) (16K17493, 15K18014) and Scientific Research (B) (17H03209, 15H03827).

References

[1] Murata S, Konagaya A, Kobayashi S, Saito H, Hagiya M 2013 Molecular robotics: A new paradigm for artifacts. *New Generat. Comput.* 31 27–45.
[2] Winfree E, Liu F, Wenzler L, Seeman N 1998 Design and self-assembly of two-dimensional DNA crystals. *Nature* 394 539–44.

[3] Yan H, Park S H, Finkelstein G, Reif J, LaBean T 2003 DNA-templated self-assembly of protein arrays and highly conductive nanowires *Science* 301-5641 1882–84.
[4] Soong R K, Bzchand G D, Neves H P, Olkhovets A G, Craighead H G, Montemagno C D 2000 Powering an inorganic nanodevice with a biomolecular motor *Science* 290 1555–8.
[5] Liu H, Schmidt J J, Bachand D G, Rizk S S, Looger L L, Hellinga H W, Montemagno C D 2002 Control of a biomolecular motor-powered nanodevice with an engineered chemical switch *Nat. Mater.* 1 173–7.
[6] Badmkc J D, Balzani V, Credi A, Silvi S, Stoddart J F 2004 A molecular elevator *Science* 303 1845–9.
[7] Hugel T, Holland N B, Cattani A, Moroder L, Seitz M, Gaub H E 2002 Single-molecule optomechanical cycle. *Science* 296 1103–6.
[8] Beebe D J, Moore J S, Bauer J M, Yu Q, Liu R H 2000 Devadoss, C.; Jo, B.H. Functional hydrogel structures for autonomous flow control inside microfluidic channels. *Nature* 404 588–90.
[9] Shepherd R F, Ilievski F, Choi W, Morin S A, Stokes A A, Mazzeo A D, Chen X, Wang M, Whitesides G M 2011 Multigait soft robot. *Proc. Natl Acad. Sci. USA* 108, 20400–3.
[10] Keplinger C, Sun J Y, Foo C C, Rothemund P, Whitesides G M, Suo Z 2013 Stretchable, transparent, ionic conductors. *Science* 341, 984–7.
[11] Larson C, Peele B, Li S, Robinson S, Totaro M, Beccai L, Mazzolai B, Shepherd R 2016 Highly stretchable electroluminescent skin for optical signaling and tactile sensing *Science* 351, 1071–4.
[12] Sidorenko A, Krupenkin T, Taylor A, Fratzl P, Aizenberg J 2007 Reversible switching of hydrogel-actuated nanostructures into complex micropatterns. *Science* 315, 487–90.
[13] Andersen E S, Dong M, Nielsen M M, Jahn K, Subramani R, Mamdouh W, Golas M M, Sander B, Stark H, Oliveira C L 2009 Pedersen, J.S. Self-assembly of a nanoscale DNA box with a controllable lid. *Nature* 459, 73–6.
[14] Douglas S, Dietz H, Liedl T, Hogberg B, Graf F, Shih W 2009 Self-assembly of DNA into Nanoscale Three-dimensional shapes. *Nature* 459, 414–8.
[15] He Y, Chen Y, Liu H, Ribbe A, Mao C 2005 Self-assembly of hexagonal DNA two-dimensional (2D) Arrays. *J. Am. Chem. Soc.* 127, 12202–3.
[16] Rothemund P 2006 Folding DNA to create nanoscale shapes and patterns. *Nature* 440, 297–302.
[17] Yan H, Park S H, Finkelstein G, Reif J, LaBean T 2003 DNA-templated self-assembly of protein arrays and highly conductive nanowires *Science* 301, 1882–4.
[18] He Y, Chen Y, Liu H, Ribbe A, Mao C 2005 Self-assembly of hexagonal DNA two-dimensional (2D) Arrays. *J. Am. Chem. Soc.* 127, 12202–3.
[19] Hamada S, Murata S 2009 Substrate-assisted assembly of interconnected single-duplex DNA nanostructures. *Angew. Chem., Int. Ed.* 48, 6820–3.
[20] Kuzuya Y, Sakai T, Yamazaki Y, Xu Y, Komiyama M 2011 Nanomechanical DNA origami 'single-molecule beacons' directly imaged by atomic force microscopy *Nat. Commun.* 2, 1–8.
[21] Hyunwoo Y, Shaoting L, Chu Ma, Mahdi T, Nicolas X F, Xuanhe Z 2017 Hydraulic hydrogel actuators and robots optically and sonically camouflaged in water. *Nature Commun.* 14230.
[22] Ma M, Guo L, Anderson D G, Langer R 2013 Bio-inspired polymer composite actuator and generator driven by water gradients. *Science* 339 186–9.
[23] Dong L, Agarwal1 K A, Beebe D J, Jiang H 2006 Adaptive liquid microlenses activated by stimuli- responsive hydrogels *Nature* 442, 551–4.
[24] Takashima Y, Hatanaka S, Otsubo M, Nakahata M, Kakuta T, Hashidzume A, Yamaguchi H, Harada A 2012 Expansion-contraction of photoresponsive artificial muscle regulated by host-guest interactions. *Nat. Commun.* 3, 1270.
[25] Sidorenko A, Krupenkin T, Taylor A, Fratzl P, Aizenberg J 2007 Reversible switching of hydrogel-actuated nanostructures into complex micropatterns. *Science* 315, 487–90.
[26] Tanaka T 1981 Gels. *Sci. Am.* 244, 110–6.

[27] Yoshida R, Uchida K, Kaneko Y, Sakai K, Kikuchi A, Sakurai Y, Okano T 1995 Comb-type grafted hydrogels with rapid de-swelling response to temperature changes. *Nature* 374, 240–2.
[28] Inadomi T, Ikeda S, Okumura Y, Kikuchi H, Miyamoto N 2014 Anomalous thermo- and photo-responsive anisotropic deformation of poly(*N*-isopropylacrylamide) gel hybridized with liquid crystalline inorganic nanosheets aligned by an electric field *Macromol. Rapid Commun.* 35, 1741–6.
[29] Kim Y S, Liu M, Ishida Y, Ebina Y, Osada M, Sasaki T, Hikima T, Takata M, Aida T 2015 Thermoresponsive actuation enabled by permittivity switching in an electrostatically anisotropic hydrogel. *Nature Mater.* 14, 1002–7.
[30] Ehrick J D, Deo S K, Browning T W, Bachas L G, Madou M J, Daunert S 2005 *Genetically engineered protein in hydrogels tailors stimuli-responsive characteristics Nat. Mater.* 4, 298–302.
[31] Matsumoto K, Kawamura A, Miyata T 2017 Conformationally regulated molecular binding and release of molecularly imprinted polypeptide hydrogels that undergo helix–coil transition. *Macromolecules* 50 2136–44.
[32] Ishiwatari T, Kawaguchi M, Mitsuishi M 1984 Oscillatory reactions in polymer systems. *J. Polym. Sci. Polym. Chem.* 22 2699–2704.
[33] Yoshida R, Sakai T, Ito S, Yamaguchi T 2002 Self-oscillation of polymer chains with rhythmical soluble-insoluble changes *J. Am. Chem. Soc.* 124 8095–8.
[34] Hara Y, Mayama H, Yamaguchi Y, Takenaka Y, Fukuda R 2013 Direct observation of periodic swelling and collapse of polymer chain induced by the Belousov-Zhabotinsky reaction *J. Phys. Chem. B* 117 14351–7.
[35] Zaikin A N, Zhabotinsky A M 1970 Concentration wave propagation in two-dimensional liquid-phase self-oscillating system *Nature* 225 535–7.
[36] Field R J, Burger M *Oscillations and Traveling Waves in Chemical Systems*; John Wiley & Sons: New York, NY, USA, 1985.
[37] Mori H, Kuramoto Y *Dissipative Structures and Chaos*; Springer-Verlag: Berlin, Germany, 1997.
[38] Gyorgyi L, Turanyi T, Field R J 1990 Mechanistic details of the oscillatory Belousov-Zhabotinskii reaction. *J. Chem. Phys.* 94, 7162–70.
[39] Reusser E J, Field R J 1979 The transition from phase waves to trigger waves in a model of the Zhabotinskii reaction. *J. Am. Chem. Soc.* 101, 1063–71.
[40] Turanyi T, Gyorgyi L, Field R J 1993 Analysis and simplification of the GTF model of the Belousov-Zhabotinsky reaction. *J. Chem. Phys.* 97, 1931–41.
[41] Hara Y, Yoshida R 2009 Damping behavior of aggregation-disaggregation self-oscillation for a polymer chain *Macromol. Rapid Commun.* 30, 1656–62.
[42] Hara Y, Yoshida R 2009 Influence of a positively charged moiety on aggregation-disaggregation self-oscillation induced by the BZ reaction *Macromol. Chem. Phys.* 210 2160–6.
[43] Hara Y, Yoshida R 2008 A viscosity self-oscillation of polymer solution induced by the BZ reaction under acid-free condition *J. Chem. Phys.* 128 224904.
[44] Hara Y, Yoshida R 2005 Self-oscillation of polymer chains induced by the Belousov-Zhabotinsky reaction under acid-free conditions *J. Phys. Chem. B* 109 9451–54.
[45] Hara Y, Sakai T, Maeda S, Hashimoto S, Yoshida R 2005 Self-oscillating soluble-insoluble changes of polymer chain including an oxidizing agent induced by the Belousov-Zhabotinsky reaction *J. Phys. Chem. B* 109, 23316–23319.
[46] Hara Y, Yoshida R 2008 Self-oscillating polymer fueled by organic acid *J. Phys. Chem. B*, 112, 8427–9.
[47] Hara Y, Yamaguchi Y, Mayama H 2014 Switching the BZ reaction with a strong-acid-free gel *The J. Phys. Chem. B* 118 634–8.
[48] Hara Y, Yoshida R 2005 Control of oscillating behavior for the self-oscillating polymer with pH-control site *Langmuir* 21 9773–6.

IOP Conf. Series: Materials Science and Engineering **242** (2017) 012095 doi:10.1088/1757-899X/242/1/012095

[49] Maeda S, Hara Y, Sakai T, Yoshida R, Hashimoto S 2007 Self-walking gel. *Adv. Mater.* 19 3480–4.

[50] Maeda S, Hara Y, Yoshida R, Hashimoto S 2008 Peristaltic motion of polymer gels. *Angew. Chem., Int. Ed.* 47 6690–3.

[51] Maeda S, Hara Y, Yoshida R, Hashimoto S 2008 Control of the dynamic motion of a gel actuator driven by the Belousov-Zhabotinsky reaction *Macromol. Rapid. Commun.* 29 401–5.

[52] Hara Y, Jahan R A 2011 Influence of initial substrate concentration of the Belouzov-Zhabotinsky reaction on transmittance self-oscillation for a nonthermoresponsive polymer chain *Polymers* 3 330–9.

[53] Hara Y 2013 Autonomous oscillation of nonthermoresponsive polymers and gels induced by the Belousov–Zhabotinsky reaction *Chemosensors* 1 3–20.

[54] Nakamaru S, Maeda S, Hara Y, Hashimoto S 2009 Control of autonomous swelling-deswelling behavior for a polymer gel *J. Phys. Chem. B* 113 4609–13.

[55] Hara Y, Fujimoto K, Mayama H 2014 Self-oscillation of polymer chains with an $Fe(bpy)_3$ catalyst induced by the Belousov–Zhabotinsky reaction *J. Phys. Chem. B* 118 608–12.

[56] Hara, Y.; Mayama, H.; Fujimoto, K. Influence of Belousov–Zhabotinsky substrate concentrations on autonomous oscillation of polymer chains with $Fe(bpy)_3$ catalyst. *J. Phys. Chem. B* **2014**, *118*, 6931–6936.

[57] Hara Y, Mayama H, Yamaguchi Y, Fujimoto K 2014 Activation energy of the Belousov–Zhabotinsky reaction in a gel with $[Fe(bpy)_3]$ catalyst *Chem. Lett.* 43, 673–5.

[58] Kuhnert L, Krug H J 1987 Kinetics of chemical waves in the acidic bromate-malonic acid-tris(bipyridine)ruthenium(2+) system in comparison with the ferroin system *J. Phys. Chem.* 91 730–3.

[59] Hara Y, Takenaka Y 2014 Autonomous oscillation of polymer chains induced by the Belousov–Zhabotinsky reaction *Sensors* 14 1497–1510.

An all-digital phase-locked loop demodulator based on FPGA

X F Gong[1], **Z D Cui**[1]

[1] School of Mechanical Engineering, University of Science & Technology Beijing, NO.30, College Road, HaiDian District, Beijing , China

2532542664@qq.com

Abstract. This paper studied the principle of analogue phase-locked loop demodulation and work process of digital phase-locked loop. It is found that the higher the reference signal frequency is, the smaller the duty ratio of the discriminator output signal is. Carrier detection is achieved by using this relationship. The experimental results indicate that the demodulator based on the principle could realize high-quality transmission of digital signals and could be an effective FM communication mode for studying wireless transmission of digital signals.

1. Introduction

The phase-locked loop (PLL) possesses a wide range of applications in modern communication systems. The all-digital phase-locked loop (ADPLL) has been widely used in digital communication and automatic control field because of the advantages of stable performance, reliable operation and so on. With the development of digital signal processing technology and programmable devices, the on-chip realization of all-digital phase-locked loop has become a hot topic [1].

In the communication system, FSK modulation method is easy to achieve, anti-noise and anti-attenuation performance is better, so it has been more widely used in the field of information transmission. When the central frequency of the ADPLL is fixed, the FSK modulation signal enters the ADPLL as an input signal, and the output of the phase detector reflects the information of the modulation signal. Then the demodulation process is completed. In this paper, the Verilog hardware description language is used to build the system, and the programmable gate array FPGA is used as the experimental platform to realize the good demodulation effect and has high engineering application value.

2. ADPLL demodulation

Phase-locked loop demodulation is known from the traditional demodulation mode. As a new demodulation method, it avoids the large number of multiplication integral operations in traditional demodulation method, which consume a large amount of digital signal processor resources. The control process is implemented as a counter and avoids the effects of system perturbations.

In addition, all-digital signal processing eliminates the impact of the environment on components, improves signal processing accuracy while simplifying the design of the system.

2.1. Digital signal modulation

The system used binary frequency shift keying (2FSK) modulation which modulate the carrier frequency according to the baseband signal. The signals used in this paper are digital signals, so it is easy to use the form of frequency control word to produce phase continuous modulation signal. The

Content from this work may be used under the terms of the Creative Commons Attribution 3.0 licence. Any further distribution of this work must maintain attribution to the author(s) and the title of the work, journal citation and DOI.
Published under licence by IOP Publishing Ltd

expression of the frequency control word is shown in $K = 2^N \bullet \frac{f_o}{f_s}$. The value of frequency control word responses to the required frequency [3].

2.2. PLL demodulation

In the analog phase-locked loop demodulator, the FM wave is amplified by the amplifier and the output of the voltage-controlled oscillator is sent to the phase detector. The output of the phase detector reflects the phase error information of the two input signals. Then the obtained voltage signal passed loop filter changes with the change of the modulation signal frequency so as to realize the mediation process. The demodulation block diagram is shown in Figure 1.

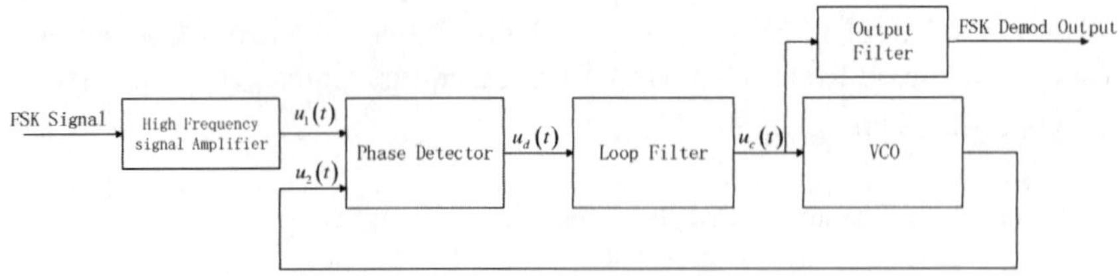

Figure 1. Analog PLL demodulator.

By the characteristics of phase detector, the average value of phase detector output signal can be described as $\overline{u_d(t)} = K_d \sin(\theta_e)$, K_d represents the gain of the phase detector, and θ_e represents the phase error of the input signals of phase detector. In order to demodulate FSK signal, the bandwidth of loop filter in the circuit should be wide enough so that all the components of the phase detector output voltage (high frequency and DC component) are able to be outputted. The instantaneous frequency of VCO reflects the change of the input FSK signal; the VCO input voltage reflects the modulation information of FSK signal.

2.3. ADPLL demodulation process

The main parts of the demodulator include not only ADPLL but also the div-phase and detect-modulate module, which are used to detect the effective input signal, control phase adjustment, and detect phase locking state and so on. The block diagram is shown in Figure 2[4].

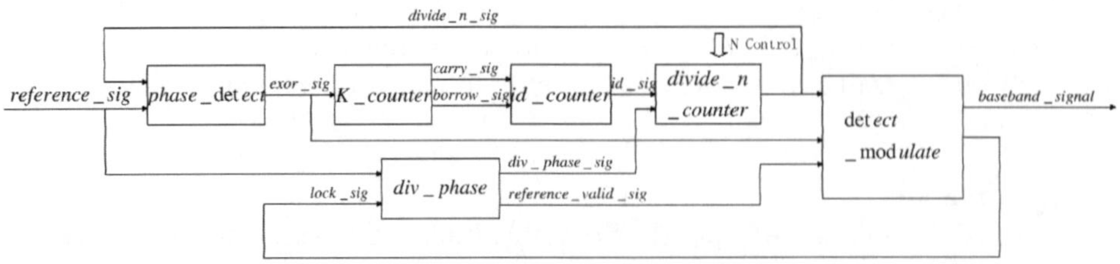

Figure 2. ADPLL demodulator.

In practical application, the data transmission is not continuous. In the absence of data transmission, the output frequency of ADPLL is center frequency. When there is data to be transmitted, the reference frequency changes. Aiming to achieve that the output frequency of ADPLL tracks reference frequency, the phase adjustment module is used to detected the jump of reference signal and the ADPLL state (whether locked). If the ADPLL is in unlocked state, the div-phase module outputs phase adjustment signal. When the frequency division counter receives the phase adjustment signal, the initial value of the id counter is adjusted immediately so that the output signal can quickly keep up with the frequency of the reference signal.

IOP Conf. Series: Materials Science and Engineering **242** (2017) 012096 doi:10.1088/1757-899X/242/1/012096

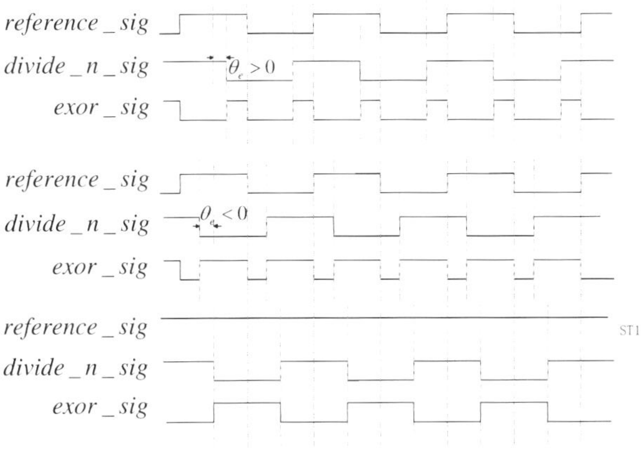

Figure 3. Detection and demodulation signal timing diagram.

The key of this paper is shown in Figure 3, it describes three situations: $\theta_e > 0$, $\theta_e < 0$ and there is no input signal. When an input signal is present, before the $divide\_n\_sig$ jump edge, $exor\_sig$ keeps low level; when there is no reference signal input, before the $divide\_n\_sig$ jump edge, the probability of $exor\_sig$ at the high and low levels is equal. So we can judge whether there is FM wave signal inputting system according to the above characteristics.

On the other hand, by the demodulation process of analog PLL, the input signal of the VCO reflects the variation of the FSK signal. Similarly, as the Figure 3 shows, the duty ratio of $exor\_sig$ reflects the carrier signal information. For example, in the case of $\theta_e > 0$, it indicates that the frequency of reference signal is greater than center frequency of system which represents that input carrier is a high frequency. On the contrary, in the case of $\theta_e < 0$, input carrier is a low frequency. In the two cases, the duty ratio of $exor\_sig$ is obviously different. After the duty ratio counted, binary data is demodulated according to the designed judge threshold.

In this paper, the 2FSK modulation is adopted. In the engineering application, the system can also implement the MSK modulation method. Equally, the demodulation process is completed according to the output signal of the phase detector.

3. System simulation and experimental verification

Before the experiment verification, ModelSim can be used for timing simulation of the system to verify the correctness of the design. The test result is shown in the Figure 4.

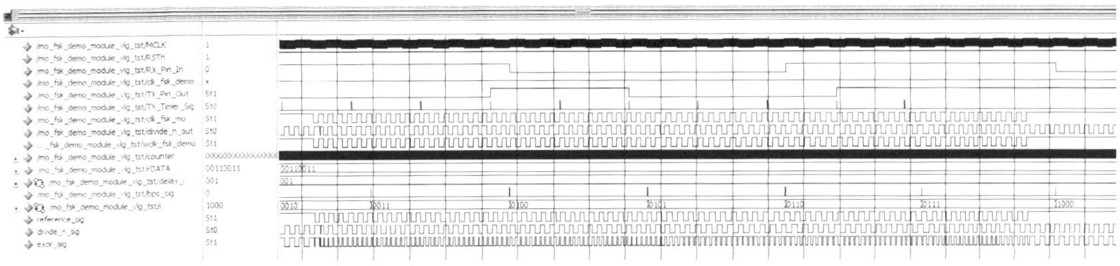

Figure 4. ModelSim simulation timing diagram.

As can be seen from the Figure 4, the relationship of $reference\_sig$, $divide\_n\_sig$ and $exor\_sig$ conforms to the analysis of Figure 3. According to this relationship, it is possible to demodulate the corresponding transmitted data information.

This experiment selects the Cyclone IV series FPGA chip EP4CE6E22I7 of Altera Company to build the experimental platform. After the program timing constraints and pin configuration, the project is downloaded to the chip for hardware verification. Two same pieces of demodulate boards will be used in the experiment. The result of experiment is realizing the asynchronous communication between the host computer and the slave computer [5].

References

[1] Xu W D 1990 *A Design of FSK modem based on phase locked loop (Water Resources Informatization* vol 2*)* pp 9-17
[2] Best R E 2003 *Phase Locked Loops: Design, Simulation, and Applications* (Beijing: Tsinghua University Press) pp 166-198
[3] Ma J R 2006 *FPGA-based waveform generator with DDS (Electronic Measurement Technology)*
[4] Du Y 2015 *MATLAB and FPGA implementation of digital modulation and demodulation technology* (Beijing: Publishing House of Electronics Industry) pp 179-244
[5] Wei W and Zhao H D 2010 *Design and simulation of UART module based on Verilog HDL (Electronic Design Engineering)*

Numerical study on fabricating rectangle microchannel in microfluidic chips by glass molding process

Tao Wang[1], Jing Chen[1] and Tianfeng Zhou[2]

[1]Peking University, No.5 Yiheyuan Road, Haidian District, Beijing, China
[2]Beijing Institute of Technology, No. 5 South Zhongguancun Street, Haidian District, Beijing, China

The corresponding author's e-mail address: j.chen@pku.edu.cn

Abstract. This paper studied the glass molding process (GMP) for fabricating a typical microstructure of glass microfluidic chips, i. e., rectangle microchannel, on soda-lime glass by finite element method. More than 100 models were established on the platform of Abaqus/Standard. The influence of parameters, i. e., temperature, aspect ratio, side wall angle and friction coefficient on deformation were studied, and the predicted morphology of the molded microchannel were presented as well. The research could provide fundamental experience for optimizing GMP process in the future.

1. Introductioon

Due to potential advantages in chemical and biomedical testing, microfluidic devices have drawn great attentions among academic and engineering communities [1]. Different kind of microfluidic devices were fabricated for wide applications, such as capillary electrophoresis [2-3], semen testing [4], electrochromatography [5] and DNA separation [6].

There are many substrate materials for fabricating microfluidic chips, such as silicon, Polydimethylsiloxane (PDMS), polymethyl methacrylate (PMMA) and glass. Among them, glass has many advantages over other materials, such as excellent optical properties, good insulating properties, high resistance to mechanical stress, high surface stability, and high solvent compatibility [7].

In order to achieve the widespread application of glass microfluidic chips, many researchers have been exploring the field of microstructure fabrication techniques in glass, and the conventional approaches can be classified into four categories: wet etching, dry etching, laser processing and mechanical processing [8]. However, the majority of conventional approaches suffer from low efficiency, high cost, poor surface quality and environmental pollution, which impede the widespread application of glass applications substantially.

Since rectangular cross-section microchannel is a fundamental structure of microfluidic devices, exploring glass molding process (GMP) to machine the microstructure is important to practical applications. This paper aims to investigate the formation process of rectangular cross-section microchannel in GMP by simulations. More than 100 models were established to study the influence of molding parameters, i. e., temperature, aspect ratio, side wall angle and friction coefficient on the mold shaped. The research could provide fundamental experience for optimizing parameters in machining rectangle microchannel on soda-lime glass in the future research.

Content from this work may be used under the terms of the Creative Commons Attribution 3.0 licence. Any further distribution of this work must maintain attribution to the author(s) and the title of the work, journal citation and DOI.
Published under licence by IOP Publishing Ltd

2. Modelling process

In terms of the glass preform material, since the soda-lime glass slides have been used by biomedical researchers to conduct experiments for decades, and it is inexpensive which can help reduce the whole fabrication cost, it is utilized as the preform materials in the paper. The physical properties of soda-lime preform are referred to [9]. According to [10], the appropriate model temperature is 620℃, the range of temperature studied in the simulation is between 610 and 660℃.

All the simulation models are established in the commercial finite element software Abaqus/Standard. Since the molding speed is relatively slow compared to other machining process, the GMP process is regarded as a quasi-static process. The GMP process consists of four stages, i.e., heating, pressing, annealing and cooling, as shown in figure 1. In this paper, the only pressing stage is simulated, and all the molds and preform are assumed to maintain a constant temperature, which is tens of degrees centigrade above its transition temperature.

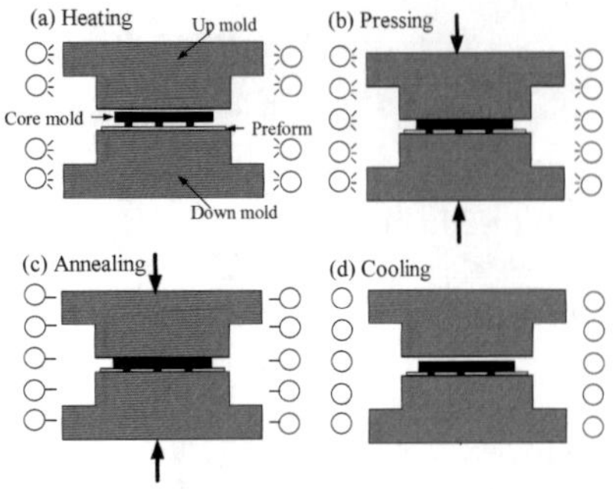

Figure 1. The GMP cycle

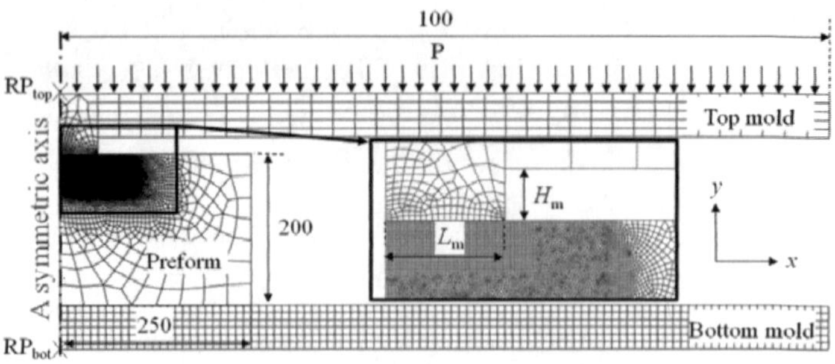

Figure 2. The established model for GMP process

At the pressing condition, the preform demonstrates high viscoelasticity, which is a time-dependent response of a material to stress or strain, and they are characterized by creep and stress relaxation behavior. In this paper, a 6-pair generalized Maxwell model is used as the constitutive model, and the specifical parameters are referred to [9]. As shown in figure 2, a rectangle specimen geometry of preform with 200μm height and 250μm width are established in the model, and it is placed on a rigid platen which is control by a reference point RPbot. All the freedoms of RPbot are constraint. A rigid platen with a micro-protrusion below is built as the top mold which is controlled by a reference point RPtop, and all of RPtop freedoms are constraint except for y direction. The height of the micro-protrusion keeps as 22μm in all the simulations. Due to the axisymmetric structure, only half of the model is built for simplicity, and the X-symmetric boundary condition is set on the left edge of preform. A uniform distributing pressure 0.2MPa is applied on the top mold. In order to alleviate the mesh distortion during pressing process, the meshes near the protrusion are densified. All contact behavior are described by the classic Coulomb friction model.

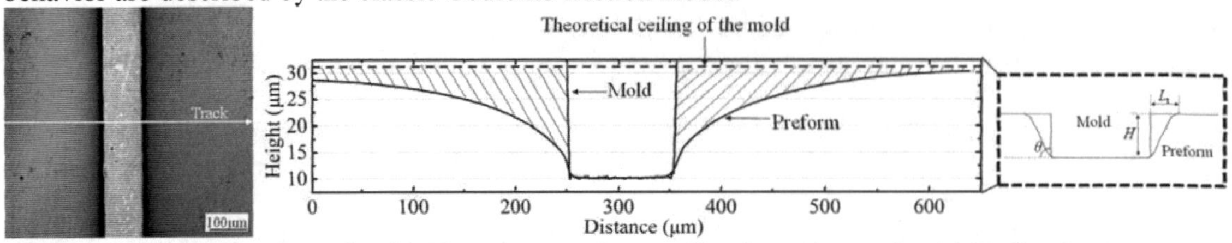

(a) The molded microchannel (b) The cross-section profile along the track (c) Defined parameters

Figure 3. A typical molded glass microchannel in experiment

Figure 3 shows a typical molded glass microchannel in experiemnt. Figure 3(a) is the actual molded microchannel obtained by Olympus Lext OLS4100, and the surface quality of the molded microchannel appears to be smooth. The cross-section profile of the microchannel is extracted along the track, and shown in figure 3(b), and it indicates that there is large area between the mold ceiling and preform, which is marked by dash area, not filled. In order to quantitatively study the influence of molding parameters on the molded morphology, the two characteristic parameter, i.e., H and L_t, are definited based on the typical deformation process, and they are shown in figure 3(c). L_t represents the distance from the side wall of core mold to the point where the preform and mold ceiling starts to separate. H stands for the height of the protrusion. Since all the height in the simulation is fixed at 22μm, a defamation factor D, which is equal to L_t divided by H, is utilized as predicted results from the simulation models. The smaller the D value, the better the shape transferability. As the glass preform is pressed at the same temperature, no heat transfer is considered in this simulation.

3. Results

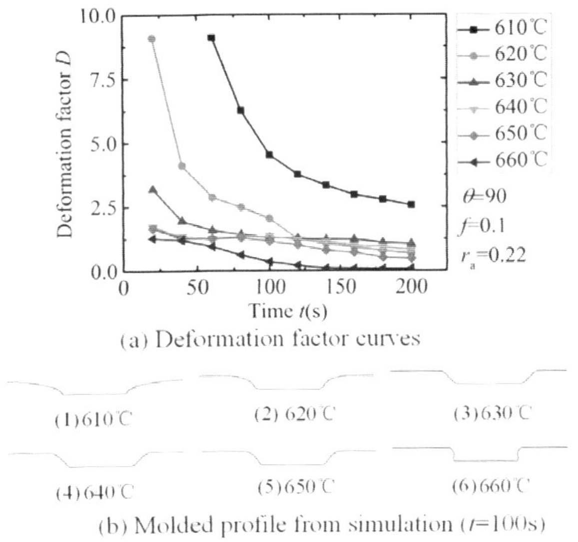

Figure 4. The effect of temperature (100s)

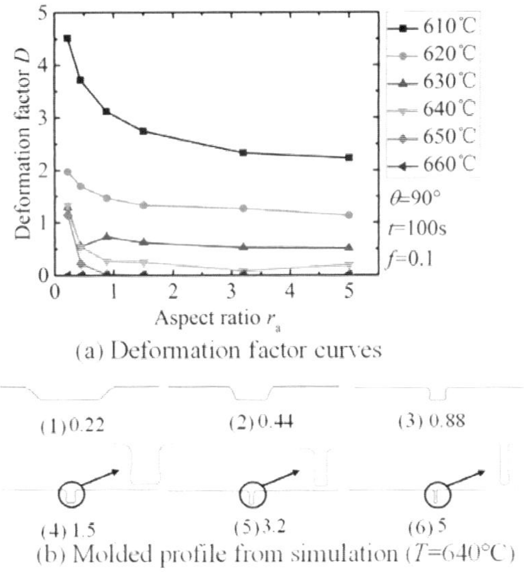

Figure 5. The effect of aspect ratio

The effect of temperature on deformation is shown in figure 4. Figure 4(a) demonstrates that the deformation factor D decreases dramatically with pressing time below 100s, and as time goes further, a slow drop of D is witnessed. It can be attributed to the glass characteristic of stress relaxation and creep under high temperature molding process. As for temperature, it is observed that D reduces with temperature as well. It is worth pointing out that it is possible to achieve almost full shape transferability when temperature is set at 660℃ and the pressing time is above 140s. In order to provide more information about the predicted shape, only the top profiles under six different temperature at 100s are extracted, as shown in figure 4(b). It confirms that high shape transferability can be achieved at high temperature.

The effect of aspect ratio on deformation factor is shown in figure 5. The horizontal axis is aspect ratio instead of pressing time, which means that each point is based on a single simulation test. D decreases dramatically when aspect ratio increases from 0.22 to 0.88; while when it increases further, a slight drop of D is witnessed. Since the height of the protrusion on the mold is fixed, the large aspect ratio represents small width of the protrusion. When aspect ratio is larger, less portion of material is supposed to experience deformation, thereby making the filling easier. As for temperature, D reduces with the increase of temperature as well. It is interesting to find that almost full shape transferability occur at both curve 650 and 660℃ when aspect ratio is equal or about 0.88. The profiles of molded shape at different aspect ratio are shown in figure 5(b). It indicates that the higher the aspect ratio, the smaller the *D*. According to figure 5(b6), the almost right angle can be achieved.

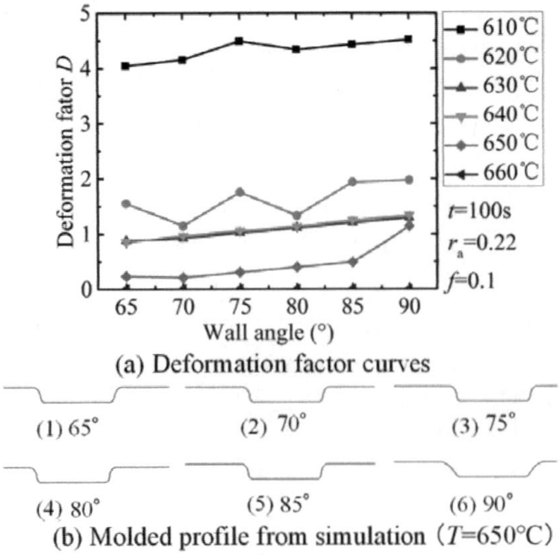

Figure 6. The effect of side wall angle

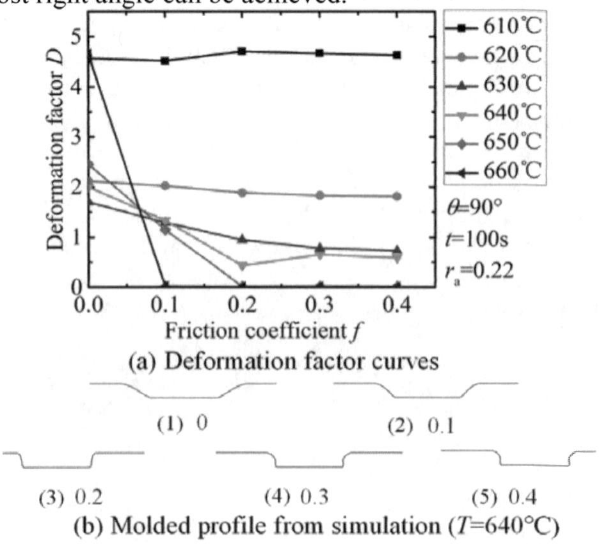

Figure 7. The effect of friction coefficient

The effect of side wall angle on deformation factor *D* is shown in figure 6. The influence of side wall angle on *D* is not significant compared with those of pressing time and aspect ratio, although a slight increasing pattern is witnessed. It can be attributed to the fact that the larger angle is supposed to make the deformation process more difficult, thereby reducing the shape transferability and increasing deformation factor *D*. The 660℃ curve can achieve almost full shape transferability at all investigated pressing condition. The profiles of molded shape at different angles are shown in figure 6(b). It is confirmed that the higher the side wall angle, the higher the *D* value.

The influence of friction coefficient on deformation factor *D* is shown in figure 9. It is interesting to find that the influence of friction coefficient on *D* is negligible when temperature is at 610 and 620℃, and its influence becomes increasingly significant as temperature increases further. Especially, *D* decreases from 4.6 to around 0 when friction coefficient increases from 0 to 0.1 at 660℃. It indicates that the temperature and friction have interactive influence on *D*. When temperature is higher, the preform is easier to deform which increases the influence of friction coefficient greatly. In addition, higher friction coefficient makes the deformation of preform harder along horizontal direction, which in the end leads to higher shape transferability. The profiles of molded shape at different angles are shown in figure 9(b). It is confirmed that the higher the friction coefficient, the smaller the D.

4. Conclusions

(1) The deformation factor *D* decreased dramatically with pressing time below 100s, and as time went further, a slow drop of *D* was witnessed. The deformation factor *D* decreased significantly when molding temperature increased.

(2) The deformation factor *D* decreased dramatically when aspect ratio increase from 0.22 to 0.88; while when it increases further, a slight drop of D is experienced.

(3) The influence of side wall angle on *D* is not significant compared with those of pressing time and aspect ratio, although a slight increasing pattern is witnessed. In addition, the temperature and friction coefficient had obvious interactive influence on *D*.

Acknowledgements
This paper is supported by National Natural Science Foundation of China (Grant No. U1537208), the National High Technology Research and Development Program of China (863) Grant No. 2015AA043601, and the National Natural Science Foundation of China (Grant No. 51375050).

References

[1] Rodriguez I, Spicar-Mihalic P, Kuyper CL, Fiorini GS and Chiu DT 2003 *Anal. Chem. Acta.* **469** 205-215
[2] Harrison D J, Manz A, Fan Z, Luedi H and Widmer HM 1992 *Anal. Chem.* **64** 1926-32
[3] Seiler K, Harrison DJ and Manz A 1993 *Anal. Chem* **65** 1481-88
[4] Kricka LJ, Faro I, Heyner S, Garside WT and Fitzpatrick G 1997 *J. Pharm. Biomed. Anal* **15** 1443-47
[5] Jacobson SC, Hergenröder R, Koutny LB and Ramsey JM 1994 *Anal. Chem* **66** 2369-73
[6] Effenhauser CS, Paulus A, Manz A and Widmer HM 1994 *Anal. Chem.* **66** 2949-53
[7] Chen Q, Li G, Jin QH, Zhao JL, Ren QS and Xu YS 2007 *J. Microelectromech. Syst.* **16** 1193-1200
[8] Kawai K, Yamaguchi F, Nakahara A and Shoji S 2010 *In 14th international conference on miniaturized systems for chemistry and life sciences,* **2** 1193-95
[9] Joshi D, 2014 Dissertation, Clemson University
[10] Huang CY, Kuo CH, and Chou CP 2010 *J. Mater. Process. Technol.* **212** 633-9

IOP Conf. Series: Materials Science and Engineering **242** (2017) 012098 doi:10.1088/1757-899X/242/1/012098

Design and performance analysis of the low-frequency and broadband piezoelectric energy harvester

Laizhao J[1,2], Rui H[1,2] and Weike W[1,2]

[1] School of Mechanical Engineering, Shandong University, Ji'nan 250061, China;
[2] Key Lab. of High Efficiency and Clean Mechanical Manufacture (Shandong University) Ministry of Education, Ji'nan 250061, China

jinglaizhao@126.com

Abstract. A novel type energy harvester composed with n masses and n beams was proposed to solve the problems of high-frequency and narrow band of cantilever energy harvester to improve its energy conversion efficiency, and broaden its application. Firstly, We taken micro beam as the research object, improved its dynamic model with size effect, then verified it by experiments. Secondly, the dynamic model was established based on the improved one, and its response and the output voltage of the external excitation was deduced formula. Finally, the performance was studied when n=2. The results show that resonance frequency was obvious reduced, and there are two frequency points less than 50Hz, 15.25 Hz and 23.08 Hz. Moreover, when the effective output voltage reached 80 mv, the effective frequency band was 20.32 Hz, which can make up for the energy conversion efficiency of single resonant frequency harvester.

1. Introduction

Since the P Curie and J Curie brothers discovered the positive / inverse piezoelectric effect of crystal by experiment, many scholars and research institutes have begun to try to apply it to real life, as a result, the piezoelectric energy conversion technology has been rapidly developed and cantilever energy harvester thereof emerges[1-3]. In 2005, Jeon[4]designed a PZT film-based collector, whose resonant frequency was tested to be 13.9KHz. Wu Xiaoming et al[5] took monocrystalline silicon as the material of cantilever beam and pasted thin piezoelectric material on its surface. At the end of the cantilever beam, they pasted mass blocks prepared by wet etching to form a silicon-based piezoelectric cantilever energy harvester, for which the resonant frequency was 1673Hz. Renaud[6] used AIN-based cantilever got the resonant frequency was 1.8KHz. And its output power was 40μW with an external excitation of 180nm. Gafforelli[7] established a three-layer structured mathematical model for cantilever beam vibration energy harvester. Combined the analytical method and finite element analysis method, he analyzed the mechanical property of harvester, and the results of two different methods were similar, namely the resonant frequency were 1kHz. Fang[8] pasted nickel masses at the end of the cantilever beam to reduce its resonant frequency, and the final value was 608Hz.

It's found by statistics that frequency of ambient vibration energy mainly distributes within a range of 300Hz, however, the resonant frequencies of harvesters in literature [4-8] are higher than 300Hz[9,10], as a result, the harvesters can't be effectively stimulated, nor can they convert mechanical vibration energy into electric energy, thus the propose of power supply fails. At the same time, these harvesters can only convert energy effectively at the resonant frequency, the working band is quite narrow. Therefore, this paper designed a new energy harvester while considering the demands of low resonant

Content from this work may be used under the terms of the Creative Commons Attribution 3.0 licence. Any further distribution of this work must maintain attribution to the author(s) and the title of the work, journal citation and DOI.
Published under licence by IOP Publishing Ltd

frequency and wide working band. The dynamic performance of the equipment under the influence of scale effect has been studied.

2. Dynamic Performance Analysis of New Energy Harvester

New energy harvester is a new structure consist of n beams and n lumped mass block, as shown in figure.1. With $m_i(i=1,2,\ldots,n)$ referring the quality, m_n locates to the end of the cantilever beam. The whole cantilever beam is divided into n sections by the location of each additional mass, the length and stiffness of each beam is marked as l_i and $E_iI_i(i=1,2,\ldots,n)$. Each straight beam is a composite beam, that is to say PZT piezoelectric coating and Cu electrode are attached on the upper and lower surfaces of Si base beam.

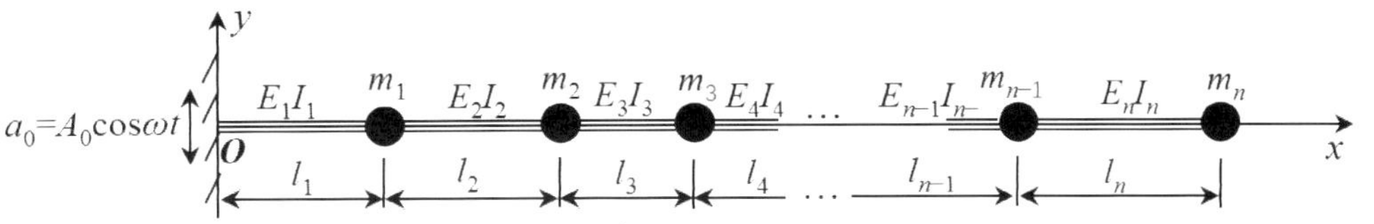

Figure 1. Schematic Diagram of Cantilever Beam With Additional Lumped Mass

It's known that the fixed end O of cantilever beam is given a ground vibration excitation, whose acceleration is $a_0=A_0\cos\omega t$. $w_i\ (x,t)\ (i=1,2,\ldots,n)$ is taken to describe the transverse vibration displacement of each beam under the excitation of acceleration. According to [11], $w_i\ (x,t)$meets the following differential equation:

$$\rho_i A_i w_{i,tt} + \left(E_i I_i + \mu A_i l^2\right) w_{i,xxxx} = 0 \quad (i=1,2,\ldots,n) \tag{1}$$

$$w_1(x_0,t) = -A_0 e^{j\omega t}/\omega^2 \qquad w_{1,x}(x_0,t)=0 \tag{2}$$

At $x_i=\sum_{k=1}^{i} l_k$ (i<n)

$$w_i(x_i,t)=w_{i+1}(x_i,t) \quad w_{i,x}(x_i,t)=w_{i+1,x}(x_i,t) \tag{3}$$

$$\left(E_iI_i+\mu A_i l^2\right) w_{i,xx}(x_i,t) = \left(E_{i+1}I_{i+1}+\mu A_{i+1} l^2\right) w_{i+1,xx}(x_i,t) \tag{4}$$

$$\left(E_iI_i+\mu A_i l^2\right) w_{i,xxx}(x_i,t) = \left(E_{i+1}I_{i+1}+\mu A_{i+1} l^2\right) w_{i+1,xxx}(x_i,t) + m_i w_{i,tt}(x_i,t) \tag{5}$$

At $x_n=\sum_{k=1}^{n} l_k$ (i<n)

$$\begin{aligned}&\left(E_nI_n+\mu A_n l^2\right) w_{n,xx}(x_n,t)=0\\&\left(E_nI_n+\mu A_n l^2\right) w_{n,xxx}(x_n,t)=m_n w_{n,tt}(x_n,t)\end{aligned} \tag{6}$$

Taking a Fourier transformation on formula (1):

$$W_{i,xxxx}(x,\omega)-\lambda_i^4 W_i(x,\omega)=0 \quad (i=1,2,\ldots,n) \tag{7}$$

The $\lambda_i^4=\rho_i A_i \omega^2 (E_i I_i+\mu A_i l^2)^{-1}$ in formula (7) has a general solution $W_i(x,\omega)$ as the following form:

$$W_i(x,\omega)=C_{i1}e^{-j\lambda_i x}+C_{i2}e^{-\lambda_i x}+C_{i3}e^{j\lambda_i x}+C_{i4}e^{\lambda_i x} \tag{8}$$

Under the harmonic excitation, the vibration displacement $w_i\ (x,t)$ of beam i can be marked as:

$$w_i(x,t)=W_i(x,\omega)e^{j\omega t}=\begin{bmatrix}e^{j(\omega t-\lambda_i x)} & e^{-\lambda_i x+j\omega t}\end{bmatrix}\begin{bmatrix}C_{i1}\\C_{i2}\end{bmatrix}+\begin{bmatrix}e^{j(\omega t+\lambda_i x)} & e^{\lambda_i x+j\omega t}\end{bmatrix}\begin{bmatrix}C_{i3}\\C_{i4}\end{bmatrix} \tag{9}$$

Front half part of right equation in formula (9) refers to the flexural wave of composite beam along the positive x axis, the latter part refers to the negative flexural wave.

As for each beam, the section corner $w_{i,x}$ is referred to as:

$$w_{i,x}(x,t)=\begin{bmatrix}-j\lambda_i e^{j(\omega t-\lambda_i x)} & -\lambda_i e^{-\lambda_i x+j\omega t}\end{bmatrix}\begin{bmatrix}C_{i1}\\C_{i2}\end{bmatrix}+\begin{bmatrix}j\lambda_i e^{j(\omega t+\lambda_i x)} & \lambda_i e^{\lambda_i x+j\omega t}\end{bmatrix}\begin{bmatrix}C_{i3}\\C_{i4}\end{bmatrix} \tag{10}$$

Bending moment of section $M_{i,x}(x,t)$is:

IOP Conf. Series: Materials Science and Engineering **242** (2017) 012098 doi:10.1088/1757-899X/242/1/012098

$$M_i(x,t)=(E_iI_i+\mu A_il^2)W_{i,xx}(x,\omega)e^{j\omega t}=(E_iI_i+\mu A_il^2)\left(\begin{bmatrix}-\lambda_i^2e^{j(\omega t-\lambda_ix)} & \lambda_i^2e^{-\lambda_ix+j\omega t}\end{bmatrix}\begin{bmatrix}C_{i1}\\C_{i2}\end{bmatrix}+\begin{bmatrix}-\lambda_i^2e^{j(\omega t+\lambda_ix)} & \lambda_i^2e^{\lambda_ix+j\omega t}\end{bmatrix}\begin{bmatrix}C_{i3}\\C_{i4}\end{bmatrix}\right) \tag{11}$$

Section shear $Q_{i,x}(x,t)$is:

$$Q_i(x,t)=(E_iI_i+\mu A_il^2)W_{i,xxx}(x,\omega)e^{j\omega t}=(E_iI_i+\mu A_il^2)\left(\begin{bmatrix}j\lambda_i^3e^{j(\omega t-\lambda_ix)} & -\lambda_i^3e^{-\lambda_ix+j\omega t}\end{bmatrix}\begin{bmatrix}C_{i1}\\C_{i2}\end{bmatrix}+\begin{bmatrix}-j\lambda_i^3e^{j(\omega t+\lambda_ix)} & \lambda_i^3e^{\lambda_ix+j\omega t}\end{bmatrix}\begin{bmatrix}C_{i3}\\C_{i4}\end{bmatrix}\right) \tag{12}$$

Defining:

$$\boldsymbol{C}_i^+=\begin{bmatrix}C_{i1}\\C_{i2}\end{bmatrix},\quad \boldsymbol{C}_i^-=\begin{bmatrix}C_{i3}\\C_{i4}\end{bmatrix},\quad \boldsymbol{e}_i^+=\begin{bmatrix}e^{-j\lambda_ix} & 0\\0 & e^{-\lambda_ix}\end{bmatrix},\quad \boldsymbol{e}_i^-=\begin{bmatrix}e^{j\lambda_ix} & 0\\0 & e^{\lambda_ix}\end{bmatrix},\quad \boldsymbol{\lambda}_{Wi}^+=\begin{bmatrix}1 & 1\\-j\lambda_i & -\lambda_i\end{bmatrix},\quad \boldsymbol{\lambda}_{Wi}^-=\begin{bmatrix}1 & 1\\j\lambda_i & \lambda_i\end{bmatrix}$$

$$\boldsymbol{\lambda}_{Fi}^+=(E_iI_i+\mu A_il^2)\lambda_i^2\begin{bmatrix}-1 & 1\\j\lambda_i & -\lambda_i\end{bmatrix},\quad \boldsymbol{\lambda}_{Fi}^-=(E_iI_i+\mu A_il^2)\lambda_i^2\begin{bmatrix}-1 & 1\\-j\lambda_i & \lambda_i\end{bmatrix}$$

Taking formula (9) and (10) into formula (2), we can get:

$$\boldsymbol{\lambda}_{W1}^+\boldsymbol{C}_i^++\boldsymbol{\lambda}_{W1}^-\boldsymbol{C}_i^-=\begin{bmatrix}-A_0/\omega^2\\0\end{bmatrix} \tag{13}$$

Taking formula (11) and (12) into formula (4) and (5), we can get:

$$\boldsymbol{\lambda}_{Fi}^+\boldsymbol{e}_i^+(x_i)\boldsymbol{C}_i^++\boldsymbol{\lambda}_{Fi}^-\boldsymbol{e}_i^-(x_i)\boldsymbol{C}_i^--\boldsymbol{\lambda}_{Fi+1}^+\boldsymbol{e}_{i+1}^+(x_i)\boldsymbol{C}_{i+1}^+-\boldsymbol{\lambda}_{Fi+1}^-\boldsymbol{e}_{i+1}^-(x_i)\boldsymbol{C}_{i+1}^-=\begin{bmatrix}0\\-\omega^2m_iW_i(x_i,\omega)\end{bmatrix} \tag{14}$$

In the derivation of formula (14), the relationship between formula (1) and (7) has been cited for the following transformation:

$$\begin{aligned}w_{i,tt}(x,t)&=-(E_iI_i+\mu A_il^2)w_{i,xxxx}(\rho_iA_i)^{-1}=-(E_iI_i+\mu A_il^2)W_{i,xxxx}(x,\omega)e^{j\omega t}(\rho_iA_i)^{-1}\\&=-(E_iI_i+\mu A_il^2)\lambda_i^4W_i(x,\omega)e^{j\omega t}(\rho_iA_i)^{-1}=-\omega^2W_i(x,\omega)e^{j\omega t}\end{aligned}$$

Taking it into formula (9) and we can get:

$$(\boldsymbol{\lambda}_{Fi}^+-\boldsymbol{\lambda}_{Fi})\boldsymbol{e}_i^+(x_i)\boldsymbol{C}_i^++(\boldsymbol{\lambda}_{Fi}^--\boldsymbol{\lambda}_{Fi})\boldsymbol{e}_i^-(x_i)\boldsymbol{C}_i^--\boldsymbol{\lambda}_{Fi+1}^+\boldsymbol{e}_{i+1}^+(x_i)\boldsymbol{C}_{i+1}^+-\boldsymbol{\lambda}_{Fi+1}^-\boldsymbol{e}_{i+1}^-(x_i)\boldsymbol{C}_{i+1}^-=\begin{bmatrix}0\\0\end{bmatrix} \tag{15}$$

Combined the formula (13), (14) and (15):

$$\begin{bmatrix}A_{11} & A_{12} & 0 & 0 & 0 & 0 & \dots & 0 & 0 & 0 & 0\\ A_{21} & A_{22} & A_{23} & A_{24} & 0 & 0 & \dots & 0 & 0 & 0 & 0\\ A_{31} & A_{32} & A_{33} & A_{34} & 0 & 0 & \dots & 0 & 0 & 0 & 0\\ 0 & 0 & A_{43} & A_{44} & A_{45} & A_{46} & \dots & 0 & 0 & 0 & 0\\ 0 & 0 & A_{53} & A_{54} & A_{55} & A_{56} & \dots & 0 & 0 & 0 & 0\\ \vdots & \vdots & \vdots & \vdots & \vdots & \vdots & \ddots & \vdots & \vdots & \vdots & \vdots\\ 0 & 0 & 0 & 0 & 0 & 0 & \dots & A_{2n-2,2n-3} & A_{2n-2,2n-2} & A_{2n-2,2n-1} & A_{2n-2,2n}\\ 0 & 0 & 0 & 0 & 0 & 0 & \dots & A_{2n-1,2n-3} & A_{2n-1,2n-2} & A_{2n-1,2n-1} & A_{2n-1,2n}\\ 0 & 0 & 0 & 0 & 0 & 0 & \dots & 0 & 0 & A_{2n,2n-1} & A_{2n,2n}\end{bmatrix}\begin{bmatrix}C_1^+\\C_1^-\\C_2^+\\C_2^-\\\vdots\\C_{n-1}^+\\C_{n-1}^-\\C_n^+\\C_n^-\end{bmatrix}=\begin{bmatrix}\boldsymbol{g}\\0\\0\\0\\0\\\vdots\\0\\0\\0\end{bmatrix} \tag{16}$$

In formula (16), $g=[-A_0\omega^{-2},0]^T$; $A_{11}=\lambda+W1$, $A_{12}=\lambda-W1$;

when i=1,2,…n-1,

$$\begin{aligned}&A_{2i,2i-1}=\lambda+Wi\times e+\ i(x_i),A_{2i,2i}=\lambda-Wi\times e-\ i(x_i)\\&A_{2i,2i+1}=-\lambda+W(i+1)\times e+\ i+1(x_i),A_{2i,2i+2}=-\lambda-W(i+1)\times e-i+1(x_i)\\&A_{2i+1,2i-1}=(\lambda+\ Fi-\lambda_{Fi})\times e+\ i(x_i),A_{2i+1,2i}=(\lambda-Fi-\lambda_{Fi})\times e-i(x_i)\\&A_{2i+1,2i+1}=-\lambda+\ F(i+1)\times e+\ i+1(x_i),A_{2i+1,2i+2}=-\lambda-F(i+1)\times e-i+1(x_i)\\&A_{2n,2n-1}=(\lambda+\ Fn-\lambda_{Fn})\times e+\ n(x_n),A_{2n,2n}=(\lambda-Fn-\lambda_{Fn})\times e-n(x_n)\end{aligned}$$

3. Calculation of Output Voltage

The straight beam in figure.1 is a composite beam, that attaches PZT piezoelectric coating and Cu electrode on the upper and lower surfaces of Si base beam. Its cross section is shown in Figure 2, in which t_b, t_p and t_e refer to the the thickness of Si beam, PZT material and Cu electrode respectively, b is the width of beam.

When the piezoelectric composite beam suffers a bending deformation, it's assumed that the composite layers are tightly bonded without relative slip, then the longitudinal normal stress in the PZT piezoelectric material will generate the piezoelectric effect, its electric displacement D_3 on the y direction is :

$$D_3 = \varepsilon_{33} E_3 + d_{31} \overline{\sigma_p} \tag{17}$$

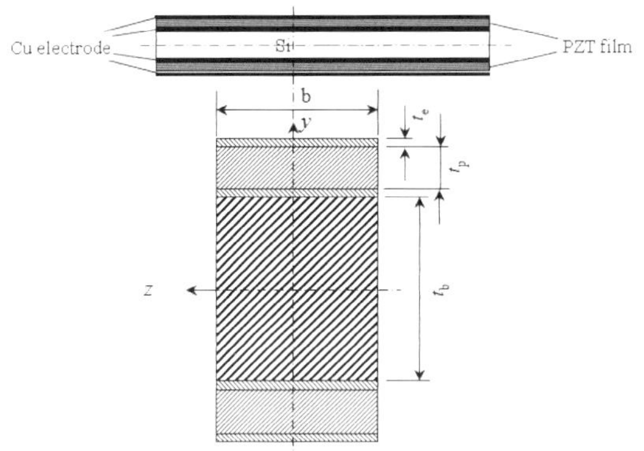

Figure 2 Layered Structure of Symmetrical Cross Section Piezoelectric Composite beam

In the formula (17), ε_{33} and d_{31} respectively refer to the dielectric constant and piezoelectric constant of PZT material, E_3 is the electric field intensity and $\overline{\sigma_p}$ refers to the mean normal stress of PZT piezoelectric material.

When PZT piezoelectric material is used for voltage output, the electric field intensity E_3 =0. Combined with formula (17), it will be:

$$D_3 = d_{31} \overline{\sigma_p} = E_p d_{31} \overline{t_p} w_{i,xx}(x,t) \tag{18}$$

In formula (18), $\overline{t_p}$ refers to the distance between the middle surface of PZT piezoelectric material to the neutrosphere.

Therefore, the Cu electrodes on the upper and lower PZT layer will generate a charge accumulation $Q_i(t)$ as following:

$$Q_i(t) = \int_{x_{i-1}}^{x_i} D_3(x,t) \cdot b \cdot dx = E_p d_{31} b \overline{t_p} \left[w_{i,x}(x_i,t) - w_{i,x}(x_{i-1},t) \right] \tag{19}$$

Taking formula (10) into formula (19), we can get:

$$Q_i(t) = E_p d_{31} b \overline{t_p} e^{j\omega t} \left\{ \begin{aligned} & \left[-j\lambda_i \left(e^{-j\lambda_i x_i} - e^{-j\lambda_i x_{i-1}} \right) \quad -\lambda_i \left(e^{-\lambda_i x_i} - e^{-\lambda_i x_{i-1}} \right) \right] \boldsymbol{C}_i^+ \\ & + \left[j\lambda_i \left(e^{j\lambda_i x_i} - e^{j\lambda_i x_{i-1}} \right) \quad \lambda_i \left(e^{-\lambda_i x_i} - e^{-\lambda_i x_{i-1}} \right) \right] \boldsymbol{C}_i^- \end{aligned} \right\} \tag{20}$$

Defining C_s as the equivalent capacitance of Cu electrodes on the upper and lower PZT layer, then $C_s = \varepsilon_{33} b(x_i - x_{i-1})/t_p$, the voltage $U_i(t)$ of Cu electrodes can be described as:

$$U_i(t) = \frac{Q_i(t)}{C_s} = \frac{E_p d_{31} t_p \overline{t_p}}{\varepsilon_{33}(x_i - x_{i-1})} e^{j\omega t} \left[(\boldsymbol{\Lambda}_i^+)^T \boldsymbol{C}_i^+ + (\boldsymbol{\Lambda}_i)^T \boldsymbol{C}_i^- \right] \quad (i=1,2) \tag{21}$$

In formula (21), $\boldsymbol{\Lambda}_i^+ = \begin{bmatrix} -j\lambda_i \left(e^{-j\lambda_i x_i} - e^{-j\lambda_i x_{i-1}} \right) \\ -\lambda_i \left(e^{-\lambda_i x_i} - e^{-\lambda_i x_{i-1}} \right) \end{bmatrix} \quad \boldsymbol{\Lambda}_i^- = \begin{bmatrix} j\lambda_i \left(e^{\lambda_i x_i} - e^{\lambda_i x_{i-1}} \right) \\ \lambda_i \left(e^{\lambda_i x_i} - e^{\lambda_i x_{i-1}} \right) \end{bmatrix}$

As for the energy harvester that has a segmentation structure in figure.1 and a cross section shape of figure 2, its total output voltage can be described as:

$$U(t) = 2\sum_{i=1}^{n} U_i(t) = \frac{2E_p d_{31}}{\varepsilon_{33}} e^{j\omega t} \sum_{i=1}^{n} \left\{ \frac{t_{pi}}{x_i - x_{i-1}} \left(\frac{t_{bi} + t_{pi}}{2} + t_e \right) \left[(\boldsymbol{\Lambda}_i^+)^T \boldsymbol{C}_i^+ + (\boldsymbol{\Lambda}_i^-)^T \boldsymbol{C}_i^- \right] \right\} \tag{22}$$

4. Performance Analysis

In order to verify the feasibility of the optimization scheme mentioned above, a new harvester whose n equals to 2 has been taken in this part for numerical analysis, as shown in figure.3. The influence of

IOP Conf. Series: Materials Science and Engineering **242** (2017) 012098 doi:10.1088/1757-899X/242/1/012098

upper and lower Cu electrodes will be ignored, namely $t_{e11}=t_{e12}=t_{e21}=t_{e22}=0$. PZT-5H is selected as the piezoelectric material that is only pasted on one side of SI beam, namely $t_{p2}=0$. With b_1、t_{b1}、t_{p11}、b_2、t_{b2}、t_{p12}respectively refer to the width and thickness of beam 1 and 2.

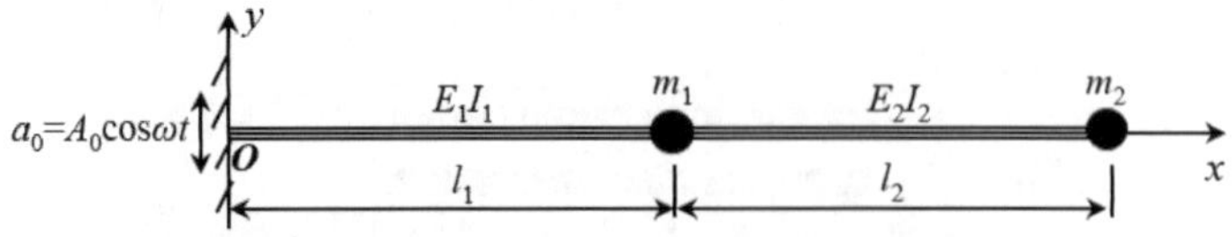

Figure 3 Diagram of Two Beams' Harvester

Table 1. Geometric Parameters of Harvester When n=2

Name	l_1 mm	l_2 mm	m_1 kg	m_2 kg	b_1 mm	b_2 mm	t_{b1} μm	t_{b2} μm	t_{p11} μm	t_{p12} μm
Value	15	6	$1*10^{-6}$	$1.7*10^{-6}$	1	1.8	8	13	3.3	6.2

According to parameters in table 1, the charge quantity and voltage of each beam generated from its deformation are taken as the research object to analyze the changes along the frequency.

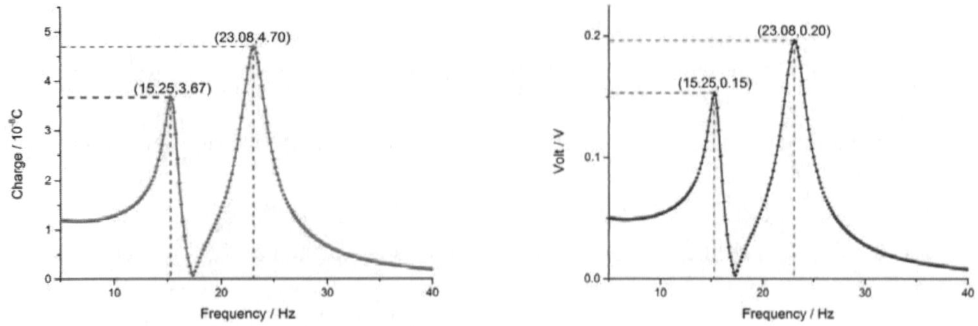

a) Charge-Frequency Curve of PZT-5H **b)** Voltage-Frequency

Figure 4 Variation Curve of Charge Along With the Voltage on Beam 1

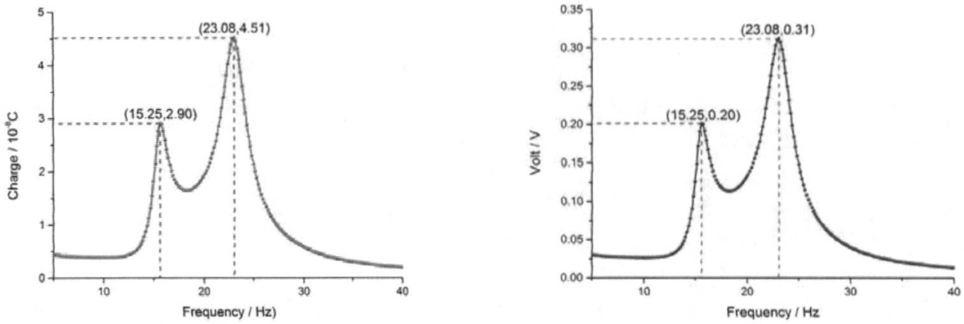

a) Charge-Frequency Curve of PZT-5H **b)** Voltage-Frequency

Figure 5 Variation Curve of Charge Along With the Voltage on Beam 2

Comparing figure 4 and 5, we know that the peak value of charge quantity and voltage for the two beams appear at the frequency 15.25Hz and 23.08Hz, namely the two frequencies are the resonant frequency of harvester. When the frequency is 15.25Hz, the charge quantity of beam 1 is $3.67*10^{-8}$C, and its output voltage is 0.15V. The charge quantity of beam 2 is $2.90*10^{-8}$C, corresponding output voltage is 0.20V. When the frequency is 23.08Hz, the charge quantity of beam 1 is $4.70*10^{-8}$C, and its output voltage is 0.2V. The charge quantity of beam 2 is $4.51*10^{-8}$C, corresponding output voltage is 0.31V.

The new harvester takes each beam as a micro harvester, and two micro harvesters are connected in series, so the charge quantity and output voltage of the whole beam equal to the sum of that on the two beams respectively. The simulation results also proved this relationship, When the frequency is 15.25Hz, the output voltage of harvester is 0.35V, when the frequency is 23.08Hz, the output voltage is 0.51V. If the effective value of output voltage is 80mV, then the energy harvester has two frequencies with in a frequency scope of 10Hz~50Hz, which are 9.63Hz and 29.95Hz. The harvester also forms an effective operating frequency band between the two frequencies, which is 20.32Hz.Thus it can be seen that the new energy harvester consisted of two beams and two mass blocks indeed meets the demand of reducing resonant frequency and broadening working band. It can match the low frequency vibration environment more perfectly and provide sufficient power for the equipment to be supplied.

5. Conclusion

This paper designed a new structure consisted of n beams and n mass blocks, that can meet the demands of reducing resonant frequency and broadening working band. Considering the scale effects, transverse vibration displacement of each beams have been solved, the calculation formula of output voltage and flexural rigidity of cross section for the new harvester have been derived with external excitation. Meanwhile, an example has been taken when n=2, the change laws of concentrated charge and output voltage for PZT-5H surface along the frequency have been analyzed.There are two resonant frequencies within 50Hz(namely 15.25Hz and 23.08Hz) that can better match the low frequency ambient vibration. When the output voltage is 80mV, the effective working band of harvester is 20.32Hz. The resonant frequency and working band of this harvester are optimized. Moreover it can adjust the parameters according to actual power demand, which greatly optimizes the performance of micro power supply and better meets the energy supply requirements of the equipment to be supplied.

References

[1] Kuehne I, Frey A Marinkovic D 2008 *Sensors and Actuators* A: *Phys* **142(1)** 263-269
[2] Roundy S, Wright P K, Rabaey J 2003 *Computer Communications Ubiquitous Computing* **26(11)**1131-1144
[3] Piezoelectric energy converter for electronic implants 1969 *Google Patents*
[4] Jeon Y B, Sood R, Jeong J H, et al. 2005 *Sensors and Actuators* A *Phys* **122(1)** 16-22
[5] WU Xiao-ming, FANG Huan-jun, LIN Jian-hui, et al. 2008 *Journal of function materials and devices* 14(2)
[6] Renaud M, Karakaya K, Sterken T, et al. 2008 *Sensors and Actuators* A: *Phys* 145 – 146, 380-386
[7] Gafforelli G, Ardito R, Corigliano A 2015 *Composite Structures* **127(0)** 369-381
[8] Fang H, Liu J, Xu Z, et al 2006 *Microelectronics Journal* **37(11)**: 1280-1284
[9] Trawiński W, Bobiński J, Tejchman J 2016 Engineering *Fracture Mechanics* 168 Part A, 204-226
[10] LIU Ying, WANG YAN-fen, LI Guang 2014 *Optics and Precision Engineering* **22(9)**:2476-2482
[11] Zhidong W, Rui H, Daokun Z 2016 *Piezoelectric and Acoustooptics* 320-323

IOP Conf. Series: Materials Science and Engineering **242** (2017) 012099 doi:10.1088/1757-899X/242/1/012099

Research on cutting path optimization of sheet metal parts based on ant colony algorithm

Z Y Wu[1], H Ling[1], L Li[2], L H Wu[1] and N B Liu[1]

[1] School of Mechanical & Electrical Engineering, Henan University of Technology, Zhengzhou, China
[2] Automobile Engineering Department, Henan Polytechnic, Zhengzhou, China

Corresponding author's e-mail: wzhaoyun@163.com

Abstract. In view of the disadvantages of the current cutting path optimization methods of sheet metal parts, a new method based on ant colony algorithm was proposed in this paper. The cutting path optimization problem of sheet metal parts was taken as the research object. The essence and optimization goal of the optimization problem were presented. The traditional serial cutting constraint rule was improved. The cutting constraint rule with cross cutting was proposed. The contour lines of parts were discretized and the mathematical model of cutting path optimization was established. Thus the problem was converted into the selection problem of contour lines of parts. Ant colony algorithm was used to solve the problem. The principle and steps of the algorithm were analyzed.

1. Introduction

In recent years, the cutting path optimization problem of sheet metal parts becomes a hot research topic in the field of computer-aided manufacturing and control. NC laser cutting machine has advantages of high speed and precision, but a large number of empty travel will be produced in the process of cutting the workpiece, which seriously affects the cutting efficiency. By the method of path optimization, shortening the distance of empty travel in the cutting process has the great practical significance to shorten the cutting time and improve machining efficiency.

In this field, the domestic and foreign scholars have made some related research. Han et al. [1] used the simulated annealing algorithm to optimize the cutting path problem of irregular parts, and considered the influence of cutting temperature. Pan et al. [2] proposed a 'punching - cutting' planning CAD/CAM system based on the knowledge base. Huixia Liu et al. [3] used the hierarchical planning three-step algorithm to solve the cutting path optimization problem. Huiping Sun et al. [4] transformed the cutting path optimization problem into the Traveling Salesman Problem by using the method of adding nodes, and used the genetic algorithm to solve the problem. Nini Li et al. [5] put forward a new laser cutting path optimization algorithm with the combination of the local search method and genetic algorithm.

These literatures studied the cutting path optimization problem from the different angles. However, their basic ideas were similar. They all transformed the problem into the TSP and used the relative algorithms to solve it according to the constraint rule of serial cutting. The constraint rule of serial cutting has shortcomings in solving the actual cutting path optimization problem.

In this paper, the cutting path optimization of sheet metal parts with no holes and public edges are taken as the research object. The cutting constraint rule and the associated optimization algorithm will

Content from this work may be used under the terms of the Creative Commons Attribution 3.0 licence. Any further distribution of this work must maintain attribution to the author(s) and the title of the work, journal citation and DOI.
Published under licence by IOP Publishing Ltd

be mainly discussed. The aim of this paper is resolving some key problems in the CNC laser cutting path optimization.

2. Description of cutting path optimization problem

As shown in figure 1, M parts with no holes and public edges in a sheet need to be cut. The laser head starts from the bottom left corner of the sheet (starting point O), in turn cuts out the contour shapes of each part, and then returns to the origin point O, thus the process of cutting all the parts is completed. In the process of cutting, the laser head is in the state of 'light-cutting' when it moves along the contour lines, and it is in the state of 'not-cutting' when it moves between the contour lines (that is empty travel). Because the length of the contours and the speed of the laser head are constant, the overall length of cutting paths can be reduced only by shortening the distance of empty travel between the contours. Therefore, the goal of cutting path optimization is to find a best laser operation path, so that the distance of empty travel is the shortest in the cutting process.

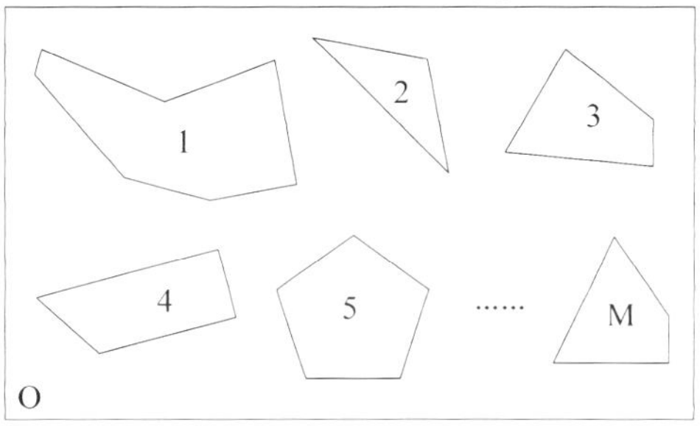

Figure 1. Description of cutting path optimization problem

3. Mathematical model of optimization problem

3.1. *Constraint rules of parts cutting*

In the most current literature, the constraint rules of serial cutting parts were adopted. The laser head must run along the contour of a part, and then cut the next part. So when the cutting path was optimized and analyzed, a point could be picked up from the contours as the punch point (cutting starting point and end point), and the point can replace the contour of parts. So the contours of parts to be cut can be simplified to a point set in the plate plane. The cutting path optimization problem was transformed into solving the shortest distance between point sets, known as the TSP [7]. Most of the literature first determined the punch points of the contours by the nearest neighbor algorithm, then used all kinds of related algorithm to solve the TSP problem. This method can make the cutting path optimization problem be resolved. But after the further study, it was found that the method could cause certain influence to the optimization results. As shown in figure 2 (a), there are two parts to be cut in a sheet. The vertices of part 1 are (P_1^1, P_2^1, P_3^1, P_4^1, P_5^1), and the vertices of part 2 are (P_1^2, P_2^2, P_3^2, P_4^2). Using the nearest neighbor algorithm, the punch points can be determined: P_1^1 and P_1^2. The complete cutting path can be expressed as: O→ P_1^1 →[P_2^1 → P_3^1 → P_4^1 → P_5^1 → P_1^1]→ P_1^2 →[P_2^2 → P_3^2 → P_4^2 → P_1^2]→O. The distance of the empty travel is: $S_e = \overline{OP_1^1} + \overline{P_1^1P_1^2} + \overline{P_1^2O}$. If not restricted by the constraint rules of serial cutting, the laser head can be allowed to cut other parts before a part is not yet finished. The cutting path was shown in figure 2 (b): O→ P_1^1 →[P_2^1 → P_3^1 → P_4^1]→ P_1^2 →[P_2^2 → P_3^2 → P_4^2 → P_1^2] → P_4^1 → [P_5^1 → P_1^1] → O. The distance of the empty travel is: $S_e' = \overline{OP_1^1} + \overline{P_4^1P_1^2} + \overline{P_1^2P_4^1} + \overline{P_1^1O} = 2\overline{OP_1^1} + 2\overline{P_4^1P_1^2}$. It is obvious that the latter is shorter. Thus, under the influence of the constraint rules of serial cutting, the traditional algorithm sometimes can not find the optimal solution. Aimed at the problem, this paper put forward the constraint rules of cross cutting. Namely in the process of cutting parts, the laser head could freely transfer between the contours of the parts. The cutting path in figure 2 (b) is an instance of the cross cutting rules. The rules can make the optimization algorithm find the better cutting path.

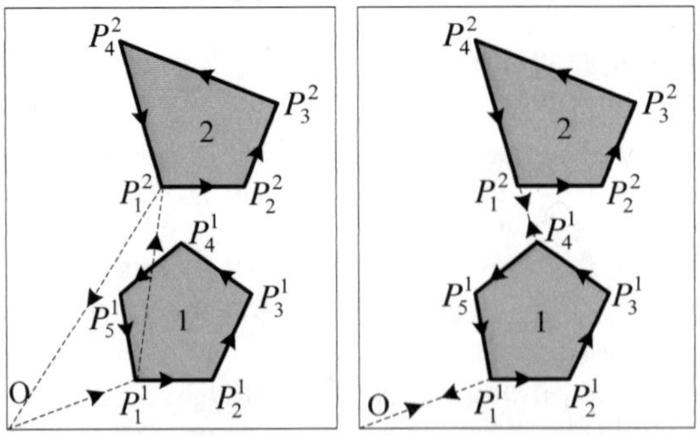

Figure 2. Comparison of two cutting path

3.2. *Mathematical model of cutting path optimization problem*

There are M parts to be cut in a sheet. The line group of part 1 is: $\{L_{11}, L_{12}, L_{13}, \cdots, L_{1m}\}$. The line group of part 2 is: $\{L_{21}, L_{22}, L_{23}, \cdots, L_{2n}\}$. The line group of part M is: $\{L_{M1}, L_{M2}, L_{M3}, \cdots, L_{Mp}\}$. According to the requirement of cutting parts, no matter what order, as long as the laser head walks along all the discretization contour lines of parts, all parts will be finished. At this point, the cutting path optimization problem was transformed into: how to plan the cutting order of the contour lines in order to make the empty travel become the shortest. According to certain rules, it is supposed that the cutting order sequence is obtained, as shown in table 1. The cutting path of laser head can be represented as: O→[P_{11}→P_{12}]→[P_{21}→P_{22}]→……→[P_{N1}→P_{N2}]→O. The empty travel path is:

$$S_e = \overline{OP_{11}} + \overline{P_{12}P_{21}} + \overline{P_{22}P_{31}} + \cdots + \overline{P_{N-1,2}P_{N1}} + \overline{P_{N2}O} = \overline{OP_{11}} + \sum_{i=1}^{N-1} \overline{P_{i,2}P_{i+1,1}} + \overline{P_{N2}O} \quad (1)$$

So, the mathematical model of cutting path optimization problem is:

$$\min S_e = \sqrt{x_{11}^2 + y_{11}^2} + \sqrt{x_{N2}^2 + y_{N2}^2} + \sum_{i=1}^{N-1} \sqrt{(x_{i+1,1} - x_{i,2})^2 + (y_{i+1,1} - y_{i,2})^2} \quad (2)$$

$$\text{s.t. } x_{ij} \geq 0 \ (i = 1,2,\cdots,N; j = 1,2)$$

$$y_{ij} \geq 0 \ (i = 1,2,\cdots,N; j = 1,2)$$

Table 1. Cutting order of contour lines

Order	Line	Point1	Point2	Point 1 X Coordinate	Point 1 Y Coordinate	Point 2 X Coordinate	Point 2 Y Coordinate
1	L_1	P_{11}	P_{12}	x_{11}	y_{11}	x_{12}	y_{12}
2	L_2	P_{21}	P_{22}	x_{21}	y_{21}	x_{22}	y_{22}
3	L_3	P_{31}	P_{32}	x_{31}	y_{31}	x_{32}	y_{32}
4	L_4	P_{41}	P_{42}	x_{41}	y_{41}	x_{42}	y_{42}
……	……	……	……	……	……	……	……
N	L_N	P_{N1}	P_{N2}	x_{N1}	y_{N1}	x_{N2}	y_{N2}

4. Optimization algorithm of cutting path

In order to solve the cutting path optimization problem, the optimization algorithm based on ant colony algorithm was proposed in this paper. The steps of the algorithm include:

Step1. Initializing variables. Set number of cycle NC, count of ant colony u, pheromone matrix $\boldsymbol{Info}$, Min-Max limited value of pheromone $\tau_{\min}$, $\tau_{\max}$, volatile coefficient of pheromone ρ, index of cycle $nc=1$, coordinate matrix of contour lines $NPLM$ generating by (3).

$$\begin{cases} NPLM(2i-1,1:4)=PLM(i,1:4) \\ NPLM(2i,1:2)=PLM(i,3:4) \\ NPLM(2i,3:4)=PLM(i,1:2) \end{cases} \tag{3}$$

Step2. Starting nc iterative calculation. $\boldsymbol{Path}$ and S are all set zero matrix, $k=1$.

Step3. Traversing the ant colony. For ant k, its location is the origin point, $x=0$, $y=0$. $Allowed$ is set 'one' matrix. $t=1$.

Step4. Selecting the lines. The matrix of heuristic factor $PM_{2N\times 1}$ is obtained by formula (4), and then it is processed by formula (5).

$$PM(i)=\begin{cases} 1\Big/\sqrt{[x-NPLM(i,1)]^2+[y-NPLM(i,2)]^2}, \\ \qquad\qquad\qquad\qquad Allowed(i)=1 \\ 0, \ Allowed(i)=0 \end{cases} \tag{4}$$

$$PM=PM/sum(PM) \tag{5}$$

The selecting probability matrix of line $PK_{2N\times 1}$ is obtained by formula (6), and then it is processed by formula (7).

$$PK(j)=Info(j,t)^{\alpha}\,PM(j)^{\beta} \tag{6}$$

$$PK=PK/sum(PK) \tag{7}$$

The index of line selected by ant is calculated by the roulette rules.

Step5. Triggering the event. Read the endpoint coordinates of selected lines.

$$\begin{cases} x_1=NPLM(h,1) \\ y_1=NPLM(h,2) \\ x_2=NPLM(h,3) \\ y_2=NPLM(h,4) \end{cases} \tag{8}$$

Calculate the distance of empty travel.

$$S(k)=S(k)+\sqrt{(x-x_1)^2+(y-y_1)^2} \tag{9}$$

Save the path: $Path(k,t)=h$. Update the coordinates of current point of the ant: $x=x_2$, $y=y_2$. Update $Allowed$: $Allowed(h)=0$.

Step6. Judge whether the ant k meet the termination conditions. If $t<N$, then set $t=t+1$, and go to the Step4. Otherwise, go to the Step7.

Step7. The ant k returns to the origin point. Put the travel returned to the origin point into the empty schedule.

$$S(k)=S(k)+\sqrt{x^2+y^2} \tag{10}$$

Step8. Judge whether all the ants have completed the search. If $k<u$, then set $k=k+1$, and go to Step3. If $k=u$, then go to Step9.

Step9. Update the global optimal solution. Find the local optimal vector $\boldsymbol{Y}'$ and the optimal path matrix $\boldsymbol{PathBest}'$, and put $\boldsymbol{Y}'$ into $\boldsymbol{RCourse}$.

Step10. Update the pheromone matrix by the formula (11).

$$Info(i,j)=(1-\rho)\cdot Info(i,j)+\Delta\tau_{ij}^{best} \tag{11}$$

Step11. Judge whether the iteration of algorithm is over. If $nc<NC$, then set $nc=nc+1$, and go to Step2. If $nc=NC$, then go to Step12.

IOP Conf. Series: Materials Science and Engineering **242** (2017) 012099 doi:10.1088/1757-899X/242/1/012099

Step12. Output the coordinates points of the optimal path. The coordinates matrix of the optimal path $PathBest\_XY$ is generated, and the calculated results is output.

5. Conclusion

The traditional CNC laser cutting path optimization method of sheet metal pieces was improved in this paper. The constraint rules of cross cutting and discretization method of contour lines were proposed. Based on these, the cutting path optimization mathematical model was established, and the optimization method was designed by the ant colony algorithm.

Acknowledgments

This work was supported by the National Nature Science Foundation of China (Grant No. 51405133), the Province Science and Technology Research Projects of Henan(Grant No. 172102210213), and the Fundamental Research Funds for the Henan Provincial Colleges and Universities in Henan University of Technology(Grant No. 2016RCJH02).

References

[1] Gukchan H and Suckjoo N 1999 *Journal of Manufacturing Processes* **1** 62-70
[2] Pan M and Rao Y 2009 *Knowledge-Based Systems* **22** 368-75
[3] Huixia L, Xiao W and Lan C 2004 *Journal of CAD & CG* **16** 660-5
[4] Huiping S, Jian L and Weigang G 2008 *Transactions of the Chinese Society for Agricultural Machinery* **39** 158-60
[5] Nini L, Zhangwei C and Shize C 2010 *Computer Engineering and Applications* **46** 234-6

IOP Conf. Series: Materials Science and Engineering **242** (2017) 012100 doi:10.1088/1757-899X/242/1/012100

Multi-scale method for ill-conditioned linear systems

Jinmei Wang[a] and Ke Wang[b,*]

[a] School of Mathematical Sciences, University of Jinan, Jinan 250022, P.R. China
[b] Department of Mathematics, Shanghai University, Shanghai 200444, P.R. China

The first author's e-mail address: ss_wangjm@ujn.edu.cn
*The corresponding author, e-mail address: kwang@shu.edu.cn

Abstract. A multi-scale method is presented for solving ill-conditioned linear systems of equations. Numerical examples are provided to show the effectiveness and efficiency of the proposed method compared with traditional iterative methods.

1. Introduction

The linear system of equations

$$Ax = b, \tag{1}$$

where $A = (a_{ij}) \in \Re^{n\times n}$ is nonsingular, $x = (x_j) \in \Re^n$ unknown and $b \in \Re^n$ known, plays an important role in science and engineering computing [1]. There are two approaches for solving (1) — direct methods and iterative methods [2]. For large sparse systems (1), the iterative methods are popular and efficient. Many iterative methods have been presented, such as Jacobi, Gauss-Seidel, SOR, Conjugate Gradient, Subspace, Preconditioned, GMRES methods [3]. However, it's difficult to develop effective methods for ill-conditioned linear systems whose coefficient matrices have large condition numbers [4-22].

In this paper a multi-scale method is suggested to solve ill-conditioned system (1). Through the numerical examples it is showed that the new method is effective and more efficient than the others.

The paper is organized as follows. First, the method is presented. Then, numerical analysis is conducted. Finally, the numerical examples are provided and the conclusion is made.

2. The multi-scale method

If the coefficient matrix A of the equations (1) is ill-conditioned, the system (1) can be solved by the following equations

$$\widetilde{A}\widetilde{x} = b, \tag{2}$$

where $\widetilde{A}$ and $\widetilde{x}$ are as follows

$$\widetilde{A} = \begin{pmatrix} \frac{1}{\|\alpha_1\|_2} a_{11} & \frac{1}{\|\alpha_2\|_2} a_{12} & \cdots & \frac{1}{\|\alpha_n\|_2} a_{1n} \\ \frac{1}{\|\alpha_1\|_2} a_{21} & \frac{1}{\|\alpha_2\|_2} a_{22} & \cdots & \frac{1}{\|\alpha_n\|_2} a_{2n} \\ \vdots & \vdots & \ddots & \vdots \\ \frac{1}{\|\alpha_1\|_2} a_{n1} & \frac{1}{\|\alpha_2\|_2} a_{n2} & \cdots & \frac{1}{\|\alpha_n\|_2} a_{nn} \end{pmatrix}, \quad \widetilde{x} = \begin{pmatrix} \|\alpha_1\|_2 x_1 \\ \|\alpha_2\|_2 x_2 \\ \vdots \\ \|\alpha_n\|_2 x_n \end{pmatrix},$$

here α_j, $j = 1,2,\cdots,n$ is the jth column of A.

Content from this work may be used under the terms of the Creative Commons Attribution 3.0 licence. Any further distribution of this work must maintain attribution to the author(s) and the title of the work, journal citation and DOI.
Published under licence by IOP Publishing Ltd

With (2), the following multi-scale method is obtained for system (1).

THE MULTI-SCALE ALGORITHM

Step 1 solve $\widetilde{A}\widetilde{x} = b$ with iterative methods;

$$Step\,2 \quad x = \begin{pmatrix} \frac{1}{\|\alpha_1\|_2}\widetilde{x}_1 \\ \frac{1}{\|\alpha_2\|_2}\widetilde{x}_2 \\ \vdots \\ \frac{1}{\|\alpha_n\|_2}\widetilde{x}_n \end{pmatrix}.$$

The following remarks show that the multi-scale method could improve the condition number of the coefficient matrix.

Remark 1. From (2), it yields

$$\widetilde{A} = \begin{pmatrix} a_{11} & a_{12} & \cdots & a_{1n} \\ a_{21} & a_{22} & \cdots & a_{2n} \\ \vdots & \vdots & \ddots & \vdots \\ a_{n1} & a_{n2} & \cdots & a_{nn} \end{pmatrix} \begin{pmatrix} \frac{1}{\|\alpha_1\|_2} & 0 & \cdots & 0 \\ 0 & \frac{1}{\|\alpha_2\|_2} & \cdots & 0 \\ \vdots & \vdots & \ddots & \vdots \\ 0 & 0 & \cdots & \frac{1}{\|\alpha_n\|_2} \end{pmatrix} \stackrel{\Delta}{=} AD.$$

Thus, for the condition number with 1-norm, denoted by

$$d_{\max} = \max\{\|\alpha_1\|_2, \|\alpha_2\|_2, \cdots, \|\alpha_n\|_2\},\ d_{\min} = \min\{\|\alpha_1\|_2, \|\alpha_2\|_2, \cdots, \|\alpha_n\|_2\},$$

the following holds

$$\begin{aligned} \mathrm{cond}(\widetilde{A}) &= \|\widetilde{A}\|\|\widetilde{A}^{-1}\| = \|AD\|\|D^{-1}A^{-1}\| \\ &\le \|A\|\|D\|\|D^{-1}\|\|A^{-1}\| = \tfrac{d_{\max}}{d_{\min}}\|A\|\|A^{-1}\| = \tfrac{d_{\max}}{d_{\min}}\mathrm{cond}(A). \end{aligned} \tag{3}$$

Therefore, if $d_{\min}$ is very close to $d_{\max}$, then the condition number cond($\widetilde{A}$) could be no more than the condition number cond(A).

Remark 2. If D is taken as an orthogonal matrix, then the following can be obtained from (3)

$$\mathrm{cond}(\widetilde{A}) \le \mathrm{cond}(A).$$

3. Numerical examples

In this section, numerical examples are given to illustrate the multi-scale method. The initial guess is 0 and the stopping criterion is

$$\frac{\|r^{(k)}\|_2}{\|r^{(0)}\|_2} < 10^{-6},$$

where $r^{(k)}$ is the residual vector after k iterations. The results are compared with classical iterative methods.

Example 1. Consider the following system

$$\begin{pmatrix} 1 & 1 & 0 \\ 0 & 100 & 1 \\ 1 & 0 & 10000 \end{pmatrix} \begin{pmatrix} x_1 \\ x_2 \\ x_3 \end{pmatrix} = \begin{pmatrix} 2 \\ 101 \\ 10001 \end{pmatrix}.$$

The coefficient matrix A is almost singular. The condition number of A is 1.0×10^4 ($\|A\|_2\|A^{-1}\|_2$). The exact solution is $(1,\ 1,\ 1)^T$. The condition number of $\widetilde{A}$ is 2.4. The numerical results are listed in Table 1 compared with classical Jacobi and Gauss-Seidel methods.

Table 1. Iterations (IT) and relative error (ERR) for Example 1.

Jacobi		Gauss-Seidel		Multi-scale	
IT	ERR	IT	ERR	IT	ERR
5	5.8e-9	4	5.8e-9	6	1.0e-12

Example 2. Consider the following system

$$\begin{pmatrix} 1 & \frac{1}{2} & \frac{1}{3} \\ \frac{1}{2} & \frac{1}{3} & \frac{1}{4} \\ \frac{1}{3} & \frac{1}{4} & \frac{1}{5} \end{pmatrix}\begin{pmatrix} x_1 \\ x_2 \\ x_3 \end{pmatrix} = \begin{pmatrix} 1 \\ 1 \\ 1 \end{pmatrix}.$$

The coefficient matrix A is almost singular. The condition number of A is 524.1 ($\|A\|_2\|A^{-1}\|_2$). The exact solution is $(3,\ -24,\ 30)^T$. The condition number of $\widetilde{A}$ is 373.7. The numerical results are listed in Table 2 compared with classical Jacobi and Gauss-Seidel methods.

Table 2. Iterations (IT) and relative error (ERR) for Example 2.

Jacobi		Gauss-Seidel		Multi-scale	
IT	ERR	IT	ERR	IT	ERR
1305	NaN	907	2.6e-8	880	4.4e-8

4. Conclusions

In this paper, a multi-scale method is presented to solve system of linear equations (1) which is ill-conditioned. The method improves the condition number of the ill-conditioned system. Numerical results show that the multi-scale method is effective and efficient for linear ill-conditioned problems.

Acknowledgement

This work was supported by National Natural Science Foundation of China (11301330), Gaoyuan Discipline of Shanghai — Mathematics, Shanghai City, China and Teaching Reform Project of Shandong Province (Z2016Z007).

References

[1] Lay D C, Lay S R and McDonald J J 2016 *Linear Algebra and Its Applications, 5th Edition* (Boston: Pearson)
[2] Demmel J W 1997 *Applied Numerical Linear Algebra* (Philadelphia : SIAM)
[3] Jin X Q and Wei Y M 2004 *Numerical Linear Algebra and Its Applications* (Beijing: Science Press).
[4] You C 1992 *Journal of Lanzhou University* (*Natural Sciences*) The successive adjustment elimination (SAE) method of solving some systems of linear equation with ill-condition **28** 35-42
[5] You C and Wei H 1998 *Journal of Gansu Sciences* Modified method of successive adjustment elimination — MSAE method **10** 35-39
[6] Volokh K Y and Vilnay O 2000 *Applied Mathematics Letters* Pin-pointing solution of ill-conditioned square systems of linear equations **13** 119-124
[7] Wu X Y, Shao R and Zhu Y R 2002 *Computers & Mathematics with Applications* New iterative improvement of a solution for an ill-conditioned system of linear equations based on a linear

dynamic system **44** 8-9
[8] Estatico C 2002 *BIT Numerical Mathematics* A class of filtering superoptimal preconditioners for highly ill-conditioned linear systems **42** 753-778
[9] Brezinski C, Redivo-Zaglia M, Rodriguez G and Seatzu S 2003 *Numerische Mathematik* Multi-parameter regularization techniques for ill-conditioned linear systems **94** 203-228
[10] Qian A L 2003 *Journal of Xianning College* The elimination by step-adjust to solve morbid equations **23** 32-33
[11] Rodriguez G, Seatzu S and Theis D 2003 *Numerical Algorithms* A new technique for ill-conditioned linear systems **33** 433-442
[12] Wu X 2007 *Applied Mathematics and Computation* Error analysis for the predictor-corrector process relating to ill-conditioned linear system of equations **186** 530-534
[13] Hoang N S and Ramm AG 2008 *Inverse Problems in Science and Engineering* Solving ill-conditioned linear algebraic systems by the dynamical systems method **16** 617-630
[14] Salahi M 2009 *Applied Mathematics & Information Sciences* On ill-conditioned linear systems and regularized GMRES method **3** 45-51
[15] Hoang N S and Ramm A G 2010 *Acta Applicandae Mathematicae* Dynamical systems gradient method for solving ill-conditioned linear algebraic systems **111** 189-204
[16] Salkuyeh D K and Fahim A 2011 *International Journal of Computer Mathematics* A new iterative refinement of the solution of ill-conditioned linear system of equations **88** 950-956
[17] Brezinski C, Novati P and Redivo-Zaglia M 2012 *Journal of Computational and Applied Mathematics* A rational Arnoldi approach for ill-conditioned linear systems **236** 2063-2077
[18] Ku C Y 2013 *CMES-Computer Modeling in Engineering & Sciences* A novel method for solving ill-conditioned systems of linear equations with extreme physical property contrasts **96** 409-434
[19] Ma Y and Li Y 2014 *Proceedings of the 4th International Conference on Intelligent System and Applied Material* Cuckoo Search Algorithm for solving ill-conditioned linear equations **1049** 1367-1370
[20] Wang D and Li Y 2015 *Proceedings of the 2015 International Symposium on Computers and Informatics* Ill-conditioned linear equations and its algorithm **13** 2216-2221
[21] Adib M 2016 *Calcolo* A Krylov subspace generalization of the improved Wilkinson's algorithm for ill-conditioned linear systems **53** 51-58
[22] Yeung M C, Douglas C C and Lee L 2017 *Journal of Computational Science* A spectral projection preconditioner for solving ill conditioned linear systems **20** 177-186

Modeling and simulation of flow field in giant magnetostrictive pump

Yapeng Zhao[a], Shiyong Ren[a], Quanguo Lu[b]

[a]The 713th Research Institute of China Shipbuilding Industry Corporation, Zhengzhou, China

[b]Institute of Micro/Nano Actuation and Control, Nanchang Institute of Technology, Nanchang, China

Abstract. Recent years, there has been significant research in the design and analysis of giant magnetostrictive pump. In this paper, the flow field model of giant magnetostrictive pump was established and the relationship between pressure loss and working frequency of piston was studied by numerical simulation method. Then, the influence of different pump chamber height on pressure loss in giant magnetostrictive pump was studied by means of flow field simulation. Finally, the fluid pressure and velocity vector distribution in giant magnetostrictive pump chamber were simulated.

1. Introduction

The hydraulic pump which converts mechanical energy into hydraulic energy is the power source and core components of hydraulic system. Traditional hydraulic pumps require the use of internal combustion engine or electromotor to provide power. However, the traditional pumps are always heavy and huge, and their energy loss is always very huge in energy transmission process. Compared with other smart materials, the rare earth giant magnetostrictive material has the features of large strain, big output force and fast response et al[1]. It provides the possibility to change Hydraulic pump drive mode. Using Terfenol-D material to drive the fluid directly, the complexity of energy transduction is reduced.

Recent years, there has been significant research activity in the development of hydraulic pump driven by Terfenol-D material. A magnetostrictive stack based hybrid pump was developed by Gerver et al[3]. This design used Terfenol-D as the driving element and employed stroke amplification to obtain high flow rates. The pump achieved a flow rate of 15 ml/s with a power input of 41 W. CSA Engineering reported the development of several hybrid actuators, including a magnetostrictive hydraulic actuator with power output exceeding 100 W[4]. The maximum no-load flow rate was 25.2 ml/s and the maximum developed pressure was 11.7 MPa.

Inertial loads, fluid viscosity and compressibility combine with loss mechanisms inherent in the active material limit the effective bandwidth of the driving pump and the total pump output power[6]. Care must be taken in the design and analysis of giant magnetostrictive pump to match the mechanical impedance characteristics of the Terfenol-D material to the fluid transmission so that maximum efficiency of operation is obtained. In this paper, we have carried out the design, modeling and simulation of the flow field in giant magnetostrictive pump to find some laws in the design method.

Content from this work may be used under the terms of the Creative Commons Attribution 3.0 licence. Any further distribution of this work must maintain attribution to the author(s) and the title of the work, journal citation and DOI.
Published under licence by IOP Publishing Ltd

2. Flow field model of giant magnetostictive pump

As shown in Figure 1, the giant magnetostrictive pump has several main components: Terfenol-D rod, piston, pump body, pump chamber, inlet tube, outlet tube and check valve. The piston is driven by a Terfenol-D rod to compress the liquid in the pump chamber. The diameter of the giant magnetostrictive pump chamber is also the diameter of the piston d_p. The height of the pump chamber is h_{ch}. The diameter of inlet tube and outlet tube both are d_t and the distance from the center axis of pump chamber to outlet tube or inlet tube is *R*.

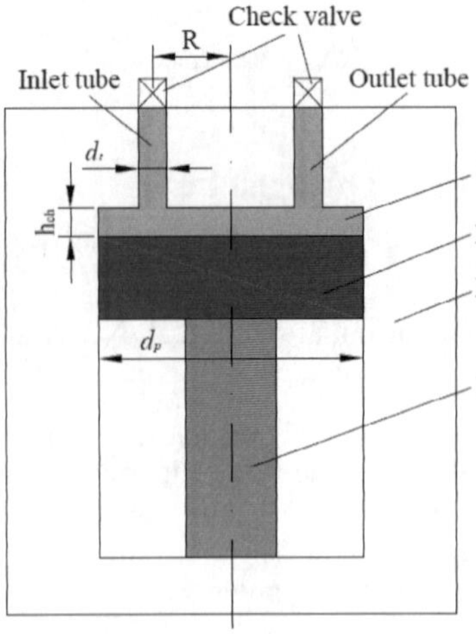

Figure 1. Structure diagram of giant magnetostrictive pump

It is assumed that the liquid in pump chamber is steady. The displacement of the piston is equal to the amount of deformation of the giant magnetostrictive rod Δl.The output flow of pump chamber is following equation:

$$Q = \Delta l \frac{\pi d_p^2}{4} f \quad (1)$$

Where, *f* is the working frequency of piston.

For the liquid in the pump chamber, ρ is a constant, v_r、 v_θ and v_z are their velocity components in the *r*、 θ and *z* directions respectively. By the law of conservation of mass, simplifying the continuity equation of fluid mechanics, we can get its expression in cylindrical coordinate system as follows:

$$\frac{v_r}{r} + \frac{\partial v_r}{\partial r} = 0 \quad (2)$$

By the law of conservation of momentum, the momentum equation of hydrodynamics is transformed into the cylindrical coordinate system, and the expression of steady state incompressible flow can be obtained[7].

$$\rho v_r \frac{\partial v_r}{\partial r} = -\frac{\partial p_{ch}}{\partial r} + \mu \frac{\partial^2 v_r}{\partial z^2} \quad (3)$$

Where ρ is the density of the liquid, μ is the dynamic viscosity of the liquid, P_{ch} is the pressure of the liquid in pump chamber.

By the law of conservation of mass, the velocity distribution of the liquid in the radial direction should be

$$v_r = \frac{C_1 z(h_{ch} - z)}{r_p} \quad (4)$$

Where C_1 is the speed coefficient, r_p is radius of piston.

Integrate z above this equation,

$$\int_0^{h_{ch}} v_r = \int_0^{h_{ch}} \frac{C_1 z(h_{ch} - z)}{r_p} = \frac{Q}{2\pi r_p} \quad (5)$$

Where Q denotes the average flow in the height direction of pump chamber.

Solve the above formula, you can get

$$C_1 = \frac{3Q}{\pi h_{ch}^3} \tag{6}$$

Substituting equations (4) and (6) into equation (3), the radial pressure gradient of the liquid in pump chamber can be obtained.

$$\frac{\partial p_{ch}}{\partial r} = \frac{9\rho Q^2}{\pi^2 h_{ch}^6 r^3} z^2 (h_{ch} - z)^2 - \frac{6Q\mu}{\pi h_{ch}^3 r} \tag{7}$$

Integrate the above equation, you can get the pressure loss in pump chamber,

$$\Delta p_{ch} = \frac{6Q\mu}{\pi h_{ch}^3} ln\left(\frac{d_p}{d_t}\right) + \frac{\rho}{60}\left(\frac{6Q}{\pi h_{ch} d_p}\right)^2 \left[\left(\frac{d_p}{d_t}\right)^2 - 1\right] \tag{8}$$

Where d_p denotes the pump chamber diameter, d_t denotes the inlet tube diameter.

The first term in equation (8) represents the pressure loss due to the viscosity of the liquid; the second term indicates the pressure loss due to the change in the radial flow area when the flow of liquid pass from pump chamber to outlet tube.

Similarly, the pump chamber is simplified, considering only its two-dimensional state, ignoring the radial area of the change, the fluid flow path as a rectangular path, you can get its two-dimensional model.

In [8], the flow expression of the slit flow is given:

$$q = \frac{bh^3 \Delta p}{12\mu l} \tag{9}$$

Where $l = d_p - d_t$, b=12.5.

Besides, q is the flow rate per unit length and its expression can be obtained by Q

$$q = \frac{Q}{\frac{1}{2}\pi(d_p+d_t)} \tag{10}$$

Combine Equation (9) and Equation (10), the equation of pump chamber pressure loss is as the following,

$$\Delta p_{ch}' = \frac{48Q\mu}{25\pi h_{ch}^3}\left(\frac{d_p-d_t}{d_p+d_t}\right) \tag{11}$$

Substitute equation (1) into equations (11), the expression of pressure loss in the two-dimensional space is obtained when the pump chamber is discharged,

$$\Delta p_{ch}' = \frac{12\mu d_p^2 f \Delta l}{25 h_{ch}^3}\left(\frac{d_p-d_t}{d_p+d_t}\right) \tag{12}$$

Substitute equation (1) into equations (8), the expression of pressure loss in the three-dimensional space is obtained when the pump chamber is discharged,

$$\Delta p_{ch} = \frac{3d_p^2 \mu f \Delta l}{2h_{ch}^3} ln\left(\frac{d_p}{d_t}\right) + \frac{3d_p^2 f^2 \Delta l^2 \rho}{80 h_{ch}^2}\left[\left(\frac{d_p}{d_t}\right)^2 - 1\right] \tag{13}$$

From the equations (12) and (13), we can get the relationship between pressure loss in pump chamber and working frequency of the piston under two-dimensional and three-dimensional conditions. The structure parameters of giant magnetostrictive pump designed in this paper are as follows: d_p=50 mm; d_t=4 mm; h_{ch}=0.5 mm; Δl=0.1 mm; Parameters of the selected hydraulic oil are as follows: ρ=850 kg/m3; ν=6.8×10-5 m2/s; μ=0.0578 Pa·s.

IOP Conf. Series: Materials Science and Engineering **242** (2017) 012101 doi:10.1088/1757-899X/242/1/012101

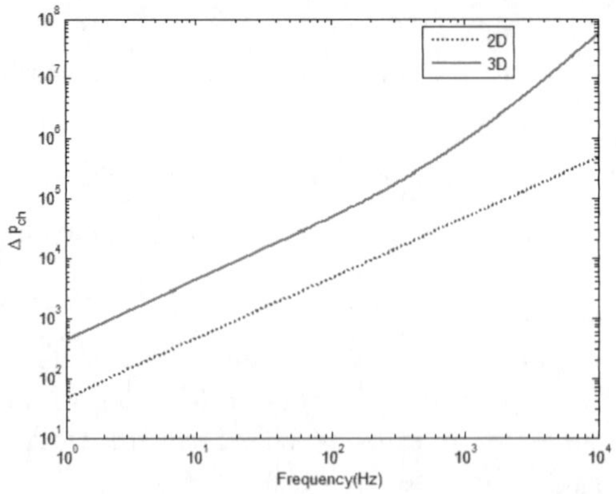

Figure 2. Relationship between the pump chamber pressure loss and the working frequency

As shown in Figure 2, we can see the relationship between the pump chamber pressure loss and the working frequency. It can be seen from the Figure 2, in the ideal state, the pump chamber pressure loss increases with increasing working frequency. With the same working frequency, the pressure loss of three-dimensional model is more than the pressure loss of the two-dimensional model, and the loss gap increases between them when their models are under high frequency. The slope of the three-dimensional model curve changes around 200 Hz, because the effect of the flow area changes on the pressure loss exceeds the effect of liquid viscosity on the pressure loss when the liquid flows radially on the piston surface. At low frequencies, the difference between two-dimensional model and three-dimensional model is the liquid viscosity term. At high frequencies, the difference becomes larger because two-dimensional model does not contain the effect of changes in the radial over-current area. In the actual flow field, the pressure loss in pump chamber increases faster with the increase of the working frequency due to some factors such as the response speed of the reed, the leakage of pump chamber at high frequencies and effect of some cavitation.

3. Influence of pump chamber height on pressure loss

The most important factor affecting the flow field in pump chamber is its height. By studying the influence of different pump chamber height on the pressure loss in the pump chamber, the optimal height parameter is determined. The two-dimensional geometric model of pump chamber was established by FLOTRAN, and the FLUID141 fluid element was used for meshing. The results are shown in Figure 3.

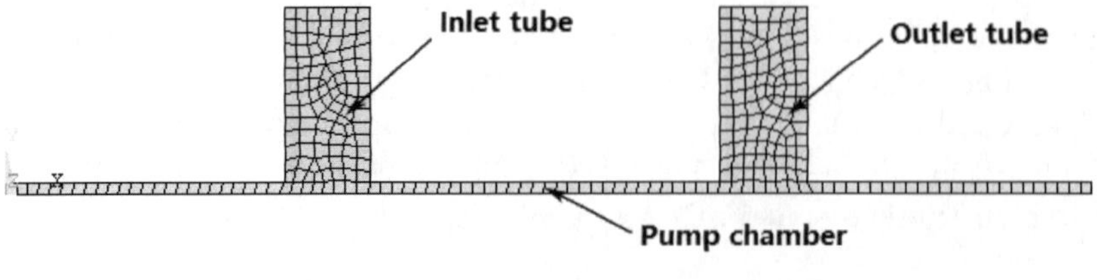

Figure 3. Geometrical model of pump chamber

The flow of liquid in pump chamber is derived from the reciprocating linear motion of piston. In the simulation analysis, the speed of piston is the excitation load. For the same pump chamber height, with different initial speed of piston to load, the relationship between the pump chamber pressure loss and the piston speed curve is obtained.

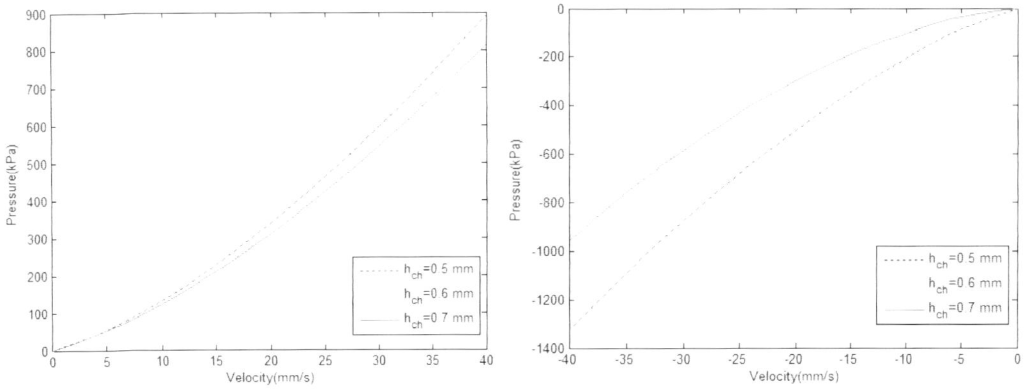

(a) Discharge (b) Intake

Figure 4. Relationship between pump chamber pressure loss and piston speed

Figure 4 shows the relationship between pump chamber pressure loss and working speed of the piston when the pump chamber height h_{ch} were 0.5 mm, 0.6 mm, 0.7 mm,. Figure 4 (a) for the discharge state, Figure 4 (b) for the intake state. It can be seen that the pressure loss in pump chamber increases non-linearly with the decrease of pump chamber height. As is well known, the smaller the pump chamber height, the greater changing rate of pump volume, the more self-priming capacity gets. For the selection of pump chamber height, it is necessary to make a trade-off between the self-priming capacity of pump and the pressure loss in pump chamber. Giant magnetostrictive pump is used to drive the hydraulic system, which requires high output pressure and good self-priming capacity, so a smaller pump chamber height should be selected.

4. Fluid pressure and velocity vector distribution in pump chamber

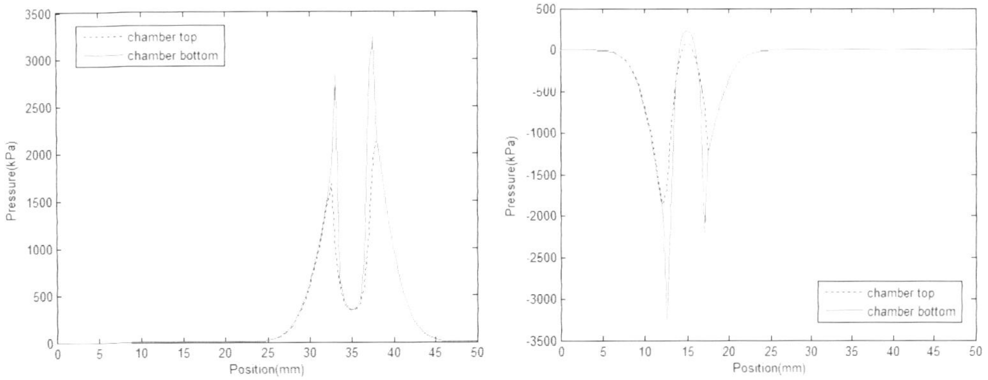

(a) Discharge (b) Intake

Figure 5. Fluid pressure distribution in pump chamber

Figure 5 shows the radial pressure distribution of the top and bottom of the bump whose chamber height is v=0.5 mm and the initial piston excitation speed is 20 mm/s. As can be seen from Figure 5, the pressure rises sharply at the inlet and outlet tube walls (x = 13 mm, 17 mm, 33 mm and 37 mm), because there is a rapid decline in flow velocity when it pass though the interface between the pump chamber and the inlet tube or outlet tube. Because the liquid flows out of pump chamber at a higher speed, there is a greater pressure loss around the center line place (x = 15 mm and 35 mm).

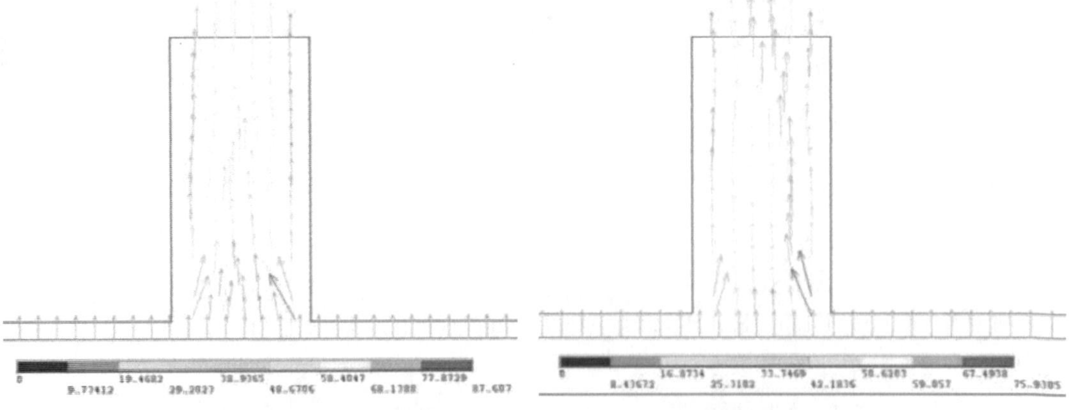

(a) h_{ch}=0.5 mm (b) h_{ch}=0.7 mm

Figure 6. Fluid velocity vector distribution in pump chamber

In Figure 6, the direction of the arrow in a certain point is the direction of liquid flow at that point. In the direction of the arrow in the connection between the pump chamber and the inlet or outlet piping, it can be seen that the direction of the pump is reduced to the horizontal, that is, the lateral flow increases. The lateral flow at the pump chamber exit will prevent the liquid from flowing out. That is the reason that when the pump chamber height is reduced, the pressure loss of the pump chamber increases.

5. Conclusion

In this paper, two-dimensional flow field model and three-dimensional flow field model of giant magnetostrictive pump are established. The numerical simulation results show that the pressure loss in pump chamber increases as the piston working frequency increases. The flow field simulation analysis of giant magnetostrictive pump chamber is carried out and it is found that with the decrease of the pump chamber height, the pressure loss in pump chamber increases. For the selection of pump chamber height, it is necessary to make a tradeoff between the self-priming capacity of pump and the pressure loss in pump chamber. Finally, the fluid pressure and velocity vector distribution in giant magnetostrictive pump chamber are simulated. The simulation results also verify the relationship between pump chamber height and pressure loss in pump chamber.

References

[1] Fang ZW, Zhang YW, Li X, Ding H, Chen LQ. 2017. Integration of a nonlinear energy sink and a giant magnetostrictive energy harvester. Journal of Sound and Vibration, 391:35-49.

[2] Zhu YC, Yang XL.2015. Dynamical model research on energy-conversion process of giant magnetstrictive materials-based electrohydrostatic actuator. *ASME/BATH 2015 Symposium on Fluid Power and Motion Control*.

[3] Gerver M.J. 1998. Magnetostrictive water pump. *SPIE Conference on Smart Structures and Integrated Systems*, 3329:694-705.

[4] Sneed R, Smith R.R, Cash M. F, Bales G. L and Anderson E. H. 2006. Development of smart material - hydraulic pumps and actuators. *In Proceedings of ASME International Mechanical Engineering Congress and Exposition*, Chicago, IL, Nov 5-10.

[5] Sneed R, Smith R.R, Cash M. F and Anderson E. H. 2007. Smart-material based hydraulic pump system for actuation of a morphing wing. *In 15th AIAA/ASME/AHS Adaptive Structures Conference*, Honululu, Hawaii, April.

[6] Lu QG, Zhao YP, Nie Q. 2016. Flow field simulation analysis and optimization of the magnetostrictive precision pump. *Materials Science Forum*, 852:943-951.

[7] Shaju J, Christopher C, Jin-Hyeong Y and Norman M. W. 2006. Application of CFD in the

Design and Analysis of a Piezoelectric Hydraulic Pump. *Journal of intelligent material systems and structures*, 17:967-979
[8] Liu HP. 2011. Hydraulic and Pneumatic Transmission. *Shanghai Jiaotong University Press.*

IOP Conf. Series: Materials Science and Engineering **242** (2017) 012102 doi:10.1088/1757-899X/242/1/012102

Simulation Research on Vehicle Active Suspension Controller Based on G1 Method

Gen Li[1], Hang Li[1], Shuaiyang Zhang[1] and Qiuhui Luo[1]

1 School of Mechatronics Engineering, He'nan University of Science & Technology, Luoyang 471003,China.

E-mail: lyligen520@163.com.

Abstract. Based on the order relation analysis method (G1 method), the optimal linear controller of vehicle active suspension is designed. The system of the main and passive suspension of the single wheel vehicle is modeled and the system input signal model is determined. Secondly, the system motion state space equation is established by the kinetic knowledge and the optimal linear controller design is completed with the optimal control theory. The weighting coefficient of the performance index coefficients of the main passive suspension is determined by the relational analysis method. Finally, the model is simulated in Simulink. The simulation results show that: the optimal weight value is determined by using the sequence relation analysis method under the condition of given road conditions, and the vehicle acceleration, suspension stroke and tire motion displacement are optimized to improve the comprehensive performance of the vehicle, and the active control is controlled within the requirements.

1. Introduction

Suspension system is one of the most important components in the vehicle, it can buffer the impact force on the vehicle that caused by the uneven ground, which have a significant effect on the ride comfort, vehicle handling and safety. The traditional passive suspension system stiffness and damping is not adjustable and its own range of changes is small, it is difficult to adapt to different road conditions. The active suspension system can adjust the suspension stiffness and damping automatically according to the different road conditions and the current vehicle movement condition, so that the vehicle can reach the optimal driving state. The determination of the weighting factor is most important when designing an active suspension using the LQG (Linear Quadratic Gaussian) controller. In the literature [2], the weighting coefficients of the performance indexes are determined by constructing judgment matrix and the original suspension controller is optimized, which improves the suspension performance to a certain extent. In literature [3], based on the genetic algorithm, the error correction factor is introduced and the optimal control rules of the weighting coefficient are determined by combining the fuzzy control strategy. The performance of the system is improved to a large extent, and the stability of the optimized system is further improved. In literature [4], a method of determining the weighting coefficient based on the local elite strategy artificial bee colony algorithm is proposed to improve the ride comfort of the vehicle.

In this study, MATLAB/Simulink is used as the simulation platform, aiming at the main passive suspension model of single-wheel vehicle, selecting the system input signal model and combining the sequence relation analysis method to determine the weight value of each performance index, using vehicle dynamics and optimal linear control theory of LQG controller design. According to the

Content from this work may be used under the terms of the Creative Commons Attribution 3.0 licence. Any further distribution of this work must maintain attribution to the author(s) and the title of the work, journal citation and DOI.
Published under licence by IOP Publishing Ltd

simulation results, the performance of the active suspension controller is analyzed to verify whether the design method is correct and effective.

2. Model Establishment

2.1 Active and passive model

After a reasonable simplification, the model of single-wheel vehicle passive suspension is established, as shown in figure 1.

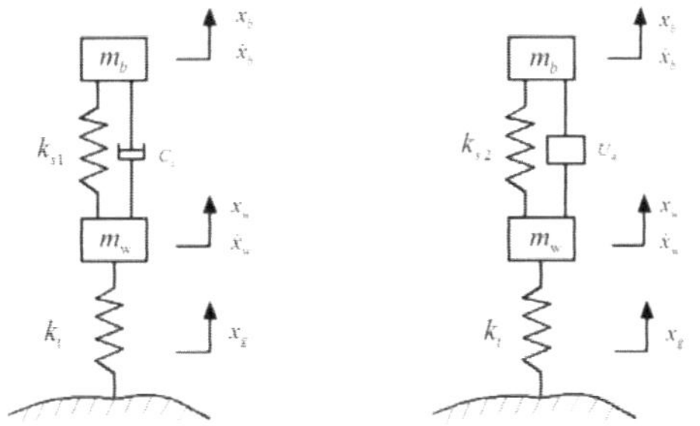

(a) Passive suspension model (b) Active suspension model

m_b - A quarter of the body quality; m_w -Tire quality; k_t -Tire stiffness;

k_{s1} -Passive suspension spring stiffness coefficient;

k_{s2} -Active suspension spring stiffness coefficient;

C_s -Damping coefficient; U_a -Active control force;

x_b -Body displacement; $\dot{x}_b$ -Body speed; x_w -Tire displacement; $\dot{x}_w$ -Tire speed;

x_g -Pavement displacement;

Figure 1. Single-wheel vehicle active suspension model

2.2 Pavement incentive model

According to the requirements of the optimal linear control theory, the filtered white noise of the Gaussian distribution is chosen as the system input [5], as in equation (1):

$$\dot{x}_g(t) = -2\pi f_0 x_g(t) + 2\pi\sqrt{G_0 U_0}\, w(t) \tag{1}$$

Where x_g is the road vertical displacement; G_0 is the road roughness coefficient; u is the vehicle speed; w is the Gaussian white noise Which mathematical expectation is zero and f_0 is the cut-off frequency.

3. LQG Controller Design

3.1 System state equation

From the kinetic knowledge we can know that:

Passive suspension system vehicle model state space equation, as shown in equation (2):

$$\begin{aligned} \dot{X}(t) &= A_1 X(t) + B_1 W(t) \\ Y &= CX(t) + DW(t) \end{aligned} \tag{2}$$

According to Newton's second law, active suspension system motion as in equation (3) (4):

$$m_b \ddot{x}_b(t) = U_a(t) - K_s[x_b(t) x_w(t)] \tag{3}$$

$$m_w\ddot{x}_w(t) = -U_a(t) + K_s[x_b(t) - x_w(t)] - K_t[x_w(t) - x_g(t)] \quad (4)$$

Obtained by the equation (1) (3) (4), the state space equation of the active suspension vehicle model system is rewritten into the form of matrix, as in equation (5):

$$\begin{aligned} \dot{X}(t) &= A_2X(t) + B_2U(t) + FW(t) \\ Y &= CX(t) \end{aligned} \quad (5)$$

Where: $X(t)$ is the system state vector; $W(t)$ is the system input matrix; $U(t)$ is the control input matrix; A_1, A_2 is passive and active system matrix; B_1, B_2 is passive and active input matrix; C is output matrix; D is direct transfer matrix.

$$A_1 = \begin{bmatrix} -\frac{C_s}{m_b} & \frac{C_s}{m_b} & -\frac{K_{s1}}{m_b} & \frac{K_{s1}}{m_b} & 0 \\ \frac{Cs}{m_w} & -\frac{Cs}{m_w} & \frac{K_{s1}}{m_w} & \frac{-K_{s1}-K_t}{m_w} & \frac{K_t}{m_w} \\ 1 & 0 & 0 & 0 & 0 \\ 0 & 1 & 0 & 0 & 0 \\ 0 & 0 & 0 & 0 & -2\pi f_0 \end{bmatrix}, \quad A_2 = \begin{bmatrix} 0 & 0 & -\frac{K_{s2}}{m_b} & \frac{K_{s2}}{m_b} & 0 \\ 0 & 0 & \frac{K_{s2}}{m_w} & \frac{-K_{s2}-K_t}{m_w} & \frac{K_t}{m_w} \\ 1 & 0 & 0 & 0 & 0 \\ 0 & 1 & 0 & 0 & 0 \\ 0 & 0 & 0 & 0 & -2\pi f_0 \end{bmatrix}$$

$$B_1 = F = \begin{bmatrix} 0 & 0 & 0 & 0 & 2\pi\sqrt{G_0U_0} \end{bmatrix}^T, \quad B_2 = \begin{bmatrix} \frac{1}{m_b} & -\frac{1}{m_w} & 0 & 0 & 0 \end{bmatrix}^T$$

$$C = \begin{bmatrix} 1 & 0 & 0 & 0 & 0 \\ 0 & 1 & 0 & 0 & 0 \\ 0 & 0 & 1 & 0 & 0 \\ 0 & 0 & 0 & 1 & 0 \\ 0 & 0 & 0 & 0 & 1 \end{bmatrix}, \quad D = \begin{bmatrix} 0 & 0 & 0 & 0 & 0 \end{bmatrix}^T$$

3.2 Controller design

In the vehicle suspension design, the main of the three factors is considered. The tire displacement reflects the ability of the tire to contact the ground; the acceleration of the body reflects the ride comfort; the suspension stroke reflects the smoothness of the vehicle. Among them, the size of the body acceleration and actuator output force is proportional to the size of the relationship.

Definition: the LQG optimization index function of the active suspension is the integral value of the weighted square sum of the tire dynamic load, the vertical acceleration of the vehicle body and the suspension stroke, as in equation (6).

$$J = \lim_{T\to\infty}\frac{1}{T}\int_0^T \{q_1[x_w(t) - x_g(t)]^2 + q_2[x_b(t) - x_w(t)]^2 + \rho\ddot{x}_b(t)\}dt \quad (6)$$

Where: q_1, q_2, ρ are the weighting factors of each performance index [6]. For the convenience of research, take the vehicle body vertical acceleration of the weighting factor $\rho = 1$.

To facilitate the calculation, the equation (6) is rewritten as a matrix form, as in equation (7):

$$J = \lim_{T\to\infty}\frac{1}{T}\int_0^T (X^TQX + U^TRU + 2X^TNU)dt \quad (7)$$

Where: X is state variable, Q is state variable weighting matrix, R is control variable weighting matrix, N is weighting matrix of the cross term.

$$Q=\begin{bmatrix}0&0&0&0&0\\0&0&0&0&0\\0&0&q_2+\frac{K_{s2}^2}{m_b^2}&-q_2-\frac{K_{s2}^2}{m_b^2}&0\\0&0&-q_2-\frac{K_{s2}^2}{m_b^2}&q_1+q_2+\frac{K_{s2}^2}{m_b^2}&-q_1\\0&0&0&-q_1&q_1\end{bmatrix},\quad N=\frac{1}{m_b^2}\begin{bmatrix}0\\0\\-K_s\\K_s\\0\end{bmatrix},\quad R=\frac{1}{m_b^2}$$

When the vehicle parameters and the weighting coefficients are determined, the optimal control feedback gain matrix can be obtained by the Riccati equation [7-8], as in equation (8).

$$AK+KA^T+Q-KBR^{-1}B^TK+FWF^T=0 \tag{8}$$

And then according to any moment of feedback state variables, according to equation (9) can be obtained at any time the optimal active control force.

$$U_a(t)=-KX(t) \tag{9}$$

3.3 Determination of weighting factor

3.3.1 Quantitative factor. As the difference between the units and value of performance indicators is too large, which need to be quantified before compared. In this paper, the LQG controller design of the vehicle active suspension is carried out on the basis of the original passive suspension [9], and the RMS (Root mean square) value of the vehicle statistic is compared with the vehicle body acceleration δ_{BA}^2, the tire moving displacement δ_{DTD}^2 and the suspension stroke δ_{SWS}^2.

The quantitative factor of the active suspension is 1, according to equation (10) to determine the quantitative factor of the tire movement and suspension of active suspension.

$$\delta_{BA}^2\bullet 1=\beta_1\bullet\delta_{DTD}^2=\beta_2\bullet\delta_{sws}^2 \tag{10}$$

3.3.2 Subjective weighting factor. Order relationship analysis method (G1 method), all indicators are sorted according to the importance of the evaluation criteria for the different degree, and then in the two comparison at the same time, given different weight values. Compared with the fuzzy comprehensive evaluation method and the tomographic analysis method, there is no need to construct the judgment matrix, and it is not necessary to carry out the consistency test. It has the advantages of small computation and intuitive results [10-11].

1. Determine the order relationship

Order relationship analysis method defined: relative to the user's subjective evaluation criteria, if the importance of indicators x_i greater than or equal to the index x_j, there are $x_i \succ x_j$.

For the index set $\{x_1,x_2,\cdots,x_m\}$, the order relation $x_1^*\succ x_2^*\succ\cdots\succ x_m^*$ is established according to the importance of each indicator relative to the evaluation target. x_i represents the evaluation index that $\{x_i\}$ is sorted by the relationship "$\succ$".

The steps to establish the order relationship are as follows:

(1) The decision maker selects one of the most important indicators in the index set $\{x_1,x_2,\cdots,x_m\}$ as x_1^*;

(2) The decision maker chooses the most important one in the remaining $\{x_1,x_2,\cdots,x_m\}$ indicators as x_2^*;

And so on, until after $m-1$ times to select the rest of the evaluation index recorded as x_m^*;

This uniquely identifies an order relationship: $x_1^* \succ x_2^* \succ \cdots \succ x_m^*$.

So we determined: body acceleration $\succ$ suspension moving stroke $\succ$ tire displacement.

2. Comparison between the relative importance of x_{k-1} and x_k.

The degree of importance of an evaluation index x_{k-1} and x_k, the rationality of the ratio ω_{k-1}/ω_k is r_k. As shown in table 1.

Table 1. Assignment table of r_k

r_k	Description
1.0	Indicator x_{k-1} are just as important as indicator x_k
1.2	Indicator x_{k-1} are slightly important than indicator x_k
1.4	Indicator x_{k-1} are obvious important than indicator x_k
1.6	Indicator x_{k-1} are more important than indicator x_k
1.8	Indicator x_{k-1} are most important than indicator x_k

Also stipulated that $r_{k-1} > 1/r_k$ must be satisfied.

According to the assignment table we give: body acceleration is more important than the tire moving displacement, $r_2=1.2$ tire moving displacement is extremely important than the suspension moving stroke, $r_3=1.8$.

3. Calculation of weight factor

For the sorting completion of the evaluation index set can be calculated according to equation (11), (12) weight coefficient.

$$\omega_m = \left(1+\sum_{k=2}^{m}\prod_{i=k}^{m} r_i\right)^{-1} \tag{11}$$

$$\omega_{k-1} = r_k\omega_k, k = m, m-1, \cdots, 3, 2 \tag{12}$$

4. Subjective weighting factor

The subjective weighting factor of the body acceleration is 1, and the subjective weighting factor x and x of the other two indicators are determined according to equation (13).

$$\frac{\omega_1}{1} = \frac{\omega_2}{\gamma_1} = \frac{\omega_3}{\gamma_2} \tag{13}$$

3.3.3 Determination of final weighting factor. Determine the final weighting factor according to equation (14).

$$q_1 = \beta_1\gamma_1$$
$$q_2 = \beta_2\gamma_2 \tag{14}$$

4. Model Simulation and Analysis

4.1nVehicle model parameters

The relevant vehicle model parameters are shown in table 2.

Table 2. The relevant vehicle model parameters

Model Parameters	Symbol	Value	Unit
1/4 Body Quality	m_b	320	Kg
Wheel Quality	m_w	40	Kg
Passive Suspension Stiffness	K_{s1}	20000	$N{\bullet}m^{-1}$
Active Suspension Stiffness	K_{s2}	22000	$N{\bullet}m^{-1}$
Tire Stiffness	K_t	200000	$N{\bullet}m^{-1}$
Passive Suspension Damping Coefficient	C_s	1000	$N{\bullet}s{\bullet}m^{-1}$
Suspension of Work Space	SWS	±100	mm
Actuator Control	U_a	±3000	N
Simulation of Road Roughness Coefficient	G_0	5.0×10^{-6}	$m^3{\bullet}cycle^{-1}$
Vehicle Speed	u	20	$m{\bullet}s^{-1}$
Cut-off Frequency	f_0	0.1	Hz

In the given road condition, the importance order of the factor is: body acceleration $\succ$ suspension moving stroke $\succ$ tire displacement. According to the Eq.（10）（11）（12）（13）（14）calculate the results:

$$q_1 = 72538.17,\quad q_2 = 4842.31$$

Call the linear quadratic optimal controller design function $[K,\ S,\ E] = LQR[A,\ B,\ Q,\ R,\ N]$ to complete design optimal active controller. The resulting optimal feedback gain matrix result is as follows:

$$K = [3826.54 \quad -1176.67 \quad 2267.75 \quad -11639.95 \quad 10926.46]$$

4. 2 Establish simulink model

According to the above design scheme, the main passive suspension model is simulated in Simulink environment [12-13], as shown in figure 2. The white noise can be called by the MATLAB function WGN (M, N, P) to produce a M-row N-column Gaussian white noise matrix. The noise intensity is determined by the P value (in unit of DB). In the simulation, take M = 10001, N = 1, P = 20, that is, take a total of 10001 sampling points, sampling time is set to 0.005s, when the speed of 20m/s, the equivalent of the simulation road length is 1000m.

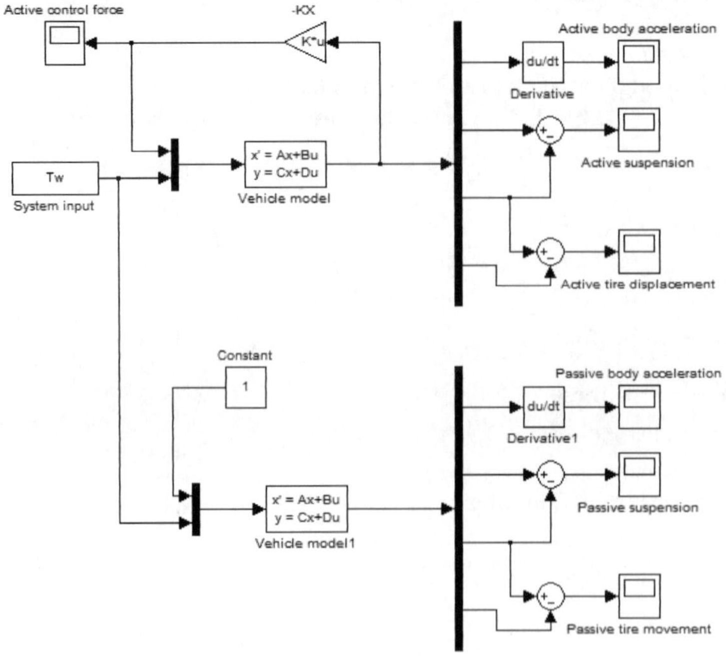

Figure 2. Schema diagram of the system simulation in Simulink environment

4. 3 Results and analysis

The simulation results are shown in figure 3 and figure 4.

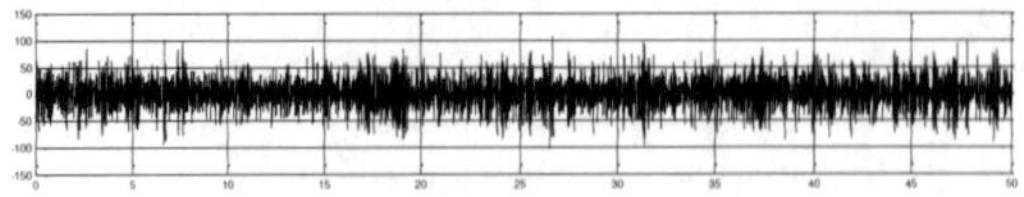

(a) Body acceleration curve

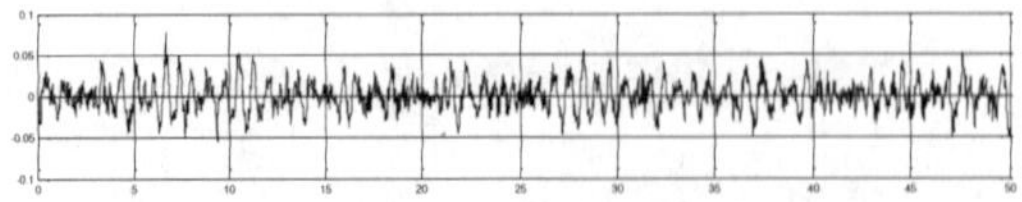

(b) Suspension stroke change curve

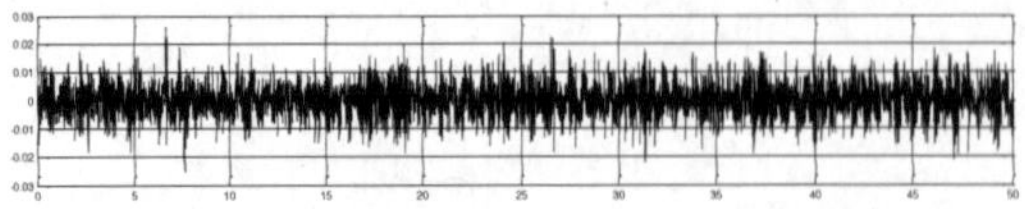

(c) Tire displacement curve

Figure 3. Passive suspension system simulation results

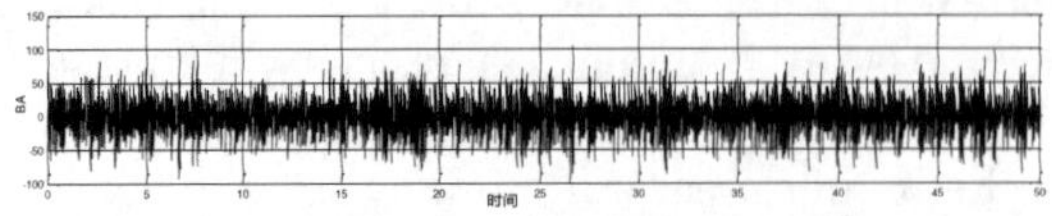

(a) Body acceleration curve

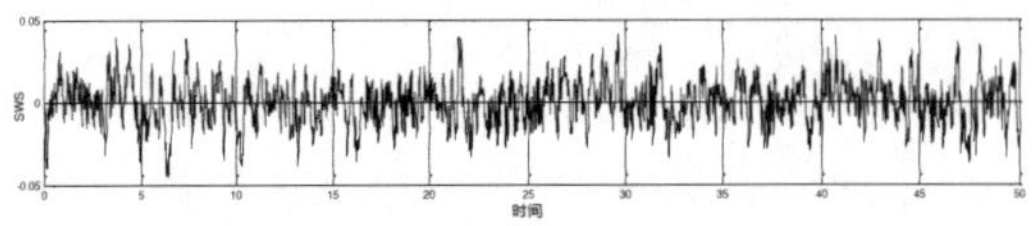

(b)Suspension stroke change curve

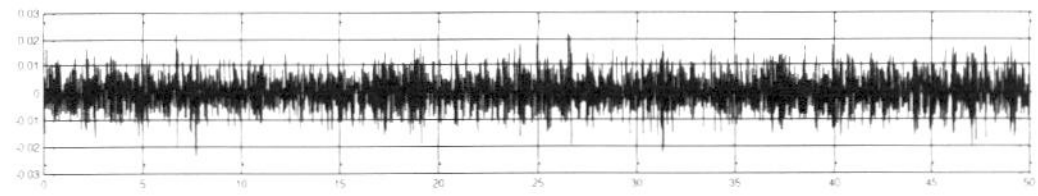

(c) Tire movement displacement curve

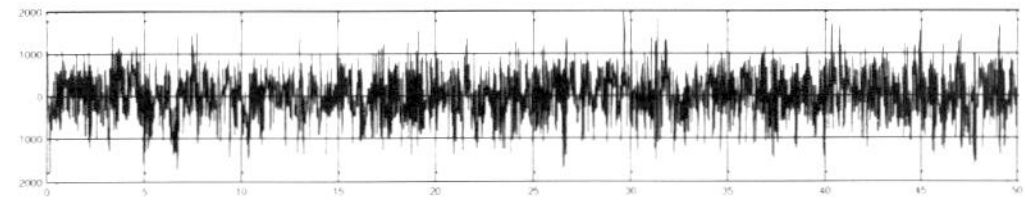

(d) Active output control curve

Figure 4. Simulation results of active suspension system

From the simulation results we can do the following conclusions:

(1) Calculate the RMS value of the performance index of the main passive suspension, as shown in table 3.

Table 3. Main passive suspension performance RMS value

	Body Acceleration /(m/s^2)	Suspension Stroke /(m)	Tire Movement /(m)
Active Suspension System	1.533	0.013366	0.005984
Passive Suspension System	1.850	0.017917	0.006350

In the case of the difference of the displacement of the tire, the optimal active suspension significantly reduces the vehicle acceleration and reduces the suspension stroke. Compared with the passive suspension, the active suspension of the LQG controller based on the G1 method, the root mean square values of each performance index were reduced respectively by 17.14%, 25.4% and 5.76%.

(2)It can be seen from figure.4(d) that the active control force is in the range of -2422.167~1947.846N, and the calculated RMS value of the active control force is 523.565N, which means that the control force of the dominant force is within the required range (±3000N).

(3) Compared with figure 3 and figure 4, it can be seen that the active suspension can smooth the change of each performance index in the adjacent time compared with the passive suspension, which is beneficial to ensure the ride comfort of the passenger and the ride comfort of the vehicle. Suspension travel is limited to the range of ±50mm, which further improve the occupants ride comfort, but also on the rational design of the suspension played a reference role.

(4) The G1 method is used to determine the weighting coefficient of each performance index, and the appropriate weighting coefficient is selected to get a better comprehensive performance. This suggests that we can design the suspension according to the needs of different users, while the focus can also make the performance indicators to achieve a balance, and then get the best design results.

5. Conclusions

The dynamic model of the main passive suspension system of the single-wheel vehicle is established. The weighting coefficient of each performance index is determined by the sequence relation analysis method, the construction of the judgment matrix is avoided, the calculation is simplified and the motion state space is established by using the knowledge of dynamics equation and combined with the optimal control theory to complete the optimal linear controller design, and Simulink in the model was simulated. The results show that: the appropriate weight value is determined by using the sequence relation analysis method, which makes the acceleration of the vehicle body, the suspension stroke and the tire displacement are optimized to a certain extent, and the overall performance of the vehicle is improved.

References

[1] Y. Bai and N. Sang 2015. Ride comfort of vehicle active suspension using active disturbance rejection control. *Journal of PLA University of Science and Technology (Natural Science Edition).* **16(6)**:575-581.

[2] X.Y. Luo and S.W.Yang 2013. Design of a LQG controller for a vehicle active suspension system based on AHP. *Journal of vibration and shock.***32(2)**:102-106.

[3] L. Zeng, D.M. Sun and B.Zhou. Design of controller of active suspension based on improved genetic algorithms. *Computer Measurement & Control.* 2016.**24(4)**:58-60,64.

[4] H.T. Zhang 2016. Design of the LQG controller for automobile active suspensions based on artificial bee colony algorithm. *NOISE AND VIBRATION CONTROL.* **36(5)**:65-69,81.

[5] Z.X. Li ,J.Y. Huang and Y.W.Liu 2016. Modeling and Simulink on white noise of road roughness in time domain. *Journal of Jiang Su University(Natural Science Edition).* **37(5)**:503-506,524.

[6] Y.B. Yin, Q. Fang and G.H. Bai 2013. Study on effect of weight coefficient to vibration of the vehicle Body with Active Suspension. *Vehicle & Power Technology.* **(1)**:24-28.

[7] Z.F. Zhang, J.L. Liu and Z.M. Xu 2012. Design of LQG controller for semi-active suspension based oil improved analytic hierarchy process. *Automotive Engineering.* **34(6)**:528-533.

[8] S. Zheng, L.Y. Zhu and J. Zhang 2015. Study of LQG control for vehicle active suspension system based on improved analytic hierarchy process. *Journal of Chinese Agricultural Mechanization,* **36(2)**:189-193.

[9] F. Yu. Vehicle dynamics and control. *Machinery Industry Press.BeiJing.*2010.

[10] M. Chen, Y.J. Guo and Z.M 2011. Yu. An improved method for rank correlation analysis and its application. *Journal of Systems & Management.***20(3)**:352-355.

[11] X.J. Wang, Y.J. Guo and T. Lan 2006. Rank correlation analysis of formation of consistent judgment matri x. *Journal of Northeastern University (Natural Science).* **27(1)**:115-118.

[12] W.J. Zhao, D.P. Zhao and Y.W. Qu 2015. The simulation research of the active suspension based on Matlab/Simulink. *Modern Manufacturing Technology & Equipment.* **(4)**:10-13.

[13] J.C. Ma, J.P. Si and Z.G.Gao 2015. Simulation analysis of semi-active suspension based on MATLAB/Simulink. *Highways & Automation Applications.* **(2)**:13-17.

A review of typical thermal fatigue failure models for solder joints of electronic components

Xiaoyan Li[1,a,*] , Ruifeng Sun[2,b] ,Yongdong Wang[3,c]

[1,2,3]China Aero-Polytechnology Establishment, Beijing, China

[a]lxylyx20100214@163.com, [b]waiguge1314@163.com, [c]orpfee83@126.com

Abstract. For electronic components, cyclic plastic strain makes it easier to accumulate fatigue damage than elastic strain. When the solder joints undertake thermal expansion or cold contraction, different thermal strain of the electronic component and its corresponding substrate is caused by the different coefficient of thermal expansion of the electronic component and its corresponding substrate, leading to the phenomenon of stress concentration. So repeatedly, cracks began to sprout and gradually extend [1]. In this paper, the typical thermal fatigue failure models of solder joints of electronic components are classified and the methods of obtaining the parameters in the model are summarized based on domestic and foreign literature research.

1. Introduction

Relevant record data indicates that the performance, service life and reliability of electronic products are greatly influenced by temperature. Because among the environmental factors that cause the failure of airborne electronic equipment, there are 45% of the failure caused by the temperature factor. In recent years, with the development of fault physics technology, reliability test based on fault physics is more and more popular, which is used to evaluate fatigue life of electronic products. Statistics show that 70% of the electronic components failure is caused by the failure of packaging and assembly and that in electronic packaging and assembly failure, solder joint failure is the main reason [2]. So this paper focuses on the research and summary of the typical thermal fatigue failure models of solder joints of electronic components.

2. Classification of thermal fatigue failure models of solder joints of electronic components

A lot of researches on the thermal fatigue failure of solder joints have been carried out, and the thermal fatigue life model of solder joints is generally classified as shown in Fig.1 [3,4]:

Content from this work may be used under the terms of the Creative Commons Attribution 3.0 licence. Any further distribution of this work must maintain attribution to the author(s) and the title of the work, journal citation and DOI.
Published under licence by IOP Publishing Ltd

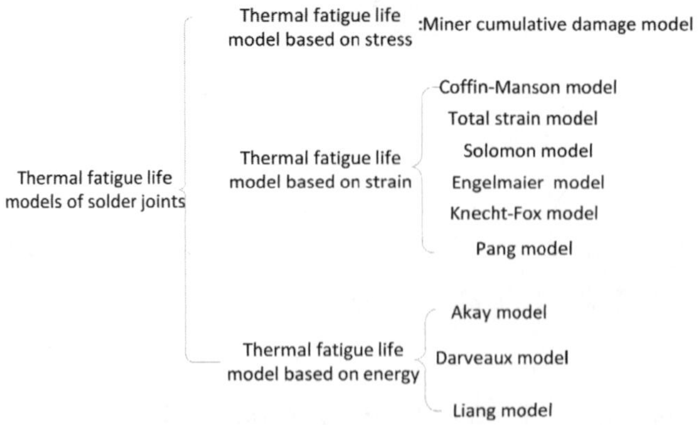

Fig. 1 Thermal fatigue life model of solder joints

2.1 Thermal fatigue life model based on stress

The application of this model is mainly due to the condition that electronic packaging devices are under thermal cycling load. Every time thermal cyclic loading will cause damage of products, and when the cumulative amount of damage exceeds the limit value of product or material, it will cause failure of the product. This model is called cumulative damage model.

In many linear damage accumulation theories, Palmgren-Miner theory is more typical, which is also referred to as Miner theory [5]. This theory has been widely used in engineering because it is very simple [6]. There are also some nonlinear cumulative deterministic theories, where many parameters need to be corrected by a large number of experimental data, leading to that its accuracy is not higher than the Miner theory. Miner theory also has its own shortcomings: it is assumed that the load state of the product and the load order don't affect the damage accumulation.

2.2 Thermal fatigue life model based on strain

In the thermal fatigue model based on strain, the strain is composed of three parts :

$$\triangle\gamma=\gamma_e+\gamma_p+\gamma_c \qquad (1)$$

γ_e—Elastic strain, which is obtained from formula calculation or finite element analysis or test;

γ_p—Plastic strain, which is obtained from formula calculation or finite element analysis or test;

γ_c—Creep strain, which is obtained from formula calculation or finite element analysis or test.

Those strain will exist at the same time in most cases, so it is not easy to distinguish strictly.

Thermal fatigue life model based on strain has the following categories:

- Coffin-Manson model
- Total strain model
- Solomon model
- Engelmaier model

The Coffin-Manson model is mainly used to calculate the low cycle fatigue life and its application is very wide. The expression is as follows:

$$\frac{\Delta\varepsilon_P}{2}=\varepsilon_f(2N_f)^c \qquad (2)$$

N_f—Number of failure cycles;

ε_P—Amplitude of plastic strain, which could be obtained from formula calculation or finite element analysis;

ε_f—Fatigue ductility coefficient, which could be obtained from 《Dictionary of materials》;

c—Fatigue ductility exponent, which could be obtained from formula calculation.

The total strain model combines Coffin-Manson model and Basquin equation effectively, not only considering the effect of elastic strain on fatigue life, and considering the influence of plastic strain [7]. The expression is as follows:

$$\frac{\Delta\varepsilon}{2}=\frac{\sigma_f}{E}(2N_f)^b+\varepsilon_f(2N_f)^c \quad (3)$$

$\Delta\varepsilon$—Strain range, which could be obtained from formula calculation or finite element analysis;

E—Elastic modulus of solder joints, which could be obtained from 《Dictionary of materials》;

σ_f—Fatigue strength coefficient; which could be obtained from 《Dictionary of materials》;

b—Fatigue strength exponent, which could be obtained by experiment;

$\varepsilon_f$and c—as mentioned before.

Solomon model only considers the effect of plastic shear strain on the fatigue life of solder joint [8]:

$$\Delta\gamma_P N_P^{\alpha}=\theta \quad (4)$$

$\Delta\gamma_P$—Plastic shear strain, which could be obtained from formula calculation or finite element analysis;

N_P—Number of failure cycles;

θ—The reciprocal of fatigue ductility coefficient, which could be obtained from formula calculation;

α—Material constant, which could be obtained from 《Dictionary of materials》.

After the Solomon model was proposed, many scholars has focused on the model and used it to the QFP package, receiving good results. However, the Solomon model didn't consider the influence of creep strain on the fatigue life of solider joint. As a result of this, the actual application is limited.

Engelmaier model, an improved version of Coffin-Manson model, takes the thermal cycle frequency, temperature effect and elastic-plastic strain into account. As a result of this, the actual application is popular. The expression is as follows:

$$N_f=\frac{1}{2}\left(\frac{\Delta\gamma}{2\varepsilon_f}\right)^{\frac{1}{c}} \quad (5)$$

N_f—Fatigue life;

ε_f—Fatigue ductility coefficient, and $\varepsilon_f=0.325$ for the eutectic solder;

c, reflecting the temperature stress, is determined by:

$$c=-0.442-0.0006T_{sj}+0.0147\left(1+\frac{360}{t_H}\right) \quad (6)$$

T_{sj}—Average temperature, obtained from formula calculation;

t_H—Time of duration of high temperature;

$\Delta\gamma$—The total shear strain range, which could be obtained from formula calculation or finite element analysis.

The fatigue model based on creep, proposed by Knecht and Fox, is relatively simple:

$$N_f=\frac{C}{\Delta\gamma_{mc}} \quad (7)$$

$$\gamma_{mc}=\int C_0\left(\frac{\tau}{\tau_0}\right)^{nMC}dt \quad (8)$$

γ_{mc}—The shear strain caused by matrix creep, which could be obtained from formula calculation or finite element analysis;

nMC, C_0, τ_0—material parameter, which could be obtained by experiment.

Some researchers think we should combine the matrix creep and grain boundary sliding to fully consider the influence of the two parts on solder joint life, and then they put forward the new model [9]:

$$N_f=\left(0.022D_{gbs}+0.063D_{mc}\right)^{-1} \quad (9)$$

In this model, D_{gbs}, the creep caused by grain boundary sliding, and D_{mc}, the matrix creep, are fully considered [10]. In addition, Syed has done a lot of research and experiments based on TSOP package, and given the conclusion: when the rate of temperature change increases gradually and the whole package stiffness changes, the creep mechanism has a gradual transition from grain boundary sliding to the matrix creep.

Pang proposed a new life model based on the plastic strain model from Solomon and the creep fatigue model from Knecht-Fox organically, including plastic strain and creep strain, as shown in Eq. 10:

$$\frac{1}{N_f}=\frac{1}{N_p}+\frac{1}{N_c} \qquad (10)$$

N_p—The fatigue life of Solomon model;

N_c—The fatigue life of Knecht-Fox model.

Based on this model, Pang predicted the life of FCOA and CBGA package and compared with the expected results of other models, finally he was in support of the use of this model.

2.3 Thermal fatigue life model based on energy

This model considers hysteresis energy and stress-strain based on volume weighted average [5]. After Akay considered the total strain energy, he proposed new fatigue model, as follows:

$$N_f=\left(\frac{\Delta W_{total}}{W_0}\right)^{\frac{1}{k}} \qquad (11)$$

N_f—Cycle number of failure;

ΔW_{total}—Total strain energy, which is obtained from formula calculation or finite element analysis;

k—fatigue quotient, from available experimental data.

Based on the above model, Akay took the pin package structure as the test object and ultimately determined the constant value: $k = -0.6342$, $W_0 = 0.1573$, but he did not further study the fatigue life of solder joints.

Taking into account the crack propagation and the plastic dissipation for crack propagation, a new fatigue model is proposed by Darveaux. The implementation of this model is as follows: Firstly, the finite element method is used to calculate the accumulated plastic work of per cycle. The accumulated plastic work is then used to calculate the number of cycles the solder joint experiences before the internal crack initiation of the solder joint and the number of cycles the solder joint experiences during the diffusion of the crack inside the solder joint. And then by adding the two results, the fatigue life can be obtained. The Darveaux model uses formula Eq.12 and Eq.13 to calculate the number.

$$N_0=k_1\Delta W^{K_2} \qquad (12)$$

$$\frac{da}{dN}=K_3\Delta W^{K_4} \qquad (13)$$

k_1, K_2, K_3, K_4—material constant.

For the internal work of the solder joints, we calculate the solution according to the average value, such as formula Eq. 14:

$$\Delta W=\frac{\sum_i \Delta W_i V_i}{\sum_i V_i} \qquad (14)$$

Then the fatigue life of solder joints is calculated according to formula Eq. 15:

$$\alpha=N_0+\frac{\alpha}{da/dN} \qquad (15)$$

Liang proposed another fatigue model based on energy, as shown in the following. The model takes into account the influence of the geometric size of solder joints and the elastic creep analysis.

$$N_f=C(W_{ss})^{-m} \qquad (16)$$

W_{ss}—Energy density under the hysteresis loop of stress and strain, determined based on hysteresis loop;

C, m—Material constant.

3. Summary

At home and abroad, the statistical data shows that about 52% of the airborne electronic equipment faults is caused by environmental factors, of which the temperature factor accounts for about 42% of the faults, the vibration factor accounts for 27%, and the humidity factor accounts for 19%. Therefore, the environmental stress, which affects the reliability of avionics equipment, is temperature, vibration and humidity in turn. So it is necessary to evaluate the fatigue life of electronic products. In recent years, with the development of fault physics technology, it is more popular to evaluate the fatigue life of electronic products by using the reliability test based on fault physics. The method ,temperature fatigue life of electronic products obtained by Fault Analysis of physical, need to understand the characteristics of electronic products and the typical thermal fatigue failure model. This article gives an investigation and summary of typical thermal fatigue models of solder joint of the electronic components and has certain engineering significance.

References

[1] Nishad Patil,Diganta Das,Michael Pecht, A prognostic approach for non-punch through and field stop IGBTs, J. Microelectronics Reliability,2012,523.

[2] Rainer Tilgner. Physics of failure for interconnect structures: an essay, J. Microsystem Technologies,2009,151.

[3] Engelmaier W. The use environments of electronic assemblies and their impact on surface mount solder attachment reliability, J. Components, Hybrids, and Manufacturing Technology, IEEE Transactions on, 1990, 13(4): 903-908.

[4] Osterman M, Stadterman T. Failure assessment software for circuit card assemblies, C. Reliability and Maintainability Symposium, 1999. Proceedings. Annual. IEEE, 1999:269-276.

[5] Han C, Han B. Board level reliability analysis of chip resistor assemblies under thermal cycling: A comparison study between SnPb and SnAgCu, J. Journal of Mechanical Science & Technology, 2014, 28(3):879-886.

[6] Barker D, Vodzak J, Dasgupta A, et al. Combined Vibrational and Thermal Solder Joint Fatigue—A Generalized Strain Versus Life Approach, J. Journal of Electronic Packaging, 1990, 112(2):129-134.

[7] Sharif I, Barker D B, Dasgupta A, et al. Fatigue Analysis of a Planarpak Surface Mount Component, J. Journal of Electronic Packaging, 1991, 113(2):194-199.

[8] Pecht M, Dasgupta A. Physics-of-failure: an approach to reliable product development, C. Integrated Reliability Workshop, 1995. Final Report., International. IEEE, 1995:1 - 4.

[9] Kotlowitz R W. Comparative compliance of representative lead designs for surface-mounted components, J. Components, Hybrids, and Manufacturing Technology, IEEE Transactions on, 1989, 12(4): 431-448.

[10] Verma S, Dasgupta A, Barker D. A Numerical Study of Fatigue Life of J-Leaded Solder Joints Using the Energy Partitioning Approach, J. Journal of Electronic Packaging, 1993, 115(4):416-423.

IOP Conf. Series: Materials Science and Engineering **242** (2017) 012104 doi:10.1088/1757-899X/242/1/012104

Analysis of blood cell particles in axial blood pump by using computational fluid dynamics/discrete element method model

Cheng Lizhi[1,2], Tan Jianping[1], Yun ZHong[1], Xu Yan[2]

[1] School of Mechanical and Electrical Engineering, Central South University Changsha 410083, China

[2] Department of Mechanical and Electrical Engineering, Changsha University Changsha 410022, China

E-mail: jptan@163.com

Abstract. Computational fluid dynamics/discrete element method (CFD-DEM) model was used to investigate the microscopic motion law of blood cells in a blood pump. The macro-hydrostatic performance indexes of the blood pump, as well as the characteristics of motion, collision. Result showed that collision among the blood cells mainly occurred at the front guide and the back guide, and the collision rate between the blood cells and the wall surface was 2.3 times the collision rate among the blood cells at the front guide vane area of the vane rotor.

1. Introduction

Blood pump as an important supplementary device for life saving has become the focus of research for scholars worldwide [1,2]. CFD technology can be used to conduct more detailed analyses of flow field to obtain the specific area where the solid-phase blood cell particles are broken and damaged. However, no technical breakthrough exists in CFD technology to study the dynamic characteristics of the motion and collision of blood cell particles at different time points and positions. Alobaid[3] built the extended CFD/discrete element method (CFD-DEM) evaluation model, which can accurately predict the motion and pressure gradient of particles. Fries[4] simulated the dynamic CFD-DEM coupling of particles on the fluid bed and studied the dynamic properties of particles. Ren[5] investigated the CFD-DEM coupling of conical gas–solid flow characteristics, the gas–solid flow structure, particle velocity, and the distribution of particle concentration. With CFD-DEM liquid–solid–gas multiphase flow coupling technology, particle distinct element method can be used to analyze the motion, collision, and breakage of molecular-scale blood cell particles, taking into consideration the true blood rheology characteristics of blood cells to lay the foundation for building the hemolytic index model for blood cell mechanical properties and blood rheology coupling.

2. Calculation model

2.1 Geometric model

This work studies the axial blood pump model (Fig. 1). The model consists of three parts, namely, the front guide, the vane wheel, and the back guide. In addition to supporting the rotor of the vane wheel, the front and back guides divert the flow of blood at the inlet and outlet. The inner diameter of the blood pump is 16 mm and the total length is 81 mm. The rotor vane is formed by two long blades and two short blades, and the clearance between the blade tip of the rotor and the inner wall of the body is 0.1 mm. To simplify the calculations, the blood cell model used was a single sphere with a radius of

Content from this work may be used under the terms of the Creative Commons Attribution 3.0 licence. Any further distribution of this work must maintain attribution to the author(s) and the title of the work, journal citation and DOI.
Published under licence by IOP Publishing Ltd

0.01 mm, as blood is mainly composed of red blood cells.

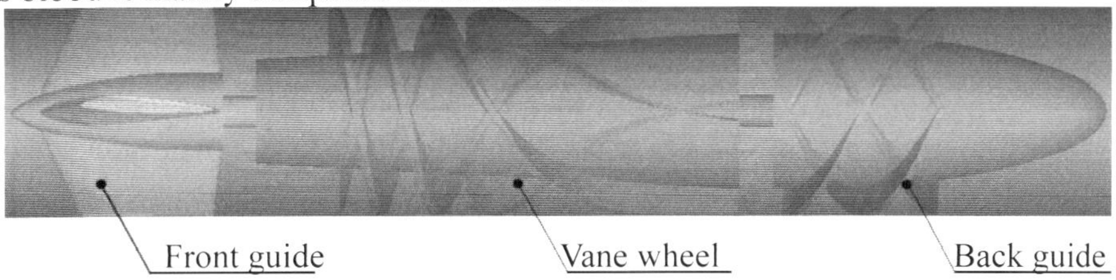

Fig. 1 Axial blood pump model

2.2 Division of meshes

As the vane wheel rotates at a high speed, the inner flow field is divided into three parts, namely, blood pump front guide fluid, vane wheel fluid, and back guide fluid, to increase the calculation accuracy. All three parts used non-structural tetrahedral meshes. The clearance fluid on the blade tip is encrypted by the mesh on the border. The space between the meshes of the front and back guide fluids is set at 0.4, and the space of some meshes of the vane wheel fluid is set at 0.1. The total number of meshes is 2 586 162. Fluent was used to import the mesh file, and the unit is converted to mm. Surface finish, exchange, and considerable refinement were conducted on the meshes, and the quality was checked.

3. Border conditions

Through multiple numerical calculations, the multiple reference frame parameter was used in Fluent to calculate the standard $k-\varepsilon$ used by the model, and pressure inlet and outlet were used to obtain stable calculation results. The other Fluent and EDEM (Engineering Discrete Element Method) coupling calculation parameters are listed in Tables 1 and 2.

Table 1. CFD calculation parameters

Parameters	**Values**
Inlet pressure p_i / Pa	1 333
Outlet pressure p_e / Pa	13 333
Viscosity μ_b / (Pa·s)	0.004 7
Simulation time t / s	0.2

Table 2. EDEM calculation parameters

Parameters	**Values**
Particle diameter d_p / μm	10
Particle density ρ_p / (kg·m^{-3})	1 050
Particle shear modulus G_p / Pa	100 000
Border shear modulus G_w / Pa	100 000
Particle Poisson ratio v_p	0.3
Border Poisson ratio v_w	0.3
Static friction coefficient μ_p	0.3

4. CFD-DEM analysis of the flow field and blood cell particles in the blood pump

4.1 Analysis of blood cell particle collision

From the number of collision times and rates of blood cell particles among themselves and between the blood cell particles and wall surface (including vane wheel surface, flow guide surface, and the inner wall surface of the pump body), the collision laws under different velocities are essentially the same, and the collision rate is quite stable (Fig. 2). When blood cell particles flow to the vane wheel area, they do not flow completely and collide with the vane wheel that rotates at high speed at a velocity of 1.2 m/s. As the linear speed of the vane wheel, particularly the blade top, is quite high, the blood cell particles are instantly driven to collide with the wall surface and with each other. With the creation of the passageway, particularly at the back guide area, the flow becomes stable and the diversion effect is quite significant.

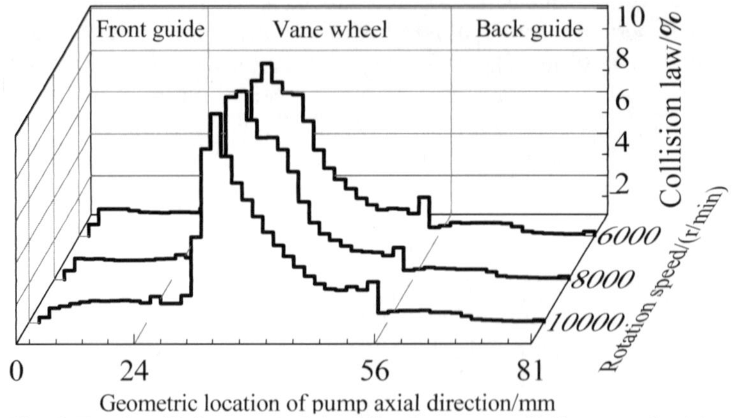

Fig. 2 Total collision rate of the blood pump at different velocities

The collision law of blood cell particles against the wall surface was thoroughly analyzed. As shown in Fig. 3, the rates of collision among the blood cell particles and between blood cell particles and the wall surface are similar, but the two types of collision rates are not the same when the fluid passes various parts of the pump body. The blood pump model shown in Fig. 1 consists mainly of the collision among blood cell particles because it only has four guide blades, and the flow is mainly streamlined. Based on the field velocity of blood fluid in Fig. 5, when the blood cell particles entered the vane wheel area with the blood fluid, the velocity of blood cell particles is significantly different from the linear velocity of the vane wheel. At this point, a large amount of blood cell particles are driven toward the wall surface, and the rate of collision between the blood cell particles and the wall surface is 2.3 times the rate of collision among the blood cell particles. With the flow diversion effect of the vane wheel blades and the back guide blades, the two types of collision rates are similar and are lower than the inlet.

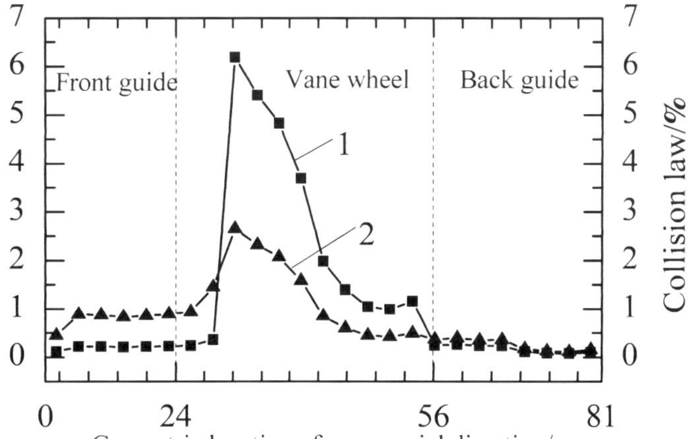

Fig. 3 Collision rate between blood cells and wall surface in the blood pump at 8 000 rpm

5. Conclusions

The liquid–solid multiphase coupling technology based on CFD-DEM used particle distinct element method to analyze the motion law of molecular-scale blood cell particles in the axial blood pump, taking into consideration the macro flow of blood cells. The convergence characteristics of blood cell particles were also studied. This axial blood pump has similar collision law at 10 000 rpm. The collision mainly occurred among the blood cell particles at the front and back guides. The collision rate between the blood cell particles and the wall surface at the front guide blade area of the vane wheel is 2.3 times the collision rate between blood cell particles. The collisions have created evident blood cell particle convergence phenomena at the clearance between the vane wheel and the back guide as well as the outlet of the back guide.

Acknowledgments

This work was supported by National Natural Science Foundation of China (Grant No. 51475477, 31670999) and Scientific Research Project of the Education Department of Hunan Province (Grant No. 13C1084, 15A019).

References

[1] Schüle CY, Thamsen B, Blümel B, et al. Experimental and Numerical Investigation of an Axial Rotary Blood Pump[J]. Artificial Organs, 2016, 40(11): 192-202.
[2] Teman N R, Demos D S, Bryner B S, et al. In Vivo Testing of a Novel Blood Pump for Short-Term Extracorporeal Life Support[J]. The Annals of Thoracic Surgery, 2014,98(1):97-102.
[3] Alobaid F, Ströhle J, Epple B. Extended CFD/DEM model for the simulation of circulating fluidized bed[J]. Advanced Powder Technology, 2013, 24(1):403-415.
[4] Fries L, Antonyuk S, Heinrich S, et al. DEM–CFD modeling of a fluidized bed spray granulator[J]. Chemical Engineering Science, 2011,66(11): 2340-2355.
[5] REN Bing, ZHONG Wenqi, JIN Baosheng, et al. DEM Simulation of Flow Behaviors in 3D Spouted Beds [J]. Journal of Engineering Thermophysics, 2012,33(08): 1341-1344 (in Chinese).

IOP Conf. Series: Materials Science and Engineering **242** (2017) 012105 doi:10.1088/1757-899X/242/1/012105

A model of airport security work flow based on petri net

Xinming Dong

International School,Beijing University of Posts and Telecommunication

No. 10, Xitucheng Road, Haidian district, Beijing, China

Email: dongxinming@bupt.edu.cn

Abstract. Extremely long lines at airports in the United States have been sharply criticized. In order to find out the bottleneck in the existing security system and put forward reasonable improvement plans and proposal, the Petri net model and the Markov Chain are introduced in this paper. This paper uses data collected by transportation Security Agency (TSA), assuming the data can represent the average level of all airports in the Unites States, to analysis the performance of security check system. By calculating the busy probabilities and the utilization probabilities, the bottleneck is found. Moreover, recommendation is given based on the parameters' modification in Petri net model.

1 Introduction

During 2016, the U.S. Transportation Security Agency (TSA) came under sharp criticism for extremely long lines, in particular at Chicago's O'Hare international airport. This phenomenon not only can easily lead to mass incidents but also brings a lot of negative impact on the air transport industry. Therefore, in order to reduce the security queue length as well as improve security efficiency, this article analyzes the airport security check workflow and its performance.

At present, researches of workflow and performance based on Petri net are common while application of Petri net in security check process is rear. In terms of passengers' service process, Guanghui Gu applied Service Model to general departure process in China and used the approach to assess working flow in different airports as well as made recommendation [1]. Besides, Naiyi Li used simulation software to simulate flight departure, reflecting the customers' leaving process dynamically [2]. When it comes to Petri net modeling and performance analysis, Xiaoqing Zhang established the Petri net model of the urban traffic emergency management system to determine its weaknesses and propose improvement measures [3].Jun Xiao extended the Petri net model to evaluate the performance of four basic model instances of the workflow in the Exsept environment [4]. Jiangtao Qin built a model of the manufacturing system using a hierarchical colouring Petri net method and parameter call technique [5].

To make feasible proposals, this article takes Chicago's O'Hare international airport's security check process as a standard and establishes a simple Petri net model, analyzing its overall performance.

2 Airport Security Process Analysis

2.1 Security Check Work Flow

The goals of security measures are to prevent passengers from hijacking or destroying aircraft and to keep all passengers safe during their travel. Chicago's O'Hare international airport's security check process can be separated into four zones.

Content from this work may be used under the terms of the Creative Commons Attribution 3.0 licence. Any further distribution of this work must maintain attribution to the author(s) and the title of the work, journal citation and DOI.
Published under licence by IOP Publishing Ltd

IOP Conf. Series: Materials Science and Engineering **242** (2017) 012105 doi:10.1088/1757-899X/242/1/012105

The work flow can be divided into two parts, one is that passengers should go through millimeter wave scanner or metal detector, and another is that all their belongings should pass X-ray screening machine.

Passengers randomly arrive at zone A and wait in a queue until a security officer can inspect their identification and boarding documents. After passengers' identifications checked, they enter zone B waiting for millimeter wave scanner or metal detector while their belongings ready for being X-ray screened. Passengers must remove shoes, belts, jackets, metal objects, electronics, and containers with liquids, placing them in a bin to be X-rayed separately; laptops and some medical equipment also need to be removed from their bags and placed in a separate bin. All of their belongings, including the bins containing the aforementioned items, are moved by conveyor belt through an X-ray machine, where some items are flagged for additional search or screening by a security officer (Zone D).Passengers that fail this step receive a pat-down inspection by a security officer (Zone D).The passengers finally proceed to the conveyor belt on the other side of the X-ray scanner to collect their belongings and depart the checkpoint area.

2.2 Assumptions

According to the working flow of security check, we make the assumptions as below:

• Collected data can represent staffs' average working efficiency in Chicago's O'Hare international airport.

• No time was observed for placing items on the belt going into x-ray.

• Transit time between process stations is contained in the waiting time.

• People can leave security check area only if he or she has retrieved their belongings with themselves and their belongings checked.

• Passengers are considered to leave the system if they enter the zone D and it will not have an effect on efficiency.

3 Mathematical Model of Airport Security Work Flow Based on Petri Net

3.1GSPN Model Establishment

According to the security checkpoint process, build up a single-channelled security checkpoint process GSPN model, that is, the channel consists of only one security officer, one X-ray scanning machine and one millimeter wave scanning machine. The GSPN model is given by Fig.1.

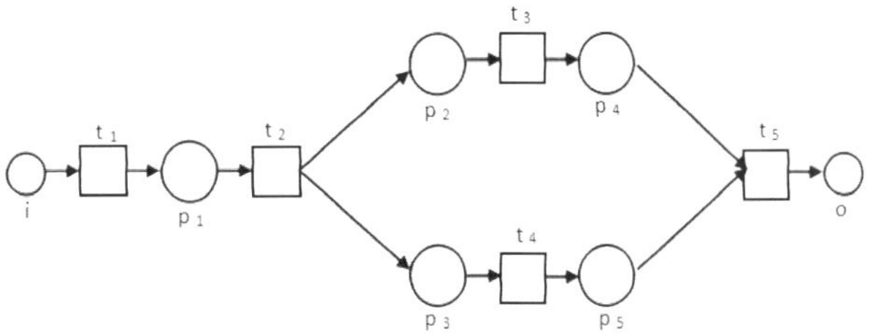

Fig. 1GSPN model that can represent a single-channelled security checkpoint

Illustration of symbols used in PSGN model is given by Table.1 and Table.2.

Table. 1 Places and their meanings

Place	*Meaning*
i	*Passengers enter the security checkpoint*
P_1	*Passengers that pass ID check process*
P_2	*Items wait for x-ray scanning*
P_3	*Passengers wait for millimeter wave scanning*
P_4	*Items wait for being retrieved*
P_5	*Passengers wait for retrieving belongings*
o	*Passengers leave the security check point*

Table. 2 Transitions and their meanings

Transition	*Meaning*
t_1	*Passengers get their identification and board documents checked*
t_2	*Passengers move to the screening line*
t_3	*Passengers' belongings move through x-ray machine*
t_4	*Passengers go through millimeter wave scanning machine*
t_5	*Passengers retrieve their belongings*

3.2Markov Chain Establishment

The MC (Markov Chain) of a SPN can be obtained from the reachability graph of the Petri net (N, M_0). The MC state space is the reachability set $R(M_0)$, and the transition rate from state M_i to state M_i is given by $q_{ij}=\lambda'_i$, the firing rate of transition ti transforming Mi into $M_j(q_{ij}=\lambda'_{i1}+\lambda'_{i2}+\cdots)$, if there are two or more transitions t_{i1},t_{i2} transforming M_i into M_j; $q_{ij}=0$ if no transitions transforming M_i into $M_j,(i\neq j)$. The square matrix $Q=[q_{ij}]$ of order $s=|R(M_0)|$ is known as the transition rate matrix [6].

(1) Passengers' arrival

The arrival of the passenger is independent of the arrival of the preceding passenger. Calculate the average arrival rate of channels.

Table. 3 Data processing of passengers' arrival

Total Time	*0:17:99.5*
Arrival	*106*
Arrival Rate (p/min)	*5.7*

(2) ID Check

Consider the machines' and staffs' handling capacity as a whole, we have assumed that the data can represent the average level of airport efficiency.

Table. 4 Data processing of ID check

Time	*00:08.0*	*00:05.3*	*00:11.1*	*00:10.0*
Time	*00:09.0*	*00:09.0*	*00:12.6*	*00:15.0*
Time	*00:14.6*	*00:11.8*	*00:14.8*	*00:20.4*
Time	*00:07.7*	*00:07.5*	*00:10.9*	*00:11.9*
	Average Time (s)	*11.225*	*Processing Rate (p/min)*	*5.35*

(3) Passengers' Arrival at security Check Area

We assume a passenger arrives at the security check area once he or she leaves the ID check service counter, so the firing rate is the same as ID check processing rate, which is 5.35 person per minute.

(4) Millimeter Wave Scanning

Table. 5 Data processing of millimeter wave scanning

Total Time	*0:07:33.8*
leaving	*40*
Leaving Rate (p/min)	*5.29*

(5) X-Ray Scanning

Table. 6 Data processing of x-ray scanning

Total Time	*0:0:26.4*
leaving	*15*
Leaving Rate (p/min)	*10.41*

(6) Passengers' Leaving of Security Check Area

People can leave security check area only if he or she has retrieved their belongings with themselves and their belongings checked. The whole scanning process' average service rate can be treated as people's average leaving rate from security area.

Table. 7 Data processing of passengers' leaving

Total Time	*0:13:50*
leaving	*29*
Leaving Rate (p/min)	*2.1*

Firing rates of each transition are yielded by Table.8 (with regular security check mechanism)

Table. 8 Transitions and their firing rates

Transition	*Average Time/s*	*Firing Rate*	*Firing Rate Value*
t1	*11.225*	$\lambda 1$	*5.35*
t2	*11.225*	$\lambda 2$	*5.35*
t3	*5.76*	$\lambda 3$	*10.41*
t4	*11.345*	$\lambda 4$	*5.29*
t5	*28.62*	$\lambda 5$	*2.1*
t6	*9.05*	λ	*5.7*

Add firing rates to the GSPN, the model can be modified into Fig.3 and the MC can be inferred (Fig.4).

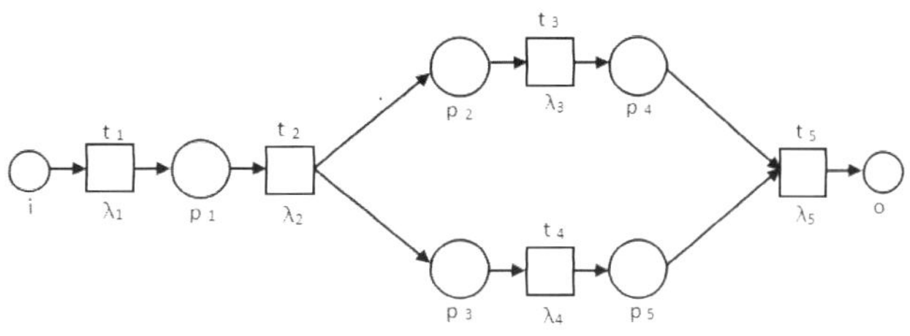

Fig. 2 Modified GSPN model

IOP Conf. Series: Materials Science and Engineering **242** (2017) 012105 doi:10.1088/1757-899X/242/1/012105

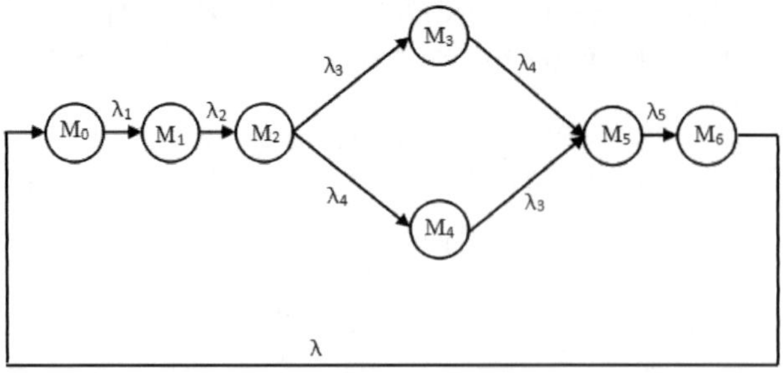

Fig. 3Markov Chain

3.3Bottlenecks Identification

Since the SPN is reversible, i.e., $m_0 \in R$ for every $m_i \in R(m_0)$, then, the SPN can generate an ergodic continuous-time MC and it is possible to compute the steady-state probability distribution X by solving the linear system

$$\begin{cases} XQ = 0 \\ \sum_{i=1}^{s} x_i = 1 \end{cases} \quad (1)$$

Transition rate metrics is given by

$$Q = \begin{bmatrix} -\lambda_1 & \lambda_1 & 0 & 0 & 0 & 0 & 0 \\ 0 & -\lambda_2 & \lambda_2 & 0 & 0 & 0 & 0 \\ 0 & 0 & -\lambda_3-\lambda_4 & \lambda_3 & \lambda_4 & 0 & 0 \\ 0 & 0 & 0 & -\lambda_4 & 0 & \lambda_4 & 0 \\ 0 & 0 & 0 & 0 & -\lambda_3 & \lambda_3 & 0 \\ 0 & 0 & 0 & 0 & 0 & -\lambda_5 & \lambda_5 \\ \lambda & 0 & 0 & 0 & 0 & 0 & -\lambda \end{bmatrix} \quad (2)$$

Assume the steady-state probability is

$$X = [x_1, x_2, x_3, x_4, x_5, x_6] \quad (3)$$

We can get the linear system

$$\begin{cases} -\lambda_1 x_0 + \lambda x_6 = 0 \\ \lambda_1 x_0 - \lambda_2 x_1 = 0 \\ \lambda_2 x_1 - (\lambda_3 + \lambda_4) x_2 = 0 \\ \lambda_3 x_2 - \lambda_4 x_3 = 0 \\ \lambda_4 x_2 - \lambda_3 x_4 = 0 \\ \lambda_4 x_3 + \lambda_3 x_4 - \lambda_5 x_5 = 0 \\ \lambda_5 x_5 - \lambda x_6 = 0 \end{cases} \quad (4)$$

Thus the steady-state probability of MC is given by

$$\begin{cases} x_0 = 0.1499 \\ x_1 = 0.1499 \\ x_2 = 0.0511 \\ x_3 = 0.1005 \\ x_4 = 0.0260 \\ x_5 = 0.3819 \\ x_6 = 0.1407 \end{cases} \quad (5)$$

The expected value of the number of tokens: Let $B(i,n)$ be the subset of $R(m_0)$ for which the number of tokens in a k-bounded place pi is n. Then, the expected value of the number of tokens in place p_i is given by

$$E[m_i] = \sum_{n=1}^{k} \{nP[B(i,n)]\} \quad (6)$$

Transition utilization probability is the sum of steady-state probabilities of states that enable the transition.

Table. 9 Expected token of each place and transition utilization probability of each transition

Place	*Expected token*	*Transition*	*Transition Utilization Probability*
i	*0.1499*	*t1*	*0.1499*
P_1	*0.1499*	*t2*	*0.1499*
P_2	*0.0771*	*t3*	*0.0771*
P_3	*0.1516*	*t4*	*0.1516*
P_4	*0.4824*	*t5*	*0.3819*
P_5	*0.4079*		
o	*0.1407*		

As can be seen from the table, there are problems at places and tokens. For one thing, token numbers at P_4 and P_5 is larger. P_4 and P_5 are all part of the extraction of baggage. Accumulation of passengers here may be caused by uncoordinated work efficiency between X-ray scanner and millimeter wave scanner. That results in either passengers waiting for their items or baggage waiting to be retrieved. For another thing, the utilization rate of change t_5 is 40.79%, which is relatively busy compared with other transitions. t_5 is also part of the extraction of baggage, this may be caused by some passengers' low packing up efficiency.

4 Improve Proposals

By analyzing the data, we conclude that $E[p_4] = x_3 + x_5$ and $E[p_5] = x_4 + x_5$ are larger than others. The part they share is x_5, so decreasing x_5 can make the values smaller, and thus reduce the congestion.

According to the analysis:

$$\begin{cases} -\lambda_1 x_0 + \lambda x_6 = 0 \\ \lambda_1 x_0 - \lambda_2 x_1 = 0 \\ \lambda_2 x_1 - (\lambda_3 + \lambda_4) x_2 = 0 \\ \lambda_3 x_2 - \lambda_4 x_3 = 0 \\ \lambda_4 x_2 - \lambda_3 x_4 = 0 \\ \lambda_4 x_3 + \lambda_3 x_4 - \lambda_5 x_5 = 0 \\ \lambda_5 x_5 - \lambda x_6 = 0 \end{cases} \quad (7)$$

Obviously, changing the relevant variable x_3 and x_4 will reduce x_5. The rate of entering is decided by the passengers, so we should improve the efficiency of services, which reduces the x_3 and x_4.

Checking of people and baggage is carried out at the same time, so the effect on the total time depends on the longer one, that is, the time spent on checking people. But after the observation of the data, we found that the total time of passengers from the beginning the check point was much more than the time of checking itself. We can easily figure out that the crux is the time spent on collecting items. So we give the corresponding feasible solution to this phenomenon.

First of all, we suggest that the airport should move the process of collecting items to another zone. The staffs help passengers transfer baggage, and passengers only need to reach the appropriate location to take their baggage. This can make the efficiency of exiting is not affected by personal efficiency, which can greatly reduce the congestion.

Secondly, the TSA can double the number of staff at checking channel, so that the checking efficiency, which is also known as the firing rate of t_4, can be doubled.

Furthermore, the TSA should add a separate zone E where is provide for passengers to pack their bags. And increase staff in the area to remove the baggage into the specific area. Thus we can reduce x_5.and increasing the leaving rate. The revised figure is:

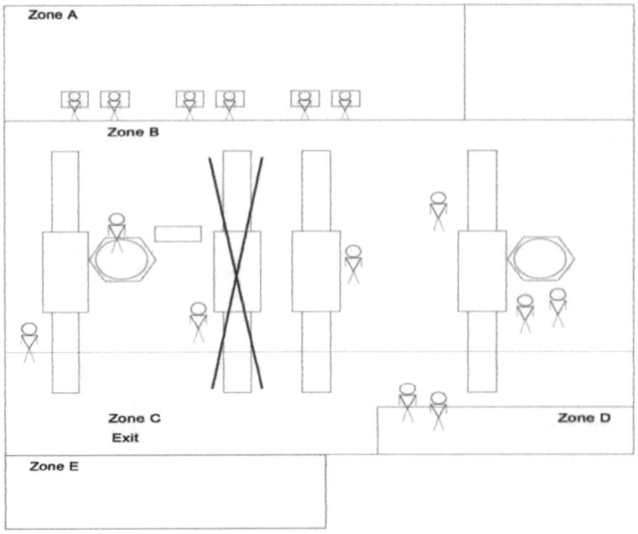

Fig. 4 Illustration of thetas security screening process after modification

Finally, TSA should also horn the efficiency of internal processes to reduce x_3 and x_4, thereby reducing the x_5. That is, the TSA should focus on increasing staff in the place where staff check baggage and people, but also should pay attention to the occupational training.

Assume that after allocating another zone for storing luggage and doubling the checking efficiency, the firing rate is changed as below:

Table. 10 Transitions and their firing rates after modification

Transition	*Average Time/s*	*Firing Rate*	*Firing Rate Value*
t1	*11.225*	$\lambda 1$	*5.35*
t2	*11.225*	$\lambda 2$	*5.35*
t3	*5.76*	$\lambda 3$	*10.41*
t4	*11.345*	$\lambda 4$	*10.58*
t5	*28.62*	$\lambda 5$	*6*
t6	*9.05*	λ	*5.7*

The steady-state probability of MC is changed into

$$\begin{cases} x_0 = 0.2176 \\ x_1 = 0.2176 \\ x_2 = 0.0555 \\ x_3 = 0.0546 \\ x_4 = 0.0564 \\ x_5 = 0.1940 \\ x_6 = 0.2042 \end{cases} \quad (8)$$

In that sense, all of the process can stay balanced. The modified token number and transition utilization probability is shown in Chart.11, in which we can see the accumulation of passengers is sharply reduced.

Table. 11 Expected token of each place and transition utilization probability of each transition after modification

Place	*Expected token*	*Transition*	*Transition Utilization Probability*
i	*0.2176*	*t1*	*0.2176*
P_1	*0.2176*	*t2*	*0.2176*
P_2	*0.1119*	*t3*	*0.1119*
P_3	*0.1101*	*t4*	*0.1101*
P_4	*0.2486*	*t5*	*0.1940*
P_5	*0.2504*		
o	*0.2042*		

5 Summary

This paper applies a simple Petri net model to American airport security system work flow, aiming at finding the bottleneck and give proposals. The calculation in Petri net model shows uncoordinated efficiency between X-ray scanner and millimeter wave scanner and passengers' low luggage extraction rate is the main crux to the long queue. In this respect, we balance the places' busy probabilities and the transitions' utilization probability, finding that the addition of millimeter scanners and temporary luggage depositary at non-security area can horn the efficiency.

References

[1] Guanghui Gu, Li Xu, and Qiangwei Xia, *Research on Simulation of Terminal Flight Departure.* Science, 2011(33): p. 91-92.(in Chinese)

[2] Naiyi Li, *Research on Dynamic Simulation of Terminal Flight Departure.*KE JI FENG, 2012(12): p. 40-41.(in Chinese)

[3] Xiaoqing Zhang and Qinming Wang, *Analysis of Urban Traffic Emergency Management System Based on Petri Net.* Western China Communications Science and Technology, 2015(6).(in Chinese)

[4] Jun Xiao, et al., *Workflow Modeling and Time Performance Analysis Based on Petri Net.* Journal of Jilin University (Information Science Edition), 2009. 27(1): p. 104-112.(in Chinese)

[5] Jiangtao Qin, *Performance Analysis of Manufacturing System Based on Petri Net Simulation.* Industrial Engineering and Management, 2014. 19(1): p. 8-15.(in Chinese)

[6] Marsan, M.A., G. Balbo, and G. Conte, *Performance models of multiprocessor systems*. 1986: MIT Pr. 403.

A survey on the modeling and applications of cellular automata theory

Yimin Gong[1, a)]

[1] Jiangsu Tianyi High School, Wuxi 214101, China

[a)] gongyimin1704@163.com

Abstract .The Cellular Automata Theory is a discrete model which is now widely used in scientific researches and simulations. The model is comprised of some cells which changes according to a specific rule over time. This paper provides a survey of the Modeling and Applications of Cellular Automata Theory, which focus on the program realization of Cellular Automata Theory and the application of Cellular Automata in each field, such as road traffic, land use, and cutting machines. Each application is further explained, and several related main models are briefly introduced. This research aims to help decision-makers formulate appropriate development plans.

1. Introduction

Cellular automata were first developed by John von Neumann as formal models of self-reproducing organisms [1-3]. An original two-dimensional cellular automata, which each cell is a small square in a large grid paper, is developed afterwards. Each cell has two possible states, black and white, which are determined by its neighborhood. In John von Neumann's theory, the neighborhood of a cell is four adjacent squares, as the gray squares in the figure shown below, which is called Von Neumann neighborhood, shown in Fig. 1 and addressed by Z^2.

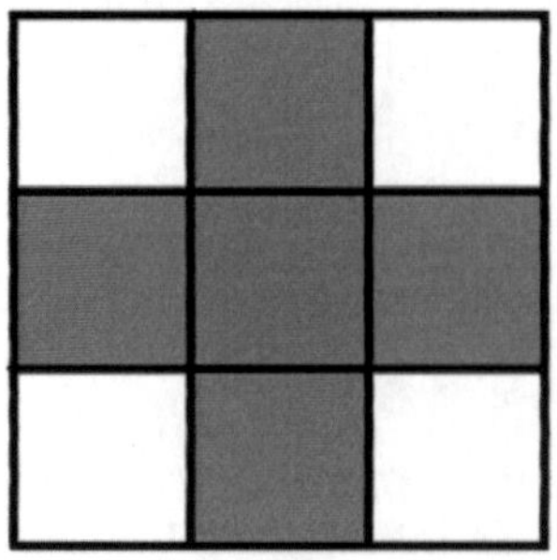

Figure 1. Von Neumann neighborhood

In 1969, computer scientist Alvy Ray Smith completed a Stanford PhD dissertation on cellular automata theory [3]. The author proved that two-dimensional CA were computation universal, introduced 1-dimensional CA, and showed that they too were computation universal, even with simple neighborhoods. It is also showed how to subsume the complex von Neumann proof of construction universality (and hence self-reproducing machines) into a consequence of computation universality in a 1-dimensional CA. In 1970s, a two-state, two-dimensional cellular automaton named Game of Life became widely known [4]. In this cellular automaton, it use a neighborhood named Moore

Content from this work may be used under the terms of the Creative Commons Attribution 3.0 licence. Any further distribution of this work must maintain attribution to the author(s) and the title of the work, journal citation and DOI.
Published under licence by IOP Publishing Ltd

neighborhood (shown in Fig. 2), which is the surrounding eight squares, as the gray squares in the figure shown below [5].

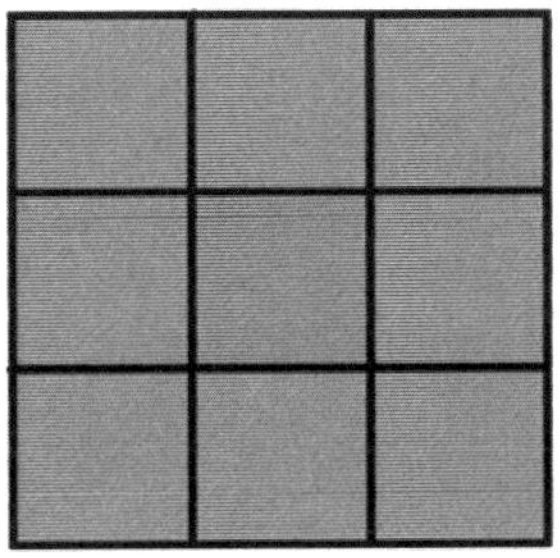

Figure 2. Moore neighborhood

Later the cellular automaton had been extended to a d-dimensional board, addressed Z^d . Recently, in modern cellular automaton theory, the lattice structure is provided by any group. This new algorithm has been referred to as the classical case in the rest of the present paper. Therefore, cellular automata have been used in various topics like group theory, but also recognition, decidability questions, computational universality, dynamical systems. The composition and rules of the game of life are as follows: cellular distribution rules in the grid division; cellular with 0 and 1 of two states, 0 represents "death", 1 represents "live"; at the present time, if a cellular state is "alive", and the eight adjacent cell in two or three the status of "alive", the next time the cellular remain "alive" , otherwise it will turn to "dead"; at the present time, if a cellular state is "dead", and the eight adjacent cell in exactly three "students", the cell will "born"(turn to "alive". Although the rules are very simple, the game of life is a cellular automaton model with the ability to generate dynamic structure. The distribution of the game of life and the initial state of the cell value is related to the initial state, given an arbitrary distribution, after several step operation, some patterns will soon disappear, but some patterns are fixed, and some repeat two or several patterns. The most famous is the glider, it will move to a fixed direction.

After that, the cellular automata theory is further developed and used in a wide range of areas. Its applications appear in biology, ecology, physics, chemistry, transportation science, computer science, information science, geography, environment science, sociology, military science and complexity science. This paper presents the state-of-the-art modeling and applications of cellular automata theory in recent years. The rest of this paper is organized as follows. In section 2, the datasets employed in experiments are reviewed briefly. Section 3 investigates the properties of the methods on modeling and applications in detail. The conclusion is presented in section 4.

2. Modeling of Cellular Automata Theory

2.1 Definition of Cellular Automata

A cellular automata is a tuple (d,S,N,f) where S is a finite set of states, $N \subseteq_{finite} Z^d$ is the finite neighborhood and $f : S^N \to S$ is the local rule of the cellular automaton. A configuration $c \in S^{Z^d}$ is a coloring Z^d of by S, shown in Fig. 3. The global map $G : S^Z \to S^{Z^d}$ applies uniformly and locally as:

$$\forall c \in S^Z,\ \forall z \in Z^d$$

$$F(c)(z) = f\left(c_{|z+N}\right). \quad (1)$$

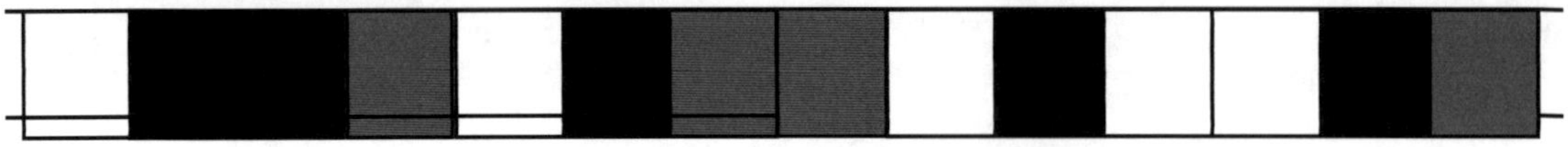

Figure 3. A configuration of cellular automata.

A space-time diagram $\Delta \in S^{Z^d \times N}$ satisfies, for all $t \in N$, there is $\Delta(t+1) = F(\Delta(t))$. Then, the von Neumann neighborhood and the Moore neighborhood can be represented as Eq. 2 and Eq. 3, respectively. Moreover, the set of configurations, S^{Z^d} is uncountable. We can consider a reasonable countable subset, for recursive configurations are useless, undecidability is everywhere (Rice theorem) [4]. First, finite configurations are with a quiescent state. Second, periodic configurations are ultimately periodic. Last, ultimately periodic configurations compromise. It is locality that can also consider partial space-time diagrams to study all configurations, used in next section.

$$N_{vN} = \{0\} \times \{-1,0,1\} \cup \{-1,0,1\} \times \{0\} . \quad (2)$$

$$N_{Moore} = \{-1,0,1\} \times \{-1,0,1\} . \quad (3)$$

2.2 Modeling of Elementary Cellular Automata

Use a two-dimensional array of $m(n,t)$ in the design of a cellular automaton data storage. In a Matlab simulation environment, put an axis into $n \times t$ box, where t means an element of each cell corresponds to the array, and n is the cellular space. I cellular in the state at t with $m(I,t)$ to represent the 0 indicates death, 1 indicates that the cell is alive. Besides, blue is also used to represent the cell of the 1 state, and white is used to represent the cell. The boundary condition of the cell is periodic, and the neighbor is a cell of the I cell, which is the neighbor length of 1, and the transformation rules are determined for a given number of elementary cellular automata. The algorithm is illustrated in Table 1.

Table 1. The algorithm of elementary cellular automata.

Steps	Definitions
1.	Initialize cell states with random numbers.
2.	Enter the conversion rules from the user interface (0-255).
3.	Process the conversion rule number makes him an algorithm subroutine. *Specific algorithm*: read the initial value from the t time as $m(1:n,t)$, by the rules to determine and generate the next moment the state of each cell $m(1:n,t+1)$.
4.	Call the subroutine from the 1-T loop.
5.	Print the user interface output.

2.3 Modeling of Game of Life

Design of "game of life", we will divide the plane into square, each square represents a cell, cellular state of 0 deaths, 1 said live, the neighborhood radius is 1, the neighborhood was of the Moose type, evolution rules:

$$S(t)=1,\ S(t+1)=\begin{cases}1 & S'=2,3\\ 0 & S'\neq 2,3\end{cases};$$
$$S(t)=0,\ S(t+1)=\begin{cases}1 & S'=3\\ 0 & S'\neq 3\end{cases}. \qquad (4)$$

Among them, $S(t)$ represents the state of t cell, and S' is the number of viable cells in 8 adjacent cells. In the program, the user uses the mouse to enter the initial state of the cell through the graphic field. After giving the definition of the neighbor and the local rules, the program can run automatically, and produce a variety of evolutionary patterns. We use two-dimensional matrix $X(m,m)$ to define the state of the cell at the moment t. The specific algorithms are illustrated as follows.

1. Define four direction vectors named adjacency matrix.

$$n=[1 \quad 1:m-1]$$
$$e=[1:m-1 \quad m]$$
$$s=[m \quad 2:m]$$
$$w=[2:m \quad 1] \qquad (5)$$

2. Sum the neighborhood matrix according to the Moore neighbor model and matrix N for east, south, west, north, southeast, southwest, northeast, northwest and the eight neighbors.

$$N=X(n,:)+X(s,:)+X(:,e)+X(:,w)+X(n,e)+$$
$$X(n,w)+X(s,e)+X(s,w) \qquad (6)$$

3. Get the algorithm of computing the state matrix of t+1 at the next t moment according to the rule of Comvay.

$$X=(X\,\&(N=2))\,|\,(N=3) \qquad (7)$$

3. Applications of Cellular Automata Theory

In this section, it provides the description of Applications of Cellular Automata Theory that are regarded as a benchmark in the current research.

3.1 Traffic Flow Based on Cellular Automaton

A traffic flow based on cellular automaton model named NaSch is a basic model describing road traffic [7]. The model is a discrete dynamical model. The Space is discrete and consists of an infinite regular grid of cells. Each cell is described by a state among a common finite set. Similarly, time is discrete, and at each clock tick cells change their state deterministically, synchronously and uniformly according to a common local update rule. This paper illustrates that the model cannot explain some phenomena because the rules of increasing and decreasing speed of cars is not similar to that in the real world. It should include the influence of the nearest car and the second nearest car. The author introduced some new variables such as the sensitivity coefficient of nearest car and the second nearest car.

3.2 Simulation of Complex Changes Based on Neural-network-based Cellular Automaton

A simulation of complex changes Based on Neural-network-based Cellular Automaton named ANN-CA model is composed of simple networks [9]. The model consists of two relatively independent modules: model correction (training) and simulation. The two modules use the same neural network. In the model correction module, the parameters of the model are obtained automatically by the training data. The structure of the entire model is very simple- the user does not have to define their own rules and parameters, which are suitable for simulating complex land use system. The network has 3 layers, the first layer is the data input layer, where the neurons correspond to the effects of land use change variables; the second layer is the hidden layer; the third layer is the output layer, which is composed of a number (N) of output neurons, and get the probability of changing between each type of land use.

3.3 Temperature Field Simulation in Cutting Process Based on Cellular Automaton

The study on temperature field simulation in cutting process based on cellular automaton takes the temperature field [8]. The field simulation milling insert produced in cutting process as the research object, carries on the grid division of the cutting tool entity with the cellular automata, establishes the cellular automata's model of the milling insert temperature field and studies the optimized mechanism of the milling insert in cutting process, it can enable the cutting value of the milling insert arrive the optimization. In this example, there is a one-to-one correspondence between the equivalence classes of construction triples and the cellular automaton. Note that it is quite common to define a cellular automaton as an equivalence class of construction triples. Many papers use this same definition without mentioning it, for it is supposed to be known, but you may still see that the map in this case is a global transition map.

4. Conclusion

The question arises whether other classical theorems on cellular automata are also true for different models of cellular automata. As the application of cellular automata theory becomes more and more widespread, the modelling of cellular automata on computers became more and more useful. In different fields, the cellular automata are used differently to suit the need. By using specially designed cellular automata, such as Neural-network-based cellular automata and traffic flow cellular automata, we can solve problems with complexity, which are hardly solved by other methods.

References

[1] Sarkar, P. (2000). A brief history of cellular automata. Acm Computing Surveys, 32(1), 80-107.
[2] T. Ceccherini-Silberstein, M. Coornaert, Cellular automata and groups, Springer Monographs in Mathematics, Springer-Verlag, Berlin (2010).
[3] E.R. Berlekamp, J.H. Conway, R.K. Guy, Winning Ways for Your Mathematical Plays, Academic Press (1982).
[4] M. Margenstern, on a characterization of cellular automata in tilings of the hyperbolic plane, Internat. J. Found. Comput. Sci. 19 (2008), no. 5, 1235–1257.
[5] J. Kari, Theory of cellular automata: A survey, Theoretical Computer Science 334 (2005), 3–33.
[6] Shun Wai Tsang, Yee Leung. A Theory-Based Cellular Automata for the Simulation of Land-Use Change. Geographical Analysis 43(2011), no. 2, 142-171.
[7] Xianjuan Kong. Research on Modelling and Characteristics Analysis of Traffic Flow Based on Cellular Automaton. Beijing Jiaotong University, 2007.
[8] Xingqin Cao. The Cellular Automata Studying of Complex System. Huazhong University of Science and Technology, 2006.
[9] Xia Li, Jiaan Ye. Neural network based cellular automata and simulating complex land use system. Geography Research (2005), no. 24, 19-27.

IOP Conf. Series: Materials Science and Engineering **242** (2017) 012107 doi:10.1088/1757-899X/242/1/012107

Fast manifold spectral clustering algorithm for intelligent traffic remote sensing image fuzzy edge

Cong Deng[1, a)], Zelin Jia[1, b)], Shen'an Li[1, c)] and Pengfei Tang[1, d)]

[1]School of Natural and Applied Sciences, Northwestern Polytechnical University, Xi'an 710072, China

a) dengconggood@mail.nwpu.edu.cn

b) 2014lishenan@mail.nwpu.edu.cn

c) jiazelin@mail.nwpu.edu.cn

d) 1430342822@ mail.nwpu.edu.cn

Abstract. Intelligent transportation system is the future development of traffic systems. Higher precision map requires a higher absolute coordinate accuracy. However, cluster labels cloud be connected to the wrong pixels near the edges of regions, and edges will be misplaced. This paper proposed a fast Manifold Spectrum Clustering Algorithm for the fuzzy-edge of intelligent traffic remote sensing image. we first construct the fuzzy edge of remote sensing image, then analyze and evaluate the influence degree of the fuzzy edge on detailed image, and proposed a spectrum clustering method based on manifold to improve the accuracy and speed to obtain the fuzzy edge of the road in the high resolution remote sensing image. The simulation environment is built by Matlab 2015a, and it is proved that this paper is superior to the existing method in clustering speed and precision performance.

1. Introduction

Intelligent transportation system is the future development of traffic systems, which is a wide-range, real-time, accurate and efficient integrated transportation management system that plays a role in all-round. Intelligent transportation system relies on the high-precision map, which is much different from the traditional map. Higher precision map requires a higher absolute coordinate accuracy, and more information of details as well as high-precision map contains more rich and detailed road traffic information elements.

This paper aims to improve the accuracy and speed of obtaining the fuzzy edges of the road by using the high-resolution remote sensing image through a new manifold spectral clustering algorithm, which provides more accurate road information for high-precision maps - road shape and width. It provides a feasible method to build high-precision map for the future.

Image clustering is an effective method for image segmentation, but it is usually very sensitive to noise. Some other researchers use texture descriptors or spectral reduction to reduce the difference between pixels in an area. But these methods are usually dependent on some important parameters, resulting in reduced robustness of clustering. To enhance the robustness of clustering, Ahmed et al. proposed bias-corrected fuzzy C-means (BCFCM). Based on the fuzzy C-means clustering algorithm (FCM), BCFCM increases the estimation of offset field and the use of spatial information, which can eliminate the influence of image offset field on segmentation. However, BCFCM has not considered the effect of noise on the estimation of offset field, so the effect of segmentation of high-noise image is poor.

Content from this work may be used under the terms of the Creative Commons Attribution 3.0 licence. Any further distribution of this work must maintain attribution to the author(s) and the title of the work, journal citation and DOI.
Published under licence by IOP Publishing Ltd

Liew et al. presented an adaptive FCM which local spatial information is applied to impose local spatial continuity, and noise suppression and ambiguity resolution are allowed. Dulyakarn and Rangsanseri added some priori spatial information to FCM, we can see that the segmentation results are improved by the spatial information from the experiments. Krinidis and Chatzis presented a robust method called fuzzy local information C-means (FLICM). FLICM is a noise insensitive method without prior knowledge. FELICM introduce the weights of pixels within local neighbor windows to reduce edge degradation. However, FELICM assumes that the label of one pixel is related to the labels of its spatial neighbors. Therefore, cluster labels cloud be connected to the wrong pixels near the edges of regions, and the edges will be misplaced.

Aiming at the above problems, this paper proposed a fast Manifold Spectrum Clustering Algorithm for the fuzzy-edge of intelligent traffic remote sensing image. By analyzing the hierarchical methods, the partition-based methods, density-based methods and the grid-based methods, which provide the theoretical and methodological basis for building a new clustering algorithm for intelligent traffic remote sensing image. In this paper, we first construct the fuzzy edge of remote sensing image, then analyze and evaluate the influence degree of fuzzy edge on detailed image, and proposed a spectrum clustering method based on manifold to improve the accuracy and speed to obtain the fuzzy edge of the road in the high resolution remote sensing image. The simulation environment is built by Matlab 2015a, and it is proved that this paper is superior to the existing method in clustering speed and precision performance.

2. Basic theory of remote sensing image clustering

2.1 Hierarchical Methods

The method has two main paths – agglomerative way (bottom-up) and divisive way (top-down). The beginning of the agglomerative way is to regard every object as a single cluster, then seek the same cluster according to the linkage and finally form clusters. The top-down approach is the opposite process, which regard all objects as one cluster, then exclude object according to the linkage and finally every object belongs to a cluster. These two kinds of path have no difference between the pros and cons. Only in the practical application, we will consider which path is faster according to the feature of the data and the number of clusters. The linkages to decide the cluster are the shortest distance linkage, the longest distance linkage, the median distance linkage and the average linkage, etc. (where the average linkage is usually regarded as the best linkage because its good monotony and its moderate degree of space expansion /concentration). BIRCH (Balanced Iterative Reducing and Clustering Using Hierarchies) is mainly used when the data volume is very large and the data type is numerical; ROCK (A Hierarchical Clustering Algorithm for Categorical Attributes) is mainly used when the data type is categorical; Chameleon (A Hierarchical Clustering Algorithm Using Dynamic Modeling) uses linkage – Knn (k-nearest-neighbor) and its effect of clustering is better than BIRCH but its computational complexity is high (O(n^2)).

2.2 Partition-based Methods

The principle is simple to understand that there is a pile of points need to cluster, the cluster goal is "the points within clusters are close enough, the points between clusters are far enough." We must first determine the number of clusters, then select a few points as the initial center point, then iterative relocate the data points according to the pre-set heuristic algorithm until it reaches the goal. According to the heuristic algorithm, it forms the k-means heuristic algorithm and its variants includes k-medoids, k-modes, k-medians and kernel k-means; k-means++, intelligent k-means, genetic k-means are not sensitive to initial values; k-medoids and k-medians are not sensitive to noise and outliers; k-modes can be used when data type is categorical; kernel k-means can be used to process the non-convex data.

2.3 Density-based Methods

K-means cannot cluster the data with irregular shape. So density-based methods are proposed to systematically solve this problem. These methods handle noise data better. The principle is like drawing circles, in which two parameters need to be defined, one is the maximum radius of the circle, the other is the smallest number of points within a single circle. Finally, the points in a same circle belong to a cluster. DBSCAN (Density-Based Spatial Clustering of Applications with Noise) is a typical one, but the two parameters of the settings are very sensitive. DBSCAN's extension called OPTICS (Ordering Points To Identify Clustering Structure) The shortcomings of DBSCAN are improved by preferentially searching for high density and then setting parameters according to high density characteristics.

2.4 Grid-based Methods

The principle is to divide the data space into grid cells, map the data object sets into the grid cells, and calculate the density of each cell. According to the pre-set threshold to determine whether each grid cell is a high-density unit and then the adjacent high-density group forms a cluster. The speed of this method is independent of the number of data objects and depends only on the number of cells in each dimension in the data space. Therefore, the advantage is that the execution efficiency is high. But the shortcomings are many, for example, this method is sensitive to the parameters, cannot deal with irregular distribution of data and may cause a dimension of disaster. STING (Statistical Information Grid) and CLIQUE (Clustering In Quest) is the representative algorithm in grid-based methods.

3. Fuzzy edge fast manifold spectral clustering theory

3.1 Fuzzy Edge of Intelligent Traffic Remote Sensing Image

In the late 1970s, scholars have begun to study the extraction of roads in remote sensing images. With the improvement of satellite remote sensing image resolution as well as urban and rural construction, more and more algorithms have been proposed to adapt to the new situation.

The existing algorithm has achieved good results for the road network in the remote sensing image with simple scenes, but few algorithms can extract the complete ones. This is because the presence of trees and buildings, vehicles on the road and a variety of cross-line will cause shadow or occlusion. Similar road feature that might be parking lots, ponds, rivers and roofs, *etc.*, not only makes the road network very messy, also causes the edges of the road in the remote sensing image to be very vague and unclear.

At present, methods to extract road edge from the remote sensing image mainly are the template matching, gradient edge operator detection and others. The template matching algorithm uses the gray scale shape of the image to match the recognition image, but the computation is large and the matching speed is low. The gradient operator edge detection finds the edge by analyzing the gray scale of the image. Commonly used gradient edge detection operators are Roberts, Sobel, Prewitt, Laplacian and Canny et al. However, these traditional operators consider less directional features and often lose some of the edge details. This paper uses the improved 5 * 5 Sobel operator with 8-direction proposed by Yuan Tan *et al.*

3.2 Neighborhood Weighting

Sobel operator is an edge detection operator based on the first-order derivative. The traditional Sobel operator consists of two 2 * 3 matrixes, respectively, the vertical gradient matrix is used to detect the horizontal edge as well as horizontal gradient matrix is used to detect the vertical edges. The advantage of the Sobel operator is the simple principle and the small computation. However, since here are only two directions, the edge detection effect is better for the horizontal and vertical direction, but the effect is not very satisfactory for the image with more complex texture and edges in more directions.

1) The Sobel Operator Used in This Paper

As the traditional Sobel operator has a limitation on the edge direction, we used the improved 5*5 Sobel operator with 8-direction to reduce the deviation and increase the detection directions of edges. Eight directions are 0 °, 22.5 °, 45 °, 67.5 °, 90 °, 112.5 °, 135 ° and 157.5 ° as the Fig.1 shows,

IOP Conf. Series: Materials Science and Engineering **242** (2017) 012107 doi:10.1088/1757-899X/242/1/012107

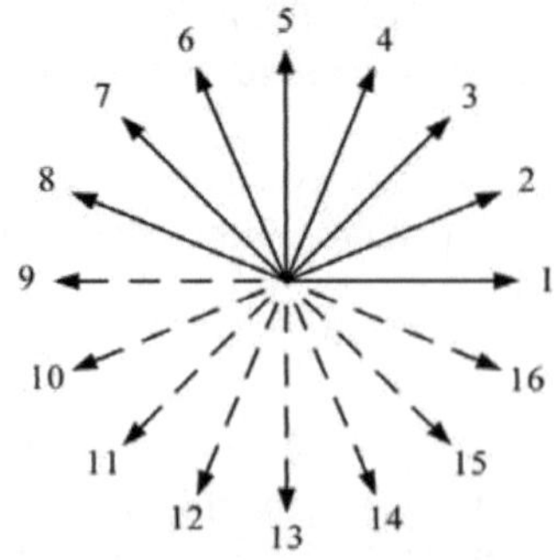

Figure 1. The improved Sobel operator

2) The Direction Templates

Traditional Sobel operator and conventional Sobel expansion operator are typically 3*3 matrix templates, we use the 5*5 matrix templates to describe the edge of the image more accurately and reduce the impact of noise on the detection results. The templates of each directions are as Fig.2 shown.

$$\begin{pmatrix} 1 & 4 & 6 & 4 & 1 \\ 2 & 8 & 12 & 8 & 2 \\ 0 & 0 & 0 & 0 & 0 \\ -2 & -8 & -12 & -8 & -2 \\ -1 & -4 & -6 & -4 & -1 \end{pmatrix} \quad \begin{pmatrix} 4 & 6 & 4 & 1 & 2 \\ 1 & 12 & 12 & 8 & 0 \\ 2 & 8 & 0 & -8 & -2 \\ 0 & -8 & -12 & -12 & -1 \\ -2 & -1 & -4 & -6 & -4 \end{pmatrix} \quad \begin{pmatrix} 6 & 4 & 1 & 2 & 0 \\ 4 & 12 & 8 & 0 & -2 \\ 1 & 8 & 0 & -8 & -1 \\ 2 & 0 & -8 & -12 & -4 \\ 0 & -2 & -1 & -4 & -6 \end{pmatrix} \quad \begin{pmatrix} 4 & 1 & 2 & 0 & -2 \\ 6 & 12 & 8 & -8 & -1 \\ 4 & 12 & 0 & -12 & -4 \\ 1 & 8 & -8 & -12 & -6 \\ 2 & 0 & -2 & -1 & -4 \end{pmatrix}$$

(a) 0° (b) 22.5° (c) 45° (d) 67.5°

$$\begin{pmatrix} 1 & 2 & 0 & -2 & -1 \\ 4 & 8 & 0 & -8 & -4 \\ 6 & 12 & 0 & -12 & -6 \\ 4 & 8 & 0 & -8 & -4 \\ 1 & 2 & 0 & -2 & -1 \end{pmatrix} \quad \begin{pmatrix} 2 & 0 & -2 & -1 & -4 \\ 1 & 8 & -8 & -12 & -6 \\ 4 & 12 & 0 & -12 & -4 \\ 6 & 12 & 8 & -8 & -1 \\ 4 & 1 & 2 & 0 & -2 \end{pmatrix} \quad \begin{pmatrix} 0 & -2 & -1 & -4 & -6 \\ 2 & 0 & -8 & -12 & -4 \\ 1 & 8 & 0 & -8 & -1 \\ 4 & 12 & 8 & 0 & -2 \\ 6 & 4 & 1 & 2 & 0 \end{pmatrix} \quad \begin{pmatrix} -2 & -1 & -4 & -6 & -4 \\ 0 & -8 & -12 & -12 & -1 \\ 2 & 8 & 0 & -8 & -2 \\ 1 & 12 & 12 & 8 & 0 \\ 4 & 6 & 4 & 1 & 2 \end{pmatrix}$$

(e) 90° (f) 112.5° (g) 135° (h) 157.5°

Figure 2. Templates of each directions

3.3 Fuzzy Edge Spectrum Clustering Based on Manifold

In order to meet the overall importance of clustering, ensure high similarity of data in the same manifold structure, and low similarity of data in the different manifold structure, manifold distance kernel measure is adopted in this paper, which is described as follows.

Firstly, we define the line length, manifold distance and manifold distance kernel.

1) Line Length on Manifold

$$L(x,y) = e^{\rho d(x,y)} - 1 \tag{1}$$

where, $d(x,y)$ presents the Euclidean distance between x and y; ρ presents the extension factor to control the line length between two data points.

2) Manifold Distance between Data Points

$$D_{i,j}^{\rho} = \frac{1}{\rho} ln(1 + d_{sp}(x_i, x_j))^2 \tag{2}$$

$$d_{sp}(x_i, x_j) = \min_{p \subset P_{ij}} \sum_{k=1}^{|p|-1} \left(e^{\rho d(p_k, p_{k+1})} - 1\right) \tag{3}$$

where, $d_{sp}(x_i, x_j)$ presents the shortest path distance between node x_i and node x_j on graph G. $d_{sp}(x_i, x_j)$ presents the shortest path Euclidean distance between x_i and node x_j on graph G. $d_{sp}(x_i, x_j)$.

3) Manifold Distance Kernel

$$K_{ij} = exp\left(-\frac{D_{i,j}^{\rho}}{2\sigma^2}\right) \tag{4}$$

Then we give the algorithmic process of fuzzy edge spectrum clustering based on manifold:

a) Input the data points $\{x_l\}_{l=1}^{n}$, the number of classes k, and the extension factor ρ.
b) Construct similarity matrix $K \in R^{n\times n}$ based on similarity measurement of manifold distance kernel where $K_{ij} = \exp\left(-\frac{D_{i,j}^{\rho}}{2\sigma^2}\right)$ presents the matrix elements (the kernel width σ is obtained by repeated experiments) ; $K_{ii} = 0, 1 \le i, j < n$ matrix main diagonal elements.
c) Construct the Laplacian matrix $\boldsymbol{L} = \boldsymbol{D}^{-1/2}\boldsymbol{K}\boldsymbol{D}^{-1/2}$ where $\boldsymbol{D}$ presents the diagonal degree matrix $D_{ii} = \sum_{j=1}^{n} k_{ij}$.
d) Solve the eigenvectors $v_1, v_2 \cdots, v_k$ corresponding to the k maximum eigenvalues of Laplacian matrix $\boldsymbol{L}$. And construct the eigenvector matrix $\boldsymbol{V} = [v_1, v_2 \cdots, v_k] \in R^{n\times k}$ where v_k is column vector.
e) Unitize the row vectors of matrix $\boldsymbol{V}$ and get matrix $\boldsymbol{Y}$ where $Y_{ij} = V_{ij}/\left(\sum_j {V_{ij}}^2\right)^{1/2}$.
f) Regard each row of matrix $\boldsymbol{V}$ as a point in the cluster space, then cluster them into k class.
g) Divide the data points of each line of the eigenvector matrix.
h) Output the result of classes $c_1, c_2, \cdots, c_k$.

4. Experiment and simulation

4.1 Simulation Environment
The experiment was carried out under the Matlab 2015a. The satellite remote sensing images were from the Quick Bird satellite and the original images were pretreated with ENVI. The purpose of pretreating the original image is to obtain the part that needs to be clustered while eliminating the influence of the vehicle on the road by cutting the remote sensing image. Secondly, the edge detection experiment is carried out on the pretreated image, and the test result is labeled on the original image and compared with the actual road edge, and the accuracy performance of the algorithm is analyzed statistically.

The result of edge detection experiment is shown in Figure 3.

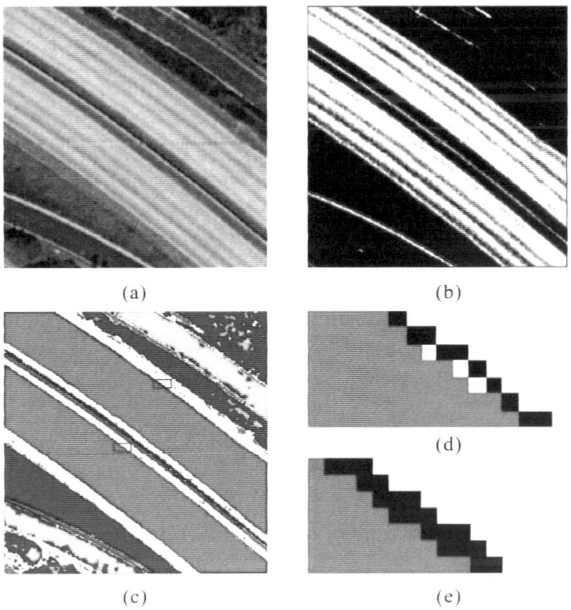

Figure 3. Edge detection experiment

4.2 Accuracy Analysis and Comparison of Simulation Results
For the complex image, many different objects have similar features, and the objects cannot be

distinguished well by only increasing the number of clusters. We compare the results of edge detection and clustering to evaluate the different clustering methods.

We use two different accuracies to quantitatively evaluate the image clustering methods: regional accuracy and edge accuracy.

First, the regional accuracy is defined as:

$$RA = \frac{\sum_{i=1}^{n} max\{A_{ik}\}}{N} \tag{5}$$

where N presents the total number of pixels in the remote sensing image, n is the number of objects in the result of edge detection, A_{ik} is the number of pixels of cluster k in object i, and $max\{\cdot\}$ is used to get the maximum number of pixels in an object.

The regional accuracy of different algorithms including FCM, MeanShift, FLICM and Ours are shown in Table 1.

Table 1. Regional accuracy

	FCM	*MeanShift*	*FLICM*	*Ours*
Accuracy	65.2%	75.3%	78.2%	83.1%

Then we use the accuracy of pixels around edges to represent the accuracies of edges. The edge accuracy can be described as:

$$EA = \frac{\sum_{i=1}^{n} A'_{ik}}{N'}, K = arg \max_{k\in\{1,c\}} A_{ik} \tag{6}$$

where N' is the number of pixels which are get by edge detection with a window, A'_{iK} is the number of pixels of cluster K in object i within the range of detection area, and K is the cluster which has the maximum number of pixels in the object.

The edge accuracy of different algorithms including FCM, MeanShift, FLICM and Ours are shown in Figure 4.

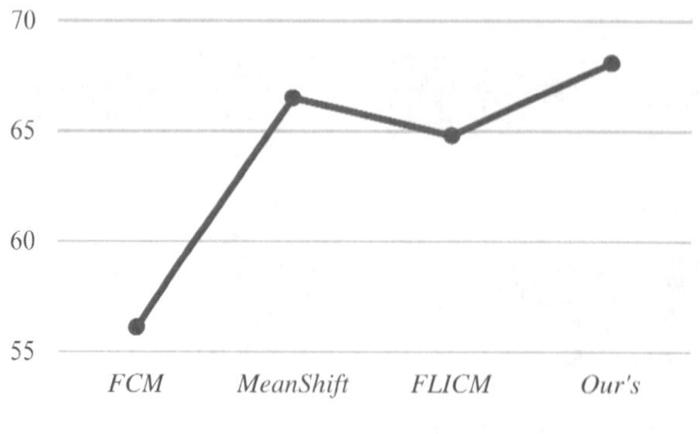

Figure 4. Performance of edge accuracy

5. Conclusion

It is an innovative practice to introduce spectral clustering method into the fuzzy edge of the road. However, the resolution of remote sensing image is limited (1 pixel equals 60 cm), and the precision of clustering cannot fully meet the requirement of high precision map. At present, our work shows that the manifold spectral clustering method is available for the map to make the road blurred edges become more accurate.

In the future, with the increase of satellite remote sensing resolution, manifold spectral clustering method can also provide more accurate road edge information.

Acknowledgments

First and foremost, I would like to show my deepest gratitude to my advisor, Dr. Lin Wei, a pleasingly affectionate rigorous and reliable scholar, have put forward many valuable opinions in the process of the thesis writing. She doesn't mind taking the trouble to give me timely help, so I can finally finish writing work. Her accurate and forward-looking academic observation benefits me in my future study.

I want to thank all the teachers in the Department of mathematics in Northwestern Polytechnical University, their selfless devotion and dedication of the attitude of academic benefits me a lot.

I would also thank Ms. Guo for all her help in the process of my research. She made valuable comments on the revision of the paper, which made my paper more perfect.

This work is supported by the National Training Programs of Innovation and Entrepreneurship for Undergraduates (No.201610699328).

Finally, I' d like to thank all my friends and my family members, especially my parents.

Reference

[1] Prabhu, S., & Tensing, D. (2015). Adapted normalized graph cut segmentation with boundary fuzzy classifier object indexing on road satellite images. International Arab Journal of Information Technology, 12(6), 597-603.

[2] Tan Y, Huang H X, Xu J M, et al. Road edge detection from remote sensing image based on improved Sobel operator[J]. Remote Sensing for Land and Resources, 2016, 28(3):7-11.

[3] Zang, Yu, et al. "Joint Enhancing Filtering for Road Network Extraction." PP.99 (2016):1-15.

[4] Shi, J, and J. Malik. "Normalized cuts and image segmentation." IEEE Transactions on Pattern Analysis & Machine Intelligence 22.8(2000):888-905.

[5] Luxburg, Ulrike. "A tutorial on spectral clustering." Statistics and Computing 17.4(2007):395-416.

[6] Ozertem, Umut, D. Erdogmus, and R. Jenssen. "Mean shift spectral clustering." Pattern Recognition 41.6(2008):1924-1938.

[7] Jung, Cheolkon, et al. "Image segmentation via manifold spectral clustering." Journal of Allergy & Clinical Immunology 127.2(2011):1-6.

IOP Conf. Series: Materials Science and Engineering **242** (2017) 012108 doi:10.1088/1757-899X/242/1/012108

A flow model for optimization of highway toll stations

Yifan Chen[1, a)]

[1]School of Mathematical Sciences, University of Chinese Academy of Sciences, Beijing 101407, China

[a)] chen-nt@outlook.com

Abstract. This paper is aimed at developing a highway toll stations route optimization model (HTSROM) for cost-effective and sustainable highway infrastructure planning and design. There are many factors for the locations of highway toll stations, including potential ridership, costs of land, construction and operation, and travel times. We developed HTSROM algorithm which used a genetic model to optimize and integrate a geographic information system for transfer of land-use, environmental, and topographic data during the optimal search process. The experimental results are discussed on the application, and more extensions of the HTSROM model remain to be studied in the future work.

1 Introduction

As road transportation brings great changes of our lives, it also brings safety and environmental problems [1-3]. The road constructions consumed uncountable resources. New transportation technologies such as connected vehicle and travel demand control, are urgently in need of reducing the negative environmental and safety impacts from transportation [2]. Barrier tolls are used as an instrument to finance new highway infrastructure. A barrier toll is a series of tollbooths which are placed across the highway, perpendicular to the traffic flow. Drivers pay their tolls when vehicles pass through the tollbooths. There are usually more tollbooths than lanes of the highway. Congestion arises when vehicles converge on the exit of the toll plaza.

A traditional mode of toll plaza on one side is illustrated in Fig.1, and the depth of color shows the willingness of drivers to go through the booth. The darker the block is, the lower that drivers are likely to pass it. The red block marks the area where congestion is mostly like to arise. Most traffic delays when an evacuation occur at intersections, so lane-based routing is important for reducing the delays [3,4]. Typically all changing lanes makes the traffic much slower, because three lines of traffic smoothly move from the left to the right, as shown in Fig.2. Suddenly a vehicle marked by a indigo dot changes its lane from the upper to the middle lane, causing a series of vehicles slowing down, marked in deep gray dots. Therefore several vehicles, marked in purple dots, decide to change lane, resulting in more deceleration, finally result in a congestion by the butterfly effect.

Content from this work may be used under the terms of the Creative Commons Attribution 3.0 licence. Any further distribution of this work must maintain attribution to the author(s) and the title of the work, journal citation and DOI.
Published under licence by IOP Publishing Ltd

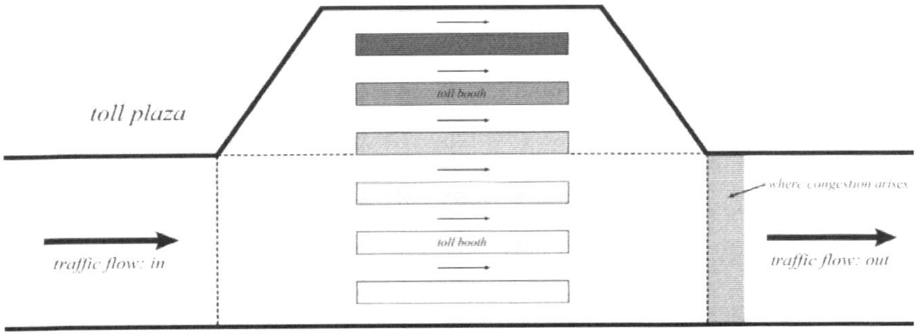

Figure 1. A traditional mode of toll plaza on one side.

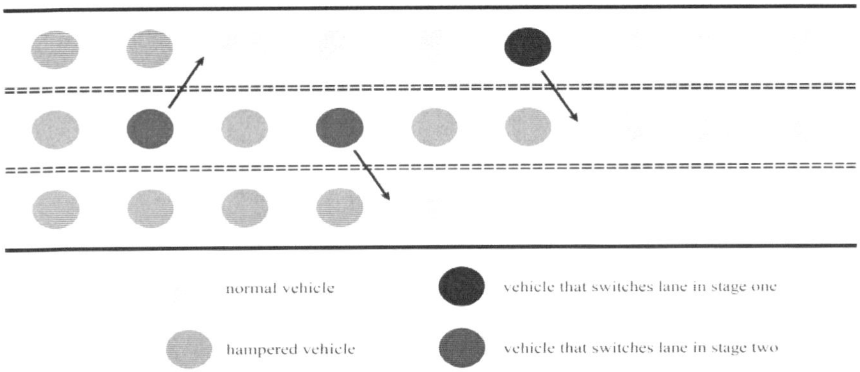

Figure 2. Typical all changing lane mode.

This paper presents a highway toll stations route optimization model (HTSROM) for cost-effective and sustainable highway infrastructure planning and design. The aim of the algorithm is that ***Average Lane-Changing Times (ALT)*** of all the vehicles involved in the system may serve as a key factor to decide the performance of the toll plaza (the lower, the better). Thereafter, another method to increase the ***Occupancy Rate of Lanes (ORL)*** is developed with a method to decrease the ***ALT***. We review the previous traditional work in prescriptive routing criteria for routing plan. The next section presents the model and formulation. Manual analysis and simulation are used to evaluate the efficiency of the HTSROM model. In the end, the paper concludes with the discussion of results and further works.

2 The proposed algorithm

As illustrated in Fig.3, if one lane has more empty places than the other lane beside, there will be a tendency for vehicles on the other lane to change lane. That is because vehicles of sparser traffic flow enjoy higher speed limit, due to safety consideration. Therefore by increasing ***ORL***, less lane-changing will happen in the traffic system, increasing the handling capacity of the toll plaza, and meanwhile ensuring a safer traffic environment.

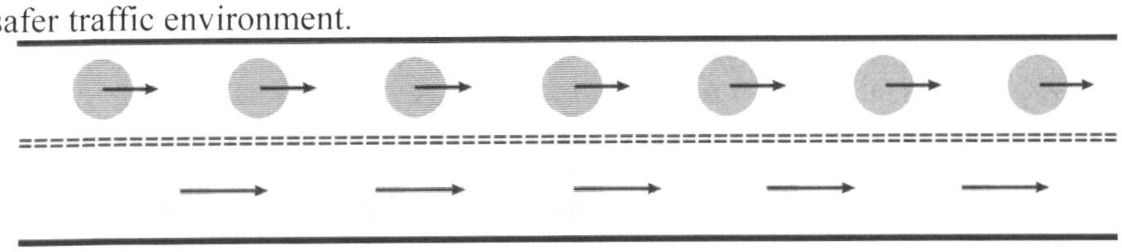

Figure 3. The basic assumption in the HTSROM mode.

Then, we choose the case when $B = 8$ and $L = 4$ as a general representation of different tolls. To simplify our discussion, we assume that the high way before and after the toll plaza contain the same number of lanes, which is always less than the number of toll booths. The table below shows some variables that we cover later in this paper.

Table 1. A brief introduction of variables in this paper.

Variables	Definitions
B	the number of tollbooths
L	the number of lanes in the highway before and after the toll plaza
plaza	the plaza matrix 1 = car, 0 = empty, -1 = forbid, -3 = booth
d_i	the distance between two vehicles
p	the probability of random deceleration of vehicles
P	the probability that a driver choose a certain booth
t	token of the booths, varies from one to eight
$v_{\max}$	maximal vehicle speed
influx	the number of vehicles that reach the toll plaza in an hour (veh/hr)
J	the number of vehicles per hour passing the area where the end of the plaza joins the ***L*** outgoing traffic lanes (veh/hr)
J_{max}	the limit of handling capacity of the toll station

2.1 CA Model

The first key of HTSROM is ***cellular automation model (CA Model)***, which is used for the simulation of specified toll plaza conditions. In the CA model, we set a main matrix to represent the toll plaza on highway. Each element of the matrix represents a cell which has four different states. The specific definition of each state is shown in Table.2. The main part of matrix ***plaza*** and the simulation diagram of CA model are shown in Fig.4. It is shown that the yellow cells are occupied by a toll barrier. Then, we mark the toll booths (yellow cells in Fig.4) with a list of numbers, shown in Fig.5.

Table 2. Four States of Cell

1	the cell is occupied by a vehicle
-1	the cell is forbidden to pass(including the road boundary)
0	the cell is unoccupied
-3	the cell is occupied by a toll barrier

$$\begin{pmatrix} -1 & 0 & 0 & 0 & 0 & -1 & -1 & -1 & -1 & -1 \\ -1 & 0 & 0 & 0 & 0 & -1 & -1 & -1 & -1 & -1 \\ \cdots & \cdots & \cdots & \cdots & \cdots & \cdots & \cdots & \cdots & \cdots & \cdots \\ -1 & 0 & 0 & 0 & 0 & 0 & 0 & 0 & 0 & -1 \\ -1 & -3 & -3 & -3 & -3 & -3 & -3 & -3 & -3 & -1 \\ -1 & 0 & 0 & 0 & 0 & 0 & 0 & 0 & 0 & -1 \\ \cdots & \cdots & \cdots & \cdots & \cdots & \cdots & \cdots & \cdots & \cdots & \cdots \\ -1 & 0 & 0 & 0 & 0 & 0 & 0 & 0 & 0 & -1 \\ -1 & 0 & 0 & 0 & 0 & 0 & 0 & 0 & 0 & -1 \end{pmatrix}$$

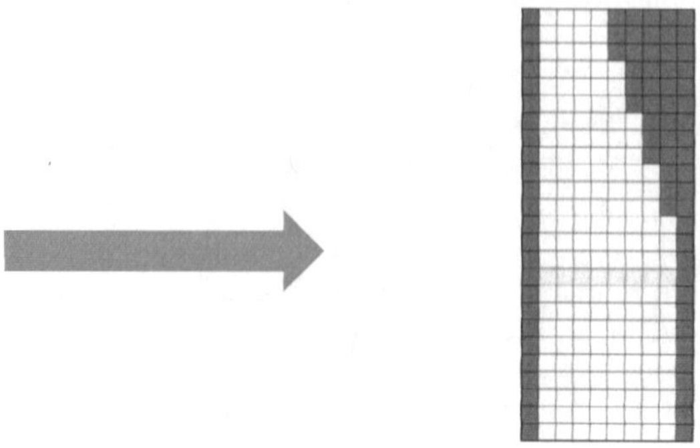

Figure 4. Matrix and simulation diagram of CA model.

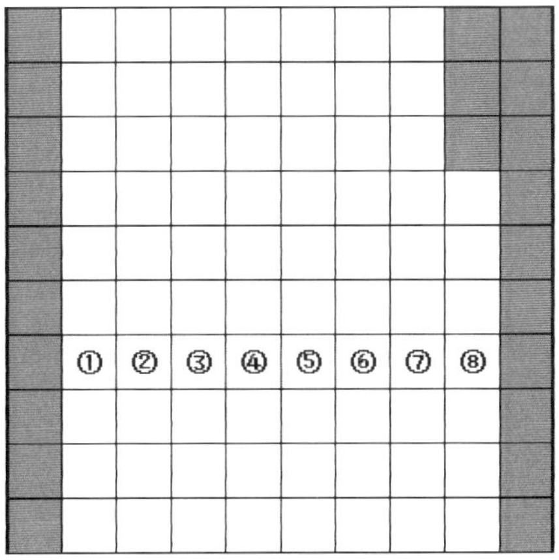

Figure 5. The simulation diagram with digital number tokens.

2.2 NS Rules

Another key of HTSROM is ***NS rules***, which is employed to initialize our CA model with NS rules that were first suggested by Nagel and Schreckenberg [4,6]. NS rules definite three states of cars-acceleration, deceleration and random moderation. And the maximum speed of cars is larger than 1. We set this CA model on the basis of NS rules as discrete model of which parameters like time, speed and place are all discrete. The set of rules is shown in Table 3.

Table 3. Rules of Vehicles Moving Forward Based on NS Rules

States	**CA code**
initial state	$v_1 = 2, v_2 = 1.v_3 = 1, v_4 = 0$
acceleration rules	$v_i(t+1) = \min(v_i(t)+1, v_{\max})$
accident prevention	$v_i = \min(v_i, d_i - 1)$
deceleration rules	$v_i(t+1) = \min(v_i(t+1), gap_i(t))$
random deceleration rules	$v_i(t+1) = \max(v_i(t+1)-1, 0) \quad rand<p$
update place rules	$x_i(t+1) = x_i(t) + v_i(t+1)$

2.3 Experimental Assumption

Besides, we assume that the number of vehicles reaching the toll plaza per time period obeys the ***poisson*** distribution. By real life experience we know that drivers tend to make a short cut and are less likely to choose lanes on the side. Therefore in the following table we define the probability for drivers to choose a certain booth.

Table 4. The probability for drivers to choose a booth.

t	**1**	**2**	**3**	**4**	**5**	**6**	**7**	**8**
P(t)	$\frac{24}{111}$	$\frac{24}{111}$	$\frac{24}{111}$	$\frac{24}{111}$	$\frac{8}{111}$	$\frac{4}{111}$	$\frac{2}{111}$	$\frac{1}{111}$

It can be seen that the probabilities of choosing the ***No.1,2,3,4*** booths are the same, that is because these booths directly face the high way. The probability significantly decreases when it comes to ***No.5*** booth, and smoothly goes down on ***No.6,7,8*** booths.

In the following discussion we will not encounter with large change of the space of toll plaza, therefore we take no consideration to the cost of land occupied by the construction. We introduce ***Average Lane-Changing Times (ALT) to*** decide the safety performance of the system. We use the handling capacity per hour J to decide the capacity performance of the system. Fig.6 shows that the handling capacity increases as the influx improves, and the limit handling capacity J_{max} of the original system is around 2290 veh/hr. In Fig. 7, the fluctuation of ***ALT*** when the influx varies. As illustrated in Fig. 6 and Fig. 7, when the influx comes to the limit, vehicles change lanes for more than 2.5 times on average, which contributes to the congestion.

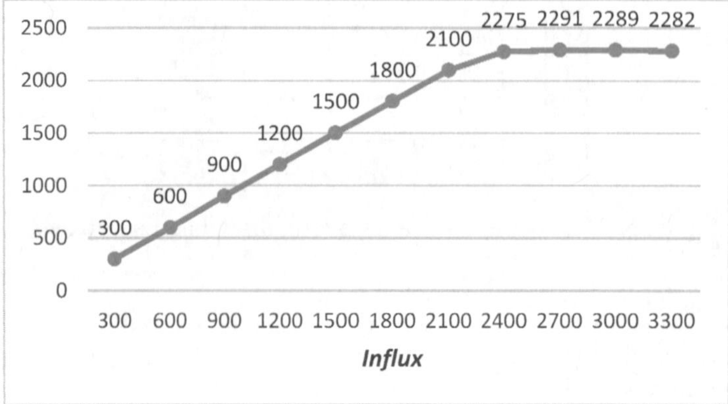

Figure 6. The handling capacity increases as the influx improves.

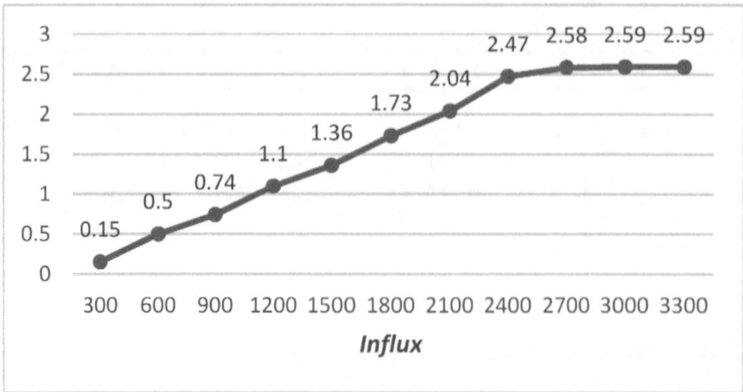

Figure 7. The fluctuation of ***ALT*** when the influx varies.

3 Optimization and analysis

3.1 Arrangement of ETC

ETC *(Electronic Toll Collection)* is nowadays the most advanced technique of collecting tolls. It allows drivers to pay their tolls without stopping or even slowing down. Apparently this technique improve the efficiency of the barrier toll, yet it to some extent guides the choice of lanes of the drivers as well. The Fig. 8 illustrates how the HTSROM is realized in program. We mark the ETC with light blue. Vehicles do not decelerate when passing through the ETC. To make our research simple, we first assume that one of eight booths is replaced by ETC booth, then which booth should we change?

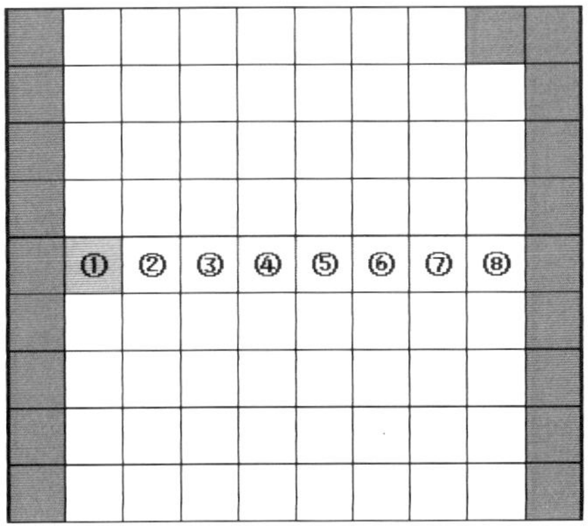

Figure 8. The simulation diagram with ETC and digital number tokens.

From above perspective views, we illustrates the fluctuation of ***J*** and ***ALT*** when the position of ETC varies in Fig. 9, respectively. Then we can see that we should replace the ***No.1,2,3,4*** booths by the ETC. It is because these four lanes directly face the high way and are the closest route for the drivers. It seems that the ***No.8*** booth is the worst choice to be replaced. However, in the discussion above we fail to take the influence of the drivers' choice of booths into account. The fact is that ETC will attract drivers to pass through because it simply saves time. Now let's reset the distribution that drivers choose the booths and see what will happen.

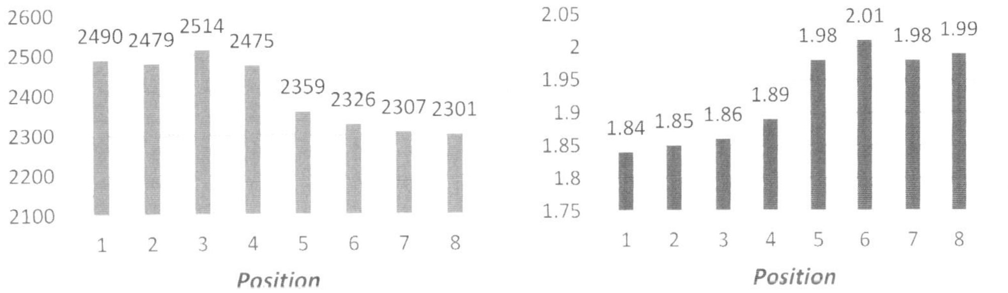

Figure 9. The fluctuation of ***J*** and ***ALT*** when the position of ETC varies, when ***influx*** is set by 4000 veh/hr. Left: fluctuation of ***J***, Right: fluctuation of ***ALT***.

The update of distribution is not remarkable if we replace the ***No.2,3,4*** booths. However the distribution significantly changes when we replace the ***No.5,6,7,8*** booths. Note that since the ***No.1*** booth has already been replaced by ETC, the probability of choosing the other booths can not exceed the probability of choosing the ***No.1*** booth. And when a new ***ETC*** is set on the ***No.8*** booth, an example of the update of distribution is shown in Table 5. Note that the ***No.1*** booth has already been replaced. Then, the new fluctuation of ***J*** and ***ALT*** when the position of ETC varies in Fig. 10, respectively. From above we can see that the ***ALT*** experiences a remarkable decrease, and the number of ***J*** increases. This inspires us the best choice of the booths to be replaced, the ***No.1,8*** booths. In common cases, we should replace the booths on the side.

Table 5. The update of distribution.

t	**1**	**2**	**3**	**4**	**5**	**6**	**7**	**8**
$P(t)$	$\frac{27}{111}$	$\frac{19}{111}$	$\frac{19}{111}$	$\frac{19}{111}$	$\frac{6}{111}$	$\frac{2}{111}$	$\frac{1}{111}$	$\frac{18}{111}$

IOP Conf. Series: Materials Science and Engineering **242** (2017) 012108 doi:10.1088/1757-899X/242/1/012108

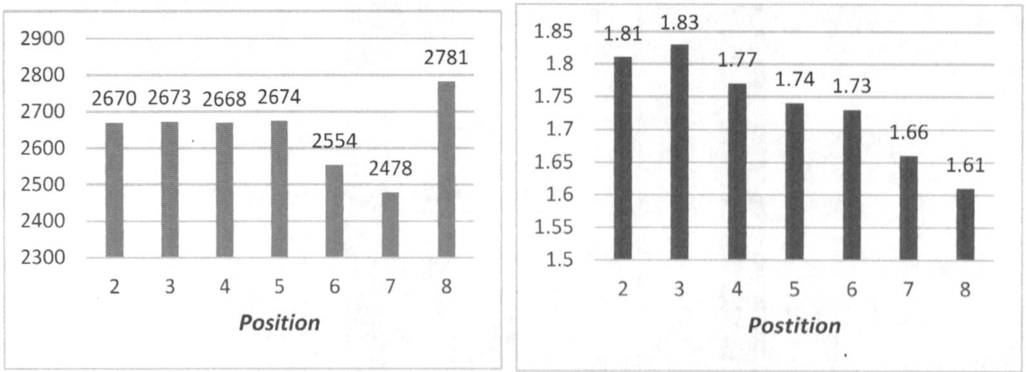

Figure 10. The fluctuation of ***J*** and ***ALT*** when the position of ETC varies. Note that the ***No.1*** booth has already been replaced by ***ETC***. The ***influx*** is set by 1200, 2400, 3600 veh/hr. Left: fluctuation of ***J***, Right: fluctuation of ***ALT***.

3.2 Rearrangement of Toll Booths

After the former discussion we have noticed that the traffic condition can be improved if there are vehicles guided to the lanes on the side. Since we know that drivers hardly see things behind the toll booths, let us analyze the new situation if we artificially "pull" the booths on the side closer to the drivers. The rearrangement of toll booths is shown in Fig. 11. By changing the distribution of toll booths we substantially change the probability of drivers choosing a certain lane, as shown in Table 6. Since we have no data on the influence to the choice of drivers, we could simply assume that the change of arrangement **do** bring the influence we except, by real life experience. We can later rearrange the booths again if we gain the data of the influence after actual operation, and then reach the most optimized rearrangement. From above, we can see that the rearrangement of toll booths successfully brings a great improve to the system.

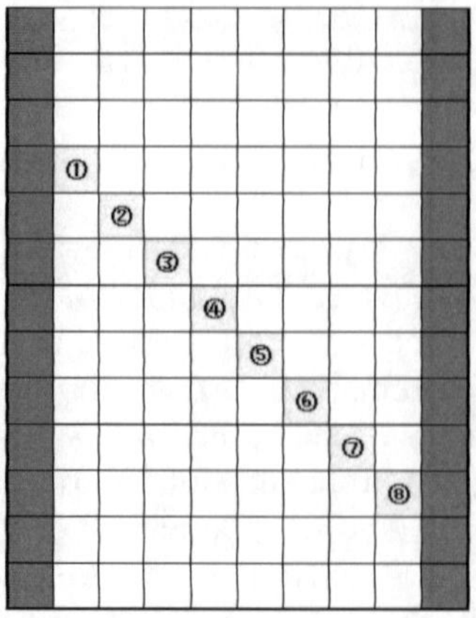

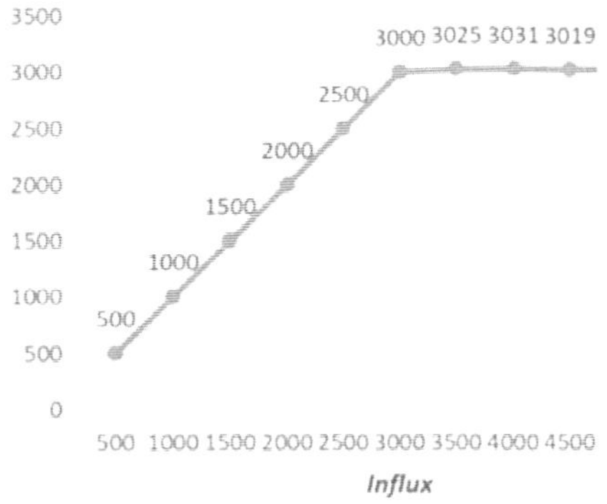

Figure 11. The rearrangement of toll booths and new fluctuation of ***J*** when the influx varies.

Table 6. A probable distribution after the rearrangement of the toll booths.

t	1	2	3	4	5	6	7	8
$P(t)$	$\frac{4}{15}$	$\frac{2}{15}$	$\frac{1}{15}$	$\frac{1}{30}$	$\frac{1}{30}$	$\frac{1}{15}$	$\frac{2}{15}$	$\frac{4}{15}$

3.3 Truck Lanes

Drivers do meet such cases when they see a better lane beside yet blocked by large trucks. Trucks significantly affect traffic efficiency due to their volume and poor steering ability. One way to handle the trouble caused by trucks is to set truck lanes. Apparently truck lanes should be set right on the side of the road. We will try to figure out whether they should be set in the ***No.1*** booth or the ***No.8*** booth. We assume that trucks are two times longer than small vehicles, which means they occupy to blocks in the ***CA*** model. We also assume that the limiting speed of trucks is 70% of small vehicles. Notice that most of the vehicles on road are small cars, therefore we assume that the proportion of trucks is no larger than 50%. What's more, trucks should only move on truck lanes, yet the route of small vehicles is not limited.

Fig. 12 (a) shows the different arrangement of truck lanes after rearranging toll booths. Truck lanes are marked in light blue. Fig. 12(b) illustrates the fluctuation of ***J*** when there is no truck lanes, when the truck lane is set before toll station ***No.1***, and when the truck lane is set before toll station ***No.8***. The proportion of trucks goes from 10% to 50%. The ***influx*** is set by 4000veh/hr. The handling capacity significantly improves after the set of truck lanes, when there is not too much trucks on the road. However, with the increase of the proportion of trucks, truck lanes loses its competitiveness. That is because too much vehicles are gathered in one certain lane, causing a great waste of space on the other lanes.

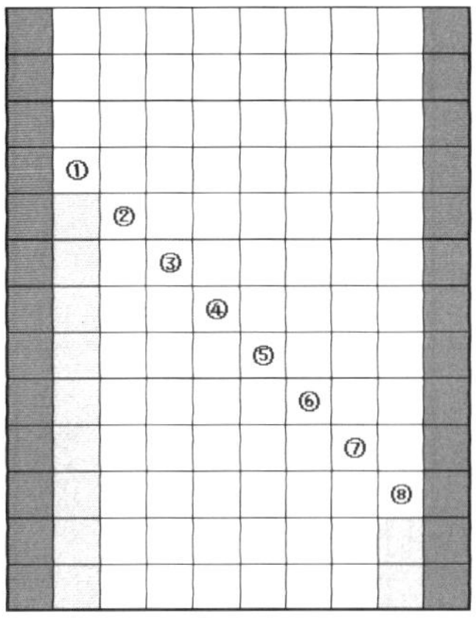

IOP Conf. Series: Materials Science and Engineering **242** (2017) 012108 doi:10.1088/1757-899X/242/1/012108

Figure 12(a): the different arrangement of truck lanes after rearranging toll booths

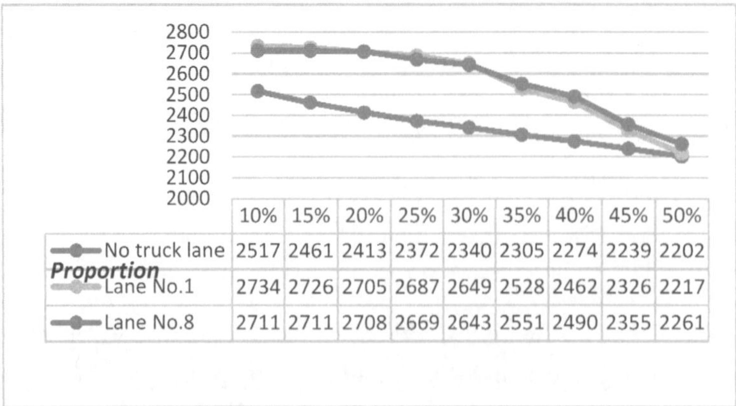

	10%	15%	20%	25%	30%	35%	40%	45%	50%
No truck lane	2517	2461	2413	2372	2340	2305	2274	2239	2202
Lane No.1	2734	2726	2705	2687	2649	2528	2462	2326	2217
Lane No.8	2711	2711	2708	2669	2643	2551	2490	2355	2261

Figure 12(b): the fluctuation of ***J***

4 Conclusion

Toll stations are always the weak spots of high way. Through the discussion above we come up with different ways of alleviating congestion. The set of ETC increases the speed of passing, and together with the rearrangement of toll booths, optimizes the utilization of road spaces. The set of Truck lanes resolves the contradiction between trucks and small vehicles. Hopefully the future technique of intelligent autonomous vehicles will get the congestion solved once and forever.

Reference

[1] Ziliaskopoulous, A.K., Mahmassani, H.S., 1996. A note on least time path computation considering delays and prohibitions for intersection movements. Transportation Research B 30, 359–367.

[2] Cova T J, Johnson J P. A network flow model for lane-based evacuation routing [J]. Transportation Research Part A-policy and Practice, 2003, 37(7): 579-604.

[3] Duranton G, Turner M A. Urban Growth and Transportation [J]. The Review of Economic Studies, 2012, 79(4): 1407-1440.

[4] Van Holle V, Deforche B, Van Cauwenberg J, et al. Relationship between the physical environment and different domains of physical activity in European adults: a systematic review [J]. BMC Public Health, 2012, 12(1): 807-807.

[5] Wu, Chaoqun, Yulong Pei, and Jingpeng Gao. Model for Estimation Urban Transportation Supply-Demand Ratio. Mathematical Problems in Engineering (2015): 1-12.

[6] Junhua Wang, Randy Iwasaki, Alexandre M Bayen, John Harvey. Future road transportation technology. http://doi.org/10.1016/j.ijtst.2016.09.001

IOP Conf. Series: Materials Science and Engineering **242** (2017) 012109 doi:10.1088/1757-899X/242/1/012109

Semantic computing and language knowledge bases

Lei Wang[1,2] Houfeng Wang[1] Shiwen Yu[1]

[1]Key Lab of Computational Linguistics of Ministry of Education, Peking University, Beijing, China

[2]School of Foreign Languages, Peking University, Beijing, China

{wangleics, wanghf, yusw}@pku.edu.cn

Abstract. As the proposition of the next-generation Web – semantic Web, semantic computing has been drawing more and more attention within the circle and the industries. A lot of research has been conducted on the theory and methodology of the subject, and potential applications have also been investigated and proposed in many fields. The progress of semantic computing made so far cannot be detached from its supporting pivot – language resources, for instance, language knowledge bases. This paper proposes three perspectives of semantic computing from a macro view and describes the current status of affairs about the construction of language knowledge bases and the related research and applications that have been carried out on the basis of these resources via a case study in the Institute of Computational Linguistics at Peking University.

1 Introduction

Semantic computing is a technology to compose information content (including software) based on meaning and vocabulary shared by people and computers. Its goal is to bridge the semantic gap through this common ground, to let people and computers cooperate more closely as in Fig. 1. The task of semantic computing is to explain the meaning of various constituents of sentences (words or phrases) in a natural language. We believe that semantic computing is a field that addresses two core problems: First, to map the semantics of user with that of content for the purpose of content retrieval, management, creation, etc.; second, to understand the meanings (semantics) of computational content of various sorts, including texts, videos or audios and expressing them in a form that can be processed by machine.

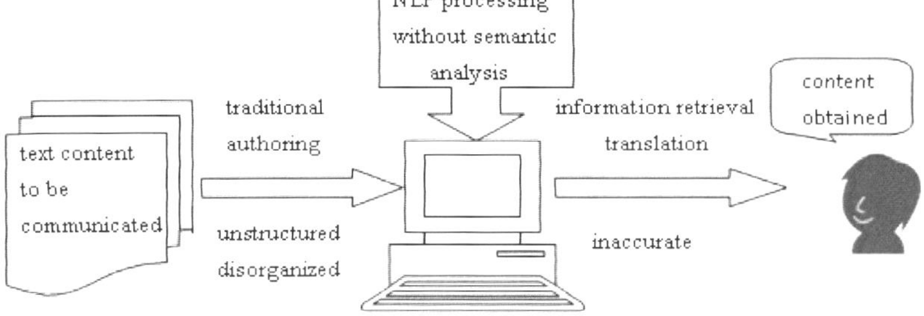

Fig. 1. Human-computer interaction is handicapped without semantic computing.

Content from this work may be used under the terms of the Creative Commons Attribution 3.0 licence. Any further distribution of this work must maintain attribution to the author(s) and the title of the work, journal citation and DOI.
Published under licence by IOP Publishing Ltd

But the way to the success of semantic computing is not even and it has taken a quite long time for researchers to make some progress in this field. The difficulties of semantic computing involve many aspects: ambiguity, polysemy, domain of quantifier, metaphor, etc. Different individuals will have different understanding of the same word or the same sentence. Research on the theory and methodology of semantic computing still has a long way to go.

2 Related Work on Semantic Computing

Semantics is an interesting but controversial topic. Many a theory has been proposed in attempt to describe what meaning really means. But up until now there has not been a theory that can describe the meaning of various language units (words, phrases and sentences) so perfectly that was accepted universally, even though Fillmore's proposition of Framework semantics [1] is successful enough. Since Gildea [2] initiated the research on automatic semantic role labeling, many evaluations have been conducted internationally, such as Senseval-3 and SemEval 2007, as well as CoNLL SRL Shared Task 2004, 2005 and 2008. And word sense disambiguation remains a popular research topic. [3][4]

In recent years, semantics-based analysis such as data and web mining, analysis of social networks and semantic system design and synthesis have begun to draw more attention from researchers. Applications using semantics such as search engines and question answering [5], content-based multimedia retrieval and editing, natural language interfaces [6] based on semantics have also been attracting attentions.

3 The Theory and Methodology of Semantic Computing

Why semantic computing (or NLU) has posed so great a challenge? We may attribute this to two major reasons: First, it is based on the knowledge of human language mechanism. If fully-developed complicated brains are often seen as a crowning achievement of biological evolution, the interpersonal communication is no simpler than human biological mechanism. Language has to be a crucial part of the evolutionary process, which has not been fully understood by scientific research. Second, in NLP research the language is both the target and the tool. Current NLP research focuses on either speech or written texts only. However, in the real world scenario, reading and interaction between humans are multi-dimensional (through different forms of information such as text, speech, or images and utilizing our different senses such as vision, hearing).

There are complex many-to-many relations between the form and the meaning of a language. Semantic computing is not only the way but also the ultimate goal of natural language understanding. Although it is hard, we should not give up. Here we propose that the main contents of semantic computing include the following three aspects: ontological, cognitive and pragmatic. As for ontologies, much progress has been made worldwide. The remarkable achievements in English include: WordNet by Princeton University, PropBank by University of Pennsylvania [7], etc. Also there are quite a number of efforts made on building ontologies in Chinese, which will be elaborated in Section 4.

As to WSD tasks on the word level, some problems can be solved when ontology is applied. But ambiguity can also appear on the syntactic level. For this, it is usually difficult for ontologies to do much, so we may seek help from language knowledge bases. The following examples of syntactic semantic analysis will illustrate how different syntactic structures will change the meaning of sentences:

Example 1

这样的电影不是垃圾是什么?	--该电影是垃圾。
zhè yàng de diàn yǐng bú shì lā jī shì shén me?	-- gāi diàn yǐng shì lā jī
If a movie as such is not rubbish, what is it?	-- It is rubbish.
这样的电影怎么能说是垃圾呢?	-- 该电影不是垃圾。
zhè yàng de diàn yǐng zěn me néng shuō shì lā jī ne?	-- gāi diàn yǐng bú shì lā jī
How can a movie as such be rubbish?	-- It is not rubbish.

With respect to semantic computing on cognitive level, we will use metaphor as an example. For a long time, NLP research has focused on ambiguity resolution. Can NLU be realized after ambiguity resolution? Metaphor, insinuation, pun, hyperbole (exaggeration), humor, personification, as well as intended word usage or sentence composing, pose a great challenge to NLU research. If the computer can deal with metaphors, it will greatly improve the ability of natural language understanding.

The linguistic function of metaphor is also important. Metaphor is the base of new word creation and polysemy production (sense evolution), for example, 垃圾箱 lā jī xiāng (recycle) and 病毒 bìng dú (virus) are used in a computer setting and words like 高峰 gāo fēng (peak), 瓶颈 píng jǐng (bottleneck) and 线索 xiàn suǒ (clue) are endowed with new meanings which have not been included in traditional Chinese dictionaries. As for the NLP tasks of metaphor computing, we can conclude that there are three tasks to be accomplished: First, metaphor recognition. For instance, how can we distinguish 知识的海洋 from 海洋资源考察 hǎi yáng zī yuán kǎo chá (investigation of ocean resources); Second, metaphor understanding and translation. For instance, 知识的海洋 actually means 知识像海洋一样丰富。zhī shí xiàng hǎi yáng yí yàng fēng fù (Knowledge is as rich as the ocean.). Third, metaphor generation. For instance, how phrases such as 信息的海洋 xìn xī de hǎi yáng (ocean of information) and 鲜花的海洋 xiān huā de hǎi yáng (ocean of flowers) can be generated successfully by computer?

Second, empirical (statistical) method i.e., providing machine with a large number of samples and training a model. From a statistical point of view, metaphor recognition can be seen as a problem to compute the conditional probability $p(m|c)$ to decide whether 海洋 is a metaphor in context c. The reversed order of two variants m and c will not change the value of unified probability of $p(m|c)$ and $p(c|m)$, while the relation between unified probability and conditional probability can be written as:

$$p(c)p(m \mid c) = p(m)p(c \mid m) \tag{1}$$

Then,

$$p(m \mid c) = p(m)p(c \mid m) / p(c) \tag{2}$$

Given c, $p(c)$ is a constant. Then,

$$p(m \mid c) \propto p(m)p(c \mid m) \tag{3}$$

Given a threshold δ, if $p(m)p(c \mid m) > \delta$, then we can deem this 海洋 is a metaphor.

Then the problem becomes how to compute $p(m)p(c \mid m)$. We can compute it based on large-scale annotated corpus and get

$$p(m) = N_m / N \tag{4}$$

N_m — the times of 海洋 as a metaphor in the corpus;

N — the total times of 海洋 in the corpus.

Then we simplify 海洋 and its context c into: $W_{-k} \ldots W_{-1}$ 海洋 $W_1 \ldots W_i$, where $W_{-k}, \ldots, W_{-1}, W_1, \ldots, W_i$ represent the n-gram of 海洋 and its syntactic and semantic attributes respectively.

$$p(c \mid m) = p(W_{-k} \mid m) \cdots p(W_{-1} \mid m) p(W_1 \mid m) \cdots p(W_i \mid m) \tag{5}$$

$$p(W_s \mid m) = N(W_s) / N_w, \quad (s = -k, \cdots, -1, 1, \cdots, i) \tag{6}$$

$N(W_s)$ stands for the times of co-occurrence of 海洋 as a metaphor and word W with designated attributes at position. Here an important hypothesis of independence is: words at different position s is not correlated with the word 海洋 as is in Fig. 2.

Last, we will discuss semantic computing on the pragmatic perspective, which is more or less unique of Chinese language. First, the change of construction in Chinese will affect the meaning of a sentence even though the words themselves are not changed. The emphasized meaning of the construction is not equal to the combination of the underlying meaning from each element in the construction. The meaning reflects the distribution of quantity of entities and the relative locations among entities. Although the underlying syntactic relationship among the main verb, the agent and the object(s) still exists, such syntactic relationship is only secondary.

IOP Conf. Series: Materials Science and Engineering **242** (2017) 012109 doi:10.1088/1757-899X/242/1/012109

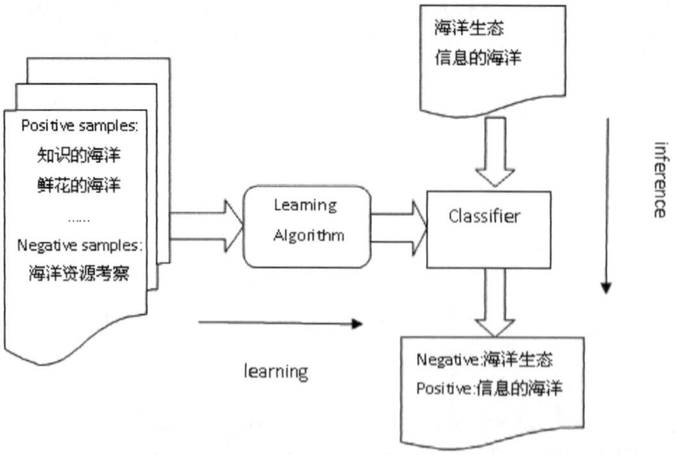

Fig. 2. Empirical (statistical) method of metaphor processing

4 Language Knowledge Bases as the Foundation of Semantic Computing

Language knowledge base is an indispensable component for NLP system, and its quality and scale determines the failure or success of the system to a great extent. For the past two decades, a number of important language knowledge bases have been built through the effort of people in Institute of Computational Linguistics (ICL) at Peking University. Among them, the Grammatical Knowledge Base of Contemporary Chinese (GKB) [8] is the most influential. Based on GKB, various research projects have been initiated. For instance, a project on the quantitative analysis of "numeral-noun" construction of Chinese was conducted by Wang [9] to further analyze the attributes of Chinese words. A project aiming at the emotion prediction of entries in CIKB was completed by Wang [10] to further understand how the compositional elements of a fossilized construct like an idiom function from the token level.

Following GKB, language knowledge bases of large scale, high quality and various type (words and texts, syntactic and semantics, multi-lingual) have been built, such as the Chinese Semantic Dictionary (CSD) for Chinese-English machine translation, the Chinese Concept Dictionary (CCD) for cross-language text processing, the multi-level Annotated Corpus of Contemporary Chinese, etc. When we want to do cross-lingual information retrieval, the two senses need to be distinguished. Hence, CCD can serve as a useful tool to complete the task for it organizes semantic knowledge from a different angle. Concepts in CCD are represented by Synsets, i.e. sets of synonyms as in Table 2. For instance, the concept 教师 is in a Synset {教师 教员 老师 先生 导师 老板 孩子王 臭老九 ...} and all the concepts form a network to associate the various semantic relations between or among the concepts: hypernym-hyponym, part-whole, antonym, cause and entailment, by which we can retrieve information in either an extensive or a contractive way so as to improve the precision or recall of a search engine. It can also provide support for WSD tasks.

Table 1. The Synset of the word 教师 jiào shī and its related Synsets

Offset	Synset	Csyncet	Hypernym	Hyponym	Definition	Cdefinition
07632177	teacher instructor	教师 教员 老师 先生 导师 孩子王 臭老九	07235322	07086332 07162304 07209465 07243767	a person whose occupation is teaching	以教学为职业的人

		…		0727965 9 0729762 2 0734117 6 0740109 8 …		

In 2009, the various knowledge bases built by ICL were integrated into the CLKB. The integration of heterogeneous knowledge bases is realized by a resolution of "a pivot of word sense". Three basic and important knowledge bases, GKB, CSD and CCD have been integrated into a unified system which includes language processing module, knowledge retrieval module and knowledge exploration module. Although there are some fundamental resources on semantic computing, it needs further improvement, updating, integration and specification to form a collective platform to perform more complicated NLP tasks. To further improve the result of semantic computing, innovative projects for new tasks should also be launched, for instance:

- metaphor knowledge base
- ultra-ontology dynamic knowledge base (generalized valence mode)
- the integration of information based on multi-lingual translation

5 Conclusion

Why semantics is so useful in the first place? Linguists and psychologists are interested in the study of word senses to shed light on important aspects of human communication, such as concept formation and language use. Lexicographers need computational aids to analyze in a more compact and extensive way word definitions in dictionaries. Computer scientists need semantics for the purpose of natural language processing and understanding. Therefore, the significance of semantic computing in NLP is obvious and more research needs to be done with this respect.

Acknowledgement

Our work is supported by National High Technology Research and Development Program of China (863 Program) (No. 2015AA015402).

References

[1] Fillmore, C. J.. Frame Semantics and the Nature of Language. In Annals of the New York Academy of Sciences: Conference on the Origin and Development of Language and Speech, pp 20-32. New York (1976).
[2] Gildea, Denial and Denial Jurafsky. Automatic Labeling of Semantic Roles. Computational Linguistics, 28(3), 245-288(2002).
[3] Ide, Nancy and Jean Véronis. Introduction to the Special Issue on Word Sense Disambiguation: The State of the Art, Computational Linguistics, 24(1), 2-40(1998).
[4] Schutze, Hinrich. Automatic Word Sense Discrimination. Computational Linguistics, 24(1), 97-124(1998).
[5] Li, Sujian, Zhang Jian, Huang Xiong and Bai Shuo. Semantic Computation in Chinese Question-Answering System, Journal of Computer Science and Technology, 17(6), 993-999(2002).
[6] Neo, Ee Sian, Takeshi Sakaguchi and Kazuhito Yokoi. A Humanoid Robot that Listens, Speaks, Sees and Manipulates in Human Environments. Proceedings of IEEE International Conference on Multisensor Fusion and Integration for Intelligent Systems Seoul, Korea, pp 419-425. Seoul (2008).
[7] Xue, Nianwen and Martha Palmer. Automatic Semantic Role Labeling for Chinese Verbs. In Proceedings of the 19th International Joint Conference on Artificial Intelligence, pp 1160-1165. Edinburgh (2005).
[8] Yu, Shiwen et al.. Introduction to Grammatical Knowledge Base of Contemporary Chinese (Second

Edition), Tsinghua University Press, Beijing, China (2003).
[9] Wang, Meng et al. Quantitative Research on Grammatical Characteristics of Noun in Contemporary Chinese. Journal of Chinese Information Processing, 22(5), 22-29(2009).
[10] Wang, L., Yu, S., Zhu, X., Li, Y.. Chinese Idiom Knowledge Base for Chinese Information Processing. In Proceedings of the Chinese Language Semantics Workshop (2012), Wuhan, pp 85-90, China (2012).

A weighted method based on Lars algorithm

Lin Chen[1, a)], Shanxiong Chen[2, b)], Chunrong Chen[3, c)] and Yuchen Zhu[3, c)]

[1,2,3] College of Computer & Information Science in Southwest University, Chongqing, China.

a)palegg_cl@126.com

b)Corresponding author: csxpml@163.com

c)cafunechen@126.com and Llumzyc@163.com

Abstract: LASSO (Least Absolute Shrinkage and Selection Operator) is mainly used to realize variable selection, simultaneously its algorithm and some improved algorithm have gotten wide attention in many fields. To improve the accuracy of LASSO problem in calculating regression coefficients, this paper proposes a new algorithm based on LASR (Least Angle Regression) algorithm to change its approximation direction, which uses two weighted method (coefficient of variation method or entropy weight method) to calculate the weight of linear relationship between the independent and the dependent variables, so we can calculate a regression coefficients set of linear regression model. Compared with LARS algorithm, it can be proved that the improved algorithm mentioned in this paper has a more precise ability for prediction.

1 Introduction

Linear model plays a significant role in modern statistics and has been wildly used in various fields [1]. The values of variable selection and regression parameter are helpful to get a precise prediction.

To decrease the deviation, we often choose plenty of independent variables into the model but a number of them exhibits little correlation or even no correlation with dependent variables, which decreases the precision of prediction [2]. Hence, it's indispensable to search out the set of independent variables related closely to dependent ones.

Robert Tibshirani proposed a method used to describe series of constrained optimization problems, called LASSO, which can make regression coefficients towards the minimum [3]. LASSO problem can improve the rationality of the model and realize variable selection simultaneously. Efron[6] proposed 'Least Angle Regression' to resolve LASSO problem, making use of correlation coefficient of current residuals and current variables to determine a new algorithm path.

We propose an improved method based on LARS algorithm in this paper. The key point of the algorithm is that using weighted algorithm to change the approximation direction and to improve the precision. Moreover, the experimental result indicates which method gets more precise performance:

- The effects on dependent variables change from sample to sample. Therefore, according to the different importance of each independent variable, we propose improved method based on LARS to change the approximation angle to direct the independent one with higher weight coefficient instead of angular bisector;
- By the experiments compared with original LARS algorithm, the weighted LARS algorithm was proved with more precise regression coefficient estimation as well as the results prediction.
-

Content from this work may be used under the terms of the Creative Commons Attribution 3.0 licence. Any further distribution of this work must maintain attribution to the author(s) and the title of the work, journal citation and DOI.
Published under licence by IOP Publishing Ltd

2 Lasso and lars algorithm

2.1 The LASSO Problem

Suppose that we have data $x_i \in R^n, i = 1,2,\cdots,m$, $y \in R^n$ who are considered to be the predictor variables and responses corresponding for the i th observation. As the linear regression problem set up:

$$y = X\beta + \alpha$$ （错误!文档中没有指定样式的文字。-1）

Where y is an observation vector of n*1, and X is a $n\times p$ full column rank matrix.

Through the linear regression model, the constraint which is less than threshold t is applied to the norm ℓ_1 of regression coefficient β , the Lasso estimate is defined as [5]:

$$(\hat{\alpha},\hat{\beta}) = \arg\min_{\beta}\left\{\sum_i (y_i - \alpha_i - \Sigma\beta_j x_{ij})^2\right\}, s.t.\sum_j\left|\beta_j\right| \le t$$ （错误!文档中没有指定样式的文字。-2）

where $\beta = \left(\hat{\beta}_1,\cdots,\hat{\beta}_p\right)^T$, and t controls the amount of regression coefficients to the minimum. With the limit of $\Sigma_{j=1}^{J}\left|\beta_j\right| \le t$, the regression coefficient shrinkage towards 0 or even to be exactly equal to 0.

Insight about the coefficients of the shrinkage can be gleaned from the orthonormal design case. X is
a $n\times p$ design matrix with $X^T X = I$ (unit matrix). Hence the solutions to the LASSO can be shown to be:

$$\hat{\beta}_j = sign(\hat{\beta}_j^0)(|\hat{\beta}_j^0| - \gamma)^+$$ （错误!文档中没有指定样式的文字。-3）

where γ is determined by the condition $\Sigma\left|\hat{\beta}_j\right| = t$.

2.2 LARS Algorithm

All coefficients start as 0, and find the predictor x_{j1} who has largest correlation with the response. Then take the largest step in the direction of selected predictor until the next predictor x_{j2} has same correlation with the current residual. At this point LARS proceeds in equiangular direction of x_{j1} and x_{j2}, repeat this step until the residual $y' = y - \overline{y}$ is equal to 0 or the predictors is x_{jk} [4].

After k steps, there are k predictor variables selected, the set can be defined as [6]:

$$A = \{j_1,\cdots,j_k\} \subseteq \{1,2,\cdots,m\}$$ （错误!文档中没有指定样式的文字。-4）

As well as obtaining the *k-1* th regression variable, define the matrix:

$$X_A = \left[s_{j1}x_{j1},\cdots,s_{jk}x_{jk}\right] \in R^{n\times k}$$ （错误!文档中没有指定样式的文字。-5）

where the $S_j = \text{sign}((Y - \mu_{k-1})' X_j)$ is described as the set of approximation direction, so the equiangular direction of next step can be defined as:

$$w_A = A_A G_A^{-1} 1_A \quad , \quad u_A = X_A w_A$$ （错误!文档中没有指定样式的文字。-6）

which defines $G_A = X_A^T X_A$ $A_A = (1_A^T G_A^{-1} 1_A)^{-\frac{1}{2}} \cdot 1_A$ is a column vector of k-dimension and all of its elements are 1. Suppose that u_A is the vector of current correlations between selected predictors and selecting predictor to update the next step.

The LASSO can make up the deficiency on variables selection, and LARS algorithm can get more precise variable section and resolve LASSO problem. However, owning to singleness of approximation angle, there may be a bigger error than expected.

3 A weighted method on the basis of Lars

3.1 The Basic Concepts of Improved Method

We've explained that LARS algorithm proceeds in a direction equiangular. However, each predictor reflects the different importance in the data set. For this main point, we can propose a new algorithm to make the approximation direction to slant the predictor variable who has higher weighted coefficient.

The approximation direction of LARS algorithm mentioned by Efron[7] is:

$$s_{jl} = \begin{cases} 1 & \left(if \left\langle x_{jl}, y \right\rangle > 0\right) \\ 0 & \left(if \left\langle x_{jl}, y \right\rangle = 0\right) \\ -1 & \left(if \left\langle x_{jl}, y \right\rangle < 0\right) \end{cases}$$ （错误!文档中没有指定样式的文字。-7）

$\left\langle x_j, y \right\rangle$ is considered as the correlation coefficient of predictor variables and responses. s_{jl} can change vector approximation direction according to the value of $\left\langle x_j, y \right\rangle$.

To change the approximation direction based on weighted method, we define:

$$S_{jl} = \begin{cases} r & (if \left\langle x_{jl}, y \right\rangle > 0) \\ 0 & (if \left\langle x_{jl}, y \right\rangle = 0) \\ -r & (if \left\langle x_{jl}, y \right\rangle < 0) \end{cases}$$ （错误!文档中没有指定样式的文字。-8）

where $r = \omega_j$ is the weight coefficient of each predictor variable in the data set.

LARS selects those independent variables just with sufficiently large value and proceeds in a direction equiangular. On the basis of LARS, with the different weight coefficient of each independent variable in the data set X_A, we can multiply corresponding weight coefficient to make approximation vector change. Through computation of weighting, we can obtain the index vector set of the data set $X = [w_j s_{jk} x_{jk}]$, and calculate the corresponding vector approximation length according to weight coefficients, finally, we can resolve LASSO.

The example of independent variables x_1, x_2 is served for illustrating in Fig.3.1. Assume that the weight coefficients of two independent variables are ω_1 、 ω_2, the $\langle x_1, y \rangle$ is bigger than zero, and $\langle x_2, y \rangle$ is smaller than zero. τ is approximation direction of LARS algorithm, while τ_1 is the new approximation direction of weighted method. The direction of vector y is required to be targeted.

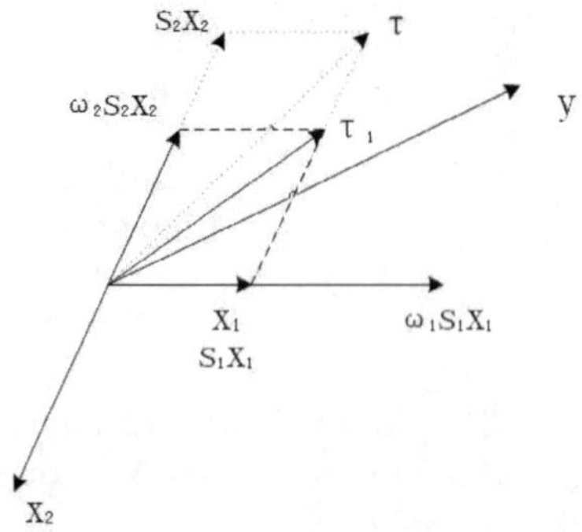

Figure 3.1 Geometric Graph of Weighted Method

In a word, the central part of proposed method is to calculate the weight of each independent variable related to dependent variable, also called weight coefficient.

3.2 Weighted Calculation

3.2.1 Coefficient of Variation Method. Calculate the mean and sample standard deviation of the data set, then calculate coefficient of variation, says CV here [8]; after that we can obtain the weight of each predictor variable by normalization processing.

$$CV = \frac{S_t}{\overline{X}_t}$$ （错误!文档中没有指定样式的文字。-9）

3.2.2 Entropy Weight Method. 1)Calculate the information entropy

Suppose that the system may be in K different indicators and the probability of each one to appear is $P_i(=1,2,\cdots,m)$; when $p_i = 1/m(i=1,2,\cdots m)$, namely when the probability of each one to appear is equal, the entropy of the system E and the maximum entropy $e_{\max}$ are [9]:

$$E = -\sum_{i=1}^{m} p_i \ln p_i , \quad e_{\max} = \ln m$$ （错误!文档中没有指定样式的文字。-10）

If there are m items waited for evaluation and n evaluation indicators, the original evaluation matrix is $R = (r_{ij})_{m\times n}$ and the information entropy of indicators is

$$e_j = -\sum_{i=1}^{m} p_{ij} \cdot \ln p_{ij} \quad \text{,where } p_{ij} = r_{ij} / \sum_{i=1}^{m} r_{ij}$$ （错误!文档中没有指定样式的文字。-11）

2) Obtain the entropy of each indicator

Calculate the information entropy of each indicators ($E_1, E_2, \cdots, E_k$), and the formula is defined:

$$W_i = \frac{1-E_i}{k-\sum E_i} \quad (i=1,2,\cdots,k)$$ （错误!文档中没有指定样式的文字。-12）

3.3 Algorithm Flows

Input: the set of predictors X，the set of responses Y;
Output: regression coefficientsβ;

I. data processing: X and Y normalization;

II. initialization: $A = \phi$, the amount of looping $k = 0$,

III. approximation direction μ_A=0, residual $\hat{y} = y - \mu$;

IV. calculating the weight coefficient: $w_j = \frac{st}{\overline{x}}(j \in A)$ or $w_j = \frac{1-E_j}{k-\sum E_j}(j \in A)$;

V. $X = [w_j s_{jk} x_{jk}]$, where $j \in A$ and $s_j = sign\{\hat{c}_j\}$;

VI. $a = X^T u_A$

VII. $\hat{\gamma} = \min_{j \notin A}^{+}\left\{\frac{C-c_j}{A_A - a_j}, \frac{C+c_j}{A_A + a_j}\right\}$

VIII. $k = k+1$, add with a new predictor, repeat the process steps 3~7;

IX. Obtain the regression coefficients β;

4 Experimental analysis

4.1 Experiment Data Processing

All requisite experiments in this paper are implemented in MATLAB. The experimental data [10] can be derived from the research which German credit classifies people as good or bad credit risks. The dataset contains 20 attribute information which can proof of his financial ability. The last column predictive classifications can be in two states, 0 or 1, namely reliable or unreliable customer.

The half of dataset is used to calculate regression coefficients, the rest of dataset is used to judge if this algorithm can get precise prediction based on obtained regression coefficients.

For the four prediction possibilities we can analyze them with ROC curves [11]:

- True positive (TP): the prediction is reliable customer, yet the fact is also reliable;
- False positive (FP): the prediction is unreliable customer, while the fact is reliable;
- True negative (TN): the prediction is unreliable customer, yet the fact is also unreliable;
- False negative (FN): the prediction is reliable customer, while the fact is reliable.

4.2 The Comparison between LARS Algorithm and Improved Algorithm

Use MATLAB to run LARS algorithm and then add improved algorithm on the basis of LARS. Each measure of two algorithms are as follows:

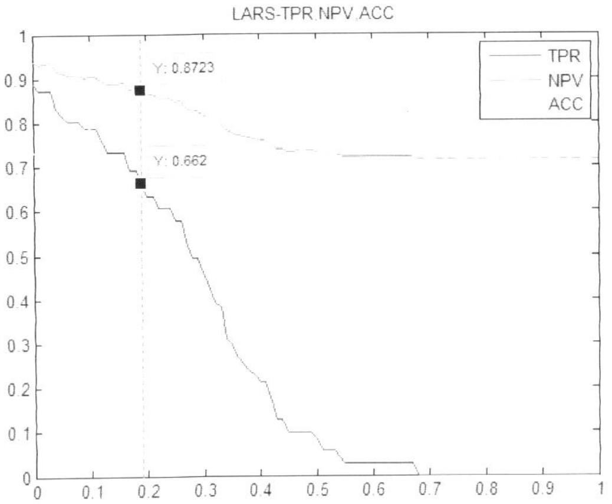

Figure 4.1 NPV and TPR of LARS Algorithm

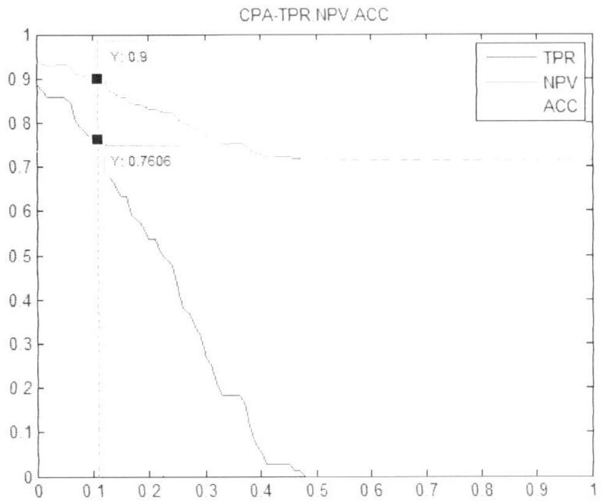

Figure 4.2 NPV and TPR of Coefficient of Variation Method

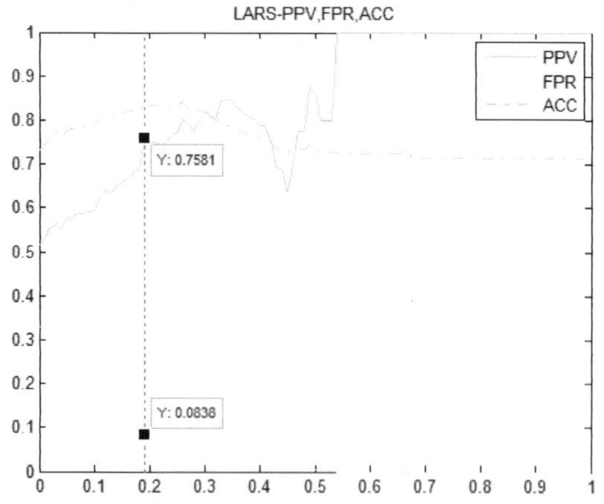

Figure 4.3 FPR and PPV of LARS Algorithm

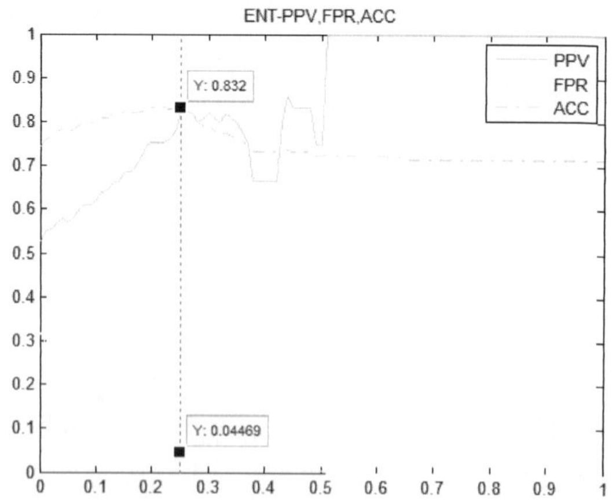

Figure 4.4 FPR and PPV of Entropy Weight Method

Along with illustrations, accuracy rate is defined as $(ACC = (TP + TN) / (P + N))$. In our experiment, the higher ACC is, the greater the ability of predicted positive condition is;

True positive rate ($TPR = TP / TP + FN$), in our experiment, it indicates the ratio of the number of true positive to all conditions positive. The higher value means the reliable positive condition detection.

Negative predictive value $(NPV = TN / (TN + FN))$ and Positive predictive value $(PPV = TP / (TP + FP))$, the higher value indicates the greater result of negative/ positive condition detection; and False positive rate $(FPR = FP / (FP + TN))$ The lower FPR means the less the misjudged condition .

It is illustrated that these indexes of the improved algorithm have improved based on LARS algorithm. It is concluded from the comparison of figure 4.1 and figure 4.2 that compared with LARS algorithm, the improved algorithm based on coefficient of variation method improves several indexes. The true positive rate was increased by 10% and negative predictive value 2.8%; while comparing from figure 4.3 and figure 4.4, the false positive rate of improved algorithm based on entropy weight method was lowered by 4% and the positive predictive value was increased by 6.4%. As for the experimental data, it is worse to predict false positive than to predict false negative. So the change of these indexes corresponds with what we have expected. Furthermore, the ability of prediction is improved by two improved algorithm.

5 Conclusion

LASSO is a method for variables selection in linear models. LARS algorithm resolves LASSO problem by proceeding in a direction equiangular of selected independent variables. The weighted method is introduced into LARS algorithm to change the approximation direction. The key to achieve this objective lies in that each independent variable has various effect on the dependent variables in linear model. We distinguish the impact of various predictors on responses by the design of algorithm weighting for the basis of LARS algorithm, which means the more precise prediction. Compared LARS algorithm with improved algorithm, the experimental verification shows that it is more effective to combine the weighted method and LARS algorithm. However, the false negative rate has no significant effect during experimental process. So the precision of variables selection and the calculation of regression coefficients are worthy to explore in the further study.

References

[1] Liu Liu, Dacheng Tao. Review on Recent Method of Solving Lasso Problem[J]. Journal of Data Acquisition and Processing. 2015(1).
[2] Wei Yang, The component selection of functional regression model[D]. Northeast Normal University, 2009.
[3] Ke Zhenlin. The optimal property of Lasso and other related methods on multiple linear regression

model[D]. Master 2011.
[4] Efron B, Hastie T, Johnstone I, et al. Least angle regression[J]. Journal of the Institute of Mathematical Statistics, 2004, 32(2): 407—499.
[5] Robert Tibshirani. (1996), Regression Shrinkage and Selection via the Lasso. Journal of the Royal Statistical Society. Series B, Vol.58, No.1.267-288.
[6] Liu GuoTian, Feng Lin, Ding Xiang. A feature selection algorithm based on LARS[C]. 2012.
[7] ZHOU Yanzhou, QIAO Hui.WU Xiao-ping. Research of Software Complexity Metric Attributes Feature Selection Based on LASSO-LARS[J]. Computer Science. 2013(11).
[8] Kai Sun, Xiaofeng Ju, Yuhua Li. Performance Evaluation of Enterprise Incubators Based on Variation Coefficient Method [J]. JOURNAL HARBIN UNIV. SCI. & TECH.2007,12(3):165-167.
[9] XIE Chi, ZHONG Zan. Entropy Method and Its Application in Comprehensive Evaluation of Bank's Performance[D]. China Soft Science.2002(9).
[10] Professor Dr. Hans Hofmann Institut f"ur Statistik und "Okonometrie Universit"at Hamburg FB Wirtschaftswissenschaften Von-Melle-Park 5 2000 Hamburg 13.
[11] https://en.wikipedia.org/wiki/Receiver_operating_characteristic

IOP Conf. Series: Materials Science and Engineering **242** (2017) 012111 doi:10.1088/1757-899X/242/1/012111

A brief history and technical review of the expert system research

Haocheng Tan[1, a)]

[1]School of Information Science and Technology, University of Science and Technology of China, Hefei 230026, China

[a)] thc1800@163.com

Abstract. The expert system is a computer system that emulates the decision-making ability of a human expert, which aims to solve complex problems by reasoning knowledge. It is an important branch of artificial intelligence. In this paper, firstly, we briefly introduce the development and basic structure of the expert system. Then, from the perspective of the enabling technology, we classify the current expert systems and elaborate four expert systems: The Rule-Based Expert System, the Framework-Based Expert System, the Fuzzy Logic-Based Expert System and the Expert System Based on Neural Network.

1. Introduction

The expert system is a computer system that emulates the decision-making ability of a human expert, which aims to solve complex problems by reasoning knowledge. In 1959, Newell, Shaw and Simon developed the General Problem Solver (GPS) [1], which is a summary of the thinking activity for people to solve problems. They found that when people solving problems, the thinking activities include three steps: 1. developing a rough plan; 2. according to the memory of the axiom, theorem and problem-solving plan, solving the problem according to plan; 3. in the implementation of the problem-solving process, constantly analyzing the method and purpose, and revising the plan.

In 1965, according to the requirements of the National Aeronautics and Space Administration, the Stanford University successfully developed the DENDRAL system [2]. By the pre-entered rule of thumb, the DENDRAL system can automatically generate molecular structures that can interpret spectral data. The program is a first successful program that uses the knowledge of the problem itself rather than the complex search technology. The DENDRAL guides artificial intelligence researchers to realize that the intelligent behavior relies not only on interference methods but also on the knowledge used in their interference. Researchers begin to build the program that uses the rules of the code to represent the knowledge to solve the input problem.

Since then, MIT has developed the MACSYMA system [3], as a mathematician's assistant, which uses heuristics to transform algebraic expressions. After continuous expansion, it can solve more than 600 kinds of mathematical problems, including calculus, matrix operation, solving equation, etc. The successful development of these systems makes the expert system widely concerned by academia and engineering. Many researchers in the process of developing expert systems have realized that knowledge representation, knowledge utilization and knowledge acquisition are three basic issues of artificial intelligence systems.

In the late 1980s, the Framework-Based Expert System began to enter people's vision. Because of its higher ability to represent descriptive and behavioral object information, a Framework-Based Expert System could handle more complex problems than the Rule-Based Expert System. At the same time,

Content from this work may be used under the terms of the Creative Commons Attribution 3.0 licence. Any further distribution of this work must maintain attribution to the author(s) and the title of the work, journal citation and DOI.
Published under licence by IOP Publishing Ltd

the study of the expert system encountered difficulties, exposing the defects of artificial intelligence systems, such as narrow application areas, knowledge acquisition difficulties, reasoning mechanism and so on. Researchers need to get rid of the dilemma by exploring the basic point of view and the use of new techniques and theories.

With the rising of neural network and statistical learning, machine learning has become a new research hotspot in artificial intelligence. Naturally, people thought that if the machine learning technology applied to the expert system, the reasoning performance of the expert system would be promoted. The fuzzy logic, as another new area, was the same applied to the expert system. Now, all kinds of different expert system blossom, expert system research has entered a new period of prosperity. Today, the demand for expert systems is further enhanced, and people are eager to use expert system to solve more complex problems.

1.1 The Basic Structure of the Expert System

The basic structure of an expert system consists of the following parts: knowledge base, working memory, reasoning machine, interpreter and human-computer interaction interface. The knowledge base is used to store expert system expertise, including facts and rules. In the process of building a knowledge base, the knowledge base should be able to acquire new knowledge, expressing and storing knowledge in a way that the computer can accomplish. The working memory is responsible for storing the input fact. The reasoning machine matches the facts in the working memory with the knowledge and gets new information. The intermediate information obtained during processing should also be stored in a storage unit. The interpreter is responsible for interpreting the results of the inference engine output, including explaining the correctness and reason of the conclusion.

2. The Classification of the Expert System

According to the working principle, the expert system can be sorted into four classes: the Rule-Based Expert System, the Frame-Based Expert System, the Fuzzy Logic-Based Expert System and the Expert System Based on Neural Network. This section will briefly introduce the model structure of these four expert systems.

2.1 The Rule-Based Expert System

The Rule-Based Expert System is the earliest type of expert systems and the most common way for researchers to build expert system. The expert system, which is completed by production rules, has been integrated into various research areas to help researchers solve all kinds of problems with pre-input knowledge.

The structure of the expert system is usually composed of three parts, including production rule, database and control strategy. Production rule is a "conditional + result" structure of the statement. For example, the condition: "if a species is a dinosaur", the result: "then the species is a reptile". If the condition in the statement is met, the result of the rule can be executed.

The database is responsible for storing the conditions and results in the production rule statement. When the production rule is executed, the corresponding condition is called from the database and the rule result is put into the database as a condition for other rules.

The role of the control strategy is to explain how to apply the rules, that is, in the process of solving the problem to select the appropriate rules. The process from selecting rules to performing operations is usually the case that a problem contains several conditions, looking for matching conditions from the database, and finding the rules for the corresponding conditions. When more than one condition is matched, you need to sort the strategy to decide which rule to use first. After the rule is selected, the operation (result) section of the rule is executed. The operation of the control strategy is usually performed in a module ("inference engine"). The inference process of the inference engine can be divided into both forward and reverse.

The Rule-based Expert System has many advantages. Using the "condition - outcome" statement to express knowledge rules is very natural for humans. At the same time, in the Rule-based Expert System, knowledge and reasoning are stored and processed separately, in line with people's daily habits. However, some shortcomings also hindered the further development of such expert system. For example, when a rule matches a condition, the expression of the statement must be written strictly in accordance with the statement in the database, even if the relative subtle differences are rejected because of the matching accuracy requirements. In addition, the speed of the Rule-based Expert System is not dominant compared to other types of expert systems, since the entire rule set in the database needs to be scanned when a rule is used.

2.2 The Framework-Based Expert System

In 1975, Minsky proposed a data structure that encodes a concept-related information and calls this data structure as "frame" [4]. The framework contains the name of the concept, the label slot of the main attribute/feature, and the possible values of each attribute, or the process of capturing procedural information about the concept. When a particular instance of the concept is encountered, the relevant eigenvalue of the instance is entered into the framework, which is called the instance.

The framework can be used for reasoning. The framework contains multi-aspect information of a concept that can be used even if the information itself is not observed. For example, whether or not we can see the roots, because the "tree" contains the "root" of the label, so we can think that the tree has root.

When looking for a framework to describe the current instance, it will often fail to match the situation. Specific examples will correspond to some frame fragments that used to match the candidate frames. The framework is usually used when most of the structure of a frame can be matched, since each frame contains information that can allow mismatches of attributes (features) to increase the fault tolerance of the frame. In addition, we can also save the framework of the search process to optimize the direction of the test until we find the most appropriate framework.

The Framework-Based Expert System uses the framework in the database to handle the specific issues of the input and output new information through the inference engine. A framework represents a concept of "class", and a framework may be a "subclass" of another framework, such as the "man" framework is a subset of the "human" framework. Some subclass-and-class relationships are semantic types, such as "I am" + "New York", "New Jersey", "Los Angeles".

The subclass framework inherits all the properties of its parent class framework, eliminating the need to reenter the properties of the process. However, we should pay attention to some special circumstances: some subclasses and their father class may be in a common property differences. When a frame belongs to multiple frames at the same time, it also inherits the properties of all these frames.

2.3 The Fuzzy Logic-Based Expert System

In 1965, American mathematician L. Zadeh first proposed the concept of fuzzy set, marking the birth of fuzzy mathematics [5]. Fuzziness refers to the indiscriminate nature of objective things or attributes, and there is a series of transitional states between them, without obvious dividing line. The fuzzy theory allows people to use mathematical tools to deal with the non-precise phenomenon in the real world.

In the Fuzzy Logic-Based Expert System, the fuzzy logic is the basis of expert system reasoning. This kind of reasoning method uses fuzzy rule as a prerequisite, and uses fuzzy language rule to deduce an approximate fuzzy judgment conclusion. The fuzzy language rules include the Generalized Modus Ponens (GMP) and the Generalized Modus Tollens (GMT).

The GMP rule can be expressed as: Prerequisite 1: x is A '. Precondition 2: If x is A, then y is B. Conclusion: y is B '. The GMT rule can be expressed as: premise 1: y is B, premise 2: If x is A, then y is B, Conclusion: x is A'. A, A', B, and B' in the above formula are fuzzy sets, and x and y are linguistic variables. The fuzzy function $A \rightarrow B$ is considered as a "if - then" statement, forming a fuzzy reasoning rule.

The result of fuzzy reasoning is a fuzzy set, but in practical, a certain output value is required. The fuzzy decision is the process of obtaining a single value representing a fuzzy set. The simplest method is the maximum membership method, which takes the maximum value of membership in all fuzzy sets as output.

The advantage of the Fuzzy Logic-Based Expert System is that it can express the high skill of expert skills and has enough robustness, and can also carry out heuristic and tentative reasoning. However, this kind of expert system has difficulty in obtaining knowledge and its reasoning depends on fuzzy knowledge base, learning ability is not strong, prone to error.

2.4 The Expert System Based on Neural Network

The neural network model is fundamentally different from the logical system described above. In the neural network, knowledge is no longer transformed into explicit rules by manual processing, but is automatically acquired by the learning algorithm and produces its own implicit rules. Compared with the traditional expert system, the neural network has more powerful function: it is more efficient than the traditional serial operation; it has a certain degree of fault tolerance; the weight of the neural network connection can be changed, etc.

The neural network acquires knowledge automatically through learning instances. Experts provide examples and expectations of the solution, neural network learning algorithm constantly modify the weight distribution of the network, achieving a stable output after training. Since the input and output of the neural network are numerical, it is necessary to encode the instance when using the neural network to acquire the knowledge.

The Expert System Based on Neural Network also has inherent weaknesses: the system performance is limited by the training sample set. In the case of improper sample set selection or too little sample, the neural network's induction reasoning ability is very poor. In addition, the neural network is unable to explain its own reasoning process and the significance of storing knowledge, because its model is based on human superficial neural activity.

At present, the most commonly used neural network model includes BP model, ART, CMAC, SOM and so on. Different models can be applied to specific requirements of the expert system.

3. Conclusions

After decades of efforts, people have gradually developed a variety of expert systems based different principles. From the Rule-Based Expert System to the Expert System Based on Neural Network, each generation of expert systems has its own advantages. Based on the rules, the framework of the expert system, although subject to their own technical constraints, but because of their relatively easy encoding, the two types of expert systems is still useful in dealing with relatively simple problems. The Fuzzy Logic-Based Expert System can reflect the fuzzy phenomenon in reality, imitating the process of human inexact reasoning to complete its work, playing a very important role in many fields. The Expert System Based on Neural Network uses neural network to further improve their ability and expand the scope to acquire knowledge, and uses large-scale parallel processing to improve the system reasoning ability. Of course, how to make the neural network to explain the process of their own reasoning is still one of the difficulties, but this does not affect such experts in the field to show their talents. In the future, with the further development of neural network technology, we can expect a more powerful expert system based on this technology. At the same time, we also look forward to the emergence of new technologies in the expert system.

References

[1] Newell, A., Shaw, J. C., & Simon, H. A. (1959, January). Report on a general problem solving program. In IFIP congress (Vol. 256, p. 64).
[2] J. Lederberg, (1965).
[3] Petrick, S. R. (1971). Proceedings of the Second Symposium on Symbolic and Algebraic Manipulation (pp. 23-25). March.
[4] Minksy, M. (1975). A framework for representing knowledge. The psychology of computer vision, 73, 211-277.

[5] Zadeh, L. A. (1965). Fuzzy sets. Information and control, 8(3), 338-353.

IOP Conf. Series: Materials Science and Engineering **242** (2017) 012112 doi:10.1088/1757-899X/242/1/012112

A new method of hybrid indoor localization with inertial sensor

Wang weigang[a), Wang wenrui[b)]

School of Electronic Science and Engineering, Nanjing University of Posts and Telecommunications, Nanjing 210003, China.

a)Corresponding author: velcon@126.com

b)2214688550@qq.com

Abstract. Indoor Localization has been an interesting and significant area of research in the recent days because of its diverse applications. However, current indoor localization method has some drawback in accuracy and it needs to be improved urgently. In this paper, we proposed a hybrid localization algorithm which combines the received signal strength indication with inertial sensors. The corresponding flow chart was also provided. Simulation results show that the proposed algorithm has made a great improvement in localization accuracy.

1. Introduction

There have recently been many studies on localization because location information is important for providing valuable applications. It is critical to several new services such as ubiquitous network systems [1, 2]. Localization techniques can be classified into several categories, such as using received signal strength (RSS), time of arrival (TOA), time difference of arrival (TDOA), angle of arrival (AOA), location fingerprint, and combinations thereof [3].

The traditional localization techniques of AOA, TDOA, and AOA was used in outdoor because of complex radio wave propagation [4]. Therefore, the location fingerprint technique, which generally uses RSSI, is attractive for indoor localization [5, 6]. The basic idea of the fingerprint approach is one kind of matching patterns based on localization technique. A multipath delay profile can also be used to locate a mobile terminal with pattern matching.

The main advantage of the fingerprinting technique is that it can provide a low-cost and high-accuracy localization solution by utilizing in-building Wi-Fi infrastructures. However, fingerprinting localization algorithm in the literature, such as the traditional *K* nearest neighbor (KNN) algorithm, may bring large error when calculating location. The most important reason for the error is that these algorithms may select the wrong reference point for localization. In order to reduce the localization error in the fingerprinting estimation due to dynamic environment, there is also some paper use the Wi-Fi tag to resolve above problems [7].

Received signal strength indication (RSSI) is very convenient to obtain in the smart phone. Since the received signal contains noise and is also infrastructure dependent [8], a hybrid system of combing RSSI fingerprint with inertial sensor is worthy of study. Hence we make use of inertial sensors embedded on smart phone such as accelerometers and magnetic compass, to detect any motion and heading direction to reduce the error compare to traditional RSSI approach.

Content from this work may be used under the terms of the Creative Commons Attribution 3.0 licence. Any further distribution of this work must maintain attribution to the author(s) and the title of the work, journal citation and DOI.
Published under licence by IOP Publishing Ltd

2. Localization based on RSSI fingerprint

The fingerprint technique was introduced in [9] as an effective approach for indoor localization. It utilized the received signal strength (RSS) in Wi-Fi circumstance. The detail procedure of RSSI localization is as following.

Suppose an indoor localization system consisting of n access points and m reference points. In off-line phase, the RSSI $\{R_k(x_i, y_j) \mid i = 1,\cdots,m, j = 1,\cdots,n, k = 1,\cdots,L\}$ is measured and recorded to the fingerprint with coordinate $\{(x_i, y_j) \mid i = 1,\cdots,m, y = 1,\cdots,n\}$.

In this model, Euclidean distance of RSSI is used to evaluate the geographical relation between off-line RSSI fingerprint and on-line RSSI. Euclidean distance D can be represented as

$$D(x_i, y_j) = \left\| \boldsymbol{R}' - \boldsymbol{R}(x_i, y_j) \right\|_2, \tag{1}$$

Where R' denotes the on-line RSSI value. According to the KNN algorithm and the minimum Euclidean distance theory, the first K coordinates of N *were* selected. Then, by matching the minimum difference to off-line database, the optimal location of smart phone can be determined.

3. Hybrid localization system

3.1 Inertial Sensor System

When pedestrian moving with a smart phone in hand, outputs of the accelerometer in smart phone usually presents obvious periodic waveform. However, the uncertain pose of the smart phone will make it difficult to determine the orientation of the accelerometer's with three axes in the navigation coordinate. As a result, the total magnitude of the three-axis acceleration is usually used as a pose-invariant signal feature. The magnitude a_m can be expressed as

$$a_m = \sqrt{a_x^{\,2} + a_y^{\,2} + a_z^{\,2}}, \tag{2}$$

Where a_x, a_y and a_z are the measurements from the three-axis accelerometer. The axes of sensor readings are based on the smart phone.

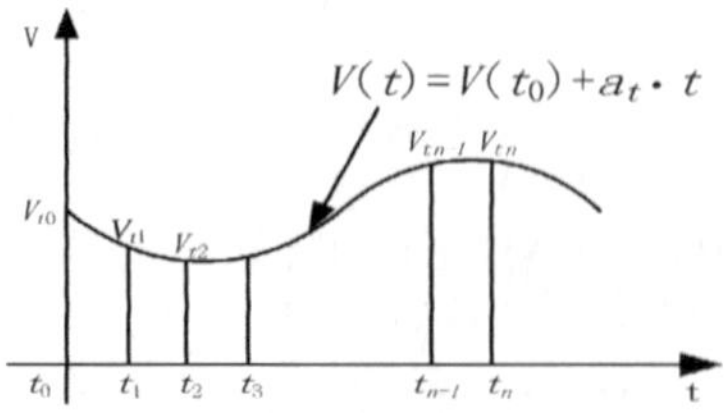

Figure 1. Velocity-Time Curve

The horizontal axes and vertical axes represent time and velocity individually in Figure 1. The $a(t)$ represents the changing acceleration values with time increase. The acceleration at any given t moment is the slope of the corresponding point on the curve. The displacement $s(t)$ and velocity $v(t)$ of an object can be obtained according to the integral principle in the continuous time from t_0 to t.

Since the output value of the accelerometer is discrete, the velocity-time curve is discredited and the graph can be decomposed into many small right angle trapeziums. There is $s(t_0) = 0$, $t_1 - t_0 = t_2 - t_1 = \ldots = t_n - t_{n-1} = \triangle t$, Where $\triangle t$ is the time interval of acceleration sensor sampling. Thus the moving distance of the target object at the n moment can be represented as

$$s_n = v_0 \cdot n \cdot \Delta t + [(n-1)\cdot a_1 + (n-2)\cdot a_2 + \ldots + a_{n-1}]\cdot \Delta t^2 + \frac{1}{4}(a_0 + a_n)\cdot \Delta t^2 \tag{3}$$

According to the equation (3), the displacement of the target can be obtained by the initial velocity v_0 and the acceleration of the accelerometer. However, when the n value is large enough, lots of mathematical operations are needed and the system memory consume too much. Therefore, iterative method is adopted to simplify the calculation process. The equation of velocity and displacement can be written as

$$v_n = v_{n-1} + \frac{a_n + a_{n-1}}{2} \cdot \Delta t, \tag{4}$$

$$s_n = s_{n-1} + \frac{v_n + v_{n-1}}{2} \cdot \Delta t = s_{n-1} + v_{n-1} \cdot \Delta t + \frac{1}{4}(a_n + a_{n-1}) \cdot \Delta t^2. \tag{5}$$

By equation (4) and (5), the current velocity v_n and displacement s_n can be obtained by calculating v_{n-1}, s_{n-1} and a_{n-1}, a_n obtained on the previous time.

3.2 Hybrid Localization Approach

The classical RSSI fingerprint localization method has disadvantage in accuracy. In inertial localization, there is also the problem of error accumulation. Although, the inertial sensor has a few drawbacks, it can estimate new location accurately if the last estimated location is offered. We turn to study on hybrid approach of localization. When RSSI fingerprint and inertial sensor are combined, they overcome each other drawbacks.

The hybrid localization approach uses the on-line RSSI value and acceleration value to determine the new location based on its previous location. If the RSSI value is accepted, the KNN algorithm can be used. Hence, the RSSI approach is reliable in the vicinity of access point. When the mobile terminal is not in the vicinity of any of access points, the inertial integral localization is used for localization. Accumulated error in inertial integral localization is reduced by re-initializing when the mobile terminal is in the vicinity of an access point again. The hybrid localization method flow chart is given as following.

If the location is determined by RSSI fingerprint only, error would increase distinctly when the RSSI value is low. If the inertial integral localization is used only, the error would increase as time goes on. Therefore, we use RSSI fingerprint and inertial sensors for hybrid localization which named inertial k-nearest neighbor (IKNN) algorithm. In our algorithm, only the difference of location by RSSI and inertial is less than 3 meters in same direction, the result is accepted.

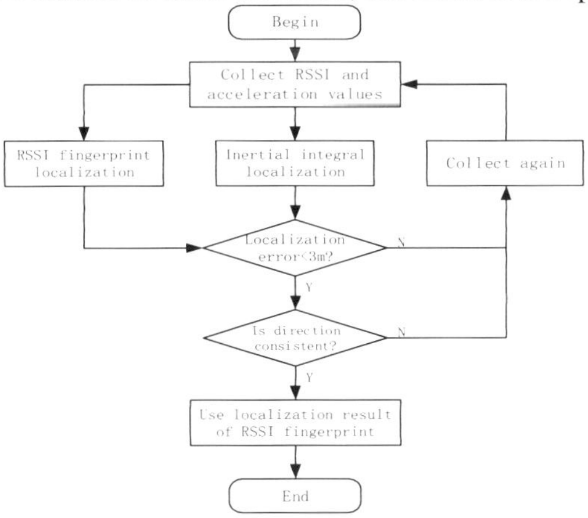

Figure 2. Hybrid localization method flow chart

4. Numerical simulation

Simulations were conducted in a zone of size 6 ×10 × 3m in 3D coordinate. The indexes of reference locations are shown in Figure 3. Adjacent receiver points are set in 1 m apart away.

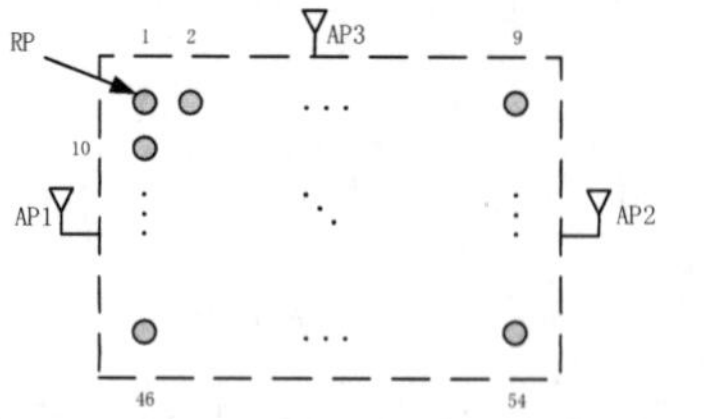

Figure 3. The indexes of reference location in database

In our experiment, the receiver point location is indicated by "RP" in Figure 3. The access points (APs) are set in 2-m tall. Monte Carlo simulations are performed to compare the performance of NN, KNN, IKNN and FKNN algorithm.

In Figure 4(a), we did the first simulation experiment which RSSI and inertial individually. When the difference between these two results is less than 3 meters in same direction, the result is accepted otherwise it should be measured again.

In Figure 4(b), the second experiment result is shown. Firstly, the discrete location points were obtained by inertial integral localization and then fitting curve was made by polynomial fitting. Secondly, the RSSI fingerprint was used to locate the target named fitting KNN (FKNN) algorithm. If the location coordinates deviated from the fitting curve more than 3 meters, the RSSI was measured again.

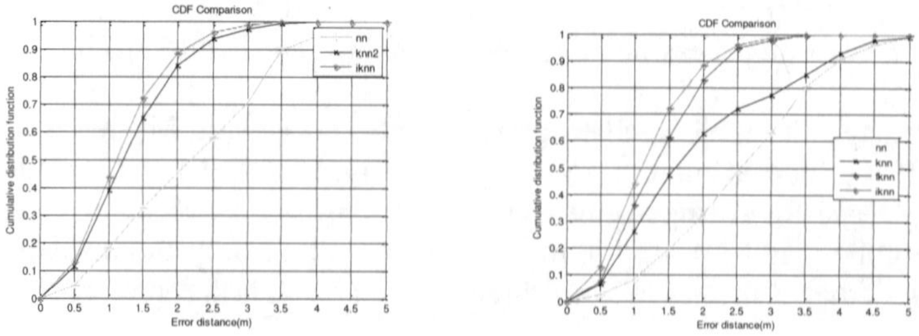

(a) (b)

Figure 4. The succeed probability of NN, KNN and IKNN algorithm

From the Figure 4, the cumulative distribution function raise when the error distance increase. We can see that the performance of the FKNN and IKNN algorithm are better than NN and KNN. The IKNN gets the best performance among of them.

Next, we did the third experiment to verify its cumulative effect of error between the inertial integral localization and hybrid localization.

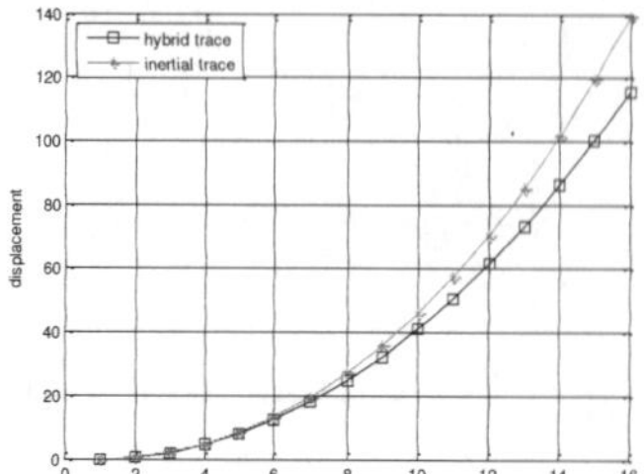

Figure 5. The contrast diagram of simulation effect

From the Figure 5, it shows the displacement raises with the sample position altering. The localization error of inertial integral localization increases sharply when displacement exceeds more than 20m. The hybrid localization can decrease this error with adjustment by RSSI fingerprint database.

5. Conclusion

In this paper, aiming to the low accuracy of indoor localization, we proposed a hybrid indoor localization method based on RSSI fingerprint database and inertial sensor. Our method can decrease the location error and adjust adaptively by referring the data from other single traditional method. The simulation results show the effectiveness of our IKNN algorithm.

Acknowledgments

The project is supported in part by the foundation of NanJing University of Posts and Telecommunications No. NY215164 and by A Project Funded by the Priority Academic Program Development of Jiangsu Higher Education Institutions and by the Key University Science Research Project of Jiangsu Province under Grant No. 14KJA510003. It also supported by the National Natural Science Foundation under grant No. 61271335 and No.61571233.

References

[1] C. M. Vikas, S. Rajendran, A. Pattar, H. S. Jamadagni and R. Budihal, "Wi-Fi RSSI and inertial sensor based indoor localisation system: A simplified hybrid approach,"*2016 International Conference on Signal and Information Processing (IConSIP)*, Vishnupuri, 2016, pp. 1-6.

[2] P. Zhang, X. Chen, D. Fang, W. Wang, Z. Tang and Y. Ma, "SmartMTra: Put the Limit of the Inertial Sensor Based Indoor Motion Trajectory Tracing Using Smartphone,"*2016 Intl IEEE Conferences on Ubiquitous Intelligence & Computing, Advanced and Trusted Computing, Scalable Computing and Communications, Cloud and Big Data Computing, Internet of People, and Smart World Congress (UIC/ATC/ScalCom/CBDCom/IoP/SmartWorld)*, Toulouse, 2016, pp. 66-73.

[3] A. Yilmaz and A. Gupta, "Indoor locationing using visual and inertial sensors," *2016 IEEE SENSORS*, Orlando, FL, 2016, pp. 1-3.

[4] A. Galov and A. Moschevikin, "Automatic signal strength map construction in indoor locationing system based on round trip time of flight measurements and inertial navigation," *2016 3rd International Symposium on Wireless Systems within the Conferences on Intelligent Data Acquisition and Advanced Computing Systems (IDAACS-SWS)*, Offenburg, 2016, pp. 25-29.

[5] Y. Li, J. Chen, Y. Shi, Y. Cheng and L. Wang, "Wi-Fi-assisted multi-floor indoor localization with inertial sensors," *2016 8th International Conference on Wireless Communications & Signal Processing (WCSP)*, Yangzhou, 2016, pp. 1-5.

[6] D. Caruso, M. Sanfourche, G. Le Besnerais and D. Vissière, "Infrastructureless indoor navigation with an hybrid magneto-inertial and depth sensor system," *2016 International Conference on Indoor Locationing and Indoor Navigation (IPIN)*, Alcala de Henares, 2016, pp. 1-8.

[7] H. Yang *et al.*, "Smartphone-Based Indoor Localization System Using Inertial Sensor and Acoustic Transmitter/Receiver," in *IEEE Sensors Journal*, vol. 16, no. 22, pp. 8051-8061, Nov.15, 2016.

[8] Y. Li, Y. Guo and H. Luo, "Accurate indoor localization based on the inertial navigation and the iBeacon," *2016 IEEE International Conference on Consumer Electronics-Taiwan (ICCE-TW)*, Nantou, 2016, pp. 1-2.

[9] C. BASRI and A. El Khadimi, "Survey on indoor localization system and recent advances of WI-FI fingerprinting technique," *2016 5th International Conference on Multimedia Computing and Systems (ICMCS)*, Marrakech, 2016, pp. 253-259.

IOP Conf. Series: Materials Science and Engineering **242** (2017) 012113 doi:10.1088/1757-899X/242/1/012113

Development of 3D browsing and interactive web system

Xiaonan Shi[1,a)] ,Jian Fu[1,b)] and Chaolin Jin[1,c)]

[1]Xi`an University of Science and Technology, Xi`an710600,China

E-mail: [a)]12606137@qq.com, [b)] Corresponding author:1046019528@qq.com, [c)]chaolinjin@gmail.com

Abstract: In the current market, users need to download specific software or plug-ins to browse the 3D model, and browsing the system may be unstable, and it cannot be 3D model interaction issues. In order to solve this problem, this paper presents a solution to the interactive browsing of the model in the server-side parsing model, and when the system is applied, the user only needs to input the system URL and upload the 3D model file to operate the browsing. The server real-time parsing 3D model, the interactive response speed, these completely follows the user to walk the minimalist idea, and solves the current market block 3D content development question.

1. Introduction

Traditional web pages can only render content through text, etc., however with the WEB technology[1] ,rendering 3D content in the web pages has become possible[2 3]. For the study of WEB 3D technology, the most authoritative country is the United States[4 5], and the research focused on perception, user interface, background software and hardware. Compared with some developed countries, there is a great disparity in the research level of China[6], but with the rapid development of computer technology, government departments and scientists have begun to pay attention to the development of WEB 3D technology. China's universities have carried out the corresponding research, such as the National Defense University of Science and Technology developed a three-dimensional Internet-based exquisite roaming environment Universe3D. As well as the Shanghai Museum of bronze Museum, the ancient scene of the simulation show and so are all in the web page to show the 3D browsing.

Open 3D model files require pass the specific PC graphics software, which is currently the biggest technical difficulty in rendering and managing 3d content in the Internet[8]. In the context of the Internet + and other national policies, the Internet system that can present and manage different formats of 3d data with a unified interface is a key role in the tide of industrial upgrading. This is also the main research content of this paper.

2. System framework

Nginx is a high-performance HTTP and reverse proxy server, and Node.js is a platform based on the V8 engine runtime[9], which is used to easily build up the network application. Node.js uses event-driven, non-blocking I/O models to be lightweight and efficient, and is ideal for running data-intensive real-time applications on distributed devices. So the system adopts the Nginx + NodeJS server architecture, as shown in Figure 1, with HTML +CSS +JavaScript as the front-end programming language, at the same time, the system adopts a browser supporting WebGL.

Content from this work may be used under the terms of the Creative Commons Attribution 3.0 licence. Any further distribution of this work must maintain attribution to the author(s) and the title of the work, journal citation and DOI.
Published under licence by IOP Publishing Ltd

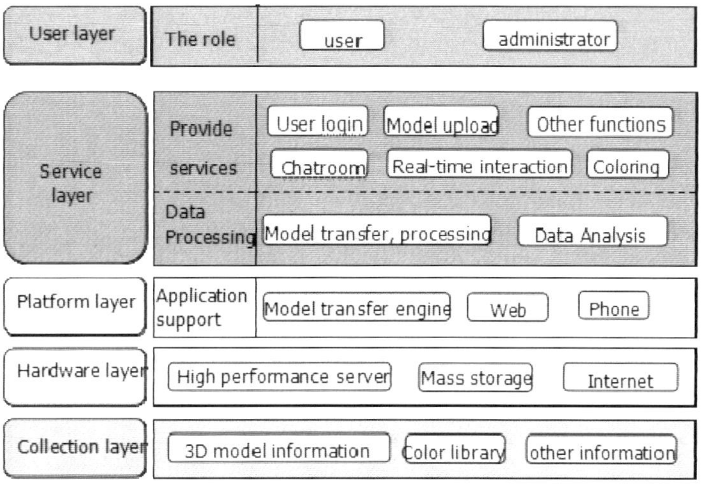

Figure 1. System architecture diagram

The system is divided into three parts: model resolution, model rendering and model interaction. By uploading the 3d model, the model storage, the analysis of the model,and through the model format conversion, ThreeJ engine and other system completes model rendering. Finally it's human-computer interaction. The System flowchart, is shown in Figure2.

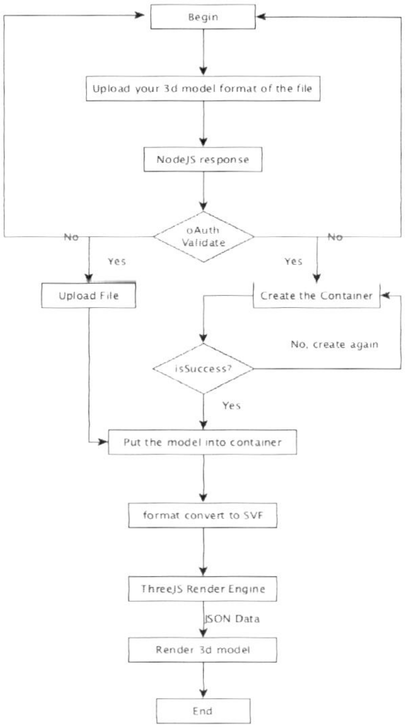

Figure 2.Flowchart

3. Key technologies

3.1 3D Model Information

The manifold triangular mesh is one of the simplest 3d model representations[10], and its external file (the type name is "SVF") describes the 3d data mainly including the geometric coordinates of the vertices and the triangular surface table information (table information of the triangles).And the SVF file in the format is composed of some text lines, according to the conventional file header and graphics data area it can be divided into two parts[11]:

(1)File header: generally fixed length, in addition to describing the type and characteristics of the file itself,mainly used to describe the geometric properties of grid elements, such as the number of vertex and surface, and their data types and other basic information.

1) File type names: convention for .svf;

2) Data line format format ASCII, using ASCII text line (without binary binary);

3) The number of vertex elements is element vertex;

4) Vertex data type property float vetrex_coord, optional float or double type;

5) The number of elements element face

6) The data type of the Surface Table property list UINT int vertex_indices, the vertex index must be an integer;Head end marker End_header;

(2)Graphic data area: is a variable length.It can be used to specifically define the 3D model of geometric data (vertex coordinates) and topological information (surface structure): x y z ... where each triples XYZ represents a three-dimensional coordinate of a vertex, separated by spaces. For example,0.605538,0.183122,0.472278 defines a vertex.Vertex_number refers to the number of vertices of the polygon surface, because the triangle mesh is used in this paper, so the vertex number of the surface is 3. A vertex index table is followed by a representation of a triangular surface. For example: 3 1 0 1 1 defines a triangular face, which is composed of 1 vertices indexed by 0, 1, 1 3 respectively.

3.2 File Parsing

The need to analyze the structural characteristics of a 3d file is the first step in completing a transformation from a 3d file to a JSON string. So there are a wide variety of tags that may appear in 3d, and you need to turn 3d files into the standard format [12] before converting to JSON.

After you get the 3d file in the standard format, split the labels and contents into an array in order of precedence. And then through an auxiliary stack to transform the conversion process during the auxiliary stack state changes, shown in Figure 3.

Table 1. Label Conversion Table

Original label format	Converted label format
<element name attributes=value>	<element name > <@attributes> < attributes> value </attributes> </@attributes>
<element name/>	<element name> </element name>
<element name attributes=value>	<element name> <@attributes> < attributes> value </ attributes> </@attributes>

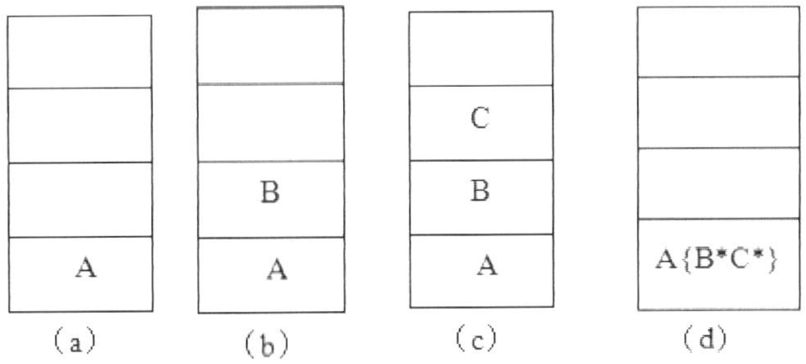

Figure 3.Auxiliary stack state diagram

Finally, the last element ejected from the stack is the converted JSON string, rendering the conversion process as shown in Figure 4 below[13].

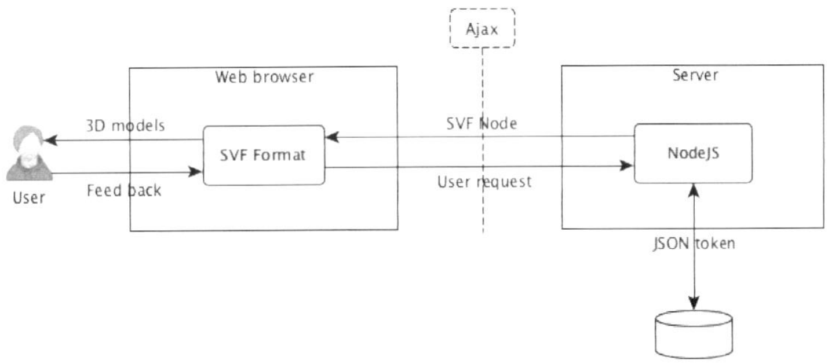

Figure 4 .Render the map

Use the browser WebGL standard to support 3d data parsing. The system uses a unified 3D file format .svf, and through AJAX interacts the front and rear data. The 3D format files are asynchronous without refreshing to parse out the SVF node, through the node to build the entire 3D model, rendering the 3d model required for the construction data stored in JSON, such as JSON rendering vector data.

```
{
"metadata": {
"version": 4,
"type": "geometry",
 "generator": "GeometryExporter"
 },
"vertices": [0,50,20,14.14213],
"normals": [0.3826834323650897,0,0.9238795325112866],
"uvs": [[0,1,0,0,0.125,1,0.125],
"faces": [56,0,9,1,0,1,2,0,1],
}
```

3.3 Research on Spatial Transformation

Computer 3D graphics and matrix are closely related, and the graphical API interface directly uses the matrix, and Transform2d/3d encapsulates the most basic transformation operations. Each transformation can be transformed into a matrix. But for matrix multiplication, it means that multiple matrix transformations are superimposed. In the matrix operation, there is a law that the matrix does not satisfy the multiplication law, which means the transformation order is different, directly resulting in the final result of a large difference[14], therefore, in the 3d transformation, every a transformations are relative to the last transformation to do, and the reference coordinate system time is changing,

IOP Conf. Series: Materials Science and Engineering **242** (2017) 012113 doi:10.1088/1757-899X/242/1/012113

regardless of 2d, 3d, so ZXY, and YXZ results in the opposite[15].

A description of any azimuth is the azimuth angle that is derived from alpha, beta, and gamma-order rotation. This description can be viewed as three rotating orthogonal matrices, and the transformed coordinates are obtained by multiplying the order.

Rotate around the Z axis to get alpha, then rotate around X to get beta, and finally rotate around the Y axis to get gamma;

There are two characteristics of restrictive Euler angles:

Value range: Alpha:0-360,beta: ± 90 ,gamma: ± 180

When beta= ± 90 , both mobile phone flip, alpha, Gamma will instantly add or subtract 180

The three orientations alpha, beta, and gamma correspond to Z, X, Y, relative to our Z-axis world coordinate system.

So the final matrix transformation formula of Euler angular azimuth is:

$$R = ZXY = \begin{bmatrix} \cos(\alpha)\cos(\gamma)-\sin(\alpha)\sin(\beta)\sin(\gamma) & -\cos(\beta)\sin(\alpha) & \cos(\gamma)\sin(\alpha)\sin(\beta)+\cos(\alpha)\sin(\gamma) \\ \cos(\lambda)\sin(\alpha)+\cos(\alpha)\sin(\beta)\sin(\gamma) & \cos(\alpha)\cos(\beta) & \sin(\alpha)\sin(\gamma)-\cos(\alpha)\cos(\gamma)\sin(\beta) \\ -\cos(\beta)\sin(\gamma) & \sin(\beta) & \cos(\beta)\cos(\gamma) \end{bmatrix} \quad 1)$$

Based on the above, the gyroscope returns the position of the handset relative to the world coordinate system, and the matrix transformation of the gyroscope is finally multiplied by ZXY. Relative to the world coordinates, on the mobile phone screen, according to the above matrix transformation, it converses to the current position, but for the screen of the object, the applied matrix transformation is relative to the screen coordinate system. The matrix transformation required by the final graphics API is the ZXY opposite direction, which is its inverse matrix. That is, ZXY multiplies the sequential upside down, and YXZ can get the inverse matrix. This ensures that the relative to the world coordinate system remains unchanged.

4. Instance

To upload the bridge model in the system as an example, Figure 5 shows the upload Model page. As shown in Figure 5, System supports .3ds, .obj, .fbx, .dwg, .stf, .3dm, .dxf ,. skp, .iges, .step and other 3d model format files upload. Figure 6 for the systems shows the main operations of the 3d models, including translation, rotation, model dismantling and other interactive operations, as well as changing color materials, viewing the model attributes (Long, Wide, High-level) and so on.

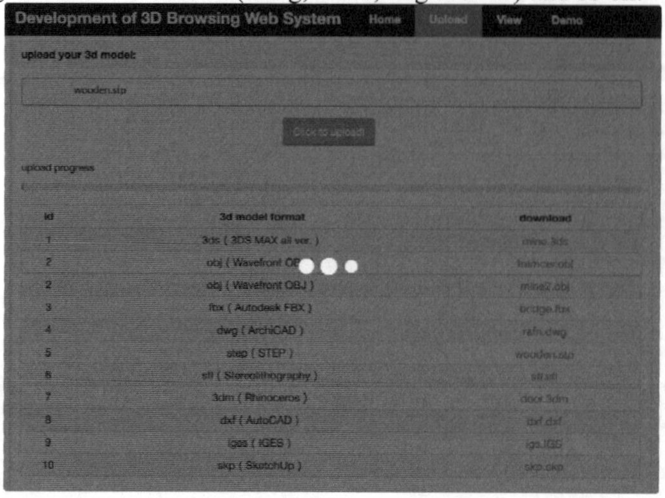

Figure 5.Upload the model

IOP Conf. Series: Materials Science and Engineering **242** (2017) 012113 doi:10.1088/1757-899X/242/1/012113

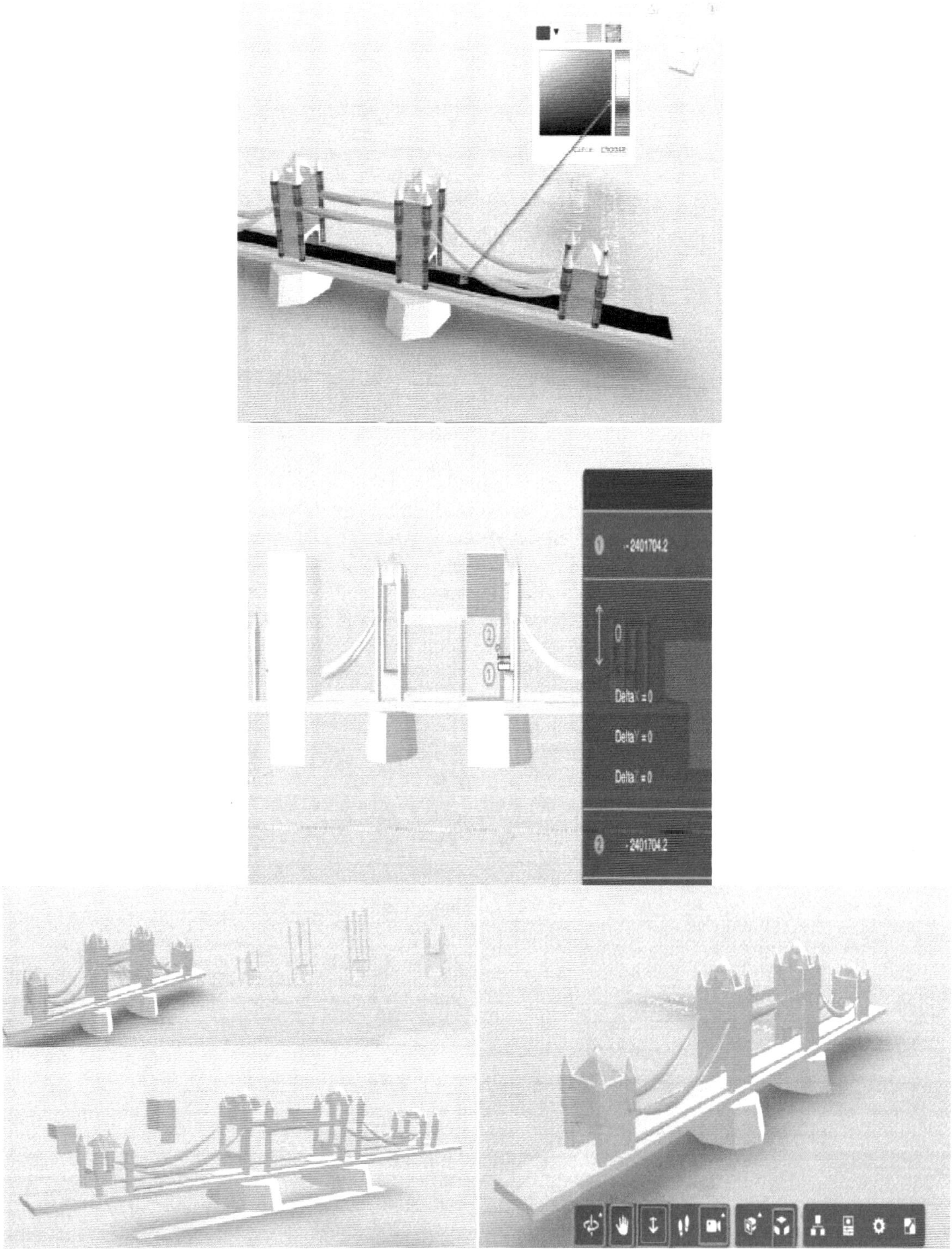

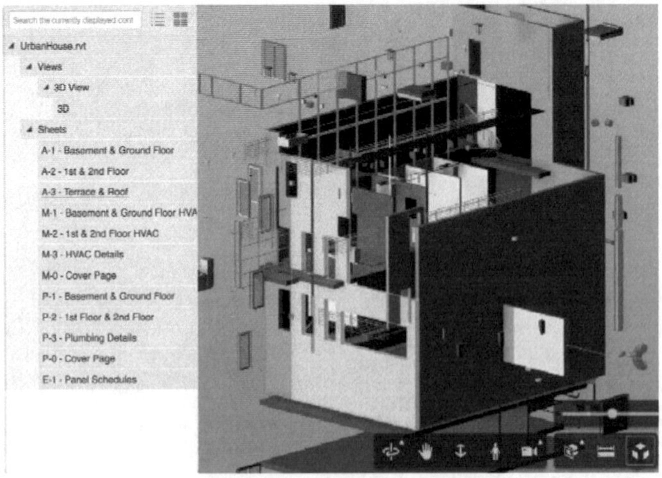

Figure 6.Basic operation display

5. Summary

It is feasible to provide interface rendering and management of 3D data in different formats. By converting the contents of different storage 3d model format files into JSON data, it can be transformed into a unified management 3d model. The basic 3D model interactive operation makes WEB 3D on the Internet have been a certain degree of promotion. And the system can be applied to online games, virtual communities, e-commerce and other fields, so that network viewers, shoppers and other multimedia users such as mobile phones to achieve better viewing entertainment.

References

[1] Han Yi. The New Development Trend In Web3D and Web Visualization—Taking WebGL and O3D as Example[J].Technology Square,2010,(05):81-86.
[2] Xiang Jun.Research Of Parallelization WebGL Graphics Rendering Based WebKit Browser[D].University of Electronic Science and Technology.2015.
[3] Dang Baosheng.Virtual reality and its development trend [J].Chinese Modern Educational Equipment,2007,(04):94-96.
[4] Cui Xingyuan,Qian Hua.Virtual reality and its evolution [J].Mechanical Engineer,2007,(04):94-96.
[5] Cui Xingyuan, Qian Hua.Virtual Reality and Its Evolution [J] .Application of Engineering Geology, 2005,(04): 14-17+ 41.
[6] Xu Wei.Current situation and development of virtual reality technology at home and abroad [j]. Modern business industry,2009,(02):279-280.
[7] Zhang Yu. Analysis on the development of 3d Web page based on Web3D technology [j]. Journal of Fuyang Teachers College (Natural Sciences Edition), 2008, (03): 57-60.
[8] Wang Weimin.Research of Web3D Technique and Compare and Choice Of SeveralVeb3D Techniques[D].Wuhan University,2004.
[9] Gao Fei,He Lili,Gao Jinbiao.Web Service Performance Study Based on Node.JS In-memory Cache[J].Industrial control computer,2015,(11):109-110.
[10] Li hua, Xiaowei.Input and processing of several 3d model files in OpenGL [J].Geophysical prospecting computing Technology. 2007, (01): 83-86.
[11] Zhou Li,Su Honggen.Research on general file formats and algorithms of 3-D models and their implementation with OpenGL[J].Computer Engineering and Design,2009,(02):433-436+439.
[12] Cheng Xiaodong, Tommy. The method of converting the X3D file to JSON [J]. Journal of Chizhou College, 2009, (03): 29-30.
[13] Dr.Javier Morales and Mr.Ashutosh Kumar.Web based 3D analysis and visualization using

HTML5 and WebGL.2014.
[14] Qiu Wei. Development and application of three-dimensional model interaction design system based on Java3D [J]. Microelectronics and Computers, 2008, (11): 195-198.
[15] Zhang Wanfang,Su Honggen.Research on LOD algoirthm of 3D models and its implementation with openGL[J],Computer Engineering and Design,2004,(10):1790-1791+1796.

IOP Conf. Series: Materials Science and Engineering **242** (2017) 012114 doi:10.1088/1757-899X/242/1/012114

An Exploration of Constructive Approach to the Higher Institutions' MOOC with Chinese Local Characteristics

Yong Liu

Personnel Department, China West Normal University, No.1 ShiDa Road, Nanchong, 637009, China

Email: liuyong007688@126.com

Abstract. Although the paradigm change led by MOOC in the field of higher education has many supporters, and MOOC has been applied widely for many years in the world. Now, more and more people begin to pay attention to the disputes about disadvantages and advantages of the MOOC. At first, the article indicates that some innate bad genes of MOOC, such as the cultural erosion tendency, the high drop out rate, the low effectiveness of the teaching evaluation and the low reliability of the course credits. And then the article faces the process of Chinese MOOC construction in different colleges and universities, and focuses on how to improve the innate bad genes of MOOC. Finally, the article gives some suggestions to explore a constructive approach to the higher institutions' MOOC with Chinese local characteristics.

1. Introduction

Although the paradigm change led by MOOC in the field of higher education has many supporters, MOOC has been applied widely for many years. Now, more and more people begin to pay attention to the disputes of the MOOC, especially in many Chinese well-known colleges and universities have built the MOOC. The people must change the bad genes of MOOC in the construction of local MOOC, to explore a constructive approach to the higher institutions' MOOC with Chinese local characteristics.

2. The Innate Bad Genes of MOOC

2.1 The Cultural Erosion Tendency

Education sovereignty is the same as economic sovereignty, political sovereignty, cultural sovereignty and information sovereignty. It is the subordinate concept of state sovereignty. The economic sovereignty has always been of concern, but the importance of education sovereignty is often overlooked. The educational sovereignty is the logical extension of the state sovereignty and cultural sovereignty [1]. Educational institutions are the place of shaping the national personality and the cohesion of the people. In essence, safeguarding educational sovereignty is the maintenance of national education right and the national mainstream ideology and cultural values. Western countries take their own ideological penetration to influence the sovereignty of education in developing countries. At the same time, MOOC is a pushing hand of the internationalization of higher education, and it will become a new recessive infiltration tool in education sovereignty.

2.2 The High Dropout Rate

The cost of drop out of MOOC is almost zero, so it accompanied by a high dropout rate from birth, which is a very normal situation. Meyer (2012) reported that the drop out of MOOC in Stanford, MIT,

Content from this work may be used under the terms of the Creative Commons Attribution 3.0 licence. Any further distribution of this work must maintain attribution to the author(s) and the title of the work, journal citation and DOI.
Published under licence by IOP Publishing Ltd

and University of California at Berkeley are between 80% and 95%. For example, 50 thousand students enrolled in "software engineering" course in University of California at Berkeley, only 7% of those who completed the learning courses of the MOOC. "social network analysis" MOOC, on the platform of Coursera, only 2% learners finish the MOOC and have the basic badges, 0.17% learners have achieved excellent certificates among them [2]。

2.3 The Low Effectiveness of Teaching Evaluation
At present, the MOOC mainly adopts online short multiple-choice with automatic scoring evaluation as a learning tool, only a handful of MOOC adopt open short answer as test method. But considering the number of learners, it is hard to imagine the MOOC teachers have the time and energy to review so many operations [3]. Therefore, MOOC often relies on the peer assessment way of learning measurements, such as s Coursera and Udacity.

The validity of this evaluation method has been questioned by many researchers. First of all, there are a large number of learners, dozens or hundreds of learners who will evaluate more than 10000 assignments. The assignments involved are very complex. Secondly, MOOC peers without teachers' participation, guidance or supervision, the quality of peer assessment is not guaranteed. Thirdly, MOOC learners are often distributed all over the world. They are different from language, culture, values and world outlook each other. Without the guidance and supervision of teachers, learners lack the sense of responsibility or effective motivation. As a result, it is difficult for them to deal with other students' assignments seriously. Therefore, the most controversial issue about peer assessment is the reliability of the results.

2.4 The Low Reliability of the Course Credits
MOOC is more controversial in the credit and certification issues. So far, most of the MOOC platforms only provide students with the certificate of the success in the course of examination, or the completion of the entire learning process, rather than the credit. The latest trend is that more and more universities began to recognize and accept the MOOC credits, because the activity can help schools increase visibility in the marketing, and thus bring more students registered in [4]. However, according to the survey of higher education in the United States, 72% of MOOC teachers think they should not be given class credits [5].

3. The Constructive approach to the higher institutions' MOOC with chinese Local Characteristics

MOOC era is the inevitable result of the development of information technology. China is completely different from the developed countries in higher education system, education idea, and mode of thinking, the level of information technology and other aspects. In view of the above MOOC controversy, we should change the innate bad genes of MOOC as much as possible, and explore a constructive approach with Chinese local characteristics in the MOOC construction course.

3.1 Highlighting the Guiding Role of Local Culture
The globalization of MOOC will inevitably bring about conflicts and integration between different cultures and ideologies. MOOC is not only a theoretical problem, but also a practical one. The rise of MOOC in the United States is, on the face of it, the advocacy and practice of individual Elite Teachers. In essence, it is jointly promoted by University, government and enterprises. It can be said that the key role is precisely the government, which has a deep purpose behind the spread of American culture.

This requires the construction of local MOOC in China and should put the national cultural safety, education safety, information security, ideology and even political stability safety into the research vision. We can carry forward the unique charm of Chinese culture through MOOC, let more people all over the world understand China, and improve China higher education in Higher Education of the world's right to speak. Therefore, Chinese MOOC should adhere to education, technology, commercial, and security in the localization process. From the perspective of education, we mainly study the problem of MOOC and educational reform, which involves the global competition of higher education and even basic education. From the perspective of technology, we mainly study the research of MOOC platform technology, which need the experts and scholars in technical field (not only educational technology) to participate the construction widely. From the perspective of business, we mainly research the operation mode of Chinese MOOC platform, to provide a solid economic foundation for Chinese MOOC, which requires the addition of economic theorists and practitioners. From the perspective of security, we should mainly study the penetration of the MOOC with western national ideology.

3.2 Enhancing the Attraction of MOOC

Although the overall dropout rate of MOOC is high, there are statistics which show that some of the MOOC in terms of teaching design, course display, or learning feedback, teaching evaluation, do some more attractive than the other MOOC, the dropout rate is much lower. Therefore, it is necessary to conduct the analysis of learners in the construction of MOOC. For example, if the learner are Chinese learners or foreigners, we should analyze their learning and cognitive style respectively. How to design an attractive MOOC is worthy of attention in the course of Chinese MOOC construction.

3.3 Improving the Effectiveness of Teaching Evaluation and Promoting the Learning Achievement Certification

The credit degree is a relatively fair evaluation standard of the learning experience and learning ability. In order to make the MOOC learning into effective learning and social recognition, the people must take practical measures to promote the effective development of teaching evaluation, and ultimately promote the MOOC learning achievement; it has been widely recognized by society.

First, we should strengthen the face-to-face interactive teaching. Through it we can eliminate the lonely feeling of MOOC learners. Face to face teaching and teaching evaluation can greatly make up for the defects of online teaching evaluation with low confidence.

Secondly, the establishment of standardized professional examination room and the improvement of the online education evaluation system can improve the effectiveness of MOOC teaching evaluation. For example, we can establish enough standardized professional examination rooms in all colleges and universities which have joined the MOOC construction union, similar to the Chinese Foreign Language Examination or Computer Professional Examination. One can participate in the teaching evaluation of the candidates through the network registration, and choose the test site and test matches freely. The evaluation can base on the results obtained from the site invigilation mode has its own higher reliability. At the same time, we can rely on the remote monitor software which can provide full supervision of the whole process of assessment and the video data. These auxiliary file materials are helpful and creditable for future queries.

Finally, we should construct an influential Chinese MOOC platform through the strong combination among colleges and universities. In July 2013, Shanghai Jiao Tong University joint with Peking University, Tsinghua University, Fudan University, Zhejiang University, Nanjing University, University of Science and Technology of China, Harbin Institute of Technology, Xi'an Jiao Tong University have completed the construction of "China C9 University alliance". Tongji University, Dalian University of Technology, and Chongqing University and other colleges and universities construct online open curriculum standard and sharing mechanism synergistically, sharing platform, sharing course, have developed a high level inter-university program for students to choose courses, join training of minor professional, and jointly implement the open online course resources to the public, expand and enjoy high-quality teaching resources in the domestic users. In the same year, both sides of the five Jiaotong University (Shanghai Jiaotong University, Xi'an Jiaotong University, Southwest Jiaotong University, Beijing Jiaotong University, Taiwan Jiaotong University) jointly established the MOOC platform, launched the online learning consortium. The joining of the famous universities in China will help to build up MOOC platform with high social influence, so that the results of the study will be of high credibility.

4. Conclusion

In general, in order to explore a constructive approach to higher institutions' MOOC with Chinese local characteristics, highlighting the guiding role of local culture, enhancing the attraction of MOOC, and improving the effectiveness of teaching evaluation and promoting the learning achievement certification are effective way to go much further.

Acknowledgments

This paper is one of the achievements of the project funded by Sichuan Province Education Development Research

Center and the project number is CJF15053.

References

[1] Liang Jia-shun. National sovereignty and educational sovereignty [J]. Academic research, 2008, (1):92-93.

[2] Weller, M.. Amnesi MOOC [EB/OL]. Blog the Ed Techie, [2015-06-13]. http://nogoodreason.typepad.co.uk/no_good_ reason/MOOC/.

[3] Yuan, L., Powell, S., Cetis, J., & Yuan, B. L.. MOOCs and Open Education: Implications for Higher Education [M]. cetis publications. http://publications.cetis.ac.uk/2013/667¥ninternal-pdf://1837/Yuan et al. - MOOCs and Open Education Implications for Higher.html, 2013.

[4] Carey, K.. Into the future with MOOC's [J]. The Chronicle of Higher Education.[2014-10-21]. http://chronicle.com/article/Into-the-Future-With-MOOCs/134080.

[5] Gregory Ferenstein. Study: Massive Online Courses Enroll an Average of 43,000 Students, 10% Completion[EB/OL].[2014-10-28]http://techcrunch.com/2014/03/03/study-massive-online-courses -enroll-an-average-of-43000-students-10-completion/.

IOP Conf. Series: Materials Science and Engineering **242** (2017) 012115 doi:10.1088/1757-899X/242/1/012115

A New Moving Object Detection Method Based on Frame-difference and Background Subtraction

Jiajia Guo [1, a)], Junping Wang[1, b)] , Ruixue Bai [1, c)], Yao Zhang[1, d)] and Yong Li [1,e)]

[1]School of Telecommunication Engineering, XIDIAN University, Xian 710071, China.
E-mail: [a)] JJG_xdu@163.com, [b)]jpwang@mail.xidian.edu.cn, [c)]674856388@qq.com, [d)]864190188@qq.com, [e)]lyiong@126.com

Abstract. Although many methods of moving object detection have been proposed, moving object extraction is still the core in video surveillance. However, with the complex scene in real world, false detection, missed detection and deficiencies resulting from cavities inside the body still exist. In order to solve the problem of incomplete detection for moving objects, a new moving object detection method combined an improved frame-difference and Gaussian mixture background subtraction is proposed in this paper. To make the moving object detection more complete and accurate, the image repair and morphological processing techniques which are spatial compensations are applied in the proposed method. Experimental results show that our method can effectively eliminate ghosts and noise and fill the cavities of the moving object. Compared to other four moving object detection methods which are GMM, VIBE, frame-difference and a literature's method, the proposed method improve the efficiency and accuracy of the detection.

1. Introduction

Moving object detection is always focused and difficult in the video sequence analysis. Many classic methods have been proposed in the long-term study of moving object detection, including inter-frame difference [1], optical flow [2] and background subtraction [3].

Recently, the combination of frame-difference and background subtraction for moving object detection has caused widespread attention[4]. An improved object detection method based on background model and inter-frame difference is proposed in [5], which reduced the amount of ghosts in the mixed Gaussian background subtraction and the serious cavities in the inter-frame difference and detected moving objects effectively under the condition of the background mutations. However, the detection is not accurate for complex environment. Literature [6] put forward a new moving object detection method, which calculated optical flow information with selecting the image representative Harris corner pixels, simplified the calculation of optical flow and three-frame difference method was introduced as a supplement, while the optical flow was more complex. Literature [7] presented a technique which improved the frame difference method by first classifying the blocks in the frame as background and others using correlation coefficient.

In this paper, a new moving object detection method is proposed. This method which combines the idea of background subtraction and frame-difference has the following new features: 1) the method proposed can greatly fill the cavities of the foreground object and eliminate the noise because it is a fusion method; 2) the algorithm of image combination is introduced; 3) image repair and morphological processing techniques were took into the method to repair the moving objects.

Content from this work may be used under the terms of the Creative Commons Attribution 3.0 licence. Any further distribution of this work must maintain attribution to the author(s) and the title of the work, journal citation and DOI.
Published under licence by IOP Publishing Ltd

2. The new moving object detection method

In this paper, we propose an improved frame-difference.The improved frame-difference as the formula 1:

$$D(x,y,\Delta t)=|f(x,y,t)-f(x,y,t-1)|\oplus|f(x,y,t+1)-f(x,y,t-1) \quad 1)$$

Where $f(x,y,t-1)$, $f(x,y,t)$ and $f(x,y,t+1)$ are three adjacent images in the image sequence, the difference image is $D(x,y,\Delta t)$. The improved frame-difference which takes full advantages of the accumulation frame-difference[8] and the discontinuous frame-difference[9] can enhance the movement target and weak the impact of environmental factors and noise.

In the GMM background model, the quality of the foreground object is highly dependent on a fixed threshold. A high threshold may cause fragmented foreground objects, while a low threshold can result in noisy pseudo-foreground objects. While selecting an appropriate threshold for different frames is very difficult and also is not impractical. In this paper, we set different thresholds to produce three images and the result is shown in Fig.1. (a) is the original video, (b) is the foreground image which has a high threshold called SP, (c) is SS1 which has a low threshold and (d) is SS2 which has a lower threshold than SS1.

(a) (b) (c) (d)

FIGURE 1. Example of difference thresholds. (a) Original video; (b) SP; (c) SS1; (d) SS2

Fig.1 shows that SP is mainly composed of important parts of the foreground and some fragment, SS1 and SS2 contain more detailed image information. In this paper, the foreground image produced by improved frame-difference is called F.

In most previous work, the combination of frame difference and background subtraction is based on the simple AND or OR operation to get the foreground objects. However, it does not suitable for some scenes because of much more noise. The follow image combination algorithm is used to solve this problem. This image combination algorithm is well summarized in Algorithm1. On account of F and SP contain important foreground information, first use the Algorithm1 to combine F and SP , as shown in the Fig.2, where FSP is the result of the combined of F and SP.

(a) (b) (c)

FIGURE 2. Combined image. (a) F; (b) SP; (c) FSP.

Algorithm1 image combination algorithm

Input: two RGB images, one used as main picture, the other used as background.
Output: combined picture
Step 1: Convert the color image to array.
(i) Convert the color image to L*a*b color.
(ii) Convert main picture to Data $X_0=\{\{x_1, x_2 ...\},\{x_k, x_{k+1} ...\}, ...\}$;the background Data $X_1=\{\{y_1, y_2, ...\},\{y_k, y_{k+1} ...\}, ...\}$;
Step 2: Initial output Data $X_2=\{\{0, 0...\}, \{0, 0...\} ...\}$.
Step3: Literal all pixels in X_0, for a white pixel, check 3*3 in X_1, if found any one not zero, set

corresponding pixel in X_2 with 255.
Step4: Literal all pixels in X_1, for a white pixel, check 3*3 in X_0, if found any one not zero, set corresponding pixel in X_2 with 255.
Step 5: Convert X_2 to image as result.

Although FSP has clearly eliminated noise and ghosts, there are still obvious cavities. Therefore, it is necessary to use the image repair to further process the foreground objects. FPS is repaired with the images which contains richer details. FPS is first improved by SS1, And then improved by SS2,then can get the repaired image. Repair processing is based on algorithm1.

There are still some cavities after the image repair, we use expansion processing of morphological technique to eliminate cavities based on the basis of no increase noise. The result is shown in Fig.3, where (a) is the repaired image and (b) is the result of expansion.

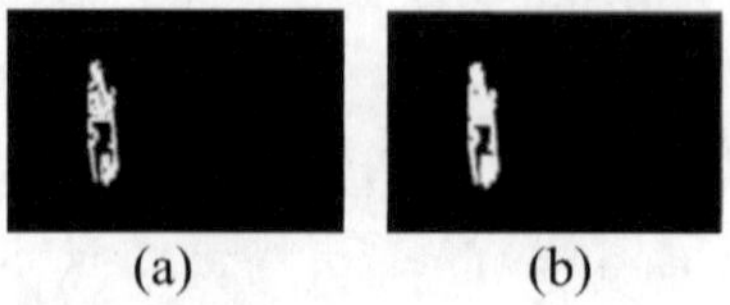

FIGURE 3 Expansion processing. (a)Repaired image; (b)The result of expansion processing.

3. Experimental results

In order to verify the effectiveness of the method, the experiment deals with four videos in the paper. The videos are taken from the Context Aware Vision using Image-based Active Recognition (CAVIAR) database [10]. We compare the proposed method against GMM, VIBE, frame-difference and literature [4]. The comparison result is shown in Fig.4, where (a) is Highway, (b) is Offers, (c) is PETS2006 and (d) is Pedestrians from CAVIAR. We chose five frames from every video to show the result. In (a), the second row is the results of frame-difference, the third row is the results of GMM, the fourth row is the results of VIBE, the fifth row is the results of [4] , the last row is the results of the proposed method, and the first row of the picture is video frame, and it also applies to (b), (c) and (d).

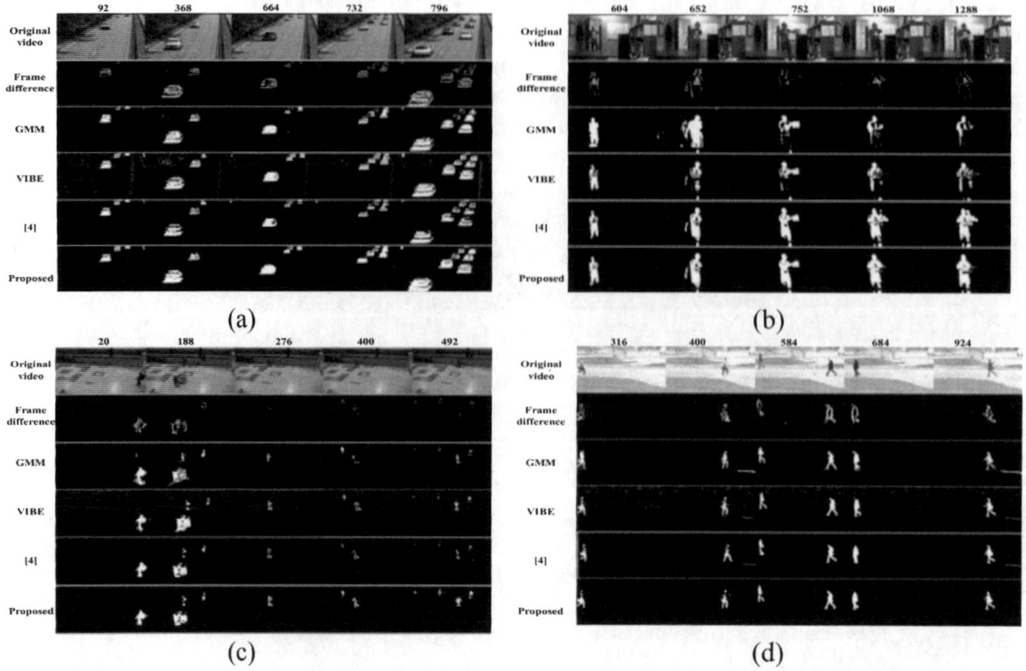

FIGURE 4. Detection result. (a) Highway; (b) Office; (c)PETS2006; (d) Pedestrians.

The Fig.4 shows that the method proposed performs better than other methods. Although GMM and VIBE can accurately detect the moving target in most cases, it is difficult to detect the object which moves slowly because some of the pixels of the object become part of the background, and the quality of the foreground object is highly dependent on a fixed threshold. The frame difference can detect the moving target better, but the cavities are more serious. [4] is only do a simple AND operation between the frame-difference and the background subtraction, this will case small noise and ghosts because of the noise corresponding to the superposition. The method proposed in this paper can not only detect the slowly moving object, but also eliminate noise and ghosts and fill the cavities effectively.

4. Conclusion

Background subtraction, optical flow, frame difference and some current improvement methods in recent years are introduced in this paper, and a new moving object detection method based on frame-difference and background subtraction is proposed, which introduces the image repair and morphological processing to the method. Finally, the results of multiple videos test show that the proposed method can eliminate noise and fill cavities quickly, and detect moving object more completely.

Acknowledgments

This work was supported in part by the National Natural Science Foundation of China under Grant 61173088, in part by the Science and Technology Program, Xi'an, China, under Grant CX12485, and in part by the Programmer of Introducing Talents of Discipline to Universities through the 111 Project under Grant B08038.

Reference

[1] M.E. Rashid, Vinu Thomas. "A Background Foreground Competitive Model for Background Subtraction in Dynamic Background. Procedia Technology, 2016:25:536-543
[2] Xin Y, Hou J, Dong L, et al. A self-adaptive optical flow method for the moving object detection in the video sequences [J]. Optik - International Journal for Light and Electron Optics, 2014, 125(19):5690-5694.
[3] Wang X, Guo L. Highway traffic video detection method [J]. Journal of Computer Applications, 2012, 32(6):1585-1588.
[4] Liu, Hongkun, et al. "Combining background subtraction and three-frame difference to detect moving object from underwater video." Oceans IEEE, 2016:1-5.
[5] Lan, Yongtian, et al. "Robot fish detection based on a combination method of three-frame-difference and background subtraction." China Control and Decision Conference, 2014:3905-3909.
[6] Chen J C, Zhang J H, Liu S J, et al. Improved Target Detection Algorithm Based on Background Modeling and Frame Difference[J]. Computer Engineering, 2011.
[7] Yuan, Guo Wu, et al. "A Moving Object Detection Algorithm Based on a Combination of Optical Flow and Three-Frame Difference." Journal of Chinese Computer Systems 34.3(2013):668-671.
[8] Ramya, P., and R. Rajeswari. "A Modified Frame Difference Method Using Correlation Coefficient for Background Subtraction." Procedia Computer Science 93(2016):478-485.
[9] Mo Lin, Liao Peng, Liu Xun. "A Motion Detection Algorithm Based on Backgroud Subtraction and Three Frame Differencing." Microcomputer Information 25.12(2009):274-276.
[10] Cavir: Context aware Vision Using image-based active recognition, Nov.2011. [Online]. Available: http://groups.inf.ed.ac.uk/Vision/CAVIAR/CAVIARDATA.

IOP Conf. Series: Materials Science and Engineering **242** (2017) 012116 doi:10.1088/1757-899X/242/1/012116

The Risk Assessment Study for Electric Power Marketing Competitiveness Based on Cloud Model and TOPSIS

Cunbin Li[1, a)], Yi Wang[1, b)] and Shuaishuai Lin[1, c)]

[1]Institute School of Economics and Management, North China Electric Power University, Beijing 102206, China

E-mail:[a)] lcb999@263.com, [b)] wyi@ncepu.edu.cn

Abstract. With the rapid development of the energy internet and the deepening of the electric power reform, the traditional marketing mode of electric power does not apply to most of electric power enterprises, so must seek a breakthrough, however, in the face of increasingly complex marketing information, how to make a quick, reasonable transformation, makes the electric power marketing competitiveness assessment more accurate and objective becomes a big problem. In this paper, cloud model and TOPSIS method is proposed. Firstly, build the electric power marketing competitiveness evaluation index system. Then utilize the cloud model to transform the qualitative evaluation of the marketing data into quantitative values and use the entropy weight method to weaken the subjective factors of evaluation index weight. Finally, by TOPSIS method the closeness degrees of alternatives are obtained. This method provides a novel solution for the electric power marketing competitiveness evaluation. Through the case analysis, the effectiveness and feasibility of this model are verified.

1. Introduction

In recent years, with the deepening of the electric power reform, the electric power marketing [1] becomes more and more important for the basic work of electric power enterprises. The effective risk assessment of electric power marketing will not only affect the stability of enterprise and customer satisfaction, but also has an important significance for long-term development of enterprise. So, how to set up the electric power marketing risk assessment index and make an effective risk assessment is the first problem of electric power enterprise, and has an important theoretical and practical significance.

At present, more and more researches have been done on power marketing. In the literature [1], the interval of the assessment level is studied by the concept of cloud measure, and the early warning model of power marketing risk assessment based on cloud measure is constructed. The fuzzy AHP and fuzzy synthesis are used in literature [2] and [3]. In order to evaluate the effectiveness of the service strategy of the power supply enterprise, the literature [4] adopts the method of analytic hierarchy process and linear weighting method to study the quality of the power supply service and the power supply enterprise's marketing work. Combining the order relation method and AHP method, Literature [5] evaluates the marketing execution ability of the power supply enterprise and selects the Yantai power supply company as an example. The literature [6] analyzes the power marketing safety through expert investigation and brainstorming risk identification, taking into account the BP neural network, factor analysis and other methods of risk analysis and evaluation.

Based on the above literatures, we can see that: 1) At present, most of the current risk research for power marketing has taken into account the combination of qualitative and quantitative, but how to

Content from this work may be used under the terms of the Creative Commons Attribution 3.0 licence. Any further distribution of this work must maintain attribution to the author(s) and the title of the work, journal citation and DOI.
Published under licence by IOP Publishing Ltd

convert the two is still a problem to be solved urgently. 2) There are many literatures on the evaluation of electric power marketing, but the research subjects rarely choose the risk assessment of power marketing competitiveness. Aiming at the above two points, this paper will establish a risk assessment model of power marketing competitiveness based on cloud model [7] and TOPSIS.

How to describe information is a major problem in the process of data mining and knowledge acquisition. The best way is to be expressed in natural language, because natural language can be better understood [8-10], while natural language is also the basis of thinking and intelligence, but many of the concepts in natural language are unknown factors, such as ambiguity, randomness, imprecision and incompleteness. In these unknown factors, randomness and fuzziness have aroused people's attention [12]. Although these uncertainties are difficult to define precisely, but its existence does not affect the exchange between people [11]. However, in order to make the computer have similar to human understanding and judgment, for the study of ambiguity and randomness is necessary [13].

Rough set and probability theory are the two most effective methods to deal with the uncertainty. However, considering the randomness of membership degree and probability, Professor Deyi Li put forward the cloud model theory on the basis of rough set and probability theory in 1995. This model is an indeterminate conversion model that deals with qualitative and quantitative descriptions. Because experts can't express their subjective professional judgments in numerical form, they are often replaced by fuzzy numbers or interval numbers, but both of the two ways have some limitations, and can't be like the natural language expression that fit the experts. On this basis, put forward the cloud model to implement the natural language evaluation transformation into quantitative values, provides a good foundation for quantitative decision.

2. Competitiveness Risk ASSESMENT Index System

Electric power marketing competitiveness assessment refers to the electric power enterprise marketing aspects of the existence of the unique, advantages, in the face of adversity can survive and develop the ability to assess. Evaluation of the competitiveness of an enterprise should pay attention to the following four points:1) The operation of the current electric power marketing strategy; 2) What are the opportunities and threats, strengths and weaknesses that companies face in marketing;3) Whether the price and cost of the enterprise are competitive; 4) Relative to other enterprises, marketing competitive position.

Considering the above four points, and in order to make the evaluation valuable, we must use the following principles in the establishment of the index system: systematic, dynamic, typical, comparable and quantifiable. This article established an evaluation index system, including three first-level indicators and 20 secondary indicators, as shown in Figure 1.

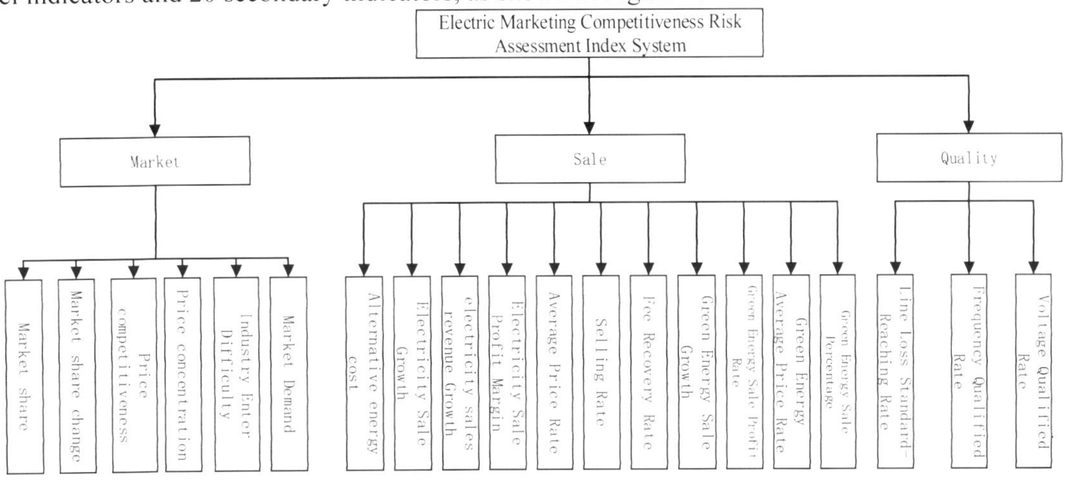

FIGURE 1. Electric marketing competitiveness risk assessment index system.

3. Cloud model introduction

3. 1 Definition of cloud model

The cloud model is a method to convert the qualitative concept and the quantitative numerical phase. The model combines the knowledge of fuzzy mathematics and statistics to reflect the uncertainty of things and the ambiguity and randomness of human subjective thinking.

Definition1 U is a field with exact values， C is a qualitative concept in U, if there is a quantitative variable x∈C, and x is a random implementation in the qualitative concept C, the certainty u(x) ∈ [0,1] for C is a random number with a stable tendency， such as equation 0， then the distribution in U of x is called cloud, each x is a cloud droplet.

$$u: U \to [0,1] \qquad \forall x \in U x \to u(x) \tag{0}$$

The cloud model has three digital features, namely, expectation value Ex, entropy En and super entropy He. In the process of the assessment, these three numerical features can be used to realize the transformation between qualitative concept and electric quantity data. Among them, the expectation value Ex is the most representative of the concept of qualitative point, and is the central information value of fuzzy concept. It is not only the center gravity of the cloud position, but also the most typical sample points of quantification for the qualitative concept. This paper selects the expected value Ex as the quantitative value for the conversion of qualitative concept. Entropy En reflects the uncertainty of qualitative concepts, including the randomness and ambiguity of qualitative concepts. In terms of randomness, entropy En mainly reflects the degree of discretization of cloud droplets, it also represents the randomness of cloud droplets. In terms of fuzziness, entropy En reflects the possibility of the qualitative concept, and it represents the value range of the droplet. In general, the greater En represents the greater uncertainty. The super entropy He is the entropy of the En, which is the uncertainty of the entropy En. It is determined by the randomness and fuzziness of the En. The super entropy He can indirectly reflect the cloud thickness of the cloud droplets in certain domain. The thickness of the cloud is not the same, close to the concept center and away from the concept of the center of the cloud thickness is small, between the cloud thickness is large, with the largest fuzzy value of the cloud thickness of the largest, as shown in Figure 2.

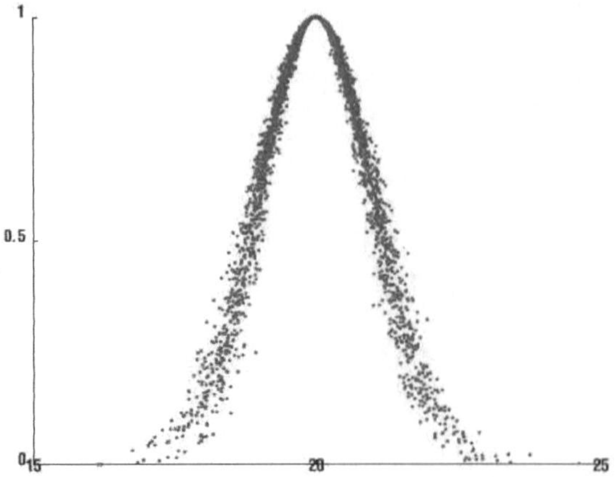

FIGURE 2. 20km membership cloud chart

Figure 2 shows the membership cloud of the concept of 20 km or so, in which $Ex = 20$, $En=1$, $He=0.1$, N (cloud droplet number) =3000. In the distance of 20 km and 20 km away, because the concept is uncertain small, cloud thickness is small; not far from 20 km near the cloud thickness increases, because the concept of uncertainty, and the subjective feelings of people is consistent.

3.2 The language evaluation value into the cloud method

Set the language of the experts to assess the scale as n (according to the actual situation arbitrarily selected), experts specify the effective domain as $[X_{min}, X_{max}]$, you can generate n clouds to express the language value. In this paper, the language used in the evaluation scale is 5, using golden segmentation to generate five clouds. Set the middle cloud as $C_0(Ex_0, En_0, He_0)$, the left and right as $C_1(Ex_1, En_1, He_1)$, $C_{-1}(Ex_{-1}, En_{-1}, He_{-1})$, $C_2(Ex_2, En_2, He_2)$, $C_{-2}(Ex_{-2}, En_{-2}, He_{-2})$ .The characteristics of these five clouds are shown in Table 1.

Table 1 Golden segmentation method to generate five cloud features digital calculation method

Cloud	*Ex*	*En*	*He*
$C_{+2}(Ex_{+2}, En_{+2}, He_{+2})$	X_{max}	$En_{+1}/0.618$	$He_{+1}/0.618$
$C_{+1}(Ex_{+1}, En_{+1}, He_{+1})$	$Ex_0 + 0.382(X_{max} + X_{min})/2$	$0.382(X_{max} - X_{min})/6$	$He_0/0.618$
$C_0(Ex_0, En_0, He_0)$	$(X_{min} + X_{max})/2$	$0.618En_{+1}$	
$C_{-1}(Ex_{-1}, En_{-1}, He_{-1})$	$Ex_0 - 0.382(X_{max} + X_{min})/2$	$0.382(X_{max} - X_{min})/6$	$He_0/0.618$
$C_{-2}(Ex_{-2}, En_{-2}, He_{-2})$	X_{min}	$En_{+1}/0.618$	$He_{-1}/0.618$

4. Research on risk assessment model of power marketing competitiveness based on cloud model and grey TOPSIS

Given m schemes and n evaluation indexes of electric power marketing competitiveness, for schema i, the evaluation value of risk factor C_j is $x_{ij}(0 < i \le m, 0 < j \le n)$, and the language evaluation matrix is $(x_{ij})_{m \times n}$. Based on the cloud model and gray correlation TOPSIS method, evaluation calculation steps are as follows:

4.1 Convert language evaluation value

According to the expert's language evaluation value, the method shown in Table 1 is transformed into the normal cloud model, and the language evaluation matrix given by the expert is transformed into the corresponding cloud model matrix.

Given the language evaluation level V ={(high/hard/strong), (higher/harder/stronger), (general), (lower/easier/weaker), (low/easy/weak)}. The valid domain of the cloud model is $[0,1]$. Let $H_{e0} = 0.005$, according to the method of Table 1, five clouds $\{C_{+2}, C_{+1}, C_0, C_{-1}, C_{-2}\}$ corresponding to the language evaluation level are generated, and the numerical characteristics of the five clouds are shown in Table 2.

Table2 the characteristic number of 5 cloud

Evaluation Level	Cloud Model	*Ex*	*En*	*He*
high/hard/strong	$C_{+2}(Ex_2, En_2, He_2)$	1	0.104	0.013
higher/harder/stronger	$C_{+1}(Ex_1, En_1, He_1)$	0.691	0.064	0.008
general	$C_0(Ex_0, En_0, He_0)$	0.5	0.040	0.005
lower/easier/weaker	$C_{-1}(Ex_{-1}, En_{-1}, He_{-1})$	0.309	0.064	0.008
low/easier/weak	$C_{-2}(Ex_{-2}, En_{-2}, He_{-2})$	0	0.104	0.013

4.2Normalize the decision matrix

Because of the existence of cost-type indicators and efficiency indicators in the listed index system, the initial decision matrix $(x_{ij})_{m \times n}$ is transformed into a standardized decision matrix $D = (d_{ij})_{m \times n}$ by normalizing the indexes using vector normalization.

$$d_{ij}=\frac{x_{ij}}{\sqrt{\sum_{i=1}^{m}d_{ij}^{2}}},i=1,2,\cdots,m;j=1,2,\cdots,n \tag{1}$$

4.3 Obtain the index weight.

As the subjective weighting method has some limitations, here we take the entropy method to find the weight of the index, $W=(w_1,w_2,\cdots,w_n)$.

1) Calculate the proportion p_{ij} of the i schema plan under the j index.

$$p_{ij}=d_{ij}\Big/\sum_{i=1}^{m}d_{ij} \tag{2}$$

2) Calculate the entropy e_j of the j index.

$$e_j=-k\sum_{i=1}^{m}p_{ij}\cdot\ln p_{ij},\quad k=1/\ln m \tag{3}$$

3) Calculate the entropy weight w_j of the j index.

$$w_j=(1-e_j)\Big/\sum_{j=1}^{n}(1-e_j) \tag{4}$$

4. 4Find the weighted normalization matrix

the formula is as follows.

$$Y=WD=\begin{bmatrix} y_1(1) & y_1(2) & \cdots & y_1(n)\\ y_2(1) & y_2(2) & \cdots & y_2(n)\\ \vdots & \vdots & \vdots & \vdots\\ y_m(1) & y_m(2) & \cdots & y_m(n)\end{bmatrix} \tag{5}$$

4. 5Determine the positive and negative ideal scheme

The positive and negative ideal solution of TOPSIS method is the optimal solution and the worst solution of the index under all schemes.

$$\begin{aligned} U_0^{+}&=\{(\max_{1\le i\le m}y_i(j),j\in J^{+}),(\min_{1\le i\le m}y_i(j),j\in J^{-})\}=\left(u_0^{+}(j)\right)\\ U_0^{-}&=\{(\min_{1\le i\le m}y_i(j),j\in J^{+}),(\max_{1\le i\le m}y_i(j),j\in J^{-})\}=\left(u_0^{-}(j)\right)\end{aligned} \tag{6}$$

J^{+} for efficiency indicator, J^{-} for cost-type indicator.

4.6Determine the solution to the positive and negative ideal solution of the distance.

By calculating the distance of each scheme to the positive and negative ideal solution, we can see the distance between each scheme and the positive and negative ideal solution.

$$\begin{aligned} D_i^{+}&=\sqrt{\sum_{j=1}^{n}\left[g_i(j)-u_0^{+}(j)\right]^2},i=1,2,\cdots,m\\ D_i^{-}&=\sqrt{\sum_{j=1}^{n}\left[g_i(j)-u_0^{-}(j)\right]^2},i=1,2,\cdots,m\end{aligned} \tag{7}$$

4.7 Determine the relative degree of the program.

According to the size of the close degree of the optimal sort, the greater the degree of close and negative ideal solution that the greater the distance, that is, the better the program.

$$C_i^* = \frac{D_i^-}{D_i^+ + D_i^-} \quad (8)$$

5. Case study

Table 3 shows the marketing competitiveness data of an electric power enterprise from July to December, which quantitative data are the actual data of the enterprise, qualitative evaluations are based on the expert evaluation. The problem is a multi-attribute decision-making problem with six decision-making schemes and 20 evaluation indexes, and to determine the order of competitiveness in the final 6 months.

Table 3 Power marketing competitiveness original data

	7	8	9	10	11	12
C_1	82.38	78.46	66.15	68.62	76.91	80.53
C_2	5.16	4.96	3.87	4.02	4.01	5.22
C_3	0.78	0.86	1.14	1.15	0.99	1.02
C_4	general	general	higher	higher	High	high
C_5	harder	hard	hard	general	hard	hard
C_6	strong	strong	stronger	general	strong	strong
C_7	higher	higher	general	lower	general	general
C_8	9.79	12.89	2.28	5.54	8.17	6.98
C_9	2.27	5.65	1.05	2.02	7.64	9.98
C_{10}	high	higher	general	lower	lower	general
C_{11}	110.56	102.43	84.62	99.04	104.56	108.11
C_{12}	4.39	6.78	5.66	8.01	5.16	6.99
C_{13}	71.22	89.45	87.66	90.21	94.35	95.12
C_{14}	6.01	6.78	3.23	2.12	2.25	3.01
C_{15}	2.57	2.43	0.69	0.22	2.79	2.98
C_{16}	796.98	799.16	800.17	801.16	815.21	811.17
C_{17}	4.95	4.92	3.28	2.98	1.31	1.09
C_{18}	109.8	112.5	100.3	99.2	96.5	95.4
C_{19}	93.21	92.11	97.06	99.67	99.82	98.16
C_{20}	96.12	96.71	97.43	98.98	99.17	99.16

1) According to Table 1 and Table 2, the qualitative language in Table 3 is transformed into a cloud model. Since the expected value is the most representative point of qualitative concept, the expected value Ex is used as the index value after conversion.

Table 4 Qualitative language into quantitative values

	7	8	9	10	11	12
C_4	0.5	0.5	0.691	0.691	1	1
C_5	0.691	1	1	0.5	1	1
C_6	1	1	0.691	0.5	1	1
C_7	0.691	0.691	0.5	0.309	0.5	0.5
C_{10}	1	0.691	0.5	0.309	0.309	0.5

2) Normalize the quantitative data in Table 3 and Table 4 according to the formula (1) to obtain the normalized matrix $D=(d_{ij})_{m\times n}$ and use the entropy method of the formula (2-4) for the matrix to obtain 20 evaluation indexes weighing, based on the known weight, we can use the formula (5) to calculate the weighted evaluation matrix. Here, due to the limited space, only the final weight value and weighted evaluation matrix are listed.

W={0. 026,0. 045, 0. 023, 0. 009, 0. 005, 0. 013, 0. 011
0. 092, 0. 085, 0. 113, 0. 068, 0. 033, 0. 053,
0. 031, 0. 024, 0. 022, 0. 026, 0. 046, 0. 137, 0. 147}

Weighted evaluation matrix:

$$Y = WD = \begin{bmatrix} 0.012 & 0.011 & 0.009 & 0.010 & 0.011 & 0.011 \\ 0.021 & 0.020 & 0.016 & 0.016 & 0.016 & 0.021 \\ 0.007 & 0.008 & 0.011 & 0.011 & 0.009 & 0.010 \\ 0.002 & 0.002 & 0.003 & 0.003 & 0.005 & 0.005 \\ 0.002 & 0.002 & 0.002 & 0.001 & 0.002 & 0.002 \\ 0.006 & 0.006 & 0.004 & 0.003 & 0.006 & 0.006 \\ 0.006 & 0.006 & 0.004 & 0.003 & 0.004 & 0.004 \\ 0.044 & 0.058 & 0.010 & 0.025 & 0.037 & 0.031 \\ 0.014 & 0.034 & 0.006 & 0.012 & 0.046 & 0.060 \\ 0.077 & 0.053 & 0.038 & 0.024 & 0.024 & 0.038 \\ 0.030 & 0.028 & 0.023 & 0.027 & 0.029 & 0.030 \\ 0.009 & 0.014 & 0.012 & 0.017 & 0.011 & 0.015 \\ 0.018 & 0.022 & 0.022 & 0.022 & 0.023 & 0.023 \\ 0.017 & 0.020 & 0.009 & 0.006 & 0.007 & 0.009 \\ 0.011 & 0.011 & 0.003 & 0.001 & 0.012 & 0.013 \\ 0.009 & 0.009 & 0.009 & 0.009 & 0.009 & 0.009 \\ 0.016 & 0.015 & 0.010 & 0.009 & 0.004 & 0.003 \\ 0.020 & 0.020 & 0.018 & 0.018 & 0.018 & 0.017 \\ 0.054 & 0.053 & 0.056 & 0.058 & 0.058 & 0.057 \\ 0.055 & 0.055 & 0.056 & 0.057 & 0.057 & 0.057 \end{bmatrix}$$

3) The distance from the scheme to the positive and negative ideal scheme is calculated by the formula (6-7) as shown below and plotted as shown in figure 3.

$$D_i^+ = \{0.049, 0.036, 0.084, 0.082, 0.062, 0.050\}$$

$$D_i^- = \{0.068, 0.067, 0.019, 0.019, 0.051, 0.062\}$$

In the figure 3, the abscissa represents 1-6 of the six schemas, the ordinate is the distance between the schema and the positive and negative ideal solution, line1 represents the distance between the scheme and the positive ideal solution, and line2 represents the distance between the schema and the negative ideal solution. From the two curves in the figure can be roughly analyzed the 6 months of the enterprise's power marketing competitiveness of the situation.

4) Finally, according to the calculated positive and negative ideal solution of the distance from the formula (8) to get the proximity of each schema.

$$C_i^* = \{0.577, 0.647, 0.181, 0.188, 0.451, 0.555\}$$

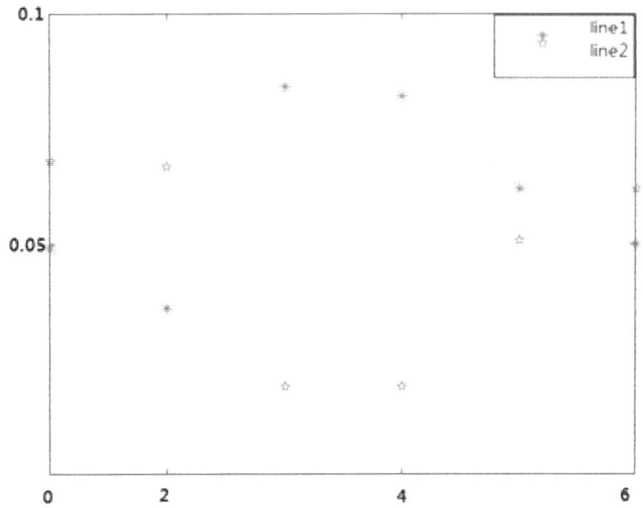

FIGURE 3. The distance between the schemes and the positive and negative ideal solutions

5) Sorting. The 6-month electricity marketing competitiveness is sorted according to the degree of proximity in step 4): August> July> December> November> October> September.

Because the electric power marketing competitive ability and profit directly related, so the profit of the six months of the comparative analysis found that the size of the profit and power marketing competitiveness of the same order, which also shows that the model built in this article has a scientific and rationality, and can provide a basis for the decision-making of electric power marketing competitiveness.

6. Conclusion

1) Based on the comprehensive consideration of market share, sales situation and product quality, this paper has established 20 comprehensive evaluation indexes system which includes qualitative and quantitative, cost-effective and cost-type. Through the study of evaluation indexes systems, it can be a deeper understanding about the status quo of the electric power marketing competitiveness.

2) In order to obtain more objective evaluation results, this paper introduces the cloud model, realizes the qualitative evaluation of the scientific and rational transformation of the qualitative evaluation language to the quantitative value, effectively solves the fuzziness and randomness of the natural language evaluation. At the same time, the cloud model and TOPSIS method are combined to propose a more scientific evaluation method. In addition, in the calculation of index weight, the use of entropy, effectively weaken the subjective factors of expert evaluation, making the evaluation results more practical and objective.

3) An example is given to analyze the marketing data of an electric power enterprise for six months. The evaluation model proposed in this paper is analyzed and verified, which provides a new method and new idea for the evaluation of power marketing competitiveness. At the same time, the model can also be used for electric power target market state analysis, electric power marketing evaluation early warning and other fields.

Acknowledgments

This article is sponsored by the National Natural Science Foundation of China, No 71271084.

References

[1] Yang Wu and Junyong Li. Research on assessment and forewarning model of power marketing condition based on cloud measure theory[J] (The Power System Protection and Control, 2013, 1) pp. 61–69.
[2] He Wang, Ming Zeng, Shan Chen, Jian Zhou and Zhixiang Zhang. *Comprehensive Evaluation*

IOP Conf. Series: Materials Science and Engineering **242** (2017) 012116 doi:10.1088/1757-899X/242/1/012116

Model for Power Supply Service Quality Based on Fuzzy Analytic Hierarchy Process[J] (The Power System Technology, 2006, 17) pp. 92–96.

[3] Wei Sun and Jingru Li. *Establishment and Evaluation Method of Performance Indicator System for Comprehensive Power Marketing of Power Utilities*[J] (Electric Power Technologic Economics, 2006, 03) pp. 54–58.

[4] Ming Zeng, Zhou Liang, Yong Ju and Guo Tian. *Evaluation of Service Strategy of Power Supply Enterprise Based on AHP [J]* (Technoeconomics & Management Research,2010,02) pp. 8–11.

[5] Tingjing Ruan. *Research on the Evaluation of Marketing Executive Power in Power Supply Companies [D],* North China Electric Power University,2014.

[6] Xiaoqian Du. *Research on marketing security risk evaluation and managerment system of electricity power enterprise[D]*, North China Electric Power University,2012.

[7] Zongze Li and Chengjun Shi. *Simulation research on PID excitation system of synchronous generator based on two-dimensional cloud model[J]* (Power System Protection and Control,2016,07) pp 19-24.

[8] K. Alexandridis, Y. Maru, Collapse and reorganization patterns of social knowledge representation in evolving semantic networks, Inform. Sci. 200(2012) 1–21.

[9] J. Dassow, F. Manea, R. Mercas, Regular languages of partial words, Inform. Sci. 268 (2014) 290–304.

[10] J.Q. Wang, P. Lu, H.Y. Zhang, X.H. Chen, Method of multi-criteria group decision-making based on cloud aggregation operators with linguistic information, Inform. Sci. 274 (2014) 177–91.

[11] Akbarzadeh-T, A qualified description of extended fuzzy logic, Inform. Sci. 244 (2013) 60–74.

[12] G.Y. Wang, C.L. Xu., Q.H. Zhang, X.R. Wang, A multi-step backward cloud generator algorithm, in: Proceeding of 8th International Conference on Rough Sets and Current Trends in Computing, LNAI, vol. 7413, 2012, pp. 313–322.

[13] K. Alexandridis, Y. Maru, Collapse and reorganization patterns of social knowledge representation in evolving semantic networks, Inform. Sci. 200(2012) 1–21.

Diagnostics Method for Analog Circuits Based on Improved KECA and Minimum Variance ELM

ZhijieYuan [1, a)], Yigang He[1,b)] and LifenYuan[1,c)]

[1]School of Electrical Engineering and Automation,Hefei University of Technology , Hefei230009,China

E-mail:[a)] *yuanzhijie@hfut.edu.cn*, [b)]Corresponding author: *18655136887@163.com*, [c)]bridge506@163.com

Abstract. Kernel entropy component analysis(KECA) is a new method for data transformation and dimensionality reduction.However it is sensitive to a single kernel radius.By analysis of the relation of statistics in the kernel feature space, improved KECA introduces two kernel radii and an adjusting factor to make KECA less sensitive to kernel radius.A method for fault diagnosis of analog circuits based on the combination of improved KECA and minimum variance extreme learning machine(ELM)is presented.Through wavelet decomposition of sampled signals,features are extracted. Improved KECA for feature dimension reduction is used.Then the fault patterns are classified by minimum variance ELM.Case studies on two analog circuits demonstrating our diagnostics method are presented.

1. Introduction

Analog circuits are used in industrial systems for conditioning signals, domestic appliance,aerospace vehicle,and more[1]. Circuit failures can affect system functionality, and can cause severe effects[2]. The research for analog circuits fault diagnosis has been deeply carried out and many efficiency methods have sprung up in the past several decades, such as the method neural network with learning by backward error propagation[3-5], support vector machine[6-10], data preprocessing techniques like wavelet decomposition and principal component analysis[11,12]and so on. However,analog circuits fault diagnosis is still full of challenges on account of tolerance of component parameters and noise etc.[13-15].The key operation of data preprocessing is feature selection and extraction. So an effective data transformation is needed to improve diagnosis efficiency.

Kernel entropy component analysis(KECA) [16,17]as a new data transformation has been used in clustering and image processing[18,19].Recently the technique was applied to analogue circuit diagnosis by Zhang *et al.*[20].Firstly the probability density function(PDF) of data set is estimated by the classical Parzen window, while window function is also as kernel function cleverly to project data onto kernel feature space.The Renyi entropy comes to squared euclidean length of the mean vector of kernel feature space by calculation.So eigendecomposition of kernel matrix is used for kernel PCA axes,which contribute most to Renyi entropy.Then data with reduction dimension is obtained by project onto those axes.

KECA is actually a principal component analysis method in the kernel feature space, which is sensitive to the kernel parameter[21] .And it has little discussion on the statistics in the space.Against that,our paper gives adequate consideration in three statistics in the kernel feature space and proposes a novel method of data-dimension reduction.This method preserves as much as possible of amount —density minus entropy(DME),which improves KECA in the way that the kernel radius in

Content from this work may be used under the terms of the Creative Commons Attribution 3.0 licence. Any further distribution of this work must maintain attribution to the author(s) and the title of the work, journal citation and DOI.
Published under licence by IOP Publishing Ltd

the first term of DME differs from the radius in the second term,and moreover there is an adjusting factor in DME.

Classifier is required to verify the performance of the feature selection and extraction method. Extreme learning machine (ELM) is a relatively new learning algorithm with fast network training and low human supervision[22,23]. Minimum Variance ELM (MVELM) algorithm extends the ELM algorithm[24],which aims at minimizing both the network output weights norm and the dispersion of the training data.Extensive evaluation on two publicly available datasets shows the outperformance of MVELM other ELM-based classification schemes.

A new method based on improved KECA and MVELM to analog circuit soft fault is proposed.Through Wavelet decomposition of the circuit's response,features are extracted for fault diagnosis.The low-dimensional data is obtained by the proposed data dimension reduction algorithm .Finally the classifier MVELM is used for fault classification.

This paper is organized as follows. Section2 outlines the principle and implementing steps of improved KECA. Section 3 briefly describes Minimum Variance ELM (MVELM) algorithm. In section 4, the proposed diagnostics method for analog circuits based on improved KECA and MVELM is put forward. The experimental results are demonstrated in Section 5 .Finally,the conclusions are drawn in Section 6.

2. Improved keca

2.1 The principle of improved KECA

KECA is a new method for dimensionality reduction with as much as possible of the Renyi entropy of the original data preserved[16]. The Renyi entropy estimator is given by

$$\hat{V}(P)=\frac{1}{N}\sum_{\mathbf{x}_u\in D}p_\sigma(\mathbf{x}_u)=\frac{1}{N}\sum_{\mathbf{x}_u\in D}\sum_{\mathbf{x}_t\in D}k_\sigma(\mathbf{x}_u,\mathbf{x}_t) \tag{1}$$

where $p_\sigma(\mathbf{x})$ is the probability density function generating the data set $D=\mathbf{x}_1,\cdots,\mathbf{x}_N$ and $k_\sigma(\mathbf{x},\mathbf{x}_t)$ is the Parzen window, or kernel function to project data onto kernel feature space. Let the map from original data set $D=\mathbf{x}_1,\cdots,\mathbf{x}_N$ to kernel feature space is ϕ ,that is $\mathbf{x}_t\rightarrow\phi(\mathbf{x}_t)$ .Under the consideration of the definition of kernel function ,we have

$k_\sigma(\mathbf{x}_u,\mathbf{x}_t)=\langle\phi(\mathbf{x}_u),\phi(\mathbf{x}_t)\rangle$

Using matrix notation,(1) is given by

$$\hat{V}(P)=\frac{1}{N^2}\mathbf{1}^T\mathbf{K}_\sigma\mathbf{1} \tag{2}$$

where element (u,t) of the $N\times N$ kernel matrix $\mathbf{K}_\sigma$ equals $k_\sigma(\mathbf{x}_u,\mathbf{x}_t)$ and $\mathbf{1}$ is an $N\times 1$ vector where each element equals one. Obviously $\mathbf{K}_\sigma$ is a positive semidefinite matrix and can be eigendecomposed as $\mathbf{K}_\sigma=\mathbf{EDE}^T$,where $\mathbf{D}$ is a diagonal matrix storing the eigenvalues $\lambda_1,\cdots,\lambda_N$ and $\mathbf{E}$ is a matrix with the corresponding eigenvectors $\mathbf{e}_1,\cdots,\mathbf{e}_N$ as columns. Rewriting(2) , we have

$$\hat{V}(p)=\frac{1}{N^2}\sum_{i=1}^{N}(\sqrt{\lambda_i}\mathbf{e}_i^T\mathbf{1})^2 \tag{3}$$

By selecting eigenvalues and eigenvectors which contribute most to the Renyi entropy estimate, KECA performs dimensionality reduction.

The paper turns attention to the statistics in the kernel feature space. First, we notice that the relationship between statistics $\mathbf{mm}^T$ and estimated entropy $\hat{V}(p)$ is

$$\hat{V}(p)=\int\phi^T(\mathbf{x})\mathbf{mm}^T\phi(\mathbf{x})d\mathbf{x} \tag{4}$$

where $\mathbf{m}=\frac{1}{N}\sum_{\mathbf{x}_i\in D}\phi(\mathbf{x}_i)$ is the mean vector vector of the kernel feature space data set[25].Concerning the common statistics, the equation illustrates the relations between them:

$$\mathbf{mm}^T=R_{\phi(\mathbf{x})}-C_{\phi(\mathbf{x})} \tag{5}$$

where $R_{\phi(\mathbf{x})}=\frac{1}{N}[\phi(\mathbf{x}_1),\cdots,\phi(\mathbf{x}_N)][\phi(\mathbf{x}_1),\cdots,\phi(\mathbf{x}_N)]^T$ is autocorrelation matrix and

$C_{\phi(\mathbf{x})}=\frac{1}{N}\sum_{\mathbf{x}_u\in D}(\phi(\mathbf{x}_u)-\mathbf{m})(\phi(\mathbf{x}_u)-\mathbf{m})^T$ is covariance matrix of data set $\phi(\mathbf{x}_1),\cdots,\phi(\mathbf{x}_N)$ respectively.

Taking (5) to (4) ,we have

$$\int\phi^T(\mathbf{x})C_{\phi(\mathbf{x})}\phi(\mathbf{x})d\mathbf{x}=\frac{1}{N}\int\left(\sum_{\mathbf{x}_t\in D}k_\sigma^2(\mathbf{x},\mathbf{x}_t)\right)d\mathbf{x}-\hat{V}(p) \tag{6}$$

Gaussian kernel function is chosen for ease of calculation , then the first term of the right side of equation(6) is given by

$$\frac{1}{N}\int\left(\sum_{\mathbf{x}_t\in D}k_\sigma^2(\mathbf{x},\mathbf{x}_t)\right)d\mathbf{x}=\frac{1}{N}\int\left(\sum_{\mathbf{x}_t\in D}k_{\sigma/\sqrt{2}}(\mathbf{x},\mathbf{x}_t)\right)d\mathbf{x}\approx\frac{1}{N}\sum_{\mathbf{x}_u\in D}\sum_{\mathbf{x}_t\in D}k_{\sigma/\sqrt{2}}(\mathbf{x}_u,\mathbf{x}_t)=\frac{1}{N}\mathbf{1}^T\mathbf{K}_{\sigma/\sqrt{2}}\mathbf{1} \tag{7}$$

Improved KECA concentrates on the following amount ,that is

$$\int\phi^T(\mathbf{x})C_{\phi(\mathbf{x})}\phi(\mathbf{x})d\mathbf{x}=\int p_{\sigma/\sqrt{2}}(\mathbf{x})d\mathbf{x}-V_\sigma(p) \tag{8}$$

$$\approx\frac{1}{N}\left(\mathbf{1}^T\mathbf{K}_{\sigma/\sqrt{2}}\mathbf{1}-\frac{1}{N}\mathbf{1}^T\mathbf{K}_\sigma\mathbf{1}\right) \tag{9}$$

The first term of (8) is the sum of the probability density , while the second term is estimate entropy.Since the kernel parameter in (9) is determined by Sigma, adjusting factor $\mu(0\le\mu\le1)$ is used to make the method be more robust.That is the method preserves as much as possible of amount $\frac{1}{N}\left(\mu\mathbf{1}^T\mathbf{K}_{\sigma/\sqrt{2}}\mathbf{1}-(1-\mu)\frac{1}{N}\mathbf{1}^T\mathbf{K}_\sigma\mathbf{1}\right)$ which is called by density minus entropy(DME).Improved KECA is a data transform to preserve as much as possible of DME.

2.2 The condition using improved KECA

In order to reduce the data dimension while trying to preserve DME,the matrix $\mathbf{K}=\mu\mathbf{K}_{\sigma/\sqrt{2}}-(1-\mu)\frac{1}{N}\mathbf{K}_\sigma$ is required to be positive semidefinite .This paper will give a sufficient condition about it. First,two lemmas on eigenvalue in the theory of matrices are invoked.

Let $\lambda_i(\cdot)$ denote the eigenvalue of a $N\times N$ matrix $\mathbf{A}$,and $\lambda_N(\cdot)\le\cdots\le\lambda_1(\cdot)$ .The following conclusion illustrates a lower bound for the minimum eigenvalue of the sum of two matrix.

Lemma1[26]:

$$\lambda_N(\mathbf{A}+\mathbf{B})\ge\lambda_N(\mathbf{A})+\lambda_N(\mathbf{B}) \tag{10}$$

Bounds for the maximum eigenvalue and minimum eigenvalue of a symmetric matrix is given by Lemma2. We denote by S_N the sets of symmetric matrices, the trace and Frobenius norm by

$tr(\cdot)$ and $\|\cdot\|_F$ respectively, then

Lemma2[27]: If $\mathbf{A} \in S_N$, then $\lambda_1(\mathbf{A}) \le \frac{tr(\mathbf{A})}{N} + (N-1)^{\frac{1}{2}} s(\lambda)$ (11)

$$\lambda_N(\mathbf{A}) \le \frac{tr(\mathbf{A})}{N} - (N-1)^{\frac{1}{2}} s(\lambda) \quad (12)$$

where $s(\lambda) = \left[\frac{\|A\|_F^2 - tr^2(A)/N}{N}\right]^{\frac{1}{2}}$ is the standard deviation of the eigenvalues .

We obtain a sufficient condition for which matrix $\mathbf{K}$ is positive semidefinite based on the two lemmas.

Theorem: If $\sigma \le \sigma^* = \left(\frac{m^*}{\ln N}\right)^{\frac{1}{2}}$,then $\mathbf{K} = \mu \mathbf{K}_{\sigma/\sqrt{2}} - (1-\mu)\frac{1}{N}\mathbf{K}_{\sigma}$ is positive semidefinite,where $m^* = \min_{\mathbf{x}_u, \mathbf{x}_t \in D} \|\mathbf{x}_u - \mathbf{x}_t\|^2$

Proof : We denote $\|\mathbf{x}_i - \mathbf{x}_j\|^2 = z_{ij}$ .Then recall the definition of kernel function ,we have

$$\mathbf{K}_\sigma(i,j) = \exp(-2z_{ij}/\sigma^2), \mathbf{K}_{\sigma/\sqrt{2}}(i,j) = \exp(-z_{ij}/\sigma^2) \quad (13)$$

Obviously,the following inequality is a consequence of Lemma 1

$$\lambda_n\left(\mu \mathbf{K}_{\sigma/\sqrt{2}} - \frac{1-\mu}{N}\mathbf{K}_\sigma\right) \ge \mu \lambda_n\left(\mathbf{K}_{\sigma/\sqrt{2}}\right) - \frac{1-\mu}{N}\lambda_1\left(\mathbf{K}_\sigma\right) \quad (14)$$

Using the fact $tr(K_\sigma) = tr(K_{\sigma/\sqrt{2}}) = N$ and lemma2,we have

$$\lambda_n(K_\sigma) \ge 1 - (N-1)^{\frac{1}{2}}\left[\left(\|K_\sigma\|_F^2 - N\right)/N\right]^{\frac{1}{2}} = 1 - (N-1)^{\frac{1}{2}}\left[\left(\sum_i \sum_{j \ne i} \exp(-4z_{ij}/\sigma^2)\right)\Big/N\right]^{\frac{1}{2}} \quad (15)$$

$$\lambda_1\left(K_{\sigma/\sqrt{2}}\right) \le 1 + (N-1)^{\frac{1}{2}}\left[\left(\|K_{\sigma/\sqrt{2}}\|_F^2 - N\right)\Big/N\right]^{\frac{1}{2}} = 1 + (N-1)^{\frac{1}{2}}\left[\left(\sum_i \sum_{j \ne i} \exp(-2z_{ij}/\sigma^2)\right)\Big/N\right]^{\frac{1}{2}} \quad (16)$$

The inequalities yields the lower bound for the minimum eigenvalue of $\mathbf{K}$,that is

$$\lambda_n\left(\mu \mathbf{K}_{\sigma/\sqrt{2}} - \frac{1-\mu}{N}\mathbf{K}_\sigma\right) \ge \left(\mu - \frac{1-\mu}{N}\right) - \mu(N-1)\exp\left(-\frac{2m^*}{\sigma^2}\right) - \frac{1-\mu}{N}(N-1)\exp\left(-\frac{m^*}{\sigma^2}\right) \quad (17)$$

where $m^* = \min_{\mathbf{x}_u, \mathbf{x}_t \in D} \|\mathbf{x}_u - \mathbf{x}_t\|^2$

Suppose $\mu \ge \frac{2}{N+1}$, (17) can be minimized by

$$\lambda_n\left(\mu \mathbf{K}_{\sigma/\sqrt{2}} - \frac{1-\mu}{N}\mathbf{K}_\sigma\right) \ge \frac{1}{N} - \mu N \exp\left(-\frac{2m^*}{\sigma^2}\right) - (1-\mu)\exp\left(-\frac{m^*}{\sigma^2}\right) \quad (18)$$

To ensure $\mathbf{K}$ be positive semidefinite,we have

$$\frac{1}{N} - \mu N \exp\left(-\frac{2m^*}{\sigma^2}\right) - (1-\mu)\exp\left(-\frac{m^*}{\sigma^2}\right) \ge 0$$

The solution of the inequality is $\sigma \le \sigma^* = \left(\frac{m^*}{\ln N}\right)^{\frac{1}{2}}$.

When Sigma satisfies the inequality above, $\mathbf{K}$ can be eigendecomposed to seek project axes ,followed by the dimension reduction of data set.

2.3 The Algorithm of Improved KECA

The procedure for improved KECA transforming data set $\mathbf{\Phi}$ to k dimensional data is summarized as:

step1: Select kernel function $k_\sigma(\mathbf{x},\mathbf{x}_t)$,we use a Gaussian kernel . Calculate kernel matrix $\mathbf{K}_\sigma$ and $\mathbf{K}_{\sigma/\sqrt{2}}$ based on the data set.Then calculate $\mathbf{K} = \mu\mathbf{K}_{\sigma/\sqrt{2}} - \frac{1-\mu}{N}\mathbf{K}_\sigma$.

step2 : Perform eigenvalue decomposition of $\mathbf{K}$, $\mathbf{K} = \mathbf{Q}^T diag(\zeta_1,\cdots,\zeta_n)\mathbf{Q}$, where $\zeta_1,\cdots,\zeta_N$ is the eigenvalue and $\mathbf{Q}$ is a orthogonal transform matrix with the corresponding eigenvector $\mathbf{q}_i$ as column, here $\zeta_i^{-\frac{1}{2}}\Phi\mathbf{q}_i$ is kernel PCA axes.

step3: Calculate the contribution of every axes to the renyi entropy $\left(\sqrt{\zeta_i}\mathbf{q}_i^T\mathbf{1}\right)^2$,select those k axes contributing most to the entropy estimate which span to U_k ,the corresponding diagonal matrix is $\mathbf{D}_k$ and orthogonal transform matrix is $\mathbf{Q}_k$.

step4: Project Φ onto U_k , $\Phi_{MECA} = P_{U_k}\Phi = D_k^{\frac{1}{2}}Q_k^T$, Φ_{MECA} is the low-dimensional data set.

3. Mvelm

MVELM theory aims to reach the smallest training error and the smallest norm of output weight as a single-hidden layer feedforward neural (SLFN) networks training algorithm.

Let us denote by $(\mathbf{x}_i,\mathbf{o}_i)_{i=1,\cdots,N}$ a training set consisting of input vector $\mathbf{x}_i = [x_{i1},\cdots,x_{iD}]^T \in R^{D\times1}$ and output vectors $\mathbf{o}_i = [o_{i1},\cdots,o_{iC}]^T \in R^{C\times1}$ .The network input weights matrix is $\mathbf{W} = [\mathbf{w}_1,\cdots,\mathbf{w}_L] \in R^{D\times L}$,where $\mathbf{w}_i \in R^{D\times1}$ is the weight vector connecting the ith hidden node and the input nodes.While the network output weights $\boldsymbol{\beta}_i = [\beta_{1i},\cdots,\beta_{Li}]^T \in R^{L\times1}$ connects the ith output node and the hidden nodes, the hidden layer bias values is $\mathbf{b} \in R^{L\times1}$.

For a given activation function for the network hidden layer $h(x)$ and a linear activation function for the network output layer, the output $\mathbf{o}_i$ of SLFN corresponding to $\mathbf{x}_i$ is calculated by:

$$\mathbf{o}_{ik} = \sum_{j=1}^{L}\beta_{jk}h(\mathbf{w}_j\cdot\mathbf{x}_i + b_j), k = 1,\cdots,C \tag{19}$$

here $\mathbf{w}_j\cdot\mathbf{x}_i$ denotes the inner product of $\mathbf{w}_j$ and $\mathbf{x}_i$.

By storing the network hidden layer outputs corresponding to the training vectors $\mathbf{x}_i, i = 1,\cdots,N$ in a matrix $\mathbf{H} = \begin{bmatrix} h(\mathbf{w}_1\cdot\mathbf{x}_1 + b_1) & \cdots & h(\mathbf{w}_1\cdot\mathbf{x}_N + b_1) \\ \vdots & & \vdots \\ h(\mathbf{w}_L\cdot\mathbf{x}_1 + b_L) & \cdots & h(\mathbf{w}_L\cdot\mathbf{x}_N + b_L) \end{bmatrix} = [\mathbf{h}_1,\cdots,\mathbf{h}_N]$,where $\mathbf{h}_i$ is the hidden layer output corresponding to $\mathbf{x}_i$. equation (19) can be expressed in a matrix form as

$$\mathbf{O} = \boldsymbol{\beta}^T\mathbf{H} \tag{20}$$

where $\mathbf{O}=[\mathbf{o}_1,\cdots,\mathbf{o}_L]\in R^{C\times N}$ is an output matrix.

MVELM not only tries to minimize the training errors and the norm of the network output weights, but also tries to minimize the variance of the training vectors . The dispersion of the training vectors can be denoted as the total scatter matrix of the training vectors.

$$\mathbf{S}=\sum_{i=1}^{N}(\mathbf{h}_i-\boldsymbol{\mu})(\mathbf{h}_i-\boldsymbol{\mu})^T \tag{21}$$

where $\boldsymbol{\mu}=\frac{1}{N}\sum_{i=1}^{N}\mathbf{h}_i$ .So $\boldsymbol{\beta}$ can be calculated by solving the following optimization problem:

$$\min L_{MVEIM}=\left\|\mathbf{S}^{\frac{1}{2}}\boldsymbol{\beta}\right\|^2+\lambda\sum_{i=1}^{N}\|\zeta_i\|_2^2 \quad S.T.\quad \boldsymbol{\beta}^T\mathbf{h}_i=\mathbf{t}_i-\zeta_i, i=1,\cdots,N \tag{22}$$

where λ is the penalty coefficient on the training errors, ζ_i is the error vector with respect to the ith training pattern, $\mathbf{t}_i$ is the target vector corresponding to $\mathbf{x}_i$ .By calculation,we obtain

$$L_{MVEIM}=\sum_{i=1}^{N}\left(\|\mathbf{o}_i-\mathbf{o}\|^2+\lambda\|\mathbf{o}_i-\mathbf{t}_i\|^2\right) \tag{23}$$

where $\mathbf{o}=\frac{1}{N}\sum_{i=1}^{N}\boldsymbol{\beta}^T\mathbf{h}_i$ denotes the mean network output.(23)shows that MVELM reaches a compromise between the training vectors dispersion in the network output and the network training error.

Based on the Karush-Kuhn_Tucker theorem and Lagrange duality, the output weights can be calculated by

$$\boldsymbol{\beta}=\left(\mathbf{H}\mathbf{H}^T+\frac{1}{\lambda}\mathbf{S}\right)^{-1}\mathbf{H}\mathbf{T}^T \tag{24}$$

where $\mathbf{T}=[\mathbf{t}_1,\cdots,\mathbf{t}_N]\in R^{C\times N}$ is a matrix containing the network target vectors.

After the calculation of the network output weights $\boldsymbol{\beta}$,the network response for a given vector $\mathbf{x}_m$ is given by $\mathbf{o}_m=\boldsymbol{\beta}^T\mathbf{h}_m$ where $\mathbf{h}_m$ is the network hidden layer output for $\mathbf{x}_m$.

4. The new method based on improved keca and mvelm

The proposed diagnostics method for analog electronic circuits can be described as follows:

1)Data preprocessing.

Firstly we use Haar wavelet transform as a preprocessor. Output voltage signal from the circuit under test(CUT) under various fault conditions are gained by Pspice simulation. Then the sampled signal are decomposed into 3 levels using Haar wavelet transform. Energies contained in the coefficients at third level are normalized to get 8-dimensional fault feature vectors.

2)Feature selection.

For the purpose of cutting down training time of the classifier, it is crucial to reduce the dimension of fault features. Thus the novel method-improved KECA is applied to obtain the low-dimension feature vectors .

3)Fault classification

The features are used to MVELM classifier and the fault patterns are classified.

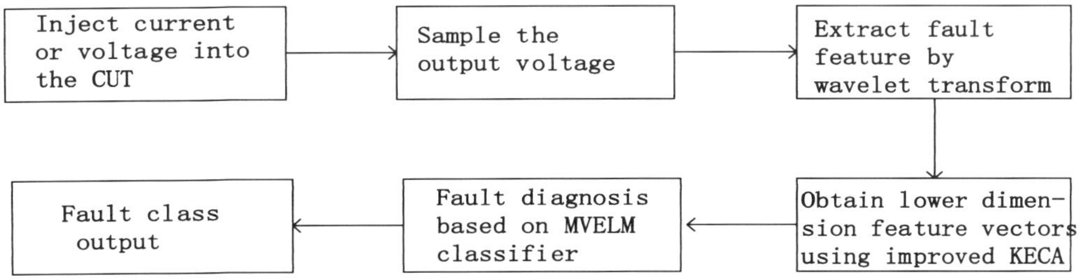

FIGURE 1. The flow chart of the proposed method

5. Fault dignosis reseults

5.1Example circuits and faults

In this section we present two circuits in order to evaluate the performance of the proposed method. In all circuits,a single 10 V pulse with 10 us duration is used as the input. Responses are acquired by sampling the CUTs'outputs voltage. The resistors and capacitors are assumed to have tolerances of 10% and 5% respectively. We assume that the components are 50% higher or lower than their nominal values .For each fault class,40 cases are generated. 20 of the cases are used to train the classifier,and the other 20 cases are used to test the method.

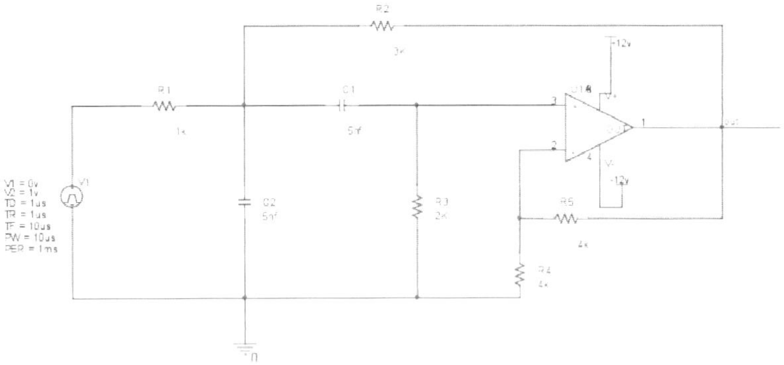

FIGURE 2.a. Sallen-Key bandpass filter

IOP Conf. Series: Materials Science and Engineering **242** (2017) 012117 doi:10.1088/1757-899X/242/1/012117

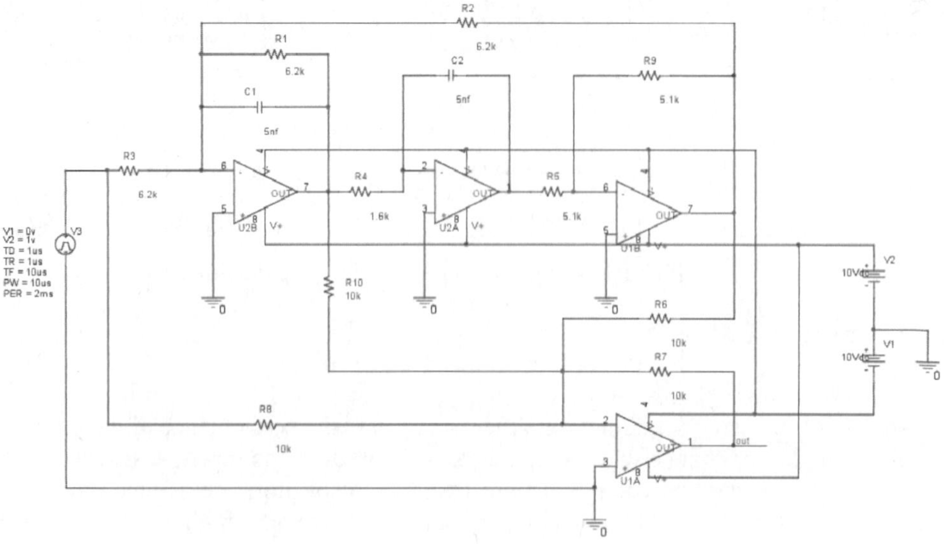

FIGURE 2.b. Four opamp biquad highpass filter

The first circuit is a Sallen-Key band-pass filter with 25 kHz central frequency in Fig2a. Fig2a also shows the nominal values of the components. The fault classes include R2↓,R2↑, R3↓, R3↑, C1↓, C1↑, C2↓,C2↑,and normal state, where ↑ and ↓ refer to values higher and lower than the nominal one , respectively.

The second circuit is more complicated,as shown in Fig2b.This is a 2-Four-Opamp Biquad High-pass Filter. The nominal values of the components are shown in the same figure.11 fault classes are considered ,including C1↑, C1↓, C2↑,C2↓, R1↑, R1↓, R2↑, R2↓ ,R3↑,R3↓ and the normal state.

TABLE 1.a. Fault classes for Sallen-Key highpass filter

Fault code	Model	Standard value	Fault value
F1	Normal	——	——
F2	R2↓	3kΩ	1.5kΩ
F3	R2↑	3kΩ	6kΩ
F4	R3↓	2kΩ	1kΩ
F5	R3↑	2kΩ	4kΩ
F6	C1↓	5nF	2.5nF
F7	C1↑	5nF	10nF
F8	C2↓	5nF	2.5nF
F9	C2↑	5nF	10nF

TABLE 1.b. Fault classes for Four opamp biquad highpass filter

Fault code	Model	Standard value	Fault value
F1	Normal	——	——
F2	C1↑	5nF	10nF
F3	C1↓	5nF	2.5nF
F4	C2↑	5nF	15nF
F5	C2↓	5nF	2.5nF
F6	R1↑	6.2kΩ	15kΩ
F7	R1↓	6.2kΩ	3kΩ
F8	R2↑	6.2kΩ	18kΩ
F9	R2↓	6.2kΩ	2kΩ
F10	R3↑	6.2kΩ	12kΩ
F11	R3↓	6.2kΩ	2.7kΩ

5.2Diagnosis results

5.2.1Features extracted by improved KECA and KECA methods. In order to verify the effectiveness of the proposed method,a comparison is conducted with the two CUTs for improved KECA and KECA[20],that is the two methods are applied to reduce the dimension of the energy features.

Using the proposed improved KECA we deprive the first two most significant component vectors which contribute most to DME for the first circuit. Two-dimensional features is displayed in Fig3.a.It is obviously that all fault classes are fallen into different 9 ambiguity groups.And there is none of asunder distributed fault classes.A 2-dimension feature set extracted by KECA for the same circuit, is shown in Fig3.b.Although all fault classes are distinct ambiguity groups,there are several flaws. Obviously F2 fault classes is divided into two parts, as well as F8 . Moreover dots from F9 fault classes distributes very separately.The results reveal that our method is an improvement of KECA and it can generate better extraction performance than KECA in this experiment.

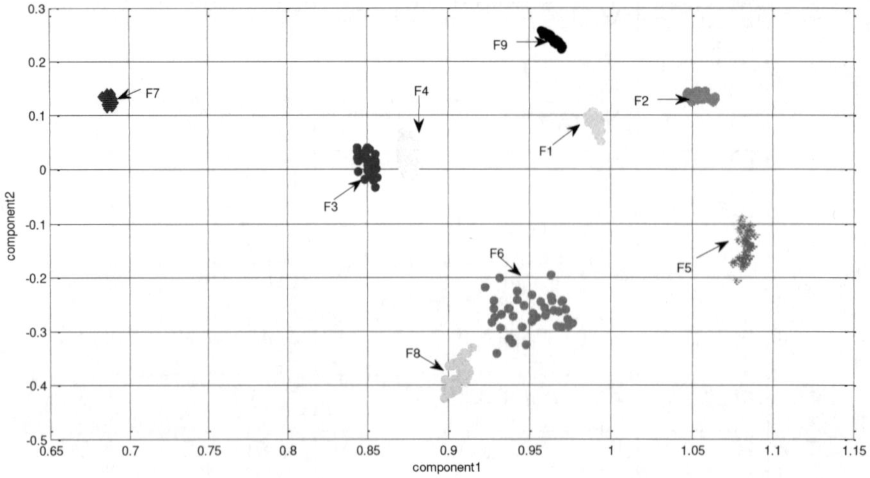

FIGURE 3.a. The two-dimensional feature of Sallen-Key filter by improved KECA

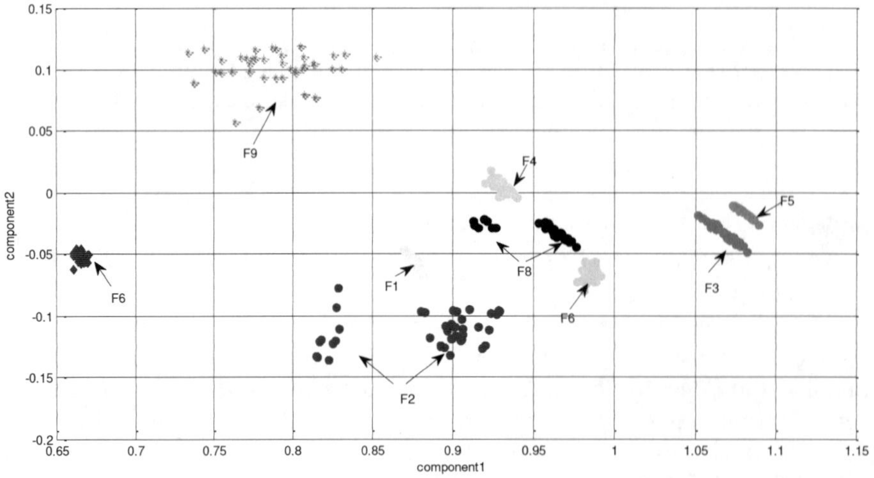

FIGURE 3.b. The two-dimensional feature of Sallen-Key filter by KECA

Three-dimensional features obtained by improved KECA and KECA is displayed in Fig4.a. and Fig4.b respectively for the second circuit. The figure Fig4.a.shows that all fault classes can be distinct. It is obviously that F2, F3, F4, F5, F6, F7, F9,F10 and F11 fault classes are different ambiguity groups, only F1and F8 fault classes are partially overlapping. Figure4.b.shows there is completely overlapping for F4 and F11 fault classes, while F2, F4 and F11 fault classes are partially overlapping ambiguity groups.Overlap between different fault classes means bad extraction performance.Besides,F3,F4and F5 are not compact fault classes.Fig4 shows improved KECA is still superior to KECA for more complicated circuit in extraction performance.

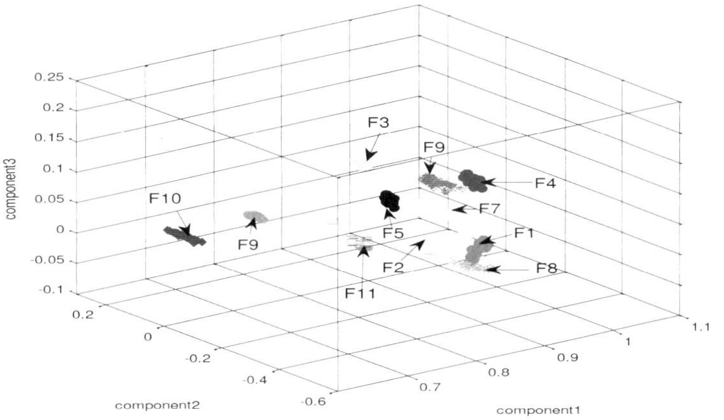

FIGURE 4.a. The three-dimensional feature of Four opamp biquad highpass filter by improved KECA

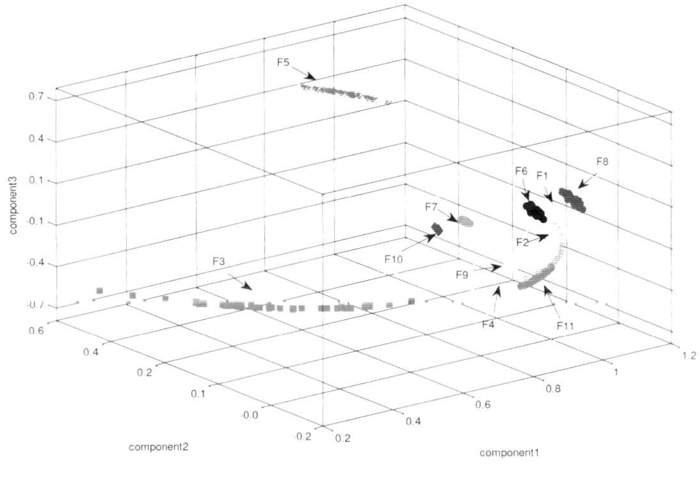

FIGURE 4.b. The three-dimensional feature of Four opamp biquad highpass filter by KECA

5.2.2 Diagnosis Accuracies. The followed table reveals the classing accuracy for two circuits based on MVELM. Table2 is the comparison of the test accuracy between improved KECA method and KECA using MVELM. From table2, it can be seen that the proposed approach is accurate(100%) in the first circuit and the accuracy is 99.09% in the second circuit in the fault diagnosis.While the method KECA have accuracies of 98.89% and a more lower 97.72% in the two circuits.From table2,it can be concluded that the proposed method has a highter accuracy than the method KECA in the case of the two CUTs.

TABLE 2. Comparison of the classification accuracy between the proposed method and KECA

method	Circuit1	Circuit2
KECA	98.33%	97.72%
Our Method	100%	99.09%

5.2.3 Sensitivity of Kernel Radius. In the section,we discuss the sensitivity of kernel radius.Cluster validity of reduced dimension of energy feature and classification accuracy of faults will be used.

From the scatter plots of fault classes by improved KECA and KECA in figure3and figure4,it is obviously that the reduced-dimensional data has been clustered.In this part ,cluster indexes is used to analyze sensitivity of kernel parameter Sigma. Dunn Validity Index(DVI) is based on geometrical considerations as a cluster validity indexes, which can provide the validation results for evaluating clustering. Suppose we have partitioned dataset X into n clusters C_i ,the Dunn Index for the set is defined as:

$$DVI = \frac{\min\limits_{1\le i\le n}\left\{\min\limits_{1\le j\le n}\left\{dist(C_i, C_j)\right\}\right\}}{\max\limits_{1\le t\le n}\left\{diam(C_t)\right\}} \tag{25}$$

,where $diam(C_t) = \max\limits_{x_u, x_v \in C_t} \|x_u - x_v\|$ and $dist(C_i, C_j) = \min\limits_{x_u \in C_i, x_v \in C_j} \|x_u - x_v\|$. $n = 9$

DVI can evaluate the effectiveness of techniques improved KECA and KECA.Fault classes are characterized by

2-dimensional features of low-pass filter by improved KECA and KECA.For this circuit, $\sigma^* = 0.178$,and here $n = 9$, C_i contains 40 feature vectors of 2-dimension.We adjust the value of μ and select the best one corresponding to every Sigma.Figure5.shows relation between DVI and kernel parameter Sigma,using data-dimension reduction techniques- improved KECA and KECA.It can be seen that when Sigma is less than 0.145,DVI increases slowly and fluctuated narrowly in the range $0.145 \le \sigma \le 0.155$,then keeps a value around DVI=0.264 using our method. The combinatorial kernel matrix and different kernel radius in MDE can be responsible to it. In the contrast,there are several suddenly smaller or larger value of DVI because of fixed kernel radius in KECA.On the other hand ,the red curve lines always above the blue curve, which partially can be explained by figure3.b,parted fault classes can cause larger distance within class.Figure 5.shows that DVI in the dataset using KECA is more sensitive than our method.We can find that improved KECA has ascendancy over KECA in robustness.

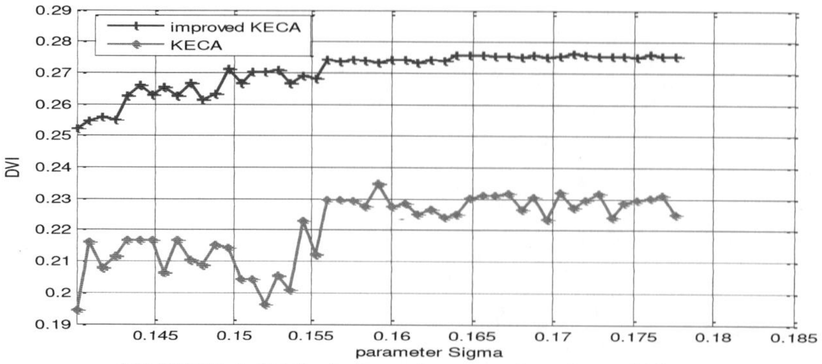

FIGURE 5. DVI of clustering as a function of Sigma

6. Conclusion

In this paper, improved KECA is presented,which is a stable method for data transformation and dimensionality reduction.The method is different from KECA in the consideration of three statistics in the kernel feature space.As an application of improved KECA,a fault diagnosis method based on improved KECA and MVELM is proposed. The reduced data and fault diagnosis efficiency is demonstrated. It is showed that the proposed method can achieve better effect .

Acknowledgements
This work was supported by the National Natural Science Foundation of China under Grant No. 51577046 ,the State Key Program of National Natural Science Foundation of China under Grant No. 51637004,the national key research and development plan "important scientific instruments and equipment development" Grant No.2016YFF010220、Anhui Provincial Science and Technology Foundation of China under Grant No. 1301022036.

References
[1] A.Vasan,L.Bing and M.Pecht, Diagnositics and prognositcs method for analog electronic circuits,IEEE Trans.Ind.Electron.60(11),pp.5277-5291,2013.
[2] J. Sikorska, M. Hodkiewicz and L. Ma.Prognostic modeling options for remaining useful life estimation by industry, Mech. Syst. Signal Process.vol. 25, no. 5, pp.1803–1836,Jul. 2011.
[3] Y.tan,Y.He and C.Cui,G.Qiu,A novel method for analog fault diagnosis based on neural networks and genetic algorithms,IEEE Trans.Instrum.Meas.57(11),pp.2631-2639,2008.
[4] M. A. El-Gama land M. D. A. Mohamed,Ensembles of Neural Networks for Fault Diagnosis in Analog Circuits.Journal of Electronic Testing.23(4),pp.323-339,2011.
[5] L. Yuan, Y. He, J. Huang, and Y. Sun, A new neural-network-based fault diagnosis approach for analog circuits by using kurtosis and entropy as a preprocessor,IEEE Trans. Instrum. Meas., vol. 59, no. 3, pp. 586–595.Mar. 2010
[6] J. Huang, X. Hu, and F. Yang , Support vector machine with genetic algorithm for machinery fault diagnosis of high voltage circuit breaker,Measurement, vol. 44, no. 6, pp. 1018–1027,Jul. 2011.
[7] B. Long, S. L. Tian, Q. Miao, and M. Pecht, Research on features for diagnostics of filtered analog circuits based on LS-SVM, IEEE AUTOTESTCON, pp. 360–366,2011.
[8] B. Long, S. Tian, and H. Wang(Jun. 2012), Diagnostics of filtered analog circuits with tolerance based on LS-SVM using frequency features,J. Electron.Test.: Theory Appl., vol. 28, no. 3, pp. 291–300.
[9] C.-W.Hsu,C.-J.Lin,A comparison of methods for multiclass support vector machines,IEEE Trans.Neural.Netw.13(2),pp.415-425,2002.
[10] A.Zhang, C.Chen and B. Jian,Analog circuit fault diagnosis based UCISVM, Neurocomputing, Volume 173, Part 3, 15 January 2016, pp.1752-1760,2016.
[11] M. Aminian and F. Aminian, Neural-network based analog circuit fault diagnosis using wavelet transform as preprocessor, IEEE Trans. Circuits Syst. II, Analog Digit. Signal Process., vol. 47, no. 2, pp. 151–156 , Feb.2000.
[12] Y. Xiao and Y. He, A novel approach for analog fault diagnosis based on neural networks and improved kernel PCA,Neurocomputing, vol. 74,no. 7, pp. 1102–1115,Mar. 2011.
[13] M. Pecht and R. Jaai, A prognostics and health management roadmap for information and electronics-rich systems, Microelectron. Reliab., vol. 50,no. 3, pp. 317–323,Mar. 2010.
[14] J. Sikorska, M. Hodkiewicz, and L. Ma, Prognostic modeling options for remaining useful life estimation by industry, Mech. Syst. Signal Process.,vol. 25, no. 5, pp. 1803–1836,Jul. 2011.
[15] D. Binu, B.S. Kariyappa, A Survey on Fault Diagnosis of Analog Circuits: Taxonomy and State of the Art,International Journal of Electronics and Communications. .73,pp.68-83,2017.
[16] JENSSEN R.Kernel entropy component analysis,IEEE Trans.Pattern Analysis and Machine Intelligence,32(5),pp.847-860,2010.
[17] J. Shi, Q.Jiang, Q. Zhang and Q. Huang, X.Li,Sparse kernel entropy component analysis for dimensionality reduction of biomedical data,Neurocomputing,vol.168,pp.930-940,2015.
[18] SHEKAR B H, SHARMILA KUMARI M and MESTETSKIY L M, et al. Face recognition using kernel entropy component analysis[J]. Neurocomputing , 74(6): pp.1053-1057,2011.
[19] XIE Z and GUAN L. Multimodal information fusion of audio emotion recognition based on kernel entropy component analysis[J]. Multimedia, IEEE International Symposium on, 41(11) : pp.1-8,2012.
[20] C. Zhang and Y. He, et..A novel approach to diagnosis of analog circuit incipient faults based of

KECA and OAO LSSVM. Metrol. Meas. Syst., 22(2),pp. 251 – 262,2015.
[21] Jonas Nordhaug Myhre.Semi-Supervised Classification using Kernel Entropy Component Analysis and the LASSO.FYS-3921 Master's thesis in Electrical Engineering December 2011 .The University of Tromsø.
[22] G.B. Huang, L. Chen, and C.K. Siew, Universal approximation using incremental constructive feedforward networks with random hidden nodes, IEEE Transactions on Neural Networks, vol. 17, no. 4, pp. 879– 892, 2006.
[23] Y. Wang, F. Cao, and Y. Yuan.A study on effectiveness of extreme learning machine,Neurocomputing, vol. 74,no. 16, pp. 2483–2490,2011.
[24] A.Iosifidis, A. Tefas, and I. Pitas, Minimum class variance extreme learning machine for human action recognition,IEEE Trans. Circuits Syst. Video Technol., vol. 23, no. 11, pp. 1968–1979,Nov. 2013.
[25] X.Yan,S.WANG,h.GUO,Feature reduction of high-order statistics based On Parzen window,CAAI Transactions On Intelligent Systems,8(1):pp.1-10,2013.
[26] R.A. Horn， C.R. Johnson. Matrix Analysis[M]. Cambridge: Cambridge University Press,1985.
[27] P.Tarazag,Eigenvalue Estimate for symmetric Matrices.Linear Algebra and its Application,135(1),pp.171-179,1990.

IOP Conf. Series: Materials Science and Engineering **242** (2017) 012118 doi:10.1088/1757-899X/242/1/012118

A Novel Local Transform Inverse S-Transform Algorithm for Statistical Filter

Baiqiang Yin[1, 2,3,a)], Zhanfeng Sun[1,b)] ,Zhong Yi[3] and Yigang He[1,c)]

[1] The School of Electrical and Automation Engineering, Hefei University of Technology, Hefei 230009, China.
[2] Key Laboratory of Nondestructive Testing,Ministry of Education, Nanchang Hangkong University,Nanchong 637009, China.
[3] Beijing Institute of Spacecraft Environment Engineering, Beijing 100094, China.

E-mail: [a)]Corresponding author: yinbaiqiang123@163.com, [b)] 1003000542 @qq.com, [c)] 18655136887 @qq.com

Abstract. S-transform (ST) is a useful tool for time-frequency filter. However, the conventional inverse S-transform (IST) algorithm suffers from time or frequency leakage. In this paper, we proposed a novel local transform inverse S-Transform (LTIST) algorithm for statistical filter. First, the matrix S-transform (MST) and MIST are derived. Then the proposed LTIST approach applies to denoising. The statistical property of stochastic noise in the MIST is discussed. The results show that the proposed MIST algorithm has better time-frequency localization in statistical filtering than the conventional methods. Illustrative examples verify the effectiveness of the proposed algorithm.

1. Introduction

Time-frequency analysis is a powerful tool for time-varying signal and many algorithms have been proposed [1]. There are two well know lincar ones: short-time Fourier transform (STFT) and continuous wavelet transform (CWT). CWT has a progressive resolution thanks to its time-scale analysis methods, which replaces the frequency shift of the STFT by the dilation of a basis function, also called mother wavelet [2]. These methods are applied to many fields including electrical and mechanical engineering, medical imaging and geophysics [3-6].

The S-transform (ST) proposed by Stockwell et al.[7], which can be viewed as an intermediate step between the STFT and the CWT. This method succeeds the advantages of STFT and it enables the use of the frequency variable as well as the multiresolution strategy of the wavelets. An adaptive Gaussian window was used to keep higher frequency resolution with narrower windows and lower frequency resolution with wider windows. Therefore, it provides a powerful performance for time-frequency analysis. Nowadays the ST has been widely applied in many areas, filtering is one of the hotter applications in the field of data processing [8, 9].

There are two factors that influence the filtering performance: IST algorithms and the pass-regions in the S-transfrom domain [10-13].To solve the former problem, some IST algorithms have been proposed with different filtering properties in the time-frequency domain. Stockwell et al.'s IST algorithm is efficient but has unsatisfactory time resolution. In order to improve this problem, Schimmel and Gallart proposed another excellent time localization IST algorithm [10]. However, their method contains some reconstruction errors in it. Fortunately the distortion can be corrected by deconvolution [11]-[13]. To second problem, Pei and Wang define the pass-regions using global

Content from this work may be used under the terms of the Creative Commons Attribution 3.0 licence. Any further distribution of this work must maintain attribution to the author(s) and the title of the work, journal citation and DOI.
Published under licence by IOP Publishing Ltd

threshold value method, but the criteria used to define the threshold value in the wideband and narrowband filters is just from empirical value [12].

In this paper, we propose a new method to determine pass-regions in the S-transfrom time-frequency domain. This method will provide more higher precise in time-frequency filtering and does not require any noise synchronous acquisition as reference.

2. S-Transform and Its Inverses

2.1 Continuous S-Transform and its inverses

The continuous ST derives from STFT, which also can be viewed as a special CWT [6]. The continuous ST of the signal $x(t)$ is defined as

$$S(\tau,f)=\int_{-\infty}^{\infty}x(t)w(\tau-t,f)e^{-i2\pi ft}dt \tag{1}$$

where $w(\tau-t,f)$ is a Gaussian window, it can be denoted as

$$w(\tau-t,f)=\frac{|f|}{k\sqrt{2\pi}}e^{\frac{-f^2(\tau-t)^2}{2k^2}},\ k>0 \tag{2}$$

where f is frequency variables, t is time variables and τ is delay and k is a scaling factor that controls time-frequency resolution.

2.2 Matrix Formulation of ST and IST

Let $x[p]=x(pT)$, $p=0,1,\cdots,N-1$ denote a discrete time series corresponding to $x(t)$ with a time sampling interval of T. Let $f_s=1/T$ be the sampling frequency and f_0 be the frequency step, and $M=f_s/f_0$ and $n=-M/2,\cdots,-M/2-1$ is the index of frequency range. The discrete ST of the time series $x[p]$ can then write as [14]

$$S[m,n]=\sum_{p=0}^{N-1}x[p]\frac{|n|}{\sqrt{2\pi}kN}e^{-(n^2(m-p)^2/2k^2N^2+i2\pi pn/N)},\ n\neq 0 \tag{3}$$

equation (3) can be rewritten in the following way [15,16]:

$$\begin{aligned}S[m,n]&=\sum_{p=0}^{N-1}x[p]\frac{|n|}{\sqrt{2\pi}kN}e^{-(n^2(m-p)^2/2k^2N^2+i2\pi pn/N)}\\&=x[0]\frac{|n|}{\sqrt{2\pi}kN}e^{-n^2m^2/2k^2N^2}+x[1]\frac{|n|}{\sqrt{2\pi}kN}e^{-(n^2(m-1)^2/2k^2N^2+i2\pi n/N)}+\\&\quad\cdots+x[p]\frac{|n|}{\sqrt{2\pi}kN}e^{-(n^2(m-p)^2/2k^2N^2+i2\pi pn/N)}+\\&\quad\cdots+x[N-1]\frac{|n|}{\sqrt{2\pi}kN}e^{-(n^2(m-(N-1))^2/2k^2N^2+i2\pi(N-1)n/N)}\\&=T_{(mn)0}x[0]+T_{(mn)1}x[1]+\cdots+T_{(mn)p}x[p]+\cdots+T_{(mn)(N-1)}x[N-1]\end{aligned} \tag{4}$$

where $T_{(mn)p}$ represent transform coefficients of ST, which is related to m, n and p. Let $s[(n-1)\times N+m]=S[m,n]$, (4) can be formulated in the form of matrices

$$\begin{bmatrix}s[1]\\s[2]\\\vdots\\s[N]\\s[N+1]\\\vdots\\s[M\times N]\end{bmatrix}=\begin{bmatrix}T_{11}&T_{12}&\cdots&T_{1N}\\T_{21}&T_{22}&\cdots&T_{2N}\\\vdots&\vdots&\cdots&\vdots\\T_{N1}&T_{N2}&\cdots&T_{NN}\\T_{(N+1)1}&T_{(N+1)2}&\cdots&T_{(N+1)N}\\\vdots&\vdots&\cdots&\vdots\\T_{(M\times N)1}&T_{(M\times N)2}&\cdots&T_{(M\times N)N}\end{bmatrix}\cdot\begin{bmatrix}x[0]\\x[1]\\\vdots\\\vdots\\\vdots\\x[N-2]\\x[N-1]\end{bmatrix} \tag{5}$$

Therefore, discrete ST in matrix form can be modeled as

$$s = Tx \tag{6}$$

where s and x are $N^2 \times 1$ matrix, T is a $N^2 \times N$ matrix respectively. Let T^+ denotes the Moore-Penrose pseudoinverse, discrete IST in the form of matrix form can be modeled as

$$x = T^+ s \tag{7}$$

where T^+ exists only if matrix T has full rank..

3. Local Ist Time-Frequency Filters

Time-frequency analysis can be applied in many areas. Extracting the instantaneous parameters of a signal and de-noising of data are the most important applications in filtering. One of the interests of the ST lies in filtering the noise in the time-frequency domain and then using inverse transform to extract useful signal.

Let $C[m,n]$ be an time-frequency equalization filter. S-transform time-frequency domain filters as follows

$$S_f[m,n] = S[m,n] * C[m,n]. \tag{8}$$

The corresponding matrix IST filters can be describe as

$$x_f = T^+ s_f \tag{9}$$

where s_f is the filtered time-frequency domain S_f, x_f represents the output signal after filter. Conventionally, filtering would multiply the pass-region by one and the others by zeros in the time-frequency domain,

Discrete IST in matrix form be modeled as (9), which can be rewritten in the following way

$$\begin{bmatrix} x_f[0] \\ x_f[1] \\ \vdots \\ \vdots \\ \vdots \\ x_f[N-2] \\ x_f[N-1] \end{bmatrix} = \begin{bmatrix} T_{11} & T_{12} & \cdots & T_{1N} \\ T_{21} & T_{22} & \cdots & T_{2N} \\ \vdots & \vdots & \cdots & \vdots \\ T_{N1} & T_{N2} & \cdots & T_{NN} \\ T_{(N+1)1} & T_{(N+1)2} & \cdots & T_{(N+1)N} \\ \vdots & \vdots & \cdots & \vdots \\ T_{(MN)1} & T_{(MN)2} & \cdots & T_{(MN)N} \end{bmatrix}^+ \cdot \begin{bmatrix} s_{f11} \\ s_{f12} \\ \vdots \\ s_{f1N} \\ s_{f21} \\ \vdots \\ s_{fMN} \end{bmatrix}. \tag{10}$$

Notice that the output discrete time series $x_f[i]$ depend on $[T_{i1}, T_{i2}, \cdots T_{iN}]$ multiplied by $[s_{11}, s_{12}, \cdots, s_{1N}]^*$. Each element of the time series is related with the whole time-frequency domain matrix and the corresponding row of the inverse transform matrix. In other words, if we want to extract only interesting time regions signal in whole time-frequency domain, it is necessary to take account of the entire time-frequency domain and corresponding row elements of inverse transform matrix. Therefore, the proposed LT IST can be formulated in matrix form as follow:

$$\begin{bmatrix} x_f[i] \\ \vdots \\ x_f[j] \end{bmatrix} = \begin{bmatrix} T_{i1} & T_{i2} & \cdots & T_{iN} \\ \vdots & \vdots & \cdots & \vdots \\ T_{j1} & T_{j2} & \cdots & T_{jN} \end{bmatrix}^+ \cdot \begin{bmatrix} s_{f11} \\ s_{f12} \\ \vdots \\ s_{f1N} \\ s_{f21} \\ \vdots \\ s_{fMN} \end{bmatrix} \tag{11}$$

$$x_f = T_f^{\ +} s \tag{12}$$

where $T_f^{\ +}$ correspond to the interesting time-frequency regions s_f in a certain period of time. LT IST filtering in the pass-region would multiply by one and other time by zeros. Unlike the conventional filtering method, LT IST has good time localization in exacting the transient signals. We notice that (12) is only an IST algorithm without filtering. It is necessary to take account of the time-frequency domain.

The spectrum of the ST in the time-frequency domain is defined as the squared modulus of $S[m,n]$, and the mean power spectrum (MPS) of the white noise can be defined as

$$\begin{aligned} E &= \{|S(\tau,f)|^2\} \\ &= \frac{|f|^2}{2\pi}\iint_{\infty} E[h(t)h^*(u)]\mathrm{e}^{-(t-\tau)^2 f^2/2-(u-\tau)^2 f^2/2} e^{-2\pi fi(t-u)}dudt \\ &= \frac{|f|^2}{2\pi}\sigma^2\iint_{\infty}\delta(t-u)\mathrm{e}^{-(t-\tau)^2 f^2/2-(u-\tau)^2 f^2/2}e^{[-2\pi fi(t-u)]}dudt \\ &= \frac{|f|}{2\sqrt{\pi}}\sigma^2 \end{aligned} \tag{13}$$

where $h(t)$ is white noise with zero mean and standard deviation is σ^2, E denotes the expectation. From (13), we know that the MPS of white noise in the time-frequency domain has linear relationship with the frequency. If white noise is normally χ distributed with one degree of freedom (DOF), it seems reasonable that the MPS of the ST in the time-frequency domain is χ^2 distributed with two DOF. The hypothesis can be proved by the Monte Carlo statistical method.

First of all, let $x(t)=so(t)+n(t)$, where $so(t)$ is signal and n(t) is white noise. The null hypothesis H_0 and the alternative hypothesis H_1 are presented. Assume that

$H_0: x(t)$ is noise,

$H_1: x(t)$ included the interesting signal .

Give a significant level at μ and the confidence level is $1-\mu$, then calculate power spectrum of the noisy signal. Let $Z(\tau,f)=|S(\tau,f)|^2/|S(\tau,f)|^2_{mean}$. If the probability $P(Z(\tau,f)\le\chi^2_\mu)$ is an occurrence, we accept H_0 and reject H_1. Because of the distribution for ST power spectrum is χ^2 with two DOF, therefore,

$$P(Z(\tau,f)\le\chi^2_\mu)=1-\mu. \tag{14}$$

The accepted domain is

$$|S(\tau,f)|^2\le\frac{1}{2}|S(\tau,f)|^2_{mean}\chi^2(\mu) \tag{15}$$

whereas the rejected domain is

$$|S(\tau,f)|^2>\frac{1}{2}|S(\tau,f)|^2_{mean}\chi^2(\mu). \tag{16}$$

4. Simulation Experiments

Previous studies demonstrate that time-frequency filter efficiency depends on the inverse transform algorithms and the area in time-frequency domain. In this section, there is an example to evaluate the performance of the proposed method. Some synthetic test signals would be used to analyze and MSE in the time domain to evaluate the time-frequency filter performance.

Removing the white noise: In this example, a signal $x(t)=\cos(10\pi t/128)$ interfered by the white noise with zero mean and standard deviation 0.1 are plotted in Fig.1 (a) and Fig.1 (b). First of all, it is important to distinguishing signal from noise in the time-frequency. Fig.1(c) shows the ST spectrum of the noisy signal and statistic characteristics of the white noise. According to the (15), if the ST amplitude of the noisy signal greater than statistic characteristics of the white noise, it is reasonable to believe that the area contains all useful characteristic signals, otherwise, there are noises. Therefore, conventionally time-frequency filters are set the interesting areas to one and the others to zeros, as illustrated in Fig.1 (d). The filtered output signals after different IST are plotted in Fig.1 (e), which shows that the four ISTs have similar performance, but the proposed LT IST has advantage over the other ones.

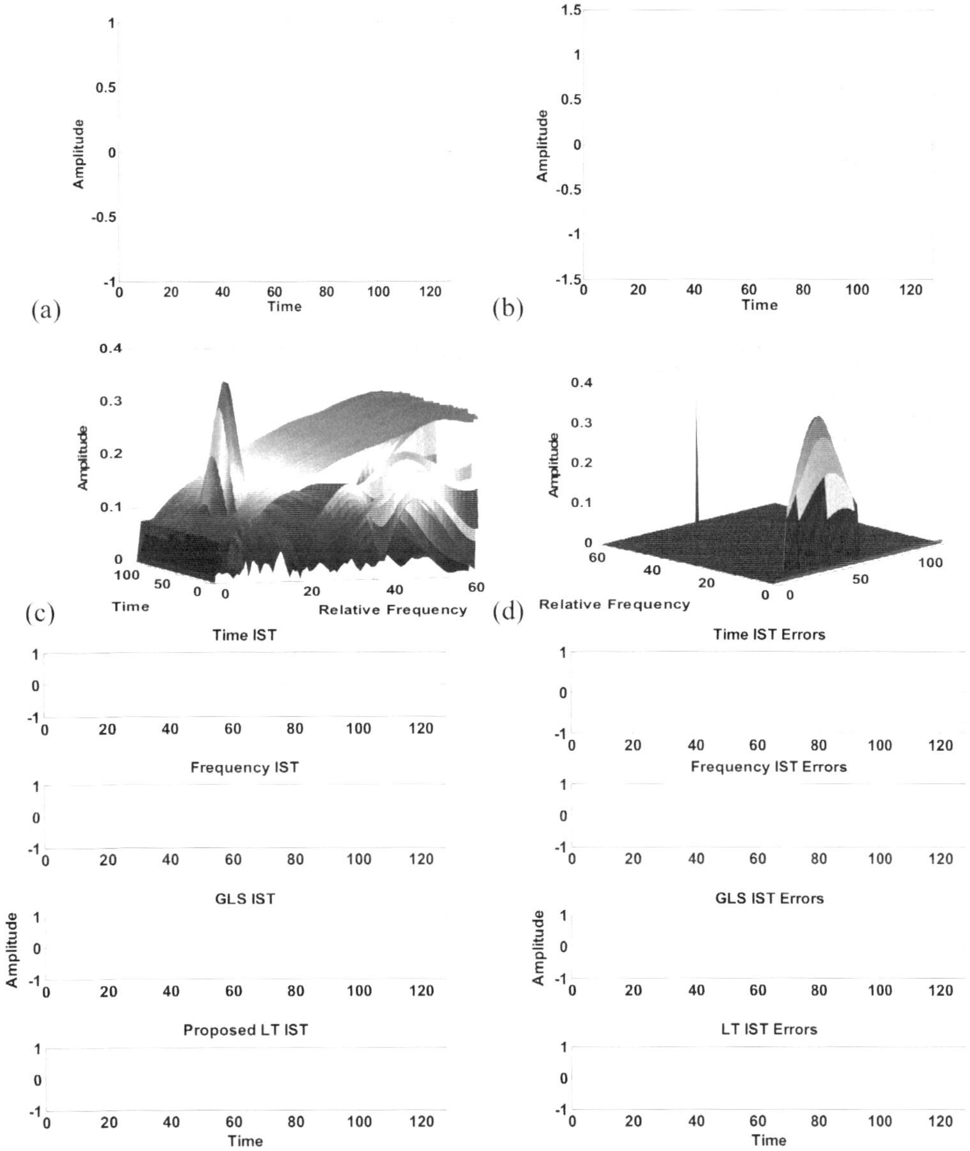

FIGURE 1. (a) The original noiseless signal. (b) The noisy signal in time domains interfered by white noise with zero mean and standard deviation 0.1. (c)Noisy signal in the S-transform time-frequency domain relative to MPS.(d) The filtered signal in the time-frequency domain.(e) The left is filtered signals after four different IST and the right is their errors.

5. Conclusion and Discuss

This paper presented a new strategy for the local matrix inverse *S*-transform. The LT IST was proposed, which can avoid restrict errors when the data processing involves a manipulation of special time in the time-frequency domain. The MPS of the white noise in ST was discussed. The results indicated that the local spectrum of the white noise follow a $0.5\chi^2$ distribution with two DOF. What's more, a filtering algorithm based on hypothesis testing was proposed, and the simulation results showed that the proposed algorithm is effective to distinguish signal from noise. In fact, during the realization process of time-frequency filters, errors will be always exist in the result because of time-frequency domain are modified and not the same as noiseless signal. The proposed LT IST algorithm is suit to extract specific signal in local time-frequency domain. If the specific frequency domain signal wants to be extracted, LT IST algorithm in matrix form can be modified. The corresponding column components of inverse transform matrix would multiply by one and others by zero.

Acknowledgments

This work was supported by the National Natural Science Foundation of China under Grant No. 61501162, No.51577046,the State Key Program of National Natural Science Foundation of China under Grant No. 51637004,the national key research and development plan "important scientific instruments and equipment development" Grant No.2016YFF0102200,China Postdoctoral Science Foundation Grant No. 2015M571926, Key Laboratory of Nondestructive Testing (Nanchang Hangkong University) Opening Foundation,Ministry of Education Grant No.JZ2015SJSY0018,CAST-BISEE Opening Foundation Grant No.W2016JSFW0066.

References

[1] J. H. Gao, W. C. Chen, Y. M. Li, and F. Tian, Generalized S-transform and seismic response analysis of thin interbeds,Chinese J. Geophys., vol. 46, no. 4, pp. 526–532(2003).
[2] C. R. Pinnegar, H. Khosravani, and P. Federico, Time-frequency phase analysis of ictal EEG recording with the S-transform,IEEE Trans. Signal Process., vol.56, no. 11, pp.2583-2593(2009).
[3] P. K. Dash, B. K. Panigrahi, and G. Panda, Power quality analysis usingS-transform,IEEE Trans. Power Deliv., vol. 18, pp. 406–411(2003).
[4] Z. Li, F. Ji, Z. G. Jie, and Z. L.Hua, Wavelet transformation for magnetocardiography signal, Chinese Physics B, vol. 54, no. 4, pp.1943-1949(2005).
[5] S. Ventosa, C. Simon, M. Schimmel, J. J. Dantildeobeitia, and A. Magravenuel, TheS-transform from a wavelet point of view,IEEE Trans. Signal Process. , vol. 56, no. 7, pp. 2771–2780(2008).
[6] I. Daubechies, The wavelet transform, time-frequency localization and signal analysis, IEEE Trans. Information Theory., vol. 36, no. 5, pp. 961-1005(1990).
[7] R.G. Stockwell, L. Mansinha, and R. P. Lowe, Localization of the complex spectrum: The transform, IEEE Trans. Signal Process., vol. 44, no.4, pp. 998–1001(1996).
[8] C. Simon, S. Ventosa, M. Schimmel, A. Helding, J. J. Danobeitia, J. Gallart, and A. Manuel, TheS-transform and its inverse: Side effects of discretizing and filtering, IEEE Trans. Signal Process., vol. 55, no.10, pp. 4928–4937(2007).
[9] C. R. Pinnegar, Comments on 'The inverseS-transform in filters with time-frequency localization', IEEE Trans. Signal Process., vol. 55, no. 10, pp. 5117–5120(2007).
[10] M. Schimmel and J. Gallart, The inverseS-transform in filters with time-frequency localization, IEEE Trans. Signal Process., vol. 53, no. 11, pp. 4417–4422(2005).
[11] M. Schimmel and J. Gallart, Author's reply to comments on 'The inverseS-transform in filters with time-frequency localization, IEEE Trans. Signal Process. , vol. 55, no. 10, pp. 5120–5121(2007).
[12] S. C. Pei and P. W. Wang, Novel inverseS-transform with equalization filter, IEEE Trans. Signal Process., vol. 57, no. 10, pp. 3858–3868(2009).
[13] M. Weishi, G. Jinghuai, Statistical denoising of signals in the S-transform domain, Computers & Geosciences., vol.35, no.6, pp. 1079-1086(2009)
[14] S. C. Pei and P. W. Wang, Discrete inverseS-transform with least square error in time-frequency filters, IEEE Trans. Signal Process., vol. 58, no. 7, pp.3557-3568(2010).
[15] Baiqiang, Yigang, Bing, YUAN, & Lifen.. An adaptive svd method for solving the pass-region problem in s-transform time-frequency filters. Chinese Journal of Electronics, 24(1), 115-123(2015)
[16] Yin B, He Y, Zhu Y. Detection and classification of power quality multi-disturbances based on

generalized S-transform and fuzzy SOM neural network. proceedings of the Chinese Society of Electrical Engineering, 35(4):866-872(2015).

Design and Implementation of Mobile Learning System for Soldiers' Vocational Skill Identification Based on Android

Jinqiang Ma

The Armed Police Academy, Langfang 065000, China;

E-mail:wjxy001@126.com

Abstract: To carry out the identification of the professional skills of the soldiers is to further promote the regularization of the needs of the fire brigade, in accordance with the "public security active forces soldiers professional skills identification implementation approach" to meet the needs of candidates for mobile learning to solve the paper learning materials bring a lot of inconvenience; This article uses the Android technology to develop a set of soldiers professional skills Identification Theory learning app, the learning software based on mobile learning, learning function is perfect, you can learn to practice, to achieve the goal of learning at any time, to enhance the soldier's post ability has a good practical value.

1. Introduction

The soldiers to carry out occupation skill appraisal is to further promote the standardization of building fire protection needs, is a new measure to strengthen the construction of public security forces personnel, to further improve the soldiers evaluation system, enhance the ability to post office soldiers, have important meaning to improve the combat effectiveness of the armed forces [1]; According to the measures for the evaluation of professional skills of soldiers in active service, the identification of soldiers ' professional skills is divided into different grades, such as primary, middle and advanced skills, technicians and senior technicians. Among them, the basic theory of fire fighters is the basis of each level, the basic theory of learning content rich, more knowledge, if the use of paper-based review materials to learn, not only waste a lot of paper, but also unable to grasp the knowledge of their correct assessment, lack of timeliness and effectiveness.

With the rapid development of mobile Internet technology, represented by smart phones and tablets enrich the mobile terminal equipment, mobile learning has gradually become a new way of learning [2]. Mobile learning is the combination of mobile computing and digital learning, it includes anytime, anywhere learning resources, powerful search capabilities, rich interactivity, strong support for effective learning and performance-based evaluation [3]. The mobile learning style is free of time and space limitation, flexible, highly targeted, and the absorption rate is higher. Android, as an open source, free mobile terminal operating system, is widely used in the education area. Here, the use of the Android development techniques developed to adapt to the mobile learning soldiers learn software professional skill appraisal theory, using the software learners can use of fragmented time learning anytime and anywhere, and can understand their own learning situation in time, improve the test efficiency.

2. System Analysis

The assessment of professional skill appraisal of firemen is divided into three parts: theoretical knowledge examination, skill performance appraisal and fitness assessment; Among them, the theoretical knowledge test took the written examination method, the percentage system was adopted, the

Content from this work may be used under the terms of the Creative Commons Attribution 3.0 licence. Any further distribution of this work must maintain attribution to the author(s) and the title of the work, journal citation and DOI.
Published under licence by IOP Publishing Ltd

above 60 is qualified, the skill operation appraisal adopts the percentage system, the above 80 is qualified, the physical examination adopts the standard system; The above three measures are certified as qualified[4]. The theoretical knowledge is divided into fire fighting occupational ethics and legal regulations, basic knowledge, fire fighting, emergency rescue and combat duty in five articles, altogether 26 chapters. The software is mainly suitable for firefighters occupation skill appraisal theory knowledge, learners can use this software to choose the relevant sections of online learning materials presented in the form of multimedia, and can record their own learning process; at the same time support online simulation test.

Table 1 The attributes of simulated test questions

Classification	Illustration
Knowledge point	According to the requirements of the outline is divided into 5 chapters, a total of 26 chapters
Question types	Single topic, Multiple choice, Judgment
Examination difficulty	Simple, Intermediate, Difficulty
Importance	Understand, General, Important

3. Development Platform

Android application development requires a specific development environment support, so the mobile learning system before the development of the function, you need to build the system development environment. The system uses integrated development tools Eclipse, the operating environment for Windows7, the program development language for the Java language.

3.1 Eclipse

Eclipse is Java based, open source, extensible development platform, Eclipse comes with a standard plug-in set, including the Java development tool (Java Development Tools, JDT). Eclipse software can be integrated software development tools of different vendors products, any development tool vendors can put their tools or components added to the Eclipse platform, users can use different tools through the same interface. Therefore, Eclipse platform has not only been supported by many developers and emerging small and medium-sized software companies, but also supported by large software companies such as IBM, Sun, Oracle and so on. Currently, Eclipse supports the Windows operating system and the Linux operating system, so you can easily switch between two operating systems.

3.2 MySQL

MySQL is a relational database management system, MySQL software because of its small size, fast, low total cost of ownership, in the development process of the system, the choice of MySQL Front database development software, it is a small application management MySQL, Key features include multi-document interface, syntax highlighting, dragging databases and tables, editable / addable / deleted fields, editable / insertable / erasable records, displayable members, executable SQL scripts, Interface with external programs, save data to CSV files, and so on.

3.3 APK

APK full name is Androgenic, is the Android installation package. When installing, use the simulator to execute the file for installation. APK format is zip, after decompression to see the dec file, is the implementation of Andrews procedures, rather than bytecode. Before the operation of the program is the need for Unzip bedding, and then directly connected to the relevant procedures.

IOP Conf. Series: Materials Science and Engineering **242** (2017) 012119 doi:10.1088/1757-899X/242/1/012119

4. System Design

4.1 Software Architecture

The software architecture shown in Figure 1. The mobile client learning system sends the HTTP network connection request to the back-end server, and the back-end server program accesses the MySQL database to obtain the required data to return to the client. The client communicates with each other through the WEB server and the database.

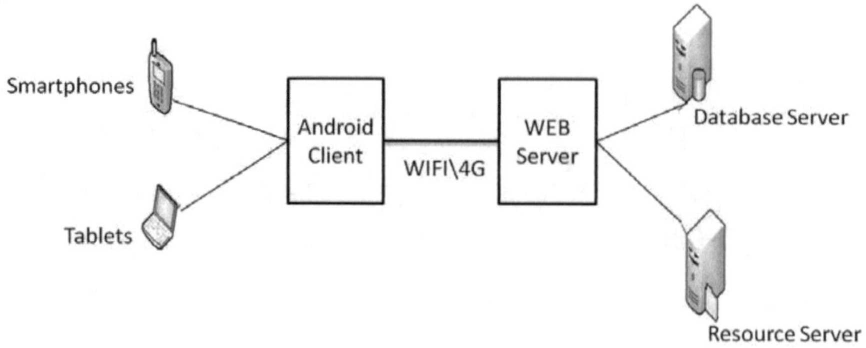

Fig 1 System architecture

4.2 System Function

The software is divided into client module and server management module, as shown in figure 2.

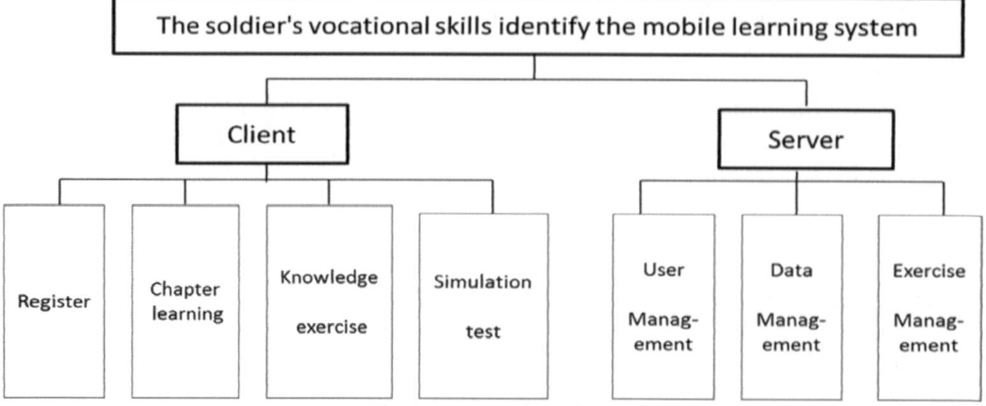

Fig 2 System function

The client user can easily retrieve the content of the resource that he needs through the system query module, while the user can collect and review the resource content and comment and evaluate the contents. The server management module is used to add, modify, delete, query, etc. At the same time, the administrator can manage the user's login user account and resource classification. Including the system user management module, the student information management module, learning information management module, the practice test module and curriculum resources management module, etc., each module can be the content of the management to add, modify, delete, and query operations. Student users through a web side or Android mobile terminal after login system, can undertake courses available information, resources, and comment on the operation, such as the administrator to the user's evaluation, curriculum resources and student information management.

4.3 Database Design

Mobile learning system needs to store a large amount of offline data and network data, and it needs to select database and data storage mode according to the corresponding requirements. The design of database plays a very important role in the development of the system, and it is also an important component of the system design. According to the function design and the type and characteristics of the data in the system, SQLite and MySQL are used as the two types of database. In order to facilitate students' learning off-line and can obtain the corresponding learning resources without a network, the learning curriculum resource system data and database data storage and exercise test in the SQLite database and server network; and the need to exchange data, including user registration and login data and interactive communication module the data is stored in the MySQL database.

4.4 Random Sampling Algorithm

In this system, an improved random sampling algorithm is designed. The random problem algorithm based on the attribute of the knowledge question bank is used to supplement the constraints of difficulty, knowledge and importance in each type. The corresponding question number in the database is extracted and stored in an array. A random numbering algorithm is called in the set of arrays of each question type to generate a random number corresponding to the number of questions requested. Road, more than 10 questions, 30 questions to determine the principle of the deposit into the corresponding array. After the random number is generated, the type of problem is generated. The process of the algorithm is similar to that of the manual process. The operation is simple and convenient. Through the control of the constraint condition, the problem of the same knowledge point can be avoided and the problem of the knowledge point Can be pumped.

4.5 Interface Design

Human computer interface is the medium for transmitting and exchanging information between human and computer. The quality of human computer interface influences the user's sense of software. It has become one of the standards to measure the usability of software, and the good man-machine interface has become an important aspect of software design. In the design of mobile learning system interface, we should take full account of the characteristics of mobile devices and screen size and other factors, in order to better presentation of learning content, and to minimize the memory space occupied by the system. At the same time, it is necessary to consider the style and beauty of the interface, so as to better attract learners' interest in learning, and thus enhance the learners' visual experience of the learning system. The interface of the learning system is divided into client and server, and the client is the course study page.

The server completes the information interaction between the client and the database, and plays an intermediary role. Management login into the background system, you can view user management, data management, chapter management, title management and other functional modules of operation.

5. Conclusion

Design and develop mobile learning system based on Android platform is humanized, responsive, powerful, rich in learning, and provides a platform for soldiers to use. The software can make better use of scattered time and learn anytime and anywhere Professional skills knowledge, improve learning efficiency.

Reference

[1] Steele,J.& Nelson.T.The Android Developer's Cookbook:Building Applications with the Android SDK[M].America： Addison-Wesley Professional， 2012
[2] Roger Ye.Embedded Programming with Android[M].America,2015.
[3] Reto Meier.Professional Android 4 Application Development[M].Canada,2012.
[4] Blanchet B. Escape analysis for object-oriented languages: application to Java[J]. Acm Sigplan Notices, 2015, 34(10):20-34.
[5] Li Y, Long Q, Lui T, et al. Partition-based method and apparatus for diagnosing memory leak in java systems: US, US8775769[P]. 2014.
[6] Marko Gargenta.Android.Learning.Android[D].Published by Oreilly Media， 2011(3):24-45.

IOP Conf. Series: Materials Science and Engineering **242** (2017) 012120 doi:10.1088/1757-899X/242/1/012120

A Sustainable City Planning Algorithm Based on TLBO and Local Search

Ke Zhang[1,b)], Li Lin[1,c)], Xuanxuan Huang [1], Yiming Liu [1], and Yonggang Zhang [1,2,a)]

[1]College of Computer Science and Technology, Jilin University, Changchun 130012, China

[2] Key Laboratory of Symbol Computation and Knowledge Engineering（Jilin University）, Ministry of Education, Changchun 130012, China

E-mail: [a)] zhangyg@jlu.edu.cn, [b)]monicazhk@163.com, [c)]xiaoliang102@vip.qq.com

Abstract. Nowadays, how to design a city with more sustainable features has become a center problem in the field of social development, meanwhile it has provided a broad stage for the application of artificial intelligence theories and methods. Because the design of sustainable city is essentially a constraint optimization problem, the swarm intelligence algorithm of extensive research has become a natural candidate for solving the problem. TLBO (Teaching-Learning-Based Optimization) algorithm is a new swarm intelligence algorithm. Its inspiration comes from the "teaching" and "learning" behavior of teaching class in the life. The evolution of the population is realized by simulating the "teaching" of the teacher and the student "learning" from each other, with features of less parameters, efficient, simple thinking, easy to achieve and so on. It has been successfully applied to scheduling, planning, configuration and other fields, which achieved a good effect and has been paid more and more attention by artificial intelligence researchers. Based on the classical TLBO algorithm, we propose a TLBO_LS algorithm combined with local search. We design and implement the random generation algorithm and evaluation model of urban planning problem. The experiments on the small and medium-sized random generation problem showed that our proposed algorithm has obvious advantages over DE algorithm and classical TLBO algorithm in terms of convergence speed and solution quality.

1 Introduction

City has been a significant space for human activity since it was formed. For gathering substances, wealth and technology among a certain range of terrain, city gradually becomes the center of economic trade. Prosperous urban also gains opportunity to develop and promote. However when human went into the 21st century, series of severe problems of city development came along[1]. Firstly, urban population exploited from 10% of world total population in 1800 to 15% in 1900, and it will be 50% or more in 2000 as expected. With the expanding of city scale, earth is bearing grand pressure. Then, interaction and accumulation of urban environment, economy and social issues of different historic period tensing the city problems that are already in serious situation. Therefore, in this sense, only when city performs sustainable, will country or even the whole world develop sustainable. In the long run, studying how to design more sustainable city is the core issue in society development realm[2,3].

Inspired by the achievements artificial intelligence acquired in solving complex engineering problems, multiple artificial intelligence theories and methods start applying to systematically study on city development and sustainable city planning design. According to data published by United Nations'

Content from this work may be used under the terms of the Creative Commons Attribution 3.0 licence. Any further distribution of this work must maintain attribution to the author(s) and the title of the work, journal citation and DOI.
Published under licence by IOP Publishing Ltd

projections, global urbanization rate will raise around 70% by 2050 [4]. As a result, crucial questions arise on how to develop conditions for a balance. Take China as an example, *The Chinese Dream: A Society Under Construction* written by Mars. N and Hornsby. A. says that China plans to annually create 20 whole new cites from now to 2020, around one million inhabitants each, to accommodate farmers in urban environments[5]. One famous instance of adopting artificial intelligence in city planning is a national research plan—SUSTAINS sponsored by French government. This project is aimed at getting urban planners and scholars in artificial intelligence together to plan newly built cities. The project focuses both on the provision of an interactive design tool which informs decision makers of the impacts of their choices, and an interactive communication tool on large tactile surfaces for public engagement.

The process of designing whole new cities is by nature a collaborative endeavor gathering urban planners and decision makers around a coarse-grain map of a territory on which to place urban shapes such as centers, industries, housings, commercial units, public equipment etc. The number of elements as well as their spatial layout needs to be strongly guided by a collection of rules related to social, economic, energy, mobility and sustainability issues. Belin from University of Nantes and his partners brings out the visualized intelligent city planning tools cover four stages [6].

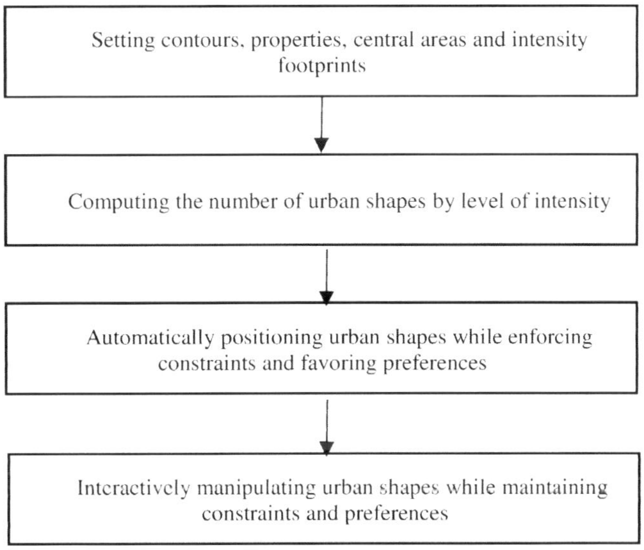

Figure 1. the visualized intelligent city planning tools cover four stages

This paper's work focuses on the third stage, automatically positioning urban shapes while enforcing constraints and favoring preferences. In order to handle this issue, we propose a sustainable city planning algorithm based on TLBO and local search. We design and implement algorithm for creating random city planning task and the solution evaluating model, on basis of which, we carry out simulation experiment. The experiment result shows for middle and small scale of random data, our algorithm has notable advantage among differential evolution algorithm and original TLBO algorithm. Loosing completeness is not an issue for our application: it is pointless to pay a high computational cost to build optimal solutions, since these solutions will be modified by hand afterwards. We thus do not need optimality proofs.

2. Fundamental introduction of basic algorithms

In this section, every algorithm will use FLUSH function to refresh the original swarm. FLUSH is to compare the new swarm calculated from the original one with the old swarm, the original one, and if a member in new swarm performs better, the individual in new swarm substitute the old one in original swarm.

FLUSH

:*f(Xi)* is the performance of *Xi*

```
function Flush(class, oldclass)
for( i = 0 ; i < N ; i++ ){
  if  f(Xi) in oldclass is better then
        Xi in class←Xi in oldclass
  end if
}
return class
end function
```

2.1 Teaching-Learning-Based-Optimization (TLBO) Algorithms

TLBO algorithms (Teaching-Learning-Based-Optimization algorithms) is a novel swarm intelligence optimization algorithm proposed by Rao and his team in 2010[7]. It simulates the teaching processes both between teachers and students and between every two students in order to perform evolution by means of the 'teaching' process by teachers/students and 'learning' process by students. TLBO algorithm has few parameters and is easy to understand, also it possesses high efficiency in calculating precise solutions with strong convergence ability. Though it's not long since TLBO being proposed, it draws messes of attention by scholars and it is being wildly applied[8].

TLBO

```
: X is an array of students, which equals class , N is the number of students in class
: f(X_i) is the performance of X_i, teacher is the student with best f(Xi) in class
: mean is the average f(Xi) in class
: depth is the iterate times
: input : class
: output : class
oldclass←class, using oldclass to record the original class
mean←∑_{i=0}^{N-1} f(Xi)/N
teacher← MAX(f(X_i))
while depth not reached do
  for(i = 0 ; i < N ; i++)
     X_i learn from teacher
  FLUSH(class, oldclass)
  for(i = 0 ; i < N ; i++)
     choose random j ( j≠i) , X_i learn from X_j
  FLUSH(class, oldclass)
  choose new teacher and calculate new mean in class
end while
return class
```

2.2 Local Search

Local Search [9](LS) algorithm is a typical greedy algorithm, it peeks a best solution from the current solution domain as its temporary answer every time until it reaches a local optimal result. This meta-heuristic takes advantage of the structure of the problem in terms of constraints and variables to guide the search. Local Search has proven its efficiency on large and various instances. The implement of Local Search can traced back to late 50s and early 60s in 20 century[9]. In former Soviet Union, research in this area started from Zhuravlev, who putted forward the algebraic theory of Local Search, while in the meantime, Rasitrigin focused on Local Search in probability. Local Search succeeded in calculating TSP (Traveling Salesman Problem) problems in early time, and later it is selected to solve location problems, network designations, scheduling problems and configuration problems. However, only with Local Search don't guarantee the optimal result in global. The practical applications of Local Search mostly are searching compound schemes or decomposition scheme originated from multiple strategies and finding approximate solutions in complex discrete problems[10].

Local Search

: *depth* is the irritate times
: f is the global cost function, f_i is the cost function for variable V_i
: s is the current configuration
: T is the adaptive tabu list
: j is the index of the variable with the worst cost
1: $s \leftarrow V_k$ // k is a random value
2: $T \leftarrow \varnothing$
3: while *depth* not reached do
4: while T is not full do
5: For all i such that $V_i \notin T$ do compute $f_i(s)$ end for
6: Select V_j a variable for which $f_j(s)$ is maximum ;
7: Compute the cost f of the configuration obtained from s by swapping V_j with another variable
8: Select s_0 the configuration for which $f(s_0)$ is minimum;
9: Update T by removing its oldest variable ;
10: if s_0 can improve current solution s then
11: $s \leftarrow s_0$
12: else
13: $T \leftarrow T \cup V_j$
14: end if
15: end while
16: end while
17: return s

The basic idea is to compute the error function for each constraint, then combine for each variable the errors of all the constraints in which it appears, thereby projecting constraint errors onto the relevant variables. This combination of errors is problem-dependent. In our case, it is a weighted sum, so that the constraints can be given different priorities.

3. Algorithm based on tlbo and local search

3.1 TLBO_LS Algorithm

In order to solve the problem of urban planning, this paper proposes a sustainable city planning algorithm based on TLBO and Local Search, hereinafter referred to as TLBO_LS. TLBO_LS is based on the demand of simulating a more realistic teaching process. According to the above, the learning steps of TLBO only includes "learning from teacher" and "learning from students", whereas the practical situation, self-study is also an indispensable step in the teaching process. It can be seen that LS is an algorithm that changes the elements in a collection and does not change relationships between elements. Thus, the improved algorithm will use LS to simulate "self-learning" to enhance the efficiency of class performance improvement, in the meantime avoid the local optimal generation, so as to get a more appropriate solution. The specific procedure of TLBO_LS is explained as follows.

TLBO_LS

: X is an array of students, which equals *class*, N is the number of students in *class*
: $f(X_i)$ is the performance of X_i, *teacher* is the student with best $f(X_i)$ in class
: *mean* is the average $f(X_i)$ in class
: *depth* is the iterate times
: input : *class*
: output : *class*
1: *oldclass* $\leftarrow$ *class*, using *oldclass* to record the original class

$mean \leftarrow \sum_{i=0}^{N-1} f(Xi)/N$
$teacher \leftarrow MAX(f(X_i))$
while *depth* not reached do
for($i = 0$; $i < N$; i++)
X_i learn from *teacher*
FLUSH(*class*, *oldclass*)
for($i = 0$; $i < N$; i++)
choose random j ($j \neq i$) , X_i learn from X_j
FLUSH(*class*, *oldclass*)
for($i = 0$; $i < N$; i++)
X_i learn by itself using Local Search
FLUSH(*class*, *oldclass*)
choose new *teacher* and calculate new *mean* in *class*
end while
return *class*

Comparing to the LS, TLBO_LS not only needs to search around the individual, but also calculate with other individuals in the population inducing the possibility to jump out of local optimality, then find a better solution. Compared with TLBO, TLBO_LS can broaden the scope of the calculation among the swarm individuals, so that the content able to be searched distributes in a larger interval. Therefore, the probability of finding a better solution will be increased. With appropriate size of swarm and iteration times will also lead TLBO_LS to find a more optimistic solution for large-scale data. Similarly, differential evolution (DE) is an algorithm for calculating the interaction between individuals within a population. However, the computational cost of DE is prohibitive. Since the computational efficiency of DE is influenced by various parameters, TLBO_LS possesses more advantages.

3.2 Constraint Model for City Planning Problem

TABLE 1 Parameters and descriptions

Parameters	**Descriptions**
City	Planning object
N*M	Planning object size.
S	Planning results
N_A	Total number of architecture types
$A_i(i=1,2,3,\ldots,N_A)$	Architectures of *i* type
n_i	Number of Ai planned to be built.

In the Table 1 listed five parameters used in our model, here are their specific explanations. In order to quantify the results clearly and straightforward, this paper makes the following assumptions about the model: the planned city can be approximated as a rectangle. In our model, in consideration of actual living conditions, common buildings are divided into 12 types: open space, residential, administrative office buildings, educational buildings, medical buildings, commercial buildings, theatres, gymnasium, hotels, traffic hubs, parks, factories. While calculating, an A_i values i. The model completes the planning task for newly built city using the parameters mentioned above in the Table 1. First, the size of the city is settled when user enters N and M. Next user enters n_i *as* the expected number of Architecture A_i according to user's demand. After the execution of the whole algorithm, the planning result ***S*** is obtained.

Assuming that the constraints are divided into two categories:

a) It is acknowledged that an architecture A_j ($i \neq j$) is not expected to be set within a certain range of architecture A_i. For example, a factory is built near the house or a business area is located near the school.

b) It is inappropriate to see that an architecture A_i does not appear in a certain area of the architecture A_j ($i \neq j$). For example, no hotel is set near the hospital or no school is established near the house.

3.3 City Planning Problems Solving

In order to model the practical problems, the specific variables will be given corresponded practical significance. ***Class*** will be a set of various planning results $X_i(i \in [1, N*M])$; $f(X_i)$ is the cost of each X_i; the most appropriate with lowest cost planning result X_t will become ***teacher***; and the ***mean*** is a plan with around average cost of the ***Class***. The total number of architectures' kind is N_A

This algorithm is based on users' needs, that is, corresponding number of various types of building A_i is already determined by customers before calculating urban planning solutions. Therefore, to compute among ***Class*** and meet the demand at the same time, this paper puts forward a method in calculating every building's location. The specific process states as follows:

I. X_i uses one-dimensional array with capacity of $N*M$ to demonstrate. The value j on a position k in this one-dimensional array means a building A_j is placed at the position k in this plan X_i.

II. X_i learns to X_j following these regulations. First, initialize temporary result X_i' as NULL. While traversing X_j, at the position k_j finds the m-th A_l of X_j. Meanwhile, in X_i search for the m-th A_l of X_i, and learn that its position is k_i. Then do learning calculations between k_i and k_j. When X_j is ***teacher***, use the computational method of *learning from teacher*. Or if X_j is a student in ***Class***, choose the *learning from students* way to compute. The new position k_i' is obtained by using the rounding and modulo calculation on the output of *learning* computing. In the temporary planning X_i', if there is an already existed building at the position k_i' in X_i', then $k_i' = k_i'+1$ (mod $N*M$) until there is no building on k_i' in X_i'. Then, A_l is settled down on the k_i' in X_i'.

a. learning from teacher

When X_i is trying to learn from ***teacher***:

(1) First, according to original TLBO, we need to define the *TF* (*teaching factor*), a teaching factor that decides the value of ***mean*** to be changed. *TF* is computed from r, a random number in the range of (0,1), and TF is chosen either 1 or 2, which is again a heuristic step and decided randomly:

$$r = \text{rand}(0,1) \tag{1}$$

$$TF = round[1 + r] \tag{2}$$

(2) Next, the *difference[k]* is required as describing the difference of the value at position k between ***teacher*** and ***mean***.

$$difference[k] = \text{r} * (\textbf{teacher}[k] - TF * \textbf{mean}[k]) \tag{3}$$

(3) This *difference[k]* modifies the existing solution according to the following expression, the total number of architectures' kind is N_A

$$X_{new,i}[k] = \text{round}\ (X_{old,i}[k] + difference[k] + N_A) \bmod N_A \tag{4}$$

b. learning from students:

When X_i is trying to learn from X_j:

(1) First, r_i is needed to simulate the learning ability of X_i:

$$r_i = \text{rand}(0\ ,\ 1) \tag{5}$$

(2) Then, X_i learn from X_j according to the follow expression:

$$X_{new,i}[k] = (\ X_{old,i}[k] + r_i * |\ X_{old,i}[k] - X_{old,j}[k]\ | + N_A\ ) \bmod N_A \tag{6}$$

III. Self-study of X_i will reference to the idea of Local Search. Firstly, initialize a temporary empty plan X_i'. Traversing X_i to place the current A_l building x on the position k', which is obtained by calculating original position k of current A_l building x using *learning by itself*. k will compute with a random number and then output the value k' within domain $\boldsymbol{\delta}$, the new position k' is obtained by rounded and modulo operation on original result k'. In the temporary planning $\boldsymbol{X}_i'$, if there is an already existed building at the position k_i' in $\boldsymbol{X}_i'$, then $k_i' = k_i'+1$ (mod $N*M$) until there is no building on k_i' in $\boldsymbol{X}_i'$. Then, A_l is settled down on the k_i' in $\boldsymbol{X}_i'$.

learning by itself:

When $\boldsymbol{X}_i$ is trying to learn by itself:

(1) First, $r_{i,k}$ is needed to ensure the new position of the architecture on $\boldsymbol{X}_{old,i}[k]$:

$$r_{i,k} = \text{rand}(\ \boldsymbol{\delta}\ ) \bmod N*M \tag{7}$$

(2) Then, $\boldsymbol{X}_i$ learn by itself according to the follow expression:

$$\boldsymbol{X}_{new,i}[r_{i,k}] = \boldsymbol{X}_{old,i}[k] \tag{8}$$

4. Experimental results and analysis

4.1 Experimental Design

In order to evaluate planning results, we designed an evaluation model applying scoring methods. Score will be accumulated according to the number of violated constraints. Specific calculating regulars based on constraints states as follows:

(1) If there is a kind of building $A_j(i{\neq}j)$ in a certain range of building A_i, it is scored from 5 to 1 according to the degree of intolerance in real life. For example, if the factory stands near the house, that is unbearable, we assume it values 5 points; if the school is established near business district, though it is unreasonable, we can tolerate it, then it weigh 3 points.

(2) If there is no building $A_j(i{\neq}j)$ in a certain range of building A_i, it is scored from 5 to 1 according to the degree of intolerance in real life. For example, if there is no hotel in the vicinity of the hospital, that is unbearable, we give it 5 points; if there is no school near the residential area, though it is unreasonable, we can tolerate it, then it weigh 3 points.

Judgment of result depends on its score, the lower the score, the better the plan. In this paper, a variety of comparative tests were performed on different sizes (10x10, 50x50) grids to assess the quality of the results and the actual computational costs. 10x10 grid represents 100 square kilometers, which is roughly the size of a county in China. 50x50 grid represents 2500 square kilometers, that is about a small and medium sized city in China[12]. Due to the planning task focuses on the initially opened up city, we do not need to consider the architectural factors already exist in city. Based on the planning data of Marne-la-Vallée from Bruno[6], we allocate all kinds of buildings and areas referring to the needs of various types of architectures. We will use the data in Table 2 to conduct experiment. In order to observe the efficiency of TLBO_LS algorithm, we will use TLBO, DE [11] to calculate the same data, and evaluate the planning effect of these three algorithms respectively. Experimental environment is i5-4210U 1.70GHz CPU, Windows 7 (32-bit) operating system and Java programming language. Experiments will test swarms of 100 population in 100 times of iteration by TLBO_LS, DE, TLBO respectively.

TABLE 2 Experimental Data

Urban Size	**Number of Architecture**
10*10 Small-scale	open space*7, residence*20, administrative office building*10, school*5, hospital*8, shopping plaza*5, theatre*5, sports center*5, hotel*5, traffic hub*5, park*20, industrial factory*5
50*50 large-scale	open space*200, residence*500, administrative office building*200, school*100, hospital*200, shopping plaza*150, theatre*100, sports center*100, hotel*150, traffic hub*100, park*500, industrial factory*200

IOP Conf. Series: Materials Science and Engineering **242** (2017) 012120 doi:10.1088/1757-899X/242/1/012120

4.2 Experimental Results and Analysis

The experimental results of DE/TLBO/TLBO_LS are evaluated by five parameters. *Initial Cost* is the cost of the initial urban planning; *Final Cost* represents the planning result's cost after using algorithms; *Time* is the total time consumed during calculation; *Reduction* is how the initial planning being improved, and the formula is:

$$Reduction = Initial\ Cost - Final\ Cost \tag{9}$$

The last parameter is *Convergence Rate* used for evaluating convergence rate of the algorithm with the expression below:

$$Convergence\ Rate = \frac{Time}{Reduction} \tag{10}$$

4.2.1 Experimental Result and Analysis of Small-Scale City

TABLE 3 Results of Testing Small-scale Data Using DE

Initial Cost	Final Cost	Time (ms)	Reduction	Convergence Rate (Time/Reduction)
377	93	23121	284	81.41
385	105	21535	280	76.91
377	76	22377	301	74.34
369	110	20758	259	80.15
364	157	20409	207	98.59
396	157	21098	239	88.28
396	210	22215	186	119.44
391	161	21371	230	92.92
384	174	21556	210	102.65
372	99	23560	273	86.3

TABLE 4 Results of Testing Small-scale Data Using TLBO

Initial Cost	Final Cost	Time (ms)	Reduction	Convergence Rate (Time/Reduction)
398	196	1036	202	5.13
375	188	1017	187	5.44
380	194	1048	186	5.63
391	192	1069	199	5.37
377	205	1055	172	6.13
370	193	1043	177	5.89
378	203	1090	175	6.23
371	189	1040	182	5.71
363	188	1025	175	5.86
388	197	1054	191	5.52

TABLE 5 Results of Testing Small-scale Data Using TLBO_LS

Initial Cost	Final Cost	Time (ms)	Reduction	Convergence Rate (Time/Reduction)
374	170	1843	204	9.03
374	162	1855	212	8.75
373	156	1846	217	8.51
390	156	1812	234	7.74
370	167	1810	203	8.92
387	169	1874	218	8.60

369	191	1827	178	10.26
379	190	1835	189	9.71
401	159	1806	242	7.46
371	172	1843	199	9.26

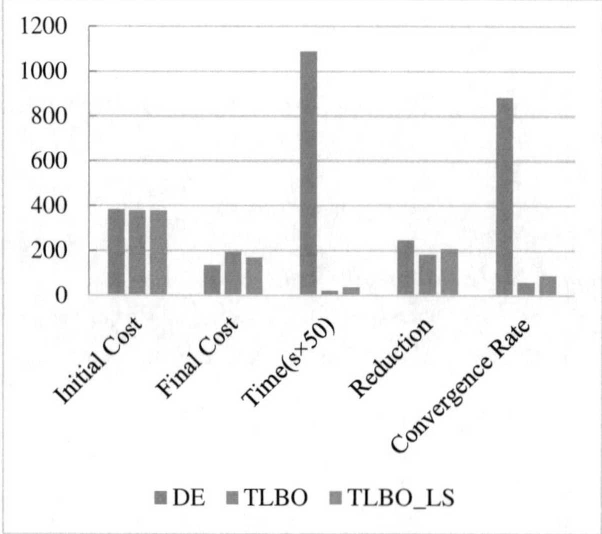

Figure 2 Comprehensive Analysis of Planning Results of Small-scale City

Referring to the Fig. 2, DE algorithm is the best one in the small-scale problem, for its promotion of the original population is significantly larger than the other two algorithms. However, it is so time-consuming that the required time reaches 10 to 20 times of the other two algorithms. As for TLBO algorithm, the quick speed and high promotion in the unit time is definitely outstanding. While the planning result of TLBO is the worst among three. TLBO_LS algorithm has the same advantages as the TLBO algorithm as similar time-consuming, and TLBO_LS can achieve as appropriate result as DE algorithms do in shorter time. In short, TLBO_LS algorithm is more suitable in solving small-scale task than the other two algorithms.

4.2.2 Experimental Result and Analysis of Large-Scale City

TABLE 6 Results of Testing big-scale data using DE

Initial Cost	Final Cost	Time (ms)	Reduction	Convergence Rate (Time/Reduction)
8243	6038	543534	2205	246.50
8290	6011	530389	2279	232.73
8300	6035	536070	2265	236.68
8279	5969	529212	2310	229.10
8282	5906	546572	2376	230.04
8330	6403	532223	1927	276.19
8357	5718	534037	2639	202.36
8333	5701	532607	2632	202.36
8175	6241	529058	1934	273.56
8366	5909	529067	2457	215.33

TABLE 7 Results of Testing big-scale data using TLBO

Initial Cost	Final Cost	Time (ms)	Reduction	Convergence Rate (Time/Reduction)
8285	5662	86890	2623	33.13
8304	5526	87194	2778	31.39
8223	5506	87375	2717	32.16
8258	5363	86565	2895	29.90
8218	5533	87344	2685	32.53
8291	5454	86845	2837	30.61
8299	5441	88817	2858	31.08
8234	5436	86598	2798	30.95
8355	5540	88777	2815	31.54
8226	5534	90193	2692	33.50

TABLE 8 Results of Testing big-scale data using TLBO_LS

Initial Cost	Final Cost	Time (ms)	Reduction	Convergence Rate (Time/Reduction)
8293	5194	350776	3099	113.19
8249	5168	342061	3081	111.02
8323	5205	341192	3118	109.43
8352	5184	322725	3168	101.87
8235	5108	336676	3127	107.67
8231	5010	341577	3221	106.05
8229	4963	360743	3266	110.45
8221	4853	364556	3368	108.24
8269	5761	326394	2508	130.14
8397	4466	340168	3931	86.53

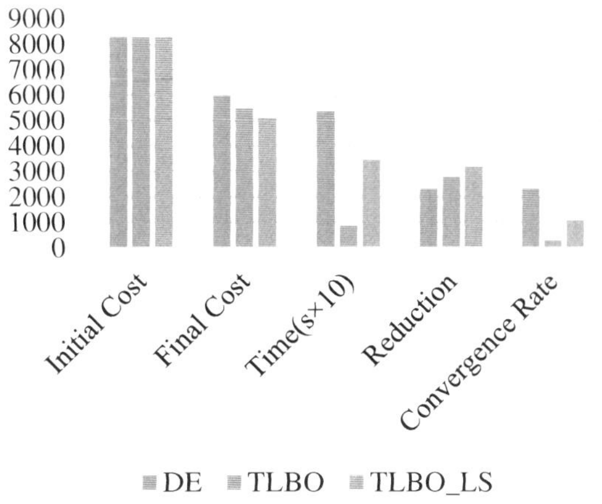

Figure 3 Comprehensive Analysis of Planning Results of Small-scale City

According to Fig 3, in the large-scale urban planning, the DE's results, which are the worst of the three planning results, are in the opposite of its performance in small-scale city. It's extremely time-consuming, and thus it is not suitable for large-scale urban planning. The advantage of TLBO's speed is obvious and it costs far less time than DE, comparing to the smaller scale situation. As a result, TLBO is the ideal planning algorithm in smaller scale. As the optimization of TLBO, TLBO_LS provides the best planning results in little time, and the planning procedure is highly efficient. To sum up, TLBO_LS is an excellent large-scale urban planning algorithm.

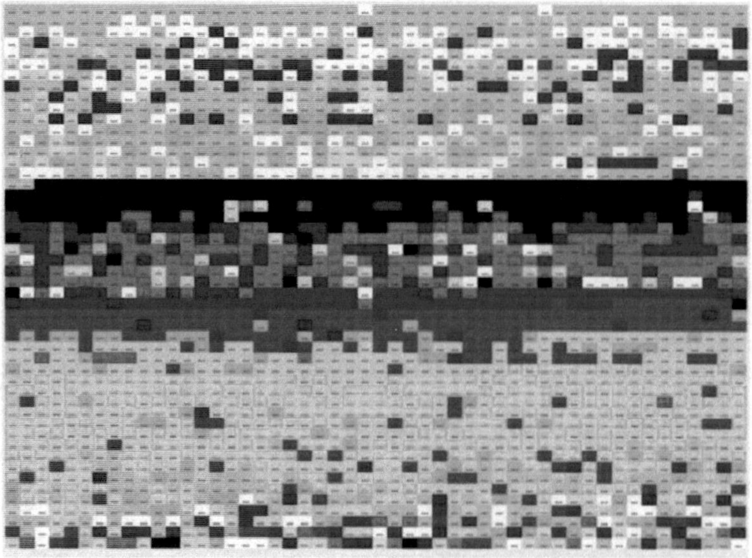

light grey: open space, grey: traffic hub, light blue: residence, dark blue: hotel,red: sports centre, dark red: administrative office building, orange: school, yellow: shopping plaza, green: park, purple: theatre, white: hospital, dark: industrial factory

Figure 4 A result of using TLBO_LS solving large-scale data instance

5. Conclusions

This paper proposes a constraint model based on city planning, and meanwhile, on the basis of the classical TLBO algorithm, this paper brings about a TLBO_LS algorithm combined with local search, designs and implements both the random generation algorithm and evaluation model of urban planning problem. The experimental results show that the proposed algorithm outperforms the best-known discrete differential evolution algorithms or TLBO and some evolutionary algorithms due to its outstanding convergence rate and quality of solutions. TLBO_LS shows promise for a reliable and efficient constraint optimization city planning problem.

Acknowledgments

This work is supported by the National Natural Science Foundation of China (61170314,61373052), the Project of Jilin Provincial Science and Technology Development(20170414004GH).

Zhang Ke, born in 1997. BSc (expected). Her major is network and information security.

Lin li, born in 1997. BSc (expected). His major is network and information security.

Huang Xuanxuan, born in 1996. BSc (expected). Her major is network and information security.

Liu Yiming, born in 1996. BSc (expected). His major is network and information security.

Zhang Yonggang, born in 1975. PhD, associate professor. Member of China Computer Federation. His main research interests include Constraint solving and Optimization.

References

[1] Ye Zhongnan. The Use Of Computer Simulation Technology In Urban Design[J]. Planners, 2014, 30(4): 40-46(in Chinese)

[2] LONG Ying, HAN Haoying, GU Yizhen, et al. Spatio-temporal Evaluation of Urban Planning Implementation[J]. PROGRESS IN GEOGRAPHY, 2011, 30(8): 967-977(in Chinese)

[3] Zhao Sidong, Ouyang Dong, Zhong Yuan. Intelligent City Planning's Effect To Urban Planning[J]. Planners, 2103, 29(2): 5-10(in Chinese)

[4] United Nations. Department of Economic and Social Affairs. World urbanization prospects: the 2007

revision. Highlights.[J]. New York New York United Nations Feb, 2007, 22(8):656-659
[5] Hornsby A. The Chinese Dream: A Society Under Construction[J]. Architectural Record, 2008, 196(7):55-55
[6] Belin B, Christie M, Truchet C. Interactive Design of Sustainable Cities with a Distributed Local Search Solver[M]// Integration of AI and OR Techniques in Constraint Programming. Springer International Publishing, 2014:104-119
[7] Rao RV, Savsani VJ, Vakharia DP. Teaching–learning-based optimization: a novel method for constrained mechanical design optimization problems[J]. Computer-Aided Design, 2011, 43: 303-315
[8] WANG Pei-chong, MA Yue, GENG Ming-yue, WANG Shen-wen. New Teaching-Learning-Based Optimization with Neighborhood Structure Based on Small World[J]. Computer Science and Exploration, 2016, 10(9): 1341-1350(in Chinese)
[9] Codognet P, Diaz D. Yet Another Local Search Method for Constraint Solving[M]// Random Algorithms: Foundations and Applications. Springer Berlin Heidelberg, 2001:73-90
[10] Caniou Y, Codognet P, Diaz D, et al. Experiments in parallel constraint-based local search[C]// European Conference on Evolutionary Computation in Combinatorial Optimization. Springer-Verlag, 2011:96-107.
[11] Storn R, Price K. DE-a simple and efficient heuristic for global optimization over continuous space[J]. Journal of Global Optimization, 1997, 114(4):341-359
[12] Wang Yang, Wang Sahojian, Qin Jin. Spatial evaluation of land urbanization level and process in Chinese cities[J]. Geographical Research, 2014, 33(12): 2228-2238(in Chinese)

IOP Conf. Series: Materials Science and Engineering **242** (2017) 012121 doi:10.1088/1757-899X/242/1/012121

Short-term data forecasting based on wavelet transformation and chaos theory

Yi Wang 1, a), Cunbin Li 1, b) and Liang Zhang1, c)

[1]Institute School of Economics and Management, North China Electric Power University, Beijing 102206, China

E-mail: [a] wyi@ncepu.edu.cn, [b] wyi@ncepu.edu.cn

Abstract. A sketch of wavelet transformation and its application was given. Concerning the characteristics of time sequence, Haar wavelet was used to do data reduction. After processing, the effect of "data nail" on forecasting was reduced. Chaos theory was also introduced, a new chaos time series forecasting flow based on wavelet transformation was proposed. The largest Lyapunov exponent was larger than zero from small data sets, it verified the data change behavior still met chaotic behavior. Based on this, chaos time series to forecast short-term change behavior could be used. At last, the example analysis of the price from a real electricity market showed that the forecasting method increased the precision of the forecasting more effectively and steadily.

1. Introduction

At present, there are many short-term data forecasting methods, such as regression model method and time series method. Not only the wavelet analysis, data mining and neural network are used in the modern forecasting methods, but also the emerging chaos theory provides a new way for the forecasting work. Literature [1-2] used the chaotic model to process the short-term power load forecasting. On the basis of chaos theory, the objective law of the data itself is used to predict the power load. literature [3] gave a short-term load forecasting method according to the phase space reconstruction of load series chaotic features. In literature [4-5], the data sequence was decomposed into low frequency part and high frequency part by wavelet decomposition principle. On this basis, different regression models were established for each sub-sequence and predicted. Literature [6] introduced the big data and data mining technology used in the power system. In literature [7], a short-term price forecasting method based on time series similarity in data mining was proposed.

The usage of data mining technology prediction often does not pay attention to the feature of data sequence itself. In the past, wavelet analysis and chaos theory predicts were mostly lack of the data preprocessing, some only make a simple adjustment of the data, and the "data nail" in the timing data reduced the accuracy of the forecast. Due to the impact of many uncertainties, sometimes data fluctuate with a high degree of non-linear and even chaos.

In this paper, wavelet transform and chaos theory are combined to propose a chaotic time series prediction method based on wavelet transformation. Firstly, according to the characteristics of the time series of data, the data loss method is used to reduce the data and then the maximum Lyapunov exponent is used. Finally, the validity and feasibility of the method is verified by analyzing the data of a real power market.

Content from this work may be used under the terms of the Creative Commons Attribution 3.0 licence. Any further distribution of this work must maintain attribution to the author(s) and the title of the work, journal citation and DOI.
Published under licence by IOP Publishing Ltd

2. Wavelet Transformation and Time Series Data Preprocessing

2.1 Wavelet Transformation and Haar wavelet

The main feature of wavelet transformation is that the non-stationary time series is transformed into a much more stationary time series than the original sequence. Haar wavelet is a popular wavelet transformation technique. Haar wavelet function is a set of functions that are orthogonal to each other. Haar wavelet is derived from this set of function sets, and it is a single rectangular wave which the support domain in the $t \in [0, 1]$ range. Because of $\int \psi(t)dt = 0$, and $\int t\psi(t)dt \neq 0$, the $\psi(\omega)$ in the point $\omega = 0$ only has one-degree zero. $\psi(t)$ constitutes a family of the simplest orthogonal normalized wavelet family in the multiresolution system of $a = 2^j$ [8], that is, $\psi(t)$ is not only orthogonal to $\psi(2^j t)$,but also orthogonal to its own integer displacement $\int \psi(t)\psi(t-k)dt = 0$ ($k \in z$). The Haar wavelet basis function is constructed as Eq. 1.

$$\psi(t) = \begin{cases} 1 & 0 \leq t \leq 0.5 \\ -1 & 0.5 \leq t \leq 1 \\ 0 & other \end{cases} \qquad (1)$$

The Haar wavelet basis function value is 1 in the range of [0, 0.5], -1 in the range of [0.5, 1], and 0 for other intervals. The original data affect the Haar wavelet function value, the sequence data value affects the same absolute value of positive and negative function value, these steps can offset the frequent repeated shocks in a very short period time of the original data, while the non-frequent shocks part of the data will not be offset and will be retained. This is the idea and principle of using Haar wavelet for time series data reduction.

2.2 Application of Wavelet Transformation in Time Series Data Preprocessing

The impact of the various risk factors in reality will inevitably be reflected in the historical data. Fig. 1 is the electricity price data from January 1 to August 31 of a power market, it can be seen that the price in some periods of great fluctuations constitute a "data nail." From the waveforms after the decomposition of the wavelet, it is observed that the abrupt changes in these prices lead to violent oscillations of the wavelet waveforms. Some of the changes are largely unpredictable due to the limited amount of information available, and the "data nails" also affect the other periods price forecast. In order to mitigate the adverse effects of these "data nails" in the forecast, it is necessary to properly preprocess the time series data.

First, statistics a period time to obtain the data interval in the probability of a certain probability; and then middle pass filter is used to compress the parts beyond the given range at an appropriate proportion. Compression uses monotonic increasing function, and does not change the size relation of data in time series.

Given $[B, T]$ is the data interval in the probability of a certain probability, P is the predicted value, then the middle pass filter function is listed as Eq. 2.

$$F(P) = \begin{cases} T + a\ln(P - T + 1) & P > T \\ P & P \in [B, T] \\ B - a\ln(B - P + 1) & P < B \end{cases} \qquad (2)$$

The filter function for P is continuous and monotonous, a is the experimental tunable parameter, and used to adjust the compression ratio of the data beyond the statistical interval part. Dealt with this filter function, the original size order relationship of the data sequence will not be destroyed, and the greater the data deviation from the statistical interval, the greater the compression ratio. With such preprocessing, the data is smoothed while maintaining the original size data relationship.

Wavelet decomposition techniques can reflect the regularity of sequences on different scales, and short-term predictions are mainly found in the history data. There are a lot of concrete and unpredictable factors in reality that work to give data rich in details. Such as sudden peaks or troughs, and these details create obstacles to reveal the general laws of data changes, and Fig. 2 is the data time series after wavelet reduction.

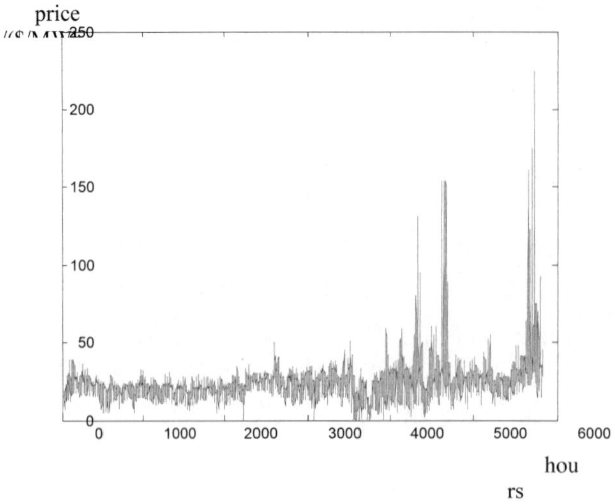

FIGURE1 Electricity price series

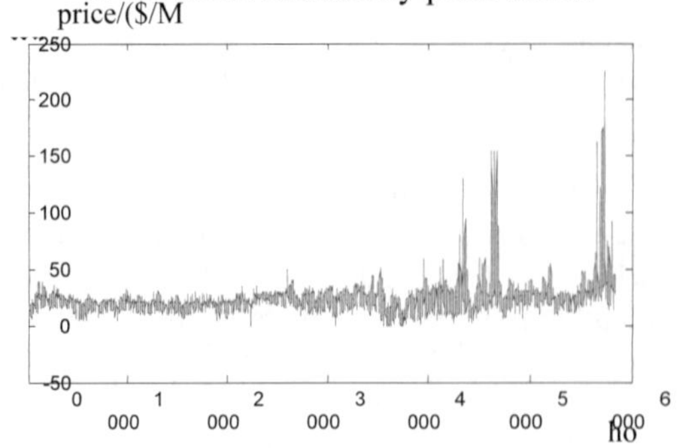

FIGURE2 data time series after wavelet reduction

3. Forecast Model of Chaotic Time Series Based on Wavelet Transformation

3.1 Phase Space Reconstruction Theory and Phase Space Reconstruction Parameters Selection

Packard et al originally proposed the theory of phase space reconstruction is to restore the chaotic attractors in high-dimensional phase space. The chaotic attractors reflect the regularity of the chaotic system, which means that the chaotic system will eventually fall into a particular trajectory. The evolution of any component of the system is determined by the other components that interact with it, so that the system itself can be predicted from the time series data of a component. The most commonly used reconstruction method is the delay coordinates vector method proposed by Packard in 1950. Takens proves that a suitable embedding dimension can be found, that is, if the dimension of the delay coordinate $m \geq 2d+1$ (d is the dimension of the power system), in this embedding dimension space, the regular trajectory can be recovered, that is, the trajectory in the reconstructed space R^m of the original power system keeps the diffeomorphism, thus laying a solid theoretical foundation for the prediction of chaotic time series [9].

a time series $x(i), i = 1,2,\cdots,N$, the appropriate time delay τ and embedding dimension m can be used to extend the time series into a phase distribution of a $M-$dimensional phase space:

$$Y_i = \{x(i), x(i+\tau), \cdots, x[i+(m-1)\tau]\} \qquad (3)$$

In Eq. 3, $i = 1,2,\cdots,M$, $M = N - (m-1)\tau$ and M is the number of phase points in the phase space.

In phase space reconstruction, the time delay τ and embedding dimension m have great significance, their values are also difficult to select. In practice, the methods of time delay are mainly autocorrelation method, complex autocorrelation method, mutual information method and C-C method. The embedded dimension is mainly obtained by GP algorithm, and the C-C method can be used to determine the time delay get the appropriate embedded dimension.

3.2 Chaotic Time Series Forecasting Method Based on Lyapunov Exponent

Chaotic time series forecasting is carried out directly according to the objective law calculated by the data sequence itself. The commonly used methods include the global forecast method, the local forecast method and the Lyapunov exponent method.

"Butterfly effect" is a classic example of chaos theory, as it describes, the orbits generated by two very close initial values are exponentially separated with time, and the Lyapunov exponent is a feature that quantitatively describes the phenomenon. It is a good predictor of the system as a feature to quantify the divergence of the initial orbit and estimate the chaotic level of the system. X_M can be set as the prediction central point, the nearest point of X_M in the phase space is X_k , the distance is $d_M(0)$, and the maximum Lyapunov exponent is λ_1 , that is:

$$d_M(0) = \min_j \left\| X_M - X_j \right\| = \left\| X_M - X_k \right\| \tag{4}$$

$$\left\| X_M - X_{M+1} \right\| = \left\| X_k - X_{k+1} \right\| e^{\lambda_1} \tag{5}$$

In the above equations, the point X_{M+1} only has one last component, so $x(t_{n+1})$ is predictable.

3.3 Forecasting of Chaotic Time Series Based on Wavelet Transformation

Before making short-term forecasting, the data is preprocessed, and the data is reduced by wavelet transformation, and it is integrated into the chaotic time series prediction process as shown in Fig. 3. The forecasting steps are as follows:

(1) Data preprocessing mainly includes data selection and data transformation. Choosing the appropriate data interval is an important step before wavelet transformation and chaos prediction.

(2) The prediction of chaotic time series is carried out after data reduction, and the time delay and embedding dimension are calculated in turn, then the correlation dimension and the largest Lyapunov exponent are used to complete the prediction.

(3) Analyze the results of the forecast and provide users with knowledge.

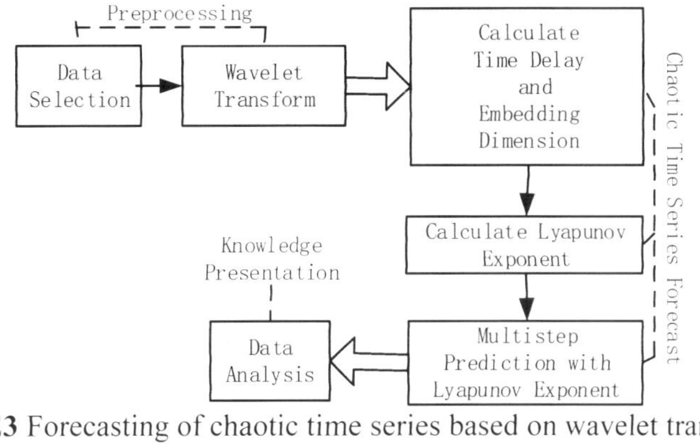

FIGURE3 Forecasting of chaotic time series based on wavelet transformation

4. Case Study

Using the data from January 1st to August 31st in a power market as a predictive data source, Fig. 4

shows the self-correlation method to determine the delay time τ =12, which is consistent with the results of the mutual information method. As shown in Fig. 5, the embedding dimension m=6 is obtained from the pseudo neighborhood point.

Under the known delay time and embedding dimension, the small data method [10] is used to find the maximum Lyapunov exponent λ =0.0921>0, and the chaotic characteristics of the data sequence are also proved. Generally, the longest forecast time is defined as $T_m = 1/\lambda$, it represents the longest time required to double the system state error. It can be used as one of the reliability indicators of short-term forecasts. By using of Eq. 5, the electricity price of a power market in September 1st is predicted in 10 steps. The forecast results are shown in Table1.

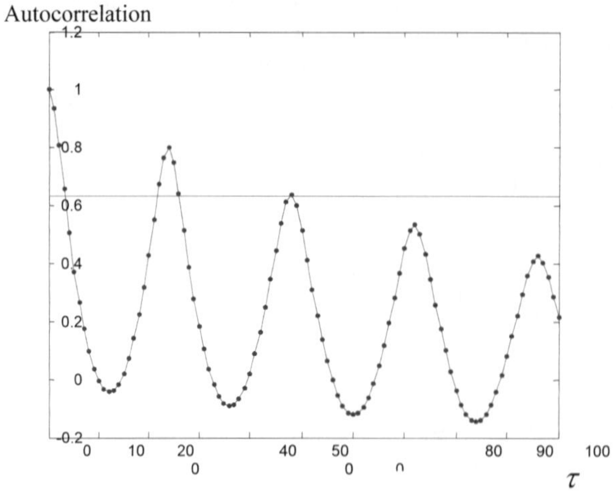

FIGURE4 the delay time τ calculated with autocorrelation method

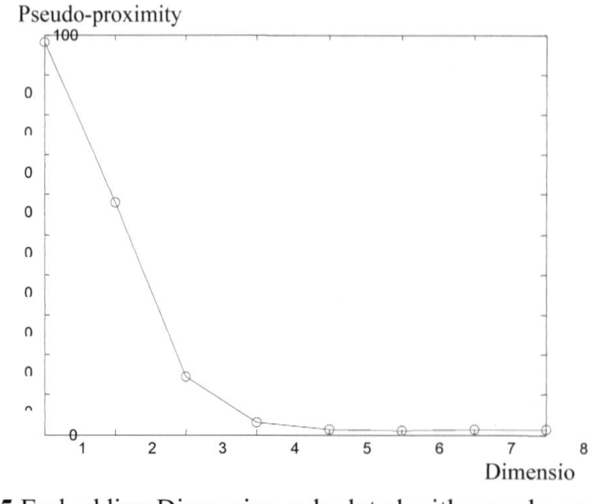

FIGURE5 Embedding Dimension calculated with pseudo-proximity

Table1 Forecasting results from 1:00 ~10:00 in September 1st

Time	Real value	Forecasting value	Relative error(%)
1	25.4453	24.4364	-3.96
…	….	…	…
10	32.999	36.062	9.28

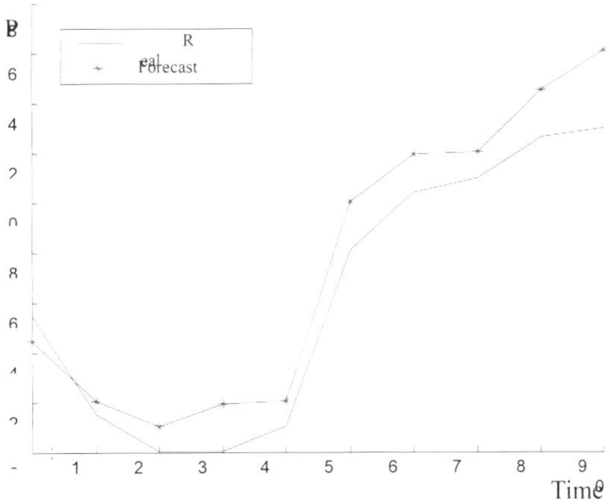

FIGURE6 The Forcast and real electric price in September 1

As can be seen from Table 1 and Fig. 6, the forcasting errors of the 10 periods are within 10%. The average error is 5.65% by calculation. Compared with the general model predicted by time series, the accuracy is very high.

5. Conclusions

In order to reduce the influence of "data nail" on the short-term forecasting, the Haar wavelet transform is used in this paper, and the data are preprocessed before forecasting. Wavelet analysis can highlight the characteristics of sequence in the frequency domain, but if only the removal of the reconstruction after details of the data as the forecasting data source, some of the missing information leads to the forecast error increases. In the process of wavelet reconstruction, the threshold of each layer of high frequency coefficients is set. The part that exceeds the threshold is defined as the singularity. The middle pass filter method is used to compress the data value at the singular point. It is found that this can effectively improve the forecasting accuracy.

In the forecasting of chaotic time series, the prediction method based on the maximum Lyapunov exponent method is adopted. The maximum Lyapunov exponent forecasting method is a very practical prediction method. It avoids the complicated calculation and error of advanced mathematics. The simple method is often successfully avoid the calculation of defects caused. At present, many scholars have put forward a lot of improved short-term forecasting methods for the maximum Lyapunov exponent, which is an incentive for the author of this paper.

Acknowledgment

This work is supported by National Natural Science Foundation of China (No.71671065) and the "Research on cascading failures and control strategies of physical integration system for power grid information" (No. 5204BB1600CP) of National Power Grid Corp science and technology project, all supports are gratefully acknowledged.

References

[1] Zhang Shuqing, Shi Rongyan, Dong Yulan, Li Pan, Ren Shuang and Jian Wanlu. 2015. Bivariate Wavelet Threshold Denoising and Improved Chaotic Forecasting Model and Its Application in Short-term Power Load Forecasting[J], Proceedings of the CSEE, 35(22) pp. 5723-5730.

[2] Li Dongdong, Qin Zishan, Lin Shunfu, Zheng Xiaoxia and Wang Tianxiang. 2015. Short-term Load Forecasting for Microgrid Based on Method of Chaotic Time Series[J], Proceedings of the CSU-EPSA, 27(5) pp. 14-18

[3] Gu Xi and Liao Zhiwei. 2017. Short-term Load Forecasting Based on Phase Space Reconstruction and Gaussian Process Regression [J], Power System Protection and Control, 45(5) pp. 73-79.

[4] Niu Lixiao, Wang Zhengfang, Cang Chuanzhi, Shang Wenli and Zhang Shengshan. 2014. Hybrid

IOP Conf. Series: Materials Science and Engineering **242** (2017) 012121 doi:10.1088/1757-899X/242/1/012121

model based on wavelet transform and ARIMA for short-term electricity price forecasting[J], Application Research of Computers, 31(3) pp.688-691.
[5] Xie Pinjie, Tan Zhongfu, Shang Jincheng, Hou Jianchao and Wang Mianbin. 2008. Short-Term Electricity Price Forecasting Based on Wavelet Transform and Generalized Autoregressive Conditional Heteroskedasticity Model Research on Short-term Electricity Price Forecasting[J], Power System Technology, 32(16) pp.96-100.
[6] Wu Keqi and Cheng Fan. 2016. Data Mining Technology and Its Application in Power System[J], Electrotechnics Electric, 6 pp.28-32.
[7] Yuan Guichuan, Cheng Li and Wang Jianquan. 2003. Electric Price Forecasting Using Data Mining, Proceedings of the EPSA,15(2) pp.19-22.
[8] Chai Yanling and Tang Zhitao. 2016. Application of Haar Wavelet Transform in Well-Logging Curve Similarity [J], Computer Technology and Development, 26(1) pp.119-123.
[9] Wu Huawen. 2014. Chaotic time series analysis and the research of its application in railway freight volumne prediction[D], China Academy of Railway Sciences.
[10] Rosenstein M T, Collins J J, and De Luca C J.A. 1993. Practical method for calculating largest Lyapunov exponents from small data sets[J]. Physica: D, 65 pp.117-134.

Efficient Grid-based Clustering Algorithm with Leaping Search and Merge Neighbors Method

Feng Liu[1)], Peng Wen[1)] and Erzhou Zhu[1, a)]

[1]School of Computer Science and Technology, Anhui University, Hefei 230601, China.

E-mail:[a)] ezzhu@ahu.edu.cn

Abstract. The increasing data size makes the research of clustering algorithm still an important topic in data mining. As one of the fastest algorithms, the grid clustering algorithm now still suffers from low precision problem. And the efficiency of the algorithm also needed improvement. In order to cope with these problems, this paper proposes an efficient grid-based clustering algorithm by using leaping search and Merge Neighborhood (LSMN). In the algorithm, the LSMN first divides the data space into a finite number of grids and determines the validity of the grid according to the threshold. Then, leaping search mechanism is used to find valid grids of the grid by retrieving all the odd columns and odd rows. Finally, if the number of valid grids is greater than the invalid grid, the invalid grids are merged together. In the algorithm, the time cost is reduced and the accuracy is improved by leaping search and re-judgment of the invalid grid mechanisms respectively. The experimental results have shown that the proposed algorithm exhibits relatively better performance when compared with some popularly used algorithms.

1. Introduction

With the advent of massive data era, it is urgently needed to accurately excavate usefulness data effectively. As a very important and useful technology in data mining, clustering technology can improve the efficiency of data mining. So, it still attracted many focuses from research and industry communities. Up to now, many clustering algorithms have been proposed. Generally, the categories of clustering algorithms can be divided into partition clustering algorithm, hierarchical clustering algorithm, density-based clustering algorithm, grid-based clustering algorithm and hybrid algorithm. Among all kinds of the clustering algorithms, the density-based clustering algorithm and the grid-based clustering algorithms are the two most important ones.

The density-based clustering algorithm is insensitive to noise points which can effectively remove noise points and process any shape of the cluster. DBSCAN [1] is a classic density-based clustering algorithm. DBSCAN is highly accurate, but the operating efficiency is low.

Generally, grid-based clustering algorithms are the most computationally efficient ones. The grid clustering algorithm first divides the data space into a finite number of grids. Then, the grids are divided into valid grids and invalid grids according to the input threshold parameters. Finally, all valid grids in the data space are searched in order and all neighborhood valid grids are merged into clusters. Since the grid clustering algorithm transforms the data points in processing into the grid, so the time complexity of the algorithm depends on the number of grids. When the number of grids is much smaller than data points, the time complexity of the algorithm will be significantly reduced. The classical grid-based clustering algorithms are STING [2] and CLIQUE [3].

Although the grid-based clustering algorithm is an important clustering algorithm, but the way that

Content from this work may be used under the terms of the Creative Commons Attribution 3.0 licence. Any further distribution of this work must maintain attribution to the author(s) and the title of the work, journal citation and DOI.
Published under licence by IOP Publishing Ltd

searching all the grids in the data space in order will seriously increase its execution time. Moreover, there could be some errors when the validity of the grid is only determined by the threshold. In order to improve the speed of grid searching and the quality of clustering, an efficient grid-based clustering algorithm, called LSMN, is proposed in this paper. By LSMN, the odd columns and odd rows in the grid are leaping search. Then the searched valid grids are taken as the starting points to merge the neighborhood valid grids. During this process, when the number of valid grids is greater than the invalid ones, invalid grids are merged together.

2. Lsmn clustering algorithm

Grid-based clustering algorithms always search for rows on a search grid, which undoubtedly reduces the efficiency of the algorithm. Aiming at remarkably reducing the grid searching time cost, this paper proposes a grid-based and density-based clustering algorithm by leaping searching and merging neighborhood (LSMN) grids. After dividing the data space into a grid according to the input parameter ε (grid step size), the input parameter Minpts (threshold) is used to determine whether the grid is valid or invalid. Then, the odd columns and odd rows among all the grids are sequentially searched. During this process, if an invalid grid is found, it is marked as a searched grid. When searching for a valid grid, its neighbor grids are put into a group, and 4 directions leaping grid searching is performed Depending on whether the grid is the valid one or been flagged, iterative searched grids are merged. In order to prevent the situation that leaping searched grids are all invalid grids, the LSMN provides an offset mechanism. By this mechanism, the directly neighbor valid grids become the center for continue grid searching. When all the odd columns and odd rows are searched, the data is clustered. The LSMN algorithm not only can reduce the grid search time but also reduce the chance that the grid is mistaken for an invalid grid.

2.1 Load Data Set and Input Parameters

At the beginning of the algorithm, the ε and Minpts parameters needed to be set. (1) Grid step size (ε): the purpose of the ε parameters is to divide the data set into non-overlapping grids. The appropriate ε parameters can improve the quality of the cluster. (2) The threshold (Minpts): the validity of the grid is determined by Minpts parameters. If the number of data points in the grid exceeds the parameter Minpts, then the grid is defined as a valid grid. Other grids are defined as invalid grid, and the points in the invalid grid are defined as noise.

2.2 Divide the Grid Space

The grid space is divided into non-overlapping grids by parameter ε, and the number of grids is the value of ε multiplies ε. Fig.1 shows all data points classified in the grids.

2.3 Filter Out Noise Points

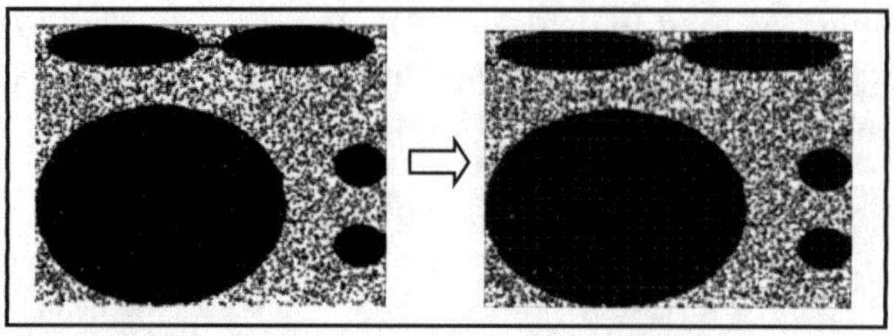

FIGURE 1. Dividing the grid space.

Fig.2 shows the original data set. The "valid" and "invalid" grids in the data space are judged according to the parameter Minpts and the point in the invalid grids is temporarily marked as noise points.

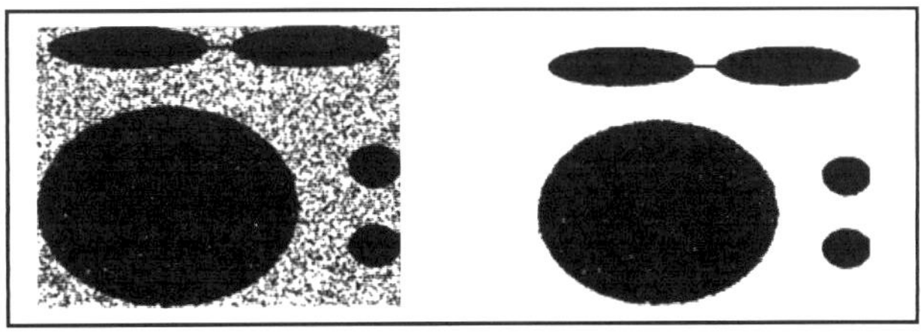

FIGURE 2. The original data sets and the data sets without noise data points.

2.4 Sequentially Search All Odd Columns and All Odd Rows of Grids

The searched grid may be valid grid or invalid grid. When the searched grid is invalid one, mark it and search for the next grid. When the searched grid is valid grid and not been marked, the neighboring all valid grids are merged with it. If the invalid grid is less than three, merge them together into the cluster. Then, leaping searching grids along four directions (as shown in Fig.3).

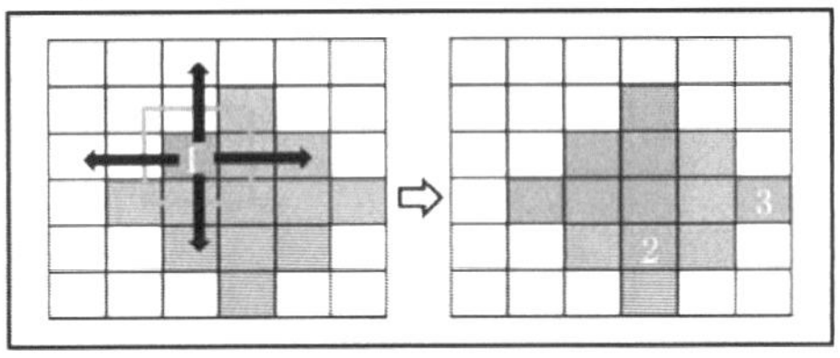

FIGURE 3. Valid grid processing.

During this process, if a valid and not marked grid is encountered, continue merging neighbor grids and leaping searches (as shown in Fig.4)

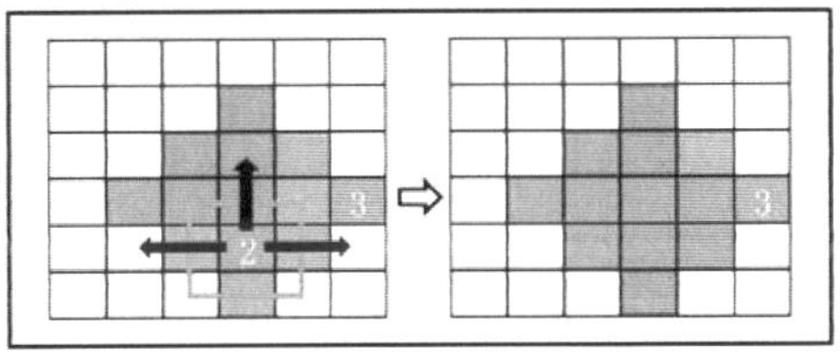

FIGURE 4. Leaping search for valid grid.

In order to prevent the situation appeared in Fig.5, the algorithm provides an offset mechanism. If four grids that leaping searched are all invalid, the directly neighbor valid grids become the center to continue to search.

2.5 Complete Clustering of Data Points

After all odd columns and all odd rows of grids are searched, clustering of data points is completed.

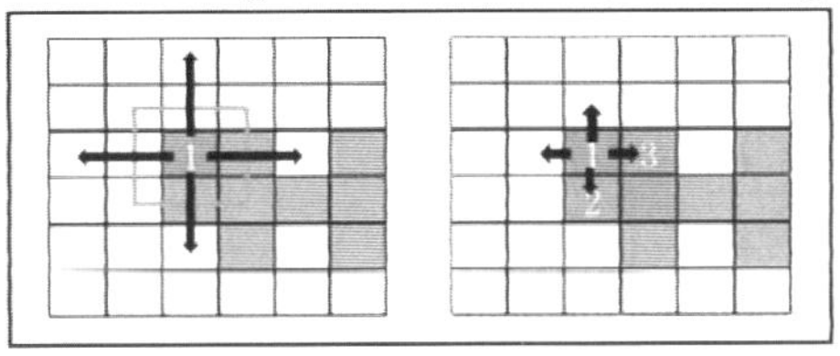

FIGURE 5. Offset operation.

3. Performance of lsmn

All experiments in this paper are carried out in JAVA program, and run on Microsoft Windows 7 desktop computer with Intel i3 CPU (3.7GHz) equipped with 4GB RAM.

The efficiency and quality of LSMN are verified by comparing the experimental results with 3 popularly used clustering algorithms (DBSCAN, CLIQUE and RTHVDS [4]).

Table 1 lists the details of the 4 teat data sets. Table 2 shows the parameter values.

Table 3 shows the experimental results. All algorithms for each dataset are tested 10 times, and average time, accuracy were calculated. DBSCAN has the highest accuracy, but as the number of data sets increases, the time consumption increases dramatically. Although LSMN is less accurate than DBSCAN, it is significantly reduced in time consumption. Compared to CLIQUE and RTHVDS, LSMN is also improved in time and precision. Fig.6 shows the original data sets for the four groups and depicts the clustering results by our LSMN algorithm using four groups of data set.

In addition, LSMN's clustering time is reduced by more than 10% compared to CLIQUE and HVDTHS. And the more the grid number, the faster the clustering time. The increase in algorithm time in the Table 3 is not obvious because the time it takes to read the data is too much. In fact, the efficiency of LSMN has been greatly improved.

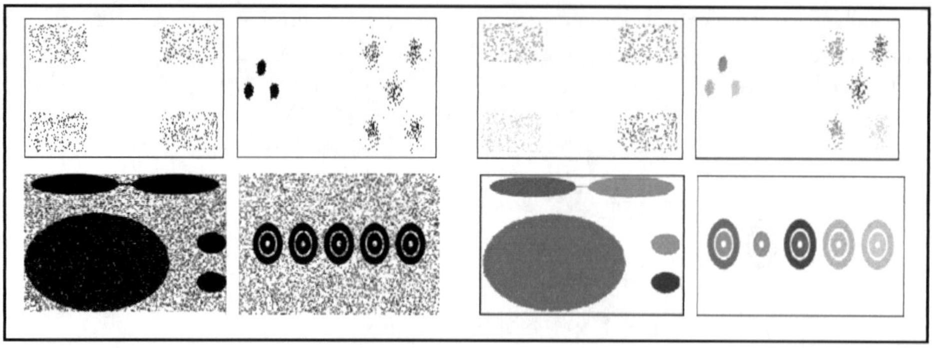

FIGURE 6. The original data set and the clustering results by LSMN.

4. Conclusion

In order to improve the efficiency and quality for clustering mass data, this paper proposed a clustering algorithm based on grid and density. By the algorithm, time cost was reduced by leaping searching; the accuracy is improved by neighbor grids merging. Compared with the DBSCAN algorithm based on density, LSMN greatly improves the running efficiency. Compared with the CLIQUE algorithm and RTHVDS algorithm, the clustering efficiency and precision are improved.

TABLE 1. Information on Four Groups of Data Set

Information	DataSet-1	DataSet-2	DataSet-3 [5]	DataSet-4 [6]
Data Source	By ourselves	M. Rezaei and P. Frnti,2016	TSAI and CHIANG,2016	TSAI and ZHANG,2012
Total number of data point	1000	6500	115000	575000
Number of noise point	0	0	15000	75000
Number of clusters	4	8	5	10

TABLE 2. Parameter Values

Parameter	DataSet-1	DataSet-2	DataSet-3	DataSet-4
ε	20	60	60	200
Minpts	2	2	20	20

TABLE 3. Experimental Results (Run-Time in Second)

Algorithm	Item	DataSet-1	DataSet-2	DataSet-3	DataSet-4
DBSCAN	Run-time	0.29	60	483	27502
	Accuracy	100.00%	100.00%	98.46%	99.46%
CLIQUE	Run-time	0.042	0.119	0.264	0.551
	Accuracy	98.50%	99.70%	98.81%	99.71%
RTHVDS	Run-time	0.039	0.104	0.256	0.545
	Accuracy	99.50%	99.78%	98.87%	99.54%
LSMN	Run-time	0.028	0.098	0.251	0.539
	Accuracy	100.00%	99.75%	98.91%	99.63%

References

[1] M. Ester, H.P. Kriegel, X. Xu, “A Density-Based Algorithm for Discovering Clusters in Large Spatial Databases with Noise”. International Conference on Knowledge Discovery & Data Mining, 1996, pp.226-231.

[2] W. Wang, J. Yang, R.R. Muntz, “STING: A statistical information grid approach to spatial data mining”. International Conference on Very Large Data Bases, 1997, pp.186-195.

[3] R. Agrawal, J. Gehrke, D. Gunopulos, et al, “Automatic subspace clustering of high dimensional data for data mining application”. Data Mining & Knowledge Discovery, 1998, pp.94-105.

[4] J.D. Zhao, Y.W. Yu, J.L. Liu, “A Data Clustering Algorithm over Real Time High-Volume Data Streams”. Journal of Beijing University of Posts and Telecommunications, 2016, pp.114-119.

[5] C.F Tsai, C. Yao, “ENHANCEMENT OF DATA CLUSTERING USING TSS-DBSCAN APPROACH FOR DATA MINING”. International Conference on Machine Learning & Cybernetics, 2016, pp.535-540.

[6] C.F Tsai, J.H. Zhang, “Grid Clustering Algorithm with Simple Leaping Search Technique”. International Symposium on Computer, 2012, pp.938-941.

IOP Conf. Series: Materials Science and Engineering **242** (2017) 012123 doi:10.1088/1757-899X/242/1/012123

Accurate Grid-based Clustering Algorithm with Diagonal Grid Searching and Merging

Feng Liu[1)], Chengcheng Ye[1)] and Erzhou Zhu[1, a)]

[1]School of Computer Science and Technology, Anhui University, Hefei 230601, China

E-mail:[a)] ezzhu@ahu.edu.cn

Abstract. Due to the advent of big data, data mining technology has attracted more and more attentions. As an important data analysis method, grid clustering algorithm is fast but with relatively lower accuracy. This paper presents an improved clustering algorithm combined with grid and density parameters. The algorithm first divides the data space into the valid meshes and invalid meshes through grid parameters. Secondly, from the starting point located at the first point of the diagonal of the grids, the algorithm takes the direction of "horizontal right, vertical down" to merge the valid meshes. Furthermore, by the boundary grid processing, the invalid grids are searched and merged when the adjacent left, above, and diagonal-direction grids are all the valid ones. By doing this, the accuracy of clustering is improved. The experimental results have shown that the proposed algorithm is accuracy and relatively faster when compared with some popularly used algorithms.

1. Introduction

As an important means in data mining, clustering algorithm is widely used in many fields such as data analysis, pattern recognition, image processing and so on. By clustering, sample points in each cluster have high similarity while different cluster sample points with low similarity.

By now, many kinds of clustering algorithms are emerged. By calculating the distance between the points and the cluster, the partition-based clustering algorithm, likes K-Means [1], is suitable for finding convex clusters. Unavailable of the number of clusters in advance, hierarchical clustering algorithm, likes BIRCH [2], is clustered by computing the distance between each cluster pair. This algorithm is suitable for aggregating multiple local clusters, but cannot deal with large-scale data properly.

By iteratively merging neighborhood high-density regions, Density-based clustering algorithms can identify arbitrary shapes data sets. But the time cost of this method is huge. DBSCAN [3] is a spatial clustering algorithm based on high density connectivity. It can find any shape of the cluster in the noisy data. However, it is very sensitive to the parameters ε (radius) and MinPts (neighborhood density threshold).

Grid-based clustering algorithms divide the data space into finite units and then cluster them. CLIQUE [4] proposed by Agrawal is a representative algorithm based on density and grid. It can automatically identify the subspace of the high dimensional data space. However, the need for finding the clustering of all the lower subspaces reduces the time and space performance of the algorithm. TSING [5] is another grid clustering algorithm with high performance but the accuracy is relatively low.

In order to elevate the accuracy of the grid clustering algorithm, this paper proposes DSM, an improved grid-based clustering algorithm with diagonal grid searching and merging. At the beginning,

Content from this work may be used under the terms of the Creative Commons Attribution 3.0 licence. Any further distribution of this work must maintain attribution to the author(s) and the title of the work, journal citation and DOI.
Published under licence by IOP Publishing Ltd

the data space is meshed by the grid parameter. If the points in the grid are greater than or equal to the threshold parameter, it is treated as a valid grid; otherwise it is considered an invalid grid. The algorithm takes the first point of the diagonal is selected as the starting grids. It also adopts horizontal-right and vertical-downward direction along the diagonal to search the valid girds. Furthermore, the valid grids that are superjacent or on the left of the valid grids located at horizontal and vertical directions of the diagonal are merged into the same groups. This search strategy can effectively reduce the time cost. During the search process, the boundary grids are judged. If it is an invalid grid, then expand searching the left, above, and diagonal meshes of this invalid grid. If they are all valid grids, then directly merging them for clustering accuracy elevates.

2. The implementation of dsm

The algorithm proposed in this paper aims at improving the accuracy and efficiency of the grid-based clustering algorithm. This method is based on dividing the grid into space and then dividing them into "valid grid" or "invalid grid" according to the grid containing at least the predetermined threshold parameter. By taking the first point of the diagonal of grids as starting grids, the effective grids are searched along the horizontal and vertical directions of the diagonal. During the searching process, if the grids adjacent to the searched valid grids are already clustered, these grids can be treated as the same group. Otherwise, they are put into a new group. Finally, if the left adjacent grid and the upper adjacent grid contain multiple groups, all groups will be merged into the same group.

```
Input: DataSet, GridSize, MinPts.          Output: Clusters.
Init(DataSet);
PartitionGrids(GridSize);
for i,j = 1 to GridColumn,GridRow do
for i1 = i to GridColumn do
   NeighborGridCheck(i1,j);
      for j1 = j+1 to GridRow do
   NeighborGridCheck(i,j1);
function NeighborGridCheck(i,j)
if CurrentGrid is valid then
        if UpGrid is valid then
           UpGridGroup.append(CurrentGrid);
        if LeftGrid is valid then
           LeftGridGroup.append(CurrentGrid);
        if UpGrid is valid && LeftGrid is valid then
GridGroupMerge(CurrentGrid,LeftGridGroup,RightGridGroup)
    else UpGrid && LeftGrid && DiagonalGrid are valid
           LeftGridGroup.append(CurrentGrid);
function GridGroupMerge(CurrentGrid,LeftGridGroup,UpGridGroup)
        LeftGridGroup.append(CurrentGrid);
        for i = 0 to UpGridGroup.size() do
```

FIGURE 1. The DSM algorithm.

In order to illustrate the detailed steps of the algorithmic method, it is needed to specify some parameters in advance: the *DataSet* represents the data set, *GridSize* represents the size of the grids, and *MinPts* represents the least data points in the grid. In this algorithm, *NeighborGridCheck()* and *GridGroupMerge()* are two important modules. *NeighborGridCheck()* will check whether the adjacent left and top of the grid are valid and have already been clustered. If so, they can be merged into the same group. If the adjacent left, top and diagonal of the grid are invalid grid, it may be treated as a new group. In addition, *GridGroupMerge()* is responsible for merging the current grid and the group where the neighbors are located, and merging all the groups into the same group. As Fig.1 shows, the proposed algorithm has the following steps:

Step1: Initialize all parameters.

Step2: Divide the grid space and distinguish between the valid grids and invalid grids according to the grid parameter and threshold parameter to. As shown in Fig.2.

IOP Conf. Series: Materials Science and Engineering **242** (2017) 012123 doi:10.1088/1757-899X/242/1/012123

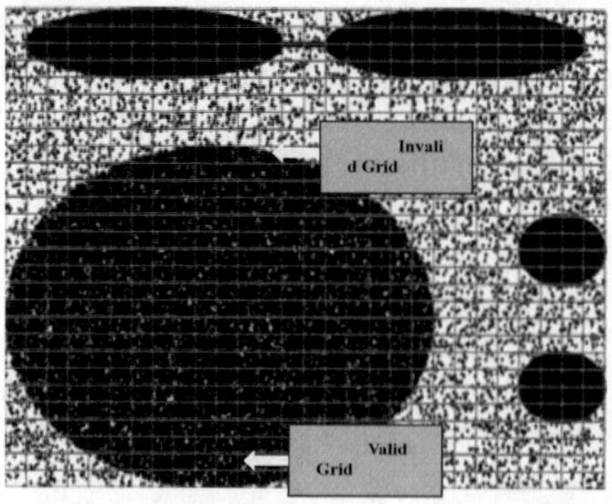

FIGURE 2. Dividing grid space into valid and invalid grids.

Step 3: Use the diagonal grids as the starting grids to search valid grid on horizontal and vertical directions. If the current grid on the left or top of the searched grid located at the diagonal is the valid one, it is merged into its adjacent left or upper group. Otherwise, all the valid grids are merged into the same group.

As shown in Fig.3, the diagonal grid 1 horizontal search to the effective grid 2. If 1 and 3 are valid grid, then 1, 2 and 3 are in the same group. Otherwise, set to a new group. Grid 1 vertical down search to grid 4. If 1 and 5 are valid grids, then the grids 1, 4 and 5 are in the same group. So 1, 2, 3, 4 and 5 are in the same group.

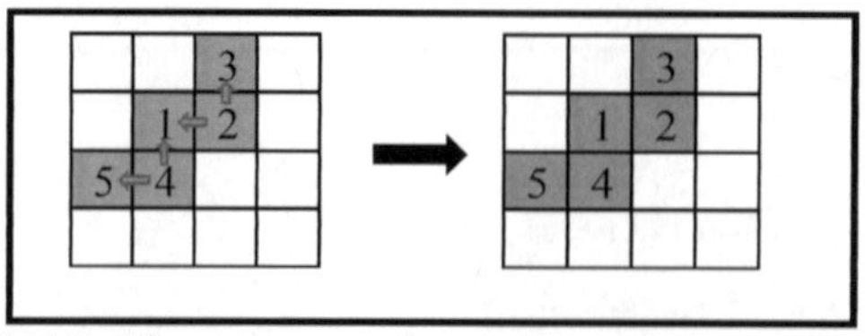

FIGURE 3. The process of searching and merging grids.

Step 4: In the search and merge process, if the left, top and diagonal adjacent grids are valid, the grid is determined as the boundary grid and merged it into the group of the left, top and diagonal grid.

As shown in Fig.4, invalid grid 2 extended search grids 1, 2 and 3. If the grids 1, 2 and 3 are valid grids, then merge together. Fig.5 shows the clustering results.

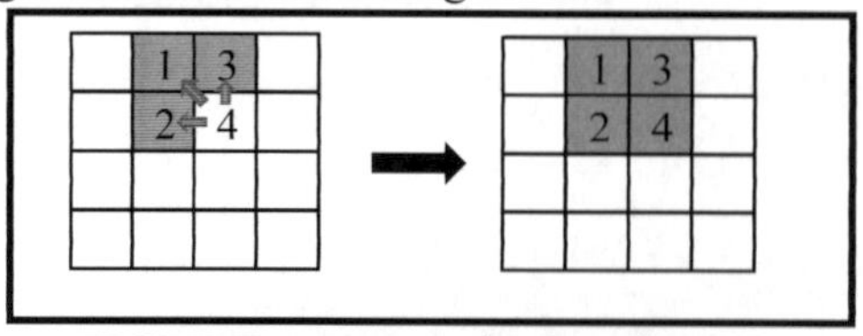

FIGURE 4. Merge the invalid grid.

FIGURE 5. Clustering result.

3. Experiment and analysis

In this work, the algorithm is implemented in Java program and executed in the desktop computer with 4GB RAM, Intel Core i3 3.7 GHz on Windows 10 operating system. The clustering results are compared with the DBSCAN, CLIQUE, and TSING methods. According to Fig.6, the data sets 1 to 4 have different distribution patterns with 575,000 data points and 75,000 noise points. Data set 1, 2, 3 and 4 has 5, 10, 3and 4 clusters respectively [6].

The data sets 1 to 4 were clustered 10 times using the proposed algorithm and the conventional method, and the average results are listed in Table 1. Fig.6 shows the clustering results. It can be seen from Table 1 that the comparison between the proposed algorithm and other methods shows that the clustering accuracy is improved by extending the boundary void grid. Meanwhile, the only search three adjacent grids per grid improves the time efficiency. DSM consumes more time but exhibits more clustering accuracy when compared with TSING.

FIGURE 6. The original datasets and clustering results of datasets 1-4.

TABLE1. Experimental results; item1 indicates run-time cost(s); item 2 indicates clustering accuracy rate(%).

Algorithm	Item	DataSet-1	DataSet-2	DataSet-3	DataSet-4
DBSCAN	1	27495	27545	27483	27479
	2	99.99%	99.98%	99.99%	99.99%
TSING	1	0.052	0.057	0.055	0.053
	2	98.62%	99.58%	99.45%	99.49%
CLIQUE	1	0.563	0.556	0.570	0.573
	2	98.85%	99.72%	99.60%	99.64%
DSM	1	0.432	0.437	0.454	0.448
	2	98.93%	99.79%	99.70%	99.80%

4. Conclusion

In this paper, a grid-based algorithm is proposed to improve the accuracy of grid clustering algorithm. It takes the diagonal grids as starting grids to horizontal and vertical search and merge valid grids. For each grid, only two adjacent grids are searched. As a result, the proposed data clustering algorithm can reduce the cost of clustering time and improve the efficiency. In addition, the processing of the boundary value is added to improve the clustering accuracy.

References

[1] J. B. McQueen, "Some Methods of Classification and Analysis of Multivariate Observations," the 5th Berkeley Symposium on Mathematical Statistics and Probability, 1967, pp. 281-297.

[2] T. Zhang and R. Ramakrishnan, "BIRCH: An efficient Data Clustering Method for Very Large Databases," ACM SIGMOD International Conference on Management of Data, 1996, pp. 103-114.

[3] M. Ester, H. P. Kriegel and X. Xu, "A Density-Based Algorithm for Discovering Clusters in Large Spatial Databases with Noise," International Conference on Knowledge Discovery and Data Mining, 1996, pp. 226-231.
[4] R. Agrawal, J. E. Gehrke, D. Gunopulos and P. Raghavan, "Automatic Subspace Clustering of High Dimensional Data for Data Mining Applications," ACM SIGMOD International Conference on Management of Data, 1998, pp. 94-105.
[5] C. F. Tsai and Y. C. Hu, "Enhancement of efficiency by thrifty search of interlocking neighbor grids approach for grid-based data clustering," International Conference on Machine Learning & Cybernetics, 2013, pp. 1279-1284.
[6] C. F. Tsai and S. C. Huang, "An effective and efficient grid-based data clustering algorithm using intuitive neighbor relationship for data mining," International Conference on Machine Learning & Cybernetics, 2015, pp. 478-483.

Corrigendum: Accurate Grid-based Clustering Algorithm with Diagonal Grid Searching and Merging

IOP Conf. Series: Materials Science and Engineering **242** (2017) 012123

Feng Liu[1)], Chengcheng Ye[1)] and Erzhou Zhu[1, a)]

[1]School of Computer Science and Technology, Anhui University, Hefei 230601, China

E-mail:[a)] ezzhu@ahu.edu.cn

1. Page 2:

In the Figure 1, the following figure appears:

```
Input: DataSet, GridSize, MinPts.            Output: Clusters.
--------------------------------------------------------------
Init(DataSet);
PartitionGrids(GridSize);
for i,j = 1 to GridColumn,GridRow do
for i1 = i to GridColumn do
   NeighborGridCheck(i1,j);
      for j1 = j+1 to GridRow do
   NeighborGridCheck(i,j1);
function NeighborGridCheck(i,j)
if CurrentGrid is valid then
         if UpGrid is valid then
            UpGridGroup.append(CurrentGrid);
         if LeftGrid is valid then
            LeftGridGroup.append(CurrentGrid);
         if UpGrid is valid && LeftGrid is valid then
GridGroupMerge(CurrentGrid,LeftGridGroup,RightGridGroup)
      else UpGrid && LeftGrid && DiagonalGrid are valid
            LeftGridGroup.append(CurrentGrid);
                                                     function
GridGroupMerge(CurrentGrid,LeftGridGroup,UpGridGroup)
         LeftGridGroup.append(CurrentGrid);
         for i = 0 to UpGridGroup.size() do
```

This should be:

```
Input: DataSet, GridSize, MinPts.            Output: Clusters.
Init(DataSet);
PartitionGrids(GridSize);
for i,j = 1 to GridColumn,GridRow do
    for i1 = i to GridColumn do
        NeighborGridCheck(i1,j);
    for j1 = j+1 to GridRow do
        NeighborGridCheck(i,j1);
function NeighborGridCheck(i,j)
    if CurrentGrid is valid then
        if UpGrid is valid then
            UpGridGroup.append(CurrentGrid);
        if LeftGrid is valid then
            LeftGridGroup.append(CurrentGrid);
        if UpGrid is valid && LeftGrid is valid then
            GridGroupMerge(CurrentGrid,LeftGridGroup,RightGridGroup)
    else UpGrid && LeftGrid && DiagonalGrid are valid
            LeftGridGroup.append(CurrentGrid);
function GridGroupMerge(CurrentGrid,LeftGridGroup,UpGridGroup)
        LeftGridGroup.append(CurrentGrid);
        for i = 0 to UpGridGroup.size() do
                LeftGridGroup.append(UpGridGroup.get(i));
```

2. Page 3:

In the Figure 2, the following figure appears:

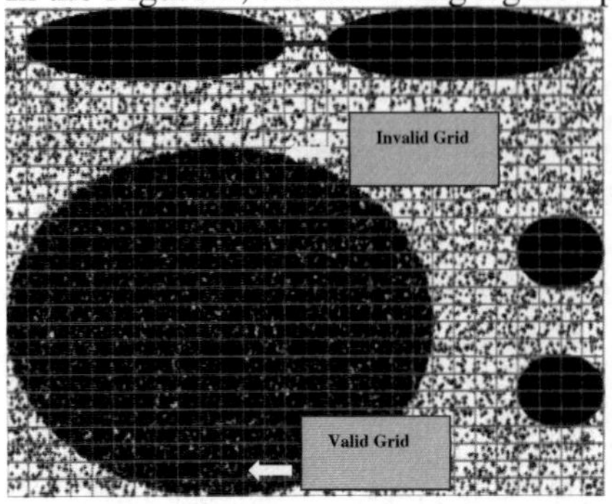

This should be:

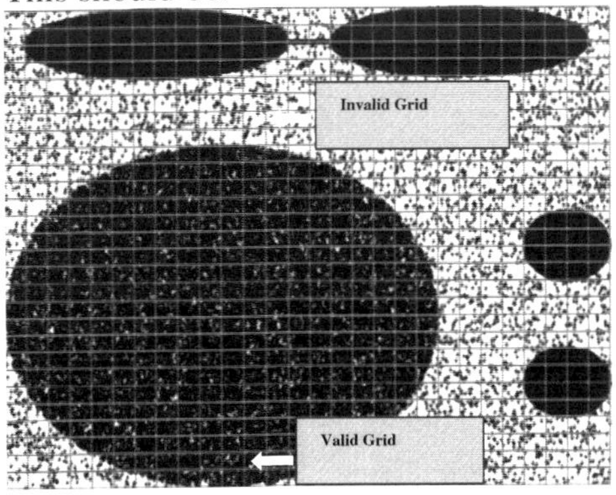

3. Page 4:

In the Table 1, the following table appears:

TABLE1. Experimental results; item1 indicates run-time cost(s); item 2 indicates clustering accuracy rate(%).

Algorithm	Item	DataSet-1	DataSet-2	DataSet-3	DataSet-4
DBSCAN	1	27495	27545	27483	27479
	2	99.99%	99.98%	99.99%	99.99%
TSING	1	0.052	0.057	0.055	0.053
	2	98.62%	99.58%	99.45%	99.49%
CLIQUE	1	0.563	0.556	0.570	0.573
	2	98.85%	99.72%	99.60%	99.64%
DSM	1	0.432	0.437	0.454	0.448
	2	98.93%	99.79%	99.70%	99.80%

This should be:

TABLE2. Experimental results; item1 indicates run-time cost(s); item 2 indicates clustering accuracy rate(%).

Algorithm	Item	DataSet-1	DataSet-2	DataSet-3	DataSet-4
DBSCAN	1	27495	27545	27483	27479
	2	99.99%	99.98%	99.99%	99.99%
TSING	1	0.052	0.057	0.055	0.053
	2	98.62%	99.58%	99.45%	99.49%
CLIQUE	1	0.563	0.556	0.570	0.573
	2	98.85%	99.72%	99.60%	99.64%
DSM	1	0.432	0.437	0.454	0.448
	2	98.93%	99.79%	99.70%	99.80%

A Shellcode Detection Method Based on Full Native API Sequence and Support Vector Machine

Yixuan Cheng[1, a)], Wenqing Fan[1, b)] , Wei Huang[1, c)] and Jing An[1, d)]

[1]A College of Computer, Communication University of China, Beijing 100024, China.

E-mail: [a)] cuc2014is_bcxq@163.com, [b)]fan.winking@gmail.com, [c)]i@huangwei.me, [d)]anjingcuc@gmail.com

Abstract. Dynamic monitoring the behavior of a program is widely used to discriminate between benign program and malware. It is usually based on the dynamic characteristics of a program, such as API call sequence or API call frequency to judge. The key innovation of this paper is to consider the full Native API sequence and use the support vector machine to detect the shellcode. We also use the Markov chain to extract and digitize Native API sequence features. Our experimental results show that the method proposed in this paper has high accuracy and low detection rate.

1. Introduction

In today's era, computer viruses like Trojans and other malicious software, especially shellcode, continue to appear, bringing a huge threat to the development of Internet. Traditional detection method is based on signature. With the development and update of computer virus technology, signature code is virtually impossible to accurately identify malicious software, especially when malware has been confused or encrypted or after other operations.

Within up-to-date analysis methods, a common way is to extract information in the PE file header as its static feature[1], such as the number of dlls and APIs. However, these features are not sufficient to represent the actual behavior of a malicious program, and its PE file header information can be easily tampered with. As the behavior of malware is mostly based on API functions while shellcode is an instruction set with specific functions, its function is limited. Therefore, to complete the more complex functions, it often calls system API.

One way to detect shellcode is to use the Hook key API, such as LoadLibraryA(W), GetProcAddressA(W), CreateFilea(W), etc. The main problem is that a malicious purpose with unknown behavior shellcode does not just use this limited number of API functions, if it uses other non-Hook API function, the detection method will be invalid.

Another common method is based on the API call sequence to determine[3]. A typical approach is to obtain the call sequence, based on a supervised heuristic learning, to determine whether it is malware[4]. It can also improve the direction of the following two aspects:

1. Compared to Win32 API, the native APIs better reflect the core nature of a program because Win32 API functions often call the underlying native API;

2. The analysis process is static. It disassembles the program, and then generate the API call sequence based on the disassembly results, but the static analysis results are not equal to its actual running results. The actual API call sequence is what we really concerned about.

Aiming at the problem of some previous methods, this paper presents a shellcode detection method based on full Native API sequence and support vector machine. Using the WinDbg tool to monitor the

Content from this work may be used under the terms of the Creative Commons Attribution 3.0 licence. Any further distribution of this work must maintain attribution to the author(s) and the title of the work, journal citation and DOI.
Published under licence by IOP Publishing Ltd

whole Native API sequence information of the shellcode, the Markov chain is used to extract the sequence features, and the support vector machine is used to automate the judgment of the shellcode. The main contributions of this paper include:

1. Take a dynamic analysis method to analyze the program. We determine the final result by dynamically analyzing the whole API sequence information that the program actually calls at execution time.

2. Use the Native API full sequence. Compared to the Win32 API sequence, we use the native API sequences that represent the actual core functionality of the program as the data source for the decision. And we monitored the full sequence of Native API calls to ensure that it can monitor its complete behavioral information.

3. Apply the support vector machine to the detection of the shellcode. We use the support vector machine to judge the characteristics of the native API full sequence and apply it to the detection and judgment of the shellcode.

The remainder of this article is organized as follows. In the second part, we introduce some of the related work in the field of malware detection. In the third part, we describe the architecture and implementation of the shellcode detection system. In the fourth part, we present our experimental results, while concluding remarks are addressed in fifth part.

2. Related work

There are two common malware detection methods: static analysis and dynamic analysis.

The static analysis method is to acquire its signature, or to extract information of its PE head or data portion as its characteristics. Hu[5] used the static method to select the PE header information and the middle information as a feature to determine. Alazab[4,6] and Sathyanarayan[7] used the IDA tool to disassemble the program and extract the API call sequence from the disassembly results for analysis. Santos[9] used the opcode sequence to calculate the similarity of the two PE executables. Static method is the current mainstream detection method, but the static methods for current malware which often uses packaging and confusing technology cannot be detected.

The dynamic method performs discriminant detection by obtaining API call information when the program is running. Sami[8] proposed API calls that combine time characteristics with spatial features to monitor samples. Wagener[10] combined with virtual machine technology, first constructed two similar sequences of program API sequences, by calculating the similarity of the classification. Tian[2] and so on recorded the program in the virtual machine to run the log information, such as API call information, to classify it. The current mainstream dynamic analysis methods mainly dynamically run the target program, record the API information for further classification or judgment. But most of the analysis process to determine the information is based on Win32 API, not the Native API which can represent the core function of a program. So in this paper, we use Native API as basic information.

3. Approach

This section describes the architecture of the entire system. As shown in Fig.1, the entire system consists of three parts.

The Native API is a lightweight program programming interface used by Windows NT and user-mode applications that is implemented by the operating system and is provided in Ntdll.dll. Win32 API main function encapsulated in Kernel32.dll. Painting related system services packaged in Gdi32.dll. Window management is mainly encapsulated in User32.dll. Windows related controls encapsulated in comctl32.dll. Win32 API function in the specific implementation of the function, will call the more underlying relevant Native API function in Ntdll.dll. If the Win32 API uses the CreateFile() function, the NtCreateFile() function in the Native API is called inside the CreateFile() function, so the Native API is better to reflect the core functionality of the program. And because the Native API can be invoked directly by the application, the detection method based on the Win32 API will fail if the malware directly calls the Native API to implement the functionality without using the Win32 API, so in the method presented in this article, we use dynamic analysis methods and the full

Native API call sequence is used as the basis for subsequent judgments.

In the first part, there are two steps:

Step 1: Use scripts builder, combined with Native API prototype, to generate WinDbg debug scripts. WinDbg through the bp and other instructions to the function under the breakpoint operation. When hit the breakpoint, the program is in an interrupt state, so we can record the Native API related information, such as the name of the Native API and other information. So according to Native API prototype, we can generate automated debugging script, and the target program will automatically log information and so on to get its dynamic runtime information.

Step 2: On the target host, use the previously generated WinDbg script, using WinDbg to debug the shellcode.

In the second part, call the WinDbg script to debug the target program. Put the shellcode into the target host and run the shellcode. WinDbg records the full Native API information based on the previously generated debug script and generates a log file.

After obtaining the log file, it is necessary to extract its digital features to carry out the upcoming classification work. In this article, we take 4-Gram to characterize. Markov hypothesis think that the appearance of a word depends only on the finite word. And Markov chain is a discrete sequence satisfying the Markov hypothesis. Combining the previous studies with the Markov Chain applications in the API sequence and the experimental results, we believe that the 4-Gram algorithm to express the API sequence can achieve better results. The extraction algorithm is as follows:

Step 1: Extract the API sequence in each log file as input to the next step 4-Gram algorithm;

Step 2: In the 4-Gram algorithm, for each function, the probability of its occurrence is only related to the previous three functions, so we choose a sliding window of size 4. Using the number of occurrences of each Gram, the number of times as the characteristics of each Gram, we get the Gram characteristics group of each program;

Step 3: The highest number of n Grams in shellcode will be used as the key Gram characteristic vector of shellcode. In this paper, we select n as 40 as the number of Gram feature vectors;

Step 4: Select the Gram vector in the Gram feature vector of the shellcode extracted from step 3 from the Gram attribute group of each program to be analyzed as the Gram characteristic vector of the program to complete the feature extraction of each program.

In the third part, we use Support Vector Machines (SVM) to classify. SVM is a kind of supervised learning model with related learning algorithm in machine learning, which can classify and analyze the data. The main idea is to find a super-plane as a training sample segmentation, to ensure the smallest classification error rate. SVM allocates a given training set to one or more categories by constructing a non-probabilistic binary linear classifier. According to the Gram feature vector extracted in the second part, 50% of the training feature is selected as the training set, and the other 50% is used as the test set, and the support vector machine is used to classify. Finally, the results are judged. SVM needs to select the kernel function. After several experiments, we think that the RBF kernel function is better for the final classification. In this paper, RBF kernel function is chosen as the kernel function of SVM.

IOP Conf. Series: Materials Science and Engineering **242** (2017) 012124 doi:10.1088/1757-899X/242/1/012124

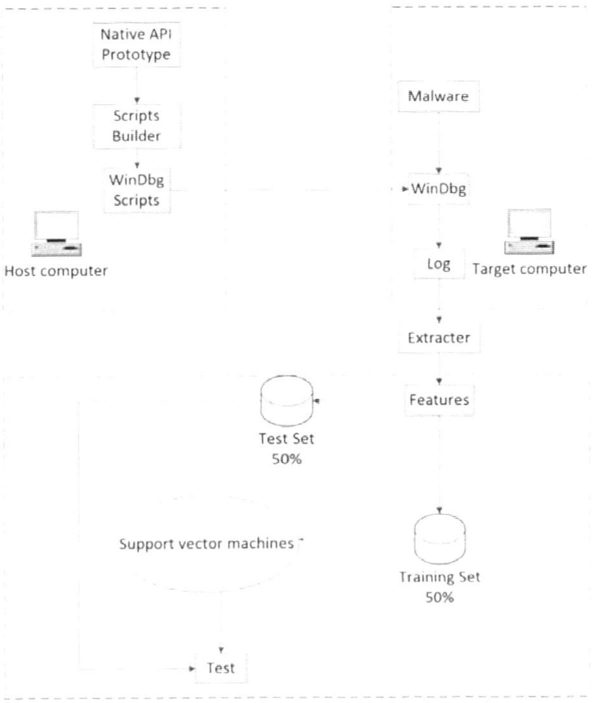

FIGURE 1. Architecture of shellcode detection system.

4. Experimental evaluation

The data sets for this article is made up of 18 shellcode programs on exploit-db and 72 PE files in authentic Windows XP system directories. In order to ensure that there is no deviation from the classifier in the course of training, we select 50% of the shellcode program and the benign program with the same number as the training set and the other part as the test set.

In the Scripts Builder, we generated the WinDbg script shown in Fig.2. We use the bp command to perform breakpoints on all functions in ntdll and output their function names to the log file as the full Native API call sequence.

```
bp ntdll!NtCreateEvent".echo Function Name :NtCreateEvent\n;g"
bp ntdll!NtCreateEventPair".echo Function Name :NtCreateEventPair\n;g"
bp ntdll!NtCreateFile".echo Function Name :NtCreateFile\n;g"
bp ntdll!NtCreateIoCompletion".echo Function Name :NtCreateIoCompletion\n;g"
bp ntdll!NtCreateJobObject".echo Function Name :NtCreateJobObject\n;g"
bp ntdll!NtCreateJobSet".echo Function Name :NtCreateJobSet\n;g"
bp ntdll!NtCreateKey".echo Function Name :NtCreateKey\n;g"
bp ntdll!NtCreateKeyedEvent".echo Function Name :NtCreateKeyedEvent\n;g"
```

FIGURE 2. Part of the script command.

In Extracter, we first process the log file into a pure API sequence, as shown in Fig.3.

```
NtQueryInformationProcess
NtSetInformationProcess
NtOpenKey
NtOpenKey
NtCreateSemaphore
NtCreateSemaphore
NtOpenKey
NtOpenKey
NtAllocateVirtualMemory
NtOpenKey
NtOpenKey
NtOpenKey
NtQueryValueKey
```

FIGURE 3. Part of the API sequence.

According to the extraction algorithm in Section 3, the API sequence is processed into a Gram feature vector and the number 1-40 is used to represent the selected 40 feature vectors, as shown in Fig.4. We also count the number of feature vectors in the program as the value of the feature vector.

IOP Conf. Series: Materials Science and Engineering **242** (2017) 012124 doi:10.1088/1757-899X/242/1/012124

```
-1 1:31 2:6 3:3 4:5 5:3 6:3 7:3 8:2 9:5 10:20 11:3 12:20 13:3 14:3 15:3 16:5 17:5 18:10 19:5 20:17 21:
-1 1:42 2:1 3:4 4:7 5:4 6:4 7:1 8:2 9:0 10:0 11:0 12:0 13:0 14:0 15:0 16:0 17:0 18:0 19:0 20:0 21:0 22
-1 1:31 2:6 3:3 4:5 5:3 6:3 7:3 8:2 9:5 10:20 11:3 12:20 13:3 14:3 15:3 16:5 17:5 18:10 19:5 20:17 21:
-1 1:30 2:3 3:3 4:5 5:3 6:3 7:0 8:2 9:2 10:0 11:2 12:0 13:2 14:2 15:2 16:0 17:0 18:0 19:0 20:0 21:0 22
-1 1:31 2:6 3:3 4:5 5:3 6:3 7:3 8:2 9:5 10:20 11:3 12:20 13:3 14:3 15:3 16:5 17:5 18:10 19:5 20:17 21:
-1 1:30 2:1 3:3 4:5 5:3 6:3 7:1 8:2 9:0 10:0 11:0 12:0 13:0 14:0 15:0 16:0 17:0 18:0 19:0 20:0 21:0 22
-1 1:31 2:6 3:3 4:5 5:3 6:3 7:3 8:2 9:5 10:20 11:3 12:20 13:3 14:3 15:3 16:5 17:5 18:10 19:5 20:17 21:
-1 1:31 2:6 3:3 4:5 5:3 6:3 7:3 8:2 9:5 10:20 11:3 12:20 13:3 14:3 15:3 16:5 17:5 18:10 19:5 20:17 21:
-1 1:30 2:28 3:0 4:0 5:0 6:0 7:51 8:0 9:0 10:28 11:0 12:28 13:0 14:0 15:0 16:31 17:180 18:104 19:30 20
+1 1:0 2:7 3:0 4:0 5:0 6:0 7:0 8:0 9:7 10:8 11:6 12:8 13:0 14:0 15:7 16:22 17:71 18:24 19:22 20:0 21:0
+1 1:0 2:0 3:0 4:0 5:0 6:0 7:0 8:0 9:0 10:0 11:0 12:0 13:0 14:0 15:0 16:0 17:50 18:0 19:0 20:0 21:0 22
+1 1:0 2:0 3:0 4:0 5:0 6:0 7:0 8:0 9:14 10:14 11:11 12:14 13:0 14:11 15:12 16:12 17:45 18:20 19:12 20:
+1 1:0 2:11 3:0 4:0 5:0 6:0 7:13 8:0 9:7 10:21 11:0 12:21 13:0 14:0 15:7 16:11 17:44 18:22 19:11 20:0
+1 1:0 2:11 3:0 4:0 5:0 6:0 7:13 8:0 9:7 10:21 11:0 12:21 13:0 14:0 15:7 16:11 17:44 18:22 19:11 20:0
+1 1:0 2:7 3:0 4:0 5:0 6:0 7:6 8:0 9:7 10:8 11:6 12:8 13:5 14:5 15:7 16:11 17:44 18:19 19:11 20:0 21:0
+1 1:0 2:8 3:0 4:0 5:0 6:0 7:0 8:0 9:7 10:9 11:6 12:9 13:0 14:0 15:7 16:11 17:44 18:20 19:11 20:0 21:0
+1 1:0 2:9 3:0 4:0 5:0 6:0 7:0 8:0 9:7 10:23 11:6 12:23 13:0 14:0 15:7 16:13 17:52 18:18 19:13 20:0 21
+1 1:0 2:9 3:0 4:0 5:0 6:0 7:0 8:0 9:7 10:9 11:6 12:9 13:0 14:0 15:7 16:11 17:44 18:17 19:11 20:0 21:0
```

FIGURE 4. Part of the training set.

We choose to use LIBSVM to achieve SVM classification. We choose RBF as the kernel function. The final judgment accuracy is 94.37%, with the false positive rate of 1.41%, and the false negative rate accounting for 44.44%.

5. Conclusion

The experimental results show that the shellcode detection method based on the full Native API sequences and the support vector machine has a high detection accuracy as well as a low false positive rate. Good results are achieved from benign programs. We still have defects: few experimental data. The reason for the high false negative rate is that the size of training sets is too small and the extracted Gram characteristic vector is not big enough. In the future improvement plan, we can increase the data sets size, especially the number of shellcode samples. We have carried out a preliminary verification of the proposed scheme. A high detection accuracy and a low false positive rate indicates that our program effect is effective.

Acknowledgments

We would like to thank Xinyuan Chen, Yue Pan and Yawen Song for their support, offering good ideas to help us finish this paper. Yixuan Cheng is with the Communication University of China.

References

[1] Schultz, M. G., Eskin, E., Zadok, F., & Stolfo, S. J. (2001). Data mining methods for detection of new malicious executables. In Security and Privacy, 2001. S&P 2001. Proceedings. 2001 IEEE Symposium on (pp. 38-49). IEEE.

[2] Tian, R., Islam, R., Batten, L., & Versteeg, S. (2010, October). Differentiating malware from cleanware using behavioural analysis. In Malicious and Unwanted Software (MALWARE), 2010 5th International Conference on (pp. 23-30). IEEE.

[3] Xu, J. Y., Sung, A. H., Chavez, P., & Mukkamala, S. (2004, December). Polymorphic malicious executable scanner by API sequence analysis. In Hybrid Intelligent Systems, 2004. HIS'04. Fourth International Conference on (pp. 378-383). IEEE.

[4] Alazab, M., Venkatraman, S., Watters, P., & Alazab, M. (2011, December). Zero-day malware detection based on supervised learning algorithms of API call signatures. In Proceedings of the Ninth Australasian Data Mining Conference-Volume 121 (pp. 171-182). Australian Computer Society, Inc..

[5] Hu, X. (2011). Large-scale malware analysis, detection, and signature generation (Doctoral dissertation, University of Michigan).

[6] Alazab, M., Venkataraman, S., & Watters, P. (2010, July). Towards understanding malware behaviour by the extraction of API calls. In Cybercrime and Trustworthy Computing Workshop (CTC), 2010 Second (pp. 52-59). IEEE.

[7] Sathyanarayan, V. S., Kohli, P., & Bruhadeshwar, B. (2008, July). Signature generation and detection of malware families. In Australasian Conference on Information Security and Privacy (pp. 336-349). Springer, Berlin, Heidelberg.

[8] Sami, A., Yadegari, B., Rahimi, H., Peiravian, N., Hashemi, S., & Hamze, A. (2010, March). Malware detection based on mining API calls. In Proceedings of the 2010 ACM symposium on

applied computing (pp. 1020-1025). ACM.
[9] Santos, I., Brezo, F., Nieves, J., Penya, Y. K., Sanz, B., Laorden, C., & Bringas, P. G. (2010, February). Idea: Opcode-sequence-based malware detection. In International Symposium on Engineering Secure Software and Systems (pp. 35-43). Springer, Berlin, Heidelberg.
[10] Wagener, G., & Dulaunoy, A. (2008). Malware behaviour analysis. Journal in computer virology, 4(4), 279-287.

ICAMMT 2017 IOP Publishing
IOP Conf. Series: Materials Science and Engineering **242** (2017) 012125 doi:10.1088/1757-899X/242/1/012125

The Study on Human-Computer Interaction Design Based on the Users' Subconscious Behavior

Lingyuan Li

New York University, New York, USA

E-mail: ll3291@nyu.edu

Abstract. Human-computer interaction is human-centered. An excellent interaction design should focus on the study of user experience, which greatly comes from the consistence between design and human behavioral habit. However, users' behavioral habits often result from subconsciousness. Therefore, it is smart to utilize users' subconscious behavior to achieve design's intention and maximize the value of products' functions, which gradually becomes a new trend in this field.

1. Introduction

Living in such an era with information explosion every day, people are heavily dependent on the electronic products, such as mobile phone, electronic reader and computer. These electronics are changing what we think and what we do, and the interaction experience of users also promotes the technology development of these electronic products. Donald Norman assumes that a good design should serve as the communication between object and user, and optimize that conduit of communication in order to make the experience of using object pleasurable[1] . How to make people feel more comfortable to use them has been a challenge to every designer. The pleasant experience of interaction with these products always makes users love them. So, users' experience is the focus that an interaction designer is always on. The behaviors of users' interacting on the screen of the products have been paid close attention by designers, who often make good use of users' conscious behaviors to design the users interface and tell them how to finish their operation. Actually, users' perception can be harnessed to dig out underlying needs for product design. Especially in the field of human computer interaction, if the interaction designer is good at utilizing users' subconscious behavior, the intention of designers and users' thoughts will be connected with each other on the same page, like the fact that two objects with same frequency will cause a resonance effect, thus the design can be made most use of to achieve greatest values.

2. The influence of users' subconscious behavior on interaction design

2.1 Traits of Subconscious Behavior

The term "subconscious" was first raised by Sigmund Freud in 1893, who believed that it is between conscious and unconscious[2]. According to Wikipedia, in a strict psychology sense, it can be defined as "operating or existing outside of consciousness"[3]. Thus subconscious behavior is described as an acting or response in a subconscious state.

Joseph Murphy points out in his book "The power of your subconscious mind" that subconsciousness has enormous power and subconscious behavior is, usually, highly effective[4]. And Brian Tracy also says our subconscious mind is 30,000 times more powerful than our conscious mind[5].

Content from this work may be used under the terms of the Creative Commons Attribution 3.0 licence. Any further distribution of this work must maintain attribution to the author(s) and the title of the work, journal citation and DOI.
Published under licence by IOP Publishing Ltd

It can not only store and retrieve data, namely pieces of memory, but also make sure that people can respond exactly in the way they are programmed.

Generally speaking, subconscious behavior happens with emotional information. Emotion exerts a huge influence on our subconsciousness, for subconscious mind is most likely to absorb emotional information. The more greatly mood fluctuates, the easier information is taken[6].

Subconscious mind is generally with weak memories and needs strong and repeating stimulations. Your subconscious mind is an unquestioning servant that works day and night to make your behavior fits a pattern consistent with your emotionalized thoughts, hopes, and desires[5].

Subconscious behavior is more susceptible to image stimuli. But it cannot tell whether the image is generated from people's real experience or from self-imagination. In other words, repeating the stimuli will lead to the final goal.

Subconscious memories cannot last very long, which means that it needs to be stimulated strongly or repeatedly. Only doing in this way can we activate the subconscious memory, namely habit. Habit is one of the most common forms of the subconsciousness.

Subconscious behavior, most of time, can be learned easily when people are relaxed because information is easier to be absorbed by subconscious mind in a relaxing state. Some scientific research shows that being relaxed and feeling well can contribute to make brain waves into Q-wave state, which is more convenient for the exchange between information and subconsciousness[7].

2.2 Subconscious Behavior in User Experience

When users interact with so many products and service in the real world, those designs that harness subconscious behavior make great achievements. One of the best examples is "a fly in urinal". It is reported that all the urinals in Amsterdam's Schiphol Airport have a vivid image of a fly[8]. It turns out that men, due to their urinal behavior, cannot resist peeing on things, especially if they look as though they might wash away. When flies were introduced into the design of urinal at Schiphol Airport, spillage rates dropped 80 percent, says manager Aad Keiboom[9]. Apparently, the designer made good use of the fact that guys love to play with their urine stream and like to aim at targets so they will pee on the fly subconsciously when they see it in the urinal. It has been proved that this idea reduces the cleaning costs, meanwhile men won't feel a sense of resistance psychology in that such a behavior is like a habit or a pattern in their comfortable zone of subconscious mind, which works better than simple warning signs.

Another successful example adopts reverse subconscious thinking. New York City's came up with a new way to clean up its garbage-strewn, rat-infested subway stations: get rid of the trashcans in the train[10]. The vast majority of subway passengers are responsible and upstanding. When passengers find a trashcan in the train, they will subconsciously throw litter. But now without any trashcan around, they are more inclined to throw their waste out before entering the subway or simply hold on to it or carry their refuse out of the system instead of dropping it on the platform or tossing it on the tracks. The designer paid attention to people's habitual behavior and found out the reason. It is because of the subconscious behavior that people drop litter when they see a trashcan that the designer decided not to install trashcans in the trains in order to decrease the possibility that litter is carried into trains. In essence, the designer utilized human beings' subconscious acting to achieve the final goal.

3. The application of users' subconscious behavior in human-computer interaction design

We've seen the power and effectiveness of subconscious behavior in daily life. For human-computer interaction, it's not difficult to imagine that its capacity will be unlimited! Some interaction design harness users' subconsciousness in a very smart way, which not only helps users operate them in the way they were intended but also let users follow their hearts.

4. Helping Bring about Users' Subconscious Behavior Based on Repeated Image Stimulation

In most websites, it is easy for users to find icons of Facebook, Twitter, Instagram and LinkedIn placed on the side of a page. If users like the webpage's content or they want to share what they're looking at to the famous social network websites, clicking the icons can achieve the goals. Users are exposed to the icon stimulation all the time, which helps them strengthen the memory of the operation and function behind the icons themselves and establish perception of icons in their subconscious minds. The strategy is effective. Almost everyone who surfs on the Internet a lot knows how to deal with the icons when they browse a website. When they want to share, they will subconsciously look for the icons to get share links, which not only saves the behavioral costs but also increases the user stickiness.

5. Taking Advantage of Users' Habitual Behavior under Subconscious Minds

A typical example close to our life is iPhone red dot issue. When there is new message or unread message of an app, there will be a red dot on the right top of the app's icon. Only if users open the app or read the message then the annoying red dot will disappear (Of course users have the option to cancel this mechanism if they don't want). In fact, everyone truly has obsessive thoughts in their subconscious minds and people tend to force themselves to do something or make things perfect. For example, some must wipe the spot on the computer screen, mobile phone screen or window as soon as they find it or they will feel uncomfortable; others think stuff must be placed in a certain rule, like according to color family, size, angle, category and alphabet. It's incredible that people will usually feel a sense of mysterious joy in deep heart after they do those things subconsciously before they realize. Even if people now are aware of the psychology, they understand its reasonable existence and are not exclusive to it. Overall, many people are inclined to regard such mental tendency as normal and natural. Designers of iPhone leverage this psychology cleverly. The red dot, a proper trigger, seems to annoy users but makes them happy in reality, which persuades users successfully to read message without too much thinking. The result lives up to the intention of the design that makes users involved in every process and helps them avoid missing important messages. For those who have a stronger tendency to be a little picky or perfectionist, they will read the unread message in time in order to get the sense of peace and joy after getting rid of red dots; for those who are not so obsessive about this, they don't care much about red dots. So red dots will not annoy them at all and user experience gets guaranteed.

Similarly, there is a tendency that social network websites choose progress bar with percentage to show how much information users have filled into the information form. People tend to make the incomplete complete in their subconscious minds. They hope to finish the form to make the bar show one hundred percent, a complete form. Actually the progress bar with percentage is to remind people of finishing the form subconsciously. In addition to the basic information such as user name and email address, websites also want to get more useful information in the same way like user preferences that is used to improve user experience and interaction design based on this kind of users' behavior.

6. Adding Information into Users' Subconscious Minds When They Are Relaxed

Google is an expert at this field. Google doodle is widely praised. Doodles are the fun, surprising, and sometimes, spontaneous changes that are made to the Google logo to celebrate holidays, anniversaries, and the lives of famous artists, pioneers, and scientists[11]. When users open the Google search engine, they will be attracted by the doodle which sometimes is in the form of game, or music or animation. It owns good interaction with users. They feel relaxed and enjoy the game, the sound or the motion design. During the relaxing process, users first become interested in the website and then rely upon it by osmosis and at last trust it. They associate Google with fun, trustworthiness, easily use and accessibility. Next time when they want to use a search engine, they are inclined to make Google as their first choice subconsciously!

These designs share one thing in common: They make users think it is themselves that make decisions and participate in every process of the product. Meanwhile, user experience is little affected and functions of the product (websites, apps, etc) play their parts.

7. Conclusion

Excellent interaction design is always human-orientated. The full-considering design for people's feeling and behavior can make the jobs on the interface done easily. Especially, the good use of users' subconscious behavior can enhance users' loyalty to the product and improves users' experience.

Human computer interaction should attach importance to the needs of human beings. The reasonable human computer interaction aims for the liberation of human nature. It is needs of human beings that result in the development of computer technology.

References

[1] Don Norman, The Design of Everyday Things (Basic Books, 2013)
[2] Sigmund Freud, The Question of Lay Analysis (W. W. Norton & Company, 1990)
[3] https://en.wikipedia.org/wiki/Subconscious
[4] Joseph Murphy, The Power of Your Subconscious Mind (Prentice Hall Press, 2008)
[5] https://www.briantracy.com/blog/general/understanding-your-subconscious-mind/
[6] D. Norman, Emotional Design (Basic Books, 2007)
[7] http://www.360doc.com/content/12/0222/10/200521_188551840.shtml
[8] https://worksthatwork.com/1/urinal-fly
[9] http://www.npr.org/templates/story/story.php?storyId=121310977
[10] https://www.wired.com/2015/08/nycs-cleaning-subway-taking-away-trash-cans/
[11] https://www.google.com/doodles/about

IOP Conf. Series: Materials Science and Engineering **242** (2017) 012126 doi:10.1088/1757-899X/242/1/012126

Design of Electronic Clock based on STM32

Jian HUANG

XiJing University, Xi'an, Shaanxi, China

E-mail:565200245@qq.com

Abstract. This paper describes the design method of electronic clock based on STM32.The hardware circuit is designed, which uses the built-in RTC module of the STM32, and the backup battery is used to supply power. When first programming, the current date and time is wrote, after that the RTC module will automatically count. It also can be used to set the alarm time. Software programming can read the current time and conversion it, then it will be displayed on the LCD screen. The test results show that the design has the advantages of accurate timing, high reliability and long service life.

1. Introduction

In production and life, people need to know the exact time. Ancient times, there are many kinds of method. Such as the ancient hourglass, burning incense, etc.. In modern society, with the development of industry, there are all kinds of mechanical clocks.

In modern times, with the development of electronic industry, there have been a variety of electronic components, such as crystal oscillator, counter module and so on. This device is not only compact, but also accurate timing. A variety of counter modules, in its periphery connected to a 32.768K crystal, can produce accurate 1s information. Other hours and minutes can be calculated according to the second.

In this paper, the use of STM32 internal RTC module timing, the time can be calculated more accurately.

2. The Working Principle of Electronic Clock

The block diagram of the RTC is shown in figure 1. The figure includes two parts, namely the APB1 bus, through which the RTC read and write operations, and the other part is the RTC register, through different settings to control the work of RTC.

Content from this work may be used under the terms of the Creative Commons Attribution 3.0 licence. Any further distribution of this work must maintain attribution to the author(s) and the title of the work, journal citation and DOI.
Published under licence by IOP Publishing Ltd

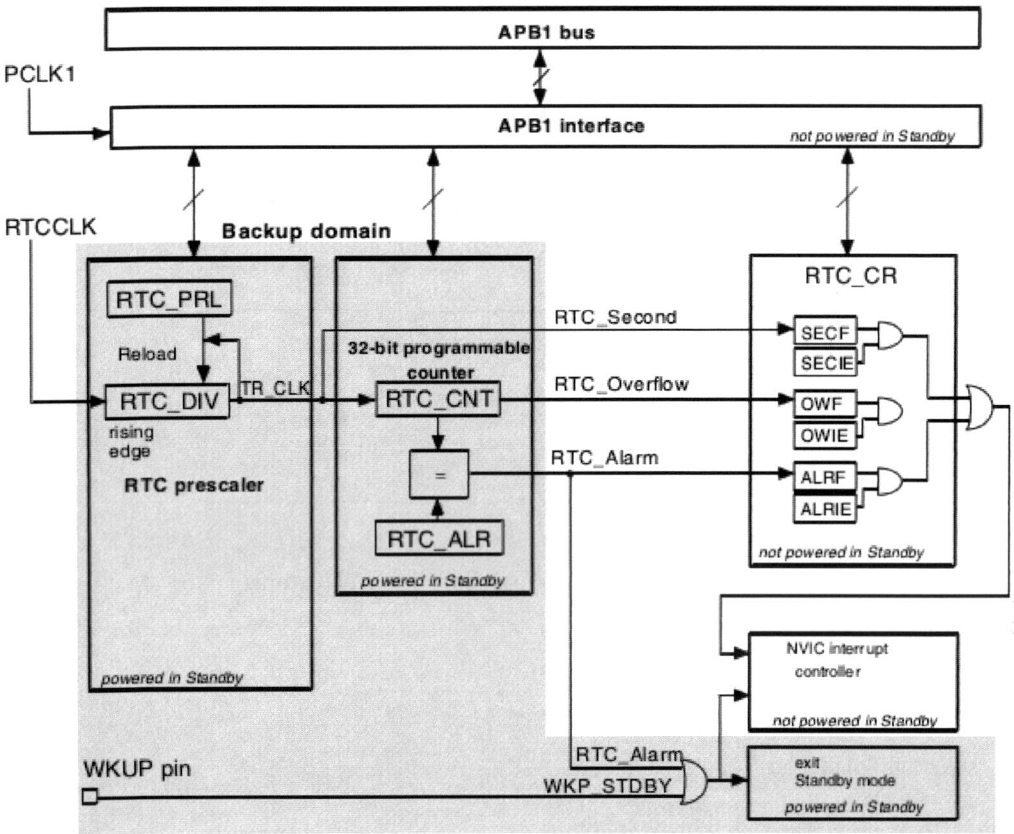

Figure 1 RTC Block Diagram

The schematic diagram of the key, as shown in Figure 2, a total of 4 keys, you can set the alarm time.

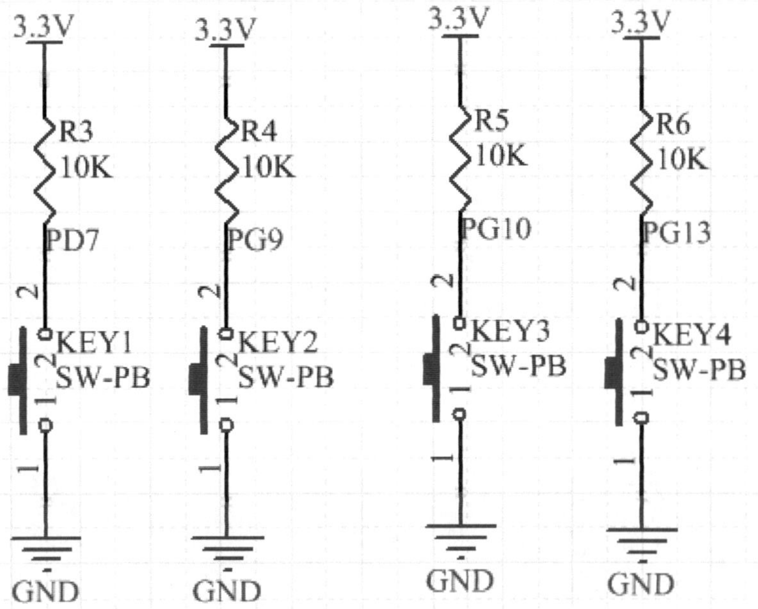

Figure 2 Key Button Circuit

IOP Conf. Series: Materials Science and Engineering **242** (2017) 012126 doi:10.1088/1757-899X/242/1/012126

3. Software Program

Programming, in KEIL 5 use the C programming language. Read and write operations on the RTC as shown in figure 3. The following part of the code:

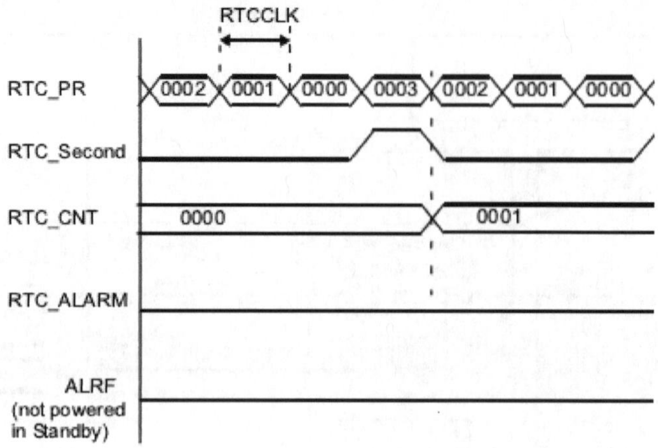

Figure 3 RTC Second and Alarm Waveform

```
u8 RTC_Init(void)
{
    u8 temp=0;
    if(BKP->DR1!=0X5050)
    {
        RCC->APB1ENR|=1<<28;
        RCC->APB1ENR|=1<<27;
        PWR->CR|=1<<8;
        RCC->BDCR|=1<<16;
        RCC->BDCR&=~(1<<16);
       RCC->BDCR|=1<<0;
       while((!(RCC->BDCR&0X02))&&temp<250)
        {
            temp++;
            delay_ms(10);
        };
        if(temp>=250)return 1;
        RCC->BDCR|=1<<8;
        RCC->BDCR|=1<<15;
        while(!(RTC->CRL&(1<<5)));
    while(!(RTC->CRL&(1<<3)));
    RTC->CRH|=0X01;
    while(!(RTC->CRL&(1<<5)));
        RTC->CRL|=1<<4;
        RTC->PRLH=0X0000;
        RTC->PRLL=32767;
        RTC_Set(2017,2,23,17,9,00);
        RTC->CRL&=~(1<<4);
        while(!(RTC->CRL&(1<<5)));
        BKP->DR1=0X5050;
        printf("FIRST TIME\n");
    }else
```

IOP Conf. Series: Materials Science and Engineering **242** (2017) 012126 doi:10.1088/1757-899X/242/1/012126

```
    {
    while(!(RTC->CRL&(1<<3)));
    RTC->CRH|=0X01;
    while(!(RTC->CRL&(1<<5)));
        printf("OK\n");
    }
    MY_NVIC_Init(2,2,RTC_IRQn,2);
    return 0;
}
```

4. Generation

This paper describes the use of STM32 design principle and method of electronic clock, with built-in RTC module. Easy to set up the initial time. In the software programming, according to the fixed time to read and write, and calculate. You can also set the alarm time by pressing the button, when the time to the clock, the driver to make a sound. The design has some practical value.

Reference

[1] Huang Jian, Zhang Shanwen, Zhou Duan. An improved algorithm of A/D sampling software filter based on [J] STM32, instrument technology and sensor.2016,3:83-85.
[2] Liu Xin, Yang Feng, AT89C52. Design of multifunctional electronic clock based on MCU [J],,.2010,36 (4): 61-64.
[3] Zhao Wenlai, Yang Junxiu, Yan Guohong, Chen Qiumei. Design and implementation of [J] electronic clock based on FPGA Journal of Zhejiang Sci-Tech University,.2010,27 (4): 590-594.
[4] Xie Jiaxing, Wang Jian,,.LED dot matrix display multifunction digital electronic clock design [J], software.2014,13 (4): 93-95.
[5] Huang Gang, Lin Li. Design of remote control electronic clock system based on single chip microcomputer [J], experimental science and technology,.2014,12 (4): 62-64.

IOP Conf. Series: Materials Science and Engineering **242** (2017) 012127 doi:10.1088/1757-899X/242/1/012127

Development of Temperature Acquisition System based on STM32

Jian HUANG

XiJing University, Xi'an, Shaanxi, China

E-mail:565200245@qq.com

Abstract: In this paper, the method of temperature acquisition by STM32 is described. The hardware circuit is designed. In order to make the measurement value accurate and reliable, the software is programmed,the temperature sample for 100 times. Remove the maximum number of twenty, the same as the minimum, with the remaining 60 data to sum, and calculate the average value as the final acquisition value. This mean filtering algorithm can effectively remove interference. The test results show that the design can accurately collect the temperature value, and has a certain practical value.

1. Introduction

In production and life, it is often to measure the natural temperature, or the surface temperature of some objects. At present, there are many ways to measure temperature. Mercury thermometer, temperature sensor, etc.. Among them, the mercury thermometer uses the principle of thermal expansion and contraction of the object to measure, the accuracy is poor.

In industrial production and experiment, the temperature is often measured by various temperature sensors to obtain the accurate temperature value. DS18B20 is often used as temperature sensor, the interface circuit is very simple, only 3 pins. One is the power supply, the other is the ground wire, the middle of the pin can output 8 or 12 digits, convenient connection with the Microcontroller. And the measurement precision is higher.

In this paper, the high performance microprocessor STM32 is used to realize the data acquisition of DS18B20, the data is filtered by software, and the results are displayed on the LCD screen.

2. The Principle of Temperature Sample

The STM32 pin diagram is shown in figure 1. PE3 is the temperature of the external input pin. Connected to the data output pin of the DS18B20.

Figure 2 is the DS18B20 connection diagram, one pin connected to the +3.3V voltage, the other pin connect to the GND. The output value of the central pin output is 12 bit digital. The STM32 is used to process the digital data and convert it into the actual temperature.

Content from this work may be used under the terms of the Creative Commons Attribution 3.0 licence. Any further distribution of this work must maintain attribution to the author(s) and the title of the work, journal citation and DOI.
Published under licence by IOP Publishing Ltd

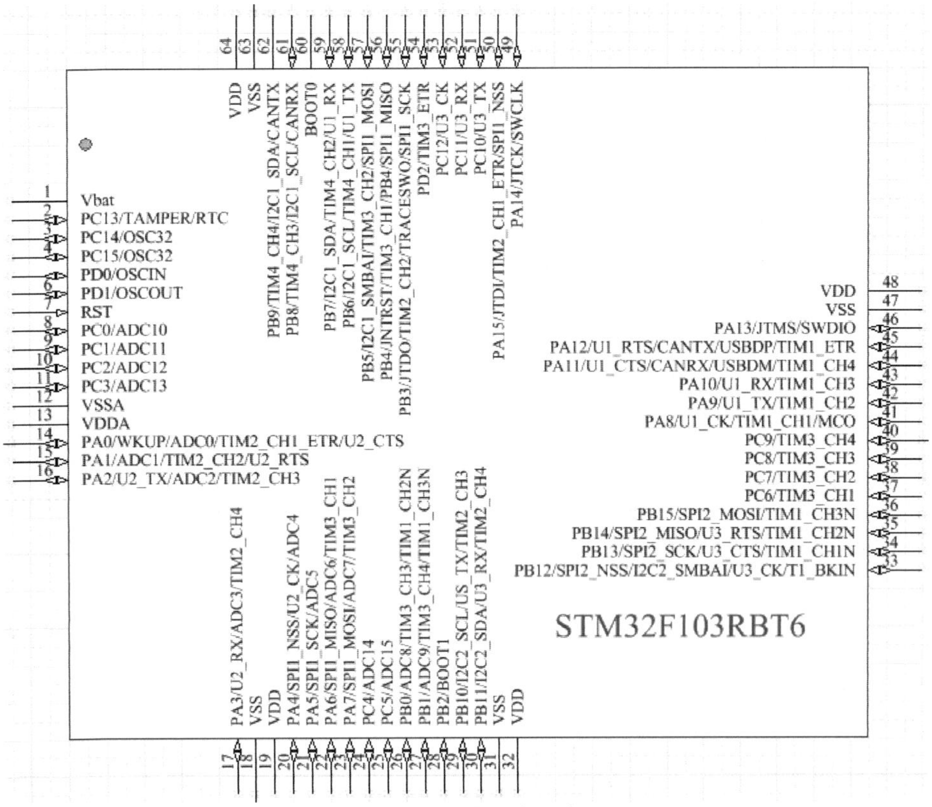

Figure 1 The Pin of STM32

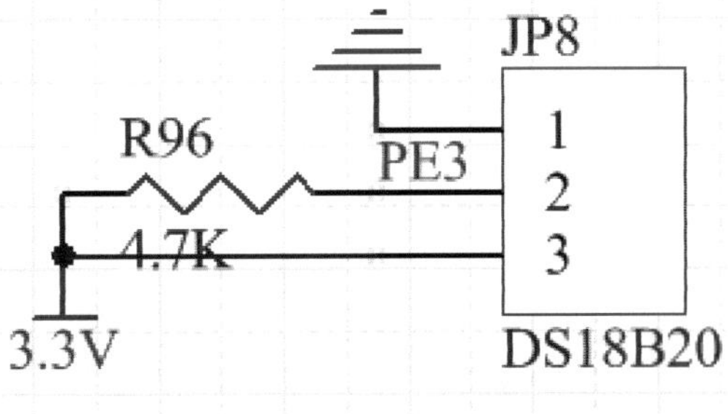

Figure 2 Temperature Sample

3. Software Program

In this design, in KEIL 5 use the C programming language. Here are some of the code, the function DS18B20_Init () to complete the initialization of the DS18B20 function. DS18B20_Read_Byte () to complete the read one byte of DS18B20 , DS18B20_Write_Byte () to complete the write one byte of DS18B20.

```
u8 DS18B20_Init(void)
{

    GPIO_InitTypeDef GPIO_InitStructure;
    RCC_APB2PeriphClockCmd(RCC_APB2Periph_GPIOE, ENABLE);
    GPIO_InitStructure.GPIO_Pin = GPIO_Pin_3;
```

```
    GPIO_InitStructure.GPIO_Mode = GPIO_Mode_Out_PP;
    GPIO_InitStructure.GPIO_Speed = GPIO_Speed_50MHz;
    GPIO_Init(GPIOE, &GPIO_InitStructure);
     DS18B20_Rst();
     return DS18B20_Check();
}
u8 DS18B20_Read_Byte(void)
{
    u8 i,j,dat;
    dat=0;
     for (i=1;i<=8;i++)
     {
        j=DS18B20_Read_Bit();
        dat=(j<<7)|(dat>>1);
    }
    return dat;
}
void DS18B20_Write_Byte(u8 dat)
 {
    u8 j;
    u8 testb;
     DS18B20_IO_OUT();//SET PA0 OUTPUT;
    for (j=1;j<=8;j++)
     {
        testb=dat&0x01;
        dat=dat>>1;
        if (testb)
        {
            DS18B20_DQ_OUT=0;// Write 1
            delay_us(2);
            DS18B20_DQ_OUT=1;
            delay_us(60);
        }
        else
        {
            DS18B20_DQ_OUT=0;// Write 0
            delay_us(60);
            DS18B20_DQ_OUT=1;
            delay_us(2);
        }
    }
}
```

4. Generation

This paper describes the working principle of temperature acquisition using STM32, and designs the interface circuit of DS18B20. In the software programming, the DS18B20 initialization, read and write operations are completed, and the data are processed by means of mean filtering algorithm. The experimental results show that the design can accurately measure the temperature value, and the precision is accurate to 0.1 degree, which can meet the needs of daily life and production.

Reference:

[1] Huang Jian, Zhang Shanwen, Zhou Duan. An improved algorithm of A/D sampling software filter based on [J] STM32, instrument technology and sensor.2016,3:83-85.
[2] Dai LAN, Jiang Yanfeng, Liu Wenkai. Research on high speed and high precision ADC measurement based on [J] Matlab, computer measurement and control.2010,18 (9): 2044-2049.
[3] Yang Junfeng, Li Zheying, Xiulimei. [J] design and error analysis of transconductance amplifier high-speed pipeline ADC Journal of electronic measurement and instrument,.2014,28 (5): 538-544.
[4] Qin Guojie, Liu Guoman, Gao Meiguo, Fu Xiongjun, Xu Peng. A time interleaved ADC time mismatch adaptive error correction method [J] Chinese Journal of scientific instrument.2013,34 (12): 1730-1735.
[5]Li Tao, Dai Yutang . A new type of high sensitivity fiber optic temperature sensor [J], photoelectron. Laser.2014,25 (4): 625-630. ()

IOP Conf. Series: Materials Science and Engineering **242** (2017) 012128 doi:10.1088/1757-899X/242/1/012128

Research Based on AMESim of Electro-hydraulic Servo Loading System

Jinlong Li[1,a)], Zhiyong Hu[1,b)]

[1] Inner Mongolia University of Technology, College of Mechanics Engineening, Hohhot, Inner Mongolia010051,China

[a)] 1404211748@qq.com
[b)] Corresponding author:lijinlong10144@163.com

Abstract. Electro-hydraulic servo loading system is a subject studied by many scholars in the field of simulation and control at home and abroad. The electro-hydraulic servo loading system is a loading device simulation of stress objects by aerodynamic moment and other force in the process of movement, its function is all kinds of gas in the lab condition to analyze stress under dynamic load of objects. The purpose of this paper is the design of AMESim electro-hydraulic servo system, PID control technology is used to configure the parameters of the control system, complete the loading process under different conditions, the optimal design parameters, optimization of dynamic performance of the loading system.

1. Introduction

The electro-hydraulic servo loading system is one of the main equipment in the aircraft design stage of semi physical simulation on the ground, with the loading process of various loads to simulate the aircraft in launching and flying in the process of the bear, in order to test the aircraft shell strength and other mechanical properties, and provide experimental data for the development of new aircraft, the performance of applications [1-2]. Because of the servo hydraulic control system is a force feedback system for simulating various loads in the simulation process, the electro-hydraulic servo loading system is a complex system which is very complex in the machine - electric - liquid, which relates to the comprehensive content of system dynamics, hydraulic transmission, control theory and computer control science.

2. Eiectro-hydraulic serve loading system principle and composition

2.1 Basic Principle of Electro-hydraulic Servo Loading System

Hydraulic servo control system is a kind of control system which is based on hydraulic cylinder mechanism. It can not only automatically quickly and accurately reproduce the input output variations, and can carry out amplification and to transform the input signal. The utility model has the advantages of small size, light weight, small inertia, high torque and high control precision [3]。

Electro hydraulic servo loading system, as a kind of hydraulic servo control system, can be divided into two categories according to the motion of the object. The first one is on the structure and material of the

Content from this work may be used under the terms of the Creative Commons Attribution 3.0 licence. Any further distribution of this work must maintain attribution to the author(s) and the title of the work, journal citation and DOI.
Published under licence by IOP Publishing Ltd

static and dynamic strength test loading system, known as static load or active loading, structure of the system is relatively simple, easy to realize correction; the other is to load the active motion of the object, called motion loading or passive loading, because this kind of system interference bearing object motion parameters, the system structure is complex, it is difficult to analyze and design.

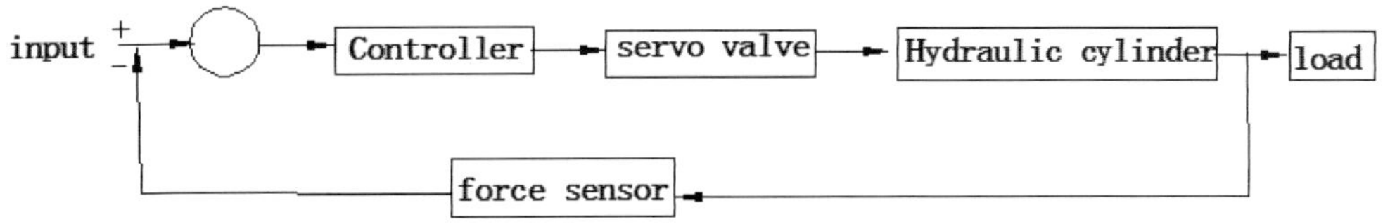

FIGURE 1. Diagram of electro hydraulic servo loading system

The design of the electro-hydraulic servo system block diagram as shown in fig 1. It mainly consists of input, controller, servo valve, hydraulic cylinder, load and sensor. Stretching through the servo valve control hydraulic cylinder, so as to realize the load control, the load force sensor through the sensor installation, incoming input in the load test data of input data and the feedback data are compared, the difference re input to the controller, to achieve control of the servo valve flow.

2.2 Composition of Electro-hydraulic Servo Loading System

Electro hydraulic servo system is generally composed of the loading system and the load object, the executive component has a hydraulic cylinder and a hydraulic motor. The design of the hydraulic system is mainly composed of hydraulic pump, hydraulic cylinder, servo valve, servo controller, relief valve, force sensor, accumulator, filter, cooler and so on. In the AMESim to establish electro-hydraulic servo loading system hydraulic model, as shown in Figure2.

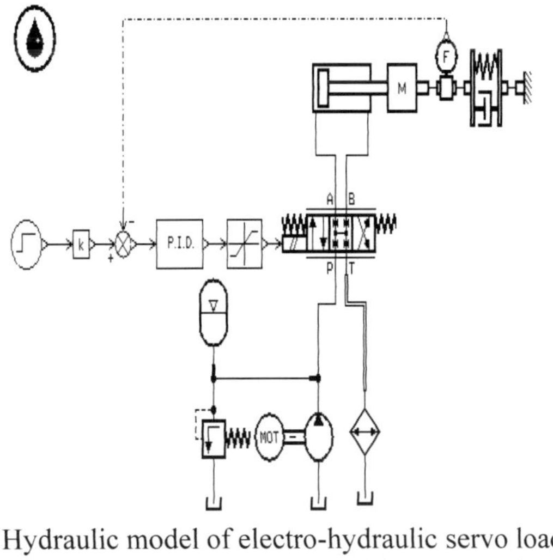

FIGURE 2. Hydraulic model of electro-hydraulic servo loading system

The electro-hydraulic servo loading system adopts closed-loop control, tension and pressure signal of You Li sensor loading hydraulic cylinder output, measured after feedback to the controller and input signal before comparing the error signal formed by PID controller after input to the servo valve, servo valve spool movement, increase or decrease the output force servo the output pressure and loading cylinder, the force sensor and the input error is reduced, and the hydraulic cylinder output force is equal to the input.

3. Simulation and analysis ofelectro hydraulic servo loading system

3.1 Simulation of Electro-hydraulic Servo Loading System

In this paper, the electro-hydraulic servo loading system, the use of AMEsim simulation software, the simulation model of the loading system the simulation analysis. According to the above calculation, the parameters of the hydraulic components are determined, the parameters of the components are set in the software, and the leakage of the oil is ignored.

The parameters of the hydraulic simulation system are set, as shown in table1.

Table 1. Parameters of electro hydraulic servo loading system

Rated current of servo value(mA)	150
Servo valve natural frequency(Hz)	50
Hydraulic cylinder rod chamber volume (cm^3)	700
Hydraulic pump speed(r/min)	1450

Set AMESim simulation time is 10s, the simulation step is 0.1s, select the following 3 groups of PID controller parameters：Kp=6，Ki=0，Kd=0；Kp=6，Ki=0.1，Kd=0；Kp=30，Ki=0，Kd=0。The hydraulic model of AMESim is simulated, and the influence of PID controller on the input signal is compared.

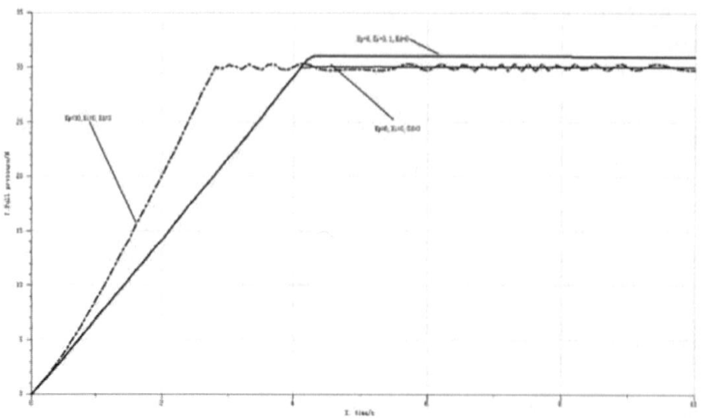

FIGURE 3. System response of different PID parameters

The response of different PID parameters to the system is shown in figure 3. You can see from Figure 3, the input pressure of 30N, when Kp=30, Ki=0, Kd=0, 2.7s in the system when the system reaches the specified value, but large overshoot and instability of the system; when Kp=6, Ki=0.1, Kd=0, 4.5s in the system is stable, and no overshoot. The error is 2N greater than the input value; when Kp=6, Ki=0, Kd=0, 4.3s in the system is stable, and no overshoot, steady-state error is 0.1N. The simulation result shows that the increase of the proportion coefficient Kp, can shorten the system response time, speed up the system response, reduce the steady-state error, but the proportion coefficient is too large, causing the overshoot will increase, not conducive to the stability of the system. Increasing the integral coefficient Ki can reduce the overshoot of the system and reduce the steady-state error of the system.

3.2 Effect of PID Value on Hydraulic Cylinder

Electro hydraulic servo loading system output force depends on the size of the hydraulic cylinder two large hydraulic pressure. The stability of the two chamber pressure of the hydraulic cylinder directly determines the response time, overshoot and steady-state error of the output force, which mainly depends on the change of the hydraulic pressure of the hydraulic cylinder in the two chamber. As shown in Figure 4, Kp=6, Ki=0, Kd=0 pressure change curve of the two cylinder. You can see from chart of hydraulic cylinder pressure changes after 4.3s, two cavity stabilized, rod chamber pressure is 2.7MP, non rod chamber pressure of 1.5MP, combined with the response of the system is shown in Figure 3, after 4.3s, the stability of output power, and no overshoot, steady-state error is sma

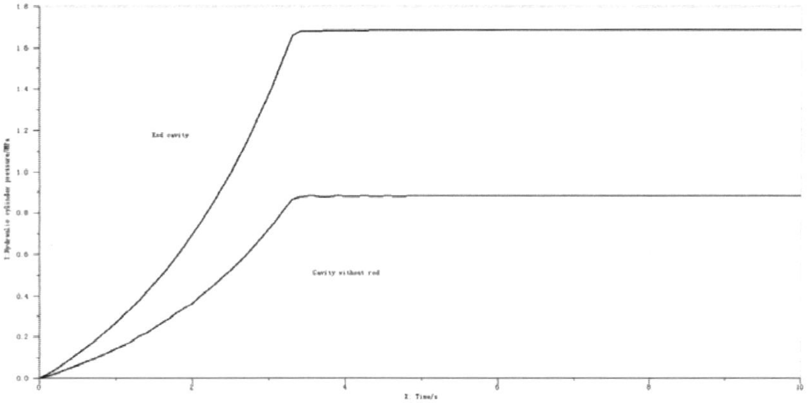

FIGURE 4. Kp=6, Ki=0, Kd=0 hydraulic cylinder pressure curve

When Kp=6, Ki=0.1, Kd=0, the pressure change of the hydraulic cylinder system is shown in figure 5. The pressure of the hydraulic cylinder is increased, the pressure of the rod cavity is increased rapidly, and the pressure of the non rod chamber is kept constant.

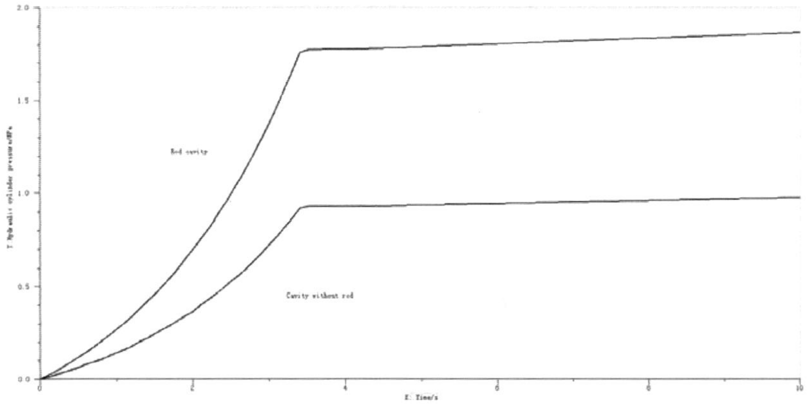

Figure 5. Kp=6, Ki=0.1, Kd=0 hydraulic cylinder pressure curve

4. Conclusion

(1)Establish the hydraulic system diagram of the electro-hydraulic servo loading system, and analyze the response of the AMESim system.

(2)Using different PID control parameters to simulate the system, the optimal control parameters are obtained, and the influence of different parameters on the force response is analyzed.

(3)Under different parameters, the size of the pressure change of the rod and the rod cavity of the hydraulic cylinder is obtained, and the optimal PID value is obtained.

IOP Conf. Series: Materials Science and Engineering **242** (2017) 012128 doi:10.1088/1757-899X/242/1/012128

References

[1] Wang Xin. Design and experimental study of electro-hydraulic servo loading table [D]. Northwestern Polytechnical University, 2003
[2] Huang Yong.Research and design of electro-hydraulic servo loading system [D]. Northwestern Polytechnical University, 2002
[3] Wei Liejiang, Han Xiao, Li Nana, Zhang Jing, Liu Wenguang. High frequency electro-hydraulic servo loading system installation frame structure optimization method of [J]. machine tool & hydraulics, 2016,05:5-8.

IOP Conf. Series: Materials Science and Engineering **242** (2017) 012129 doi:10.1088/1757-899X/242/1/012129

Research on Design of MUH Attitude Stability Augmentation Control System

Shigang Fan[1, a)]

[1]High-tech Institute,Fan Gong-ting South Street on the 12th, QingZhou 262500 , China.

Abstract. Attitude stability augmentation control system with a lower cost need to be designed so that MUH (Mini Unmanned Helicopter) can adapt to different types of geographic environment and fly steadily although the weather may be bad. Attitude feedback was calculated mainly by filtering estimation within attitude acquisition module in this system. Stability augmentation can be improved mainly by PI. This paper will depict running principle and designing process of MUH attitude stability augmentation control system and algorithm that is considered as an important part in this system.

1. Introduction

With the advance of science and technology, research on MUH (unmanned aerial vehicle) has become a hot point within intelligent control area. Researchers versatilely study flying stability control of MUH and application of MUH[1]. With the development of MUH technology, nowadays MUH has been applied widely in many areas in our country and MUH technology has made much more progress. If someone wants to use this kind of technology, he/she must have a systematic, complete and full cognition on this technology.

MUH (Mini Unmanned Helicopter) has a very important value in many areas including civil use, military use and science area in our country. MUH has the characteristic of high performance and low-cost. The model of human-machine separation plays an important role in investigating situation, aerial imaging, site monitoring of disaster and other aspects. Stability of MUH is a prerequisite for all kinds of complex tasks. If flying state can not be controlled well, MUH would jitter and fly unsteadily, even crash. As a result, MUH can not achieve flying tasks[2].

2. Running Principle of MUH Attitude Stability Augmentation Control System

Flight control system is the main part of MUH (Mini Unmanned Helicopter). MUH attitude stability augmentation control system is an effective way to guarantee MUH to accomplish tasks smoothly. Figure 1 is a frame structure diagram of the whole MUH (Mini Unmanned Helicopter) attitude stability augmentation control system. According to the diagram, it can be known that MUH attitude stability augmentation control system is composed by six parts that are should be regarded as a complete and unified system. The six parts are remote sensor, ground station, sensor, receiver, controller and actuator. In this system, these machines are assembled separately into two big parts which are ground and airborne. MUH attitude stability augmentation control system can voluntarily tell the current state of MUH (Mini Unmanned Helicopter) through the coordination of these two big parts. If so, the stability of MUH (Mini Unmanned Helicopter) can be increased in flying process and other running process so that operation difficulty, which is caused by complex composition can be decreased further. Flight operators can be assisted to control stability and accuracy of MUH (Mini Unmanned Helicopter) manually so that MUH (Mini Unmanned Helicopter) can be in a stable flying state.

Content from this work may be used under the terms of the Creative Commons Attribution 3.0 licence. Any further distribution of this work must maintain attribution to the author(s) and the title of the work, journal citation and DOI.
Published under licence by IOP Publishing Ltd

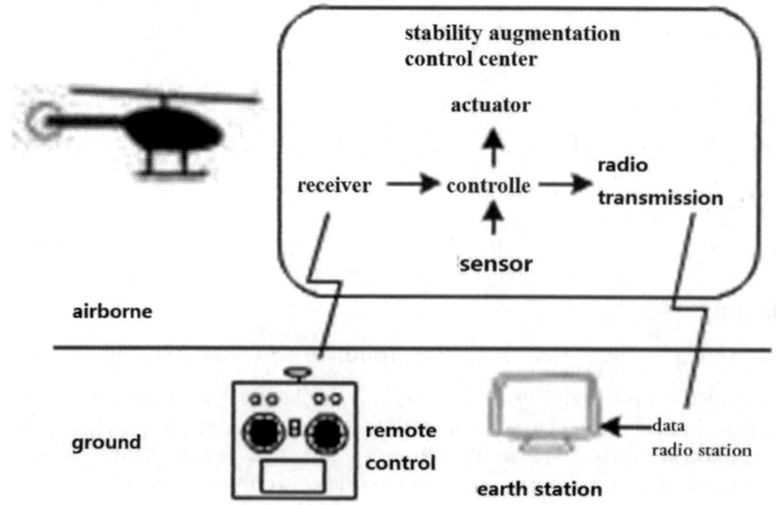

FIGURE 1. Structure of MUH (Mini Unmanned Helicopter) Attitude Stability Augmentation Control System

3. Design of MUH Attitude Stability Augmentation Control System

3.1 Hardware Designing of MUH Attitude Stability Augmentation Control System
Attitude stability augmentation control system of MUH (Mini Unmanned Helicopter) mainly adopted the model and method of internal loop control. Meanwhile, there are also other primary different models: flying attitude maintaining and flying attitude controlling of MUH and flying speed controlling of MUH. Based on triaxial gyro and tilt sensor, the two things are used to acquire data and information about flying attitude of MUH. And then, the elevator and the rudder can be controlled so that MUH can keep stable and controlled during the flying process. As a basic part in outer loop control of MUH, the internal loop control plays a very important role in keeping MUH's flying stable and safe. Internal loop is a attitude stability augmentation controlling loop that consists of triaxial gyro and tilt sensor[3]. Internal loop interface connection is shown in figure 2.

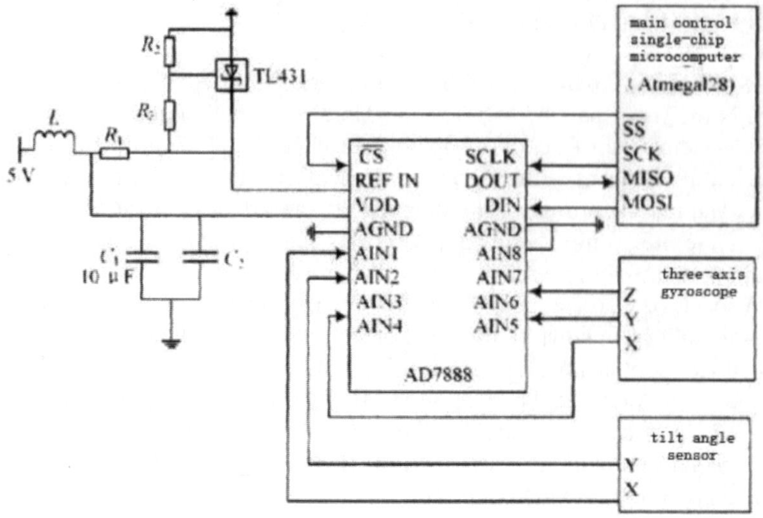

FIGURE 2. Internal Loop Interface Connection
As shown in figure 2, AD7888 is an ADC(analog-digital converter) with high calculating speed and

IOP Conf. Series: Materials Science and Engineering **242** (2017) 012129 doi:10.1088/1757-899X/242/1/012129

low energy consumption. Its transferring rate reaches up to 125KSPS and AD7888 possesses multipath analog input channel. External voltage can guarantee that AD7888 has stable energy during running process. In order to keep stability of reference voltage, TL431 will be usually used as reference voltage. As a result, loop is composed of above parts. The advantage of TL431 is to regulate the three ends of the shunt stably. The single chip of TL431 and AD7888 are connected and equipped with the host and the slave. The function of triaxial gyro is to measure the angular velocity during MUH flying process, such as pitch angle velocity, roll angle velocity and yaw angle velocity. SCA100T is a kind of tilt sensor that is made by a Finland company. The advantage of the tilt sensor is its small volume and light mass. All the above hardware compose jointly hardware designing of MUH attitude stability augmentation control system.

3.2 Software Designing of MUH Attitude Stability Control System

Flying data and numerical value collected by the sensor is input values and inputs of flying control system in the whole flying process. Inputs in this system is the control variable of the equation when MUH is in flying state, so from a certain view, flying system is a control system that possesses many channels in essence. Flying system can also be called control system with much input and little output. Within this system, as a core part in flying control system, manipulating loop is commonly said as internal loop by the public. Feedback information collected by pitch angle velocity or roll angle velocity is the basis to form flying control system. And then feedback information composes a more safe flying attitude and loop that is controlled by flying system during MUH is in flying state.

In the design of the internal loop, external loop must be introduced so that the whole design can be more reasonable and flying state of MUH can be stable. Because of the symmetry of the longitudinal plane of MUH, flying control of MUH can be divided into two mutually independent controlling channels, which are transverse controlling channel and longitudinal controlling channel[4]. Firstly, the transverse controlling channel is used to control feedback of roll angle and roll angle speed, which is caused by rolling and yaw motion of MUH. The roll angle and roll angle speed jointly compose core loop of roll channel which is also called internal loop. Secondly, longitudinal controlling channel is a controlling channel that can make MUH do pitch and lift motion by manipulating elevators of MUH when MUH is doing longitudinal motion or in pitch and lift process. Composing principle of longitudinal controlling channel is similar to transverse controlling channel's. Longitudinal controlling channel is a core controlling loop, which is also called internal loop. It is composed by feedback of pitch angle and pitch angle's speed. Content in figure 3 is the framework sketches of longitudinal controlling pitch channel.

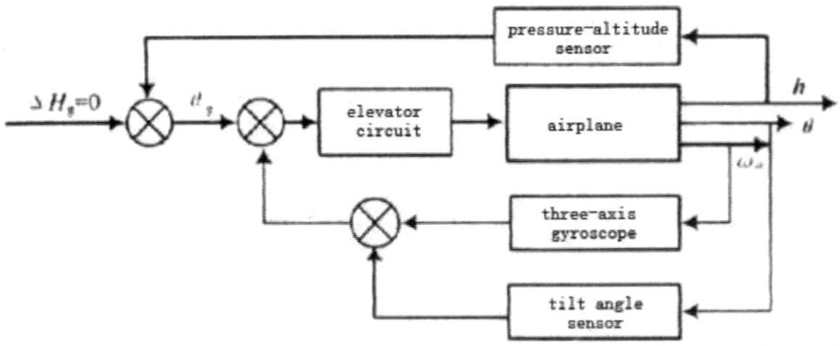

FIGURE 3. Framework Sketches of Longitudinal Controlling Pitch Angle Channel

Flying process of MUH can be divided into five independent phases. They are taking off phase, flying phase during flight, turning phase, hovering phase and landing phase. Within the six phases, tasks at every phase are independent and different from each other. Every phase has their own characteristic and speciality. Tasks belonging to different phases should coordinate with each other so that they can compose tasks of a complete flying phase. It is just like all the flight phases should blend so that they can compose a complete flying phase. They are closely related to each other. Complex flying state at every phase can be divided into three basic flying tasks. They are flying during flight, turning and elevating.

The flying state can connect with transverse controlling channel and longitudinal controlling channel so that all of them can accomplish flying tasks[5].

4. Conclusion

MUH attitude stability augmentation control system is given a lot of attention and has become a hot point and a key point because of its complexity. Attitude stability augmentation control system with remarkable features is a basic and prominent system that can guarantee MUH to accomplish tasks smoothly and efficiently. Only if learning and mastering fully above attitude stability augmentation control system, people can construct and use attitude stability augmentation control system so that this system can promote social development.

Reference

[1] Z.H.Cheng: The Design of PID Controller for Attitude Angle Stabilization Based on Four-rotor Dynamic Model[J], Machine Design and Manufacturing Engineering, 2015,44(12)：47-51.
[2] C.W.Mei, H.Liu and H.L.Pei: Research and Design of Engine Control System for Small-scaled Unmanned Helicopter[J], Computer Engineering and Design, 2017(3):749-755.
[3] Z.H.Ma, J.Chen, L.P.Chen and W.T.Fan: Gyroscope Achieves the Four Rotor UAV Attitude Stabilization Control[J], Computer Simulation, 2016,33(10):32-38.
[4] Y.L.Yu, J.K.Zu, Z.L.Liao, etc: The Research on Attitude Determination Algorithm Based on Kalman Filter for Unmanned Helicopter[J], Electronic Measurement Technology, 2015, 38(5):106-110.
[5] P.B.Luo, B.Han, K.Yu and P.Li: Design and Realization of Attitude Stability Augmentation Control System for Mini Unmanned Helicopter[J], Computer Engineering, 2012, 38(4):212-214.

A Filter-Filled Function Method for Global Optimization with Box Constraints

Hongxia Ji [a]**, Zhiyuan Tian and Yunxia Gu**

School of Mathematics and Statistics, Qingdao University, Qingdao 266071,China.

E-mail:991826157@qq.com

Abstract. A single parameter filled function for solving global optimization problems with box constraints is proposed and its properties are investigated. The filter technique avoids the difficulty of selecting the penalty parameter and has a good numerical results in the local optimization algorithm. The filter is introduced and combined with the filled function to form the filter-filled function algorithm in this paper. The numerical results illustrate that the algorithm is reliable and effective.

1. Intronduction

Consider the global optimization problem as follows :

$$(P) \quad \min_{x \in X \subset R^n} f(x) \qquad (1)$$

Where $f: R^n \to R$ is a continuous differentiable function. $X \subset R^n$ is a closed bounded box, where $L \le X \le U, L, U \subset R^n$.

The global optimization problems are widespread in various fields of engineering, technology, finance as well as social science. Ge[1]first proposed the definition of the filled function, many scholars have made the further research on the filled function during last few decades[2-5].The filter technique is proposed by Leyffer and Fletcher[6].In the process of minimizing the filled function, filter is introduced to supervise the iteration and determine whether to accept the current iteration point.

The framework of this paper is as follows: Some assumptions and definitions are given in section2. In sect.3, a new filled function is proposed and its properties are investigated. In Sect.4, we give a new filter filled function algorithm. In Sect.5, some experiments and numerical results are reported. The final section gives the summary.

2. Assumptions and definitions

Assumption 1. $f(x)$ is a Lipschitz continuously differentiable over $X \subset R^n$,i.e., there exists a constant $L > 0$ such that $|f(x) - f(y)| \le L\|x - y\|$ holds for all x, $y \in X$, X is a closed bounded box.

Assumption 2. Suppose $f(x)$ has no infinite number of local minima in X .

Definition 3. A point x^k takes priority over another point x^l if and only if $f(x^k) \le f(x^l)$ and $T(x^k) \le T(x^l)$.

Definition 4. A iteration point is added in the filter set if and only if the point is not dominated by any points in the filter set and the points controlled by the current iteration point are removed. The process is called the update of the filter set. Let filter set is F_{k-1} at present, the new point is added in the filter set, the

Content from this work may be used under the terms of the Creative Commons Attribution 3.0 licence. Any further distribution of this work must maintain attribution to the author(s) and the title of the work, journal citation and DOI.
Published under licence by IOP Publishing Ltd

filter set is updated to

$$F_k = F_{k-1} \bigcup \{x_k\} / \{x_l \in F_{k-1} \mid \{f(x_k) < f(x_l), T(x_k) < T(x_l)\}\}. \quad (2)$$

3. A filter filled function and its properties

The single parameter filled function for problem (1):

$$T(x, x_1^*, \rho) = -\frac{1}{\rho}\|x - x_1^*\|^2 (f(x) - f(x_1^*)) - \rho\|x - x_1^*\|^3. \quad (3)$$

Where $\rho > 0, x_1^* \in L(P)$, all local minima of $f(x)$ are contained in $L(P)$. The filled function is shown by the following theorems, when $\rho > \sqrt{L} > 0$ is satisfactorily large.

Theorem 5. If $x_1^* \in L(P)$, then x_1^* is a strictly local maximizer of $T(x, x_1^*, \rho)$.

Proof. Since $x_1^* \in L(P)$, there exists a neighborhood $U(x_1^*, \delta)$ of x_1^* with $\delta > 0$, such that $f(x) \geq f(x_1^*)$, for all $x \in U(x_1^*, \delta) \bigcap X$ and $x \neq x_1^*$.

$$T(x, x_1^*, \rho) = -\frac{1}{\rho}\|x - x_1^*\|^2 (f(x) - f(x_1^*)) - \rho\|x - x_1^*\|^3 < 0 = T(x_1^*, x_1^*, \rho). \quad (4)$$

Hence, x_1^* is a strictly local maximizer of $T(x, x_1^*, \rho)$.

Theorem 6. Let x_1^* is a local minimizer of (P), the function $T(x, x_1^*, \rho)$ has no stationary point in the region $S_1 = \{x \mid f(x) \geq f(x_1^*), x \in X \setminus \{x_1^*\}\}$, when $\rho > 0$ is satisfactorily large.

Proof. Let $x \in S_1$, *i.e.*, $f(x) \geq f(x_1^*)$, it follows that

$$\nabla T(x, x_1^*, \rho) = -\frac{1}{\rho}\left[2(x - x_1^*)(f(x) - f(x_1^*)) + \|x - x_1^*\|^2 \nabla f(x)\right] - 3\rho\|x - x_1^*\|^2 \frac{x - x_1^*}{\|x - x_1^*\|}. (5)$$

$$\frac{(x - x_1^*)^T}{\|x - x_1^*\|}\nabla T(x, x_1^*, \rho) = -\frac{1}{\rho}\left[2\|x - x_1^*\|(f(x) - f(x_1^*)) + \|x - x_1^*\|^2 \frac{(x - x_1^*)^T \nabla f(x)}{\|x - x_1^*\|}\right] - 3\rho\|x - x_1^*\|^2$$

$$\leq -\frac{1}{\rho}\|x - x_1^*\|^2 \frac{(x - x_1^*)^T \nabla f(x)}{\|x - x_1^*\|} - 3\rho\|x - x_1^*\|^2. \quad (6)$$

By assumption 1, $\frac{|(x - x_1^*)^T \nabla f(x)|}{\|x - x_1^*\|} \leq \nabla f(x)$ is bounded in X. Furthermore,

we have $-\frac{1}{\rho}\|x - x_1^*\|^2 \frac{(x - x_1^*)^T \nabla f(x)}{\|x - x_1^*\|}$ is bounded. Therefore, when $\rho > 0$ is sufficiently large, we have $\frac{(x - x_1^*)^T}{\|x - x_1^*\|}\nabla T(x, x_1^*, \rho) < 0$, that indicates $T(x, x_1^*, \rho)$ has no stationary point in S_1.

Theorem 7. If $x_1^* \in L(P)$ but $x_1^* \notin G(P)$, then there is a minimizer of $T(x, x_1^*, \rho)$ at least in the region $S_2 = \{x \mid f(x) < f(x_1^*), x \in X\}$, when $\rho > \sqrt{L} > 0$ is sufficiently large.

Proof. If $x_1^* \in L(P)$ but $x_1^* \notin G(P)$, then there are a point $x_G \in L(P)$ such that $f(x_G) < f(x_1^*)$. By assumption 1, we have

$$T(x_G, x_1^*, \rho) = -\|x_G - x_1^*\|^2 (f(x_G) - f(x_1^*)) / \rho - \rho\|x_G - x_1^*\|^3$$

$$\le -\|x_G - x_1^*\|^2(-L\|x_G - x_1^*\|)/\rho - \rho\|x_G - x_1^*\|^3 = L\|x_G - x_1^*\|^3/\rho - \rho\|x_G - x_1^*\|^3$$
$$< \sqrt{L}\|x_G - x_1^*\|^3 - \sqrt{L}\|x_G - x_1^*\|^3 = 0. \quad (7)$$

Since $T(x, x_1^*, \rho)$ is continuous in X , there must be a minimizer $\bar{x}$ of $T(x, x_1^*, \rho)$, such that $T(\bar{x}, x_1^*, \rho) \le T(x_G, x_1^*, \rho) < 0$ .It implies that

$$T(\bar{x}, x_1^*, \rho) = -\|\bar{x} - x_1^*\|^2(f(\bar{x}) - f(x_1^*))/\rho - \rho\|\bar{x} - x_1^*\|^3 < 0. \quad (8)$$

Therefore, $\rho > \sqrt{-(f(\bar{x}) - f(x_1^*))/\|\bar{x} - x_1^*\|}$. Since $\rho > 0$, hence, $f(\bar{x}) - f(x_1^*) < 0$.

It implies that $\bar{x} \in S_2(x_1^*)$.

4. Filter filled function algorithm

Based on the assumptions and theorems in the above sections, a filter filled function method with one parameter is proposed as follow:

Step0 Initialization. Choose a large positive number M and a initial point $x_0 \in X; \rho = 1; G = 200$;

Step1 Minimize $f(x)$ starting from $x_0 \in X$, and obtain a minimizer $x_1^*, i = 1$;

Step2 Initialize the filter set to $F_0 = \{x_1^*\}$;

Step3 Construct the new filled function

$$T(x, x_1^*, \rho) = -\frac{1}{\rho}\|x - x_1^*\|^2(f(x) - f(x_1^*)) - \rho\|x - x_1^*\|^3 . j = 1;$$

Step 4 Choose a iteration point randomly $x_1 \in X/U(x_1^*, \sigma), \sigma > 0$;

Step 5 If $f(x_1) < f(x_1^*)$ then $x_0 = x_1$, turn to step1, else go to step6;

Step 6 Minimize $T(x, x_1^*, \rho)$ starting from x_1 ,In the process of minimizing $T(x, x_1^*, \rho)$, if finding a $x_k \in X$ and the point is accepted by the filter set, then update the filter set

$$F_k = F_{k-1} \cup \{x_k\}/\{x_l \in F_{k-1} \mid \{f(x_k) < f(x_l), T(x_k) < T(x_l)\}\}.$$

$x_0 = x_k$, then go to Step1;else go to Step7;

Step 7 If $j \le G$,then $j = j + 1$,go to Step4;else go to Step 8;

Step 8 If $\rho \le M$, then add $\rho : \rho = \rho \times 10$, go to Step 3; otherwise, stop, $x^* = x_1^*$ as the global minimizer of (P);

5. Numerical experiments

In this section, we apply the proposed algorithm to two test examples. Numerical results prove that the method is efficient. Some symbols used in the following tables1 and 2 are given as follows:

i: The iterations in finding the ith local minimizer.

x_i^0 : The ith initial point for the objective function.

x_i^* : The ith local minimizer. F_i : The ith filter set.

$f(x_i^*)$: The evaluation of function of the ith local minimizer.

Problem1. $\min f(x) = 4x_1^2 - 2.1x_1^4 + x_1^6/3 - x_1x_2 - 4x_2^2 + 4x_2^4, s.t. -3 \le x_1, x_2 \le 3$.

Problem2. $\min f(x) = x_1^2 + x_2^2 - \cos(18x_1) - \cos(18x_2), s.t. -1 \le x_1, x_2 \le 1$.

TABLE1. The numerical results for problem 1

i	x_i^0	x_i^*	$f(x_i^*)$	F_i
1	(-2;-1)	(-1.703606;-0.796083)	-0.215463	(-0.563866;-0.730267)
2	(-0.563866;-0.730267)	(-0.089842;-0.712656)	-1.031628	(-0.089842;-0.712656)

TABLE2. The numerical results for problem 2

i	x_i^0	x_i^*	$f(x_i^*)$	F_i
1	$\begin{pmatrix}1.0000000000\\1.0000000000\end{pmatrix}$	$\begin{pmatrix}1.0000000000\\1.0000000000\end{pmatrix}$	0.6793665835	$\begin{pmatrix}0.1903221267\\-0.2972792932\end{pmatrix}$
2	$\begin{pmatrix}0.1903221267\\-0.2972792932\end{pmatrix}$	$\begin{pmatrix}0.6938434921\\-0.6938423271\end{pmatrix}$	-1.0312074241	$\begin{pmatrix}0.0457380489\\0.3754688632\end{pmatrix}$
3	$\begin{pmatrix}0.0457380489\\0.3754688632\end{pmatrix}$	$\begin{pmatrix}0.0000002446\\0.3469228475\end{pmatrix}$	-1.8789006513	$\begin{pmatrix}0.1077380489\\0.0001546886\end{pmatrix}$
4	$\begin{pmatrix}0.1077380489\\0.0001546886\end{pmatrix}$	$\begin{pmatrix}0.0000004850\\-0.0000002273\end{pmatrix}$	-1.9999999999	$\begin{pmatrix}0.0000004850\\-0.0000002273\end{pmatrix}$

6. Conclusions

A new filled function is proposed and its properties are investigated in this paper. In the process of the filter-filled function algorithm, the initial points are chosen randomly and the filter is introduced to supervise the iteration, whether the current point is acceptable. The numerical results illustrate that the proposed algorithm is reliable and effective.

References

[1] R. P. Ge. A filled function method for finding a global minimizer of a function of several variables. Math. Program. 46:191-204(1990).
[2] Y. J. Yang and Y. L. Shang. A new filled function method for unconstrained global optimization.[J]. Comp. Appl. Math.173:501-512(2006).
[3] H. W. Lin, Y. P. Wang and Lei Fan. A filled function method with one parameter for unconstrained global optimization.[J]. Comp. Appl. Math.218:3776-3785(2011).
[4] X. S. Zhang. A filter-filled function method for solving nonlinear programming problem.[D].Shang hai: East China University Of Science And Technology.19-21(2015).
[5] Q.Hu, W. Wang. A filled function method based on filter for global optimization with box constraints. [J].Operation Research Transaction. 20:57-67. (2016).
[6] Fletcher R, Leyffer S. Nonlinear programming without a penalty function[J]. Math. Program.91: 239-269(2002).

Design and Implementation of BDS RTCM SSR Message

Xia Xiao[1, 2,3,a)], Fang Cheng[1, 2,b)] , Feng Liu [1,2, c)] and XiaoChun Lu[1, 2,d)]

[1]I National Time Service Centre, Chinese Academy of Sciences, Xi'an 710600, China.

[2] Key Laboratory of Precision Navigation and Timing Technology, National Time Service Center Chinese Academy of Sciences, Xi'an 710600, China.

[3] University of Chinese Academy of Sciences, Beijing 100049, China.

E-mail:[a)] xiaoxia@ntsc.ac.cn, [b)] chengfang@ntsc.ac.cn, [c)] liufeng@ntsc.ac.cn, [d)] luxc@ntsc.ac.cn

Abstract. Based on the real-time differential positioning system of Beidou wide area, a set of Beidou high-precision differential information transmission protocol is studied and designed. After the system generates satellite precise orbit and clock difference products, it needs to be sent to the users in a certain data format through C band GEO communication satellite. The transport protocol should be designed according to the data volume and update rate of the system. At the same time, it is necessary to combine the signal system, satellite resources and the restriction and restriction of the receiver implementation technology. It refers to the relevant fields and international similar systems of technology and standards to design. This paper establishes a set of technical indicators which can measure the pros and cons of differential message organization and broadcasting strategy. Through modeling, simulation and actual test, this paper makes a comprehensive comparison of different strategies. Finally, a transmission protocol design scheme which has good performance and meets the system service requirements is presented.

1. Introduction

In order to realize the harmonization of different receiver differential data formats to facilitate the exchange and processing of differential data, the International Maritime Radio Technical Committee (RTCM) has developed a differential global navigation system service standard. According to the age of development, it is divided into two stages, second and third[1]. Each stage consists of a series of message structures and content standards similar to the content of the message (referred to as the RTCM 10402.X series and the RTCM 10403.X series). It has been widely used in the field of satellite navigation [4]. RTCM 10403.x and RTCM 10402.x in the message structure, teletext length, calibration methods are not the same, the latest standard is RTCM 10403.2 It supports high bandwidth, broadcast and point to multipoint communications, and also supports the following high precision land, sea and air applications [2]:

1. Single frequency, dual frequency and multi frequency RTK applications. Can get better than the sub-level real-time positioning accuracy;

2. RTK applications in single GNSS mode or multiple GNSS modes. Can get better than the sub-level real-time positioning accuracy;

3.Network RTK applications in single GNSS mode or multiple GNSS modes. Can be better than the

Content from this work may be used under the terms of the Creative Commons Attribution 3.0 licence. Any further distribution of this work must maintain attribution to the author(s) and the title of the work, journal citation and DOI.
Published under licence by IOP Publishing Ltd

sub-level positioning accuracy;

4. SSR (state space representation) applications in single GNSS mode or multiple GNSS modes. You can get real-time positioning accuracy of the sub-meter level.

In the satellite navigation and positioning, satellite orbit and clock error, ionospheric delay and other errors are independent of the location of users. These error sources together form the state space of satellite navigation and positioning.

Precision single point positioning (PPP) mainly uses accurate satellite orbit and clock parameters (calculated by the global IGS tracking network data) and the global atmospheric model to achieve single station precision positioning. By providing the user with the above-mentioned state space information, the observation error is corrected to improve the user positioning accuracy. On a global scale, the use of dual-band receivers can achieve the accuracy of the sub-meter or even sub-meter scale. This method is called in the RTCM standard state space representation, also known as wide-area difference.

2. BDS RTCM SSR transmission protocol

In this paper, the definition of the Beidou System State Space Information (BDS SSR) is based on the RTCM V3.2[3]. Data is defined in order to better broadcast, so in the definition of BDS SSR message format first need to determine the message type.

Beidou wide area differential message broadcast information includes the following five types: orbit correction, clock correction, code deviation, user ranging accuracy and fast clock correction. The data volume and update interval for each type of information are shown in Table 1.

TABLE 1. BDS SSR message type table.

Message Type	Message Name	The number of bytes	Bytes Update interval (s)	Description
1235	orbit correction	4.125+16.625*Ns=236.875	30	
1236	clock correction	4.125+9.25*Ns=133.625	10	
1237	code deviation	4.125+0.875*Ns+2.125*Ns*6=194.875	1800	Ns=Number of satellites
1238	ura	4.125+1.25*Ns=21.625	2	
1239	High speed clock correction	4.125+3.25*Ns=49.625	2	

To 14 satellites, each satellite 6 signal calculation.

1) The orbit correction information is: 4.125 + 16.625 * 14 = 236.875 bytes, rounded 237 bytes, or 1896 bits.

2) The clock correction information is: 4.125 + 9.25 * 14 = 133.625 bytes, rounded up to 134 bytes, that is, 1072 bits

3) The code offset is 4.125 + 0.875 * 14 + 2.125 * 14 * 6 = 194.875 bytes, rounded up to 195 bytes, or 1560 bits.

4) URA information is occupied as: 4.125 + 1.25 * 14 = 21.625 bytes, rounded 22 bytes, that is, 176 bits.

5) The High speed clock correction information occupies: 4.125 + 3.25 * 14 = 49.625 bytes, rounds up to 50 bytes, that is, 400 bits.

According to the parameters of the update rate can be calculated in one second at the same time broadcast all the parameters of the amount of 744bit information.

3. Coding scheme

Beidou wide area real-time differential positioning system uses GEO satellite to broadcast differential information, while receiving signals can also be received by a large number of common terminals. Therefore, the transmission scheme is designed to require an original differential information rate of up to 744bps and a bit error rate of less than 10^{-7}.The downlink constraints must be taken into account. Corresponding major system performance constraints can be summarized as shown in Table 2.

TABLE 2. The main control parameters of the system.

Message Type	Restrictions
Carrier synchronization signal - to - noise ratio requirements	≥4dB
Satellite EIRP	39dBW
Satellite output back	4dB
Terminal antenna gain	0dBi
Polarization Mismatch Loss	3dB
Power stability	0.5dB/Day

The quality of the satellite downlink is a bottleneck that constrains system performance. By satellite equivalent EIRP and terrestrial terminal reception capacity constraints, satellite downlink quality is poor. To meet the system performance indicators, it need to use a special error correction coding technology. Based on the construction theory of cyclic permutation matrix, the QC-LDPC code with code length of 992 and bit rate of 3/4 is constructed according to the requirements of the actual system. The main method is to construct multiplexing the cyclic permutation matrix A new parity check matrix is obtained, and the required QC-LDPC code is reconstructed[6].

For QC-LDPC codes with a code length of 992, we take L = 62, and the parity check matrix is

$$\mathbf{H} = \begin{bmatrix} I(p_{1,1}) & I(p_{1,2}) & \cdots & I(p_{1,16}) \\ I(p_{2,1}) & I(p_{2,2}) & \cdots & I(p_{2,16}) \\ \vdots & \vdots & \ddots & \vdots \\ I(p_{4,1}) & I(p_{4,2}) & \cdots & I(p_{4,16}) \end{bmatrix} \quad 1)$$

We follow the following steps to eliminate the ring in H [7].

1. Initialize the cyclic shift value of all the 1 elements in the base matrix X corresponding to its check H_b, that is, set it to zero.

2. In the base matrix H_b, from the first column, up to the 16th column, select the current column one by one. In the current column, from the line 1 to the last, select the current 1 element, the order of one by one optimization. That is, by adjusting the shift value of the current 1 element while keeping the shift value of 1 element has been traversed, the loop with any length of 2g contained in the current 1 element satisfies the loop constraint condition. The current column divides the matrix into left and right parts. The current 1 element divides the current column into upper and lower parts.

3. Find the ring contained in the current 1 element in the range above the current column and the current 1 element and create a ring constraint list.

4. Eliminate the extended loop by setting the shift value of the current element. The value of the shift value corresponding to the current 1 element is [0,61] [8]. I If the current 1 element corresponds to the shift value, the contained ring cannot satisfy the constraint condition, The shift value corresponding to the current 1 element is increased at tolerance 1 until the current 1 element corresponds to the shift value. So that all the rings contained satisfy the ring constraint.

After the above steps we can get the check matrix H. We use Matlab to simulate its performance. Through the AWGN channel, the log domain and product decoding algorithm are used [11]. The decoding is iterated 30 times. The bit error rate performance shown in Figure 1.

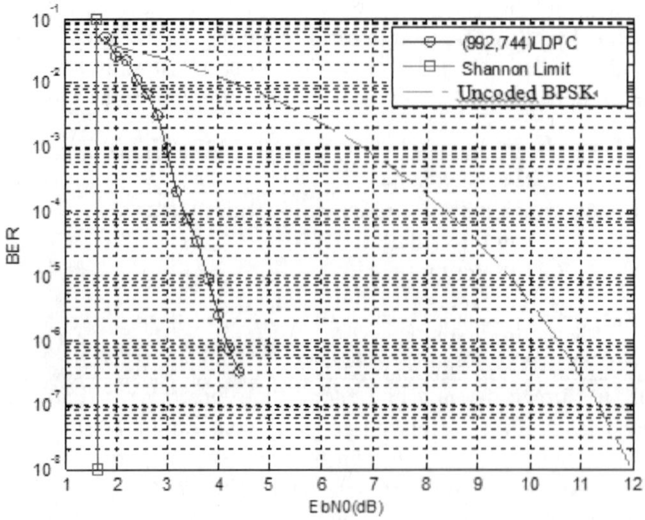

FIGURE 1. (992,744)LDPC coding bit error rate performance curve.

As can be seen from the above figure, the (992, 744) LDPC coded bit error rate performance can meet the bit error rate requirements. When the bit signal to noise ratio exceeds 5 dB, it is possible to obtain a bit error rate of less than 10^{-7}.

The bit-to-noise ratio and the bit-to-noise ratio and coding gain required for coding the BPSK system at different bit error rates are shown in Table 3.

TABLE 3. The coding gain at different bit error levels ((992,744) LDPC codes).

Bit Error Rate Level	Not Encoded Required Signal to Noise Ratio (dB)	Required Signal to Noise Ratio (dB)	Encoded Gain (dB)
10^{-5}	9.59	3.79	5.8
10^{-6}	10.53	4.15	6.38
10^{-7}	11.31	4.6	6.71

As can be seen from the above table, LDPC coding using 3/4 rate can achieve a coding gain of 6.5 dB or more at a bit error rate of 10^{-7}, at a cost of 4/3 times the transmit power. If LDPC encoding is not used, there is no coding redundancy, and the gain is only 1.25dB gain due to 4/3 power, which is much smaller than the gain caused by LDPC coding.

In fact, due to the relatively poor situation of the downlink, it is impossible to achieve the desired original information rate close to 750 bps (744 bps) without using the LDPC encoding. As can be seen in the link budget described below, the receiver can obtain an encoded bit signal to noise ratio of about 5 dB at which the LDPC decoder can achieve an error rate level of less than 10^{-7}. In this link condition, if the LDPC coding is not used, the receiver can obtain the bit signal to noise ratio of about 8dB. The bit signal to noise ratio can only support about 2×10^{-4} bit error rate level, can not reach the system Claim.

4. Frame structure design

In this paper, the 3/4 rate LDPC code is used as the error control code, and the LDPC code belongs to the block code. The decoder at the receiving end needs to determine the head of a coded packet to start the decoding process [5]. Therefore, it is necessary to add a header flag for each coded bit group.

Since the system requires the original differential information rate of 744bps, and from the system to

IOP Conf. Series: Materials Science and Engineering **242** (2017) 012131 doi:10.1088/1757-899X/242/1/012131

meet the bit error rate requirements.This paper also selected 3/4 rate of LDPC coding. Therefore, this time within 1 second 744 original differential information bits Encoding, get 992 encoded bits. For the sake of convenience, it is desirable that the final frame length is 1000 bits, and therefore 8 bits are available as the header indication of the coded packet[10]. Due to the basis of frame synchronization LDPC decoding, the header indication flag must have a very high detection accuracy rate. In order to achieve this goal, this paper uses a good self-correlation characteristics of the Barker code as a coded packet header. The 7-bit Barker code of the currently known maximum length is used to search for the packet header by correlating the received signal with the local signal. And further adds one padding bit to form a data frame having a length of 1000 bits.

5. BDS RTCM SSR message structure

Based on the above research basis, the final message structure is designed as follows.

C1 frequency of the I-channel broadcast fast clock correction and code offset information, the information rate of 250 bps, 2 seconds into the message once, each injection 500 bits.C2 frequency I branch broadcast track correction and URA information, information rate 200 bps, 2 seconds into the message once, each time 400bits. In the case of the design of the message, the orbit correction, code offset, URA, fast clock correction information were 30 seconds, 128 seconds, 2 seconds, 2 seconds a complete broadcast time,.It fully meet the requirements of data updates.

The detailed message is as follows.

1. C1 I message

C1 I Road message broadcast Fast clock correction with code offset information. The message structure is a frame containing 64 subframes. Each subframe 500 bits, 2 seconds to generate a subframe, a total of 32000 bits, with 128 seconds to complete the broadcast. The information rate of 250bps. The structure of each message subframe is consistent and the fast clock correction information appears once every subframe. The code offset information is broadcast once through 64 subframes.

2. C2 I message

C2 I message broadcast track revision and URA information. The complete data unit is a frame containing 15 subframes. Each subframe 400 bits, 2 seconds to generate a subframe, a total of 6000 bits with 30 seconds to complete the broadcast. The information rate of 200bps. The structure of each message subframe is consistent and the URA information is presented once every subframe. The track correction information is broadcast once through 14 subframes (the corresponding position of the 15th subframe is left blank).

6. Conclusions

In order to adapt to the development of new systems (especially BDS and Galileo), there have been system upgrades (such as new L2C and L5 signals) and regional wide area differential enhancement systems[9]. In this paper BDS RTCM and SSR messages are designed according to the standard of data volume and update rate of Beidou wide area real-time difference system. It refers to the technology and RTCM standards of international similar systems. It also combines the C band signal system, satellite resources and the restriction and restriction of the receiver implementation technology.

The new message protocol design provides a strong support for the construction of stable and reliable high-precision enhanced information broadcasting means. The use of space-based broadcast will provide high-precision differential products to provide high-precision users to use, covering the "one side" area of advanced performance, Continuous and reliable, reliable service Beidou satellite navigation system High-precision civilian service system, and other domestic service system for the " One belt One road " area to provide seamless BeiDou high-precision navigation location services.

Acknowledgments
The author wishes to thank the IEEE for providing this template and all colleagues who previously provided technical support. The authors also gratefully acknowledge National Natural Science Foundation of China (Grant No. 41504006 and No. 41674034) for the financial.

References
[1] Li Sichao, Ye Tianchun and Xu Jianhua, "Brief introduction of DGPS RTCM data format and its decoding algorithm,"in Electronic Measurement Technology, 2008, 12.
[2] Liu Zhimin, Lin Wenjie and Tang Weiming, " Journal of Guilin Institute of Technology,"in Journal of Guilin Institute of Technology, 2004, 24 (2): 188-191.
[3] HUANG Yunqian, CHENG Pengfei and CAI Yanhui, "Study on pre-definition of RTCM message types for Beidou satellite navigation system"in Bulletin of Surveying and Mapping, 2013 (5): 5-7.
[4] GUO Hong-tao, REN Chao, "Transmission GPS Data Communication Format RTCM3.0 and Application Development,"in Global Positioning System, 2010, 35 (003): 63-65.
[5] LIU Wen-tao, QUAN Lu-lu and LIANG Hong, et al, "Synchronous design and performance simulation of navigation bit in satellite-based augmented system receiver, "in Telecommunications Technology, 2008, 7: 54-56.
[6] Marc P. C. Fossorier, "Quasi-Cyclic Low-Density Parity-Check Codes From Circulant Permutation Matrices,"in IEEE Transactions on Information Theory. 2004, 50(8):1788-1793.
[7] Jun Xu, Lei Chen and Shu Lin, "Construction of Regular and Irregular LDPC Codes: Geometry Decomposition and Masking,"in IEEE Transactions on Information Theory. 2007, 53(1):121-134.
[8] Zongwang Li, Lei Chen and Lingqi Zeng, "Efficient Encoding of Quasi-Cyclic Low-Density Parity-Check Codes,"in IEEE Transactions on Communications. 2006, 54(1):71-81.
[9] WANG Quan; MEI Jin-zhong; GUAN Shou-zhou and FANG Shu-shan, "YUANG Gan;New Generation International Standard RTCM V3.2 and Its Application CHEN Zhen,"in Journal of Navigation and Location,2014-12-20.
[10] Hua C X and Oksman J, "A new algorithm to optimize Barker code sidelobe suppression filters,"in IEEE Transactions on Aerospace and Electronic Systems, 1990, 26(4): 673-677.
[11] Declercq D and Fossorier M, "Decoding algorithms for nonbinary LDPC codes over GF $(q) $,"in IEEE Transactions on Communications, 2007, 55(4): 633-643.

IOP Conf. Series: Materials Science and Engineering **242** (2017) 012132 doi:10.1088/1757-899X/242/1/012132

Acoustic Analysis of Nasal Vowels in Monguor Language

Hanbin Zhang

School of Foreign Languages, Northwest Minzu University, Lanzhou730030, China

E-mail:zhanghb06@163.com

Abstract. The purpose of the study is to analyze the spectrum characteristics and acoustic features for the nasal vowels [ɑ̃] and [ɔ̃] in Monguor language. On the base of acoustic parameter database of the Monguor speech, the study finds out that there are five main zero-pole pairs appearing for the nasal vowel [ɑ̃] and two zero-pole pairs appear for the nasal vowel [ɔ̃]. The results of regression analysis demonstrate that the duration of the nasal vowel [ɑ̃] or the nasal vowel [ɔ̃] can be predicted by its F1, F2 and F3 respectively.

1. Introduction

When a vowel is pronounced, if the soft palate lowers down and oral and nasal passages open at the same time, the vowel will take nasal feature. Under the circumstance, the vowel is called nasal vowel [1]. A nasal vowel is shown with an additional symbol "˜" over a vowel in the international phonetic alphabet. When a nasal vowel is pronounced, the breath flows from the mouth and nose at the same time. Here are some examples of various kinds of language. In mandarin, [ɔ̃] in "DengEr" [tɔ̃˞55] is a nasal vowel, but [ɤ] in "Deng" [tɤŋ55] is not a nasal vowel. Polish ¥ę¥ is the nasalized sound of [e]. French letter combinations "an", "am", "en" and "em" are nasal vowels in the general case. Portuguese letter combinations "am" and "in" are also nasal vowels.

Studies on nasal vowels based on acoustic analysis are mainly concerning French, Portuguese, Hindi, Japanese, and some Chinese dialects. For example, the study on lingual differences in Brazilian Portuguese oral and nasal vowels found that the lingual configurations are in line with the effects of vowel nasalization on formant frequencies [2]. The investigation of the role of lingual and pharyngeal articulation in the production of the nasal vowel system of French showed that lingual and pharyngeal articulatory configurations enhance formant-frequency-related acoustic effects associated with nasalization [3]. A research on the oral articulations of French nasal vowels suggests that many of the oral articulatory configurations of Northern Metropolitan French nasal vowels enhance the acoustic effect of velo-pharyngeal coupling on F1 and F2 frequencies [4].

In Monguor language, when "a" and "o" appear before velar nasal sound [ŋ], "a" and "o" will become nasal vowels [ɑ̃] and [ɔ̃] respectively [5]. For example, while Monguor word "hairghan" (cute) is pronounced as [xaɪʐɢᴀn], "rghang" (power) is pronounced as [ʐɢɑ̃ŋ]; while "manqog" (cockscomb) as [mᴀntɕʰɔk], "manglii" (forehead) as [mɑ̃ŋli:]; while "jog" (aspect) as [tɕɔk], "jong" (hundred) as [tɕɔ̃ŋ]; while "rzogla" (overturn) as [ʐtsɔklᴀ], "zongna" (honey) as [tsɔ̃ŋna:]. In Monguor language, the appearance of nasal vowels [ɑ̃] and [ɔ̃] has their particular conditions and rules, but nasal vowels [ɑ̃] and [ɔ̃] have no the function of distinguishing the meaning of words.

This paper analyzes the nasal vowels [ɑ̃] and [ɔ̃] in Monguor language from the respect of acoustic features, such as duration and formants, especially the first formant (F1), the second formant (F2) and the third formant (F3), in which F1 corresponds to the height of the tongue, F2 the front or back of the tongue, and F3 the lip rounding [1].

Content from this work may be used under the terms of the Creative Commons Attribution 3.0 licence. Any further distribution of this work must maintain attribution to the author(s) and the title of the work, journal citation and DOI.
Published under licence by IOP Publishing Ltd

2. Method

2.1 Source of Corpus

An acoustic parameter database of the Monguor speech is built for this study. The Monguor words recorded for this study are just the words containing the nasal vowels [ɑ̃] and [ɔ̃]. There are 34 Monguor words with the nasal vowel [ɑ̃] and 37 Monguor words with the nasal vowel [ɔ̃]. These words are composed of monosyllable, disyllable, trisyllable, and quadrisyllable words.

2.2 Speech Signal Collection

A Dell Notebook, a Behringer recording microphone, and a sound card of YAMAHA Steinberg are used in this study as the recording equipment. The recording sampling rate is 44.1 kHz with 16 bits resolution ratio. The sounds are saved with *.wav files. The sounds are recorded in a recording studio. Each word is read three times.

2.3 Speakers

One speaker, male, aged 45 reads all the selected words. The speaker serves as a teacher in a local primary school. His speech belongs to Tianzhu dialect, similar to the Huzhu dialect.

2.4 Research Questions

The questions addressed in this study are:

(1) What are the spectrum characteristics for the nasal vowels [ɑ̃] and [ɔ̃]?

(2) What are the acoustic features for the nasal vowels [ɑ̃] and [ɔ̃]?

3. Results And Discussion

3.1 Spectrum Characteristics

When a sound comes from a branching acoustic tube, it has not only zero point but also pole point on its spectrogram [1]. When a nasal vowel is pronounced, the nasal cavity coupled with the oral cavity comes into being a branching acoustic tube. Figure 1 shows the spectrums of /a/ for [ᴀ] before nasalization in Fig. 1 (A), and for [ɑ̃] after nasalization in Fig. 1 (B).

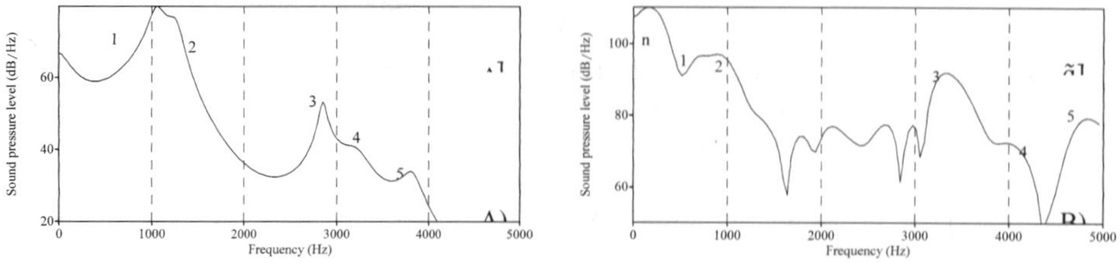

FIGURE 1. Spectrums of [ᴀ] and [ɑ̃]

In Fig. 1 (A), F1, F2, F3, F4 and F5 of the oral vowel [ᴀ] descend successively in the sound pressure level. However, in Fig. 1 (B), a zero point appears on the left of F1 of the nasal vowel [ɑ̃]. At the point of around 200Hz, a new formant comes into being, which is called nasal formant (Fn). Between F2 and F3, a group of zero points appear, and between F4 and F5, there also exists a zero point. Figure 1 (B) shows there are five main zero-pole pairs appearing for the nasal vowel [ɑ̃]. For the first zero-pole pair, the zero point is about 530Hz and the pole point is about 160Hz. The zero point is about 1630Hz and the pole point is about 1450Hz for the second zero-pole pair. The zero point is about 2840Hz and the pole point is about 2670Hz for the third zero-pole pair. The zero point is about 3060Hz and the pole point is about 2970Hz for the fourth zero-pole pair. The zero point is about 4380Hz and the pole point is about 4070Hz for the fifth zero-pole pair.

In Fig. 2 (A), F1, F2, F3 and F4 of the oral vowel [ɔ] are almost the same in the sound pressure level. However, in Fig. 2 (B), F1, F2, F3 and F4 of the nasal vowel [ɔ̃] vary greater. Two zero-pole pairs appear on the left of F1 of the nasal vowel [ɔ̃]. At the point of around 400Hz, a nasal formant (Fn) comes into being. The zero point is about 230Hz and the pole point is about 40Hz For the first zero-pole pair. The zero point is about 950Hz and the pole point is about 400Hz for the second zero-pole pair.

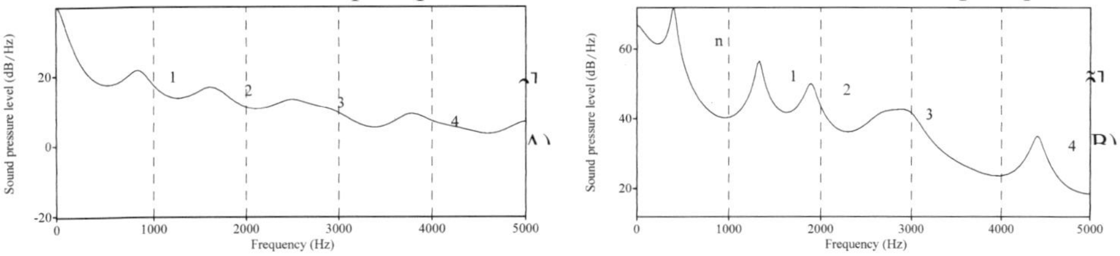

FIGURE 2. Spectrums of [ɔ] and [ɔ̃]

3.2 Acoustic Features

Acoustic features mainly include formants and duration in this study. Figure 3 shows the formants for [ɑ̃] and [ɔ̃]. F1 and F2 stay close for both [ɑ̃] and [ɔ̃], while F3, F4 and F5 stay close for [ɑ̃], and F3 and F4 stay close for [ɔ̃].

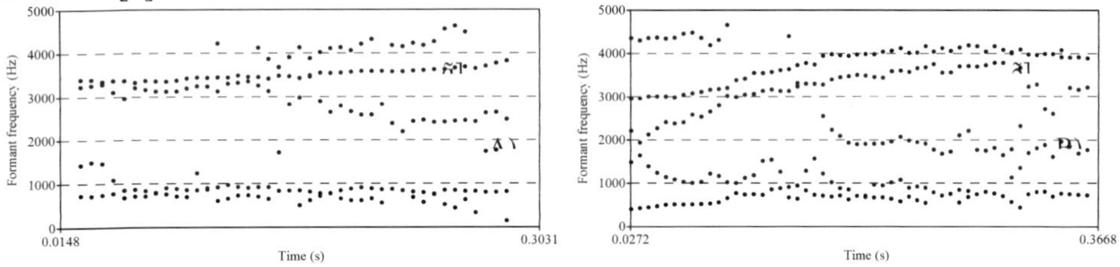

FIGURE 3. Formants for [ɑ̃] and [ɔ̃]

Table 1 shows the descriptive statistics of formants and duration for [ɑ̃]. F1, F2 and F3 of [ɑ̃] are 712Hz, 1442Hz, and 2769Hz respectively. The mean duration of [ɑ̃] is 276 milliseconds (ms).

TABLE 1. Descriptive statistics for [ɑ̃]

	Mean	**Std. Deviation**	**Std. Error**	**Range**
F1	712	126	22	552
F2	1442	211	36	717
F3	2769	205	35	734
Duration	276	88	15	357

The standard deviation for F1, F2 and F3 of [ɑ̃] is 126Hz, 211Hz, and 205Hz respectively. The standard deviation for duration is 88ms. The standard error for F1, F2 and F3 of [ɑ̃] is 22Hz, 36Hz, and 35Hz respectively. The standard error for duration is 15ms. The range for F1, F2 and F3 of [ɑ̃] is 552Hz, 717Hz, and 734Hz respectively. The range for duration is 357ms.

The result of ANOVA shows that there exists a regression relationship among F1, F2, F3 and duration of the nasal vowel [ɑ̃] ($F = 135$, $p = 0.000 < 0.05$). The result of regression analysis demonstrates that the relationship among F1, F2, F3 and duration of the nasal vowel [ɑ̃] can be the equation:

$$D = -0.212F1 + 0.153F2 + 0.1F3 \qquad (1)$$

The equation (1) shows that the duration of the nasal vowel [ɑ̃] can be predicted by F1, F2 and F3. The coefficient of F1 is -0.212, the coefficient of F2 0.153, and the coefficient of F3 0.1. F1, F2 and F3 are the independent variables, and duration is the dependent variable. In Fig. 4 (A), the scatterplot verifies the prediction.

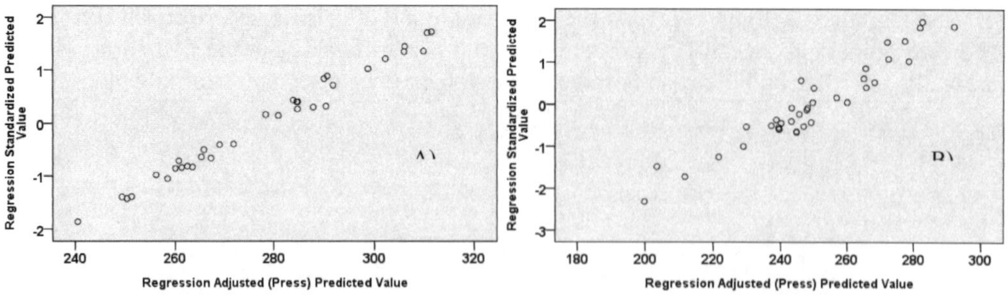

FIGURE 4. Scatterplots of formants and duration for [ɑ̃] and [ɔ̃]

Table 2 shows the descriptive statistics of formants and duration for [ɔ̃]. F1, F2 and F3 of [ɔ̃] are 720Hz, 1558Hz, and 2907Hz respectively. The mean duration of [ɔ̃] is 249 milliseconds (ms). The standard deviation for F1, F2 and F3 of [ɔ̃] is 142Hz, 276Hz, and 206Hz respectively. The standard deviation for duration is 75ms. The standard error for F1, F2 and F3 of [ɔ̃] is 23Hz, 45Hz, and 34Hz respectively. The standard error for duration is 12ms. The range for F1, F2 and F3 of [ɔ̃] is 513Hz, 1023Hz, and 818Hz respectively. The range for duration is 316ms.

TABLE 2. Descriptive statistics for [ɔ̃]

	Mean	**Std. Deviation**	**Std. Error**	**Range**
F1	720	142	23	513
F2	1588	276	45	1023
F3	2907	206	34	818
Duration	249	75	12	316

The result of ANOVA shows that there also exists a regression relationship among F1, F2, F3 and duration of the nasal vowel [ɔ̃] ($F = 151$, $p = 0.000 < 0.05$). The result of regression analysis demonstrates that the relationship among F1, F2, F3 and duration of the nasal vowel [ɔ̃] can be the equation:

$$D = -0.072F1 + 0.017F2 + 0.096F3 \quad (2)$$

The equation (2) shows that the duration of the nasal vowel [ɔ̃] can also be predicted by F1, F2 and F3. The coefficient of F1 is -0.072, the coefficient of F2 0.017, and the coefficient of F3 0.095. F1, F2 and F3 are the independent variables, and duration is the dependent variable. In Fig. 4 (B), the scatterplot also verifies the prediction.

4. Conclusion

On the base of acoustic parameter database of the Monguor speech, the study discusses the spectrum characteristics and the acoustic features for the nasal vowels [ɑ̃] and [ɔ̃]. The study has found that there are five main zero-pole pairs appearing for the nasal vowel [ɑ̃]. A zero point appears on the left of F1. A new formant, a nasal formant, comes into being at the point of around 200Hz. Between F2 and F3, three obvious zero-pole pairs appear, and the fifth zero-pole pair emerges between F4 and F5. Two zero-pole pairs appear on the left of F1 of the nasal vowel [ɔ̃]. A new nasal formant comes into being at the point of around 400Hz. F1, F2 and F3 of [ɑ̃] are 712Hz, 1442Hz, and 2769Hz respectively. The mean duration of [ɑ̃] is 276ms. F1, F2 and F3 of [ɔ̃] are 720Hz, 1558Hz, and 2907Hz respectively. The mean duration of [ɔ̃] is 249ms.

The results of regression analysis demonstrate that the duration of the nasal vowel [ɑ̃] or the nasal vowel [ɔ̃] can be predicted by F1, F2 and F3. The relationship among F1, F2, F3 and duration for the nasal vowels [ɑ̃] and [ɔ̃] can be the equations: [ɑ̃]: $D = -0.212F1 + 0.153F2 + 0.1F3$; [ɔ̃]: $D = -0.072F1 + 0.017F2 + 0.096F3$.

Monguor language is disappearing nowadays. As professor Wang Yuanxin [6] says, a certain minority language is not only the historical creation of the nation, but also the excellent culture of the nation. It is also the common cultural wealth of all mankind. The main significance of the study lies in language application and protection. The disappearance of a large number of languages and a large number of unique cultures, without a doubt, is a great loss of human civilization. Therefore, the purpose of the study is for the preservation of the precious minority language. The results of the study can be applied in speech recognition and speech synthesis.

Acknowledgments
This research was supported by Fund Yxm2015014 to Northwest Minzu University.

Refrences

[1] H. Q. Bao and M. C. Lin, *Essentials of Experimental Phonetics* (Peking University Press, Beijing, 2014).
[2] M. Barlaz, M. J. Fu, Z. P. Liang, S. Ryan, S. Brad, "Lingual Differences in Brazilian Portuguese Oral and Nasal Vowels: An MRI Study," In Proceedings of the 18th International Congress of Phonetic Sciences (Glasgow, Scotland, UK, 2015).
[3] C. Christopher, K. S. Ryan, M. J. Fu, Z. P. Liang, P. S. Bradley, "A real-time MRI investigation of the role of lingual and pharyngeal articulation in the production of the nasal vowel system of French," Journal of Phonetics 50, 34–51(2015).
[4] C. Christopher, "An acoustic and articulatory examination of the "oral" in "nasal": The oral articulations of French nasal vowels are not arbitrary," Journal of Phonetics 46, 23–33(2014).
[5] Chinggeltei. *The Monguor language and Mongolian,* (Inner Mongolia people's publishing house, Hohhot, 1991).
[6] Y. X. Wang, "The investigation and protection of the vulnerable and endangered language of the less populous ethnic groups," Journal of Xinjiang Normal University, (2008).

Group Multilateral Relation Analysis Based on Large Data

Qiang LIU[1, a)], Guo-min ZHOU[1, b)], Guang-xuan CHEN[1, c)] and Yong XU[2, a)]

[1]Big data Key Laboratory of the M.P.S, Zhejiang Police College, Hangzhou 310053, China

[2]Hangzhou Hezhong Data Technology Co., Ltd, Hangzhou 310052, China

E-mail: [a)]Corresponding author: navylq@126.com, [b)]zhouguomin@zjjcxy.cn, c)chenguangxuan@zjjcxy.cn

Abstract. Massive, multi-source, heterogeneous police data and social data brings challenges to the current police work. The existing massive data resources are studied as the research object to excavate the group of multilateral relations by using large data technology for data archiving. The results of the study could provide technical support to police enforcement departments for fighting crime and preventing crime.

1. Introduction

In recent years, with the continuous application of public security information systems, a variety of public security business applications are gradually built and put into practical application, accumulated a large number of valuable data resources [1-2]. Outside of police department, there are also accumulated a wealth of data resources in the information systems from all walks of life. The massive data resources from police data and social data may reflect the attributes, trajectories and trends of the society and the individual, but the information is disorganized and the relationship can not be straightened out [3-4]. Now, ETL technology has been more mature, fusion heterogeneous, multi-source data is no longer a problem.

For large data processing, it is an important subject to dig interpersonal relationships by using massive & multi-source data. You should organize the information into a structured, centralized archival data, according to the classification of the category, and the information with the plane/train trips, hotel accommodation, Internet cafes online data, online shopping consumption, communication, household registration information, case and so on. Then you can use reality and virtual network activities to visualize the direct or indirect relationship among people by the whole, automatic and deep mining analysis.

Based on the existing large data technology represented by Hadoop, this paper uses distributed database technology, Map-Reduce, full-text search and other technical means to organize the massive data resources inside and outside the public into organized and centralized archival data, according to the classification of the main categories, then analyzes the visually interpersonal relationships by using the data described above.

2. Super dimension aggregation method

The information in the public security information system is build up based on a specific needs, for example, the hotel system maily including accommodation records, the driver management system maily including the driver basic information and his car's illegal information. It is difficult to grasp the

Content from this work may be used under the terms of the Creative Commons Attribution 3.0 licence. Any further distribution of this work must maintain attribution to the author(s) and the title of the work, journal citation and DOI.
Published under licence by IOP Publishing Ltd

whole information of people or cars through a single system. However, if we pile up the information purely, a large number of bare data (original file) may be produced, and the data can not form archival data which can be be used effectively. The traditional way is to let the police carry out multi-library, multi-table correlation query in the massive data and spend a lot of time, effort to carry out a variety of complex relationship analysis. So that it's not efficient and unrealistic.

So we can effectively aggregate a large number of scattered bare data for truly combat services, by data archiving and a variety of archives for effective correlation. Data archiving mainly refers to organize the chaotic, scattered data into a structured, centralized archival data, by using the data relevance in the data object. Each file can be analyzed automatically, and all other data with a specific association of the current data object may be summarized into the file of the current data object. As shown in Fig.1 and Fig.2, there are personnel files, vehicle files, case files, virtual identity files, organization files and other files in the aggregative classification data.

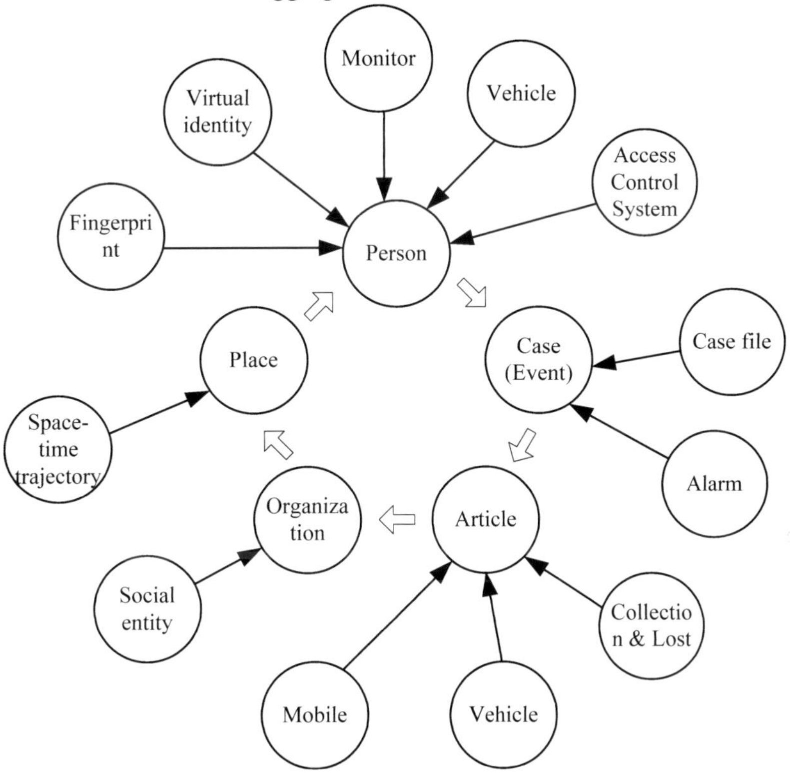

FIGURE 1. The display information of file data (1)

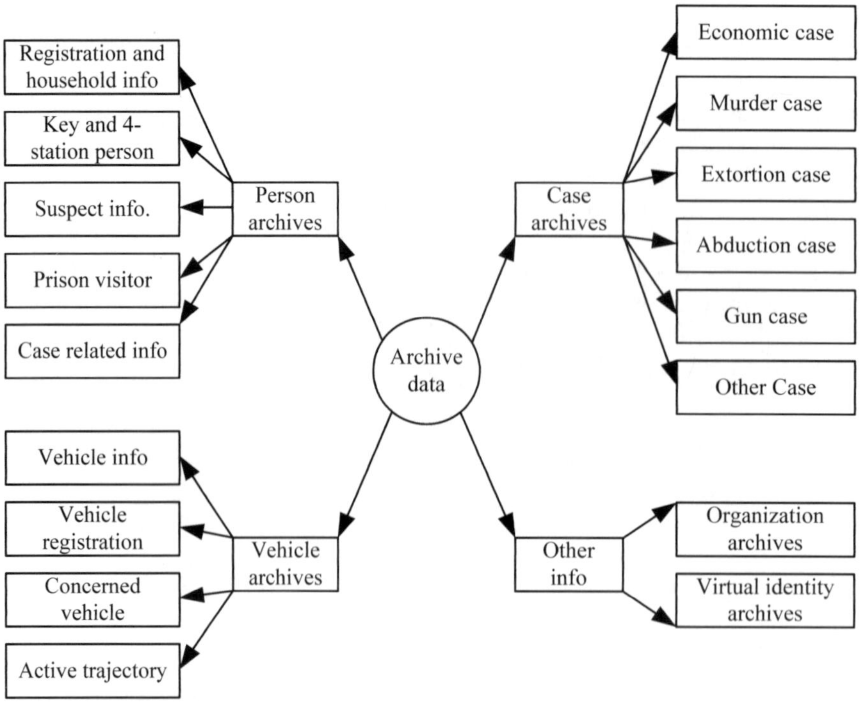

FIGURE 2. The display information of file data (2)

3. Super dimension aggregation method

The data files formed by the hyper-dimension aggregation collation are stored in the NoSQL database. The NoSQL database supports the storage and management for massive structure, semi-structured, unstructured data, also supports search query, analysis and mining, is very suitable for massive data query, full text search, offline analysis, offline transaction computing and other scenes, especially for non-format data query, retrieval [5-7]. The data which is scattered in each information system can be easily analyzed and archived according to NoSQL database, such as personnel, case, article, organization and other data.

First of all, based on the NoSQL database, the public security data is described on the basis of the super dimension aggregation method. Then, you can perform unilateral relational analysis, relationship analysis, or multi-party relationship by using the high-performance batch computing framework MapReduce. Finally, the relationships between the different feature objects can be automatically discovered and displayed graphically, thus visually revealing the interrelationships between objects, including hidden or indirect links.

Let {OD} denote a object which contains attributes of various elements, OD={M, T, L, G, OR}. Where M is the personal element, M={m1, m2, m3, …, mi}. T is the time element, T={t1, t2, t3, …, ti}. L is the location element, L={l1, l2, l3, …, li}. G is the articles element, G={g1, g2, g3, …, gi}. OR is the organization element, OR={or1, or2, or3, …, ori}. The principle of social relation self-discovery is shown in Fig.3.

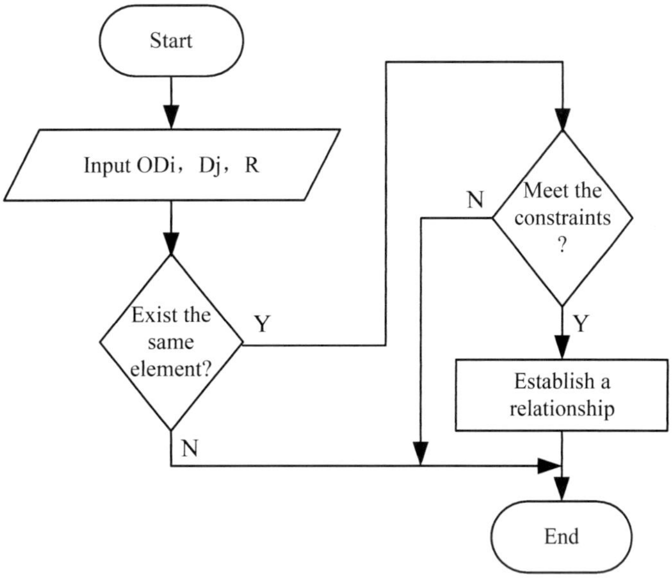

FIGURE 3. Social relation self-discovery

Here, ODi and ODj are two ODs, R is the constraint condition. When the two kinds of data ODi and ODj exist with the same characteristics of the property, the system may compare each other according to the constraints of the alignment. If the result of the comparison satisfies the constraints R, a relationship may be established. Normally, the constraints R are based on public security work experience. The system will automatically tap the social relations of known persons, automatically dig peer, living, Internet, relatives, social ties and other relations. With the deepening of the integration of information resources, a new further social relation may be generated.

After the social relations dig out, the relationship between the various data, and the information represented by the data will form a social relationship network. Based on the six-dimensional theory of space, the platform provides an indirect correlation analysis function through the network analysis technology. The platform could specify the path or use the shortest path, and provide iterative mining function, then a infinite drill analysis of social relations.

4. Conclusions

The massive data resources from police data and social data are used to study the interpersonal relations of a group. The use of super-dimensional aggregation classification methods and social relations self-discovery technology may provide police a new method to find an important clue for investigate and solve a case. The police may easily deal with diversity, complexity and concealment of social relations.

Acknowledgements

This work was financially supported by the NSFC-Zhejiang Joint Fund for the Integration of Industrialization and Informatization (U1509219), Science and Technology Program of Ministry of Public Security of China (2015GABJC41) and Scientific Research Fund of Zhejiang Provincial Education Department (Y201533572).

References

[1] Q. Shen, and B. Chen, *Electronic Design Engineering*, Vol. 24 (2016), pp. 25-27.
[2] D. Zhang, K. Y. Qi, Z. Y. Su and J. Zhang, *Information Technology & Standardization*, No. 7 (2015), pp. 18-22.
[3] W. Li, L. Q. Sun, and S. L. Li, *Journal of Chinese People's Public Security University(Science and Technology)*, No. 4 (2013), pp. 20-25.

IOP Conf. Series: Materials Science and Engineering **242** (2017) 012133 doi:10.1088/1757-899X/242/1/012133

[4] J. Liu, and L. H. Zhang, *Police Technology*, No. 3 (2015), pp. 4-7.
[5] J. P. Meng, C. Chen, and J. J. Meng, *Microcomputer ITS Application*, Vol. 33 (2014), pp. 19-22.
[6] W. Li, *Information & Communication*, No. 7 (2014), pp. 72-73.
[7] Y. B. Wu, Q. Xue, D. W. Xiang and Y. H. Mai, *Modern Electronics Technique*, Vol. 39 (2016), pp. 44-47.